D0143063

STUDENT'S SOLUTIONS MANUAL

DAVID DUBRISKE
University of Arkansas for Medical Sciences

CALCULUS AND ITS APPLICATIONS EXPANDED VERSION

Marvin L. Bittinger
Indiana University Purdue University Indianapolis

David J. Ellenbogen
Community College of Vermont

Scott A. Surgent
Arizona State University

PEARSON

Boston Columbus Indianapolis New York San Francisco Upper Saddle River
Amsterdam Cape Town Dubai London Madrid Milan Munich Paris Montreal Toronto
Delhi Mexico City São Paulo Sydney Hong Kong Seoul Singapore Taipei Tokyo

The author and publisher of this book have used their best efforts in preparing this book. These efforts include the development, research, and testing of the theories and programs to determine their effectiveness. The author and publisher make no warranty of any kind, expressed or implied, with regard to these programs or the documentation contained in this book. The author and publisher shall not be liable in any event for incidental or consequential damages in connection with, or arising out of, the furnishing, performance, or use of these programs.

Reproduced by Pearson from electronic files supplied by the author.

Copyright © 2014 Pearson Education, Inc.
Publishing as Pearson, 75 Arlington Street, Boston, MA 02116.

All rights reserved. No part of this publication may be reproduced, stored in a retrieval system, or transmitted, in any form or by any means, electronic, mechanical, photocopying, recording, or otherwise, without the prior written permission of the publisher. Printed in the United States of America.

ISBN-13: 978-0-321-84417-0
ISBN-10: 0-321-84417-3

1 2 3 4 5 6 EBM 16 15 14 13

www.pearsonhighered.com

TABLE OF CONTENTS

Chapter R
Functions, Graphs, and Models

Exercise Set R.1

1. Graph $y = x + 4$.

We choose some x-values and calculate the corresponding y-values to find some ordered pairs that are solutions of the equation. Then we plot the points and connect them with a smooth curve.

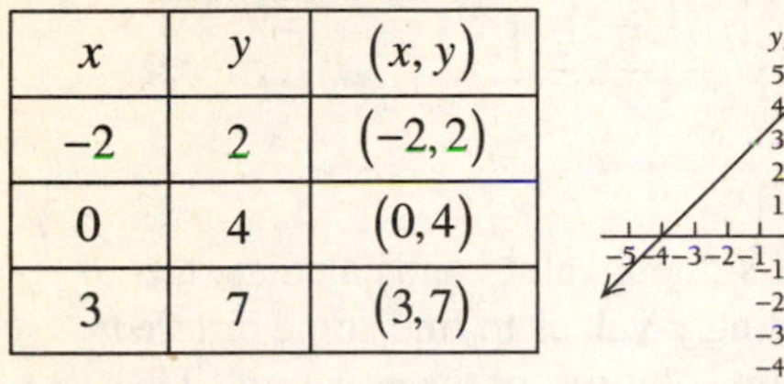

x	y	(x, y)
-2	2	$(-2, 2)$
0	4	$(0, 4)$
3	7	$(3, 7)$

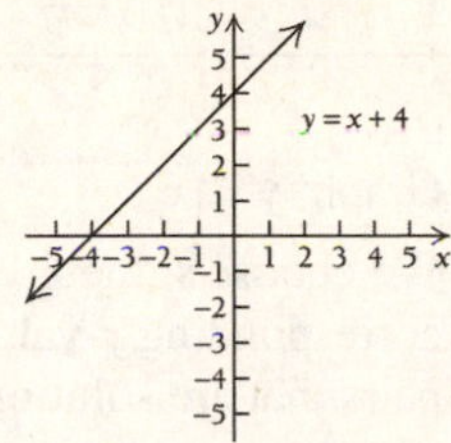

3. Graph $y = -3x$.

We choose some x-values and calculate the corresponding y-values to find some ordered pairs that are solutions of the equation. Then we plot the points and connect them with a smooth curve.

x	y	(x, y)
-1	3	$(-1, 3)$
0	0	$(0, 0)$
2	-6	$(2, -6)$

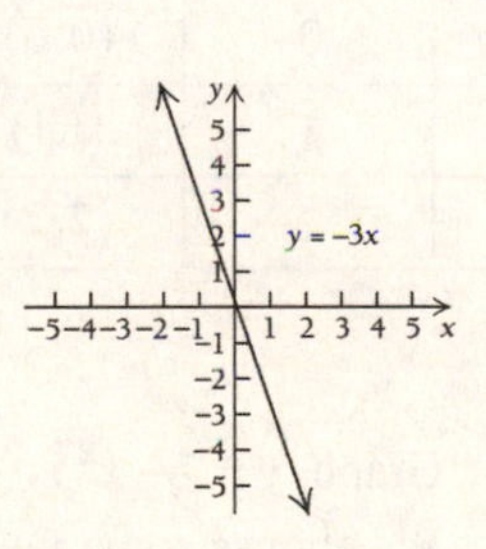

5. Graph $y = \frac{2}{3}x - 4$.

We choose some x-values and calculate the corresponding y-values to find some ordered pairs that are solutions of the equation. Then we plot the points and connect them with a smooth curve.

x	y	(x, y)
-3	-6	$(-3, -6)$
0	-4	$(0, -4)$
3	-2	$(3, -2)$

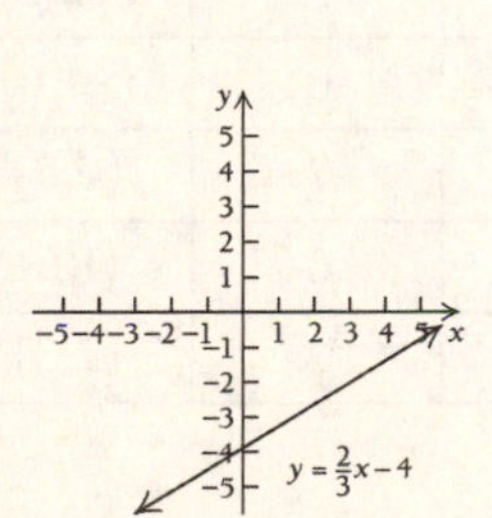

7. Graph $x + y = 5$.

We solve for y first.

$x + y = 5$

$y = 5 - x$ subtract x from both sides

$y = -x + 5$ commutative property

Next, we choose some x-values and calculate the corresponding y-values to find some ordered pairs that are solutions of the equation. Then we plot the points and connect them with a smooth curve.

x	y	(x, y)
-1	6	$(-1, 6)$
0	5	$(0, 5)$
2	3	$(2, 3)$

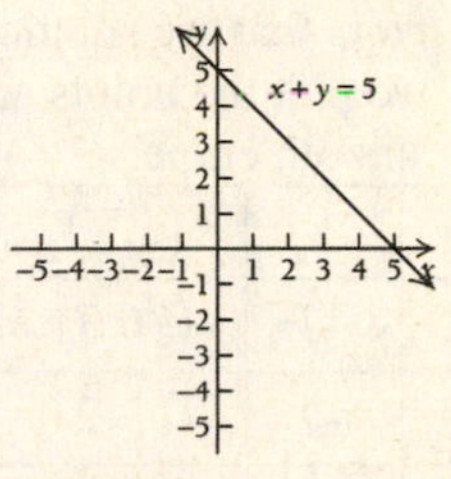

9. Graph $8y - 2x = 4$.

We solve for y first.

$8y - 2x = 4$

$8y = 2x + 4$ add $2x$ to both sides

$y = \frac{2}{8}x + \frac{4}{8}$ divide both sides by 8

$y = \frac{1}{4}x + \frac{1}{2}$

Next, we choose some x-values and calculate the corresponding y-values to find some ordered pairs that are solutions of the equation. Then we plot the points and connect them with a smooth curve.

x	y	(x, y)
-2	0	$(-2, 0)$
2	1	$(2, 1)$
6	2	$(6, 2)$

Copyright © 2014 Pearson Education, Inc. Publishing as Addison-Wesley.

11. Graph $5x-6y=12$.

We solve for y first.

$5x-6y=12$

$-6y=12-5x$ subtract $5x$ from both sides

$y=\frac{1}{-6}(12-5x)$ divide both sides by -6

$y=-2+\frac{5}{6}x$

$y=\frac{5}{6}x-2$

We choose some x-values and calculate the corresponding y-values to find some ordered pairs that are solutions of the equation. Then we plot the points and connect them with a smooth curve.

x	y	(x,y)
−6	−7	(−6,7)
0	−2	(0,−2)
6	3	(6,3)

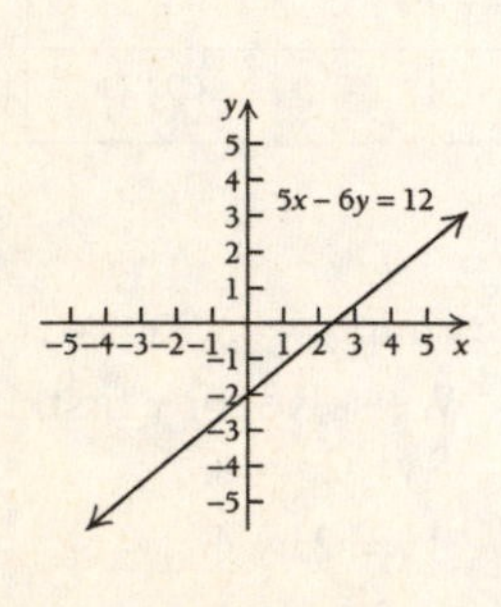

13. Graph $y=x^2-5$.

We choose some x-values and calculate the corresponding y-values to find some ordered pairs that are solutions of the equation. Then we plot the points and connect them with a smooth curve.

x	y	(x,y)
−2	−1	(−2,−1)
−1	−4	(−1,−4)
0	−5	(0,−5)
1	−4	(1,−4)
2	−1	(2,−1)

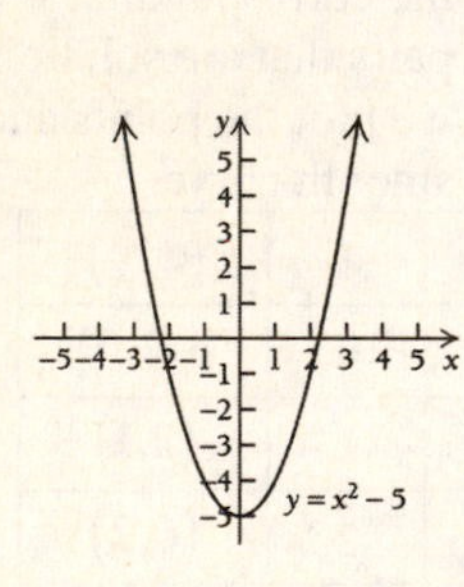

15. Graph $x=2-y^2$.

Since x is expressed in terms of y we first choose values for y and then compute x. Then we plot the points that are found and connect them with a smooth curve.

x	y	(x,y)
−2	−2	(−2,−2)
1	−1	(1,−1)
2	0	(2,0)
−1	1	(−1,1)
−2	2	(−2,2)

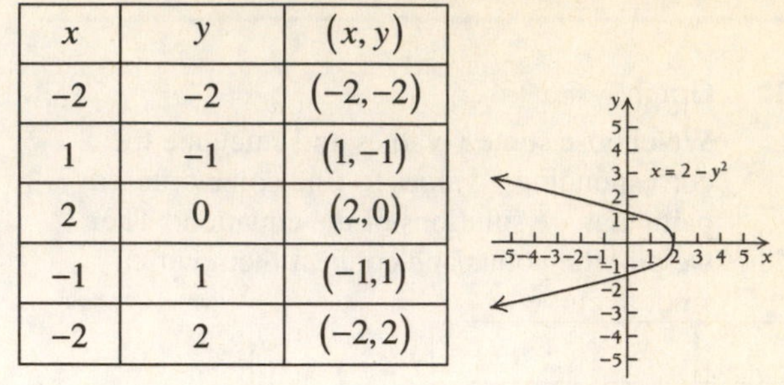

17. Graph $y=|x|$.

We choose some x-values and calculate the corresponding y-values to find some ordered pairs that are solutions of the equation. Then we plot the points and connect them with a smooth curve.

x	y	(x,y)
−2	2	(−2,2)
−1	1	(−1,1)
0	0	(0,0)
1	1	(1,1)
2	2	(2,2)

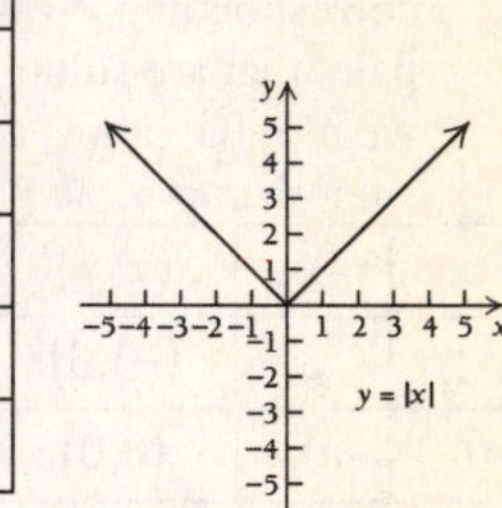

19. Graph $y=7-x^2$.

We choose some x-values and calculate the corresponding y-values to find some ordered pairs that are solutions of the equation. Then we plot the points and connect them with a smooth curve.

x	y	(x,y)
−2	3	(−2,3)
−1	6	(−1,6)
0	7	(0,7)
1	6	(1,6)
2	3	(2,3)

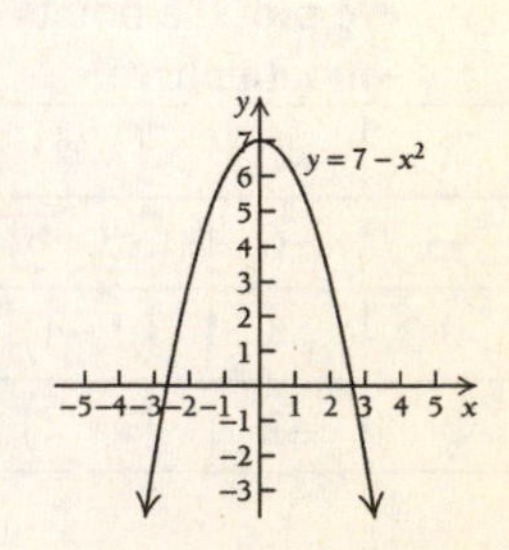

Copyright © 2014 Pearson Education, Inc. Publishing as Addison-Wesley.

21. Graph $y+1=x^3$.

First we solve for y.

$y+1=x^3$

$y=x^3-1$

Next, we choose some x-values and calculate the corresponding y-values to find some ordered pairs that are solutions of the equation. Then we plot the points and connect them with a smooth curve.

x	y	(x,y)
-2	-9	$(-2,-9)$
-1	-2	$(-1,-2)$
0	-1	$(0,-1)$
1	0	$(1,0)$
2	7	$(2,7)$

$y+1=x^3$

23. $R=-0.00582x+15.3476$

We substitute 1954 in for x to get

$R=-0.00582(1954)+15.3476$

$=3.97532$

According to this model, the world record for the mile in 1954 is approximately 3.97532 minutes. To convert this to traditional minutes-seconds we multiply the decimal part by 60 seconds.

$0.97532(60)=58.5192$

Therefore the world record for the mile in 1954 was 3:58.5.

Likewise, we substitute 2008 in for x to get

$R=-0.00582(2008)+15.3476$

$=3.66104$

According to the model, the world record for the mile in 2008 will be approximately 3.66104 minutes or 3:39.7.

Finally, we substitute 2012 in for x to get

$R=-0.00582(2012)+15.3476$

$=3.63776$

According to the model, the world record for the mile in 2012 will be approximately 3.63776 minutes or 3:38.3.

25. $v(t)=10.9t$

We substitute 2.5 in for t to get

$v(2.5)=10.9(2.5)$

$=27.25$

White was traveling at 27.25 miles per hour when he reentered the half pipe.

27. a) Locate 20 on the horizontal axis and go directly up to the graph. Then move left to the vertical axis and read the value there. We estimate the number of hearing-impaired Americans of age 20 is about 1.8 million.

Follow the same process for 40, 50, and 60 to determine the number of hearing-impaired Americans at each of those ages.

We estimate the number of hearing-impaired Americans of age 40 is about 3.7 million.

We estimate the number of hearing-impaired Americans of age 50 is about 4.4 million.

We estimate the number of hearing-impaired Americans of age 60 is about 4.5 million.

b) Locate 4 on the vertical axis and move horizontally across to the graph. There are two x-values that correspond to the y-value of 4. They are 44 and 70, so there are approximately 4 million Americans age 44 who are hearing-impaired and approximately 4 million Americans age 70 who are hearing-impaired.

c) The highest point on the graph appears to correspond to the x-value of 58. Therefore, age 58 appears to be the age at which the greatest number of Americans are hearing-impaired.

d) ✎

29. a) $A=P(1+i)^t$

$A=100{,}000(1+0.028)^1$

$=100{,}000(1.028)$

$=102{,}800$

At the end of 1 year, the investment is worth \$102,800.

b) $A=P\left(1+\frac{i}{n}\right)^{nt}$

$A=100{,}000\left(1+\frac{0.028}{2}\right)^{2\cdot 1}$

$=100{,}000(1+0.014)^2$

$=100{,}000(1.014)^2$

$A=100{,}000(1.028196)$

$=102{,}819.60$

At the end of 1 year, the investment is worth \$102,819.60.

Copyright © 2014 Pearson Education, Inc. Publishing as Addison-Wesley.

c) $A = P\left(1+\frac{i}{n}\right)^{nt}$

$$A = 100{,}000\left(1+\frac{0.028}{4}\right)^{4\cdot 1}$$
$$= 100{,}000(1+0.07)^4$$
$$= 100{,}000(1.07)^4$$
$$= 100{,}000(1.0282953744)$$
$$= 102{,}829.537$$
$$\approx 102{,}829.54$$

At the end of 1 year, the investment is worth $102,829.54.

d) $A = P\left(1+\frac{i}{n}\right)^{nt}$

$$A = 100{,}000\left(1+\frac{0.028}{365}\right)^{365\cdot 1}$$
$$= 100{,}000(1+0.00076712329)^{365}$$
$$= 100{,}000(1.00076712329)^{365}$$
$$= 100{,}000(1.02839458002)$$
$$= 102{,}839.458002$$
$$\approx 102{,}839.46$$

At the end of 1 year, the investment is worth $102,839.46.

e) There are $24 \cdot 365 = 8760$ hours in one year.

$$A = P\left(1+\frac{i}{n}\right)^{nt}$$
$$A = 100{,}000\left(1+\frac{0.028}{8760}\right)^{8760\cdot 1}$$
$$= 100{,}000(1+0.00000196347)^{8760}$$
$$= 100{,}000(1.00000196347)^{8760}$$
$$= 100{,}000(1.02839563811)$$
$$= 102{,}839.563811$$
$$\approx 102{,}839.56$$

At the end of 1 year, the investment is worth $102,839.56.

31. a) $A = P(1+i)^t$

$$A = 30{,}000(1+0.04)^1$$
$$= 30{,}000(1.04)^1$$
$$= 31{,}200.00$$

At the end of 1 year, the investment is worth $30,200.00.

b) $A = P\left(1+\frac{i}{n}\right)^{nt}$

$$A = 30{,}000\left(1+\frac{0.04}{2}\right)^{2\cdot 1}$$
$$= 30{,}000(1.02)^2$$
$$= 30{,}000(1.0404)$$
$$= 31{,}212.00$$

At the end of 1 year, the investment is worth $30,212.00.

c) $A = P\left(1+\frac{i}{n}\right)^{nt}$

$$A = 30{,}000\left(1+\frac{0.04}{4}\right)^{4\cdot 1}$$
$$= 30{,}000(1.01)^4$$
$$= 30{,}000(1.04060401)$$
$$= 31{,}218.1203$$
$$\approx 31{,}218.12$$

At the end of 1 year, the investment is worth $30,218.12.

d) $A = P\left(1+\frac{i}{n}\right)^{nt}$

$$A = 30{,}000\left(1+\frac{0.04}{365}\right)^{365\cdot 1}$$
$$= 30{,}000(1.000109589)^{365}$$
$$= 30{,}000(1.04080847752)$$
$$= 31{,}224.2543257$$
$$\approx 31{,}224.25$$

At the end of 1 year, the investment is worth $30,224.25.

e) There are $24 \cdot 365 = 8760$ hours in one year.

$$A = P\left(1+\frac{i}{n}\right)^{nt}$$
$$A = 30{,}000\left(1+\frac{0.04}{8760}\right)^{8760\cdot 1}$$
$$= 30{,}000(1.000004566210046)^{8760}$$
$$= 30{,}000(1.04081067873)$$
$$= 31{,}224.3203618$$
$$\approx 31{,}224.32$$

At the end of 1 year, the investment is worth $30,224.32.

Copyright © 2014 Pearson Education, Inc. Publishing as Addison-Wesley.

33. Using the formula:

$$M = P\left[\frac{\frac{i}{12}\left(1+\frac{i}{12}\right)^n}{\left(1+\frac{i}{12}\right)^n - 1}\right]$$

We substitute 18,000 for P, 0.064 $(6.4\% = 0.064)$ for i, and 36 $(3\cdot 12 = 36)$ for n. Then we use a calculator to perform the computation.

$$M = 18,000\left[\frac{\frac{0.064}{12}\left(1+\frac{0.064}{12}\right)^{36}}{\left(1+\frac{0.064}{12}\right)^{36} - 1}\right]$$

$$\approx 550.86$$

The monthly payment on the loan will be approximately \$550.86.

35. $W = P\left[\frac{(1+i)^n - 1}{i}\right]$

We substitute 3000 for P, 0.0657 $(6.57\% = 0.0657)$ for i, and 18 for n.

$$W = 3000\left[\frac{(1+0.0657)^{18} - 1}{0.0657}\right]$$

$$\approx 97,881.97$$

Rounded to the nearest cent, the annuity will be worth \$97,881.97 after 18 years.

37. a) Locate 250,000 on the vertical axis and then think of horizontal lines extending across the graph from this point. The years for which the graph lies above this line are the years for which the deer population was at or above 250,000. Those time periods are 1996 – 2000, and then a brief time between 2001 and 2002..

b) Locate 200,000 on the vertical axis and then think of a horizontal line extending across the graph from this point. The years for which the graph touches this line are the years for which the population was at 200,000. Those time periods are 1987 and 1990.

c) Locate the highest point on the graph and extend a line vertically to the horizontal axis. The year which the deer population was the highest was 1999.

d) Locate the lowest point on the graph and extend a line vertically to the horizontal axis. In this case there are two points that are exactly at 200,000. The years when the deer population was the lowest are 1987 and 1990.

39. a) Using the formula $W = P\left[\frac{(1+i)^n - 1}{i}\right]$ we substitute 1200 for P, 0.08 $(8\% = 0.08)$ for i and 35 for n.

$$W = 1200\left[\frac{(1+0.08)^{35} - 1}{0.08}\right]$$

$$\approx 206,780.16$$

Sally will have approximately \$206,780.16 in her account when she retires.

b) Sally invested \$1200 per year for 35 years. Therefore, the total amount of her original payments is: $\$1200 \cdot 35 = \$42,000$. Since the total amount in the account was \$206,780, the interest earned over the 35 years is:

$\$206,780.16 - \$42,000 = \$164,780.16$

Therefore, \$42,000 was the total amount of Sally's payments and \$164,780.16 was the total amount of her interest.

41. Graph $y = x - 150$

We use the following window.

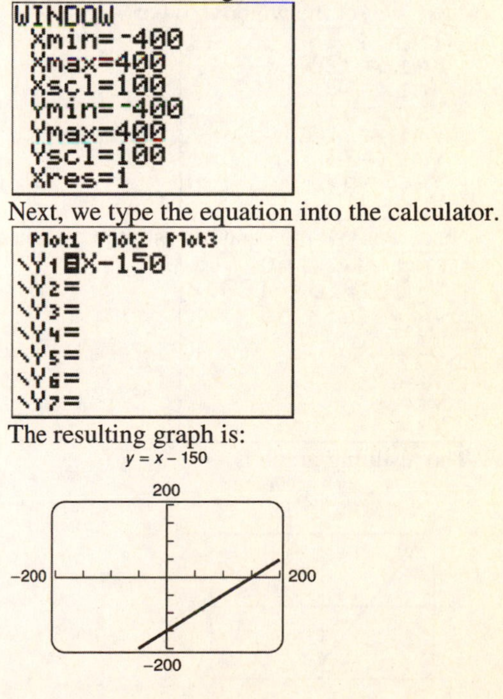

Copyright © 2014 Pearson Education, Inc. Publishing as Addison-Wesley.

43. Graph $y = x^3 + 2x^2 - 4x - 13$

We use the following window:

```
WINDOW
 Xmin=-10
 Xmax=10
 Xscl=1
 Ymin=-20
 Ymax=20
 Yscl=5
 Xres=1
```

Next, we type the equation in to the calculator.

```
Plot1 Plot2 Plot3
\Y1=X^3+2X^2-4X-
13
\Y2=
\Y3=
\Y4=
\Y5=
\Y6=
```

The resulting graph is:

$y = x^3 + 2x^2 - 4x - 13$

15

−10 10

−15

45. Graph $9.6x + 4.2y = -100$.

First, we solve for y.

$$9.6x + 4.2y = -100$$

$$4.2y = -100 - 9.6x \quad \text{subtract } 9.6x \text{ from both sides}$$

$$y = \frac{-9.6x - 100}{4.2}$$

Next, we set the window to be:

```
WINDOW
 Xmin=-20
 Xmax=10
 Xscl=5
 Ymin=-40
 Ymax=10
 Yscl=5
 Xres=1
```

Next, we type the equation into the calculator:

```
Plot1 Plot2 Plot3
\Y1=(-9.6X-100)/
4.2
\Y2=
\Y3=
\Y4=
\Y5=
\Y6=
```

The resulting graph is:

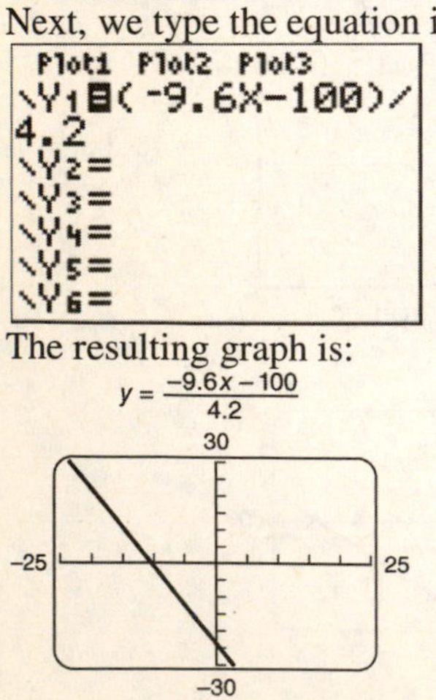

47. Graph $x = 4 + y^2$.

First we solve for y.

$$x = 4 + y^2$$

$$x - 4 = y^2 \quad \text{subtracting 4 from both sides}$$

$$\pm\sqrt{x-4} = y \quad \text{taking the square root of both sides}$$

Next, we set the window to the standard window:

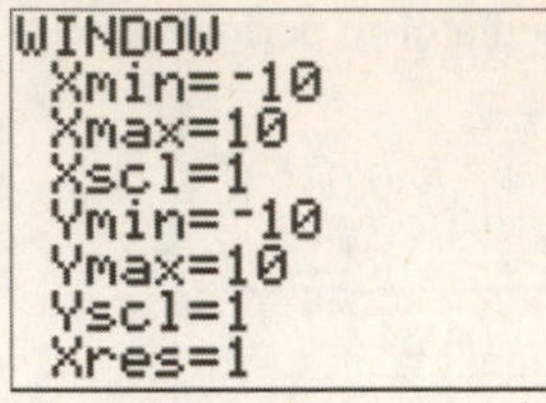

```
WINDOW
 Xmin=-10
 Xmax=10
 Xscl=1
 Ymin=-10
 Ymax=10
 Yscl=1
 Xres=1
```

It is important to remember that we must graph both the positive root and the negative root. We type both equations into the calculator:

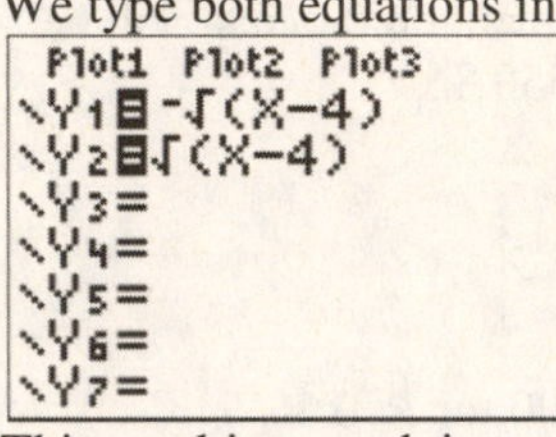

```
Plot1 Plot2 Plot3
\Y1=-√(X-4)
\Y2=√(X-4)
\Y3=
\Y4=
\Y5=
\Y6=
\Y7=
```

This resulting graph is:

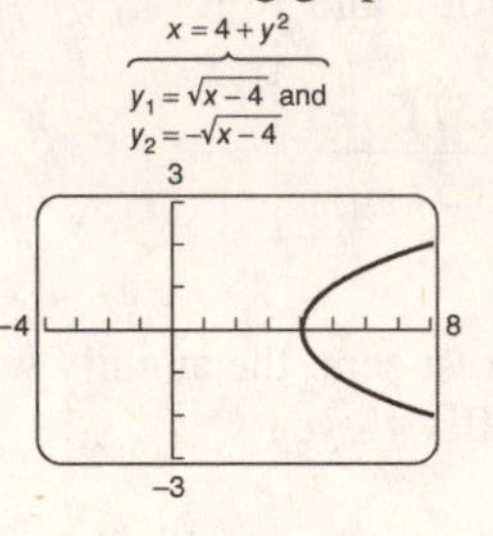

Copyright © 2014 Pearson Education, Inc. Publishing as Addison-Wesley.

Exercise Set R.2

1. The correspondence is a function because each member of the domain corresponds to only one member of the range.

3. The correspondence is a function because each member of the domain corresponds to only one member of the range.

5. The correspondence is a function because each member of the domain corresponds to only one member of the range, even though two members of the domain, Quarter Pounder with Cheese ® and Big N' Tasty with cheese ® correspond to $3.20.

7. The correspondence is a function because each iPod has exactly one amount of memory.

9. The correspondence is a function because each iPod has exactly one number of songs at any given time.

11. The correspondence is a function because any number squared and then increased by 8, corresponds to exactly one number greater than or equal to 8.

13. The correspondence is a function because every female has exactly one biological mother.

15. This correspondence is *not* a function, because it is reasonable to assume at least one avenue is intersected by more than one cross street.

17. The correspondence is a function because each shape has exactly one area.

19. a) $f(x)=4x-3$

$$f(5.1)=4(5.1)-3=17.4$$
$$f(5.01)=4(5.01)-3=17.04$$
$$f(5.001)=4(5.001)-3=17.004$$
$$f(5)=4(5)-3=17$$

x	5.1	5.01	5.001	5
$f(x)$	17.4	17.04	17.004	17

b) $f(x)=4x-3$

$$f(4)=4(4)-3=13$$
$$f(3)=4(3)-3=9$$
$$f(-2)=4(-2)-3=-11$$
$$f(k)=4(k)-3=4k-3$$
$$f(1+t)=4(1+t)-3=4+4t-3=4t+1$$
$$f(x+h)=4(x+h)-3=4x+4h-3$$

21. $g(x)=x^2-3$

$$g(-1)=(-1)^2-3=1-3=-2$$
$$g(0)=(0)^2-3=0-3=-3$$
$$g(1)=(1)^2-3=1-3=-2$$
$$g(5)=(5)^2-3=25-3=22$$
$$g(u)=(u)^2-3=u^2-3$$
$$g(a+h)=(a+h)^2-3=a^2+2ah+h^2-3$$
$$\frac{g(a+h)-g(a)}{h}=\frac{(a+h)^2-3-\left[(a)^2-3\right]}{h}$$
$$=\frac{a^2+2ah+h^2-3-\left[a^2-3\right]}{h}$$
$$=\frac{2ah+h^2}{h}$$
$$=\frac{h(2a+h)}{h}$$
$$=2a+h$$

23. $f(x)=\dfrac{1}{(x+3)^2}$

a) $f(4)=\dfrac{1}{((4)+3)^2}=\dfrac{1}{(7)^2}=\dfrac{1}{49}$

$f(-3)=\dfrac{1}{((-3)+3)^2}=\dfrac{1}{(0)^2}$, Output is undefined.

$f(0)=\dfrac{1}{((0)+3)^2}=\dfrac{1}{(3)^2}=\dfrac{1}{9}$

$f(a)=\dfrac{1}{((a)+3)^2}=\dfrac{1}{(a+3)^2}$

The solution is continued on the next page.

Copyright © 2014 Pearson Education, Inc. Publishing as Addison-Wesley.

$$f(t+4)=\frac{1}{((t+4)+3)^2}=\frac{1}{(t+7)^2}$$

$$f(x+h)=\frac{1}{((x+h)+3)^2}=\frac{1}{(x+h+3)^2}$$

$$\frac{f(x+h)-f(x)}{h}$$

$$=\frac{\frac{1}{(x+h+3)^2}-\frac{1}{(x+3)^2}}{h}$$

$$=\frac{\frac{(x+3)^2}{(x+h+3)^2(x+3)^2}-\frac{(x+h+3)^2}{(x+h+3)^2(x+3)^2}}{h}$$

$$=\frac{x^2+6x+9-\left(x^2+2hx+6x+h^2+6h+9\right)}{h(x+h+3)^2(x+3)^2}$$

$$=\frac{-2hx-h^2-6h}{h(x+h+3)^2(x+3)^2}$$

$$=\frac{h(-2x-h-6)}{h(x+h+3)^2(x+3)^2}$$

$$=\frac{-2x-h-6}{(x+h+3)^2(x+3)^2},\quad h\neq 0$$

b) The function squares the input, then it adds six times the input, then it adds 9 and then it takes the reciprocal of the result.

25. Graph $f(x)=2x-5$.

First, we choose some values for x and compute the values for $f(x)$, in order to form the ordered pairs that we will plot on the graph.

$f(-1)=2(-1)-5=-7$

$f(0)=2(0)-5=-5$

$f(1)=2(1)-5=-3$

$f(2)=2(2)-5=-1$

x	$f(x)$	$(x,f(x))$
−1	−7	(−1,−7)
0	−5	(0,−5)
1	−3	(1,−3)
2	−1	(2,−1)

Next we plot the input – output pairs from the table and, in this case, draw the line to complete the graph.

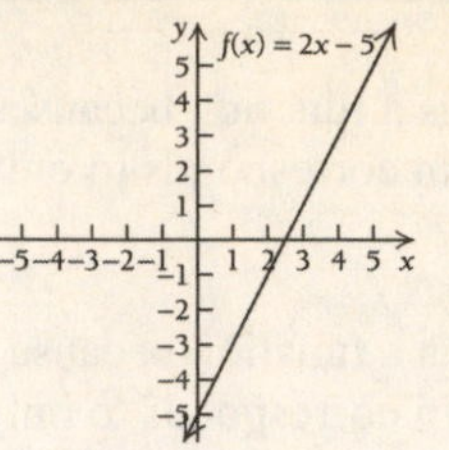

27. Graph $g(x)=-4x$.

First, we choose some values for x and compute the values for $g(x)$, in order to form the ordered pairs that we will plot on the graph.

$g(-1)=-4(-1)=4$

$g(0)=-4(0)=0$

$g(1)=-4(1)=-4$.

x	$g(x)$	$(x,g(x))$
−1	4	(−1,4)
0	0	(0,0)
1	−4	(0,−4)

Next we plot the input – output pairs from the table and, in this case, draw the line to complete the graph.

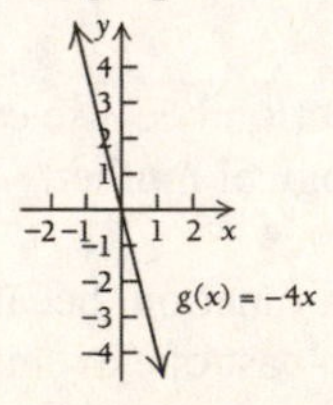

29. Graph $f(x)=x^2-2$.

First, we choose some values for x and compute the values for $f(x)$, in order to form the ordered pairs that we will plot on the graph.

$f(-2)=(-2)^2-2=2$

$f(-1)=(-1)^2-2=-1$

$f(0)=(0)^2-2=-2$ T

$f(1)=(1)^2-2=-1$

$f(2)=(2)^2-2=2$

The input – output table and the graph are shown at the top of the next page.

Copyright © 2014 Pearson Education, Inc. Publishing as Addison-Wesley.

We organized the function values from the previous page into an input – output table.

x	$f(x)$	$(x, f(x))$
-2	2	$(-2,2)$
-1	-1	$(-1,-1)$
0	-2	$(0,-2)$
1	-1	$(1,-1)$
2	2	$(2,2)$

Next we plot the input – output pairs from the table and, in this case, draw the curve to complete the graph.

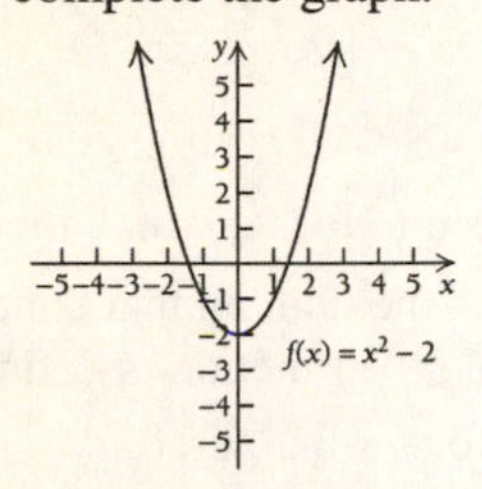

31. Graph $f(x) = 6 - x^2$.
First, we choose some values for x and compute the values for $f(x)$, in order to form the ordered pairs that we will plot on the graph.

$f(-2) = 6 - (-2)^2 = 2$

$f(-1) = 6 - (-1)^2 = 5$

$f(0) = 6 - (0)^2 = 6$

$f(1) = 6 - (1)^2 = 5$

$f(2) = 6 - (2)^2 = 2$

x	$f(x)$	$(x, f(x))$
-2	2	$(-2,2)$
-1	5	$(-1,5)$
0	6	$(0,6)$
1	5	$(1,5)$
2	2	$(2,2)$

Next we plot the input – output pairs from the table and, in this case, draw the curve to complete the graph.

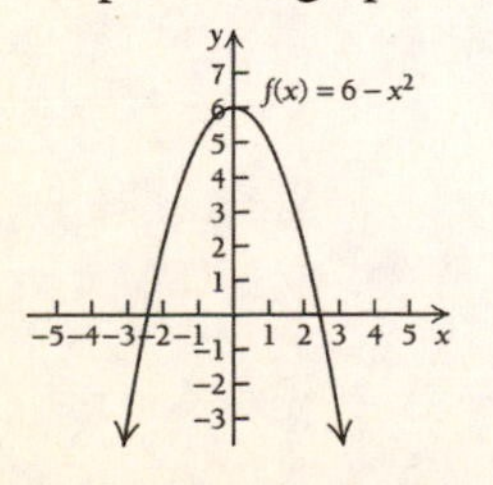

33. Graph $g(x) = x^3$.
First, we choose some values for x and compute the values for $g(x)$, in order to form the ordered pairs that we will plot on the graph.

$g(-2) = (-2)^3 = -8$

$g(-1) = (-1)^3 = -1$

$g(0) = (0)^3 = 0$

$g(1) = (1)^3 = 1$

$g(2) = (2)^3 = 8$

x	$f(x)$	$(x, f(x))$
-2	-8	$(-2,-8)$
-1	-1	$(-1,-1)$
0	0	$(0,0)$
1	1	$(1,1)$
2	8	$(2,8)$

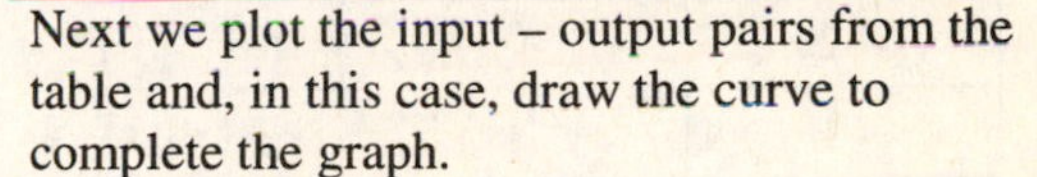
Next we plot the input – output pairs from the table and, in this case, draw the curve to complete the graph.

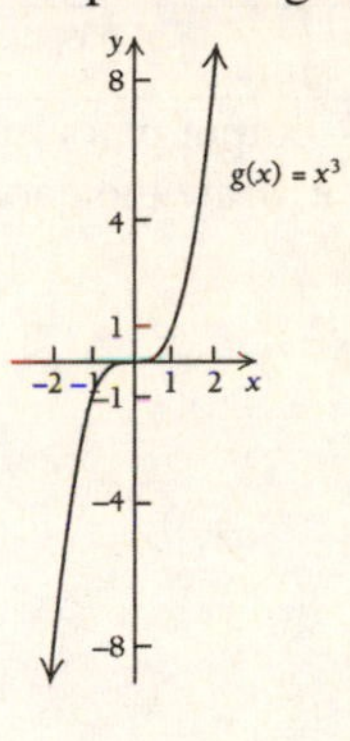

35. The graph is a function, it is impossible to draw a vertical line that intersects the graph more than once.

37. The graph is a function, it is impossible to draw a vertical line that intersects the graph more than once.

39. The graph is not that of a function. A vertical line can intersect the graph more than once.

41. The graph is not that of a function. A vertical line can intersect the graph more than once.

43. The graph is a function, it is impossible to draw a vertical line that intersects the graph more than once.

Copyright © 2014 Pearson Education, Inc. Publishing as Addison-Wesley.

45. The graph is a function, it is impossible to draw a vertical line that intersects the graph more than once.

47. Graph $x = y^2 - 2$.

a) First, we choose some values for y (since x is expressed in terms of y) and compute the values for x, in order to form the ordered pairs that we will plot on the graph.

For $y = -2; x = (-2)^2 - 2 = 2$

For $y = -1; x = (-1)^2 - 2 = -1$

For $y = 0; x = (0)^2 - 2 = -2$

For $y = 1; x = (1)^2 - 2 = -1$

For $y = 2; x = (2)^2 - 2 = 2$

x	y	(x, y)
2	−2	$(2,-2)$
−1	−1	$(-1,-1)$
−2	0	$(-2,0)$
−1	1	$(-1,1)$
2	2	$(2,2)$

Next we plot the input – output pairs from the table and, in this case, draw the curve to complete the graph.

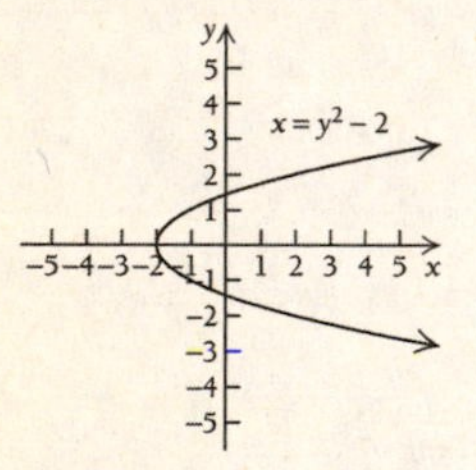

b) The graph is not that of a function. A vertical line can intersect the graph more than once.

49. $f(x) = x^2 - 3x$

$$\frac{f(x+h) - f(x)}{h}$$

$$= \frac{(x+h)^2 - 3(x+h) - \left[x^2 - 3x\right]}{h}$$

$$= \frac{x^2 + 2xh + h^2 - 3x - 3h - \left[x^2 - 3x\right]}{h}$$

$$= \frac{2xh + h^2 - 3h}{h} \quad \text{combining like terms}$$

$$= \frac{h(2x + h - 3)}{h} \quad \text{Factoring}$$

$$= 2x + h - 3, \quad h \neq 0$$

51. To find $f(-1)$ we need to locate which piece defines the function on the domain that contains $x = -1$. When $x = -1$, the function is defined by $f(x) = -2x + 1;$ for $x < 0$; therefore,

$f(-1) = -2(-1) + 1 = 2 + 1 = 3$.

To find $f(1)$ we need to locate which piece defines the function on the domain that contains $x = 1$. When $x = 1$, the function is defined by $f(x) = x^2 - 3;$ for $0 < x < 4$; therefore,

$f(1) = (1)^2 - 3 = 1 - 3 = -2$.

53. To find $f(0)$ we need to locate which piece defines the function on the domain that contains $x = 0$.

When $x = 0$, the function is defined by $f(x) = 17;$ for $x = 0$; therefore,

$f(0) = 17$.

To find $f(10)$ we need to locate which piece defines the function on the domain that contains $x = 10$. When $x = 10$, the function is defined by $f(x) = \frac{1}{2}x + 1;$ for $x \geq 4$; therefore,

$f(10) = \frac{1}{2}(10) + 1 = 5 + 1 = 6$.

Copyright © 2014 Pearson Education, Inc. Publishing as Addison-Wesley.

55. Graph $f(x)=\begin{cases}1 & \text{for } x<0\\ -1 & \text{for } x\geq 0\end{cases}$.

First, we graph $f(x)=1$ for inputs less than 0. We note for any x-value less than 0, the graph is the horizontal line $y=1$. Note that for $f(x)=1$

$f(-2)=1$

$f(-1)=1$

The open circle indicates that $(0,1)$ is not part of the graph.

Next, we graph $f(x)=-1$ for inputs greater than or equal to 0. We note for any x-value less than 0, the graph is the horizontal line $y=-1$.

Note that for $f(x)=-1$.

$f(0)=-1$

$f(1)=-1$

$f(2)=-1$

The solid dot indicates that $(0,-1)$ is part of the graph.

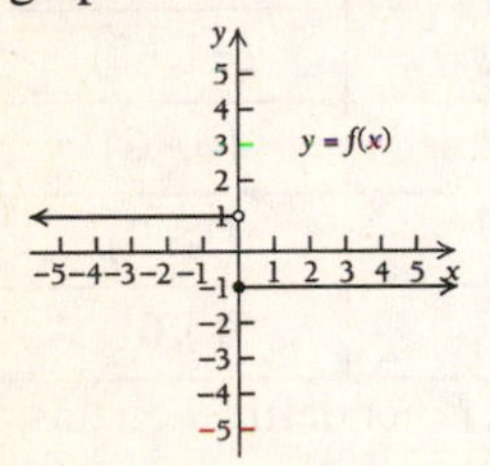

57. Graph $f(x)=\begin{cases}6, & \text{for } x=-2\\ x^2, & \text{for } x\neq -2\end{cases}$.

First, we graph $f(x)=6$ for $x=-2$.

This graph consists of only one point, $(-2,6)$.

The solid dot indicates that $(-2,6)$ is part of the graph.

Next, we graph $f(x)=x^2$ for inputs $x\neq -2$.

Note that for $f(x)=x^2$

$f(-3)=(-3)^2=9$

$f(-1)=(-1)^2=1$

$f(0)=(0)^2=0$

$f(1)=(1)^2=1$

$f(2)=(2)^2=4$

Creating the input – output table, we have:

x	$f(x)$	$(x, f(x))$
−3	9	(−3,9)
−1	1	(−1,1)
0	0	(0,0)
1	1	(1,1)
2	4	(2,4)

Since the input $x=-2$ is not defined on this part of the graph, the point $(-2,4)$ is not part of the graph. The open circle indicates that $(-2,4)$ is not part of the graph.

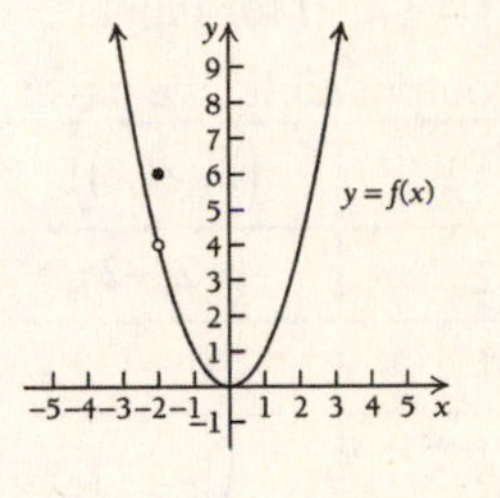

59. Graph $g(x)=\begin{cases}-x, & \text{for } x<0\\ 4, & \text{for } x=0\\ x+2, & \text{for } x>0\end{cases}$.

First, we graph $g(x)=-x$ for inputs $x<0$.

Creating the input – output table, we have:

x	$g(x)$	$(x, g(x))$
−3	3	(−3,3)
−2	2	(−2,2)
−1	1	(−1,1)

The open circle indicates that $(0,0)$ is not part of the graph.

Next, we graph $g(x)=4$ for $x=0$. This part of the graph consists of a single point. The solid dot indicates that $(0,4)$ is part of the graph.

Next, we graph $g(x)=x+2$ for inputs $x>0$.

Creating the input – output table, we have:

x	$g(x)$	$(x, g(x))$
1	3	(1,3)
2	4	(2,4)
3	5	(3,5)

The open circle indicates that $(0,2)$ is not part of the graph. The graph is shown on the next page.

Copyright © 2014 Pearson Education, Inc. Publishing as Addison-Wesley.

Using the information from the previous page, we sketch a graph of the function.

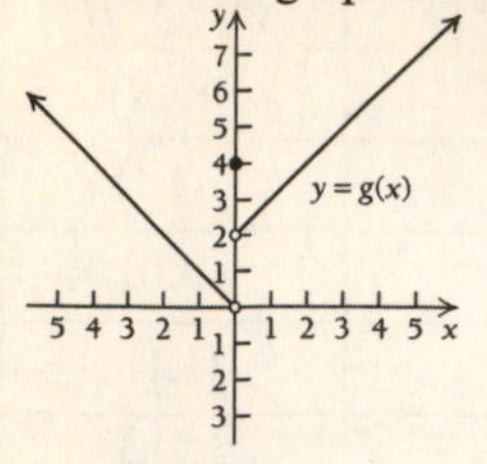

61. Graph $g(x)=\begin{cases}\frac{1}{2}x-1, & \text{for } x<2\\ -4, & \text{for } x=2\\ x-3, & \text{for } x>2\end{cases}$.

First, we graph $g(x)=\frac{1}{2}x-1$ for inputs $x<2$.

Creating the input – output table, we have:

x	$g(x)$	$(x, g(x))$
-2	-2	$(-2,-2)$
0	-1	$(0,-1)$
1	$-\frac{1}{2}$	$(1,-\frac{1}{2})$

The open circle indicates that $(2,0)$ is not part of the graph.

Next, we graph $g(x)=-4$ for $x=2$. This part of the graph consists of a single point. The solid dot indicates that $(2,-4)$ is part of the graph.

Next, we graph $g(x)=x-3$ for inputs $x>2$.

Creating the input – output table, we have:

x	$g(x)$	$(x, g(x))$
3	0	$(3,0)$
4	1	$(4,1)$
5	2	$(5,2)$

The open circle indicates that $(2,-1)$ is not part of the graph.

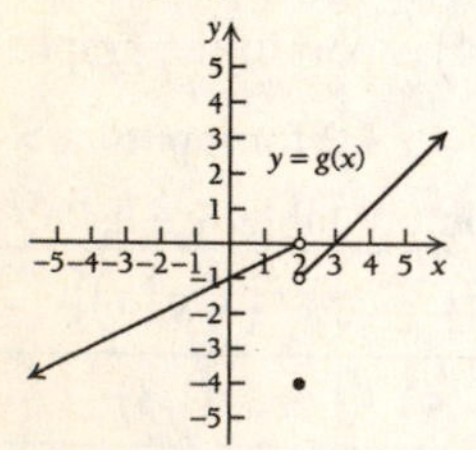

63. Graph $f(x)=\begin{cases}-7, & \text{for } x=2\\ x^2-3, & \text{for } x\neq 2\end{cases}$.

First, we graph $f(x)=-7$ for $x=2$.

This graph consists of only one point, $(2,-7)$.

The solid dot indicates that $(2,-7)$ is part of the graph.

Next, we graph $f(x)=x^2-3$ for inputs $x\neq-2$.

Note that for $f(x)=x^2-3$

$f(-3)=(-3)^2-3=6$

$f(-1)=(-1)^2-3=-2$

$f(0)=(0)^2-3=-3$

$f(1)=(1)^2-3=-2$

$f(3)=(3)^2-3=6$

Creating the input – output table, we have:

x	$f(x)$	$(x, f(x))$
-3	6	$(-3,6)$
-1	-2	$(-1,-2)$
0	-3	$(0,-3)$
1	-2	$(1,-2)$
3	6	$(3,6)$

Since the input $x=2$ is not defined on this part of the graph, the point $(2,1)$ is not part of the graph. The open circle indicates that $(2,1)$ is not part of the graph.

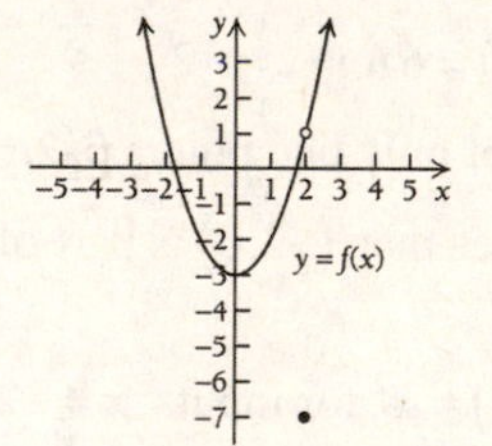

Copyright © 2014 Pearson Education, Inc. Publishing as Addison-Wesley.

65. $A(t)=P\left(1+\frac{0.06}{4}\right)^{4t}$

We substitute 500 in for P and 2 in for t:

$$A(t)=500\left(1+\frac{0.06}{4}\right)^{4\cdot 2}$$
$$=500(1.015)^8$$
$$=500(1.126492587)$$
$$=563.2462933$$
$$\approx 563.25$$

The investment will be worth approximately \$563.25 after 2 years.

67. $s=\sqrt{\frac{hw}{3600}}$

a) We substitute 170 for h and 70 for w.

$$s=\sqrt{\frac{(170)(70)}{3600}}\approx 1.818$$

The patient's approximate surface area is $1.818\,m^2$

b) We substitute 170 for h and 100 for w.

$$s=\sqrt{\frac{(170)(100)}{3600}}\approx 2.173$$

The patient's approximate surface area is $2.173\,m^2$

c) We substitute 170 for h and 50 for w.

$$s=\sqrt{\frac{(170)(50)}{3600}}\approx 1.537$$

The patient's approximate surface area is $1.537\,m^2$

69. a) Yes, the table represents a function. Each event is assigned exactly one scale of impact number.

b) The inputs are the events; the outputs are the scale of impact numbers.

71. First we solve the equation for y.

$2y^2+3x=4x+5$

$2y^2=x+5$ subtract $3x$ from both sides

$y^2=\frac{x+5}{2}$ divide both sides by 2

$y=\pm\sqrt{\frac{x+5}{2}}$ take the square root of both sides

We sketch a graph of the equation.

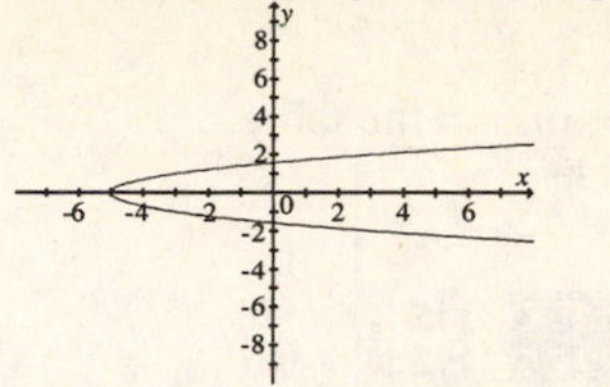

We can see that a vertical line will intersect the graph more than once; therefore, this is not a function.

73. First, we solve the equation for y.

$$\left(3y^{3/2}\right)^2=72x$$
$$9y^3=72x$$
$$y^3=8x$$
$$y=\sqrt[3]{8x}$$
$$y=2\sqrt[3]{x}\qquad [y\geq 0]$$

Note: since y must be non-negative to satisfy the original equation, we only graph the points for which y is non-negative.

Next, we sketch a graph of the equation:

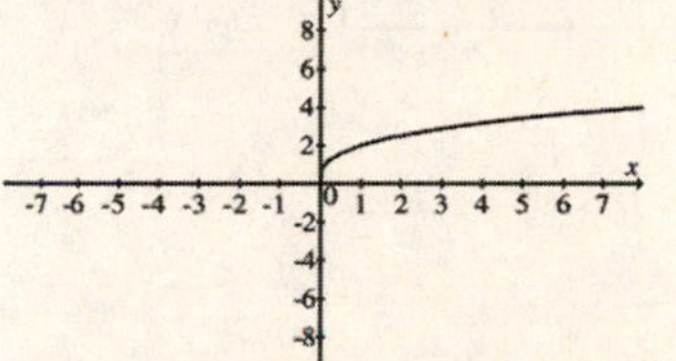

No vertical line meets the graph more than once. Thus, the equation represents a function.

75. 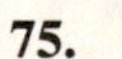

Copyright © 2014 Pearson Education, Inc. Publishing as Addison-Wesley.

77. $f(x)=\dfrac{3}{x^2-4}$

We begin by setting up the table:

```
TABLE SETUP
 TblStart=-3
 ΔTbl=1
Indpnt: Auto Ask
Depend: Auto Ask
```

Next, we will type in the equation into the graphing editor.

```
 Plot1 Plot2 Plot3
\Y1=3/(X^2-4)
\Y2=
\Y3=
\Y4=
\Y5=
\Y6=
\Y7=
```

Now, we are able to look at the table:

X	Y1
-3	.6
-2	ERROR
-1	-1
0	-.75
1	-1
2	ERROR
3	.6

X=-3

79. Each graph is shown below.

In order to graph $f(x)=x^3+2x^2-4x-13$, we use the window:

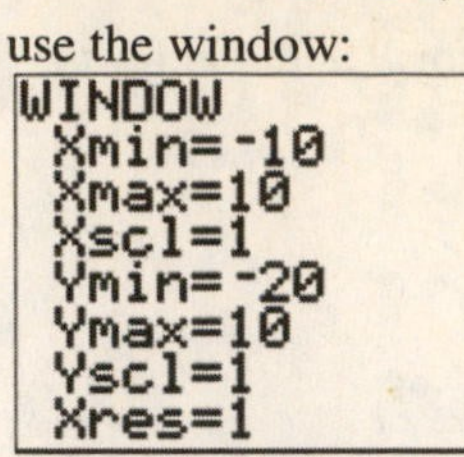

After entering the function into the graphing editor, we get:

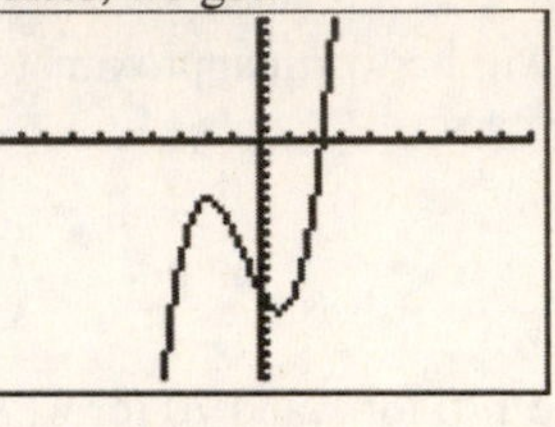

In order to graph $f(x)=\dfrac{3}{x^2-4}$, we use the standard window:

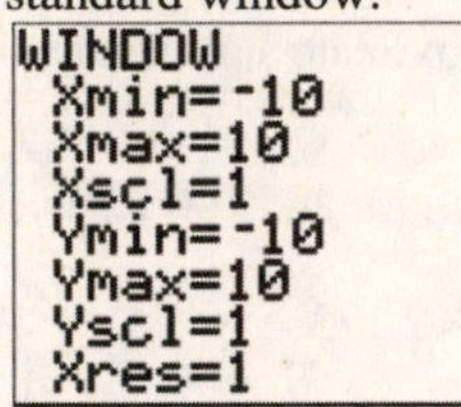

After entering the function into the graphing editor, we get:

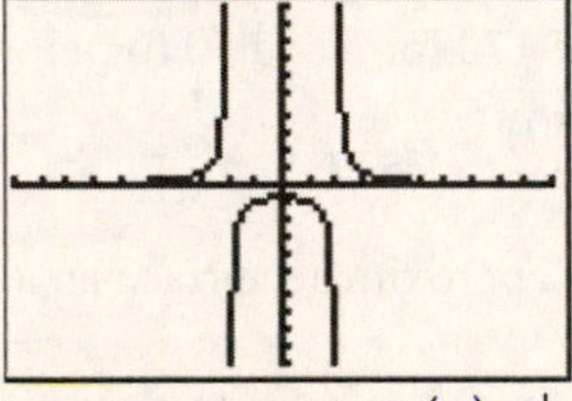

In order to graph $f(x)=|x-2|+|x+1|-5$, we use the standard window.

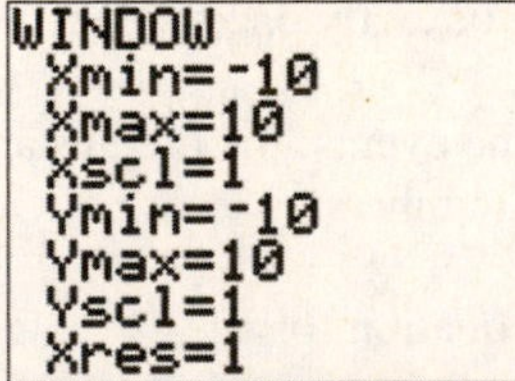

After entering the function into the graphing editor, we get:

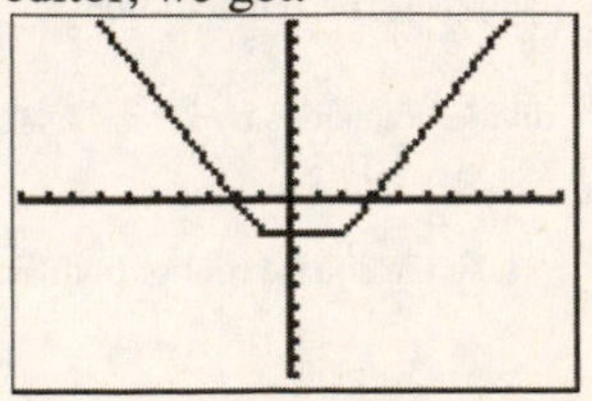

Copyright © 2014 Pearson Education, Inc. Publishing as Addison-Wesley.

Exercise Set R.3

1. $[-2,4]$

3. $(0,5)$

5. $[-9,-4)$

7. $[x,x+h]$

9. (p,∞)

11. $[-2,2]$

-5 -4 -3 -2 -1 0 1 2 3 4 5

13. $[-4,-1)$

-5 -4 -3 -2 -1 0 1 2 3 4 5

15. $(-\infty,-2]$

-5 -4 -3 -2 -1 0 1 2 3 4 5

17. $(-2,3]$

-5 -4 -3 -2 -1 0 1 2 3 4 5

19. $(-\infty,12.5)$

-10 -5 0 10 12.5 15

21. a) First, we locate 1 on the horizontal axis and then we look vertically to find the point on the graph for which 1 is the first coordinate. From that point, we look to the vertical axis to find the corresponding y-coordinate, 3. Thus, $f(1)=3$.

 b) The domain is the set of all x-values of the points on the graph. The domain is $\{-3,-1,1,3,5\}$.

 c) First, we locate 2 on the vertical axis and then we look horizontally to find any points on the graph for which 2 is the second coordinate. One such point exists, $(3,2)$. Thus the x-value for which $f(x)=2$ is $x=3$.

 d) The range is the set of all y-values of the points on the graph. The range is $\{-2,0,2,3,4\}$.

23. a) First, we locate 1 on the horizontal axis and then we look vertically to find the point on the graph for which 1 is the first coordinate. From that point, we look to the vertical axis to find the corresponding y-coordinate, 4. Thus, $f(1)=4$.

 b) The domain is the set of all x-values of the points on the graph. The domain is $\{-5,-3,1,2,3,4,5\}$.

 c) First, we locate 2 on the vertical axis and then we look horizontally to find any points on the graph for which 2 is the second coordinate. Three such point exists, $(-5,2)$; $(-3,2)$; and $(4,2)$. Thus the x-values for which $f(x)=2$ are $\{-5,-3,4\}$.

 d) The range is the set of all y-values of the points on the graph. The range is $\{-3,2,4,5\}$

25. a) First, we locate 1 on the horizontal axis and then we look vertically to find the point on the graph for which 1 is the first coordinate. From that point, we look to the vertical axis to find the corresponding y-coordinate, -1. Thus, $f(1)=-1$.

 b) The domain is the set of all x-values of the points on the graph. These extend from -2 to 4. Thus, the domain is $\{x\,|\,-2\le x\le 4\}$, or in interval notation $[-2,4]$.

 c) First, we locate 2 on the vertical axis and then we look horizontally to find any points on the graph for which 2 is the second coordinate. One such point exists, $(3,2)$. Thus the x-value for which $f(x)=2$ is $x=3$.

 d) The range is the set of all y-values of the points on the graph. These extend from -3 to 3. Thus, the range is $\{y\,|\,-3\le y\le 3\}$, or, in interval notation $[-3,3]$.

27. a) First, we locate 1 on the horizontal axis and then we look vertically to find the point on the graph for which 1 is the first coordinate. From that point, we look to the vertical axis to find the corresponding y-coordinate, -2. Thus, $f(1)=-2$.

Copyright © 2014 Pearson Education, Inc. Publishing as Addison-Wesley.

b) The domain is the set of all x-values of the points on the graph. These extend from -4 to 2. Thus, the domain is $\{x|-4 \le x \le 2\}$, or, in interval notation $[-4,2]$.

c) First, we locate 2 on the vertical axis and then we look horizontally to find any points on the graph for which 2 is the second coordinate. One such point exists, $(-2,2)$. Thus the x-value for which $f(x)=2$ are $x=-2$.

d) The range is the set of all y-values of the points on the graph. These extend from -3 to 3. Thus, the range is $\{y|-3 \le y \le 3\}$, or in interval notation $[-3,3]$.

29. a) First, we locate 1 on the horizontal axis and then we look vertically to find the point on the graph for which 1 is the first coordinate. From that point, we look to the vertical axis to find the corresponding y-coordinate, 3. Thus, $f(1)=3$.

b) The domain is the set of all x-values of the points on the graph. These extend from -3 to 3. Thus, the domain is $\{x|-3 \le x \le 3\}$, or, in interval notation $[-3,3]$.

c) First, we locate 2 on the vertical axis and then we look horizontally to find any points on the graph for which 2 is the second coordinate. Two such point exists, $(-1.4,2)$ and $(1.4,2)$. Thus the x-values for which $f(x)=2$ are $\{-1.4,1.4\}$.

d) The range is the set of all y-values of the points on the graph. These extend from -5 to 4. Thus, the range is $\{y|-5 \le y \le 4\}$, or in interval notation $[-5,4]$.

31. a) First, we locate 1 on the horizontal axis and then we look vertically to find the point on the graph for which 1 is the first coordinate. From that point, we look to the vertical axis to find the corresponding y-coordinate, 1. Thus, $f(1)=1$.

b) The domain is the set of all x-values of the points on the graph. These extend from -5 to 5. However, the open circle at the point $(5,2)$ indicates that 5 is not in the domain. Thus, the domain is $\{x|-5 \le x < 5\}$, or in interval notation $[-5,5)$.

c) First, we locate 2 on the vertical axis and then we look horizontally to find any points on the graph for which 2 is the second coordinate. We notice all the points with x-values in the set $\{x|3 \le x < 5\}$ Thus the x-values for which $f(x)=2$ are $\{x|3 \le x < 5\}$, or $[3,5)$.

d) The range is the set of all y-values of the points on the graph. The range is $\{-2,-1,0,1,2\}$.

33. $f(x)=\dfrac{6}{2-x}$

Since the function value cannot be calculated when the denominator is equal to 0, we solve the following equation to find those real numbers that must be excluded from the domain of f.

$2-x=0$ setting the denominator equal to 0

$2=x$ adding x to both sides

Thus, 2 is not in the domain of f, while all other real numbers are. The domain of f is $\{x|x \text{ is a real number and } x \ne 2\}$; or, in interval notation, $(-\infty,2)\cup(2,\infty)$.

35. $f(x)=\sqrt{2x}$

Since the function value cannot be calculated when the radicand is negative, the domain is all real numbers for which $2x \ge 0$. We find them by solving the inequality.

$2x \ge 0$ setting the radicand ≥ 0

$x \ge 0$ dividing both sides by 2

The domain of f is $\{x|x \text{ is a real number and } x \ge 0\}$; or, in interval notation, $[0,\infty)$.

37. $f(x)=x^2-2x+3$

We can calculate the function value for all values of x, so the domain is the set of all real numbers $\mathbb{R}$.

Copyright © 2014 Pearson Education, Inc. Publishing as Addison-Wesley.

39. $f(x)=\frac{x-2}{6x-12}$

Since the function value cannot be calculated when the denominator is equal to 0, we solve the following equation to find those real numbers that must be excluded from the domain of *f*.

$6x-12=0$ setting the denominator equal to 0

$6x=12$ adding 12 to both sides

$x=2$ dividing both sides by 6

Thus, 2 is not in the domain of *f*, while all other real numbers are. The domain of *f* is $\{x \mid x \text{ is a real number and } x \neq 2\}$; or, in interval notation, $(-\infty,2)\cup(2,\infty)$.

41. $f(x)=|x-4|$

We can calculate the function value for all values of *x*, so the domain is the set of all real numbers $\mathbb{R}$.

43. $f(x)=\frac{3x-1}{7-2x}$

Since the function value cannot be calculated when the denominator is equal to 0, we solve the following equation to find those real numbers that must be excluded from the domain of *f*.

$7-2x=0$ setting the denominator equal to 0

$7=2x$ adding $2x$ to both sides

$\frac{7}{2}=x$ dividing both sides by 2

Thus, $\frac{7}{2}$ is not in the domain of *f*, while all other real numbers are. The domain of *f* is $\left\{x \mid x \text{ is a real number and } x \neq \frac{7}{2}\right\}$; or, in interval notation, $\left(-\infty,\frac{7}{2}\right)\cup\left(\frac{7}{2},\infty\right)$.

45. $g(x)=\sqrt{4+5x}$

Since the function value cannot be calculated when the radicand is negative, the domain is all real numbers for which $4+5x\geq 0$. We find them by solving the inequality.

$4+5x\geq 0$ setting the radicand ≥ 0

$5x\geq -4$ subtracting 4 from both sides

$x\geq -\frac{4}{5}$ dividing both sides by 5

The domain of *g* is $\left\{x \mid x \text{ is a real number and } x\geq -\frac{4}{5}\right\}$; or, in interval notation, $\left[-\frac{4}{5},\infty\right)$.

47. $g(x)=x^2-2x+1$

We can calculate the function value for all values of *x*, so the domain is the set of all real numbers $\mathbb{R}$.

49. $g(x)=\frac{2x}{x^2-25}$

Since the function value cannot be calculated when the denominator is equal to 0, we solve the following equation to find those real numbers that must be excluded from the domain of *g*.

$x^2-25=0$ setting the denominator equal to 0

$x^2=25$ adding 25 to both sides

$x=\pm\sqrt{25}$ taking the square root or both sides

$x=\pm 5$

Thus, -5 and 5 are not in the domain of *g*, while all other real numbers are. The domain of *g* is $\{x \mid x \text{ is a real number and } x\neq -5,\ x\neq 5\}$; or, in interval notation, $(-\infty,-5)\cup(-5,5)\cup(5,\infty)$.

51. $g(x)=|x|+1$

We can calculate the function value for all values of *x*, so the domain is the set of all real numbers $\mathbb{R}$.

Copyright © 2014 Pearson Education, Inc. Publishing as Addison-Wesley.

53. $g(x)=\dfrac{2x-6}{x^2-6x+5}$

Since the function value cannot be calculated when the denominator is equal to 0, we solve the following equation to find those real numbers that must be excluded from the domain of *f*.

$x^2-6x+5=0$ setting the denominator equal to 0

$(x-5)(x-1)=0$ factoring the quadratic equation

$x-5=0$ or $x-1=0$ Using the principle of zero products

$x=5$ or $x=1$

Thus, 1 and 5 are not in the domain of *g*, while all other real numbers are. The domain of *g* is $\{x \mid x \text{ is a real number and } x\neq 1, x\neq 5\}$; or, in interval notation, $(-\infty,1)\cup(1,5)\cup(5,\infty)$.

55. The graph of *f* lies on or below the *x*-axis when $f(x)\leq 0$, so we scan the graph from left to right looking for the values of *x* for which the graph lies on or below the *x* axis. Those values extend from -1 to 2. So the set of *x*-values for which $f(x)\leq 0$ is $\{x \mid -1\leq x\leq 2\}$, or, in interval notation, $[-1,2]$.

57. a) We use the compound interest formula from Theorem 2 in section R.1 and substitute 5000 for *P*, 2 for *n* and 0.08 (8%) for *i*. Using the information, we derive the function.

$$A=P\left(1+\frac{i}{n}\right)^{nt}$$

$$A=5000\left(1+\frac{0.08}{2}\right)^{2t}$$

$$A=5000(1.04)^{2t}$$

b) The independent variable *t* is the time in years the principal has been invested in the account. It would not make sense to have time be a negative number in this case. Therefore, the domain is the set of all non-negative real numbers. $\{t \mid 0\leq t<\infty\}$.

59. a) The graph extends from $x=0$ to $x=84.7$, so the domain, in interval notation, of the function *N* is $[0,84.7]$.

b) The graph extends from $N(x)=0$ to $N(x)=4.6$ million. Therefore, the range, in interval notation, of the function *N* is $[0, 4,600,000]$.

c) ✎

61. a) The graph extends from $t=0$ to $t=70$, so the domain, in interval notation, of the function *L* is $[0,70]$.

b) The graph extends from $L(t)=8$ to $L(t)=75$, so the range, in interval notation, of the function *L* is $[8,75]$.

63. ✎

65. ✎

67. The range in interval notation for each function is shown below.

Exercise 34: $(-\infty,0)\cup(0,\infty)$

Exercise 36: $[0,\infty)$

Exercise 48: $\mathbb{R}$

Exercise 51: $[1,\infty)$

Exercise 54: $\mathbb{R}$

Copyright © 2014 Pearson Education, Inc. Publishing as Addison-Wesley.

Exercise Set R.4

1. Graph $x = 3$.
The graph consists of all ordered pairs whose first coordinate is 3. This results in a vertical line whose x-intercept is the point $(3,0)$

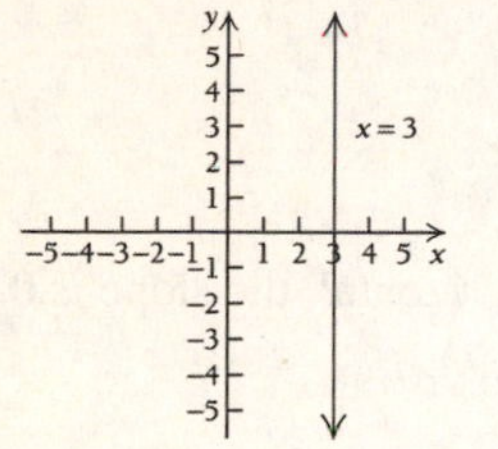

3. Graph $y = -2$.
The graph consists of all ordered pairs whose second coordinate is -2. This results in a horizontal line whose y-intercept is the point $(0,-2)$.

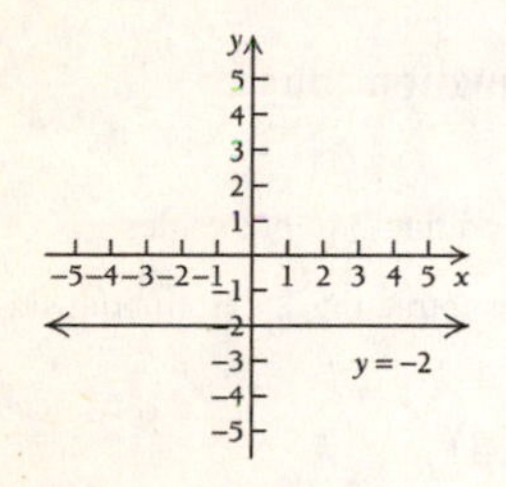

5. Graph $x = -4.5$.
The graph consists of all ordered pairs whose first coordinate is – 4.5. This results in a vertical line whose x-intercept is the point $(-4.5,0)$.

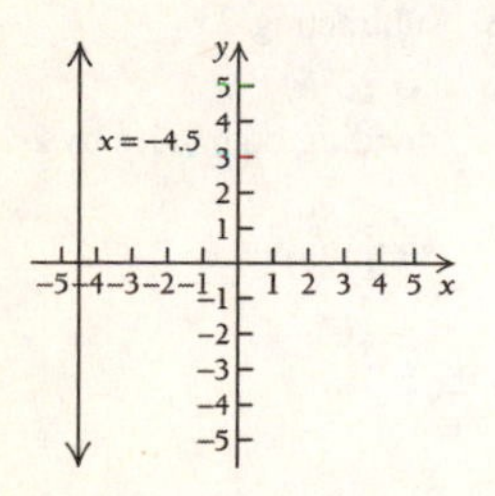

7. Graph $y = 3.75$.
The graph consists of all ordered pairs whose second coordinate is 3.75. This results in a horizontal line whose y-intercept is the point $(0,3.75)$.

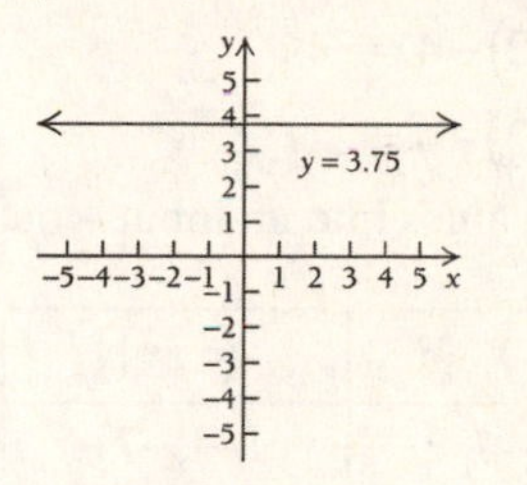

9. Graph $y = -2x$.
Using Theorem 4, The graph of y is the straight line through the origin $(0,0)$ and the point $(1,-2)$. We plot these two points and connect them with a straight line.

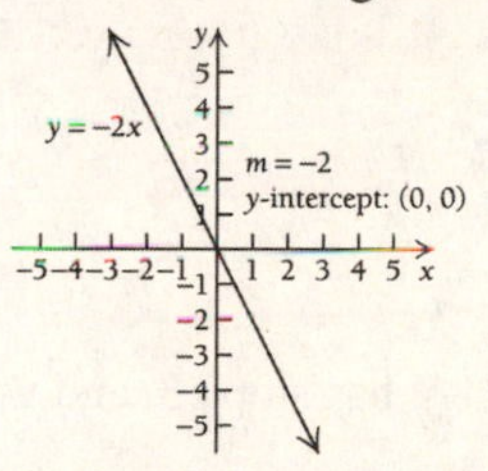

The function $y = -2x$ has slope –2, and y-intercept $(0,0)$.

11. Graph $f(x) = 0.5x$.
Using Theorem 4, The graph of $f(x)$ is the straight line through the origin $(0,0)$ and the point $(1,0.5)$. We plot these two points and connect them with a straight line.

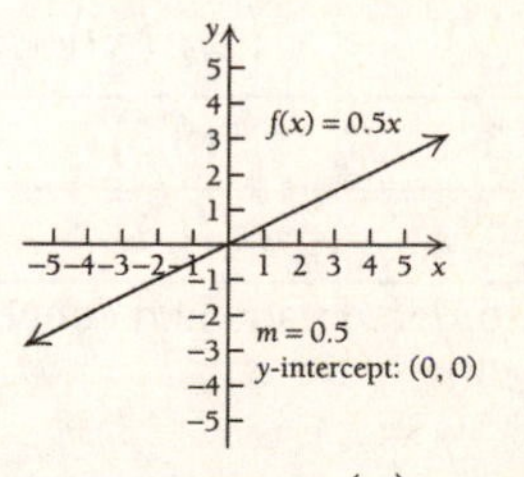

The function $f(x) = 0.5x$ has slope 0.5, and y-intercept $(0,0)$.

Copyright © 2014 Pearson Education, Inc. Publishing as Addison-Wesley.

13. Graph $y = 3x - 4$

First, we make a table of values. We choose any number for x and then determine y by substitution.

When $x = -1, y = 3(-1) - 4 = -7$.

When $x = 0, y = 3(0) - 4 = -4$.

When $x = 2, y = 3(2) - 4 = 2$.

We organize these values into an input – output table.

x	y	(x, y)
-1	-7	$(-1,-7)$
0	-4	$(0,-4)$
2	2	$(2,2)$

Next, we plot these ordered pairs and connect them with a straight line.

The function $y = 3x - 4$ has slope 3, and y-intercept $(0,-4)$.

15. Graph $g(x) = -x + 3$.

First, we make a table of values. We choose any number for x and then determine y by substitution.

When $x = 0, g(0) = -(0) + 3 = 3$.

When $x = 1, g(1) = -(1) + 3 = 2$.

We organize these values into a table.

x	$g(x)$	$(x, g(x))$
0	3	$(0,3)$
1	2	$(1,2)$

Next, we plot these ordered pairs and connect them with a straight line.

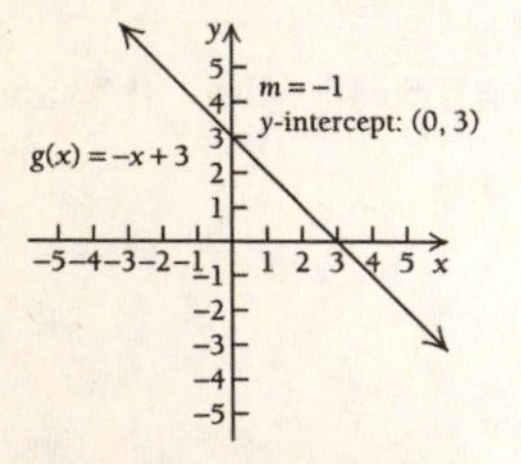

The function $g(x) = -x + 3$ has slope -1, and y-intercept $(0,3)$.

17. Graph $y = 7$.

The graph consists of all ordered pairs whose second coordinate is 7. This results in a horizontal line, whose y-intercept is the point $(0,7)$.

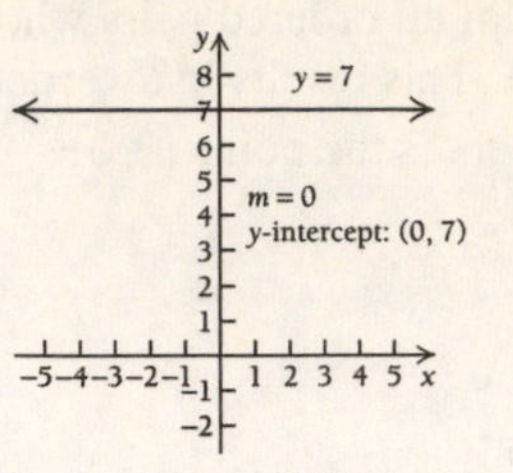

Since the graph is horizontal, the slope is 0 and the y-intercept is $(0,7)$.

19. First, we solve the equation for y.

$y - 3x = 6$

$y = 3x + 6$ adding $3x$ to both sides

The slope is 3.

The y-intercept is $(0,6)$.

21. First, we solve the equation for y.

$2x + y - 3 = 0$

$2x + y = 3$ adding 3 to both sides

$y = -2x + 3$ subtracting $2x$ from both sides

The slope is -2.

The y-intercept is $(0,3)$.

23. First, we solve the equation for y.

$2x + 2y + 8 = 0$

$2x + 2y = -8$ subtracting 8

$2y = -2x - 8$ subtracting $2x$

$y = \dfrac{-2x - 8}{2}$ dividing both sides by 2

$y = -x - 4$ simplifying

The slope is -1.

The y-intercept is $(0,-4)$.

25. First, we solve the equation for y.

$x = 3y + 7$

$3y + 7 = x$ commutative property of equality

$3y = x - 7$ subtracting 7

$y = \dfrac{x - 7}{3}$ dividing by 3

$y = \dfrac{1}{3}x - \dfrac{7}{3}$ simplifying

The slope is $\dfrac{1}{3}$. The y-intercept is $\left(0, -\dfrac{7}{3}\right)$.

Copyright © 2014 Pearson Education, Inc. Publishing as Addison-Wesley.

27.
$$\begin{aligned} y-y_1&=m(x-x_1) \\ y-(-3)&=-5(x-(-2)) && \text{Substituting} \\ y+3&=-5(x+2) && \text{Simplifying} \\ y+3&=-5x-10 \\ y&=-5x-13 && \text{Subtracting 3} \end{aligned}$$

29.
$$\begin{aligned} y-y_1&=m(x-x_1) \\ y-(3)&=-2(x-(2)) && \text{Substituting} \\ y-3&=-2x+4 \\ y&=-2x+7 && \text{Adding 3} \end{aligned}$$

31.
$$\begin{aligned} y-y_1&=m(x-x_1) \\ y-(0)&=2(x-(3)) && \text{Substituting} \\ y&=2x-6 \end{aligned}$$

33.
$$\begin{aligned} y&=mx+b \\ y&=\frac{1}{2}x+(-6) && \text{Substituting} \\ y&=\frac{1}{2}x-6 && \text{Simplifying} \end{aligned}$$

35.
$$\begin{aligned} y-y_1&=m(x-x_1) \\ y-(3)&=0(x-(2)) && \text{Substituting} \\ y-3&=0 && \text{Simplifying} \\ y&=3 && \text{Adding 3} \end{aligned}$$

37. $m=\dfrac{y_2-y_1}{x_2-x_1}$

Substituting, we have:

$$\begin{aligned} m&=\frac{1-(-3)}{-2-5} \\ &=\frac{1+3}{-2-5} \\ &=\frac{4}{-7} \\ &=-\frac{4}{7} \end{aligned}$$

39. $m=\dfrac{y_2-y_1}{x_2-x_1}$

Substituting, we have:

$$\begin{aligned} m&=\frac{-4-(-3)}{-1-2} \\ &=\frac{-4+3}{-1-2} \\ &=\frac{-1}{-3} \\ &=\frac{1}{3} \end{aligned}$$

41. $m=\dfrac{y_2-y_1}{x_2-x_1}$

Substituting, we have:

$$\begin{aligned} m&=\frac{-9-(-7)}{3-3} \\ &=\frac{-9+7}{3-3} \\ &=\frac{-2}{0} \end{aligned}$$

Since we cannot divide by 0, the slope is undefined.

43. $m=\dfrac{y_2-y_1}{x_2-x_1}$

Substituting, we have:

$$\begin{aligned} m&=\frac{\frac{2}{5}-(-3)}{\frac{1}{2}-\frac{4}{5}} \\ &=\frac{\frac{2}{5}-\left(-\frac{15}{5}\right)}{\frac{5}{10}-\frac{8}{10}} && \text{finding a common denominator} \\ &=\frac{\frac{17}{5}}{-\frac{3}{10}} && \text{adding fractions} \\ &=\frac{17}{5}\cdot\left(-\frac{10}{3}\right) && \text{Multiplying by the reciprical} \\ &=-\frac{34}{3} \end{aligned}$$

Copyright © 2014 Pearson Education, Inc. Publishing as Addison-Wesley.

45. $m = \dfrac{y_2 - y_1}{x_2 - x_1}$

Substituting, we have:

$$m = \frac{3-(3)}{-1-2} = \frac{0}{-3} = 0$$

47. $m = \dfrac{y_2 - y_1}{x_2 - x_1}$

Substituting, we have:

$$m = \frac{3(x+h)-(3x)}{(x+h)-x} = \frac{3x+3h-3x}{x+h-x} = \frac{3h}{h} = 3$$

49. $m = \dfrac{y_2 - y_1}{x_2 - x_1}$

Substituting, we have:

$$m = \frac{[2(x+h)+3]-(2x+3)}{(x+h)-x} = \frac{2x+2h+3-(2x+3)}{x+h-x} = \frac{2h}{h} = 2$$

51. From Exercise 37, we know that the slope is $-\frac{4}{7}$. Using the point $(5,-3)$, we substitute into the point-slope equation.

$$y-(-3) = -\frac{4}{7}(x-5)$$
$$y+3 = -\frac{4}{7}x+\frac{20}{7}$$
$$y = -\frac{4}{7}x+\frac{20}{7}-3$$
$$y = -\frac{4}{7}x-\frac{1}{7}$$

Note: We could have found the equation of the line using the point $(-2,1)$ as follows

$$y-(1) = -\frac{4}{7}(x-(-2))$$
$$y-1 = -\frac{4}{7}x-\frac{8}{7}$$
$$y = -\frac{4}{7}x-\frac{8}{7}+1$$
$$y = -\frac{4}{7}x-\frac{1}{7}$$

53. From Exercise 39, we know that the slope is $\frac{1}{3}$. Using the point $(2,-3)$, we substitute into the point-slope equation.

$$y-(-3) = \frac{1}{3}(x-2)$$
$$y+3 = \frac{1}{3}x-\frac{2}{3}$$
$$y = \frac{1}{3}x-\frac{2}{3}-3$$
$$y = \frac{1}{3}x-\frac{11}{3}$$

Note: We could have found the equation of the line using the point $(-1,-4)$ as follows

$$y-(-4) = \frac{1}{3}(x-(-1))$$
$$y+4 = \frac{1}{3}x+\frac{1}{3}$$
$$y = \frac{1}{3}x+\frac{1}{3}-4$$
$$y = \frac{1}{3}x-\frac{11}{3}$$

55. From Exercise 41, we know that the slope is undefined. The graph is a line which contains all ordered pairs whose first coordinate is 3. The equation of the line is $x = 3$

Copyright © 2014 Pearson Education, Inc. Publishing as Addison-Wesley.

57. From Exercise 43, we know that the slope is $-\frac{34}{3}$. Using the point $\left(\frac{4}{5},-3\right)$, we substitute into the point-slope equation.

$$y-(-3)=-\frac{34}{3}\left(x-\frac{4}{5}\right)$$
$$y+3=-\frac{34}{3}x+\frac{136}{15}$$
$$y=-\frac{34}{3}x+\frac{136}{15}-3$$
$$y=-\frac{34}{3}x+\frac{91}{15}$$

Note: We could have found the equation of the line using the point $\left(\frac{1}{2},\frac{2}{5}\right)$.

$$y-\left(\frac{2}{5}\right)=-\frac{34}{3}\left(x-\frac{1}{2}\right)$$
$$y-\frac{2}{5}=-\frac{34}{3}x+\frac{17}{3}$$
$$y=-\frac{34}{3}x+\frac{17}{3}+\frac{2}{5}$$
$$y=-\frac{34}{3}x+\frac{91}{15}$$

59. From Exercise 45, we know that the slope is 0. The line is horizontal, thus the equation is $y=3$.

61. $\text{Slope}=\frac{2\text{ ft}}{5\text{ ft}}=\frac{2}{5}=0.4.$
The slope of the skateboard ramp is 0.4.

63. $\text{Slope}=\frac{43.33\text{ ft}}{1238\text{ ft}}=\frac{43.33}{1238}=\frac{7}{200}=0.035.$
Expressing the slope as a percentage, we find the head of the river is 3.5%.

65. a) If I, the number of inkjet cartridges required each year, is directly proportional to the number of students, s. Then there exists a constant m such that
$I=ms$
To find m, we substitute 16 for I and 2800 for s into $I=ms$ and solve for m.

$$16=m\cdot 2800$$
$$\frac{16}{2800}=m$$
$$0.0057=m$$

The constant of variation is $m=0.0057$. The equation of variation is $I=0.0057s$.

b) $I=0.0057s$
$=0.0057\cdot 3100$
$=17.67\approx 18.$
The university would need 18 inkjet cartridges if 3100 students were enrolled.

67. a) Total costs = Variables costs + Fixed Costs
To produce x skis, it costs \$80 dollars per ski, that is the variable costs are $80x$. In addition to the variable costs the fixed costs are \$45,000. The total cost is
$C(x)=80x+45{,}000.$
The graph is shown on the next page.

b) Revenue = Price times Quantity.
The price of a pair of skis is \$225, therefore, the revenue from selling x pairs of skis is given by $R(x)=255x.$

c) Profit = Revenue – Cost.
Using the Cost and Revenue functions found in part (a) and (b) we have:

$$P(x)=R(x)-C(x)$$
$$P(x)=255x-(80x+45{,}000)$$
$$=175x-45{,}000.$$

The graph is shown below

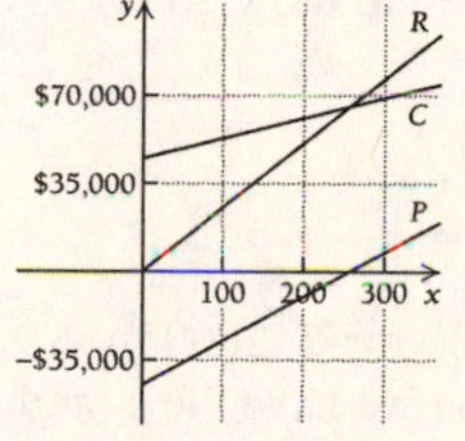

d) Substituting 3000 for x into the profit equation, we have
$P(3000)=175(3000)-45{,}000=480{,}000.$
Total profit will be \$480,000 if they sell the expected 3000 pairs of skis.

e) The break even point occurs when $P(x)=0$. Therefore, we set the profit function equal to 0 and solve for x.

$$P(x)=0$$
$$175x-45{,}000=0$$
$$175x=45{,}000$$
$$x=\frac{45{,}000}{175}$$
$$x=257.14$$
$$x\approx 258.$$

They will need to sell 258 pairs of skis in order to break even.

Copyright © 2014 Pearson Education, Inc. Publishing as Addison-Wesley.

69. a) Total costs = Variables costs + Fixed Costs
The gas and maintenance cost are $4 dollars per lawn, that is the variable costs are $4x$.
In addition to the variable costs the initial fixed cost is $250. The total cost is
$C(x)=4x+250$.

b) $P(x)=R(x)-C(x)$
Substituting the profit in cost functions into the equation we have:

$$9x-250=R(x)-(4x+250)$$
$$9x-250=R(x)-4x-250$$
$$13x=R(x).$$

Since Revenue is equal to price times quantity. Jimmy charges $13 per lawn.

c) The break even point occurs when $P(x)=0$. Therefore, we set the profit function equal to 0 and solve for x.
Solve $P(x)=0$

$$9x-250=0$$
$$9x=250$$
$$x=27.77\overline{7}$$
$$x\approx 28.$$

Jimmy will need to mow 28 lawns in order to break even.

71. $V(t)=C-t\left(\frac{C-S}{N}\right)$

Substituting 60,000 for *C*, 2000 for *S*, and 5 for *N* into the equation we have the straight line depreciation $V(t)$:

$$V(t)=60{,}000-t\left(\frac{60{,}000-2000}{5}\right)$$
$$=60{,}000-11{,}600t$$
$$V(3)=60{,}000-11{,}600(3)=25{,}200.$$

The computer system will have a book value of $25,200 after 3 years.

73. Slope is rise over run. The maximum riser on a stair is 8.25 inches, and the minimum run for a tread is 9 inches. Therefore:

$$\text{Slope}=\frac{\text{height of riser}}{\text{length of run}}=\frac{8.25\text{ in}}{9\text{ in}}=\frac{11}{12}=0.91\overline{6}.$$

The maximum grade of stairs in North Carolina is 91%. Don't round up for legal reasons!

75. The average rate of change can be found using the coordinates of any two points on the line. We use the given coordinates
$(2005, 4024)$ and $(2009, 4824)$.

$$\text{Rate of Change}=\frac{4824-4024}{2009-2005}$$
$$=\frac{800}{4}=200.$$

The average rate of change in the annual premium for a single person is $200 per year.

77. The average rate of change can be found using the coordinates of any two points on the line. We use the given coordinates
$(2004,\ \$26{,}450)$ and $(2008,\ \$19{,}580)$.

$$\text{Rate of Change}=\frac{19{,}580-26{,}450}{2009-2005}$$
$$=\frac{-6870}{4}\approx -1717.5.$$

The average rate of change of the cost of a formal wedding is approximately negative $1717.50 per year.

79. The equation of variation is $D=293t$. The distance the impulse has to travel is 6 ft. So we solve:

$$6=293t$$
$$\frac{6}{293}=t$$
$$0.0204778157=t$$
$$0.02\approx t.$$

It would take an impulse 0.02 seconds to travel from the toes of the person to the brain.

81. a) Since the brain weight *B* is directly proportional to the person's body weight *W*, there is a positive constant *m* such that
$B=mW$
To find *m*, we substitute 3 for *B* and 120 for *W* into $B=mW$ solve for *m*.

$$3=m\cdot 120$$
$$0.025=m.$$

The constant of variation is $m=0.025$. The equation of variation is $B=0.025W$.

b) The constant of variation $m=0.025$ is equivalent to 2.5%, so we have $B=2.5\%W$. The weight, *B*, of a human's brain is 2.5% of the persons total body weight, *W*.

Copyright © 2014 Pearson Education, Inc. Publishing as Addison-Wesley.

c) Substituting 160 for W into the equation of variation we have:
$B = 0.025(160)$
$= 4$
A person weighing 160 pounds has a brain that weighs 4 pounds.

83. a) $D(r) = \dfrac{11r+5}{10}$

$D(5) = \dfrac{11(5)+5}{10} = 6$
When traveling 5 miles per hour, the car's reaction distance is 6 feet.

$D(10) = \dfrac{11(10)+5}{10} = 11.5$
When traveling 10 miles per hour, the car's reaction distance is 11.5 feet.

$D(20) = \dfrac{11(20)+5}{10} = 22.5$
When traveling 20 miles per hour, the car's reaction distance is 22.5 feet.

$D(50) = \dfrac{11(50)+5}{10} = 55.5$
When traveling 50 miles per hour, the car's reaction distance is 55.5 feet.

$D(65) = \dfrac{11(65)+5}{10} = 72$
When traveling 65 miles per hour, the car's reaction distance is 72 feet.

b) Plotting the points found in part (a) and connecting the points with a smooth curve we have:

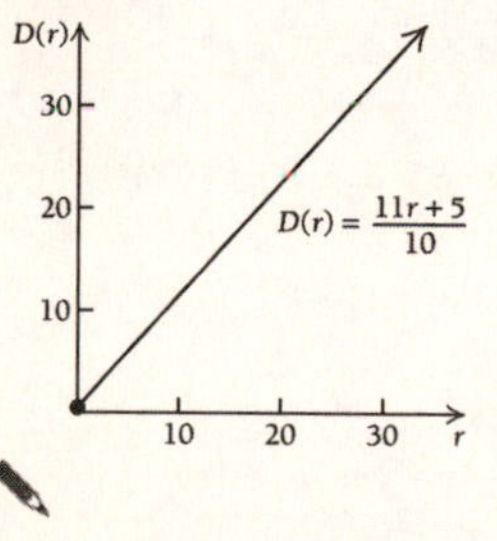

c) ✎

85. a) We find the equation of the line that contains the points $(2000, 72)$ and $(2009, 92)$. First we find the slope of the line.
$m = \dfrac{92-72}{2009-2000} = \dfrac{20}{9} \approx 2.22$
Next, we use the point-slope equation. We will use the point $(2000, 72)$.

$y - y_1 = m(x - x_1)$
$y - 72 = \frac{20}{9}(x - 2000)$
$y - 72 = \frac{20}{9}x - 4444.44$
$y = \frac{20}{9}x - 4372.44.$

b) Using the equation and substituting 2010 in for x we have:
$y = \frac{20}{9}(2010) - 4372.44 \approx 94.23$
In 2010, approximately 94.23% of adults used the internet.

c) Using the equation and substituting 100 in for y we have:
$100 = \frac{20}{9}x - 4372.44$
$4472.44 = \frac{20}{9}x$
$4472.44\left(\frac{9}{20}\right) = x$
$2012.598 = x$
According to the equation, 100% of adults will be using the internet in 2012.

d) ✎

87. a) We know that $N = P + 0.02P = 1.02P$. Therefore the equation of variation is $N = 1.02P$.

b) Substituting 200,000 for P we have $N = 1.02(200{,}000) = 204{,}000$.
The new population is 204,000 after a growth of 2%.

c) Substituting 367,200 for N we have
$367{,}200 = 1.02P$
$\dfrac{367{,}200}{1.02} = P$
$360{,}000 = P$
The previous population was 360,000.

89. ✎

91. a) Graph III is appropriate, because it shows the rate before January 1 is approximately \$3000 per month, and the rate after January 1 is approximately \$2000 per month.

b) Graph IV is appropriate, because it shows the rate before January 1 is approximately \$3000 per month, and the rate after January 1 is approximately –\$4000 per month.

Copyright © 2014 Pearson Education, Inc. Publishing as Addison-Wesley.

c) Graph I is appropriate, because it shows the rate before January 1 is approximately \$1000 per month, and the rate after January 1 is approximately \$2000 per month.

d) Graph II is appropriate, because it shows the rate before January 1 is approximately \$4000 per month, and the rate after January 1 is approximately –\$2000 per month.

93. Answers may vary. Regions of profit will occur when total revenue is greater than total cost, or when total profit is above the *x*-axis. Regions of loss will occur when total revenue is less than total cost, or when total profit is below the *x*-axis.

Copyright © 2014 Pearson Education, Inc. Publishing as Addison-Wesley.

Exercise Set R.5

1. Graph $y=\frac{1}{2}x^2$ and $y=-\frac{1}{2}x^2$.

Starting with $y=\frac{1}{2}x^2$, we first find the vertex or the turning point. The x-coordinate of the vertex is

$$x=-\frac{b}{2a}$$
$$=-\frac{0}{2\left(\frac{1}{2}\right)}$$
$$=0$$

Substituting 0 for x into the equation, we find the second coordinate of the vertex:

$y=\frac{1}{2}(0)^2=0$.

The vertex is $(0,0)$. The y-axis (The vertical line $x=0$.) is the axis of symmetry. Next, we choose some x-values on each side of the vertex and compute the y-values.

When $x=1$, $y=\frac{1}{2}(1)^2=\frac{1}{2}\cdot 1=\frac{1}{2}$

When $x=2$, $y=\frac{1}{2}(2)^2=\frac{1}{2}\cdot 4=2$

When $x=-1$, $y=\frac{1}{2}(-1)^2=\frac{1}{2}\cdot 1=\frac{1}{2}$

When $x=-2$, $y=\frac{1}{2}(-2)^2=\frac{1}{2}\cdot 4=2$

x	y
0	0
1	$\frac{1}{2}$
2	2
−1	$\frac{1}{2}$
−2	2

We plot these points and connect them with a smooth curve on the axis below.

Next, we graph $y=-\frac{1}{2}x^2$. First, we find the vertex or the turning point. The x-coordinate of the vertex is

$$x=-\frac{b}{2a}$$
$$x=-\frac{0}{2\left(-\frac{1}{2}\right)}=0.$$

Substituting 0 for x into the equation, we find the second coordinate of the vertex:

$y=-\frac{1}{2}(0)^2=0$.

The vertex is $(0,0)$. The y-axis (The vertical line $x=0$.) is the axis of symmetry. Next, we choose some x-values on each side of the vertex and compute the y-values.

When $x=1$, $y=-\frac{1}{2}(1)^2=-\frac{1}{2}\cdot 1=-\frac{1}{2}$

When $x=2$, $y=-\frac{1}{2}(2)^2=-\frac{1}{2}\cdot 4=-2$

When $x=-1$, $y=-\frac{1}{2}(-1)^2=-\frac{1}{2}\cdot 1=-\frac{1}{2}$

When $x=-2$, $y=-\frac{1}{2}(-2)^2=-\frac{1}{2}\cdot 4=-2$

x	y
0	0
1	$-\frac{1}{2}$
2	−2
−1	$-\frac{1}{2}$
−2	−2

We plot these points and connect them with a smooth curve on the axis below.

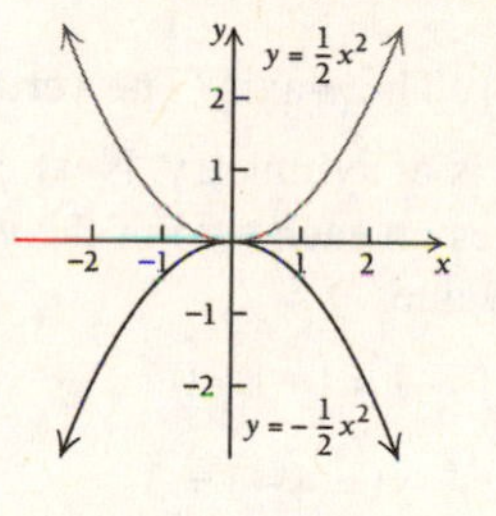

3. Graph $y=x^2$ and $y=x^2-1$.

Starting with $y=x^2$, we first find the vertex or the turning point. The x-coordinate of the vertex is

$$x=-\frac{b}{2a}=-\frac{0}{2(1)}=0.$$

Substituting 0 for x into the equation, we find the second coordinate of the vertex:

$y=(0)^2=0$.

The vertex is $(0,0)$. The y-axis (The vertical line $x=0$.) is the axis of symmetry.
Next, we choose some x-values on each side of the vertex and compute the y-values. The calculations are shown at the top of the next page.

Copyright © 2014 Pearson Education, Inc. Publishing as Addison-Wesley.

When $x = 1, y = (1)^2 = 1$

When $x = 2, y = (2)^2 = 4$

When $x = -1, y = (-1)^2 = 1$

When $x = -2, y = (-2)^2 = 4$

x	y
0	0
1	1
2	4
−1	1
−2	4

We plot these points and connect them with a smooth curve on the axis below.

Next, we graph $y = x^2 - 1$. First, we find the vertex or the turning point. The x-coordinate of the vertex is

$$x = -\frac{b}{2a}$$
$$= -\frac{0}{2(1)}$$
$$= 0.$$

Substituting 0 for x into the equation, we find the second coordinate of the vertex:

$y = (0)^2 - 1 = -1$.

The vertex is $(0,-1)$. The y-axis (The vertical line $x = 0$.) is the axis of symmetry. Next, we choose some x-values on each side of the vertex and compute the y-values.

When $x = 1, y = (1)^2 - 1 = 1 - 1 = 0$

When $x = 2, y = (2)^2 - 1 = 4 - 1 = 3$

When $x = -1, y = (-1)^2 - 1 = 1 - 1 = 0$

When $x = -2, y = (-2)^2 - 1 = 4 - 1 = 3$

x	y
0	−1
1	0
2	3
−1	0
−2	3

We plot these points and connect them with a smooth curve on the axis below.

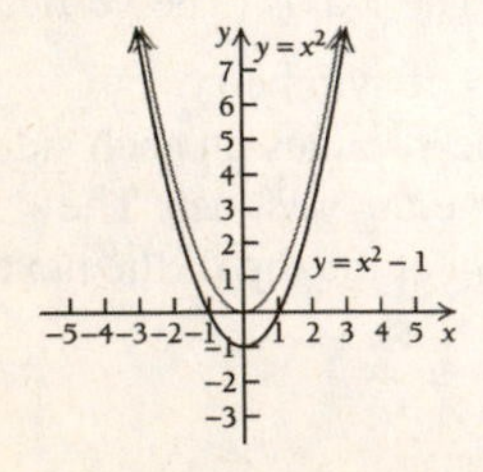

5. Graph $y = -2x^2$ and $y = -2x^2 + 1$.

Starting with $y = -2x^2$, we first find the vertex or the turning point. The x-coordinate of the vertex is:

$$x = -\frac{b}{2a}$$
$$= -\frac{0}{2(-2)}$$
$$= 0.$$

Substituting 0 for x into the equation, we find the second coordinate of the vertex:

$y = -2(0)^2 = 0$.

The vertex is $(0,0)$. The y-axis (The vertical line $x = 0$.) is the axis of symmetry. Next, we choose some x-values on each side of the vertex and compute the y-values.

When $x = 1, y = -2(1)^2 = -2\cdot 1 = -2$

When $x = 2, y = -2(2)^2 = -2\cdot 4 = -8$

When $x = -1, y = -2(-1)^2 = -2\cdot 1 = -2$

When $x = -2, y = -2(-2)^2 = -2\cdot 4 = -8$

x	y
0	0
1	−2
2	−8
−1	−2
−2	−8

We plot these points and connect them with a smooth curve on the next page.

Next, we graph $y = -2x^2 + 1$. First, we find the vertex or the turning point. The x-coordinate of the vertex is

$$x = -\frac{b}{2a}$$
$$= -\frac{0}{2(-2)}$$
$$= 0.$$

Substituting 0 for x into the equation, we find the second coordinate of the vertex:

$y = -2(0)^2 + 1 = 1$.

The vertex is $(0,1)$. The y-axis (The vertical line $x = 0$.) is the axis of symmetry. Next, we choose some x-values on each side of the vertex and compute the y-values. The calculations are shown at the top of the next page.

Copyright © 2014 Pearson Education, Inc. Publishing as Addison-Wesley.

When $x = 1, y = -2(1)^2 + 1 = -2 + 1 = -1$

When $x = 2, y = -2(2)^2 + 1 = -8 + 1 = -7$

When $x = -1, y = -2(-1)^2 + 1 = -2 + 1 = -1$

When $x = -2, y = -2(-2)^2 + 1 = -8 + 1 = -7$

x	y
0	1
1	−1
2	−7
−1	−1
−2	−7

We plot these points and connect them with a smooth curve on the axis below.

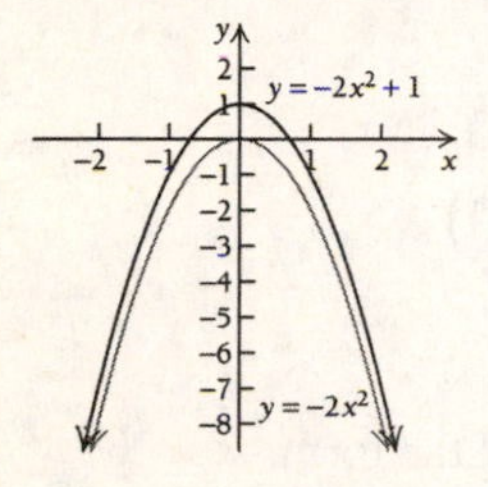

7. Graph $y = |x|$ and $y = |x - 3|$.

Starting with $y = |x|$, we choose some x-values and compute the y-values to make a table of points.

When $x = -2, y = |-2| = -(-2) = 2$

When $x = -1, y = |-1| = -(-1) = 1$

When $x = 0, y = |0| = 0$

When $x = 1, y = |1| = 1$

When $x = 2, y = |2| = 2$

x	y
−2	2
−1	1
0	0
1	1
2	2

We plot these points and connect them with a smooth curve on the axis below.

Next, we graph $y = |x - 3|$. We choose some x-values and compute the y-values to make a table of points.

When $x = -2, y = |-2 - 3| = |-5| = -(-5) = 5$

When $x = 0, y = |0 - 3| = |-3| = -(-3) = 3$

When $x = 3, y = |3 - 3| = |0| = 0$

When $x = 6, y = |6 - 3| = |3| = 3$

When $x = 8, y = |8 - 3| = |5| = 5$

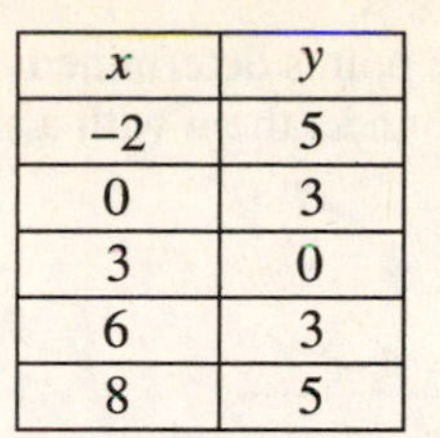

x	y
−2	5
0	3
3	0
6	3
8	5

We plot these points and connect them with a smooth curve on the axis below.

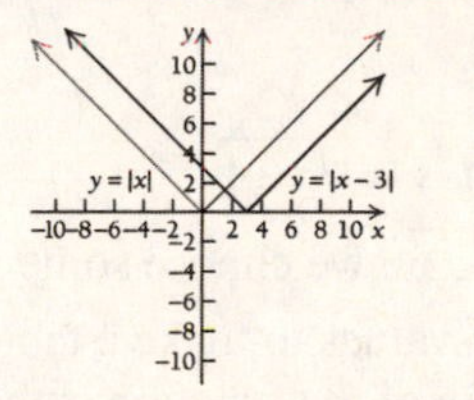

9. Graph $y = x^3$ and $y = x^3 + 2$.

Starting with $y = x^3$, we choose some x-values and compute the y-values to make a table of points.

When $x = -2, y = (-2)^3 = -8$

When $x = -1, y = (-1)^3 = -1$

When $x = 0, y = (0)^3 = 0$

When $x = 1, y = (1)^3 = 1$

When $x = 2, y = (2)^3 = 8$

x	y
−2	−8
−1	−1
0	0
1	1
2	8

We plot these points and connect them with a smooth curve on the axis on the next page.

Next, we graph $y = x^3 + 2$. We choose some x-values and compute the y-values to make a table of points.

When $x = -2, y = (-2)^3 + 2 = -8 + 2 = -6$

When $x = -1, y = (-1)^3 + 2 = -1 + 2 = 1$

When $x = 0, y = (0)^3 + 2 = 0 + 2 = 2$

When $x = 1, y = (1)^3 + 2 = 1 + 2 = 3$

When $x = 2, y = (2)^3 + 2 = 8 + 2 = 10$

x	y
−2	−6
−1	1
0	2
1	3
2	10

The solution is continued on the next page.

Copyright © 2014 Pearson Education, Inc. Publishing as Addison-Wesley.

We plot the points determined on the previous page and connect them with a smooth curve.

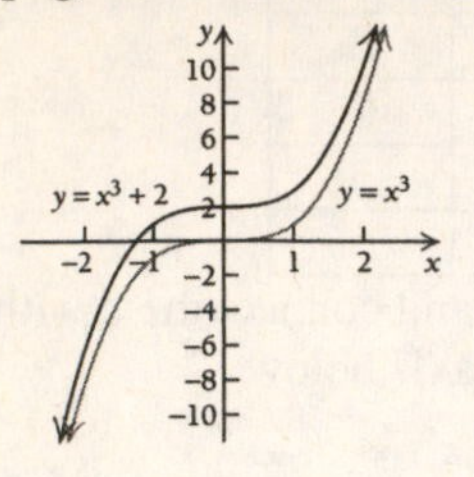

11. Graph $y=\sqrt{x}$ and $y=\sqrt{x-1}$.

Starting with $y=\sqrt{x}$, we choose some x-values and compute the y-values to make a table of points. The domain of the function is the set of all nonnegative real numbers, so we choose x-values that are in the set $[0,\infty)$.

When $x=0, y=\sqrt{0}=0$

When $x=1, y=\sqrt{1}=1$

When $x=4, y=\sqrt{4}=2$

When $x=9, y=\sqrt{9}=3$

x	y
0	0
1	1
4	2
9	3

We plot these points and connect them with a smooth curve on the axis below.

Next, we graph $y=\sqrt{x-1}$. we choose some x-values and compute the y-values to make a table of points. The domain of the function is the set of all positive real numbers greater than or equal to 1, so we choose x-values that are in the set $[1,\infty)$

When $x=1, y=\sqrt{1-1}=\sqrt{0}=0$

When $x=2, y=\sqrt{2-1}=\sqrt{1}=1$

When $x=5, y=\sqrt{5-1}=\sqrt{4}=2$

When $x=10, y=\sqrt{10-1}=\sqrt{9}=3$

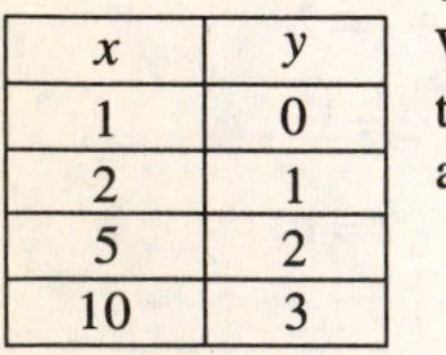

x	y
1	0
2	1
5	2
10	3

We plot these points and connect them with a smooth curve on the axis below.

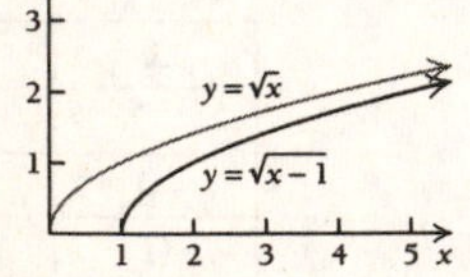

13. $y=x^2+4x-7$

This function is of the form $y=ax^2+bx+c,\ a\neq 0$, so its graph is a parabola.

We have $a=1$ and $b=4$, so the first coordinate of the vertex is

$$x=-\frac{b}{2a}$$
$$=-\frac{4}{2(1)}=-2.$$

Substituting -2 into the equation, we find the second coordinate of the vertex:

$$y=(-2)^2+4(-2)-7$$
$$=4-8-7$$
$$=-11.$$

The vertex is $(-2,-11)$.

15. $y=2x^4-4x^2-3$

The function is not of the form $y=ax^2+bx+c,\ a\neq 0$, so its graph is not a parabola.

17. Graph $y=x^2-4x+3$.

First, we should recognize that this function is a quadratic function. We find the vertex or the turning point. The x-coordinate of the vertex is

$$x=-\frac{b}{2a}$$
$$=-\frac{-4}{2(1)}$$
$$=2.$$

Substituting 2 for x into the equation, we find the second coordinate of the vertex:

$$y=(2)^2-4(2)+3$$
$$=4-8+3=-1.$$

The vertex is $(2,-1)$. The vertical line $x=2$ is the axis of symmetry. Next, we choose some x-values on each side of the vertex and compute the y-values.

When $x=-1, y=(-1)^2-4(-1)+3=8$

When $x=\ 0, y=(0)^2-4(0)+3=3$

When $x=\ 1, y=(1)^2-4(1)+3=0$

When $x=\ 3, y=(3)^2-4(3)+3=0$

When $x=\ 4, y=(4)^2-4(4)+3=3$

When $x=\ 5, y=(5)^2-4(5)+3=8$

Copyright © 2014 Pearson Education, Inc. Publishing as Addison-Wesley.

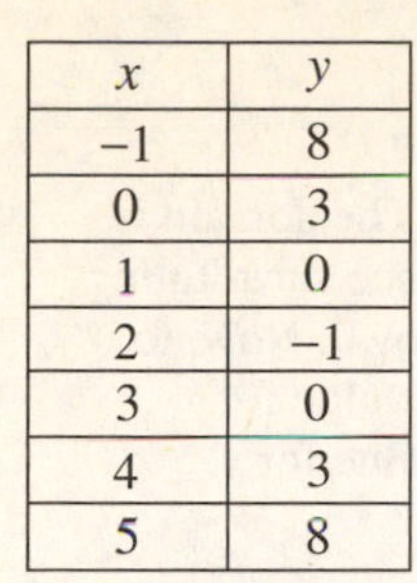

x	y
−1	8
0	3
1	0
2	−1
3	0
4	3
5	8

Using the information from the previous page, we plot the points in the table and connect them with a smooth curve on the axis below.

19. Graph $y=-x^2+2x-1$.

First, we should recognize that this function is a quadratic function. We find the vertex or the turning point. The x-coordinate of the vertex is

$$x=-\frac{b}{2a}=-\frac{2}{2(-1)}=1.$$

Substituting 1 for x into the equation, we find the second coordinate of the vertex:

$$y=-(1)^2+2(1)-1$$
$$=-1+2-1=0.$$

The vertex is $(1,0)$. The vertical line $x=1$ is the axis of symmetry. Next, we choose some x-values on each side of the vertex and compute the y-values.

When $x=-1$, $y=-(-1)^2+2(-1)-1=-4$

When $x=\ 0$, $y=-(0)^2+2(0)-1=-1$

When $x=\ 2$, $y=-(2)^2+2(2)-1=-1$

When $x=\ 3$, $y=-(3)^2+2(3)-1=-4$

x	y
−1	−4
0	−1
1	0
2	−1
3	−4

We plot these points and connect them with a smooth curve on the axis below.

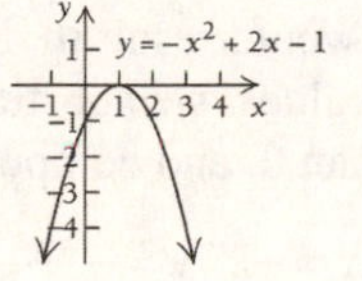

21. Graph $f(x)=2x^2-6x+1$.

First, we should recognize that this function is a quadratic function. We find the vertex or the turning point. The x-coordinate of the vertex is

$$x=-\frac{b}{2a}=-\frac{-6}{2(2)}=\frac{3}{2}.$$

Substituting $3/2$ for x into the equation, we find the second coordinate of the vertex at the top of the next column.

$$f\left(\frac{3}{2}\right)=2\left(\frac{3}{2}\right)^2-6\left(\frac{3}{2}\right)+1$$
$$=2\left(\frac{9}{4}\right)-9+1$$
$$=\frac{9}{2}-8$$
$$=-\frac{7}{2}.$$

The vertex is $\left(\frac{3}{2},-\frac{7}{2}\right)$. The vertical line $x=\frac{3}{2}$ is the axis of symmetry.

Next, we choose some x-values on each side of the vertex and compute the y-values.

When $x=0$, $y=2(0)^2-6(0)+1=1$

When $x=1$, $y=2(1)^2-6(1)+1=-3$

When $x=2$, $y=2(2)^2-6(2)+1=-3$

When $x=3$, $y=2(3)^2-6(3)+1=1$

x	y
0	1
1	−3
$\frac{3}{2}$	$-\frac{7}{2}$
2	−3
3	1

We plot these points and connect them with a smooth curve on the axis below.

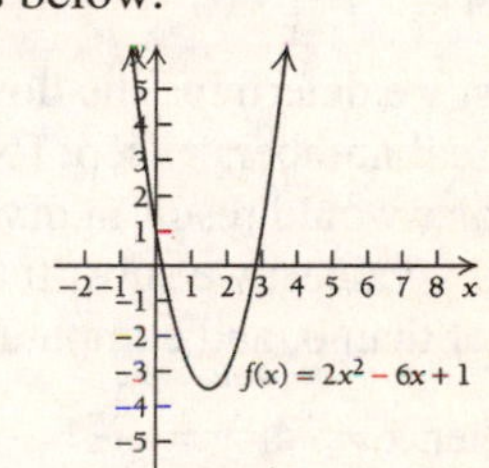

23. Graph $g(x)=-3x^2-4x+5$.

First, we should recognize that this function is a quadratic function. We find the vertex or the turning point. The x-coordinate of the vertex is

$$x=-\frac{b}{2a}=-\frac{-4}{2(-3)}=-\frac{2}{3}.$$

Substituting $-2/3$ for x into the equation, we find the second coordinate of the vertex.

$$f\left(-\frac{2}{3}\right)=-3\left(-\frac{2}{3}\right)^2-4\left(-\frac{2}{3}\right)+5$$
$$=-3\left(\frac{4}{9}\right)+\frac{8}{3}+5=$$
$$=-\frac{4}{3}+\frac{8}{3}+5=\frac{19}{3}.$$

The vertex is $\left(-\frac{2}{3},\frac{19}{3}\right)$.

The solution is continued on the next page.

Copyright © 2014 Pearson Education, Inc. Publishing as Addison-Wesley.

The vertical line $x=-\frac{2}{3}$ is the axis of symmetry. Next, we choose some x-values on each side of the vertex and compute the y-values.

When $x=-2$, $y=-3(-2)^2-4(-2)+5=1$

When $x=-1$, $y=-3(-1)^2-4(-1)+5=6$

When $x=0$, $y=-3(0)^2-4(0)+5=5$

When $x=1$, $y=-3(1)^2-4(1)+5=-2$

x	y
-2	1
-1	6
$-\frac{2}{3}$	$\frac{19}{3}$
0	5
1	-2

We plot these points and connect them with a smooth curve on the axis below.

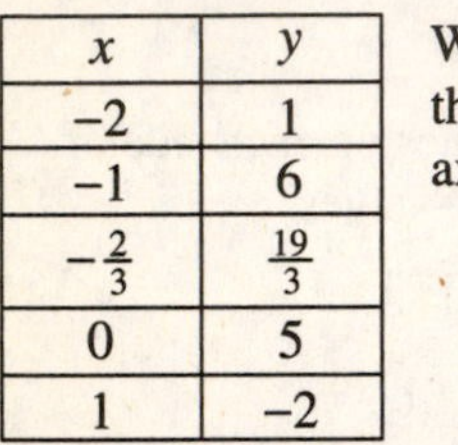

25. Graph $y=\frac{2}{x}$.

First we determine the domain. The domain is all real numbers except for 0, since substituting 0 for x would result in division by 0. Now, to find y-values, we substitute any value for x other than 0, and compute the value for y.

When $x=-4$, $y=\frac{2}{-4}=-\frac{1}{2}$

When $x=-2$, $y=\frac{2}{-2}=-1$

When $x=-1$, $y=\frac{2}{-1}=-2$

When $x=\ 1$, $y=\frac{2}{1}=2$

When $x=\ 2$, $y=\frac{2}{2}=1$

When $x=\ 4$, $y=\frac{2}{4}=\frac{1}{2}$

x	y
-4	$-\frac{1}{2}$
-2	-1
-1	-2
1	2
2	1
4	$\frac{1}{2}$

We plot these points and connect them with a smooth curve on the axis below.

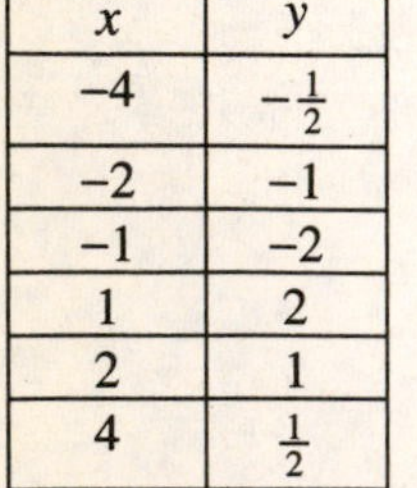

27. Graph $y=-\frac{2}{x}$.

First we determine the domain. The domain is all real numbers except for 0, since substituting 0 for x would result in division by 0. Now, to find y-values, we substitute any value for x other than 0, and compute the value for y.

When $x=-4$, $y=-\frac{2}{-4}=\frac{1}{2}$

When $x=-2$, $y=-\frac{2}{-2}=1$

When $x=-1$, $y=-\frac{2}{-1}=2$

When $x=\ 1$, $y=-\frac{2}{1}=-2$

When $x=\ 2$, $y=-\frac{2}{2}=-1$

When $x=\ 4$, $y=-\frac{2}{4}=-\frac{1}{2}$

x	y
-4	$\frac{1}{2}$
-2	1
-1	2
1	-2
2	-1
4	$-\frac{1}{2}$

We plot these points and connect them with a smooth curve on the axis below.

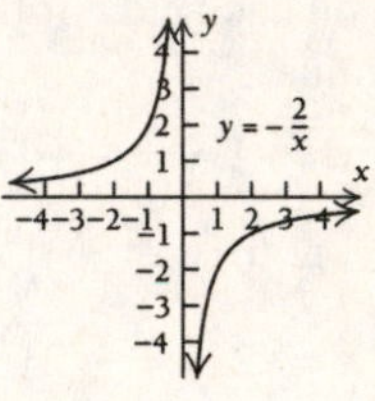

29. Graph $y=\frac{1}{x^2}$.

First we determine the domain. The domain is all real numbers except for 0, since substituting 0 for x would result in division by 0. Now, to find y-values, we substitute any value for x other than 0, and compute the value for y.

When $x=-2$, $y=\frac{1}{(-2)^2}=\frac{1}{4}$

When $x=-1$, $y=\frac{1}{(-1)^2}=1$

When $x=-\frac{1}{2}$, $y=\frac{1}{\left(-\frac{1}{2}\right)^2}=\frac{1}{\frac{1}{4}}=4$

When $x=\ \frac{1}{2}$, $y=\frac{1}{\left(\frac{1}{2}\right)^2}=\frac{1}{\frac{1}{4}}=4$

The solution is continued on the next page.

Copyright © 2014 Pearson Education, Inc. Publishing as Addison-Wesley.

When $x = 1$, $y = \frac{1}{(1)^2} = 1$

When $x = 2$, $y = \frac{1}{(2)^2} = 4$

x	y
-2	$\frac{1}{4}$
-1	1
$-\frac{1}{2}$	4
$\frac{1}{2}$	4
1	1
2	$\frac{1}{4}$

We plot these points and connect them with a smooth curve on the axis below.

31. Graph $y = \sqrt[3]{x}$.

First we determine the domain of the function. Since the index of the radicand is odd (3) the domain is all real numbers. We are free to choose any number for x and compute the value for y.

When $x = -8$, $y = \sqrt[3]{-8} = -2$

When $x = -1$, $y = \sqrt[3]{-1} = -1$

When $x = 0$, $y = \sqrt[3]{0} = 0$

When $x = 1$, $y = \sqrt[3]{1} = 1$

When $x = 8$, $y = \sqrt[3]{8} = 2$

x	y
-8	-2
-1	-1
0	0
1	1
8	2

We plot these points and connect them with a smooth curve on the axis below.

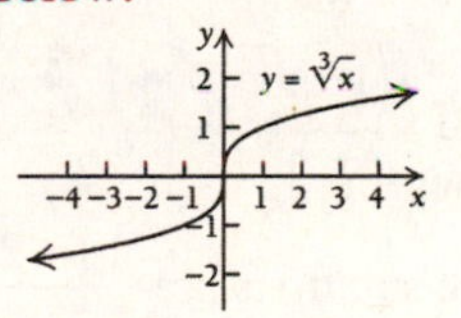

33. Graph $f(x) = \frac{x^2 + 5x + 6}{x + 3}$.

First, we simplify the function, by factoring the numerator and removing a factor of 1 as follows:

$$f(x) = \frac{x^2 + 5x + 6}{x + 3}$$
$$= \frac{(x+3)(x+2)}{x+3}$$
$$f(x) = \frac{x+3}{x+3} \cdot \frac{x+2}{1}$$
$$= x + 2, \quad x \neq -3$$

The simplification assumes that x is not -3. The number -3 is not in the domain of the original function because it would result in division by zero. Thus we can express the function as follows:

$y = f(x) = x + 2, \quad x \neq -3.$

To find function values, we substitute any value for x other than -3 and calculate the y-values.

x	y
-5	-3
-4	-2
-2	0
-1	1
0	2
1	3
2	4

We plot these points and draw the graph. The open circle at the point $(-3, -1)$ indicates that it is not part of the graph.

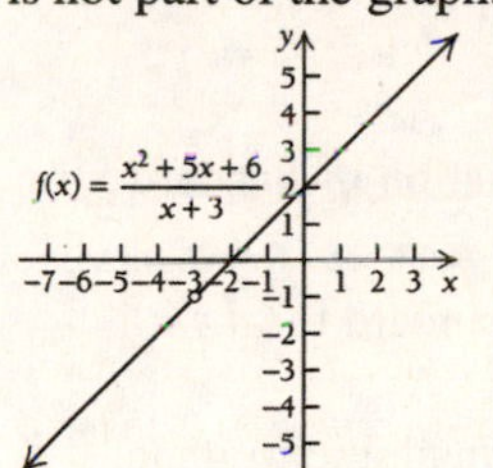

35. Graph $f(x) = \frac{x^2 - 1}{x - 1}$.

First, we simplify the function, by factoring the numerator and removing a factor of 1 as follows:

$$f(x) = \frac{x^2 - 1}{x - 1}$$
$$= \frac{(x-1)(x+1)}{x-1}$$
$$f(x) = \frac{x-1}{x-1} \cdot \frac{x+1}{1}$$
$$= x + 1, \quad x \neq 1$$

The simplification assumes that x is not 1. The number 1 is not in the domain of the original function because it would result in division by zero. Thus we can express the function as follows:

$y = f(x) = x + 1, \quad x \neq 1.$

The solution is continued on the next page.

Copyright © 2014 Pearson Education, Inc. Publishing as Addison-Wesley.

To find function values, we substitute any value for x other than 1 and calculate the y-values.

x	y
−2	−1
−1	0
0	1
2	3
3	4
4	5

We plot these points and draw the graph. The open circle at the point $(1,2)$ indicates that it is not part of the graph.

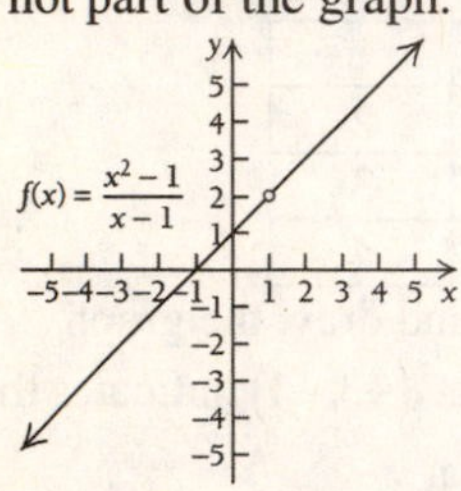

37. Solve $x^2-2x=2$.
First, we put the equation in standard form.
$x^2-2x-2=0$ subtract 2 from both sides
The equation is in standard form with
$a=1,\ b=-2,\ c=-2$
Next, we apply the quadratic formula.

$$x=\frac{-b\pm\sqrt{b^2-4ac}}{2a}$$

Substituting the values for a, b, and c, we get:

$$x=\frac{-(-2)\pm\sqrt{(-2)^2-4(1)(-2)}}{2(1)}$$
$$=\frac{2\pm\sqrt{4+8}}{2}=\frac{2\pm\sqrt{12}}{2}$$
$$=\frac{2\pm2\sqrt{3}}{2} \qquad \left(\sqrt{12}=\sqrt{4\cdot3}=2\sqrt{3}\right)$$
$$=\frac{2\left(1\pm\sqrt{3}\right)}{2\cdot1}=1\pm\sqrt{3}$$

The solutions are $1+\sqrt{3}$ and $1-\sqrt{3}$.

39. Solve $x^2+6x=1$.
First, we put the equation in standard form.
$x^2+6x-1=0$ subtract 1 from both sides
The equation is in standard form with
$a=1,\ b=6,\ c=-1$
Next, we apply the quadratic formula.

$$x=\frac{-b\pm\sqrt{b^2-4ac}}{2a}$$

Substituting the values for a, b, and c, we get:

$$x=\frac{-(6)\pm\sqrt{(6)^2-4(1)(-1)}}{2(1)}$$
$$=\frac{-6\pm\sqrt{36+4}}{2}=\frac{-6\pm\sqrt{40}}{2}$$
$$=\frac{-6\pm2\sqrt{10}}{2} \qquad \left(\sqrt{40}=\sqrt{4\cdot10}=2\sqrt{10}\right)$$
$$=\frac{2\left(-3\pm\sqrt{10}\right)}{2\cdot1}=-3\pm\sqrt{10}$$

The solutions are $-3+\sqrt{10}$ and $-3-\sqrt{10}$.

41. Solve $4x^2=4x+1$.
First, we put the equation in standard form.
$4x^2-4x-1=0$ subtract $4x$ and 1 from both sides
The equation is in standard form with
$a=4,\ b=-4,\ c=-1$
Next, we apply the quadratic formula.

$$x=\frac{-b\pm\sqrt{b^2-4ac}}{2a}$$

Substituting the values for a, b, and c, we have

$$x=\frac{-(-4)\pm\sqrt{(-4)^2-4(4)(-1)}}{2(4)}$$
$$=\frac{4\pm\sqrt{16+16}}{8}$$
$$=\frac{4\pm\sqrt{32}}{8}$$
$$x=\frac{4\pm4\sqrt{2}}{8} \qquad \left(\sqrt{32}=\sqrt{16\cdot2}=4\sqrt{2}\right)$$
$$=\frac{4\left(1\pm\sqrt{2}\right)}{4\cdot2}=\frac{1\pm\sqrt{2}}{2}$$

The solutions are $\frac{1+\sqrt{2}}{2}$ and $\frac{1-\sqrt{2}}{2}$.

43. Solve $3y^2+8y+2=0$.
The equation is in standard form with
$a=3,\ b=8,\ c=2$
Next, we apply the quadratic formula.

$$y=\frac{-b\pm\sqrt{b^2-4ac}}{2a}$$

The solution is continued on the next page.

Copyright © 2014 Pearson Education, Inc. Publishing as Addison-Wesley.

Substituting the values for *a, b,* and *c*, we get:

$$y=\frac{-(8)\pm\sqrt{(8)^2-4(3)(2)}}{2(3)}$$

$$=\frac{-8\pm\sqrt{64-24}}{6}=\frac{-8\pm\sqrt{40}}{6}$$

$$=\frac{-8\pm2\sqrt{10}}{6}\qquad\left(\sqrt{40}=\sqrt{4\cdot10}=2\sqrt{10}\right)$$

$$=\frac{2\left(-4\pm\sqrt{10}\right)}{2\cdot3}=\frac{-4\pm\sqrt{10}}{3}$$

The solutions are $\frac{-4+\sqrt{10}}{3}$ and $\frac{-4-\sqrt{10}}{3}$.

45. Solve $x+7+\frac{9}{x}=0$.

Multiplying both sides by *x*, we get:

$$x\cdot\left(x+7+\frac{9}{x}\right)=0\cdot x$$

$$x^2+7x+9=0.$$

This is a quadratic equation in standard form with $a=1, b=7$, and $c=9$.

Next, we apply the quadratic formula.

$$x=\frac{-b\pm\sqrt{b^2-4ac}}{2a}$$

Substituting the values for *a, b,* and *c*, we get:

$$x=\frac{-(7)\pm\sqrt{(7)^2-4(1)(9)}}{2(1)}$$

$$=\frac{-7\pm\sqrt{49-36}}{2}=\frac{-7\pm\sqrt{13}}{2}$$

The solutions are $\frac{-7+\sqrt{13}}{2}$ and $\frac{-7-\sqrt{13}}{2}$.

47. $\sqrt{x^3}=x^{3/2}\qquad\left(\text{The index is 2; }\sqrt[n]{a^m}=a^{m/n}\right)$

49. $\sqrt[5]{a^3}=a^{3/5}\qquad\left(\sqrt[n]{a^m}=a^{m/n}\right)$

51. $\sqrt[7]{t}=t^{1/7}$

53. $\sqrt[4]{x^{12}}=x^{12/4}=x^3$

55. $\frac{1}{\sqrt{t^5}}=\frac{1}{t^{5/2}}\qquad\left(\sqrt[n]{a^m}=a^{m/n}\right)$

$$=t^{-5/2}\qquad\left(\frac{1}{a^n}=a^{-n}\right)$$

57. $\frac{1}{\sqrt{x^2+7}}=\frac{1}{\left(x^2+7\right)^{1/2}}\qquad\left(\sqrt[n]{a^m}=a^{m/n}\right)$

$$=\left(x^2+7\right)^{-1/2}\qquad\left(\frac{1}{a^n}=a^{-n}\right)$$

59. $x^{1/5}=\sqrt[5]{x^1}=\sqrt[5]{x}\qquad\left(a^{m/n}=\sqrt[n]{a^m}\right)$

61. $y^{2/3}=\sqrt[3]{y^2}\qquad\left(a^{m/n}=\sqrt[n]{a^m}\right)$

63. $t^{-2/5}=\frac{1}{t^{2/5}}\qquad\left(a^{-n}=\frac{1}{a^n}\right)$

$$=\frac{1}{\sqrt[5]{t^2}}\qquad\left(a^{m/n}=\sqrt[n]{a^m}\right)$$

65. $b^{-1/3}=\frac{1}{b^{1/3}}\qquad\left(a^{-n}=\frac{1}{a^n}\right)$

$$=\frac{1}{\sqrt[3]{b}}\qquad\left(a^{m/n}=\sqrt[n]{a^m}\right)$$

67. $e^{-17/6}=\frac{1}{e^{17/6}}\qquad\left(a^{-n}=\frac{1}{a^n}\right)$

$$=\frac{1}{\sqrt[6]{e^{17}}}\qquad\left(a^{m/n}=\sqrt[n]{a^m}\right)$$

69. $\left(x^2-3\right)^{-1/2}=\frac{1}{\left(x^2-3\right)^{1/2}}\qquad\left(a^{-n}=\frac{1}{a^n}\right)$

$$=\frac{1}{\sqrt{x^2-3}}\qquad\left(a^{m/n}=\sqrt[n]{a^m}\right)$$

71. $\frac{1}{t^{2/3}}=\frac{1}{\sqrt[3]{t^2}}\qquad\left(a^{m/n}=\sqrt[n]{a^m}\right)$

73. $9^{3/2}$

$$=\left(9^{1/2}\right)^3\qquad\left(\tfrac{3}{2}=\tfrac{1}{2}\cdot3;\ a^{m\cdot n}=\left(a^m\right)^n\right)$$

$$=\left(\sqrt{9}\right)^3\qquad\left(a^{1/n}=\sqrt[n]{a}\right)$$

$$=(3)^3=27$$

Copyright © 2014 Pearson Education, Inc. Publishing as Addison-Wesley.

75. $64^{2/3}$

$= \left(64^{1/3}\right)^2 \qquad \left(\frac{2}{3} = \frac{1}{3}\cdot 2;\ a^{m\cdot n} = \left(a^m\right)^n\right)$

$= \left(\sqrt[3]{64}\right)^2 \qquad \left(a^{1/n} = \sqrt[n]{a}\right)$

$= (4)^2 \qquad \left(\sqrt[3]{64} = 4\right)$

$= 16$

77. $16^{3/4}$

$= \left(16^{1/4}\right)^3 \qquad \left(\frac{3}{4} = \frac{1}{4}\cdot 3;\ a^{m\cdot n} = \left(a^m\right)^n\right)$

$= \left(\sqrt[4]{16}\right)^3 \qquad \left(a^{1/n} = \sqrt[n]{a}\right)$

$= (2)^3 \qquad \left(\sqrt[4]{16} = 2\right)$

$= 8$

79. The domain of a rational function is restricted to those input values that do not result in division by 0. To determine the domain of

$$f(x) = \frac{x^2 - 25}{x - 5}$$

we set the denominator equal to zero and solve:

$x - 5 = 0$

$x = 5.$

Therefore, 5 is not in the domain. The domain of f consists of all real numbers except 5.

81. The domain of a rational function is restricted to those input values that do not result in division by 0.
To determine the domain of

$$f(x) = \frac{x^3}{x^2 - 5x + 6}$$

we set the denominator equal to zero and solve:

$x^2 - 5x + 6 = 0$

$(x-2)(x-3) = 0$ factoring

$x - 2 = 0$ or $x - 3 = 0$ Principle of Zero Products

$x = 2$ or $x = 3.$

Therefore, 2 and 3 are not in the domain. The domain of f consists of all real numbers except 2 and 3.

83. The domain of the radical function $f(x) = \sqrt{5x+4}$ is restricted to those input values that result in the value of the radicand being greater than or equal to 0. In other words, the domain will be the set of real numbers that satisfy the inequality $5x + 4 \geq 0$.

To find the domain, we solve the inequality:

$5x + 4 \geq 0$

$5x \geq -4$

$x \geq -\frac{4}{5}.$

Therefore, the domain of f consists of all real numbers greater than or equal to $-\frac{4}{5}$, or in interval notation $\left[-\frac{4}{5}, \infty\right)$.

85. The domain of the radical function $f(x) = \sqrt[4]{7-x}$ is restricted to those input values that result in the value of the radicand being greater than or equal to 0. In other words, the domain will be the set of real numbers that satisfy the inequality $7 - x \geq 0$.
To find the domain, we solve the inequality:

$7 - x \geq 0$

$7 \geq x.$

Therefore, the domain of f consists of all real numbers less than or equal to 7, or in interval notation $(-\infty, 7]$.

87. We set the demand equation equal to the supply equation and solve for x.

$1000 - 10x = 250 + 5x$

$750 = 15x$

$50 = x.$

Thus, the equilibrium price is \$50. To find the equilibrium quantity, we substitute 50 for x into either the demand equation or supply equation. We use the demand equation.

$q = 1000 - 10(50)$

$q = 1000 - 500 = 500$

The equilibrium quantity is 500 units. The equilibrium point is $(50, 500)$.

89. We set the demand equation equal to the supply equation and solve for x.

$\frac{5}{x} = \frac{x}{5}$

$25 = x^2$ multiply both sides by $5x$

$\sqrt{25} = \sqrt{x^2}$ take the square root of both sides

$\pm 5 = x$

Since it is not appropriate to have a negative price, the equilibrium price is 5 hundred dollars or \$500. We find the equilibrium on the next page.

Copyright © 2014 Pearson Education, Inc. Publishing as Addison-Wesley.

In order to find the equilibrium quantity, we substitute 5 for x into either the demand equation or supply equation. We use the demand equation.

$$q=\frac{5}{(5)}=1$$

The equilibrium quantity is 1 thousand units or 1000 units. The equilibrium point is $(5,1)$.

91. We set the demand equation equal to the supply equation and solve for x.

$$(x-3)^2=x^2+2x+1$$
$$x^2-6x+9=x^2+2x+1$$
$$-6x+9=2x+1 \quad \text{subtracting } x^2 \text{ from both sides}$$
$$8=8x$$
$$1=x$$

The equilibrium price is \$1.

To find the equilibrium quantity, we substitute 1 for x into either the demand equation or supply equation. We use the demand equation.

$$q=(1-3)^2$$
$$=(-2)^2=4$$

The equilibrium quantity is 4 hundred units or 400 units. The equilibrium point is $(1,4)$.

93. We set the demand equation equal to the supply equation and solve for x.

$$5-x=\sqrt{x+7} \qquad 0\le x\le 5$$
$$(5-x)^2=\left(\sqrt{x+7}\right)^2 \quad \text{Squaring both sides}$$
$$25-10x+x^2=x+7$$
$$18-11x+x^2=0$$
$$(9-x)(2-x)=0 \quad \text{Factoring the quadratic}$$
$$9-x=0 \text{ or } 2-x=0 \quad \text{Principle of Zero Products}$$
$$9=x \text{ or } 2=x$$

Since 9 is not in the domain of the demand function, the equilibrium price is 2 thousand dollars or \$2000.

To find the equilibrium quantity, we substitute 2 for x into either the demand equation or supply equation. We use the demand equation.

$$q=5-(2)=3$$

The equilibrium quantity is 3 thousand units or 3000 units. The equilibrium point is $(2,3)$.

95. If the price per share S is inversely proportional to the prime rate R, then we have:

$$S=\frac{k}{R}.$$

We find the constant of variation by substituting 205.93 for S and 0.0325 (3.25%) for R.

$$205.93=\frac{k}{0.0325}$$
$$6.692725=k$$

The equation of variation is $S=\frac{6.692725}{R}$.

If the prime rate rose to 4.75% we find S by substituting 0.0475 in for R.

$$S=\frac{6.692725}{0.0475}$$
$$=140.899\approx 140.90$$

The price per share would be approximately \$140.90 if the assumption of inverse proportionality is correct.

97. a) $R(x)=11.74x^{0.25}$

$$R(40{,}000)=11.74(40{,}000)^{0.25}$$
$$=11.74(14.14213562)$$
$$=166.0286722\approx 166$$

The maximum range will be approximately 166 miles when the peak power is 40,000 watts.

$$R(50{,}000)=11.74(50{,}000)^{0.25}$$
$$=11.74(14.95348781)$$
$$=175.5539469\approx 176$$

The maximum range will be approximately 176 miles when the peak power is 50,000 watts.

$$R(60{,}000)=11.74(60{,}000)^{0.25}$$
$$=11.74(15.6508458)$$
$$=183.7409297\approx 184$$

The maximum range will be approximately 184 miles when the peak power is 60,000 watts.

b) Plotting the points found in part (a) and connecting them with a smooth curve we see:

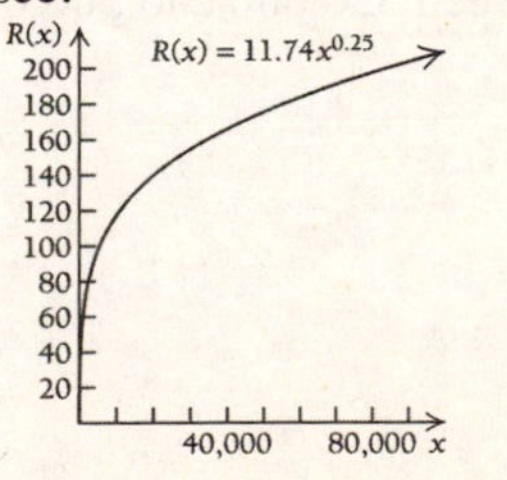

Copyright © 2014 Pearson Education, Inc. Publishing as Addison-Wesley.

99. a) In 2005, $t = 2005 - 1970 = 35$.

$P = 1000(35)^{5/4} + 14{,}000 \approx 99{,}130$

In 2005, average pollution was approximately 99,130 particles per cubic centimeter.

In 2008, $t = 2008 - 1970 = 38$.

$P = 1000(38)^{5/4} + 14{,}000 \approx 108{,}347$

In 2008, average pollution will be approximately 108,347 particles per cubic centimeter.

In 2014, $t = 2014 - 1970 = 44$.

$P = 1000(44)^{5/4} + 14{,}000 \approx 127{,}322$

In 2014, average pollution will be approximately 127,322 particles per cubic centimeter.

b) Plot the points above and others, if necessary, and draw the graph.

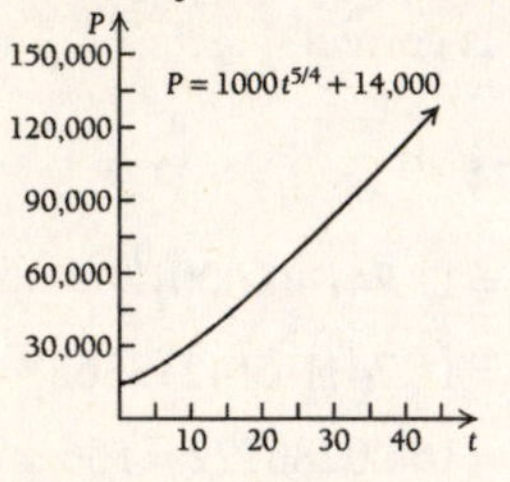

101. The number of cities N with a population greater than S is inversely proportional to S.

$$N = \frac{k}{S}$$

$$52 = \frac{k}{350{,}000} \quad \text{Substituting}$$

$18{,}200{,}000 = k$

The equation of variation is

$$N = \frac{18{,}200{,}000}{S}.$$

We find N when S is 500,000.

$$N = \frac{18{,}200{,}000}{500{,}000} = 36.4 \approx 36$$

Using the fact that there were 52 cities with a population greater than 350,000 and 36 cities with a population of 500,000 or greater, we estimate that there are $52 - 36 = 16$ cities with a population between 350,000 and 500,000.

To estimate the number of cities with a population between 300,000 and 600,000 we find N when S is 300,000 and when S is 600,000.

$$N = \frac{18{,}200{,}000}{300{,}000} \approx 60.667 \approx 61$$

$$N = \frac{18{,}200{,}000}{600{,}000} \approx 30.33 \approx 30$$

There are 61 cities with a population greater than 300,000 and 30 cities with a population greater than 600,000, so there are $61 - 30 = 31$ cities with a population between 300,000 and 600,000.

103. ✎

105. $f(x) = 2x^3 - x^2 - 14x - 10$

Enter the function into your calculator.

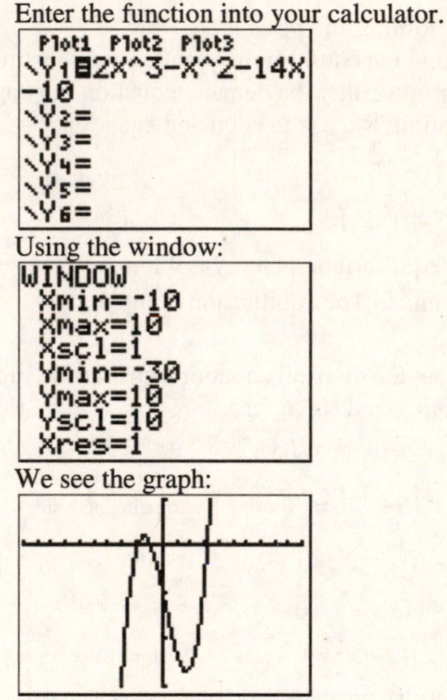

Now using the ZERO feature on the calculator, we approximate the zeros. The zeros are -1.831, -0.856, 3.188.

107. $f(x) = x^4 + 4x^3 - 36x^2 - 160x + 300$

Enter the function into your calculator.

The solution is continued on the next page.

Copyright © 2014 Pearson Education, Inc. Publishing as Addison-Wesley.

Using the window:

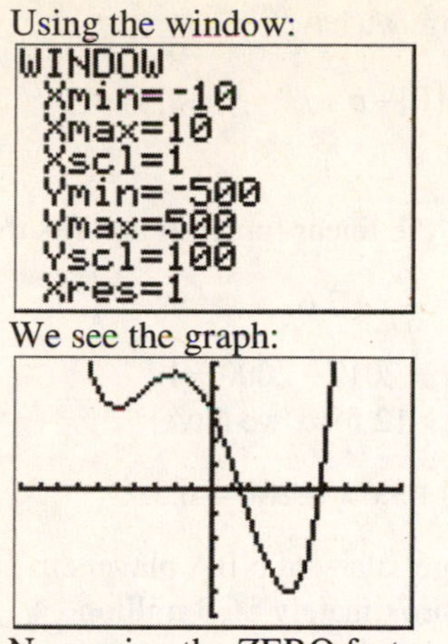

We see the graph:

Now using the ZERO feature on the calculator, we approximate the zeros. The zeros are 1.489 and 5.673.

109. $f(x)=|x+1|+|x-2|-5$

Enter the function into your calculator.

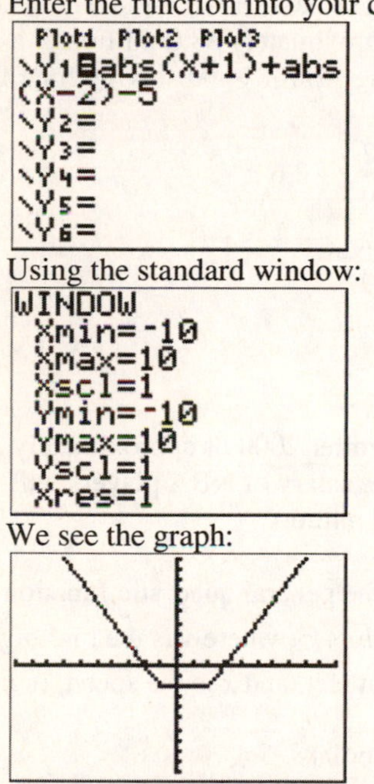

Using the standard window:

We see the graph:

Now using the ZERO feature on the calculator, we approximate the zeros. The zeros are –2 and 3.

111. $f(x)=|x+1|+|x-2|-3$

Enter the function into your calculator.

```
Plot1 Plot2 Plot3
\Y1=abs(X+1)+abs
(X-2)-3
\Y2=
\Y3=
\Y4=
\Y5=
\Y6=
```

Using the standard window:

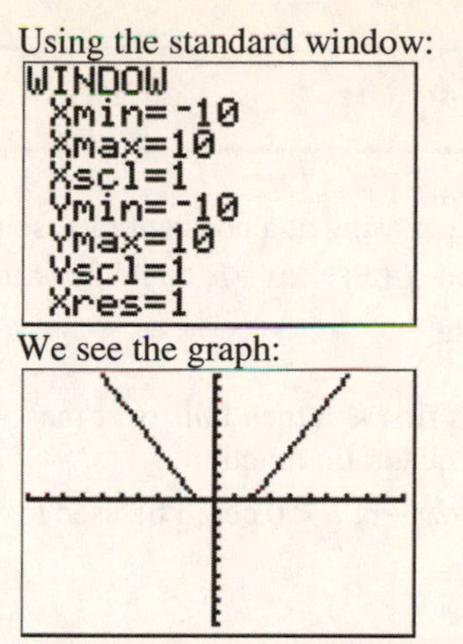

We see the graph:

We see that the graph intersects the x-axis between $-1\le x\le 2$. The zeros of this function are all real numbers in $[-1,2]$

113. We enter the demand and supply equations into the graphing editor on the calculator.

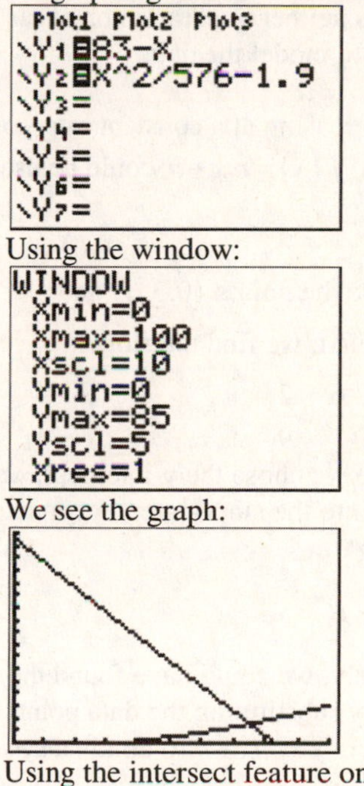

Using the window:

We see the graph:

Using the intersect feature on the calculator we find the intersection to be:

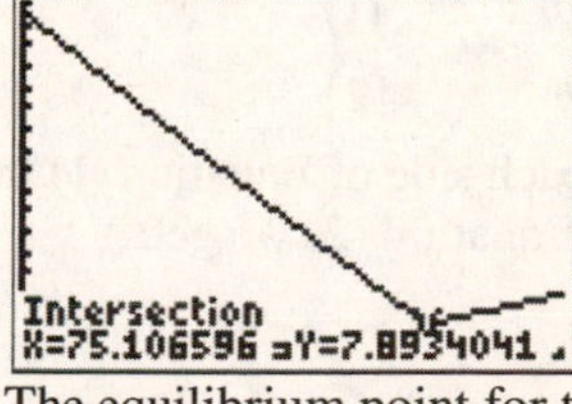

The equilibrium point for this market is $(75.11, 7.893)$. In other words, 7893 units will be sold at a price of \$75.11.

Copyright © 2014 Pearson Education, Inc. Publishing as Addison-Wesley.

Exercise Set R.6

1. The data is decreasing at a constant rate, so a linear function $f(x) = mx + b$ could be used to model the data.

3. The data rises first and then falls over the domain, so a quadratic function $f(x) = ax^2 + bx + c,\ a < 0$ could be used to model the data.

5. The data is increasing at a constant rate, so a linear function $f(x) = mx + b$ could be used to model the data.

7. The data rises and falls several times over the domain. This implies that a polynomial function that is neither quadratic nor linear would be best to model the data.

9. The data is decreasing at a constant rate, so a linear function $f(x) = mx + b$ could be used to model the data.

11. a) We will use the points $(0, 3.6)$ and $(9, 5.6)$. First, we find the slope:

$$m = \frac{5.6 - 3.6}{9 - 0} = \frac{2}{9}$$

Next, since we chose the y-intercept, we can substitute into the slope-intercept equation.

$$y = mx + b$$

$$y = \frac{2}{9}x + 3.6$$

Alternatively, we could have found the equation by substituting the data points in to the equation $y = mx + b$ to obtain a system of equations.

$$3.6 = m \cdot 0 + b \qquad (1)$$

$$5.6 = m \cdot 9 + b \qquad (2)$$

Subtracting each side of Equation (1) from each side of Equation (2) we get:

$$2 = 9m$$

$$\frac{2}{9} = m$$

Now substitute $m = \frac{2}{9}$ in for either Equation (1) or (2) and solve for b. At the top of the next column, we use Equation (1).

Substituting, we have

$$3.6 = \left(\frac{2}{9}\right)(0) + b$$

$$3.6 = b.$$

Therefore, the linear function that fits the data is $y = \frac{2}{9}x + 3.6$

b) In 2012, $x = 2012 - 2000 = 12$

Substituting 12 for x we have:

$$y = \frac{2}{9}(12) + 3.6 = 6.26\overline{6} \approx 6.3.$$

The average salary of NBA players in 2012 will be approximately \$6.3 million.

In 2020, $x = 2020 - 2000 = 20$

Substituting 12 for x we have:

$$y = \frac{2}{9}(20) + 3.6 = 8.04\overline{4} \approx 8.0.$$

The average salary of NBA players in 2020 will be approximately \$8.0 million.

c) Substituting 9.0 for y and solving for x we have:

$$9.0 = \frac{2}{9}x + 3.6$$

$$5.4 = \frac{2}{9}x$$

$$\frac{9}{2}(5.4) = x$$

$$24.3 \approx x.$$

24.3 years after 2000 or approximately 2024 the average salary of NBA players will reach \$9.0 million.

13. a) Consider the general quadratic function $y = ax^2 + bx + c$, where y is the braking distance, in feet, and x is the speed, in miles per hour.

Using the points $(20, 25), (40, 105)$, and $(60, 300)$, we substitute each point into the general quadratic function obtain the following system of equations:

$$25 = a \cdot 20^2 + b \cdot 20 + c$$

$$105 = a \cdot 40^2 + b \cdot 40 + c$$

$$300 = a \cdot 60^2 + b \cdot 60 + c$$

or,

$$25 = 400a + 20b + c$$

$$105 = 1600a + 40b + c$$

$$300 = 3600a + 60b + c$$

The solution is continued on the next page.

Copyright © 2014 Pearson Education, Inc. Publishing as Addison-Wesley.

Solving the system of equations from the previous page, we get:
$a = 0.14375$, $b = -4.625$, and $c = 60$.

Therefore, the function is

$y = 0.144x^2 - 4.63x + 60$.

b) We substitute 50 for x and compute the value of y.

$y = 0.144(50)^2 - 4.63(50) + 60 = 188.5$

The breaking distance of a car traveling at 50 mph is about 188.5 ft.

c) No, the function does not make sense for speeds less than 15 mph. The vertex of the parabola occurs around 16 mph. Thus for speeds less than 16 mph the breaking distance starts increasing as the speed decreases.

15. a) Answers will vary depending on which points are used to find the function.
We will use the points $(30, 1.4)$ and $(70, 53.0)$. We substitute in to the equation $y = mx + b$ to obtain a system of equations.

$1.4 = m \cdot 30 + b \qquad (1)$

$53.0 = m \cdot 70 + b \qquad (2)$

Subtracting each side of Equation (1) from each side of Equation (2) we get:

$51.6 = 40m$

$1.29 = m$.

Now substitute $m = 1.29$ in for either Equation (1) or (2) and solve for b. We use Equation (1).

$1.4 = (1.29)(30) + b$

$1.4 = 38.7 + b$

$-37.3 = b$..

Therefore, the linear function that fits the data is $y = 1.29x - 37.3$.

b) We plot the points in the table in the text and then graph the function found in part (a) on the same set of axes.

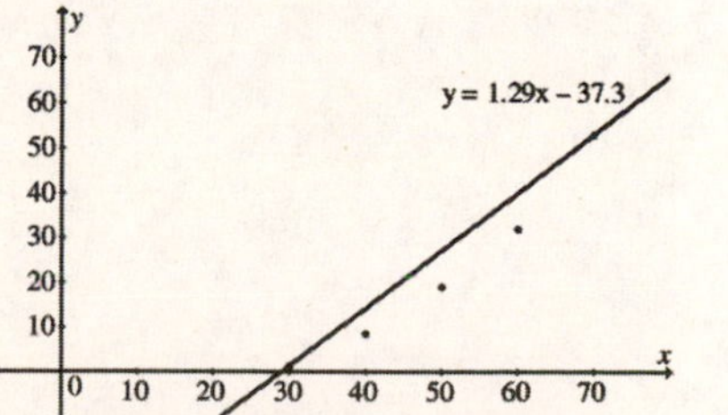

c) Substituting in 55 for x we have:

$y = 1.29(55) - 37.3 = 33.65$.

Approximately 33.65% of 55-yr-old women have high blood pressure.

17.

19.

21. a) First, we enter the data into the statistic editor on the calculator, letting L_1 be the values for x and L_2 be the values for y.

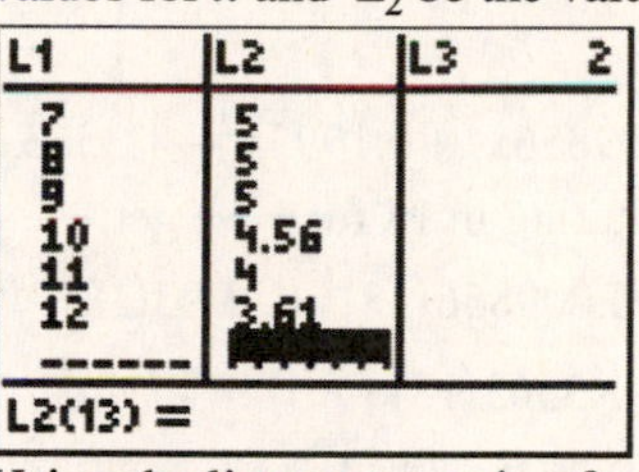

Using the linear regression feature we get:

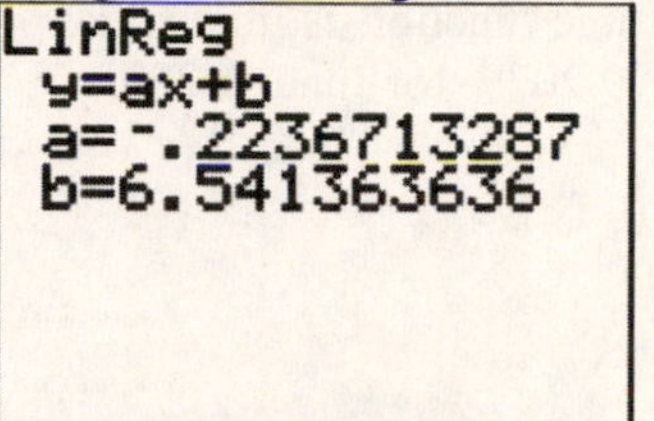

The linear function that fits the data is $y = -0.224x + 6.5414$

b) Substituting 18 in for x we get:

$y = -0.224(18) + 6.5414$

≈ 2.509

≈ 2.51

The prime rate in June 2009 will be approximately 2.51%

c) The actual prime rate in June of 2009 was 3.25%. So the linear regression on the calculator gave us a closer approximation to the actual prime rate for that month. The regression answer seems more plausible; it uses all the data.

Copyright © 2014 Pearson Education, Inc. Publishing as Addison-Wesley.

d) Using the cubic regression feature on the calculator results in:

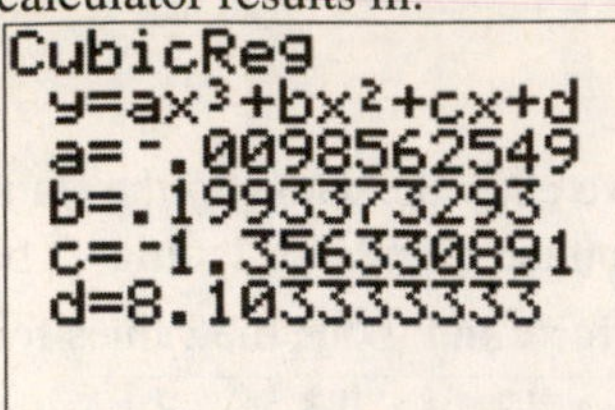

$y = -0.009856x^3 + 0.1993x^2 - 1.3563x + 8.103$

Substituting in 18 for x we get:

$$y = -0.009856(18)^3 + .1993(18)^2 - 1.3563(18) + 8.103$$
$$\approx -9.207$$

The cubic function estimated the prime rate to be -9.207% for June of 2005.

e) ✎

Copyright © 2014 Pearson Education, Inc. Publishing as Addison-Wesley.

Chapter 1

Differentiation

Exercise Set 1.1

1. We select x-values closer and closer to 3, and observe the corresponding output.

x	2.99	2.999	3	3.001	3.01
$2x+5$	10.98	10.998	$\to 11 \leftarrow$	11.002	11.02

As x approaches 3, the value of $2x+5$ approaches 11.

3. We solve the equation:

$-3x = 6$

$x = -2$

Therefore, As x approaches -2, the value of $-3x$ approaches 6.

5. The notation $\lim_{x\to 4} f(x)$ is read "the limit, as x approaches 4, of $f(x)$."

7. The notation $\lim_{x\to 5^-} F(x)$ is read "the limit, as x approaches 5 from the left, of $F(x)$."

9. The notation $\lim_{x\to 2^+}$ is read "the limit, as x approaches 2 from the right."

11. As inputs x approach 3 from the right, outputs $f(x)$ approach 2. Thus the limit from the right is 2. That is,

$\lim_{x\to 3^+} f(x) = 2.$

13. As inputs x approach -1 from the left, outputs $f(x)$ approach -3. Thus the limit from the left is -3. That is,

$\lim_{x\to -1^-} f(x) = -3.$

15. From Exercise (11) and (12) we k now that $\lim_{x\to 3^-} f(x) = 1$ and $\lim_{x\to 3^+} f(x) = 2$. Since the limit from the left, 1, is not the same as the limit from the right, 2, $\lim_{x\to 3} f(x)$ *does not exist.*

17. As inputs x approach 4 from the left, outputs $f(x)$ approach 3. Thus the limit from the left is 3. That is,

$\lim_{x\to 4^-} f(x) = 3.$

As inputs x approach 4 from the right, outputs $f(x)$ approach 3. Thus the limit from the right is 4. That is,

$\lim_{x\to 4^+} f(x) = 3.$

Since the limit from the left, 3, is the same as the limit from the right, 3, we have

$\lim_{x\to 4} f(x) = 3.$

19. As inputs x approach -2 from the left, outputs $g(x)$ approach 4. Thus the limit from the left is 4. That is,

$\lim_{x\to -2^-} g(x) = 4.$

21. As inputs x approach 4 from the right, outputs $g(x)$ approach -1. Thus the limit from the right is -1. That is,

$\lim_{x\to 4^+} g(x) = -1.$

23. Since the limit from the left, -1, is the same as the limit from the right, -1, we have

$\lim_{x\to 4} g(x) = -1.$

25. As inputs x approach 2 from the left, outputs $g(x)$ approach 0. Thus the limit from the left is 0. That is,

$\lim_{x\to 2^-} g(x) = 0.$

As inputs x approach 2 from the right, outputs $g(x)$ approach 0. Thus the limit from the right is 0. That is,

$\lim_{x\to 2^+} g(x) = 0.$

Since the limit from the left, 0, is the same as the limit from the right, 0, we have

$\lim_{x\to 2} g(x) = 0.$

Copyright © 2014 Pearson Education, Inc. Publishing as Addison-Wesley.

27. As inputs x approach -3 from the left, outputs $F(x)$ approach 5. Thus the limit from the left is 5. That is,
$\lim_{x\to-3^-} F(x)=5$.
As inputs x approach -3 from the right, outputs $F(x)$ approach 5. Thus the limit from the right is 5. That is,
$\lim_{x\to-3^+} F(x)=5$.
Since the limit from the left, 5, is the same as the limit from the right, 5, we have
$\lim_{x\to-3} F(x)=5$.

29. As inputs x approach -2 from the left, outputs $F(x)$ approach 4. Thus the limit from the left is 4. That is,
$\lim_{x\to-2^-} F(x)=4$.
As inputs x approach -2 from the right, outputs $F(x)$ approach 2. Thus the limit from the right is 2. That is,
$\lim_{x\to-2^+} F(x)=2$.
Since the limit from the left, 4, is not the same as the limit from the right, 2, we have
$\lim_{x\to-2} F(x)$ *does not exist.*

31. As inputs x approach 4 from the left, outputs $F(x)$ approach 2. Thus the limit from the left is 2. That is,
$\lim_{x\to4^-} F(x)=2$.
As inputs x approach 4 from the right, outputs $F(x)$ approach 2. Thus the limit from the right is 2. That is,
$\lim_{x\to4^+} F(x)=2$.
Since the limit from the left, 2, is the same as the limit from the right, 2, we have
$\lim_{x\to4} F(x)=2$.

33. As inputs x approach -2 from the right, outputs $F(x)$ approach 2. Thus the limit from the right is 2. That is,
$\lim_{x\to-2^+} F(x)=2$.

35. As inputs x approach -2 from the left, outputs $G(x)$ approach 1. Thus the limit from the left is 1. That is,
$\lim_{x\to-2^-} G(x)=1$.
As inputs x approach -2 from the right, outputs $G(x)$ approach 1. Thus the limit from the right is 1. That is,
$\lim_{x\to-2^+} G(x)=1$.
Since the limit from the left, 1, is the same as the limit from the right, 1, we have
$\lim_{x\to-2} G(x)=1$.

37. As inputs x approach 1 from the left, outputs $G(x)$ approach 4. Thus the limit from the left is 4. That is, $\lim_{x\to1^-} G(x)=4$.

39. As inputs x approach 1 from the left, outputs $G(x)$ approach 4. Thus the limit from the left is 4. That is,
$\lim_{x\to1^-} G(x)=4$.
As inputs x approach 1 from the right, outputs $G(x)$ approach -1. Thus the limit from the right is -1. That is,
$\lim_{x\to1^+} G(x)=-1$.
Since the limit from the left, 4, is not the same as the limit from the right, -1, we have
$\lim_{x\to1} G(x)$ *does not exist.*

41. As inputs x approach 3 from the right, outputs $G(x)$ approach 0. Thus the limit from the right is 0. That is, $\lim_{x\to3^+} G(x)=0$.

43. As inputs x approach -3 from the left, outputs $H(x)$ approach 0. Thus the limit from the left is 0. That is,
$\lim_{x\to-3^-} H(x)=0$.
As inputs x approach -3 from the right, outputs $H(x)$ approach 0. Thus the limit from the right is 0. That is,
$\lim_{x\to-3^+} H(x)=0$.
Since the limit from the left, 0, is the same as the limit from the right, 0, we have
$\lim_{x\to-3} H(x)=0$.

Copyright © 2014 Pearson Education, Inc. Publishing as Addison-Wesley.

45. As inputs x approach -2 from the right, outputs $H(x)$ approach 1. Thus the limit from the right is 1. That is,
$\lim_{x\to -2^+} H(x)=1$.

47. As inputs x approach 1 from the left, outputs $H(x)$ approach 4. Thus the limit from the left is 4. That is,
$\lim_{x\to 1^-} H(x)=4$.

49. As inputs x approach 1 from the left, outputs $H(x)$ approach 4. Thus the limit from the left is 4. That is,
$\lim_{x\to 1^-} H(x)=4$.
The solution is continued on the next page.
As inputs x approach 1 from the right, outputs $H(x)$ approach 2. Thus the limit from the right is 2. That is,
$\lim_{x\to 1^+} H(x)=2$.
Since the limit from the left, 4, is not the same as the limit from the right, 2, we have
$\lim_{x\to 1} H(x)$ *does not exist.*

51. As inputs x approach 3 from the right, outputs $H(x)$ approach 1. Thus the limit from the right is 1. That is,
$\lim_{x\to 3^+} H(x)=1$.

53. As inputs x approach -1 from the left, outputs $f(x)$ approach 1. Thus the limit from the left is 1. That is,
$\lim_{x\to -1^-} f(x)=1$.
As inputs x approach -1 from the right, outputs $f(x)$ approach 1. Thus the limit from the right is 1. That is,
$\lim_{x\to -1^+} f(x)=1$.
Since the limit from the left, 1, is the same as the limit from the right, 1, we have
$\lim_{x\to -1} f(x)=1$.

55. As inputs x approach -3 from the left, outputs $f(x)$ increase without bound. We say that the limit from the left is infinity. That is
$\lim_{x\to -3^-} f(x)=\infty$.
As inputs x approach -3 from the right, outputs $f(x)$ decrease without bound. We say that limit from the right is negative infinity. That is,
$\lim_{x\to -3^+} f(x)=-\infty$.
Since the function values as $x \to 3$ from the left increase without bound, and the function values as $x \to 3$ from the right decrease without bound, the limit does not exist. We have,
$\lim_{x\to -3} f(x)$ *does not exist.*

57. As inputs x approach 3 from the left, outputs $f(x)$ approach 0. Thus the limit from the left is 0. That is,
$\lim_{x\to 3^-} f(x)=0$.
As inputs x approach 3 from the right, outputs $f(x)$ approach 0.
Thus the limit from the right is 0. That is,
$\lim_{x\to 3^+} f(x)=0$.
Since the limit from the left, 0, is the same as the limit from the right, 0, we have
$\lim_{x\to 3} f(x)=0$.

59. As inputs x approach -4 from the left, outputs $f(x)$ approach 3. Thus the limit from the left is 3. That is,
$\lim_{x\to -4^-} f(x)=3$.
As inputs x approach -4 from the right, outputs $f(x)$ approach 3. Thus the limit from the right is 3. That is,
$\lim_{x\to -4^+} f(x)=3$.
Since the limit from the left, 3, is the same as the limit from the right, 3, we have
$\lim_{x\to -4} f(x)=3$.

61. As inputs x get larger and larger, outputs $f(x)$ get closer and closer to 2. We have
$\lim_{x\to \infty} f(x)=2$.

Copyright © 2014 Pearson Education, Inc. Publishing as Addison-Wesley.

63. Defining $f(x)=|x|$ as a piecewise defined function we have:

$$f(x)=\begin{cases}-x, & x<0\\ x, & x\geq 0\end{cases}.$$

We graph the function by creating an input-output table.

x	-2	-1	0	1	2
$f(x)$	2	1	0	1	2

Next, we plot the points from the table and draw the graph.

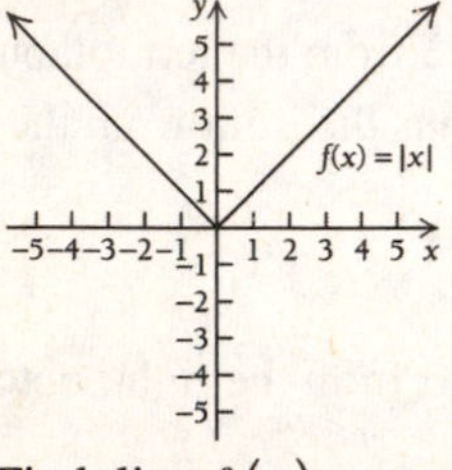

Find $\lim_{x\to 0} f(x)$.

As inputs x approach 0 from the left, outputs $f(x)$ approach 0. We have,

$\lim_{x\to 0^-} f(x)=0$.

As inputs x approach 0 from the right, outputs $f(x)$ approach 0. We have,

$\lim_{x\to 0^+} f(x)=0$

Since the limit from the left is the same as the limit from the right, we have

$\lim_{x\to 0} f(x)=0$.

Find $\lim_{x\to -2} f(x)$.

As inputs x approach -2 from the left, outputs $f(x)$ approach 2. We have,

$\lim_{x\to -2^-} f(x)=2$.

As inputs x approach -2 from the right, outputs $f(x)$ approach 2. We have,

$\lim_{x\to -2^+} f(x)=2$

Since the limit from the left is the same as the limit from the right, we have

$\lim_{x\to -2} f(x)=2$.

65. $g(x)=x^2-5$

We graph the function by creating an input-output table.

x	-2	-1	0	1	2
$g(x)$	-1	-4	-5	-4	-1

Next, we plot the points from the table and draw the graph.

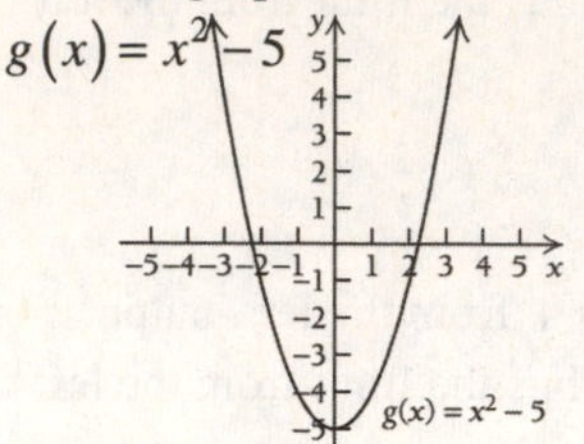

Find $\lim_{x\to 0} g(x)$.

As inputs x approach 0 from the left, outputs $g(x)$ approach -5. We have,

$\lim_{x\to 0^-} g(x)=-5$.

As inputs x approach 0 from the right, outputs $g(x)$ approach -5. We have,

$\lim_{x\to 0^+} g(x)=-5$

Since the limit from the left is the same as the limit from the right, we have

$\lim_{x\to 0} g(x)=-5$.

Find $\lim_{x\to -1} g(x)$.

As inputs x approach -1 from the left, outputs $g(x)$ approach -4. We have,

$\lim_{x\to -1^-} g(x)=-4$.

As inputs x approach -1 from the right, outputs $g(x)$ approach -4. We have,

$\lim_{x\to -1^+} g(x)=-4$

Since the limit from the left is the same as the limit from the right, we have

$\lim_{x\to -1} g(x)=-4$.

67. $F(x)=\dfrac{1}{x-3}$

Since $x=3$ makes the denominator zero, we exclude the value 3 from the domain. Creating an input-output table we have

x	1	2	2.5	2.9	3.1	3.5	4	5
$F(x)$	$-\frac{1}{2}$	-1	-2	-10	10	2	1	$\frac{1}{2}$

The solution is continued on the next page.

Copyright © 2014 Pearson Education, Inc. Publishing as Addison-Wesley.

Next we plot the points from the previous page and draw the graph.

Find $\lim\limits_{x \to 3} F(x)$.

As inputs x approach 3 from the left, outputs $F(x)$ decrease without bound . We have,

$\lim\limits_{x \to 3^-} F(x) = -\infty$.

As inputs x approach 3 from the right, outputs $F(x)$ increase without bound. We have,

$\lim\limits_{x \to 3^+} F(x) = \infty$

Since the function values as $x \to 3$ from the left decrease without bound, and the function values as $x \to 3$ from the right increase without bound, the limit does not exist. We have,

$\lim\limits_{x \to 3} F(x)$ *does not exist.*

Find $\lim\limits_{x \to 4} F(x)$.

As inputs x approach 4 from the left, outputs $F(x)$ approach 1. We have,

$\lim\limits_{x \to 4^-} F(x) = 1$.

As inputs x approach 4 from the right, outputs $F(x)$ approach 1. We have,

$\lim\limits_{x \to 4^+} F(x) = 1$

Since the limit from the left is the same as the limit from the right, we have

$\lim\limits_{x \to 4} F(x) = 1$.

69. $f(x) = \dfrac{1}{x} - 2$

Since $x = 0$ makes the denominator zero, we exclude the value 0 from the domain. Creating an input-output table we have

x	-1	-0.5	-0.1	0.1	0.5	1
$f(x)$	-3	-4	-12	8	0	-1

Next we plot the points and draw the graph.

Find $\lim\limits_{x \to \infty} f(x)$.

As inputs x get larger and larger, outputs $f(x)$ get closer and closer to -2. We have

$\lim\limits_{x \to \infty} f(x) = -2$.

Find $\lim\limits_{x \to 0} f(x)$.

As inputs x approach 0 from the left, outputs $f(x)$ decrease without bound. We have,

$\lim\limits_{x \to 0^-} f(x) = -\infty$.

As inputs x approach 0 from the right, outputs $f(x)$ increase without bound. We have,

$\lim\limits_{x \to 0^+} f(x) = \infty$

Since the function values as $x \to 0$ from the left decrease without bound, and the function values as $x \to 0$ from the right increase without bound, the limit does not exist. We have,

$\lim\limits_{x \to 0} f(x)$ *does not exist.*

71. $g(x) = \dfrac{1}{x+2} + 4$

Since $x = -2$ makes the denominator zero, we exclude the value -2 from the domain. Creating an input-output table we have

x	-3	-2.5	-2.1	-1.9	-1.5	-1	0
$g(x)$	3	2	-6	14	6	5	$\frac{9}{2}$

Next we plot the points and draw the graph.

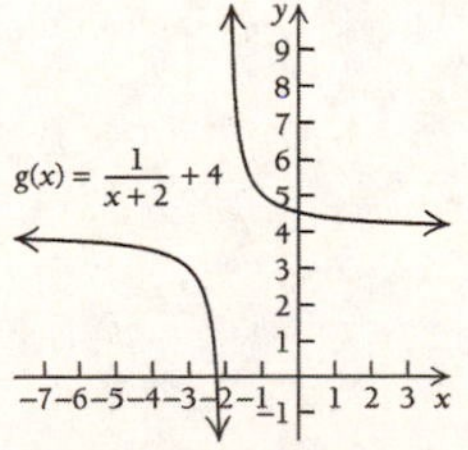

The solution is continued on the next page.

Copyright © 2014 Pearson Education, Inc. Publishing as Addison-Wesley.

Find $\lim_{x\to\infty} g(x)$.

As inputs x get larger and larger, outputs $g(x)$ get closer and closer to 4. We have $\lim_{x\to\infty} g(x) = 4$.

Find $\lim_{x\to -2} g(x)$.

As inputs x approach -2 from the left, outputs $g(x)$ decrease without bound. We have, $\lim_{x\to -2^-} g(x) = -\infty$.

As inputs x approach -2 from the right, outputs $g(x)$ increase without bound. We have, $\lim_{x\to -2^+} g(x) = \infty$.

Since the function values as $x \to 2$ from the left decrease without bound, and the function values as $x \to 2$ from the right increase without bound, the limit does not exist. We have, $\lim_{x\to -2} g(x)$ *does not exist*.

73. $F(x) = \begin{cases} 2x+1, & \text{for } x < 1 \\ x, & \text{for } x \geq 1. \end{cases}$

We create an input-output table for each piece of the function.

For $x < 1$

x	-1	0	0.9
$F(x)$	-1	1	2.8

We plot the points and draw the graph. Notice we draw an open circle at the point $(1,3)$ to indicate that the point is not part of the graph.

For $x \geq 1$

x	1	2	3
$F(x)$	1	2	3

We plot the points and draw the graph. Notice we draw a solid circle at the point $(1,1)$ to indicate that the point is part of the graph.

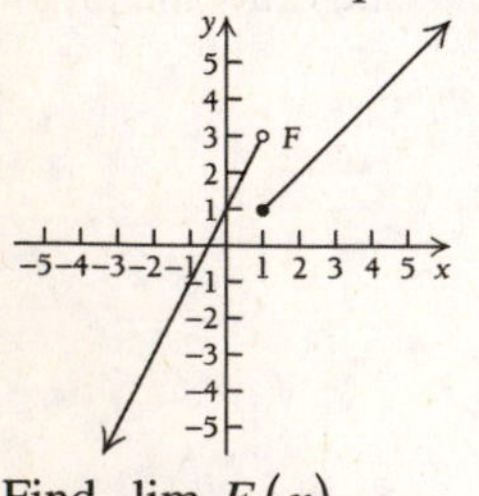

Find $\lim_{x\to 1^-} F(x)$.

As inputs x approach 1 from the left, outputs $F(x)$ approach 3. That is, $\lim_{x\to 1^-} F(x) = 3$.

Find $\lim_{x\to 1^+} F(x)$.

As inputs x approach 1 from the right, outputs $F(x)$ approach 1. That is, $\lim_{x\to 1^+} F(x) = 1$.

Find $\lim_{x\to 1} F(x)$

Since the limit from the left, 3, is not the same as the limit from the right, 1, we have $\lim_{x\to 1} F(x)$ *does not exist.*

75. $g(x) = \begin{cases} -x+4, & \text{for } x < 3 \\ x-3, & \text{for } x > 3. \end{cases}$

We create an input-output table for each piece of the function.

For $x < 3$

x	0	1	2	2.9
$g(x)$	4	3	2	1.1

We plot the points from the table draw the graph. Notice we draw an open circle at the point $(3,1)$ to indicate that the point is not part of the graph.

For $x > 3$

x	3.1	4	5	6
$g(x)$	0.1	1	2	3

We plot the points and draw the graph. Notice we draw an open circle at the point $(3,0)$ to indicate that the point is not part of the graph.

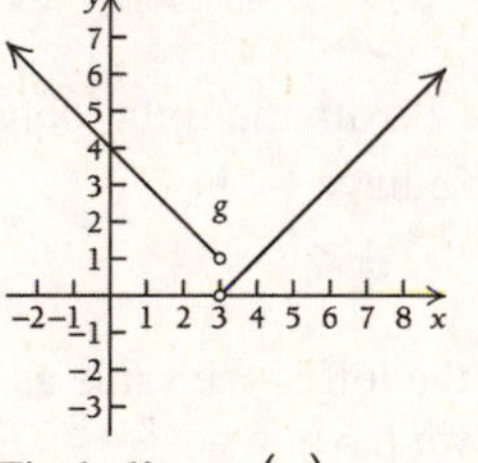

Find $\lim_{x\to 3^-} g(x)$.

As inputs x approach 3 from the left, outputs $g(x)$ approach 1. That is, $\lim_{x\to 3^-} g(x) = 1$.

Find $\lim_{x\to 3^+} g(x)$.

As inputs x approach 3 from the right, outputs $g(x)$ approach 0. That is, $\lim_{x\to 3^+} g(x) = 0$.

Find $\lim_{x\to 3} g(x)$

Since the limit from the left, 1, is not the same as the limit from the right, 0, we have $\lim_{x\to 3} g(x)$ *does not exist.*

Copyright © 2014 Pearson Education, Inc. Publishing as Addison-Wesley.

77. $G(x)=\begin{cases} x^2, & \text{for } x<-1 \\ x+2, & \text{for } x>-1. \end{cases}$

We create an input-output table for each piece of the function.

For $x<-1$

x	-3	-2	-1.1
$G(x)$	9	4	-1.21

We plot the points and draw the graph. Notice we draw an open circle at the point $(-1,1)$ to indicate that the point is not part of the graph.

For $x>-1$

x	-0.9	0	1
$G(x)$	1.1	2	3

We plot the points and draw the graph. Notice we draw an open circle at the point $(-1,1)$ to indicate that the point is not part of the graph.

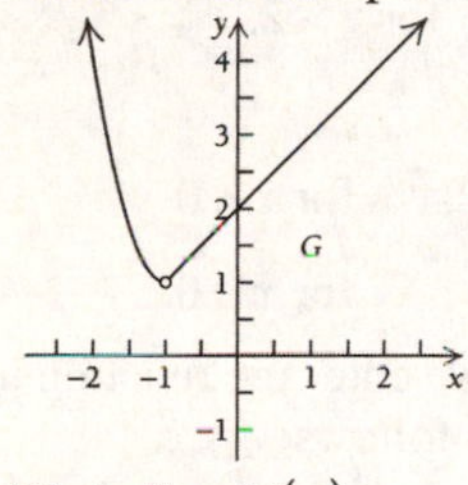

Find $\lim_{x\to -1} G(x)$.

As inputs x approach -1 from the left, outputs $G(x)$ approach 1. We have,

$\lim_{x\to -1^-} G(x)=1$.

As inputs x approach -1 from the right, outputs $G(x)$ approach 1. We have,

$\lim_{x\to -1^+} G(x)=1$

Since the limit from the left is the same as the limit from the right, we have

$\lim_{x\to -1} G(x)=1$.

79. $H(x)=\begin{cases} x+1, & \text{for } x<0 \\ 2, & \text{for } 0\le x<1 \\ 3-x, & \text{for } x\ge 1. \end{cases}$

We create an input-output table for each piece of the function.

For $x<0$

x	-1	-0.5	-0.1
$H(x)$	0	0.5	0.9

We plot the points and draw the graph. Notice we draw an open circle at the point $(0,1)$ to indicate that the point is not part of the graph.

For $0\le x<1$, the function has value of 2. We draw a solid circle at the point $(0,2)$ to indicate the point is part of the graph and we draw an open circle at $(1,2)$ to indicate that the point is not part of the graph.

For $x\ge 1$

x	1	2	3
$H(x)$	2	1	-1

We plot the points and draw the graph. Notice we draw a solid circle at the point $(1,2)$ to indicate that the point is part of the graph.

Find $\lim_{x\to 0} H(x)$

As inputs x approach 0 from the left, outputs $H(x)$ approach 1. That is,

$\lim_{x\to 0^-} H(x)=1$.

As inputs x approach 0 from the right, outputs $H(x)$ approach 2. That is,

$\lim_{x\to 0^+} H(x)=2$.

Since the limit from the left, 1, is not the same as the limit from the right, 2, we have

$\lim_{x\to 0} H(x)$ *does not exist.*

Find $\lim_{x\to 1} H(x)$

As inputs x approach 1 from the left, outputs $H(x)$ approach 2. That is,

$\lim_{x\to 1^-} H(x)=2$.

As inputs x approach 1 from the right, outputs $H(x)$ approach 2. That is,

$\lim_{x\to 1^+} H(x)=2$.

Since the limit from the left, 2, is the same as the limit from the right, 2, we have

$\lim_{x\to 0} H(x)=2$.

81. $\lim_{x\to 0.25^-} C(x)=\$3.30$

$\lim_{x\to 0.25^+} C(x)=\$3.30$

$\lim_{x\to 0.25} C(x)=\$3.30$

Copyright © 2014 Pearson Education, Inc. Publishing as Addison-Wesley.

83. $\lim_{x\to0.6^-} C(x) = \$3.70.$

$\lim_{x\to0.6^+} C(x) = \$4.10.$

$\lim_{x\to0.6} C(x)$ *does not exist.*

85. $\lim_{x\to2^-} p(x) = \$1.10.$

$\lim_{x\to2^+} p(x) = \$1.30.$

$\lim_{x\to2} p(x)$ *does not exist.*

87. $\lim_{x\to3^-} p(x) = \$1.30.$

$\lim_{x\to3^+} p(x) = \$1.50.$

$\lim_{x\to3} p(x)$ *does not exist.*

89. $\lim_{t\to1.5^-} p(t) = 11.0$ hundred deer.

$\lim_{t\to1.5^+} p(t) = 12.0$ hundred deer.

$\lim_{t\to1.5} p(t)$ *does not exist.*

91. ✎

93. $\lim_{t\to0.8^-} p(t) = 35$ bears.

$\lim_{t\to0.8^+} p(t) = 34$ bears.

$\lim_{t\to0.8} p(t)$ *does not exist.*

95. As inputs x approach 2 from the right, outputs $f(x)$ approach 4. We have,

$\lim_{x\to2^+} f(x) = 4$. In order for $\lim_{x\to2} f(x)$ to exist we need $\lim_{x\to2^-} f(x) = 4$. We will use the letter c for the unknown in the equation; therefore,

$\lim_{x\to2^-} \frac{1}{2}(x) + c = 4$.

Substitute 2 in for x to get the equation:

$\frac{1}{2}(2) + c = 4$ and solving for c we get

$1 + c = 4$

$c = 3.$

Therefore, in order for the limit to exist as x approaches 2, the function must be:

$$f(x) = \begin{cases} \frac{1}{2}x + \underline{3} & \text{for } x < 2 \\ -x + 6 & \text{for } x > 2. \end{cases}$$

97. As inputs x approach 2 from the left, outputs $f(x)$ approach -5. We have,

$\lim_{x\to2^-} f(x) = -5$. In order for $\lim_{x\to2} f(x)$ to exist we need $\lim_{x\to2^+} f(x) = -5$. We will use the letter c for the unknown in the equation and this gives us

$\lim_{x\to2^+} \left(-x^2 + c\right) = -5.$

Substitute 2 in for x to get the equation:

$-(2)^2 + c = -5$ and solving for c we get

$-(4) + c = -5$

$c = -1.$

Therefore, in order for the limit to exist as x approaches 2, the function must be:

$$f(x) = \begin{cases} x^2 - 9 & \text{for } x < 2 \\ -x^2 + \underline{-1} & \text{for } x > 2. \end{cases}$$

99. Graph $f(x) = \begin{cases} x^2 - 2, & \text{for } x < 0 \\ 2 - x^2, & \text{for } x \geq 0. \end{cases}$

Using the calculator we enter the function into the graphing editor as follows:

When you select the table feature you get:

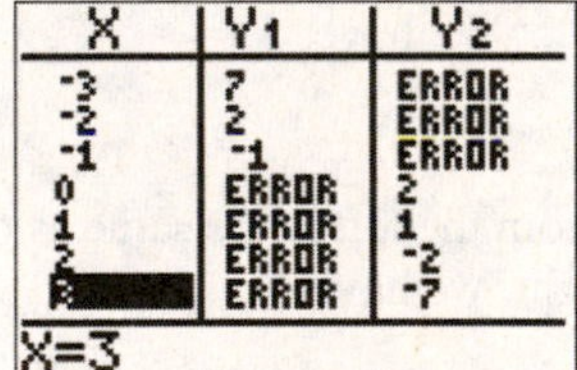

X	Y1	Y2
-3	7	ERROR
-2	2	ERROR
-1	-1	ERROR
0	ERROR	2
1	ERROR	1
2	ERROR	-2
■	ERROR	-7

X=3

The calculator graphs the function

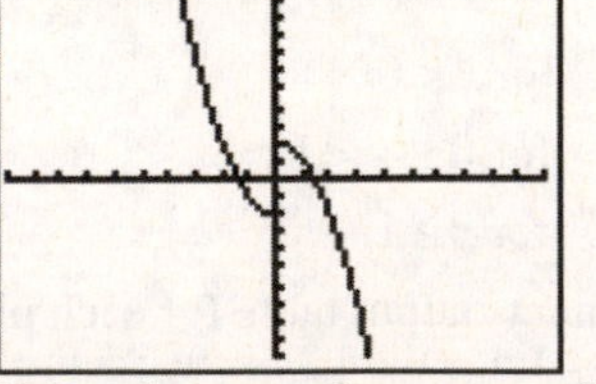

Using the trace feature, we find the limits.

Find $\lim_{x\to0} f(x)$.

The solution is continued on the next page.

Copyright © 2014 Pearson Education, Inc. Publishing as Addison-Wesley.

From the previous page, we see as inputs x approach 0 from the left, outputs $f(x)$ approach -2. We have,

$\lim_{x\to 0^-} f(x) = -2$.

As inputs x approach 0 from the right, outputs $f(x)$ approach 2. We have,

$\lim_{x\to 0^+} f(x) = 2$

Since the limit from the left, -2, is not the same as the limit from the right, 2, we have

$\lim_{x\to 0} f(x)$ *does not exist.*

Find $\lim_{x\to -2} f(x)$.

As inputs x approach -2 from the left, outputs $f(x)$ approach 2. We have,

$\lim_{x\to -2^-} f(x) = 2$.

As inputs x approach -2 from the right, outputs $f(x)$ approach 2. We have,

$\lim_{x\to -2^+} f(x) = 2$

Since the limit from the left is the same as the limit from the right, we have

$\lim_{x\to -2} f(x) = 2$.

101. Graph $f(x) = \dfrac{1}{x^2 - 4x - 5}$

Using the calculator we enter the function into the graphing editor.

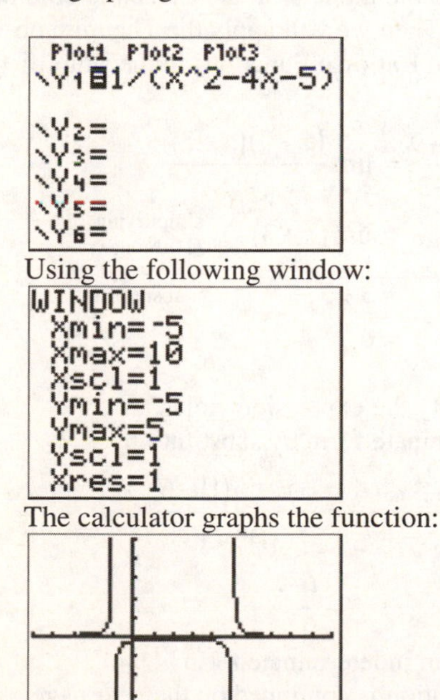

Using the following window:

The calculator graphs the function:

Using the trace feature on the calculator we find the limits.

Find $\lim_{x\to -1} f(x)$.

As inputs x approach -1 from the left, outputs $f(x)$ increase without bound. We have,

$\lim_{x\to -1^-} f(x) = \infty$.

As inputs x approach -1 from the right, outputs $f(x)$ decrease without bound. We have,

$\lim_{x\to -1^+} f(x) = -\infty$

Since the function values as $x \to -1$ from the left increase without bound, and the function values as $x \to -1$ from the right decrease without bound, the limit does not exist.

$\lim_{x\to -1} f(x)$ *does not exist.*

Find $\lim_{x\to 5} f(x)$.

As inputs x approach 5 from the left, outputs $f(x)$ decrease without bound. We have,

$\lim_{x\to 5^-} f(x) = -\infty$.

As inputs x approach 5 from the right, outputs $f(x)$ increase without bound. We have,

$\lim_{x\to 5^+} f(x) = \infty$

Since the function values as $x \to 5$ from the left decrease without bound, and the function values as $x \to 5$ from the right increase without bound, the limit does not exist. We have,

$\lim_{x\to 5} f(x)$ *does not exist.*

Copyright © 2014 Pearson Education, Inc. Publishing as Addison-Wesley.

Exercise Set 1.2

1. By limit principle L1,
$\lim_{x\to 3} 7 = 7.$
Therefore, the statement is a true.

3. By limit principle L2,
$\lim_{x\to 1}[g(x)]^2 = \left[\lim_{x\to 1} g(x)\right]^2 = [5]^2 = 25.$
Therefore, the statement is false.

5. By the definition of continuity, in order for *f* to be continuous at $x = 2$, $f(2)$ must exist. Therefore, the statement is true.

7. This statement is false. If $\lim_{x\to 4} F(x)$ exists but is not equal to $F(4)$, then F is not continuous.

9. It follows from the Theorem on Limits of Rational Functions that we can find the limit by substitution:

$$\begin{aligned}\lim_{x\to 1}(3x+2) &= 3(1)+2\\ &= 5.\end{aligned}$$

11. It follows from the Theorem on Limits of Rational Functions that we can find the limit by substitution:

$$\begin{aligned}\lim_{x\to -1}\left(x^2-4\right) &= (-1)^2-4\\ &= 1-4\\ &= -3.\end{aligned}$$

13. It follows from the Theorem on Limits of Rational Functions that we can find the limit by substitution:

$$\begin{aligned}\lim_{x\to 3}\left(x^2-4x+7\right) &= (3)^2-4(3)+7\\ &= 9-12+7\\ &= 4.\end{aligned}$$

15. It follows from the Theorem on Limits of Rational Functions that we can find the limit by substitution:

$$\begin{aligned}&\lim_{x\to 2}\left(2x^4-3x^3+4x-1\right)\\ &= 2(2)^4-3(2)^3+4(2)-1\\ &= 2(16)-3(8)+8-1\\ &= 32-24+8-1\\ &= 15.\end{aligned}$$

17. It follows from the Theorem on Limits of Rational Functions that we can find the limit by substitution:

$$\begin{aligned}\lim_{x\to 3}\left(\frac{x^2-8}{x-2}\right) &= \frac{(3)^2-8}{3-2}\\ &= \frac{9-8}{3-2}\\ &= 1.\end{aligned}$$

19. We verify the expression yields an indeterminate form by substitution:

$$\begin{aligned}\lim_{x\to 3}\frac{x^2-9}{x-3} &= \frac{(3)^2-9}{(3)-3}\\ &= \frac{0}{0}.\end{aligned}$$

This is an indeterminate form. In order to find the limit we will simplify the function by factoring the numerator and canceling common factors. Then we will apply the Theorem on Limits of Rational Functions to the simplified function.

$$\begin{aligned}\lim_{x\to 3}\frac{x^2-9}{x-3} &= \lim_{x\to 3}\frac{(x-3)(x+3)}{x-3}\\ &= \lim_{x\to 3}(x+3) && \text{simplifying, assuming } x\neq -3\\ &= 3+3 && \text{substitution}\\ &= 6.\end{aligned}$$

21. We verify the expression yields an indeterminate form by substitution:

$$\begin{aligned}\lim_{x\to 1}\frac{x^2+5x-6}{x^2-1} &= \frac{(1)^2+5(1)-6}{(1)^2-1}\\ &= \frac{0}{0}.\end{aligned}$$

This is an indeterminate form.
The solution is continued on the next page.

Copyright © 2014 Pearson Education, Inc. Publishing as Addison-Wesley.

In order to find the limit we will simplify the function by factoring the numerator and denominator and canceling common factors. Then we will apply the Theorem on Limits of Rational Functions to the simplified function.

$$\begin{aligned}\lim_{x\to1}\frac{x^2+5x-6}{x^2-1}&=\lim_{x\to1}\frac{(x-1)(x+6)}{(x-1)(x+1)}\\&=\lim_{x\to1}\frac{(x+6)}{(x+1)}\quad\text{simplifying, assuming } x\neq1\\&=\frac{1+6}{1+1}\quad\text{substitution}\\&=\frac{7}{2}.\end{aligned}$$

23. We verify the expression yields an indeterminate form by substitution:

$$\begin{aligned}\lim_{x\to2}\frac{3x^2+x-14}{x^2-4}&=\frac{3(2)^2+(2)-14}{(2)^2-4}\\&=\frac{0}{0}.\end{aligned}$$

This is an indeterminate form. In order to find the limit we will simplify the function by factoring the numerator and denominator and canceling common factors. Then we will apply the Theorem on Limits of Rational Functions to the simplified function.

$$\begin{aligned}\lim_{x\to2}\frac{3x^2+x-14}{x^2-4}&=\lim_{x\to2}\frac{(3x+7)(x-2)}{(x+2)(x-2)}\\&=\lim_{x\to2}\frac{(3x+7)}{(x+2)}\quad\text{simplifying, assuming } x\neq2\\&=\frac{3(2)+7}{2+2}\quad\text{substitution}\\&=\frac{13}{4}.\end{aligned}$$

25. We verify the expression yields an indeterminate form by substitution:

$$\begin{aligned}\lim_{x\to1}\frac{x^3-1}{x-1}&=\frac{(1)^3-1}{(1)-1}\\&=\frac{0}{0}.\end{aligned}$$

This is an indeterminate form. In order to find the limit we will simplify the function by factoring the numerator and canceling common factors. Then we will apply the Theorem on Limits of Rational Functions to the simplified function.

$$\begin{aligned}\lim_{x\to1}\frac{x^3-1}{x-1}&=\lim_{x\to1}\frac{(x-1)(x^2+x+1)}{(x-1)}\\&=\lim_{x\to1}(x^2+x+1)\quad\text{simplifying, assuming } x\neq1\\&=(1)^2+(1)+1\quad\text{substitution}\\&=3.\end{aligned}$$

27. We verify the expression yields an indeterminate form by substitution:

$$\begin{aligned}\lim_{x\to25}\frac{\sqrt{x}-5}{x-25}&=\frac{\sqrt{25}-5}{(25)-25}\\&=\frac{0}{0}.\end{aligned}$$

This is an indeterminate form. In order to find the limit we will simplify the function by factoring the denominator and canceling common factors. Then we will apply the Theorem on Limits of Rational Functions to the simplified function.

$$\begin{aligned}\lim_{x\to25}\frac{\sqrt{x}-5}{x-25}&=\lim_{x\to25}\frac{\sqrt{x}-5}{(\sqrt{x}-5)(\sqrt{x}+5)}\\&=\lim_{x\to25}\frac{1}{(\sqrt{x}+5)}\quad\text{simplifying, assuming } x\neq25\\&=\frac{1}{\sqrt{25}+5}\quad\text{substitution}\\&=\frac{1}{10}.\end{aligned}$$

29. We verify the expression yields an indeterminate form by substitution:

$$\begin{aligned}\lim_{x\to-1}\frac{x^2+5x+4}{x^2+2x+1}&=\frac{(-1)^2+5(-1)+4}{(-1)^2+2(-1)+1}\\&=\frac{0}{0}.\end{aligned}$$

This is an indeterminate form. In order to find the limit we will simplify the function by factoring the numerator and denominator and canceling common factors. Then we will apply the Theorem on Limits of Rational Functions to the simplified function.

The solution is continued on the next page.

Copyright © 2014 Pearson Education, Inc. Publishing as Addison-Wesley.

$$\lim_{x\to-1}\frac{x^2+5x+4}{x^2+2x+1}=\lim_{x\to-1}\frac{(x+1)(x+4)}{(x+1)(x+1)}$$
$$=\lim_{x\to-1}\left(\frac{x+4}{x+1}\right) \quad \text{simplifying, assuming } x\neq-1$$
$$=\frac{-1+4}{-1+1} \quad \text{substitution}$$
$$=\frac{3}{0}.$$

Substitution yields division by zero. Therefore,

$\lim_{x\to-1}\frac{x^2+5x+4}{x^2+2x+1}$ *does not exist.*

31. $\lim_{x\to4}\sqrt{x^2-9}$

By limit principle L2,

$$\lim_{x\to4}\sqrt{x^2-9}=\sqrt{\lim_{x\to4}\left(x^2-9\right)}$$
$$=\sqrt{\lim_{x\to4}x^2-\lim_{x\to4}9} \quad \text{By L3}$$
$$=\sqrt{\left(\lim_{x\to4}x\right)^2-9} \quad \text{By L2 and L1}$$
$$=\sqrt{(4)^2-9}$$
$$=\sqrt{16-9}$$
$$=\sqrt{7}.$$

33. $\lim_{x\to2}\sqrt{x^2-9}$

By limit principle L2,

$$\lim_{x\to2}\sqrt{x^2-9}=\sqrt{\lim_{x\to2}\left(x^2-9\right)}$$
$$=\sqrt{\lim_{x\to2}x^2-\lim_{x\to2}9} \quad \text{By L3}$$
$$=\sqrt{\left(\lim_{x\to2}x\right)^2-9} \quad \text{By L2 and L1}$$
$$=\sqrt{(2)^2-9}$$
$$=\sqrt{4-9}$$
$$=\sqrt{-5}.$$

Therefore, $\lim_{x\to2}\sqrt{x^2-9}$ *does not exist.*

35. $\lim_{x\to3^+}\sqrt{x^2-9}$

By limit principle L2,

$$\lim_{x\to3^+}\sqrt{x^2-9}=\sqrt{\lim_{x\to3^+}\left(x^2-9\right)}$$
$$=\sqrt{\lim_{x\to3^+}x^2-\lim_{x\to3^+}9} \quad \text{By L3}$$
$$=\sqrt{\left(\lim_{x\to3^+}x\right)^2-9} \quad \text{By L2 and L1}$$
$$=\sqrt{(3)^2-9}$$
$$=\sqrt{9-9}$$
$$=\sqrt{0}$$
$$=0.$$

37. The function is not continuous over the interval, because $f(x)$ is not continuous at $x=1$. As x approaches 1 from the left, $f(x)$ approaches 2. However, as x approaches 1 from the right $f(x)$ approaches -1. Therefore, $f(x)$ is not continuous at 1.

39. The function is not continuous over the interval, because $k(x)$ is not continuous at $x=-1$. The function is not defined at $x=-1$, in other words $k(-1)$ does not exist. Therefore, $k(x)$ is not continuous at -1.

41. The function is not continuous over the interval, because it is not continuous at $x=-2$. The limit does not exist as x approaches -2, furthermore, $t(-2)$ does not exist. Therefore the function is not continuous at $x=-2$.

43. a) As inputs x approach 1 from the right, outputs $g(x)$ approach -2. Thus, the limit from the right is -2. $\lim_{x\to1^+}g(x)=-2$.
As inputs x approach 1 from the left, outputs $g(x)$ approach -2. Thus, the limit from the left is -2. $\lim_{x\to1^-}g(x)=-2$.
Since the limit from the left, -2, is the same as the limit from the right, -2, we have. $\lim_{x\to1}g(x)=-2$.

b) When the input is 1, the output $g(1)$ is -2. That is $g(1)=-2$.

Copyright © 2014 Pearson Education, Inc. Publishing as Addison-Wesley.

c) The function $g(x)$ is continuous at $x=1$, because
1) $g(1)$ exists, $g(1)=-2$
2) $\lim_{x\to 1} g(x)$ exists, $\lim_{x\to 1} g(x)=-2$, and
3) $\lim_{x\to 1} g(x)=-2=g(1)$.

d) As inputs x approach -2 from the right, outputs $g(x)$ approach -3. Thus, the limit from the right is -3. $\lim_{x\to -2^+} g(x)=-3$.
As inputs x approach -2 from the left, outputs $g(x)$ approach 4. Thus, the limit from the left is 4. $\lim_{x\to -2^-} g(x)=4$.
Since the limit from the left, 4, is not the same as the limit from the right, -3, we say $\lim_{x\to -2} g(x)$ does not exist.

e) When the input is -2, the output $g(-2)$ is -3. That is $g(-2)=-3$.

f) Since the limit of $g(x)$ as x approaches -2 does not exist, the function is not continuous at $x=-2$.

45. a) As inputs x approach 1 from the right, outputs $h(x)$ approach 2. Thus, the limit from the right is 2. $\lim_{x\to 1^+} h(x)=2$.
As inputs x approach 1 from the left, outputs $h(x)$ approach 2. Thus, the limit from the left is 2. $\lim_{x\to 1^-} h(x)=2$.
Since the limit from the left, 2, is the same as the limit from the right, 2, we have. $\lim_{x\to 1} h(x)=2$.

b) When the input is 1, the output $h(1)$ is 2. That is $h(1)=2$.

c) The function $h(x)$ is continuous at $x=1$, because
1) $h(1)$ exists, $\mathrm{h}(1)=2$
2) $\lim_{x\to 1} h(x)$ exists, $\lim_{x\to 1} h(x)=2$, and
3) $\lim_{x\to 1} h(x)=2=h(1)$.

d) As inputs x approach -2 from the right, outputs $h(x)$ approach 0. Thus, the limit from the right is 0. $\lim_{x\to -2^+} h(x)=0$.
As inputs x approach -2 from the left, outputs $h(x)$ approach 0. Thus, the limit from the left is 0. $\lim_{x\to -2^-} h(x)=0$.
Since the limit from the left, 0, is the same as the limit from the right, 0, we say $\lim_{x\to -2} h(x)=0$.

e) When the input is -2, the output $h(-2)$ is 0. That is $h(-2)=0$.

f) The function $h(x)$ is continuous at $x=-2$, because
1) $h(-2)$ exists, $h(-2)=0$
2) $\lim_{x\to -2} h(x)$ exists, $\lim_{x\to -2} h(x)=0$, and
3) $\lim_{x\to -2} h(x)=0=h(-2)$.

47. a) As inputs x approach 3 from the right, outputs $G(x)$ approach 3. Thus, $\lim_{x\to 3^+} G(x)=3$.

b) As inputs x approach 3 from the left, outputs $G(x)$ approach 1. Thus, $\lim_{x\to 3^-} G(x)=1$.

c) Since the limit from the left, 1, is not the same as the limit from the right, 3, the limit does not exist. $\lim_{x\to 3} G(x)$ does not exist.

d) $G(3)=1$

e) The function $G(x)$ is not continuous at $x=3$ because the limit does not exist as x approaches 3.

f) The function $G(x)$ is continuous at $x=0$, because
1) $G(0)$ exists,
2) $\lim_{x\to 0} G(x)$ exists, and
3) $\lim_{x\to 0} G(x)=G(0)$.

g) The function $G(x)$ is continuous at $x=2.9$, because
1) $G(2.9)$ exists,
2) $\lim_{x\to 2.9} G(x)$ exists, and
3) $\lim_{x\to 2.9} G(x)=G(2.9)$.

Copyright © 2014 Pearson Education, Inc. Publishing as Addison-Wesley.

49. First we find the function value when $x = 5$.
$f(5) = 3(5) - 2 = 13$. Hence, $f(5)$ exists.
Next, we find the limit as x approaches 5. It follows from the Theorem on Limits of Rational Functions that we can find the limit by substitution: $\lim_{x\to 5} f(x) = 3(5) - 2 = 13$
Therefore, $\lim_{x\to 5} f(x) = 13 = f(5)$ and the function is continuous at $x = 5$.

51. The function $G(x) = \frac{1}{x}$ is not continuous at $x = 0$ because $G(0) = \frac{1}{0}$ is undefined.

53. First we find the function value when $x = 3$.
$g(3) = \frac{1}{3}(3) + 4 = 1 + 4 = 5$. Hence, $g(3)$ exists.
Next, we find the limit as x approaches 3. As the inputs x approach 3 from the right, the outputs $g(x)$ approach 5, that is,

$\lim_{x\to 3^-} g(x) = \frac{1}{3}(3) + 4 = 5$.

As the inputs x approach 3 from the left, the outputs $g(x)$ approach 5, that is,

$\lim_{x\to 3^+} g(x) = 2(3) - 1 = 5$.

Since the limit from the left, 5, is the same as the limit from the right, 5. The limit exists. We have:

$\lim_{x\to 3} g(x) = 5$.

Therefore, we have

$\lim_{x\to 3} g(x) = 5 = g(3)$.

Thus the function is continuous at $x = 3$.

55. The function is not continuous at $x = 3$ because the limit does not exist as x approaches 3. To verify this we take the limit as x approaches 3 from the left and the limit as x approaches 3 from the right.
As x approaches 3 from the left we have

$\lim_{x\to 3^-} F(x) = \frac{1}{3}(3) + 4 = 5$.

As x approaches 3 from the right we have

$\lim_{x\to 3^+} F(x) = 2(3) - 5 = 1$.

Since the limit from the left, 5, is not the same as the limit from the right, 1, the limit does not exist. $\lim_{x\to 3} F(x)$ does not exist.

57. First we find the function value when $x = 3$.
$f(3) = 2(3) - 1 = 5$. Hence, $f(3)$ exists.
Next, we find the limit as x approaches 3. As the inputs x approach 3 from the left, the outputs $f(x)$ approach 5, that is,

$\lim_{x\to 3^-} f(x) = \frac{1}{3}(3) + 4 = 5$.

As the inputs x approach 3 from the left, the outputs $f(x)$ approach 5, that is,

$\lim_{x\to 3^+} f(x) = 2(3) - 1 = 5$.

Since the limit from the left, 5, is the same as the limit from the right, 5. The limit exists. We have:

$\lim_{x\to 3} f(x) = 5$.

Therefore, we have

$\lim_{x\to 3} f(x) = 5 = f(3)$.

Thus the function is continuous at $x = 3$.

59. The function is not continuous at $x = 2$. To verify this, we take the limit as x approaches 2. Using the Theorem on Limits of Rational Functions, we simplify the function near 2 by factoring the numerator and canceling common factors.

$$\begin{aligned}\lim_{x\to 2} G(x) &= \lim_{x\to 2} \frac{x^2 - 4}{x - 2} \\ &= \lim_{x\to 2} \frac{(x-2)(x+2)}{x-2} \\ &= \lim_{x\to 2}(x+2) \\ &= 2 + 2 = 4\end{aligned}$$

Therefore,

$\lim_{x\to 2} G(x) = 4$.

However, when $x = 2$, the output $G(2)$ is defined to be 5. That is, $G(2) = 5$. Therefore, $\lim_{x\to 2} G(x) = 4 \neq 5 = G(2)$. Thus the function is not continuous at $x = 2$.

61. First we find the function value when $x = 5$.
$f(5) = (5) + 1 = 6$, $f(5)$ exists.
Next we find the limit as x approaches 5.

The solution is continued on the next page.

Copyright © 2014 Pearson Education, Inc. Publishing as Addison-Wesley.

To find the limit as x approaches 5 from the left, we first simplify the rational function by factoring the numerator and canceling common factors.

$$\begin{aligned}\lim_{x\to5^-} f(x) &= \lim_{x\to5^-} \frac{x^2-4x-5}{x-5}\\ &= \lim_{x\to5^-} \frac{(x-5)(x+1)}{x-5}\\ &= \lim_{x\to5^-}(x+1)\\ &= 5+1\\ &= 6\end{aligned}$$

To find the limit as x approaches 5 from the right, we can use substitution.

$$\begin{aligned}\lim_{x\to5^+} f(x) &= \lim_{x\to5^+} x+1\\ &= 5+1\\ &= 6\end{aligned}$$

Therefore, the limit exists.

$\lim_{x\to5} f(x) = 6$.

Thus we have,

$\lim_{x\to5} f(x) = 6 = f(5)$.

Therefore, the function in continuous at $x = 5$.

63. The function is not continuous at $x = 5$ because $g(5)$ does not exist.

$$\begin{aligned}g(5) &= \frac{1}{(5)^2-7(5)+10}\\ &= \frac{1}{25-35+10}\\ &= \frac{1}{0}\end{aligned}$$

65. First we find the function value when $x = 4$.

$$\begin{aligned}F(4) &= \frac{1}{(4)^2-7(4)+10}\\ &= \frac{1}{16-28+10}\\ &= \frac{1}{-2}\\ &= -\frac{1}{2}\end{aligned}$$

Hence, $F(4)$ exists.

Next we find the limit. Applying the Theorem on Limits of Rational Functions we have:

$$\lim_{x\to4} F(x) = \frac{1}{(4)^2-7(4)+10} = -\frac{1}{2}.$$

We now have,

$$\lim_{x\to4} F(x) = \frac{1}{(4)^2-7(4)+10} = -\frac{1}{2} = F(4).$$

Therefore, the function is continuous at $x = 4$.

67. Yes, the function is continuous over the interval $(-4,4)$. Since the function is defined for every value in the interval, the Theorem on Limits of Rational Functions tells us $\lim_{x\to a} g(x) = g(a)$ for all values a in the interval. Thus $g(x)$ is continuous over the interval.

69. No, the function is not continuous over the interval $(-7,7)$ because the function does not exist at $x = 0$. $f(0) = \frac{1}{0}+3$, which is undefined.

71. Yes, the function is continuous on $\mathbb{R}$. The function is defined for all real numbers, so by the Theorem on Limits of Rational Functions, $\lim_{x\to a} g(x) = g(a)$ for all a in $\mathbb{R}$.

73. a) In order for the function to be continuous at $x = 20$ the limit as x approaches 20 from the left must equal the limit as x approaches 20 from the right. The limit from the left of the function is:

$\lim_{x\to20^-}(1.5x) = 1.50(20) = 30$. Therefore, the limit of $p(x)$ as x approaches 20 from the right must be equal to 30. We set the right hand limit equal to 30:

$\lim_{x\to20^+}(1.25x+k) = 30$.

We allow x to approach 20. By Limit Property L3, we have:

$1.25(20)+k = 30$

Solving this equation for k yields:

$$\begin{aligned}25+k &= 30\\ k &= 5.\end{aligned}$$

Therefore, k must equal 5 in order for the price function to be continuous at $x = 20$.

b) It is preferred to have continuity at $x = 20$ so there will not be a possibility that customers can pay less for quantities greater than 20 pounds of candy then they do for quantities less than 20 pounds of candy resulting in lost revenue for the seller.

Copyright © 2014 Pearson Education, Inc. Publishing as Addison-Wesley.

75. As the inputs x approach 0 from the left, the outputs approach -1. We see this by looking at a table:

x	-0.1	-0.01	-0.001
$\frac{\|x\|}{x}$	-1	-1	-1

$\lim_{x\to 0^-} \frac{|x|}{x} = -1$.

As the inputs x approach 0 from the right, the outputs approach 1. We see this by looking at a table:

x	0.001	0.01	0.1
$\frac{\|x\|}{x}$	1	1	1

$\lim_{x\to 0^+} \frac{|x|}{x} = 1.$

Since the limit from the left, -1, is not the same as the limit from the right, 1, the limit does not exist. $\lim_{x\to 0} \frac{|x|}{x}$ does not exist.

77. 6

79. -0.2887 or $-\frac{1}{2\sqrt{3}}$

81. 0.75 or $\frac{3}{4}$

83. 0.25 or $\frac{1}{4}$

Copyright © 2014 Pearson Education, Inc. Publishing as Addison-Wesley.

Exercise Set 1.3

1. a) $f(x)=4x^2$

so,

$f(x+h)=4(x+h)^2$ substituing $x+h$ for x

$=4(x^2+2xh+h^2)$

$=4x^2+8xh+4h^2$

Then

$\dfrac{f(x+h)-f(x)}{h}$ Difference quotient

$=\dfrac{(4x^2+8xh+4h^2)-4x^2}{h}$ Substituting

$=\dfrac{8xh+4h^2}{h}$

$=\dfrac{h(8x+4h)}{h}$ Factoring the numerator

$=\dfrac{h}{h}\cdot(8x+4h)$ Removing a factor = 1.

$=8x+4h,$ Simplified difference quotient

or $4(2x+h)$

b) The difference quotient column in the table can be completed using the simplified difference quotient.

$8x+4h$ Simplified difference quotient

$=8(5)+4(2)=48$

substitiuting 5 for x and 2 for h;

$=8(5)+4(1)=44$

substitiuting 5 for x and 1 for h;

$=8(5)+4(0.1)=40.4$

substitiuting 5 for x and 0.1 for h;

$=8(5)+4(0.01)=40.04$

substitiuting 5 for x and 0.01 for h.

The completed table is:

x	h	$\dfrac{f(x+h)-f(x)}{h}$
5	2	48
5	1	44
5	0.1	40.4
5	0.01	40.04

3. a) $f(x)=-4x^2$

so,

$f(x+h)=-4(x+h)^2$ substituing $x+h$ for x

$=-4(x^2+2xh+h^2)$

$=-4x^2-8xh-4h^2$

Then

$\dfrac{f(x+h)-f(x)}{h}$ Difference quotient

$=\dfrac{(-4x^2-8xh-4h^2)-(-4x^2)}{h}$ Substituting

$=\dfrac{-8xh-4h^2}{h}$

$=\dfrac{h(-8x-4h)}{h}$ Factoring the numerator

$=\dfrac{h}{h}\cdot(-8x-4h)$ Removing a factor = 1.

$=-8x-4h,$ Simplified difference quotient

or $-4(2x+h)$

b) The difference quotient column in the table can be completed using the simplified difference quotient.

$-8x-4h$ Simplified difference quotient

$=-8(5)-4(2)=-48$

substitiuting 5 for x and 2 for h;

$=-8(5)-4(1)=-44$

substitiuting 5 for x and 1 for h;

$=-8(5)-4(0.1)=-40.4$

substitiuting 5 for x and 0.1 for h;

$=-8(5)-4(0.01)=-40.04$

substitiuting 5 for x and 0.01 for h.

The completed table is:

x	h	$\dfrac{f(x+h)-f(x)}{h}$
5	2	−48
5	1	−44
5	0.1	−40.4
5	0.01	−40.04

Copyright © 2014 Pearson Education, Inc. Publishing as Addison-Wesley.

5. a) $f(x)=x^2+x$

We substitute $x+h$ for x

$$f(x+h)=(x+h)^2+(x+h)$$
$$=(x^2+2xh+h^2)+x+h$$
$$=x^2+2xh+h^2+x+h$$

Then

$\dfrac{f(x+h)-f(x)}{h}$ Difference quotient

$$=\frac{(x^2+2xh+h^2+x+h)-(x^2+x)}{h}$$

$$=\frac{2xh+h^2+h}{h}$$

$=\dfrac{h(2x+h+1)}{h}$ Factoring the numerator

$=\dfrac{h}{h}\cdot(2x+h+1)$ Removing a factor = 1.

$=2x+h+1$ Simplified difference quotient

b) The difference quotient column in the table can be completed using the simplified difference quotient.

$2x+h+1$ Simplified difference quotient

$=2(5)+(2)+1=13$

substitiuting 5 for x and 2 for h;

$=2(5)+(1)+1=12$

substitiuting 5 for x and 1 for h;

$=2(5)+(0.1)+1=11.1$

substitiuting 5 for x and 0.1 for h;

$=2(5)+(0.01)+1=11.01$

substitiuting 5 for x and 0.01 for h.

The completed table is:

x	h	$\dfrac{f(x+h)-f(x)}{h}$
5	2	13
5	1	12
5	0.1	11.1
5	0.01	11.01

7. a) $f(x)=\dfrac{2}{x}$

We substitute $x+h$ for x

$$f(x+h)=\frac{2}{x+h}$$

Then

$\dfrac{f(x+h)-f(x)}{h}$ Difference quotient

$$=\frac{\left(\frac{2}{x+h}\right)-\left(\frac{2}{x}\right)}{h}$$

$=\dfrac{\left(\frac{2}{x+h}\cdot\frac{x}{x}\right)-\left(\frac{2}{x}\cdot\frac{(x+h)}{(x+h)}\right)}{h}$ multiplying by 1

$$=\frac{\left(\frac{2x}{x(x+h)}\right)-\left(\frac{2(x+h)}{x(x+h)}\right)}{h}$$

$=\dfrac{\frac{-2h}{x(x+h)}}{h}$ adding fractions

$=\dfrac{\frac{-2h}{x(x+h)}}{\frac{h}{1}}$ $h=\dfrac{h}{1}$

$=\dfrac{-2h}{x(x+h)}\cdot\dfrac{1}{h}$ multiplying by the reciprocal

$=\dfrac{h}{h}\cdot\left(\dfrac{-2}{x(x+h)}\right)$ Removing a factor = 1.

$=\dfrac{-2}{x(x+h)}$ Simplified difference quotient

b) The difference quotient column in the table can be completed using the simplified difference quotient.

$-\dfrac{2}{x(x+h)}$ Simplified difference quotient

$=-\dfrac{2}{5(5+2)}=-\dfrac{2}{35}$

substitiuting 5 for x and 2 for h;

$=-\dfrac{2}{5(5+1)}=-\dfrac{2}{30}=-\dfrac{1}{15}$

substitiuting 5 for x and 1 for h;

The solution is continued on the next page.

Copyright © 2014 Pearson Education, Inc. Publishing as Addison-Wesley.

$= -\dfrac{2}{5(5+0.1)} = -\dfrac{2}{25.5} = -\dfrac{4}{51}$

substitiuting 5 for x and 0.1 for h;

$= -\dfrac{2}{5(5+0.01)} = -\dfrac{2}{25.05} = -\dfrac{40}{501}$

substitiuting 5 for x and 0.01 for h.

The completed table is:

x	h	$\frac{f(x+h)-f(x)}{h}$
5	2	$-\frac{2}{35}$
5	1	$-\frac{1}{15}$
5	0.1	$-\frac{4}{51}$
5	0.01	$-\frac{40}{501}$

9. a) $f(x) = -2x+5$

We substitute $x+h$ for x

$f(x+h) = -2(x+h)+5$

$-2x-2h+5$

Then

$\dfrac{f(x+h)-f(x)}{h}$ Difference quotient

$= \dfrac{(-2x-2h+5)-(-2x+5)}{h}$

$= \dfrac{-2h}{h}$

$= -2$ Simplified difference quotient

b) The difference quotient is -2 for all values of x and h. Therefore, the completed table is

x	h	$\frac{f(x+h)-f(x)}{h}$
5	2	-2
5	1	-2
5	0.1	-2
5	0.01	-2

11. a) $f(x) = 1-x^3$

so,

$f(x+h) = 1-(x+h)^3$ substituing $x+h$ for x

$= 1-\left(x^3+3x^2h+3xh^2+h^3\right)$

$= 1-x^3-3x^2h-3xh^2-h^3$

Then

$\dfrac{f(x+h)-f(x)}{h}$ Difference quotient

$= \dfrac{\left(1-x^3-3x^2h-3xh^2-h^3\right)-\left(1-x^3\right)}{h}$

$= \dfrac{-3x^2h-3xh^2-h^3}{h}$

$= \dfrac{h\left(-3x^2-3xh-h^2\right)}{h}$ Factoring the numerator

$= \dfrac{h}{h}\cdot\left(-3x^2-3xh-h^2\right)$ Removing a factor = 1.

$= -3x^2-3xh-h^2$

Simplified difference quotient

b) The difference quotient column in the table can be completed using the simplified difference quotient.

$-3x^2-3xh-h^2$

$= -3(5)^2-3(5)(2)-(2)^2 = -109$

substitiuting 5 for x and 2 for h;

$= -3(5)^2-3(5)(1)-(1)^2 = -91$

substitiuting 5 for x and 1 for h;

$= -3(5)^2-3(5)(0.1)-(0.1)^2 = -76.51$

substitiuting 5 for x and 0.1 for h;

$= -3(5)^2-3(5)(0.01)-(0.01)^2 = -75.1501$

substitiuting 5 for x and 0.01 for h.

The completed table is:

x	h	$\frac{f(x+h)-f(x)}{h}$
5	2	-109
5	1	-91
5	0.1	-76.51
5	0.01	-75.1501

Copyright © 2014 Pearson Education, Inc. Publishing as Addison-Wesley.

13. a) $f(x)=x^2-3x$

We substitute $x+h$ for x

$$f(x+h)=(x+h)^2-3(x+h)$$
$$=(x^2+2xh+h^2)-3x-3h$$
$$=x^2+2xh+h^2-3x-3h$$

Then

$$\frac{f(x+h)-f(x)}{h} \quad \text{Difference quotient}$$
$$=\frac{(x^2+2xh+h^2-3x-3h)-(x^2-3x)}{h}$$
$$=\frac{2xh+h^2-3h}{h}$$
$$=\frac{h(2x+h-3)}{h} \quad \text{Factoring the numerator}$$
$$=\frac{h}{h}\cdot(2x+h-3) \quad \text{Removing a factor = 1.}$$
$$=2x+h-3 \quad \text{Simplified difference quotient}$$

b) The difference quotient column in the table can be completed using the simplified difference quotient.

$2x+h-3$ Simplified difference quotient

$=2(5)+(2)-3=9$

substitiuting 5 for x and 2 for h;

$=2(5)+(1)-3=8$

substitiuting 5 for x and 1 for h;

$=2(5)+(0.1)-3=7.1$

substitiuting 5 for x and 0.1 for h;

$=2(5)+(0.01)-3=7.01$

substitiuting 5 for x and 0.01 for h.

Using the values from the previous page, the completed table is:

x	h	$\frac{f(x+h)-f(x)}{h}$
5	2	9
5	1	8
5	0.1	7.1
5	0.01	7.01

15. a) $f(x)=x^2+4x-3$

We substitute $x+h$ for x

$$f(x+h)=(x+h)^2+4(x+h)-3$$
$$=(x^2+2xh+h^2)+4x+4h-3$$
$$=x^2+2xh+h^2+4x+4h-3$$

Then

$$\frac{f(x+h)-f(x)}{h} \quad \text{Difference quotient}$$
$$=\frac{(x^2+2xh+h^2+4x+4h-3)-(x^2+4x-3)}{h}$$
$$=\frac{2xh+h^2+4h}{h}$$
$$=\frac{h(2x+h+4)}{h} \quad \text{Factoring the numerator}$$
$$=\frac{h}{h}\cdot(2x+h+4) \quad \text{Removing a factor = 1}$$
$$=2x+h+4 \quad \text{Simplified difference quotient}$$

b) The difference quotient column in the table can be completed using the simplified difference quotient.

$2x+h+4$ Simplified difference quotient

$=2(5)+(2)+4=16$

substitiuting 5 for x and 2 for h;

$=2(5)+(1)+4=15$

substitiuting 5 for x and 1 for h;

$=2(5)+(0.1)+4=14.1$

substitiuting 5 for x and 0.1 for h;

$=2(5)+(0.01)+4=14.01$

substitiuting 5 for x and 0.01 for h.

The completed table is:

x	h	$\frac{f(x+h)-f(x)}{h}$
5	2	16
5	1	15
5	0.1	14.1
5	0.01	14.01

Copyright © 2014 Pearson Education, Inc. Publishing as Addison-Wesley.

17. a) To find the average rate of change from 2000 to 2005, we locate the corresponding points $(2000,0)$ and $(2005,1.5)$. Using these two points we calculate the average rate of change

$$\frac{1.5-0}{2005-2000}=\frac{1.5}{5}=0.3$$

The average rate of change of total employment from 2000 to 2005 increased approximately 0.3% per year.

b) To find the average rate of change from 2005 to 2009, we locate the corresponding points $(2005,1.5)$ and $(2009,-0.6)$. Using these two points we calculate the average rate of change

$$\frac{-0.6-1.5}{2009-2004}=\frac{-2.1}{4}=-0.525\approx-0.5$$

The average rate of change of total employment from 2005 to 2009 increased approximately -0.5% per year.

c) To find the average rate of change from 2000 to 2009, we locate the corresponding points $(2000,0)$ and $(2009,-0.6)$. Using these two points we calculate the average rate of change

$$\frac{-0.6-0}{2009-2000}=\frac{-0.6}{9}\approx-0.07$$

The average rate of change of total employment from 2000 to 2009 increased approximately -0.07% per year.

19. a) To find the average rate of change from 2000 to 2005, we locate the corresponding points $(2000,0)$ and $(2005,1.75)$. Using these two points we calculate the average rate of change

$$\frac{1.75-0}{2005-2000}=\frac{1.75}{5}=0.35$$

The average rate of change of employment for professional services from 2000 to 2005 increased approximately 0.35% per year.

b) To find the average rate of change from 2005 to 2009, we locate the corresponding points $(2005,1.75)$ and $(2009,-0.5)$. Using these two points we calculate the average rate of change

$$\frac{-0.5-1.75}{2009-2004}=\frac{-2.25}{4}=-0.5625\approx-0.56$$

The average rate of change of employment for professional services from 2005 to 2009 increased approximately -0.56% per year.

c) To find the average rate of change from 2000 to 2009, we locate the corresponding points $(2000,0)$ and $(2009,-0.5)$. Using these two points we calculate the average rate of change

$$\frac{-0.5-0}{2009-2000}=\frac{-0.5}{9}\approx-0.06$$

The average rate of change of employment for professional services from 2000 to 2009 increased approximately -0.06% per year.

21. a) To find the average rate of change from 2000 to 2005, we locate the corresponding points $(2000,0)$ and $(2005,18.6)$. Using these two points we calculate the average rate of change

$$\frac{18.6-0}{2005-2000}=\frac{18.6}{5}=3.72\approx3.7$$

The average rate of change of education employment from 2000 to 2005 increased approximately 3.7% per year.

b) To find the average rate of change from 2005 to 2009, we locate the corresponding points $(2005,18.6)$ and $(2009,29.25)$. Using these two points we calculate the average rate of change

$$\frac{29.25-18.6}{2009-2005}=\frac{10.65}{4}=2.6625\approx2.7$$

The average rate of change of education employment from 2000 to 2005 increased approximately 2.7% per year.

c) To find the average rate of change from 2000 to 2009, we locate the corresponding points $(2000,0)$ and $(2009,29.25)$. Using these two points we calculate the average rate of change

$$\frac{29.25-0}{2009-2000}=\frac{29.25}{9}=3.25$$

The average rate of change of education employment from 2000 to 2009 increased approximately 3.5% per year.

23. a) To find the average rate of change from 2000 to 2005, we locate the corresponding points $(2000,0)$ and $(2005,4.84)$. Using these two points we calculate the average rate of change

$$\frac{4.84-0}{2005-2000}=\frac{4.84}{5}=0.968\approx0.97$$

The average rate of change of natural resources employment from 2000 to 2005 increased approximately 0.97% per year.

Copyright © 2014 Pearson Education, Inc. Publishing as Addison-Wesley.

b) To find the average rate of change from 2005 to 2009, we locate the corresponding points $(2005, 4.84)$ and $(2009, 16.86)$.
Using these two points we calculate the average rate of change
$$\frac{16.86-4.84}{2009-2004}=\frac{12.02}{4}=3.0005\approx 3$$
The average rate of change of natural resources employment from 2005 to 2009 increased approximately 3% per year.

c) To find the average rate of change from 2000 to 2009, we locate the corresponding points $(2000, 0)$ and $(2009, 16.86)$. Using these two points we calculate the average rate of change
$$\frac{16.86-0}{2009-2000}=\frac{16.86}{9}\approx 1.8733\approx 1.9$$
The average rate of change of natural resources employment from 2000 to 2009 increased approximately 1.9% per year.

25. In order to find the average rate of change from 1970 to 1980, we use the data points $(1970, 67.84)$ and $(1980, 78.29)$. The average rate of change is
$$\frac{78.29-67.84}{1980-1970}=\frac{10.45}{10}=1.045.$$
Between the years 1970 to 1980 the average rate of change in U.S. energy consumption increased about 1.045 quadrillion BTUs per year.

In order to find the average rate of change from 1980 to 1990, we use the data points $(1980, 78.29)$ and $(1990, 84.67)$. The average rate of change is
$$\frac{84.67-78.29}{1990-1980}=\frac{6.38}{10}=0.638.$$
Between the years 1980 to 1990 the average rate of change in U.S. energy consumption increased about 0.638 quadrillion BTUs per year.
In order to find the average rate of change from 2000 to 2010, we use the data points $(2000, 98.90)$ and $(2010, 98.01)$. The average rate of change is
$$\frac{98.01-98.90}{2010-2000}=\frac{-0.89}{10}=-0.089.$$
Between the years 2000 to 2010 the average rate of change in U.S. energy consumption increased about -0.089 quadrillion BTUs per year.

27. a) From 0 units to 1 unit the average rate of change is
$$\frac{70-0}{1-0}=70 \text{ pleasure units per unit.}$$
From 1 unit to 2 units the average rate of change is
$$\frac{109-70}{2-1}=\frac{39}{1}=39 \text{ pleasure units per unit.}$$
From 2 units to 3 units the average rate of change is
$$\frac{138-109}{3-2}=\frac{29}{1}=29 \text{ pleasure units per unit.}$$
From 3 units to 4 units the average rate of change is
$$\frac{161-138}{4-3}=\frac{23}{1}=23 \text{ pleasure units per unit.}$$
b) ✎

29. $p(x)=0.03x^2+0.56x+8.63$

a) $p(4)=0.03(4)^2+0.56(4)+8.63=11.35$

b) $p(17)=0.03(17)^2+0.56(17)+8.63=26.82$

c) $P(17)-p(4)=26.82-11.35=15.47$

d) $\dfrac{p(17)-p(4)}{17-4}=\dfrac{15.47}{13}=1.19$

This result implies that the average price of a ticket between 1995 ($x=4$) and 2008 ($x=14$) grew at an average rate of \$1.19 per year.

31. $A(t)=5000(1.14)^t$

$A(3)=5000(1.14)^3=7407.72$

$A(2)=5000(1.14)^2=6498.00$

$$\frac{A(3)-A(2)}{3-2}=\frac{7407.72-6498.00}{3-2}=909.72$$
If unpaid, the debt on the credit card will be growing at an average rate of \$909.72 per year between the 2nd and 3rd year.

33. $C(x)=-0.05x^2+50x$

First substitute 301 for x.
$$\begin{aligned}C(301)&=-0.05(301)^2+50(301)\\&=-4530.05+15{,}050\\&=10{,}519.95\end{aligned}$$
The total cost of producing 301 units is \$10,519.95.
The solution is continued on the next page.

Copyright © 2014 Pearson Education, Inc. Publishing as Addison-Wesley.

Next substitute 300 for x.

$$C(300) = -0.05(300)^2 + 50(300)$$
$$= -4500 + 15{,}000$$
$$= 10{,}500.00$$

The total cost of producing 300 units is \$10,500.00.

Now we can substitute to find the average rate of change.

$$\frac{C(301) - C(300)}{301 - 300} = \frac{10{,}519.95 - 10{,}500}{301 - 300}$$
$$= \frac{19.95}{1}$$
$$= 19.95$$

Total cost will increase \$19.95 if the company produces the 301st unit.

35. Note: Answers will vary according to the values estimated from the graph.

a) Locate the points $(0,8)$ and $(12,20)$ on the girls growth median weight chart. Using these points we calculate the average growth rate.

$$\frac{20-8}{12-0} = \frac{12}{12} = 1.$$

The average growth rate of a girl during her first 12 months is 1 pound per month.

b) Locate the points $(12,20)$ and $(24,26.5)$ on the girls growth median weight chart. Using these points we calculate the average growth rate.

$$\frac{26.5-20}{24-12} = \frac{6.5}{12} = 0.541\overline{6} \approx 0.54.$$

The average growth rate of a girl during her second 12 months is approximately 0.54 pounds per month.

c) Locate the points $(0,8)$ and $(24,26.5)$ on the girls growth median weight chart. Using these points we calculate the average growth rate.

$$\frac{26.5-8}{24-0} = \frac{18.5}{24} = 0.7708\overline{3} \approx 0.77.$$

The average growth rate of a girl during her first 24 months is approximately 0.77 pounds per month.

d) We estimate the growth rate of a 12 month old girl to be approximately 0.67 pounds per month. This answer will vary depending upon your tangent line.

e) The graph indicates that the growth rate is fastest during the first 3 months.

37. $H(w) = 0.11w^{1.36}$

a) First we substitute 500 and 700 in for w to find the home range at the respective weights.

$$H(500) = 0.11(500)^{1.36}$$
$$= 0.11(4683.809314)$$
$$= 515.2190246$$
$$\approx 515.22$$
$$H(700) = 0.11(700)^{1.36}$$
$$= 0.11(7401.731628)$$
$$= 814.1904791$$
$$\approx 814.19$$

Next we use the function values to find the average rate at which the mammal's home range will increase

$$\frac{H(700) - H(500)}{700 - 500} = \frac{814.19 - 515.22}{700 - 500}$$
$$= \frac{298.97}{200}$$
$$\approx 1.49485$$

The average rate at which a carnivorous mammal's home range increases as the animal's weight grows from 500 g to 700 g is approximately 1.49 hectares per gram.

b) First we substitute 200 and 300 in for w to find the home range at the respective weights.

$$H(200) = 0.11(200)^{1.36}$$
$$= 0.11(1347.102971)$$
$$= 148.1813269$$
$$\approx 148.18$$
$$H(300) = 0.11(300)^{1.36}$$
$$= 0.11(2338.217499)$$
$$= 257.2039249$$
$$\approx 257.20$$

Next we use the function values to find the average rate at which the mammal's home range will increase

$$\frac{H(300) - H(200)}{300 - 200} = \frac{257.20 - 148.18}{300 - 200}$$
$$= \frac{109.02}{100}$$
$$\approx 1.0902$$

The average rate at which a carnivorous mammal's home range increases as the animal's weight grows from 200 g to 300 g is approximately 1.09 hectares per gram.

Copyright © 2014 Pearson Education, Inc. Publishing as Addison-Wesley.

39. a) We locate the points $(0,0)$ and $(8,10)$ on the graph and use them to calculate the average rate of change.

$$\frac{10-0}{8-0}=\frac{10}{8}=\frac{5}{4}=1.25$$

The average rate of change is 1.25 words per minute.

We locate the points $(8,10)$ and $(16,20)$ on the graph and use them to calculate the average rate of change.

$$\frac{20-10}{16-8}=\frac{10}{8}=\frac{5}{4}=1.25$$

The average rate of change is 1.25 words per minute.

We locate the points $(16,20)$ and $(24,25)$ on the graph and use them to calculate the average rate of change.

$$\frac{25-20}{24-16}=\frac{5}{8}=0.625$$

The average rate of change is 0.625 words per minute.

We locate the points $(24,25)$ and $(32,25)$ on the graph and use them to calculate the average rate of change.

$$\frac{25-25}{32-24}=\frac{0}{8}=0$$

The average rate of change is 0 words per minute.

We locate the points $(32,25)$ and $(36,25)$ on the graph and use them to calculate the average rate of change.

$$\frac{25-25}{36-32}=\frac{0}{4}=0$$

The average rate of change is 0 words per minute.

b) ✎

41. $s(t)=16t^2$

a) First, we find the function values by substituting 3 and 5 in for t respectively.

$$s(3)=16(3)^2=16(9)=144$$

$$s(5)=16(5)^2=16(25)=400$$

Next we subtract the function values.

$$s(5)-s(3)=400-144=256$$

The object will fall 256 feet in the two second time period between $t=3$ and $t=5$.

b) The average rate of change is calculated as

$$\frac{s(5)-s(3)}{5-3}=\frac{400-144}{5-3}=\frac{256}{2}=128$$

The average velocity of the object during the two second time period from $t=3$ to $t=5$ is 128 feet per second.

43. a) For each curve, as t changes from 0 to 4, $P(t)$ changes from 0 to 500. Thus, the average growth rate for each country is

$$\frac{500-0}{4-0}=\frac{500}{4}=125.$$

The average growth rate for each country is approximately 125 million people per year.

b) ✎

c) For Country A:

As t changes from 0 to 1, $P(t)$ changes from 0 to 290. Thus the average growth rate is $\frac{290-0}{1-0}=290$ million people per year.

As t changes from 1 to 2, $P(t)$ changes from 290 to 250. Thus the average growth rate is

$\frac{250-290}{2-1}=-40$ million people per year.

As t changes from 2 to 3, $P(t)$ changes from 250 to 200. Thus the average growth rate is

$\frac{200-250}{3-2}=-50$ million people per year.

As t changes from 3 to 4, $P(t)$ changes from 200 to 500. Thus the average growth rate is

$\frac{500-200}{4-3}=300$ million people per year.

For Country B:

As t changes from 0 to 1, $P(t)$ changes from 0 to 125. Thus the average growth rate is $\frac{125-0}{1-0}=125$ million people per year.

The solution is continued on the next page.

Copyright © 2014 Pearson Education, Inc. Publishing as Addison-Wesley.

As t changes from 1 to 2, $P(t)$ changes from 125 to 250. Thus the average growth rate is

$$\frac{250-125}{2-1}=125 \text{ million people per year.}$$

As t changes from 2 to 3, $P(t)$ changes from 250 to 375. Thus the average growth rate is

$$\frac{375-250}{3-2}=125 \text{ million people per year.}$$

As t changes from 3 to 4, $P(t)$ changes from 375 to 500. Thus the average growth rate is

$$\frac{500-375}{4-3}=125 \text{ million people per year.}$$

d) ✎

45. a) Tracing along the 4-year private school graph, we see that the largest increase in costs occurred in the 1985-86 year.

b) Tracing along the 4-year public school graph, we see that the largest increases in costs occurred during the 1975-76 year, the 2003-04 year and the 2004-05 year.

c) For the 4-year public school, the cost in 2005 dollars is approximately \$6000. To find out what the cost was in 1975 dollars assuming a 3% inflation rate over the 30 years we create the following equation using the simple compound interest formula from section R1 $C(1+r)^t=A$.

$$C(1+.03)^{30}=6000$$
$$C(1.03)^{30}=6000$$
$$C=\frac{6000}{(1.03)^{30}}$$
$$C=2471.92$$
$$C\approx 2472$$

The cost of attending a 4 year public school in 1975 was \$2472 in 1975 dollars. Likewise, for the 4-year private school, the cost in 2005 dollars is approximately \$13,000.
To find out what the cost was in 1975 dollars assuming a 3% inflation rate over the 30 years we create the equation at the top of the next column using the simple compound interest formula from section R1.

$$C(1+r)^t=A.$$
$$C(1+.03)^{30}=13{,}000$$
$$C(1.03)^{30}=13{,}000$$
$$C=\frac{13{,}000}{(1.03)^{30}}$$
$$C=5355.83$$
$$C\approx 5356$$

The cost of attending a 4 year private school in 1975 was \$5356 in 1975 dollars.

47. $f(x)=ax^2+bx+c$

Substituting $x+h$ for x we have,

$$f(x+h)=a(x+h)^2+b(x+h)+c$$
$$=a\left(x^2+2xh+h^2\right)+bx+bh+c$$
$$=ax^2+2axh+ah^2+bx+bh+c$$

Thus,

$$\frac{f(x+h)-f(x)}{h} \quad \text{Difference quotient}$$
$$=\frac{\left(ax^2+2axh+ah^2+bx+bh+c\right)-\left(ax^2+bx+c\right)}{h}$$
$$=\frac{2axh+ah^2+bh}{h}$$
$$=\frac{h(2ax+ah+b)}{h} \quad \text{Factoring the numerator}$$
$$=2ax+ah+b \quad \text{Simplified difference quotient}$$

49. $f(x)=x^4$

Substituting $x+h$ for x we have,

$$f(x+h)=(x+h)^4$$
$$=x^4+4x^3h+6x^2h^2+4xh^3+h^4$$

Thus,

$$\frac{f(x+h)-f(x)}{h} \quad \text{Difference quotient}$$
$$=\frac{\left(x^4+4x^3h+6x^2h^2+4xh^3+h^4\right)-\left(x^4\right)}{h}$$
$$=\frac{4x^3h+6x^2h^2+4xh^3+h^4}{h}$$
$$=\frac{h\left(4x^3+6x^2h+4xh^2+h^3\right)}{h} \quad \text{Factoring the numerator}$$
$$=4x^3+6x^2h+4xh^2+h^3 \quad \text{Simplified difference quotient}$$

Copyright © 2014 Pearson Education, Inc. Publishing as Addison-Wesley.

51. $f(x) = ax^5 + bx^4$

Substituting $x+h$ for x we have,

$$f(x+h) = a(x+h)^5 + b(x+h)^4$$
$$= ax^5 + 5ax^4h + 10ax^3h^2 + 10ax^2h^3 + 5axh^4 + ah^5 + bx^4 + 4bx^3h + 6bx^2h^2 + 4bxh^3 + bh^4$$

Thus,

$$\frac{f(x+h) - f(x)}{h} \quad \text{Difference quotient}$$

$$= \frac{ax^5 + 5ax^4h + 10ax^3h^2 + 10ax^2h^3 + 5axh^4 + ah^5 + bx^4 + 4bx^3h + 6bx^2h^2 + 4bxh^3 + bh^4 - (ax^5 + bx^4)}{h}$$

$$= \frac{5ax^4h + 10ax^3h^2 + 10ax^2h^3 + 5axh^4 + ah^5 + 4bx^3h + 6bx^2h^2 + 4bxh^3 + bh^4}{h}$$

$$= \frac{h(5ax^4 + 10ax^3h + 10ax^2h^2 + 5axh^3 + ah^4 + 4bx^3 + 6bx^2h^1 + 4bxh^2 + bh^3)}{h} \quad \text{Factoring the numerator}$$

$$= 5ax^4 + 10ax^3h + 10ax^2h^2 + 5axh^3 + ah^4 + 4bx^3 + 6bx^2h^1 + 4bxh^2 + bh^3 \quad \text{Simplified difference quotient}$$

53. $f(x) = \dfrac{1}{1-x}$

$$\frac{f(x+h) - f(x)}{h}$$

$$= \frac{\dfrac{1}{1-(x+h)} - \left(\dfrac{1}{1-x}\right)}{h}$$

$$= \frac{\dfrac{1}{1-x-h} \cdot \dfrac{1-x}{1-x} - \left(\dfrac{1}{1-x} \cdot \dfrac{1-x-h}{1-x-h}\right)}{h}$$

$$= \frac{\dfrac{1-x}{(1-x-h)(1-x)} - \dfrac{1-x-h}{(1-x-h)(1-x)}}{h}$$

$$= \frac{\dfrac{h}{(1-x-h)(1-x)}}{h}$$

$$= \frac{h}{(1-x-h)(1-x)} \cdot \frac{1}{h}$$

$$= \frac{1}{(1-x-h)(1-x)}$$

Copyright © 2014 Pearson Education, Inc. Publishing as Addison-Wesley.

55. $f(x)=\sqrt{2x+1}$

$$\frac{f(x+h)-f(x)}{h} \quad \text{Difference quotient}$$

$$=\frac{\sqrt{2(x+h)+1}-\sqrt{2x+1}}{h}$$

$$=\frac{\sqrt{2(x+h)+1}-\sqrt{2x+1}}{h}\cdot\frac{\sqrt{2(x+h)+1}+\sqrt{2x+1}}{\sqrt{2(x+h)+1}+\sqrt{2x+1}}$$

$$=\frac{2(x+h)+1-(2x+1)}{h\left(\sqrt{2(x+h)+1}+\sqrt{2x+1}\right)}$$

$$=\frac{2x-2h+1-2x-1}{h\left(\sqrt{2(x+h)+1}+\sqrt{2x+1}\right)}$$

$$=\frac{-2h}{h\left(\sqrt{2(x+h)+1}+\sqrt{2x+1}\right)}$$

$$=\frac{-2}{\sqrt{2(x+h)+1}+\sqrt{2x+1}}$$

Copyright © 2014 Pearson Education, Inc. Publishing as Addison-Wesley.

Exercise Set 1.4

1. $f(x)=\frac{3}{2}x^2$

a), b)

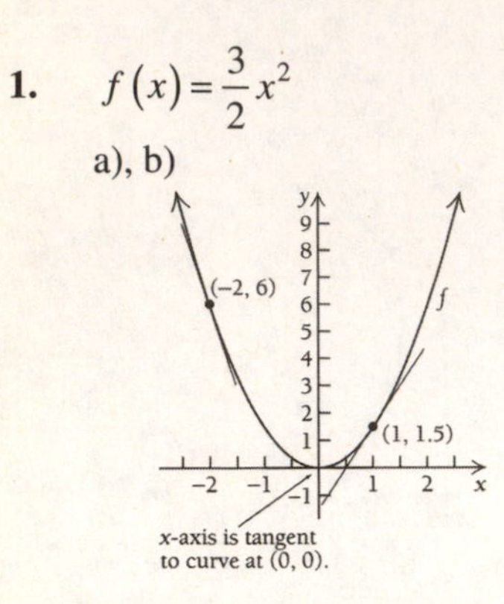

c) Find the simplified difference quotient first.

$$\frac{f(x+h)-f(x)}{h}$$
$$=\frac{\frac{3}{2}(x+h)^2-\frac{3}{2}x^2}{h}$$
$$=\frac{\frac{3}{2}(x^2+2xh+h^2)-\frac{3}{2}x^2}{h}$$
$$=\frac{\frac{3}{2}x^2+3xh+\frac{3}{2}h^2-\frac{3}{2}x^2}{h}$$
$$=\frac{3xh+\frac{3}{2}h^2}{h}$$
$$=\frac{h\left(3x+\frac{3}{2}h\right)}{h}$$
$$=3x+\frac{3}{2}h \quad \text{Simplified difference quotient}$$

Now we will find the limit of the difference quotient as $h\to 0$ using the simplified difference quotient.

$$\lim_{h\to 0}\frac{f(x+h)-f(x)}{h}=\lim_{h\to 0}\left(3x+\frac{3}{2}h\right)$$
$$=3x$$

Thus, $f'(x)=3x$.

d) Find the values of the derivative by making the appropriate substitutions.

$f'(-2)=3(-2)=-6$ Substituting -2 for x

$f'(0)=3(0)=0$ Substituting 0 for x

$f'(1)=3(1)=3$ Substituting 1 for x

3. $f(x)=-2x^2$

a), b)

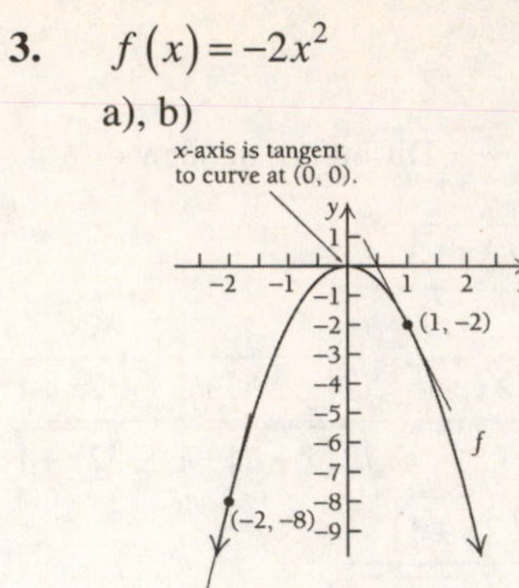

c) Find the simplified difference quotient first.

$$\frac{f(x+h)-f(x)}{h}$$
$$=\frac{-2(x+h)^2-(-2x^2)}{h}$$
$$=\frac{-2(x^2+2xh+h^2)-(-2x^2)}{h}$$
$$=\frac{-2x^2-4xh-2h^2+2x^2}{h}$$
$$=\frac{-4xh-2h^2}{h}$$
$$=\frac{h(-4x-2h)}{h}$$
$$=-4x-2h \quad \text{Simplified difference quotient}$$

Now we will find the limit of the difference quotient as $h\to 0$ using the simplified difference quotient.

$$\lim_{h\to 0}\frac{f(x+h)-f(x)}{h}=\lim_{h\to 0}(-4x-2h)$$
$$=-4x$$

Thus, $f'(x)=-4x$.

d) Find the values of the derivative by making the appropriate substitutions.

$f'(-2)=-4(-2)=8$ Substituting -2 for x

$f'(0)=-4(0)=0$ Substituting 0 for x

$f'(1)=-4(1)=-4$ Substituting 1 for x

5. $f(x)=x^3$

a), b)

(1, 1)

x-axis is tangent to curve at (0, 0).

(−2, −8)

Copyright © 2014 Pearson Education, Inc. Publishing as Addison-Wesley.

c) Find the simplified difference quotient first.

$$\frac{f(x+h)-f(x)}{h}$$

$$=\frac{(x+h)^3-(x^3)}{h}$$

$$=\frac{(x^3+3x^2h+3xh^2+h^3)-(x^3)}{h}$$

$$=\frac{3x^2h+3xh^2+h^3}{h}$$

$$=\frac{h(3x^2+3xh+h^2)}{h}$$

$=3x^2+3xh+h^2$ Simplified difference quotient

Now we will find the limit of the difference quotient as $h \to 0$ using the simplified difference quotient.

$$\lim_{h\to 0}\frac{f(x+h)-f(x)}{h}=\lim_{h\to 0}(3x^2+3xh+h^2)$$

$$=3x^2$$

Thus, $f'(x)=3x^2$.

d) Find the values of the derivative by making the appropriate substitutions.

$f'(-2)=3(-2)^2=12$ Substituting -2 for x

$f'(0)=3(0)^2=0$ Substituting 0 for x

$f'(1)=3(1)^2=3$ Substituting 1 for x

7. $f(x)=2x+3$

a), b)

c) Find the simplified difference quotient first.

$$\frac{f(x+h)-f(x)}{h}$$

$$=\frac{2(x+h)+3-(2x+3)}{h}$$

$$=\frac{2x+2h+3-2x-3}{h}$$

$$=\frac{2h}{h}$$

$=2$ Simplified difference quotient

Now we will find the limit of the difference quotient as $h \to 0$ using the simplified difference quotient.

$$\lim_{h\to 0}\frac{f(x+h)-f(x)}{h}=\lim_{h\to 0}(2)$$

$$=2$$

Thus, $f'(x)=2$.

d) Since the derivative is a constant, the value of the derivative will be 2 regardless of the value of x.

$f'(-2)=2$ Substituting -2 for x

$f'(0)=2$ Substituting 0 for x

$f'(1)=2$ Substituting 1 for x

9. $f(x)=\frac{1}{2}x-3$

a), b)

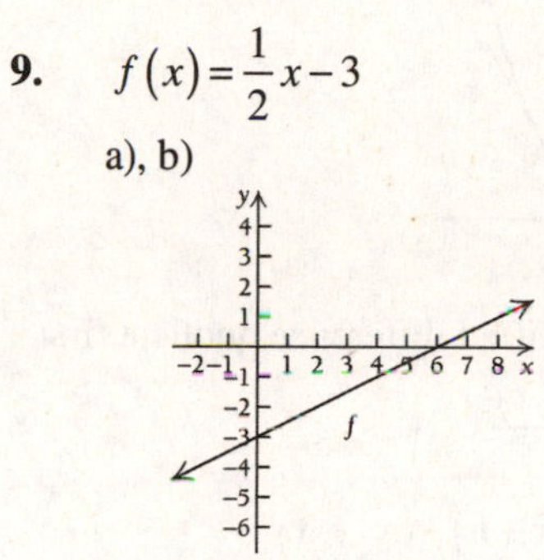

c) Find the simplified difference quotient first.

$$\frac{f(x+h)-f(x)}{h}$$

$$=\frac{\frac{1}{2}(x+h)-3-\left(\frac{1}{2}x-3\right)}{h}$$

$$=\frac{\frac{1}{2}x+\frac{1}{2}h-3-\frac{1}{2}x+3}{h}$$

$$=\frac{\frac{1}{2}h}{h}$$

$=\frac{1}{2}$ Simplified difference quotient

Now we will find the limit of the difference quotient as $h \to 0$ using the simplified difference quotient.

$$\lim_{h\to 0}\frac{f(x+h)-f(x)}{h}=\lim_{h\to 0}\left(\frac{1}{2}\right)$$

$$=\frac{1}{2}$$

Thus, $f'(x)=\frac{1}{2}$.

Copyright © 2014 Pearson Education, Inc. Publishing as Addison-Wesley.

d) Since the derivative is a constant, the value of the derivative will be $\frac{1}{2}$ regardless of the value of x.

$f'(-2)=\frac{1}{2}$ Substituting -2 for x

$f'(0)=\frac{1}{2}$ Substituting 0 for x

$f'(1)=\frac{1}{2}$ Substituting 1 for x

11. $f(x)=x^2+x$

a), b)

c) Find the simplified difference quotient first.

$$\frac{f(x+h)-f(x)}{h}$$

$$=\frac{(x+h)^2+(x+h)-(x^2+x)}{h}$$

$$=\frac{x^2+2xh+h^2+x+h-x^2-x}{h}$$

$$=\frac{2xh+h^2+h}{h}$$

$$=\frac{h(2x+h+1)}{h}$$

$=2x+h+1$ Simplified difference quotient

Now we will find the limit of the difference quotient as $h\to 0$ using the simplified difference quotient.

$$\lim_{h\to 0}\frac{f(x+h)-f(x)}{h}=\lim_{h\to 0}(2x+h+1)$$

$$=2x+1$$

Thus, $f'(x)=2x+1$.

d) Find the values of the derivative by making the appropriate substitutions.

$f'(-2)=2(-2)+1=-3$

$f'(0)=2(0)+1=1$

$f'(1)=2(1)+1=3$

13. $f(x)=2x^2+3x-2$

a), b)

c) Find the simplified difference quotient first.

$$\frac{f(x+h)-f(x)}{h}$$

$$=\frac{\left(2(x+h)^2+3(x+h)-2\right)-\left(2x^2+3x-2\right)}{h}$$

$$=\frac{2\left(x^2+2xh+h^2\right)+3x+3h-2-2x^2-3x+2}{h}$$

$$=\frac{2x^2+4xh+2h^2+3x+3h-2-2x^2-3x+2}{h}$$

$$=\frac{4xh+2h^2+3h}{h}$$

$$=\frac{h(4x+2h+3)}{h}$$

$=4x+2h+3$ Simplified difference quotient

Now we will find the limit of the difference quotient as $h\to 0$ using the simplified difference quotient.

$$\lim_{h\to 0}\frac{f(x+h)-f(x)}{h}=\lim_{h\to 0}(4x+2h+3)$$

$$=4x+3$$

Thus, $f'(x)=4x+3$.

d) Find the values of the derivative by making the appropriate substitutions.

$f'(-2)=4(-2)+3=-5$

$f'(0)=4(0)+3=3$

$f'(1)=4(1)+3=7$

15. $f(x)=\frac{1}{x}$

a), b)

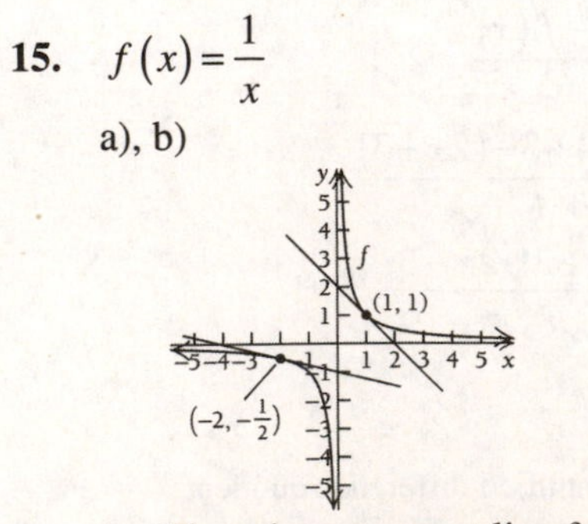

There is no tangent line for $x=0$.

Copyright © 2014 Pearson Education, Inc. Publishing as Addison-Wesley.

c) Find the simplified difference quotient first.

$$\frac{f(x+h)-f(x)}{h}$$

$$=\frac{\left(\frac{1}{x+h}\right)-\left(\frac{1}{x}\right)}{h}$$

$$=\frac{\left(\frac{1}{x+h}\cdot\frac{x}{x}\right)-\left(\frac{1}{x}\cdot\frac{(x+h)}{(x+h)}\right)}{h}$$

$$=\frac{\left(\frac{x}{x(x+h)}\right)-\left(\frac{(x+h)}{x(x+h)}\right)}{h}$$

$$=\frac{\frac{-h}{x(x+h)}}{h}$$

$$=\frac{-h}{x(x+h)}\cdot\frac{1}{h}$$

$$=\frac{-1}{x(x+h)} \quad \text{Simplified difference quotient}$$

Now we will find the limit of the difference quotient as $h \to 0$ using the simplified difference quotient.

$$\lim_{h\to 0}\frac{f(x+h)-f(x)}{h}=\lim_{h\to 0}\left(\frac{-1}{x(x+h)}\right)$$

$$=\frac{-1}{x(x+0)}$$

$$=\frac{-1}{x^2}$$

Thus, $f'(x)=\frac{-1}{x^2}$.

d) Find the values of the derivative by making the appropriate substitutions.

$$f'(-2)=\frac{-1}{(-2)^2}=-\frac{1}{4}$$

$$f'(0)=\frac{-1}{(0)^2}; \text{Thus, } f'(0) \text{ does not exist.}$$

$$f'(1)=\frac{-1}{(1)^2}=-1$$

17. a) From Example 1 we know that $f'(x)=2x$. $f'(3)=2(3)=6$, so the slope of the line tangent to the curve at $(3,9)$ is 6. We substitute the point and the slope into the point-slope equation to find the equation of the tangent line.

$$y-y_1=m(x-x_1)$$
$$y-9=6(x-3)$$
$$y-9=6x-18$$
$$y=6x-9$$

b) $f'(-1)=2(-1)=-2$, so the slope of the line tangent to the curve at $(-1,1)$ is -2. We substitute the point and the slope into the point-slope equation to find the equation of the tangent line.

$$y-y_1=m(x-x_1)$$
$$y-1=-2(x-(-1))$$
$$y-1=-2x-2$$
$$y=-2x-1$$

c) $f'(10)=2(10)=20$, so the slope of the line tangent to the curve at $(10,100)$ is 20. We substitute the point and the slope into the point-slope equation to find the equation of the tangent line.

$$y-y_1=m(x-x_1)$$
$$y-100=20(x-10)$$
$$y-100=20x-200$$
$$y=20x-100$$

19. From Exercise 16 we know that $f'(x)=\frac{-2}{x^2}$.

a) $f'(1)=\frac{-2}{(1)^2}=-2$, so the slope of the line tangent to the curve at $(1,2)$ is -2. We substitute the point and the slope into the point-slope equation to find the equation of the tangent line.

$$y-y_1=m(x-x_1)$$
$$y-2=-2(x-1)$$
$$y-2=-2x+2$$
$$y=-2x+4$$

Copyright © 2014 Pearson Education, Inc. Publishing as Addison-Wesley.

b) $f'(-1)=\frac{-2}{(-1)^2}=-2$, so the slope of the line tangent to the curve at $(-1,2)$ is -2. We substitute the point and the slope into the point-slope equation to find the equation of the tangent line.

$$y-y_1=m(x-x_1)$$
$$y-(-2)=-2(x-(-1))$$
$$y+2=-2x-2$$
$$y=-2x-4$$

c) $f'(100)=\frac{-2}{(100)^2}=-0.0002$, so the slope of the line tangent to the curve at $(100,0.02)$ is –0.002. We substitute the point and the slope into the point-slope equation to find the equation of the tangent line.

$$y-y_1=m(x-x_1)$$
$$y-0.02=-0.0002(x-100)$$
$$y-0.02=-0.0002x+0.02$$
$$y=-0.0002x+0.04$$

21. First, we find $f'(x)$:

$$\frac{f(x+h)-f(x)}{h}$$
$$=\frac{\left(4-(x+h)^2\right)-\left(4-x^2\right)}{h}$$
$$=\frac{4-\left(x^2+2xh+h^2\right)-4+x^2}{h}$$
$$=\frac{4-x^2-2xh-h^2-4+x^2}{h}$$
$$=\frac{-2xh-h^2}{h}$$
$$=\frac{h(-2x-h)}{h}$$
$$=-2x-h \quad \text{Simplified difference quotient}$$

$$f'(x)=\lim_{h\to 0}\frac{f(x+h)-f(x)}{h}$$
$$=\lim_{h\to 0}(-2x-h)$$
$$=-2x$$

a) $f'(-1)=-2(-1)=2$, so the slope of the line tangent to the curve at $(-1,3)$ is 2. We substitute the point and the slope into the point-slope equation to find the equation of the tangent line.

$$y-y_1=m(x-x_1)$$
$$y-3=2(x-(-1))$$
$$y-3=2x+2$$
$$y=2x+5$$

b) $f'(0)=-2(0)=0$, so the slope of the line tangent to the curve at $(0,4)$ is 0. We substitute the point and the slope into the point-slope equation to find the equation of the tangent line

$$y-y_1=m(x-x_1)$$
$$y-4=0(x-0)$$
$$y-4=0$$
$$y=4$$

c) $f'(5)=-2(5)=-10$, so the slope of the line tangent to the curve at $(5,-21)$ is -10. We substitute the point and the slope into the point-slope equation to find the equation of the tangent line.

$$y-y_1=m(x-x_1)$$
$$y-(-21)=-10(x-5)$$
$$y+21=-10x+50$$
$$y=-10x+29$$

23. Find the simplified difference quotient for $f(x)=mx+b$ first.

$$\frac{f(x+h)-f(x)}{h}$$
$$=\frac{m(x+h)+b-(mx+b)}{h}$$
$$=\frac{mx+mh+b-mx-b}{h}$$
$$=\frac{mh}{h}=m \quad \text{Simplified difference quotient}$$

Now we will find the limit of the difference quotient as $h\to 0$ using the simplified difference quotient.

$$\lim_{h\to 0}\frac{f(x+h)-f(x)}{h}=\lim_{h\to 0}(m)$$
$$=m$$

Thus, $f'(x)=m$.

Copyright © 2014 Pearson Education, Inc. Publishing as Addison-Wesley.

25. If a function has a "corner," it will not be differentiable at that point. Thus, the function is not differentiable at x_3, x_4, x_6. The function has a vertical tangent at x_{12}. Vertical lines have undefined slope, hence the function is not differentiable at x_{12}. Also, if a function is discontinuous at some point a, then it is not differentiable at a. The function is discontinuous at the point x_0, thus it is not differentiable at x_0.
Therefore, the graph is not differentiable at the points $x_0, x_3, x_4, x_6, x_{12}$.

27. If a function has a "corner," it will not be differentiable at that point. Thus, the function is not differentiable at x_3. The function has a vertical tangent at x_1. Vertical lines have undefined slope, hence the function is not differentiable at x_1. Also, if a function is discontinuous at some point a, then it is not differentiable at a. The function is discontinuous at the point x_1, x_2, x_4, thus it is not differentiable at x_1, x_2, x_4.
Therefore, the graph is not differentiable at the points x_1, x_2, x_3, x_4.

29. The following graph is continuous but not differentiable, at $x = 3$.

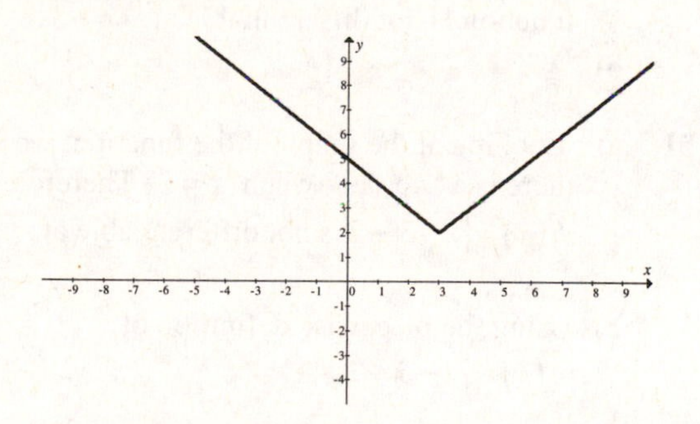

31. The following graph has horizontal tangent line at $x = 0, x = 2,$ and $x = 4$.

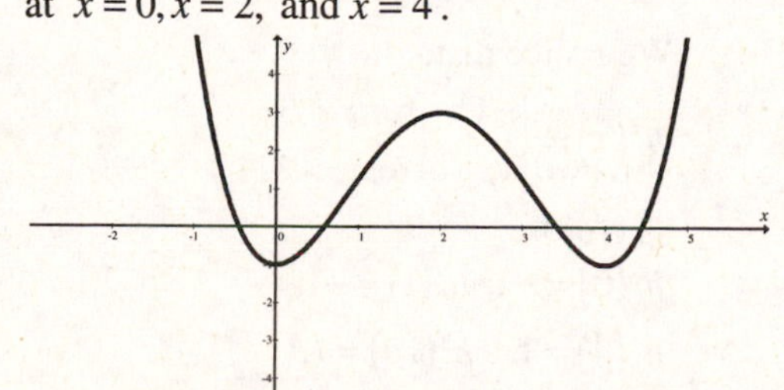

33. The following graph is smooth for all x but is not differentiable at $x = 1$.

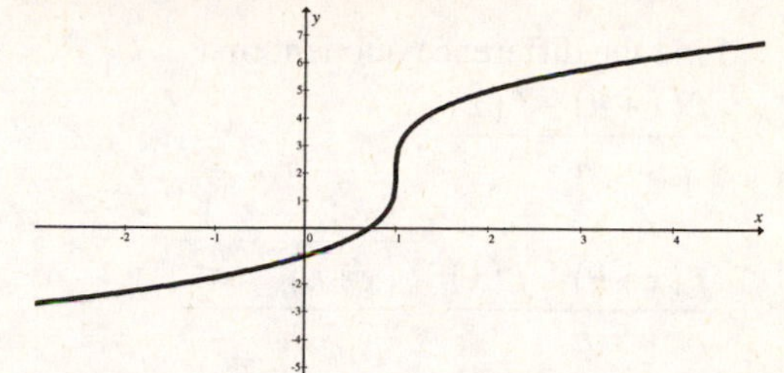

35. The postage function does not have any "corners" nor does it have any vertical tangents. However it is discontinuous at all natural numbers. Therefore the postage function is not differentiable for 1, 2, 3, 4, and so on.

37. The ticket price function is continuous everywhere on its domain. It also does not have any "corners" nor does it have any vertical tangents. Thus, the function is differentiable everywhere on it domain.

39. The graph is not differentiable at each of the points $x = 4, 5, 6, 7, 8, 9, 10, 11$ because there is a "corner" at each of the values.

41. ✎

43. $f(x) = \dfrac{1}{1-x}$

We found the simplified difference quotient in Exercise 53 of Exercise Set 1.3. We find the limit of the difference quotient as $h \to 0$.

$$\lim_{h\to 0} \frac{f(x+h)-f(x)}{h} = \lim_{h\to 0} \frac{1}{(1-x)(1-x-h)}$$

$$= \frac{1}{(1-x)(1-x-0)}$$

$$= \frac{1}{(1-x)^2}$$

Thus, $f'(x) = \dfrac{1}{(1-x)^2}$.

Copyright © 2014 Pearson Education, Inc. Publishing as Addison-Wesley.

45. $f(x)=\frac{1}{x^2}$

Find the difference quotient first.

$$\frac{f(x+h)-f(x)}{h}$$

$$\frac{f(x+h)-f(x)}{h}=\frac{\frac{1}{(x+h)^2}-\frac{1}{x^2}}{h}$$

$$=\frac{\frac{1}{(x+h)^2}\cdot\frac{x^2}{x^2}-\frac{1}{x^2}\cdot\frac{(x+h)^2}{(x+h)^2}}{h}$$

$$=\frac{\frac{x^2-(x+h)^2}{x^2(x+h)^2}}{h}$$

$$=\frac{x^2-x^2-2xh-h^2}{x^2(x+h)^2}\cdot\frac{1}{h}$$

$$=\frac{-2xh-h^2}{x^2(x+h)^2}\cdot\frac{1}{h}$$

$$=\frac{-2x-h}{x^2(x+h)^2}$$

Next, we will find the limit of the difference quotient as $h\to 0$.

$$f'(x)=\lim_{h\to 0}\frac{f(x+h)-f(x)}{h}$$

$$=\lim_{h\to 0}\frac{-2x-h}{x^2(x+h)^2}$$

$$=\frac{-2x}{x^2(x+0)^2}$$

$$=\frac{-2x}{x^4}$$

$$=\frac{-2}{x^3}$$

Thus, $f'(x)=\frac{-2}{x^3}$.

47. $f(x)=\sqrt{2x+1}$

We found the simplified difference quotient in Exercise 55 of Exercise Set 1.3. We now find the limit of the difference quotient as $h\to 0$.

$$\lim_{h\to 0}\frac{f(x+h)-f(x)}{h}$$

$$=\lim_{h\to 0}\frac{2}{\sqrt{2(x+h)+1}+\sqrt{2x+1}}$$

$$=\frac{2}{\sqrt{2(x+0)+1}+\sqrt{2x+1}}$$

$$=\frac{2}{\sqrt{2x+1}+\sqrt{2x+1}}$$

$$=\frac{2}{2\sqrt{2x+1}}$$

$$=\frac{1}{\sqrt{2x+1}}$$

Thus, $f'(x)=\frac{1}{\sqrt{2x+1}}$.

49. a) The domain of the rational function is restricted to those input values that do not result in division by 0. The domain for $f(x)=\frac{x^2-9}{x+3}$ consists of all real numbers except -3. Since $f(-3)$ does not exist, the function is not continuous at -3. Thus, the function is not differentiable at $x=-3$.

b) ✎

51. a) Looking at the graph of the function, we see there is a "corner" when $x=3$. Therefore, $h(x)=|x-3|+2$ is not differentiable at $x=3$.

b) Using the piecewise definition of

$$h(x)=|x-3|+2$$

$$=\begin{cases}-(x-3)+2, & \text{for } x<3\\ (x-3)+2, & \text{for } x\geq 3\end{cases}$$

We notice that:

$$h'(x)=\begin{cases}-1, & \text{for } x<3\\ 1, & \text{for } x>3.\end{cases}$$

Therefore,

$h'(0)=-1;\ h'(1)=-1;$

$h'(4)=1;\ \ h'(10)=1.$

The "shortcut" is noticing that this function is a linear function with slope $m=-1$ for $x<3$ and slope $m=1$ for $x>3$.

Copyright © 2014 Pearson Education, Inc. Publishing as Addison-Wesley.

53. The error was made when the student did not determine the implied domain of the function.

$$f(x)=\frac{x^2+4x+3}{x+1}$$

is undefined at $x=-1$. Therefore, $f(x)$ is not differentiable at $x=-1$. Once the domain is properly defined, the student can find the derivative of the function.

55. a) The function $F(x)$ is continuous at $x=2$, because

1) $F(2)$ exists, $F(2)=5$

2) $\lim_{x\to 2^-} F(x)=5$ and $\lim_{x\to 2^+} F(x)=5$,
Therefore,
$\lim_{x\to 2} F(x)=5$

3) $\lim_{x\to 2} F(x)=5=F(2)$.

b) The function $F(x)$ is not differentiable at $x=2$ because there is a "corner" at $x=2$.

57. In order for $H(x)$ to be differentiable at $x=3$. $H(x)$ must be continuous at $x=3$. It must also be "smooth" at $x=3$ which means the slope as x approaches 3 from the left must equal the slope as x approaches 3 from the right.
For $x\le 3$
$H'(3)=4(3)-1=11$.
Using this information, we know
$m=11$.
We also know:
$\lim_{x\to 3^+} H(x)=15$ in order for $H(x)$ to be continuous.
Using the above information and substituting $m=11$ we have:

$$\lim_{x\to 3^+} 11(x)+b=15$$
$$11(3)+b=15$$
$$33+b=15$$
$$b=-18.$$

Therefore, the values $m=11$ and $b=-18$ will make $H(x)$ differentiable at $x=3$.

59-63. Left to the student.

65. There is a vertical tangent at $x=5$, therefore, $f'(x)$ does not exist at $x=5$.

Copyright © 2014 Pearson Education, Inc. Publishing as Addison-Wesley.

Exercise Set 1.5

1. $y = x^7$

$$\frac{dy}{dx} = \frac{d}{dx}x^7$$
$$= 7x^{7-1} \quad \text{Theorem 1}$$
$$= 7x^6$$

3. $y = -3x$

$$\frac{dy}{dx} = \frac{d}{dx}(-3x)$$
$$= -3\frac{d}{dx}x \quad \text{Theorem 3}$$
$$= -3\left(1x^{1-1}\right) \quad \text{Theorem 1}$$
$$= -3\left(x^0\right)$$
$$= -3 \quad \left[a^0 = 1\right]$$

5. $y = 12$ Constant function

$$\frac{dy}{dx} = \frac{d}{dx}12$$
$$= 0 \quad \text{Theorem 2}$$

7. $y = 2x^{15}$

$$\frac{dy}{dx} = \frac{d}{dx}\left(2x^{15}\right)$$
$$= 2\frac{d}{dx}\left(x^{15}\right) \quad \text{Theorem 3}$$
$$= 2\left(15x^{15-1}\right) \quad \text{Theorem 1}$$
$$= 30x^{14}$$

9. $y = x^{-6}$

$$\frac{dy}{dx} = \frac{d}{dx}x^{-6}$$
$$= -6x^{-6-1} \quad \text{Theorem 1}$$
$$= -6x^{-7}$$

11. $y = 4x^{-2}$

$$\frac{dy}{dx} = \frac{d}{dx}\left(4x^{-2}\right)$$
$$= 4\frac{d}{dx}\left(x^{-2}\right) \quad \text{Theorem 3}$$
$$= 4\left(-2x^{-2-1}\right) \quad \text{Theorem 1}$$
$$= -8x^{-3}$$

13. $y = x^3 + 3x^2$

$$\frac{dy}{dx} = \frac{d}{dx}\left(x^3 + 3x^2\right)$$
$$= \frac{d}{dx}x^3 + \frac{d}{dx}3x^2 \quad \text{Theorem 4}$$
$$= \frac{d}{dx}x^3 + 3\frac{d}{dx}x^2 \quad \text{Theorem 3}$$
$$= 3x^{3-1} + 3\left(2x^{2-1}\right) \quad \text{Theorem 1}$$
$$= 3x^2 + 6x$$

15. $y = 8\sqrt{x} = 8x^{1/2}$

$$\frac{dy}{dx} = \frac{d}{dx}8x^{1/2}$$
$$= 8\frac{d}{dx}x^{1/2} \quad \text{Theorem 3}$$
$$\frac{dy}{dx} = 8\left(\frac{1}{2}x^{1/2-1}\right) \quad \text{Theorem 1}$$
$$= 4x^{-1/2}$$
$$= \frac{4}{x^{1/2}} = \frac{4}{\sqrt{x}} \quad \text{Properties of exponents}$$

17. $y = x^{0.9}$

$$\frac{dy}{dx} = \frac{d}{dx}x^{0.9}$$
$$= 0.9x^{0.9-1} \quad \text{Theorem 1}$$
$$= 0.9x^{-0.1}$$

19. $y = \frac{1}{2}x^{4/5}$

$$\frac{dy}{dx} = \frac{d}{dx}\left(\frac{1}{2}x^{4/5}\right)$$
$$= \frac{1}{2}\cdot\frac{d}{dx}\left(x^{4/5}\right) \quad \text{Theorem 3}$$
$$= \frac{1}{2}\left(\frac{4}{5}x^{4/5-1}\right) \quad \text{Theorem 1}$$
$$= \frac{2}{5}x^{-1/5}$$

Copyright © 2014 Pearson Education, Inc. Publishing as Addison-Wesley.

21. $y = \frac{7}{x^3} = 7x^{-3}$

$\frac{dy}{dx} = \frac{d}{dx}\left(7x^{-3}\right)$

$= 7\frac{d}{dx}\left(x^{-3}\right)$ Theorem 3

$= 7\left(-3x^{-3-1}\right)$ Theorem 1

$= -21x^{-4}$

$= -\frac{21}{x^4}$ Properties of exponents

23. $y = \frac{4x}{5} = \frac{4}{5}x$

$\frac{dy}{dx} = \frac{d}{dx}\left(\frac{4}{5}x\right)$

$= \frac{4}{5}\cdot\frac{d}{dx}(x)$ Theorem 3

$= \frac{4}{5}\cdot\left(1x^{1-1}\right)$ Theorem 1

$= \frac{4}{5}$

25. $\frac{d}{dx}\left(\sqrt[4]{x} - \frac{3}{x}\right)$

$= \frac{d}{dx}\sqrt[4]{x} - \frac{d}{dx}\frac{3}{x}$ Theorem 4

$= \frac{d}{dx}x^{1/4} - \frac{d}{dx}3x^{-1}$ Properties of exponents

$= \frac{d}{dx}x^{1/4} - 3\frac{d}{dx}x^{-1}$ Theorem 3

$= \frac{1}{4}x^{1/4-1} - 3\left(-1x^{-1-1}\right)$ Theorem 1

$= \frac{1}{4}x^{-3/4} + 3x^{-2}$

$= \frac{1}{4x^{3/4}} + \frac{3}{x^2}$

$= \frac{1}{4\sqrt[4]{x^3}} + \frac{3}{x^2}$

27. $\frac{d}{dx}\left(\sqrt{x} - \frac{2}{\sqrt{x}}\right)$

$= \frac{d}{dx}\sqrt{x} - \frac{d}{dx}\frac{2}{\sqrt{x}}$ Theorem 4

$= \frac{d}{dx}x^{1/2} - \frac{d}{dx}2x^{-1/2}$ Properties of exponents

$= \frac{d}{dx}x^{1/2} - 2\frac{d}{dx}x^{-1/2}$ Theorem 3

$= \frac{1}{2}x^{1/2-1} - 2\left(-\frac{1}{2}x^{-1/2-1}\right)$ Theorem 1

$= \frac{1}{2}x^{-1/2} + x^{-3/2}$

$= \frac{1}{2x^{1/2}} + \frac{1}{x^{3/2}}$

$= \frac{1}{2\sqrt{x}} + \frac{1}{\sqrt{x^3}}$

$= \frac{1}{2\sqrt{x}} + \frac{1}{x\sqrt{x}}$

29. $\frac{d}{dx}\left(-2\sqrt[3]{x^5}\right)$

$= -2\frac{d}{dx}\left(\sqrt[3]{x^5}\right)$ Theorem 3

$= -2\frac{d}{dx}\left(x^{5/3}\right)$

$= -2\left(\frac{5}{3}x^{5/3-1}\right)$ Theorem 1

$= -\frac{10}{3}x^{2/3} = -\frac{10\sqrt[3]{x^2}}{3}$

31. $\frac{d}{dx}\left(5x^2 - 7x + 3\right)$

$= \frac{d}{dx}5x^2 - \frac{d}{dx}7x + \frac{d}{dx}3$ Theorem 4

$= 5\frac{d}{dx}x^2 - 7\frac{d}{dx}x + \frac{d}{dx}3$ Theorem 3

$= 5\left(2x^{2-1}\right) - 7\left(1x^{1-1}\right) + 0$ Theorems 1 and 2

$= 10x - 7$

Copyright © 2014 Pearson Education, Inc. Publishing as Addison-Wesley.

33. $f(x)=0.6x^{1.5}$

$$\begin{aligned} f'(x)&=\frac{d}{dx}0.6x^{1.5}\\ &=0.6\frac{d}{dx}x^{1.5} && \text{Theorem 3}\\ &=0.6\left(1.5x^{1.5-1}\right) && \text{Theorem 1}\\ &=0.9x^{0.5}\end{aligned}$$

35. $f(x)=\frac{2x}{3}=\frac{2}{3}x$

$$\begin{aligned} f'(x)&=\frac{d}{dx}\left(\frac{2}{3}x\right)\\ &=\frac{2}{3}\frac{d}{dx}(x)\\ &=\frac{2}{3}\left(1x^{1-1}\right)\\ &=\frac{2}{3}\end{aligned}$$

37. $f(x)=\frac{4}{7x^3}=\frac{4x^{-3}}{7}=\frac{4}{7}x^{-3}$

$$\begin{aligned} f'(x)&=\frac{d}{dx}\left(\frac{4}{7}x^{-3}\right)\\ &=\frac{4}{7}\frac{d}{dx}\left(x^{-3}\right)\\ &=\frac{4}{7}\left(-3x^{-3-1}\right)\\ &=\frac{-12}{7}x^{-4}\\ &=-\frac{12}{7x^4}\end{aligned}$$

39. $f(x)=\frac{5}{x}-x^{2/3}=5x^{-1}-x^{2/3}$

$$\begin{aligned} f'(x)&=\frac{d}{dx}\left(5x^{-1}-x^{2/3}\right)\\ &=\frac{d}{dx}\left(5x^{-1}\right)-\frac{d}{dx}\left(x^{2/3}\right)\\ &=5\frac{d}{dx}\left(x^{-1}\right)-\frac{d}{dx}\left(x^{2/3}\right)\\ &=5\left(-1x^{-1-1}\right)-\frac{2}{3}x^{2/3-1}\\ &=-5x^{-2}-\frac{2}{3}x^{-1/3}\\ &=-\frac{5}{x^2}-\frac{2}{3}x^{-1/3}\end{aligned}$$

41. $f(x)=4x-7$

$$\begin{aligned} f'(x)&=\frac{d}{dx}(4x-7)\\ &=\frac{d}{dx}(4x)-\frac{d}{dx}(7)\\ f'(x)&=4\frac{d}{dx}(x)-\frac{d}{dx}(7)\\ &=4\left(1x^{1-1}\right)-0\\ &=4\end{aligned}$$

43. $f(x)=\frac{x^{4/3}}{4}=\frac{1}{4}x^{4/3}$

$$\begin{aligned} f'(x)&=\frac{d}{dx}\left(\frac{1}{4}x^{4/3}\right)\\ &=\frac{1}{4}\frac{d}{dx}\left(x^{4/3}\right)\\ &=\frac{1}{4}\left(\frac{4}{3}x^{4/3-1}\right)\\ &=\frac{1}{3}x^{1/3}, \text{ or } \frac{\sqrt[3]{x}}{3}\end{aligned}$$

45. $f(x)=-0.01x^2-0.5x+70$

$$\begin{aligned} f'(x)&=\frac{d}{dx}\left(-0.01x^2-0.5x+70\right)\\ &=\frac{d}{dx}\left(-0.01x^2\right)-\frac{d}{dx}(0.5x)+\frac{d}{dx}(70)\\ &=-0.01\frac{d}{dx}\left(x^2\right)-0.5\frac{d}{dx}(x)+\frac{d}{dx}(70)\\ &=-0.01\left(2x^{2-1}\right)-0.5\left(1x^{1-1}\right)+0\\ &=-0.02x-0.5\end{aligned}$$

Copyright © 2014 Pearson Education, Inc. Publishing as Addison-Wesley.

47. $y = 3x^{-2/3} + x^{3/4} + x^{6/5} + \dfrac{8}{x^3}$

$y = 3x^{-2/3} + x^{3/4} + x^{6/5} + 8x^{-3}$

$$y' = \frac{d}{dx}\left(3x^{-2/3} + x^{3/4} + x^{6/5} + 8x^{-3}\right)$$

$$= \frac{d}{dx}\left(3x^{-2/3}\right) + \frac{d}{dx}\left(x^{3/4}\right) + \frac{d}{dx}\left(x^{6/5}\right) + \frac{d}{dx}\left(8x^{-3}\right)$$

$$= 3\left(\frac{-2}{3}x^{-2/3-1}\right) + \left(\frac{3}{4}x^{3/4-1}\right) + \left(\frac{6}{5}x^{6/5-1}\right) + 8\left(-3x^{-3-1}\right)$$

$$= -2x^{-5/3} + \frac{3}{4}x^{-1/4} + \frac{6}{5}x^{1/5} - 24x^{-4}$$

$$= -2x^{-5/3} + \frac{3}{4}x^{-1/4} + \frac{6}{5}x^{1/5} - \frac{24}{x^4}$$

49. $y = \dfrac{2}{x} - \dfrac{x}{2} = 2x^{-1} - \dfrac{1}{2}x$

$$y' = \frac{d}{dx}\left(2x^{-1} - \frac{1}{2}x\right)$$

$$= \frac{d}{dx}\left(2x^{-1}\right) - \frac{d}{dx}\left(\frac{1}{2}x\right)$$

$$= 2\frac{d}{dx}\left(x^{-1}\right) - \frac{1}{2}\frac{d}{dx}(x)$$

$$= 2\left(-1x^{-1-1}\right) - \frac{1}{2}\left(1x^{1-1}\right)$$

$$= -2x^{-2} - \frac{1}{2}$$

$$= -\frac{2}{x^2} - \frac{1}{2}$$

51. $f(x) = x^2 + 4x - 5$

First, we find $f'(x)$

$$f'(x) = \frac{d}{dx}\left(x^2 + 4x - 5\right)$$

$$= \frac{d}{dx}\left(x^2\right) + 4\frac{d}{dx}(x) - \frac{d}{dx}5$$

$$= \left(2x^{2-1}\right) + 4\left(1x^{1-1}\right) - 0$$

$$= 2x + 4$$

Therefore,

$$f'(10) = 2(10) + 4$$

$$= 24.$$

53. $y = \dfrac{4}{x^2} = 4x^{-2}$

Find $\dfrac{dy}{dx}$ first.

$$\frac{dy}{dx} = \frac{d}{dx}\left(4x^{-2}\right)$$

$$= 4\frac{d}{dx}\left(x^{-2}\right)$$

$$= 4\left(-2x^{-2-1}\right)$$

$$= -8x^{-3}$$

$$= -\frac{8}{x^3}$$

Therefore,

$$\left.\frac{dy}{dx}\right|_{x=-2} = -\frac{8}{(-2)^3}$$

$$= -\frac{8}{(-8)}$$

$$= 1.$$

55. $y = x^3 + 2x - 5$

Find $\dfrac{dy}{dx}$ first.

$$\frac{dy}{dx} = \frac{d}{dx}\left(x^3 + 2x - 5\right)$$

$$= \frac{d}{dx}\left(x^3\right) + 2\frac{d}{dx}(x) - 5$$

$$= 3x^{3-1} + 2\left(x^{1-1}\right) - 0$$

$$= 3x^2 + 2$$

Therefore,

$$\left.\frac{dy}{dx}\right|_{x=-2} = 3(-2)^2 + 2$$

$$= 3(4) + 2$$

$$= 14.$$

Copyright © 2014 Pearson Education, Inc. Publishing as Addison-Wesley.

57. $y=\frac{1}{3x^4}=\frac{1}{3}x^{-4}$

Find $\frac{dy}{dx}$ first.

$$\frac{dy}{dx}=\frac{d}{dx}\left(\frac{1}{3}x^{-4}\right)$$
$$=\frac{1}{3}\frac{d}{dx}\left(x^{-4}\right)$$
$$=\frac{1}{3}\left(-4x^{-4-1}\right)$$
$$=-\frac{4}{3}x^{-5}$$
$$=-\frac{4}{3x^5}$$

Therefore,

$$\left.\frac{dy}{dx}\right|_{x=-1}=-\frac{4}{3(-1)^5}$$
$$=-\frac{4}{(-3)}$$
$$=\frac{4}{3}.$$

59. We will need the derivative to find the slope of the tangent line at each of the indicated points. We find the derivative first.

$$f(x)=x^3-2x+1$$
$$f'(x)=\frac{d}{dx}\left(x^3-2x+1\right)$$
$$=\frac{d}{dx}\left(x^3\right)-\frac{d}{dx}(2x)+\frac{d}{dx}(1)$$
$$=\left(3x^{3-1}\right)-2\left(1x^{1-1}\right)+0$$
$$=3x^2-2$$

a) Using the derivative, we find the slope of the line tangent to the curve at point $(2,5)$ by evaluating the derivative at $x=2$.

$f'(2)=3(2)^2-2=10$. Therefore the slope of the tangent line is 10. We use the point-slope equation to find the equation of the tangent line.

$$y-y_1=m(x-x_1)$$
$$y-5=10(x-2)$$
$$y-5=10x-20$$
$$y=10x-15$$

b) Using the derivative, we find the slope of the line tangent to the curve at point $(-1,2)$ by evaluating the derivative at $x=-1$.

$f'(-1)=3(-1)^2-2=1$. Therefore the slope of the tangent line is 1. We use the point-slope equation to find the equation of the tangent line.

$$y-y_1=m(x-x_1)$$
$$y-2=1(x-(-1))$$
$$y-2=x+1$$
$$y=x+3$$

c) Using the derivative, we find the slope of the line tangent to the curve at point $(0,1)$ by evaluating the derivative at $x=0$.

$f'(0)=3(0)^2-2=-2$. Therefore the slope of the tangent line is -2. We use the point-slope equation to find the equation of the tangent line.

$$y-y_1=m(x-x_1)$$
$$y-1=-2(x-0)$$
$$y-1=-2x$$
$$y=-2x+1$$

61. We will need the derivative to find the slope of the tangent line at each of the indicated points. We find the derivative first.

$$f(x)=\frac{1}{x^2}=x^{-2}$$
$$f'(x)=\frac{d}{dx}\left(x^{-2}\right)$$
$$=-2x^{-2-1}$$
$$=-2x^{-3}$$
$$=-\frac{2}{x^3}$$

a) Using the derivative, we find the slope of the line tangent to the curve at point $(1,1)$ by evaluating the derivative at $x=1$.

$f'(1)=-\frac{2}{(1)^3}=-2$. Therefore the slope of the tangent line is -2.

The solution is continued on the next page.

Copyright © 2014 Pearson Education, Inc. Publishing as Addison-Wesley.

We use the information from the previous page and the point-slope equation to find the equation of the tangent line.

$$y - y_1 = m(x - x_1)$$
$$y - 1 = -2(x - 1)$$
$$y - 1 = -2x + 2$$
$$y = -2x + 3$$

b) Using the derivative, we find the slope of the line tangent to the curve at point $\left(3, \frac{1}{9}\right)$ by evaluating the derivative at $x = 3$.

$f'(3) = -\dfrac{2}{(3)^3} = -\dfrac{2}{27}$. Therefore the slope of the tangent line is $-\frac{2}{27}$. We use the point-slope equation to find the equation of the tangent line.

$$y - y_1 = m(x - x_1)$$
$$y - \frac{1}{9} = -\frac{2}{27}(x - 3)$$
$$y - \frac{1}{9} = -\frac{2}{27}x + \frac{2}{9}$$
$$y = -\frac{2}{27}x + \frac{1}{3}$$

c) Using the derivative, we find the slope of the line tangent to the curve at point $\left(-2, \frac{1}{4}\right)$ by evaluating the derivative at $x = -2$.

$f'(-2) = -\dfrac{2}{(-2)^3} = \dfrac{1}{4}$. Therefore the slope of the tangent line is $\frac{1}{4}$.

We use the point-slope equation to find the equation of the tangent line.

$$y - y_1 = m(x - x_1)$$
$$y - \frac{1}{4} = \frac{1}{4}(x - (-2))$$
$$y - \frac{1}{4} = \frac{1}{4}x + \frac{1}{2}$$
$$y = \frac{1}{4}x + \frac{3}{4}$$

63. $y = x^2 - 3$

A horizontal tangent line has slope equal to 0, so we first find the values of x that make $\dfrac{dy}{dx} = 0$.

First, we find the derivative.

$$\frac{dy}{dx} = \frac{d}{dx}\left(x^2 - 3\right)$$
$$\frac{dy}{dx} = \frac{d}{dx}x^2 - \frac{d}{dx}3$$
$$= 2x - 0$$
$$= 2x$$

Next, we set the derivative equal to zero and solve for x.

$$\frac{dy}{dx} = 0$$
$$2x = 0$$
$$x = \frac{0}{2} = 0$$

So the horizontal tangent will occur when $x = 0$. Next we find the point on the graph. For , so there is a horizontal tangent at the point

$x = 0,\ y = (0)^2 - 3 = -3\ (0, -3)$.

65. $y = -x^3 + 1$

A horizontal tangent line has slope equal to 0, so we first find the values of x that make $\dfrac{dy}{dx} = 0$.

First, we find the derivative.

$$\frac{dy}{dx} = \frac{d}{dx}\left(-x^3 + 1\right)$$
$$= -\frac{d}{dx}x^3 + \frac{d}{dx}1$$
$$= -3x^2 - 0$$
$$= -3x^2$$

Next, we set the derivative equal to zero and solve for x.

$$\frac{dy}{dx} = 0$$
$$-3x^2 = 0$$
$$x^2 = 0$$
$$x = 0$$

So the horizontal tangent will occur when $x = 0$. Next we find the point on the graph. For $x = 0,\ y = -(0)^3 + 1 = 1$, so there is a horizontal tangent at the point $(0, 1)$.

Copyright © 2014 Pearson Education, Inc. Publishing as Addison-Wesley.

67. $y = 3x^2 - 5x + 4$

A horizontal tangent line has slope equal to 0, so we need to find the values of x that make $\frac{dy}{dx} = 0$.

First, we find the derivative.

$$\frac{dy}{dx} = \frac{d}{dx}\left(3x^2 - 5x + 4\right)$$
$$= \frac{d}{dx}3x^2 - \frac{d}{dx}5x + \frac{d}{dx}4$$
$$= 3\left(2x^{2-1}\right) - 5\left(1x^{1-1}\right) + 0$$
$$= 6x - 5$$

Next, we set the derivative equal to zero and solve for x.

$$\frac{dy}{dx} = 0$$
$$6x - 5 = 0$$
$$6x = 5$$
$$x = \frac{5}{6}$$

So the horizontal tangent will occur when $x = \frac{5}{6}$. Next we find the point on the graph.

For $x = \frac{5}{6}$,

$$y = 3\left(\frac{5}{6}\right)^2 - 5\left(\frac{5}{6}\right) + 4$$
$$= 3\left(\frac{25}{36}\right) - \frac{25}{6} + 4$$
$$= \frac{25}{12} - \frac{25}{6} + 4$$
$$= \frac{25}{12} - \frac{25}{6} \cdot \frac{2}{2} + \frac{4}{1} \cdot \frac{12}{12}$$
$$= \frac{25}{12} - \frac{50}{12} + \frac{48}{12}$$
$$= \frac{25 - 50 + 48}{12}$$
$$= \frac{23}{12}$$

Therefore, there is a horizontal tangent at the point $\left(\frac{5}{6}, \frac{23}{12}\right)$.

69. $y = -0.01x^2 - 0.5x + 70$

A horizontal tangent line has slope equal to 0, so we need to find the values of x that make $\frac{dy}{dx} = 0$.

First, we find the derivative.

$$\frac{dy}{dx} = \frac{d}{dx}\left(-0.01x^2 - 0.5x + 70\right)$$
$$= -0.02x - 0.5 \qquad \text{See Exercise 45.}$$

Next, we set the derivative equal to zero and solve for x.

$$\frac{dy}{dx} = 0$$
$$-0.02x - 0.5 = 0$$
$$-0.02x = 0.5$$
$$x = \frac{0.5}{-0.02}$$
$$x = -25$$

So the horizontal tangent will occur when $x = -25$. Next we find the point on the graph.

For $x = -25$,

$$y = -0.01(-25)^2 - 0.5(-25) + 70$$
$$= -0.01(625) + 12.5 + 70$$
$$= -6.25 + 12.5 + 70$$
$$= 76.25$$

Therefore, there is a horizontal tangent at the point $(-25, 76.25)$.

71. $y = 2x + 4$ Linear function

$\frac{dy}{dx} = 2$ Slope is 2

There are no values of x for which $\frac{dy}{dx} = 0$, so there are no points on the graph at which there is a horizontal tangent.

73. $y = 4$ Constant Function

$\frac{dy}{dx} = 0$ Theorem 2

$\frac{dy}{dx} = 0$ for all values of x, so the tangent line is horizontal for all points on the graph.

Copyright © 2014 Pearson Education, Inc. Publishing as Addison-Wesley.

75. $y=-x^3+x^2+5x-1$

A horizontal tangent line has slope equal to 0, so we need to find the values of x that make $\frac{dy}{dx}=0$.

First, we find the derivative.

$$\frac{dy}{dx}=\frac{d}{dx}\left(-x^3+x^2+5x-1\right)$$
$$=-\frac{d}{dx}\left(x^3\right)+\frac{d}{dx}\left(x^2\right)+\frac{d}{dx}(5x)-\frac{d}{dx}(1)$$
$$=-3x^2+2x+5$$

Next, we set the derivative equal to zero and solve for x.

$$\frac{dy}{dx}=0$$
$$-3x^2+2x+5=0$$
$$3x^2-2x-5=0 \quad \text{Multiply both sides by -1.}$$
$$(3x-5)(x+1)=0 \quad \text{Factor the left hand side.}$$
$$3x-5=0 \quad \text{or} \quad x+1=0$$
$$3x=5 \quad \text{or} \quad x=-1$$
$$x=\frac{5}{3} \quad \text{or} \quad x=-1$$

There are two horizontal tangents. One at $x=\frac{5}{3}$ and one at $x=-1$.

Next we find the points on the graph where the horizontal tangents occur.

For $x=-1$

$$y=-(-1)^3+(-1)^2+5(-1)-1$$
$$y=-(-1)+(1)-5-1$$
$$y=-4$$

For $x=\frac{5}{3}$

$$y=-\left(\frac{5}{3}\right)^3+\left(\frac{5}{3}\right)^2+5\left(\frac{5}{3}\right)-1$$
$$y=-\left(\frac{125}{27}\right)+\left(\frac{25}{9}\right)+\frac{25}{3}-1$$
$$y=-\frac{125}{27}+\frac{75}{27}+\frac{225}{27}-\frac{27}{27}$$
$$y=\frac{148}{27}=5\tfrac{13}{27}$$

Therefore, there are horizontal tangents at the points $\left(\frac{5}{3},5\tfrac{13}{27}\right)$ and $(-1,-4)$.

77. $y=\frac{1}{3}x^3-3x+2$

A horizontal tangent line has slope equal to 0, so we need to find the values of x that make $\frac{dy}{dx}=0$.

First, we find the derivative.

$$\frac{dy}{dx}=\frac{d}{dx}\left(\frac{1}{3}x^3-3x+2\right)$$
$$=\frac{d}{dx}\left(\frac{1}{3}x^3\right)-\frac{d}{dx}(3x)+\frac{d}{dx}(2)$$
$$=x^2-3$$

Next, we set the derivative equal to zero and solve for x.

$$\frac{dy}{dx}=0$$
$$x^2-3=0$$
$$x^2=3$$
$$x=\pm\sqrt{3}$$

There are two horizontal tangents. One at $x=-\sqrt{3}$ and one at $x=\sqrt{3}$. Next we find the points on the graph where the horizontal tangents occur.

For $x=-\sqrt{3}$

$$y=\frac{1}{3}\left(-\sqrt{3}\right)^3-3\left(-\sqrt{3}\right)+2$$
$$=\frac{1}{3}\left(-3\sqrt{3}\right)-3\left(-\sqrt{3}\right)+2$$
$$=-\sqrt{3}+3\sqrt{3}+2$$
$$=2+2\sqrt{3}$$

For $x=\sqrt{3}$

$$y=\frac{1}{3}\left(\sqrt{3}\right)^3-3\left(\sqrt{3}\right)+2$$
$$=\frac{1}{3}\left(3\sqrt{3}\right)-3\left(\sqrt{3}\right)+2$$
$$=+\sqrt{3}-3\sqrt{3}+2$$
$$=2-2\sqrt{3}$$

Therefore, there are horizontal tangents at the points $\left(-\sqrt{3},2+2\sqrt{3}\right)$ and $\left(\sqrt{3},2-2\sqrt{3}\right)$.

Copyright © 2014 Pearson Education, Inc. Publishing as Addison-Wesley.

79. $y=\frac{1}{3}x^3+\frac{1}{2}x^2-2$

A horizontal tangent line has slope equal to 0, so we need to find the values of x that make $\frac{dy}{dx}=0$.

First, we find the derivative.

$$\frac{dy}{dx}=\frac{d}{dx}\left(\frac{1}{3}x^3+\frac{1}{2}x^2-2\right)$$
$$=\frac{d}{dx}\left(\frac{1}{3}x^3\right)+\frac{d}{dx}\left(\frac{1}{2}x^2\right)-\frac{d}{dx}(2)$$
$$=x^2+x$$

Next, we set the derivative equal to zero and solve for x.

$$\frac{dy}{dx}=0$$
$$x^2+x=0$$
$$x(x+1)=0$$
$$x=0 \quad \text{or} \quad x+1=0$$
$$x=0 \quad \text{or} \quad x=-1$$

There are two horizontal tangents. One at $x=0$ and one at $x=-1$.

Next we find the points on the graph where the horizontal tangents occur.

For $x=0$

$$y=\frac{1}{3}(0)^3+\frac{1}{2}(0)^2-2$$
$$y=-2$$

For $x=-1$

$$y=\frac{1}{3}(-1)^3+\frac{1}{2}(-1)^2-2$$
$$=-\frac{1}{3}+\frac{1}{2}-2$$
$$=-\frac{2}{6}+\frac{3}{6}-\frac{12}{6}$$
$$=-\frac{11}{6}$$

Therefore, there are horizontal tangents at the points $(0,-2)$ and $\left(-1,-\frac{11}{6}\right)$.

81. $y=20x-x^2$

To find the tangent line that has slope equal to 1, so we need to find the values of x that make $\frac{dy}{dx}=1$.

First, we find the derivative.

$$\frac{dy}{dx}=\frac{d}{dx}\left(20x-x^2\right)$$
$$=\frac{d}{dx}20x-\frac{d}{dx}x^2$$
$$=20-2x$$

Next, we set the derivative equal to 1 and solve for x.

$$\frac{dy}{dx}=1$$
$$20-2x=1$$
$$-2x=1-20$$
$$-2x=-19$$
$$x=\frac{19}{2}$$

So the tangent will occur when $x=\frac{19}{2}$.

Next we find the point on the graph.

For $x=\frac{19}{2}$,

$$y=20\left(\frac{19}{2}\right)-\left(\frac{19}{2}\right)^2$$
$$y=190-\left(\frac{361}{4}\right)$$
$$=\frac{760}{4}-\frac{361}{4}$$
$$=\frac{399}{4}$$

The tangent line has slope 1 at the point $\left(\frac{19}{2},\frac{399}{4}\right)$.

83. $y=-0.025x^2+4x$

To find the tangent line that has slope equal to 1, we need to find the values of x that make $\frac{dy}{dx}=1$.

First, we find the derivative.

$$\frac{dy}{dx}=\frac{d}{dx}\left(-0.025x^2+4x\right)$$
$$=\frac{d}{dx}-0.025x^2+\frac{d}{dx}4x$$
$$=-0.025(2x)+4$$
$$=-0.05x+4$$

The solution is continued on the next page.

Copyright © 2014 Pearson Education, Inc. Publishing as Addison-Wesley.

Next, we set the derivative from the previous page equal to 1 and solve for x.

$$\frac{dy}{dx}=1$$
$$-0.05x+4=1$$
$$-0.05x=-3$$
$$x=\frac{-3}{-0.05}$$
$$x=60$$

So the tangent will occur when $x=60$. Next we find the point on the graph.
For $x=60$,

$$y=-0.025(60)^2+4(60)$$
$$=-0.025(3600)+240$$
$$=-90+240$$
$$=150$$

The tangent line has slope 1 at the point $(60,150)$.

85. $y=\frac{1}{3}x^3+2x^2+2x$

To find the tangent line that has slope equal to 1, we need to find the values of x that make $\frac{dy}{dx}=1$.

First, we find the derivative.

$$\frac{dy}{dx}=\frac{d}{dx}\left(\frac{1}{3}x^3+2x^2+2x\right)$$
$$=\frac{d}{dx}\left(\frac{1}{3}x^3\right)+\frac{d}{dx}2x^2+\frac{d}{dx}2x$$
$$=x^2+4x+2$$

Next, we set the derivative equal to 1 and solve for x.

$$\frac{dy}{dx}=1$$
$$x^2+4x+2=1$$
$$x^2+4x+1=0$$

This is a quadratic equation, not readily factorable, so we use the quadratic formula where $a=1, b=4,$ and, $c=1$.

Applying the quadratic formula, we have:

$$x=\frac{-b\pm\sqrt{b^2-4ac}}{2a}$$
$$x=\frac{-(4)\pm\sqrt{(4)^2-4(1)(1)}}{2(1)} \quad \text{Substituting}$$
$$=\frac{-4\pm\sqrt{12}}{2}$$
$$=\frac{-4\pm2\sqrt{3}}{2} \quad \left[\sqrt{12}=\sqrt{4\cdot3}=2\sqrt{3}\right]$$
$$=\frac{2\left(-2\pm\sqrt{3}\right)}{2}$$
$$=-2\pm\sqrt{3}.$$

We know there are two tangent lines that have slope equal to 1. The first one occurs at $x=-2+\sqrt{3}$ and the second one occurs at $x=-2-\sqrt{3}$. We use the original equation to find the point on the graph.

For $x=-2+\sqrt{3}$,

$$y=\frac{1}{3}\left(-2+\sqrt{3}\right)^3+2\left(-2+\sqrt{3}\right)^2+2\left(-2+\sqrt{3}\right)$$
$$=\frac{1}{3}\left(-26+15\sqrt{3}\right)+2\left(7-4\sqrt{3}\right)-4+2\sqrt{3}$$
$$=-\frac{26}{3}+5\sqrt{3}+14-8\sqrt{3}-4+2\sqrt{3}$$
$$=\frac{4}{3}-\sqrt{3}$$

For $x=-2-\sqrt{3}$,

$$y=\frac{1}{3}\left(-2-\sqrt{3}\right)^3+2\left(-2-\sqrt{3}\right)^2+2\left(-2-\sqrt{3}\right)$$
$$=\frac{1}{3}\left(-26-15\sqrt{3}\right)+2\left(7+4\sqrt{3}\right)-4-2\sqrt{3}$$
$$=-\frac{26}{3}-5\sqrt{3}+14+8\sqrt{3}-4-2\sqrt{3}$$
$$=\frac{4}{3}+\sqrt{3}$$

The tangent lines have slope 1 at the points $\left(-2+\sqrt{3},\frac{4}{3}-\sqrt{3}\right)$ and $\left(-2-\sqrt{3},\frac{4}{3}+\sqrt{3}\right)$.

Copyright © 2014 Pearson Education, Inc. Publishing as Addison-Wesley.

87. a) In order to find the rate of change of the area with respect to the radius, we must find the derivative of the function with respect to *r*.

$$A'(r) = \frac{d}{dr}\left(3.14r^2\right)$$
$$= 3.14\left(2r^{2-1}\right)$$
$$= 6.28r$$

b) ✎

89. $w(t) = 8.15 + 1.82t - 0.0596t^2 + 0.000758t^3$

a) In order to find the rate of change of weight with respect to time, we take the derivative of the function with respect to *t*.

$$w'(t)$$
$$= \frac{d}{dt}\left(8.15 + 1.82t - 0.0596t^2 + 0.000758t^3\right)$$
$$= 0 + 1.82 - 0.0596(2t) + 0.000758\left(3t^2\right)$$
$$= 1.82 - 0.1192t + 0.002274t^2$$

Therefore, the rate of change of weight with respect to time is given by:

$$w'(t) = 1.82 - 0.1192t + 0.002274t^2$$

b) The weight of the baby at age 10 months can be found by evaluating the function when $t = 10$.

$$w(10) = 8.15 + 1.82(10) - 0.0596(10)^2 + 0.000758(10)^3$$
$$\approx 21.148 \quad \text{Using a calculator}$$

Therefore, a 10 month old boy weighs approximately 21.148 pounds.

c) The rate of change of the baby's weight with respect to time at age of 10 months can be found by evaluating the derivative when $t = 10$.

$$w'(10)$$
$$= 1.82 - 0.1192(10) + 0.002274(10)^2$$
$$\approx 0.8554$$

A 10 month old boys weight will be increasing at a rate of 0.86 pounds per month.

91. $R(v) = \frac{6000}{v} = 6000v^{-1}$

a) Using the power rule, we take the derivative of *R* with respect to *v*.

$$R'(v) = 6000\left(-1v^{-1-1}\right)$$
$$= -6000v^{-2}$$
$$= -\frac{6000}{v^2}$$

The rate of change of heart rate with respect to the output per beat is

$$R'(v) = -\frac{6000}{v^2}.$$

b) To find the heart rate at $v = 80$ ml per beat, we evaluate the function $R(v)$ when $v = 80$.

$$R(80) = \frac{6000}{80} = 75.$$

The heart rate is 75 beats per minute when the output per beat is 80 ml per beat.

c) To find the rate of change of the heart beat at $v = 80$ ml per beat, we evaluate the derivative $R'(v)$ at $v = 80$.

$$R'(80) = -\frac{6000}{80^2}$$
$$R'(80) = -\frac{15}{16}$$
$$= -0.9375$$

The heart rate is decreasing at a rate of 0.94 beats per minute when the output per beat is 80 mL per beat.

93. a) Using the power rule, we find the growth rate $\frac{dP}{dt}$.

$$\frac{dP}{dt} = \frac{d}{dt}\left(100{,}000 + 2000t^2\right)$$
$$= 0 + 2000(2t)$$
$$= 4000t$$

b) Evaluate the function *P* when $t = 10$.

$$P(10) = 100{,}000 + 2000(10)^2$$
$$= 100{,}000 + 2000(100)$$
$$= 300{,}000$$

The population of the city will be 300,000 people after 10 years.

Copyright © 2014 Pearson Education, Inc. Publishing as Addison-Wesley.

c) Evaluate the derivative $P'(t)$ when $t = 10$.

$$\left.\frac{dP}{dt}\right|_{t=10} = P'(10) = 4000(10) = 40,000$$

The population's growth rate after 10 years is 40,000 people per year.

d) ✎

95. $V = 1.22\sqrt{h} = 1.22h^{1/2}$

a) Use the power rule to find the derivative.

$$\frac{dV}{dh} = \frac{d}{dh}\left(1.22h^{1/2}\right)$$
$$= 1.22\left(\frac{1}{2}h^{1/2-1}\right)$$
$$= 0.61h^{-1/2}$$
$$= \frac{0.61}{h^{1/2}} = \frac{0.61}{\sqrt{h}}$$

b) Evaluate the function V when $h = 40,000$.

$$V = 1.22\sqrt{40,000}$$
$$= 244$$

A person would be able to see 244 miles to the horizon from a height of 40,000 feet.

c) Evaluate the derivative $\frac{dV}{dh}$ when $h = 40,000$.

$$\left.\frac{dV}{dh}\right|_{h=40,000} = \frac{0.61}{\sqrt{40,000}}$$
$$= \frac{0.61}{200}$$
$$= 0.00305$$

The rate of change at $h = 40,000$ is 0.0031 miles per foot.

d) ✎

97. $f(x) = x^2 - 4x + 1$

The derivative is positive when $f'(x) > 0$.

Find $f'(x)$.

$$f'(x) = \frac{d}{dx}\left(x^2 - 4x + 1\right)$$
$$= 2x - 4$$

Next, we solve the inequality

$$f'(x) > 0$$
$$2x - 4 > 0$$
$$2x > 4$$
$$x > 2$$

Therefore, the interval for which $f'(x)$ is positive is $(2, \infty)$.

99. $f(x) = \frac{1}{3}x^3 - x^2 - 3x + 5$

The derivative is positive when $f'(x) > 0$.

Find $f'(x)$.

$$f'(x) = \frac{d}{dx}\left(\frac{1}{3}x^3 - x^2 - 3x + 5\right)$$
$$= x^2 - 2x - 3$$

Next, we solve the inequality.

$$f'(x) > 0$$
$$x^2 - 2x - 3 > 0$$

First we find where the quadratic is equal to zero, in order to determine the intervals that we will need to test.

$$x^2 - 2x - 3 = 0$$
$$(x-3)(x+1) = 0$$
$$x + 1 = 0 \quad \text{or} \quad x - 3 = 0$$
$$x = -1 \quad \text{or} \quad x = 3$$

Now we will test a value to the left of -1, between -1 and 3 and to the right of 3 to determine where the quadratic is positive or negative. We choose the values $x = -2$, $x = 0$, and, $x = 4$ to test.

When $x = -2$, the derivative $f'(x)$ is

$$f'(-2) = (-2)^2 - 2(-2) - 3 = 5.$$

When $x = 0$, the derivative $f'(x)$ is

$$f'(0) = (0)^2 - 2(0) - 3 = -3.$$

When $x = 4$, the derivative $f'(x)$ is

$$f'(4) = (4)^2 - 2(4) - 3 = 5.$$

We organize the results in the table below.

Test point x	Test $x = -2$	-1	Test $x = 0$	3	Test $x = 4$
$f'(x)$	$f'(-2) = 5$	0	$f'(0) = -3$	0	$f'(4) = 5$

From the table, we can see that $f'(x)$ is positive on the interval $(-\infty, -1)$ and the interval $(3, \infty)$.

Copyright © 2014 Pearson Education, Inc. Publishing as Addison-Wesley.

101. $y = 2x^6 - x^4 - 2$

A horizontal tangent line has slope equal to 0, so we need to find the values of x that make $\frac{dy}{dx} = 0$.

First, we find the derivative.

$$\frac{dy}{dx} = \frac{d}{dx}\left(2x^6 - x^4 - 2\right)$$
$$= \frac{d}{dx}2x^6 - \frac{d}{dx}x^4 - \frac{d}{dx}2$$
$$= 2\left(6x^{6-1}\right) - \left(4x^{4-1}\right) + 0$$
$$= 12x^5 - 4x^3$$

Next, we set the derivative equal to zero and solve for x.

$$\frac{dy}{dx} = 0$$
$$12x^5 - 4x^3 = 0$$
$$4x^3\left(3x^2 - 1\right) = 0$$
$$4x^3 = 0 \quad \text{or} \quad 3x^2 - 1 = 0$$
$$x = 0 \quad \text{or} \quad 3x^2 = 1$$
$$x = 0 \quad \text{or} \quad x^2 = \frac{1}{3}$$
$$x = 0 \quad \text{or} \quad x = \pm\sqrt{\frac{1}{3}} = \pm\frac{1}{\sqrt{3}}$$

So the horizontal tangent will occur when $x = 0$, $x = \frac{1}{\sqrt{3}}$, and $x = -\frac{1}{\sqrt{3}}$. Next we find the points on the graph.

For $x = 0$,

$$y = 2(0)^6 - (0) - 2$$
$$= -2$$

For $x = \frac{1}{\sqrt{3}}$,

$$y = 2\left(\frac{1}{\sqrt{3}}\right)^6 - \left(\frac{1}{\sqrt{3}}\right)^4 - 2$$
$$= 2\left(\frac{1}{27}\right) - \frac{1}{9} - 2$$
$$= \frac{2}{27} - \frac{3}{27} - \frac{54}{27}$$
$$= -\frac{55}{27}$$

For $x = -\frac{1}{\sqrt{3}}$,

$$y = 2\left(-\frac{1}{\sqrt{3}}\right)^6 - \left(-\frac{1}{\sqrt{3}}\right)^4 - 2$$
$$= 2\left(\frac{1}{27}\right) - \frac{1}{9} - 2$$
$$= \frac{2}{27} - \frac{3}{27} - \frac{54}{27}$$
$$= -\frac{55}{27}$$

Therefore, there are horizontal tangents at the points $(0,-2)$, $\left(\frac{1}{\sqrt{3}}, -\frac{55}{27}\right)$, and $\left(-\frac{1}{\sqrt{3}}, -\frac{55}{27}\right)$.

103. $f(x) = x^3 + 2x$

Taking the derivative we have:

$$f'(x) = \frac{d}{dx}\left(x^3 + 2x\right)$$
$$= \frac{d}{dx}x^3 + 2\frac{d}{dx}x$$
$$= 3x^{3-1} + 2x^{1-1}$$
$$= 3x^2 + 2$$

Notice that $f'(x) \geq 0$ for all values of x.

Therefore, $f(x)$ is always increasing.

105. $f(x) = \sqrt{x}, \quad x \geq 0$

Taking the derivative we have:

$$f'(x) = \frac{d}{dx}\left(\sqrt{x}\right)$$
$$= \frac{d}{dx}\left(x^{\frac{1}{2}}\right)$$
$$= \frac{1}{2}x^{\frac{1}{2}-1}$$
$$= \frac{1}{2}x^{-\frac{1}{2}}$$
$$= \frac{1}{2\sqrt{x}}$$

Notice that $f'(x) \geq 0$ for all values of $x \geq 0$

Therefore, $f(x)$ is always increasing over the interval $x \geq 0$.

107. $y = (x+3)(x-2) = x^2 + x - 6$

$$\frac{dy}{dx} = 2x + 1$$

Copyright © 2014 Pearson Education, Inc. Publishing as Addison-Wesley.

109. $y=\dfrac{x^5-x^3}{x^2}$

First, we separate the fraction.

$$y=\frac{x^5}{x^2}-\frac{x^3}{x^2}$$
$$=x^{5-2}-x^{3-2} \qquad \left[\frac{a^m}{a^n}=a^{m-n}\right]$$
$$=x^3-x^1$$

Therefore,

$$\frac{dy}{dx}=\frac{d}{dx}\left(x^3-x\right)$$
$$=\frac{d}{dx}x^3-\frac{d}{dx}x$$
$$=3x^2-1$$

111. $y=\dfrac{x^5+x}{x^2}$

First, we separate the fraction.

$$y=\frac{x^5}{x^2}+\frac{x}{x^2}$$
$$=x^{5-2}+x^{1-2}$$
$$=x^3+x^{-1}$$

Therefore,

$$\frac{dy}{dx}=\frac{d}{dx}\left(x^3+x^{-1}\right)$$
$$=3x^2+\left(-x^{-1-1}\right)$$
$$=3x^2-x^{-2}$$
$$=3x^2-\frac{1}{x^2}$$

113. $y=(-4x)^3$

$$y=(-4)^3\cdot x^3=-64x^3$$

Therefore,

$$\frac{dy}{dx}=-64\left(3x^{3-1}\right)=-192x^2$$

115. $y=\sqrt[3]{8x}=(8x)^{1/3}=8^{1/3}\cdot x^{1/3}=2x^{1/3}$

$$\frac{dy}{dx}=\frac{2}{3}x^{-\frac{2}{3}}=\frac{2}{3x^{2/3}}=\frac{2}{3\sqrt[3]{x^2}}$$

117. $y=\left(\sqrt{x}-\dfrac{1}{\sqrt{x}}\right)^2$

$$y=\left(x^{1/2}-x^{-1/2}\right)^2$$
$$=\left(x^{1/2}-x^{-1/2}\right)\left(x^{1/2}-x^{-1/2}\right)$$
$$=x-2x^0+x^{-1}$$
$$=x+x^{-1}-2$$
$$\frac{dy}{dx}=1-2x^{-2}$$
$$=1-\frac{2}{x^2}$$

119. $y=(x+1)^3=x^3+3x^2+3x+1$

$$\frac{dy}{dx}=3x^2+6x+3$$

121. ✎

123. $f(x)=x^4-3x^2+1$

First we enter the equation into the graphing editor on the calculator.

```
Plot1 Plot2 Plot3
\Y1 = X^4-3X^2+1
\Y2=
\Y3=
\Y4=
\Y5=
\Y6=
\Y7=
```

Using the window:

```
WINDOW
 Xmin=-2
 Xmax=2
 Xscl=.5
 Ymin=-2
 Ymax=2
 Yscl=.5
 Xres=1
```

We get the graph:

$y=x^4-3x^2+1$

2, −2, 2, −2

We estimate the x-values at which the tangent lines are horizontal are $x=-1.225,\ x=0,$ and, $x=1.225$.

Copyright © 2014 Pearson Education, Inc. Publishing as Addison-Wesley.

125. $f(x)=10.2x^4-6.9x^3$

First we enter the equation into the graphing editor on the calculator.

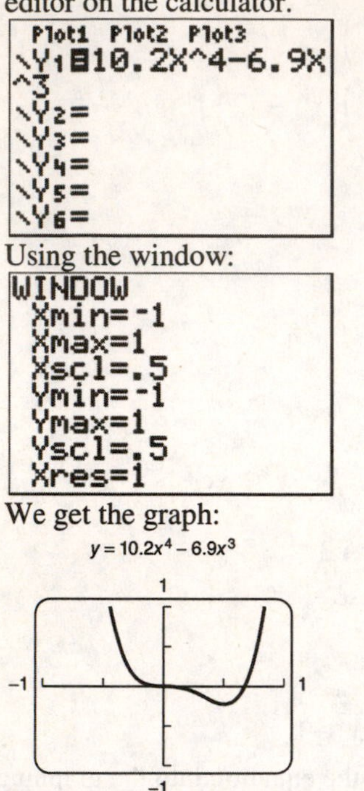

Using the window:

We get the graph:

We estimate the x-values at which the tangent lines are horizontal are $x=0$ and $x=0.507$.

127. $f(x)=20x^3-3x^5$

Using the calculator, we graph the function and the derivative in the same window. We can use the nDeriv feature to graph the derivative without actually calculating the derivative.

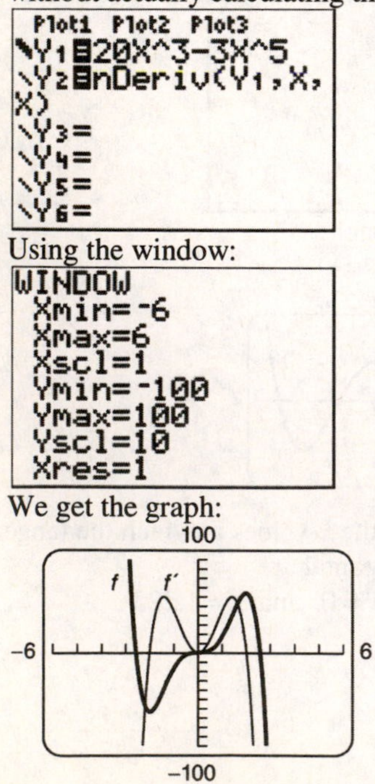

Using the window:

We get the graph:

Note, the function $f(x)$ is the thicker graph.

Using the calculator, we can find the derivative of the function when $x=1$.

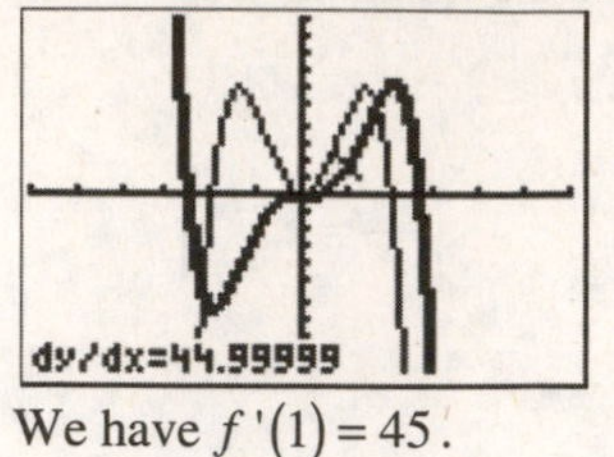

We have $f'(1)=45$.

129. $f(x)=x^3-2x-2$

Using the calculator, we graph the function and the derivative in the same window. We can use the nDeriv feature to graph the derivative without actually calculating the derivative.

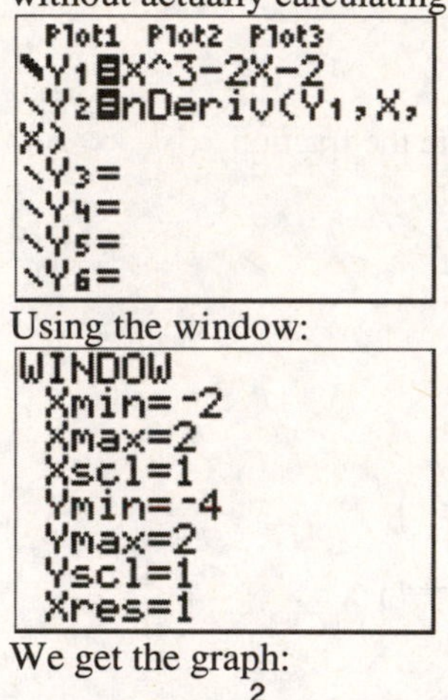

Using the window:

We get the graph:

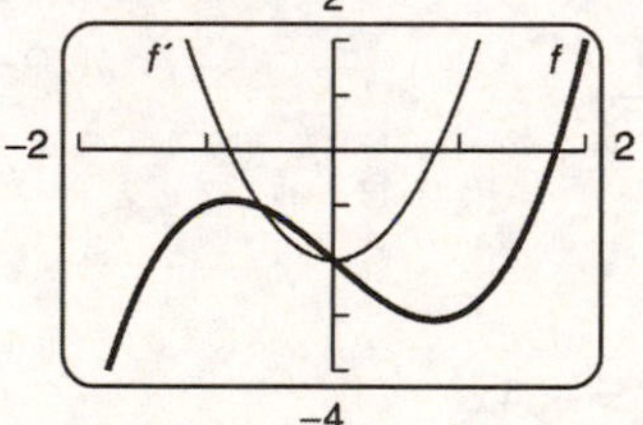

Note, the function $f(x)$ is the thicker graph.

Using the calculator, we can find the derivative of the function when $x=1$.

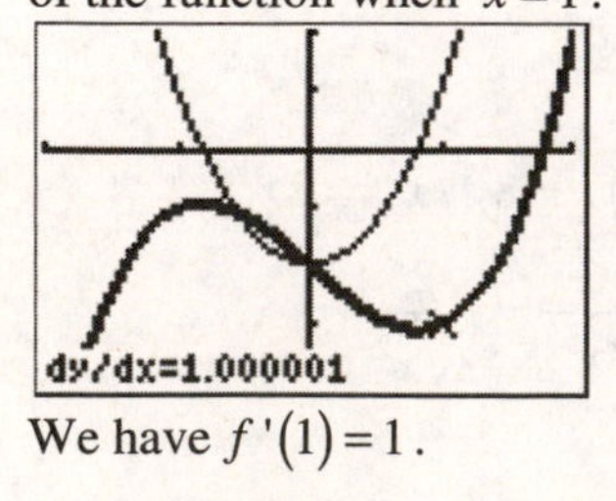

We have $f'(1)=1$.

Copyright © 2014 Pearson Education, Inc. Publishing as Addison-Wesley.

131. $f(x)=\dfrac{4x}{x^2+1}$

Using the calculator, we graph the function and the derivative in the same window. We can use the nDeriv feature to graph the derivative without actually calculating the derivative.

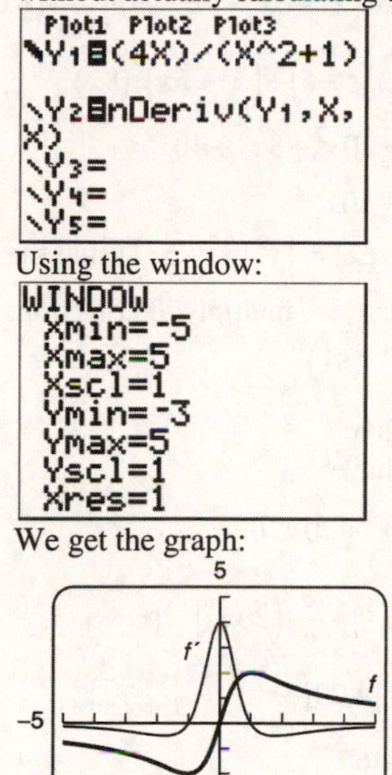

Using the window:

We get the graph:

Note, the function $f(x)$ is the thicker graph. Using the calculator, we can find the derivative of the function when $x=1$.

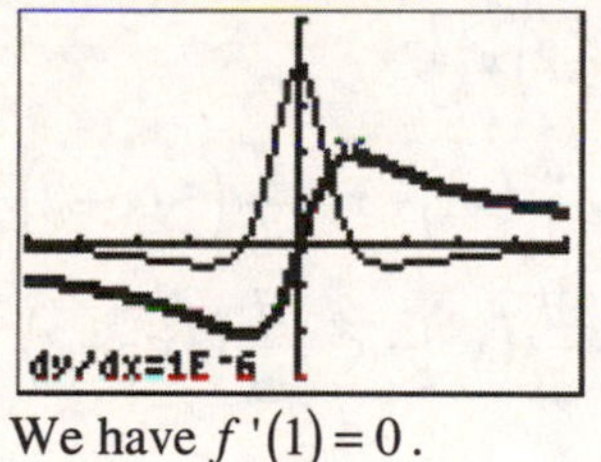

We have $f'(1)=0$.

133. a) $f(x)=x^3-x^2$

Zooming in on the calculator we see the picture:

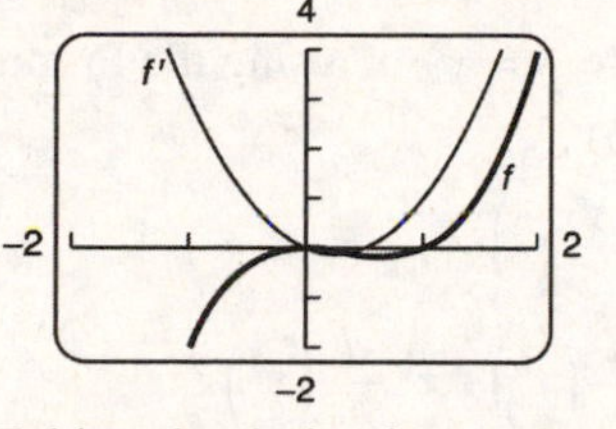

b) Taking the derivative we have:

$$f'(x)=\frac{d}{dx}\left(x^3-x^2\right)$$
$$=3x^2-2x$$

In order to find the horizontal tangents, we set the derivative equal to zero and solve for x.

$$3x^2-2x=0$$
$$x(3x-2)=0$$
$$x=0 \quad \text{or} \quad 3x-2=0$$
$$x=0 \quad \text{or} \quad x=\frac{2}{3}$$

The graph will have horizontal tangent lines at $x=0$ and $x=\dfrac{2}{3}$.

c) Using the result from part (b) we conclude that $f(x)$ is decreasing over the interval $0<x<\dfrac{2}{3}$.

d) No, $f'(x)\geq 0$ for all other values of x. Therefore, $f(x)$ cannot be decreasing on any other interval.

Copyright © 2014 Pearson Education, Inc. Publishing as Addison-Wesley.

Exercise Set 1.6

1. Differentiate $y = x^5 \cdot x^6$ using the Product Rule (Theorem 5).

$$\frac{dy}{dx} = \frac{d}{dx}\left(x^5 \cdot x^6\right)$$
$$= x^5 \cdot \frac{d}{dx}\left(x^6\right) + x^6 \frac{d}{dx}\left(x^5\right)$$
$$= x^5 \cdot 6x^5 + x^6 \cdot 5x^4$$
$$= 6x^{10} + 5x^{10}$$
$$= 11x^{10}$$

Differentiate $y = x^5 \cdot x^6 = x^{11}$ using the Power Rule (Theorem 1).

$$\frac{dy}{dx} = \frac{d}{dx}x^{11}$$
$$= 11x^{11-1}$$
$$= 11x^{10}$$

The two results are equivalent.

3. Differentiate $f(x) = (2x+5)(3x-4)$ using the Product Rule (Theorem 5).

$$f'(x) = \frac{d}{dx}\left[(2x+5)(3x-4)\right]$$
$$= (2x+5)\cdot\frac{d}{dx}(3x-4) + (3x-4)\cdot\frac{d}{dx}(2x+5)$$
$$= (2x+5)\cdot 3 + (3x-4)\cdot 2$$
$$= 6x + 15 + 6x - 8$$
$$= 12x + 7$$

Differentiate $f(x) = (2x+5)(3x-4)$ using the Power Rule (Theorem 1). First, we multiply the binomial terms in the function.

$$f(x) = (2x+5)(3x-4)$$
$$= 6x^2 + 7x - 20$$

Therefore, by Theorem 1 and Theorem 4 we have:

$$f'(x) = \frac{d}{dx}\left(6x^2 + 7x - 20\right)$$
$$= \frac{d}{dx}\left(6x^2\right) + \frac{d}{dx}(7x) - \frac{d}{dx}(20) \quad \text{Theorem 4}$$
$$= 12x + 7 \quad \text{Theorem 1}$$

The two results are equivalent.

5. Differentiate $G(x) = 4x^2\left(x^3 + 5x\right)$ using the Product Rule.

$$G'(x) = \frac{d}{dx}\left[4x^2\left(x^3+5x\right)\right]$$
$$= 4x^2 \cdot \frac{d}{dx}\left(x^3+5x\right) + \left(x^3+5x\right)\cdot\frac{d}{dx}\left(4x^2\right)$$
$$= 4x^2 \cdot \left(3x^2+5\right) + \left(x^3+5x\right)\cdot(8x)$$
$$= 12x^4 + 20x^2 + 8x^4 + 40x^2$$
$$= 20x^4 + 60x^2$$

Differentiate $G(x) = 4x^2\left(x^3 + 5x\right)$ using the Power Rule. First, we multiply the function.

$$G(x) = 4x^2\left(x^3+5x\right)$$
$$= 4x^5 + 20x^3$$

Therefore, we have:

$$G'(x) = \frac{d}{dx}\left(4x^5 + 20x^3\right)$$
$$= \frac{d}{dx}\left(4x^5\right) + \frac{d}{dx}\left(20x^3\right) \quad \text{Theorem 4}$$
$$= 4\left(5x^4\right) + 20\left(3x^2\right) \quad \text{Theorem 1, Theorem 3}$$
$$= 20x^4 + 60x^2$$

The two results are equivalent.

7. Differentiate $y = \left(3\sqrt{x}+2\right)x^2$ using the Product Rule.

$$\frac{dy}{dx} = \frac{d}{dx}\left[\left(3\sqrt{x}+2\right)x^2\right]$$
$$= \left(3\sqrt{x}+2\right)\cdot\frac{d}{dx}\left(x^2\right) + x^2\cdot\frac{d}{dx}\left(3\sqrt{x}+2\right)$$
$$= \left(3x^{1/2}+2\right)\cdot\frac{d}{dx}\left(x^2\right) + x^2\cdot\frac{d}{dx}\left(3x^{1/2}+2\right)$$
$$= \left(3x^{1/2}+2\right)\cdot 2x + x^2\cdot\frac{3}{2}x^{-1/2} \quad \text{Theorem 1}$$
$$= 6x^{3/2} + 4x + \frac{3}{2}x^{3/2}$$
$$= \frac{15}{2}x^{3/2} + 4x$$

The solution is continued on the next page.

Copyright © 2014 Pearson Education, Inc. Publishing as Addison-Wesley.

Differentiate $y=\left(3\sqrt{x}+2\right)x^2$ using the Power Rule. First, we multiply the function.

$$y=\left(3\sqrt{x}+2\right)x^2$$
$$=3x^{1/2+2}+2x^2$$
$$=3x^{5/2}+2x^2$$

Therefore, we have:

$$\frac{dy}{dx}=\frac{d}{dx}\left(3x^{5/2}+2x^2\right)$$
$$=\frac{d}{dx}\left(3x^{5/2}\right)+\frac{d}{dx}\left(2x^2\right) \quad \text{Theorem 4}$$
$$=3\left(\frac{5}{2}x^{5/2-1}\right)+2\left(2x^1\right) \quad \text{Theorem 1, Theorem 3}$$
$$=\frac{15}{2}x^{3/2}+4x$$

The two results are equivalent.

9. Differentiate $g(x)=(4x-3)\left(2x^2+3x+5\right)$ using the Product Rule.

$$g'(x)=\frac{d}{dx}\left[(4x-3)\left(2x^2+3x+5\right)\right]$$
$$=(4x-3)\cdot\frac{d}{dx}\left(2x^2+3x+5\right)+$$
$$\left(2x^2+3x+5\right)\cdot\frac{d}{dx}(4x-3)$$
$$=(4x-3)\cdot(4x+3)+\left(2x^2+3x+5\right)\cdot 4$$
$$=16x^2-9+8x^2+12x+20$$
$$=24x^2+12x+11$$

Differentiate $g(x)=(4x-3)\left(2x^2+3x+5\right)$ using the Power Rule. First, we multiply the terms in the function.

$$g(x)=(4x-3)\left(2x^2+3x+5\right)$$
$$=8x^3+6x^2+11x-15$$

Therefore, we have:

$$g'(x)=\frac{d}{dx}\left(8x^3+6x^2+11x-15\right)$$
$$=\frac{d}{dx}\left(8x^3\right)+\frac{d}{dx}\left(6x^2\right)+$$
$$\frac{d}{dx}(11x)-\frac{d}{dx}(15)$$
$$=24x^2+12x+11$$

The two results are equivalent.

11. Differentiate $F(t)=\left(\sqrt{t}+2\right)\left(3t-4\sqrt{t}+7\right)$ using the Product Rule.

$$F'(t)=\frac{d}{dt}\left[\left(\sqrt{t}+2\right)\left(3t-4\sqrt{t}+7\right)\right]$$
$$=\left(t^{1/2}+2\right)\cdot\frac{d}{dt}\left(3t-4t^{1/2}+7\right)+$$
$$\left(3t-4t^{1/2}+7\right)\cdot\frac{d}{dt}\left(t^{1/2}+2\right) \quad \left[\sqrt{t}=t^{1/2}\right]$$
$$=\left(t^{1/2}+2\right)\cdot\left(3-4\left(\frac{1}{2}t^{-1/2}\right)\right)+$$
$$\left(3t-4t^{1/2}+7\right)\cdot\left(\frac{1}{2}t^{-1/2}\right)$$
$$F'(t)=\left(t^{1/2}+2\right)\cdot\left(3-2t^{-1/2}\right)+$$
$$\left(3t-4t^{1/2}+7\right)\cdot\left(\frac{1}{2}t^{-1/2}\right)$$
$$=3t^{1/2}-2+6-4t^{-1/2}+\frac{3}{2}t^{1/2}-2+\frac{7}{2}t^{-1/2}$$
$$=\frac{9}{2}t^{1/2}-\frac{1}{2}t^{-1/2}+2$$
$$=\frac{9\sqrt{t}}{2}-\frac{1}{2\sqrt{t}}+2$$

Differentiate $F(t)=\left(\sqrt{t}+2\right)\left(3t-4\sqrt{t}+7\right)$ using the Power Rule

$$F(t)=\left(\sqrt{t}+2\right)\left(3t-4\sqrt{t}+7\right)$$
$$=3t^{3/2}-4t+7t^{1/2}+6t-8t^{1/2}+14$$
$$=3t^{3/2}-t^{1/2}+2t+14$$

Therefore, we have:

$$F'(t)=\frac{d}{dt}\left(3t^{3/2}-t^{1/2}+2t+14\right)$$
$$=\frac{d}{dt}\left(3t^{3/2}\right)-\frac{d}{dt}\left(t^{1/2}\right)+\frac{d}{dt}(2t)+\frac{d}{dt}(14)$$
$$=\frac{9t^{1/2}}{2}-\frac{1}{2}t^{-1/2}+2$$
$$=\frac{9\sqrt{t}}{2}-\frac{1}{2\sqrt{t}}+2$$

The two results are equivalent.

Copyright © 2014 Pearson Education, Inc. Publishing as Addison-Wesley.

13. Differentiate $y=\frac{x^7}{x^3}$ using the Quotient Rule (Theorem 6).

$$\frac{dy}{dx}=\frac{d}{dx}\left(\frac{x^7}{x^3}\right)$$
$$=\frac{x^3\frac{d}{dx}\left(x^7\right)-x^7\frac{d}{dx}\left(x^3\right)}{\left(x^3\right)^2}$$
$$=\frac{x^3\left(7x^6\right)-x^7\left(3x^2\right)}{x^6}$$
$$\frac{dy}{dx}=\frac{7x^9-3x^9}{x^6}$$
$$=\frac{4x^9}{x^6}$$
$$=4x^3,\quad \text{for } x\neq 0$$

Differentiate $y=\frac{x^7}{x^3}=x^4$ using the Power Rule.

$$\frac{dy}{dx}=\frac{d}{dx}x^4$$
$$=4x^{4-1}$$
$$=4x^3,\quad \text{for } x\neq 0$$

The two results are equivalent.

15. Differentiate $f(x)=\frac{2x^5+x^2}{x}$ using the Quotient Rule.

$$f'(x)=\frac{d}{dx}\left(\frac{2x^5+x^2}{x}\right)$$
$$=\frac{x\frac{d}{dx}\left(2x^5+x^2\right)-\left(2x^5+x^2\right)\frac{d}{dx}(x)}{(x)^2}$$
$$f'(x)=\frac{x\left(10x^4+2x\right)-\left(2x^5+x^2\right)(1)}{x^2}$$
$$=\frac{10x^5+2x^2-2x^5-x^2}{x^2}$$
$$=\frac{8x^5+x^2}{x^2}$$
$$=\frac{x^2\left(8x^3+1\right)}{x^2}$$
$$=8x^3+1,\quad \text{for } x\neq 0$$

Differentiate $f(x)=\frac{2x^5+x^2}{x}$ using the Power Rule. First, factor the numerator and divide the common factors.

$$f(x)=\frac{2x^5+x^2}{x}$$
$$=\frac{x\left(2x^4+x\right)}{x}$$
$$=2x^4+x$$
$$f'(x)=\frac{d}{dx}\left(2x^4+x\right)$$
$$=8x^3+1,\quad \text{for } x\neq 0$$

The two results are equivalent.

17. Differentiate $G(x)=\frac{8x^3-1}{2x-1}$ using the Quotient Rule.

$$G'(x)=\frac{d}{dx}\left(\frac{8x^3-1}{2x-1}\right)$$
$$=\frac{(2x-1)\frac{d}{dx}\left(8x^3-1\right)-\left(8x^3-1\right)\frac{d}{dx}(2x-1)}{(2x-1)^2}$$
$$=\frac{(2x-1)\left(24x^2\right)-\left(8x^3-1\right)(2)}{(2x-1)^2}$$
$$=\frac{(2x-1)\left(24x^2\right)-(2x-1)\left(4x^2+2x+1\right)(2)}{(2x-1)^2}$$
$$=\frac{(2x-1)\left[\left(24x^2\right)-\left(4x^2+2x+1\right)(2)\right]}{(2x-1)^2}$$
$$=\frac{(2x-1)\left[\left(24x^2\right)-\left(8x^2+4x+2\right)\right]}{(2x-1)^2}$$
$$=\frac{\left[16x^2-4x-2\right]}{(2x-1)}$$
$$=\frac{(2x-1)(8x+2)}{(2x-1)}$$
$$=8x+2;\quad x\neq \tfrac{1}{2}$$

The solution is continued on the next page.

Copyright © 2014 Pearson Education, Inc. Publishing as Addison-Wesley.

Differentiate $G(x)=\frac{8x^3-1}{2x-1}$ using the Power Rule. First, factor the numerator and divide the common factors.

$$G(x)=\frac{8x^3-1}{2x-1}$$
$$=\frac{(2x-1)(4x^2+2x+1)}{2x-1} \quad \text{Difference of cubes}$$
$$=4x^2+2x+1$$
$$G'(x)=\frac{d}{dx}(4x^2+2x+1)$$
$$=8x+2; \qquad x\neq\tfrac{1}{2}$$

The two results are equivalent.

19. Differentiate $y=\frac{t^2-16}{t+4}$ using the Quotient Rule.

$$\frac{dy}{dt}=\frac{d}{dt}\left(\frac{t^2-16}{t+4}\right)$$
$$=\frac{(t+4)\frac{d}{dt}(t^2-16)-(t^2-16)\frac{d}{dt}(t+4)}{(t+4)^2}$$
$$=\frac{(t+4)(2t)-(t^2-16)(1)}{(t+4)^2}$$
$$=\frac{2t^2+8t-t^2+16}{(t+4)^2}$$
$$=\frac{t^2+8t+16}{(t+4)^2}$$
$$\frac{dy}{dt}=\frac{(t+4)^2}{(t+4)^2}$$
$$=1; \qquad t\neq-4$$

Differentiate $y=\frac{t^2-16}{t+4}$ using the Power Rule. First, factor the numerator and divide the common factors.

$$y=\frac{t^2-16}{t+4}$$
$$=\frac{(t+4)(t-4)}{t+4} \quad \text{Difference of squares}$$
$$=t-4$$
$$\frac{dy}{dt}=\frac{d}{dt}(x-4)$$
$$=1, \qquad \text{for } t\neq-4$$

The two results are equivalent.

21. $f(x)=(3x^2-2x+5)(4x^2+3x-1)$

Using the Product Rule, we have:

$$f'(x)=\frac{d}{dx}\left[(3x^2-2x+5)(4x^2+3x-1)\right]$$
$$=(3x^2-2x+5)\cdot\frac{d}{dx}(4x^2+3x-1)+$$
$$(4x^2+3x-1)\cdot\frac{d}{dx}(3x^2-2x+5)$$
$$=(3x^2-2x+5)\cdot(8x+3)+$$
$$(4x^2+3x-1)\cdot(6x-2)$$

Simplifying, we get

$$=(24x^3-7x^2+34x+15)+$$
$$(24x^3+10x^2-12x+2)$$
$$=48x^3+3x^2+22x+17$$

23. $y=\frac{5x^2-1}{2x^3+3}$

Using the Quotient Rule.

$$\frac{dy}{dx}=\frac{d}{dx}\left(\frac{5x^2-1}{2x^3+3}\right)$$
$$=\frac{(2x^3+3)\frac{d}{dx}(5x^2-1)-(5x^2-1)\frac{d}{dx}(2x^3+3)}{(2x^3+3)^2}$$
$$=\frac{(2x^3+3)(10x)-(5x^2-1)(6x^2)}{(2x^3+3)^2}$$
$$=\frac{20x^4+30x-30x^4+6x^2}{(2x^3+3)^2}$$
$$=\frac{-10x^4+6x^2+30x}{(2x^3+3)^2}$$
$$=\frac{-2x(5x^3-3x-15)}{(2x^3+3)^2}$$

Copyright © 2014 Pearson Education, Inc. Publishing as Addison-Wesley.

25. $G(x)=\left(8x+\sqrt{x}\right)\left(5x^2+3\right)$

$G(x)=\left(8x+x^{1/2}\right)\left(5x^2+3\right) \quad \left[\sqrt{x}=x^{1/2}\right]$

Using the Product Rule, we calculate the derivative.

$$G'(x)=\frac{d}{dx}\left[\left(8x+x^{1/2}\right)\left(5x^2+3\right)\right]$$
$$=\left(8x+x^{1/2}\right)\cdot\frac{d}{dx}\left(5x^2+3\right)+\left(5x^2+3\right)\cdot\frac{d}{dx}\left(8x+x^{1/2}\right)$$
$$=\left(8x+x^{1/2}\right)\cdot(10x)+\left(5x^2+3\right)\cdot\left(8+\frac{1}{2}x^{-1/2}\right)$$

Simplifying, we get

$$=\left(80x^2+10x^{3/2}\right)+\left(40x^2+\frac{5}{2}x^{3/2}+\frac{3}{2}x^{-1/2}+24\right)$$
$$=120x^2+\frac{25}{2}x^{3/2}+\frac{3}{2}x^{-1/2}+24$$

27. $g(t)=\frac{t}{3-t}+5t^3$

Differentiating we have:

$$g'(t)=\frac{d}{dt}\left(\frac{t}{3-t}+5t^3\right)$$
$$=\frac{d}{dt}\left(\frac{t}{3-t}\right)+\frac{d}{dt}\left(5t^3\right)$$

We will apply the Quotient Rule to the first term, and the Power Rule to the second term.

$$g'(t)=\underbrace{\frac{(3-t)\cdot\frac{d}{dt}(t)-t\cdot\frac{d}{dt}(3-t)}{(3-t)^2}}_{\text{Quotient Rule}}+15t^2$$
$$=\frac{(3-t)(1)-t(-1)}{(3-t)^2}+15t^2$$
$$=\frac{3}{(3-t)^2}+15t^2$$

29. $F(x)=(x+3)^2=(x+3)(x+3)$

Using the Product Rule, we have

$$F'(x)=\frac{d}{dx}\left[(x+3)(x+3)\right]$$
$$=(x+3)\cdot\frac{d}{dx}(x+3)+(x+3)\cdot\frac{d}{dx}(x+3)$$
$$=(x+3)\cdot(1)+(x+3)\cdot(1)$$
$$=2x+6$$
$$=2(x+3)$$

31. $y=\left(x^3-4x\right)^2=\left(x^3-4x\right)\left(x^3-4x\right)$

Using the Product Rule, we have

$$\frac{dy}{dx}=\frac{d}{dx}\left[\left(x^3-4x\right)\left(x^3-4x\right)\right]$$
$$=\left(x^3-4x\right)\cdot\frac{d}{dx}\left(x^3-4x\right)+\left(x^3-4x\right)\cdot\frac{d}{dx}\left(x^3-4x\right)$$
$$=\left(x^3-4x\right)\cdot\left(3x^2-4\right)+\left(x^3-4x\right)\cdot\left(3x^2-4\right)$$
$$=2\left(x^3-4x\right)\left(3x^2-4\right)$$
$$=2x\left(x^2-4\right)\left(3x^2-4\right)$$

33. $g(x)=5x^{-3}\left(x^4-5x^3+10x-2\right)$

Using the Product Rule:

$$g'(x)=5x^{-3}\frac{d}{dx}\left(x^4-5x^3+10x-2\right)+\left(x^4-5x^3+10x-2\right)\frac{d}{dx}\left(5x^{-3}\right)$$
$$=\left(5x^{-3}\right)\left(4x^3-15x^2+10\right)+\left(x^4-5x^3+10x-2\right)\left(-15x^{-4}\right)$$

Simplifying, we get

$$=20-75x^{-1}+50x^{-3}-15+75x^{-1}-150x^{-3}+30x^{-4}$$
$$=5-100x^{-3}+30x^{-4}$$

Copyright © 2014 Pearson Education, Inc. Publishing as Addison-Wesley.

35. $F(t)=\left(t+\frac{2}{t}\right)\left(t^2-3\right)=\left(t+2t^{-1}\right)\left(t^2-3\right)$

Using the Product Rule, we have:

$$f'(t)=\frac{d}{dt}\left[\left(t+2t^{-1}\right)\left(t^2-3\right)\right]$$
$$=\left(t+2t^{-1}\right)\cdot\frac{d}{dt}\left(t^2-3\right)+$$
$$\left(t^2-3\right)\cdot\frac{d}{dt}\left(t+2t^{-1}\right)$$
$$f'(t)=\left(t+2t^{-1}\right)\cdot(2t)+\left(t^2-3\right)\cdot\left(1-2t^{-2}\right)$$

Simplifying, we get

$$=2t^2+4+\left(t^2-2-3+6t^{-2}\right)$$
$$=3t^2-1+6t^{-2}$$
$$=3t^2-1+\frac{6}{t^2}$$

37. $y=\frac{x^2+1}{x^3-1}-5x^2$

Differentiating we have:

$$\frac{dy}{dx}=\frac{d}{dx}\left(\frac{x^2+1}{x^3-1}-5x^2\right)$$
$$=\frac{d}{dx}\left(\frac{x^2+1}{x^3-1}\right)-\frac{d}{dx}\left(5x^2\right)$$

We will apply the Quotient Rule to the first term, and the Power Rule to the second term.

$$\frac{dy}{dx}=\underbrace{\frac{\left(x^3-1\right)\cdot(2x)-\left(x^2+1\right)\cdot\left(3x^2\right)}{\left(x^3-1\right)^2}}_{\text{Quotient Rule}}-10x$$

Simplifying, we get

$$=\frac{2x^4-2x-\left(3x^4+3x^2\right)}{\left(x^3-1\right)^2}-10x$$
$$=\frac{-x^4-3x^2-2x}{\left(x^3-1\right)^2}-10x$$

39. $y=\frac{\sqrt[3]{x}-7}{\sqrt{x}+3}=\frac{x^{1/3}-7}{x^{1/2}+3}$

Using the Quotient Rule.

$$\frac{dy}{dx}=\frac{d}{dx}\left(\frac{x^{1/3}-7}{x^{1/2}+3}\right)$$
$$=\frac{\left(x^{1/2}+3\right)\frac{d}{dx}\left(x^{1/3}-7\right)-\left(x^{1/3}-7\right)\frac{d}{dx}\left(x^{1/2}+3\right)}{\left(x^{1/2}+3\right)^2}$$
$$=\frac{\left(x^{1/2}+3\right)\left(\frac{1}{3}x^{-2/3}\right)-\left(x^{1/3}-7\right)\left(\frac{1}{2}x^{-1/2}\right)}{\left(x^{1/2}+3\right)^2}$$

Note, the previous derivative can be simplified as follows

$$\frac{dy}{dx}=\frac{\left(x^{1/2}+3\right)\left(\frac{1}{3}x^{-2/3}\right)-\left(x^{1/3}-7\right)\left(\frac{1}{2}x^{-1/2}\right)}{\left(x^{1/2}+3\right)^2}$$
$$=\frac{\frac{1}{3}x^{-1/6}+x^{-2/3}-\frac{1}{2}x^{-1/6}+\frac{7}{2}x^{-1/2}}{\left(x^{1/2}+3\right)^2}$$
$$=\frac{x^{-2/3}-\frac{1}{6}x^{-1/6}+\frac{7}{2}x^{-1/2}}{\left(x^{1/2}+3\right)^2}\cdot\frac{6x^{2/3}}{6x^{2/3}}$$
$$=\frac{6-\sqrt{x}+21x^{1/6}}{6x^{2/3}\left(\sqrt{x}+3\right)^2}$$

41. $f(x)=\frac{x}{x^{-1}+1}$

Using the Quotient Rule, we have

$$f'(x)=\frac{d}{dx}\left(\frac{x}{x^{-1}+1}\right)$$
$$=\frac{\left(x^{-1}+1\right)\frac{d}{dx}(x)-(x)\frac{d}{dx}\left(x^{-1}+1\right)}{\left(x^{-1}+1\right)^2}$$
$$=\frac{\left(x^{-1}+1\right)(1)-(x)\left(-1x^{-2}\right)}{\left(x^{-1}+1\right)^2}$$
$$=\frac{x^{-1}+1+x^{-1}}{\left(x^{-1}+1\right)^2}$$
$$=\frac{2x^{-1}+1}{\left(x^{-1}+1\right)^2},\qquad \text{for } x\neq 0 \text{ and } x\neq -1$$

The solution is continued on the next page.

Copyright © 2014 Pearson Education, Inc. Publishing as Addison-Wesley.

Note, the derivative on the previous page could be simplified as follows:

$$f'(x)=\frac{2x^{-1}+1}{\left(x^{-1}+1\right)^2}$$

$$=\frac{\frac{2}{x}+1}{\left(\frac{1}{x}+1\right)^2}$$

$$=\frac{\frac{2+x}{x}}{\left(\frac{1+x}{x}\right)^2}$$

$$f'(x)=\frac{2+x}{x}\cdot\frac{x^2}{(1+x)^2}$$

$$=\frac{x(x+2)}{(1+x)^2} \quad \text{for } x\neq 0 \text{ and } x\neq -1$$

43. $F(t)=\frac{1}{t-4}$

Using the Quotient Rule, we have

$$F'(t)=\frac{d}{dt}\left(\frac{1}{t-4}\right)$$

$$=\frac{(t-4)\frac{d}{dt}(1)-(1)\frac{d}{dt}(t-4)}{(t-4)^2}$$

$$=\frac{(t-4)(0)-(1)(1)}{(t-4)^2}$$

$$=\frac{-1}{(t-4)^2}$$

45. $f(x)=\frac{3x^2+2x}{x^2+1}$

Using the Quotient Rule, we have

$$f'(x)=\frac{d}{dx}\left(\frac{3x^2+2x}{x^2+1}\right)$$

$$=\frac{\left(x^2+1\right)\frac{d}{dx}\left(3x^2+2x\right)-\left(3x^2+2x\right)\frac{d}{dx}\left(x^2+1\right)}{\left(x^2+1\right)^2}$$

$$=\frac{\left(x^2+1\right)(6x+2)-\left(3x^2+2x\right)(2x)}{\left(x^2+1\right)^2}$$

$$=\frac{6x^3+2x^2+6x+2-\left(6x^3+4x^2\right)}{\left(x^2+1\right)^2}$$

$$=\frac{-2x^2+6x+2}{\left(x^2+1\right)^2}$$

$$=\frac{-2\left(x^2-3x-1\right)}{\left(x^2+1\right)^2}$$

47. $g(t)=\frac{-t^2+3t+5}{t^2-2t+4}$

Using the Quotient Rule, we have

$$g'(t)=\frac{d}{dt}\left(\frac{-t^2+3t+5}{t^2-2t+4}\right)$$

$$=\frac{\left(t^2-2t+4\right)\frac{d}{dt}\left(-t^2+3t+5\right)}{\left(t^2-2t+4\right)^2}-\frac{\left(-t^2+3t+5\right)\frac{d}{dt}\left(t^2-2t+4\right)}{\left(t^2-2t+4\right)^2}$$

$$=\frac{\left(t^2-2t+4\right)(-2t+3)-\left(-t^2+3t+5\right)(2t-2)}{\left(t^2-2t+4\right)^2}$$

Note, the previous derivative could be simplified as follows:

$$g'(t)=\frac{-2t^3+7t^2-14t+12-\left(-2t^3+8t^2+4t-10\right)}{\left(t^2-2t+4\right)^2}$$

$$=\frac{-t^2-18t+22}{\left(t^2-2t+4\right)^2}$$

Copyright © 2014 Pearson Education, Inc. Publishing as Addison-Wesley.

49. – 95. Left to the student.

97. $y=\frac{8}{x^2+4}$

$$\frac{dy}{dx}=\frac{(x^2+4)(0)-8(2x)}{(x^2+4)^2}$$

$$\frac{dy}{dx}=\frac{-16x}{(x^2+4)^2}$$

a) When $x=0$, $\frac{dy}{dx}=\frac{-16(0)}{(0^2+4)^2}=0$, so the slope of the tangent line at $(0,2)$ is 0. The equation of the horizontal line passing through $(0,2)$ is $y=2$.

b) When $x=-2$,

$\frac{dy}{dx}=\frac{-16(-2)}{((-2)^2+4)^2}=\frac{32}{64}=\frac{1}{2}$, so the slope of the tangent line at $(-2,1)$ is $\frac{1}{2}$. Using the point-slope equation, we have:

$$y-y_1=m(x-x_1)$$
$$y-1=\frac{1}{2}(x-(-2))$$
$$y-1=\frac{1}{2}x+1$$
$$y=\frac{1}{2}x+2$$

99. $y=x^2+\frac{3}{x-1}$

$$\frac{dy}{dx}=2x+\frac{(x-1)(0)-3(1)}{(x-1)^2}$$
$$=2x-\frac{3}{(x-1)^2}$$

a) When $x=2$, $y=(2)^2+\frac{3}{2-1}=4+3=7$,

and $\frac{dy}{dx}=2(2)-\frac{3}{(2-1)^2}=4-3=1$.

Therefore, the slope of the tangent line at $(2,7)$ is 1.

Using the point-slope equation, we have:

$$y-y_1=m(x-x_1)$$
$$y-7=1(x-2)$$
$$y-7=x-2$$
$$y=x+5$$

b) When $x=3$, $y=(3)^2+\frac{3}{3-1}=9+\frac{3}{2}=\frac{21}{2}$,

and $\frac{dy}{dx}=2(3)-\frac{3}{(3-1)^2}=6-\frac{3}{4}=\frac{21}{4}$.

Therefore, the slope of the tangent line at $\left(3,\frac{21}{2}\right)$ is $\frac{21}{4}$.

Using the point-slope equation, we have:

$$y-y_1=m(x-x_1)$$
$$y-\frac{21}{2}=\frac{21}{4}(x-3)$$
$$y-\frac{21}{2}=\frac{21}{4}x-\frac{63}{4}$$
$$y=\frac{21}{4}x-\frac{21}{4}$$

101. The average cost of producing x items is

$A_C(x)=\frac{C(x)}{x}$. Therefore,

$A_C(x)=\frac{950+15\sqrt{x}}{x}$.

Next, we take the derivative using the Quotient Rule to find the rate at which average cost is changing.

$$A_C'(x)=\frac{d}{dx}\left(\frac{950+15x^{1/2}}{x}\right)$$

$$A_C'(x)=\frac{x\left(\frac{15}{2}x^{-1/2}\right)-\left(950+15x^{1/2}\right)(1)}{(x)^2}$$

$$=\frac{\frac{15}{2}x^{1/2}-950-15x^{1/2}}{x^2}$$

$$=\frac{-\frac{15}{2}x^{1/2}-950}{x^2}$$

$$=\frac{-\frac{15}{2}\sqrt{x}-950}{x^2}$$

The solution is continued on the next page.

Copyright © 2014 Pearson Education, Inc. Publishing as Addison-Wesley.

Substituting 400 for x, we have

$$A_C'(400)=\frac{-\frac{15}{2}\sqrt{400}-950}{(400)^2}$$
$$=\frac{-150-950}{160{,}000}$$
$$\approx -0.006875$$

Therefore, when 400 jackets have been produced, average cost is changing at a rate of -0.0069 dollars per jacket.

103. The average revenue of producing x items is $A_R(x)=\frac{R(x)}{x}$. Therefore,

$$A_R(x)=\frac{85\sqrt{x}}{x}=\frac{85}{x^{1/2}}.$$

Next, we take the derivative using the Quotient Rule to find the rate at which average revenue is changing.

$$A_R'(x)=\frac{d}{dx}\left(\frac{85}{\sqrt{x}}\right)$$
$$=\frac{x^{1/2}(0)-(85)\left(\frac{1}{2}x^{-1/2}\right)}{\left(x^{1/2}\right)^2}$$
$$=\frac{-\frac{85}{2}x^{-1/2}}{x}$$
$$=-\frac{85}{2x^{3/2}}$$

Substituting 400 for x, we have

$$A_R'(400)=-\frac{85}{2(400)^{3/2}}$$
$$=-\frac{85}{16000}$$
$$=-0.00531250$$

Therefore, when 400 jackets have been produced, average revenue is changing at a rate of -0.0053 dollars per jacket.

105. $A_P(x)=\frac{P(x)}{x}=\frac{R(x)-C(x)}{x}$

From Exercises 101 and 103, we know that

$$A_P(x)=\frac{85x^{1/2}-\left(950+15x^{1/2}\right)}{x}=\frac{70x^{1/2}-950}{x}$$

Using the Quotient Rule to take the derivative, we have:

$$A_P'(x)=\frac{x\left(\frac{70}{2}x^{-1/2}\right)-\left(70x^{1/2}-950\right)(1)}{(x)^2}$$
$$=\frac{35x^{1/2}-70x^{1/2}+950}{x^2}$$
$$=\frac{-35x^{1/2}+950}{x^2}$$

Substituting 400 for x, we have:

$$A_P'(400)=\frac{-35(400)^{1/2}+950}{(400)^2}$$
$$=\frac{-700+950}{16{,}000}$$
$$=\frac{250}{16{,}000}$$
$$\approx 0.015625$$

When 400 jackets have been produced and sold, the average profit is changing at a rate of 0.0156 dollars per jacket.

Alternatively, we could have used the information in Exercises 101 and 103 to find the rate of change of average profit when 400 jackets are produced and sold. Notice that

$$A_P'(x)=A_R'(x)-A_C'(x)$$
$$=-0.0053125-(-0.006875)$$
$$=0.00156250$$

107. The average profit of producing x items is $A_P(x)=\frac{R(x)-C(x)}{x}$. Therefore,

$$A_P(x)=\frac{65x^{0.9}-\left(4300+2.1x^{0.6}\right)}{x}$$
$$=\frac{65x^{0.9}-2.1x^{0.6}-4300}{x}.$$

Using the Quotient Rule to take the derivative, we have

$$A_P'(x)$$
$$=\frac{x\left(65\left(0.9x^{-0.1}\right)-2.1\left(0.6x^{-0.4}\right)\right)-\left(65x^{0.9}-2.1x^{0.6}-4300\right)(1)}{(x)^2}$$
$$=\frac{58.5x^{0.9}-1.26x^{0.6}-\left(65x^{0.9}-2.1x^{0.6}-4300\right)}{x^2}$$
$$=\frac{-6.5x^{0.9}+0.84x^{0.6}+4300}{x^2}$$

The solution is continued on the next page.

Copyright © 2014 Pearson Education, Inc. Publishing as Addison-Wesley.

Substituting 50 for x, we have

$$A_P'(50) = \frac{-6.5(50)^{0.9} + 0.84(50)^{0.6} + 4300}{(50)^2}$$
$$= \frac{4089.00428745}{2500}$$
$$= 1.63560171$$
$$\approx 1.64$$

Therefore, when 50 vases have been produced and sold, the average profit is changing at rate of 1.64 dollars per vase.

109. $P(t) = 567 + t\left(36t^{0.6} - 104\right)$

a) Using the Product Rule and remembering that the derivative of a constant is 0, we have

$$P'(t) = 0 + t\left(36\left(0.6t^{-0.4}\right)\right) + \left(36t^{0.6} - 104\right)(1)$$
$$= 21.6t^{0.6} + 36t^{0.6} - 104$$
$$= 57.6t^{0.6} - 104$$

b) Substituting 45 for t, we have:

$$P'(45) = 57.6(45)^{0.6} - 104$$
$$= 565.39243291 - 104$$
$$= 461.39243291$$

c) ✎

111. $T(t) = \frac{4t}{t^2+1} + 98.6$

a) $$T'(t) = \frac{\left(t^2+1\right)(4) - (4t)(2t)}{\left(t^2+1\right)^2} + 0$$
$$= \frac{4t^2 + 4 - 8t^2}{\left(t^2+1\right)^2}$$
$$= \frac{-4t^2 + 4}{\left(t^2+1\right)^2}$$
$$= \frac{-4\left(t^2-1\right)}{\left(t^2+1\right)^2}$$

b) $$T(2) = \frac{4(2)}{(2)^2+1} + 98.6 = \frac{8}{5} + 98.6 = 100.2$$

After 2 hours, the temperature of the ill person is approximately 100.2 degrees Fahrenheit.

c) $$T'(2) = \frac{-4\left((2)^2-1\right)}{\left((2)^2+1\right)^2} = \frac{-12}{25} = -0.48$$

After 2 hours, the person's temperature is changing at rate of -0.48 degrees per hour.

113. $y(t) = 5t(t-1)(2t+3)$

First, group the factors of $y(t)$ in order to apply the product rule.

$$y(t) = \left[5t(t-1)\right]\cdot(2t+3)$$

Now we calculate the derivative.

Notice that when we take the derivative of the first term, $\left[5t(t-1)\right]$ we will have to apply the Product Rule again.

$$y'(t) = \left[5t(t-1)\right](2) + (2t+3)\left[\underbrace{(5t)(1) + (t-1)(5)}_{\text{Product Rule for }[5t(t-1)]}\right]$$
$$= 10t(t-1) + (2t+3)[5t + 5t - 5]$$
$$= 10t(t-1) + (2t+3)(10t-5)$$

The previous derivative can be simplified as follows:

$$y'(t) = 10t^2 - 10t + 20t^2 + 30t - 10t - 15$$
$$= 30t^2 + 10t - 15$$

115. $g(x) = \left(x^3-8\right)\cdot\frac{x^2+1}{x^2-1}$

We will begin by applying the Product Rule.

$$g'(x) = \left(x^3-8\right)\frac{d}{dx}\left(\frac{x^2+1}{x^2-1}\right) + \frac{x^2+1}{x^2-1}\cdot\frac{d}{dx}\left(x^3-8\right)$$

Notice, that we will have to apply the Quotient Rule to take the derivative of $\frac{x^2+1}{x^2-1}$.

$$g'(x) = \left(x^3-8\right)\frac{\left(x^2-1\right)(2x) - \left(x^2+1\right)(2x)}{\left(x^2-1\right)^2} + \left(\frac{x^2+1}{x^2-1}\right)\cdot\left(3x^2\right)$$
$$= \left(x^3-8\right)\frac{-4x}{\left(x^2-1\right)^2} + \left(\frac{x^2+1}{x^2-1}\right)\cdot\left(3x^2\right)$$

The solution is continued on the next page.

Copyright © 2014 Pearson Education, Inc. Publishing as Addison-Wesley.

The derivative from the previous page can be simplified as follows.

$$g'(x)=\frac{-4x\left(x^3-8\right)}{\left(x^2-1\right)^2}+\frac{3x^2\left(x^2+1\right)}{x^2-1}$$

$$=\frac{-4x^4+32x}{\left(x^2-1\right)^2}+\frac{3x^4+3x^2}{x^2-1}\cdot\frac{x^2-1}{x^2-1}$$

$$=\frac{-4x^4+32x+3x^6-3x^2}{\left(x^2-1\right)^2}$$

$$=\frac{3x^6-4x^4-3x^2+32x}{\left(x^2-1\right)^2}$$

117. $f(x)=\dfrac{(x-1)\left(x^2+x+1\right)}{x^4-3x^3-5}$

First we will group the numerator, to apply the Quotient Rule. Remember that we will have to apply the Product Rule when taking the derivative of the numerator.

$$f(x)=\frac{\left[(x-1)\left(x^2+x+1\right)\right]}{x^4-3x^3-5}$$

We calculate the derivative.

$$f'(x)$$

$$=\frac{\left(x^4-3x^3-5\right)\left[(x-1)(2x+1)+\left(x^2+x+1\right)(1)\right]}{\left(x^4-3x^3-5\right)^2}-\frac{\left[(x-1)\left(x^2+x+1\right)\right]\left(4x^3-9x^2\right)}{\left(x^4-3x^3-5\right)^2}$$

$$=\frac{\left(x^4-3x^3-5\right)\left[2x^2-x-1+x^2+x+1\right]}{\left(x^4-3x^3-5\right)^2}-\frac{\left[x^3-1\right]\left(4x^3-9x^2\right)}{\left(x^4-3x^3-5\right)^2}$$

$$=\frac{\left(x^4-3x^3-5\right)\left[3x^2\right]-\left[x^3-1\right]\left(4x^3-9x^2\right)}{\left(x^4-3x^3-5\right)^2}$$

$$=\frac{3x^6-9x^5-15x^2-4x^6+9x^5+4x^3-9x^2}{\left(x^4-3x^3-5\right)^2}$$

$$=\frac{-x^6+4x^3-24x^2}{\left(x^4-3x^3-5\right)^2}$$

119. $f(x)=\dfrac{x^2}{x^2-1}$ and $g(x)=\dfrac{1}{x^2-1}$

a) $f'(x)=\dfrac{\left(x^2-1\right)(2x)-x^2(2x)}{\left(x^2-1\right)^2}=\dfrac{-2x}{\left(x^2-1\right)^2}$

b) $g'(x)=\dfrac{\left(x^2-1\right)(0)-1(2x)}{\left(x^2-1\right)^2}=\dfrac{-2x}{\left(x^2-1\right)^2}$

c)

121.

123. a) Definition of the derivative.

b) Adding and subtracting the same quantity is the same as adding 0.

c) The limit of a sum is the sum of the limits.

d) Factoring common factors.

e) The limit of a product is the product of the limits and $\lim\limits_{h\to 0} f(x+h)=f(x)$.

f) Definition of the derivative.

g) Using Leibniz's notation.

125. The break-even point occurs when $P(x)=0$.

$$P(x)=R(x)-C(x)$$

$$=85x^{1/2}-\left(950+15x^{1/2}\right)$$

$$=70x^{1/2}-950$$

Using the window:

```
WINDOW
 Xmin=0
 Xmax=500
 Xscl=100
 Ymin=-500
 Ymax=500
 Yscl=100
 Xres=1
```

We graph the profit function on the calculator.

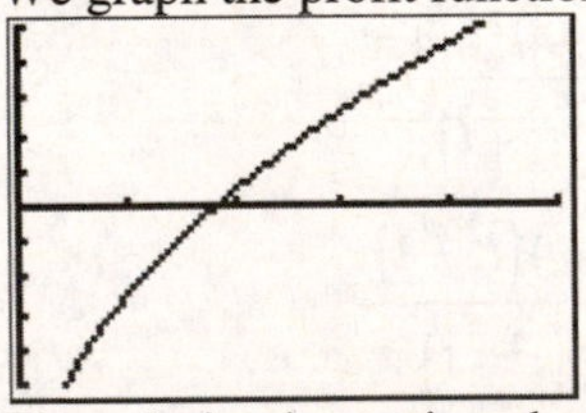

The solution is continued on the next page.

Copyright © 2014 Pearson Education, Inc. Publishing as Addison-Wesley.

The break even point will be the zero of the function. Using the zero finder on the calculator we have:

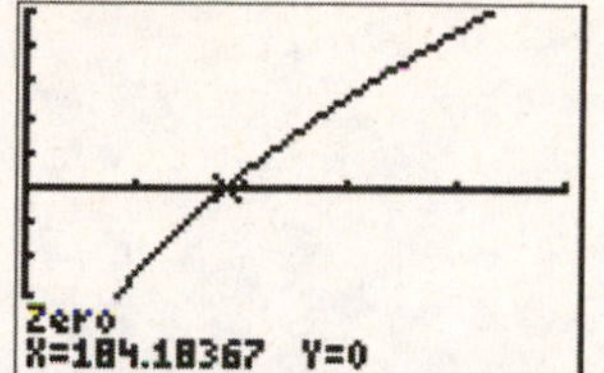

We see that the break-even point occurs at $x = 184$ jackets.

The profit is changing at rate of

$$P'(x) = 70\left(\frac{1}{2}x^{-1/2}\right) = \frac{70}{2x^{1/2}} = \frac{35}{\sqrt{x}}$$

Substituting 184 for x we have:

$$P'(184) = \frac{35}{\sqrt{184}}$$
$$= 2.580234233$$
$$\approx 2.58$$

Therefore, at the break-even point, profit is increasing at a rate of 2.58 dollars per jacket.

From Exercise 105 we know that:

$$A_P'(x) = \frac{-35x^{1/2} + 950}{x^2}$$

Substituting 184 for x we get:

$$A_P'(184) = \frac{-35(184)^{1/2} + 950}{(184)^2}$$
$$\approx 0.014037$$
$$\approx 0.014$$

At the break-even point, average profit is changing at a rate of 0.014 dollars per jacket.

127. $f(x) = \left(x + \frac{2}{x}\right)\left(x^2 - 3\right)$

Using the calculator, we graph the function and the derivative in the same window. We can use the nDeriv feature to graph the derivative without actually calculating the derivative.

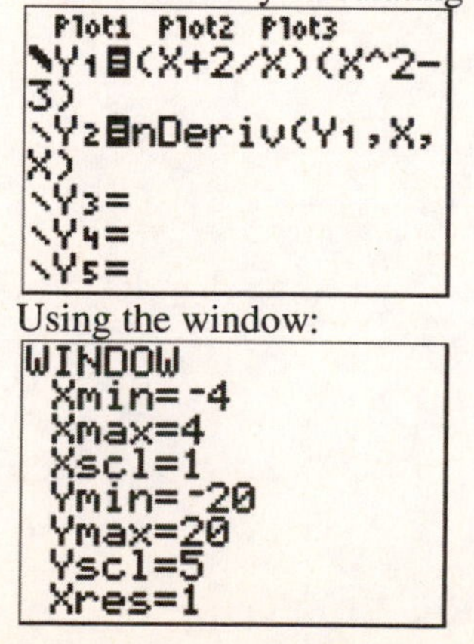

We get the graph:

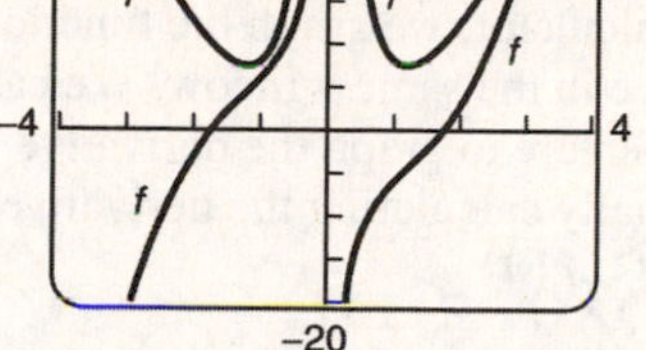

Note, the function $f(x)$ is the thicker graph.

The horizontal tangents occur at the turning points of this function, or at the x-intercepts of the derivative. We can see that the derivative never intersects the x-axis, therefore, there are no points at which the tangent line is horizontal.

129. $f(x) = \frac{0.3x}{0.04 + x^2}$

Using the calculator, we graph the function and the derivative in the same window. We can use the nDeriv feature to graph the derivative without actually calculating the derivative.

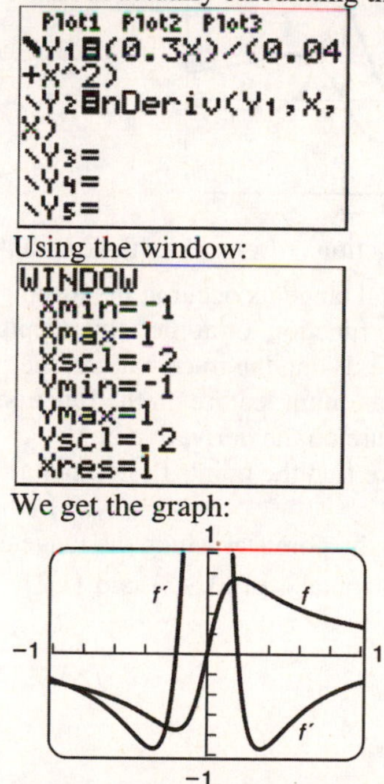

Note, the function $f(x)$ is the thicker graph.

The horizontal tangents occur at the turning points of this function, or at the x-intercepts of the derivative. Using the calculator, we find the points of horizontal tangency.

We estimate the points at which the tangent lines are horizontal are $(-0.2, -0.75)$ and $(0.2, 0.75)$.

Copyright © 2014 Pearson Education, Inc. Publishing as Addison-Wesley.

131. $f(x)=\dfrac{4x}{x^2+1}$

Using the calculator, we graph the function and the derivative in the same window. We can use the nDeriv feature to graph the derivative without actually calculating the derivative.

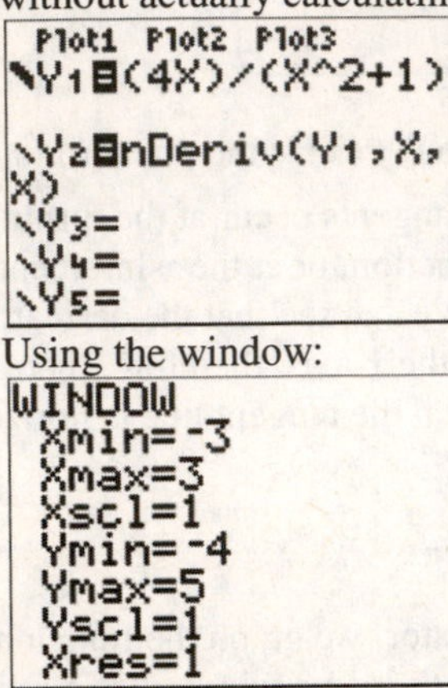

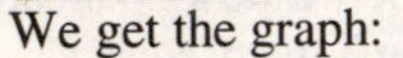

We get the graph:

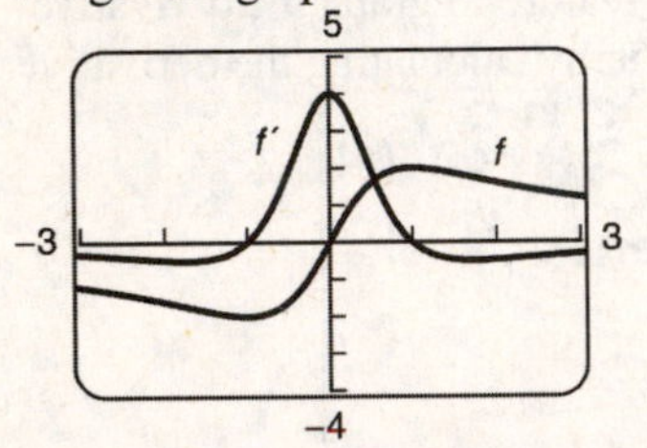

Note, the function $f(x)$ is the thicker graph.

The horizontal tangents occur at the turning points of this function, or at the x-intercepts of the derivative. Using the trace feature, the minimum/maximum feature on the function, or the zero feature on the derivative on the calculator, we find the points of horizontal tangency.

We estimate the points at which the tangent lines are horizontal are $(-1,-2)$ and $(1,2)$.

Copyright © 2014 Pearson Education, Inc. Publishing as Addison-Wesley.

Exercise Set 1.7

1. $y=(2x+1)^2$

Using the Extended Power Rule:

$$\frac{dy}{dx}=\frac{d}{dx}\left[(2x+1)^2\right]$$
$$=2(2x+1)^{2-1}\cdot\frac{d}{dx}(2x+1)$$
$$=2(2x+1)(2)$$
$$=8x+4.$$

Simplifying the function first, we have:

$$y=(2x+1)^2$$
$$=(2x+1)(2x+1)$$
$$=4x^2+4x+1.$$

Now we take the derivative using the Power Rule.

$$\frac{dy}{dx}=\frac{d}{dx}(4x^2+4x+1)$$
$$=\frac{d}{dx}(4x^2)+\frac{d}{dx}(4x)+\frac{d}{dx}(1)$$
$$=8x+4$$

The results are the same.

3. $y=(7-x)^{55}$

Using the Extended Power Rule:

$$\frac{dy}{dx}=\frac{d}{dx}\left[(7-x)^{55}\right]$$
$$=55(7-x)^{55-1}\cdot\frac{d}{dx}(7-x)$$
$$\frac{dy}{dx}=55(7-x)^{54}(-1)$$
$$=-55(7-x)^{54}.$$

5. $y=\sqrt{1+8x}=(1+8x)^{1/2}$

Using the Extended Power Rule

$$\frac{dy}{dx}=\frac{d}{dx}\left[(1+8x)^{1/2}\right]$$
$$=\frac{1}{2}(1+8x)^{-1/2}\frac{d}{dx}(1+8x)$$
$$=\frac{1}{2(1+8x)^{1/2}}\cdot(8)$$
$$=\frac{4}{\sqrt{1+8x}}.$$

7. $y=\sqrt{3x^2-4}=(3x^2-4)^{1/2}$

Using the Extended Power Rule

$$\frac{dy}{dx}=\frac{d}{dx}\left[(3x^2-4)^{1/2}\right]$$
$$=\frac{1}{2}(3x^2-4)^{-1/2}\frac{d}{dx}(3x^2-4)$$
$$=\frac{1}{2(3x^2-4)^{1/2}}\cdot(6x)$$
$$=\frac{3x}{\sqrt{3x^2-4}}.$$

9. $y=(8x^2-6)^{-40}$

Using the Extended Power Rule

$$\frac{dy}{dx}=\frac{d}{dx}\left[(8x^2-6)^{-40}\right]$$
$$=-40(8x^2-6)^{-40-1}\frac{d}{dx}(8x^2-6)$$
$$=-40(8x^2-6)^{-41}\cdot(16x)$$
$$=-640x(8x^2-6)^{-41}$$
$$=\frac{-640x}{(8x^2-6)^{41}}.$$

11. $y=(x-4)^8(2x+3)^6$

Using the Product Rule, we have

$$\frac{dy}{dx}=\frac{d}{dx}\left[(x-4)^8(2x+3)^6\right]$$
$$=(x-4)^8\frac{d}{dx}(2x+3)^6+(2x+3)^6\frac{d}{dx}(x-4)^8.$$

Next, we will apply the Extended Power Rule.

$$\frac{dy}{dx}=(x-4)^8\left[6(2x+3)^{6-1}\frac{d}{dx}(2x+3)\right]+(2x+3)^6\left[8(x-4)^{8-1}\frac{d}{dx}(x-4)\right]$$
$$=(x-4)^8\left[6(2x+3)^5(2)\right]+(2x+3)^6\left[8(x-4)^7(1)\right]$$
$$=12(x-4)^8(2x+3)^5+8(2x+3)^6(x-4)^7.$$

The solution is continued on the next page.

Copyright © 2014 Pearson Education, Inc. Publishing as Addison-Wesley.

Factoring out common factors from the derivative on the previous page, we have:

$$\frac{dy}{dx} = 4(x-4)^7(2x+3)^5\left[3(x-4)+2(2x+3)\right]$$
$$= 4(x-4)^7(2x+3)^5\left[3x-12+4x+6\right]$$
$$= 4(x-4)^7(2x+3)^5(7x-6).$$

13. $y = \dfrac{1}{(3x+8)^2} = (3x+8)^{-2}$

Using the Extended Power Rule

$$\frac{dy}{dx} = \frac{d}{dx}\left[(3x+8)^{-2}\right]$$
$$= -2(3x+8)^{-2-1}\frac{d}{dx}(3x+8)$$
$$= -2(3x+8)^{-3}\cdot(3)$$
$$= -6(3x+8)^{-3}$$
$$= \frac{-6}{(3x+8)^3}.$$

15. $y = \dfrac{4x^2}{(7-5x)^3}$

First, we use the Quotient Rule.

$$\frac{dy}{dx} = \frac{d}{dx}\left[\frac{4x^2}{(7-5x)^3}\right]$$
$$= \frac{(7-5x)^3\frac{d}{dx}(4x^2) - 4x^2\frac{d}{dx}(7-5x)^3}{\left((7-5x)^3\right)^2}.$$

Next, using the Extended Power Rule, we have:

$$\frac{dy}{dx} = \frac{(7-5x)^3(8x) - 4x^2\left[3(7-5x)^2(-5)\right]}{(7-5x)^6}$$
$$= \frac{8x(7-5x)^3 + 60x^2(7-5x)^2}{\left((7-5x)^3\right)^2}$$
$$= \frac{(7-5x)^2\left[8x(7-5x)+60x^2\right]}{(7-5x)^6} \quad \text{Factoring}$$
$$= \frac{56x - 40x^2 + 60x^2}{(7-5x)^4} \quad \text{Dividing common factors}$$
$$= \frac{20x^2+56x}{(7-5x)^4}$$
$$= \frac{4x(5x+14)}{(7-5x)^4}.$$

17. $f(x) = (1+x^3)^3 - (2+x^8)^4$

Using the Difference Rule and then the Extended Power Rule we have:

$$f'(x) = \frac{d}{dx}\left[(1+x^3)^3 - (2+x^8)^4\right]$$
$$= \frac{d}{dx}(1+x^3)^3 - \frac{d}{dx}(2+x^8)^4$$
$$= 3(1+x^3)^{3-1}\left(\frac{d}{dx}(1+x^3)\right) - 4(2+x^8)^{4-1}\left(\frac{d}{dx}(2+x^8)\right)$$
$$= 3(1+x^3)^2(3x^2) - 4(2+x^8)^3(8x^7)$$
$$= 9x^2(1+x^3)^2 - 32x^7(2+x^8)^3.$$

19. $f(x) = x^2 + (200-x)^2$

Using the Sum Rule and the Extended Power Rule, we have:

$$f'(x) = \frac{d}{dx}\left[x^2 + (200-x)^2\right]$$
$$= \frac{d}{dx}(x^2) + \frac{d}{dx}(200-x)^2$$
$$= 2x + 2(200-x)^{2-1}\left[\frac{d}{dx}(200-x)\right]$$
$$= 2x + 2(200-x)(-1)$$
$$= 2x + 2x - 400$$
$$= 4x - 400.$$

21. $g(x) = \sqrt{x} + (x-3)^3 = x^{1/2} + (x-3)^3$

Using the Sum Rule and the Extended Power Rule, we have:

$$g'(x) = \frac{d}{dx}\left[x^{1/2} + (x-3)^3\right]$$
$$= \frac{d}{dx}\left(x^{1/2}\right) + \frac{d}{dx}(x-3)^3$$
$$= \frac{1}{2}x^{1/2-1} + 3(x-3)^{3-1}\left[\frac{d}{dx}(x-3)\right]$$
$$= \frac{1}{2}x^{-1/2} + 3(x-3)^2(1)$$
$$= \frac{1}{2x^{1/2}} + 3(x-3)^2$$
$$= \frac{1}{2\sqrt{x}} + 3(x-3)^2.$$

Copyright © 2014 Pearson Education, Inc. Publishing as Addison-Wesley.

23. $f(x) = -5x(2x-3)^4$

Using the Product Rule, we have

$$f'(x) = \frac{d}{dx}\left[-5x(2x-3)^4\right]$$
$$= -5x\frac{d}{dx}\left[(2x-3)^4\right] + (2x-3)^4\frac{d}{dx}(-5x).$$

Using the Extended Power Rule, we have

$$f'(x) = -5x\left[4(2x-3)^3\left(\frac{d}{dx}(2x-3)\right)\right] + (2x-3)^4(-5)$$
$$= -5x\left[4(2x-3)^3(2)\right] + (2x-3)^4(-5)$$
$$= -40x(2x-3)^3 - 5(2x-3)^4$$
$$= -5(2x-3)^3\left[8x + (2x-3)\right] \quad \text{Factoring}$$
$$= -5(2x-3)^3(10x-3).$$

25. $g(x) = (3x-1)^7(2x+1)^5$

Using the Product Rule and the Extended Power Rule, we have

$$g'(x) = \frac{d}{dx}\left[(3x-1)^7(2x+1)^5\right]$$
$$= (3x-1)^7\frac{d}{dx}(2x+1)^5 + (2x+1)^5\frac{d}{dx}(3x-1)^7$$
$$= (3x-1)^7\left[5(2x+1)^4(2)\right] + (2x+1)^5\left[7(3x-1)^6(3)\right]$$
$$= 10(3x-1)^7(2x+1)^4 + 21(2x+1)^5(3x-1)^6$$
$$= (3x-1)^6(2x+1)^4\left[10(3x-1) + 21(2x+1)\right]$$
$$= (3x-1)^6(2x+1)^4\left[30x - 10 + 42x + 21\right]$$
$$= (3x-1)^6(2x+1)^4(72x+11).$$

27. $f(x) = x^2\sqrt{4x-1} = x^2(4x-1)^{1/2}$

Using the Product Rule and the Extended Power Rule, we have

$$f'(x) = \frac{d}{dx}\left[x^2(4x-1)^{1/2}\right]$$
$$= x^2\left[\frac{1}{2}(4x-1)^{-1/2}(4)\right] + (4x-1)^{1/2}(2x)$$
$$f'(x) = \frac{2x^2}{(4x-1)^{1/2}} + 2x(4x-1)^{1/2}$$
$$= \frac{2x^2}{\sqrt{(4x-1)}} + 2x\sqrt{(4x-1)}.$$

The derivative can be simplified as follows:

$$f'(x) = \frac{2x^2}{\sqrt{4x-1}} + \frac{2x\sqrt{4x-1}}{1}\cdot\frac{\sqrt{4x-1}}{\sqrt{4x-1}}$$
$$= \frac{2x^2}{\sqrt{4x-1}} + \frac{2x(4x-1)}{\sqrt{4x-1}}$$
$$= \frac{2x^2}{\sqrt{4x-1}} + \frac{8x^2-2x}{\sqrt{4x-1}}$$
$$= \frac{10x^2-2x}{\sqrt{4x-1}}$$
$$= \frac{2x(5x-1)}{\sqrt{4x-1}}.$$

29. $G(x) = \sqrt[3]{x^5+6x} = (x^5+6x)^{1/3}$

Using the Extended Power Rule, we have

$$G'(x) = \frac{d}{dx}\left[(x^5+6x)^{1/3}\right]$$
$$= \frac{1}{3}(x^5+6x)^{1/3-1}\frac{d}{dx}(x^5+6x)$$
$$= \frac{1}{3}(x^5+6x)^{-2/3}(5x^4+6)$$
$$= \frac{5x^4+6}{3(x^5+6x)^{2/3}}$$
$$= \frac{5x^4+6}{3\cdot\sqrt[3]{(x^5+6x)^2}}.$$

31. $f(x) = \left(\frac{3x-1}{5x+2}\right)^4$

Using the Extended Power Rule, we have

$$f'(x) = 4\left(\frac{3x-1}{5x+2}\right)^3\frac{d}{dx}\left[\frac{3x-1}{5x+2}\right]$$

Using the Quotient Rule, we have

$$f'(x) = 4\left(\frac{3x-1}{5x+2}\right)^3\left[\frac{(5x+2)(3)-(3x-1)(5)}{(5x+2)^2}\right]$$
$$= 4\left(\frac{3x-1}{5x+2}\right)^3\left[\frac{15x+6-15x+5}{(5x+2)^2}\right]$$
$$= 4\left(\frac{3x-1}{5x+2}\right)^3\left[\frac{11}{(5x+2)^2}\right]$$
$$= \frac{44(3x-1)^3}{(5x+2)^5}.$$

Copyright © 2014 Pearson Education, Inc. Publishing as Addison-Wesley.

33. $g(x)=\sqrt{\dfrac{4-x}{3+x}}=\left(\dfrac{4-x}{3+x}\right)^{1/2}$

Using the Extended Power Rule, we have

$$g'(x)=\frac{1}{2}\left(\frac{4-x}{3+x}\right)^{1/2-1}\frac{d}{dx}\left[\frac{4-x}{3+x}\right].$$

Using the Quotient Rule, we have

$$g'(x)=\frac{1}{2}\left(\frac{4-x}{3+x}\right)^{-1/2}\left[\frac{(3+x)(-1)-(4-x)(1)}{(3+x)^2}\right]$$
$$=\frac{1}{2}\left(\frac{4-x}{3+x}\right)^{-1/2}\left[\frac{-3-x-4+x}{(3+x)^2}\right]$$
$$=\frac{1}{2}\left(\frac{3+x}{4-x}\right)^{1/2}\left[\frac{-7}{(3+x)^2}\right]$$
$$=\frac{-7}{2(3+x)^{3/2}(4-x)^{1/2}}$$
$$=\frac{-7}{2\sqrt{(3+x)^3}\cdot\sqrt{4-x}}.$$

35. $f(x)=\left(2x^3-3x^2+4x+1\right)^{100}$

Using the Extended Power Rule, we have

$$f'(x)=100\left(2x^3-3x^2+4x+1\right)^{99}\left(6x^2-6x+4\right)$$
$$=200\left(2x^3-3x^2+4x+1\right)^{99}\left(3x^2-3x+2\right).$$

37. $g(x)=\left(\dfrac{2x+3}{5x-1}\right)^{-4}=\left(\dfrac{5x-1}{2x+3}\right)^{4}$

Using the Extended Power Rule, we have

$$g'(x)=\frac{d}{dx}\left[\left(\frac{5x-1}{2x+3}\right)^4\right]$$
$$=4\left(\frac{5x-1}{2x+3}\right)^{4-1}\left[\frac{d}{dx}\left(\frac{5x-1}{2x+3}\right)\right]$$

Next, using the Quotient Rule, we have

$$g'(x)=4\left(\frac{5x-1}{2x+3}\right)^3\left[\frac{(2x+3)(5)-(5x-1)(2)}{(2x+3)^2}\right]$$
$$=4\left(\frac{5x-1}{2x+3}\right)^3\left[\frac{10x+15-10x+2}{(2x+3)^2}\right]$$
$$=4\left(\frac{5x-1}{2x+3}\right)^3\left[\frac{17}{(2x+3)^2}\right]$$
$$=\frac{68(5x-1)^3}{(2x+3)^5}.$$

39. $f(x)=\sqrt{\dfrac{x^2+x}{x^2-x}}=\left(\dfrac{x^2+x}{x^2-x}\right)^{1/2}$

Using the Extended Power Rule, we have

$$f'(x)=\frac{1}{2}\left(\frac{x^2+x}{x^2-x}\right)^{1/2-1}\frac{d}{dx}\left[\frac{x^2+x}{x^2-x}\right].$$

Using the Quotient Rule, we have

$$f'(x)$$
$$=\frac{1}{2}\left(\frac{x^2+x}{x^2-x}\right)^{-1/2}\left[\frac{\left(x^2-x\right)(2x+1)-\left(x^2+x\right)(2x-1)}{\left(x^2-x\right)^2}\right]$$
$$=\frac{1}{2}\left(\frac{x^2+x}{x^2-x}\right)^{-1/2}\left[\frac{2x^3-x^2-x-2x^3-x^2+x}{\left(x^2-x\right)^2}\right]$$
$$=\frac{1}{2}\left(\frac{x^2-x}{x^2+x}\right)^{1/2}\left(\frac{-2x^2}{\left(x^2-x\right)^2}\right)$$
$$=\frac{-x^2}{\left(x^2-x\right)^{3/2}\left(x^2+x\right)^{1/2}}.$$

The derivative can be simplified as follows:

$$f'(x)=\frac{-x^2}{\left(x^2-x\right)^{3/2}\left(x^2+x\right)^{1/2}}$$
$$=\frac{-x^2}{x^{3/2}(x-1)^{3/2}x^{1/2}(x+1)^{1/2}}\quad\text{Factoring}$$
$$=\frac{-x^2}{x^2(x-1)^{3/2}(x+1)^{1/2}}$$
$$=\frac{-1}{(x-1)^{3/2}(x+1)^{1/2}}.$$

Copyright © 2014 Pearson Education, Inc. Publishing as Addison-Wesley.

41. $f(x)=\dfrac{(2x+3)^4}{(3x-2)^5}$

Using the Quotient Rule and the Extended Power Rule, we have:

$$f'(x)=\frac{d}{dx}\left[\frac{(2x+3)^4}{(3x-2)^5}\right]$$

$$=\frac{(3x-2)^5\left[4(2x+3)^3(2)\right]-(2x+3)^4\left[5(3x-2)^4(3)\right]}{\left((3x-2)^5\right)^2}$$

$$=\frac{8(2x+3)^3(3x-2)^5-15(2x+3)^4(3x-2)^4}{(3x-2)^{10}}$$

$$=\frac{(2x+3)^3(3x-2)^4\left[8(3x-2)-15(2x+3)\right]}{(3x-2)^{10}}$$

$$=\frac{(2x+3)^3\left[24x-16-30x-45\right]}{(3x-2)^6}$$

$$=\frac{(2x+3)^3\left[-6x-61\right]}{(3x-2)^6}$$

$$=\frac{-(2x+3)^3\left[6x+61\right]}{(3x-2)^6}$$

$$f'(x)=\frac{-(2x+3)^3\left[6x+61\right]}{(3x-2)^6}.$$

43. $f(x)=12(2x+1)^{2/3}(3x-4)^{5/4}$

Using the Product Rule and the Extended Power Rule, we have:

$$f'(x)=12\frac{d}{dx}\left[(2x+1)^{2/3}(3x-4)^{5/4}\right]$$

$$=12\left[(2x+1)^{2/3}\left[\frac{5}{4}(3x-4)^{1/4}(3)\right]\right]+12\left[(3x-4)^{5/4}\left[\frac{2}{3}(2x+1)^{-1/3}(2)\right]\right]$$

$$=12\left[\frac{15}{4}(2x+1)^{2/3}(3x-4)^{1/4}\right]+12\left[\frac{4}{3}(3x-4)^{5/4}(2x+1)^{-1/3}\right]$$

$$=45(2x+1)^{2/3}(3x-4)^{1/4}+\frac{16(3x-4)^{5/4}}{(2x+1)^{1/3}}.$$

We simplify the derivative at the top of the column.

$$f'(x)$$

$$=\frac{45(2x+1)^{2/3}(3x-4)^{1/4}}{1}\cdot\frac{(2x+1)^{1/3}}{(2x+1)^{1/3}}+\frac{16(3x-4)^{5/4}}{(2x+1)^{1/3}}$$

$$=\frac{45(2x+1)(3x-4)^{1/4}}{(2x+1)^{1/3}}+\frac{16(3x-4)^{5/4}}{(2x+1)^{1/3}}$$

$$=\frac{45(2x+1)(3x-4)^{1/4}+16(3x-4)^{5/4}}{(2x+1)^{1/3}}$$

$$=\frac{(3x-4)^{1/4}\left[45(2x+1)+16(3x-4)^{4/4}\right]}{(2x+1)^{1/3}}$$

$$=\frac{(3x-4)^{1/4}\left[45(2x+1)+16(3x-4)\right]}{(2x+1)^{1/3}}$$

$$=\frac{(3x-4)^{1/4}\left[90x+45+48x-64\right]}{(2x+1)^{1/3}}$$

$$=\frac{(3x-4)^{1/4}\left[138x-19\right]}{(2x+1)^{1/3}}$$

$$f'(x)=\frac{(3x-4)^{1/4}(138x-19)}{(2x+1)^{1/3}}.$$

45. $y=\sqrt{u}=u^{1/2}$ and $u=x^2-1$

$$\frac{dy}{du}=\frac{1}{2}u^{1/2-1}=\frac{1}{2}u^{-1/2}=\frac{1}{2\sqrt{u}}$$

$$\frac{du}{dx}=2x^{2-1}=2x$$

Applying the Chain Rule, we have:

$$\frac{dy}{dx}=\frac{dy}{du}\cdot\frac{du}{dx}$$

$$=\frac{1}{2\sqrt{u}}\cdot 2x$$

$$=\frac{2x}{2\sqrt{x^2-1}} \quad \text{Subtituting } x^2-1 \text{ for } u.$$

$$=\frac{x}{\sqrt{x^2-1}}. \quad \text{Simplifying}$$

Copyright © 2014 Pearson Education, Inc. Publishing as Addison-Wesley.

47. $y = u^{50}$ and $u = 4x^3 - 2x^2$

$$\frac{dy}{du} = 50u^{50-1} = 50u^{49}$$

$$\frac{du}{dx} = 4\left(3x^{3-1}\right) - 2\left(2x^{2-1}\right) = 12x^2 - 4x$$

$$\frac{dy}{dx} = \frac{dy}{du} \cdot \frac{du}{dx}$$

$$= 50u^{49} \cdot \left(12x^2 - 4x\right)$$

Subtituting $4x^3 - 2x^2$ for u.

$$= 50\left(4x^3 - 2x^2\right)^{49} \cdot \left(12x^2 - 4x\right)$$

$$= 200x(3x-1)\left(4x^3 - 2x^2\right)^{49}.$$ Simplifying

49. $y = u(u+1)$ and $u = x^3 - 2x$

$$\frac{dy}{du} = u(1) + (u+1)(1)$$ Product Rule

$$= 2u + 1$$

$$\frac{du}{dx} = 3x^2 - 2$$

$$\frac{dy}{dx} = \frac{dy}{du} \cdot \frac{du}{dx}$$

$$= (2u+1) \cdot \left(3x^2 - 2\right)$$

Subtituting $x^3 - 2x$ for u.

$$= \left(2\left(x^3 - 2x\right) + 1\right) \cdot \left(3x^2 - 2\right)$$

$$= \left(2x^3 - 4x + 1\right) \cdot \left(3x^2 - 2\right).$$ Simplifying

51. $y = 5u^2 + 3u$ and $u = x^3 + 1$

$$\frac{dy}{du} = 10u + 3$$

$$\frac{du}{dx} = 3x^2$$

Applying the Chain Rule, we have:

$$\frac{dy}{dx} = \frac{dy}{du} \cdot \frac{du}{dx}$$

$$= (10u + 3) \cdot \left(3x^2\right)$$

$$= \left(10\left(x^3 + 1\right) + 3\right) \cdot \left(3x^2\right)$$ Substituting for u.

$$= 3x^2\left(\left(10x^3 + 10\right) + 3\right)$$

$$= 3x^2\left(10x^3 + 13\right).$$

53. $y = \sqrt[3]{2u+5} = (2u+5)^{1/3}$ and $u = x^2 - x$

$$\frac{dy}{du} = \frac{1}{3}(2u+5)^{-2/3}(2)$$ Extended Power Rule

$$= \frac{2}{3(2u+5)^{2/3}}$$

$$\frac{du}{dx} = 2x - 1.$$

We apply the Chain Rule.

$$\frac{dy}{dx} = \frac{dy}{du} \cdot \frac{du}{dx}$$

$$= \left(\frac{2}{3(2u+5)^{2/3}}\right) \cdot (2x-1)$$

$$= \left(\frac{2}{3\left(2\left(x^2 - x\right) + 5\right)^{2/3}}\right) \cdot (2x-1)$$ Substituting

$$= \frac{2(2x-1)}{3\left(2x^2 - 2x + 5\right)^{2/3}}.$$

55. $y = \dfrac{1}{u^2 + u}$ and $u = 5 + 3t$

$$\frac{dy}{du} = \frac{\left(u^2 + u\right)(0) - (1)(2u+1)}{\left(u^2 + u\right)^2}$$ Quotient Rule

$$= \frac{-(2u+1)}{\left(u^2 + u\right)^2}$$

$$\frac{du}{dt} = 3$$

The solution is continued on the next page.

Copyright © 2014 Pearson Education, Inc. Publishing as Addison-Wesley.

Using the information from the previous page, we apply the Chain Rule.

$$\frac{dy}{dt}=\frac{dy}{du}\cdot\frac{du}{dt}$$

$$=\left(\frac{-(2u+1)}{(u^2+u)^2}\right)\cdot(3)$$

$$=\left(\frac{-(2(5+3t)+1)}{\left((5+3t)^2+(5+3t)\right)^2}\right)\cdot(3) \quad \text{Substituting}$$

$$=\frac{-3(10+6t+1)}{\left((5+3t)^2+(5+3t)\right)^2}$$

$$=\frac{-3(6t+11)}{(5+3t)^2\left((5+3t)+1\right)^2} \quad \text{Factoring}$$

$$=\frac{-3(6t+11)}{(5+3t)^2(6+3t)^2}$$

57. $y=\sqrt{x^2+3x}=(x^2+3x)^{1/2}$

First, we find the derivative using the Extended Power Rule.

$$\frac{dy}{dx}=\frac{1}{2}(x^2+3x)^{1/2-1}(2x+3)$$

$$=\frac{2x+3}{2\sqrt{x^2+3x}}$$

When $x=1$,

$$\frac{dy}{dx}=\frac{2(1)+3}{2\sqrt{(1)^2+3(1)}}=\frac{5}{2\sqrt{4}}=\frac{5}{2\cdot 2}=\frac{5}{4}.$$

Thus, the slope of the tangent line at $(1,2)$ is $\frac{5}{4}$.

Using the point-slope equation, we find the equation of the tangent line.

$$y-y_1=m(x-x_1)$$

$$y-2=\frac{5}{4}(x-1)$$

$$y-2=\frac{5}{4}x-\frac{5}{4}$$

$$y=\frac{5}{4}x+\frac{3}{4}$$

59. $y=x\sqrt{2x+3}=x(2x+3)^{1/2}$

First, we find the derivative using the Product Rule and the Extended Power Rule.

$$\frac{dy}{dx}=x\left(\frac{1}{2}(2x+3)^{1/2-1}(2)\right)+(2x+3)^{1/2}(1)$$

$$=\frac{x}{\sqrt{2x+3}}+\sqrt{2x+3}$$

When $x=3$,

$$\frac{dy}{dx}=\frac{(3)}{\sqrt{2(3)+3}}+\sqrt{2(3)+3}$$

$$=\frac{3}{\sqrt{9}}+\sqrt{9}$$

$$=\frac{3}{3}+3$$

$$=1+3$$

$$=4.$$

Thus, the slope of the tangent line at $(3,9)$ is 4.

Using the point-slope equation, we find the equation of the tangent line.

$$y-y_1=m(x-x_1)$$

$$y-9=4(x-3)$$

$$y-9=4x-12$$

$$y=4x-3$$

61. $f(x)=\frac{x^2}{(1+x)^5}$

a) Using the Quotient Rule and the Extended Power Rule, we have:

$$f'(x)=\frac{(1+x)^5(2x)-x^2\left[5(1+x)^4(1)\right]}{\left((1+x)^5\right)^2}$$

$$=\frac{2x(1+x)^5-5x^2(1+x)^4}{(1+x)^{10}}$$

$$=\frac{(1+x)^4\left(2x(1+x)-5x^2\right)}{(1+x)^{10}} \quad \text{Factoring}$$

$$=\frac{2x+2x^2-5x^2}{(1+x)^6}$$

$$=\frac{2x-3x^2}{(1+x)^6}$$

$$=\frac{x(2-3x)}{(1+x)^6}.$$

Copyright © 2014 Pearson Education, Inc. Publishing as Addison-Wesley.

b) Using the Product Rule and the Extended Power Rule on $f(x)=x^2(1+x)^{-5}$, we have

$$f'(x)=x^2\left(-5(1+x)^{-6}(1)\right)+(1+x)^{-5}(2x)$$
$$=\frac{-5x^2}{(1+x)^6}+\frac{2x}{(1+x)^5}$$
$$=\frac{-5x^2}{(1+x)^6}+\frac{2x}{(1+x)^5}\cdot\frac{(1+x)}{(1+x)}$$
$$=\frac{-5x^2+2x+2x^2}{(1+x)^6}$$
$$=\frac{2x-3x^2}{(1+x)^6}$$
$$=\frac{x(2-3x)}{(1+x)^6}.$$

c) The results are the same.

63. $h(x)=(3x^2-7)^5$

Let $f(x)=x^5$ and $g(x)=3x^2-7$.

$$(f\circ g)(x)=f(g(x))=f(3x^2-7)=(3x^2-7)^5$$

Thus, $h(x)=(f\circ g)(x)$.

Answers may vary.

65. $h(x)=\frac{x^3+1}{x^3-1}$

Let $f(x)=\frac{x+1}{x-1}$ and $g(x)=x^3$.

$$(f\circ g)(x)=f(g(x))=f(x^3)=\frac{x^3+1}{x^3-1}$$

Thus, $h(x)=(f\circ g)(x)$.

Answers may vary.

67. Using the Chain Rule:

$f(u)=u^3,\ g(x)=u=2x^4+1$

First find $f'(u)$ and $g'(x)$.

$$f'(u)=3u^2$$
$$f'(g(x))=3(2x^4+1)^2 \quad \text{Substituting } g(x) \text{ for } u.$$
$$g'(x)=8x^3$$

The Chain Rule states

$$(f\circ g)'(x)=f'(g(x))\cdot g'(x)$$

Substituting, we have:

$$(f\circ g)'(x)=3(2x^4+1)^2\cdot(8x^3)$$
$$=24x^3(2x^4+1)^2.$$

Therefore,

$$(f\circ g)'(-1)=24(-1)^3\left(2(-1)^4+1\right)^2$$
$$=-24(2+1)^2$$
$$=-24\cdot 9$$
$$=-216.$$

Finding $f(g(x))$ first, we have:

$$f(g(x))=f(2x^4+1)=(2x^4+1)^3.$$

By the Extended Power Rule:

$$f'(g(x))=3(2x^4+1)^2(8x^3)$$
$$=24x^3(2x^4+1)^2.$$

Therefore, $f'(g(-1))=-216$ as above.

69. Using the Chain Rule:

$f(u)=\sqrt[3]{u}=u^{1/3},\ g(x)=u=1+3x^2.$

First find $f'(u)$ and $g'(x)$.

$$f'(u)=\frac{1}{3}u^{-2/3}=\frac{1}{3\cdot\sqrt[3]{u^2}}$$
$$f'(g(x))=\frac{1}{3\cdot\sqrt[3]{(1+3x^2)^2}} \quad \text{Substituting } g(x) \text{ for } u.$$
$$g'(x)=6x$$

The Chain Rule states

$$(f\circ g)'(x)=f'(g(x))\cdot g'(x).$$

Substituting, we have:

$$(f\circ g)'(x)=\frac{1}{3\cdot\sqrt[3]{(1+3x^2)^2}}\cdot(6x)$$
$$=\frac{2x}{\sqrt[3]{(1+3x^2)^2}}.$$

The solution is continued on the next page.

Copyright © 2014 Pearson Education, Inc. Publishing as Addison-Wesley.

Evaluating the derivative from the previous page yields:

$$(f \circ g)'(2) = \frac{2(2)}{\sqrt[3]{\left(1+3(2)^2\right)^2}} = \frac{4}{\sqrt[3]{(13)^2}} \approx 0.72348760.$$

Finding $f\left(g(x)\right)$ first, we have:

$$f\left(g(x)\right) = f\left(1+3x^2\right) = \left(1+3x^2\right)^{1/3}.$$

By the Extended Power Rule:

$$f'\left(g(x)\right) = \frac{1}{3}\left(1+3x^2\right)^{-2/3}(6x) = \frac{2x}{\left(1+3x^2\right)^{2/3}}.$$

Therefore $(f \circ g)'(2) = \dfrac{4}{\sqrt[3]{(13)^2}} \approx 0.72348760$.

71. $f(x) = \left(2x^3 + (4x-5)^2\right)^6$

Letting $u = 2x^3 + (4x-5)^2$ and applying the Chain Rule, we have:

$$f'(x) = 6\left(2x^3+(4x-5)^2\right)^{6-1} \cdot \frac{d}{dx}\left(2x^3+(4x-5)^2\right).$$

We will have to apply the chain rule again to find $\dfrac{d}{dx}\left(2x^3+(4x-5)^2\right)$.

Applying the chain rule again, we have:

$$\frac{d}{dx}\left(2x^3+(4x-5)^2\right) = 2\frac{d}{dx}x^3 + \frac{d}{dx}(4x-5)^2 = 2\left(3x^2\right) + 2(4x-5)^{2-1} \cdot \frac{d}{dx}(4x-5) = 6x^2 + 2(4x-5) \cdot 4 = 6x^2 + 32x - 40.$$

Therefore, the derivative is:

$$f'(x) = 6\left(2x^3+(4x-5)^2\right)^5\left(6x^2+32x-40\right).$$

73. $f(x) = \sqrt{x^2+\sqrt{1-3x}} = \left(x^2+(1-3x)^{\frac{1}{2}}\right)^{\frac{1}{2}}$

Applying the Chain Rule, we have

$$f'(x) = \frac{1}{2}\left(x^2+(1-3x)^{\frac{1}{2}}\right)^{\frac{1}{2}-1} \cdot \frac{d}{dx}\left(x^2+(1-3x)^{\frac{1}{2}}\right) = \frac{1}{2}\left(x^2+(1-3x)^{\frac{1}{2}}\right)^{-\frac{1}{2}} \cdot \frac{d}{dx}\left(x^2+(1-3x)^{\frac{1}{2}}\right).$$

Applying the chain rule again, we have

$$\frac{d}{dx}\left(x^2+(1-3x)^{\frac{1}{2}}\right) = \frac{d}{dx}x^2 + \frac{d}{dx}(1-3x)^{\frac{1}{2}} = 2x^{2-1} + \frac{1}{2}(1-3x)^{\frac{1}{2}-1} \cdot \frac{d}{dx}(1-3x) = 2x + \frac{1}{2\sqrt{1-3x}} \cdot (-3) = 2x - \frac{3}{2\sqrt{1-3x}}.$$

Therefore, the derivative is:

$$f'(x) = \frac{1}{2}\left(x^2+(1-3x)^{\frac{1}{2}}\right)^{-\frac{1}{2}}\left(2x - \frac{3}{2\sqrt{1-3x}}\right) = \frac{1}{2\sqrt{x^2+\sqrt{1-3x}}}\left(2x - \frac{3}{2\sqrt{1-3x}}\right).$$

75. $R(x) = 1000\sqrt{x^2-0.1x} = 1000\left(x^2-0.1x\right)^{1/2}$

Using the Extended Power Rule, we have

$$R'(x) = 1000\left[\frac{1}{2}\left(x^2-0.1x\right)^{-1/2}(2x-0.1)\right] = 500\left[\frac{2x-0.1}{\left(x^2-0.1x\right)^{1/2}}\right] = \frac{500(2x-0.1)}{\sqrt{x^2-0.1x}}.$$

Substituting 20 for x, we have

$$R'(20) = \frac{500(2(20)-0.1)}{\sqrt{(20)^2-0.1(20)}} = 1000.00314070 \approx 1000.$$

When 20 items have been sold, revenue is changing at a rate of 1,000 thousand of dollars per item, or 1,000,000 dollars per item.

Copyright © 2014 Pearson Education, Inc. Publishing as Addison-Wesley.

77. $P(x)=R(x)-C(x)$ and
$P'(x)=R'(x)-C'(x)$
Since we are trying to find the rate at which total profit is changing as a function of x, we can use the derivatives found in Exercise 71 and 72 to find the derivative of the profit function. There is no need to find the Profit function first and then take the derivative.

$$P'(x)=R'(x)-C'(x)$$
$$=\frac{500(2x-0.1)}{\sqrt{x^2-0.1x}}-\frac{4000x}{3(x^2+2)^{2/3}}$$

79. Let x be the number of years since 1995.

$$C(x)=0.21x^4-5.92x^3+50.53x^2-18.92x+1114.93$$

a) $$\frac{dC}{dx}=0.21(4x^3)-5.92(3x^2)+50.53(2x)-18.92$$
$$=0.84x^3-17.76x^2+101.06x-18.92$$

b) ✎

c) Substitute 15 for x. $[2010-1995=15]$

$$\frac{dC}{dx}=0.84(15)^3-17.76(15)^2+101.06(15)-18.92$$
$$=2835-3996+1515.9-18.92$$
$$=335.98$$

Outstanding consumer credit will be rising approximately at a rate of 336 billion of dollars per year in 2010.

81. $A=1000(1+i)^3$

a) Using the Extended Power Rule, we have

$$\frac{dA}{di}=1000(3)(1+i)^2(1)$$
$$=3000(1+i)^2.$$

b) ✎

83. $D(p)=\frac{80,000}{p}$ and $p=1.6t+9$

a) Substitute $1.6t+9$ in for p in the demand function.

$$D(t)=\frac{80,000}{1.6t+9}$$

b) Using the Quotient Rule, we have

$$D'(t)=\frac{(1.6t+9)(0)-(80,000)(1.6)}{(1.6t+9)^2}$$
$$=\frac{-128,000}{(1.6t+9)^2}.$$

Substituting 100 for t into the derivative, we have:

$$D'(100)=\frac{-128,000}{(1.6(100)+9)^2}$$
$$=\frac{-128,000}{28,561}$$
$$\approx -4.48163580.$$

After 100 days, quantity demanded is changing -4.482 units per day.

85. $D=0.85A(c+25)$ and $c=(140-y)\frac{w}{72x}$

a) Substituting 5 for A we have:

$$D(c)=0.85(5)(c+25)$$
$$=4.25(c+25)$$
$$=4.25c+106.25.$$

Substituting 0.6 for x, 45 for y, we have:

$$c(w)=(140-45)\frac{w}{72(0.6)}$$
$$=95\frac{w}{43.2}$$
$$=\frac{95w}{43.2}\approx 2.199w.$$

b) $$\frac{dD}{dc}=\frac{d}{dc}(4.25c+106.25)=4.25$$

The dosage changes at a rate of 4.25 mg per unit of creatine clearance.

c) $$\frac{dc}{dw}=\frac{d}{dw}\left(\frac{95w}{43.2}\right)=\frac{95}{43.2}\approx 2.199$$

The creatine clearance changes at a rate of 2.199 unit of creatine clearance per kilogram.

d) By the Chain Rule:

$$\frac{dD}{dw}=\frac{dD}{dc}\cdot\frac{dc}{dw}=(4.25)\left(\frac{95}{43.2}\right)\approx 9.346$$

The dosage changes at a rate of 9.35 milligrams per kilogram.

e) ✎

Copyright © 2014 Pearson Education, Inc. Publishing as Addison-Wesley.

87. $f(x) = x + \sqrt{x}$

Note that $f'(x) = 1 + \frac{1}{2\sqrt{x}}$ and

$f'(f(x)) = 1 + \frac{1}{2\sqrt{x+\sqrt{x}}}$.

Applying the Chain Rule to the iterated function, we have

$$\frac{d}{dx}[(f \circ f)(x)]$$
$$= \frac{d}{dx}[f(f(x))]$$
$$= f'(f(x)) \cdot f'(x)$$
$$= \left[1 + \frac{1}{2\sqrt{x+\sqrt{x}}}\right] \cdot \left[1 + \frac{1}{2\sqrt{x}}\right]$$
$$= 1 + \frac{1}{2\sqrt{x}} + \left(\frac{1}{2\sqrt{x+\sqrt{x}}}\right)\left(1 + \frac{1}{2\sqrt{x}}\right).$$

89. $f(x) = \sqrt[3]{x} = x^{\frac{1}{3}}$

Note that $f'(x) = \frac{1}{3}x^{-\frac{2}{3}}$,

$f'(f(x)) = \frac{1}{3}x^{-\frac{2}{9}}$, and

$f'(f(f(x))) = \frac{1}{3}x^{-\frac{2}{27}}$.

Applying the Chain Rule to the iterated function, we have

$$\frac{d}{dx}[(f \circ f \circ f)(x)]$$
$$= \frac{d}{dx}[f(f(f(x)))]$$
$$= f'(f(f(x))) \cdot f'(f(x)) \cdot f'(x)$$
$$= \frac{1}{3}x^{-\frac{2}{27}} \cdot \frac{1}{3}\left(x^{-\frac{2}{9}}\right) \cdot \frac{1}{3}\left(x^{-\frac{2}{3}}\right)$$
$$= \frac{1}{27}x^{-\frac{2}{27}-\frac{2}{9}-\frac{2}{3}}$$
$$= \frac{1}{27}x^{-\frac{26}{27}}.$$

91. $y = \sqrt[3]{x^3 + 6x + 1} \cdot x^5 = \left(x^3 + 6x + 1\right)^{1/3} \cdot x^5$

Using the Product Rule and the Extended Power Rule, we have

$$\frac{dy}{dx} = \left(x^3 + 6x + 1\right)^{1/3} \cdot \left(5x^4\right) + x^5\left[\frac{1}{3}\left(x^3 + 6x + 1\right)^{-2/3}\left(3x^2 + 6\right)\right]$$
$$= 5x^4\left(x^3 + 6x + 1\right)^{1/3} + \frac{3x^5\left(x^2 + 2\right)}{3\left(x^3 + 6x + 1\right)^{2/3}}.$$

The derivative can be further simplified by finding a common denominator and combining the fractions.

$$\frac{dy}{dx} = \frac{\left(x^3 + 6x + 1\right)^{2/3}}{\left(x^3 + 6x + 1\right)^{2/3}} \cdot \frac{5x^4\left(x^3 + 6x + 1\right)^{1/3}}{1} + \frac{x^5\left(x^2 + 2\right)}{\left(x^3 + 6x + 1\right)^{2/3}}$$
$$= \frac{5x^4\left(x^3 + 6x + 1\right)}{\left(x^3 + 6x + 1\right)^{2/3}} + \frac{x^5\left(x^2 + 2\right)}{\left(x^3 + 6x + 1\right)^{2/3}}$$
$$= \frac{5x^7 + 30x^5 + 5x^4 + x^7 + 2x^5}{\left(x^3 + 6x + 1\right)^{2/3}}$$
$$= \frac{6x^7 + 32x^5 + 5x^4}{\left(x^3 + 6x + 1\right)^{2/3}}.$$

Copyright © 2014 Pearson Education, Inc. Publishing as Addison-Wesley.

93. $y=\left(\dfrac{x}{\sqrt{x-1}}\right)^3$

Using the Extended Power Rule and the Quotient Rule, we have:

$$\frac{dy}{dx}=3\left(\frac{x}{(x-1)^{1/2}}\right)^{3-1}\frac{d}{dx}\left(\frac{x}{(x-1)^{1/2}}\right)$$

$$=3\left(\frac{x}{\sqrt{x-1}}\right)^2\cdot\left(\frac{(x-1)^{1/2}(1)-x\left(\frac{1}{2}(x-1)^{-1/2}\cdot 1\right)}{\left(\sqrt{x-1}\right)^2}\right)$$

$$=\frac{3x^2}{x-1}\left(\frac{(x-1)^{1/2}-\dfrac{x}{2(x-1)^{1/2}}}{x-1}\right)$$

$$=\frac{3x^2}{x-1}\left(\frac{\dfrac{2x-2}{2(x-1)^{1/2}}-\dfrac{x}{2(x-1)^{1/2}}}{x-1}\right)$$

$$=\frac{3x^2}{(x-1)}\left(\frac{x-2}{2(x-1)^{3/2}}\right)$$

$$=\frac{3x^2(x-2)}{2(x-1)^{5/2}}.$$

95. $y=\dfrac{\sqrt{1-x^2}}{1-x}=\dfrac{(1-x^2)^{1/2}}{1-x}$

Using the Quotient Rule and the Extended Power Rule, we have:

$$\frac{dy}{dx}=\frac{(1-x)\left(\frac{1}{2}(1-x^2)^{-1/2}(-2x)\right)-(1-x^2)^{1/2}(-1)}{(1-x)^2}$$

$$\frac{dy}{dx}=\frac{\dfrac{x(x-1)}{(1-x^2)^{1/2}}+(1-x^2)^{1/2}}{(1-x)^2}$$

$$=\frac{\dfrac{x^2-x}{(1-x^2)^{1/2}}+\dfrac{(1-x^2)^{1/2}}{1}\cdot\dfrac{(1-x^2)^{1/2}}{(1-x^2)^{1/2}}}{(1-x)^2}$$

$$=\frac{\dfrac{x^2-x}{(1-x^2)^{1/2}}+\dfrac{(1-x^2)}{(1-x^2)^{1/2}}}{(1-x)^2}$$

$$=\frac{1-x}{(1-x^2)^{1/2}(1-x)^2}$$

$$=\frac{1}{(1-x)\sqrt{1-x^2}}.$$

97. $y=\left(\dfrac{x^2-x-1}{x^2+1}\right)^3$

Using the Extended Power Rule and the Quotient Rule, we have

$$\frac{dy}{dx}$$

$$=3\left(\frac{x^2-x-1}{x^2+1}\right)^2\left[\frac{(x^2+1)(2x-1)-(x^2-x-1)(2x)}{(x^2+1)^2}\right]$$

$$=3\left(\frac{x^2-x-1}{x^2+1}\right)^2\left[\frac{x^2+4x-1}{(x^2+1)^2}\right]$$

$$=\frac{3(x^2-x-1)^2(x^2+4x-1)}{(x^2+1)^4}.$$

Copyright © 2014 Pearson Education, Inc. Publishing as Addison-Wesley.

99. $f(t)=\sqrt{3t+\sqrt{t}}=\left(3t+t^{1/2}\right)^{1/2}$

Using the Extended Power Rule, we have

$$f'(t)=\frac{1}{2}\left(3t+t^{1/2}\right)^{-1/2}\left(3+\frac{1}{2}t^{-1/2}\right)$$

$$=\frac{3+\frac{1}{2\sqrt{t}}}{2\sqrt{3t+\sqrt{t}}}$$

$$=\frac{\frac{3}{1}\cdot\frac{2\sqrt{t}}{2\sqrt{t}}+\frac{1}{2\sqrt{t}}}{2\sqrt{3t+\sqrt{t}}}.$$

Simplifying, we have:

$$f'(t)=\frac{\frac{6\sqrt{t}+1}{2\sqrt{t}}}{2\sqrt{3t+\sqrt{t}}}$$

$$=\frac{6\sqrt{t}+1}{4\sqrt{t}\sqrt{3t+\sqrt{t}}}.$$

101. ✎

103. $f(x)=1.68x\sqrt{9.2-x^2};\quad [-3,3]$

Using the calculator, we graph the function and the derivative in the same window. We can use the nDeriv feature to graph the derivative without actually calculating the derivative.

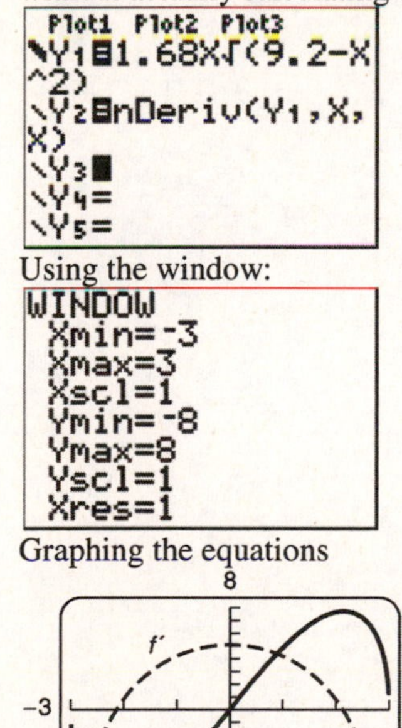

Graphing the equations

8

f′

−3 3

f

−8

Note, the function $f(x)$ is the solid graph. The horizontal tangents occur at the turning points of this function, or at the x-intercepts of the derivative. Using the trace feature, the minimum/maximum feature on the function, or the zero feature on the derivative on the calculator, we find the points of horizontal tangency. We estimate the points at which the tangent lines are horizontal are $(-2.14476,-7.728)$ and $(2.14476,7.728)$.

105. $f(x)=x\sqrt{4-x^2}=x\left(4-x^2\right)^{1/2}$

$$f'(x)=x\left[\frac{1}{2}\left(4-x^2\right)^{-1/2}(-2x)\right]+\left(4-x^2\right)^{1/2}(1)$$

$$=\frac{-x^2}{\left(4-x^2\right)^{1/2}}+\left(4-x^2\right)^{1/2}$$

$$=\frac{-x^2}{\left(4-x^2\right)^{1/2}}+\frac{\left(4-x^2\right)}{\left(4-x^2\right)^{1/2}}$$

$$=\frac{4-2x^2}{\left(4-x^2\right)^{1/2}}$$

$$=\frac{4-2x^2}{\sqrt{4-x^2}}$$

Using the window:

WINDOW
Xmin=-2
Xmax=2
Xscl=1
Ymin=-5
Ymax=5
Yscl=1
Xres=1

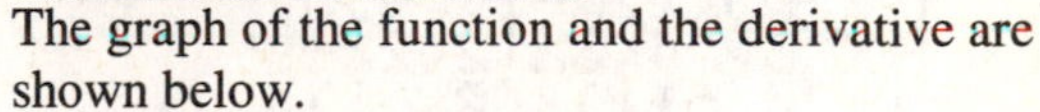

The graph of the function and the derivative are shown below.

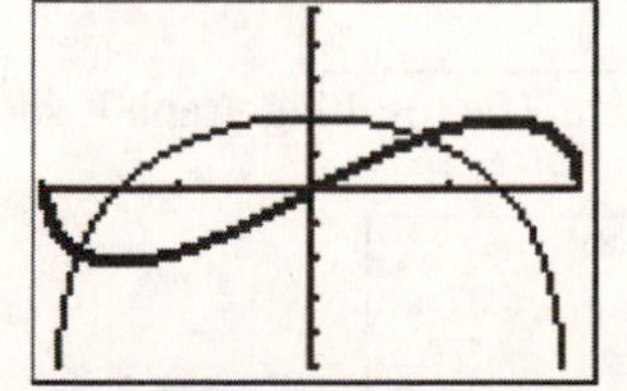

Note: The graph of the function is the thicker graph.

Copyright © 2014 Pearson Education, Inc. Publishing as Addison-Wesley.

107. $f(x)=\left(\sqrt{2x-1}+x^3\right)^5=\left((2x-1)^{\frac{1}{2}}+x^3\right)^5$

$$f'(x)=5\left((2x-1)^{\frac{1}{2}}+x^3\right)^4\left[\frac{1}{2}(2x-1)^{-\frac{1}{2}}(2)+3x^2\right]$$

$$=5\left((2x-1)^{\frac{1}{2}}+x^3\right)^4\left[\frac{1}{(2x-1)^{\frac{1}{2}}}+3x^2\right]$$

Simplifying, we have:

$$f'(x)=5\left((2x-1)^{\frac{1}{2}}+x^3\right)^4\left[\frac{3x^2(2x-1)^{\frac{1}{2}}+1}{(2x-1)^{\frac{1}{2}}}\right]$$

$$=\frac{5\left(\sqrt{2x-1}+x^3\right)^4\left(3x^2\sqrt{2x-1}+1\right)}{\sqrt{2x-1}}.$$

Using the window:

```
WINDOW
 Xmin=-1
 Xmax=3
 Xscl=1
 Ymin=-40
 Ymax=40
 Yscl=10
 Xres=1
```

First we enter the function

$Y_1=\left((2x-1)^{0.5}+x^3\right)^5$ into the graphing editor.

Next, we enter $Y_2=nDeriv\left(y_1,x,x\right)$ into the graphing editor. Finally, we enter

$$Y_3=\frac{5\left((2x-1)^{0.5}+x^3\right)^4\left(3x^2(2x-1)^{0.5}+1\right)}{(2x-1)^{0.5}}.$$

The screen shot is shown below:

We graph Y_2 first. The resulting graph is shown below:

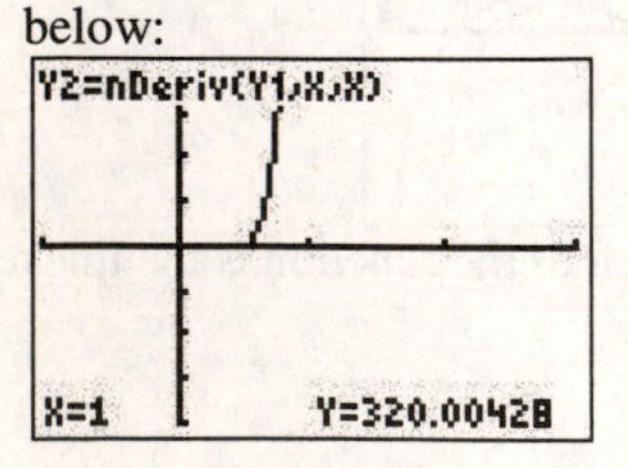

Next we graph Y_3. The resulting graph is shown below:

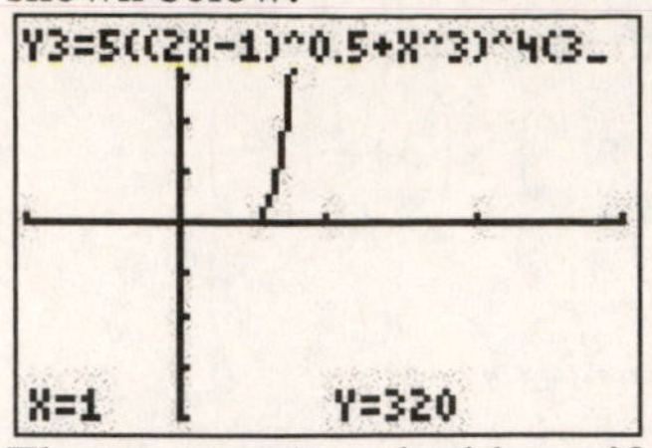

The two graphs coincide verifying the result.

Copyright © 2014 Pearson Education, Inc. Publishing as Addison-Wesley.

Exercise Set 1.8

1. $y = x^5 + 9$

$\frac{dy}{dx} = 5x^{5-1} = 5x^4$ First Derivative

$\frac{d^2y}{dx^2} = 5\left(4x^{4-1}\right) = 20x^3$ Second Derivative

3. $y = 2x^4 - 5x$

$\frac{dy}{dx} = 2\left(4x^3\right) - 5 =$

$= 8x^3 - 5$ First Derivative

$\frac{d^2y}{dx^2} = 8\left(3x^2\right)$

$= 24x^2$ Second Derivative

5. $y = 4x^2 + 3x - 1$

$\frac{dy}{dx} = 4\left(2x^{2-1}\right) + 3 - 0$

$= 8x + 3$ First Derivative

$\frac{d^2y}{dx^2} = 8$ Second Derivative

7. $y = 7x + 2$

$\frac{dy}{dx} = 7$ First Derivative

$\frac{d^2y}{dx^2} = 0$ Second Derivative

9. $y = \frac{1}{x^2} = x^{-2}$

$\frac{dy}{dx} = -2x^{-2-1}$

$= -2x^{-3}$

$= \frac{-2}{x^3}$ First Derivative

$\frac{d^2y}{dx^2} = -2\left(-3x^{-3-1}\right)$

$= 6x^{-4}$

$= \frac{6}{x^4}$ Second Derivative

11. $y = \sqrt{x} = x^{1/2}$

$\frac{dy}{dx} = \frac{1}{2}x^{1/2-1}$

$= \frac{1}{2}x^{-1/2}$

$= \frac{1}{2x^{1/2}} = \frac{1}{2\sqrt{x}}$ First Derivative

$\frac{d^2y}{dx^2} = \frac{1}{2}\cdot\left(-\frac{1}{2}x^{-1/2-1}\right)$

$= -\frac{1}{4}x^{-3/2}$

$= -\frac{1}{4x^{3/2}} = -\frac{1}{4\sqrt{x^3}}$ Second Derivative

13. $f(x) = x^4 + \frac{3}{x}$

$f'(x) = 4x^{4-1} + 3\left(-1x^{-1-1}\right)$

$= 4x^3 - 3x^{-2}$

$= 4x^3 - \frac{3}{x^2}$ First Derivative

$f''(x) = 4\left(3x^{3-1}\right) - 3\left(-2x^{-2-1}\right)$

$= 12x^2 + 6x^{-3}$

$= 12x^2 + \frac{6}{x^3}$ Second Derivative

15. $f(x) = x^{1/5}$

$f'(x) = \frac{1}{5}x^{1/5-1}$

$= \frac{1}{5}x^{-4/5}$

$= \frac{1}{5x^{4/5}}$ First Derivative

$f''(x) = \frac{1}{5}\left(-\frac{4}{5}\right)x^{-4/5-1}$

$= -\frac{4}{25}x^{-9/5}$

$= -\frac{4}{25x^{9/5}}$ Second Derivative

Copyright © 2014 Pearson Education, Inc. Publishing as Addison-Wesley.

17. $f(x)=4x^{-3}$

$f'(x)=4\left(-3x^{-3-1}\right)$

$=-12x^{-4}$

$=-\dfrac{12}{x^4}$ First Derivative

$f''(x)=-12\left(-4x^{-4-1}\right)$

$=48x^{-5}$

$=\dfrac{48}{x^5}$ Second Derivative

19. $f(x)=\left(x^2+3x\right)^7$

$f'(x)=7\left(x^2+3x\right)^{7-1}(2x+3)$ Theorem 7

$=7(2x+3)\left(x^2+3x\right)^6$ First Derivative

$f''(x)=7(2x+3)\left(6\left(x^2+3x\right)^{6-1}(2x+3)\right)+$

$7\left(x^2+6x\right)^6(2)$ Theorem 5

$=42(2x+3)^2\left(x^2+3x\right)^5+14\left(x^2+3x\right)^6$

We can simplify the second derivative by factoring out common factors.

$f''(x)$

$=14\left(x^2+3x\right)^5\left[3(2x+3)^2+\left(x^2+3x\right)\right]$

$=14\left(x^2+3x\right)^5\left[3\left(4x^2+12x+9\right)+\left(x^2+3x\right)\right]$

$=14\left(x^2+3x\right)^5\left[12x^2+36x+27+x^2+3x\right]$

$=14\left(x^2+3x\right)^5\left(13x^2+39x+27\right)$ Second Derivative

21. $f(x)=\left(2x^2-3x+1\right)^{10}$

$f'(x)=10\left(2x^2-3x+1\right)^{10-1}(4x-3)$ Theorem 7

$=10\left(2x^2-3x+1\right)^9(4x-3)$ First Derivative

$f''(x)=10\left(2x^2-3x+1\right)^9(4)+$ Theorem 5

$10(4x-3)\cdot 9\left(2x^2-3x+1\right)^8(4x-3)$

$=40\left(2x^2-3x+1\right)^9+$

$90(4x-3)^2\left(2x^2-3x+1\right)^8$

$=10\left(2x^2-3x+1\right)^8\left(4\left(2x^2-3x+1\right)+\right.$

$\left.9\left(16x^2-24x+9\right)\right)$

$=10\left(2x^2-3x+1\right)^8\left(8x^2-12x+4+\right.$

$\left.144x^2-216x+81\right)$

$=10\left(2x^2-3x+1\right)^8\left(152x^2-228x+85\right)$

Second Derivative

23. $f(x)=\sqrt[4]{\left(x^2+1\right)^3}=\left(x^2+1\right)^{3/4}$

$f'(x)=\dfrac{3}{4}\left(x^2+1\right)^{-1/4}(2x)$ Theorem 7

$=\dfrac{3}{2}x\left(x^2+1\right)^{-1/4}$ First Derivative

$f''(x)=\dfrac{3}{2}x\cdot\dfrac{-1}{4}\left(x^2+1\right)^{-5/4}(2x)+$

$\dfrac{3}{2}\left(x^2+1\right)^{-1/4}(1)$ Theorem 5

$=-\dfrac{3}{4}x^2\left(x^2+1\right)^{-5/4}+\dfrac{3}{2}\left(x^2+1\right)^{-1/4}$

$=\dfrac{-3x^2}{4\left(x^2+1\right)^{5/4}}+\dfrac{3}{2\left(x^2+1\right)^{1/4}}$

We can simplify the second derivative by finding a common denominator and combining the fractions.

$f''(x)=\dfrac{-3x^2}{4\left(x^2+1\right)^{5/4}}+\dfrac{3}{2\left(x^2+1\right)^{1/4}}\cdot\dfrac{2\left(x^2+1\right)}{2\left(x^2+1\right)}$

$=\dfrac{-3x^2}{4\left(x^2+1\right)^{5/4}}+\dfrac{6\left(x^2+1\right)}{4\left(x^2+1\right)^{5/4}}$

$=\dfrac{3x^2+6}{4\left(x^2+1\right)^{5/4}}$

$=\dfrac{3\left(x^2+2\right)}{4\left(x^2+1\right)^{5/4}}$

Copyright © 2014 Pearson Education, Inc. Publishing as Addison-Wesley.

25.
$$y = x^{2/3} + 4x$$
$$y' = \frac{2}{3}x^{2/3-1} + 4$$
$$= \frac{2}{3}x^{-1/3} + 4 \quad \text{First Derivative}$$
$$y'' = \frac{2}{3}\cdot\frac{-1}{3}x^{-1/3-1}$$
$$= -\frac{2}{9}x^{-4/3}$$
$$= -\frac{2}{9x^{4/3}} \quad \text{Second Derivative}$$

27.
$$y = \left(x^3 - x\right)^{3/4}$$
$$y' = \frac{3}{4}\left(x^3 - x\right)^{3/4-1}\left(3x^2 - 1\right) \quad \text{Theorem 7}$$
$$= \frac{3}{4}\left(x^3 - x\right)^{-1/4}\left(3x^2 - 1\right) \quad \text{First Derivative}$$
$$y'' = \frac{3}{4}\left(x^3 - x\right)^{-1/4}(6x) + \quad \text{Theorem 5}$$
$$\frac{3}{4}\left(3x^2 - 1\right)\cdot\frac{-1}{4}\left(x^3 - x\right)^{-1/4-1}\left(3x^2 - 1\right)$$
$$= \frac{9}{2}x\left(x^3 - x\right)^{-1/4} +$$
$$\frac{-3}{16}\left(3x^2 - 1\right)^2\left(x^3 - x\right)^{-5/4}$$
$$= \frac{9x}{2\left(x^3 - x\right)^{1/4}} - \frac{3\left(3x^2 - 1\right)^2}{16\left(x^3 - x\right)^{5/4}}$$

The second derivative can be simplified by finding a common denominator and combining the fractions.

$$y'' = \frac{9x}{2\left(x^3 - x\right)^{1/4}}\cdot\frac{8\left(x^3 - x\right)}{8\left(x^3 - x\right)} - \frac{3\left(3x^2 - 1\right)^2}{16\left(x^3 - x\right)^{5/4}}$$
$$= \frac{72x^4 - 72x^2}{16\left(x^3 - x\right)^{5/4}} - \frac{3\left(9x^4 - 6x^2 + 1\right)}{16\left(x^3 - x\right)^{5/4}}$$
$$= \frac{72x^4 - 72x^2 - 27x^4 + 18x^2 - 3}{16\left(x^3 - x\right)^{5/4}}$$
$$= \frac{45x^4 - 54x^2 - 3}{16\left(x^3 - x\right)^{5/4}} \quad \text{Second Derivative}$$

29.
$$y = 2x^{5/4} + x^{1/2}$$
$$y' = 2\cdot\frac{5}{4}x^{5/4-1} + \frac{1}{2}x^{1/2-1}$$
$$= \frac{5}{2}x^{1/4} + \frac{1}{2}x^{-1/2} \quad \text{First Derivative}$$
$$y'' = \frac{5}{2}\cdot\frac{1}{4}x^{1/4-1} + \frac{1}{2}\cdot\frac{-1}{2}x^{-1/2-1}$$
$$= \frac{5}{8}x^{-3/4} - \frac{1}{4}x^{-3/2}$$
$$= \frac{5}{8x^{3/4}} - \frac{1}{4x^{3/2}} \quad \text{Second Derivative}$$

31.
$$y = \frac{2}{x^3} + \frac{1}{x^2} = 2x^{-3} + x^{-2}$$
$$y' = 2\left(-3x^{-3-1}\right) + \left(-2x^{-2-1}\right)$$
$$= -6x^{-4} - 2x^{-3} \quad \text{First Derivative}$$
$$y'' = -6\left(-4x^{-4-1}\right) - 2\left(-3x^{-3-1}\right)$$
$$= 24x^{-5} + 6x^{-4}$$
$$= \frac{24}{x^5} + \frac{6}{x^4} \quad \text{Second Derivative}$$

33.
$$y = \left(x^2 + 3\right)(4x - 1)$$
$$y' = \left(x^2 + 3\right)(4) + (4x - 1)(2x) \quad \text{Theorem 5}$$
$$= 4x^2 + 12 + 8x^2 - 2x$$
$$= 12x^2 - 2x + 12 \quad \text{First Derivative}$$
$$y'' = 12\left(2x^{2-1}\right) - 2$$
$$= 24x - 2 \quad \text{Second Derivative}$$

35.
$$y = \frac{3x + 1}{2x - 3}$$
$$y' = \frac{(2x - 3)(3) - (3x + 1)(2)}{(2x - 3)^2} \quad \text{Theorem 6}$$
$$= \frac{6x - 9 - 6x - 2}{(2x - 3)^2}$$
$$= \frac{-11}{(2x - 3)^2} \quad \text{First Derivative}$$

The solution is continued on the next page.

Copyright © 2014 Pearson Education, Inc. Publishing as Addison-Wesley.

We continue the solution from the previous page.

$$y''=\frac{(2x-3)^2(0)-(-11)\left(2(2x-3)^{2-1}(2)\right)}{\left((2x-3)^2\right)^2}$$

Theorem 6 and Theorem 7

$$=\frac{44(2x-3)}{(2x-3)^4}$$

$$=\frac{44}{(2x-3)^3} \quad \text{Second Derivative}$$

37. $y=x^4$

$$\frac{dy}{dx}=4x^{4-1}=4x^3 \quad \text{First Derivative}$$

$$\frac{d^2y}{dx^2}=4\left(3x^{3-1}\right)=12x^2 \quad \text{Second Derivative}$$

$$\frac{d^3y}{dx^3}=12\left(2x^{2-1}\right)=24x \quad \text{Third Derivative}$$

$$\frac{d^4y}{dx^4}=24 \quad \text{Fourth Derivative}$$

39. $y=x^6-x^3+2x$

$$\frac{dy}{dx}=6x^{6-1}-3x^{3-1}+2$$

$$=6x^5-3x^2+2 \quad \text{First Derivative}$$

$$\frac{d^2y}{dx^2}=6\left(5x^{5-1}\right)-3\left(2x^{2-1}\right)$$

$$=30x^4-6x \quad \text{Second Derivative}$$

$$\frac{d^3y}{dx^3}=30\left(4x^{4-1}\right)-6$$

$$=120x^3-6 \quad \text{Third Derivative}$$

$$\frac{d^4y}{dx^4}=120\left(3x^{3-1}\right)$$

$$=360x^2 \quad \text{Fourth Derivative}$$

$$\frac{d^5y}{dx^5}=360\left(2x^{2-1}\right)$$

$$=720x \quad \text{Fifth Derivative}$$

41. $f(x)=x^{-2}-x^{1/2}$

$$f'(x)=-2x^{-2-1}-\frac{1}{2}x^{1/2-1}$$

$$=-2x^{-3}-\frac{1}{2}x^{-1/2} \quad \text{First Derivative}$$

$$f''(x)=-2\left(-3x^{-3-1}\right)-\frac{1}{2}\cdot\frac{-1}{2}x^{-1/2-1}$$

$$=6x^{-4}+\frac{1}{4}x^{-3/2} \quad \text{Second Derivative}$$

$$f'''(x)=6\left(-4x^{-4-1}\right)+\frac{1}{4}\cdot\frac{-3}{2}x^{-3/2-1}$$

$$=-24x^{-5}-\frac{3}{8}x^{-5/2} \quad \text{Third Derivative}$$

$$f^{(4)}(x)=-24\left(-5x^{-5-1}\right)-\frac{3}{8}\cdot\frac{-5}{2}x^{-5/2-1}$$

$$=120x^{-6}+\frac{15}{16}x^{-7/2} \quad \text{Fourth Derivative}$$

43. $g(x)=x^4-3x^3-7x^2-6x+9$

$$g'(x)=4x^{4-1}-3\left(3x^{3-1}\right)-7\left(2x^{2-1}\right)-6$$

$$=4x^3-9x^2-14x-6 \quad \text{First Derivative}$$

$$g''(x)=4\left(3x^{3-1}\right)-9\left(2x^{2-1}\right)-14$$

$$=12x^2-18x-14 \quad \text{Second Derivative}$$

$$g'''(x)=12\left(2x^{2-1}\right)-18$$

$$=24x-18 \quad \text{Third Derivative}$$

$$g^{(4)}(x)=24 \quad \text{Fourth Derivative}$$

$$g^{(5)}(x)=0 \quad \text{Fifth Derivative}$$

$$g^{(6)}(x)=0 \quad \text{Sixth Derivative}$$

45. $s(t)=t^3+t$

a) $v(t)=s'(t)=3t^2+1$

b) $a(t)=v'(t)=s''(t)=6t$

c) When $t=4$

$v(4)=3(4)^2+1=49$

$a(4)=6(4)=24$

After 4 seconds, the velocity is 49 feet per second, and the acceleration is 24 feet per second squared.

47. $s(t)=3t+10$

a) $v(t)=s'(t)=3$

b) $a(t)=v'(t)=s''(t)=0$

Copyright © 2014 Pearson Education, Inc. Publishing as Addison-Wesley.

c) When $t = 2$

$v(2) = 3$

$a(2) = 0$

After 2 hours, the velocity is 3 miles per hour, and the acceleration is 0 miles per hour squared.

d) ✎

49. $s(t) = 16t^2$

a) When $t = 3$, $s(3) = 16(3)^2 = 144$.

The hammer falls 144 feet in 3 seconds.

b) $v(t) = s'(t) = 32t$

When $t = 3$, $v(3) = 32(3) = 96$

the hammer is falling at 96 feet per second after 3 seconds.

c) $a(t) = v'(t) = s''(t) = 32$

When $t = 3$, $a(3) = 32$

the hammer is accelerating at 32 feet per second squared after 3 seconds.

51. $s(t) = 4.905t^2$

The velocity and acceleration are given by:

$v(t) = s'(t) = 9.81t$

$a(t) = v'(t) = s''(t) = 9.81$

After 2 seconds, we have

$v(2) = 9.81(2) = 19.62$

The stone is falling at 19.62 meters per second.

$a(2) = 9.81$

The stone is accelerating at 9.81 meters per second squared.

53. a) The plane's velocity is greater at $t = 20$ seconds. We know this because the slope of the tangent line is greater at $t = 20$ then it is at $t = 6$.

b) The plane's acceleration is positive, since the velocity (slope of the tangent lines) is increasing over time.

55. $S(t) = 2t^3 - 40t^2 + 220t + 160$

a) $S'(t) = 6t^2 - 80t + 220$

When $t = 1$,

$S'(1) = 6(1)^2 - 80(1) + 220 = 146$

After 1 month, sales are increasing at 146 thousand (146,000) dollars per month.

When $t = 2$,

$S'(2) = 6(2)^2 - 80(2) + 220 = 84$

After 2 month, sales are increasing at 84 thousand (84,000) dollars per month.

When $t = 4$,

$S'(4) = 6(4)^2 - 80(4) + 220 = -4$

After 4 months, sales are changing at a rate of -4 thousand (-4000) dollars per month.

b) $S''(t) = 12t - 80$

When $t = 1$, $S''(1) = 12(1) - 80 = -68$

After 1 month, the rate of change of sales are changing at a rate of -68 thousand $(-68,000)$ dollars per month squared.

When $t = 2$,

$S''(2) = 12(2) - 80 = -56$

After 2 months, the rate of change of sales are changing at a rate of -56 thousand $(-56,000)$ dollars per month squared.

When $t = 4$, $S''(4) = 12(4) - 80 = -32$

After 4 months, the rate of change of sales are changing at a rate of -32 thousand $(-32,000)$ dollars per month squared.

c) ✎

57. a) $p(t) = \dfrac{2000t}{4t + 75}$

First find the derivative of the population function. Using the quotient rule, we have:

$$p'(t)$$

$$= \frac{d}{dt}\left(\frac{2000t}{4t + 75}\right)$$

$$= \frac{(4t + 75)\frac{d}{dt}(2000t) - (2000t)\frac{d}{dt}(4t + 75)}{(4t + 75)^2}$$

$$= \frac{(4t + 75)(2000) - (2000t)(4)}{(4t + 75)^2}$$

$$= \frac{8000t + 150,000 - 8000t}{(4t + 75)^2}$$

$$= \frac{150,000}{(4t + 75)^2}.$$

The solution is continued on the next page.

Copyright © 2014 Pearson Education, Inc. Publishing as Addison-Wesley.

Next we substitute the appropriate values in for t.

$$p'(10)=\frac{150,000}{(4(10)+75)^2}=\frac{150,000}{13,225}\approx 11.34.$$

$$p'(50)=\frac{150,000}{(4(50)+75)^2}=\frac{150,000}{75,625}\approx 1.98.$$

$$p'(100)=\frac{150,000}{(4(100)+75)^2}=\frac{150,000}{225,625}\approx 0.665.$$

b) First find the second derivative.

$$p''(t)=\frac{d}{dt}\left(\frac{150,000}{(4t+75)^2}\right)=\frac{d}{dt}\left(150,000(4t+75)^{-2}\right)=-300,000(4t+75)^{-3}\cdot 4=-\frac{1,200,000}{(4t+75)^{-3}}$$

Next we substitute the appropriate values in for t.

$$p''(10)=-\frac{1,200,000}{(4(10)+75)^3}=-\frac{1,200,000}{1,520,875}\approx -0.789.$$

$$p''(50)=-\frac{1,200,000}{(4(50)+75)^3}=-\frac{1,200,000}{20,796,875}\approx -0.0577.$$

$$p''(100)=-\frac{1,200,000}{(4(100)+75)^3}=-\frac{1,200,000}{107,171,875}\approx -0.0112.$$

c) ✎

59. $y=\frac{1}{(1-x)}=(1-x)^{-1}$

$$y'=-1(1-x)^{-1-1}(-1)=1(1-x)^{-2}$$
$$y''=-2(1-x)^{-2-1}(-1)=2(1-x)^{-3}$$
$$y'''=2(-3)(1-x)^{-3-1}(-1)=6(1-x)^{-4}$$

Therefore,

$$y'''=\frac{6}{(1-x)^4}$$

61. $y=\frac{1}{\sqrt{2x+1}}=(2x+1)^{-1/2}$

$$y'=\frac{-1}{2}(2x+1)^{-1/2-1}(2)=-(2x+1)^{-3/2}$$
$$y''=-\frac{-3}{2}(2x+1)^{-3/2-1}(2)=3(2x+1)^{-5/2}$$
$$y'''=3\left(\frac{-5}{2}\right)(2x+1)^{-5/2-1}(2)=-15(2x+1)^{-7/2}=-\frac{15}{(2x+1)^{7/2}}$$

63. $y=\frac{\sqrt{x}+1}{\sqrt{x}-1}=\frac{x^{1/2}+1}{x^{1/2}-1}$

$$y'=\frac{(x^{1/2}-1)\left(\frac{1}{2}x^{-1/2}\right)-(x^{1/2}+1)\left(\frac{1}{2}x^{-1/2}\right)}{(x^{1/2}-1)^2}=\frac{\frac{1}{2}-\frac{1}{2}x^{-1/2}-\frac{1}{2}-\frac{1}{2}x^{-1/2}}{(x^{1/2}-1)^2}=\frac{-x^{-1/2}}{(x^{1/2}-1)^2}$$

The solution is continued on the next page.

Copyright © 2014 Pearson Education, Inc. Publishing as Addison-Wesley.

$$y'' = \frac{\left(x^{1/2}-1\right)^2\left(\frac{1}{2}x^{-3/2}\right)-\left(-x^{-1/2}\right)\left[2\left(x^{1/2}-1\right)\frac{1}{2}x^{-1/2}\right]}{\left(\left(x^{1/2}-1\right)^2\right)^2}$$

$$= \frac{\frac{1}{2}x^{-3/2}\left(x^{1/2}-1\right)^2+x^{-1}\left(x^{1/2}-1\right)}{\left(x^{1/2}-1\right)^4}$$

$$= \frac{\left(x^{1/2}-1\right)\left(\frac{1}{2}x^{-3/2}\left(x^{1/2}-1\right)+x^{-1}\right)}{\left(x^{1/2}-1\right)^4}$$

$$= \frac{\frac{1}{2}x^{-1}-\frac{1}{2}x^{-3/2}+x^{-1}}{\left(x^{1/2}-1\right)^3}$$

$$= \frac{\left(\frac{3}{2}x^{-1}-\frac{1}{2}x^{-3/2}\right)}{\left(x^{1/2}-1\right)^3}$$

$$= \frac{\frac{3}{2x}-\frac{1}{2x^{3/2}}}{\left(x^{1/2}-1\right)^3}$$

$$= \frac{\frac{3x^{1/2}}{2x^{3/2}}-\frac{1}{2x^{3/2}}}{\left(x^{1/2}-1\right)^3}$$

$$= \frac{3x^{1/2}-1}{2x^{3/2}\left(x^{1/2}-1\right)^3}$$

65. $y = x^k$

$$\frac{dy}{dx} = kx^{k-1}$$

$$\frac{d^2y}{dx^2} = k(k-1)x^{k-2}$$

$$\frac{d^3y}{dx^3} = k(k-1)(k-2)x^{k-3}$$

$$\frac{d^4y}{dx^4} = k(k-1)(k-2)(k-3)x^{k-4}$$

$$\frac{d^5y}{dx^5} = k(k-1)(k-2)(k-3)(k-4)x^{k-5}$$

67. $f(x) = \frac{x-1}{x+2}$

$$f'(x) = \frac{(x+2)(1)-(x-1)(1)}{(x+2)^2}$$

$$= \frac{3}{(x+2)^2}$$

Notice, $f'(x) = \frac{3}{(x+2)^2} = 3(x+2)^{-2}$

$$f''(x) = 3(-2)(x+2)^{-2-1}$$

$$= -6(x+2)^{-3}$$

$$= -\frac{6}{(x+2)^3}$$

Notice, $f''(x) = -\frac{6}{(x+2)^3} = -6(x+2)^{-3}$

$$f'''(x) = -6(-3)(x+2)^{-3-1}(1)$$

$$= 18(x+2)^{-4}$$

$$= \frac{18}{(x+2)^4}$$

Notice, $f'''(x) = \frac{18}{(x+2)^4} = 18(x+2)^{-4}$

$$f^{(4)}(x) = 18(-4)(x+2)^{-4-1}(1)$$

$$= -72(x+2)^{-5}$$

$$= -\frac{72}{(x+2)^5}$$

Copyright © 2014 Pearson Education, Inc. Publishing as Addison-Wesley.

69. $s(t)=16t^2$

Find the velocity function, $v(t)=s'(t)=32t$.

This velocity function is in feet per second, we need to convert 50 miles per hour to feet per second. There are 5280 feet in one mile and 3600 seconds in one hour. Therefore,

$$\left(50\frac{\text{mi}}{\text{hr}}\right)\left(\frac{5280\text{ ft}}{1\text{ mi}}\right)\left(\frac{1\text{ hr}}{3600\text{ sec}}\right)=\frac{220}{3}\frac{\text{ft}}{\text{sec}}$$

Now we solve the equation:

$$v(t)=\frac{220}{3}$$
$$32t=\frac{220}{3}$$
$$t=\frac{220}{3}\cdot\frac{1}{32}$$
$$t=\frac{55}{24}\approx 2.29$$

The ball will have to fall approximately 2.29 seconds to reach a speed of 50 mi/hr.

71. a) $s(2)=0.81(2)^2\approx 3.24$

The object has fallen 3.24 meters after 2 seconds.

b) To determine the speed, we find the first derivative of the position function, which is $s'(t)=1.62t$.

Therefore,
$s'(2)=1.62(2)=3.24$.

The object is traveling at 3.24 meters per second after 2 seconds.

c) To determine the acceleration, we find the second derivative of the position function, which is:
$s''(t)=1.62$.

The object is accelerating at 1.62 meters per second squared.

d) The second derivative represents the acceleration due to gravity on the moon. It is a constant $1.62\frac{m}{\text{sec}^2}$.

73. First we must find out how long it took to fall 28ft. Solving the equation

$$s(t)=28$$
$$16t^2=28$$
$$t^2=1.75$$
$$t=\pm 1.3229$$

We find out it took about 1.3229 seconds to fall to the ramp.

Next, we find the velocity function which is the first derivative of the position function. The first derivative is
$s'(t)=32t$.

Substituting 1.3229 in for t, we have
$s'(1.32)=32(1.3229)=42.33$

Danny Way was traveling around 42.33 feet per second when he touched down on the ramp.

75. $s(t)=-t^3+3t;\qquad [-3,3]$

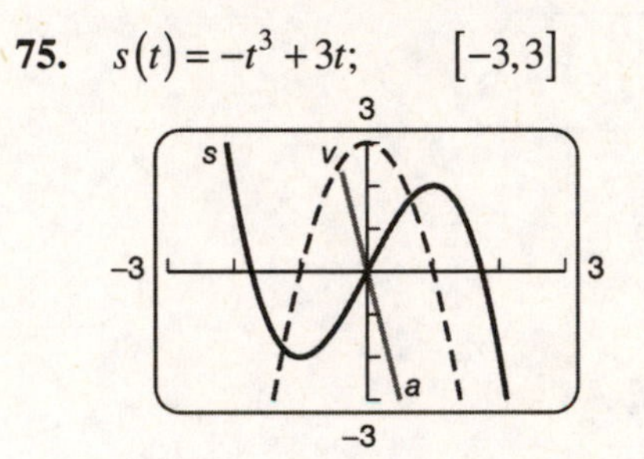

From the graph we see that $v(t)$ switches at $t=0$.

77. $s(t)=t^3-3t^2+2;\qquad [-2,4]$

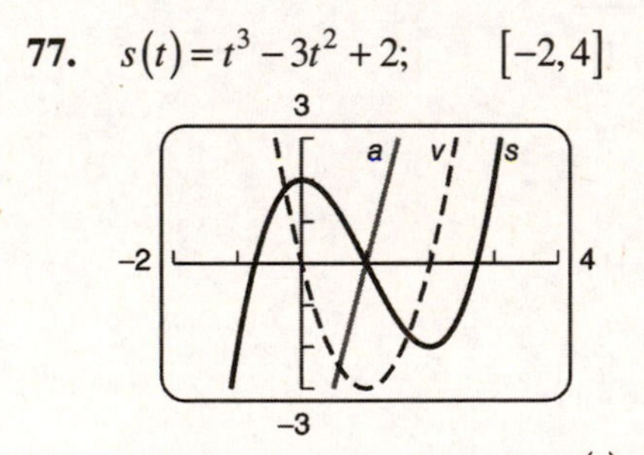

From the graph we see that $v(t)$ switches at $t=1$.

Copyright © 2014 Pearson Education, Inc. Publishing as Addison-Wesley.

Chapter 2

Applications of Differentiation

Exercise Set 2.1

1. $f(x)=x^2+4x+5$

First, find the critical points.

$f'(x)=2x+4$

$f'(x)$ exists for all real numbers. We solve

$f'(x)=0$

$2x+4=0$

$2x=-4$

$x=-2.$

The only critical value is -2. We use -2 to divide the real number line into two intervals, A: $(-\infty,-2)$ and B: $(-2,\infty)$:

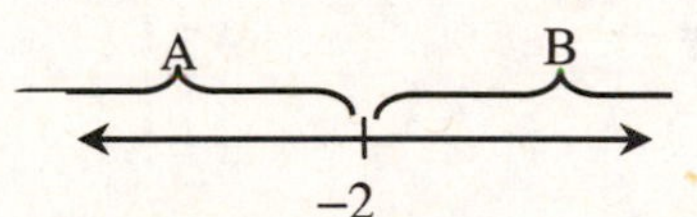

We use a test value in each interval to determine the sign of the derivative in each interval.

A: Test $-3, f'(-3)=2(-3)+4=-2<0$

B: Test 0, $\quad f'(0)=2(0)+4=4>0$

We see that $f(x)$ is decreasing on $(-\infty,-2)$ and increasing on $(-2,\infty)$, and the change from decreasing to increasing indicates that a relative minimum occurs at $x=-2$. We substitute into the original equation to find $f(-2)$:

$f(-2)=(-2)^2+4(-2)+5=1.$

Thus, there is a relative minimum at $(-2,1)$. We use the information obtained to sketch the graph. Other function values are listed below.

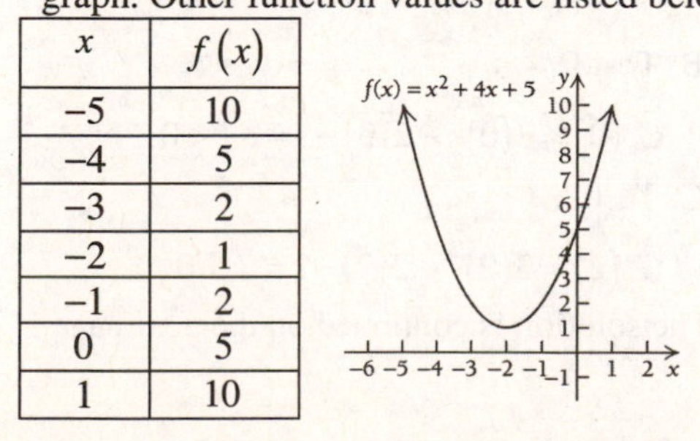

x	$f(x)$
−5	10
−4	5
−3	2
−2	1
−1	2
0	5
1	10

3. $f(x)=5-x-x^2$

First, find the critical points.

$f'(x)=-1-2x$

$f'(x)$ exists for all real numbers. We solve

$f'(x)=0$

$-1-2x=0$

$-2x=1$

$x=-\frac{1}{2}.$

The only critical value is $-\frac{1}{2}$. We use $-\frac{1}{2}$ to divide the real number line into two intervals,

A: $\left(-\infty,-\frac{1}{2}\right)$ and B: $\left(-\frac{1}{2},\infty\right)$:

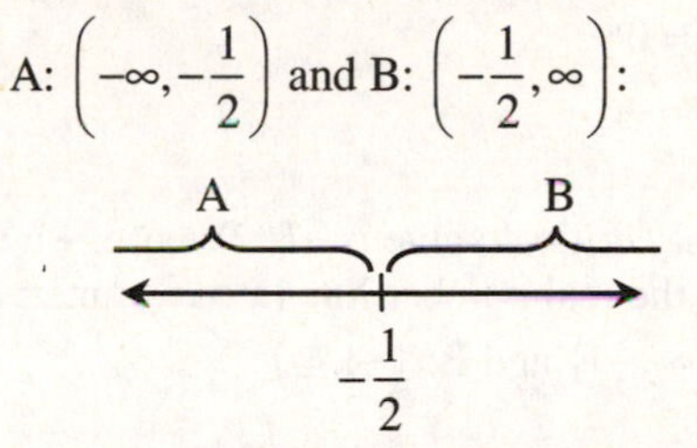

We use a test value in each interval to determine the sign of the derivative in each interval.

A: Test $-1, f'(-1)=-1-2(-1)=1>0$

B: Test 0, $\quad f'(0)=-1-2(0)=-1<0$

We see that $f(x)$ is increasing on $\left(-\infty,-\frac{1}{2}\right)$ and decreasing on $\left(-\frac{1}{2},\infty\right)$, and the change from increasing to decreasing indicates that a relative maximum occurs at $x=-\frac{1}{2}$. We substitute into the original equation to find $f\left(-\frac{1}{2}\right)$:

$f\left(-\frac{1}{2}\right)=5-\left(-\frac{1}{2}\right)-\left(-\frac{1}{2}\right)^2=\frac{21}{4}.$

Thus, there is a relative maximum at $\left(-\frac{1}{2},\frac{21}{4}\right)$.

The solution is continued on the next page.

Copyright © 2014 Pearson Education, Inc. Publishing as Addison-Wesley.

We use the information obtained on the previous page to sketch the graph. Other function values are listed below.

x	$f(x)$
−3	−1
−2	3
−1	5
$-\frac{1}{2}$	$\frac{21}{4}$
0	5
1	3
2	−1

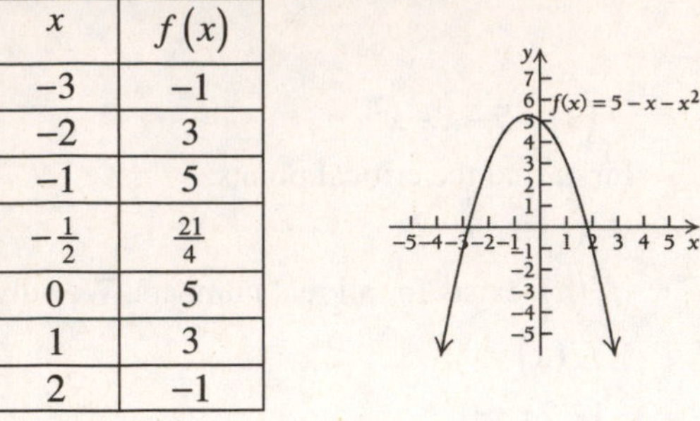

5. $g(x) = 1 + 6x + 3x^2$

First, find the critical points.

$g'(x) = 6 + 6x$

$g'(x)$ exists for all real numbers. We solve:

$g'(x) = 0$

$6 + 6x = 0$

$6x = -6$

$x = -1.$

The only critical value is -1. We use -1 to divide the real number line into two intervals, A: $(-\infty, -1)$ and B: $(-1, \infty)$:

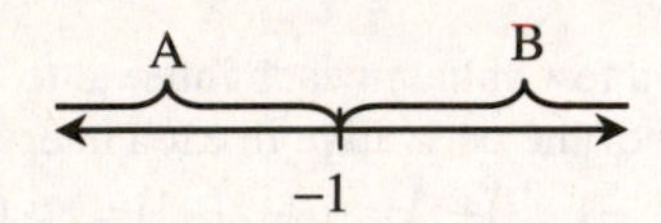

We use a test value in each interval to determine the sign of the derivative in each interval.

A: Test -2, $g'(-2) = 6 + 6(-2) = -6 < 0$

B: Test 0, $g'(0) = 6 + 6(0) = 6 > 0$

We see that $g'(x)$ is decreasing on $(-\infty, -1)$ and increasing on $(-1, \infty)$, and the change from decreasing to increasing indicates that a relative minimum occurs at $x = -1$. We substitute into the original equation to find $g(-1)$:

$g(-1) = 1 + 6(-1) + 3(-1)^2 = -2.$

Thus, there is a relative minimum at $(-1, -2)$.

We use the information obtained to sketch the graph. Other function values are listed below.

x	$g(x)$
−4	25
−3	10
−2	1
−1	−2
0	1
1	10
2	25

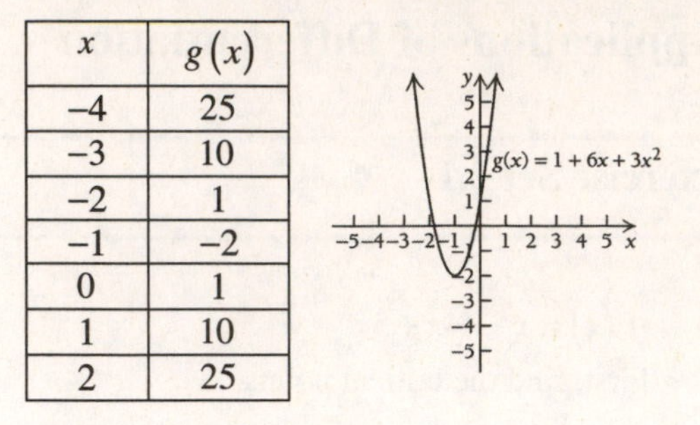

7. $G(x) = x^3 - x^2 - x + 2$

First, find the critical points.

$G'(x) = 3x^2 - 2x - 1$

$G'(x)$ exists for all real numbers. We solve

$G'(x) = 0$

$3x^2 - 2x - 1 = 0$

$(3x + 1)(x - 1) = 0$

$3x + 1 = 0$ or $x - 1 = 0$

$3x = -1$ or $x = 1$

$x = -\frac{1}{3}$ or $x = 1.$

The critical values are $-\frac{1}{3}$ and 1. We use them to divide the real number line into three intervals,

A: $\left(-\infty, -\frac{1}{3}\right)$, B: $\left(-\frac{1}{3}, 1\right)$, and C: $(1, \infty)$.

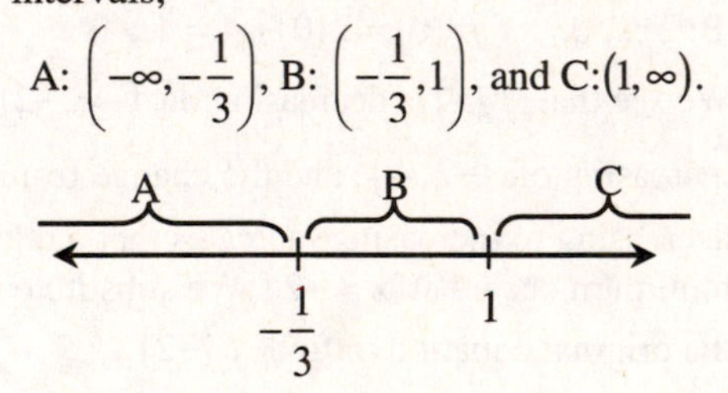

We use a test value in each interval to determine the sign of the derivative in each interval.

A: Test -1,

$G'(-1) = 3(-1)^2 - 2(-1) - 1 = 4 > 0$

B: Test 0,

$G'(0) = 3(0)^2 - 2(0) - 1 = -1 < 0$

C: Test 2,

$G'(2) = 3(2)^2 - 2(2) - 1 = 7 > 0$

The solution is continued on the next page.

Copyright © 2014 Pearson Education, Inc. Publishing as Addison-Wesley.

On the previous page, we see that $G(x)$ is increasing on $\left(-\infty,-\frac{1}{3}\right)$, decreasing on $\left(-\frac{1}{3},1\right)$, and increasing on $(1,\infty)$. So there is a relative maximum at $x=-\frac{1}{3}$ and a relative minimum at $x=1$.

We find $G\left(-\frac{1}{3}\right)$:

$$G\left(-\frac{1}{3}\right)=\left(-\frac{1}{3}\right)^3-\left(-\frac{1}{3}\right)^2-\left(-\frac{1}{3}\right)+2$$
$$=-\frac{1}{27}-\frac{1}{9}+\frac{1}{3}+2$$
$$=\frac{59}{27}.$$

Then we find $G(1)$:

$$G(1)=(1)^3-(1)^2-(1)+2$$
$$=1-1-1+2$$
$$=1.$$

There is a relative maximum at $\left(-\frac{1}{3},\frac{59}{27}\right)$, and there is a relative minimum at $(1,1)$. We use the information obtained to sketch the graph. Other function values are listed below.

x	$G(x)$
−2	−8
−1	1
0	2
2	4
3	17

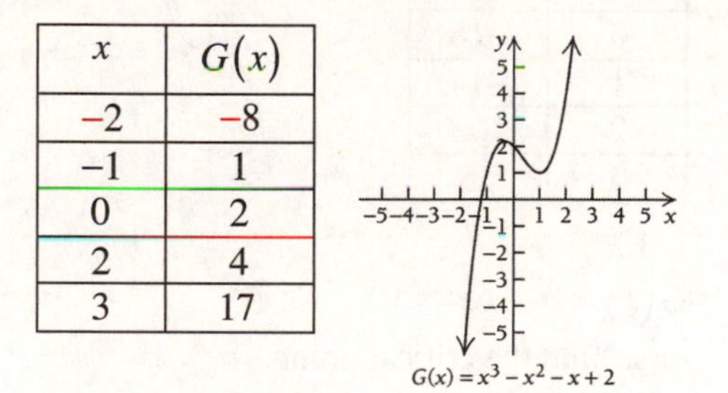

9. $f(x)=x^3-3x+6$

First, find the critical points.

$f'(x)=3x^2-3$

$f'(x)$ exists for all real numbers. We solve

$$f'(x)=0$$
$$3x^2-3=0$$
$$3x^2=3$$
$$x^2=1$$
$$x=\pm 1.$$

The critical values are -1 and 1. We use them to divide the real number line into three intervals,

A: $(-\infty,-1)$, B: $(-1,1)$, and C:$(1,\infty)$.

A B C

−1 1

We use a test value in each interval to determine the sign of the derivative in each interval.

A: Test -3, $f'(-3)=3(-3)^2-3=24>0$

B: Test 0, $f'(0)=3(0)^2-3=-3<0$

C: Test 2, $f'(2)=3(2)^2-3=9>0$

We see that $f(x)$ is increasing on $(-\infty,-1)$, decreasing on $(-1,1)$, and increasing on $(1,\infty)$. So there is a relative maximum at $x=-1$ and a relative minimum at $x=1$.

We find $f(-1)$:

$$f(-1)=(-1)^3-3(-1)+6=-1+3+6=8.$$

Then we find $f(1)$:

$$f(1)=(1)^3-3(1)+6=1-3+6=4.$$

There is a relative maximum at $(-1,8)$, and there is a relative minimum at $(1,4)$. We use the information obtained to sketch the graph. Other function values are listed below.

x	$f(x)$
−3	−12
−2	4
0	6
2	8
3	24

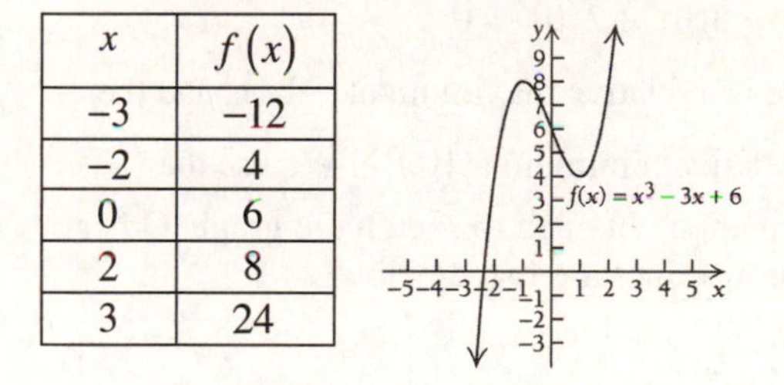

11. $f(x)=3x^2+2x^3$

First, find the critical points.

$f'(x)=6x+6x^2$

$f'(x)$ exists for all real numbers. We solve

$$f'(x)=0$$
$$6x+6x^2=0$$
$$6x(1+x)=0$$
$$6x=0 \quad \text{or} \quad x+1=0$$
$$x=0 \quad \text{or} \quad x=-1.$$

The solution is continued on the next page.

Copyright © 2014 Pearson Education, Inc. Publishing as Addison-Wesley.

From the previous page, we know the critical values are -1 and 0. We use them to divide the real number line into three intervals,
A: $(-\infty,-1)$, B: $(-1,0)$, and C:$(0,\infty)$.

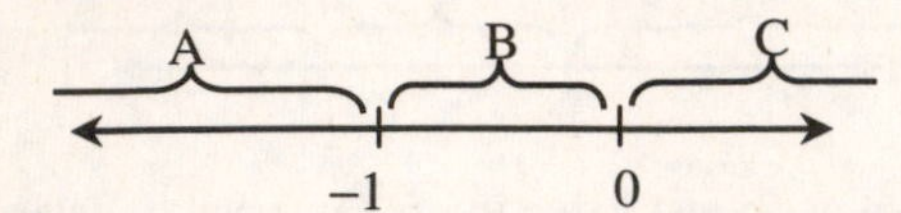

We use a test value in each interval to determine the sign of the derivative in each interval.

A: Test -2,

$$f'(-2)=6(-2)+6(-2)^2=12>0$$

B: Test $-\frac{1}{2}$,

$$f'\left(-\frac{1}{2}\right)=6\left(-\frac{1}{2}\right)+6\left(-\frac{1}{2}\right)^2=-\frac{3}{2}<0$$

C: Test 1,

$$f'(1)=6(1)+6(1)^2=12>0$$

We see that $f(x)$ is increasing on $(-\infty,-1)$, decreasing on $(-1,0)$, and increasing on $(0,\infty)$. So there is a relative maximum at $x=-1$ and a relative minimum at $x=0$.

We find $f(-1)$:

$$f(-1)=3(-1)^2+2(-1)^3=1.$$

Then we find $f(0)$:

$$f(0)=3(0)^2+2(0)^3=0.$$

There is a relative maximum at $(-1,1)$, and there is a relative minimum at $(0,0)$. We use the information obtained to sketch the graph. Other function values are listed below.

x	$f(x)$
-3	-27
-2	-4
$\frac{1}{2}$	1
2	28

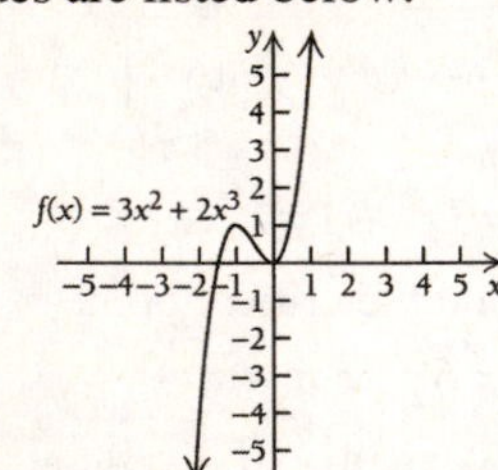

13. $g(x)=2x^3-16$

First, find the critical points.

$g'(x)=6x^2$

$g'(x)$ exists for all real numbers. We solve

$$g'(x)=0$$
$$6x^2=0$$
$$x=0.$$

The only critical value is 0. We use 0 to divide the real number line into two intervals, A: $(-\infty,0)$, and B: $(0,\infty)$.

We use a test value in each interval to determine the sign of the derivative in each interval.

A: Test -1, $g'(-1)=6(-1)^2=6>0$

B: Test 1, $g'(1)=6(1)^2=6>0$

We see that $g(x)$ is increasing on $(-\infty,0)$ and increasing on $(0,\infty)$, so the function has no relative extema. We use the information obtained to sketch the graph. Other function values are listed below.

x	$g(x)$
-2	-32
-1	-18
0	-16
1	-14
2	0
3	38

15. $G(x)=x^3-6x^2+10$

First, find the critical points.

$G'(x)=3x^2-12x$

$G'(x)$ exists for all real numbers. We solve

$$G'(x)=0$$

$x^2-4x=0$ Dividing by 3

$x(x-4)=0$

$x=0$ or $x-4=0$

$x=0$ or $x=4.$

The critical values are 0 and 4.

The solution is continued on the next page.

Copyright © 2014 Pearson Education, Inc. Publishing as Addison-Wesley.

We use the critical values determined on the previous page to divide the real number line into three intervals,

A: $(-\infty,0)$, B: $(0,4)$, and C: $(4,\infty)$.

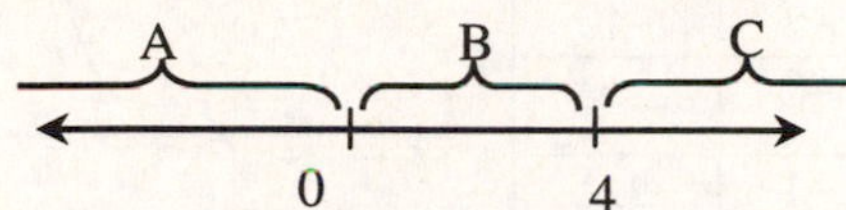

We use a test value in each interval to determine the sign of the derivative in each interval.

A: Test -1, $G'(-1)=3(-1)^2-12(-1)=15>0$

B: Test 1, $G'(1)=3(1)^2-12(1)=-9<0$

C: Test 5, $G'(5)=3(5)^2-12(5)=15>0$

We see that $G(x)$ is increasing on $(-\infty,0)$, decreasing on $(0,4)$, and increasing on $(4,\infty)$. So there is a relative maximum at $x=0$ and a relative minimum at $x=4$.

We find $G(0)$:

$$G(0)=(0)^3-6(0)^2+10$$
$$=10.$$

Then we find $G(4)$:

$$G(4)=(4)^3-6(4)^2+10$$
$$=64-96+10$$
$$=-22.$$

There is a relative maximum at $(0,10)$, and there is a relative minimum at $(4,-22)$. We use the information obtained to sketch the graph. Other function values are listed below.

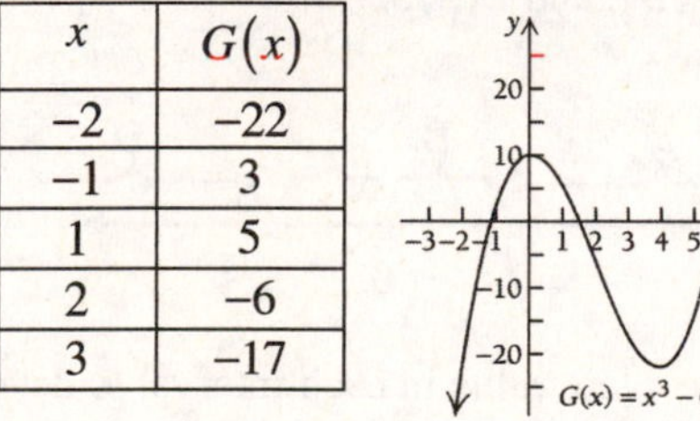

x	$G(x)$
-2	-22
-1	3
1	5
2	-6
3	-17

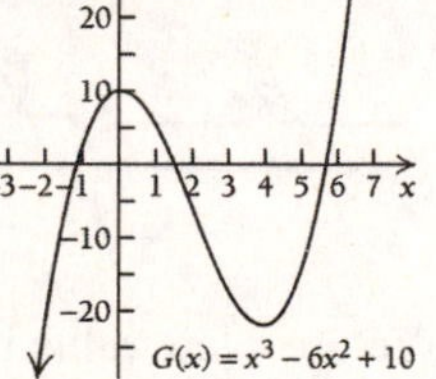

17. $g(x)=x^3-x^4$

First, find the critical points.

$g'(x)=3x^2-4x^3$

$g'(x)$ exists for all real numbers. We solve

$$g'(x)=0$$
$$3x^2-4x^3=0$$
$$x^2(3-4x)=0$$
$$x^2=0 \quad \text{or} \quad 3-4x=0$$
$$x=0 \quad \text{or} \quad -4x=-3$$
$$x=0 \quad \text{or} \quad x=\frac{3}{4}.$$

The critical values are 0 and $\frac{3}{4}$.

We use the critical values to divide the real number line into three intervals,

A: $(-\infty,0)$, B: $\left(0,\frac{3}{4}\right)$, and C: $\left(\frac{3}{4},\infty\right)$.

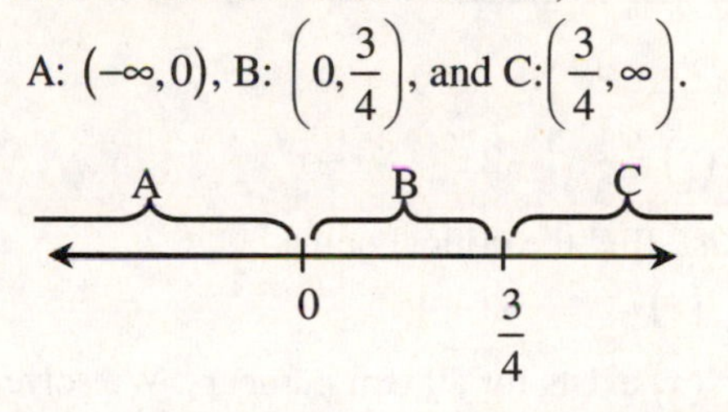

We use a test value in each interval to determine the sign of the derivative in each interval.

A: Test -1, $g'(-1)=3(-1)^2-4(-1)^3=7>0$

B: Test $\frac{1}{2}$, $g'\left(\frac{1}{2}\right)=3\left(\frac{1}{2}\right)^2-4\left(\frac{1}{2}\right)^3$

$$=3\left(\frac{1}{4}\right)-4\left(\frac{1}{8}\right)=\frac{1}{4}>0$$

C: Test 1, $g'(1)=3(1)^2-4(1)^3=-1<0$

We see that $g(x)$ is increasing on $(-\infty,0)$ and $\left(0,\frac{3}{4}\right)$, and is decreasing on $\left(\frac{3}{4},\infty\right)$. So there is no relative extrema at $x=0$ but there is a relative maximum at $x=\frac{3}{4}$.

The solution is continued on the next page.

Copyright © 2014 Pearson Education, Inc. Publishing as Addison-Wesley.

We find $g\left(\frac{3}{4}\right)$:

$$g\left(\frac{3}{4}\right)=\left(\frac{3}{4}\right)^3-\left(\frac{3}{4}\right)^4=\frac{27}{64}-\frac{81}{256}=\frac{27}{256}$$

There is a relative maximum at $\left(\frac{3}{4},\frac{27}{256}\right)$. We use the information obtained to sketch the graph. Other function values are listed below.

x	$g(x)$
-2	-24
-1	-2
0	0
$\frac{1}{2}$	$\frac{1}{16}$
1	0
2	-8

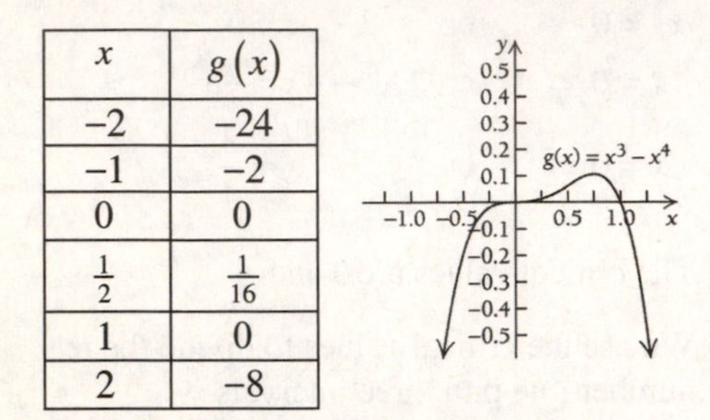

19. $f(x)=\frac{1}{3}x^3-2x^2+4x-1$

First, find the critical points.

$f'(x)=x^2-4x+4$

$f'(x)$ exists for all real numbers. We solve

$$f'(x)=0$$
$$x^2-4x+4=0$$
$$(x-2)^2=0$$
$$x-2=0$$
$$x=2.$$

The only critical value is 2.
We divide the real number line into two intervals,

A: $(-\infty,2)$ and B: $(2,\infty)$.

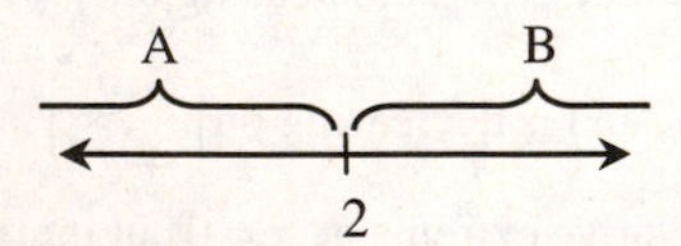

We use a test value in each interval to determine the sign of the derivative in each interval.

A: Test 0, $f'(0)=(0)^2-4(0)+4=4>0$

B: Test 3, $f'(3)=(3)^2-4(3)+4=1>0$

We see that $f(x)$ is increasing on both $(-\infty,2)$ and $(2,\infty)$. Therefore, there are no relative extrema.

We use the information obtained to sketch the graph. Other function values are listed below.

x	$f(x)$
-3	-40
-2	$-\frac{59}{3}$
-1	$-\frac{22}{3}$
0	-1
1	$\frac{4}{3}$
2	$\frac{5}{3}$
3	2

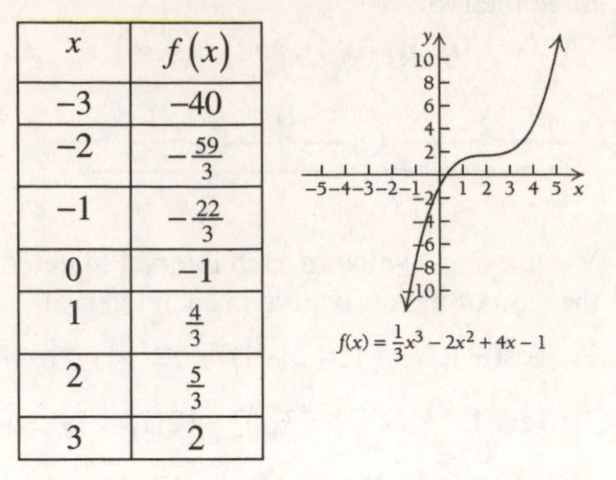

21. $g(x)=2x^4-20x^2+18$

First, find the critical points.

$g'(x)=8x^3-40x$

$g'(x)$ exists for all real numbers. We solve

$$g'(x)=0$$
$$8x^3-40x=0$$
$$8x\left(x^2-5\right)=0$$
$$8x=0 \quad \text{or} \quad x^2-5=0$$
$$x=0 \quad \text{or} \quad x^2=5$$
$$x=0 \quad \text{or} \quad x=\pm\sqrt{5}.$$

The critical values are 0, $\sqrt{5}$ and $-\sqrt{5}$. We use them to divide the real number line into four intervals,

A: $(-\infty,-\sqrt{5})$, B: $(-\sqrt{5},0)$,

C: $(0,\sqrt{5})$, and D: $(\sqrt{5},\infty)$.

A B C D
$-\sqrt{5}$ 0 $\sqrt{5}$

We use a test value in each interval to determine the sign of the derivative in each interval.

The solution is continued on the next page.

Copyright © 2014 Pearson Education, Inc. Publishing as Addison-Wesley.

Using the information from the previous page, we test a value in each interval.

A: Test -3,

$$g'(-3)=8(-3)^3-40(-3)=-96<0$$

B: Test -1,

$$g'(-1)=8(-1)^3-40(-1)=32>0$$

C: Test 1,

$$g'(1)=8(1)^3-40(1)=-32<0$$

D: Test 3,

$$g'(3)=8(3)^3-40(3)=96>0$$

We see that $g(x)$ is decreasing on $(-\infty,-\sqrt{5})$, increasing on $(-\sqrt{5},0)$, decreasing again on $(0,\sqrt{5})$, and increasing again on $(\sqrt{5},\infty)$.

Thus, there is a relative minimum at $x=-\sqrt{5}$, a relative maximum at $x=0$, and another relative minimum at $x=\sqrt{5}$.

We find $g(-\sqrt{5})$:

$$g(-\sqrt{5})=2(-\sqrt{5})^4-20(-\sqrt{5})^2+18=-32.$$

Then we find $g(0)$:

$$g(0)=2(0)^4-20(0)^2+18=18.$$

Then we find $g(\sqrt{5})$

$$g(\sqrt{5})=2(\sqrt{5})^4-20(\sqrt{5})^2+18=-32.$$

There are relative minima at $(-\sqrt{5},-32)$ and $(\sqrt{5},-32)$. There is a relative maximum at $(0,18)$ We use the information obtained to sketch the graph. Other function values are listed below.

x	$g(x)$
-4	210
-3	0
-1	0
1	0
3	0
4	210

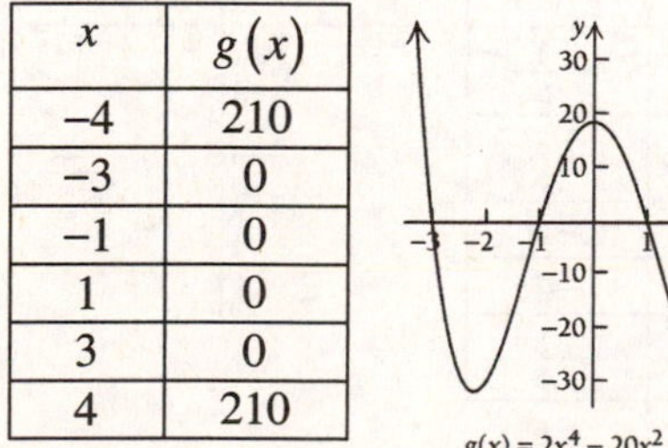
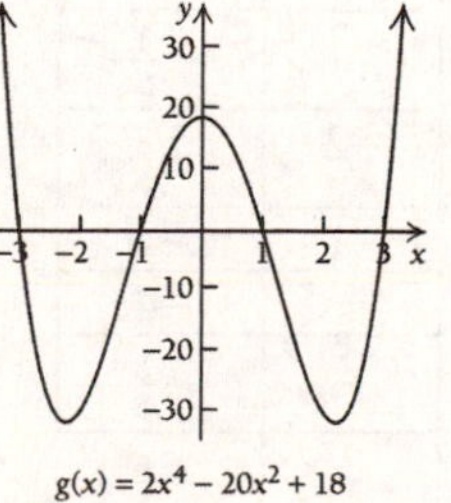

23. $F(x)=\sqrt[3]{x-1}=(x-1)^{1/3}$

First, find the critical points.

$$F'(x)=\frac{1}{3}(x-1)^{-2/3}(1)$$
$$=\frac{1}{3(x-1)^{2/3}}$$

$F'(x)$ does not exist when $3(x-1)^{2/3}=0$, which means that $F'(x)$ does not exist when $x=1$. The equation $F'(x)=0$ has no solution, therefore, the only critical value is $x=1$.

We use 1 to divide the real number line into two intervals,

A: $(-\infty,1)$ and B: $(1,\infty)$:

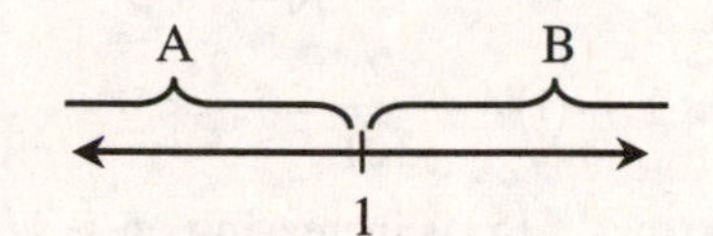

We use a test value in each interval to determine the sign of the derivative in each interval.

A: Test 0, $F'(0)=\dfrac{1}{3(0-1)^{2/3}}=\dfrac{1}{3}>0$

B: Test 2, $F'(2)=\dfrac{1}{3(2-1)^{2/3}}=\dfrac{1}{3}>0$

We see that $F(x)$ is increasing on both $(-\infty,1)$ and $(1,\infty)$. Thus, there are no relative extrema for $F(x)$. We use the information obtained to sketch the graph. Other function values are listed.

x	$F(x)$
-7	-2
0	-1
1	0
2	1
9	2

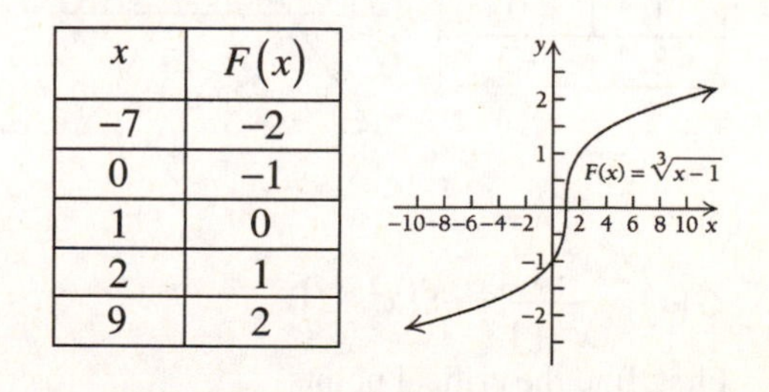

25. $f(x)=1-x^{2/3}$

First, find the critical points.

$$f'(x)=\frac{-2}{3}x^{-1/3}$$
$$=\frac{-2}{3\sqrt[3]{x}}$$

The solution is continued on the next page

Copyright © 2014 Pearson Education, Inc. Publishing as Addison-Wesley.

From the previous page, we see that $f'(x)$ does not exist when $3\sqrt[3]{x}=0$, which means that $f'(x)$ does not exist when $x=0$. The equation $f'(x)=0$ has no solution, therefore, the only critical value is $x=0$.

We use 0 to divide the real number line into two intervals,

A: $(-\infty,0)$ and B: $(0,\infty)$:

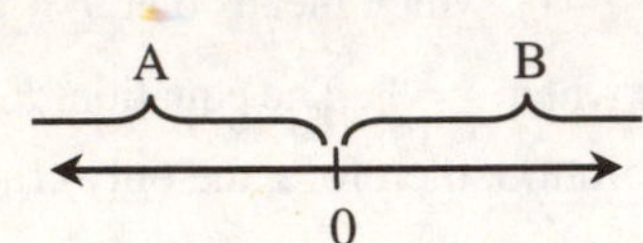

We use a test value in each interval to determine the sign of the derivative in each interval.

A: Test $-1, f'(-1)=-\frac{2}{3\sqrt[3]{-1}}=\frac{2}{3}>0$

B: Test $1, f'(1)=-\frac{2}{3\sqrt[3]{1}}=-\frac{2}{3}<0$

We see that $f(x)$ is increasing on $(-\infty,0)$ and decreasing on $(0,\infty)$. Thus, there is a relative maximum at $x=0$.

We find $f(0)$:

$f(0)=1-(0)^{2/3}=1$.

Therefore, there is a relative maximum at $(0,1)$.

We use the information obtained to sketch the graph. Other function values are listed below.

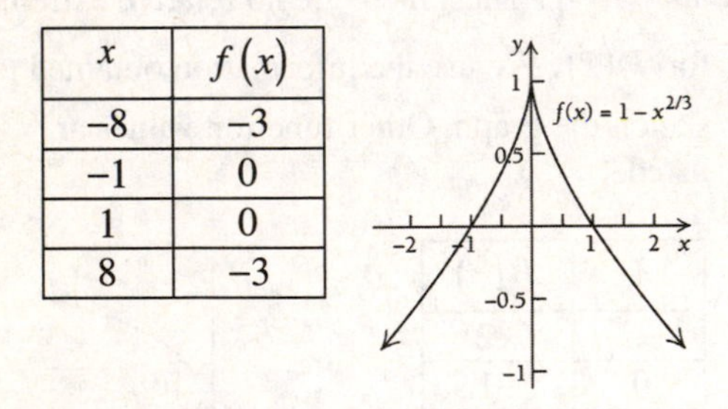

x	$f(x)$
-8	-3
-1	0
1	0
8	-3

27. $G(x)=\frac{-8}{x^2+1}=-8\left(x^2+1\right)^{-1}$

First, find the critical points.

$G'(x)=-8(-1)\left(x^2+1\right)^{-2}(2x)$

$=\frac{16x}{\left(x^2+1\right)^2}$

$G'(x)$ exists for all real numbers.

We set the derivative equal to zero and solve the equation.

$G'(x)=0$

$\frac{16x}{\left(x^2+1\right)^2}=0$

$16x=0$

$x=0$

The only critical value is 0.

We use 0 to divide the real number line into two intervals,

A: $(-\infty,0)$ and B: $(0,\infty)$:

A B

0

We use a test value in each interval to determine the sign of the derivative in each interval.

A: Test $-1, G'(-1)=\frac{16(-1)}{\left((-1)^2+1\right)^2}=\frac{-16}{4}=-4<0$

B: Test 1, $G'(1)=\frac{16(1)}{\left((1)^2+1\right)^2}=\frac{16}{4}=4>0$

We see that $G(x)$ is decreasing on $(-\infty,0)$ and increasing on $(0,\infty)$. Thus, a relative minimum occurs at $x=0$.

We find $G(0)$:

$G(0)=\frac{-8}{(0)^2+1}=-8.$

Thus, there is a relative minimum at $(0,-8)$.

We use the information obtained to sketch the graph. Other function values are listed below.

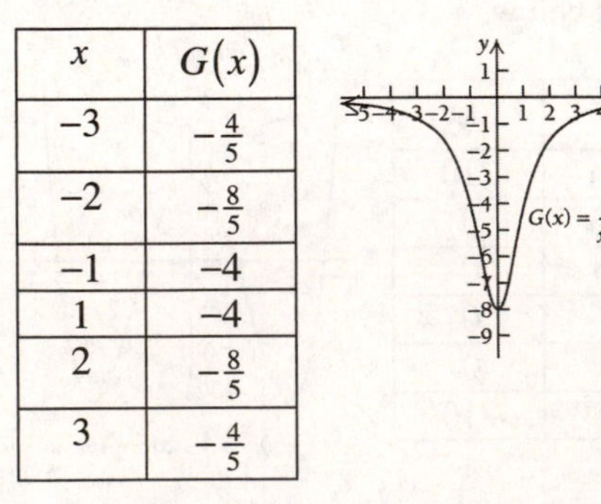

x	$G(x)$
-3	$-\frac{4}{5}$
-2	$-\frac{8}{5}$
-1	-4
1	-4
2	$-\frac{8}{5}$
3	$-\frac{4}{5}$

Copyright © 2014 Pearson Education, Inc. Publishing as Addison-Wesley.

29. $g(x)=\dfrac{4x}{x^2+1}$

First, find the critical points.

$$g'(x)=\frac{(x^2+1)(4)-4x(2x)}{(x^2+1)^2}\quad \text{Quotient Rule}$$
$$=\frac{4x^2+4-8x^2}{(x^2+1)^2}$$
$$=\frac{4-4x^2}{(x^2+1)^2}$$

$g'(x)$ exists for all real numbers. We solve

$$g'(x)=0$$
$$\frac{4-4x^2}{(x^2+1)^2}=0$$
$$4-4x^2=0 \quad \text{Multiplying by } (x^2+1)^2$$
$$x^2-1=0 \quad \text{Dividing by } -4$$
$$x^2=1$$
$$x=\pm\sqrt{1}$$
$$x=\pm 1.$$

The critical values are -1 and 1. We use them to divide the real number line into three intervals,

A: $(-\infty,-1)$, B: $(-1,1)$, and C:$(1,\infty)$.

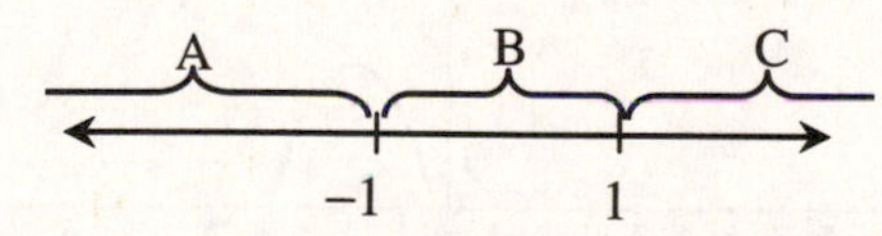

We use a test value in each interval to determine the sign of the derivative in each interval.

A: Test -2, $g'(-2)=\dfrac{4-4(-2)^2}{((-2)^2+1)^2}=-\dfrac{12}{25}<0$

B: Test 0, $g'(0)=\dfrac{4-4(0)^2}{((0)^2+1)^2}=4>0$

C: Test 2, $g'(2)=\dfrac{4-4(2)^2}{((2)^2+1)^2}=-\dfrac{12}{25}<0$

We see that $g(x)$ is decreasing on $(-\infty,-1)$, increasing on $(-1,1)$, and decreasing again on $(1,\infty)$. So there is a relative minimum at $x=-1$ and a relative maximum at $x=1$.

We find $g(-1)$:

$$g(-1)=\frac{4(-1)}{(-1)^2+1}=\frac{-4}{2}=-2.$$

Then we find $g(1)$:

$$g(1)=\frac{4(1)}{(1)^2+1}=\frac{4}{2}=2.$$

There is a relative minimum at $(-1,-2)$, and there is a relative maximum at $(1,2)$. We use the information obtained to sketch the graph. Other function values are listed below.

x	$g(x)$
−3	$-\frac{6}{5}$
−2	$-\frac{8}{5}$
0	0
2	$\frac{8}{5}$
3	$\frac{6}{5}$

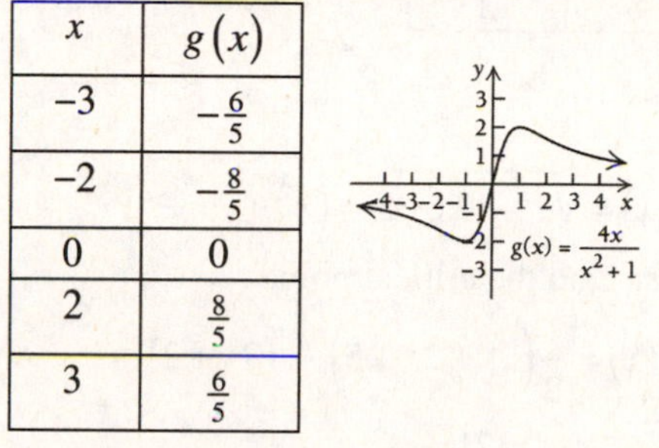

31. $f(x)=\sqrt[3]{x}=(x)^{1/3}$

First, find the critical points.

$$f'(x)=\frac{1}{3}(x)^{-2/3}$$
$$=\frac{1}{3(x)^{2/3}}=\frac{1}{3\cdot\sqrt[3]{x^2}}$$

$f'(x)$ does not exist when $x=0$. The equation $f'(x)=0$ has no solution, therefore, the only critical value is $x=0$.

We use 0 to divide the real number line into two intervals,

A: $(-\infty,0)$ and B: $(0,\infty)$:

A B

0

We use a test value in each interval to determine the sign of the derivative in each interval.

A: Test -1, $f'(-1)=\dfrac{1}{3\sqrt[3]{(-1)^2}}=\dfrac{1}{3}>0$

B: Test 1, $f'(1)=\dfrac{1}{3\left(\sqrt[3]{(1)^2}\right)}=\dfrac{1}{3}>0$

The solution is continued on the next page.

Copyright © 2014 Pearson Education, Inc. Publishing as Addison-Wesley.

On the previous page, we see that $f(x)$ is increasing on both $(-\infty,0)$ and $(0,\infty)$. Thus, there are no relative extrema for $f(x)$. We use the information obtained to sketch the graph. Other function values are listed below.

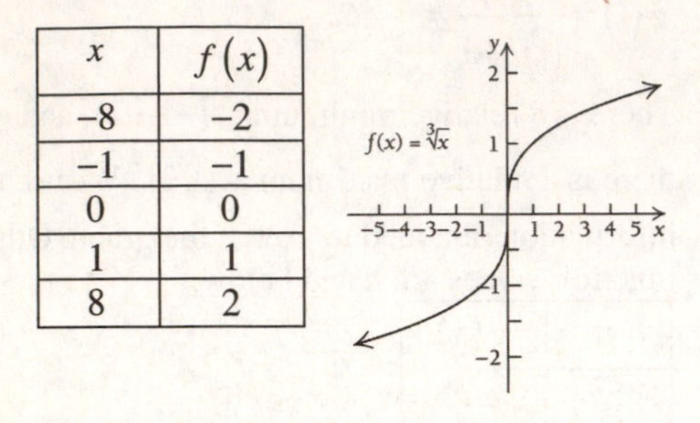

x	$f(x)$
−8	−2
−1	−1
0	0
1	1
8	2

33. $g(x)=\sqrt{x^2+2x+5}=\left(x^2+2x+5\right)^{1/2}$

First, find the critical points.

$$g'(x)=\frac{1}{2}\left(x^2+2x+5\right)^{-1/2}(2x+2)$$
$$=\frac{2(x+1)}{2\left(x^2+2x+5\right)^{1/2}}$$
$$=\frac{x+1}{\sqrt{x^2+2x+5}}$$

The equation $x^2+2x+5=0$ has no real-number solution, so $g'(x)$ exists for all real numbers. Next we find out where the derivative is zero. We solve

$$g'(x)=0$$
$$\frac{x+1}{\sqrt{x^2+2x+5}}=0$$
$$x+1=0$$
$$x=-1.$$

The only critical value is -1. We use -1 to divide the real number line into two intervals, A: $(-\infty,-1)$ and B: $(-1,\infty)$:

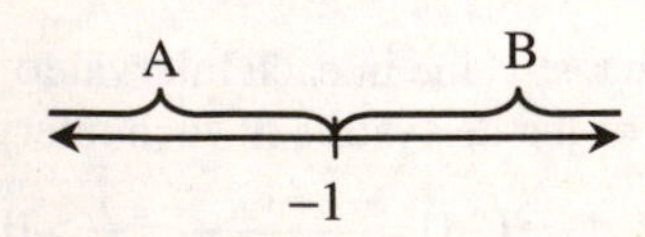

We use a test value in each interval to determine the sign of the derivative in each interval.

A: Test -2,

$$g'(-2)=\frac{(-2)+1}{\sqrt{(-2)^2+2(-2)+5}}=\frac{-1}{\sqrt{5}}<0$$

B: Test 0,

$$g'(0)=\frac{(0)+1}{\sqrt{(0)^2+2(0)+5}}=\frac{1}{\sqrt{5}}>0$$

We see that $g(x)$ is decreasing on $(-\infty,-1)$ and increasing on $(-1,\infty)$, and the change from decreasing to increasing indicates that a relative minimum occurs at $x=-1$. We substitute into the original equation to find $g(-1)$:

$$g(-1)=\sqrt{(-1)^2+2(-1)+5}=\sqrt{4}=2.$$

Thus, there is a relative minimum at $(-1,2)$. We use the information obtained to sketch the graph. Other function values are listed below.

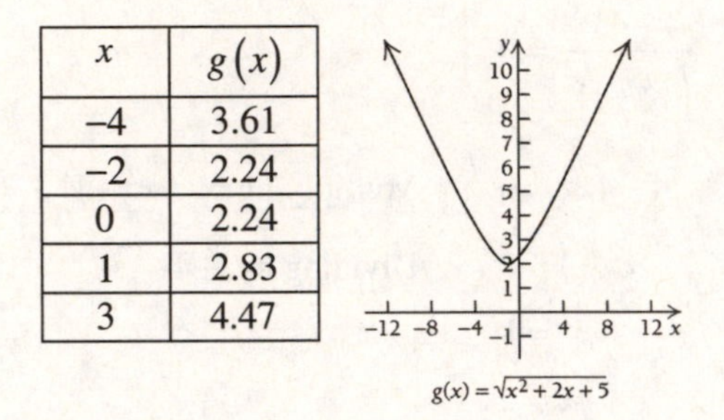

x	$g(x)$
−4	3.61
−2	2.24
0	2.24
1	2.83
3	4.47

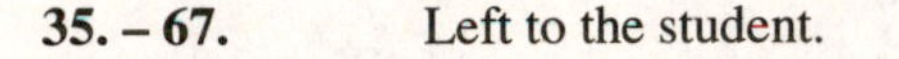

35. – 67. Left to the student.

69. Answers may vary, one such graph is:

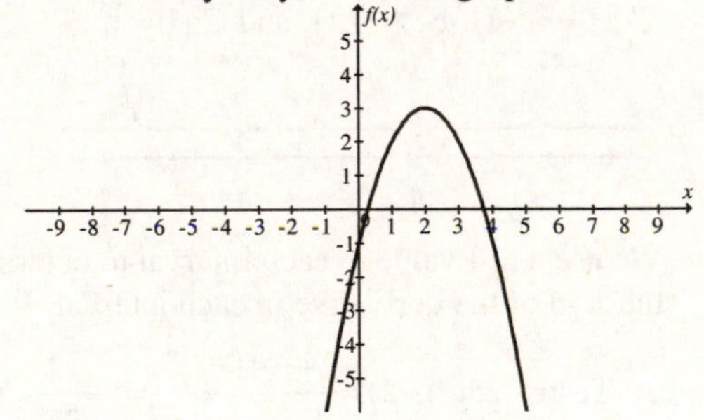

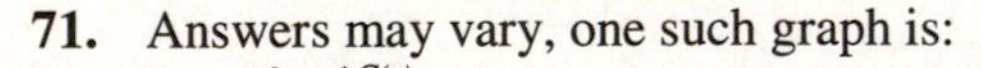

71. Answers may vary, one such graph is:

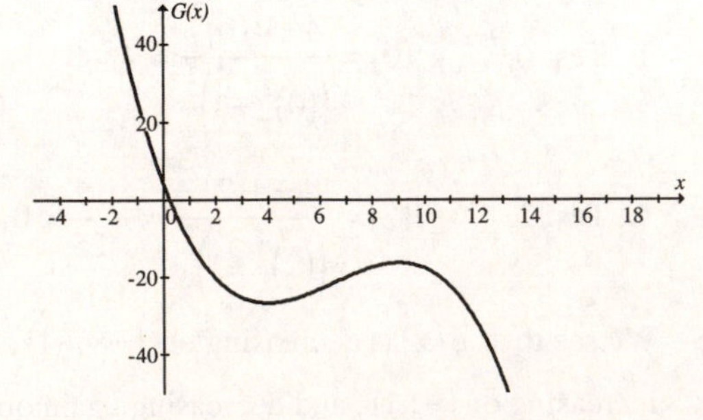

Copyright © 2014 Pearson Education, Inc. Publishing as Addison-Wesley.

73. Answers may vary, one such graph is:

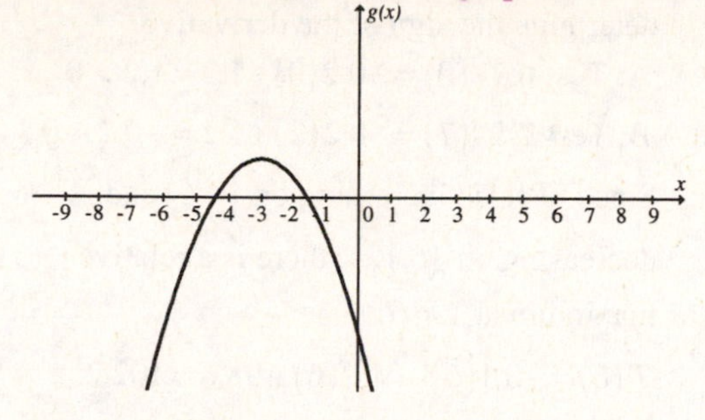

75. Answers may vary, one such graph is:

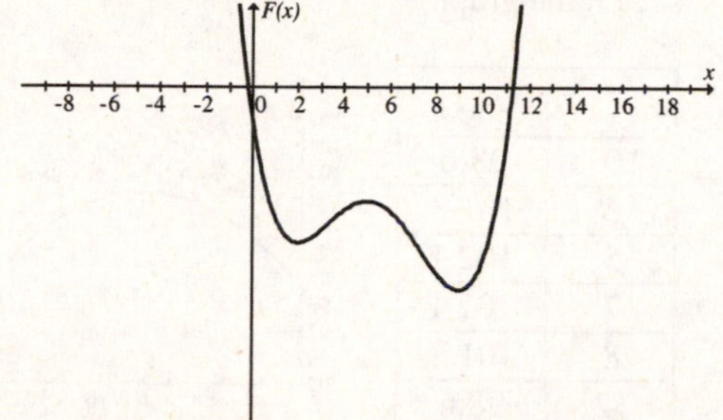

77. Answers may vary, one such graph is:

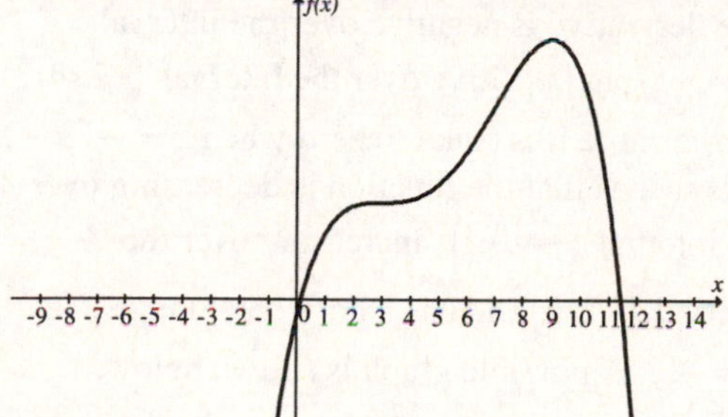

79. Answers may vary, one such graph is:

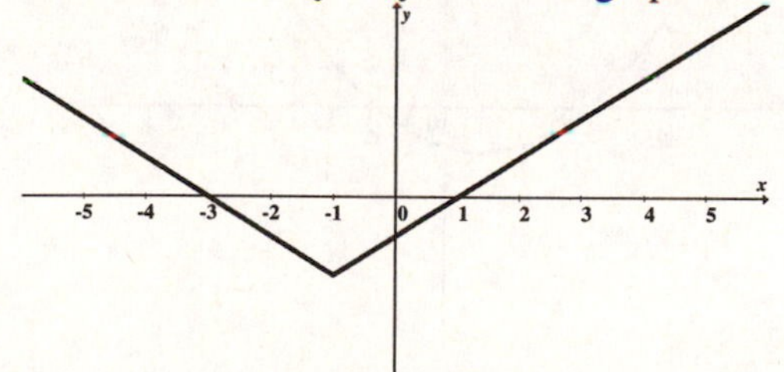

81. Answers may vary, one such graph is:

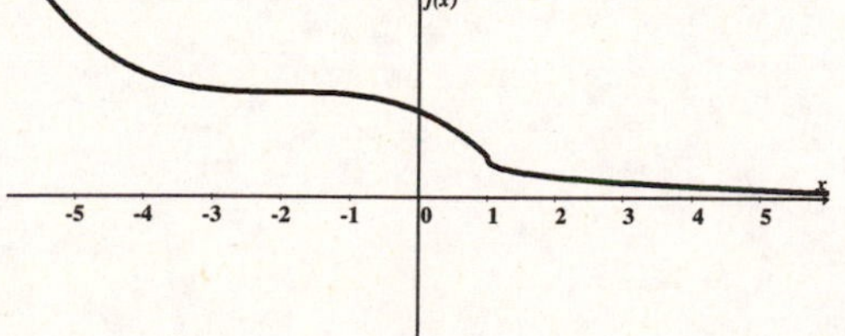

83. Answers may vary, one such graph is:

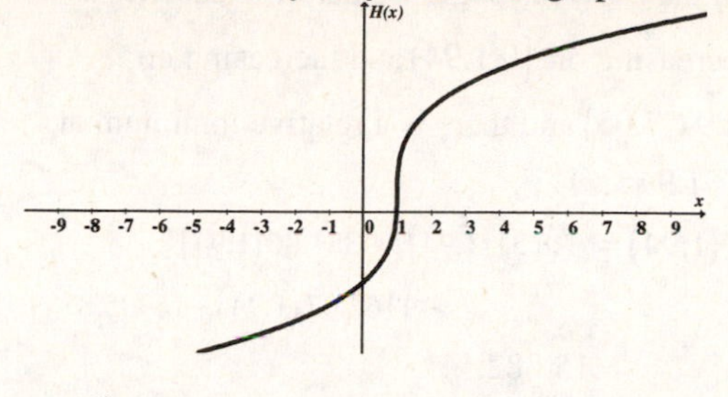

85.

87. Letting t be years since 2000 and E be thousand of employees, we have the function:

$E(t) = -28.31t^3 + 381.86t^2 - 1162.07t + 16905.87$

First, we find the critical points.

$E'(t) = -84.93t^2 + 763.72t - 1162.07$

$E'(t)$ exists for all real numbers. Solve

$$E'(t) = 0$$

$-84.93t^2 + 763.72t - 1162.07 = 0.$

Using the quadratic formula, we have:

$$t = \frac{-763.72 \pm \sqrt{(763.72)^2 - 4(-84.93)(-1162.07)}}{2(-84.93)}$$

$$= \frac{-763.72 \pm \sqrt{188,489.818}}{-169.86}$$

$t \approx 1.94$ or $t \approx 7.05$.

There are two critical values. We use them to divide the interval $[0, \infty)$ into three intervals:

A: $[0, 1.94)$ B: $(1.94, 7.05)$, and C: $(7.05, \infty)$

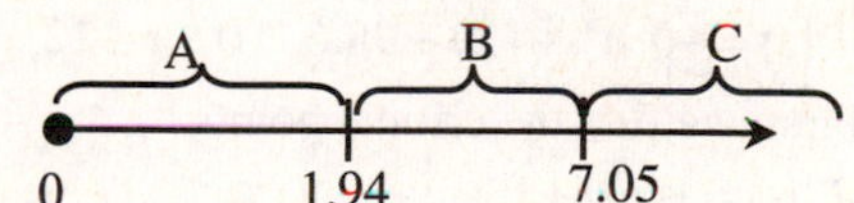

Next, we test a point in each interval to determine the sign of the derivative.

A: Test 1,

$$E'(1) = -84.93(1)^2 + 763.72(1) - 1162.07$$
$$= -483.28 < 0$$

B: Test 2,

$$E'(2) = -84.93(2)^2 + 763.72(2) - 1162.07$$
$$= 25.65 > 0$$

C: Test 8,

$$E'(8) = -84.93(8)^2 + 763.72(8) - 1162.07$$
$$= -487.83 < 0$$

The solution is continued on the next page.

Copyright © 2014 Pearson Education, Inc. Publishing as Addison-Wesley.

On the previous page, we determined $E(t)$ is decreasing on $[0,1.94)$ and increasing on $(1.94,7.05)$ and there is a relative minimum at $t=1.94$.

$$E(1.94)=-28.31(1.94)^3+381.86(1.94)^2 -1162.07(1.94)+16905.87 \approx 15,882.$$

There is a relative minimum at $(1.94,\ 15,882)$.

Since, $E(t)$ is increasing on $(1.94,7.05)$ and decreasing $[7.05,\infty)$ on and there is a relative maximum at $t=7.05$.

$$E(7.05)=-28.31(7.05)^3+381.86(7.05)^2 -1162.07(7.05)+16905.87 \approx 17,773.$$

There is a relative maxnimum at $(7.05,\ 17,773)$.

We sketch the graph.

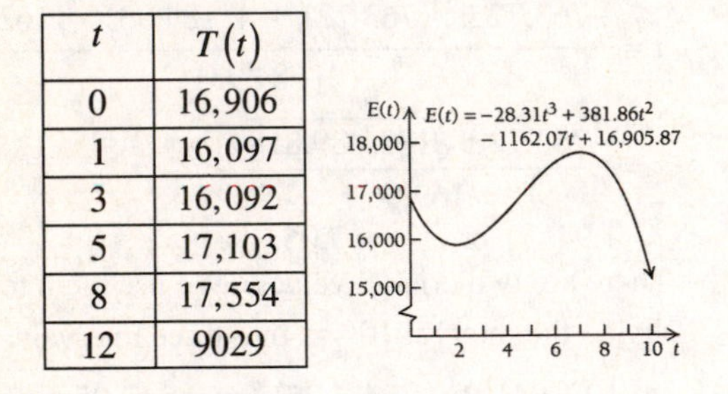

t	$T(t)$
0	16,906
1	16,097
3	16,092
5	17,103
8	17,554
12	9029

89. $T(t)=-0.1t^2+1.2t+98.6, \quad 0\le t\le 12$

First, we find the critical points.

$T'(t)=-0.2t+1.2$

$T'(t)$ exists for all real numbers. Solve

$$T'(t)=0$$
$$-0.2t+1.2=0$$
$$-0.2t=-1.2$$
$$t=6.$$

The only critical value is 6. We use it to divide the interval $[0,12]$ into two intervals:

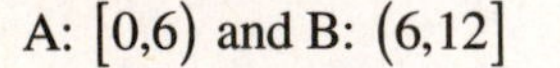

A: $[0,6)$ and B: $(6,12]$

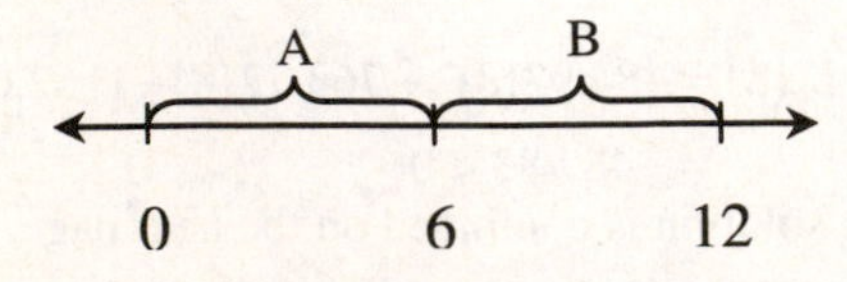

Next, we test a point in each interval to determine the sign of the derivative.

A: Test 0, $T'(0)=-0.2(0)+1.2=1.2>0$

B: Test 7, $T'(7)=-0.2(7)+1.2=-0.2<0$

Since, $T(t)$ is increasing on $[0,6)$ and decreasing on $(6,12]$, there is a relative maximum at $t=6$.

$$T(6)=-0.1(6)^2+1.2(6)+98.6=102.2.$$

There is a relative maximum at $(6,102.2)$. We sketch the graph.

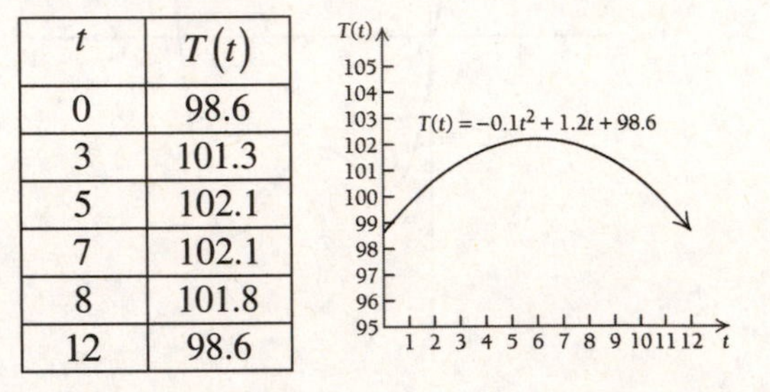

t	$T(t)$
0	98.6
3	101.3
5	102.1
7	102.1
8	101.8
12	98.6

91. The derivative is negative over the interval $(-\infty,-1)$ and positive over the interval $(-1,\infty)$. Furthermore it is equal to zero when $x=-1$. This means that the function is decreasing over the interval $(-\infty,-1)$, increasing over the interval $(-1,\infty)$ and has a horizontal tangent at $x=-1$. A possible graph is shown below.

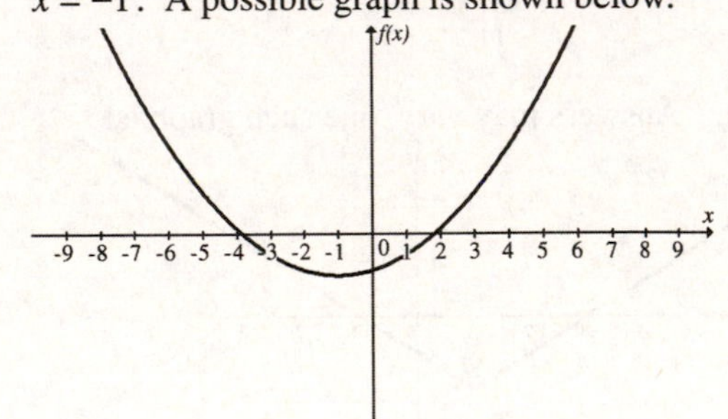

Copyright © 2014 Pearson Education, Inc. Publishing as Addison-Wesley.

93. The derivative is positive over the interval $(-\infty,1)$ and negative over the interval $(1,\infty)$. Furthermore it is equal to zero when $x=1$. This means that the function is increasing over the interval $(-\infty,1)$, decreasing over the interval $(1,\infty)$ and has a horizontal tangent at $x=1$. A possible graph is shown below.

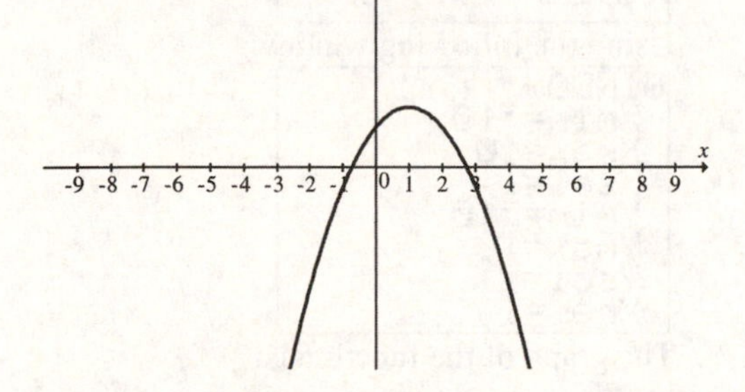

95. The derivative is positive over the interval $(-4,2)$ and negative over the intervals $(-\infty,-4)$ and $(2,\infty)$. Furthermore it is equal to zero when $x=-4$ and $x=2$. This means that the function is decreasing over the interval $(-\infty,-4)$, then increasing over the interval $(-4,2)$, and then decreasing again over the interval $(2,\infty)$. The function has horizontal tangents at $x=-4$ and $x=2$. A possible graph is:

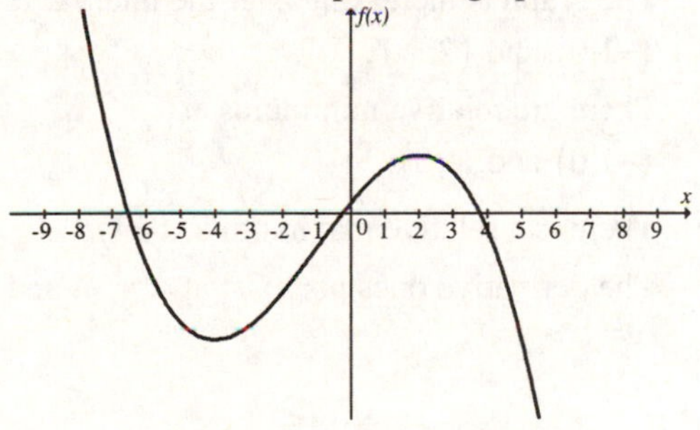

97. $f(x)=-x^6-4x^5+54x^4+160x^3-641x^2$
$-828x+1200$

Using the calculator we enter the function into the graphing editor as follows:

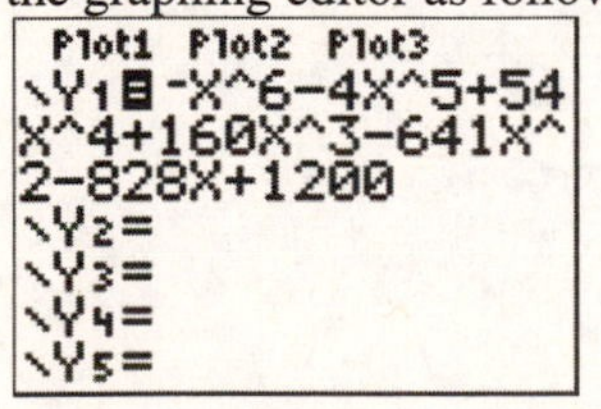

```
Plot1 Plot2 Plot3
\Y1■-X^6-4X^5+54
X^4+160X^3-641X^
2-828X+1200
\Y2=
\Y3=
\Y4=
\Y5=
```

Using the following window:

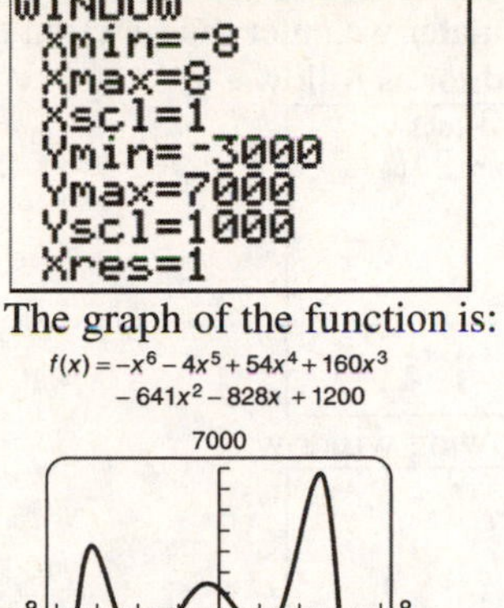

```
WINDOW
 Xmin=-8
 Xmax=8
 Xscl=1
 Ymin=-3000
 Ymax=7000
 Yscl=1000
 Xres=1
```

The graph of the function is:

$f(x)=-x^6-4x^5+54x^4+160x^3$
$-641x^2-828x+1200$

7000
−8 8
−3000

We find the relative extrema using the minimum/maximum feature on the calculator. There are relative minima at $(-3.683,-2288.03)$ and $(2.116,-1083.08)$.

There are relative maxima at $(-6.262,3213.8)$, $(-0.559,1440.06)$, and $(5.054,6674.12)$.

99. $f(x)=\sqrt[3]{|4-x^2|}+1$

We enter the function into the graphing editor.

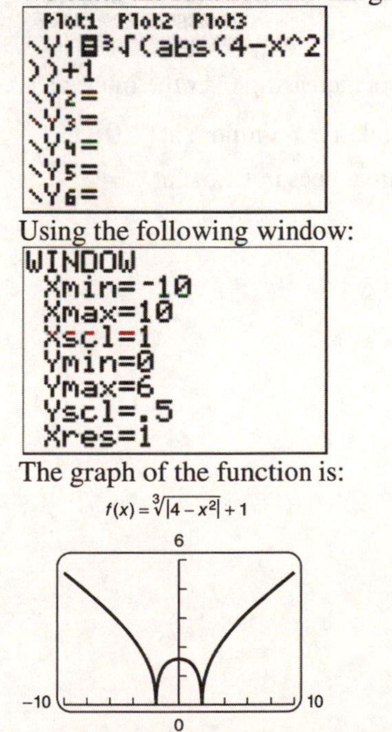

```
Plot1 Plot2 Plot3
\Y1■³√(abs(4-X^2
))+1
\Y2=
\Y3=
\Y4=
\Y5=
\Y6=
```

Using the following window:

```
WINDOW
 Xmin=-10
 Xmax=10
 Xscl=1
 Ymin=0
 Ymax=6
 Yscl=.5
 Xres=1
```

The graph of the function is:

$f(x)=\sqrt[3]{|4-x^2|}+1$

6
−10 10
0

We find the relative extrema using the minimum/maximum feature on the calculator. There are relative minima at $(-2,1)$ and $(2,1)$.

There is a relative maximum at $(0,2.587)$.

Copyright © 2014 Pearson Education, Inc. Publishing as Addison-Wesley.

101. $f(x)=|x-2|$

Using the calculator we enter the function into the graphing editor as follows:

```
Plot1 Plot2 Plot3
\Y1=abs(X-2)
\Y2=
\Y3=
\Y4=
\Y5=
\Y6=
\Y7=
```

Using the following window:

```
WINDOW
 Xmin=-10
 Xmax=10
 Xscl=1
 Ymin=-10
 Ymax=10
 Yscl=1
 Xres=1
```

The graph of the function is:

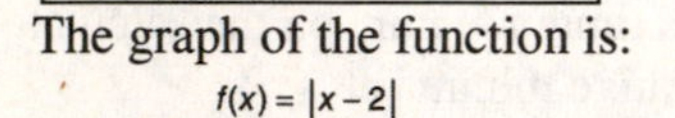

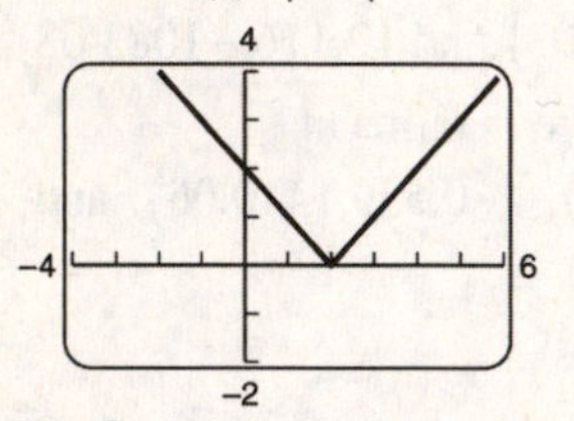

We find the relative extrema using the minimum/maximum feature on the calculator. The graph is decreasing over the interval $(-\infty,2)$.

The graph is increasing over the interval $(2,\infty)$.

There is a relative minimum at $(2,0)$.

The derivative does not exist at $x=2$.

103. $f(x)=|x^2-1|$

Using the calculator we enter the function into the graphing editor as follows:

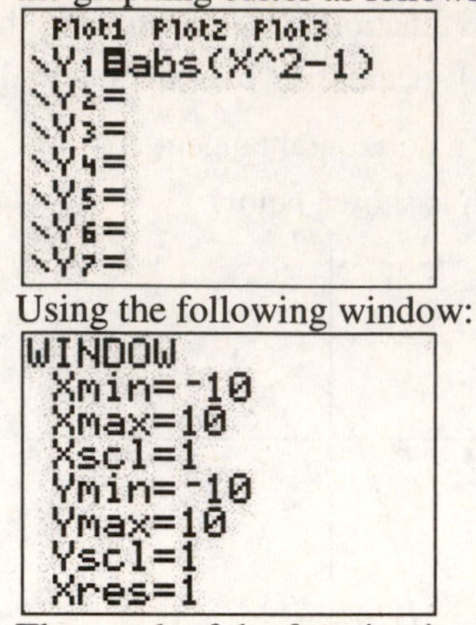

Using the following window:

The graph of the function is:

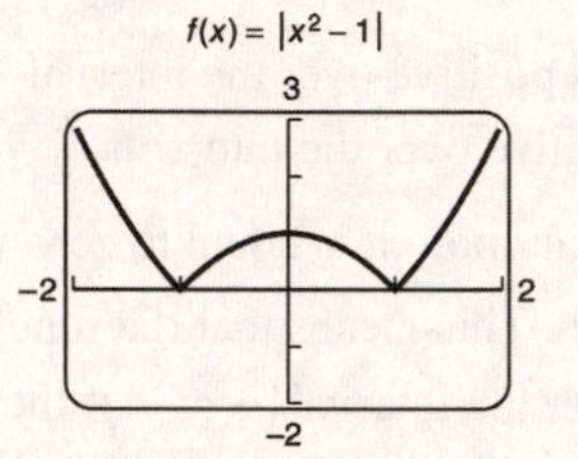

We find the relative extrema using the minimum/maximum feature on the calculator. The graph is decreasing over the interval $(-\infty,-1)$ and $(0,1)$.

The graph is increasing over the interval $(-1,0)$ and $(2,\infty)$.

There are relative minimums at $(-1,0)$ and $(1,0)$.

There is a relative maximum at $(0,1)$.

The derivative does not exist at $x=-1$ and $x=1$.

Copyright © 2014 Pearson Education, Inc. Publishing as Addison-Wesley.

105. $f(x)=\left|9-x^2\right|$

Using the calculator we enter the function into the graphing editor as follows:

```
Plot1 Plot2 Plot3
\Y1■abs(9-X^2)
\Y2=
\Y3=
\Y4=
\Y5=
\Y6=
\Y7=
```

Using the following window:

```
WINDOW
 Xmin=-10
 Xmax=10
 Xscl=1
 Ymin=-10
 Ymax=10
 Yscl=1
 Xres=1
```

The graph of the function is:

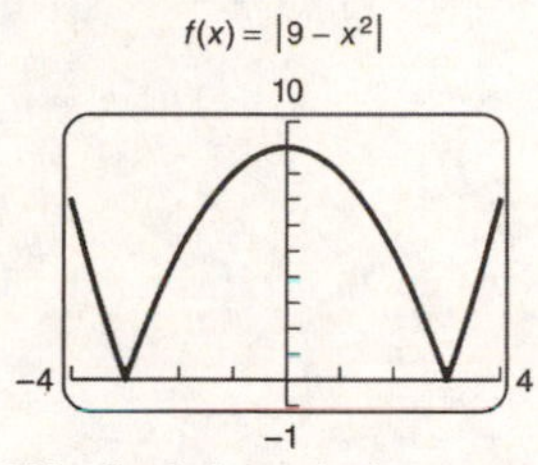

We find the relative extrema using the minimum/maximum feature on the calculator. The graph is decreasing over the interval $(-\infty,-3)$ and $(0,3)$.

The graph is increasing over the interval $(-3,0)$ and $(3,\infty)$.

There are relative minimums at $(-3,0)$ and $(3,0)$.

There is a relative maximum at $(0,9)$.

The derivative does not exist at $x=-3$ and $x=3$.

107. $f(x)=\left|x^3-1\right|$

Using the calculator we enter the function into the graphing editor as follows:

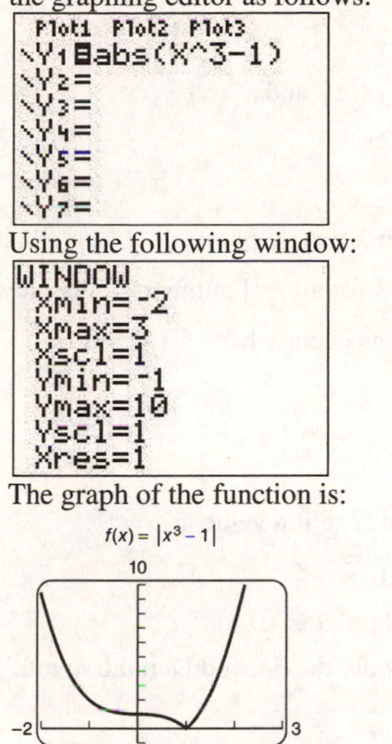

We find the relative extrema using the minimum/maximum feature on the calculator. The graph is decreasing over the interval $(-\infty,-1)$.

The graph is increasing over the interval $(1,\infty)$.

There is a relative minimum at $(1,0)$.

The derivative does not exist at $x=1$.

109. 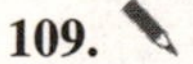

111.

Copyright © 2014 Pearson Education, Inc. Publishing as Addison-Wesley.

Exercise Set 2.2

1. $f(x)=5-x^2$

First, find $f'(x)$ and $f''(x)$.

$f'(x)=-2x$

$f''(x)=-2$

Next, find the critical points of $f(x)$. Since $f'(x)$ exists for all real numbers x, the only critical points occur when $f'(x)=0$.

$f'(x)=0$

$-2x=0$

$x=0$

We find the function value at $x=0$.

$f(0)=5-(0)^2=5$.

The critical point is $(0,5)$.

Next, we apply the Second Derivative test.

$f''(x)=-2$

$f''(0)=-2<0$

Therefore, $f(0)=5$ is a relative maximum.

3. $f(x)=x^2-x$

First, find $f'(x)$ and $f''(x)$.

$f'(x)=2x-1$

$f''(x)=2$

Next, find the critical points of $f(x)$. Since $f'(x)$ exists for all real numbers x, the only critical points occur when $f'(x)=0$.

$f'(x)=0$

$2x-1=0$

$2x=1$

$x=\frac{1}{2}$

We find the function value at $x=\frac{1}{2}$.

$f\left(\frac{1}{2}\right)=\left(\frac{1}{2}\right)^2-\left(\frac{1}{2}\right)=\frac{1}{4}-\frac{1}{2}=-\frac{1}{4}$.

The critical point is $\left(\frac{1}{2},-\frac{1}{4}\right)$.

Next, we apply the Second Derivative test.

$f''(x)=2$

$f''\left(\frac{1}{2}\right)=2>0$

Therefore, $f\left(\frac{1}{2}\right)=-\frac{1}{4}$ is a relative minimum.

5. $f(x)=-5x^2+8x-7$

First, find $f'(x)$ and $f''(x)$.

$f'(x)=-10x+8$

$f''(x)=-10$

Next, find the critical points of $f(x)$. Since $f'(x)$ exists for all real numbers x, the only critical points occur when $f'(x)=0$.

$f'(x)=0$

$-10x+8=0$

$-10x=-8$

$x=\frac{-8}{-10}$

$x=\frac{4}{5}$

We find the function value at $x=\frac{4}{5}$.

$$f\left(\frac{4}{5}\right)=-5\left(\frac{4}{5}\right)^2+8\left(\frac{4}{5}\right)-7$$
$$=-5\left(\frac{16}{25}\right)+\frac{32}{5}-\frac{7}{1}$$
$$=-\frac{16}{5}+\frac{32}{5}-\frac{35}{5}$$
$$=-\frac{19}{5}$$

The critical point is $\left(\frac{4}{5},-\frac{19}{5}\right)$.

Next, we apply the Second Derivative test.

$f''(x)=-10$

$f''\left(\frac{4}{5}\right)=-10<0$

Therefore, $f\left(\frac{4}{5}\right)=-\frac{19}{5}$ is a relative maximum.

Copyright © 2014 Pearson Education, Inc. Publishing as Addison-Wesley.

7. $f(x)=8x^3-6x+1$

First, find $f'(x)$ and $f''(x)$.

$f'(x)=24x^2-6$

$f''(x)=48x$

Next, find the critical points of $f(x)$. Since $f'(x)$ exists for all real numbers x, the only critical points occur when $f'(x)=0$.

$f'(x)=0$

$24x^2-6=0$

$4x^2-1=0$ Dividing by 6

$x^2=\frac{1}{4}$

$x=\pm\sqrt{\frac{1}{4}}$

$x=\pm\frac{1}{2}$

There are two critical values $x=-\frac{1}{2}$ and $x=\frac{1}{2}$.

We find the function value at $x=-\frac{1}{2}$

$$\begin{aligned} f\left(-\frac{1}{2}\right) &= 8\left(-\frac{1}{2}\right)^3-6\left(-\frac{1}{2}\right)+1 \\ &= 8\left(-\frac{1}{8}\right)+3+1 \\ &= -1+3+1 \\ &= 3 \end{aligned}$$

The critical point is $\left(-\frac{1}{2},3\right)$.

Next, we apply the Second Derivative test.

$f''(x)=48x$

$f''\left(-\frac{1}{2}\right)=48\left(-\frac{1}{2}\right)=-24<0$

Therefore, $f\left(-\frac{1}{2}\right)=3$ is a relative maximum.

Now, we find the function value at $x=\frac{1}{2}$.

$$\begin{aligned} f\left(\frac{1}{2}\right) &= 8\left(\frac{1}{2}\right)^3-6\left(\frac{1}{2}\right)+1 \\ &= 8\left(\frac{1}{8}\right)-3+1 \\ &= 1-3+1 \\ &= -1 \end{aligned}$$

The critical point is $\left(\frac{1}{2},-1\right)$.

Next, we apply the Second Derivative test.

$f''(x)=48x$

$f''\left(\frac{1}{2}\right)=48\left(\frac{1}{2}\right)=24>0$

Therefore, $f\left(\frac{1}{2}\right)=-1$ is a relative minimum.

9. $f(x)=x^3-12x$

a) Find $f'(x)$ and $f''(x)$.

$f'(x)=3x^2-12$

$f''(x)=6x$

The domain of f is $\mathbb{R}$.

b) Find the critical points of $f(x)$. Since $f'(x)$ exists for all real numbers x, the only critical points occur when $f'(x)=0$.

$f'(x)=0$

$3x^2-12=0$

$3x^2=12$

$x^2=4$

$x=\pm 2$

There are two critical values $x=-2$ and $x=2$.

We find the function value at $x=-2$

$f(-2)=(-2)^3-12(-2)=16$.

The critical point on the graph is $(-2,16)$.

Next, we find the function value at $x=2$.

$f(2)=(2)^3-12(2)=-16$.

The critical point on the graph is $(2,-16)$.

c) Apply the Second Derivative test to the critical points.

For $x=-2$

$f''(x)=6x$

$f''(-2)=6(-2)=-12<0$

The critical point $(-2,16)$ is a relative maximum.

The solution is continued on the next page.

Copyright © 2014 Pearson Education, Inc. Publishing as Addison-Wesley.

For $x=2$

$f''(x)=6x$

$f''(2)=6(2)=12>0$

The critical point $(2,-16)$ is a relative minimum.

If we use the critical values $x=-2$ and $x=2$ to divide the real line into three intervals, $(-\infty,-2),(-2,2)$, and $(2,\infty)$, we know from the extrema above, that $f(x)$ is increasing over the interval $(-\infty,-2)$, decreasing over the interval $(-2,2)$ and then increasing again over the interval $(2,\infty)$.

d) Find the points of inflection. $f''(x)$ exists for all real numbers, so we solve the equation

$f''(x)=0$

$6x=0$

$x=0$

Therefore, a possible inflection point occurs at $x=0$.

$f(0)=(0)^3-12(0)=0$.

This gives the point $(0,0)$ on the graph.

e) To determine concavity, we use the possible inflection point to divide the real number line into two intervals $A:(-\infty,0)$ and $B:(0,\infty)$. We test a point in each interval

A: Test $-1: f''(-1)=6(-1)=-6<0$

B: Test $1: \quad f''(1)=6(1)=6>0$

Then, $f(x)$ is concave down on the interval $(-\infty,0)$ and concave up on the interval $(0,\infty)$, so $(0,0)$ is an inflection point.

f) Finally, we use the preceding information to sketch the graph of the function.

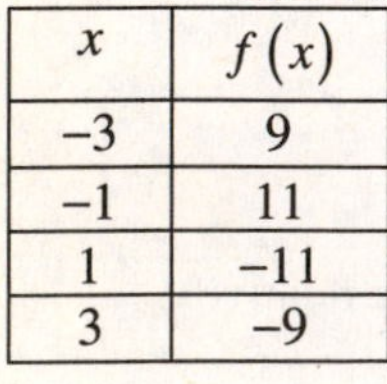

x	$f(x)$
−3	9
−1	11
1	−11
3	−9

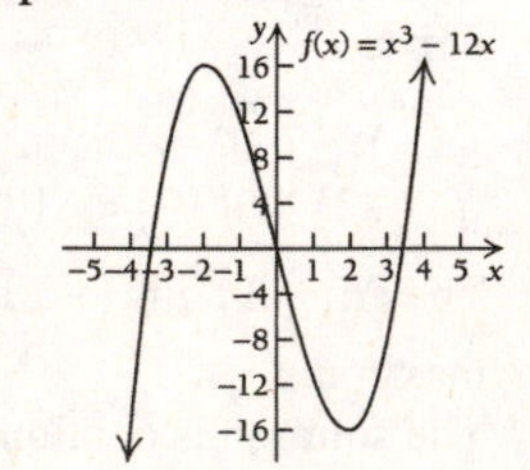

11. $f(x)=3x^3-36x-3$

a) First, find $f'(x)$ and $f''(x)$.

$f'(x)=9x^2-36$

$f''(x)=18x$

The domain of f is $\mathbb{R}$.

b) Find the critical points of $f(x)$. Since $f'(x)$ exists for all real numbers x, the only critical points occur when $f'(x)=0$.

$f'(x)=0$

$9x^2-36=0$

$9x^2=36$

$x^2=4$

$x=\pm 2$

There are two critical values $x=-2$ and $x=2$.

We find the function value at $x=-2$

$f(-2)=3(-2)^3-36(-2)-3=45$.

The critical point on the graph is $(-2,45)$.

Next, we find the function value at $x=2$.

$f(2)=3(2)^3-36(2)-3=-51$.

The critical point on the graph is $(2,-51)$.

c) Apply the Second Derivative test to the critical points.

For $x=-2$

$f''(x)=18x$

$f''(-2)=18(-2)=-36<0$

The critical point $(-2,45)$ is a relative maximum.

For $x=2$

$f''(x)=18x$

$f''(2)=18(2)=36>0$

The critical point $(2,-51)$ is a relative minimum.

c) We use the critical values $x=-2$ and $x=2$ to divide the real line into three intervals, A:$(-\infty,-2)$, B: $(-2,2)$, and C: $(2,\infty)$, we know from the extrema above, that $f(x)$ is increasing over the interval $(-\infty,-2)$, decreasing over the interval $(-2,2)$ and then increasing again over the interval $(2,\infty)$.

Copyright © 2014 Pearson Education, Inc. Publishing as Addison-Wesley.

d) Find the points of inflection. $f''(x)$ exists for all real numbers, so we solve the equation

$$f''(x)=0$$
$$18x=0$$
$$x=0$$

Therefore, a possible inflection point occurs at $x=0$.

$f(0)=3(0)^3-36(0)-3=-3$.

This gives the point $(0,-3)$ on the graph.

e) To determine concavity, we use the possible inflection point to divide the real number line into two intervals

$A:(-\infty,0)$ and $B:(0,\infty)$. We test a point in each interval

A: Test -1: $f''(-1)=18(-1)=-18<0$

B: Test 1: $f''(1)=18(1)=18>0$

Then, $f(x)$ is concave down on the interval $(-\infty,0)$ and concave up on the interval $(0,\infty)$, so $(0,-3)$ is an inflection point.

f) We use the preceding information to sketch the graph of the function. Additional function values can also be calculated as needed.

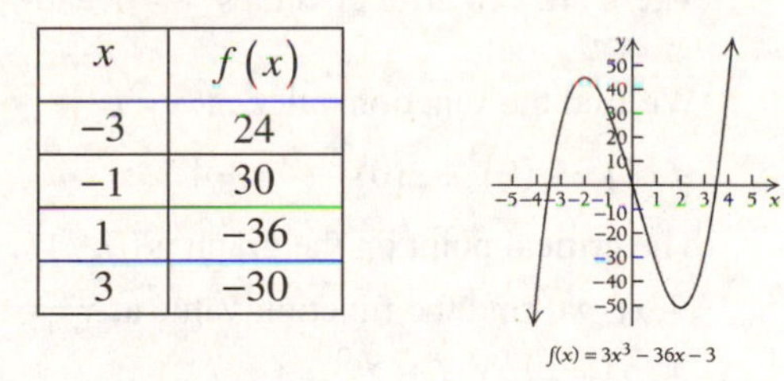

x	$f(x)$
-3	24
-1	30
1	-36
3	-30

$f(x)=3x^3-36x-3$

13. $f(x)=\frac{8}{3}x^3-2x+\frac{1}{3}$

a) Find $f'(x)$ and $f''(x)$.

$$f'(x)=8x^2-2$$
$$f''(x)=16x$$

The domain of f is $\mathbb{R}$.

b) Next, find the critical points of $f(x)$.

Since $f'(x)$ exists for all real numbers x, the only critical points occur when $f'(x)=0$.

We set the first derivative equal to zero and solve for x at the top of the next column.

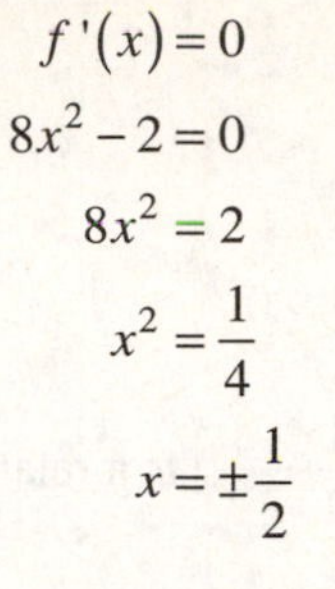

$$f'(x)=0$$
$$8x^2-2=0$$
$$8x^2=2$$
$$x^2=\frac{1}{4}$$
$$x=\pm\frac{1}{2}$$

There are two critical values $x=-\frac{1}{2}$ and $x=\frac{1}{2}$.

We find the function value at $x=-\frac{1}{2}$

$$f\left(-\frac{1}{2}\right)=\frac{8}{3}\left(-\frac{1}{2}\right)^3-2\left(-\frac{1}{2}\right)+\frac{1}{3}$$
$$=\frac{8}{3}\left(-\frac{1}{8}\right)+1+\frac{1}{3}$$
$$=-\frac{1}{3}+1+\frac{1}{3}$$
$$=1$$

The critical point on the graph is $\left(-\frac{1}{2},1\right)$.

Next, we find the function value at $x=\frac{1}{2}$.

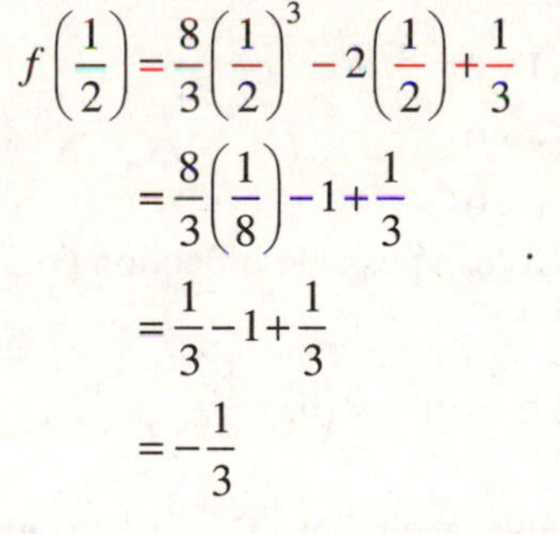

$$f\left(\frac{1}{2}\right)=\frac{8}{3}\left(\frac{1}{2}\right)^3-2\left(\frac{1}{2}\right)+\frac{1}{3}$$
$$=\frac{8}{3}\left(\frac{1}{8}\right)-1+\frac{1}{3}$$
$$=\frac{1}{3}-1+\frac{1}{3}$$
$$=-\frac{1}{3}$$

The critical point on the graph is $\left(\frac{1}{2},-\frac{1}{3}\right)$.

c) Apply the Second Derivative test to the critical points.

For $x=-\frac{1}{2}$

$$f''(x)=16x$$
$$f''\left(-\frac{1}{2}\right)=16\left(-\frac{1}{2}\right)=-8<0$$

The critical point $\left(-\frac{1}{2},1\right)$ is a relative maximum.

The solution is continued on the next page.

Copyright © 2014 Pearson Education, Inc. Publishing as Addison-Wesley.

For $x=\frac{1}{2}$

$f''(x)=16x$

$f''\left(\frac{1}{2}\right)=16\left(\frac{1}{2}\right)=8>0$

The critical point $\left(\frac{1}{2},-\frac{1}{3}\right)$ is a relative minimum.

We use the critical values $x=-\frac{1}{2}$, and $x=\frac{1}{2}$ to divide the real line into three intervals,

A: $\left(-\infty,-\frac{1}{2}\right)$, B: $\left(-\frac{1}{2},\frac{1}{2}\right)$, and C: $\left(\frac{1}{2},\infty\right)$,

we know from the extrema above, that $f(x)$ is increasing over the interval $\left(-\infty,-\frac{1}{2}\right)$, decreasing over the interval $\left(-\frac{1}{2},\frac{1}{2}\right)$ and then increasing again over the interval $\left(\frac{1}{2},\infty\right)$.

d) We find the points of inflection. $f''(x)$ exists for all real numbers, so we solve the equation

$$f''(x)=0$$
$$16x=0$$
$$x=0$$

Therefore, a possible inflection point occurs at $x=0$.

$f(0)=\frac{8}{3}(0)^3-2(0)+\frac{1}{3}=\frac{1}{3}$.

This gives the point $\left(0,\frac{1}{3}\right)$ on the graph.

e) To determine concavity, we use the possible inflection point to divide the real number line into two intervals $A:(-\infty,0)$ and $B:(0,\infty)$. We test a point in each interval

A: Test -1: $f''(-1)=16(-1)=-16<0$

B: Test 1: $f''(1)=16(1)=16>0$

Then, $f(x)$ is concave down on the interval $(-\infty,0)$ and concave up on the interval $(0,\infty)$, so $\left(0,\frac{1}{3}\right)$ is an inflection point.

f) We use the preceding information to sketch the graph of the function.

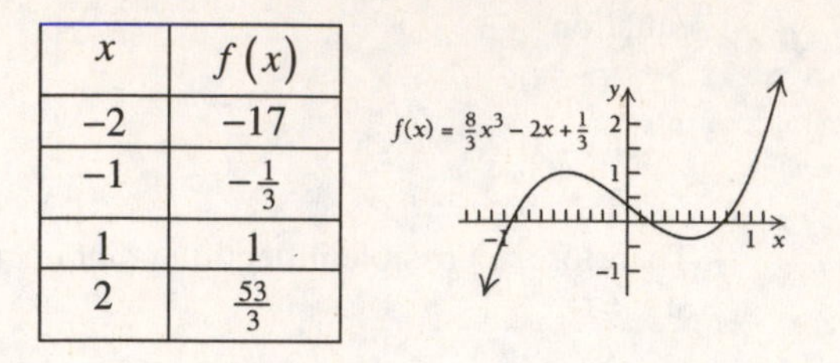

x	$f(x)$
-2	-17
-1	$-\frac{1}{3}$
1	1
2	$\frac{53}{3}$

15. $f(x)=-x^3+3x^2-4$

a) First, find $f'(x)$ and $f''(x)$.

$$f'(x)=-3x^2+6x$$
$$f''(x)=-6x+6$$

The domain of f is $\mathbb{R}$.

b) Find the critical points of $f(x)$. Since $f'(x)$ exists for all real numbers x, the only critical points occur when $f'(x)=0$.

$$f'(x)=0$$
$$-3x^2+6x=0$$
$$-3x(x-2)=0$$
$$-3x=0 \quad \text{or} \quad x-2=0$$
$$x=0 \quad \text{or} \quad x=2$$

There are two critical values $x=0$ and $x=2$.

We find the function value at $x=0$

$f(0)=-(0)^3+3(0)^2-4=-4$

The critical point on the graph is $(0,-4)$.

Next, we find the function value at $x=6$.

$$f(2)=-(2)^3+3(2)^2-4$$
$$=-8+12-4$$
$$=0$$

The critical point on the graph is $(2,0)$.

c) Apply the Second Derivative test to the critical points.

For $x=0$

$f''(x)=-6x+6$

$f''(0)=-6(0)+6=6>0$

The critical point $(0,-4)$ is a relative minimum.

The solution is continued on the next page.

Copyright © 2014 Pearson Education, Inc. Publishing as Addison-Wesley.

For $x = 2$

$f''(x) = -6x + 6$

$f''(2) = -6(2) + 6 = -6 < 0$

The critical point $(2,0)$ is a relative maximum.

We use the critical values $x = 0$ and $x = 2$ to divide the real line into three intervals, $A:(-\infty,0)$, B: $(0,2)$, and C: $(2,\infty)$, we know from the extrema above, that $f(x)$ is decreasing over the interval $(-\infty,0)$, increasing over the interval $(0,2)$ and then decreasing again over the interval $(2,\infty)$.

d) Find the points of inflection. $f''(x)$ exists for all real numbers, so we solve the equation

$f''(x) = 0$

$-6x + 6 = 0$

$-6x = -6$

$x = 1$

Therefore, a possible inflection point occurs at $x = 1$.

$f(1) = -(1)^3 + 3(1)^2 - 4$

$= -1 + 3 - 4$

$= -2$

This gives the point $(1,-2)$ on the graph.

e) To determine concavity, we use the possible inflection point to divide the real number line into two intervals $A:(-\infty,1)$ and B:$(1,\infty)$. We test a point in each interval

A: Test 0: $f''(0) = -6(0) + 6 = 6 > 0$

B: Test 2: $f''(2) = -6(2) + 6 = -2 < 0$

Then, $f(x)$ is concave up on the interval $(-\infty,1)$ and concave down on the interval $(1,\infty)$, so $(1,-2)$ is an inflection point.

f) We use the preceding information to sketch the graph of the function. Additional function values can also be calculated as needed.

x	$f(x)$
-2	16
-1	0
3	-4
4	-20

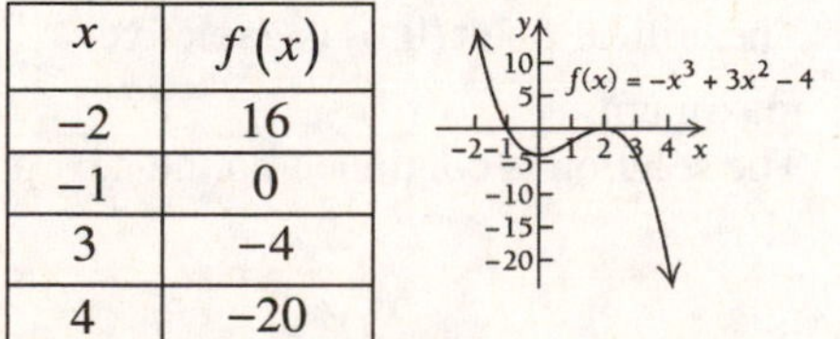

17. $f(x) = 3x^4 - 16x^3 + 18x^2$

a) First, find $f'(x)$ and $f''(x)$.

$f'(x) = 12x^3 - 48x^2 + 36x$

$f''(x) = 36x^2 - 96x + 36$

The domain of f is $\mathbb{R}$.

b) Next, find the critical points of $f(x)$. Since $f'(x)$ exists for all real numbers x, the only critical points occur when $f'(x) = 0$.

$f'(x) = 0$

$12x^3 - 48x^2 + 36x = 0$

$12x(x^2 - 4x + 3) = 0$

$12x(x-1)(x-3) = 0$

$12x = 0$ or $x - 1 = 0$ or $x - 3 = 0$

$x = 0$ or $x = 1$ or $x = 3$

There are three critical values $x = 0$, $x = 1$, and $x = 3$.

Then

$f(0) = 3(0)^4 - 16(0)^3 + 18(0)^2 = 0$

$f(1) = 3(1)^4 - 16(1)^3 + 18(1)^2 = 5$

$f(3) = 3(3)^4 - 16(3)^3 + 18(3)^2 = -27$

Thus, the critical points $(0,0)$, $(1,5)$, and $(3,-27)$ are on the graph.

c) Apply the Second Derivative test to the critical points.

$f''(0) = 36(0)^2 - 96(0) + 36 = 36 > 0$

The critical point $(0,0)$ is a relative minimum.

$f''(1) = 36(1)^2 - 96(1) + 36 = -24 < 0$

The critical point $(1,5)$ is a relative maximum.

$f''(3) = 36(3)^2 - 96(3) + 36 = 72 > 0$

The critical point $(3,-27)$ is a relative minimum.

We use the critical values $0, 1,$ and 3 to divide the real line into four intervals, $A:(-\infty,0)$, B:$(0,1)$, C:$(1,3)$ and D:$(3,\infty)$, we know from the extrema above, that $f(x)$ is decreasing over the intervals $(-\infty,0)$ and $(1,3)$ and $f(x)$ increasing over the intervals $(0,1)$ and $(3,\infty)$.

Copyright © 2014 Pearson Education, Inc. Publishing as Addison-Wesley.

d) Find the points of inflection. $f''(x)$ exists for all real numbers, so we solve the equation $f''(x)=0$.

$$f''(x)=0$$
$$36x^2-96x+36=0$$
$$12(3x^2-8x+3)=0$$
$$3x^2-8x+3=0$$

Using the quadratic formula, we find that $x=\frac{4\pm\sqrt{7}}{3}$, so $x\approx 0.451$ or $x\approx 2.215$ are possible inflection points.

$$f(0.451)\approx 2.321$$
$$f(2.215)\approx -13.358$$

So, $(0.451,2.321)$ and $(2.215,-13.358)$ are two more points on the graph.

e) To determine concavity, we use the possible inflection point to divide the real number line into three intervals $A:(-\infty,0.451)$, $B:(0.451,2.215)$, and $C:(2.215,\infty)$.

We test a point in each interval to determine the sign of the second derivative.

A: Test 0:

$$f''(0)=36(0)^2-96(0)+36=36>0$$

B: Test 1:

$$f''(1)=36(1)^2-96(1)+36=-24<0$$

C: Test 3:

$$f''(3)=36(3)^2-96(3)+36=72>0$$

Then, $f(x)$ is concave up on the interval $(-\infty,0.451)$ and concave down on the interval $(0.451,2.215)$ and concave up on the interval $(2.215,\infty)$, so $(0.451,2.321)$ and $(2.215,-13.358)$ are inflection points.

f) We use the preceding information to sketch the graph of the function.

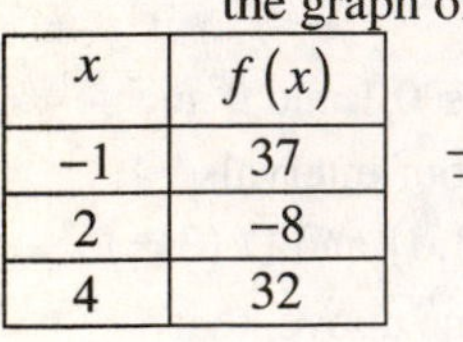

x	$f(x)$
−1	37
2	−8
4	32

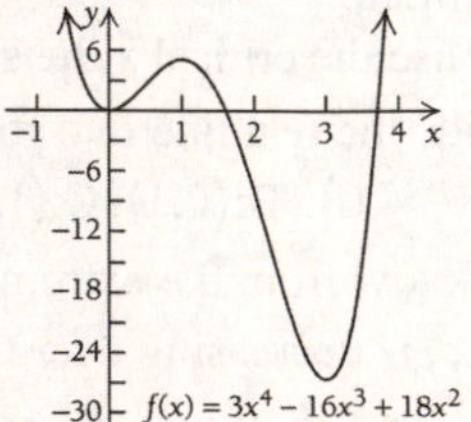

19. $f(x)=x^4-6x^2$

a) First, find $f'(x)$ and $f''(x)$.

$$f'(x)=4x^3-12x$$
$$f''(x)=12x^2-12$$

The domain of f is $\mathbb{R}$.

b) Find the critical points of $f(x)$. Since $f'(x)$ exists for all real numbers x, the only critical points occur when $f'(x)=0$.

$$f'(x)=0$$
$$4x^3-12x=0$$
$$4x(x^2-3)=0$$
$$4x=0 \quad \text{or} \quad x^2-3=0$$
$$x=0 \quad \text{or} \quad x=\pm\sqrt{3}$$

There are three critical values $-\sqrt{3}$, 0, and $\sqrt{3}$.

Then

$$f(-\sqrt{3})=(-\sqrt{3})^4-6(-\sqrt{3})^2 =9-6(3) =-9$$
$$f(0)=(0)^4-6(0)^2=0$$
$$f(\sqrt{3})=(\sqrt{3})^4-6(\sqrt{3})^2 =9-6(3) =-9$$

Thus, the critical points $(-\sqrt{3},-9)$, $(0,0)$, $(\sqrt{3},-9)$ and are on the graph.

c) Apply the Second Derivative test to the critical points.

$$f''(-\sqrt{3})=12(-\sqrt{3})^2-12 =12(3)-12=24>0$$

The critical point $(-\sqrt{3},-9)$ is a relative minimum.

$$f''(0)=12(0)^2-12=-12<0$$

The critical point $(0,0)$ is a relative maximum.

The solution is continued on the next page.

Copyright © 2014 Pearson Education, Inc. Publishing as Addison-Wesley.

Applying the Second Derivative test to the next critical point, we have:

$$f''(\sqrt{3}) = 12(\sqrt{3})^2 - 12$$
$$= 12(3) - 12 = 24 > 0$$

The critical point $(\sqrt{3}, -9)$ is a relative minimum.

If we use the critical values $-\sqrt{3}$, 0, and $\sqrt{3}$ to divide the real line into four intervals, A: $(-\infty, -\sqrt{3})$, B: $(-\sqrt{3}, 0)$, C: $(0, \sqrt{3})$, and D: $(\sqrt{3}, \infty)$

Then $f(x)$ is decreasing over the intervals $(-\infty, -\sqrt{3})$ and $(0, \sqrt{3})$, and $f(x)$ increasing over the intervals $(-\sqrt{3}, 0)$ and $(\sqrt{3}, \infty)$.

d) Find the points of inflection. $f''(x)$ exists for all real numbers, so we solve the equation

$$f''(x) = 0$$
$$12x^2 - 12 = 0$$
$$x^2 - 1 = 0$$
$$x^2 = 1$$
$$x = \pm 1$$

So $x = -1$ or $x = 1$ are possible inflection points.

$$f(-1) = (-1)^4 - 6(-1)^2 = 1 - 6 = -5$$
$$f(1) = (1)^4 - 6(1)^2 = 1 - 6 = -5$$

So, $(-1, -5)$ and $(1, -5)$ are two more points on the graph.

e) To determine concavity, we use the possible inflection point to divide the real number line into three intervals $A: (-\infty, -1)$, B: $(-1, 1)$, and C: $(1, \infty)$.

We test a point in each interval

A: Test -2:

$$f''(-2) = 12(-2)^2 - 12 = 36 > 0$$

B: Test 0:

$$f''(0) = 12(0)^2 - 12 = -12 < 0$$

C: Test 2:

$$f''(2) = 12(2)^2 - 12 = 36 > 0$$

Then, $f(x)$ is concave up on the intervals $(-\infty, -1)$ and $(1, \infty)$ and concave down on the interval $(-1, 1)$, so $(-1, -5)$ and $(1, -5)$ are inflection points.

f) We use the preceding information to sketch the graph of the function. Additional function values can also be calculated as needed.

x	$f(x)$
−3	27
−2	−8
2	−8
3	27

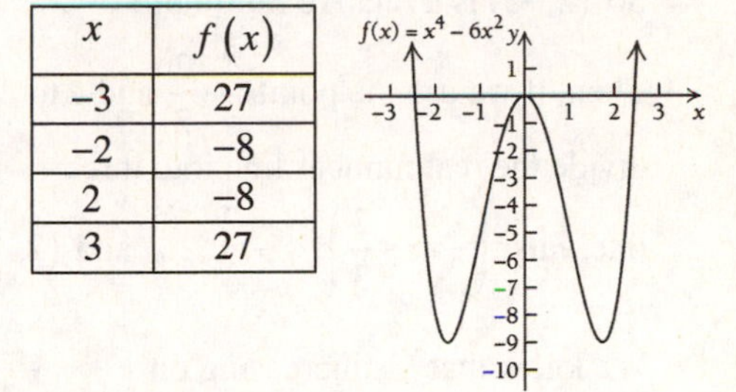

21. $f(x) = x^3 - 2x^2 - 4x + 3$

a) $f'(x) = 3x^2 - 4x - 4$

$f''(x) = 6x - 4$

The domain of f is $\mathbb{R}$.

b) $f'(x)$ exists for all values of x, so the only critical points of f are where $f'(x) = 0$.

$$3x^2 - 4x - 4 = 0$$
$$(3x + 2)(x - 2) = 0$$
$$3x + 2 = 0 \quad \text{or} \quad x - 2 = 0$$
$$x = -\frac{2}{3} \quad \text{or} \quad x = 2$$

The critical values are $-\frac{2}{3}$ and 2.

We determine the function values for each critical value.

$$f\left(-\frac{2}{3}\right) = \left(-\frac{2}{3}\right)^3 - 2\left(-\frac{2}{3}\right)^2 - 4\left(-\frac{2}{3}\right) + 3$$
$$= -\frac{8}{27} - \frac{8}{9} + \frac{8}{3} + 3$$
$$= -\frac{8}{27} - \frac{24}{27} + \frac{72}{27} + \frac{81}{27}$$
$$= \frac{121}{27}$$

$$f(2) = (2)^3 - 2(2)^2 - 4(2) + 3$$
$$= 8 - 8 - 8 + 3$$
$$= -5$$

The critical points $\left(-\frac{2}{3}, \frac{121}{7}\right)$ and $(2, -5)$ are on the graph.

Copyright © 2014 Pearson Education, Inc. Publishing as Addison-Wesley.

c) Applying the Second Derivative Test, we have:

$$f''\left(-\frac{2}{3}\right)=6\left(-\frac{2}{3}\right)-4=-4-4=-8<0$$

So $\left(-\frac{2}{3},\frac{121}{7}\right)$ is a relative maximum.

$f''(2)=6(2)-4=12-4=8>0$

So $(2,-5)$ is a relative minimum.

Then, if we use the points $-\frac{2}{3}$ and 2 to divide the real number line into three intervals, $\left(-\infty,-\frac{2}{3}\right)$, $\left(-\frac{2}{3},2\right)$, and $(2,\infty)$.

We know that f is increasing on $\left(-\infty,-\frac{2}{3}\right)$ and on $(2,\infty)$ and f is decreasing on $\left(-\frac{2}{3},2\right)$.

d) Find the points of inflection. $f''(x)$ exists for all values of x, so the only possible inflection points occur when $f''(x)=0$.

$$6x-4=0$$
$$6x=4$$
$$x=\frac{4}{6}=\frac{2}{3}$$

There is a possible inflection point at $x=\frac{2}{3}$.

$$f\left(\frac{2}{3}\right)=\left(\frac{2}{3}\right)^3-2\left(\frac{2}{3}\right)^2-4\left(\frac{2}{3}\right)+3$$
$$=\frac{8}{27}-\frac{8}{9}-\frac{8}{3}+3$$
$$=\frac{8}{27}-\frac{24}{27}-\frac{72}{27}+\frac{81}{27}$$
$$=-\frac{7}{27}$$

Another point on the graph is $\left(\frac{2}{3},-\frac{7}{27}\right)$.

e) To determine concavity we use $\frac{2}{3}$ to divide the real number line into two intervals, A: $\left(-\infty,\frac{2}{3}\right)$ and B: $\left(\frac{2}{3},\infty\right)$. Then test a point in each interval.

A: Test 0, $f''(0)=6(0)-4=-4<0$

B: Test 1, $f''(1)=6(1)-4=2>0$

We see that f is concave down on $\left(-\infty,\frac{2}{3}\right)$ and concave up on $\left(\frac{2}{3},\infty\right)$, so $\left(\frac{2}{3},-\frac{7}{27}\right)$ is an inflection point.

f) We sketch the graph using the preceding information. Additional function values may also be calculated as necessary.

x	$f(x)$
−2	−5
−1	4
0	3
1	−2
3	0
4	19

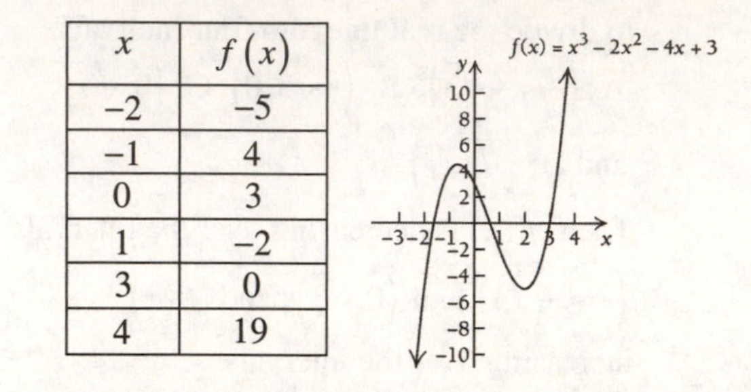

23. $f(x)=3x^4+4x^3$

a) $f'(x)=12x^3+12x^2$

$f''(x)=36x^2+24x$

The domain of f is $\mathbb{R}$.

b) $f'(x)$ exists for all values of x, so the only critical points of f are where $f'(x)=0$.

$$12x^3+12x^2=0$$
$$12x^2(x+1)=0$$
$$12x^2=0 \quad \text{or} \quad x+1=0$$
$$x=0 \quad \text{or} \quad x=-1$$

The critical values are -1 and 0.

$f(-1)=3(-1)^4+4(-1)^3=-1$

$f(0)=3(0)^4+4(0)^3=0$

The critical points $(-1,-1)$ and $(0,0)$ are on the graph.

c) Applying the Second Derivative Test, we have:

$$f''(-1)=36(-1)^2+24(-1)=36-24$$
$$=12>0$$

So $(-1,-1)$ is a relative minimum.

$f''(0)=36(0)^2+24(0)=0$

The test fails. We will use the First Derivative Test.

The solution is continued on the next page.

Copyright © 2014 Pearson Education, Inc. Publishing as Addison-Wesley.

We use 0 to divide the interval $(-1,\infty)$ into two intervals, A: $(-1,0)$ and B: $(0,\infty)$, and test a point in each interval.

A: Test $-\frac{1}{2}$,

$$f'\left(-\frac{1}{2}\right)=12\left(-\frac{1}{2}\right)^3+12\left(-\frac{1}{2}\right)^2$$
$$=\frac{3}{2}>0$$

B: Test 2,

$$f'(2)=12(2)^3+12(2)^2=144>0$$

f is increasing on both intervals $(-1,0)$ and $(0,\infty)$. Therefore, $(0,0)$ is not a relative extremum. Since $(-1,-1)$ is a relative minimum, we know that f is decreasing on $(-\infty,-1)$.

d) Find the points of inflection. $f''(x)$ exists for all values of x, so the only possible inflection points occur when $f''(x)=0$.

$$f''(x)=0$$
$$36x^2+24x=0$$
$$12x(3x+2)=0$$
$$12x=0 \quad \text{or} \quad 3x+2=0$$
$$x=0 \quad \text{or} \quad x=-\frac{2}{3}$$

There are a possible inflection points at $x=-\frac{2}{3}$ and $x=0$.

$$f\left(-\frac{2}{3}\right)=3\left(-\frac{2}{3}\right)^4+4\left(-\frac{2}{3}\right)^3$$
$$=3\left(\frac{16}{81}\right)+4\left(-\frac{8}{27}\right)$$
$$=\frac{16}{27}-\frac{32}{27}$$
$$=-\frac{16}{27}$$

$$f(0)=3(0)^4+4(0)^3=0$$

This gives one additional point $\left(-\frac{2}{3},-\frac{16}{27}\right)$ on the graph.

e) To determine concavity we use $-\frac{2}{3}$ and 0 to divide the real number line into three intervals, A: $\left(-\infty,-\frac{2}{3}\right)$, B: $\left(-\frac{2}{3},0\right)$, and C: $(0,\infty)$. Then test a point in each interval.

A: Test -1,

$$f''(-1)=36(-1)^2+24(-1)=12>0$$

B: Test $-\frac{1}{2}$,

$$f''\left(-\frac{1}{2}\right)=36\left(-\frac{1}{2}\right)^2+24\left(-\frac{1}{2}\right)$$
$$=-3<0$$

C: Test 1,

$$f''(1)=36(1)^2+24(1)=60>0$$

We see that f is concave up on the intervals $\left(-\infty,-\frac{2}{3}\right)$ and $(0,\infty)$, and concave down on the interval $\left(-\frac{2}{3},0\right)$, so both $\left(-\frac{2}{3},-\frac{16}{27}\right)$ and $(0,0)$ are inflection points.

f) We sketch the graph using the preceding information. Additional function values may also be calculated as necessary.

x	$f(x)$
-2	16
1	7
2	80

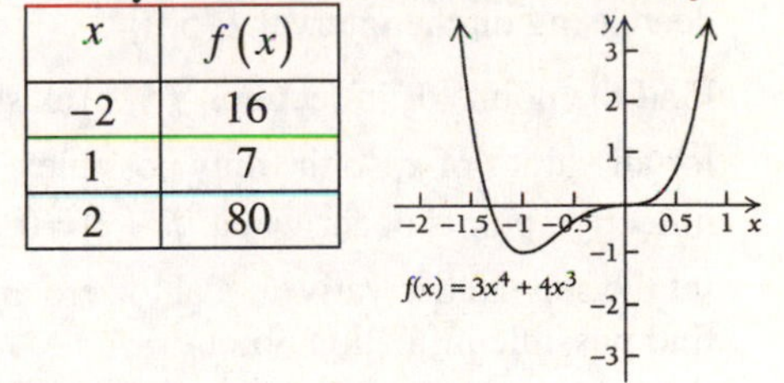

25. $f(x)=x^3-6x^2-135x$

a) $f'(x)=3x^2-12x-135$

$f''(x)=6x-12$

The domain of f is $\mathbb{R}$.

b) $f'(x)$ exists for all values of x, so the only critical points of f are where $f'(x)=0$.

$$3x^2-12x-135=0$$
$$x^2-4x-45=0$$
$$(x-9)(x+5)=0$$
$$x-9=0 \quad \text{or} \quad x+5=0$$
$$x=9 \quad \text{or} \quad x=-5$$

The solution is continued on the next page.

Copyright © 2014 Pearson Education, Inc. Publishing as Addison-Wesley.

On the previous page, we determined the critical values are -5 and 9.

$f(-5)=(-5)^3-6(-5)^2-135(-5)$
$=-125-150+675$
$=400$

$f(9)=(9)^3-6(9)^2-135(9)$
$=729-486-1215$
$=-972$

The critical points $(-5,400)$ and $(9,-972)$ are on the graph.

c) Applying the Second Derivative Test, we have:

$f''(-5)=6(-5)-12=-30-12$
$=-42<0$

The critical point $(-5,400)$ is a relative maximum.

$f''(9)=6(9)-12=54-12$
$=42>0$

The critical point $(9,-972)$ is a relative minimum.
If we use the points -5 and 9 to divide the real number line into three intervals $(-\infty,-5)$, $(-5,9)$, and $(9,\infty)$ we see that $f(x)$ is increasing on the intervals $(-\infty,-5)$ and $(9,\infty)$ and $f(x)$ is decreasing on the interval $(-5,9)$.

d) Find the points of inflection. $f''(x)$ exists for all values of x, so the only possible inflection points occur when $f''(x)=0$. We set the second derivative equal to zero and find possible inflection points.
We set the second derivative equal to zero and solve for x.

$6x-12=0$
$6x=12$
$x=2$

The only possible inflection point is 2.

$f(2)=(2)^3-(2)^2-135(2)$
$=8-24-270$
$=-286$

The point $(2,-286)$ is a possible inflection point on the graph.

e) To determine concavity we use 2 to divide the real number line into two intervals, $A:(-\infty,2)$ and B: $(2,\infty)$, Then test a point in each interval.

A: Test 0, $f''(0)=6(0)-12=-12<0$
B: Test 3, $f''(3)=6(3)-12=6>0$

We see that f is concave down on the interval $(-\infty,2)$ and concave up on the interval $(2,\infty)$, Therefore $(2,-286)$ is an inflection point.

f) We sketch the graph using the preceding information. Additional function values may also be calculated as necessary.

x	$f(x)$
−11	−572
−10	−250
−3	324
0	0
2	−286
5	−700
15	0
16	400

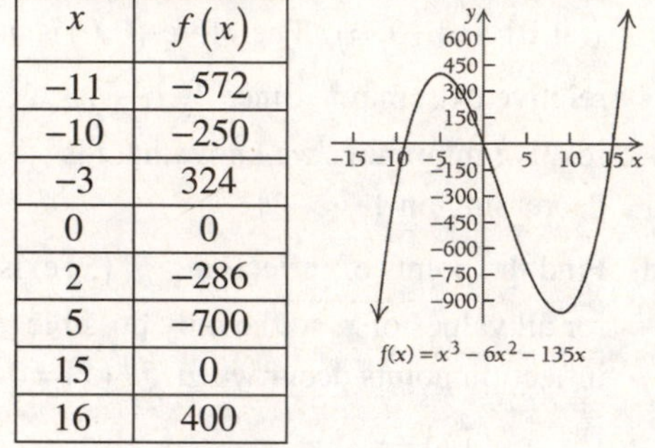

27. $f(x)=x^4-4x^3+10$

a) $f'(x)=4x^3-12x^2$
$f''(x)=12x^2-24x$
The domain of f is $\mathbb{R}$.

b) $f'(x)$ exists for all values of x, so the only critical points of f are where $f'(x)=0$.

$4x^3-12x^2=0$
$4x^2(x-3)=0$
$4x^2=0 \quad \text{or} \quad x-3=0$
$x=0 \quad \text{or} \quad x=3$

The critical values are 0 and 3.

$f(0)=(0)^4-4(0)^3+10=10$
$f(3)=(3)^4-4(3)^3+10=-17$

The critical points $(0,10)$ and $(3,-17)$ are on the graph.

c) Applying the Second Derivative Test, we have:

$f''(0)=12(0)^2-24(0)=0$

The test fails, we will use the First Derivative Test.
The solution is continued on the next page.

Copyright © 2014 Pearson Education, Inc. Publishing as Addison-Wesley.

Divide $(-\infty,3)$ into two intervals,

A: $(-\infty,0)$ and B: $(0,3)$, and test a point in each interval.

A: Test -1,

$$f'(-1)=4(-1)^3-12(-1)^2=-16<0$$

B: Test 1, $f'(1)=4(1)^3-12(1)^2=-8<0$

Since, f is decreasing on both intervals, $(0,10)$ is not a relative extremum.

We use the Second Derivative Test for $x=3$.

$$f''(3)=12(3)^2-24(3)=36>0$$

The critical point $(3,-17)$ is a relative minimum.

When we applied the First Derivative Test, we saw that $f(x)$ was decreasing on the intervals $(-\infty,0)$ and $(0,3)$. Since $(3,-17)$ is a relative minimum, we know that $f(x)$ is increasing on $(3,\infty)$.

d) Find the points of inflection. $f''(x)$ exists for all values of x, so the only possible inflection points occur when $f''(x)=0$.

$$12x^2-24x=0$$
$$12x(x-2)=0$$
$$12x=0 \quad \text{or} \quad x-2=0$$
$$x=0 \quad \text{or} \quad x=2$$
$$f(0)=(0)^4-4(0)^3+10=10$$
$$f(2)=(2)^4-4(2)^3+10=-6$$

The points $(0,10)$ and $(2,-6)$ are possible inflection points on the graph.

e) To determine concavity we use 0 and 2 to divide the real number line into three intervals,

A: $(-\infty,0)$, B: $(0,2)$, and C: $(2,\infty)$, Then test a point in each interval.

A: Test -1,

$$f''(-1)=12(-1)^2-24(-1)=36>0$$

B: Test 1,

$$f''(1)=12(1)^2-24(1)=-12<0$$

C: Test 3,

$$f''(3)=12(3)^2-24(3)=36>0$$

We see that f is concave up on the intervals $(-\infty,0)$ and $(2,\infty)$ and concave down on the interval $(0,2)$. Therefore both $(0,10)$ and $(2,-6)$ are inflection points.

f) We sketch the graph using the preceding information. Additional function values may also be calculated as necessary.

x	$f(x)$
-2	58
-1	15
1	7
4	10
5	135

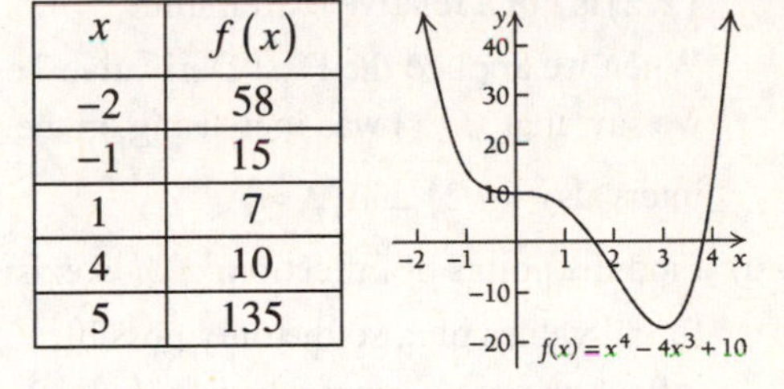

29. $f(x)=x^3-6x^2+12x-6$

a) $f'(x)=3x^2-12x+12$

$f''(x)=6x-12$

The domain of f is $\mathbb{R}$.

b) $f'(x)$ exists for all values of x, so the only critical points of f are where $f'(x)=0$.

$$3x^2-12x+12=0$$
$$x^2-4x+4=0 \quad \text{Dividing by 3}$$
$$(x-2)^2=0$$
$$x-2=0$$
$$x=2$$

The critical value is 2.

$$f(2)=(2)^3-6(2)^2+12(2)-6=2$$

The critical point $(2,2)$ is on the graph.

c) Applying the Second Derivative Test, we have:

$$f''(2)=6(2)-12=0$$

The test fails, we will use the First Derivative Test.

Divide the real line into two intervals,

A: $(-\infty,2)$ and B: $(2,\infty)$, and test a point in each interval.

The solution is continued on the next page.

Copyright © 2014 Pearson Education, Inc. Publishing as Addison-Wesley.

We test a point in each interval.
A: Test 0,

$$f'(0)=3(0)^2-12(0)+12=12>0$$

B: Test 3,

$$f'(3)=3(3)^2-12(3)+12=3>0$$

Since, f is increasing on both intervals, $(2,2)$ is not a relative extremum.

When we applied the First Derivative Test, we saw that $f(x)$ was increasing on the intervals $(-\infty,2)$ and $(2,\infty)$.

d) Find the points of inflection. $f''(x)$ exists for all values of x, so the only possible inflection points occur when $f''(x)=0$.

$$6x-12=0$$
$$6x=12$$
$$x=2$$

We have already seen that $f(2)=2$, so the point $(2,2)$ is a possible inflection point on the graph.

e) To determine concavity we use 2 to divide the real number line into two intervals, A: $(-\infty,2)$ and B: $(2,\infty)$, Then test a point in each interval.
A: Test 0, $f''(0)=6(0)-12=-12<0$
B: Test 3, $f''(3)=6(3)-12=6>0$
We see that $f(x)$ is concave down on the interval $(-\infty,2)$ and concave up on the interval $(2,\infty)$. Therefore, the point $(2,2)$ is an inflection point.

f) We sketch the graph using the preceding information. Additional function values may also be calculated as necessary.

x	$f(x)$
−1	−25
0	−6
1	1
3	3
4	10

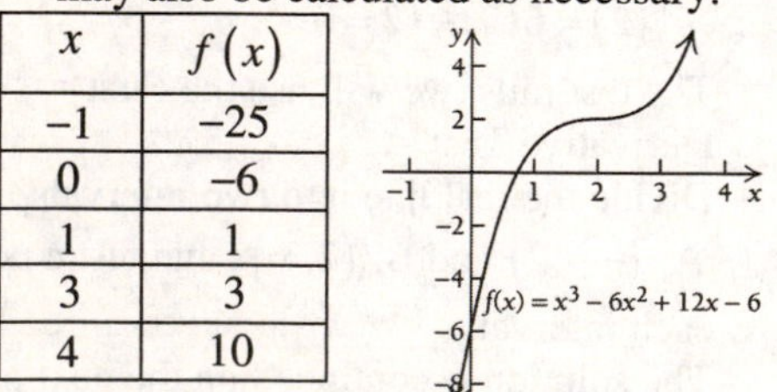

31. $f(x)=5x^3-3x^5$

a) $f'(x)=15x^2-15x^4$

$f''(x)=30x-60x^3$

The domain of f is $\mathbb{R}$.

b) $f'(x)$ exists for all values of x, so the only critical points of f are where $f'(x)=0$.

$$15x^2-15x^4=0$$
$$15x^2(1-x^2)=0$$
$$15x^2=0 \quad \text{or} \quad 1-x^2=0$$
$$x=0 \quad \text{or} \quad x=\pm 1$$

We deteremined the critical values are −1, 0, and 1.

$$f(-1)=5(-1)^3-3(-1)^5=-2$$
$$f(0)=5(0)^3-3(0)^5=0$$
$$f(1)=5(1)^3-3(1)^5=2$$

The critical points $(-1,-2)$, $(0,0)$ and $(1,2)$ are on the graph.

c) Applying the Second Derivative Test, we have:

$$f''(-1)=30(-1)-60(-1)^3=30>0$$

So, the critical point $(-1,-2)$ is a relative minimum.

$$f''(0)=30(0)-60(0)^3=0$$

The test fails, we will use the First Derivative Test.
Divide $(-1,1)$ into two intervals, A: $(-1,0)$ and B: $(0,1)$, and test a point in each interval.

A: Test $-\frac{1}{2}$,

$$f'\left(-\frac{1}{2}\right)=15\left(-\frac{1}{2}\right)^2-15\left(-\frac{1}{2}\right)^4 =\frac{45}{16}>0$$

B: Test $\frac{1}{2}$,

$$f'\left(\frac{1}{2}\right)=15\left(\frac{1}{2}\right)^2-15\left(\frac{1}{2}\right)^4 =\frac{45}{16}>0$$

Since, f is increasing on both intervals, $(0,0)$ is not a relative extremum.

The solution is continued on the next page.

Copyright © 2014 Pearson Education, Inc. Publishing as Addison-Wesley.

We use the Second Derivative Test for $x = 1$.

$f''(1) = 30(1) - 60(1)^3 = -30 < 0$

The critical point $(1, 2)$ is a relative maximum.

When we applied the First Derivative Test, we saw that $f(x)$ was increasing on the intervals $(-1, 0)$ and $(0, 1)$. Since $(-1, -2)$ is a relative minimum, we know that $f(x)$ is decreasing on $(-\infty, -1)$.

Since $(1, 2)$ is a relative maximum, we know that $f(x)$ is decreasing on $(1, \infty)$.

d) Find the points of inflection. $f''(x)$ exists for all values of x, so the only possible inflection points occur when $f''(x) = 0$.

$$30x - 60x^3 = 0$$
$$30x(1 - 2x^2) = 0$$
$$30x = 0 \quad \text{or} \quad 1 - 2x^2 = 0$$
$$x = 0 \quad \text{or} \quad x^2 = \frac{1}{2}$$
$$x = 0 \quad \text{or} \quad x = \pm\sqrt{\frac{1}{2}} = \pm\frac{1}{\sqrt{2}}$$

$$f\left(-\frac{1}{\sqrt{2}}\right) = 5\left(-\frac{1}{\sqrt{2}}\right)^3 - 3\left(-\frac{1}{\sqrt{2}}\right)^5 = -1.237$$

$$f(0) = 5(0)^3 - 3(0)^5 = 0$$

$$f\left(\frac{1}{\sqrt{2}}\right) = 5\left(\frac{1}{\sqrt{2}}\right)^3 - 3\left(\frac{1}{\sqrt{2}}\right)^5 = 1.237$$

The points $\left(-\frac{1}{\sqrt{2}}, -1.237\right)$, $(0, 0)$ and $\left(\frac{1}{\sqrt{2}}, 1.237\right)$ are possible inflection points on the graph.

e) To determine concavity we use $-\frac{1}{\sqrt{2}}$, 0, and $\frac{1}{\sqrt{2}}$ to divide the real number line into four intervals, A: $\left(-\infty, -\frac{1}{\sqrt{2}}\right)$, B: $\left(-\frac{1}{\sqrt{2}}, 0\right)$, C: $\left(0, \frac{1}{\sqrt{2}}\right)$, and D: $\left(\frac{1}{\sqrt{2}}, \infty\right)$.

We test a point in each interval.

A: Test -1, $f''(-1) = 30(-1) - 60(-1)^3 = 30 > 0$

B: Test $-\frac{1}{2}$,

$$f''\left(-\frac{1}{2}\right) = 30\left(-\frac{1}{2}\right) - 60\left(-\frac{1}{2}\right)^3 = -\frac{15}{2} < 0$$

C: Test $\frac{1}{2}$,

$$f''\left(\frac{1}{2}\right) = 30\left(\frac{1}{2}\right) - 60\left(\frac{1}{2}\right)^3 = \frac{15}{2} > 0$$

D: Test 1,

$$f''(1) = 30(1) - 60(1)^3 = -30 < 0$$

We see that f is concave up on the intervals $\left(-\infty, -\frac{1}{\sqrt{2}}\right)$ and $\left(0, \frac{1}{\sqrt{2}}\right)$ and concave down on the intervals $\left(-\frac{1}{\sqrt{2}}, 0\right)$ and $\left(\frac{1}{\sqrt{2}}, \infty\right)$. Therefore, the points $\left(-\frac{1}{\sqrt{2}}, -1.237\right)$, $(0, 0)$ and $\left(\frac{1}{\sqrt{2}}, 1.237\right)$ are inflection points.

Copyright © 2014 Pearson Education, Inc. Publishing as Addison-Wesley.

e) We sketch the graph using the preceding information. Additional function values may also be calculated as necessary.

x	$f(x)$
-2	56
$-\frac{1}{2}$	$-\frac{17}{32}$
$\frac{1}{2}$	$\frac{17}{32}$
2	-56

$f(x)=5x^3-3x^5$

33. $f(x)=x^2(3-x)^2$

$=x^2(9-6x+x^2)$

$=9x^2-6x^3+x^4$

a) $f'(x)=18x-18x^2+4x^3$

$f''(x)=18-36x+12x^2$

The domain of f is $\mathbb{R}$.

b) $f'(x)$ exists for all values of x, so the only critical points of f are where $f'(x)=0$.

$18x-18x^2+4x^3=0$

$2x(9-9x+2x^2)=0$

$2x(3-2x)(3-x)=0$

$2x=0$ or $3-2x=0$ or $3-x=0$

$x=0$ or $x=\frac{3}{2}$ or $x=3$

The critical values are 0, $\frac{3}{2}$, and 3.

$f(0)=(0)^2(3-(0))^2=0$

$f\left(\frac{3}{2}\right)=\left(\frac{3}{2}\right)^2\left(3-\left(\frac{3}{2}\right)\right)^2=\frac{81}{16}$

$f(3)=(3)^2(3-(3))^2=0$

The critical points $(0,0)$, $\left(\frac{3}{2},\frac{81}{16}\right)$, and $(3,0)$ are on the graph.

c) Applying the Second Derivative Test, we have:

$f''(0)=18-36(0)+12(0)^2=18>0$

So, the critical point $(0,0)$ is a relative minimum.

$f''\left(\frac{3}{2}\right)=18-36\left(\frac{3}{2}\right)+12\left(\frac{3}{2}\right)^2=-9<0$

So, the critical point $\left(\frac{3}{2},\frac{81}{16}\right)$ is a relative maximum.

$f''(3)=18-36(3)+12(3)^2=18>0$

So, the critical point $(3,0)$ is a relative minimum.

We use the points 0, $\frac{3}{2}$, and 3 to divide the real number line into four intervals, $(-\infty,0)$, $\left(0,\frac{3}{2}\right)$, $\left(\frac{3}{2},3\right)$, and $(3,\infty)$, we know that $f(x)$ is decreasing on the intervals $(-\infty,0)$ and $\left(\frac{3}{2},3\right)$, and $f(x)$ is increasing on the intervals $\left(0,\frac{3}{2}\right)$ and $(3,\infty)$.

d) Find the points of inflection. $f''(x)$ exists for all values of x, so the only possible inflection points occur when $f''(x)=0$.

$18-36x+12x^2=0$

$3-6x+2x^2=0$ Dividing by 6

Using the quadratic formula we have:

$x=\frac{3\pm\sqrt{3}}{2}$

$x\approx 0.634$ or $x\approx 2.366$

$f(0.634)\approx 2.250$

$f(2.366)\approx 2.250$

The points, $(0.634,2.250)$ and $(2.366,2.250)$ are possible inflection points on the graph.

e) To determine concavity we use 0.634 and 2.366 to divide the real number line into three intervals,

A: $(-\infty,0.634)$, B: $(0.634,2.366)$, and C: $(2.366,\infty)$

The solution is continued on the next page.

Copyright © 2014 Pearson Education, Inc. Publishing as Addison-Wesley.

Then test a point in each interval.

A: Test 0,

$$f''(0)=18-36(0)+12(0)^2$$
$$=18>0$$

B: Test 1,

$$f''(1)=18-36(1)+12(1)^2$$
$$=-6<0$$

C: Test 3,

$$f''(3)=18-36(3)+12(3)^2$$
$$=18>0$$

We see that f is concave up on the intervals $(-\infty,0.634)$ and $(2.366,\infty)$ and concave down on the interval $(0.634,2.366)$.

Therefore, the points $(0.634,2.250)$ and $(2.366,2.250)$ are inflection points.

f) We sketch the graph using the preceding information. Additional function values may also be calculated as necessary.

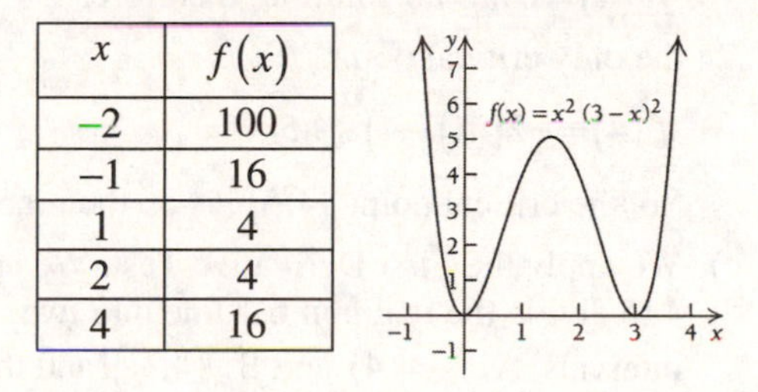

x	$f(x)$
−2	100
−1	16
1	4
2	4
4	16

35. $f(x)=(x+1)^{2/3}$

a) $f'(x)=\frac{2}{3}(x+1)^{-1/3}=\frac{2}{3(x+1)^{1/3}}$

$$f''(x)=-\frac{2}{9}(x+1)^{-4/3}=-\frac{2}{9(x+1)^{4/3}}$$

The domain of f is $\mathbb{R}$.

b) $f'(x)$ does not exist for $x=-1$. The equation $f'(x)=0$ has no solution, therefore, $x=-1$ is the only critical value.

$f(-1)=(-1+1)^{2/3}=0$.

So, the critical point, $(-1,0)$, is on the graph.

c) We apply the First Derivative Test. We use -1 to divide the real number line into two intervals $A:(-\infty,-1)$ and B: $(-1,\infty)$ and then we test a point in each interval.

A: Test -2,

$$f'(-2)=\frac{2}{3((-2)+1)^{1/3}}=-\frac{2}{3}<0$$

B: Test 0,

$$f'(0)=\frac{2}{3((0)+1)^{1/3}}=\frac{2}{3}>0$$

Thus, $(-1,0)$ is a relative minimum. We also know that $f(x)$ is decreasing on the interval $(-\infty,-1)$ and increasing on the interval $(-1,\infty)$.

d) Find the points of inflection. $f''(x)$ does not exist when $x=-1$. The equation $f''(x)=0$ has no solution, so $x=-1$ is the only possible inflection point. We know that $f(-1)=0$.

e) To determine concavity, we divide the real number line into two intervals, $A:(-\infty,-1)$ and B: $(-1,\infty)$ and then we test a point in each interval.

A: Test -2,

$$f''(-2)=-\frac{2}{9((-2)+1)^{4/3}}=-\frac{2}{9}<0$$

B: Test 0,

$$f''(0)=-\frac{2}{9((0)+1)^{4/3}}=-\frac{2}{9}<0$$

Thus, $f(x)$ is concave down on the interval $(-\infty,-1)$ and on the interval $(-1,\infty)$. Therefore, the point $(-1,0)$ is not an inflection point.

f) We sketch the graph using the preceding information. Additional function values may also be calculated as necessary.

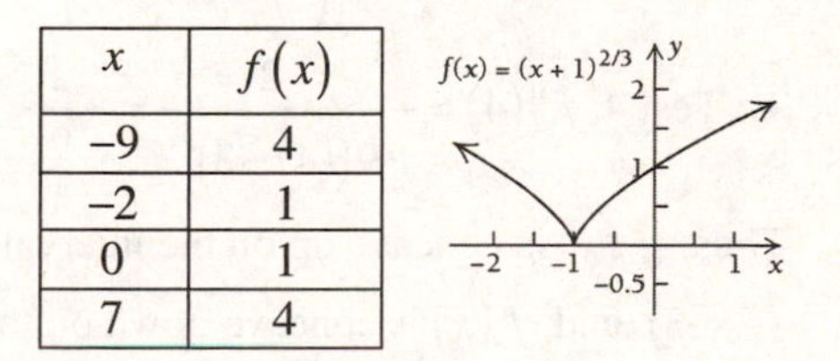

x	$f(x)$
−9	4
−2	1
0	1
7	4

Copyright © 2014 Pearson Education, Inc. Publishing as Addison-Wesley.

37. $f(x)=(x-3)^{1/3}-1$

a) $f'(x)=\frac{1}{3}(x-3)^{-2/3}=\frac{1}{3(x-3)^{2/3}}$

$f''(x)=-\frac{2}{9}(x-3)^{-5/3}=-\frac{2}{9(x-3)^{5/3}}$

The domain of f is $\mathbb{R}$.

b) $f'(x)$ does not exist for $x=3$. The equation $f'(x)=0$ has no solution, therefore, $x=3$ is the only critical value.

$f(3)=((3)-3)^{1/3}-1=-1$.

So, the critical point, $(3,-1)$ is on the graph.

c) We apply the First Derivative Test. We use 3 to divide the real number line into two intervals $A:(-\infty,3)$ and B: $(3,\infty)$ and then we test a point in each interval.

A: Test 2, $f'(2)=\frac{1}{3((2)-3)^{2/3}}=\frac{1}{3}>0$

B: Test 4, $f'(4)=\frac{1}{3((4)-3)^{2/3}}=\frac{1}{3}>0$

$f(x)$ is increasing on both intervals $(-\infty,3)$ and $(3,\infty)$, therefore $(3,-1)$ is not a relative extremum.

d) Find the points of inflection. $f''(x)$ does not exist when $x=3$. The equation $f''(x)=0$ has no solution, so at $x=3$ is the only possible inflection point. We know that $f(3)=-1$.

e) To determine concavity, we divide the real number line into two intervals, $A:(-\infty,3)$ and B: $(3,\infty)$ and then we test a point in each interval.

A: Test 2, $f''(2)=-\frac{2}{9((2)-3)^{5/3}}=\frac{2}{9}>0$

B: Test 4, $f''(4)=-\frac{2}{9((4)-3)^{5/3}}=-\frac{2}{9}<0$

Thus, $f(x)$ is concave up on the interval $(-\infty,3)$ and $f(x)$ is concave down on the interval $(3,\infty)$. Therefore, the point $(3,-1)$ is an inflection point.

f) We sketch the graph using the preceding information. Additional function values may also be calculated as necessary.

x	$f(x)$
–5	–3
2	–2
4	0
11	1

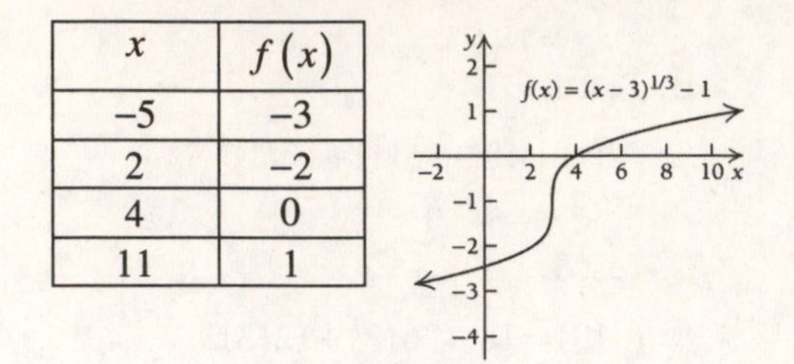

39. $f(x)=-2(x-4)^{2/3}+5$

a) $f'(x)=-\frac{4}{3}(x-4)^{-1/3}=-\frac{4}{3(x-4)^{1/3}}$

$f''(x)=\frac{4}{9}(x-4)^{-4/3}=\frac{4}{9(x-4)^{4/3}}$

The domain of f is $\mathbb{R}$.

b) $f'(x)$ does not exist for $x=4$. The equation $f'(x)=0$ has no solution, therefore, $x=4$ is the only critical point.

$f(4)=-2((4)-4)^{2/3}+5=5$.

So, the critical point $(4,5)$, is on the graph.

c) We apply the First Derivative Test. We use 4 to divide the real number line into two intervals $A:(-\infty,4)$ and B: $(4,\infty)$ and then we test a point in each interval.

A: Test 3, $f'(3)=-\frac{4}{3((3)-4)^{1/3}}=\frac{4}{3}>0$

B: Test 5, $f'(5)=-\frac{4}{3((5)-4)^{1/3}}=-\frac{4}{3}<0$

Thus, $(4,5)$ is a relative maximum. We also know that $f(x)$ is increasing on the interval $(-\infty,4)$ and decreasing on the interval $(4,\infty)$.

d) Find the points of inflection. $f''(x)$ does not exist when $x=4$. The equation $f''(x)=0$ has no solution, so $x=4$ is the only possible inflection point. We know that $f(4)=5$.

Copyright © 2014 Pearson Education, Inc. Publishing as Addison-Wesley.

e) To determine concavity, we divide the real number line into two intervals, $A:(-\infty,4)$ and $B:(4,\infty)$ and then we test a point in each interval.

$$\text{A: Test } 3, f''(3)=\frac{4}{9((3)-4)^{4/3}}=\frac{4}{9}>0$$

$$\text{B: Test } 5, f''(5)=-\frac{4}{9((5)-4)^{4/3}}=\frac{4}{9}>0$$

Thus, $f(x)$ is concave up on both intervals $(-\infty,4)$ and $(4,\infty)$ Therefore, the point $(4,5)$ is not an inflection point.

f) We sketch the graph using the preceding information. Additional function values may also be calculated as necessary.

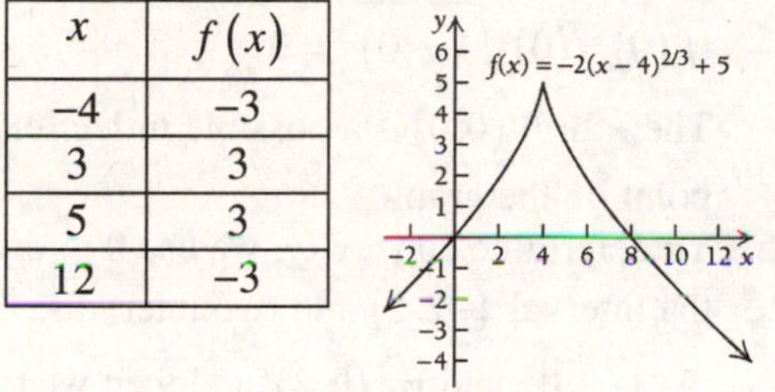

x	$f(x)$
−4	−3
3	3
5	3
12	−3

41. $f(x)=x\sqrt{4-x^2}=x(4-x^2)^{1/2}$

a) $$f'(x)=x\cdot\frac{1}{2}(4-x^2)^{-1/2}(-2x)+(4-x^2)^{1/2}\cdot(1)$$

Next, we simplify the derivative.

$$f'(x)=\frac{-x^2}{(4-x^2)^{1/2}}+(4-x^2)^{1/2}$$

$$=\frac{-x^2+4-x^2}{(4-x^2)^{1/2}}$$

$$=\frac{4-2x^2}{(4-x^2)^{1/2}}$$

$$=(4-2x^2)(4-x^2)^{-1/2}$$

We determine the second derivative at the top of the next column.

$$f''(x)$$

$$=(4-2x^2)\left(-\frac{1}{2}\right)(4-x^2)^{-3/2}(-2x)+(4-x^2)^{-1/2}(-4x)$$

$$=\frac{x(4-2x^2)}{(4-x^2)^{3/2}}-\frac{4x}{(4-x^2)^{1/2}}$$

$$=\frac{4x-2x^3-4x(4-x^2)}{(4-x^2)^{3/2}}$$

$$=\frac{4x-2x^3-16x+4x^3}{(4-x^2)^{3/2}}$$

$$=\frac{2x^3-12x}{(4-x^2)^{3/2}}$$

The domain of $f(x)$ is $[-2,2]$.

b) First, we find the critical points.

$f'(x)$ does not exist when $4-x^2=0$.

Solve:

$$4-x^2=0$$
$$x^2=4$$
$$x=\pm\sqrt{4}$$
$$x=\pm 2$$

Since the domain of $f(x)$ is $[-2,2]$, relative extrema cannot occur at $x=-2$ or $x=2$ because there is not an open interval containing -2 or 2 on which the function is defined. For this reason, we do not consider -2 or 2 in our discussion of relative extrema.

The other critical points occur where

$$f'(x)=0$$
$$\frac{4-2x^2}{\sqrt{4-x^2}}=0$$
$$4-2x^2=0$$
$$2x^2=4$$
$$x^2=2$$
$$x=\pm\sqrt{2}$$

The critical values are $-\sqrt{2}$ and $\sqrt{2}$.

The solution is continued on the next page.

Copyright © 2014 Pearson Education, Inc. Publishing as Addison-Wesley.

We determine the critical points by evaluating the function at the critical values.

$$f\left(-\sqrt{2}\right)=-\sqrt{2}\sqrt{4-\left(-\sqrt{2}\right)^2}=-\sqrt{2}\sqrt{2}=-2$$

$$f\left(\sqrt{2}\right)=\sqrt{2}\sqrt{4-\left(\sqrt{2}\right)^2}=\sqrt{2}\sqrt{2}=2$$

Therefore, $\left(-\sqrt{2},-2\right)$ and $\left(\sqrt{2},2\right)$ are critical points on the graph.

c) We use the Second Derivative Test.

$$f''\left(-\sqrt{2}\right)=\frac{2\left(-\sqrt{2}\right)^3-12\left(-\sqrt{2}\right)}{\left[4-\left(-\sqrt{2}\right)^2\right]^{3/2}}$$

$$=\frac{-4\sqrt{2}+12\sqrt{2}}{2^{3/2}}=\frac{8\sqrt{2}}{2\sqrt{2}}=4>0$$

The critical point $\left(-\sqrt{2},-2\right)$ is a relative minimum.

$$f''\left(\sqrt{2}\right)=\frac{2\left(\sqrt{2}\right)^3-12\left(\sqrt{2}\right)}{\left[4-\left(\sqrt{2}\right)^2\right]^{3/2}}$$

$$=\frac{4\sqrt{2}-12\sqrt{2}}{2^{3/2}}=\frac{-8\sqrt{2}}{2\sqrt{2}}=-4<0$$

The critical point $\left(\sqrt{2},2\right)$ is a relative maximum.

If we use the points $-\sqrt{2}$ and $\sqrt{2}$ to divide the interval $[-2,2]$ into three intervals $\left[-2,-\sqrt{2}\right)$, $\left(-\sqrt{2},\sqrt{2}\right)$, and $\left(\sqrt{2},2\right]$, we see that $f(x)$ is decreasing on the intervals $\left(-2,-\sqrt{2}\right)$ and $\left(\sqrt{2},2\right)$ and $f(x)$ is increasing on the interval $\left(-\sqrt{2},\sqrt{2}\right)$.

d) Find the points of inflection. $f''(x)$ does not exist where $4-x^2=0$. We know that this occurs at $x=-2$ and $x=2$. However, just as relative extrema cannot occur at $(-2,0)$ and $(2,0)$, they cannot be inflection points either. Inflection points could occur where $f''(x)=0$. We set the second derivative equal to zero and solve for the possible inflection points.

We set the second derivative equal to zero and solve for x.

$$f''(x)=0$$

$$\frac{2x^3-12x}{\left(4-x^2\right)^{3/2}}=0$$

$$2x^3-12x=0$$

$$2x\left(x^2-6\right)=0$$

$$2x=0 \quad \text{or} \quad x^2-6=0$$

$$x=0 \quad \text{or} \quad x^2=6$$

$$x=0 \quad \text{or} \quad x=\pm\sqrt{6}$$

Note that $f(x)$ is not defined for $x=\pm\sqrt{6}$. Therefore, the only possible inflection point is $x=0$.

$$f(0)=(0)\sqrt{4-(0)^2}=0.$$

Therefore, $(0,0)$ is a possible inflection point on the graph.

e) To determine concavity, we use 0 to divide the interval $(-2,2)$ into two intervals, A: $(-2,0)$ and B: $(0,2)$ and then we test a point in each interval.

A: Test -1,

$$f''(-1)=\frac{2(-1)^3-12(-1)}{\left[4-(-1)^2\right]^{3/2}}=\frac{10}{3^{3/2}}>0$$

B: Test 1,

$$f''(1)=\frac{2(1)^3-12(1)}{\left[4-(1)^2\right]^{3/2}}=\frac{-10}{3^{3/2}}<0$$

Thus, $f(x)$ is concave up on the interval $(-2,0)$ and $f(x)$ is concave down on the interval $(0,2)$. Therefore, the point $(0,0)$ is an inflection point.

f) We sketch the graph using the preceding information. Additional function values may also be calculated as necessary.

x	$f(x)$
-1	$-\sqrt{3}$
1	$\sqrt{3}$

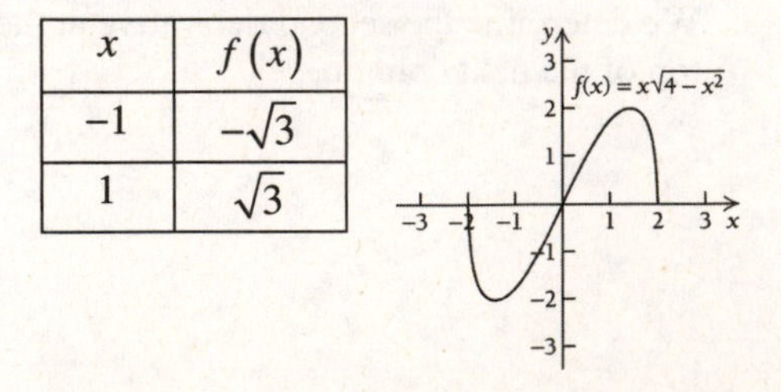

Copyright © 2014 Pearson Education, Inc. Publishing as Addison-Wesley.

43. $f(x)=\dfrac{x}{x^2+1}$

a) $f'(x)=\dfrac{(x^2+1)(1)-x(2x)}{(x^2+1)^2}$ Quotient Rule

$=\dfrac{x^2+1-2x^2}{(x^2+1)^2}$

$=\dfrac{1-x^2}{(x^2+1)^2}$

$f''(x)$

$=\dfrac{(x^2+1)^2(-2x)-(1-x^2)\left[2(x^2+1)^1(2x)\right]}{\left((x^2+1)^2\right)^2}$

$=\dfrac{(x^2+1)\left[-2x(x^2+1)-4x(1-x^2)\right]}{(x^2+1)^4}$

$=\dfrac{-2x^3-2x-4x+4x^3}{(x^2+1)^3}$

$=\dfrac{2x^3-6x}{(x^2+1)^3}$

The domain of f is $\mathbb{R}$.

b) Since $f'(x)$ exists for all real numbers, the only critical values are where $f'(x)=0$.

$\dfrac{1-x^2}{(x^2+1)^2}=0$

$1-x^2=0$ Multiplying by $(x^2+1)^2$

$x^2=1$

$x=\pm\sqrt{1}=\pm1$

The two critical values are $x=-1$ and $x=1$.

$f(-1)=\dfrac{-1}{(-1)^2+1}=-\dfrac{1}{2}$

$f(1)=\dfrac{1}{(1)^2+1}=\dfrac{1}{2}$

The critical points $\left(-1,-\dfrac{1}{2}\right)$ and $\left(1,\dfrac{1}{2}\right)$ are on the graph.

c) We use the Second Derivative Test.

$f''(-1)=\dfrac{2(-1)^3-6(-1)}{\left[(-1)^2+1\right]^3}=\dfrac{4}{8}=\dfrac{1}{2}>0$

So the point $\left(-1,-\dfrac{1}{2}\right)$ is a relative minimum.

$f''(1)=\dfrac{2(1)^3-6(1)}{\left[(1)^2+1\right]^3}=\dfrac{-4}{8}=-\dfrac{1}{2}<0$

So the point $\left(1,\dfrac{1}{2}\right)$ is a relative maximum.

We use -1 and 1 to divide the real number line into three intervals $(-\infty,-1)$, $(-1,1)$, and $(1,\infty)$. $f(x)$ is decreasing on the intervals $(-\infty,1)$ and $(1,\infty)$, and $f(x)$ is increasing on the interval $[-1,1]$.

d) Find the points of inflection. $f''(x)$ exists for all real numbers, so the only possible points of inflection occur when $f''(x)=0$.

$\dfrac{2x^3-6x}{(x^2+1)^3}=0$

$2x^3-6x=0$

$2x(x^2-3)=0$

$2x=0$ or $x^2-3=0$

$x=0$ or $x^2=3$

$x=0$ or $x=\pm\sqrt{3}$

There are three possible inflection points at $x=-\sqrt{3}, 0,$ and $\sqrt{3}$.

$f(-\sqrt{3})=\dfrac{-\sqrt{3}}{(-\sqrt{3})^2+1}=-\dfrac{\sqrt{3}}{4}$

$f(0)=\dfrac{\sqrt{0}}{(\sqrt{0})^2+1}=\dfrac{0}{1}=0$

$f(\sqrt{3})=\dfrac{\sqrt{3}}{(\sqrt{3})^2+1}=\dfrac{\sqrt{3}}{4}$

The points $\left(-\sqrt{3},-\dfrac{\sqrt{3}}{4}\right)$, $(0,0)$, and $\left(\sqrt{3},\dfrac{\sqrt{3}}{4}\right)$ are three possible inflection points on the graph.

Copyright © 2014 Pearson Education, Inc. Publishing as Addison-Wesley.

e) To determine concavity we use $-\sqrt{3}, 0,$ and $\sqrt{3}$ to divide the real number line into four intervals,

$A: \left(-\infty, -\sqrt{3}\right)$, B: $\left(-\sqrt{3}, 0\right)$, C: $\left(0, \sqrt{3}\right)$, and D: $\left(\sqrt{3}, \infty\right)$.

We test a point in each interval.

A: Test -2, $f''(-2) = -\frac{4}{125} < 0$

B: Test -1, $f''(-1) = \frac{1}{2} > 0$

C: Test 1, $f''(1) = -\frac{1}{2} < 0$

D: Test 2, $f''(2) = \frac{4}{125} > 0$

We see that f is concave down on the intervals $\left(-\infty, -\sqrt{3}\right)$ and $\left(0, \sqrt{3}\right)$ and concave up on the intervals $\left(-\sqrt{3}, 0\right)$ and $\left(\sqrt{3}, \infty\right)$. Therefore the points $\left(-\sqrt{3}, -\frac{\sqrt{3}}{4}\right)$, $(0,0)$, and $\left(\sqrt{3}, \frac{\sqrt{3}}{4}\right)$ are inflection points.

f) We sketch the graph using the preceding information. Additional function values may also be calculated as necessary.

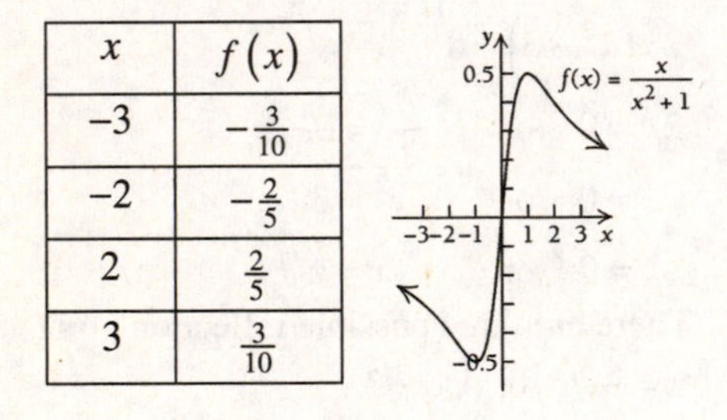

x	$f(x)$
-3	$-\frac{3}{10}$
-2	$-\frac{2}{5}$
2	$\frac{2}{5}$
3	$\frac{3}{10}$

45. $f(x) = \frac{3}{x^2+1} = 3\left(x^2+1\right)^{-1}$

a) $f'(x) = 3(-1)\left(x^2+1\right)^{-2}(2x)$

$= -6x\left(x^2+1\right)^{-2}$

$= \frac{-6x}{\left(x^2+1\right)^2}$

$f''(x)$

$= \frac{\left(x^2+1\right)^2(-6) - (-6x)\left(2\left(x^2+1\right)(2x)\right)}{\left(\left(x^2+1\right)^2\right)^2}$

$= \frac{\left(x^2+1\right)\left[\left(x^2+1\right)(-6) - (-6x)(2)(2x)\right]}{\left(x^2+1\right)^4}$

$f''(x) = \frac{-6x^2 - 6 + 24x^2}{\left(x^2+1\right)^3}$

$= \frac{18x^2 - 6}{\left(x^2+1\right)^3}$

The domain of f is $\mathbb{R}$.

b) Since $f'(x)$ exists for all real numbers, the only critical values are where $f'(x) = 0$.

$\frac{-6x}{\left(x^2+1\right)^2} = 0$

$-6x = 0$ Multiplying by $\left(x^2+1\right)^2$

$x = 0$

The critical value is $x = 0$.

$f(0) = \frac{3}{(0)^2+1} = 3$

The critical point $(0,3)$ is on the graph.

c) We use the Second Derivative Test.

$f''(0) = \frac{18(0) - 6}{\left((0)^2+1\right)^2} = \frac{-6}{1} = -6 < 0$

So the point $(0,3)$ is a relative maximum.

We use 0 to divide the real number line into two intervals $(-\infty, 0)$ and $(0, \infty)$. $f(x)$ is increasing on the interval $(-\infty, 0]$, and $f(x)$ is decreasing on the interval $[0, \infty)$.

Copyright © 2014 Pearson Education, Inc. Publishing as Addison-Wesley.

d) Find the points of inflection. $f''(x)$ exists for all real numbers, so the only possible points of inflection occur when $f''(x)=0$.

$$\frac{18x^2-6}{\left(x^2+1\right)^3}=0$$
$$18x^2-6=0$$
$$18x^2=6$$
$$x^2=\frac{1}{3}$$
$$x=\pm\frac{1}{\sqrt{3}}$$

There are two possible inflection points at $x=-\frac{1}{\sqrt{3}}$ and $\frac{1}{\sqrt{3}}$. We determine the possible inflection points.

$$f\left(-\frac{1}{\sqrt{3}}\right)=\frac{3}{\left(-\frac{1}{\sqrt{3}}\right)^2+1}=\frac{3}{\frac{1}{3}+1}=\frac{3}{\frac{4}{3}}=\frac{9}{4}$$

$$f\left(\frac{1}{\sqrt{3}}\right)=\frac{3}{\left(\frac{1}{\sqrt{3}}\right)^2+1}=\frac{3}{\frac{1}{3}+1}=\frac{3}{\frac{4}{3}}=\frac{9}{4}$$

The points $\left(-\frac{1}{\sqrt{3}},\frac{9}{4}\right)$ and $\left(\frac{1}{\sqrt{3}},\frac{9}{4}\right)$ are possible inflection points on the graph.

e) To determine concavity we use $-\frac{1}{\sqrt{3}}$ and $\frac{1}{\sqrt{3}}$ to divide the real number line into three intervals,

A: $\left(-\infty,-\frac{1}{\sqrt{3}}\right)$, B: $\left(-\frac{1}{\sqrt{3}},\frac{1}{\sqrt{3}}\right)$, and

C: $\left(\frac{1}{\sqrt{3}},\infty\right)$.

Then we test a point in each interval.

A: Test -1, $f''(-1)=\frac{18(-1)^2-6}{\left((-1)^2+1\right)^3}=\frac{3}{2}>0$

B: Test 0, $f''(0)=\frac{18(0)^2-6}{\left((0)^2+1\right)^3}=-6<0$

C: Test 1, $f''(1)=\frac{18(1)^2-6}{\left((1)^2+1\right)^3}=\frac{3}{2}>0$

We see that f is concave up on the intervals $\left(-\infty,-\frac{1}{\sqrt{3}}\right)$ and $\left(\frac{1}{\sqrt{3}},\infty\right)$ and concave down on the interval $\left(-\frac{1}{\sqrt{3}},\frac{1}{\sqrt{3}}\right)$.

Therefore the points $\left(-\frac{1}{\sqrt{3}},\frac{9}{4}\right)$ and $\left(\frac{1}{\sqrt{3}},\frac{9}{4}\right)$ are inflection points.

f) We sketch the graph using the preceding information. Additional function values may also be calculated as necessary.

x	$f(x)$
-3	$\frac{3}{10}$
-1	$\frac{3}{2}$
1	$\frac{3}{2}$
3	$\frac{3}{10}$

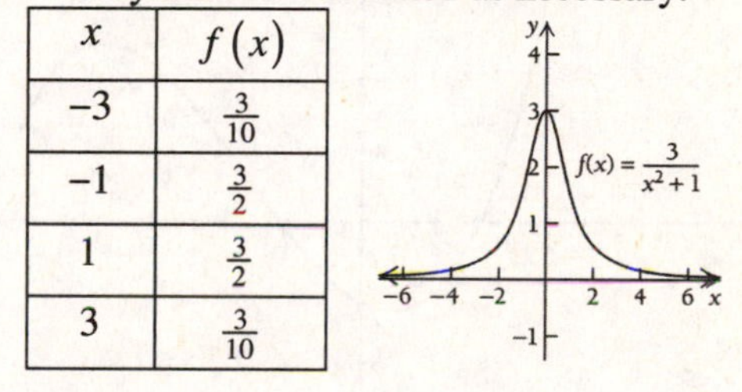

47. Answers may vary, one possible graph is:

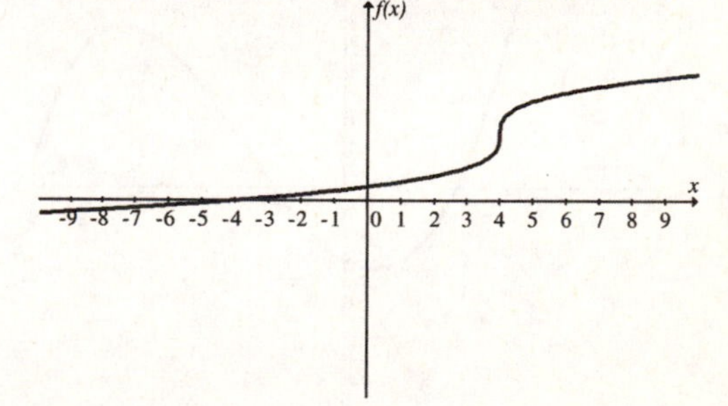

Copyright © 2014 Pearson Education, Inc. Publishing as Addison-Wesley.

49. Answers may vary, one possible graph is:

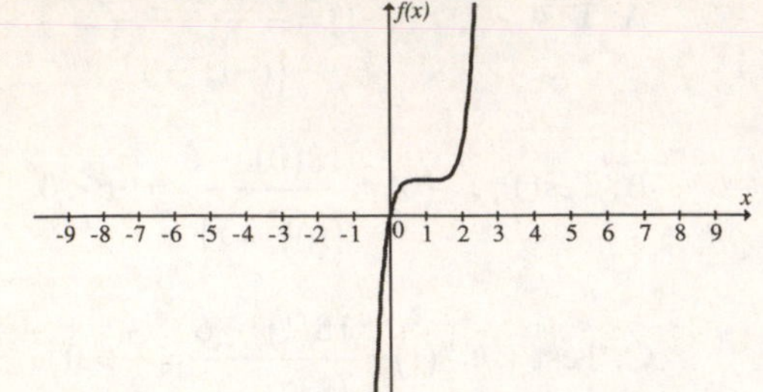

51. Answers may vary, one possible graph is:

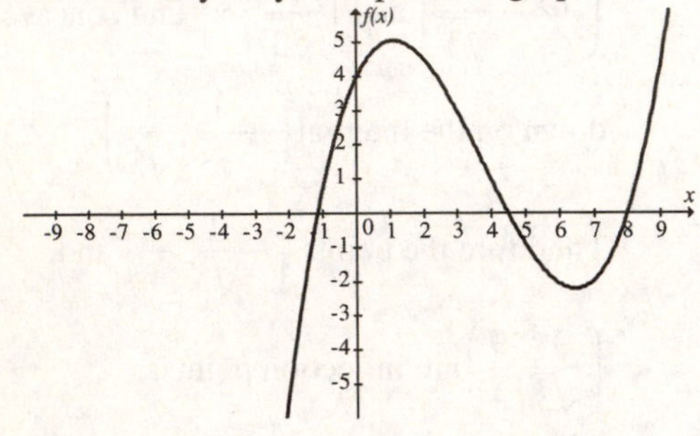

53. Answers may vary, one possible graph is:

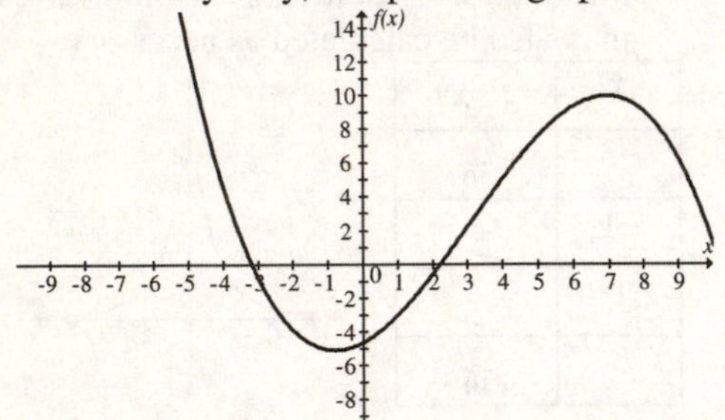

55. Answers may vary, one possible graph is:

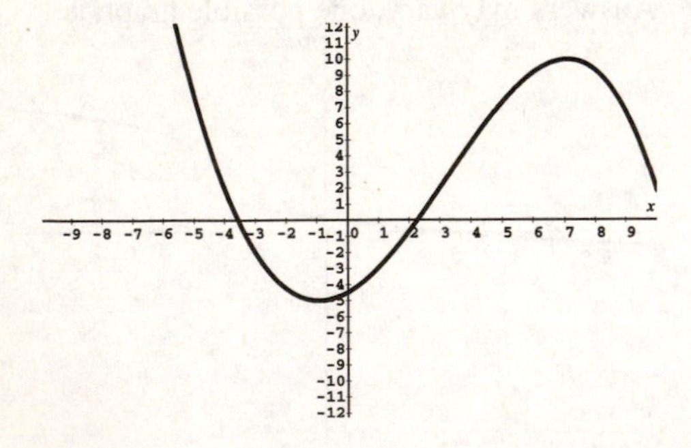

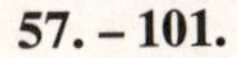

57. – 101. Left to the Student.

103. $R(x) = 50x - 0.5x^2$

$C(x) = 4x + 10$

$$P(x) = R(x) - C(x)$$
$$= (50x - 0.5x^2) - (4x + 10)$$
$$= -0.5x^2 + 46x - 10$$

We will restrict the domains of all three functions to $x \geq 0$ since a negative number of units cannot be produced and sold.

First graph $R(x) = 50x - 0.5x^2$

$R'(x) = 50 - x$

$R''(x) = -1$

Since $R'(x)$ exists for all $x \geq 0$, the only critical points are where $R'(x) = 0$.

$50 - x = 0$

$50 = x$ Critical Value

Find the function value at $x = 50$.

$$R(50) = 50(50) - 0.5(50)^2$$
$$= 2500 - 1250$$
$$= 1250$$

This critical point $(50,1250)$ is on the graph.

We use the Second Derivative Test:

$R''(50) = -1 < 0$

The point $(50,1250)$ is a relative maximum.

We use 50 to divide the interval $[0,\infty)$ into two intervals, $[0,50)$ and $(50,\infty)$, we know that R is increasing on $(0,50)$ and decreasing on $(50,\infty)$.

Next, find the inflection points. Since $R''(x)$ exists for all $x \geq 0$, and $R''(x) = -1$, there are no possible inflection points.

Furthermore, since $R''(x) < 0$ for all $x \geq 0$, R is concave down over the interval $(0,\infty)$.

Sketch the graph using the preceding information. The x-intercepts of R are found by solving $R(x) = 0$.

$$50x - 0.5x^2 = 0$$
$$0.5x(100 - x) = 0$$
$$0.5x = 0 \text{ or } 100 - x = 0$$
$$x = 0 \text{ or } 100 = x$$

The x-intercepts are $(0,0)$ and $(100,0)$.

The solution is continued on the next page.

Copyright © 2014 Pearson Education, Inc. Publishing as Addison-Wesley.

Next, we graph $C(x)=4x+10$. This is a linear function with slope 4 and y-intercept $(0,10)$.

$C(x)$ is increasing over the entire domain $x\geq 0$ and has no relative extrema or points of inflection.

Finally, we graph $P(x)=-0.5x^2+46x-10$

$P'(x)=-x+46$

$P''(x)=-1$

Since $P'(x)$ exists for all $x\geq 0$, the only critical points occur when $P'(x)=0$.

$-x+46=0$

$46=x$ Critical Value

Find the function value at $x=46$.

$P(46)=-0.5(46)^2+46(46)-10$

$=-1058+2116-10$

$=1048$

The critical point $(46,1048)$ is on the graph.

We use the Second Derivative Test:

$P''(46)=-1<0$

The point $(46,1048)$ is a relative maximum.

We use 46 to divide the interval $[0,\infty)$ into two intervals, $[0,46)$ and $(46,\infty)$, we know that P is increasing on $[0,46]$ and decreasing on $[46,\infty)$.

Next, find the inflection points. Since $P''(x)$ exists for all $x\geq 0$, and $P''(x)=-1$, there are no possible inflection points. Furthermore, since $P''(x)<0$ for all $x\geq 0$, P is concave down over the interval $(0,\infty)$.

Sketch the graph using the preceding information.

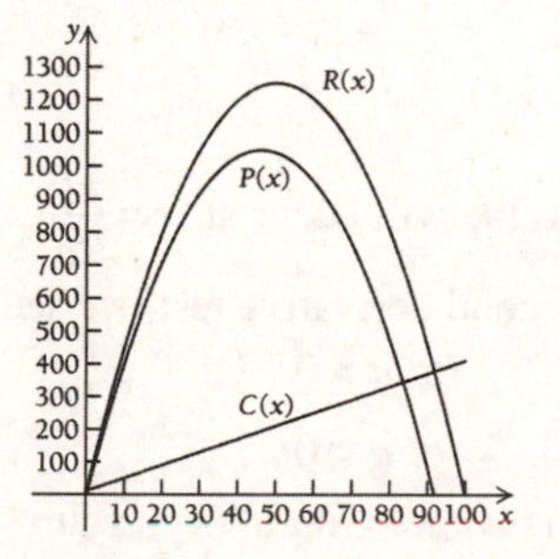

105. $p(x)=\dfrac{13x^3-240x^2-2460x+585{,}000}{75{,}000}$

$p'(x)=\dfrac{39x^2-480x-2460}{75{,}000}$

$p''(x)=\dfrac{78x-480}{75{,}000}$

Since $p'(x)$ exists for all real numbers, the only critical points are where $p'(x)=0$.

$\dfrac{39x^2-480x-2460}{75{,}000}=0$

$39x^2-480x-2460=0$

Using the quadratic formula, we have:

$x=\dfrac{-b\pm\sqrt{b^2-4ac}}{2a}$

$=\dfrac{-(-480)\pm\sqrt{(-480)^2-4(39)(-2460)}}{2(39)}$

$=\dfrac{480\pm\sqrt{614{,}160}}{78}$

$x\approx -3.89$ or $x\approx 16.20$ Critical values

Since the domain of the function is $0\leq x\leq 40$, we consider only $x\approx 16.20$

$p(16.20)$

$=\dfrac{13(16.20)^3-240(16.20)^2-2460(16.20)-585{,}000}{75{,}000}$

≈ 7.17

The critical point $(16.20,7.17)$ is on the graph.

We use the Second Derivative Test:

$p''(x)=\dfrac{78(16.20)-480}{75{,}000}\approx 0.01>0$

The point $(16.20,7.17)$ is a relative minimum.

If we use the point 16.20 to divide the domain into two intervals, $[0,16.20)$ and $(16.20,40]$, we know that p is decreasing on $(0,16.20)$ and increasing on $(16.20,40)$.

The solution is continued on the next page.

Copyright © 2014 Pearson Education, Inc. Publishing as Addison-Wesley.

Next, we find the inflection points. $p''(x)$ exists for all real numbers, so the only possible inflection points are where $p''(x)=0$

$$\frac{78x-480}{75{,}000}=0$$
$$78x-480=0$$
$$78x=480$$
$$x\approx 6.15$$

$$p(6.15)$$
$$=\frac{13(6.15)^3-240(6.15)^2-2460(6.15)-585{,}000}{75{,}000}$$
$$\approx 7.52$$

The point $(6.15,7.52)$ is a possible inflection point.

To determine concavity, we use 6.15 to divide the domain into two intervals

A: $[0,6.15)$ and B: $(6.15,40]$ and test a point in each interval.

A: Test 1, $p''(1)=\dfrac{78(1)-480}{75{,}000}=-0.005<0$

B: Test 7, $p''(7)=\dfrac{78(7)-480}{75{,}000}=0.00088>0$

Then p is concave down on $(0,6.15)$ and concave up on $(6.15,40)$ and the point $(6.15,7.52)$ is a point of inflection.

Sketch the graph for $0\le x\le 40$ using the preceding information. Additional function values may be calculated if necessary.

x	$p(x)$
0	7.8
8	7.42
12	7.25
20	7.25
24	7.57
32	9.15
40	12.46

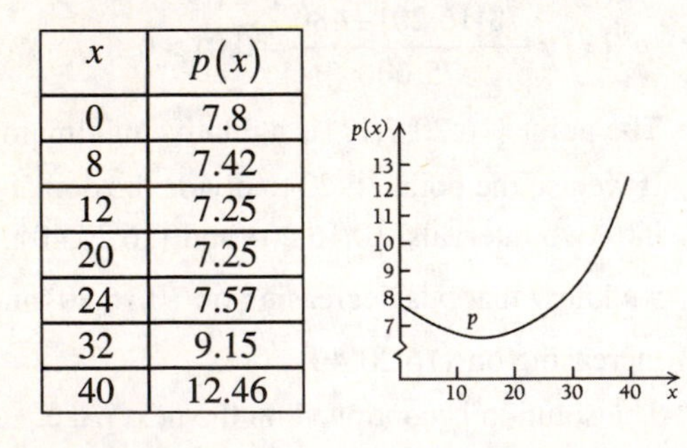

107. $V(r)=k\left(20r^2-r^3\right),\qquad 0\le r\le 20$

$$V'(r)=k\left(40r-3r^2\right)$$
$$V''(r)=k(40-6r)$$

$V'(r)$ exists for all r in $[0,20]$, so the only critical points occur where $V'(r)=0$.

$$V'(r)=0$$
$$k\left(40r-3r^2\right)=0$$
$$40r-3r^2=0$$
$$r(40-3r)=0$$
$$r=0 \text{ or } 40-3r=0$$
$$r=0 \text{ or } 40=3r$$
$$r=0 \text{ or } \frac{40}{3}=r$$

Using the Second Derivative Test:

$$V''(0)=k(40-6(0))=40k>0 \qquad [k>0]$$
$$V''\left(\frac{40}{3}\right)=k\left(40-6\left(\frac{40}{3}\right)\right)=-40k<0$$

Since $V''\left(\frac{40}{3}\right)<0$, we know that there is a relative maximum at $x=\frac{40}{3}$. Thus, for an object whose radius is $\frac{40}{3}$ mm or 13.33 mm, the maximum velocity is needed to remove the object.

109. ✎

111. ✎

113. $f(x)=ax^2+bx+c,\qquad a\ne 0$

$$f'(x)=2ax+b$$
$$f''(x)=2a$$

Since $f'(x)$ exists for all real numbers, the only critical points occur when $f'(x)=0$. We solve:

$$2ax+b=0$$
$$2ax=-b$$
$$x=\frac{-b}{2a}$$

So the critical value will occur at $x=\frac{-b}{2a}$.

Applying the second derivative test, we see that

$f''(x)=2a>0,\quad$ for $a>0$

$f''(x)=2a<0,\quad$ for $a<0$

Therefore, a relative maximum occurs at $x=\frac{-b}{2a}$ when $a<0$ and a relative minimum occurs at $x=\frac{-b}{2a}$ when $a>0$.

Copyright © 2014 Pearson Education, Inc. Publishing as Addison-Wesley.

115. True

117. True

119. False, The function could have a point of inflection at a critical value, but it does not have to have one.

121. True

123. $f(x) = 4x - 6x^{2/3}$

Graphing the function on the calculator we have:

$f(x) = 4x - 6x^{2/3}$

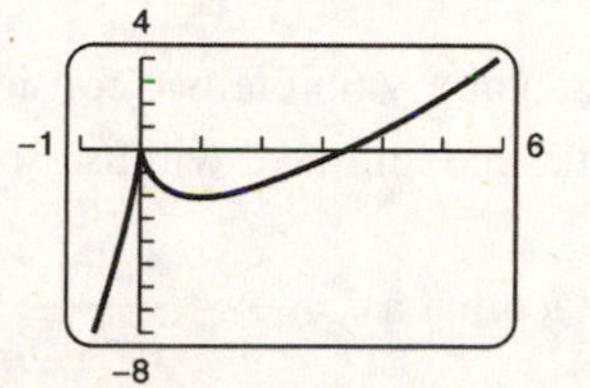

Using the minimum/maximum feature on the calculator, we estimate a relative maximum at $(0,0)$ and a relative minimum at $(1,-2)$.

125. $f(x) = x^2(1-x)^3$

Graphing the function on the calculator we have:

$f(x) = x^2(1-x)^3$

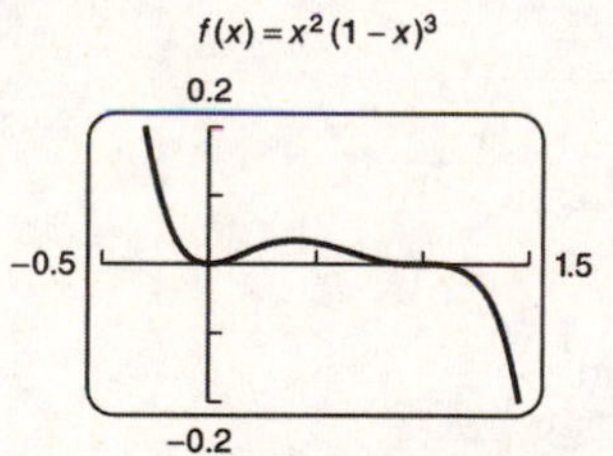

Using the minimum/maximum feature on the calculator, we estimate a relative maximum at $(0.4, 0.035)$ and a relative minimum at $(0,0)$.

127. $f(x) = (x-1)^{2/3} - (x+1)^{2/3}$

Graphing the function on the calculator we have:

$f(x) = (x-1)^{2/3} - (x+1)^{2/3}$

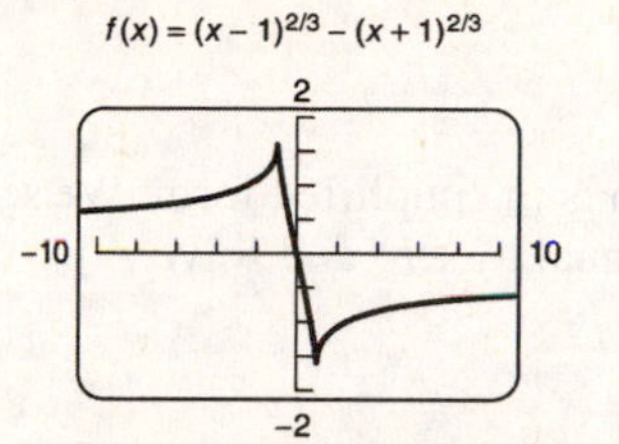

Using the minimum/maximum feature on the calculator, we estimate a relative maximum at $(-1, 1.587)$ and a relative minimum at $(1, -1.587)$.

Copyright © 2014 Pearson Education, Inc. Publishing as Addison-Wesley.

Exercise Set 2.3

1. $f(x)=\dfrac{2x-3}{x-5}$

The expression is in simplified form. We set the denominator equal to zero and solve.

$x-5=0$

$x=5$

The vertical asymptote is the line $x=5$.

3. $f(x)=\dfrac{3x}{x^2-9}$

First, we write the function in simplified form.

$f(x)=\dfrac{3x}{(x-3)(x+3)}$

Once the expression is in simplified form, we set the denominator equal to zero and solve.

$(x-3)(x+3)=0$

$x-3=0$ or $x+3=0$

$x=3$ or $x=-3$

The vertical asymptotes are the lines $x=-3$ and $x=3$.

5. $f(x)=\dfrac{x+2}{x^3-6x^2+8x}$

First, we write the function in simplified form.

$f(x)=\dfrac{x+2}{x(x^2-6x+8)}$ Factor out x.

$=\dfrac{x+2}{x(x-4)(x-2)}$

Once the expression is in simplified form, we set the denominator equal to zero and solve.

$x(x-2)(x-4)=0$

$x=0$ or $x-2=0$ or $x-4=0$

$x=0$ or $x=2$ or $x=4$

The vertical asymptotes are the lines $x=0$, $x=2$, and $x=4$.

7. $f(x)=\dfrac{x+6}{x^2+7x+6}$

First, we write the function in simplified form.

$f(x)=\dfrac{x+6}{(x+6)(x+1)}$

$=\dfrac{1}{x+1}$ Dividing common terms

Once the expression is in simplified form, we set the denominator equal to zero and solve.

$x+1=0$

$x=-1$

The vertical asymptote is the line $x=-1$.

9. $f(x)=\dfrac{6}{x^2+36}$

The function is in simplified form. The equation $x^2+36=0$ has no real solution; therefore, the function does not have any vertical asymptotes.

11. $f(x)=\dfrac{6x}{8x+3}$

To find the horizontal asymptote, we consider $\lim\limits_{x\to\infty} f(x)$. To find the limit, we will use some algebra and the fact that as $x\to\infty$, $\dfrac{b}{ax^n}\to 0$ for any positive integer n.

Calculating the limit, we have:

$$\lim_{x\to\infty} f(x)=\lim_{x\to\infty}\frac{6x}{8x+3}$$

$$=\lim_{x\to\infty}\frac{6x}{8x+3}\cdot\frac{\frac{1}{x}}{\frac{1}{x}} \quad \text{Multiplying by a form of 1}$$

$$=\lim_{x\to\infty}\frac{\frac{6x}{x}}{\frac{8x}{x}+\frac{3}{x}}$$

$$=\lim_{x\to\infty}\frac{6}{8+\frac{3}{x}}$$

$$=\frac{6}{8+0} \quad \left[\text{as } x\to\infty,\ \frac{b}{ax^n}\to 0\right]$$

$$=\frac{6}{8}=\frac{3}{4}.$$

In a similar manner, it can be shown that

$$\lim_{x\to-\infty} f(x)=\frac{3}{4}.$$

The horizontal asymptote is the line $y=\dfrac{3}{4}$.

Copyright © 2014 Pearson Education, Inc. Publishing as Addison-Wesley.

13. $f(x)=\frac{4x}{x^2-3x}$

To find the horizontal asymptote, we consider $\lim_{x\to\infty} f(x)$. To find the limit, we will use some algebra and the fact that as $x\to\infty, \frac{b}{ax^n}\to 0$ for any positive integer *n*.

$$\lim_{x\to\infty} f(x)=\lim_{x\to\infty}\frac{4x}{x^2-3x}$$

$$=\lim_{x\to\infty}\frac{4x}{x^2-3x}\cdot\frac{\frac{1}{x^2}}{\frac{1}{x^2}} \quad \text{Multiplying by a form of 1}$$

$$=\lim_{x\to\infty}\frac{\frac{4x}{x^2}}{\frac{x^2}{x^2}-\frac{3}{x^2}}$$

$$=\lim_{x\to\infty}\frac{\frac{1}{x}}{1+\frac{3}{x^2}}$$

$$=\frac{0}{1+0} \quad \left[\text{as } x\to\infty, \frac{b}{ax^n}\to 0\right]$$

$$=0.$$

In a similar manner, it can be shown that $\lim_{x\to-\infty} f(x)=0$.

The horizontal asymptote is the line $y=0$.

15. $f(x)=5-\frac{3}{x}$

To find the horizontal asymptote, we consider $\lim_{x\to\infty} f(x)$. To find the limit, we will use some algebra and the fact that as $x\to\infty, \frac{b}{ax^n}\to 0$ for any positive integer *n*.

$$\lim_{x\to\infty} f(x)=\lim_{x\to\infty} 5-\frac{3}{x}$$

$$=5-0 \quad \left[\text{as } x\to\infty, \frac{b}{ax^n}\to 0\right]$$

$$=5.$$

In a similar manner, it can be shown that $\lim_{x\to-\infty} f(x)=5$.

The horizontal asymptote is the line $y=5$.

17. $f(x)=\frac{8x^4-5x^2}{2x^3+x^2}$

To find the horizontal asymptote, we consider $\lim_{x\to\infty} f(x)$.

$$\lim_{x\to\infty} f(x)=\lim_{x\to\infty}\frac{8x^4-5x^2}{2x^3+x^2}$$

$$=\lim_{x\to\infty}\frac{8x^4-5x^2}{2x^3+x^2}\cdot\frac{\frac{1}{x^3}}{\frac{1}{x^3}} \quad \text{Multiplying by a form of 1}$$

$$=\lim_{x\to\infty}\frac{8x-\frac{5}{x}}{2-\frac{1}{x^2}}$$

$$=\frac{\lim_{x\to\infty} 8x-\frac{5}{x}}{2-0}$$

$$=\infty$$

In a similar manner, it can be shown that $\lim_{x\to-\infty} f(x)=-\infty$.

The function increases without bound as $x\to\infty$ and decreases without bound as $x\to-\infty$. Therefore, the function does not have a horizontal asymptote.

19. $f(x)=\frac{6x^4+4x^2-7}{2x^5-x+3}$

To find the horizontal asymptote, we consider $\lim_{x\to\infty} f(x)$.

$$\lim_{x\to\infty} f(x)=\lim_{x\to\infty}\frac{6x^4+4x^2-7}{2x^5-x+3}$$

$$=\lim_{x\to\infty}\frac{6x^4+4x^2-7}{2x^5-x+3}\cdot\frac{\frac{1}{x^5}}{\frac{1}{x^5}}$$

$$=\lim_{x\to\infty}\frac{\frac{6}{x}+\frac{4}{x^3}-\frac{7}{x^5}}{2-\frac{1}{x^4}+\frac{3}{x^5}}$$

$$=\frac{0}{2+0} \quad \left[\text{as } x\to\infty, \frac{b}{ax^n}\to 0\right]$$

$$=0$$

In a similar manner, it can be shown that $\lim_{x\to-\infty} f(x)=0$.

The horizontal asymptote is the line $y=0$.

Copyright © 2014 Pearson Education, Inc. Publishing as Addison-Wesley.

21. $f(x)=\dfrac{2x^3-4x+1}{4x^3+2x-3}$

To find the horizontal asymptote, we consider $\lim\limits_{x\to\infty} f(x)$. To find the limit, we will use some algebra and the fact that as $x\to\infty$, $\dfrac{b}{ax^n}\to 0$ for any positive integer n.

$$\lim_{x\to\infty} f(x)=\lim_{x\to\infty}\frac{2x^3-4x+1}{4x^3+2x-3}$$

$$=\lim_{x\to\infty}\frac{2x^3-4x+1}{4x^3+2x-3}\cdot\frac{\frac{1}{x^3}}{\frac{1}{x^3}}$$

$$=\lim_{x\to\infty}\frac{2-\frac{4}{x^2}+\frac{1}{x^3}}{4+\frac{2}{x^2}-\frac{3}{x^3}}$$

$$=\frac{2-0+0}{4+0-0}\quad\left[\text{as } x\to\infty,\ \frac{b}{ax^n}\to 0\right]$$

$$=\frac{2}{4}=\frac{1}{2}$$

In a similar manner, it can be shown that $\lim\limits_{x\to-\infty} f(x)=\dfrac{1}{2}$.

The horizontal asymptote is the line $y=\dfrac{1}{2}$.

23. $f(x)=-\dfrac{5}{x}=-5x^{-1}$

a) *Intercepts.* Since the numerator is the constant -5, there are no *x*-intercepts. The number 0 is not in the domain of the function, so there are no *y*-intercepts.

b) *Asymptotes.*
Vertical. The denominator is 0 for $x=0$, so the line $x=0$ is a vertical asymptote.
Horizontal. The degree of the numerator is less than the degree of the denominator, so $y=0$ is the horizontal asymptote.
Slant. There is no slant asymptote since the degree of the numerator is not one more than the degree of the denominator.

c) *Derivatives and Domain.*

$$f'(x)=5x^{-2}=\frac{5}{x^2}$$

$$f''(x)=-10x^{-3}=-\frac{10}{x^3}$$

The domain of f is $(-\infty,0)\cup(0,\infty)$ as determined in step (b).

d) *Critical Points.* $f'(x)$ exists for all values of x except 0, but 0 is not in the domain of the function, so $x=0$ is not a critical value. The equation $f'(x)=0$ has no solution, so there are no critical points.

e) *Increasing, decreasing, relative extrema.* We use 0 to divide the real number line into two intervals A: $(-\infty,0)$ and B: $(0,\infty)$, and we test a point in each interval.

A: Test -1, $f'(-1)=\dfrac{5}{(-1)^2}=5>0$

B: Test 1, $f'(1)=\dfrac{5}{(1)^2}=5>0$

Then $f(x)$ is increasing on both intervals. Since there are no critical points, there are no relative extrema.

f) *Inflection points.* $f''(x)$ does not exist at 0, but because 0 is not in the domain of the function, there cannot be an inflection point at 0. The equation $f''(x)=0$ has no solution; therefore, there are no inflection points.

g) *Concavity.* We use 0 to divide the real number line into two intervals A: $(-\infty,0)$ and B: $(0,\infty)$, and we test a point in each interval.

A: Test -1, $f''(-1)=-\dfrac{10}{(-1)^3}=10>0$

B: Test 1, $f''(1)=-\dfrac{10}{(1)^3}=-10<0$

Therefore, $f(x)$ is concave up on $(-\infty,0)$ and concave down on $(0,\infty)$.

h) *Sketch.* Use the preceding information to sketch the graph. Compute additional function values as needed.

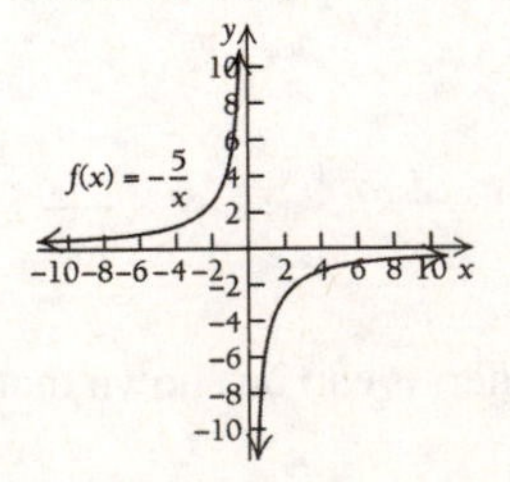

Copyright © 2014 Pearson Education, Inc. Publishing as Addison-Wesley.

25. $f(x)=\frac{1}{x-5}=(x-5)^{-1}$

a) *Intercepts.* Since the numerator is the constant 1, there are no *x*-intercepts. To find the *y*-intercepts we compute $f(0)$

$$f(0)=\frac{1}{(0)-5}=-\frac{1}{5}$$

The point $\left(0,-\frac{1}{5}\right)$ is the *y*-intercept.

b) *Asymptotes.*
Vertical. The denominator is 0 for $x=5$, so the line $x=5$ is a vertical asymptote.
Horizontal. The degree of the numerator is less than the degree of the denominator, so $y=0$ is the horizontal asymptote.
Slant. There is no slant asymptote since the degree of the numerator is not one more than the degree of the denominator.

c) *Derivatives and Domain.*

$$f'(x)=-(x-5)^{-2}=\frac{-1}{(x-5)^2}$$

$$f''(x)=2(x-5)^{-3}=\frac{2}{(x-5)^3}$$

The domain of f is $(-\infty,5)\cup(5,\infty)$ as determined in step (b).

d) *Critical Points.* $f'(x)$ exists for all values of x except 5, but 5 is not in the domain of the function, so $x=5$ is not a critical value. The equation $f'(x)=0$ has no solution, so there are no critical points.

e) *Increasing, decreasing, relative extrema.* We use 5 to divide the real number line into two intervals A: $(-\infty,5)$ and B: $(5,\infty)$, and we test a point in each interval.

A: Test 4, $f'(4)=\frac{-1}{(4-5)^2}=-\frac{1}{1}=-1<0$

B: Test 6, $f'(6)=\frac{-1}{(6-5)^2}=-\frac{1}{1}=-1<0$

Then $f(x)$ is decreasing on both intervals. Since there are no critical points, there are no relative extrema.

f) *Inflection points.* $f''(x)$ does not exist at 5, but because 5 is not in the domain of the function, there cannot be an inflection point at 5. The equation $f''(x)=0$ has no solution; therefore, there are no inflection points.

g) *Concavity.* We use 5 to divide the real number line into two intervals A: $(-\infty,5)$ and B: $(5,\infty)$, and we test a point in each interval.

A: Test 4, $f''(4)=\frac{2}{(4-5)^3}=\frac{2}{-1}=-2<0$

B: Test 6, $f''(6)=\frac{2}{(6-5)^3}=\frac{2}{1}=2>0$

Therefore, $f(x)$ is concave down on $(-\infty,5)$ and concave up on $(5,\infty)$.

h) *Sketch.* Use the preceding information to sketch the graph. Compute additional function values as needed.

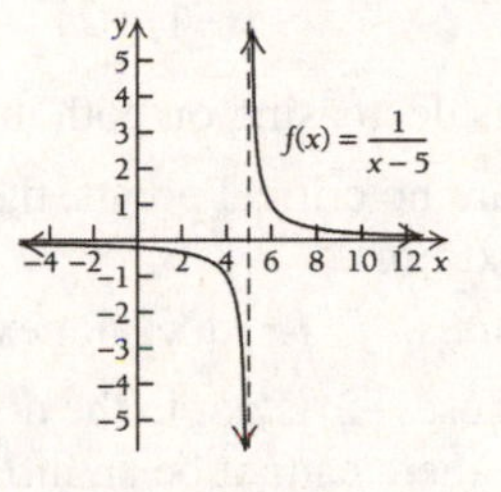

27. $f(x)=\frac{1}{x+2}=(x+2)^{-1}$

a) *Intercepts.* Since the numerator is the constant 1, there are no *x*-intercepts. To find the *y*-intercepts we compute $f(0)$

$$f(0)=\frac{1}{(0)+2}=\frac{1}{2}$$

The point $\left(0,\frac{1}{2}\right)$ is the *y*-intercept.

b) *Asymptotes.*
Vertical. The denominator is 0 for $x=-2$, so the line $x=-2$ is a vertical asymptote.
Horizontal. The degree of the numerator is less than the degree of the denominator, so $y=0$ is the horizontal asymptote.
Slant. There is no slant asymptote since the degree of the numerator is not one more than the degree of the denominator.

c) *Derivatives and Domain.*

$$f'(x)=-(x+2)^{-2}=\frac{-1}{(x+2)^2}$$

$$f''(x)=2(x+2)^{-3}=\frac{2}{(x+2)^3}$$

The domain of f is $(-\infty,-2)\cup(2,\infty)$ as determined in step (b).

Copyright © 2014 Pearson Education, Inc. Publishing as Addison-Wesley.

d) *Critical Points.* $f'(x)$ exists for all values of x except -2, but -2 is not in the domain of the function, so $x=-2$ is not a critical value. The equation $f'(x)=0$ has no solution, so there are no critical points.

e) *Increasing, decreasing, relative extrema.* We use -2 to divide the real number line into two intervals A: $(-\infty,-2)$ and B: $(-2,\infty)$, and we test a point in each interval.

A: Test -3, $f'(-3)=\dfrac{-1}{((-3)+2)^2}=-1<0$

B: Test -1, $f'(-1)=\dfrac{-1}{((-1)+2)^2}=-1<0$

Then $f(x)$ is decreasing on both intervals. Since there are no critical points, there are no relative extrema.

f) *Inflection points.* $f''(x)$ does not exist at -2, but because -2 is not in the domain of the function, there cannot be an inflection point at -2. The equation $f''(x)=0$ has no solution; therefore, there are no inflection points.

g) *Concavity.* We use -2 to divide the real number line into two intervals A: $(-\infty,-2)$ and B: $(-2,\infty)$, and we test a point in each interval.

A: Test -3, $f''(-3)=\dfrac{2}{((-3)+2)^3}=-2<0$

B: Test -1, $f''(-1)=\dfrac{2}{((-1)+2)^3}=2>0$

Therefore, $f(x)$ is concave down on $(-\infty,-2)$ and concave up on $(-2,\infty)$.

h) *Sketch.* Use the preceding information to sketch the graph. Compute additional function values as needed.

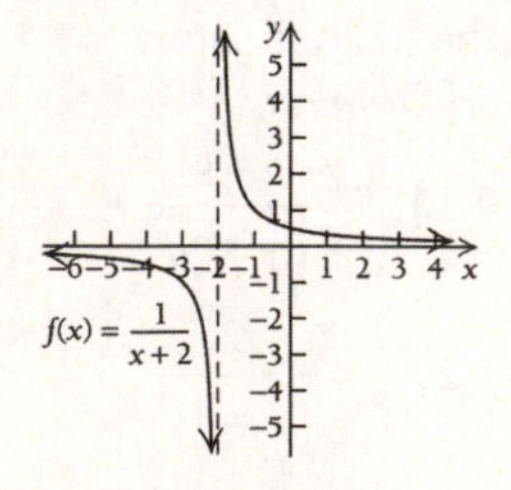

29. $f(x)=\dfrac{-3}{x-3}=-3(x-3)^{-1}$

a) *Intercepts.* Since the numerator is the constant -3, there are no x-intercepts. To find the y-intercepts we compute $f(0)$

$$f(0)=\frac{-3}{(0)-3}=\frac{3}{3}=1$$

The point $(0,1)$ is the y-intercept.

b) *Asymptotes.*
Vertical. The denominator is 0 for $x=3$, so the line $x=3$ is a vertical asymptote.
Horizontal. The degree of the numerator is less than the degree of the denominator, so $y=0$ is the horizontal asymptote.
Slant. There is no slant asymptote since the degree of the numerator is not one more than the degree of the denominator.

c) *Derivatives and Domain.*

$$f'(x)=3(x-3)^{-2}=\frac{3}{(x-3)^2}$$

$$f''(x)=-6(x-3)^{-3}=\frac{-6}{(x-3)^3}$$

The domain of f is $(-\infty,3)\cup(3,\infty)$ as determined in step (b).

d) *Critical Points.* $f'(x)$ exists for all values of x except 3, but 3 is not in the domain of the function, so $x=3$ is not a critical value. The equation $f'(x)=0$ has no solution, so there are no critical points.

e) *Increasing, decreasing, relative extrema.* We use 3 to divide the real number line into two intervals A: $(-\infty,3)$ and B: $(3,\infty)$, and we test a point in each interval.

A: Test 2, $f'(2)=\dfrac{3}{((2)-3)^2}=3>0$

B: Test 4, $f'(4)=\dfrac{3}{((4)-3)^2}=3>0$

Then $f(x)$ is increasing on both intervals. Since there are no critical points, there are no relative extrema.

f) *Inflection points.* $f''(x)$ does not exist at 3, but because 3 is not in the domain of the function, there cannot be an inflection point at 3. The equation $f''(x)=0$ has no solution; therefore, there are no inflection points.

Copyright © 2014 Pearson Education, Inc. Publishing as Addison-Wesley.

g) *Concavity.* We use 3 to divide the real number line into two intervals
A: $(-\infty,3)$ and B: $(3,\infty)$, and we test a point in each interval.

A: Test 2, $f''(2)=\dfrac{-6}{((2)-3)^3}=6>0$

B: Test 4, $f''(4)=\dfrac{-6}{((4)-3)^3}=-6<0$

Therefore, $f(x)$ is concave up on $(-\infty,3)$ and concave down on $(3,\infty)$.

h) *Sketch.* Use the preceding information to sketch the graph. Compute additional function values as needed.

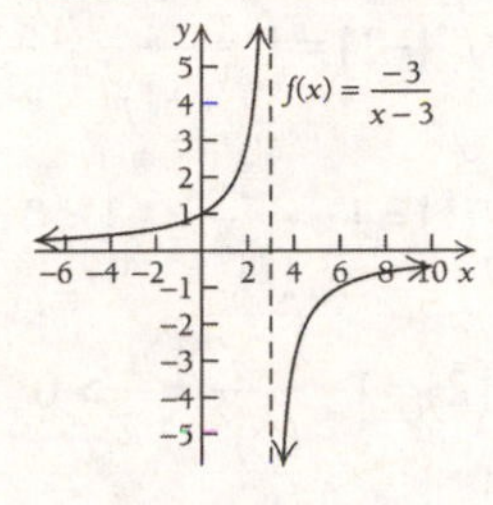

31. $f(x)=\dfrac{3x-1}{x}$

a) *Intercepts.* To find the x-intercepts, solve $f(x)=0$.

$$\frac{3x-1}{x}=0$$
$$3x-1=0$$
$$3x=1$$
$$x=\frac{1}{3}$$

Since $x=\frac{1}{3}$ does not make the denominator 0, the x-intercept is $\left(\frac{1}{3},0\right)$

The number 0 is not in the domain of $f(x)$ so there are no y-intercepts.

b) *Asymptotes.*
Vertical. The denominator is 0 for $x=0$, so the line $x=0$ is a vertical asymptote.
Horizontal. The numerator and the denominator have the same degree, so $y=\frac{3}{1}$, or $y=3$ is the horizontal asymptote.
Slant. There is no slant asymptote since the degree of the numerator is not one more than the degree of the denominator.

c) *Derivatives and Domain.*

$$f'(x)=\frac{1}{x^2}$$
$$f''(x)=-2x^{-3}=-\frac{2}{x^3}$$

The domain of f is $(-\infty,0)\cup(0,\infty)$ as determined in step (b).

d) *Critical Points.* $f'(x)$ exists for all values of x except 0, but 0 is not in the domain of the function, so $x=0$ is not a critical value. The equation $f'(x)=0$ has no solution, so there are no critical points.

e) *Increasing, decreasing, relative extrema.* We use 0 to divide the real number line into two intervals A: $(-\infty,0)$ and B: $(0,\infty)$, and we test a point in each interval.

A: Test -1, $f'(-1)=\dfrac{1}{(-1)^2}=1>0$

B: Test 1, $f'(1)=\dfrac{1}{(-1)^2}=1>0$

Then $f(x)$ is increasing on both intervals. Since there are no critical points, there are no relative extrema.

f) *Inflection points.* $f''(x)$ does not exist at 0, but because 0 is not in the domain of the function, there cannot be an inflection point at 0. The equation $f''(x)=0$ has no solution; therefore, there are no inflection points.

g) *Concavity.* We use 0 to divide the real number line into two intervals
A: $(-\infty,0)$ and B: $(0,\infty)$, and we test a point in each interval.

A: Test -1, $f''(-1)=\dfrac{-2}{(-1)^3}=2>0$

B: Test 1, $f''(1)=\dfrac{-2}{(1)^3}=-2<0$

Therefore, $f(x)$ is concave up on $(-\infty,0)$ and concave down on $(0,\infty)$.

Copyright © 2014 Pearson Education, Inc. Publishing as Addison-Wesley.

h) *Sketch.* Use the preceding information to sketch the graph. Compute additional function values as needed.

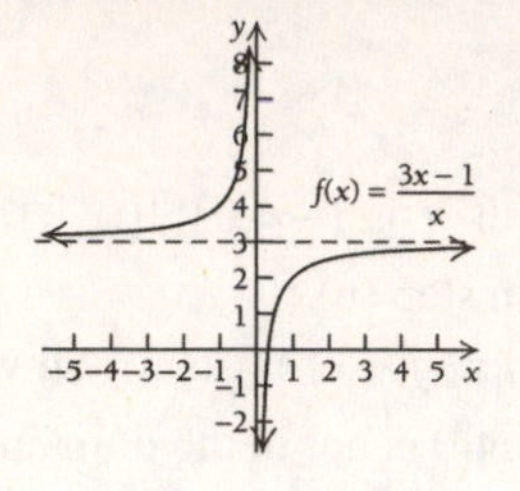

33. $f(x) = x + \frac{2}{x} = \frac{x^2+2}{x}$

a) *Intercepts.* The equation $f(x) = 0$ has no real solutions, so there are no x-intercepts. The number 0 is not in the domain of $f(x)$ so there are no y-intercepts.

b) *Asymptotes.*
Vertical. The denominator is 0 for $x = 0$, so the line $x = 0$ is a vertical asymptote.
Horizontal. The degree of the numerator is greater than the degree of the denominator, so there are no horizontal asymptotes.
Slant. The degree of the numerator is exactly one greater than the degree of the denominator. As $|x|$ gets very large, $f(x) = x + \frac{2}{x}$ approaches x. Therefore, $y = x$ is the slant asymptote.

c) *Derivatives and Domain.*

$$f'(x) = 1 - 2x^{-2} = 1 - \frac{2}{x^2}$$

$$f''(x) = 4x^{-3} = \frac{4}{x^3}$$

The domain of f is $(-\infty, 0) \cup (0, \infty)$ as determined in step (b).

d) *Critical Points.* $f'(x)$ exists for all values of x except 0, but 0 is not in the domain of the function, so $x = 0$ is not a critical value. The critical points will occur when $f'(x) = 0$.

$$1 - \frac{2}{x^2} = 0$$

$$1 = \frac{2}{x^2}$$

$$x^2 = 2$$

$$x = \pm\sqrt{2}$$

Thus, $-\sqrt{2}$ and $\sqrt{2}$ are critical values. $f(-\sqrt{2}) = -2\sqrt{2}$ and $f(\sqrt{2}) = 2\sqrt{2}$, so the critical points $(-\sqrt{2}, -2\sqrt{2})$ and $(\sqrt{2}, 2\sqrt{2})$ are on the graph.

e) *Increasing, decreasing, relative extrema.* We use $-\sqrt{2}$, 0, and $\sqrt{2}$ to divide the real number line into four intervals
A: $(-\infty, -\sqrt{2})$ B: $(-\sqrt{2}, 0)$, C: $(0, \sqrt{2})$, and D: $(\sqrt{2}, \infty)$.

A: Test -2, $f'(-2) = 1 - \frac{2}{(-2)^2} = \frac{1}{2} > 0$

B: Test -1, $f'(-1) = 1 - \frac{2}{(-1)^2} = -1 < 0$

C: Test 1, $f'(1) = 1 - \frac{2}{(1)^2} = -1 < 0$

D: Test 2, $f'(2) = 1 - \frac{2}{(2)^2} = \frac{1}{2} > 0$

Then $f(x)$ is increasing on $(-\infty, -\sqrt{2}]$ and $[\sqrt{2}, \infty)$ and is decreasing on $[-\sqrt{2}, 0)$ and $(0, \sqrt{2}]$. Therefore, $(-\sqrt{2}, -2\sqrt{2})$ is a relative maximum, and $(\sqrt{2}, 2\sqrt{2})$ is a relative minimum.

f) *Inflection points.* $f''(x)$ does not exist at 0, but because 0 is not in the domain of the function, there cannot be an inflection point at 0. The equation $f''(x) = 0$ has no solution; therefore, there are no inflection points.

g) *Concavity.* We use 0 to divide the real number line into two intervals
A: $(-\infty, 0)$ and B: $(0, \infty)$, and we test a point in each interval.

A: Test -1, $f''(-1) = \frac{4}{(-1)^3} = -4 < 0$

B: Test 1, $f''(1) = \frac{4}{(1)^3} = 4 > 0$

Therefore, $f(x)$ is concave down on $(-\infty, 0)$ and concave up on $(0, \infty)$.

Copyright © 2014 Pearson Education, Inc. Publishing as Addison-Wesley.

h) *Sketch.* Use the preceding information to sketch the graph. Compute additional function values as needed.

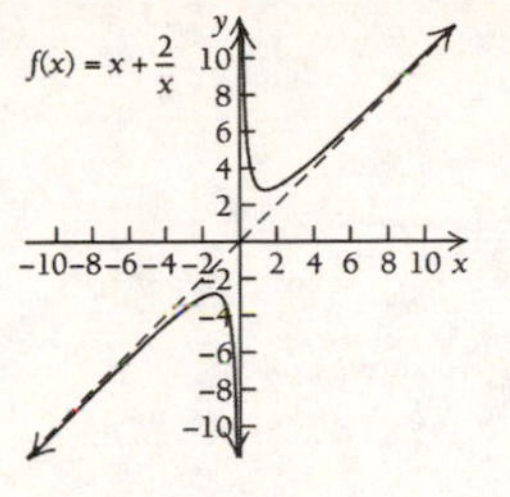

35. $f(x)=\frac{-1}{x^2}=-x^{-2}$

a) *Intercepts.* Since the numerator is the constant -1, there are no x-intercepts. The number 0 is not in the domain of the function, so there are no y-intercepts.

b) *Asymptotes.*
Vertical. The denominator is 0 for $x=0$, so the line $x=0$ is a vertical asymptote.
Horizontal. The degree of the numerator is less than the degree of the denominator, so $y=0$ is the horizontal asymptote.
Slant. There is no slant asymptote since the degree of the numerator is not one more than the degree of the denominator.

c) *Derivatives and Domain.*

$$f'(x)=2x^{-3}=\frac{2}{x^3}$$

$$f''(x)=-6x^{-4}=-\frac{6}{x^4}$$

The domain of f is $(-\infty,0)\cup(0,\infty)$ as determined in step (b).

d) *Critical Points.* $f'(x)$ exists for all values of x except 0, but 0 is not in the domain of the function, so $x=0$ is not a critical value. The equation $f'(x)=0$ has no solution, so there are no critical points.

e) *Increasing, decreasing, relative extrema.* We use 0 to divide the real number line into two intervals A: $(-\infty,0)$ and B: $(0,\infty)$, and we test a point in each interval.

A: Test -1, $f'(-1)=\frac{2}{(-1)^3}=-2<0$

B: Test 1, $f'(1)=\frac{2}{(1)^3}=2>0$

Then $f(x)$ is decreasing on $(-\infty,0)$ and is increasing on $(0,\infty)$. Since there are no critical points, there are no relative extrema.

f) *Inflection points.* $f''(x)$ does not exist at 0, but because 0 is not in the domain of the function, there cannot be an inflection point at 0. The equation $f''(x)=0$ has no solution; therefore, there are no inflection points.

g) *Concavity.* We use 0 to divide the real number line into two intervals A: $(-\infty,0)$ and B: $(0,\infty)$, and we test a point in each interval.

A: Test -1, $f''(-1)=-\frac{6}{(-1)^4}=-6<0$

B: Test 1, $f''(1)=-\frac{6}{(1)^4}=-6<0$

Therefore, $f(x)$ is concave down on both intervals.

h) *Sketch.* Use the preceding information to sketch the graph. Compute additional function values as needed.

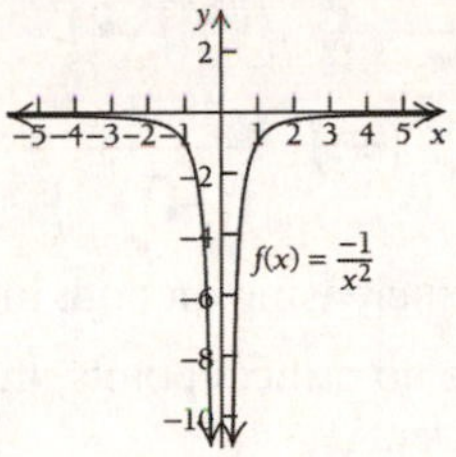

37. $f(x)=\frac{x}{x+2}$

a) *Intercepts.* To find the x-intercepts, solve $f(x)=0$.

$$\frac{x}{x+2}=0$$

$$x=0$$

Since $x=0$ does not make the denominator 0, the x-intercept is $(0,0)$. $f(0)=0$, so the y-intercept is $(0,0)$ also.

b) *Asymptotes.*
Vertical. The denominator is 0 for $x=-2$, so the line $x=-2$ is a vertical asymptote.
Horizontal. The numerator and the denominator have the same degree, so $y=\frac{1}{1}$, or $y=1$ is the horizontal asymptote.
Slant. There is no slant asymptote since the degree of the numerator is not one more than the degree of the denominator.

Copyright © 2014 Pearson Education, Inc. Publishing as Addison-Wesley.

c) *Derivatives and Domain.*

$$f'(x)=\frac{2}{(x+2)^2}$$

$$f''(x)=-4(x+2)^{-3}=-\frac{4}{(x+2)^3}$$

The domain of f is $(-\infty,-2)\cup(2,\infty)$ as determined in step (b).

d) *Critical Points.* $f'(x)$ exists for all values of x except -2, but -2 is not in the domain of the function, so $x=-2$ is not a critical value. The equation $f'(x)=0$ has no solution, so there are no critical points.

e) *Increasing, decreasing, relative extrema.* We use -2 to divide the real number line into two intervals A: $(-\infty,-2)$ and B: $(-2,\infty)$, and we test a point in each interval.

A: Test -3, $f'(-3)=\frac{2}{((-3)+2)^2}=2>0$

B: Test -1, $f'(-1)=\frac{2}{((-1)+2)^2}=2>0$

Then $f(x)$ is increasing on both intervals. Since there are no critical points, there are no relative extrema.

f) *Inflection points.* $f''(x)$ does not exist at -2, but because -2 is not in the domain of the function, there cannot be an inflection point at -2. The equation $f''(x)=0$ has no solution; therefore, there are no inflection points.

g) *Concavity.* We use -2 to divide the real number line into two intervals A: $(-\infty,-2)$ and B: $(-2,\infty)$, and we test a point in each interval.

A: Test -3, $f''(-3)=-\frac{4}{((-3)+2)^3}=4>0$

B: Test -1, $f''(-1)=-\frac{4}{((-1)+2)^3}=-4<0$

Therefore, $f(x)$ is concave up on $(-\infty,-2)$ and concave down on $(-2,\infty)$.

h) *Sketch.* Use the preceding information to sketch the graph. Compute additional function values as needed.

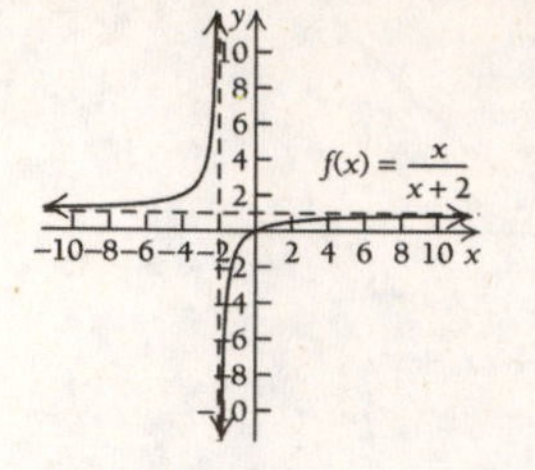

39. $f(x)=\frac{-1}{x^2+2}=-(x^2+2)^{-1}$

a) *Intercepts.* Since the numerator is the constant -1, there are no x-intercepts.

$f(0)=\frac{-1}{(0)^2+2}=-\frac{1}{2}$, so the y-intercept is $\left(0,-\frac{1}{2}\right)$.

b) *Asymptotes.*

Vertical. $x^2+2=0$ has no real solution, so there are no vertical asymptotes.

Horizontal. The degree of the numerator is less than the degree of the denominator, so $y=0$ is the horizontal asymptote.

Slant. There is no slant asymptote since the degree of the numerator is not one more than the degree of the denominator.

c) *Derivatives and Domain.*

$$f'(x)=2x(x^2+2)^{-2}=\frac{2x}{(x^2+2)^2}$$

$$f''(x)=\frac{-6x^2+4}{(x^2+2)^3}$$

The domain of f is $\mathbb{R}$.

d) *Critical Points.* $f'(x)$ exists for all real numbers. Solve $f'(x)=0$

$$\frac{2x}{(x^2+2)^2}=0$$
$$2x=0$$
$$x=0$$

The critical value is 0. From step (a) we found $\left(0,-\frac{1}{2}\right)$ is on the graph.

Copyright © 2014 Pearson Education, Inc. Publishing as Addison-Wesley.

e) *Increasing, decreasing, relative extrema.* We use 0 to divide the real number line into two intervals A: $(-\infty,0)$ and B: $(0,\infty)$, and we test a point in each interval.

A: Test -1, $f'(-1)=-\frac{2}{9}<0$

B: Test 1, $f'(1)=\frac{2}{9}>0$

Then $f(x)$ is decreasing on $(-\infty,0]$ and is increasing on $[0,\infty)$. Thus $\left(0,-\frac{1}{2}\right)$ is a relative minimum.

f) *Inflection points.* $f''(x)$ exists for all real numbers. Solve $f''(x)=0$.

$$\frac{-6x^2+4}{\left(x^2+2\right)^3}=0$$

$$-6x^2+4=0$$

$$-6x^2=-4$$

$$x^2=\frac{2}{3}$$

$$x=\pm\sqrt{\frac{2}{3}}$$

$f\left(-\sqrt{\frac{2}{3}}\right)=-\frac{3}{8}$ and $f\left(\sqrt{\frac{2}{3}}\right)=-\frac{3}{8}$

So, $\left(-\sqrt{\frac{2}{3}},-\frac{3}{8}\right)$ and $\left(\sqrt{\frac{2}{3}},-\frac{3}{8}\right)$ are possible points of inflection.

g) *Concavity.* We use $-\sqrt{\frac{2}{3}}$ and $\sqrt{\frac{2}{3}}$ to divide the real number line into three intervals

A: $\left(-\infty,-\sqrt{\frac{2}{3}}\right)$ B: $\left(-\sqrt{\frac{2}{3}},\sqrt{\frac{2}{3}}\right)$,

and C: $\left(\sqrt{\frac{2}{3}},\infty\right)$

A: Test -1, $f''(-1)=-\frac{2}{27}<0$

B: Test 0, $f''(0)=\frac{1}{2}>0$

C: Test 1, $f''(1)=-\frac{2}{27}<0$

Therefore, $f(x)$ is concave down on $\left(-\infty,-\sqrt{\frac{2}{3}}\right)$ and $\left(\sqrt{\frac{2}{3}},\infty\right)$ and concave up on $\left(-\sqrt{\frac{2}{3}},\sqrt{\frac{2}{3}}\right)$. Therefore the points $\left(-\sqrt{\frac{2}{3}},-\frac{3}{8}\right)$ and $\left(\sqrt{\frac{2}{3}},-\frac{3}{8}\right)$ are points of inflection.

h) *Sketch.* Use the preceding information to sketch the graph. Compute additional function values as needed.

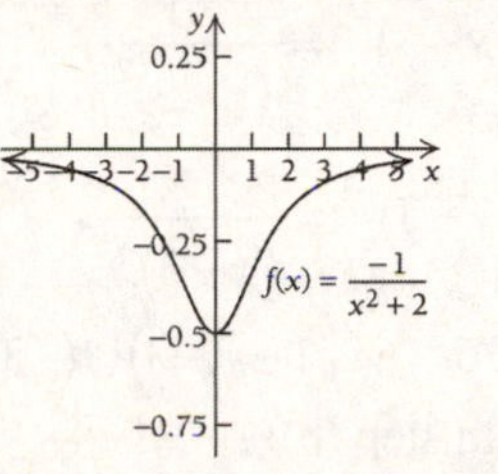

41. $f(x)=\frac{x+3}{x^2-9}=\frac{x+3}{(x+3)(x-3)}=\frac{1}{x-3},\ x\neq\pm3$

We write the expression in simplified form noting that the domain is restricted to all real numbers except for $x=\pm3$.

a) *Intercepts.* $f(x)=0$ has no solution. $x=-3$ is not in the domain of the function. Therefore, there are no x-intercepts. To find the y-intercepts we compute $f(0)$

$$f(0)=\frac{1}{(0)-3}=-\frac{1}{3}$$

The point $\left(0,-\frac{1}{3}\right)$ is the y-intercept.

b) *Asymptotes.*
Vertical. In the original function, the denominator is 0 for $x=-3$ or $x=3$, however, $x=-3$ also made the numerator equal to 0. We look at the limits to determine if there are vertical asymptotes at these points.

$$\lim_{x\to-3}\frac{x+3}{x^2-9}=\lim_{x\to-3}\frac{1}{x-3}=\frac{1}{-3-3}=-\frac{1}{6}.$$

Because the limit exists, the line $x=-3$ is not a vertical asymptote. Instead, we have a removable discontinuity, or a "hole" at the point $\left(-3,-\frac{1}{6}\right)$.

The solution is continued on the next page.

Copyright © 2014 Pearson Education, Inc. Publishing as Addison-Wesley.

An open circle is drawn at $\left(-3,-\frac{1}{6}\right)$ to show that it is not part of the graph.
The denominator is 0 for $x=3$ and the numerator is not 0 at this value, so the line $x=3$ is a vertical asymptote.
Horizontal. The degree of the numerator is less than the degree of the denominator, so $y=0$ is the horizontal asymptote.
Slant. There is no slant asymptote since the degree of the numerator is not one more than the degree of the denominator.

c) *Derivatives and Domain.*

$$f'(x)=-(x-3)^{-2}=\frac{-1}{(x-3)^2}$$

$$f''(x)=2(x-3)^{-3}=\frac{2}{(x-3)^3}$$

The domain of f is $(-\infty,-3)\cup(-3,\infty)$ as determined in step (b).

d) *Critical Points.* $f'(x)$ exists for all values of x except 3, but 3 is not in the domain of the function, so $x=3$ is not a critical value. The equation $f'(x)=0$ has no solution, so there are no critical points.

e) *Increasing, decreasing, relative extrema.* We use -3 and 3 to divide the real number line into three intervals A: $(-\infty,-3)$ B: $(-3,3)$ and C: $(3,\infty)$.
We notice that $f'(x)<0$ for all real numbers, $f(x)$ is decreasing on all three intervals $(-\infty,-3)$, $(-3,3)$, and $(3,\infty)$.
Since there are no critical points, there are no relative extrema.

f) *Inflection points.* $f''(x)$ does not exist at 3, but because 3 is not in the domain of the function, there cannot be an inflection point at 3. The equation $f''(x)=0$ has no solution; therefore, there are no inflection points.

g) *Concavity.* We use -3 and 3 to divide the real number line into three intervals A: $(-\infty,-3)$ B: $(-3,3)$ and C: $(3,\infty)$ and we test a point in each interval.

A: Test -4, $f''(-4)=\frac{2}{((-4)-3)^3}=-\frac{2}{343}<0$

B: Test 2, $f''(2)=\frac{2}{((2)-3)^3}=-2<0$

C: Test 4, $f''(4)=\frac{2}{((4)-3)^3}=2>0$

Therefore, $f(x)$ is concave down on $(-\infty,-3)$ and $(-3,3)$ and concave up on $(3,\infty)$.

h) *Sketch.* Use the preceding information to sketch the graph. Compute additional function values as needed.

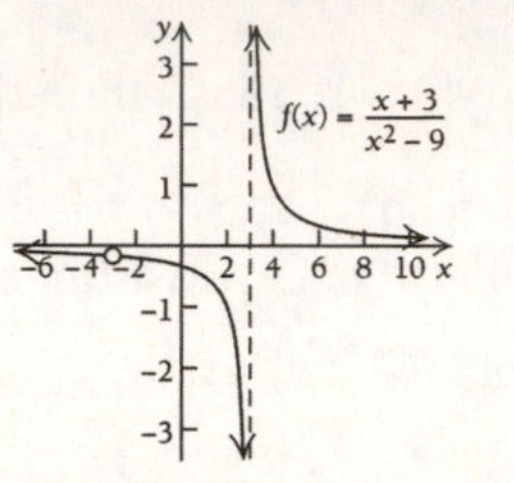

43. $f(x)=\frac{x-1}{x+2}$

a) *Intercepts.* To find the x-intercepts, solve $f(x)=0$.

$$\frac{x-1}{x+2}=0$$
$$x=1$$

Since $x=1$ does not make the denominator 0, the x-intercept is $(1,0)$.

$f(0)=\frac{0-1}{0+2}=-\frac{1}{2}$, so the y-intercept is $\left(0,-\frac{1}{2}\right)$.

b) *Asymptotes.*
Vertical. The denominator is 0 for $x=-2$, so the line $x=-2$ is a vertical asymptote.
Horizontal. The numerator and the denominator have the same degree, so $y=\frac{1}{1}$, or $y=1$ is the horizontal asymptote.
Slant. There is no slant asymptote since the degree of the numerator is not one more than the degree of the denominator.

c) *Derivatives and Domain.*

$$f'(x)=\frac{3}{(x+2)^2}$$

$$f''(x)=-\frac{6}{(x+2)^3}$$

The domain of f is $(-\infty,-2)\cup(-2,\infty)$ as determined in part (b).

Copyright © 2014 Pearson Education, Inc. Publishing as Addison-Wesley.

d) *Critical Points.* $f'(x)$ exists for all values of x except -2, but -2 is not in the domain of the function, so $x=-2$ is not a critical value. The equation $f'(x)=0$ has no solution, so there are no critical points.

e) *Increasing, decreasing, relative extrema.* We use -2 to divide the real number line into two intervals
A: $(-\infty,-2)$ and B: $(-2,\infty)$, and we test a point in each interval.
A: Test -3, $f'(-3)=3>0$
B: Test -1, $f'(-1)=3>0$
Then $f(x)$ is increasing on both intervals. Since there are no critical points, there are no relative extrema.

f) *Inflection points.* $f''(x)$ does not exist at -2, but because -2 is not in the domain of the function, there cannot be an inflection point at -2. The equation $f''(x)=0$ has no solution; therefore, there are no inflection points.

g) *Concavity.* We use -2 to divide the real number line into two intervals
A: $(-\infty,-2)$ and B: $(-2,\infty)$, and we test a point in each interval.
A: Test -3, $f''(-3)=6>0$
B: Test -1, $f''(-1)=-6<0$
Therefore, $f(x)$ is concave up on $(-\infty,-2)$ and concave down on $(-2,\infty)$.

h) *Sketch.* Use the preceding information to sketch the graph. Compute additional function values as needed.

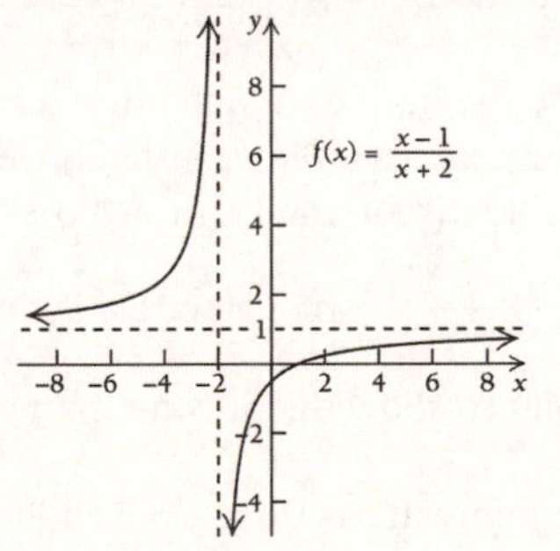

45. $f(x)=\dfrac{x^2-4}{x+3}$

a) *Intercepts.* To find the x-intercepts, solve $f(x)=0$.

$$\frac{x^2-4}{x+3}=0$$
$$x^2-4=0$$
$$x=\pm 2$$

Neither of these values make the denominator 0, so the x-intercepts are $(-2,0)$ and $(2,0)$. $f(0)=\dfrac{0^2-4}{0+3}=-\dfrac{4}{3}$, so the y-intercept is $\left(0,-\dfrac{4}{3}\right)$.

b) *Asymptotes.*
Vertical. The denominator is 0 for $x=-3$, so the line $x=-3$ is a vertical asymptote.
Horizontal. The degree of the numerator is greater than the degree of the denominator, so there are no horizontal asymptotes.
Slant. Divide the numerator by the denominator.

$$\begin{array}{r|l} & x \quad -3 \\ x+3 & x^2 \qquad -4 \\ & \underline{x^2+3x} \\ & \quad -3x-4 \\ & \quad \underline{-3x-9} \\ & \qquad\quad 5 \end{array}$$

$$f(x)=x-3+\frac{5}{x+3}$$

As $|x|$ gets very large, $f(x)$ approaches $x-3$, so $y=x-3$ is the slant asymptote.

c) *Derivatives and Domain.*

$$f'(x)=\frac{x^2+6x+4}{(x+3)^2}$$

$$f''(x)=\frac{10}{(x+3)^3}$$

The domain of f is $(-\infty,-3)\cup(-3,\infty)$ as determined in part (b).

d) *Critical Points.* $f'(x)$ exists for all values of x except -3, but -3 is not in the domain of the function, so $x=-3$ is not a critical value.
The solution is continued on the next page.

Copyright © 2014 Pearson Education, Inc. Publishing as Addison-Wesley.

Solve $f'(x)=0$.

$$\frac{x^2+6x+4}{(x+3)^2}=0$$

$$x^2+6x+4=0$$

$$x=-3\pm\sqrt{5} \quad \text{Using the Quadratic Formula}$$

$x\approx-5.236$ or $x\approx-0.764$

$f(-5.236)\approx-10.472$ and

$f(-0.764)\approx-1.528$, so $(-5.236,-10.472)$ and $(-0.764,-1.528)$ are on the graph.

e) *Increasing, decreasing, relative extrema.* We use -5.236, -3, and -0.764 to divide the real number line into four intervals

A: $(-\infty,-5.236)$, B: $(-5.236,-3)$,

C: $(-3,-0.764)$, and D: $(-0.764,\infty)$

We test a point in each interval.

A: Test -6, $f'(-6)=\frac{4}{9}>0$

B: Test -4, $f'(-4)=-4<0$

C: Test -2, $f'(-2)=-4<0$

D: Test 0, $f'(0)=\frac{4}{9}>0$

Then $f(x)$ is increasing on the intervals $(-\infty,-5.236]$ and $[-0.764,\infty)$, and is decreasing on the intervals $[-5.226,-3)$ and $(-3,-0.764]$. Therefore, $(-5.236,-10.472)$ is a relative maximum and $(-0.764,-1.528)$ is a relative minimum.

f) *Inflection points.* $f''(x)$ does not exist at -3, but because -3 is not in the domain of the function, there cannot be an inflection point at -3. The equation $f''(x)=0$ has no solution; therefore, there are no inflection points.

g) *Concavity.* We use -3 to divide the real number line into two intervals

A: $(-\infty,-3)$ and B: $(-3,\infty)$, and we test a point in each interval.

A: Test -4, $f''(-4)=-10<0$

B: Test -2, $f''(-2)=10>0$

Therefore, $f(x)$ is concave down on $(-\infty,-3)$ and concave up on $(-3,\infty)$.

h) *Sketch.* Use the preceding information to sketch the graph. Compute additional function values as needed.

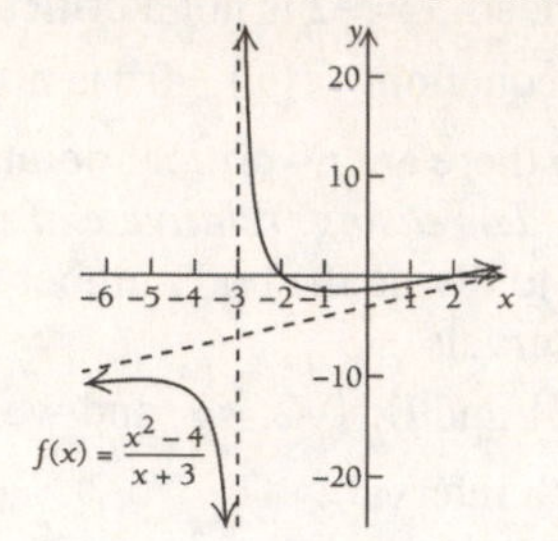

47. $f(x)=\frac{x+1}{x^2-2x-3}=\frac{x+1}{(x-3)(x+1)}=\frac{1}{x-3}$, $x\neq-1$

We write the expression in simplified form noting that the domain is restricted to all real numbers except for $x=-1$ and $x=3$

a) *Intercepts.* $f(x)=0$ has no solution. $x=-1$ is not in the domain of the function. Therefore, there are no *x*-intercepts. To find the *y*-intercepts we compute $f(0)$

$$f(0)=\frac{0+1}{(0)^2-2(0)-3}=-\frac{1}{3}$$

The point $\left(0,-\frac{1}{3}\right)$ is the *y*-intercept.

b) *Asymptotes.*
Vertical. In the original function, the denominator is 0 for $x=-1$ or $x=3$, however, $x=-1$ also made the numerator equal to 0. We look at the limits to determine if there are vertical asymptotes at these points.

$$\lim_{x\to-1}\frac{x+1}{x^2-2x-3}=\lim_{x\to-1}\frac{1}{x-3}=\frac{1}{-1-3}=-\frac{1}{4}$$

Because the limit exists, the line $x=-1$ is not a vertical asymptote. Instead, we have a removable discontinuity, or a "hole" at the point $\left(-1,-\frac{1}{4}\right)$. An open circle is drawn at this point to show that it is not part of the graph.

The denominator is 0 for $x=3$ and the numerator is not 0 at this value, so the line $x=3$ is a vertical asymptote.

Horizontal. The degree of the numerator is less than the degree of the denominator, so $y=0$ is the horizontal asymptote.

The solution is continued on the next page.

Copyright © 2014 Pearson Education, Inc. Publishing as Addison-Wesley.

Slant. There is no slant asymptote since the degree of the numerator is not one more than the degree of the denominator.

c) *Derivatives and Domain.*

$$f'(x) = -(x-3)^{-2} = \frac{-1}{(x-3)^2}$$

$$f''(x) = 2(x-3)^{-3} = \frac{2}{(x-3)^3}$$

The domain of f is $(-\infty, 3) \cup (3, \infty)$ as determined in part (b).

d) *Critical Points.* $f'(x)$ exists for all values of x except 3, but 3 is not in the domain of the function, so $x = 3$ is not a critical value. The equation $f'(x) = 0$ has no solution, so there are no critical points.

e) *Increasing, decreasing, relative extrema.* We use -1 and 3 to divide the real number line into three intervals

A: $(-\infty, -1)$ B: $(-1, 3)$ and C: $(3, \infty)$.

We notice that $f'(x) < 0$ for all real numbers, so $f(x)$ is decreasing on all three intervals $(-\infty, -1)$, $(-1, 3)$, and $(3, \infty)$. Since there are no critical points, there are no relative extrema.

f) *Inflection points.* $f''(x)$ does not exist at 3, but because 3 is not in the domain of the function, there cannot be an inflection point at 3. The equation $f''(x) = 0$ has no solution; therefore, there are no inflection points.

g) *Concavity.* We use -1 and 3 to divide the real number line into three intervals

A: $(-\infty, -1)$ B: $(-1, 3)$ and C: $(3, \infty)$, and we test a point in each interval.

A: Test -2, $f''(-2) = -\frac{2}{125} < 0$

B: Test 2, $f''(2) = -2 < 0$

C: Test 4, $f''(4) = 2 > 0$

Therefore, $f(x)$ is concave down on $(-\infty, -1)$ and $(-1, 3)$ and concave up on $(3, \infty)$.

h) *Sketch.* Use the preceding information to sketch the graph. Compute additional function values as needed.

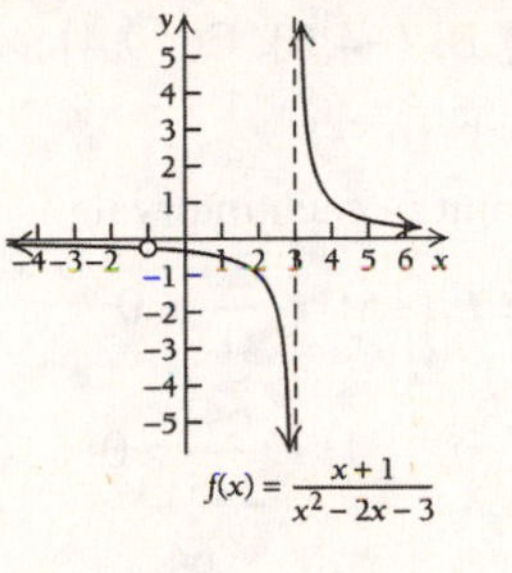

49. $f(x) = \frac{2x^2}{x^2 - 16}$

a) *Intercepts.* The numerator is 0 for $x = 0$ and this value does not make the denominator 0, the x-intercept is $(0, 0)$

$f(0) = 0$, so the y-intercept is $(0, 0)$ also.

b) *Asymptotes.*

Vertical. The denominator is 0 when

$$x^2 - 16 = 0$$
$$x^2 = 16$$
$$x = \pm 4$$

So the lines $x = -4$ and $x = 4$ are vertical asymptotes.

Horizontal. The numerator and the denominator have the same degree, so $y = \frac{2}{1}$, or $y = 2$ is the horizontal asymptote.

Slant. There is no slant asymptote since the degree of the numerator is not one more than the degree of the denominator.

c) *Derivatives and Domain.*

$$f'(x) = -\frac{64x}{(x^2 - 16)^2}$$

$$f''(x) = \frac{192x^2 + 1024}{(x^2 - 16)^3}$$

The domain of f is $(-\infty, -4) \cup (-4, 4) \cup (4, \infty)$ as determined in part (b).

d) *Critical Points.* $f'(x)$ exists for all values of x except $x = -4$ and $x = 4$, but -4 and 4 are not in the domain of the function, so $x = -4$ and $x = 4$ are not critical values. $f'(x) = 0$ for $x = 0$ so, $(0, 0)$ is the only critical point.

Copyright © 2014 Pearson Education, Inc. Publishing as Addison-Wesley.

e) *Increasing, decreasing, relative extrema.* We use -4, 0, and 4 to divide the real number line into four intervals
A: $(-\infty,-4)$ B: $(-4,0)$, C: $(0,4)$, and D: $(4,\infty)$
We test a point in each interval.
A: Test -5, $f'(-5)=\frac{320}{81}>0$
B: Test -1, $f'(-1)=\frac{64}{225}>0$
C: Test 1, $f'(1)=-\frac{64}{225}<0$
D: Test 5, $f'(5)=-\frac{320}{81}<0$
Then $f(x)$ is increasing on the intervals $(-\infty,-4)$ and $(-4,0]$, and is decreasing on the intervals $[0,4)$ and $(4,\infty)$. Thus, there is a relative maximum at $(0,0)$.

f) *Inflection points.* $f''(x)$ does not exist at -4 and 4, but because -4 and 4 are not in the domain of the function, there cannot be an inflection point at -4 or 4. The equation $f''(x)=0$ has no real solution; therefore, there are no inflection points.

g) *Concavity.* We use -4 and 4 to divide the real number line into three intervals
A: $(-\infty,-4)$ B: $(-4,4)$ and C: $(4,\infty)$, and we test a point in each interval.
A: Test -5, $f''(-5)=\frac{5824}{729}>0$
B: Test 0, $f''(0)=-\frac{1}{4}<0$
C: Test 5, $f''(5)=\frac{5824}{729}>0$
Therefore, $f(x)$ is concave up on the intervals $(-\infty,-4)$ and $(4,\infty)$ and concave down on the interval $(-4,4)$.

h) *Sketch.* Use the preceding information to sketch the graph. Compute additional function values as needed.

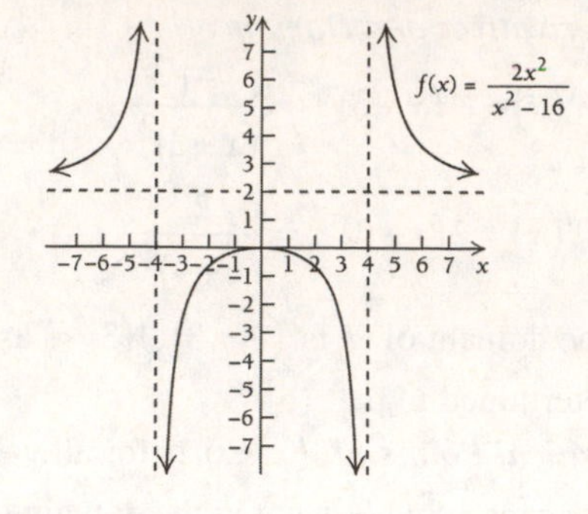

51. $f(x)=\frac{1}{x^2-1}$

a) *Intercepts.* Since the numerator is a constant 1, there are no x-intercepts.
$$f(0)=\frac{1}{(0)^2-1}=-1$$
The point $(0,-1)$ is the y-intercept.

b) *Asymptotes.*
Vertical. The denominator $x^2-1=(x-1)(x+1)$ is 0 for $x=-1$ or $x=1$, so the lines $x=-1$ and $x=1$ are vertical asymptotes.
Horizontal. The degree of the numerator is less than the degree of the denominator, so $y=0$ is the horizontal asymptote.
Slant. There is no slant asymptote since the degree of the numerator is not one more than the degree of the denominator.

c) *Derivatives and Domain.*
$$f'(x)=\frac{-2x}{(x^2-1)^2}$$
$$f''(x)=\frac{2(3x^2+1)}{(x^2-1)^3}$$
The domain of f is $(-\infty,-1)\cup(-1,1)\cup(1,\infty)$ as determined in part (b).

d) *Critical Points.* $f'(x)$ exists for all values of x except -1 and 1, but these values are not in the domain of the function, so $x=-1$ and $x=1$ are not critical values. $f'(x)=0$ for $x=0$. From step (a) we know $f(0)=-1$, so the critical point is $(0,-1)$.

Copyright © 2014 Pearson Education, Inc. Publishing as Addison-Wesley.

e) *Increasing, decreasing, relative extrema.* We use -1, 0, and 1 to divide the real number line into four intervals A: $(-\infty,-1)$, B: $(-1,0)$, C: $(0,1)$, and D: $(1,\infty)$, and we test a point in each interval.

A: Test -2, $f'(-2)=\frac{4}{9}>0$

B: Test $-\frac{1}{2}$, $f'\left(-\frac{1}{2}\right)=\frac{16}{9}>0$

C: Test $\frac{1}{2}$, $f'\left(\frac{1}{2}\right)=-\frac{16}{9}<0$

D: Test 2, $f'(2)=-\frac{4}{9}<0$

We see that $f(x)$ is increasing on the intervals $(-\infty,-1)$ and $(-1,0]$, and is decreasing on the intervals $[0,1)$ and $(1,\infty)$. Therefore, $(0,-1)$ is a relative maximum.

f) *Inflection points.* $f''(x)$ does not exist at -1 and 1, but because these values are not in the domain of the function, there cannot be an inflection point at -1 or 1. The equation $f''(x)=0$ has no real solution; therefore, there are no inflection points.

g) *Concavity.* We use -1 and 1 to divide the real number line into three intervals A: $(-\infty,-1)$ B: $(-1,1)$ and C: $(1,\infty)$. and we test a point in each interval.

A: Test -2, $f''(-2)=\frac{26}{27}>0$

B: Test 0, $f''(0)=-2<0$

C: Test 2, $f''(2)=\frac{26}{27}>0$

Therefore, $f(x)$ is concave up on $(-\infty,-1)$ and $(1,\infty)$, and concave down on $(-1,1)$.

h) *Sketch.* Use the preceding information to sketch the graph. Compute additional function values as needed.

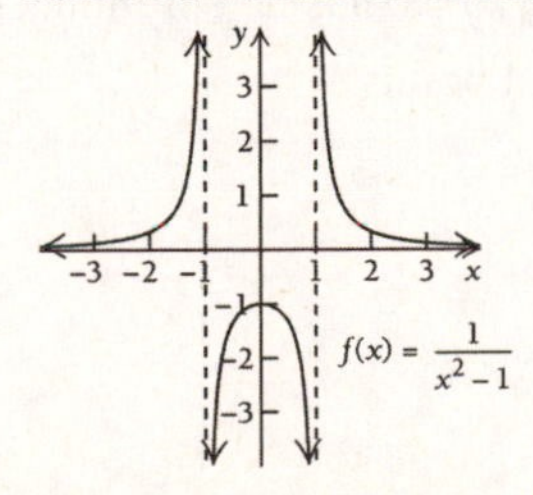

53. $f(x)=\frac{x^2+1}{x}$

a) *Intercepts.* The equation $f(x)=0$ has no real solutions, so there are no x-intercepts. The number 0 is not in the domain of $f(x)$ so there are no y-intercepts.

b) *Asymptotes.*
Vertical. The denominator is 0 for $x=0$, so the line $x=0$ is a vertical asymptote.
Horizontal. The degree of the numerator is greater than the degree of the denominator, so there are no horizontal asymptotes.
Slant. The degree of the numerator is exactly one greater than the degree of the denominator. When we divide the numerator by the denominator we have

$f(x)=\frac{x^2+1}{x}=x+\frac{1}{x}$. As $|x|$ gets very large, $f(x)=x+\frac{1}{x}$ approaches x. Therefore, $y=x$ is the slant asymptote.

c) *Derivatives and Domain.*

$$f'(x)=\frac{x^2-1}{x^2}$$

$$f''(x)=\frac{2}{x^3}$$

The domain of f is $(-\infty,-0)\cup(0,\infty)$ as determined in part (b).

d) *Critical Points.* $f'(x)$ exists for all values of x except 0, but 0 is not in the domain of the function, so $x=0$ is not a critical value. The critical points will occur when $f'(x)=0$.

$$\frac{x^2-1}{x^2}=0$$

$$x^2-1=0$$

$$x^2=1$$

$$x=\pm1$$

We found that -1 and 1 are critical values, thus $f(-1)=-2$ and $f(1)=2$, so the critical points $(-1,-2)$ and $(1,2)$ are on the graph.

Copyright © 2014 Pearson Education, Inc. Publishing as Addison-Wesley.

e) *Increasing, decreasing, relative extrema.* We use $-1, 0,$ and 1 to divide the real number line into four intervals A: $(-\infty,-1)$ B: $(-1,0)$, C: $(0,1)$, and D: $(1,\infty)$. We test a point in each interval.

A: Test $-2, f'(-2)=\frac{3}{4}>0$

B: Test $-\frac{1}{2},\ f'\left(-\frac{1}{2}\right)=-3<0$

C: Test $\frac{1}{2},\ f'\left(\frac{1}{2}\right)=-3<0$

D: Test $2, f'(2)=\frac{3}{4}>0$

Then $f(x)$ is increasing on $(-\infty,-1]$ and $[1,\infty)$ and is decreasing on $[-1,0)$ and $(0,1]$. Therefore, $(-1,-2)$ is a relative maximum, and $(1,2)$ is a relative minimum.

f) *Inflection points.* $f''(x)$ does not exist at 0, but because 0 is not in the domain of the function, there cannot be an inflection point at 0. The equation $f''(x)=0$ has no solution; therefore, there are no inflection points.

g) *Concavity.* We use 0 to divide the real number line into two intervals A: $(-\infty,0)$ and B: $(0,\infty)$, and we test a point in each interval.

A: Test $-1, f''(-1)=-2<0$

B: Test $1,\ f''(1)=2>0$

Therefore, $f(x)$ is concave down on $(-\infty,0)$ and concave up on $(0,\infty)$.

h) *Sketch.* Use the preceding information to sketch the graph. Compute additional function values as needed.

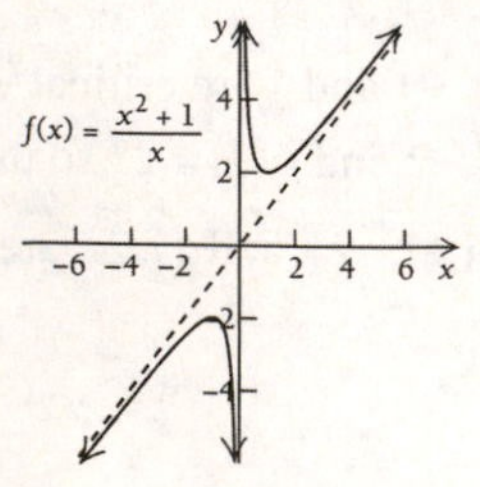

55. $f(x)=\frac{x^2-9}{x-3}$

We write the expression in simplified form:

$$f(x)=\frac{(x-3)(x+3)}{x-3}=x+3,\quad x\neq 3$$

Note that the domain is restricted to all real numbers except for $x=3$.

a) *Intercepts.* The numerator is 0 when $x=-3$ or $x=3$ however, $x=3$ is not in the domain of $f(x)$, so the x-intercept is $(-3,0)$.

$$f(0)=\frac{0^2-9}{(0)-3}=\frac{-9}{-3}=3$$

The point $(0,3)$ is the y-intercept.

b) *Asymptotes.*
In simplified form $f(x)=x+3$, a linear function everywhere except $x=3$. So there are no asymptotes of any kind.
In the original function, the denominator is 0 for $x=3$, however, $x=3$ also made the numerator equal to 0. We look at the limits to determine if there is a vertical asymptote at this point.

$$\lim_{x\to 3}\frac{x^2-9}{x-3}=\lim_{x\to 3}x+3=6.$$

We found the limit exists; therefore, the line $x=3$ is not a vertical asymptote. Instead, we have a removable discontinuity, or a "hole" at the point $(3,6)$. An open circle is drawn at this point to show that it is not part of the graph.

c) *Derivatives and Domain.*

$f'(x)=1,\ x\neq 3$

$f''(x)=0$

The domain of f is $(-\infty,3)\cup(3,\infty)$ as determined in part (b).

d) *Critical Points.* $f'(x)$ exists for all values of x except 3, but 3 is not in the domain of the function, so $x=3$ is not a critical value. The equation $f'(x)=0$ has no solution, so there are no critical points.

Copyright © 2014 Pearson Education, Inc. Publishing as Addison-Wesley.

e) *Increasing, decreasing, relative extrema.* We use 3 to divide the real number line into two intervals A: $(-\infty,3)$ and B: $(3,\infty)$. We notice that $f'(x)>0$ for all real numbers, $f(x)$ is increasing on both intervals. Since there are no critical points, there are no relative extrema.

f) *Inflection points.* $f''(x)$ is constant; therefore, there are no points of inflection.

g) *Concavity.* $f''(x)$ is 0; therefore, there is no concavity.

h) *Sketch.* Use the preceding information to sketch the graph.

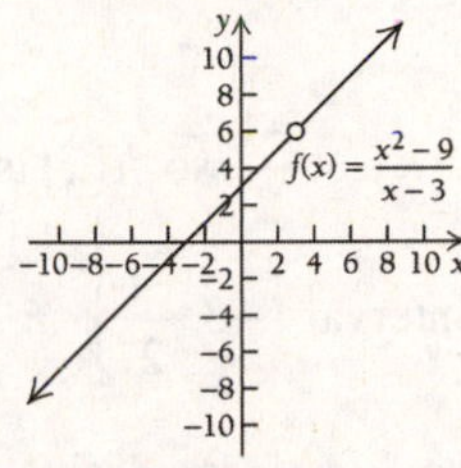

Note: In the preceding problem, we could have noticed that the graph of $f(x)=\frac{x^2-9}{x-3}$ is the graph of $f(x)=x+3$ with the exception of the point $(3,6)$ which is a removable discontinuity. We simply need to graph $f(x)=x+3$ with a hole at the point $(3,6)$ and determine all other aspects of the graph of $f(x)$ from the linear graph.

57. One possible rational function would be $f(x)=\frac{-2x}{x-2}$.

Using the techniques in this section, we sketch the graph.

Intercepts. $f(x)=0$ when $x=0$. It turns out the x-intercept and the y-intercept is $(0,0)$

Asymptotes. $x=2$ is the vertical asymptote. The degree of the numerator equals that of the denominator, the line $y=-2$ is the horizontal asymptote.

Increasing, decreasing, relative extrema.

$f'(x)=\frac{4}{(x-2)^2}$. $f'(x)$ is not defined for $x=2$, however that value is outside the domain of the function. $f'(x)>0$ so $f(x)$ is increasing on the interval $(-\infty,2)$ and $(2,\infty)$. There are no relative extrema.

Inflection points, concavity.

$f''(x)=\frac{-12}{(x-2)^3}$ does not exist when $x=2$.

The equation $f''(x)=0$ has no real solution, so there are no possible points of inflection. Furthermore, $f''(x)<0$ for all x in $(-\infty,2)$, so $f(x)$ is concave up on $(-\infty,2)$, and , $f''(x)<0$ for all x in $(2,\infty)$, so $f(x)$ is concave down on $(2,\infty)$.

We use this information to sketch the graph. Additional values may be computed as necessary.

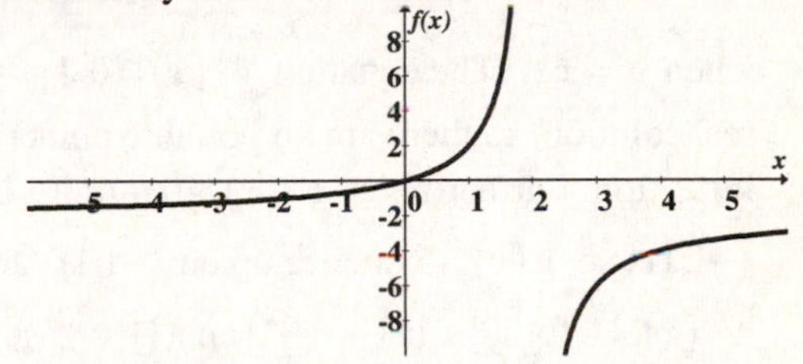

59. One possible rational function would be $g(x)=\frac{x^2-2}{x^2-1}$.

Using the techniques in this section, we sketch the graph.

Intercepts. $g(x)=0$ when $x=\pm\sqrt{2}$. The x-intercepts are $(-\sqrt{2},0)$ and $(\sqrt{2},0)$ When $x=0$, $g(x)=2$, so the y-intercept is $(0,2)$.

Asymptotes. $x=-1$ and $x=1$ are the vertical asymptote.

The degree of the numerator equals that of the denominator, the line $y=1$ is the horizontal asymptote.

Copyright © 2014 Pearson Education, Inc. Publishing as Addison-Wesley.

Increasing, decreasing, relative extrema.

$g'(x)=\frac{2x}{(x^2-1)^2}$. $g'(x)$ is not defined for $x=\pm1$, however that value is outside the domain of the function. $g'(x)=0$, when $x=0$, so there is a critical value at $(0,2)$.

$g'(x)<0$ when $x<-1$ so $g(x)$ is decreasing on the interval $(-\infty,-1)$. $g'(x)>0$ when $1<x$ so $g(x)$ is increasing on the interval $(1,\infty)$. $g'(x)<0$ when $-1<x<0$ so $g(x)$ is decreasing on the interval $(-1,0)$. $g'(x)>0$ when $0<x<1$ so $g(x)$ is increasing on the interval $(0,1)$. There are is a relative minimum at $(0,2)$.

Inflection points, concavity.

$g''(x)=\frac{-2(3x^2+1)}{(x^2-1)^3}$. $g''(x)$ does not exist when $x=\pm1$. The equation $g''(x)=0$ has no real solution, so there are no possible points of inflection. Furthermore, $g''(x)>0$ for all x in $(-1,1)$, so $g(x)$ is concave up on $(-1,1)$, and, $g''(x)<0$ for all x in $(-\infty,-1)$ and $(1,\infty)$, so $g(x)$ is concave down on $(-\infty,-1)$ and $(1,\infty)$.

We use this information to sketch the graph. Additional values may be computed as necessary.

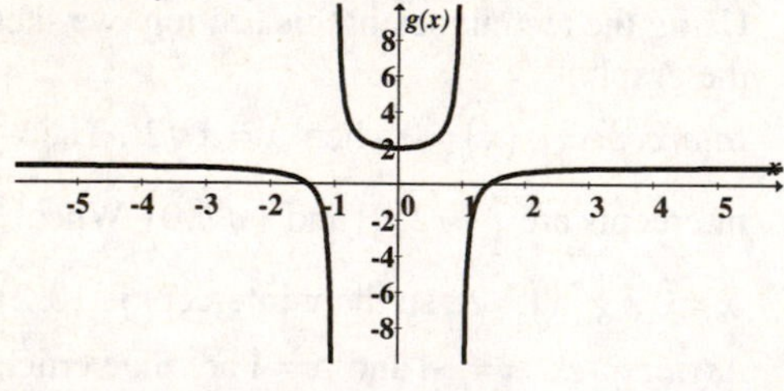

61. One possible rational function would be $h(x)=\frac{-8}{x^2+x-6}$.

Using the techniques in this section, we sketch the graph.

Intercepts. $h(x)=0$ has no real solution. The are no x-intercepts. When $x=0$, $h(x)=\frac{4}{3}$, so the y-intercept is $\left(0,\frac{4}{3}\right)$.

Asymptotes. $x=-3$ and $x=2$ are the vertical asymptote.

The degree of the numerator is less than that of the denominator, the line $y=0$ is the horizontal asymptote.

Increasing, decreasing, relative extrema.

$h'(x)=\frac{8(2x+1)}{(x-2)^2(x+3)^2}$. $h'(x)$ is not defined for $x=-3$ or $x=2$, however those values are outside the domain of the function. $h'(x)=0$, when $x=\frac{-1}{2}$ so there is a critical value at $\left(\frac{-1}{2},\frac{32}{25}\right)$

$h'(x)<0$ when $-3<x<\frac{-1}{2}$ so $h(x)$ is decreasing on the interval $\left(-3,\frac{-1}{2}\right)$. $h'(x)>0$ when $\frac{-1}{2}<x<2$ so $h(x)$ is increasing on the interval $\left(\frac{-1}{2},2\right)$. There is a relative minimum at $\left(\frac{-1}{2},2\right)$. Furthermore, $h'(x)<0$ when $x<-3$ so $h(x)$ is decreasing on the interval $(-\infty,-3)$ $h'(x)>0$ when $2<x$ so $h(x)$ is increasing on the interval $(2,\infty)$.

Inflection points, concavity.

$h''(x)=\frac{-16(3x^2+3x+7)}{(x-2)^3(x+3)^3}$. $h''(x)$ does not exist when $x=-2$ and $x=0$. The equation $h''(x)=0$ has no real solution. Therefore $h(x)$ has no points of inflection. $h''(x)<0$ when $x<-3$ and $x>2$. Therefore, $h(x)$ is concave down on the intervals $(-\infty,-3)$ and $(2,\infty)$. $h''(x)>0$ when $-3<x<2$, so $h(x)$ is concave up on $(-3,2)$.

The solution is continued on the next page.

Copyright © 2014 Pearson Education, Inc. Publishing as Addison-Wesley.

We use the information on the previous page to sketch the graph. Additional values may be computed as necessary.

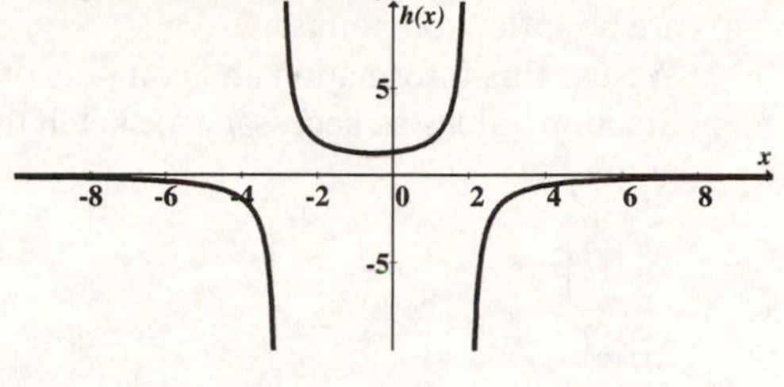

63. $V(t)=50-\dfrac{25t^2}{(t+2)^2}$

a) $V(0)=50-\dfrac{25(0)^2}{((0)+2)^2}=50-0=50$

The inventory's value after 0 months is \$50.

$V(5)=50-\dfrac{25(5)^2}{((5)+2)^2}=50-\dfrac{625}{49}\approx 37.24$

The inventory's value after 5 months is \$37.24.

$V(10)=50-\dfrac{25(10)^2}{((10)+2)^2}=50-\dfrac{2500}{144}\approx 32.64$

The inventory's value after 10 months is \$32.64.

$V(70)=50-\dfrac{25(70)^2}{((70)+2)^2}=50-\dfrac{122,500}{5184}\approx 26.37$

The inventory's value after 70 months is \$26.37.

b) Find $V'(t)$ and $V''(t)$.

$$V'(t)=-\frac{(t+2)^2(50t)-25t^2(2(t+2)(1))}{\left((t+2)^2\right)^2}$$
$$=-\frac{(t+2)\left[(t+2)(50t)-25t^2(2)\right]}{(t+2)^4}$$
$$=-\frac{50t^2-100t-50t^2}{(t+2)^3}$$
$$=-\frac{100t}{(t+2)^3}$$

$$V''(t)=-\frac{(t+2)^3(100)-100t\left[3(t+2)^2(1)\right]}{\left((t+2)^3\right)^2}$$
$$=-\frac{(t+2)^2\left[100(t+2)-100t(3)\right]}{(t+2)^6}$$
$$=-\frac{100t+200-300t}{(t+2)^4}$$
$$=-\frac{-200t+200}{(t+2)^4}$$
$$=\frac{200t-200}{(t+2)^4}$$

$V'(t)$ exists for all values of t in $[0,\infty)$.

Solve $V'(t)=0$.

$$-\frac{100t}{(t+2)^3}=0$$
$$-100t=0$$
$$t=0$$

Since $t=0$ is an endpoint of the domain, there cannot be a relative extrema at $t=0$. We notice that $V'(t)<0$ for all x in the domain, therefore, $V(t)$ is decreasing over the interval $[0,\infty)$. Since $V(t)$ is decreasing, the *absolute* maximum value of the inventory will be \$50 when $t=0$.

c) Using the techniques of this section, we find the following additional information:
Intercepts. There are no t-intercepts in $[0,\infty)$. The V-intercept is the point $(0,50)$
Asymptotes. There are no vertical asymptotes in $[0,\infty)$.
The line $V=25$ is a horizontal asymptote. There are no slant asymptotes.
Increasing, decreasing, relative extrema. We have already seen that $V(t)$ is decreasing over the interval $[0,\infty)$. There are no relative extrema.
Inflection points, concavity. $V''(t)$ exist for all values of t in $[0,\infty)$. $V''(t)=0$, when $t=1$. We use this to split the domain into two intervals. A: $(0,1)$ and B: $(1,\infty)$.
The solution is continued on the next page.

Copyright © 2014 Pearson Education, Inc. Publishing as Addison-Wesley.

Testing points in each interval, we see $V(t)$ is concave down on $(0,1)$ and concave up on $(1,\infty)$.

We use this information to sketch the graph. Additional values may be computed as necessary.

d)

65. $C(p)=\dfrac{48,000}{100-p}$

a) $C(0)=\dfrac{48,000}{100-0}=\dfrac{48,000}{100}=480$

The cost of removing 0% of the pollutants from a chemical spill is $480.

$C(20)=\dfrac{48,000}{100-(20)}=\dfrac{48,000}{80}=600$

The cost of removing 20% of the pollutants from a chemical spill is $600.

$C(80)=\dfrac{48,000}{100-(80)}=\dfrac{48,000}{20}=2400$

The cost of removing 80% of the pollutants from a chemical spill is $2400.

$C(90)=\dfrac{48,000}{100-(90)}=\dfrac{48,000}{10}=4800$

The cost of removing 90% of the pollutants from a chemical spill is $4800.

b) The domain of C is $0\le p<100$ since it is not possible to remove less than 0% or more than 100% of the pollutants, and $C(p)$ is not defined for $p=100$.

c) Using the techniques of this section we find the following additional information.

Intercepts. No p-intercepts. The point $(0,480)$ is the C-intercept.

Asymptotes. *Vertical.* $p=100$

Horizontal. $C=0$

Slant. None

Increasing, decreasing, relative extrema.

$C(p)$ is increasing over the interval $[0,100)$.

There are no relative extrema.

Inflection points, concavity. $C(p)$ is concave up on the interval $(0,100)$. There are no inflection points.

We use this information and compute other function values as necessary to sketch the graph.

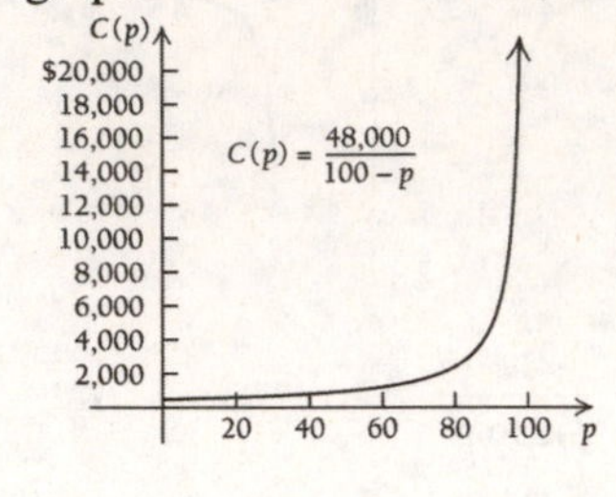

d)

67. $P(x)=\dfrac{2.632}{1+0.116x}$

a) $P(10)=\dfrac{2.632}{1+0.116(10)}=1.21851852$

In 1980, the purchasing power of a dollar was $1.22.

$P(20)=\dfrac{2.632}{1+0.116(20)}=0.79277108$

In 1990, the purchasing power of a dollar was $0.79.

$P(40)=\dfrac{2.632}{1+0.116(40)}=0.4666667$

In 2010, the purchasing power of a dollar was $0.47.

b) Solve $P(x)=0.50$

$$\frac{2.632}{1+0.116x}=0.50$$

$$2.632=0.50(1+0.116x)$$

$$2.632=0.50+0.058x$$

$$2.132=0.058x$$

$$36.7586=x$$

36.8 years after 1970, or in 2006, the purchasing power of a dollar will be $0.50.

Copyright © 2014 Pearson Education, Inc. Publishing as Addison-Wesley.

c) Find $\lim_{x\to\infty} P(x)$.

$$\lim_{x\to\infty} P(x) = \lim_{x\to\infty} \frac{2.632}{1+0.116x}$$

$$= \lim_{x\to\infty} \frac{2.632}{1+0.116x} \cdot \frac{\frac{1}{x}}{\frac{1}{x}}$$

$$= \lim_{x\to\infty} \frac{\frac{2.632}{x}}{\frac{1}{x}+0.116}$$

$$= \frac{0}{0+0.116} = 0$$

$\lim_{x\to\infty} P(x) = 0$.

69. $E(n) = 9 \cdot \frac{4}{n}$

a) Calculate each value for the given n.

$E(9) = 9 \cdot \frac{4}{9} = 4.00$

$E(6) = 9 \cdot \frac{4}{6} = 6.00$

$E(3) = 9 \cdot \frac{4}{3} = 12.00$

$E(1) = 9 \cdot \frac{4}{1} = 36.00$

$E\left(\frac{2}{3}\right) = 9 \cdot \frac{4}{\frac{2}{3}} = 9\left(4 \cdot \frac{3}{2}\right) = 54.00$

$E\left(\frac{1}{3}\right) = 9 \cdot \frac{4}{\frac{1}{3}} = 9\left(4 \cdot \frac{3}{1}\right) = 108.00$

We complete the table.

Innings Pitched (n)	Earned-Run average (E)
9	4.00
6	6.00
3	12.00
1	36.00
$\frac{2}{3}$	54.00
$\frac{1}{3}$	108.00

b) i) $\lim_{n\to 0} E(n) = \lim_{n\to 0} 9 \cdot \frac{4}{n} = \lim_{n\to 0} \frac{36}{n} = \infty$

ii) If the pitcher gives up one or more runs but gets no one out, the pitcher would be credited with zero innings pitched.

c) Evaluating the function, we would have:

$E(18) = 9 \cdot \frac{4}{18} = 2$.

Thus, the pitchers ERA over the two games is 2.00. Which means the pitcher gives up 2 runs per 9 innings.

71. ✎

73. $\lim_{x\to 0} \frac{|x|}{x}$

Using $|x| = \begin{cases} -x, & \text{for } x < 0 \\ x, & \text{for } x \ge 0 \end{cases}$, we have

$\lim_{x\to 0^-} \frac{|x|}{x} = -1$ and $\lim_{x\to 0^+} \frac{|x|}{x} = 1$.

Therefore, $\lim_{x\to 0} \frac{|x|}{x}$ does not exist.

75. We divide the numerator and denominator by x^2 the highest power of x in the denominator.

$$\lim_{x\to\infty} \frac{-6x^3+7x}{2x^2-3x-10} = \lim_{x\to\infty} \frac{-6x+\frac{7}{x}}{2-\frac{3}{x}-\frac{10}{x^2}}$$

$$= \frac{\lim_{x\to\infty}(-6x)+0}{2-0-0}$$

$$= -\infty$$

77. $$\lim_{x\to 1} \frac{x^3-1}{x^2-1} = \lim_{x\to 1} \frac{(x-1)(x^2+x+1)}{(x-1)(x+1)}$$

$$= \lim_{x\to 1} \frac{(x^2+x+1)}{(x+1)}$$

$$= \frac{(1)^2+(1)+1}{(1)+1}$$

$$= \frac{3}{2}$$

79. $$\lim_{x\to -\infty} \frac{2x^4+x}{x+1} = \lim_{x\to -\infty} \frac{2x^3+1}{1+\frac{1}{x}}$$

$$= \frac{\lim_{x\to -\infty} 2x^3+1}{1+0}$$

$$= -\infty$$

Copyright © 2014 Pearson Education, Inc. Publishing as Addison-Wesley.

81. $f(x)=\dfrac{x}{\sqrt{x^2+1}}$

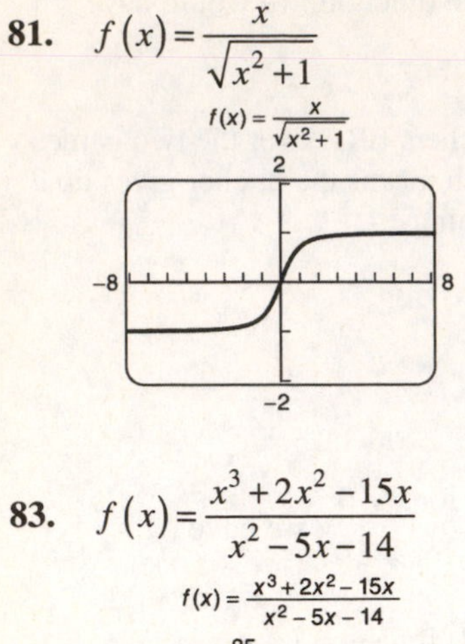

83. $f(x)=\dfrac{x^3+2x^2-15x}{x^2-5x-14}$

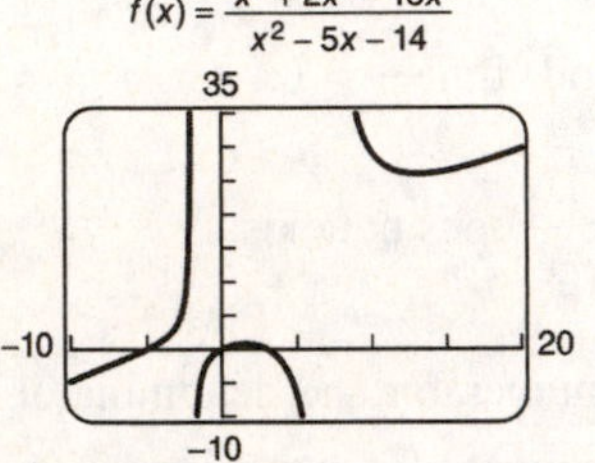

85. $f(x)=\left|\dfrac{1}{x}-2\right|$

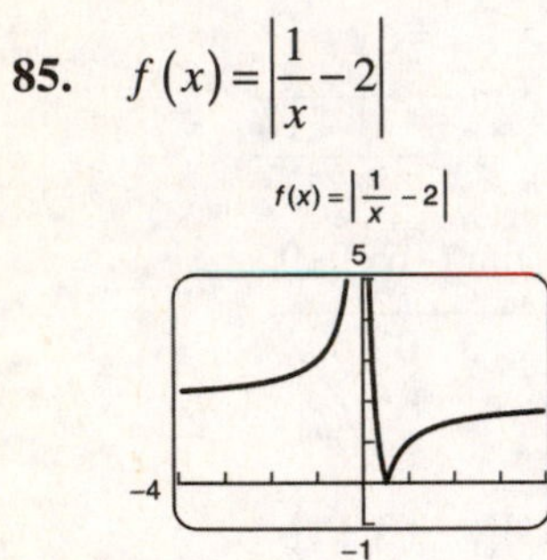

87. ✎

89. a) Visually inspecting the graph, there appears to be relative maxiumum at $\left(0,\dfrac{1}{6}\right)$.

However, noticing the graph dips below the horizontal asymptote around $x=5$, we would think there is a relative minimum for some value $x>5$. This graph does not give us enough detail to visually determine that point.

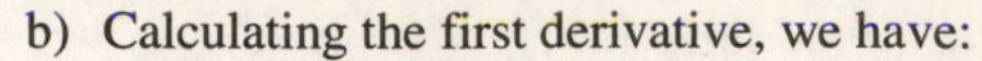

b) Calculating the first derivative, we have:

$$f'(x)=\frac{x^2-10x+1}{\left(x^2+x-6\right)}.$$

$f'(x)$ does not exist when $x=-3$ or $x=2$, however, those values are not in the domain of $f(x)$. So we set $f'(x)=0$ and solve for x.

$$\frac{x^2-10x+1}{\left(x^2+x-6\right)^2}=0$$

$$x^2-10x+1=0$$

By the quadratic formula, we have:

$$x=\frac{-(-10)\pm\sqrt{(-10)^2+4(1)(1)}}{2(1)}$$

$$=\frac{10\pm\sqrt{100+4}}{2}$$

$$=\frac{10\pm\sqrt{104}}{2}.$$

Using the calculator, $f'(x)=0$ when $x\approx 0.10$ and $x\approx 9.90$.

Substititing these values back into the function, we find the critical values occur approximately at $(0.101,0.168)$ and $(9.899,0.952)$.

c) Entering the graph in to the calculator and using the maximum feature we see:

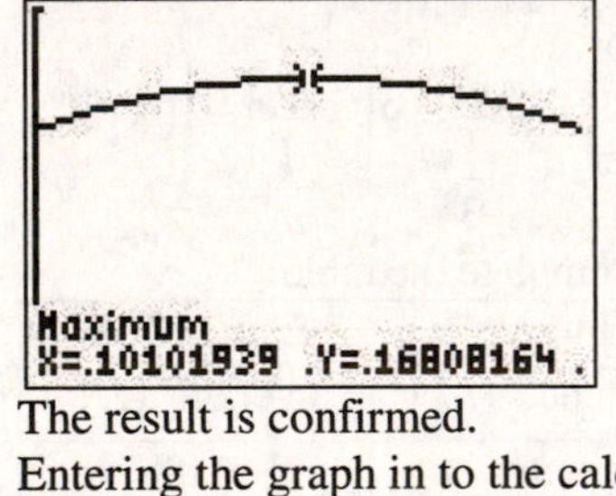

The result is confirmed.

d) Entering the graph in to the calculator and using the minimum feature we see:

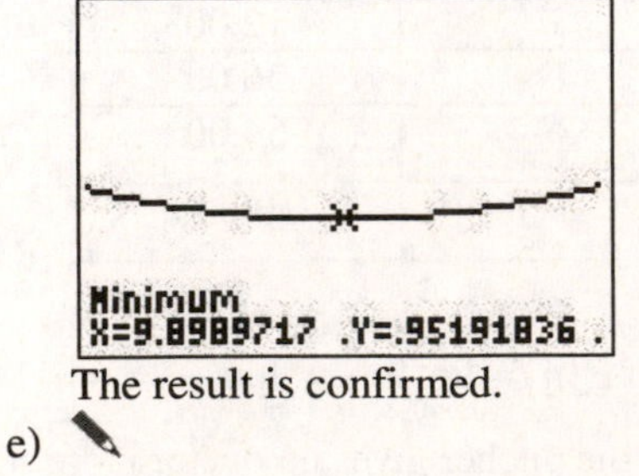

The result is confirmed.

e) ✎

Copyright © 2014 Pearson Education, Inc. Publishing as Addison-Wesley.

Exercise Set 2.4

1. a) The absolute maximum gasoline mileage is obtained at a speed of 55 mph.
b) The absolute minimum gasoline mileage is obtained at a speed of 5 mph.
c) At 70 mph, the fuel economy is 25 mpg.

3. $f(x)=5+x-x^2;\ [0,2]$

a) Find $f'(x)$

$f'(x)=1-2x$

b) Find the Critical Values. The derivative exists for all real numbers. Thus, we solve

$$f'(x)=0$$
$$1-2x=0$$
$$1=2x$$
$$\frac{1}{2}=x$$

c) List the critical values and endpoints. These values are $0,\ \frac{1}{2}$, and 2.

d) Evaluate $f(x)$ at each value in step (c).

$$f(0)=5+(0)-(0)^2=5$$
$$f\left(\frac{1}{2}\right)=5+\left(\frac{1}{2}\right)-\left(\frac{1}{2}\right)^2=\frac{21}{4}=5.25$$
$$f(2)=5+(2)-(2)^2=3$$

The largest of these values, $\frac{21}{4}$, is the absolute maximum, it occurs at $x=\frac{1}{2}$. The smallest of these values, 3, is the absolute minimum, it occurs at $x=2$.

5. $f(x)=x^3-x^2-x+2;\quad [-1,2]$

a) Find $f'(x)$

$f'(x)=3x^2-2x-1$

b) Find the Critical Values. The derivative exists for all real numbers. Thus, we solve

$$f'(x)=0$$
$$3x^2-2x-1=0$$
$$(3x+1)(x-1)=0$$
$$3x+1=0 \quad \text{or} \quad x-1=0$$
$$3x=-1 \quad \text{or} \quad x=1$$
$$x=-\frac{1}{3} \quad \text{or} \quad x=1$$

c) List the critical values and endpoints. These values are $-1,\ -\frac{1}{3},\ 1$, and 2.

d) Evaluate $f(x)$ at each value in step (c).

$$f(-1)=(-1)^3-(-1)^2-(-1)+2=1$$
$$f\left(\frac{-1}{3}\right)=\left(\frac{-1}{3}\right)^3-\left(\frac{-1}{3}\right)^2-\left(\frac{-1}{3}\right)+2=\frac{59}{27}\approx 2.2$$
$$f(1)=(1)^3-(1)^2-(1)+2=1$$
$$f(2)=(2)^3-(2)^2-(2)+2=4$$

The largest of these values, 4,is the absolute maximum, it occurs at $x=2$. The smallest of these values, 1, is the absolute minimum, it occurs at $x=-1$ and $x=1$.

7. $f(x)=x^3-x^2-x+3;\quad [-1,0]$

a) Find $f'(x)$

$f'(x)=3x^2-2x-1$

b) Find the Critical Values. The derivative exists for all real numbers. Thus, we solve

$$f'(x)=0$$
$$3x^2-2x-1=0$$
$$(3x+1)(x-1)=0$$
$$3x+1=0 \quad \text{or} \quad x-1=0$$
$$3x=-1 \quad \text{or} \quad x=1$$
$$x=-\frac{1}{3} \quad \text{or} \quad x=1$$

c) List the critical values and endpoints. The critical value $x=1$ is not in the interval, so we exclude it. We will test the values $-1,\ -\frac{1}{3}$, and 0.

Copyright © 2014 Pearson Education, Inc. Publishing as Addison-Wesley.

d) Evaluate $f(x)$ at each value in step (c).

$f(-1)=(-1)^3-(-1)^2-(-1)+3=2$

$f\left(\frac{-1}{3}\right)=\left(\frac{-1}{3}\right)^3-\left(\frac{-1}{3}\right)^2-\left(\frac{-1}{3}\right)+3=\frac{86}{27}\approx 3.2$

$f(0)=(0)^3-(0)^2-(0)+3=3$

On the interval $[-1,0]$, the absolute maximum is $\frac{86}{27}$, which occurs at $x=-\frac{1}{3}$. The absolute minimum is 2, which occurs at $x=-1$.

9. $f(x)=5x-7;\qquad [-2,3]$

a) Find $f'(x)$

$f'(x)=5$

b) and c)
The derivative exists and is 5 for all real numbers. Note that the derivative is never 0. Thus, there are no critical values for $f(x)$, and the absolute maximum and absolute minimum will occur at the endpoints of the interval.

d) Evaluate $f(x)$ at the endpoints.

$f(-2)=5(-2)-7=-17$

$f(3)=5(3)-7=8$

On the interval $[-2,3]$, the absolute maximum is 8, which occurs at $x=3$. The absolute minimum is -17, which occurs at $x=-2$.

11. $f(x)=7-4x;\qquad [-2,5]$

a) Find $f'(x)$

$f'(x)=-4$

b) and c)
The derivative exists and is -4 for all real numbers. Note that the derivative is never 0. Thus, there are no critical values for $f(x)$, and the absolute maximum and absolute minimum will occur at the endpoints of the interval.

d) Evaluate $f(x)$ at the endpoints.

$f(-2)=7-4(-2)=15$

$f(5)=7-4(5)=-13$

On the interval $[-2,5]$, the absolute maximum is 15, which occurs at $x=-2$. The absolute minimum is -13, which occurs at $x=5$.

13. $f(x)=-5;\qquad [-1,1]$

Note for all values of x, $f(x)=-5$. Thus, the absolute maximum is -5 for $-1\le x\le 1$ and the absolute minimum is -5 for $-1\le x\le 1$.

15. $f(x)=x^2-6x-3;\qquad [-1,5]$

a) $f'(x)=2x-6$

b) $f'(x)$ exists for all real numbers. Solve:

$2x-6=0$

$2x=6$

$x=3$

c) The critical value and the endpoints are -1, 3, and 5.

d) Evaluate $f(x)$ for each value in (c)

$f(-1)=(-1)^2-6(-1)-3=4$

$f(3)=(3)^2-6(3)-3=-12$

$f(5)=(5)^2-6(5)-3=-8$

On the interval $[-1,5]$, the absolute maximum is 4, which occurs at $x=-1$. The absolute minimum is -12, which occurs at $x=3$.

17. $f(x)=3-2x-5x^2;\qquad [-3,3]$

a) $f'(x)=-2-10x$

b) $f'(x)$ exists for all real numbers. Solve:

$-2-10x=0$

$-10x=2$

$x=-\frac{1}{5}$

c) The critical value and the endpoints are -3, $-\frac{1}{5}$, and 3.

Copyright © 2014 Pearson Education, Inc. Publishing as Addison-Wesley.

d) Evaluate $f(x)$ for each value in (c)

$$f(-3)=3-2(-3)-5(-3)^2=-36$$

$$f\left(-\frac{1}{5}\right)=3-2\left(-\frac{1}{5}\right)-5\left(-\frac{1}{5}\right)^2=\frac{16}{5}=3.2$$

$$f(3)=3-2(3)-5(3)^2=-48$$

On the interval $[-3,3]$, the absolute maximum is $\frac{16}{5}$, which occurs at $x=-\frac{1}{5}$. The absolute minimum is -48, which occurs at $x=3$.

19. $f(x)=x^3-3x^2;\quad [0,5]$

a) $f'(x)=3x^2-6x$

b) $f'(x)$ exists for all real numbers. Solve:

$$3x^2-6x=0$$
$$3x(x-2)=0$$
$$3x=0 \quad \text{or} \quad x-2=0$$
$$x=0 \quad \text{or} \quad x=2$$

c) The critical value and the endpoints are 0, 2, and 5. Note, since 0 is an endpoint of the interval, $x=0$ is included in this list as an endpoint, as well as a critical value.

d) Evaluate $f(x)$ for each value in (c)

$$f(0)=(0)^3-3(0)^2=0$$
$$f(2)=(2)^3-3(2)^2=-4$$
$$f(5)=(5)^3-3(5)^2=50$$

On the interval $[0,5]$, the absolute maximum is 50, which occurs at $x=5$. The absolute minimum is -4, which occurs at $x=2$.

21. $f(x)=x^3-3x;\quad [-5,1]$

a) $f'(x)=3x^2-3$

b) $f'(x)$ exists for all real numbers. Solve:

$$3x^2-3=0$$
$$x^2-1=0$$
$$x=\pm 1$$

c) The critical value and the endpoints are -5, -1, and 1. Note, since 1 is an endpoint of the interval, $x=1$ is included in this list as an endpoint, not a critical value.

d) Evaluate $f(x)$ for each value in (c)

$$f(-5)=(-5)^3-3(-5)=-110$$
$$f(-1)=(-1)^3-3(-1)=2$$
$$f(1)=(1)^3-3(1)=-2$$

On the interval $[-5,1]$, the absolute maximum is 2, which occurs at $x=-1$. The absolute minimum is -110, which occurs at $x=-5$.

23. $f(x)=1-x^3;\quad [-8,8]$

a) $f'(x)=-3x^2$

b) $f'(x)$ exists for all real numbers. Solve:

$$-3x^2=0$$
$$x=0$$

c) The critical value and the endpoints are -8, 0, and 8.

d) Evaluate $f(x)$ for each value in (c)

$$f(-8)=1-(-8)^3=513$$
$$f(0)=1-(0)^3=1$$
$$f(8)=1-(8)^3=-511$$

On the interval $[-8,8]$, the absolute maximum is 513, which occurs at $x=-8$. The absolute minimum is -511, which occurs at $x=8$.

25. $f(x)=12+9x-3x^2-x^3;\quad [-3,1]$

a) $f'(x)=9-6x-3x^2$

b) $f'(x)$ exists for all real numbers. Solve:

$$9-6x-3x^2=0$$
$$x^2+2x-3=0 \qquad \text{Divide by } -3.$$
$$(x+3)(x-1)=0$$
$$x+3=0 \quad \text{or} \quad x-1=0$$
$$x=-3 \quad \text{or} \quad x=1$$

c) The critical values and the endpoints are -3 and 1. Note, since the possible critical values are the endpoints of the interval, they are included in this list as endpoints, not as critical values.

Copyright © 2014 Pearson Education, Inc. Publishing as Addison-Wesley.

d) Evaluate $f(x)$ for each value in (c)

$$f(-3)=12+9(-3)-3(-3)^2-(-3)^3=-15$$
$$f(1)=12+9(1)-3(1)^2-(1)^3=17$$

On the interval $[-3,1]$, the absolute maximum is 17, which occurs at $x=1$. The absolute minimum is -15, which occurs at $x=-3$.

27. $f(x)=x^4-2x^3; \qquad [-2,2]$

a) $f'(x)=4x^3-6x^2$

b) $f'(x)$ exists for all real numbers. Solve:

$$4x^3-6x^2=0$$
$$2x^2(2x-3)=0$$
$$2x^2=0 \quad \text{or} \quad 2x-3=0$$
$$x=0 \quad \text{or} \quad x=\frac{3}{2}$$

c) The critical values and the endpoints are -2, 0, $\frac{3}{2}$, and 2.

d) Evaluate $f(x)$ for each value in (c)

$$f(-2)=(-2)^4-2(-2)^3=32$$
$$f(0)=(0)^4-2(0)^3=0$$
$$f\left(\frac{3}{2}\right)=\left(\frac{3}{2}\right)^4-2\left(\frac{3}{2}\right)^3=-\frac{27}{16}=-1.6875$$
$$f(2)=(2)^4-2(2)^3=0$$

On the interval $[-2,2]$, the absolute maximum is 32, which occurs at $x=-2$. The absolute minimum is $-\frac{27}{16}$, which occurs at $x=\frac{3}{2}$.

29. $f(x)=x^4-2x^2+5; \qquad [-2,2]$

a) $f'(x)=4x^3-4x$

b) $f'(x)$ exists for all real numbers. Solve:

$$4x^3-4x=0$$
$$4x\left(x^2-1\right)=0$$
$$4x=0 \quad \text{or} \quad x^2-1=0$$
$$x=0 \quad \text{or} \quad x=\pm 1$$

c) The critical values and the endpoints are -2, -1, 0, 1, and 2.

d) Evaluate $f(x)$ for each value in (c)

$$f(-2)=(-2)^4-2(-2)^2+5=13$$
$$f(-1)=(-1)^4-2(-1)^2+5=4$$
$$f(0)=(0)^4-2(0)^2+5=5$$
$$f(1)=(1)^4-2(1)^2+5=4$$
$$f(2)=(2)^4-2(2)^2+5=13$$

On the interval $[-2,2]$, the absolute maximum is 13, which occurs at $x=-2$ and $x=2$. The absolute minimum is 4, which occurs at $x=-1$ and $x=1$.

31. $f(x)=(x+3)^{2/3}-5; \qquad [-4,5]$

a) $f'(x)=\frac{2}{3}(x+3)^{-1/3}=\frac{2}{3(x+3)^{1/3}}$

b) $f'(x)$ does not exist for $x=-3$. The equation $f'(x)=0$ has no solution, so $x=-3$ is the only critical value.

c) The critical values and the endpoints are -4, -3, and 5.

d) Evaluate $f(x)$ for each value in (c)

$$f(-4)=((-4)+3)^{2/3}-5=-4$$
$$f(-3)=((-3)+3)^{2/3}-5=-5$$
$$f(5)=((5)+3)^{2/3}-5=-1$$

On the interval $[-4,5]$, the absolute maximum is -1, which occurs at $x=5$. The absolute minimum is -5, which occurs at $x=-3$.

33. $f(x)=x+\frac{1}{x}; \qquad [1,20]$

a) $f'(x)=1-x^{-2}=1-\frac{1}{x^2}$

Copyright © 2014 Pearson Education, Inc. Publishing as Addison-Wesley.

b) $f'(x)$ does not exist for $x=0$. However, $x=0$ is not in the interval. Solve $f'(x)=0$.

$$1-\frac{1}{x^2}=0$$
$$1=\frac{1}{x^2}$$
$$x^2=1$$
$$x=\pm 1$$

The critical value $x=-1$ is not in the interval, and the other critical value is an endpoint.

c) The critical values and the endpoints are 1 and 20.

d) Evaluate $f(x)$ for each value in (c)

$$f(1)=1+\frac{1}{1}=2$$
$$f(20)=20+\frac{1}{20}=\frac{401}{20}=20.05$$

On the interval $[1,20]$, the absolute maximum is 20.05, which occurs at $x=20$. The absolute minimum is 2, which occurs at $x=1$.

35. $f(x)=\frac{x^2}{x^2+1}$; $[-2,2]$

a) $$f'(x)=\frac{(x^2+1)(2x)-x^2(2x)}{(x^2+1)^2} \quad \text{Quotient Rule}$$
$$=\frac{2x^3+2x-2x^3}{(x^2+1)^2}$$
$$=\frac{2x}{(x^2+1)^2}$$

b) $f'(x)$ exists for all real numbers. Solve:

$$f'(x)=0$$
$$\frac{2x}{(x^2+1)^2}=0$$
$$2x=0$$
$$x=0$$

c) The critical values and the endpoints are -2, 0, and 2.

d) Evaluate $f(x)$ for each value in (c)

$$f(-2)=\frac{(-2)^2}{(-2)^2+1}=\frac{4}{5}$$
$$f(0)=\frac{(0)^2}{(0)^2+1}=0$$
$$f(2)=\frac{(2)^2}{(2)^2+1}=\frac{4}{5}$$

On the interval $[-2,2]$, the absolute maximum is $\frac{4}{5}$, which occurs at $x=-2$ and $x=2$. The absolute minimum is 0, which occurs at $x=0$.

37. $f(x)=(x+1)^{1/3}$; $[-2,26]$

a) $$f'(x)=\frac{1}{3}(x+1)^{-2/3}=\frac{1}{3(x+1)^{2/3}}$$

b) $f'(x)$ does not exist for $x=-1$. The equation $f'(x)=0$ has no solution, so $x=-1$ is the only critical value.

c) The critical values and the endpoints are -2, -1, and 26.

d) Evaluate $f(x)$ for each value in (c)

$$f(-2)=((-2)+1)^{1/3}=-1$$
$$f(-1)=((-1)+1)^{1/3}=0$$
$$f(26)=((26)+1)^{1/3}=3$$

On the interval $[-2,26]$, the absolute minimum is -1, which occurs at $x=-2$. The absolute maximum is 3, which occurs at $x=26$.

39. – 47. Left to the student.

49. $f(x)=12x-x^2$

When no interval is specified, we use the real line $(-\infty,\infty)$.

a) Find $f'(x)$

$$f'(x)=12-2x$$

Copyright © 2014 Pearson Education, Inc. Publishing as Addison-Wesley.

b) Find the critical values. The derivative exists for all real numbers. Thus, we solve

$f'(x)=0$.

$12-2x=0$

$12=2x$

$6=x$

The only critical value is $x=6$.

c) Since there is only one critical value, we can apply Max-Min Principle 2. First we find $f''(x)$.

$f''(x)=-2$.

The second derivative is constant, so $f''(6)=-2$. Since the second derivative is negative at 6, we have a maximum at $x=6$. Next, we find the function value at $x=6$.

$f(6)=12(6)-(6)^2=36$

Therefore, the absolute maximum is 36, which occurs at $x=6$. The function has no minimum value.

51. $f(x)=2x^2-40x+270$

When no interval is specified , we use the real line $(-\infty,\infty)$.

a) Find $f'(x)$

$f'(x)=4x-40$

b) Find the critical values. The derivative exists for all real numbers. Thus, we solve

$f'(x)=0$.

$4x-40=0$

$4x=40$

$x=10$

The only critical value is $x=10$. .

c) Since there is only one critical value, we can apply Max-Min Principle 2. First we find $f''(x)$.

$f''(x)=4$.

The second derivative is constant, so $f''(10)=4$. Since the second derivative is positive at 10, we have a minimum at $x=10$. Next, we find the function value at $x=10$.

$f(10)=2(10)^2-40(10)+270=70$

Therefore, the absolute minimum is 70, which occurs at $x=10$. The function has no maximum value.

53. $f(x)=x-\frac{4}{3}x^3$; $(0,\infty)$

a) Find $f'(x)$

$f'(x)=1-4x^2$

b) Find the critical values. The derivative exists for all real numbers. Thus, we solve

$f'(x)=0$.

$1-4x^2=0$

$-4x^2=-1$

$x^2=\frac{1}{4}$

$x=\pm\sqrt{\frac{1}{4}}$

$x=\pm\frac{1}{2}$

There are two critical values; however, the only critical value in $(0,\infty)$ is $x=\frac{1}{2}$.

c) Since there is only one critical value in the interval, we can apply Max-Min Principle 2. First we find $f''(x)$.

$f''(x)=-8x$.

The second derivative is constant, so $f''\left(\frac{1}{2}\right)=-4$. Since the second derivative is negative at $\frac{1}{2}$, we have a maximum at $x=\frac{1}{2}$.

To find the maximum value, we find the function value at $x=\frac{1}{2}$.

$f\left(\frac{1}{2}\right)=\left(\frac{1}{2}\right)-\frac{4}{3}\left(\frac{1}{2}\right)^3=\frac{1}{3}$

Therefore, the absolute maximum is $\frac{1}{3}$, which occurs at $x=\frac{1}{2}$. The function has no minimum value.

55. $f(x)=x(60-x)=60x-x^2$

When no interval is specified , we use the real line $(-\infty,\infty)$.

a) Find $f'(x)$

$f'(x)=60-2x$

Copyright © 2014 Pearson Education, Inc. Publishing as Addison-Wesley.

b) Find the critical values. The derivative exists for all real numbers. Thus, we solve
$f'(x)=0$.
$60-2x=0$
$60=2x$
$30=x$
The only critical value is $x=30$.

c) Since there is only one critical value, we can apply Max-Min Principle 2. First we find $f''(x)$.
$f''(x)=-2$.
The second derivative is constant, so $f''(30)=-2$. Since the second derivative is negative at 30, we have a maximum at $x=30$. Next, we find the function value at $x=30$.
$f(30)=30(60-30)=900$
Therefore, the absolute maximum is 900, which occurs at $x=30$. The function has no minimum value.

57. $f(x)=\frac{1}{3}x^3-3x;\qquad [-2,2]$

a) Find $f'(x)$
$f'(x)=x^2-3$

b) Find the critical values. The derivative exists for all real numbers. Thus, we solve
$f'(x)=0$.
$x^2-3=0$
$x^2=3$
$x=\pm\sqrt{3}\approx\pm1.732$
Both critical values are in the interval $[-2,2]$.

c) The interval is closed and there is more than one critical value, so we use Max-Min Principle 1.
The critical points and the endpoints are
$-2,\ -\sqrt{3},\ \sqrt{3}$, and 2.
Next, we find the function values at these points.

$$f(-2)=\frac{1}{3}(-2)^3-3(-2)=\frac{10}{3}=3.3\overline{3}$$

$$f(-\sqrt{3})=\frac{1}{3}(-\sqrt{3})^3-3(-\sqrt{3})$$
$$=-\sqrt{3}+3\sqrt{3}=2\sqrt{3}\approx3.464$$

$$f(\sqrt{3})=\frac{1}{3}(\sqrt{3})^3-3(\sqrt{3})$$
$$=\sqrt{3}-3\sqrt{3}=-2\sqrt{3}\approx-3.464$$

$$f(2)=\frac{1}{3}(2)^3-3(2)=-\frac{10}{3}=-3.3\overline{3}$$

The largest of these values, $2\sqrt{3}$, is the maximum. It occurs at $x=-\sqrt{3}$. The smallest of these values, $-2\sqrt{3}$, is the minimum. It occurs at $x=\sqrt{3}$.
Thus, the absolute maximum over the interval $[-2,2]$, is $2\sqrt{3}$, which occurs at $x=-\sqrt{3}$, and the absolute minimum over $[-2,2]$ is $-2\sqrt{3}$, which occurs at $x=\sqrt{3}$.

59. $f(x)=-0.001x^2+4.8x-60$
When no interval is specified, we use the real line $(-\infty,\infty)$.

a) Find $f'(x)$
$f'(x)=-0.002x+4.8$

b) Find the critical values. The derivative exists for all real numbers. Thus, we solve
$f'(x)=0$.
$-0.002x+4.8=0$
$-0.002x=-4.8$
$x=2400$
The only critical value is $x=2400$.

c) Since there is only one critical value, we can apply Max-Min Principle 2. First we find $f''(x)$.
$f''(x)=-0.002$.
The second derivative is constant, so $f''(2400)=-0.002$.
Since the second derivative is negative at 2400, we have a maximum at $x=2400$.
Next, we find the function value at $x=2400$.
$$f(2400)=-0.001(2400)^2+4.8(2400)-60$$
$$=5700$$
Therefore, the absolute maximum is 5700, which occurs at $x=2400$. The function has no minimum value.

Copyright © 2014 Pearson Education, Inc. Publishing as Addison-Wesley.

61. $f(x)=-\frac{1}{3}x^3+6x^2-11x-50;\quad (0,3)$

a) Find $f'(x)$

$f'(x)=-x^2+12x-11$

b) Find the critical values. The derivative exists for all real numbers. Thus, we solve $f'(x)=0$.

$-x^2+12x-11=0$

$x^2-12x+11=0$

$(x-11)(x-1)=0$

$x-11=0 \quad \text{or} \quad x-1=0$

$x=11 \quad \text{or} \quad x=1.$

c) The interval $(0,3)$ is not closed. The only critical value in the interval is $x=1$. Therefore, we can apply Max-Min Principle 2. First, we find the second derivative.

$f''(x)=-2x+12$

Evaluating the second derivative at $x=1$, we have:

$f''(1)=-2(1)+12=10>0$.

Since the second derivative is positive when $x=1$, there is a minimum at $x=1$.
Next, find the function value at $x=1$.

$f(1)=-\frac{1}{3}(1)^3+6(1)^2-11(1)-50$

$=-\frac{166}{3}=-55.33\overline{3}$

Thus, the absolute minimum over the interval $(0,3)$ is $-\frac{166}{3}$, which occurs at $x=1$.

63. $f(x)=15x^2-\frac{1}{2}x^3;\quad [0,30]$

a) Find $f'(x)$

$f'(x)=30x-\frac{3}{2}x^2$

b) Find the critical values. The derivative exists for all real numbers. Thus, we solve $f'(x)=0$.

$30x-\frac{3}{2}x^2=0$

$60x-3x^2=0$

$3x(20-x)=0$

$3x=0 \quad \text{or} \quad 20-x=0$

$x=0 \quad \text{or} \quad x=20.$

The critical values are in the interval $[0,30]$.

c) Since the interval is closed and there is more than one critical value, we apply the Max-Min Principle 1.
The critical values and the endpoints are 0, 20, and 30.
Next, we find the function values.

$f(0)=15(0)^2-\frac{1}{2}(0)^3=0$

$f(20)=15(20)^2-\frac{1}{2}(20)^3=2000$

$f(30)=15(30)^2-\frac{1}{2}(30)^3=0$

The largest of these values, 2000, is the maximum. It occurs at $x=20$. The smallest of these values, 0, is the minimum. It occurs at $x=0$ and $x=30$.
Thus, the absolute maximum over the interval $[0,30]$, is 2000, which occurs at $x=20$, and the absolute minimum over $[0,30]$ is 0, which occurs at $x=0$ and $x=30$.

65. $f(x)=2x+\frac{72}{x};\quad (0,\infty)$

$f(x)=2x+72x^{-1}$

a) Find $f'(x)$

$f'(x)=2-72x^{-2}=2-\frac{72}{x^2}$

b) Find the critical values. $f'(x)$ does not exist for $x=0$; however, 0 is not in the interval $(0,\infty)$. Therefore, we solve

$f'(x)=0$

$2-\frac{72}{x^2}=0$

$2=\frac{72}{x^2}$

$2x^2=72$ Multiplying by x^2, since $x\neq 0$.

$x^2=36$

$x=\pm 6$

Copyright © 2014 Pearson Education, Inc. Publishing as Addison-Wesley.

c) The interval $(0,\infty)$ is not closed. The only critical value in the interval is $x=6$. Therefore, we can apply Max-Min Principle 2. First, we find the second derivative.

$$f''(x)=144x^{-3}=\frac{144}{x^3}$$

Evaluating the second derivative at $x=6$, we have:

$$f''(6)=\frac{144}{(6)^3}=\frac{2}{3}>0.$$

Since the second derivative is positive when $x=6$, there is a minimum at $x=6$.

Next, find the function value at $x=6$.

$$f(6)=2(6)+\frac{72}{6}=24$$

Thus, the absolute minimum over the interval $(0,\infty)$ is 24, which occurs at $x=6$. The function has no maximum value over the interval $(0,\infty)$.

67. $f(x)=x^2+\frac{432}{x}$; $(0,\infty)$

$f(x)=x^2+432x^{-1}$

a) Find $f'(x)$

$$f'(x)=2x-432x^{-2}=2x-\frac{432}{x^2}$$

b) Find the critical values. $f'(x)$ does not exists for $x=0$; however, 0 is not in the interval $(0,\infty)$.

Therefore, we solve

$$f'(x)=0$$
$$2x-\frac{432}{x^2}=0$$
$$2x=\frac{432}{x^2}$$
$$2x^3=432 \quad \text{Multiplying by } x^2, \text{ since } x\neq 0.$$
$$x^3=216$$
$$x=6$$

c) The interval $(0,\infty)$ is not closed. The only critical value in the interval is $x=6$. Therefore, we can apply Max-Min Principle 2. First, we find the second derivative.

$$f''(x)=2+864x^{-3}=2+\frac{864}{x^3}$$

Evaluating the second derivative at $x=6$, we have:

$$f''(6)=2+\frac{864}{(6)^3}=6>0.$$

Since the second derivative is positive when $x=6$, there is a minimum at $x=6$. Next, find the function value at $x=6$.

$$f(6)=(6)^2+\frac{432}{6}=108$$

Thus, the absolute minimum over the interval $(0,\infty)$ is 108, which occurs at $x=6$. The function has no maximum value over the interval $(0,\infty)$.

69. $f(x)=2x^4-x$; $[-1,1]$

a) Find $f'(x)$

$$f'(x)=8x^3-1$$

b) Find the critical values. The derivative exists for all real numbers. Thus, we solve

$$f'(x)=0.$$
$$8x^3-1=0$$
$$8x^3=1$$
$$x^3=\frac{1}{8}$$
$$x=\frac{1}{2}$$

The only critical value $x=\frac{1}{2}$ is in the interval $[-1,1]$.

c) The interval is closed, and we are looking for both the absolute maximum and absolute minimum values, so we use Max-Min Principle 1.
The critical values and the endpoints are -1, $\frac{1}{2}$, and 1.

Next, we find the function values at these points.

$$f(-1)=2(-1)^4-(-1)=3$$
$$f\left(\frac{1}{2}\right)=2\left(\frac{1}{2}\right)^4-\left(\frac{1}{2}\right)=-\frac{3}{8}$$
$$f(1)=2(1)^4-(1)=1$$

The largest of these values, 3, is the maximum. It occurs at $x=-1$.
The solution is continued on the next page.

Copyright © 2014 Pearson Education, Inc. Publishing as Addison-Wesley.

The smallest of these values, $-\frac{3}{8}$, is the minimum. It occurs at $x = \frac{1}{2}$.

Thus, the absolute maximum over the interval $[-1,1]$, is 3, which occurs at $x = -1$, and the absolute minimum over $[-1,1]$ is $-\frac{3}{8}$, which occurs at $x = \frac{1}{2}$.

71. $f(x) = \sqrt[3]{x} = x^{1/3}$; $[0,8]$

a) Find $f'(x)$

$$f'(x) = \frac{1}{3}x^{-2/3} = \frac{1}{3\cdot\sqrt[3]{x^2}}$$

b) Find the critical values. $f'(x)$ does not exist for $x = 0$. The equation $f'(x) = 0$ has no solution, so the only critical value is 0, which is also an endpoint.

c) The interval is closed, and we are looking for both the absolute maximum and absolute minimum values, so we use Max-Min Principle 1.
The only critical value is an endpoint. The endpoints are 0 and 8.
Next, we find the function values at these points.

$$f(0) = \sqrt[3]{0} = 0$$
$$f(8) = \sqrt[3]{8} = 2$$

The largest of these values, 2, is the maximum. It occurs at $x = 8$. The smallest of these values, 0, is the minimum. It occurs at $x = 0$.
Thus, the absolute maximum over the interval $[0,8]$, is 2, which occurs at $x = 8$, and the absolute minimum over $[0,8]$ is 0, which occurs at $x = 0$.

73. $f(x) = (x+1)^3$

When no interval is specified, we use the real line $(-\infty,\infty)$.

a) Find $f'(x)$

$$f'(x) = 3(x+1)^2$$

b) Find the critical values. The derivative exists for all real numbers. Thus, we solve $f'(x) = 0$.

$$3(x+1)^2 = 0$$
$$x+1 = 0$$
$$x = -1$$

The only critical value is $x = -1$.

c) Since there is only one critical value, we can apply Max-Min Principle 2. First we find $f''(x)$.

$$f''(x) = 6(x+1).$$

Now,
$f''(-1) = 6((-1)+1) = 0$. So the Max-Min Principle 2 fails. We cannot use Max-Min Principle 1, because there are no endpoints.
We note that $f'(x) = 3(x+1)^2$ is always positive, except at $x = -1$. Thus, $f(x)$ is increasing everywhere except at $x = -1$.
Therefore, the function has no maximum or minimum over the interval $(-\infty,\infty)$.

Notice

$$f''(-2) = -6 < 0$$
$$f''(0) = 6 > 0$$

and

$$f(-1) = 0$$

Therefore, there is a point of inflection at $(-1,0)$.

75. $f(x) = 2x - 3$; $[-1,1]$

a) Find $f'(x)$

$$f'(x) = 2$$

b) and c)
The derivative exists and is 2 for all real numbers. Therefore, $f'(x)$ is never 0. Thus, there are no critical values. We apply the Max-Min Principle 1. The endpoints are -1 and 1. We find the function values at the endpoints.

$$f(-1) = 2(-1) - 3 = -5$$
$$f(1) = 2(1) - 3 = -1$$

Therefore, the absolute maximum over the interval $[-1,1]$ is -1, which occurs at $x = 1$, and the absolute minimum over the interval $[-1,1]$ is -5, which occurs at $x = -1$.

Copyright © 2014 Pearson Education, Inc. Publishing as Addison-Wesley.

77. $f(x)=2x-3;\quad [-1,5)$

a) Find $f'(x)$

$f'(x)=2$

b) and c)
The derivative exists and is 2 for all real numbers. Therefore, $f'(x)$ is never 0. Thus, there are no critical values. We apply the Max-Min Principle 1. There is only one endpoint, $x=-1$. We find the function value at the endpoint.

$f(-1)=2(-1)-3=-5$

We know $f'(x)>0$ over the interval, so the function is increasing over the interval $[-1,5)$. Therefore, the minimum value will be the left hand endpoint. The absolute minimum over the interval $[-1,5)$ is -5, which occurs at $x=-1$. Since the right endpoint is not included in the interval, the function has no maximum value over the interval $[-1,5)$.

79. $f(x)=x^{2/3};\quad [-1,1]$

a) Find $f'(x)$

$$f'(x)=\frac{2}{3}x^{-1/3}=\frac{2}{3\cdot\sqrt[3]{x}}$$

b) Find the critical values. $f'(x)$ does not exist for $x=0$. The equation $f'(x)=0$ has no solution, so the only critical value is 0.

c) The interval is closed, and we are looking for both the absolute maximum and absolute minimum values, so we use Max-Min Principle 1.
The critical value and the endpoints are -1, 0, and 1.
Next, we find the function values at these points.

$f(-1)=(-1)^{2/3}=1$

$f(0)=(0)^{2/3}=0$

$f(1)=(1)^{2/3}=1$

The largest of these values, 1, is the maximum. It occurs at $x=-1$ and $x=1$. The smallest of these values, 0, is the minimum. It occurs at $x=0$.

Thus, the absolute maximum over the interval $[-1,1]$, is 1, which occurs at $x=-1$ and $x=1$, and the absolute minimum over $[-1,1]$ is 0, which occurs at $x=0$.

81. $f(x)=\frac{1}{3}x^3-x+\frac{2}{3}$

When no interval is specified , we use the real line $(-\infty,\infty)$.

a) Find $f'(x)$

$f'(x)=x^2-1$

b) Find the critical values. The derivative exists for all real numbers. Thus, we solve $f'(x)=0$.

$x^2-1=0$

$x^2=1$

$x=\pm1$

There are two critical values -1 and 1.

c) The interval $(-\infty,\infty)$ is not closed, so the Max-Min Principle 1 does not apply. Since there is more than one critical value, the Max-Min Principle 2 does not apply.
A quick sketch of the graph will help us determine whether absolute or relative extrema occur at the critical values.

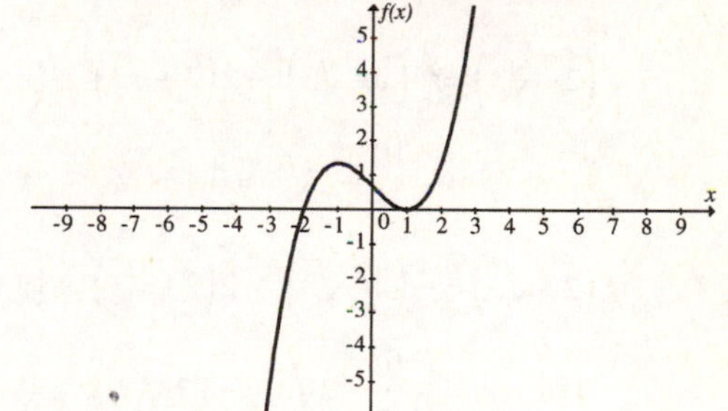

We determine that the function has no absolute extrema over the interval $(-\infty,\infty)$.

83. $f(x)=\frac{1}{3}x^3-2x^2+x;\quad [0,4]$

a) Find $f'(x)$

$f'(x)=x^2-4x+1$

Copyright © 2014 Pearson Education, Inc. Publishing as Addison-Wesley.

b) Find the critical values. The derivative exists for all real numbers. Thus, we solve $f'(x)=0$.

$x^2-4x+1=0$

We use the quadratic formula to solve the equation.

$$x=\frac{-(-4)\pm\sqrt{(-4)^2-4(1)(1)}}{2(1)}$$

$$=\frac{4\pm\sqrt{12}}{2}$$

$$=\frac{4\pm2\sqrt{3}}{2}$$

$$=2\pm\sqrt{3}$$

Both critical values $x=2-\sqrt{3}\approx0.27$ and $x=2+\sqrt{3}\approx3.73$ are in the closed interval $[0,4]$.

c) The interval is closed, and there is more than one critical value in the interval, so we use Max-Min Principle 1.

The critical values and the endpoints are $0,\ 2-\sqrt{3},\ 2+\sqrt{3}$, and 4.

Next, we find the function values at these points.

$$f(0)=\frac{1}{3}(0)^3-2(0)^2+(0)=0$$

$$f(2-\sqrt{3})=\frac{1}{3}(2-\sqrt{3})^3-2(2-\sqrt{3})^2+(2-\sqrt{3})$$

$$=-\frac{10}{3}+2\sqrt{3}\approx0.131$$

$$f(2+\sqrt{3})=\frac{1}{3}(2+\sqrt{3})^3-2(2+\sqrt{3})^2+(2+\sqrt{3})$$

$$=-\frac{10}{3}-2\sqrt{3}\approx-6.797$$

$$f(4)=\frac{1}{3}(4)^3-2(4)^2+(4)$$

$$=-\frac{20}{3}\approx-6.66\overline{6}$$

The largest of these values, $-\frac{10}{3}+2\sqrt{3}$, is the maximum. It occurs at $x=2-\sqrt{3}$ The smallest of these values, $-\frac{10}{3}-2\sqrt{3}$, is the minimum. It occurs at $x=2+\sqrt{3}$

Thus, the absolute maximum over the interval $[0,4]$, is $-\frac{10}{3}+2\sqrt{3}$, which occurs at $x=2-\sqrt{3}$, and the absolute minimum over $[0,4]$ is $-\frac{10}{3}-2\sqrt{3}$, which occurs at $x=2+\sqrt{3}$.

85. $t(x)=x^4-2x^2$

When no interval is specified, we use the real line $(-\infty,\infty)$.

a) Find $t'(x)$

$t'(x)=4x^3-4x$

b) Find the critical values. The derivative exists for all real numbers. Thus, we solve $t'(x)=0$.

$4x^3-4x=0$

$4x(x^2-1)=0$

$x=0$ or $x^2-1=0$

$x=0$ or $x=\pm1$

There are three critical values -1, 0, and 1.

c) The interval $(-\infty,\infty)$ is not closed, so the Max-Min Principle 1 does not apply. Since there is more than one critical value, the Max-Min Principle 2 does not apply.

A quick sketch of the graph will help us determine whether absolute or relative extrema occur at the critical values.

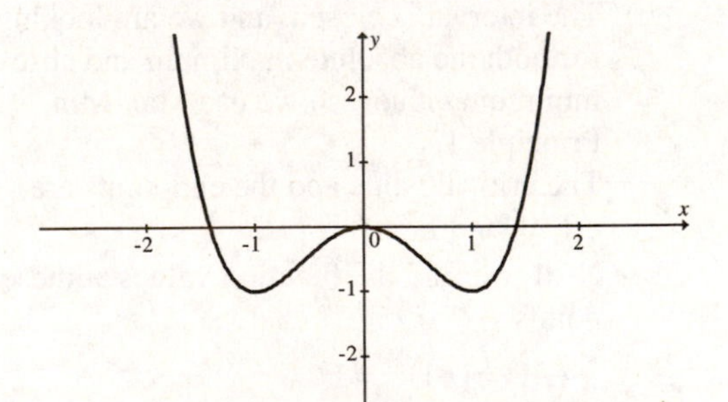

We determine that the function has no absolute maximum over the interval $(-\infty,\infty)$. The function's absolute minimum is -1, which occurs at $x=-1$ and $x=1$.

87. – 95. Left to the Student.

Copyright © 2014 Pearson Education, Inc. Publishing as Addison-Wesley.

97. $M(t) = -2t^2 + 100t + 180, \quad 0 \le t \le 40$

a) $M'(t) = -4t + 100$

b) $M'(t)$ exists for all real numbers. We solve $M'(t) = 0$.

$$-4t + 100 = 0$$
$$4t = 100$$
$$t = 25$$

c) Since there is only one critical value, we apply the Max-Min Principle 2. First, we find the second derivative.

$f''(t) = -4$. The second derivative is negative for all values of t in the interval, therefore, a maximum occurs at $t = 25$.

$$f(25) = -2(25)^2 + 100(25) + 180 = 1430$$

The maximum productivity for $0 \le t \le 40$ is 1430 units per month, which occurs at $t = 25$ years of service.

99. $p(x) = \dfrac{13x^3 - 240x^2 - 2460x + 585{,}000}{75{,}000}$

$$p(x) = \frac{1}{75{,}000}\left(13x^3 - 240x^2 - 2460x + 585{,}000\right)$$

We restrict our attention to the years 1970 to 2000. That is, we will look at the x-values $0 \le x \le 30$.

a) Find $p'(x)$.

$$p'(x) = \frac{1}{75{,}000}\left(39x^2 - 480x - 2460\right)$$

b) Find the critical values. $p'(x)$ exists for all real numbers. Therefore, we solve $p'(x) = 0$.

$$\frac{1}{75{,}000}\left(39x^2 - 480x - 2460\right) = 0$$
$$39x^2 - 480x - 2460 = 0$$

Using the quadratic formula, we have:

$$x = \frac{-(-480) \pm \sqrt{(-480)^2 - 4(39)(-2460)}}{2(39)}$$
$$= \frac{480 \pm \sqrt{614{,}160}}{78}$$

$x \approx -3.89$ or $x \approx 16.20$

Only $x \approx 16.20$ is in the interval $[0, 30]$.

c) The critical values and the endpoints are 0, 16.20, and 30.

d) Using a calculator, we find the function values.

$$p(0) \approx 7.8$$
$$p(16.20) \approx 7.17$$
$$p(30) \approx 8.62$$

From 1970 to 2000 the minimum percentage of U.S. national income generated by nonfarm proprietors was 7.17 percent. This occurred about 16.20 years after 1970, or in 1986.

101. $P(t) = 0.000008533t^4 - 0.001685t^3 + 0.090t^2 - 0.687t + 4.00, \quad 0 \le t \le 90$

a) Find $P'(t)$.

$$P'(t) = 0.00003413t^3 - 0.005055t^2 + 0.18t - 0.687$$

b) $P'(t)$ exists for all real numbers. Solve

$$P'(t) = 0$$
$$0.00003413t^3 - 0.005055t^2 + 0.18t - 0.687 = 0$$

Using a graphing calculator, we approximate the zeros of $P'(t)$. We find the solutions:

$$t \approx 4.3$$
$$t \approx 49.2$$
$$t \approx 94.6$$

Two of the critical values are in the interval.

c) The critical values and the endpoints are: 0, 4.3, 49.2, and 90.

d) Find the function values.

$$P(0) = 4.0$$
$$P(4.3) \approx 2.6$$
$$P(49.2) \approx 37.4$$
$$P(90) \approx 2.7$$

The absolute maximum production of world wide oil was 37.4 billion barrels. The world achieved this production 49.2 years after 1950, or in the year 1999.

103. $C(x) = 5000 + 600x$

$$R(x) = -\frac{1}{2}x^2 + 1000x, \; 0 \le x \le 600$$

a) $P(x) = R(x) - C(x)$

$$= -\frac{1}{2}x^2 + 1000x - (5000 + 600x)$$
$$= -\frac{1}{2}x^2 + 400x - 5000$$

Copyright © 2014 Pearson Education, Inc. Publishing as Addison-Wesley.

b) First, we find the critical values.

$P'(x) = -x + 400$

$P'(x)$ exists for all real numbers. Solve:

$$P'(x) = 0$$
$$-x + 400 = 0$$
$$x = 400.$$

The critical value is 400 and the endpoints are 0 and 600. Using the Max-Min Principle 1, we find the maximum profit.

$$P(0) = -\frac{1}{2}(0)^2 + 400(0) - 5000 = -5000$$

$$P(400) = -\frac{1}{2}(400)^2 + 400(400) - 5000 = 75{,}000$$

$$P(600) = -\frac{1}{2}(600)^2 + 400(600) - 5000 = 55{,}000$$

The total profit is maximized when 400 items are produced.

105. $B(x) = 305x^2 - 1830x^3, \quad 0 \le x \le 0.16$

a) $B'(x) = 610x - 5490x^2$

b) $B'(x)$ exists for all real numbers. Solve:

$$B'(x) = 0$$
$$610x - 5490x^2 = 0$$
$$610x(1 - 9x) = 0$$

$$x = 0 \quad \text{or} \quad 1 - 9x = 0$$

$$x = 0 \quad \text{or} \quad x = \frac{1}{9} \approx 0.11.$$

c) The critical points and the endpoints are $0, \frac{1}{9}$, and 0.16.

d) We find the function values.

$$B(0) = 305(0)^2 - 1830(0)^3 = 0$$

$$B\left(\frac{1}{9}\right) = 305\left(\frac{1}{9}\right)^2 - 1830\left(\frac{1}{9}\right)^3 = \frac{305}{243} \approx 1.255$$

$$B(0.16) = 305(0.16)^2 - 1830(0.16)^3 \approx 0.312.$$

The maximum blood pressure is approximately 1.255, which occurs at a dose of $x = \frac{1}{9}$ cc, or about 0.11 cc of the drug.

107. We look at the derivative on each piece of the Function to determine any critical values. For $-3 < x < 1$, $f'(x) = 2$ so there are no critical values for this part of the function. For $1 < x \le 2$, $f'(x) = -2x$. $f'(x) = 0$ when $x = 0$, which is outside the domain of this piece of the function. Therefore, there are no critical values of $f(x)$. The absolute extrema will occur at one of the endpoints. The function values are:

$$f(-3) = 2(-3) + 1 = -5$$
$$f(1) = 2(1) + 1 = 3$$
$$f(2) = 4 - (2)^2 = 0.$$

On the interval $[-3, 2]$, the absolute minimum is -5, which occurs at $x = -3$. The absolute maximum is 3, which occurs at $x = 1$.

A sketch of the graph is shown below.

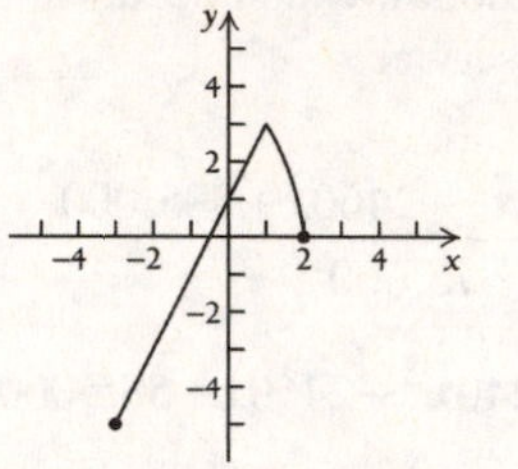

109. We look at the derivative on each piece of the Function to determine any critical values. For $-4 \le x < 0$, $h'(x) = -2x$. $h'(x) = 0$ when $x = 0$, which is also the endpoint of this part of the domain. For $0 < x \le 1$, $h'(x) = -1$. Therefore, there are no critical values of $h(x)$ on this part of the domain. For $1 \le x \le 2$, $h'(x) = 1$. Therefore, there are no critical values of $h(x)$ on this part of the domain. The absolute extrema will occur at one of the endpoints. The function values are:

$$h(-4) = 1 - (-4)^2 = -15$$
$$h(0) = 1 - (0) = 1$$
$$h(1) = (1) - 1 = 0$$
$$h(2) = (2) - 1 = 1.$$

On the interval $[-4, 2]$, the absolute minimum is -15, which occurs at $x = -4$. The absolute maximum is 1, which occurs at $x = 0$ and $x = 2$.

The solution is continued on the next page.

Copyright © 2014 Pearson Education, Inc. Publishing as Addison-Wesley.

A sketch of the graph is shown below.

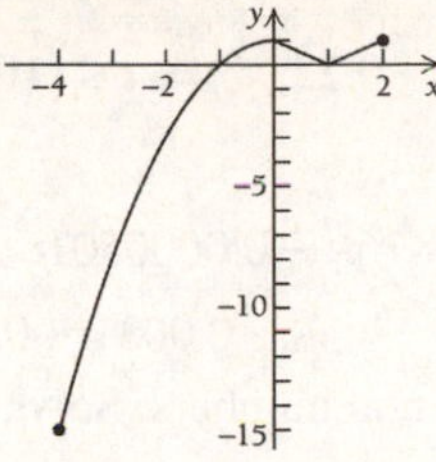

111. a) The sketch of the graph is shown below:

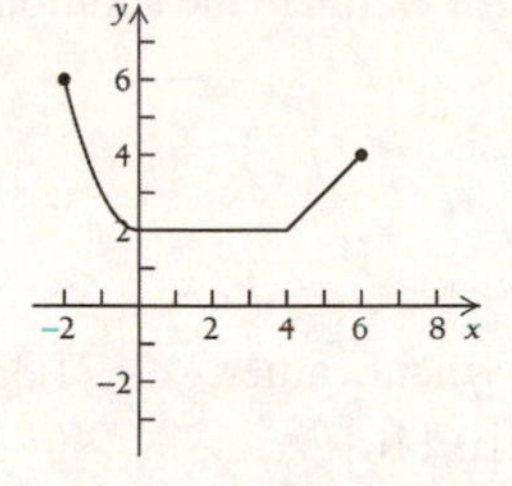

b) From the graph, the absolute maximum is 6 and occurs at $x=-2$.

c) The absolute minimum value for this function is 2. This value occurs over the range $0 \le x \le 4$.

113. $g(x)=x\sqrt{x+3}; \qquad [-3,3]$

a) Find $g'(x)$.

$$g'(x)=x\left[\frac{1}{2}(x+3)^{-1/2}(1)\right]+(1)(x+3)^{1/2}$$

$$=\frac{x}{2(x+3)^{1/2}}+(x+3)^{1/2}$$

$$=\frac{x}{2(x+3)^{1/2}}+\frac{(x+3)^{1/2}}{1}\cdot\frac{2(x+3)^{1/2}}{2(x+3)^{1/2}}$$

Multiplying by a form of 1

$$=\frac{x}{2(x+3)^{1/2}}+\frac{2(x+3)}{2(x+3)^{1/2}}$$

$$=\frac{3x+6}{2(x+3)^{1/2}}, \text{ or } \frac{3x+6}{2\sqrt{x+3}}$$

b) Find the critical values. $g'(x)$ exists for all values in $[-3,3]$ except -3. This is a critical value as well as an end point. To find the other critical values, we solve $g'(x)=0$.

$$g'(x)=0$$

$$\frac{3x+6}{2\sqrt{x+3}}=0$$

$$3x+6=0$$

$$3x=-6$$

$$x=-2$$

The second critical value on the interval is -2.

c) On a closed interval, the Max-Min Principle 1 can always be used. The critical values and the endpoints are $-3,\ -2$, and 3.

d) Find the function value at each value in (c).

$$g(-3)=(-3)\sqrt{(-3)+3}=0$$

$$g(-2)=(-2)\sqrt{(-2)+3}=-2$$

$$g(3)=(3)\sqrt{(3)+3}=3\sqrt{6}$$

Thus, the absolute maximum over the interval $[-3,3]$ is $3\sqrt{6}$, which occurs at $x=3$, and the absolute minimum is -2, which occurs at $x=-2$.

115. $C(x)=(2x+4)+\left(\frac{2}{x-6}\right), \qquad x>6$

$$=2x+4+2(x-6)^{-1}$$

a) Find $C'(x)$.

$$C'(x)=2-2(x-6)^{-2}(1)$$

$$=2-\frac{2}{(x-6)^2}$$

Copyright © 2014 Pearson Education, Inc. Publishing as Addison-Wesley.

b) Find the critical values.
$C'(x)$ does not exist for $x = 6$; however, this value is not in the domain interval, so it is not a critical value.
Solve $C'(x) = 0$.

$$2 - \frac{2}{(x-6)^2} = 0$$

$$2 = \frac{2}{(x-6)^2}$$

$$2(x-6)^2 = 2 \quad \text{Multiplying by } (x-6)^2 \text{ Since } x \neq 6.$$

$$(x-6)^2 = 1$$

$$x - 6 = \pm 1 \quad \text{Taking the square root of both sides.}$$

$$x = 6 \pm 1$$

$$x = 5 \quad \text{or} \quad x = 7$$

The only critical value in $(6, \infty)$ is 7.

c) Since there is only one critical value, we apply the Max-Min Principle 2.

$$C''(x) = 4(x-6)^{-3} = \frac{4}{(x-6)^3}$$

$$C''(7) = \frac{4}{(7-6)^3} = 4 > 0$$

Therefore, since $C''(7) > 0$, there is a minimum at $x = 7$.
The firm should use 7 “quality units” to minimize its total cost of service.

117. 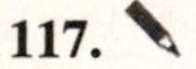

119. $P(t) = 0.0000000219t^4 - 0.0000167t^3 + 0.00155t^2 + 0.002t + 0.22, \quad 0 \le t \le 110$

a) Find $P'(t)$.

$$P'(t) = 0.0000000876t^3 - 0.0000501t^2 + 0.0031t + 0.002$$

$P'(t)$ exists for all real numbers. Solve $P'(t) = 0$. We use a calculator to find the zeros of $P'(t)$. We estimate the solutions to be:

$$x \approx -0.639$$

$$x \approx 71.333$$

$$x \approx 501.223.$$

Only one of the critical values, $x \approx 71.333$, is in the interval $[0, 110]$.
We apply the Max-Min Principle 1, to find the absolute maximum.
The critical values and the endpoints are 0, 71.333, and 110.
The function values at these points are

$$P(0) = 0.22$$

$$P(71.333) \approx 2.755$$

$$P(110) \approx 0.174.$$

Thus, the absolute maximum oil production for the U.S. after 1910 was 2.755 billion barrels per year. This production level occurred 71.333 year after 1910, or in 1981.

b) In 2004, $t = 2004 - 1910 = 96$. We plug this value into the first derivative to obtain:

$$P'(94)$$

$$= 0.0000000876(94)^3 - 0.0000501(94)^2 + 0.0031(94) + 0.002$$

$$\approx -0.1485.$$

The rate of oil was declining at approximately 0.1485 billion of barrels per year.
In 2010, $t = 2010 - 1910 = 100$. We plug this value into the first derivative to obtain:

$$P'(100)$$

$$= 0.0000000876(100)^3 - 0.0000501(100)^2 + 0.0031(100) + 0.002$$

$$\approx -0.1881.$$

The rate of oil was declining at approximately 0.1881 billion of barrels per year.

Copyright © 2014 Pearson Education, Inc. Publishing as Addison-Wesley.

121. $f(x)=\frac{3}{4}\left(x^2-1\right)^{2/3}; \qquad \left[\frac{1}{2},\infty\right)$

Using a calculator, we enter the equation into the graphing editor:

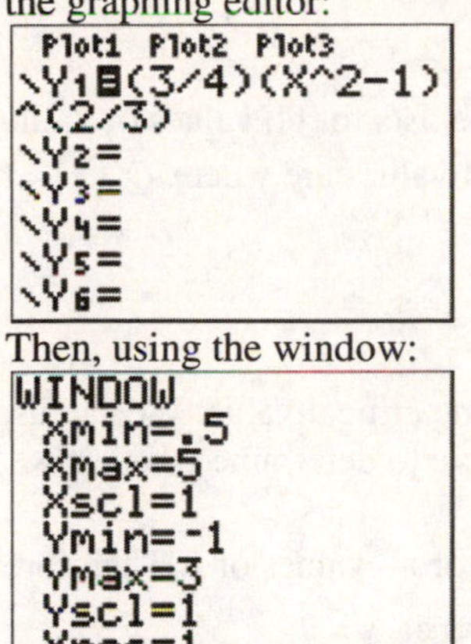

The graph of the function is shown below:

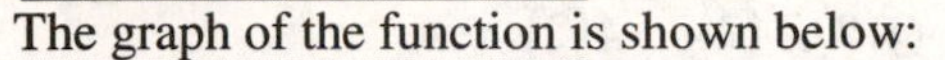

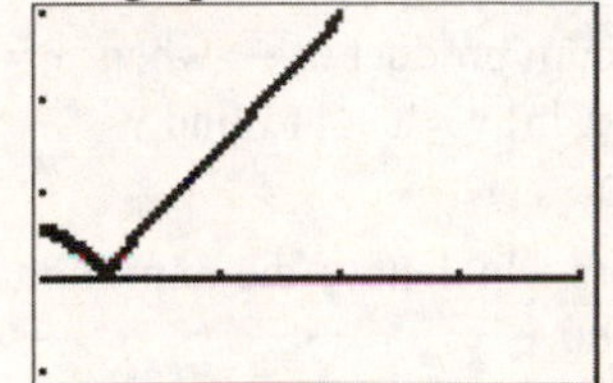

Using the table feature, we locate the extrema. We estimate the absolute minimum to be 0, which occurs at $x=1$. There is no absolute maximum.

123. a) Using a graphing calculator, we fit the linear equation $y=x+8.857$. This corresponds to the model $P(t)=t+8.857$. Where P is the pressure of the contractions and t is the time in minutes. We substitute 7 for t to find the pressure at 7 minutes.

$P(7)=7+8.857=15.857$.

The pressure at 7 minutes is 15.857 mm of Hg.

b) Rounding the coefficients to 3 decimal places, we find the quartic regression

$y=0.117x^4-1.520x^3+6.193x^2-$
$7.018x+10.009.$

Changing the variables we get the model

$P(t)=0.117t^4-1.520t^3+6.193t^2-$
$7.018t+10.009.$

Using the table feature, when $x=7$, $y=24.857$. So the pressure at 7 minutes is 24.86 mm of mercury. (If we use the rounded coefficients above, we get 23.897 mm of Hg.)

Using the trace feature, we estimate the smallest contraction on the interval $[0,10]$ was about 7.62 mm of Hg. This occurred when $x\approx 0.765$ min.

Copyright © 2014 Pearson Education, Inc. Publishing as Addison-Wesley.

Exercise Set 2.5

1. Express $Q = xy$ as a function of one variable. First, we solve $x + y = 50$ for y.

$x + y = 50$

$y = 50 - x$

Next, we substitute $50 - x$ for y in $Q = xy$.

$Q = xy$

$Q = x(50 - x)$ Substituting

$= 50x - x^2$

Now that Q is a function of one variable we can find the maximum. First, we find the critical values.

$Q'(x) = 50 - 2x$. Since $Q'(x)$ exists for all real numbers, the only critical value will occur when $Q'(x) = 0$. We solve:

$50 - 2x = 0$

$50 = 2x$

$25 = x$

There is only one critical value. We use the second derivative to determine if the critical value is a maximum. Note that:

$Q''(x) = -2 < 0$. The second derivative is negative for all values of x. Therefore, a maximum occurs at $x = 25$.

Now,

$Q(25) = 50(25) - (25)^2 = 625$.

Therefore, the maximum product is 625, which occurs when $x = 25$. If $x = 25$, then $y = 50 - 25 = 25$. The two numbers are 25 and 25.

3. ✎

5. Let x be one number and y the other number. Since the difference of the two numbers must be 4, we have $x - y = 4$.

The product, Q, of the two numbers is given by $Q = xy$, so our task is to minimize $Q = xy$, where $x - y = 4$.

First, we express $Q = xy$ as a function of one variable.

Solving $x - y = 4$ for y, we have:

$x - y = 4$

$-y = 4 - x$

$y = x - 4$.

Next, we substitute $x - 4$ for y in $Q = xy$.

$Q = x(x - 4) = x^2 - 4x$

$Q(x) = x^2 - 4x$

Find $Q'(x) = 2x - 4$

The derivative exists for all values of x; thus, the only critical values are where $Q'(x) = 0$.

$2x - 4 = 0$

$2x = 4$

$x = 2$

There is only one critical value. We can use the second derivative to determine whether we have a maximum.

$Q''(x) = 2 > 0$ for all values of x. Therefore, a minimum occurs at $x = 2$.

$Q(2) = (2)^2 - 4(2) = -4$

Thus, the minimum product is -4 when $x = 2$. Substitute 2 for x in $y = x - 4$ to find y.

$y = 2 - 4 = -2$.

The two numbers which have the minimum product are 2 and -2.

7. Maximize $Q = xy^2$, where x and y are positive numbers such that $x + y^2 = 1$.

Express $Q = xy^2$ as a function of one variable.

First, we solve $x + y^2 = 1$ for y^2.

$x + y^2 = 1$

$y^2 = 1 - x$

Next, we substitute $1 - x$ for y^2 in $Q = xy^2$.

$Q = xy^2$

$Q = x(1 - x)$ Substituting

$= x - x^2$

Now that Q is a function of one variable we can find the maximum. First, we find the critical values.

$Q'(x) = 1 - 2x$. Since $Q'(x)$ exists for all real numbers, the only critical value will occur when $Q'(x) = 0$. We solve;

$1 - 2x = 0$

$1 = 2x$

$\frac{1}{2} = x$

There is only one critical value.

The solution is continued on the next page.

Copyright © 2014 Pearson Education, Inc. Publishing as Addison-Wesley.

We use the second derivative to determine if the critical value is a maximum. Note that: $Q''(x) = -2 < 0$. The second derivative is negative for all values of x. Therefore, a maximum occurs at $x = \frac{1}{2}$.

Now,

$$Q\left(\frac{1}{2}\right) = \left(\frac{1}{2}\right) - \left(\frac{1}{2}\right)^2 = \frac{1}{4}.$$

Substitute $\frac{1}{2}$ in for x in $x + y^2 = 1$ and solve for y.

$$\frac{1}{2} + y^2 = 1$$
$$y^2 = \frac{1}{2}$$
$$y = \pm\sqrt{\frac{1}{2}} = \pm\frac{1}{\sqrt{2}}$$
$$y = \frac{1}{\sqrt{2}}, \quad x \text{ and } y \text{ must be positive}$$

Therefore, the maximum value of Q is $\frac{1}{4}$ when $x = \frac{1}{2}$ and $y = \frac{1}{\sqrt{2}}$.

9. Minimize $Q = 2x^2 + 3y^2$, where $x + y = 5$. Express Q as a function of one variable. First, solve $x + y = 5$ for y.

$$x + y = 5$$
$$y = 5 - x$$

Then substitute $5 - x$ for y in $Q = 2x^2 + 3y^2$.

$$Q = 2x^2 + 3(5 - x)^2$$
$$= 2x^2 + 3(25 - 10x + x^2)$$
$$= 2x^2 + 75 - 30x + 3x^2$$
$$= 5x^2 - 30x + 75$$

Find $Q'(x)$, where $Q(x) = 5x^2 - 30x + 75$.

$$Q'(x) = 10x - 30$$

This derivative exists for all values of x; thus the only critical values are where

$$Q'(x) = 0$$
$$10x - 30 = 0$$
$$10x = 30$$
$$x = 3.$$

Since there is only one critical value, we can use the second derivative to determine whether we have a minimum. Note that: $Q''(x) = 10$, which is positive for all real numbers. Thus $Q''(3) > 0$, so a minimum occurs when $x = 3$. The value of Q is

$$Q(3) = 2(3)^2 + 3(5 - 3)^2$$
$$= 18 + 12$$
$$= 30.$$

Substitute 3 for x in $y = 5 - x$ to find y.

$$y = 5 - x$$
$$y = 5 - 3 = 2$$

Thus, the minimum value of Q is 30 when $x = 3$ and $y = 2$.

11. Maximize $Q = xy$, where x and y are positive numbers such that $\frac{4}{3}x^2 + y = 16$.

Express Q as a function of one variable. First, solve $\frac{4}{3}x^2 + y = 16$ for y.

$$\frac{4}{3}x^2 + y = 16$$
$$y = 16 - \frac{4}{3}x^2$$

Then substitute $16 - \frac{4}{3}x^2$ for y in $Q = xy$.

$$Q = x\left(16 - \frac{4}{3}x^2\right)$$
$$= 16x - \frac{4}{3}x^3$$

Find $Q'(x)$, where $Q(x) = 16x - \frac{4}{3}x^3$.

$$Q'(x) = 16 - 4x^2$$

This derivative exists for all values of x; thus the only critical values are where

$$Q'(x) = 0$$
$$16 - 4x^2 = 0$$
$$-4x^2 = -16$$
$$x^2 = 4$$
$$x = \pm 2$$
$$x = 2. \quad x \text{ must be positive}$$

Since there is only one critical value, we can use the second derivative to determine whether we have a maximum. The solution is continued on the next page.

Copyright © 2014 Pearson Education, Inc. Publishing as Addison-Wesley.

Note that:
$Q''(x) = -8x$
and
$Q''(2) = -8(2) = -16 < 0$.
Since $Q''(2)$ is negative, a maximum occurs at $x = 2$.

$$Q(2) = 16(2) - \frac{4}{3}(2)^3$$
$$= 32 - \frac{32}{3}$$
$$= \frac{64}{3}$$

Substitute 2 for *x* in $y = 16 - \frac{4}{3}x^2$ to find *y*.

$$y = 16 - \frac{4}{3}x^2$$
$$y = 16 - \frac{4}{3}(2)^2$$
$$y = 16 - \frac{16}{3}$$
$$y = \frac{32}{3}$$

Thus, the maximum value of Q is $\frac{64}{3}$ when $x = 2$ and $y = \frac{32}{3}$.

13. Let *x* represent the length and *y* represent the width of the swimming area. It is helpful to draw a picture.

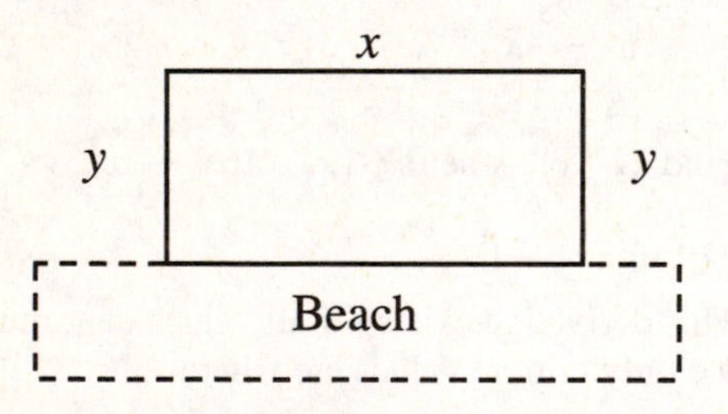

Since the life guard has 180 yd of rope and floats, the perimeter of the swimming area is $x + 2y = 180$. Solving this equation for *x*, we have $x = 180 - 2y$

The objective is to maximize area, which is given by
$A = l \cdot w$
Substituting $x = 180 - 2y$ for the length and *y* for the width, we have:
$A = (180 - 2y)y = 180y - 2y^2$.

We will maximize the area over the interval $0 < y < 90$, because *y* is the length of one side, and cannot be negative. Furthermore, since there is only 180 yards of rope, and we need two sides, *y* cannot be greater than 90 yards. If $y = 90$, then the length of the swimming area would be 0.

We now must find $A'(y)$, where
$A(y) = 180y - 2y^2$.
$A'(y) = 180 - 4y$
This derivative exists for all values of *y* in $(0, 90)$. Thus the only critical values are where

$$A'(y) = 0$$
$$180 - 4y = 0$$
$$-4y = -180$$
$$y = 45$$

Since there is only one critical value in the interval, we use the second derivative to determine whether we have a maximum. Note, $A''(y) = -4 < 0$ for all values of *y*, so there is a maximum at $y = 45$.

Next, find the dimensions and the area.
When $y = 45$, we have
$x = 180 - 2(45) = 90$
and
$A(45) = 180(45) - 2(45)^2 = 4050$.
Therefore, the maximum area is 4050 yd^2 when the overall dimensions are 45 yd by 90 yd.

15. Let *x* represent the length and *y* represent the width. It is helpful to draw a picture.

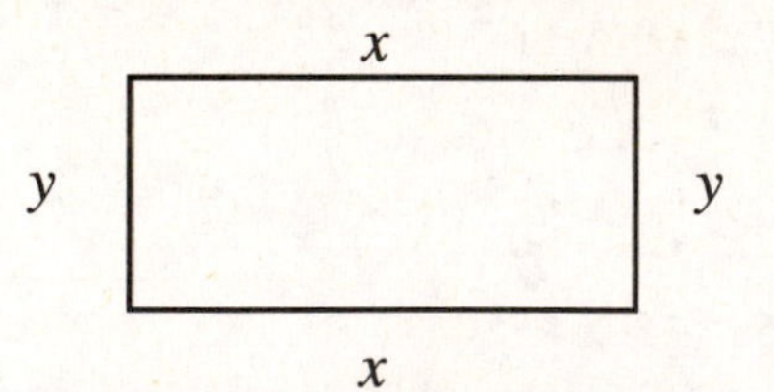

The perimeter is found by adding up the length of the sides. Since it is fixed at 54 feet, the equation of the perimeter is $2x + 2y = 54$.

The area is given by $A = xy$.

First, we solve the perimeter equation for *y*.

$$2x + 2y = 54$$
$$2y = 54 - 2x$$
$$y = 27 - x$$

The solution is continued on the next page.

Copyright © 2014 Pearson Education, Inc. Publishing as Addison-Wesley.

Then we substitute for y into the area formula.

$$A = xy$$
$$= x(27-x) = 27x - x^2$$

We want to maximize the area on the interval $(0,27)$. We consider this interval because x is the length of the shed and cannot be negative. Since the perimeter cannot exceed 54 feet, x cannot be greater than 27, also if x is 27 feet, the width of the shed would be 0 feet. We begin by finding $A'(x)$.

$$A'(x) = 27 - 2x.$$

This derivative exists for all values of x in $(0,27)$. Thus, the only critical values occur where $A'(x) = 0$. We solve the equation.

$$27 - 2x = 0$$
$$-2x = -27$$
$$x = \frac{27}{2} = 13.5$$

Since there is only one critical value in the interval, we can use the second derivative to determine whether we have a maximum. Note that

$A''(x) = -2 < 0$ for all values of x. Thus, $A''(13.5) < 0$, so a maximum occurs at $x = 13.5$.

Now,

$$A(x) = 27x - x^2$$
$$A(13.5) = 27(13.5) - (13.5)^2$$
$$= 182.25.$$

The maximum area is 182.25 ft^2.

Note: when $x = 13.5$, $y = 27 - 13.5 = 13.5$, so the overall dimensions that will achieve the maximum area are 13.5 ft by 13.5 ft.

17. When squares of length h on a side are cut out of the corners, we are left with a square base of length x. A picture will help.

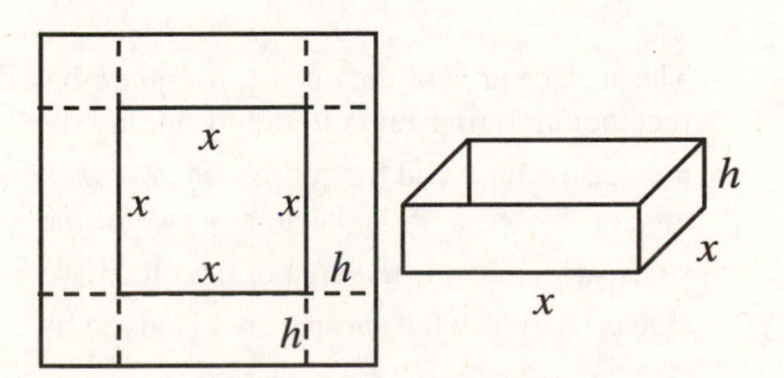

The resulting volume of the box is

$V = lwh = x \cdot x \cdot h = x^2 h$.

We want to express V in terms of one variable. Note that the overall length of a side of the aluminum is 50 cm. We see from the drawing, that $h + x + h = 50$, *or* $x + 2h = 50$. Solving for h we get:

$$2h = 50 - x$$
$$h = \frac{1}{2}(50 - x) = 25 - \frac{1}{2}x.$$

Substituting h into the volume equation, we have:

$V = x^2\left(25 - \frac{1}{2}x\right) = 25x^2 - \frac{1}{2}x^3$. The objective is to maximize $V(x)$ on the interval $(0,50)$.

First, we find the derivative.

$$V'(x) = 50x - \frac{3}{2}x^2$$

This derivative exists for all x in the interval $(0,50)$, so the critical values will occur when $V'(x) = 0$.

Solving this equation, we have:

$$50x - \frac{3}{2}x^2 = 0$$
$$x\left(50 - \frac{3}{2}x\right) = 0$$
$$x = 0 \quad \text{or} \quad 50 - \frac{3}{2}x = 0$$
$$x = 0 \quad \text{or} \quad -\frac{3}{2}x = -50$$
$$x = 0 \quad \text{or} \quad x = \frac{100}{3} \approx 33\tfrac{1}{3}.$$

The only critical value in $(0,50)$ is $\frac{100}{3}$, or about 33.33. Therefore, we can use the second derivative $V''(x) = 50 - 3x$ to determine if we have a maximum. We have

$$V''\left(\frac{100}{3}\right) = 50 - 3\left(\frac{100}{3}\right) = -50 < 0.$$

Therefore, there is a maximum at $\frac{100}{3}$.

The solution is continued on the next page.

Copyright © 2014 Pearson Education, Inc. Publishing as Addison-Wesley.

Evaluating the function at the critical value, we have:

$$V\left(\frac{100}{3}\right) = 25\left(\frac{100}{3}\right)^2 - \frac{1}{2}\left(\frac{100}{3}\right)^3$$
$$= \frac{250{,}000}{27} = 9259\tfrac{7}{27}.$$

Now, we find the height of the box.

$$h = 25 - \frac{1}{2}\left(\frac{100}{3}\right) = \frac{25}{3} = 8\tfrac{1}{3}.$$

Therefore, a box with dimensions $33\frac{1}{3}$ cm. by $33\frac{1}{3}$ cm. by $8\frac{1}{3}$ cm. will yield a maximum volume of $9259\frac{7}{27}$ cm^3.

19. First, we make a drawing.

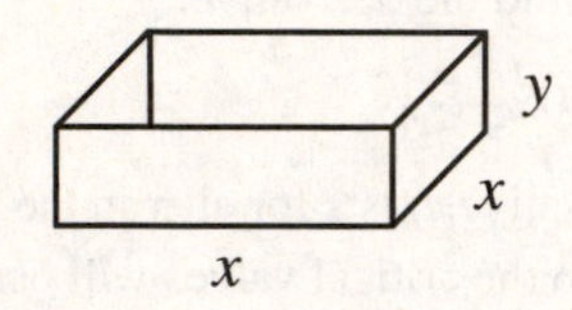

The surface area of the open-top, square-based, rectangular box is found by adding the area of the base and the four sides. x^2 is the area of the base, xy is the area of one of the sides and there are four sides, therefore the surface area is given by $S = x^2 + 4xy$.

The volume must by 62.5 cubic inches, and is given by $V = l \cdot w \cdot h = x^2 y = 62.5$.

To express S in terms of one variable, we solve $x^2 y = 62.5$ for y:

$$y = \frac{62.5}{x^2}.$$

Then

$$S(x) = x^2 + 4x\left(\frac{62.5}{x^2}\right)$$
$$= x^2 + \frac{250}{x} = x^2 + 250x^{-1}.$$

Now S is defined only for positive numbers, so we minimize S on the interval $(0,\infty)$.

First, we find $S'(x)$.

$$S'(x) = 2x - 250x^{-2}$$
$$= 2x - \frac{250}{x^2}.$$

Since $S'(x)$ exists for all x in $(0,\infty)$, the only critical values are where $S'(x) = 0$.

We solve the following equation:

$$2x - \frac{250}{x^2} = 0$$
$$2x = \frac{250}{x^2}$$
$$x^3 = 125$$
$$x = 5.$$

Since there is only one critical value, we use the second derivative to determine whether we have a minimum. Note that

$$S''(x) = 2 + 500x^{-3} = 2 + \frac{500}{x^3}.$$

$S''(5) = 2 + \frac{500}{5^3} = 6 > 0$. Since the second derivative is positive, we have a minimum at $x = 5$. We find y when $x = 5$.

$$y = \frac{62.5}{x^2}$$
$$= \frac{62.5}{5^2}$$
$$= 2.5$$

The surface area is minimized when $x = 5$ in. and $y = 2.5$ in. We find the minimum surface area by substituting these values into the surface area equation.

$$S = x^2 + 4xy$$
$$= (5)^2 + 4(5)(2.5)$$
$$= 25 + 50$$
$$= 75$$

The minimum surface area is 75 in^2 when the dimensions are 5 in. by 5 in. by 2.5 in.

21. First, we make a drawing.

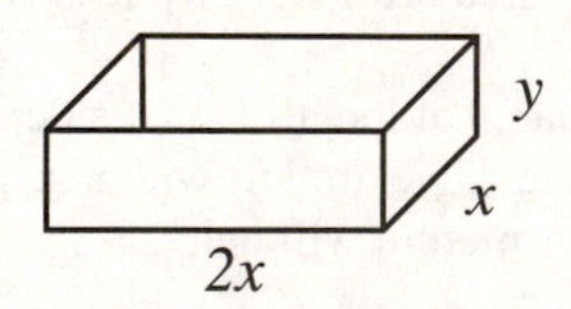

The surface area of the open-top, square-based, rectangular dumpster is found by adding the area of the base and the four sides. $2x^2$ is the area of the base, xy is the area of two of the sides, while $2xy$ is the area of the other two sides. Therefore the surface area is given by $S = 2x^2 + 2xy + 2(2xy) = 2x^2 + 6xy$.

The solution is continued on the next page.

Copyright © 2014 Pearson Education, Inc. Publishing as Addison-Wesley.

The volume must by 12 cubic yards, and is given by $V = l \cdot w \cdot h = 2x \cdot x \cdot y = 2x^2y = 12$. To express S in terms of one variable, we solve $2x^2y = 12$ for y:

$$y = \frac{6}{x^2}.$$

Then

$$S(x) = 2x^2 + 6x\left(\frac{6}{x^2}\right)$$
$$= 2x^2 + \frac{36}{x} = 2x^2 + 36x^{-1}.$$

Now S is defined only for positive numbers, so we minimize S on the interval $(0, \infty)$.

First, we find $S'(x)$.

$$S'(x) = 4x - 36x^{-2}$$
$$= 4x - \frac{36}{x^2}$$

Since $S'(x)$ exists for all x in $(0, \infty)$, the only critical values are where $S'(x) = 0$. We solve the following equation:

$$4x - \frac{36}{x^2} = 0$$
$$4x = \frac{36}{x^2}$$
$$x^3 = 9$$
$$x = \sqrt[3]{9} \approx 2.08.$$

Since there is only one critical value, we use the second derivative to determine whether we have a minimum. Note that

$$S''(x) = 4 + 72x^{-3} = 4 + \frac{72}{x^3}.$$

$S''(\sqrt[3]{9}) = 4 + \frac{72}{(\sqrt[3]{9})^3} = 12 > 0$. Since the second derivative is positive, we have a minimum at $x = \sqrt[3]{9} \approx 2.08$. The width is 2.08 yd.; therefore, the length is $2(2.08) \approx 4.16$. We find the height y

$$y = \frac{6}{x^2}$$
$$= \frac{6}{(2.08)^2}$$
$$\approx 1.387.$$

The overall dimensions of the dumpster that will minimize surface area are 2.08 yd. by 4.16 yd. by 1.387 yd.

23. $R(x) = 50x - 0.5x^2$; $C(x) = 4x + 10$

Profit is equal to revenue minus cost.

$$P(x) = R(x) - C(x)$$
$$= 50x - 0.5x^2 - (4x + 10)$$
$$= -0.5x^2 + 46x - 10$$

Because x is the number of units produced and sold, we are only concerned with the non-negative values of x. Therefore, we will find the maximum of $P(x)$ on the interval $[0, \infty)$.

First, we find $P'(x)$.

$$P'(x) = -x + 46$$

The derivative exists for all values of x in $[0, \infty)$.

Thus, we solve $P'(x) = 0$.

$$-x + 46 = 0$$
$$-x = -46$$
$$x = 46$$

There is only one critical value. We can use the second derivative to determine whether we have a maximum.

$$P''(x) = -1 < 0$$

The second derivative is less than zero for all values of x. Thus, a maximum occurs at $x = 46$.

$$P(46) = -0.5(46)^2 + 46(46) - 10$$
$$= -1058 + 2116 - 10$$
$$= 1048$$

The maximum profit is \$1048 when 46 units are produced and sold.

25. $R(x) = 2x$; $C(x) = 0.01x^2 + 0.6x + 30$

Profit is equal to revenue minus cost.

$$P(x) = R(x) - C(x)$$
$$= 2x - (0.01x^2 + 0.6x + 30)$$
$$= -0.01x^2 + 1.4x - 30$$

Because x is the number of units produced and sold, we are only concerned with the non-negative values of x. Therefore, we will find the maximum of $P(x)$ on the interval $[0, \infty)$.

First, we find $P'(x)$.

$$P'(x) = -0.02x + 1.4$$

The derivative exists for all values of x in $[0, \infty)$.

The solution is continued on the next page.

Copyright © 2014 Pearson Education, Inc. Publishing as Addison-Wesley.

Thus, we solve $P'(x)=0$.

$-0.02x+1.4=0$

$-0.02x=-1.4$

$x=70$

There is only one critical value. We can use the second derivative to determine whether we have a maximum.

$P''(x)=-0.02<0$

The second derivative is less than zero for all values of x. Thus, a maximum occurs at $x=70$.

$P(70)=-0.01(70)^2+1.4(70)-30$

$=-49+98-30$

$=19$

The maximum profit is \$19 when 70 units are produced and sold.

27. $R(x)=9x-2x^2$

$C(x)=x^3-3x^2+4x+1$

$R(x)$ and $C(x)$ are in thousands of dollars and x is in thousands of units.

Profit is equal to revenue minus cost.

$P(x)=R(x)-C(x)$

$=9x-2x^2-(x^3-3x^2+4x+1)$

$=-x^3+x^2+5x-1$

Because x is the number of units produced and sold, we are only concerned with the non-negative values of x. Therefore, we will find the maximum of $P(x)$ on the interval $[0,\infty)$.

First, we find $P'(x)$.

$P'(x)=-3x^2+2x+5$

The derivative exists for all values of x in $[0,\infty)$.

Thus, we solve $P'(x)=0$.

$-3x^2+2x+5=0$

$3x^2-2x-5=0$

$(3x-5)(x+1)=0$

$3x-5=0$ or $x+1=0$

$3x=5$ or $x=-1$

$x=\frac{5}{3}$ or $x=-1$

There is only one critical value in the interval $[0,\infty)$. We can use the second derivative to determine whether we have a maximum.

$P''(x)=-6x+2$

Therefore,

$P''\left(\frac{5}{3}\right)=-6\left(\frac{5}{3}\right)+2=-10+2=-8<0$

The second derivative is less than zero for $x=\frac{5}{3}$. Thus, a maximum occurs at $x=\frac{5}{3}$.

$P\left(\frac{5}{3}\right)=-\left(\frac{5}{3}\right)^3+\left(\frac{5}{3}\right)^2+5\left(\frac{5}{3}\right)-1$

$=-\frac{125}{27}+\frac{25}{9}+\frac{25}{3}-1$

$=-\frac{125}{27}+\frac{75}{27}+\frac{225}{27}-\frac{27}{27}$

$=\frac{148}{27}$

Note that $x=\frac{5}{3}$ thousand is approximately 1.667 thousand or 1667 units, and that $\frac{148}{27}$ thousand is approximately 5.481 thousand or 5481.

Thus, the maximum profit is approximately \$5481 when approximately 1667 units are produced and sold.

29. $p=150-0.5x$ Price per suit.

$C(x)=4000+0.25x^2$ Cost per suit.

a) Revenue is price times quantity. Therefore, revenue can be found by multiplying the number of suits sold, x, by the price of the suit, p. Substituting $150-0.5x$ for p, we have:

$R(x)=x\cdot p$

$=x(150-0.5x)$

$R(x)=150x-0.5x^2$

b) Profit is revenue minus cost. Therefore,

$P(x)=R(x)-C(x)$

$=150x-0.5x^2-(4000+0.25x^2)$

$=-0.75x^2+150x-4000.$

Since x is the number of suits produced and sold, we will restrict the domain to the interval $0\le x<\infty$.

Copyright © 2014 Pearson Education, Inc. Publishing as Addison-Wesley.

c) To determine the number of suits required to maximize profit, we first find $P'(x)$.

$P'(x) = -1.5x + 150$.

The derivative exists for all real numbers in the interval $[0, \infty)$. Thus, we solve

$$P'(x) = 0$$
$$-1.5x + 150 = 0$$
$$-1.5x = -150$$
$$x = 100.$$

Since there is only one critical value, we can use the second derivative to determine whether we have a maximum.

$P''(x) = -1.5 < 0$

The second derivative is negative for all values of x; therefore, a maximum occurs at $x = 100$.

Raggs, Ltd. must sell 100 suits to maximize profit.

d) The maximum profit is found by substituting 100 for x in the profit function.

$$P(100) = -0.75(100)^2 + 150(100) - 4000$$
$$= -7500 + 15,000 - 4000$$
$$= 3500$$

The maximum profit is \$3500.

e) The price per suit is given by:

$p = 150 - 0.5x$.

Substituting 100 for x, we have:

$p = 150 - 0.5(100) = 150 - 50 = 100$.

The price per suit will be \$100.

31. Let x be the amount by which the price of \$18 should be decreased (if x is negative, the price would be increased to maximize revenue). First, we express total revenue R as a function of x. There are two sources of revenue, revenue from tickets and revenue from concessions.

$$R(x) = \begin{pmatrix}\text{Revenue from}\\ \text{tickets}\end{pmatrix} + \begin{pmatrix}\text{Revenue from}\\ \text{concessions}\end{pmatrix}$$
$$= \begin{pmatrix}\text{Number of}\\ \text{People}\end{pmatrix} \cdot \begin{pmatrix}\text{Ticket}\\ \text{Price}\end{pmatrix} + \begin{pmatrix}\text{Number of}\\ \text{People}\end{pmatrix} \cdot 4.50$$

Note, the increase in ticket sales is $10,000\,x$, when price drops $3x$ dollars. Therefore, the increase in ticket sales is $\frac{10,000}{3}x$ when price drops x dollars.

$$R(x) = \left(40,000 + \frac{10,000}{3}x\right)(18 - x) + \left(40,000 + \frac{10,000}{3}x\right)(4.50)$$
$$= -\frac{10,000}{3}x^2 + 20,000x + 720,000 + 180,000 + 15,000x$$
$$= -\frac{10,000}{3}x^2 + 35,000x + 900,000$$

Therefore, the total revenue function is

$$R(x) = -\frac{10,000}{3}x^2 + 35,000x + 900,000.$$

To find x such that $R(x)$ is a maximum, we first find $R'(x)$:

$$R'(x) = -\frac{20,000}{3}x + 35,000.$$

This derivative exists for all real numbers x. thus, the only critical values are where $R'(x) = 0$; so we solve that equation:

$$-\frac{20,000}{3}x + 35,000 = 0$$
$$-\frac{20,000}{3}x = -35,000$$
$$x = 5.25.$$

Since this is the only critical value, we can use the second derivative,

$$R''(x) = -\frac{20,000}{3} < 0$$

to determine whether we have a maximum. since $R''(5.25)$ is negative, $R(5.25)$ is a maximum. Therefore, in order to maximize revenue, the university should charge $\$18 - \$5.25$, or \$12.75. Since \$12.75 is \$5.25 less than \$18, we can find the attendance using

$$40,000 + \frac{10,000}{3}(5.25) = 57,500.$$

The average attendance when ticket price is \$12.75 is 57,500 people.

33. Let x equal the number of additional trees per acre which should be planted. Then the number of trees planted per acre is represented by $(20 + x)$ and the yield per tree by $(30 - x)$. The total yield per acre is equal to the yield per tree times the number of trees so, we have:

$$Y(x) = (30 - x)(20 + x)$$
$$= 600 + 10x - x^2.$$

The solution is continued on the next page.

Copyright © 2014 Pearson Education, Inc. Publishing as Addison-Wesley.

To find x such that $Y(x)$ is a maximum, we first find $Y'(x)$:

$Y'(x) = 10 - 2x.$

This derivative exists for all real numbers x. Thus, the only critical values are where $Y'(x) = 0$; so we solve that equation:

$$10 - 2x = 0$$
$$-2x = -10$$
$$x = 5.$$

This corresponds to planting 5 trees.
Since this is the only critical value, we can use the second derivative,

$R''(x) = -2 < 0,$

to determine whether we have a maximum.
Since $R''(5)$ is negative, $R(5)$ is a maximum.
Therefore, in order to maximize yield, the apple farm should plant 20 + 5, or 25 trees per acre.

35. a) When $x = 25$, $q = 2.13$. When $x = 25 + 1$, or 26, then $q = 2.13 - 0.04 = 2.09$. We use the points $(25, 2.13)$ and $(26, 2.09)$ to find the linear demand function $q(x)$. First, we find the slope:

$$m = \frac{2.13 - 2.09}{25 - 26} = \frac{0.04}{-1} = -0.04.$$

Next, we use the point-slope equation:

$$q - 2.13 = -0.04(x - 25)$$
$$q - 2.13 = -0.04x + 1$$
$$q = -0.04x + 3.13.$$

Therefore, the linear demand function is:

$q(x) = -0.04x + 3.13.$

b) Revenue is price times quantity; therefore, the revenue function is:

$$R(x) = x \cdot q(x)$$
$$= x(-0.04x + 3.13)$$
$$= -0.04x^2 + 3.13x.$$

To find x such that $R(x)$ is a maximum, we first find $R'(x)$:

$R'(x) = -0.08x + 3.13.$

This derivative exists for all values of x. So the only critical values occur when $R'(x) = 0$; so we solve that equation:

$$-0.08x + 3.13 = 0$$
$$-0.08x = -3.13$$
$$x = 39.125.$$

Since this is the only critical value, we can use the second derivative,

$R''(x) = -0.08 < 0,$

to determine whether we have a maximum.
Since $R''(39.125)$ is negative, $R(39.125)$ is a maximum.
In order to maximize revenue, the State of Maryland should charge \$39.125 or rounding up to \$39.13 per license plate.

37. The volume of the box is given by
$V = x \cdot x \cdot y = x^2 y = 320.$

The area of the base is x^2. The cost of the base is $15x^2$ cents.

The area of the top is x^2. The cost of the top is $10x^2$ cents.

The area of each side is xy. The total area for the four sides is $4xy$. The cost of the four sides is $2.5(4xy)$ cents.

The total costs in cents is given by
$C = 15x^2 + 10x^2 + 2.5(4xy) = 25x^2 + 10xy.$

To express C in terms of one variable, we solve $x^2 y = 320$ for y:

$$y = \frac{320}{x^2}.$$

Then,

$$C(x) = 25x^2 + 10x\left(\frac{320}{x^2}\right)$$
$$= 25x^2 + \frac{3200}{x}.$$

The function is defined only for positive numbers, and the problem dictates that the length x must be positive, so we are minimizing C on the interval $(0, \infty)$.

First, we find $C'(x)$.

$$C'(x) = 50x - 3200x^{-2} = 50x - \frac{3200}{x^2}$$

Since $C'(x)$ exists for all x in $(0, \infty)$, the only critical values are where $C'(x) = 0$.

The solution is continued on the next page.

Copyright © 2014 Pearson Education, Inc. Publishing as Addison-Wesley.

Thus, we solve the following equation:

$$C'(x) = 0$$

$$50x - \frac{3200}{x^2} = 0$$

$$50x = \frac{3200}{x^2}$$

$$50x^3 = 3200$$

$$x^3 = 64$$

$$x = 4.$$

This is the only critical value, so we can use the second derivative to determine whether we have a minimum.

$$C''(x) = 50 + 6400x^{-3} = 50 + \frac{6400}{x^3}$$

Note that the second derivative is positive for all positive values of x, therefore we have a minimum at $x = 4$. We find y when $x = 4$.

$$y = \frac{320}{x^2} = \frac{320}{(4)^2} = \frac{320}{16} = 20$$

The cost is minimized when the dimensions are 4 ft by 4 ft by 20 ft.

39. Let x equal the lot size. Now the inventory costs are given by:

$$C(x) = \begin{matrix}\text{Yearly Carrying}\\ \text{Cost}\end{matrix} + \begin{matrix}\text{Yearly reorder}\\ \text{Cost}\end{matrix}$$

We consider each cost separately.

Yearly carrying costs, $C_c(x)$: Can be found by multiplying the cost to store the items by the number of items in storage. The average amount held in stock is $\frac{x}{2}$, and it cost \$20 per pool table for storage. Thus:

$$C_c(x) = 20 \cdot \frac{x}{2}$$

$$= 10x.$$

Yearly reorder costs, $C_r(x)$: Can be found by multiplying the cost of each order by the number of reorders. The cost of each order is $40 + 16x$, and the number of orders per year is $\frac{100}{x}$. Therefore,

$$C_r(x) = (40 + 16x)\left(\frac{100}{x}\right)$$

$$= \frac{4000}{x} + 1600.$$

Hence, the total inventory cost is:

$$C(x) = C_c(x) + C_r(x)$$

$$= 10x + \frac{4000}{x} + 1600.$$

We want to find the minimum value of C on the interval $[1,100]$. First, we find $C'(x)$:

$$C'(x) = 10 - 4000x^{-2} = 10 - \frac{4000}{x^2}.$$

The derivative exists for all x in $[1,100]$, so the only critical values are where $C'(x) = 0$.

We set the derivative equal to zero and solve the equation:

$$C'(x) = 0$$

$$10 - \frac{4000}{x^2} = 0$$

$$10 = \frac{4000}{x^2}$$

$$10x^2 = 4000$$

$$x^2 = 400$$

$$x = \pm 20.$$

The only critical value in the interval is $x = 20$, so we can use the second derivative to determine whether we have a minimum.

$$C''(x) = 8000x^{-3} = \frac{8000}{x^3}$$

Notice that $C''(x)$ is positive for all values of x in $[1,100]$, we have a minimum at $x = 20$. Thus, to minimize inventory costs, the store should order pool tables $\frac{100}{20} = 5$ times per year. The lot size will be 20 tables.

41. Let x equal the lot size. Now the inventory costs are given by:

$$C(x) = \begin{matrix}\text{Yearly Carrying}\\ \text{Cost}\end{matrix} + \begin{matrix}\text{Yearly reorder}\\ \text{Cost}\end{matrix}$$

We consider each cost separately.

Yearly carrying costs, $C_c(x)$: Can be found by multiplying the cost to store the items by the number of items in storage. The average amount held in stock is $\frac{x}{2}$, and it cost \$2 per calculator for storage. Thus:

$$C_c(x) = 2 \cdot \frac{x}{2}$$

$$= x.$$

The solution is continued on the next page.

Copyright © 2014 Pearson Education, Inc. Publishing as Addison-Wesley.

Yearly reorder costs, $C_r(x)$*:* Can be found by multiplying the cost of each order by the number of reorders. The cost of each order is $5+2.50x$, and the number of orders per year is $\frac{720}{x}$. Therefore,

$$C_r(x)=(5+2.50x)\left(\frac{720}{x}\right)$$
$$=\frac{3600}{x}+1800.$$

Hence, the total inventory cost is:

$$C(x)=C_c(x)+C_r(x)$$
$$=x+\frac{3600}{x}+1800.$$

We want to find the minimum value of C on the interval $[1,720]$. First, we find $C'(x)$:

$$C'(x)=1-3600x^{-2}=1-\frac{3600}{x^2}.$$

The derivative exists for all x in $[1,720]$, so the only critical values are where $C'(x)=0$. We solve that equation:

$$1-\frac{3600}{x^2}=0$$
$$1=\frac{3600}{x^2}$$
$$x^2=3600$$
$$x=\pm 60.$$

We determined the only critical value in the interval is $x=60$, so we can use the second derivative to determine whether we have a minimum.

$$C''(x)=7200x^{-3}=\frac{7200}{x^3}$$

Notice that $C''(x)$ is positive for all values of x in $[1,720]$, we have a minimum at $x=60$. Thus, to minimize inventory costs, the store should order calculators $\frac{720}{60}=12$ times per year. The lot size will be 60 calculators.

43. Let x equal the lot size.
Yearly carrying costs:

$$C_c(x)=2\cdot\frac{x}{2}$$
$$=x.$$

Yearly reorder costs:

$$C_r(x)=(4+2.50x)\left(\frac{256}{x}\right)$$
$$=\frac{1024}{x}+640$$

Hence, the total inventory cost is:

$$C(x)=C_c(x)+C_r(x)$$
$$=x+\frac{1024}{x}+640.$$

We want to find the minimum value of C on the interval $[1,256]$. First, we find $C'(x)$:

$$C'(x)=1-1024x^{-2}=1-\frac{1024}{x^2}.$$

The derivative exists for all x in $[1,256]$, so the only critical values are where $C'(x)=0$. We solve that equation:

$$1-\frac{1024}{x^2}=0$$
$$1=\frac{1024}{x^2}$$
$$x^2=1024$$
$$x=\pm 32.$$

The only critical value in the interval is $x=32$, so we can use the second derivative to determine whether we have a minimum.

$$C''(x)=2048x^{-3}=\frac{2048}{x^3}$$

Notice that $C''(x)$ is positive for all values of x in $[1,256]$, so we have a minimum at $x=32$. Thus, to minimize inventory costs, the store should order calculators $\frac{256}{32}=8$ times per year. The lot size will be 32 calculators.

45. The volume of the container must be 250 in^3. Therefore, we use formula for the volume cylinder to obatain $250=\pi r^2 h$.
Solving for h we have:

$$h=\frac{250}{\pi r^2}.$$

The container consists of a circular top and a circular bottom the area for each of the top and bottom of the conainer is given by $A=\pi r^2$. The side material that when laid out is a rectangle of height h and whose length is the same as the circumference of the circular ends, $2\pi r$.
The solution is continued on the next page.

Copyright © 2014 Pearson Education, Inc. Publishing as Addison-Wesley.

Therfore, the surface area of the side material is $A = 2\pi rh$. Therefore, the total surface area is the sum of the area of the top and the bottom plus the the side material:

$A = 2\pi r^2 + 2\pi rh$.

Substituting in for h, we have area as a function of the radius, r.

$$A(r) = 2\pi r^2 + 2\pi r\left(\frac{250}{\pi r^2}\right)$$

$$A(r) = 2\pi r^2 + \frac{500}{r}$$

The nature of this problem requires $r > 0$. We differentiate the area function with respect to r:

$$A'(r) = 4\pi r - \frac{500}{r^2}.$$

We find the critical values by setting the derivative equal to zero and solving for r. Remember, $r > 0$.

$$A'(r) = 0$$

$$4\pi r - \frac{500}{r^2} = 0$$

$$4\pi r = \frac{500}{r^2}$$

$$4\pi r^3 = 500$$

$$r^3 = \frac{500}{4\pi}$$

$$r = \sqrt[3]{\frac{125}{\pi}}$$

$$r \approx 3.414$$

This is the only critical value in the interval $r > 0$. We will calculate the second derivative to determine the concavity of the function.

$$A''(r) = 4\pi + \frac{1000}{r^3}.$$

Evaluating the second derivative at the critical value, we have:

$$A''(3.41) = 4\pi + \frac{1000}{(3.414)^3}$$

$$\approx 37.697 > 0.$$

Since the second derivative is positive at the critical value, the critical value represents a relative minimum. We determine the height of the container by substituting back into

$$h = \frac{250}{\pi r^2}.$$

$$h = \frac{250}{\pi(3.414)^2} \approx 6.828.$$

Therefore, the dimensions of the container that will minimize the surface area are a height of 6.828 inches and a radius of 3.414 inches.

47. The cost function for the container would be is given by:

$$C = 0.005\begin{pmatrix}\text{area of the} \\ \text{top and bottom}\end{pmatrix} + 0.003\begin{pmatrix}\text{area of the} \\ \text{side material}\end{pmatrix}$$

Using the information from problem 45, we know the total area for the top and bottom of the can was given by $2\pi r^2$ and the area of the side material was given by $\frac{500}{r^r}$. Therefore, the cost function is:

$$C(r) = 0.005(2\pi r^2) + 0.003\left(\frac{500}{r}\right)$$

$$= 0.01\pi r^2 + \frac{1.5}{r}.$$

The nature of the problem still requires $r > 0$. Calculating the derivative of the cost function we have:

$$C'(r) = \frac{d}{dr}\left[0.01(\pi r^2) + \frac{1.5}{r}\right]$$

$$= \frac{d}{dr}\left[0.01(\pi r^2)\right] + \frac{d}{dr}\left[1.5r^{-1}\right]$$

$$= 0.01(2\pi r^{2-1}) + 1.5(-1r^{-1-1})$$

$$= 0.02\pi r - \frac{1.5}{r^2}.$$

Find the critical values by setting the derivative equal to zero and solving for r.

$$C'(r) = 0$$

$$0.02\pi r - \frac{1.5}{r^2} = 0$$

$$0.02\pi r = \frac{1.5}{r^2}$$

$$r^3 = \frac{1.5}{0.02\pi}$$

$$r^3 = \frac{75}{\pi}$$

$$r = \sqrt[3]{\frac{75}{\pi}}$$

$$r \approx 2.879$$

This is the only critical value in the interval $r > 0$.

The solution is continued on the next page.

Copyright © 2014 Pearson Education, Inc. Publishing as Addison-Wesley.

Next, we find the second derivative:

$$C''(r) = \frac{d}{dr}\left(2\pi r - \frac{1.5}{r^2}\right)$$
$$= 2\pi + \frac{3}{r^3}.$$

Substituting the critical value into the second derivative, we have:

$$C''(2.879) = 2\pi + \frac{3}{(2.879)^3}$$
$$\approx 6.4 > 0.$$

Since the second derivative is positive at the critical value, the critical value represents a relative minimum. We substitute the critical value back into $h = \frac{250}{\pi r^2}$ to determine the value for *h*. Notice, we substitute before the rounding of *r*.

$$h = \frac{250}{\pi(2.879411)^2}$$
$$\approx 9.598$$

The dimensions of the container that will minimize the cost of the container are a radius of 2.879 *in* and a height of 9.589 *in*.

49. Case I.
If *y* is the length, the girth is $x + x + x + x$, or $4x$.
Case II.
If *x* is the length, the girth is $x + y + x + y$, or $2x + 2y$.
Case I.
The combine length and girth is $y + 4x = 84$.
The volume is $V = x \cdot x \cdot y = x^2 y$.
We want express *V* in terms of one variable.
We solve $y + 4x = 84$, for *y*.

$$y + 4x = 84$$
$$y = 84 - 4x$$

Thus,
$V = x^2(84 - 4x) = 84x^2 - 4x^3$.
To maximize $V(x)$ we first find $V'(x)$.
$V'(x) = 168x - 12x^2$
This derivative exists for all *x*, so the critical values will occur when $V'(x) = 0$

We solve the equation

$$V'(x) = 0$$
$$168x - 12x^2 = 0$$
$$12x(14 - x) = 0$$
$$12x = 0 \quad \text{or} \quad 14 - x = 0$$
$$x = 0 \quad \text{or} \quad x = 14.$$

Since $x \neq 0$, the only critical value is $x = 14$.
We can use the second derivative,
$V'(x) = 168 - 24x$,
to determine whether we have a maximum.
$V'(14) = 168 - 24(14) = -168 < 0$
Therefore, we have a maximum at $x = 14$.
If $x = 14$, then $y = 84 - 4(14) = 28$.
Therefore, the dimensions that will maximize the volume of the package are 14 in. by 14 in. by 28 in. The volume is $14 \times 14 \times 28 = 5488 \text{ in}^3$.
Case II.
The combine length and girth is $x + 2x + 2y = 3x + 2y = 84$.
The volume is $V = x \cdot x \cdot y = x^2 y$.
We want express *V* in terms of one variable.
We solve $3x + 2y = 84$, for *y*.

$$3x + 2y = 84$$
$$2y = 84 - 3x$$
$$y = 42 - \frac{3}{2}x$$

Thus,

$$V = x^2\left(42 - \frac{3}{2}x\right) = 42x^2 - \frac{3}{2}x^3.$$

To maximize $V(x)$ we first find $V'(x)$.

$$V'(x) = 84x - \frac{9}{2}x^2$$

This derivative exists for all *x*, so the critical values will occur when $V'(x) = 0$; therefore, we solve that equation.

$$84x - \frac{9}{2}x^2 = 0$$
$$3x\left(28 - \frac{3}{2}x\right) = 0$$
$$3x = 0 \quad \text{or} \quad 28 - \frac{3}{2}x = 0$$
$$x = 0 \quad \text{or} \quad -\frac{3}{2}x = -28$$
$$x = 0 \quad \text{or} \quad x = \frac{56}{3}$$

The solution is continued on the next page.

Copyright © 2014 Pearson Education, Inc. Publishing as Addison-Wesley.

Since $x \neq 0$, the only critical value is $x = \frac{56}{3}$.

We can use the second derivative,
$V''(x) = 84 - 9x$,
to determine whether we have a maximum.
$V''\left(\frac{56}{3}\right) = 84 - 9\left(\frac{56}{3}\right) = -84 < 0$

Therefore, we have a maximum at $x = \frac{56}{3}$.

If $x = \frac{56}{3} \approx 18.67$, then $y = 42 - \frac{3}{2}\left(\frac{56}{3}\right) = 14$.
Therefore, the dimensions that will maximize the volume of the package are 18.67 in. by 18.67 in. by 14 in. The volume is
$\frac{56}{3} \times \frac{56}{3} \times 14 \approx 4878.2 \text{ in}^3$.
Comparing Case I and Case II, we see that the maximum volume is 5488 in^3 when the dimensions are 14 in. by 14 in. by 28 in.

51. Use the figure in the text book. Since the radius of the window is *x*, the diameter of the window is $2x$, which is also the length of the base of the window.
The circumference of a circle whose radius is *x* is given by:
$C = 2\pi x. \qquad (C = 2\pi r)$
Therefore, the perimeter of the semicircle is
$\frac{1}{2}C = \frac{2\pi x}{2} = \pi x.$
The perimeter of the three sides of the rectangle which form the remaining part of the total perimeter of the window is given by:
$2x + y + y = 2x + 2y.$
The total perimeter of the window is:
$\pi x + 2x + 2y = 24$.
Maximizing the amount of light is the same as maximizing the area of the window. The area of the circle with radius *x* is:
$A = \pi x^2, \qquad (A = \pi r^2).$
Therefore, the area of the semicircle is:
$\frac{1}{2}A = \frac{1}{2}\pi x^2.$
The area of the rectangle is $2x \cdot y$.
The total area of the Norman window is
$A = \frac{1}{2}\pi x^2 + 2xy.$

To express *A* in terms of one variable, we solve $\pi x + 2x + 2y = 24$ for *y*:
$$\pi x + 2x + 2y = 24$$
$$2y = 24 - 2x - \pi x$$
$$y = 12 - x - \frac{\pi}{2}x.$$
Then,
$$A(x) = \frac{1}{2}\pi x^2 + 2x\left(12 - x - \frac{\pi}{2}x\right)$$
$$= \frac{1}{2}\pi x^2 + 24x - 2x^2 - \pi x^2$$
$$= \left(-\frac{1}{2}\pi - 2\right)x^2 + 24x.$$
We maximize *A* on the interval $(0, 24)$. We first find $A'(x)$.
$A'(x) = (-\pi - 4)x + 24$.
Since $A'(x)$ exists for all *x* in $(0, 24)$, the only critical points are where $A'(x) = 0$. Thus, we solve the following equation:
$$A'(x) = 0$$
$$(-\pi - 4)x + 24 = 0$$
$$(-\pi - 4)x = -24$$
$$x = \frac{-24}{(-\pi - 4)}$$
$$x = \frac{-24}{-(\pi + 4)}$$
$$x = \frac{24}{\pi + 4} \approx 3.36.$$
This is the only critical value, so we can use the second derivative to determine whether we have a maximum.
$A''(x) = -\pi - 4 < 0$
Since $A''(x)$ is negative for all values of *x*, we have a maximum at $x = \frac{24}{\pi + 4}$.

The solution is continued on the next page.

Copyright © 2014 Pearson Education, Inc. Publishing as Addison-Wesley.

We find y when $x=\frac{24}{\pi+4}$:

$$y=12-\frac{\pi}{2}x-x$$
$$=12-\frac{\pi}{2}\left(\frac{24}{\pi+4}\right)-\frac{24}{\pi+4}$$
$$=12\left(\frac{\pi+4}{\pi+4}\right)-\frac{12\pi}{\pi+4}-\frac{24}{\pi+4}$$
$$=\frac{12\pi+48-12\pi-24}{\pi+4}$$
$$=\frac{24}{\pi+4}\approx 3.36.$$

To maximize the amount of light through the window, the dimensions must be $x=\frac{24}{\pi+4}$ ft and $y=\frac{24}{\pi+4}$ ft, or approximately $x\approx 3.36$ ft and $y\approx 3.36$ ft.

53. Let x represent a positive number. Then, $\frac{1}{x}$ is the reciprocal of the number, and x^2 is the square of the number. The sum, S, of the reciprocal and five times the square is given by:

$$S(x)=\frac{1}{x}+5x^2.$$

We want to minimize $S(x)$ on the interval $(0,\infty)$. First, we find $S'(x)$

$$S'(x)=-x^{-2}+10x=-\frac{1}{x^2}+10x.$$

Since $S'(x)$ exists for all values of x in $(0,\infty)$, the only critical values occur when $S'(x)=0$. We solve the following equation:

$$-\frac{1}{x^2}+10x=0$$
$$10x=\frac{1}{x^2}$$
$$10x^3=1$$
$$x^3=\frac{1}{10}$$
$$x=\sqrt[3]{\frac{1}{10}}=\frac{1}{\sqrt[3]{10}}.$$

Since there is only one critical value, we use the second derivative,

$$S''(x)=2x^{-3}+10=\frac{2}{x^3}+10,$$

to determine whether it is a minimum. The second derivative is positive for all x in $(0,\infty)$; therefore, the sum is a minimum when

$$x=\frac{1}{\sqrt[3]{10}}.$$

55. Let A represent the amount deposited in savings accounts and i represent the interest rate paid on the money deposited. If A is directly proportional to i, then there is some positive constant k such that $A=ki$. The interest earned by the bank is represented by $18\%A$, or $0.18A$. The interest paid by the bank is represented by iA. Thus the profit received by the bank is given by

$$P=0.18A-iA.$$

We express P as a function of the interest the bank pays on the money deposited, i, by substituting ki for A.

$$P=0.18(ki)-i(ki)$$
$$=0.18ki-ki^2$$

We maximize P on the interval $(0,\infty)$. First, we find $P'(i)$.

$$P'(i)=0.18k-2ki$$

Since $P'(i)$ exists for all i in $(0,\infty)$, the only critical values are where $P'(i)=0$.
We solve the following equation:

$$0.18k-2ki=0$$
$$-2ki=-0.18k$$
$$i=\frac{-0.18k}{-2k}$$
$$i=0.09.$$

Since there is only one critical point, we can use the second derivative to determine whether we have a minimum. Notice that $P''(i)=-2k$, which is a negative constant $(k>0)$. Thus, $P''(0.09)$ is negative, so $P(0.09)$ is a maximum. To maximize profit, the bank should pay 9% on its savings accounts.

Copyright © 2014 Pearson Education, Inc. Publishing as Addison-Wesley.

57. Using the drawing in the text, we write a function that gives the cost of the power line. The length of the power line on the land is given by $4-x$, so the cost of laying the power line underground is given by:

$$C_L(x)=3000(4-x)=12{,}000-3000x.$$

The length of the power line that will be under water is $\sqrt{1+x^2}$, so the cost of laying the power line underwater is given by:

$$C_W(x)=5000\sqrt{1+x^2}.$$

Therefore, the total cost of laying the power line is:

$$C(x)=C_L(x)+C_W(x)$$
$$=12{,}000-3000x+5000\sqrt{1+x^2}.$$

We want to minimize $C(x)$ over the interval $0\le x\le 4$. First, we find the derivative.

$$C'(x)=-3000+5000\left(\frac{1}{2}\right)\left(1+x^2\right)^{-1/2}(2x)$$
$$=-3000+5000x\left(1+x^2\right)^{-1/2}$$
$$=-3000+\frac{5000x}{\sqrt{1+x^2}}$$

Since the derivative exists for all x, we find the critical values by solving the equation:

$$C'(x)=0$$
$$-3000+\frac{5000x}{\sqrt{1+x^2}}=0$$
$$-3000\sqrt{1+x^2}+5000x=0$$
$$5000x=3000\sqrt{1+x^2}$$
$$\frac{5}{3}x=\sqrt{1+x^2}$$
$$\left(\frac{5}{3}x\right)^2=\left(\sqrt{1+x^2}\right)^2$$
$$\frac{25}{9}x^2=1+x^2$$
$$\frac{16}{9}x^2=1$$
$$x^2=\frac{9}{16}$$
$$x=\pm\sqrt{\frac{9}{16}}$$
$$x=\pm\frac{3}{4}.$$

The only critical value in the interval $[0,4]$ is $x=\frac{3}{4}$, so we can use the second derivative to determine if we have a minimum.

$$C''(x)=\frac{\left(1+x^2\right)^{1/2}(5000)-5000x\left[\frac{1}{2}\left(1+x^2\right)^{-1/2}(2x)\right]}{\left[\left(1+x^2\right)^{1/2}\right]^2}$$
$$=\frac{5000\sqrt{1+x^2}-\frac{5000x^2}{\sqrt{1+x^2}}}{\left(1+x^2\right)}$$
$$=\frac{5000}{\sqrt{1+x^2}}-\frac{5000x^2}{\left(1+x^2\right)^{3/2}}$$
$$=\frac{5000\left(1+x^2\right)-5000x^2}{\left(1+x^2\right)^{3/2}}$$
$$=\frac{5000}{\left(1+x^2\right)^{3/2}}$$

$C''(x)$ is positive for all x in $[0,4]$; therefore, a minimum occurs at $x=\frac{3}{4}$. When $x=\frac{3}{4}$,

$$4-\frac{3}{4}=\frac{13}{4}=3.25.$$

Therefore, S should be 3.25 miles down shore from the power station.

Note: since we are minimizing cost over a closed interval, we could have used Max-Min Principle 1 to determine the minimum. The critical value and the endpoints are $0, \frac{3}{4}$, and 4. The function values at these three points are:

$$C(0)=12{,}000-3000(0)+5000\left(\sqrt{1+(0)^2}\right)$$
$$=17{,}000$$
$$C\left(\frac{3}{4}\right)=12{,}000-3000\left(\frac{3}{4}\right)+5000\left(\sqrt{1+\left(\frac{3}{4}\right)^2}\right)$$
$$=16{,}000$$
$$C(4)=12{,}000-3000(4)+5000\left(\sqrt{1+(4)^2}\right)$$
$$\approx 20{,}615.53.$$

Therefore, the minimum occurs when $x=\frac{3}{4}$, or when S is 3.25 miles down shore from the power station.

Copyright © 2014 Pearson Education, Inc. Publishing as Addison-Wesley.

59. Using the drawing in the text, we write a function which gives the total distance between the cities.
The distance from C_1 to the bridge can be given by $\sqrt{a^2+(p-x)^2}$. The distance over the bridge is r. The distance from the bridge to C_2 can be given by $\sqrt{b^2+x^2}$. Therefore, the total distance between the two cities is given by:

$$D(x)=\sqrt{a^2+(p-x)^2}+r+\sqrt{x^2+b^2}$$

To minimize the distance, we find the derivative of the function first.

$$D'(x)=\frac{1}{2}\left[a^2+(p-x)^2\right]^{-1/2}\cdot 2(p-x)(-1)+\frac{1}{2}\left[b^2+x^2\right]^{-1/2}(2x)$$

$$=\frac{x-p}{\sqrt{a^2+(p-x)^2}}+\frac{x}{\sqrt{b^2+x^2}}$$

The derivative exists for all values of x in the interval $[0,p]$. Therefore, the only critical values occur when $D'(x)=0$. We solve this equation.

$$\frac{x-p}{\sqrt{a^2+(p-x)^2}}+\frac{x}{\sqrt{b^2+x^2}}=0.$$

The solution to this equation is

$x=\frac{bp}{b-a}$ or $x=\frac{bp}{b+a}$.

Only $x=\frac{bp}{b+a}$ is in $[0,p]$.

Since there is only one critical value, we can use the second derivative to determine if there is a minimum.
The second derivative is given by:

$$D''(x)=\frac{a^2}{\left[a^2+(p-x)^2\right]^{3/2}}+\frac{b^2}{\left[x^2+b^2\right]^{3/2}}.$$

$D''(x)>0$ for all values of x; therefore, a minimum occurs at $x=\frac{bp}{b+a}$.

The bridge should be located such that the distance x is $\frac{bp}{b+a}$ units.

61. $A(x)=\frac{C(x)}{x}$

a) Taking the derivative of $A(x)$ we have:

$$A'(x)=\frac{d}{dx}\left[\frac{C(x)}{x}\right]$$

$$=\frac{x\cdot C'(x)-C(x)\cdot 1}{x^2} \quad \text{Quotient Rule}$$

$$=\frac{x\cdot C'(x)-C(x)}{x^2}$$

b) The derivative exists for all x in $(0,\infty)$, therefore, the critical values will occur when $A'(x_0)=0$. We solve the equation:

$$\frac{x_0\cdot C'(x_0)-C(x_0)}{x_0^2}=0$$

$$x_0\cdot C'(x_0)-C(x_0)=0 \quad \text{Multiplying by } x_0^2\neq 0.$$

$$x_0\cdot C'(x_0)=C(x_0)$$

$$C'(x_0)=\frac{C(x_0)}{x_0}=A(x_0)$$

63. Express Q as a function of one variable. First, solve $x^2+y^2=2$ for y. We have:

$$y^2=2-x^2$$

$$y=\pm\sqrt{2-x^2}$$

y is a real number of x in the interval $\left[-\sqrt{2},\sqrt{2}\right]$.

If $y=-\sqrt{2-x^2}$, we substitute for y to get:

$$Q=3x+y^3$$

$$Q=3x+\left(-\sqrt{2-x^2}\right)^3$$

$$=3x-\left(2-x^2\right)^{3/2}.$$

Next, we find $Q'(x)$.

$$Q'(x)=3-\left(\frac{3}{2}\right)\left(2-x^2\right)^{1/2}(-2x)$$

$$=3+3x\left(2-x^2\right)^{1/2}$$

$$=3+3x\sqrt{2-x^2}.$$

The derivative exists for all values of x in the interval $\left[-\sqrt{2},\sqrt{2}\right]$; thus, the only critical values are where $Q'(x)=0$.

The solution is continued on the next page.

Copyright © 2014 Pearson Education, Inc. Publishing as Addison-Wesley.

We solve:

$$Q'(x)=0$$
$$3+3x\sqrt{2-x^2}=0$$
$$3=-3x\sqrt{2-x^2}$$
$$1=-x\sqrt{2-x^2}$$
$$1^2=\left(-x\sqrt{2-x^2}\right)^2$$
$$1=x^2\left(2-x^2\right)$$
$$1=2x^2-x^4$$
$$x^4-2x^2+1=0$$
$$\left(x^2-1\right)^2=0$$
$$x^2-1=0$$
$$x^2=1$$
$$x=\pm 1.$$

We notice that $x=1$ is an extraneous solution which does not work.

$$3+3(1)\sqrt{2-(1)^2}=3+3=6\neq 0.$$

Therefore, the only critical value is $x=-1$. The critical point and the endpoints are $-\sqrt{2},\ -1$, and $\sqrt{2}$.

$$Q\left(-\sqrt{2}\right)=3\left(-\sqrt{2}\right)-\left(2-\left(-\sqrt{2}\right)^2\right)^{3/2}=-3\sqrt{2}$$

$$Q(-1)=3(-1)-\left(2-(-1)^2\right)^{3/2}=-4$$

$$Q\left(\sqrt{2}\right)=3\left(\sqrt{2}\right)-\left(2-\left(\sqrt{2}\right)^2\right)^{3/2}=3\sqrt{2}.$$

The minimum value of Q is $-3\sqrt{2}$ and occurs when $x=-\sqrt{2}$ and $y=-\sqrt{2-\left(-\sqrt{2}\right)^2}=0.$

Next, we repeat the process for $y=\sqrt{2-x^2}$. We notice that:

$$Q=3x+y^3$$
$$Q=3x+\left(\sqrt{2-x^2}\right)^3$$
$$=3x+\left(2-x^2\right)^{3/2}.$$

Next, we find $Q'(x)$.

$$Q'(x)=3+\left(\frac{3}{2}\right)\left(2-x^2\right)^{1/2}(-2x)$$
$$=3-3x\left(2-x^2\right)^{1/2}$$
$$=3-3x\sqrt{2-x^2}.$$

The derivative exists for all values of x in the interval $\left[-\sqrt{2},\sqrt{2}\right]$; thus, the only critical values are where $Q'(x)=0$. We solve the equation:

$$3-3x\sqrt{2-x^2}=0$$
$$3=3x\sqrt{2-x^2}$$
$$1=x\sqrt{2-x^2}$$
$$1^2=\left(x\sqrt{2-x^2}\right)^2$$
$$1=x^2\left(2-x^2\right)$$
$$1=2x^2-x^4$$
$$x^4-2x^2+1=0$$
$$\left(x^2-1\right)^2=0$$
$$x^2-1=0$$
$$x^2=1$$
$$x=\pm 1.$$

We notice that $x=-1$ is an extraneous solution which does not work.

$$3-3(-1)\sqrt{2-(-1)^2}=3+3=6\neq 0.$$

Therefore, the only critical value is $x=1$. The critical point and the endpoints are $-\sqrt{2},\ 1$, and $\sqrt{2}$.

$$Q\left(-\sqrt{2}\right)=3\left(-\sqrt{2}\right)+\left(2-\left(-\sqrt{2}\right)^2\right)^{3/2}=-3\sqrt{2}$$

$$Q(1)=3(1)+\left(2-(1)^2\right)^{3/2}=4$$

$$Q\left(\sqrt{2}\right)=3\left(\sqrt{2}\right)+\left(2-\left(\sqrt{2}\right)^2\right)^{3/2}=3\sqrt{2}.$$

The minimum value of Q is $-3\sqrt{2}$ occurs when $x=-\sqrt{2}$ and $y=-\sqrt{2-\left(-\sqrt{2}\right)^2}=0$.

Regardless of what value of y we chose, we see that the minimum of Q, is $-3\sqrt{2}$, when $x=-\sqrt{2}$ and $y=0.$

Copyright © 2014 Pearson Education, Inc. Publishing as Addison-Wesley.

65. From Exercise 60, we know that the store should order a lot size of $\sqrt{\frac{2bQ}{a}}$ units, $\sqrt{\frac{aQ}{2b}}$ times per year.
When $Q = 2500,\ a = 10,\ b = 20,\ c = 9$, the store should order:
$$\sqrt{\frac{aQ}{2b}} = \sqrt{\frac{10(2500)}{2(20)}} = 25 \text{ times per year.}$$
The lot size of each order should be:
$$\sqrt{\frac{2(20)(2500)}{10}} = 100 \text{ units.}$$

Copyright © 2014 Pearson Education, Inc. Publishing as Addison-Wesley.

Exercise Set 2.6

1. $R(x)=5x;\ C(x)=0.001x^2+1.2x+60$

a) Total profit is revenue minus cost.

$$P(x)=R(x)-C(x)$$
$$=5x-\left(0.001x^2+1.2x+60\right)$$
$$=5x-0.001x^2-1.2x-60$$
$$=-0.001x^2+3.8x-60$$

b) Substituting 100 for x into the three functions, we have:

$R(100)=5(100)=500.$

The total revenue from the sale of the first 50 units is $500.

$C(100)=0.001(100)^2+1.2(100)+60=190$

The total cost of producing the first 100 units is $190.

$$P(100)=R(100)-C(100)$$
$$=500-190$$
$$=310$$

The total profit is $310 when the first 100 units are produced and sold.
Note, we could have also used the profit function, $P(x)$, from part (a) to find the profit.

$P(100)=-0.001(100)^2+3.8(100)-60=310$

c) Finding the derivative for each of the functions, we have:

$R'(x)=5$

$C'(x)=0.002x+1.2$

$P'(x)=-0.002x+3.8.$

d) Substituting 100 for x in each of the three marginal functions, we have:

$R'(100)=5.$

Once 100 units have been sold, the approximate revenue for the 51st unit is $5.

$C'(100)=1.4$

Once 100 units have been produced, the approximate cost for the 101st unit is $1.40.

$P'(100)=-0.002(100)+3.8=3.6$

Once 100 units have been produced and sold, the approximate profit from the sale of the 101st unit is $3.60.

e) ✎

3. $C(x)=0.001x^3+0.07x^2+19x+700$

a) Substituting 25 for x into the cost function, we have:

$$C(25)=0.001(25)^3+0.07(25)^2+19(25)+700$$
$$=1234.375.$$

The current monthly cost of producing 25 chairs is $1234.38.

b) In order to find the additional cost of producing 26 chairs monthly, we first find the total cost of producing 26 chairs in a month.

$$C(26)=0.001(26)^3+0.07(26)^2+19(26)+700$$
$$=1258.896$$

Next, we subtract the cost of producing 25 chairs monthly found in part (a) from the cost of producing 26 chairs monthly.

$$C(26)-C(25)=1258.896-1234.375$$
$$=24.521$$

The additional cost of increasing production to 26 chairs monthly is $24.52.

c) First, we find the marginal cost function,

$C'(x)=0.003x^2+0.14x+19.$

Next, we substituting 25 for x, we have:

$$C'(25)=0.003(25)^2+0.14(25)+19$$
$$=24.375.$$

The marginal cost when 25 chairs have been produced is $24.38.

d) Using the marginal cost from part (c), the additional cost required to produce 2 additional chairs monthly is:

$2(24.375)=48.75.$

Therefore, the difference in cost between producing 25 and 27 chairs per month is approximately $48.75.

e) In part (a) we found that it cost 1234.38 to produce 25 chairs per month. In part (d) we found that the difference in cost between 25 chairs and 27 chairs per month was $48.75. Therefore, the approximate total cost of producing 27 chairs per month is

$C(27)\approx 1234.38+48.75=1283.13.$

We predict the cost of producing 27 chairs monthly will be $1283.13.

Copyright © 2014 Pearson Education, Inc. Publishing as Addison-Wesley.

5. $R(x) = 0.005x^3 + 0.01x^2 + 0.5x$

a) Substituting 70 for x, we have:

$$R(70) = 0.005(70)^3 + 0.01(70)^2 + 0.5(70)$$
$$= 1715 + 49 + 35$$
$$= 1799.$$

The currently daily revenue from selling 70 lawn chairs per day is \$1799.

b) Substituting 73 for x, we have:

$$R(73) = 0.005(73)^3 + 0.01(73)^2 + 0.5(73)$$
$$= 2034.875$$
$$= 2034.88.$$

Therefore, the increase in revenue from increasing sales to 73 chairs per day is:

$$R(73) - R(70) = 2034.88 - 1799$$
$$= 235.88.$$

Revenue will increase \$235.88 per day if the number of chairs sold increases to 73 per day.

c) First we find the marginal revenue function by finding the derivative of the revenue function.

$$R'(x) = 0.015x^2 + 0.02x + 0.5$$

Substituting 70 for x, we have:

$$R'(70) = 0.015(70)^2 + 0.02(70) + 0.5$$
$$= 75.40.$$

The marginal revenue when 70 lawn chairs are sold daily is \$75.40.

d) In part (a) we found that selling 70 lawn chairs per day resulted in a revenue of \$1799. In part (c) we found that the marginal revenue when 70 chairs were sold is \$75.40. Using these two numbers, we estimate the daily revenue generated by selling 71 chairs is

$$R(71) \approx R(70) + R'(70)$$
$$= \$1799 + \$75.40 = \$1874.40.$$

Similarly, the daily revenue generated by selling 72 chairs, or 2 additional chairs, daily is approximately

$$R(72) \approx R(70) + 2 \cdot R'(70)$$
$$\approx \$1799 + 2(\$75.40) \approx \$1949.80.$$

The daily revenue generated by selling 73 chairs, or 3 additional chairs, daily is approximately

$$R(73) \approx R(70) + 3 \cdot R'(70)$$
$$\approx \$1799 + 3(\$75.40) \approx \$2025.20.$$

7. $P(x) = -0.004x^3 - 0.3x^2 + 600x - 800$

a) Substituting 9 for x, we have:

$$P(9) = -0.004(9)^3 - 0.3(9)^2 + 600(9) - 800$$
$$= -2.916 - 24.30 + 5400 - 800$$
$$= 4572.784.$$

The currently weekly profit is \$4572.78.

b) First, we find the total weekly profit of selling 8 laptops per week.

$$P(8) = -0.004(8)^3 - 0.3(8)^2 + 600(8) - 800$$
$$= 3978.752.$$

The difference in weekly profit from selling 8 laptops and 9 laptops per week is

$$P(9) - P(8) = 4572.78 - 3978.75$$
$$= 594.03.$$

Therefore, Crawford Computing would lose \$594.03 each week if 8 laptops were sold each week instead of 9.

c) First, we find the marginal profit function by taking the derivative of the profit function.

$$P'(x) = -0.012x^2 - 0.6x + 600$$

Substituting 9 for x, we have:

$$P'(9) = -0.012(9)^2 - 0.6(9) + 600$$
$$= 593.628.$$

The marginal profit is \$593.63 when 9 laptops are sold weekly.

d) From part (a), we know that when 9 laptops are built and sold, total weekly profit is \$4572.78. From part (c), we know that when 9 laptops are built and sold, marginal profit is \$593.63. Therefore, we estimate:

$$P(10) \approx P(9) + P'(9)$$
$$\approx 4572.78 + 593.63$$
$$\approx 5166.41.$$

The total weekly profit is approximately \$5166.41 when 10 laptops are built and sold weekly.

9. $N(1000) = 500{,}000$ means that 500,000 computers will be sold annually when the price of the computer is \$1000. $N'(1000) = -100$ means that when the price is increased \$1 to \$1001, sales will decrease by 100 computers per year.

Copyright © 2014 Pearson Education, Inc. Publishing as Addison-Wesley.

11. $C(x)=0.01x^2+0.6x+30$

$\Delta C=C(x+\Delta x)-C(x)$

Substituting $x=70$, and $\Delta x=1$ we have

$$\Delta C=C(70+1)-C(70)$$
$$=C(71)-C(70)$$
$$=0.01(71)^2+0.6(71)+30-\left[0.01(70)^2+0.6(70)+30\right]$$
$$=2.01.$$

The additional cost of producing the 71st unit is \$2.01.

Finding the derivative of $C(x)$ we have:

$C'(x)=0.02x+0.6$

Substituting 70 for x, we have:

$C'(70)=0.02(70)+0.6=2.00$

The marginal cost when 70 units are produced is \$2.00.

13. $R(x)=2x$

$\Delta R=R(x+\Delta x)-R(x)$

Substituting $x=70$, and $\Delta x=1$ we have

$$\Delta R=R(70+1)-R(70)$$
$$=R(71)-R(70)$$
$$=2(71)-[2(70)]$$
$$=2.$$

The additional cost of producing the 71st unit is \$2.00.

Finding the derivative of $R(x)$ we have:

$R'(x)=2$.

The derivative is constant; therefore, $R'(70)=2$.

The marginal cost when 70 units are produced is \$2.00.

15. $C(x)=0.01x^2+0.6x+30;\ R(x)=2x$

a) Finding the profit function we have:

$$P(x)=R(x)-C(x)$$
$$=2x-\left(0.01x^2+0.6x+30\right)$$
$$=-0.01x^2+1.4x-30.$$

b) $\Delta P=P(x+\Delta x)-P(x)$

Substituting $x=70$, and $\Delta x=1$ we have

$$\Delta P=P(70+1)-P(70)$$
$$=P(71)-P(70)$$
$$=-0.01(71)^2+1.4(71)-30-\left[-0.01(70)^2+1.4(70)-30\right]$$
$$=-0.01.$$

The additional profit of producing and selling the 71st unit is −\$0.01.

Finding the derivative of $P(x)$ we have:

$P'(x)=-0.02x+1.4$

Substituting 70 for x, we have:

$P'(70)=-0.02(70)+1.4=0.00$

The marginal profit when 70 units are produced and sold is \$0.00.

17. $D=0.007p^3-0.5p^2+150p$

a) We take the derivative of the demand function with respect to price.

$$\frac{dD}{dp}=0.021p^2-p+150$$

b) Substituting 25 for p in the demand function we have:

$$D=0.007(25)^3-0.5(25)^2+150(25)$$
$$=109.375-312.50+3750$$
$$=3546.875.$$

Consumers will want to buy 3547 units when price is \$25 per unit.

c) ✎

d) ✎

19. $A(x)=\dfrac{13x+100}{x}$

To estimate the change in average cost as production goes from 100 to 101 units, we establish that $x=100$ and $\Delta x=1$. Next, we find the derivative of $A(x)$.

$$A'(x)=\frac{x(13)-(13x+100)(1)}{x^2}$$
$$=\frac{13x-13x-100}{x^2}$$
$$=-\frac{100}{x^2}.$$

The solution is continued on the next page.

Copyright © 2014 Pearson Education, Inc. Publishing as Addison-Wesley.

Therefore,

$$\begin{aligned}\Delta A &\approx A'(x)\Delta x\\ &\approx A'(100)\Delta x \qquad (x=100)\\ &\approx -\frac{100}{(100)^2}\Delta x\\ &\approx -\frac{100}{(100)^2}(1) \qquad (\Delta x=1)\\ &\approx -0.01.\end{aligned}$$

The average cost changes by about $-\$0.01$. (We see an approximate decrease in average cost of one cent.)

21. $P(x)=567+x\left(36x^{0.6}-104\right)$

$$=567+36x^{1.6}-104x$$

x is the number of years since 1960; therefore, the year 2009 corresponds to $x=2009-1960=49$, and the year 2010 corresponds to $x=2010-1960=50$. To estimate the increase in gross domestic product from 2009 to 2010, we establish that $x=49$ and $\Delta x=1$. Next, we find the derivative of $P(x)$:

$P'(x)=36(1.6)x^{0.6}-104=57.6x^{0.6}-104$.

Therefore,

$$\begin{aligned}\Delta P &\approx P'(x)\Delta x\\ &\approx P'(49)\Delta x \qquad [x=49]\\ &\approx \left(57.6(49)^{0.6}-104\right)\Delta x\\ &\approx (491.03173875)(1) \qquad [\Delta x=1]\\ &\approx 491.03.\end{aligned}$$

The gross domestic product should increase about \$491.03 billion between 2009 and 2010.

23. ✎

25. Alan's marginal tax rate is 21%, therefore for each additional dollar he earns, he will have to pay \$0.21 in taxes. If he earns another \$2000, dollars, we will pay an additional $\$2000(0.21)=\$420$ in taxes.

27. $y=f(x)=x^2$, $x=2$, and $\Delta x=0.01$

$$\begin{aligned}\Delta y &= f(x+\Delta x)-f(x)\\ &= f(2+0.01)-f(2) \quad \text{Substituting 2 for } x \text{ and } 0.01 \text{ for } \Delta x.\\ &= f(2.01)-f(2)\\ &= (2.01)^2-(2)^2\\ &= 0.0401\end{aligned}$$

$$f'(x)\Delta x=2x\cdot\Delta x \qquad \left[f(x)=x^2; f'(x)=2x\right]$$

$$f'(2)\Delta x=2(2)\cdot(0.01) \quad \text{Substituting 2 for } x \text{ and } 0.01 \text{ for } \Delta x.$$

$$\begin{aligned}&=4(0.01)\\ &=0.04\end{aligned}$$

29. $y=f(x)=x+x^2$, $x=3$, and $\Delta x=0.04$

$$\begin{aligned}\Delta y &= f(x+\Delta x)-f(x)\\ &= f(3+0.04)-f(3) \quad \text{Substituting 3 for } x \text{ and } 0.04 \text{ for } \Delta x.\\ &= f(3.04)-f(3)\\ &= \left[(3.04)+(3.04)^2\right]-\left[(3)+(3)^2\right]\\ &= [12.2816]-[12]\\ &= 0.2816\end{aligned}$$

$$f'(x)\Delta x=(1+2x)\cdot\Delta x \qquad \begin{bmatrix}f(x)=x+x^2;\\ f'(x)=1+2x\end{bmatrix}$$

$$f'(3)\Delta x=\left[1+2(3)\right]\cdot(0.04) \quad \text{Substituting 3 for } x \text{ and } 0.04 \text{ for } \Delta x.$$

$$\begin{aligned}&=[7](0.04)\\ &=0.28\end{aligned}$$

31. $y=f(x)=\frac{1}{x^2}=x^{-2}$, $x=1$, and $\Delta x=0.5$

$$\begin{aligned}\Delta y &= f(x+\Delta x)-f(x)\\ &= f(1+0.5)-f(1) \quad \text{Substituting 1 for } x \text{ and } 0.5 \text{ for } \Delta x.\\ &= f(1.5)-f(1)\\ &= \left[\frac{1}{(1.5)^2}\right]-\left[\frac{1}{(1)^2}\right]\\ &= \left[\frac{1}{2.25}\right]-[1]\\ &= -0.5556\end{aligned}$$

The solution is continued on the next page.

Copyright © 2014 Pearson Education, Inc. Publishing as Addison-Wesley.

Taking the derivative, we have:

$$f'(x)\Delta x = -2x^{-3}\cdot\Delta x \quad \begin{bmatrix} f(x)=x^{-2}; \\ f'(x)=-2x^{-3} \end{bmatrix}$$

$$f'(1)\Delta x = \left[-2(1)^{-3}\right]\cdot(0.5) \quad \text{Substituting 1 for } x \text{ and 0.5 for } \Delta x.$$

$$= [-2](0.5)$$
$$= -1.$$

33. $y = f(x) = 3x-1,\ x=4, \text{ and } \Delta x = 2$

$$\Delta y = f(x+\Delta x) - f(x)$$
$$= f(4+2) - f(4) \quad \text{Substituting 4 for } x \text{ and 2 for } \Delta x.$$
$$= f(6) - f(4)$$
$$= [3(6)-1] - [3(4)-1]$$
$$= [17] - [11]$$
$$= 6$$

$$f'(x)\Delta x = (3)\cdot\Delta x \quad [f(x)=3x-1; f'(x)=3]$$
$$f'(4)\Delta x = (3)\cdot(2) \quad \text{Substituting 4 for } x \text{ and 2 for } \Delta x.$$
$$= 6$$

35. We first think of the number closest to 26 that is a perfect square. This is 25. What we will do is approximate how $y = \sqrt{x}$, changes when x changes from 25 to 26. Let

$$y = f(x) = \sqrt{x} = x^{1/2}$$

Then $f'(x) = \frac{1}{2}x^{-1/2} = \frac{1}{2\sqrt{x}}$.

Using, $\Delta y \approx f'(x)\Delta x$, we have

$$\Delta y \approx f'(x)\Delta x$$
$$\approx \frac{1}{2\sqrt{x}}\cdot\Delta x.$$

We are interested in Δy as x changes from 25 to 26, so

$$\Delta y \approx \frac{1}{2\sqrt{x}}\cdot\Delta x$$
$$\approx \frac{1}{2\sqrt{25}}\cdot 1 \quad \text{Replacing } x \text{ with 25 and } \Delta x \text{ with 1}$$
$$\approx \frac{1}{2\cdot 5}$$
$$\approx \frac{1}{10} = 0.1.$$

We can now approximate $\sqrt{26}$;

$$\sqrt{26} = \sqrt{25} + \Delta y$$
$$= 5 + \Delta y$$
$$\approx 5 + 0.1$$
$$\approx 5.1.$$

To five decimal places $\sqrt{26} = 5.09902$. Thus, our approximation is reasonably accurate.

37. We first think of the number closest to 102 that is a perfect square. This is 100. What we will do is approximate how $y = \sqrt{x}$, changes when x changes from 100 to 102. Let

$$y = f(x) = \sqrt{x} = x^{1/2}$$

Then $f'(x) = \frac{1}{2}x^{-1/2} = \frac{1}{2\sqrt{x}}$

Using, $\Delta y \approx f'(x)\Delta x$, we have

$$\Delta y \approx f'(x)\Delta x$$
$$\approx \frac{1}{2\sqrt{x}}\cdot\Delta x.$$

We are interested in Δy as x changes from 100 to 102, so

$$\Delta y \approx \frac{1}{2\sqrt{x}}\cdot\Delta x$$
$$\approx \frac{1}{2\sqrt{100}}\cdot 2 \quad \text{Replacing } x \text{ with 100 and } \Delta x \text{ with 2}$$
$$\approx \frac{1}{2\cdot 10}\cdot 2$$
$$\approx \frac{1}{10} = 0.1.$$

We can now approximate $\sqrt{102}$;

$$\sqrt{102} = \sqrt{100} + \Delta y$$
$$= 10 + \Delta y$$
$$\approx 10 + 0.1$$
$$\approx 10.1.$$

To five decimal places $\sqrt{26} = 10.09950$. Thus, our approximation is reasonably accurate.

39. We first think of the number closest to 1005 that is a perfect cube. This is 1000. What we will do is approximate how $y = \sqrt[3]{x}$, changes when x changes from 1000 to 1005. Let

$$y = f(x) = \sqrt[3]{x} = x^{1/3}.$$

Then $f'(x) = \frac{1}{3}x^{-2/3} = \frac{1}{2\sqrt[3]{x^2}}$.

The solution is continued on the next page.

Copyright © 2014 Pearson Education, Inc. Publishing as Addison-Wesley.

Using, $\Delta y \approx f'(x)\Delta x$, we have

$$\Delta y \approx f'(x)\Delta x \approx \frac{1}{2\sqrt[3]{x^2}} \cdot \Delta x.$$

We are interested in Δy as x changes from 1000 to 1005, so

$$\Delta y \approx \frac{1}{3\sqrt[3]{x^2}} \cdot \Delta x$$

Replacing x with 1000 and Δx with 5, we have

$$\approx \frac{1}{3 \cdot \sqrt[3]{(1000)^2}} \cdot 5 \approx \frac{1}{3 \cdot 100} \cdot 5 \approx \frac{1}{60} = 0.017.$$

We can now approximate $\sqrt[3]{1005}$;

$$\sqrt[3]{1005} = \sqrt[3]{1000} + \Delta y = 10 + \Delta y \approx 10 + 0.017 \approx 10.017.$$

To five decimal places $\sqrt[3]{1000} = 10.01664$ Thus, our approximation is reasonably accurate.

41. $y = \sqrt{x+1} = (x+1)^{1/2}$

First, we find $\frac{dy}{dx}$:

$$\frac{dy}{dx} = \frac{1}{2}(x+1)^{-1/2}(1) = \frac{1}{2\sqrt{x+1}}.$$

Then

$$dy = \frac{1}{2\sqrt{x+1}} dx.$$

Note that the expression for dy contains two variables x and dx.

43. $y = \left(2x^3+1\right)^{3/2}$

First, we find $\frac{dy}{dx}$:

$$\frac{dy}{dx} = \frac{3}{2}\left(2x^3+1\right)^{1/2}\left(6x^2\right) \quad \text{By the extended power rule}$$

$$= 9x^2\left(2x^3+1\right)^{1/2} = 9x^2\sqrt{2x^3+1}.$$

Then

$$dy = 9x^2\sqrt{2x^3+1}\, dx.$$

Note that the expression for dy contains two variables x and dx.

45. $y = \sqrt[5]{x+27} = (x+27)^{1/5}$

First, we find $\frac{dy}{dx}$. By the extended power rule we have:

$$\frac{dy}{dx} = \frac{1}{5}(x+27)^{-4/5}(1) = \frac{1}{5 \cdot \sqrt[5]{(x+27)^4}}.$$

Then

$$dy = \frac{1}{5 \cdot \sqrt[5]{(x+27)^4}} dx.$$

Note that the expression for dy contains two variables x and dx.

47. $y = x^4 - 2x^3 + 5x^2 + 3x - 4$

First, we find $\frac{dy}{dx}$:

$$\frac{dy}{dx} = 4x^3 - 6x^2 + 10x + 3.$$

Then

$$dy = \left(4x^3 - 6x^2 + 10x + 3\right) dx.$$

Note that the expression for dy contains two variables x and dx.

49. From Exercise 47, we know:

$$dy = \left(4x^3 - 6x^2 + 10x + 3\right) dx.$$

When $x = 2$ and $dx = 0.1$ we have:

$$dy = \left(4(2)^3 - 6(2)^2 + 10(2) + 3\right)(0.1) = (32 - 24 + 20 + 3)(0.1) = (31)(0.1) = 3.1.$$

51. $y = (3x-10)^5$

First, we find $\frac{dy}{dx}$:

$$\frac{dy}{dx} = 5(3x-10)^4(3) \quad \text{By the extended power rule}$$

$$= 15(3x-10)^4.$$

Then

$$dy = 15(3x-10)^4\, dx.$$

The solution is continued on the next page.

Copyright © 2014 Pearson Education, Inc. Publishing as Addison-Wesley.

When $x = 4$ and $dx = 0.03$ we have:

$$dy = 15(3(4)-10)^4(0.03)$$
$$= 15(2)^4(0.03)$$
$$= 7.2.$$

53. Let $y = f(x) = x^4 - x^2 + 8$

First we find $f'(x)$:

$f'(x) = 4x^3 - 2x$.

Then

$$dy = f'(x)dx$$
$$= (4x^3 - 2x)dx.$$

To approximate $f(5.1)$, we will use $x = 5$ and $dx = 0.1$ to determine the differential dy.

Substituting 5 for x and 0.1 for dx we have:

$$dy = f'(5)dx$$
$$= (4(5)^3 - 2(5))(0.1)$$
$$= (4(125) - 10)(0.1)$$
$$= (500 - 10)(0.1)$$
$$= (490)(0.1)$$
$$= 49.$$

Next, we find

$$f(5) = (5)^4 - (5)^2 + 8$$
$$= 625 - 25 + 8$$
$$= 608.$$

Now,

$$f(5.1) \approx f(5) + f'(5)dx$$
$$\approx 608 + 49$$
$$\approx 657.$$

55. $S = 0.02235h^{0.42246}w^{0.51456}$

We begin by noticing that we are wanting to estimate the change in surface area due to a change in weight w; therefore, we will first find $\frac{dS}{dw}$. Since $h = 160$, we have:

$$S = 0.02235(160)^{0.42246}w^{0.51456}$$
$$= 0.02235(8.53399783)w^{0.51456}$$
$$= 0.19073485w^{0.51456}.$$

Now we can take the derivative of S with respect to w.

$$\frac{dS}{dw} = 0.19073485(0.51456)w^{-0.48544}$$
$$= 0.09814452w^{-0.48544}$$

Therefore,

$$dS = \left(0.09814452w^{-0.48544}\right)dw.$$

Now that we have the differential, we can use her weight of 60 kg to approximate how much her surface area changes when her weight drops 1 kg.

We substitute 60 for w and -1 for dw to get:

$$dS \approx \left(0.09814452(60)^{-0.48544}\right)(-1)$$
$$\approx -0.01345.$$

The patient's surface area will change by $-0.01345\ \text{m}^2$.

57. $N(t) = \frac{0.8t + 1000}{5t + 4}$

First we find $N'(t)$ by the quotient rule.

$$N'(t) = \frac{(5t+4)(0.8) - (0.8t+1000)(5)}{(5t+4)^2}$$
$$= \frac{4t + 3.2 - 4t - 5000}{(5t+4)^2}$$
$$= -\frac{4996.8}{(5t+4)^2}$$

The differential is:

$$dN = N'(t)dt$$
$$= -\frac{4996.8}{(5t+4)^2} \cdot dt.$$

We approximate the change in bodily concentration from 1.0 hr to 1.1 hr by using 1.0 for t and 0.1 for dt.

$$dN = -\frac{4996.8}{(5(1.0)+4)^2} \cdot (0.1)$$
$$= -\frac{4996.8}{(9)^2}(0.1)$$
$$= -\frac{4996.8}{81}(0.1)$$
$$\approx -61.6889(0.1)$$
$$\approx -6.16889$$

Next, we approximate the change in bodily concentration from 2.8 hr to 2.9 hr by using 2.8 for t and 0.1 for dt.

The solution is continued on the next page.

Copyright © 2014 Pearson Education, Inc. Publishing as Addison-Wesley.

Substituting, we have:

$$dN = -\frac{4996.8}{(5(2.8)+4)^2}\cdot(0.1)$$
$$= -\frac{4996.8}{(18)^2}(0.1)$$
$$= -\frac{4996.8}{324}(0.1)$$
$$\approx -15.4222(0.1)$$
$$\approx -1.54222.$$

The concentration changes more from 1.0 hr to 1.1 hr.

59. The circumference of the earth, which is the original length of the rope, is given by $C(r) = 2\pi r$, where r is the radius of the earth. We need to find the change in the length of the radius, Δr, when the length of the rope is increased 10 feet.

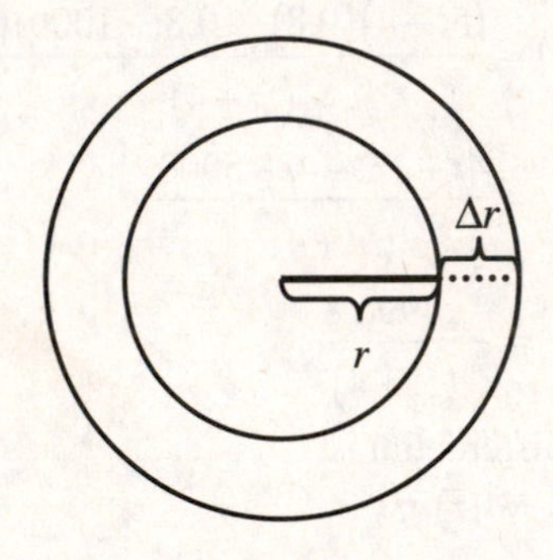

Using differentials, $\Delta C \approx C'(r)\Delta r$ represents the change in the length of the rope. Therefore, $\Delta C = 10$, and we have:

$$10 = C'(r)\Delta r$$

Noticing that $C'(r) = 2\pi$, we have:

$$10 = 2\pi \cdot \Delta r$$
$$\frac{10}{2\pi} = \Delta r.$$

Therefore, the rope is raised approximately $\Delta r = \frac{5}{\pi} \approx 1.59$ feet above the earth.

61. a) The area of the water tank is given by $A = 2\pi r^2$.
Calculating the differential we have:

$$dA = A'(r)dr$$
$$= 4\pi r dr.$$

The tolerance is in feet is ± 0.5.
Substituting, we have:

$$dA = 4(3.14)(100)(\pm 0.5)$$
$$\approx \pm 628.$$

The approximate difference in surface area when the tolerance is taken into consideration is ± 628 square feet.

b) The possible extra area is 628 square feet. Since each additional can will cover 300 square feet, they will need to bring 3 extra cans to account for any extra area.

c) 3 additional cans will cost the painters \$90.

63. $p = 100 - \sqrt{x}$

Since revenue is price times quantity, the revenue function is given by:

$$R(x) = p\cdot x$$
$$= (100 - \sqrt{x})x$$
$$= 100x - x^{3/2}.$$

To find the marginal revenue, we take the derivative of the revenue function. Thus:

$$R'(x) = 100 - \frac{3}{2}x^{1/2}$$
$$= 100 - \frac{3\sqrt{x}}{2}.$$

65. $p = 500 - x$

Since revenue is price times quantity, the revenue function is given by:

$$R(x) = p\cdot x$$
$$= (500 - x)x$$
$$= 500x - x^2$$

To find the marginal revenue, we take the derivative of the revenue function. Thus:

$$R'(x) = 500 - 2x.$$

67. $p = \frac{3000}{x} + 5$

Since revenue is price times quantity, the revenue function is given by:

$$R(x) = p\cdot x$$
$$= \left(\frac{3000}{x} + 5\right)x$$
$$= 3000 + 5x.$$

To find the marginal revenue, we take the derivative of the revenue function. Thus:

$$R'(x) = 5.$$

69. ✎

Copyright © 2014 Pearson Education, Inc. Publishing as Addison-Wesley.

Exercise Set 2.7

1. Differentiate implicitly to find $\frac{dy}{dx}$.

We have

$x^3 + 2y^3 = 6.$

Differentiating both sides with respect to x yields:

$$\frac{d}{dx}\left(x^3 + 2y^3\right) = \frac{d}{dx}(6)$$

$$\frac{d}{dx}x^3 + 2\frac{d}{dx}y^3 = \frac{d}{dx}6$$

$$3x^2 + 2\cdot 3y^2 \cdot \frac{dy}{dx} = 0.$$

Next, we isolate $\frac{dy}{dx}$

$$6y^2 \cdot \frac{dy}{dx} = -3x^2$$

$$\frac{dy}{dx} = \frac{-3x^2}{6y^2}$$

$$\frac{dy}{dx} = \frac{-x^2}{2y^2}.$$

Find the slope of the tangent line to the curve at $(2,-1)$.

$$\frac{dy}{dx} = \frac{-x^2}{2y^2}$$

Replacing x with 2 and y with -1, we have:

$$\frac{dy}{dx} = \frac{-(2)^2}{2(-1)^2} = \frac{-4}{2} = -2.$$

The slope of the tangent line to the curve at $(2,-1)$ is -2.

3. Differentiate implicitly to find $\frac{dy}{dx}$.

We have

$2x^2 - 3y^3 = 5.$

Differentiating both sides with respect to x yields:

$$\frac{d}{dx}\left(2x^2 - 3y^3\right) = \frac{d}{dx}(5)$$

$$\frac{d}{dx}2x^2 - 3\frac{d}{dx}y^3 = \frac{d}{dx}5$$

$$2\cdot 2x - 3\cdot 3y^2 \cdot \frac{dy}{dx} = 0.$$

Next, we isolate $\frac{dy}{dx}$

$$-9y^2 \cdot \frac{dy}{dx} = -4x$$

$$\frac{dy}{dx} = \frac{-4x}{-9y^2}$$

$$\frac{dy}{dx} = \frac{4x}{9y^2}.$$

Find the slope of the tangent line to the curve at $(-2,1)$.

$$\frac{dy}{dx} = \frac{4x}{9y^2}$$

Replacing x with -2 and y with 1, we have:

$$\frac{dy}{dx} = \frac{4(-2)}{9(1)^2} = \frac{-8}{9} = -\frac{8}{9}.$$

The slope of the tangent line to the curve at $(-2,1)$ is $-\frac{8}{9}$.

5. Differentiate implicitly to find $\frac{dy}{dx}$.

We have

$x^2 - y^2 = 1.$

Differentiating both sides with respect to x yields:

$$\frac{d}{dx}\left(x^2 - y^2\right) = \frac{d}{dx}(1)$$

$$\frac{d}{dx}x^2 - \frac{d}{dx}y^2 = \frac{d}{dx}1$$

$$2x - 2y\cdot\frac{dy}{dx} = 0$$

$$-2y\cdot\frac{dy}{dx} = -2x$$

$$\frac{dy}{dx} = \frac{-2x}{-2y}$$

$$\frac{dy}{dx} = \frac{x}{y}.$$

Find the slope of the tangent line to the curve at $\left(\sqrt{3},\sqrt{2}\right)$.

$$\frac{dy}{dx} = \frac{x}{y}$$

Replacing x with $\sqrt{3}$ and y with $\sqrt{2}$, we have:

$$\frac{dy}{dx} = \frac{\sqrt{3}}{\sqrt{2}} = \sqrt{\frac{3}{2}}.$$

The slope of the tangent line to the curve at $\left(\sqrt{3},\sqrt{2}\right)$ is $\sqrt{\frac{3}{2}}$.

Copyright © 2014 Pearson Education, Inc. Publishing as Addison-Wesley.

7. Differentiate implicitly to find $\frac{dy}{dx}$.

We have

$3x^2y^4 = 12.$

Differentiating both sides with respect to x yields:

$$\frac{d}{dx}\left(3x^2y^4\right) = \frac{d}{dx}(12)$$

$$3x^2\frac{d}{dx}y^4 + y^4\frac{d}{dx}3x^2 = \frac{d}{dx}12 \quad \text{Product Rule}$$

$$3x^2\left(4y^3\cdot\frac{dy}{dx}\right) + y^4\cdot(3\cdot 2x) = 0$$

$$12x^2y^3\cdot\frac{dy}{dx} + 6xy^4 = 0$$

$$12x^2y^3\cdot\frac{dy}{dx} = -6xy^4$$

$$\frac{dy}{dx} = \frac{-6xy^4}{12x^2y^3}$$

$$\frac{dy}{dx} = -\frac{y}{2x}.$$

Find the slope of the tangent line to the curve at $(2,-1)$.

$$\frac{dy}{dx} = -\frac{y}{2x}$$

Replacing x with 2 and y with -1, we have:

$$\frac{dy}{dx} = -\frac{(-1)}{2(2)} = \frac{1}{4}.$$

The slope of the tangent line to the curve at $(2,-1)$ is $\frac{1}{4}$.

9. Differentiate implicitly to find $\frac{dy}{dx}$.

We have

$x^3 - x^2y^2 = -9.$

Differentiating both sides with respect to x yields:

$$\frac{d}{dx}\left(x^3 - x^2y^2\right) = \frac{d}{dx}(-9)$$

$$\frac{d}{dx}x^3 - \frac{d}{dx}x^2y^2 = \frac{d}{dx}(-9)$$

$$3x^2 - \left[x^2\left(2y\cdot\frac{dy}{dx}\right) + y^2(2x)\right] = 0$$

$$3x^2 - 2x^2y\cdot\frac{dy}{dx} - 2xy^2 = 0.$$

Isolate $\frac{dy}{dx}$ at the top of the next column.

$$-2x^2y\cdot\frac{dy}{dx} = 2xy^2 - 3x^2$$

$$\frac{dy}{dx} = \frac{2xy^2 - 3x^2}{-2x^2y}$$

$$\frac{dy}{dx} = \frac{-x\left(3x - 2y^2\right)}{-x(2xy)}$$

$$\frac{dy}{dx} = \frac{3x - 2y^2}{2xy}$$

Find the slope of the tangent line to the curve at $(3,-2)$.

$$\frac{dy}{dx} = \frac{3x - 2y^2}{2xy}$$

Replacing x with 3 and y with -2, we have:

$$\frac{dy}{dx} = \frac{3(3) - 2(-2)^2}{2(3)(-2)} = \frac{9-8}{-12} = -\frac{1}{12}.$$

The slope of the tangent line to the curve at $(3,-2)$ is $-\frac{1}{12}$.

11. Differentiate implicitly to find $\frac{dy}{dx}$.

We have

$xy - x + 2y = 3.$

Differentiating both sides with respect to x yields:

$$\frac{d}{dx}(xy - x + 2y) = \frac{d}{dx}(3)$$

$$\frac{d}{dx}xy - \frac{d}{dx}x + 2\frac{d}{dx}y = \frac{d}{dx}(3)$$

$$\left[x\left(\frac{dy}{dx}\right) + y(1)\right] - 1 + 2\cdot\frac{dy}{dx} = 0$$

$$x\cdot\frac{dy}{dx} + y - 1 + 2\cdot\frac{dy}{dx} = 0$$

$$x\cdot\frac{dy}{dx} + 2\cdot\frac{dy}{dx} = 1 - y$$

$$(x+2)\cdot\frac{dy}{dx} = 1 - y$$

$$\frac{dy}{dx} = \frac{1-y}{x+2}.$$

Find the slope of the tangent line to the curve at $\left(-5, \frac{2}{3}\right)$.

$$\frac{dy}{dx} = \frac{1-y}{x+2}$$

The solution is continued on the next page.

Copyright © 2014 Pearson Education, Inc. Publishing as Addison-Wesley.

Replacing x with -5 and y with $\frac{2}{3}$, we have:

$$\frac{dy}{dx}=\frac{1-\left(\frac{2}{3}\right)}{(-5)+2}=\frac{\frac{1}{3}}{-3}=-\frac{1}{9}.$$

The slope of the tangent line to the curve at $\left(-5,\frac{2}{3}\right)$ is $-\frac{1}{9}$.

13. Differentiate implicitly to find $\frac{dy}{dx}$.

We have

$x^2y-2x^3-y^3+1=0.$

Differentiating both sides with respect to x yields:

$$\frac{d}{dx}\left(x^2y-2x^3-y^3+1\right)=\frac{d}{dx}(0)$$

$$\frac{d}{dx}x^2y-\frac{d}{dx}2x^3-\frac{d}{dx}y^3+\frac{d}{dx}1=0$$

$$\left[x^2\left(\frac{dy}{dx}\right)+y(2x)\right]-2\left(3x^2\right)-\left(3y^2\right)\cdot\frac{dy}{dx}=0$$

$$x^2\cdot\frac{dy}{dx}+2xy-6x^2-3y^2\cdot\frac{dy}{dx}=0$$

$$x^2\cdot\frac{dy}{dx}-3y^2\cdot\frac{dy}{dx}=6x^2-2xy$$

$$\left(x^2-3y^2\right)\cdot\frac{dy}{dx}=6x^2-2xy$$

$$\frac{dy}{dx}=\frac{6x^2-2xy}{x^2-3y^2}.$$

Find the slope of the tangent line to the curve at $(2,-3)$.

$$\frac{dy}{dx}=\frac{6x^2-2xy}{x^2-3y^2}$$

Replacing x with 2 and y with -3, we have:

$$\frac{dy}{dx}=\frac{6(2)^2-2(2)(-3)}{(2)^2-3(-3)^2}=\frac{24+12}{4-27}=\frac{36}{-23}=-\frac{36}{23}.$$

The slope of the tangent line to the curve at $(2,-3)$ is $-\frac{36}{23}$.

15. Differentiate implicitly to find $\frac{dy}{dx}$.

We have

$2xy+3=0.$

Differentiating both sides with respect to x yields:

$$\frac{d}{dx}(2xy+3)=\frac{d}{dx}(0)$$

$$2\cdot\frac{d}{dx}xy+\frac{d}{dx}3=\frac{d}{dx}(0)$$

$$2\cdot\left[x\left(\frac{dy}{dx}\right)+y(1)\right]+0=0$$

$$2x\cdot\frac{dy}{dx}+2y=0$$

$$2x\cdot\frac{dy}{dx}=-2y$$

$$\frac{dy}{dx}=\frac{-2y}{2x}$$

$$\frac{dy}{dx}=-\frac{y}{x}.$$

17. Differentiate implicitly to find $\frac{dy}{dx}$.

We have

$x^2-y^2=16.$

Differentiating both sides with respect to x yields:

$$\frac{d}{dx}\left(x^2-y^2\right)=\frac{d}{dx}(16)$$

$$\frac{d}{dx}x^2-\frac{d}{dx}y^2=\frac{d}{dx}(16)$$

$$2x-2y\cdot\frac{dy}{dx}=0$$

$$-2y\cdot\frac{dy}{dx}=-2x$$

$$\frac{dy}{dx}=\frac{-2x}{-2y}$$

$$\frac{dy}{dx}=\frac{x}{y}.$$

19. Differentiate implicitly to find $\frac{dy}{dx}$.

We have

$y^5=x^3.$

Differentiating both sides with respect to x yields:

$$\frac{d}{dx}\left(y^5\right)=\frac{d}{dx}\left(x^3\right)$$

$$5y^4\cdot\frac{dy}{dx}=3x^2$$

$$\frac{dy}{dx}=\frac{3x^2}{5y^4}.$$

Copyright © 2014 Pearson Education, Inc. Publishing as Addison-Wesley.

21. Differentiate implicitly to find $\frac{dy}{dx}$.

We have

$x^2y^3 + x^3y^4 = 11.$

Differentiating both sides with respect to x yields:

$$\frac{d}{dx}\left(x^2y^3 + x^3y^4\right) = \frac{d}{dx}(11)$$

$$\frac{d}{dx}\left(x^2y^3\right) + \frac{d}{dx}\left(x^3y^4\right) = 0.$$

Notice:

$$\frac{d}{dx}\left(x^2y^3\right) = x^2\left(3y^2\cdot\frac{dy}{dx}\right) + y^3(2x)$$
$$= 3x^2y^2\cdot\frac{dy}{dx} + 2xy^3$$

and

$$\frac{d}{dx}\left(x^3y^4\right) = x^3\left(4y^3\cdot\frac{dy}{dx}\right) + y^4\left(3x^2\right)$$
$$= 4x^3y^3\cdot\frac{dy}{dx} + 3x^2y^4.$$

Therefore,

$$\frac{d}{dx}\left(x^2y^3\right) + \frac{d}{dx}\left(x^3y^4\right) = 0$$

$$3x^2y^2\cdot\frac{dy}{dx} + 2xy^3 + 4x^3y^3\cdot\frac{dy}{dx} + 3x^2y^4 = 0.$$

Isolating $\frac{dy}{dx}$, we have:

$$\left(4x^3y^3 + 3x^2y^2\right)\cdot\frac{dy}{dx} = -3x^2y^4 - 2xy^3$$

$$\frac{dy}{dx} = \frac{-3x^2y^4 - 2xy^3}{4x^3y^3 + 3x^2y^2}$$

$$\frac{dy}{dx} = \frac{xy^2\left(-3xy^2 - 2y\right)}{xy^2\left(4x^2y + 3x\right)}$$

$$\frac{dy}{dx} = -\frac{2y + 3xy^2}{4x^2y + 3x}.$$

23. Differentiate implicitly to find $\frac{dp}{dx}$.

We have

$p^3 + p - 3x = 50.$

Differentiating both sides with respect to x yields:

$$\frac{d}{dx}\left(p^3 + p - 3x\right) = \frac{d}{dx}(50)$$

$$3p^2\cdot\frac{dp}{dx} + \frac{dp}{dx} - 3\cdot 1 = 0$$

$$\left(3p^2 + 1\right)\cdot\frac{dp}{dx} = 3$$

$$\frac{dp}{dx} = \frac{3}{3p^2 + 1}.$$

25. Differentiate implicitly to find $\frac{dp}{dx}$.

$xp^3 = 24$

Differentiating both sides with respect to x yields:

$$\frac{d}{dx}\left(xp^3\right) = \frac{d}{dx}(24)$$

$$x\left(3p^2\cdot\frac{dp}{dx}\right) + p^3(1) = 0 \quad \text{Product Rule}$$

$$3xp^2\cdot\frac{dp}{dx} = -p^3$$

$$\frac{dp}{dx} = \frac{-p^3}{3xp^2}$$

$$\frac{dp}{dx} = -\frac{p}{3x}.$$

27. Differentiate implicitly to find $\frac{dp}{dx}$.

$$\frac{xp}{x+p} = 2$$

First, we multiply both sides by $x + p$ to clear the fraction.

$$(x+p)\left(\frac{xp}{x+p}\right) = (2)(x+p)$$

$$xp = 2x + 2p$$

Next, differentiating both sides with respect to x yields:

$$\frac{d}{dx}(xp) = \frac{d}{dx}(2x + 2p)$$

$$x\cdot\frac{dp}{dx} + p\cdot 1 = 2\cdot 1 + 2\cdot\frac{dp}{dx}$$

$$x\cdot\frac{dp}{dx} - 2\cdot\frac{dp}{dx} = 2 - p$$

$$(x-2)\cdot\frac{dp}{dx} = 2 - p$$

$$\frac{dp}{dx} = \frac{2-p}{x-2}.$$

Copyright © 2014 Pearson Education, Inc. Publishing as Addison-Wesley.

29. Differentiate implicitly to find $\frac{dp}{dx}$.

$(p+4)(x+3)=48$

Expanding the left hand side of the equation we have:

$$px+3p+4x+12=48$$

$$px+3p+4x=36.$$

Differentiating both sides with respect to x yields:

$$\frac{d}{dx}(px+3p+4x)=\frac{d}{dx}(36)$$

$$p\cdot 1+x\cdot\frac{dp}{dx}+3\cdot\frac{dp}{dx}+4\cdot 1=0$$

$$(x+3)\cdot\frac{dp}{dx}=-p-4$$

$$\frac{dp}{dx}=\frac{-p-4}{x+3}.$$

31. $A^3+B^3=9$

We differentiate both sides with respect to t.

$$\frac{d}{dt}\left(A^3+B^3\right)=\frac{d}{dt}(9)$$

$$3A^2\cdot\frac{dA}{dt}+3B^2\cdot\frac{dB}{dt}=0$$

$$3A^2\cdot\frac{dA}{dt}=-3B^2\cdot\frac{dB}{dt}$$

$$\frac{dA}{dt}=\frac{-3B^2}{3A^2}\cdot\frac{dB}{dt}$$

We find B when $A=2$:

$$A^3+B^3=9$$

$$(2)^3+B^3=9$$

$$8+B^3=9$$

$$B^3=1$$

$$B=1.$$

Next, we substitute 2 for A, 1 for B, and 3 for $\frac{dB}{dt}$ into the formula for $\frac{dA}{dt}$:

$$\frac{dA}{dt}=\frac{-3B^2}{3A^2}\cdot\frac{dB}{dt}$$

$$=\frac{-3(1)^2}{3(2)^2}\cdot(3)$$

$$=\frac{-3}{12}\cdot 3$$

$$=-\frac{3}{4}.$$

33. $R(x)=50x-0.5x^2$

Differentiating with respect to time we have:

$$\frac{d}{dt}R(x)=\frac{d}{dt}\left(50x-0.5x^2\right)$$

$$\frac{dR}{dt}=50\cdot\frac{dx}{dt}-x\cdot\frac{dx}{dt}$$

$$\frac{dR}{dt}=(50-x)\cdot\frac{dx}{dt}.$$

Next, we substitute 30 for x and 20 for dx/dt.

$$\frac{dR}{dt}=(50-30)\cdot 20=(20)\cdot 20=400.$$

The rate of change of total revenue with respect to time is \$400 per day.

$C(x)=4x+10$

Differentiating with respect to time we have:

$$\frac{d}{dt}C(x)=\frac{d}{dt}(4x+10)$$

$$\frac{dC}{dt}=4\cdot\frac{dx}{dt}.$$

Next, we substitute 30 for x and 20 for dx/dt.

$$\frac{dC}{dt}=4\cdot(20)=80.$$

The rate of change of total cost with respect to time is \$80 per day.

Profit is revenue minus cost. Therefore;

$$P(x)=R(x)-C(x)$$

$$=50x-0.5x^2-(4x+10)$$

$$=-0.5x^2+46x-10.$$

Differentiating with respect to time we have:

$$\frac{d}{dt}P(x)=\frac{d}{dt}\left(-0.5x^2+46x-10\right)$$

$$\frac{dP}{dt}=-x\cdot\frac{dx}{dt}+46\frac{dx}{dt}$$

$$\frac{dP}{dt}=(46-x)\cdot\frac{dx}{dt}.$$

Next, we substitute 30 for x and 20 for dx/dt.

$$\frac{dP}{dt}=(46-(30))\cdot(20)=(16)(20)=320.$$

The rate of change of total profit with respect to time is \$320 per day.

35. $R(x)=2x$

Differentiating with respect to time we have:

$$\frac{d}{dt}R(x)=\frac{d}{dt}(2x)$$

$$\frac{dR}{dt}=2\cdot\frac{dx}{dt}.$$

The solution is continued on the next page.

Copyright © 2014 Pearson Education, Inc. Publishing as Addison-Wesley.

Next, we substitute 20 for x and 8 for dx/dt.

$$\frac{dR}{dt}=2\cdot 8=16$$

The rate of change of total revenue with respect to time is $16 per day.

$$C(x)=0.01x^2+0.6x+30$$

Differentiating with respect to time we have:

$$\frac{d}{dt}C(x)=\frac{d}{dt}\left(0.01x^2+0.6x+30\right)$$
$$\frac{dC}{dt}=0.02x\cdot\frac{dx}{dt}+0.6\cdot\frac{dx}{dt}$$
$$\frac{dC}{dt}=(0.02x+0.6)\cdot\frac{dx}{dt}.$$

Next, we substitute 20 for x and 8 for dx/dt.

$$\frac{dC}{dt}=(0.02(20)+0.6)\cdot 8=8.$$

The rate of change of total cost with respect to time is $8 per day.

Profit is revenue minus cost. Therefore;

$$P(x)=R(x)-C(x)$$
$$=2x-\left(0.01x^2+0.6x+30\right)$$
$$=-0.01x^2+1.4x-30$$

Differentiating with respect to time we have:

$$\frac{d}{dt}P(x)=\frac{d}{dt}\left(-0.01x^2+1.4x-30\right)$$
$$\frac{dP}{dt}=-0.02x\cdot\frac{dx}{dt}+1.4\frac{dx}{dt}$$
$$\frac{dP}{dt}=(1.4-0.02x)\cdot\frac{dx}{dt}.$$

Next, we substitute 20 for x and 8 for dx/dt.

$$\frac{dP}{dt}=(1.4-0.02(20))\cdot(8)=8.$$

The rate of change of total profit with respect to time is $8 per day.

37. $5p+4x+2px=60$

First, we take the derivative of both sides of the equation with respect to t.

$$\frac{d}{dt}[5p+4x+2px]=\frac{d}{dt}[60]$$

$$5\frac{dp}{dt}+4\frac{dx}{dt}+2\left(\underbrace{p\cdot\frac{dx}{dt}+\frac{dp}{dt}\cdot x}_{\text{Product Rule}}\right)=0$$

$$5\frac{dp}{dt}+4\frac{dx}{dt}+2p\cdot\frac{dx}{dt}+2x\cdot\frac{dp}{dt}=0$$

Next, we solve for $\frac{dx}{dt}$.

$$4\frac{dx}{dt}+2p\cdot\frac{dx}{dt}=-5\frac{dp}{dt}-2x\cdot\frac{dp}{dt}$$
$$(4+2p)\frac{dx}{dt}=-(5+2x)\cdot\frac{dp}{dt}$$
$$\frac{dx}{dt}=\frac{-(5+2x)}{(4+2p)}\cdot\frac{dp}{dt}$$

Substituting 3 for x, 5 for p, and 1.5 for $\frac{dp}{dt}$, we have:

$$\frac{dx}{dt}=\frac{-(5+2(3))}{(4+2(5))}\cdot(1.5)$$
$$=\frac{-(11)}{14}\cdot(1.5)$$
$$=\frac{-16.5}{14}$$
$$\approx -1.18.$$

Sales are changing at a rate of -1.18 sales per day.

39. $A=\pi r^2$

To find the rate of change of the area of the Arctic ice cap with respect to time, we take the derivative of both sides of the equation with respect to t.

$$\frac{d}{dt}A=\frac{d}{dt}\left[\pi r^2\right]$$
$$\frac{dA}{dt}=\pi\frac{d}{dt}\left[r^2\right] \qquad \text{Constant Multiple Rule}$$
$$\frac{dA}{dt}=\pi\left[2r\cdot\frac{dr}{dt}\right] \qquad \text{Chain Rule}$$
$$\frac{dA}{dt}=2\pi r\cdot\frac{dr}{dt}$$

In 2005, the r was 808 miles, and $\frac{dr}{dt}=-4.3$ miles per year. Substituting these values into the derivative, we have:

$$\frac{dA}{dt}=2\pi(808)(-4.3)$$
$$\approx -21{,}830.2990313$$
$$\approx -21{,}830.$$

Therefore, in 2005 the Arctic ice cap was changing at a rate of $-21{,}830\text{ mi}^2$ per year. Another way of stating this is to say that the Arctic ice cap was *shrinking* at a rate of 21,830 mi^2/yr.

Copyright © 2014 Pearson Education, Inc. Publishing as Addison-Wesley.

41. $S=\frac{\sqrt{hw}}{60}$

First, we substitute 165 for h, and then we take the derivative of both sides with respect to t.

$$S=\frac{\sqrt{165w}}{60}=\frac{\sqrt{165}}{60}\cdot w^{1/2}$$

$$\frac{d}{dt}[S]=\frac{d}{dt}\left[\frac{\sqrt{165}}{60}\cdot w^{1/2}\right]$$

$$\frac{dS}{dt}=\frac{\sqrt{165}}{60}\cdot\frac{1}{2}w^{-1/2}\cdot\frac{dw}{dt}$$

$$=\frac{\sqrt{165}}{120}\frac{1}{w^{1/2}}\cdot\frac{dw}{dt}$$

$$=\frac{\sqrt{165}}{120\sqrt{w}}\cdot\frac{dw}{dt}$$

Now, we will substitute 70 for w and -2 for $\frac{dw}{dt}$.

$$\frac{dS}{dt}=\frac{\sqrt{165}}{120\sqrt{70}}\cdot(-2)$$

$$\approx -0.0256$$

Therefore, Tom's surface area is changing at a rate of $-0.0256\ \text{m}^2$/month. We could also say that Tom's surface area is *decreasing* by 0.0256 m^2/month.

43. $V=\frac{p}{4Lv}\left(R^2-r^2\right)$

We assume that r, p, L and v are constants

a) Taking the derivative of both sides with respect to t, we have:

$$\frac{dV}{dt}=\frac{d}{dt}\left[\frac{p}{4Lv}\left(R^2-r^2\right)\right]$$

$$=\frac{p}{4Lv}\left[\frac{d}{dt}R^2-\frac{d}{dt}r^2\right]$$

$$=\frac{p}{4Lv}\left[2R\cdot\frac{dR}{dt}-0\right]$$

$$=\frac{pR}{2Lv}\cdot\frac{dR}{dt}.$$

Substituting 70 for L, 400 for p and 0.003 for v, we have:

$$\frac{dV}{dt}=\frac{400R}{2(70)(0.003)}\cdot\frac{dR}{dt}$$

$$=\frac{400R}{0.42}\cdot\frac{dR}{dt}$$

$$\approx 952.38R\cdot\frac{dR}{dt}.$$

b) Using the derivative in part (a), we substitute 0.00015 for dR/dt and 0.1 for R to get:

$$\frac{dV}{dt}=952.38(0.1)\cdot(0.00015)$$

$$\approx 0.0143.$$

The speed of the person's blood will be increasing at a rate of 0.0143 mm/sec^2.

45. Since the ladder forms a right triangle with the wall and the ground, we know that:

$$x^2+y^2=26^2$$

$$x^2+y^2=676.$$

We are looking for $\frac{dy}{dt}$.

Differentiating both sides with respect to t, we have:

$$\frac{d}{dt}\left[x^2+y^2\right]=\frac{d}{dt}[676]$$

$$2x\frac{dx}{dt}+2y\frac{dy}{dt}=0$$

$$2y\frac{dy}{dt}=-2x\frac{dx}{dt} \quad \text{Subtracting}$$

$$\frac{dy}{dt}=\frac{-2x}{2y}\cdot\frac{dx}{dt} \quad \text{Dividing by } 2y$$

$$\frac{dy}{dt}=\frac{-x}{y}\cdot\frac{dx}{dt}.$$

The lower end of the wall is being pulled away from the wall at a rate of 5 feet per second; therefore, $\frac{dx}{dt}=5$. When the lower end is 10 feet away from the wall, $x=10$, and

$$(10)^2+y^2=676$$

$$100+y^2=676$$

$$y^2=676-100$$

$$y^2=576$$

$$y=\pm\sqrt{576}$$

$$y=\pm 24$$

$$y=24. \quad \text{Since } y \text{ must be positive.}$$

The solution is continued on the next page.

Copyright © 2014 Pearson Education, Inc. Publishing as Addison-Wesley.

We substitute 10 for x, 24 for y and 5 for $\frac{dx}{dt}$ into the derivative to get:

$$\frac{dy}{dt} = \frac{-x}{y} \cdot \frac{dx}{dt}$$
$$= -\frac{(10)}{(24)} \cdot (5)$$
$$= -\frac{25}{12}$$
$$= -2\tfrac{1}{12}.$$

When the lower end of the ladder is 10 feet from the wall, the top of the ladder is moving down the wall at a rate of $-2\frac{1}{12}$ feet per second.

47. $V = \frac{4}{3}\pi r^3$

Differentiating both sides with respect to t, we have:

$$\frac{dV}{dt} = \frac{d}{dt}\left[\frac{4}{3}\pi r^3\right]$$
$$= \frac{4}{3}\pi \cdot \frac{d}{dt}\left[r^3\right]$$
$$= \frac{4}{3}\pi\left[3r^2\frac{dr}{dt}\right]$$
$$= 4\pi r^2 \cdot \frac{dr}{dt}.$$

Next, substituting 0.7 for dr/dt and 7.5 for r, we have:

$$\frac{dV}{dt} = 4\pi(7.5)^2(0.7)$$
$$= 4\pi(56.25)(0.7)$$
$$= 157.5\pi$$
$$\approx 494.8.$$

The cantaloupe's volume is changing at the rate of 494.8 cm^3/week.

49. $\frac{1}{x^2} + \frac{1}{y^2} = 5$

$x^{-2} + y^{-2} = 5$

We differentiating both sides with respect to x at the top of the next column.

Differentiating both sides with respect to x, we have:

$$\frac{d}{dx}\left[x^{-2}\right] + \frac{d}{dx}\left[y^{-2}\right] = \frac{d}{dx}[5]$$
$$-2x^{-3} - 2y^{-3}\frac{dy}{dx} = 0$$
$$-2y^{-3}\frac{dy}{dx} = 2x^{-3}$$
$$\frac{dy}{dx} = -\frac{x^{-3}}{y^{-3}}$$
$$\frac{dy}{dx} = -\frac{y^3}{x^3}.$$

51. $y^2 = \frac{x^2 - 1}{x^2 + 1}$

Differentiating both sides with respect to x, we have:

$$\frac{d}{dx}\left[y^2\right] = \frac{d}{dx}\left[\frac{x^2-1}{x^2+1}\right]$$
$$2y\frac{dy}{dx} = \frac{(x^2+1)(2x) - (x^2-1)(2x)}{(x^2+1)^2}$$
$$2y\frac{dy}{dx} = \frac{2x^3 + 2x - 2x^3 + 2x}{(x^2+1)^2}$$
$$2y\frac{dy}{dx} = \frac{4x}{(x^2+1)^2}$$
$$\frac{dy}{dx} = \frac{4x}{2y(x^2+1)^2}$$
$$\frac{dy}{dx} = \frac{2x}{y(x^2+1)^2}.$$

53. $(x-y)^3 + (x+y)^3 = x^5 + y^5$

Differentiating both sides with respect to x, we have:

$$\frac{d}{dx}\left[(x-y)^3 + (x+y)^3\right] = \frac{d}{dx}\left[x^5 + y^5\right]$$
$$\frac{d}{dx}(x-y)^3 + \frac{d}{dx}(x+y)^3 = \frac{d}{dx}x^5 + \frac{d}{dx}y^5.$$

The solution is continued on the next page.

Copyright © 2014 Pearson Education, Inc. Publishing as Addison-Wesley.

Differentiating, we have:

$$3(x-y)^2 \cdot \frac{d}{dx}(x-y) + 3(x+y)^2 \frac{d}{dx}(x+y) = 5x^4 + 5y^4 \frac{dy}{dx}$$

$$3(x-y)^2\left(1-\frac{dy}{dx}\right) + 3(x+y)^2\left(1+\frac{dy}{dx}\right) = 5x^4 + 5y^4 \frac{dy}{dx}$$

$$3(x-y)^2 - 3(x-y)^2\frac{dy}{dx} + 3(x+y)^2 + 3(x+y)^2\frac{dy}{dx} = 5x^4 + 5y^4\frac{dy}{dx}$$

$$\left[3(x+y)^2 - 3(x-y)^2 - 5y^4\right]\frac{dy}{dx} = 5x^4 - 3(x-y)^2 - 3(x+y)^2$$

$$\frac{dy}{dx} = \frac{5x^4 - 3(x-y)^2 - 3(x+y)^2}{3(x+y)^2 - 3(x-y)^2 - 5y^4}.$$

Simplification will yield:

$$\frac{dy}{dx} = \frac{5x^4 - 6x^2 - 6y^2}{12xy - 5y^4}.$$

55. $y^2 - xy + x^2 = 5$

Differentiate implicitly to find $\frac{dy}{dx}$

$$\frac{d}{dx}\left[y^2 - xy + x^2\right] = \frac{d}{dx}[5]$$

$$2y\frac{dy}{dx} - \left[x\frac{dy}{dx} + y \cdot 1\right] + 2x = 0$$

$$(2y-x)\frac{dy}{dx} - y + 2x = 0$$

$$(2y-x)\frac{dy}{dx} = y - 2x$$

$$\frac{dy}{dx} = \frac{y-2x}{2y-x}.$$

Differentiate $\frac{dy}{dx}$ implicitly to find $\frac{d^2y}{dx^2}$

$$\frac{d}{dx}\left[\frac{dy}{dx}\right] = \frac{d}{dx}\left[\frac{y-2x}{2y-x}\right]$$

$$\frac{d^2y}{dx^2} = \frac{(2y-x)\left(\frac{dy}{dx}-2\right) - (y-2x)\left(2\frac{dy}{dx}-1\right)}{(2y-x)^2}.$$

Simplifying the numerator we have:

$$\frac{d^2y}{dx^2} = \left[\left(2y\frac{dy}{dx} - 4y - x\frac{dy}{dx} + 2x\right) - \left(2y\frac{dy}{dx} - y - 4x\frac{dy}{dx} + 2x\right)\right] \div (2y-x)^2$$

$$\frac{d^2y}{dx^2} = \frac{-3y + 3x\frac{dy}{dx}}{(2y-x)^2}$$

Substituting $\frac{y-2x}{2y-x}$ for $\frac{dy}{dx}$

$$\frac{d^2y}{dx^2} = \frac{-3y + 3x \cdot \frac{y-2x}{2y-x}}{(2y-x)^2}$$

$$= \frac{-3y\frac{2y-x}{2y-x} + 3x \cdot \frac{y-2x}{2y-x}}{(2y-x)^2}$$

$$= \frac{-6y^2 + 3xy + 3xy - 6x^2}{(2y-x)^3}$$

$$= \frac{-6y^2 + 6xy - 6x^2}{(2y-x)^3}$$

$$= \frac{-6\left(y^2 - xy + x^2\right)}{(2y-x)^3}.$$

57. $x^3 - y^3 = 8$

Differentiate implicitly to find $\frac{dy}{dx}$

$$\frac{d}{dx}\left[x^3 - y^3\right] = \frac{d}{dx}[8]$$

$$3x^2 - 3y^2\frac{dy}{dx} = 0$$

$$-3y^2\frac{dy}{dx} = -3x^2$$

$$\frac{dy}{dx} = \frac{-3x^2}{-3y^2}$$

$$\frac{dy}{dx} = \frac{x^2}{y^2}.$$

The solution is continued on the next page.

Copyright © 2014 Pearson Education, Inc. Publishing as Addison-Wesley.

Differentiate $\frac{dy}{dx}$ implicitly to find $\frac{d^2y}{dx^2}$

$$\frac{d^2y}{dx^2} = \frac{d}{dx}\left[\frac{x^2}{y^2}\right]$$

$$= \frac{y^2 \cdot (2x) - x^2 \cdot 2y\frac{dy}{dx}}{\left(y^2\right)^2}$$

$$= \frac{2xy^2 - 2x^2y\left[\frac{x^2}{y^2}\right]}{y^4} \quad \text{Substituting for } \frac{dy}{dx}$$

$$= \frac{2xy^2 - \frac{2x^4}{y}}{y^4}$$

$$= \frac{\frac{2xy^2}{1}\cdot\frac{y}{y} - \frac{2x^4}{y}}{y^4}$$

$$= \frac{\frac{2xy^3 - 2x^4}{y}}{y^4}$$

$$= \frac{2xy^3 - 2x^4}{y^5}$$

$$= \frac{2x\left(y^3 - x^3\right)}{y^5}.$$

59. ✎

61. Using the calculator, we have:

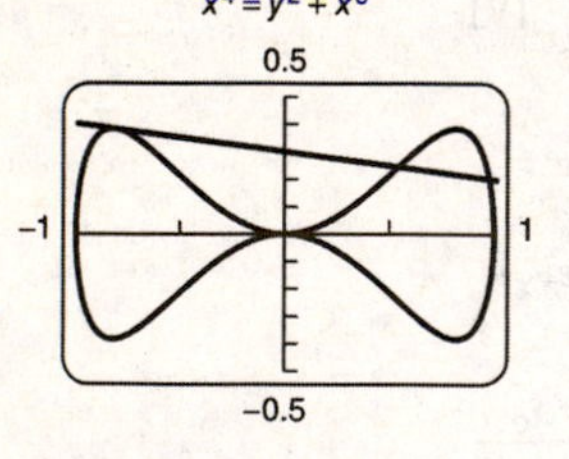

63. Using the calculator, we have:

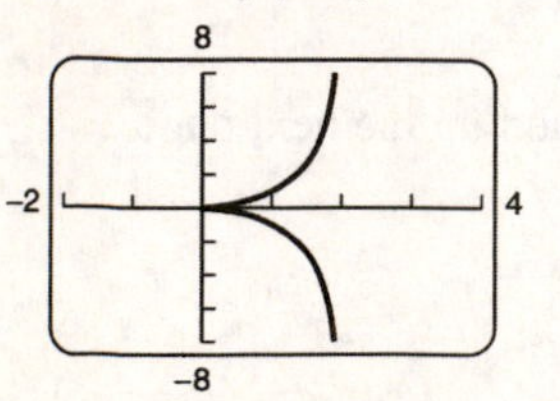

Copyright © 2014 Pearson Education, Inc. Publishing as Addison-Wesley.

Chapter 3

Exponential and Logarithmic Functions

Exercise Set 3.1

1. Graph: $y = 4^x$

First, we find some function values.

$x = -2,\ y = 4^{-2} = \dfrac{1}{4^2} = \dfrac{1}{16} = 0.0625$

$x = -1,\ y = 4^{-1} = \dfrac{1}{4} = 0.25$

$x = 0,\ \ y = 4^0 = 1$

$x = 1,\ \ y = 4^1 = 4$

$x = 2,\ \ y = 4^2 = 16$

x	y
-2	0.0625
-1	0.25
0	1
1	4
2	16

Next, we plot the points and connect them with a smooth curve.

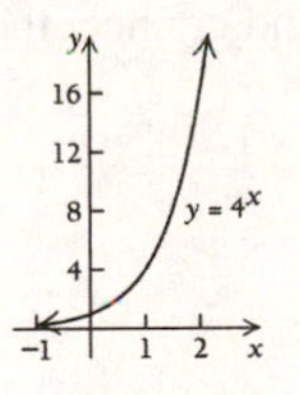

3. Graph: $y = (0.25)^x$

First note that:

$$y = (0.25)^x$$
$$= \left(\frac{1}{4}\right)^x$$
$$= \left(4^{-1}\right)^x$$
$$= 4^{-x}.$$

This will ease our work in calculating function values.

$x = -2,\ y = 4^{-(-2)} = 4^2 = 16$

$x = -1,\ y = 4^{-(-1)} = 4^1 = 4$

$x = 0,\ \ y = 4^{-(0)} = 1$

$x = 1,\ \ y = 4^{-(1)} = \dfrac{1}{4} = 0.25$

$x = 2,\ \ y = 4^{-(2)} = \dfrac{1}{4^2} = \dfrac{1}{16} = 0.0625$

x	y
-2	16
-1	4
0	1
1	0.25
2	0.0625

Next, we plot the points and connect them with a smooth curve.

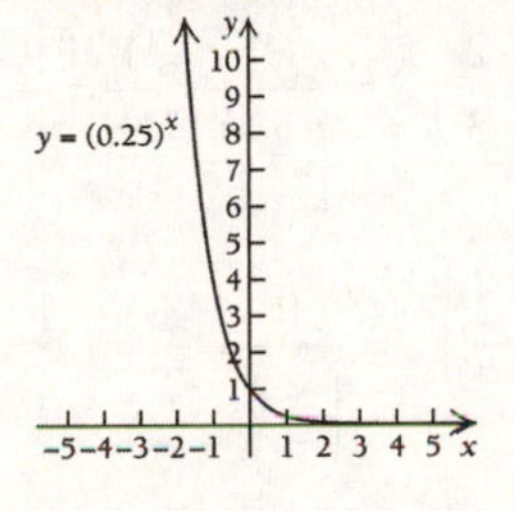

5. Graph: $f(x) = \left(\dfrac{3}{2}\right)^x$

First, we find some function values.

$x = -2,\ f(-2) = \left(\dfrac{3}{2}\right)^{-2} = \dfrac{1}{\left(\dfrac{3}{2}\right)^2} = \dfrac{1}{\left(\dfrac{9}{4}\right)} = \dfrac{4}{9} = .44\overline{4}$

$x = -1,\ f(-1) = \left(\dfrac{3}{2}\right)^{-1} = \dfrac{1}{\left(\dfrac{3}{2}\right)} = \dfrac{2}{3} = .66\overline{6}$

$x = 0,\ \ f(0) = \left(\dfrac{3}{2}\right)^0 = 1$

$x = 1,\ \ f(1) = \left(\dfrac{3}{2}\right)^1 = \dfrac{3}{2} = 1.5$

$x = 2,\ \ f(2) = \left(\dfrac{3}{2}\right)^2 = \dfrac{9}{4} = 2.25$

The solution is continued on the next page.

Copyright © 2014 Pearson Education, Inc. Publishing as Addison-Wesley.

We organize the information from the previous page into a table.

x	$f(x)$
-2	0.444
-1	0.666
0	1
1	1.5
2	2.25

Next, we plot the points and connect them with a smooth curve.

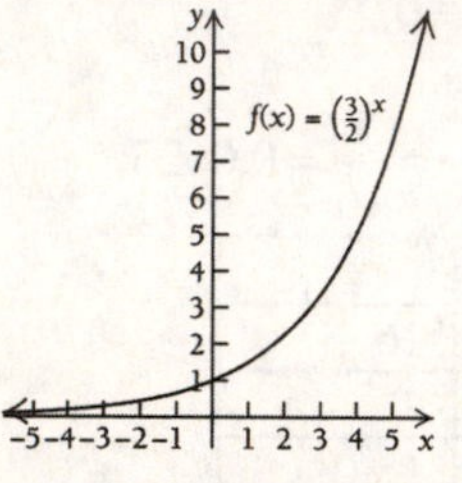

7. Graph: $g(x)=\left(\frac{2}{3}\right)^x$

First, we find some function values.

$$x=-2,\ g(-2)=\left(\frac{2}{3}\right)^{-2}=\frac{1}{\left(\frac{2}{3}\right)^2}=\frac{1}{\frac{4}{9}}=\frac{9}{4}=2.25$$

$$x=-1,\ g(-1)=\left(\frac{2}{3}\right)^{-1}=\frac{1}{\left(\frac{2}{3}\right)}=\frac{3}{2}=1.5$$

$$x=0,\quad g(0)=\left(\frac{2}{3}\right)^{0}=1$$

$$x=1,\quad g(1)=\left(\frac{2}{3}\right)^{1}=0.66\overline{6}$$

$$x=2,\quad g(2)=\left(\frac{2}{3}\right)^{2}=\frac{4}{9}=0.44\overline{4}$$

x	$g(x)$
-2	2.25
-1	1.5
0	1
1	0.666
2	0.444

Next, we plot the points and connect them with a smooth curve.

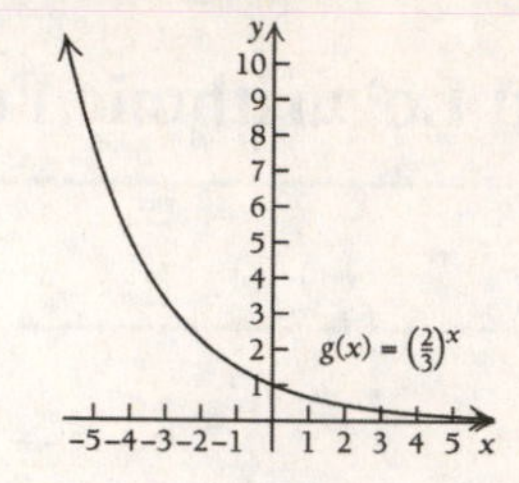

9. Graph: $f(x)=(2.5)^x$

First, we find some function values.

$$x=-2,\ f(-2)=(2.5)^{-2}=\frac{1}{(2.5)^2}=\frac{1}{6.25}=0.16$$

$$x=-1,\ f(-1)=(2.5)^{-1}=\frac{1}{2.5}=0.4$$

$$x=0,\quad f(0)=(2.5)^0=1$$

$$x=1,\quad f(1)=(2.5)^1=2.5$$

$$x=2,\quad f(2)=(2.5)^2=6.25$$

x	$f(x)$
-2	0.16
-1	0.4
0	1
1	2.5
2	6.25

Next, we plot the points and connect them with a smooth curve.

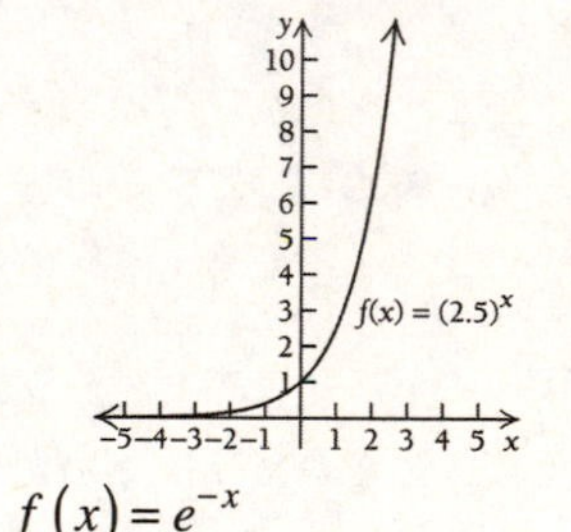

11. $f(x)=e^{-x}$

Using the chain rule $\frac{d}{dx}e^{f(x)}=f'(x)e^{f(x)}$, we have $f'(x)=e^{-x}\cdot(-1)=-e^{-x}$.

$f'(x)=-e^{-x}$

13. $g(x)=e^{3x}$

Using the chain rule $\frac{d}{dx}e^{f(x)}=f'(x)e^{f(x)}$, we have $g'(x)=e^{3x}\cdot 3=3e^{3x}$.

$g'(x)=3e^{3x}$

Copyright © 2014 Pearson Education, Inc. Publishing as Addison-Wesley.

15. $f(x) = 6e^x$

Using $\frac{d}{dx}[c \cdot f(x)] = c \cdot f'(x)$, we have:

$f'(x) = 6e^x$.

17. $F(x) = e^{-7x}$

$F'(x) = -7e^{-7x}$

19. $G(x) = 2e^{4x}$

We use $\frac{d}{dx}[c \cdot f(x)] = c \cdot f'(x)$:

$G'(x) = \frac{d}{dx}\left[2 \cdot e^{4x}\right] = 2 \cdot \frac{d}{dx}e^{4x}$.

Next, we use the Chain Rule:

$\frac{d}{dx}e^{f(x)} = f'(x)e^{f(x)}$; therefore,

$G'(x) = 2 \cdot 4e^{4x}$

$G'(x) = 8e^{4x}$.

21. $f(x) = -3e^{-x}$

$$f'(x) = \frac{d}{dx}\left[-3 \cdot e^{-x}\right] = -3 \cdot \frac{d}{dx}e^{-x}$$
$$= -3 \cdot e^{-x} \cdot (-1)$$
$$= 3e^{-x}$$

23. $g(x) = \frac{1}{2}e^{-5x}$

$$g'(x) = \frac{d}{dx}\left[\frac{1}{2} \cdot e^{-5x}\right] = \frac{1}{2} \cdot \frac{d}{dx}e^{-5x}$$
$$= \frac{1}{2}e^{-5x} \cdot (-5)$$
$$= -\frac{5}{2}e^{-5x}$$

25. $F(x) = -\frac{2}{3}e^{x^2}$

$$F'(x) = \frac{d}{dx}\left[-\frac{2}{3} \cdot e^{x^2}\right] = -\frac{2}{3} \cdot \frac{d}{dx}e^{x^2}$$
$$= -\frac{2}{3} \cdot e^{x^2} \cdot (2x)$$
$$= -\frac{4x}{3}e^{x^2}$$

27. $G(x) = 7 + 3e^{5x}$

Using the Sum Rule, we have

$$G'(x) = \frac{d}{dx}\left[7 + 3e^{5x}\right]$$
$$= \frac{d}{dx}7 + 3\frac{d}{dx}e^{5x}.$$

To differentiate the two terms remember that the derivative of a constant is zero. All that is left is to apply the Chain Rule.

$$G'(x) = 0 + 3 \cdot e^{5x} \cdot (5)$$
$$= 15e^{5x}$$

29. $f(x) = x^5 - 2e^{6x}$

First, we use the Difference Rule

$$f'(x) = \frac{d}{dx}x^5 - 2\frac{d}{dx}e^{6x}.$$

Next, we use the Power Rule on the first term and the Chain Rule on the second term

$$f'(x) = 5x^4 - 2 \cdot e^{6x} \cdot (6)$$
$$= 5x^4 - 12e^{6x}.$$

31. $g(x) = x^5 e^{2x}$

We use the Product Rule and the Chain Rule.

$$g'(x) = x^5 \cdot \left[\frac{d}{dx}e^{2x}\right] + \left[\frac{d}{dx}x^5\right] \cdot e^{2x}$$
$$= x^5 \cdot e^{2x} \cdot (2) + 5x^4 \cdot e^{2x}$$
$$= 2x^5 e^{2x} + 5x^4 e^{2x}$$
$$= (2x+5)x^4 e^{2x} \quad \text{Factoring}$$

33. $F(x) = \frac{e^{2x}}{x^4}$

We use the Quotient Rule and the Chain Rule.

$$F'(x) = \frac{x^4 \cdot (2)e^{2x} - e^{2x} \cdot 4x^3}{x^8}$$
$$= \frac{2x^3 e^{2x}(x-2)}{x^3 \cdot x^5} \quad \text{Factoring}$$
$$= \frac{x^3}{x^3} \cdot \frac{2e^{2x}(x-2)}{x^5} \quad \text{Remove the factor equal to 1.}$$
$$= \frac{2e^{2x}(x-2)}{x^5}$$

Copyright © 2014 Pearson Education, Inc. Publishing as Addison-Wesley.

35. $f(x)=(x^2+3x-9)e^x$

$f'(x)=(2x+3)e^x+(x^2+3x-9)e^x$ Product Rule

$=(2x+3+x^2+3x-9)e^x$ Factoring

$=(x^2+5x-6)e^x$

37. $f(x)=\frac{e^x}{x^4}$

$f'(x)=\frac{x^4\cdot e^x-4x^3\cdot e^x}{x^8}$ By the Quotient Rule

$=\frac{x^3e^x(x-4)}{x^3\cdot x^5}$ Factoring

$=\frac{x^3}{x^3}\cdot\frac{e^x(x-4)}{x^5}$ Remove the factor equal to 1.

$=\frac{e^x(x-4)}{x^5}$

39. $f(x)=e^{-x^2+7x}$

$f'(x)=e^{-x^2+7x}\cdot\left[\frac{d}{dx}(-x^2+7x)\right]$ By the Chain Rule

$=^{-x^2+7x}\cdot[-2x+7]$

$=(-2x+7)\cdot e^{-x^2+7x}$

41. $f(x)=e^{-x^2/2}$

$f'(x)=e^{-x^2/2}\cdot\left[\frac{d}{dx}\left(-\frac{x^2}{2}\right)\right]$

$=e^{-x^2/2}\left[\frac{-1}{2}(2x)\right]$

$=-x\cdot e^{-x^2/2}$

43. $y=e^{\sqrt{x-7}}$

First note that

$y=e^{\sqrt{x-7}}=e^{(x-7)^{1/2}}$.

Using the Chain Rule, we have

$\frac{dy}{dx}=e^{\sqrt{x-7}}\cdot\left[\frac{d}{dx}(x-7)^{1/2}\right]$.

Using the Chain Rule again

$\frac{dy}{dx}=e^{\sqrt{x-7}}\cdot\left[\frac{1}{2}(x-7)^{-1/2}\right]$

$=e^{\sqrt{x-7}}\cdot\frac{1}{2(x-7)^{1/2}}$ Properties of exponents

$=\frac{e^{\sqrt{x-7}}}{2\sqrt{x-7}}$.

45. $y=\sqrt{e^x-1}$

$y=\sqrt{e^x-1}=(e^x-1)^{1/2}$

$\frac{dy}{dx}=\frac{1}{2}(e^x-1)^{-1/2}\cdot\frac{d}{dx}(e^x-1)$ By the Chain Rule

$=\frac{1}{2}(e^x-1)^{-1/2}\cdot(e^x-0)$

$=\frac{e^x}{2(e^x-1)^{1/2}}$

$=\frac{e^x}{2\sqrt{e^x-1}}$

47. $y=xe^{-2x}+e^{-x}+x^3$

Differentiate the function term by term. We apply the Product Rule to the first term, the Chain Rule to the second term, and the Power Rule to the third term.

$\frac{dy}{dx}=\underbrace{x(-2)e^{-2x}+1\cdot e^{-2x}}_{\text{Product Rule}}+\underbrace{(-1)e^{-x}}_{\text{Chain Rule}}+\underbrace{3x^2}_{\text{Power Rule}}$

Now we simplify:

$\frac{dy}{dx}=-2xe^{-2x}+e^{-2x}-e^{-x}+3x^2$

$=(1-2x)e^{-2x}-e^{-x}+3x^2$.

49. $y=1-e^{-x}$

$\frac{dy}{dx}=0-e^{-x}\cdot(-1)$

$=e^{-x}$

51. $y=1-e^{-kx}$

We use the Chain Rule, remembering that k is a constant:

$\frac{dy}{dx}=0-e^{-kx}\cdot(-k)$

$=ke^{-kx}$.

Copyright © 2014 Pearson Education, Inc. Publishing as Addison-Wesley.

53. $g(x)=\left(4x^2+3x\right)e^{x^2-7x}$

We use the Product Rule first, and apply the Chain Rule when taking the derivative of the exponential term.

$$g'(x)=\left(4x^2+3x\right)\underbrace{(2x-7)e^{x^2-7x}}_{\text{Chain Rule}}+(8x+3)e^{x^2-7x}$$

Factoring out the common term, we have

$$g'(x)=\left[\left(4x^2+3x\right)(2x-7)+(8x+3)\right]e^{x^2-7x}.$$

Simplifying inside the bracket, yields

$$g'(x)=\left[\left(8x^3-22x^2-21x\right)+(8x+3)\right]e^{x^2-7x}$$
$$=\left[8x^3-22x^2-13x+3\right]e^{x^2-7x}.$$

55. Graph: $f(x)=e^{2x}$

Using a calculator, we first find some function values.

$f(-2)=e^{2(-2)}=e^{-4}\approx 0.0183$

$f(-1)=e^{2(-1)}=e^{-2}\approx 0.1353$

$f(0)=e^{2(0)}=e^{0}=1$

$f(1)=e^{2(1)}=e^{2}\approx 7.3891$

$f(2)=e^{2(2)}=e^{4}\approx 54.598$

x	$f(x)$
-2	0.0183
-1	0.1353
0	1
1	7.3891
2	54.598

Next, we plot the points and connect them with a smooth curve.

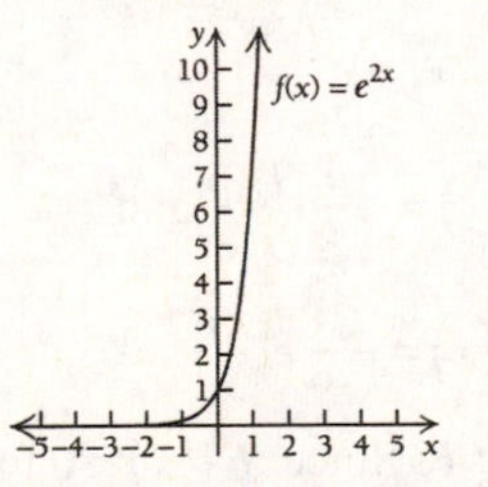

Derivatives. $f'(x)=2e^{2x}$ and $f''(x)=4e^{2x}$.

Critical values of f. Since $f'(x)>0$ for all real numbers x, we know that the derivative exists for all real numbers and there is no solution to the equation $f'(x)=0$. There are no critical values and therefore no maximum or minimum values.

Increasing. Since $f'(x)>0$ for all real numbers x, the function f is increasing over the entire real line.

Inflection points. Since $f''(x)>0$ for all real numbers x, the equation $f''(x)=0$ has no solution and there are no points of inflection.

Concavity. Since $f''(x)>0$ for all real numbers x, the function f' is increasing over the entire real line and the graph is concave up over the entire real line.

57. Graph: $g(x)=e^{(1/2)x}$

First, we find some function values.

$g(-2)=e^{(1/2)(-2)}=e^{-1}\approx 0.3679$

$g(-1)=e^{(1/2)(-1)}=e^{-1/2}\approx 0.6065$

$g(0)=e^{(1/2)(0)}=e^{0}=1$

$g(1)=e^{(1/2)(1)}=e^{1/2}\approx 1.6487$

$g(2)=e^{(1/2)(2)}=e^{1}\approx 2.7183$

x	$g(x)$
-2	0.3679
-1	0.6065
0	1
1	1.6487
2	2.7183

Next, we plot the points and connect them with a smooth curve.

Derivatives. $g'(x)=\frac{1}{2}e^{(1/2)x}$ and $g''(x)=\frac{1}{4}e^{(1/2)x}$.

Critical values of g. Since $g'(x)>0$ for all real numbers x, we know that the derivative exists for all real numbers and there is no solution to the equation $g'(x)=0$. There are no critical values and therefore no maximum or minimum values.

Increasing. Since $g'(x)>0$ for all real numbers x, the function g is increasing over the entire real line. Continued on the next page.

Copyright © 2014 Pearson Education, Inc. Publishing as Addison-Wesley.

Inflection points. Since $g''(x) > 0$ for all real numbers x, the equation $g''(x) = 0$ has no solution and there are no points of inflection.
Concavity. Since $g''(x) > 0$ for all real numbers x, the function g' is increasing over the entire real line and the graph is concave up over the entire real line.

59. Graph: $f(x) = \frac{1}{2}e^{-x}$
First, we find some function values.

$f(-2) = \frac{1}{2}e^{-(-2)} = \frac{1}{2}e^{2} \approx 3.6945$

$f(-1) = \frac{1}{2}e^{-(-1)} = \frac{1}{2}e^{1} \approx 1.3591$

$f(0) = \frac{1}{2}e^{-(0)} = \frac{1}{2}e^{0} = 0.5$

$f(1) = \frac{1}{2}e^{-(1)} = \frac{1}{2}e^{-1} \approx 0.1839$

$f(2) = \frac{1}{2}e^{-(2)} = \frac{1}{2}e^{-2} \approx 0.06767$

x	$f(x)$
-2	3.6945
-1	1.3591
0	0.5
1	0.1839
2	0.06767

Next, we plot the points and connect them with a smooth curve.

Derivatives. $f'(x) = -\frac{1}{2}e^{-x}$ and

$f''(x) = \frac{1}{2}e^{-x}$.

Critical values of f. Since $f'(x) < 0$ for all real numbers x, we know that the derivative exists for all real numbers and there is no solution to the equation $f'(x) = 0$. There are no critical values and therefore no maximum or minimum values.

Decreasing. Since $f'(x) < 0$ for all real numbers x, the function f is decreasing over the entire real line.
Inflection points. Since $f''(x) > 0$ for all real numbers x, the equation $f''(x) = 0$ has no solution and there are no points of inflection.
Concavity. Since $f''(x) > 0$ for all real numbers x, the function f' is increasing over the entire real line and the graph is concave up over the entire real line.

61. Graph: $F(x) = -e^{(1/3)x}$
First, we find some function values.

$F(-2) = -e^{(1/3)(-2)} = -e^{-2/3} \approx -0.5134$

$F(-1) = -e^{(1/3)(-1)} = -e^{-1/3} \approx -0.7165$

$F(0) = -e^{(1/3)(0)} = -e^{0} = -1$

$F(1) = -e^{(1/3)(1)} = -e^{1/3} \approx -1.3956$

$F(2) = -e^{(1/3)(2)} = -e^{2/3} \approx -1.9477$

x	$F(x)$
-2	- 0.5134
-1	- 0.7165
0	- 1
1	- 1.3956
2	- 1.9477

Next, we plot the points and connect them with a smooth curve.

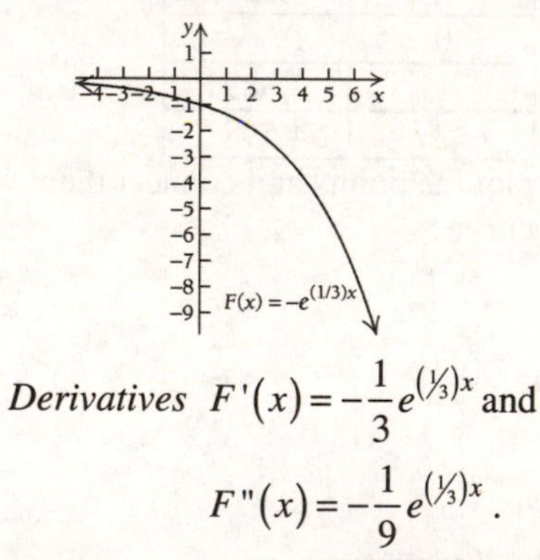

Derivatives $F'(x) = -\frac{1}{3}e^{(1/3)x}$ and

$F''(x) = -\frac{1}{9}e^{(1/3)x}$.

Critical values of F. Since $F'(x) < 0$ for all real numbers x, we know that the derivative exists for all real numbers and there is no solution to the equation $F'(x) = 0$. There are no critical values and therefore no maximum or minimum values. The solution is continued on the next page.

Copyright © 2014 Pearson Education, Inc. Publishing as Addison-Wesley.

Decreasing. Since $F'(x)<0$ for all real numbers x, the function F is decreasing over the entire real line.

Inflection points. Since $F''(x)<0$ for all real numbers x, the equation $F''(x)=0$ has no solution and there are no points of inflection..

Concavity. Since $F''(x)<0$ for all real numbers x, the function F' is decreasing over the entire real line and the graph is concave down over the entire real line.

63. Graph: $g(x)=2\left(1-e^{-x}\right)$, for $x\geq 0$

First, we find some function values:

$g(0)=2\left(1-e^{-0}\right)=0$

$g(1)=2\left(1-e^{-1}\right)\approx 1.2642$

$g(2)=2\left(1-e^{-2}\right)\approx 1.7293$

$g(3)=2\left(1-e^{-3}\right)\approx 1.9004$

$g(4)=2\left(1-e^{-4}\right)\approx 1.9634$

$g(5)=2\left(1-e^{-5}\right)\approx 1.9865$

x	$g(x)$
0	0
1	1.2642
2	1.7293
3	1.9004
4	1.9634
5	1.9865

Next, we plot the points and connect them with a smooth curve.

Derivatives. $g'(x)=2e^{-x}$ and

$g''(x)=-2e^{-x}$.

Critical values of g. Since $2e^{-x}>0$ for all real numbers x, we know that there is no solution to the equation $g'(x)=0$. There are no critical values on $(0,\infty)$.

Increasing. Since $2e^{-x}>0$ for all real numbers x, we know that g is increasing over its entire domain $[0,\infty)$.

Inflection points. Since $-2e^{-x}<0$ for all real numbers x, the equation $g''(x)=0$ has no solution and there are no points of inflection.

Concavity. Since $-2e^{-x}<0$ for all real numbers, the function g' is decreasing and the graph is concave down over the interval $(0,\infty)$.

65. From exercise 55 we know that:

$f(x)=e^{2x}$

$f'(x)=2e^{2x}$

$f''(x)=4e^{2x}$

Enter each function into your graphing calculator.

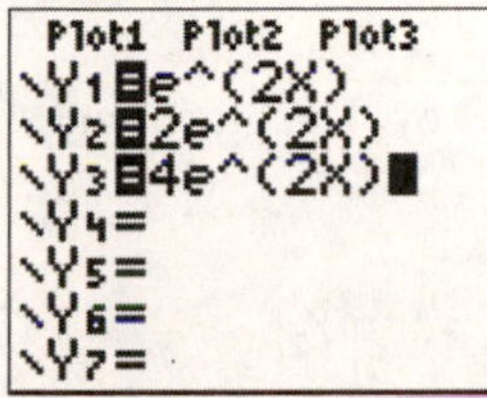

Set the window on the calculator by pressing the Window button and changing the dimensions of the window.

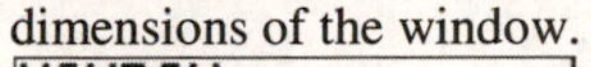

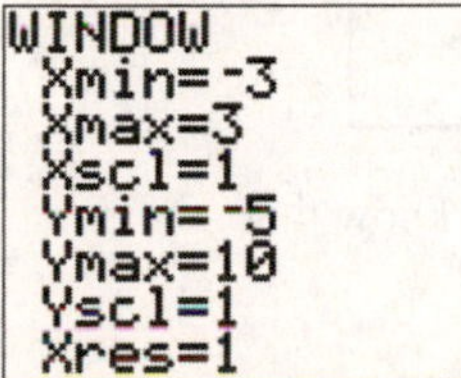

Now press the graph key. The graphs are shown in the screen shot.

$Y_3\,Y_2\,Y_1$

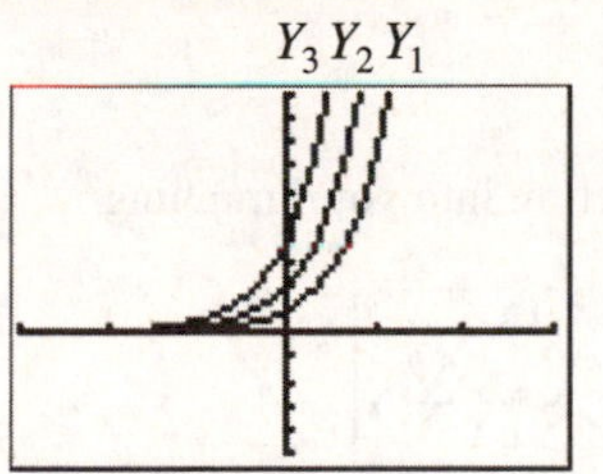

67. From Exercise 57 we know that:

$g(x)=e^{(1/2)x}$

$g'(x)=\frac{1}{2}e^{(1/2)x}$

$g''(x)=\frac{1}{4}e^{(1/2)x}$

The solution is continued on the next page.

Copyright © 2014 Pearson Education, Inc. Publishing as Addison-Wesley.

Enter each function into your graphing calculator.

```
Plot1 Plot2 Plot3
\Y1=e^((1/2)X)
\Y2=(1/2)e^((1/2
)X)
\Y3=(1/4)e^((1/2
)X)
\Y4=
\Y5=
```

Set the window on the calculator by pressing the Window button and changing the dimensions of the window.

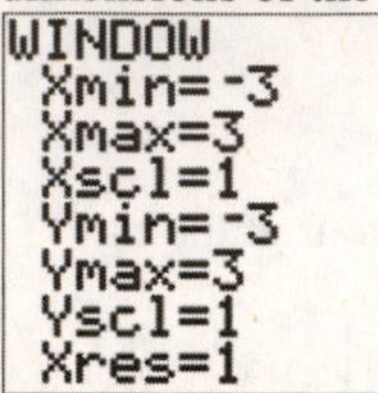

Now press the graph key. The graphs are shown in the screen shot.

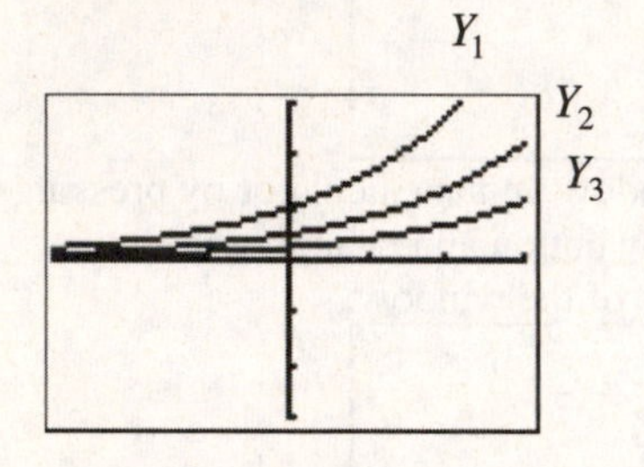

69. From Exercise 59 we know that:

$$f(x)=\frac{1}{2}e^{-x}$$

$$f'(x)=-\frac{1}{2}e^{-x}$$

$$f''(x)=\frac{1}{2}e^{-x}$$

Enter each function into your graphing calculator.

```
Plot1 Plot2 Plot3
\Y1=(1/2)e^(-X)
\Y2=(-1/2)e^(-X)
\Y3=(1/2)e^(-X)
\Y4=
\Y5=
\Y6=
```

Set the window on the calculator by pressing the Window button and changing the dimensions of the window.

```
WINDOW
 Xmin=-3
 Xmax=3
 Xscl=1
 Ymin=-3
 Ymax=3
 Yscl=1
 Xres=1
```

Now press the graph key. The graphs are shown in the screen shot.

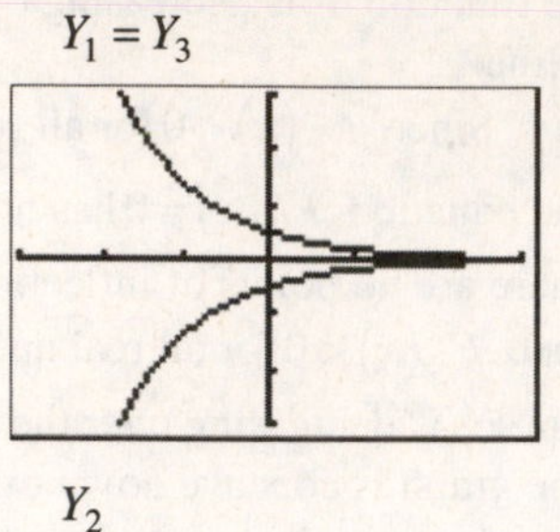

71. From Exercise 61 we know that:

$$F(x)=-e^{(1/3)x}$$

$$F'(x)=-\frac{1}{3}e^{(1/3)x}$$

$$F''(x)=-\frac{1}{9}e^{(1/3)x}$$

Enter each function into your graphing calculator.

Set the window on the calculator by pressing the Window button and changing the dimensions of the window.

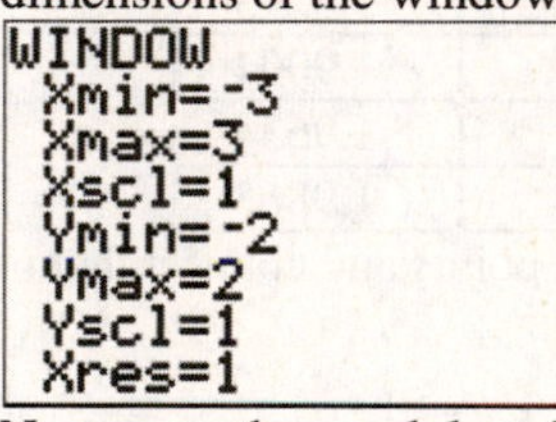

Now press the graph key. The graphs are shown in the screen shot.

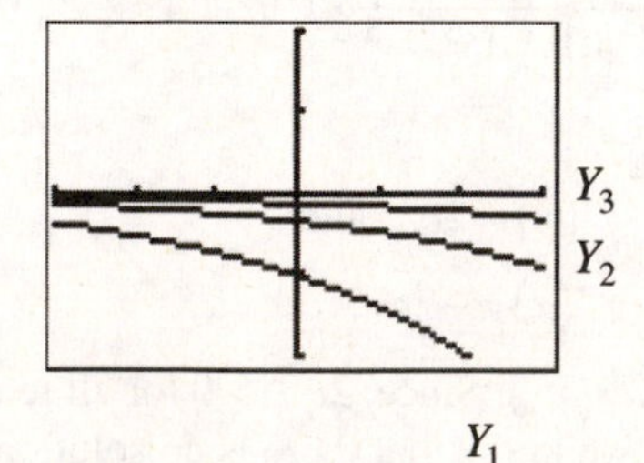

Copyright © 2014 Pearson Education, Inc. Publishing as Addison-Wesley.

73. From Exercise 63 we know that:

$g(x) = 2\left(1 - e^{-x}\right); \quad x \geq 0$

$g'(x) = 2e^{-x}; \qquad x \geq 0$

$g''(x) = -2e^{-x}; \qquad x \geq 0$

Enter each function into your graphing calculator. Notice how you can specify the domain in your calculator. Consult your owner's manual for a detailed explanation on how to do this.

```
Plot1 Plot2 Plot3
\Y1=2(1-e^(-X))/
(X≥0)
\Y2=2e^(-X)/(X≥0
)
\Y3=-2e^(-X)/(X≥
0)
\Y4=
```

Set the window on the calculator by pressing the Window button and changing the dimensions of the window.

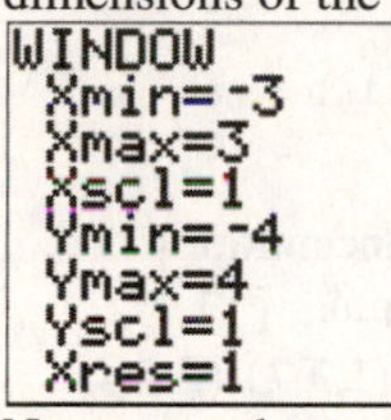

Now press the graph key. The graphs are shown in the screen shot.

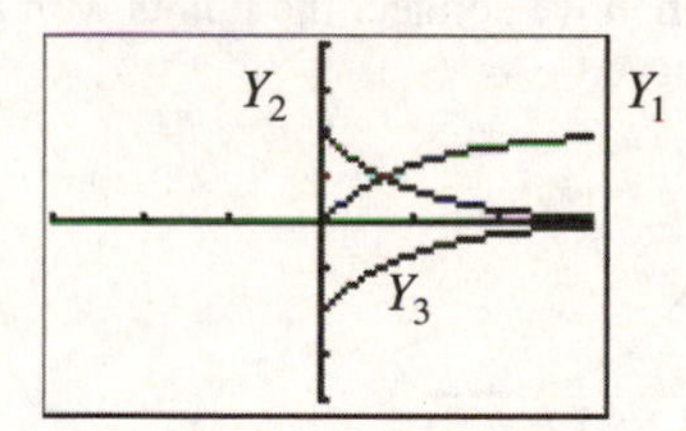

75. First find the derivative of the function.

$f(x) = e^x$

$f'(x) = e^x$

Now, evaluate the derivative at $x = 0$.

$f'(0) = e^0 = 1$

The slope of the tangent line at the point $(0,1)$ is 1.

77. First find the slope of the tangent line at $(0,1)$, by evaluating the derivative at $x = 0$.

$g(x) = e^{-x}$

$g'(x) = -e^{-x}$

$g'(0) = -e^{-0} = -1$

Then we find the equation of the line with slope -1 and containing the point $(0,1)$.

$y - y_1 = m(x - x_1)$ Point-slope equation

$y - 1 = -1(x - 0)$

$y - 1 = -x$

$y = -x + 1$

79. Enter the function and the tangent line found in problem 77 into the calculator.

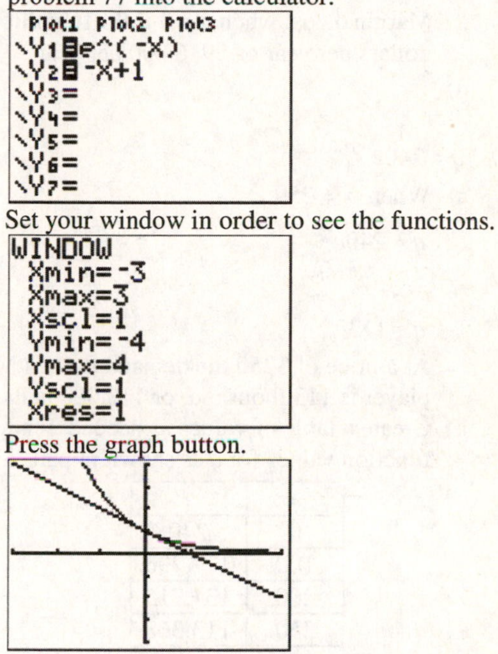

Set your window in order to see the functions.

Press the graph button.

81. a) 2009 correspond to $t = 0$. Evaluate $V(0)$.

$V(0) = 1.6e^{0.0046(0)} = 1.6$

The value of U.S. exports in 2009 is approximately \$1.6 billion.

2020 corresponds to $t = 11$. Evaluate $V(11)$.

$V(11) = 1.6e^{0.0046(11)} \approx 2.7$

The value of U.S. exports in 2020 is approximately \$2.7 billion.

b) Calculate how long it will take for $V(t)$ to increase from 1.6 billion to 3.2 billion.

$$3.2 = 1.6e^{0.0046t}$$

$$2 = e^{0.0046t}$$

$$\ln 2 = \ln\left(e^{0.0046t}\right)$$

$$\ln 2 = 0.0046t$$

$$\frac{\ln 2}{0.0046} = t$$

$$15 \approx t$$

The doubling time of U.S. exports is approximately 15 years.

Copyright © 2014 Pearson Education, Inc. Publishing as Addison-Wesley.

83. a) $C'(t) = 0 - 50(-1)e^{-t}$

$C'(t) = 50e^{-t}$

b) $C'(0) = 50e^{-0} = 50 \cdot 1 = 50$

Marginal cost when $t = 0$ is 50 million dollars per year.

c) $C'(4) = 50e^{-4} \approx 0.916$

Marginal cost when $t = 4$ is 0.916 million dollars per year or $916,000 per year.

d) ✎

85. $q = 240e^{-0.003x}$

a) When $x = 250$

$q = 240e^{-0.003(250)}$

$= 240e^{-0.75}$

$q \approx 113$

At a price of $250 the demand for this MP3 player is 113 thousand, or 113,000 units.

b) Create a table of values as needed. Find the function values for q as shown in part 'a'.

x	q
0	240
100	177.796
200	131.715
250	113.368
300	97.577
400	72.287

Next plot the points and connect the lines with a smooth curve.

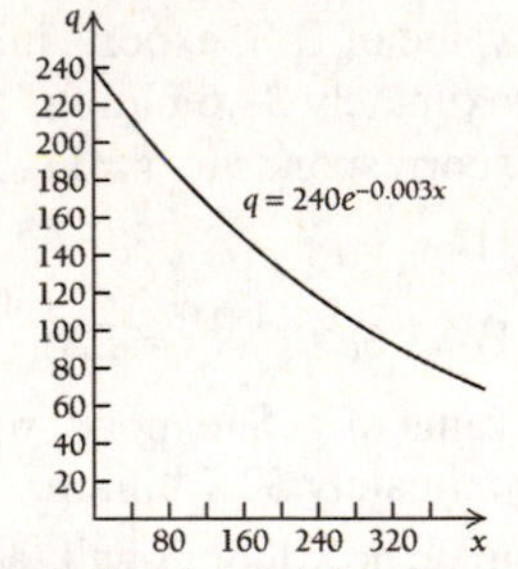

c) Take the derivative of the demand function with respect to x.

$q'(x) = 240(-0.003)e^{-0.003x}$

$= -0.72e^{-0.003x}$

d) ✎

87. a) $C(0) = 10 \cdot 0^2 e^{-0} = 0$

The initial concentration is 0 parts per million (ppm).

$C(1) = 10 \cdot 1^2 e^{-1}$

$\approx 10(0.367879)$

≈ 3.68

After 1 hour, the concentration is approximately 3.7 ppm.

$C(2) = 10 \cdot 2^2 e^{-2}$

$\approx 40(0.135335)$

≈ 5.41

After 2 hours, the concentration is approximately 5.4 ppm.

$C(3) = 10 \cdot 3^2 e^{-3}$

$\approx 90(0.049787)$

≈ 4.48

After 3 hours, the concentration is approximately 4.5 ppm.

$C(10) = 10 \cdot 10^2 e^{-10}$

$\approx 1000(0.000045)$

≈ 0.05

After 10 hours, the concentration is approximately 0.05 ppm.

b) Plot the points $(0,0);(1,3.7);(2,5.4);$ $(3,4.5)$ and $(10,0.05)$ and other points as needed. Then we connect the points with a smooth curve.

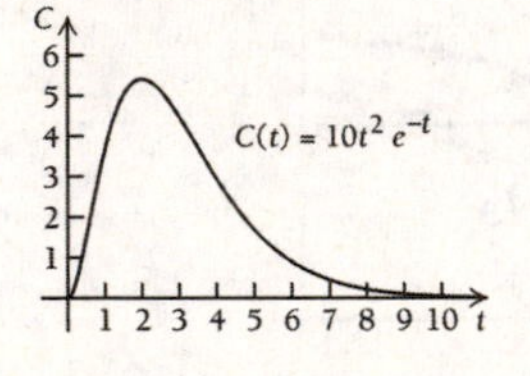

c) $C'(t) = 10t^2(-1)e^{-t} + 20te^{-t}$

$= (20t - 10t^2)e^{-t}$

d) Since $C'(x)$ exists for all values of $x \geq 0$, the only critical values are where $C'(x) = 0$. Solve:

$C'(x) = 0$

$(20t - 10t^2)e^{-t} = 0$

$20t - 10t^2 = 0 \quad (e^{-t} \neq 0)$

$10t(2 - t) = 0$

$t = 0 \ or \ t = 2$

The are two critical values. The critical value $t = 2$ is the obvious maximum.

The solution is continued on the next page.

Copyright © 2014 Pearson Education, Inc. Publishing as Addison-Wesley.

Find $C(2)$.

$$C(2)=10\cdot 2^2 e^{-2}$$
$$\approx 40(0.135335)$$
$$\approx 5.41$$

The maximum value of the concentration will be 5.41 parts per million and will occur 2 hours after the drug has been administered.

e)

89. $y=\left(e^{3x}+1\right)^5$

Using the Extended Power Rule

$$\frac{dy}{dx}=5\left(e^{3x}+1\right)^4\cdot 3e^{3x}$$
$$=15e^{3x}\left(e^{3x}+1\right)^4.$$

91. $y=\dfrac{e^{3t}-e^{7t}}{e^{4t}}$

Simplify the expression.

$$y=\frac{e^{3t}\left(1-e^{4t}\right)}{e^{3t}\cdot e^{t}}$$
$$=\frac{1-e^{4t}}{e^{t}}$$

Using the Quotient Rule.

$$\frac{dy}{dt}=\frac{e^{t}\left(-4e^{4t}\right)-e^{t}\left(1-e^{4t}\right)}{\left(e^{t}\right)^2}$$
$$=\frac{e^{t}\left(-4e^{4t}-1+e^{4t}\right)}{e^{t}\cdot e^{t}}$$
$$=\frac{-1-3e^{4t}}{e^{t}}$$
$$\frac{dy}{dt}=-e^{-t}-3e^{3t}.$$

93. $y=\dfrac{e^{x}}{x^2+1}$

$$\frac{dy}{dx}=\frac{\left(x^2+1\right)e^{x}-e^{x}(2x)}{\left(x^2+1\right)^2}\quad \text{Quotient Rule}$$
$$=\frac{e^{x}\left(x^2+1-2x\right)}{\left(x^2+1\right)^2}\quad \text{Factoring the Numerator}$$
$$=\frac{e^{x}\left(x^2-2x+1\right)}{\left(x^2+1\right)^2}$$
$$=\frac{e^{x}(x-1)^2}{\left(x^2+1\right)^2}$$

95. $f(x)=e^{\sqrt{x}}+\sqrt{e^{x}}$

$$=e^{x^{1/2}}+e^{x/2}$$
$$f'(x)=e^{x^{1/2}}\cdot\frac{1}{2}x^{-1/2}+\frac{1}{2}e^{x/2}\quad \text{Chain Rule}$$
$$=\frac{e^{\sqrt{x}}}{2\sqrt{x}}+\frac{\sqrt{e^{x}}}{2}$$

97. $f(x)=e^{x/2}\cdot\sqrt{x-1}$

First, we note that

$$f(x)=e^{x/2}\cdot(x-1)^{1/2}.$$

Next, we use the Product Rule.

$$f'(x)=e^{x/2}\cdot\frac{1}{2}(x-1)^{-1/2}+\frac{1}{2}e^{x/2}(x-1)^{1/2}$$
$$=\frac{1}{2}e^{x/2}\left[(x-1)^{-1/2}+(x-1)^{1/2}\right]\quad \text{Factoring}$$
$$=\frac{1}{2}e^{x/2}\left[\frac{1}{\sqrt{x-1}}+\sqrt{x-1}\right]$$
$$=\frac{1}{2}e^{x/2}\left[\frac{1}{\sqrt{x-1}}+\sqrt{x-1}\cdot\frac{\sqrt{x-1}}{\sqrt{x-1}}\right]\quad \text{Multiplying by 1}$$
$$=\frac{1}{2}e^{x/2}\left[\frac{1}{\sqrt{x-1}}+\frac{x-1}{\sqrt{x-1}}\right]$$
$$=\frac{1}{2}e^{x/2}\left[\frac{x}{\sqrt{x-1}}\right]\quad \text{Adding Fractions}$$
$$=e^{x/2}\left[\frac{x}{2\sqrt{x-1}}\right]$$

Copyright © 2014 Pearson Education, Inc. Publishing as Addison-Wesley.

99. $f(x)=\dfrac{e^x-e^{-x}}{e^x+e^{-x}}$

Using the Quotient Rule

$$f'(x)$$
$$=\frac{(e^x+e^{-x})(e^x+e^{-x})-(e^x-e^{-x})(e^x-e^{-x})}{(e^x+e^{-x})^2}$$
$$=\frac{(e^{2x}+e^0+e^0+e^{-2x})-(e^{2x}-e^0-e^0+e^{-2x})}{(e^x+e^{-x})^2}$$
$$=\frac{e^{2x}+e^0+e^0+e^{-2x}-e^{2x}+e^0+e^0-e^{-2x}}{(e^x+e^{-x})^2}$$
$$=\frac{4}{(e^x+e^{-x})^2}.$$

101. $e=\lim_{t\to 0} f(t);\ f(t)=(1+t)^{1/t}$

$$f(1)=(1+1)^{1/1}=2^1=2$$
$$f(0.5)=(1+0.5)^{1/0.5}=1.5^2=2.25$$
$$f(0.2)=(1+0.2)^{1/0.2}=1.2^5=2.48832$$
$$f(0.1)=(1+0.1)^{1/0.1}=1.1^{10}\approx 2.59374$$
$$f(0.001)=(1+0.001)^{1/0.001}=1.001^{1000}\approx 2.71692$$

103. $f(x)=x^2e^{-x};\ [0,4]$

$$f'(x)=x^2\cdot(-1)e^{-x}+2xe^{-x}$$
$$=-x^2e^{-x}+2xe^{-x}$$

Since $f'(x)$ exists for all values of x in $[0,4]$, the only critical values are where $f'(x)=0$.

$$f'(x)=0$$
$$-x^2e^{-x}+2xe^{-x}=0$$
$$-e^{-x}(x^2-2x)=0$$
$$x^2-2x=0 \qquad (-e^{-x}\neq 0)$$
$$x(x-2)=0$$
$$x=0 \text{ or } x=2$$

Use the Max-Min Principle 1. Find the function values at $x=0, x=2$, and $x=4$.

$$f(x)=x^2e^{-x}$$
$$f(0)=0^2\cdot e^{-0}=0\cdot 1=0$$
$$f(2)=2^2\cdot e^{-2}=4\cdot(0.135335)\approx 0.5413$$
$$f(4)=4^2\cdot e^{-4}=16(0.018315)\approx 0.2930$$

The maximum value of $f(x)$ is $4e^{-2}$ or approximately 0.5413 when $x=2$.

105. ✎

107. Graph: $y=x^2e^{-x}$

From Exercise 107 we see that here is a relative maximum at $(2,4e^{-2})$, or approximately $(2,0.5413)$.

Using our work from Exercise 107 and the First-Derivative Test we find that there is a relative minimum at $(0,0)$. Also observe that $y\geq 0$ for all x. We will calculate additional function values.

For:

$$x=-3,\ y=(-3)^2e^{-(-3)}=9e^3\approx 180.77$$
$$x=-2,\ y=(-2)^2e^{-(-2)}=4e^2\approx 29.56$$
$$x=-1,\ y=(-1)^2e^{-(-1)}=1e^1\approx 2.72$$
$$x=0,\ y=(0)^2e^{-(0)}=0\cdot 1=0$$
$$x=1,\ y=(1)^2e^{-(1)}=1e^{-1}\approx 0.37$$
$$x=2,\ y=(2)^2e^{-(2)}=4e^{-2}\approx 0.54$$
$$x=3,\ y=(3)^2e^{-(3)}=9e^{-3}\approx 0.45$$
$$x=4,\ y=(4)^2e^{-(4)}=16e^{-4}\approx 0.29$$
$$x=5,\ y=(5)^2e^{-(5)}=25e^{-5}\approx 0.17$$
$$x=10,\ y=(10)^2e^{-(10)}=100e^{-10}\approx 0.005$$

Next, we plot these points and connect them with a smooth curve.

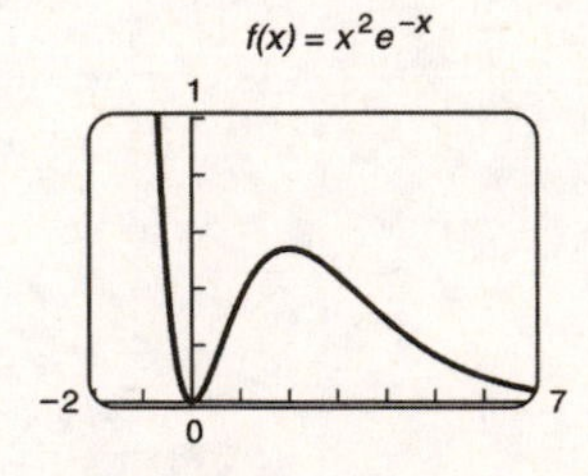

Copyright © 2014 Pearson Education, Inc. Publishing as Addison-Wesley.

109. See page 310 in the text. The graphs of $f(x), f'(x)$, and $f''(x)$ are all the graph of $y = e^x$.

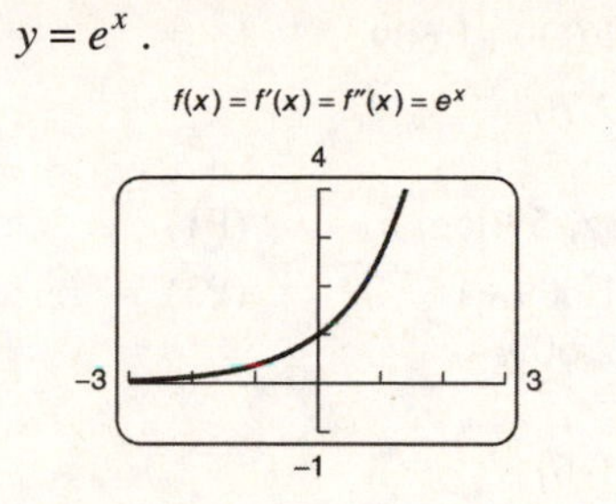

111. $f(x) = 2e^{0.3x}$

$f'(x) = 0.6e^{0.3x}$

$f''(x) = 0.18e^{0.3x}$

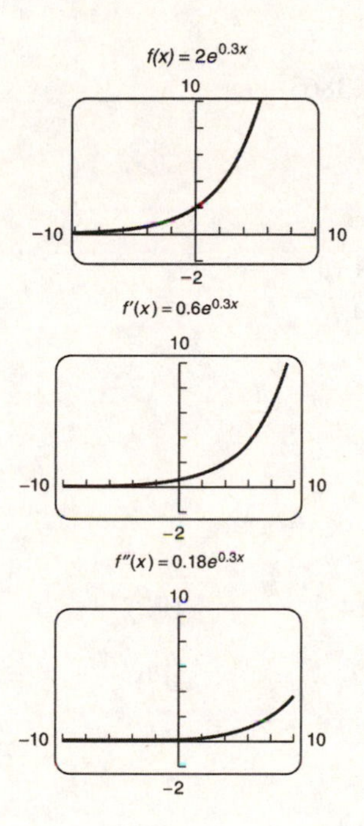

113.

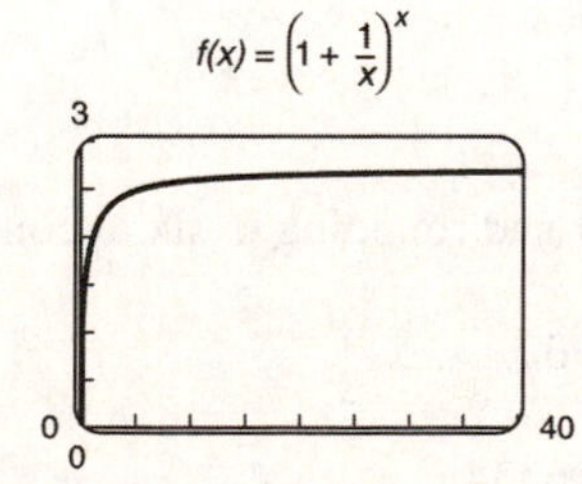

Copyright © 2014 Pearson Education, Inc. Publishing as Addison-Wesley.

Exercise Set 3.2

1. $\log_2 8 = 3$ Logarithmic equation.

$2^3 = 8$ Exponential equation; 2 is the base, 3 is the exponent.

3. $\log_8 2 = \frac{1}{3}$ Logarithmic equation.

$8^{1/3} = 2$ Exponential equation; 8 is the base, 3 is the exponent.

5. $\log_a K = J$ Logarithmic equation.

$a^J = K$ Exponential equation; a is the base, J is the exponent.

7. $-\log_{10} h = p$ Logarithmic equation.

$\log_{10} h = -p$ Multiply by -1.

$10^{-p} = h$ Exponential equation; 10 is the base, $-p$ is the exponent.

9. $e^M = b$ Exponential equation; e is the base, M is the exponent.

$\log_e b = M$ Logarithmic equation.

or $\ln b = M$ $\ln b$ is the abbreviation for $\log_e b$.

11. $10^2 = 100$ Exponential equation; 10 is the base, 2is the exponent.

$\log_{10} 100 = 2$ Logarithmic equation.

13. $10^{-1} = 0.1$ Exponential equation; 10 is the base, -1is the exponent.

$\log_{10} 0.1 = -1$ Logarithmic equation.

15. $M^p = V$ Exponential equation; M is the base, p is the exponent.

$\log_M V = p$ Logarithmic equation.

17. $\log_b\left(\frac{5}{3}\right) = \log_b 5 - \log_b 3$ (P2)

$= 1.609 - 1.099$

$= 0.51$

19. $\log_b 15 = \log_b(3\cdot 5)$

$= \log_b 3 + \log_b 5$ (P1)

$= 1.099 + 1.609$

$= 2.708$

21. $\log_b 5b = \log_b 5 + \log_b b$ (P1)

$= 1.609 + 1$ (P5)

$= 2.609$

23. $\ln 20 = \ln(4\cdot 5)$

$= \ln 4 + \ln 5$ (P1)

$= 1.3863 + 1.6094$

$= 2.9957$

25. $\ln\left(\frac{5}{4}\right) = \ln 5 - \ln 4$

$= 1.6094 - 1.3863$

$= 0.2231$

27. $\ln(5e) = \ln 5 + \ln e$

$= 1.6094 + 1$

$= 2.6094$

29. $\ln\sqrt{e^6} = \ln e^{6/2}$

$= \ln e^3$

$= 3$

31. $\ln\left(\frac{1}{4}\right) = \ln 1 - \ln 4$ (P2)

$= 0 - 1.3863$ (P6)

$= -1.3863$

33. $\ln\left(\frac{e}{5}\right) = \ln e - \ln 5$

$= 1 - 1.6094$

$= -0.6094$

35. Using a calculator and rounding to six decimal places, we have $\ln 5894 \approx 8.681690$.

37. $\ln 0.0182 \approx -4.006334$

39. $\ln 8100 \approx 8.999619$

Copyright © 2014 Pearson Education, Inc. Publishing as Addison-Wesley.

41. $e^t = 80$

$\ln e^t = \ln 80$ Taking the natural log on both sides

$t = \ln 80$ (P5)

$t \approx 4.382027$ Using a calculator

$t \approx 4.382$

43. $e^{2t} = 1000$

$\ln e^{2t} = \ln 1000$ Taking the natural log on both sides

$2t = \ln 1000$ (P5)

$t = \frac{\ln 1000}{2}$

$t \approx \frac{6.907755}{2}$ Using a calculator

$t \approx 3.453878$

$t \approx 3.454$

45. $e^{-t} = 0.1$

$\ln e^{-t} = \ln 0.1$ Taking the natural log on both sides

$-t = \ln 0.1$ (P5)

$t = -\ln 0.1$

$t \approx 2.302585$ Using a calculator

$t \approx 2.303$

47. $e^{-0.02t} = 0.06$

$\ln e^{-0.02t} = \ln 0.06$ Taking the natural log on both sides

$-0.02t = \ln 0.06$ (P5)

$t = \frac{\ln 0.06}{-0.02}$

$t \approx \frac{-2.813411}{-0.02}$ Using a calculator

$t \approx 140.671$

49. $y = -8\ln x$

$\frac{dy}{dx} = -8 \cdot \frac{1}{x}$ $\left[\frac{d}{dx}[c \cdot f(x)] = c \cdot f'(x)\right]$

$\frac{dy}{dx} = \frac{-8}{x}$

51. $y = x^4 \ln x - \frac{1}{2}x^2$

Differentiate this function term by term. We apply the Product Rule to the first term, and the Power Rule to the second term.

$$\frac{dy}{dx} = \underbrace{x^4 \cdot \frac{1}{x} + 4x^3 \cdot \ln x}_{\text{Product Rule}} - \underbrace{\frac{1}{2} \cdot 2x}_{\text{Power Rule}}$$

$$\frac{dy}{dx} = x^3 + 4x^3 \ln x - x$$

53. $f(x) = \ln(6x)$

Using Theorem 7, we have

$$\frac{d}{dx}\ln f(x) = \frac{1}{f(x)} \cdot f'(x) = \frac{f'(x)}{f(x)}$$

$$f'(x) = \frac{1}{6x} \cdot 6$$

$$= \frac{1}{x}$$

55. $g(x) = x^2 \ln(7x)$

We apply the Product Rule. Remember to use Theorem 7 when taking the derivative of the logarithm.

$$g'(x) = \underbrace{x^2 \cdot \underbrace{\frac{1}{7x} \cdot 7}_{\text{Theorem 7}} + 2x \cdot \ln(7x)}_{\text{Product Rule}}$$

$$g'(x) = x^2 \cdot \frac{1}{x} + 2x\ln(7x)$$

$$= x + 2x\ln(7x)$$

57. $y = \frac{\ln x}{x^4}$

Using the Quotient Rule, we have

$$\frac{dy}{dx} = \frac{x^4 \cdot \frac{1}{x} - 4x^3 \cdot \ln x}{x^8}$$

$$= \frac{x^3 - 4x^3 \ln x}{x^8}$$

$$= \frac{x^3(1 - 4\ln x)}{x^3 \cdot x^5} \quad \text{Factoring}$$

$$= \frac{x^3}{x^3} \cdot \frac{1 - 4\ln x}{x^5} \quad \text{Removing a factor equal to 1}$$

$$= \frac{1 - 4\ln x}{x^5}$$

59. $y = \ln\left(\frac{x^2}{4}\right)$

$y = \ln x^2 - \ln 4$ (P2)

$\frac{dy}{dx} = \frac{1}{x^2} \cdot 2x - 0$ Theorem 7

$= \frac{2}{x}$

Copyright © 2014 Pearson Education, Inc. Publishing as Addison-Wesley.

61. $y = \ln\left(3x^2 + 2x - 1\right)$

Using Theorem 7, we note:

$$\frac{d}{dx}\ln g(x) = \frac{g'(x)}{g(x)}$$

$$g(x) = 3x^2 + 2x - 1$$

$$g'(x) = 6x + 2.$$

Therefore,

$$\frac{dy}{dx} = \frac{6x+2}{3x^2+2x-1}$$

$$\frac{dy}{dx} = \frac{2(3x+1)}{3x^2+2x-1}.$$

63. $f(x) = \ln\left(\frac{x^2-7}{x}\right)$

Using the Theorem 7.

$$f'(x) = \frac{\left(\frac{d}{dx}\left[\frac{x^2-7}{x}\right]\right)}{\frac{x^2-7}{x}}$$

$$= \left(\frac{d}{dx}\left[\frac{x^2-7}{x}\right]\right)\cdot\frac{x}{x^2-7} \quad \text{Dividing fractions}$$

We apply the Quotient Rule to take the derivative of the inside function to get

$$\frac{d}{dx}\left[\frac{x^2-7}{x}\right] = \frac{x\cdot 2x - \left(x^2-7\right)\cdot 1}{x^2}$$

$$= \frac{2x^2 - x^2 + 7}{x^2}$$

$$= \frac{x^2+7}{x^2}.$$

Now we substitute back into the original derivative.

$$f'(x) = \frac{x^2+7}{x^2}\cdot\frac{x}{x^2-7}$$

$$= \frac{x^2+7}{x\left(x^2-7\right)}.$$

An alternate solution to this problem involves using property P2.

$$f(x) = \ln\left(\frac{x^2-7}{x}\right)$$

$$= \ln\left(x^2-7\right) - \ln x \quad \text{(P2)}$$

$$= \frac{2x}{\left(x^2-7\right)} - \frac{1}{x} \quad \text{Theorem 7}$$

$$= \frac{2x}{x^2-7} - \frac{1}{x}$$

Combine the fractions by finding a common denominator.

$$f'(x) = \frac{2x}{x^2-7}\cdot\frac{x}{x} - \frac{1}{x}\cdot\frac{x^2-7}{x^2-7} \quad \text{Multiply by 1}$$

$$= \frac{2x^2}{x\left(x^2-7\right)} - \frac{x^2-7}{x\left(x^2-7\right)}$$

$$= \frac{x^2+7}{x\left(x^2-7\right)}$$

65. $g(x) = e^x \ln x^2$

$$g'(x) = e^x \cdot \underbrace{\frac{2x}{x^2}}_{\text{Theorem 7}} + e^x \cdot \ln x^2 \quad \text{By the Product Rule}$$

$$= e^x \cdot \frac{2}{x} + e^x \ln x^2$$

$$= \frac{2e^x}{x} + 2e^x \ln x \quad \text{(P3)}$$

67. $f(x) = \ln\left(e^x + 1\right)$

$$f'(x) = \frac{\left(e^x + 0\right)}{e^x + 1} \quad \text{Theorem 7}$$

$$= \frac{e^x}{e^x+1}$$

69. $g(x) = (\ln x)^4$

Using the Extended Power Rule, we have

$$g'(x) = 4(\ln x)^3 \cdot \frac{1}{x}$$

$$= \frac{4(\ln x)^3}{x}.$$

Copyright © 2014 Pearson Education, Inc. Publishing as Addison-Wesley.

71. $f(x)=\ln(\ln(8x))$

First we apply Theorem 7.

$$f'(x)=\frac{1}{\ln(8x)}\cdot\left(\frac{d}{dx}\ln(8x)\right)$$

Using Theorem 7 again, we have

$$f'(x)=\frac{1}{\ln(8x)}\cdot\frac{1}{8x}\cdot 8$$
$$=\frac{1}{x\ln(8x)}.$$

73. $g(x)=\ln(5x)\cdot\ln(3x)$

Using the Product Rule along with Theorem 7, we have

$$g'(x)=\ln(5x)\cdot\underbrace{\left(\frac{3}{3x}\right)}_{\text{Theorem 7}}+\underbrace{\left(\frac{5}{5x}\right)}_{\text{Theorem 7}}\cdot\ln(3x)$$
$$=\ln(5x)\frac{1}{x}+\frac{1}{x}\ln(3x)$$
$$=\frac{\ln(5x)+\ln(3x)}{x}.$$

If we wanted to simplify the expression, we could use Property 1 to combine the logarithms.

$$g'(x)=\frac{\ln(5x\cdot 3x)}{x}=\frac{\ln(15x^2)}{x}$$

75. First, we find the point of tangency by evaluating the function at $x=2$.

$x=2;$

$$y=(2^2-2)\ln(6\cdot 2)=2\ln(12)\approx 4.9698$$

Point of tangency: $(2, 4.9698)$

Now, we find the slope of the function at $x=2$ by taking the derivative.

$$\frac{dy}{dx}=(x^2-x)\left(6\cdot\frac{1}{6x}\right)+(2x-1)\ln(6x)$$
$$=(x^2-x)\left(\frac{1}{x}\right)+(2x-1)\ln(6x)$$
$$=x-1+(2x-1)\ln(6x)$$

Next, we evaluate the derivative at $x=2$.

$$\left.\frac{dy}{dx}\right|_{x=2}=2-1+(2\cdot 2-1)\ln(6\cdot 2)$$
$$=1+3\ln(12)\approx 8.4547.$$

Now we have the slope and a point on the tangent line.

We use the Point-Slope formula to find the equation of the tangent line.

$$y-y_1=m(x-x_1)$$
$$y-4.970=8.455(x-2)$$
$$y-4.970=8.455x-16.91$$
$$y=8.455x-11.94$$

77. First, we find the point of tangency by evaluating the function at $x=3$.

$x=3;$

$$y=(\ln 3)^2\approx 1.207$$

Point of tangency: $(3, 1.207)$

Now, we find the slope of the function at $x=3$ by taking the derivative of the function using the Extended Power Rule.

$$\frac{dy}{dx}=2(\ln(x))\cdot\frac{1}{x}$$
$$=\frac{2\ln x}{x}$$

Next, we evaluate the derivative at $x=3$.

$$\left.\frac{dy}{dx}\right|_{x=3}=\frac{2\ln(3)}{3}\approx 0.732$$

Now we have the slope and a point on the tangent line. We use the Point-Slope formula to find the equation of the tangent line.

$$y-y_1=m(x-x_1)$$
$$y-1.207=0.732(x-3)$$
$$y-1.207=0.732x-2.196$$
$$y=0.732x-0.989$$

79. $N(a)=2000+500\ln a,\ a\geq 1$

a) We substitute 1 in for a.

$$N(1)=2000+500\cdot\ln 1$$
$$=2000+500\cdot 0$$
$$=2000$$

Thus, 2000 units were sold after spending $1000 on advertising.

b) Taking the derivative with respect to a, we have

$$N'(a)=0+500\cdot\frac{1}{a}$$
$$=\frac{500}{a}.$$

Therefore,

$$N'(10)=\frac{500}{10}=50.$$

Copyright © 2014 Pearson Education, Inc. Publishing as Addison-Wesley.

c) $N'(a) > 0$ for all $a \geq 1$. Thus $N(a)$ is an increasing function and has a minimum value of 2000 when $a = 1$ thousand dollars. There is no maximum, because there is not an upper limit on the advertising budget.

d) ✎

81. How long should the campaign last in order to maximize profit? We need to find the revenue and cost functions in order to find the profit. We find the revenue function first.

$$R(t) = \begin{pmatrix} \text{Price} \\ \text{Per} \\ \text{Unit} \end{pmatrix} \cdot \begin{pmatrix} \text{Target} \\ \text{Market} \end{pmatrix} \cdot \begin{pmatrix} \text{Percentage} \\ \text{Buying} \end{pmatrix}$$

$$R(t) = 0.50(1{,}000{,}000)\left(1 - e^{-0.04t}\right)$$
$$= 500{,}000 - 500{,}000e^{-0.04t}$$

Next, we find the cost function.

$$C(t) = \begin{pmatrix} \text{Advertising cost} \\ \text{per day} \end{pmatrix} \cdot \begin{pmatrix} \text{Number of} \\ \text{days} \end{pmatrix}$$

$C(t) = 2000 \cdot t.$

Now, we find the profit function.

$$P(t) = R(t) - C(t)$$
$$= 500{,}000 - 500{,}000e^{-0.04t} - 2000t.$$

We take the derivative of the profit function.

$$P'(t) = 0 - 500{,}000(-0.04)e^{-0.04t} - 2000$$
$$= 20{,}000e^{-0.04t} - 2000.$$

Next, we set the derivative of the profit function equal to zero and solve for t to find the critical values.

$$P'(t) = 0$$
$$20{,}000e^{-0.04t} - 2000 = 0$$
$$20{,}000e^{-0.04t} = 2000$$
$$e^{-0.04t} = \frac{2000}{20{,}000}$$
$$e^{-0.04t} = 0.1$$
$$\ln\left(e^{-0.04t}\right) = \ln(0.1)$$
$$-0.04t = -2.30258$$
$$t \approx \frac{-2.30258}{-0.04} \approx 57.565$$

Rounding up the value, we see that the only critical value is 58 days. We will apply the second derivative test to see if we have a maximum.

The second derivative of the profit function is $P''(t) = -800e^{-0.04t}$.

Evaluating this at the critical value we get $P''(58) = -800e^{-0.04 \cdot 58} < 0$ so we have a maximum.

The length of the advertising campaign must be 58 days to result in maximum profit.

83. $V(t) = 58\left(1 - e^{-1.1t}\right) + 20$

Where $V(t)$ is the value of the stock after time t, in months.

a) $$V(1) = 58\left(1 - e^{-1.1(1)}\right) + 20$$
$$= 58(1 - 0.332871) + 20$$
$$= 58(0.667129) + 20$$
$$= 38.693482 + 20$$
$$= 58.693482 \approx 58.69$$

The value of the stock one month after purchase is \$58.69

$$V(12) = 58\left(1 - e^{-1.1(12)}\right) + 20$$
$$= 58(1 - 0.000002) + 20$$
$$= 58(0.999998) + 20$$
$$= 57.999884 + 20$$
$$= 77.999884 \approx 78.00$$

The value of the stock 12 months after purchase is \$78.00

b) $$V'(t) = 58\left(0 - (-1.1)e^{-1.1t}\right) + 0$$
$$= 58\left(1.1e^{-1.1t}\right)$$
$$V'(t) = 63.8e^{-1.1t}$$

Copyright © 2014 Pearson Education, Inc. Publishing as Addison-Wesley.

c) Solve: $V(t)=75$

$$58\left(1-e^{-1.1t}\right)+20=75$$
$$58\left(1-e^{-1.1t}\right)=55$$
$$\left(1-e^{-1.1t}\right)=\frac{55}{58}$$
$$1-e^{-1.1t}=0.948276$$
$$-e^{-1.1t}=-0.051724$$
$$e^{-1.1t}=0.051724$$
$$\ln e^{-1.1t}=\ln(0.051724)$$
$$-1.1t=-2.96183$$
$$t=\frac{-2.96183}{-1.1}$$
$$t\approx 2.6926.$$

The stock will first reach $75 approximately 2.7 months after the purchase.

d) ✎

85. a) Taking the derivative of the profit function $P(x)=2x-0.3x\ln x$ will give us marginal profit.

$$P'(x)=2-\left[0.3x\left(\frac{1}{x}\right)+0.3\ln x\right]$$
$$=2-0.3-0.3\ln x$$
$$=1.7-0.3\ln x$$

b) ✎

c) Since $P'(x)$ is defined for all values of $x>0$. We find the critical values by setting $P'(x)=0$.

$$1.7-0.3\ln x=0$$
$$-0.3\ln x=-1.7$$
$$\ln x=\frac{-1.7}{-0.3}$$
$$\ln x=5.666667$$
$$x=e^{5.66667}$$
$$x\approx 289.069$$

Notice that:

$$P''(x)=\frac{-0.3}{x}<0 \text{ for all values of } x>0.$$

The critical value results in a maximum. Therefore, 289,069 mechanical pencils should be sold to maximize profit.

87. $S(t)=68-20\ln(t+1),\quad t\geq 0$

a) $S(0)=68-20\ln(0+1)$

$$=68-20\ln(1)$$
$$=68-20\cdot 0$$
$$=68$$

The average score when they initially took the exam was 68%.

b) $S(4)=68-20\ln(4+1)$

$$=68-20\ln(5)$$
$$=68-20(1.609438)$$
$$=68-32.18876$$
$$\approx 35.8$$

The average score 4 months after they took the exam was 35.8%.

c) $S(24)=68-20\ln(24+1)$

$$=68-20\ln(25)$$
$$=68-20(3.218876)$$
$$=68-64.37752$$
$$\approx 3.6$$

The average score 24 months after they took the exam was 3.6%.

d) First, we reword the question: "3.6 (the average score after 24 months) is what percent of 68 (the average score on the initial exam)?

Next, we translate the sentence into an equation and solve:

$$3.6=x\cdot 68$$
$$\frac{3.6}{68}=x$$
$$0.052941=x$$
$$5.3\%\approx x.$$

The students retained approximately 5.3% of their original answers after 2 years.

e) $S(t)=68-20\ln(t+1)$

$$S'(t)=0-20(1)\left(\frac{1}{t+1}\right)$$
$$=-\frac{20}{t+1}$$

f) $S'(t)<0$ for all values of $t\geq 0$. Therefore, $S(t)$ is a decreasing function and has a maximum value of 68% when $t=0$. The function does not have a minimum value.

g) ✎

Copyright © 2014 Pearson Education, Inc. Publishing as Addison-Wesley.

89. $v(p)=0.37\ln p+0.05$

a) $v(571)=0.37\ln 571+0.05$
$$=0.37(6.347389)+0.05$$
$$=2.398534$$
$$\approx 2.4$$
The average walking speed of a person living in Seattle is 2.40 feet per second.

b) $v(8100)=0.37\ln 8100+0.05$
$$=0.37(8.999619)+0.05$$
$$=3.379859$$
$$\approx 3.4$$
The average walking speed of a person living in New York is 3.4 feet per second.

c) $v'(p)=0.37\frac{1}{p}+0=\frac{0.37}{p}$

d) ✎

91. $P=P_0e^{kt}$

$\frac{P}{P_0}=e^{kt}$ Divide by P_0

$\ln\left(\frac{P}{P_0}\right)=\ln\left(e^{kt}\right)$ Take the Natural Log on both sides

$\ln\left(\frac{P}{P_0}\right)=kt$ (P5)

$\frac{\ln\left(\frac{P}{P_0}\right)}{k}=t$ Divide by k.

93. $f(t)=\ln\left(t^2-t\right)^7$

$f(t)=7\ln\left(t^2-t\right)$ (P3)

$f'(t)=7\left(\frac{(2t-1)}{t^2-t}\right)$ By Theorem 7

$f'(t)=\frac{7(2t-1)}{t^2-t}$ Simplifying

95. $f(x)=\ln\left[\ln\left(\ln(3x)\right)\right]$

Differentiating this function will require several applications of the Chain Rule. We will start on the outside and work our way in to the middle.

$$f'(x)=\left(\frac{1}{\ln(\ln(3x))}\right)\cdot\frac{d}{dx}\left(\ln(\ln(3x))\right)$$
$$=\left(\frac{1}{\ln(\ln(3x))}\right)\cdot\left[\left(\frac{1}{\ln(3x)}\right)\cdot\frac{d}{dx}\left(\ln(3x)\right)\right]$$
$$=\left(\frac{1}{\ln(\ln(3x))}\right)\cdot\left(\frac{1}{\ln(3x)}\right)\left[\frac{1}{3x}\cdot 3\right]$$
$$=\frac{1}{x\ln(3x)\cdot\ln(\ln(3x))}$$

97. $f(t)=\ln\frac{1-t}{1+t}$

$f(t)=\ln(1-t)-\ln(1+t)$ (P2)

$f'(t)=\frac{(-1)}{1-t}-\frac{(1)}{1+t}$ By Theorem 7

$=\frac{-1}{1-t}-\frac{1}{1+t}$

$=\frac{1}{(-1)(1-t)}-\frac{1}{1+t}$ Properties of fractions

$=\frac{1}{(t-1)}-\frac{1}{t+1}$ Distribution

Next, we find a common denominator.

$=\frac{t+1}{(t-1)(t+1)}-\frac{t-1}{(t-1)(t+1)}$

$f'(t)=\frac{2}{t^2-1}$

99. $f(x)=\log_5 x$

Using (P7), the change-of-base formula.

$f(x)=\frac{\ln x}{\ln 5}$

Remember that $\ln 5$ is a constant.

$f'(x)=\frac{1}{\ln 5}\cdot\frac{1}{x}$

$=\frac{1}{x\ln 5}$

101. $y=\ln\sqrt{5+x^2}$

$y=\ln\left(5+x^2\right)^{1/2}$

$y=\frac{1}{2}\ln\left(5+x^2\right)$ (P3)

$\frac{dy}{dx}=\frac{1}{2}\cdot\frac{2x}{5+x^2}$ Theorem 7

$\frac{dy}{dx}=\frac{x}{5+x^2}$

Copyright © 2014 Pearson Education, Inc. Publishing as Addison-Wesley.

103. $f(x)=\frac{1}{5}x^5\left(\ln x-\frac{1}{5}\right)$

Using the Product Rule, we have:

$$f'(x)=\frac{1}{5}x^5\left(\frac{1}{x}-0\right)+\frac{1}{5}\left(5x^4\right)\left(\ln x-\frac{1}{5}\right)$$
$$=\frac{1}{5}x^4+x^4\ln x-\frac{1}{5}x^4$$
$$=x^4\ln x.$$

105. $f(x)=\ln\frac{1+\sqrt{x}}{1-\sqrt{x}}$

$f(x)=\ln\left(1+\sqrt{x}\right)-\ln\left(1-\sqrt{x}\right)$ (P2)

We use theorem 7, and the Chain Rule to differentiate both logarithms.

Note: $\frac{d}{dx}\left(1+\sqrt{x}\right)=\frac{1}{2}x^{-1/2}$

$\frac{d}{dx}\left(1-\sqrt{x}\right)=-\frac{1}{2}x^{-1/2}$

$$f'(x)=\frac{\left(\frac{1}{2}\right)x^{-1/2}}{1+\sqrt{x}}-\frac{\left(-\frac{1}{2}\right)x^{-1/2}}{1-\sqrt{x}}$$

Simplifying

$$f'(x)=\frac{1}{1+\sqrt{x}}\cdot\frac{1}{2x^{1/2}}+\frac{1}{1-\sqrt{x}}\cdot\frac{1}{2x^{1/2}}$$
$$=\frac{1}{2\sqrt{x}\left(1+\sqrt{x}\right)}+\frac{1}{2\sqrt{x}\left(1-\sqrt{x}\right)}.$$

Now, we combine the fractions.

$f'(x)$
$$=\frac{1}{2\sqrt{x}\left(1+\sqrt{x}\right)}\cdot\frac{1-\sqrt{x}}{1-\sqrt{x}}+\frac{1}{2\sqrt{x}\left(1-\sqrt{x}\right)}\cdot\frac{1+\sqrt{x}}{1+\sqrt{x}}$$
$$=\frac{1-\sqrt{x}}{2\sqrt{x}\left(1+\sqrt{x}\right)\left(1-\sqrt{x}\right)}+\frac{1+\sqrt{x}}{2\sqrt{x}\left(1+\sqrt{x}\right)\left(1-\sqrt{x}\right)}$$
$$=\frac{2}{2\sqrt{x}\left(1+\sqrt{x}\right)\left(1-\sqrt{x}\right)}$$
$$=\frac{1}{\sqrt{x}\left(1+\sqrt{x}\right)\left(1-\sqrt{x}\right)}$$

or, if we multiply the two binomials

$$f'(x)=\frac{1}{\sqrt{x}(1-x)}.$$

107. Let $X=\log_a M$ and $Y=\log_a N$

Proof of Property 1:

$M=a^X$ and $N=a^Y$ <u>Definition of Logarithm</u>

So,

$MN=a^Xa^Y=a^{X+Y}$ <u>Product Rule for Exponents</u>

Thus,

$\log_a(MN)=X+Y$ <u>Definition of Logarithm</u>

$=\log_a M+\log_a N$ <u>Substitution</u>

109. Proof of Property 3:

$M=a^X$ <u>Definition of Logarithm</u>

So,

$M^k=\left(a^X\right)^k$

$\left[a=b\Rightarrow a^c=b^c\right]$

$=a^{X\cdot k}$ <u>Power Rule for Exponents</u>

Thus,

$\log_a\left(M^k\right)=X\cdot k$ <u>Definition of Logarithm</u>

$=k\cdot\log_a M$ <u>Substitution</u>

111. $\lim_{h\to0}\frac{\ln(1+h)}{h}$

Note: $\ln(1+h)$ exists only for $h>-1$.

Since the function is not continuous at $h=0$, we will use input-output tables.

First, we look as h approaches 0 from the left.

h	$\frac{\ln(1+h)}{h}$
-0.9	2.56
-0.5	1.39
-0.1	1.05
-0.01	1.01
-0.001	1.001

From the table we observe that

$$\lim_{h\to0^-}\frac{\ln(1+h)}{h}=1.$$

Now, we look as h approaches 0 from the right.

h	$\frac{\ln(1+h)}{h}$
0.9	0.71
0.5	0.81
0.1	0.95
0.01	0.995
0.001	0.9995

The solution is continued on the next page.

Copyright © 2014 Pearson Education, Inc. Publishing as Addison-Wesley.

From the table on the previous page we observe:

$$\lim_{h\to 0^+} \frac{\ln(1+h)}{h} = 1.$$

Thus,

$$\lim_{h\to 0} \frac{\ln(1+h)}{h} = 1.$$

113. We consider the function $y = \frac{\ln x}{x}$. It can be shown that this function has a maximum at $x = e$. Thus:

$$\frac{\ln e}{e} > \frac{\ln \pi}{\pi} \quad \text{Definition of Maximum}$$

$$\pi \cdot \ln e > e \cdot \ln \pi$$

$$\ln e^{\pi} > \ln \pi^{e} \quad \text{(P3)}$$

$$e^{\pi} > \pi^{e}.$$

Therefore; e^{π} is larger.

115. Find $\lim_{x\to 1} \ln x$

First, we observe as x approaches 1 from the left.

x	$\ln x$
0.1	-2.30
0.4	-0.92
0.7	-0.36
0.9	-0.11
0.99	-0.01
0.999	-0.001

From the table we observe:

$$\lim_{x\to 1^-} \ln x = 0$$

Now, we observe as x approaches 1 from the right.

x	$\ln x$
1.9	0.64
1.7	0.53
1.4	0.34
1.1	0.10
1.01	0.01
1.001	0.001

From the table we observe:

$$\lim_{x\to 1^+} \ln x = 0$$

Thus, $\lim_{x\to 1} \ln x = 0$.

117. Using the window:

```
WINDOW
  Xmin=-2
  Xmax=10
  Xscl=1
  Ymin=-5
  Ymax=5
  Yscl=1
  Xres=1
```

We graph the functions using a calculator.

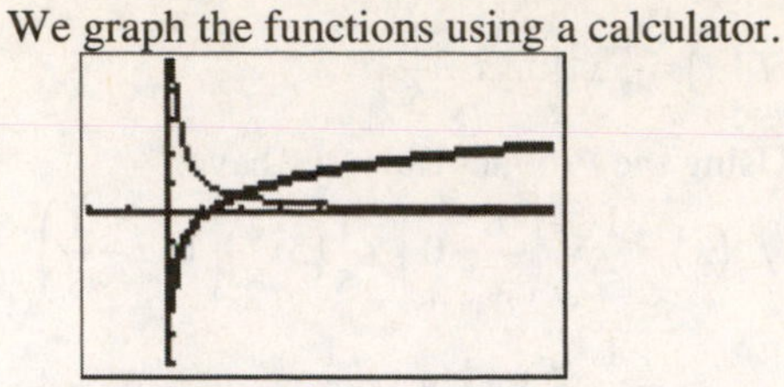

Notice the function is the thicker graph.

119. Using the window:

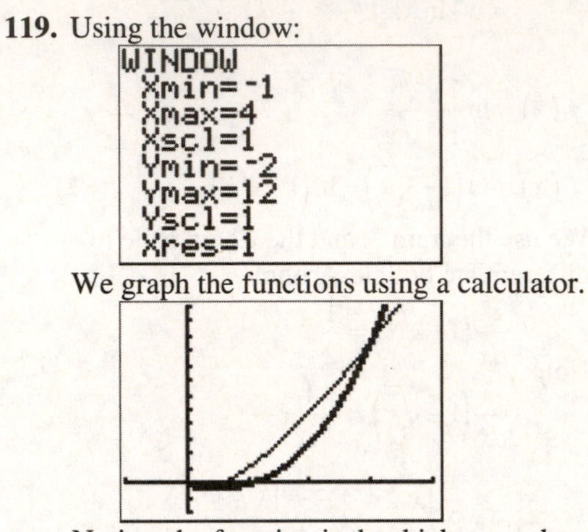

We graph the functions using a calculator.

Notice, the function is the thicker graph.

121. $f(x) = x \ln x$

$$f'(x) = x \cdot \frac{1}{x} + \ln x = 1 + \ln x$$

$f'(x)$ exists for all x in the domain of $f(x)$.

Solve:

$$f'(x) = 0$$

$$1 + \ln x = 0$$

$$\ln x = -1$$

$$x = e^{-1}.$$

The critical value occurs when $x = e^{-1}$.

Use the Second-Derivative test to determine if the critical value represents a relative maximum or a relative minimum.

$$f''(x) = \frac{1}{x}$$

$$f''(e^{-1}) = \frac{1}{e^{-1}} = e > 0$$

The function is concave up and there is a relative minimum at the critical value $x = e^{-1}$.

$$f(e^{-1}) = e^{-1} \ln e^{-1}$$

$$= e^{-1} \cdot (-1) = -\frac{1}{e} \approx -0.368$$

The minimum value that occurs at $x = e^{-1}$ is approximately -0.368.

Copyright © 2014 Pearson Education, Inc. Publishing as Addison-Wesley.

Exercise Set 3.3

1. Using Theorem 8, the general form of f that satisfies the equation $f'(x) = 4 \cdot f(x)$ is $f(x) = ce^{4x}$ for some constant c.

3. Using Theorem 8, the general form of A that satisfies the equation $\frac{dA}{dt} = -9 \cdot A$ is $A = ce^{-9t}$, or $A(t) = ce^{-9t}$ for some constant c.
Note: If we were to let the initial population, the population when $t = 0$, be represented by $A(0) = A_0$, then we have $A_0 = A(0) = ce^{-9 \cdot 0} = ce^0 = c$. Thus, $A_0 = c$, and we can express $A(t) = A_0 e^{-9t}$.

5. Using Theorem 8, the general form of Q that satisfies the equation $\frac{dQ}{dt} = k \cdot Q$ is $Q = ce^{kt}$, or $Q(t) = ce^{kt}$ for some constant c.
Note: If we were to let the initial population, the population when $t = 0$, be represented by $Q(0) = Q_0$, then we have $Q_0 = Q(0) = ce^{k \cdot 0} = ce^0 = c$. Thus, $Q_0 = c$, and we can express $Q(t) = Q_0 e^{kt}$.

7. a) Using Theorem 8, the general form of N that satisfies the equation $N'(t) = 0.046 \cdot N(t)$ is $N(t) = ce^{0.046t}$ for some constant c.
Allowing $t = 0$ to correspond to 1980 when approximately 112,000 patent applications were received, gives us the initial condition $N(0) = 112,000$. Therefore:

$$112,000 = ce^{0.046 \cdot 0}$$
$$112,000 = ce^0$$
$$112,000 = c$$

Substituting this value for c. We get: $N(t) = 112,000e^{0.046t}$.

b) In 2010, $t = 2010 - 1980 = 30$.

$$\begin{aligned} N(30) &= 112,000e^{0.046(30)} \\ &= 112,000e^{1.38} \\ &= 112,000(3.974902) \\ &= 445,188.98 \\ &\approx 445,189 \end{aligned}$$

There will be approximately 445,189 patent applications received in 2010.

c) From Theorem 9, the doubling time T is given by $T = \frac{\ln 2}{k}$.

$$T = \frac{\ln 2}{0.046} \approx 15.1$$

It will take approximately 15.1 years for the number of patent applications to double. This means that in 1995, the number of patents will have doubled.

9. a) Using Theorem 8, the general form of P that satisfies the equation $\frac{dP}{dt} = 0.059 \cdot P(t)$ is $P(t) = P_0 e^{0.059t}$ for some initial principal P_0.

b) If \$1000 is invested, then $P_0 = 1000$ and $P(t) = 1000e^{0.059t}$.

$$\begin{aligned} P(1) &= 1000e^{0.059(1)} \\ &= 1000e^{0.059} \\ &= 1000(1.06077524074) \\ &\approx 1060.78 \end{aligned}$$

The balance after 1 year is \$1060.78.

$$\begin{aligned} P(2) &= 1000e^{0.059(2)} \\ &= 1000e^{0.118} \\ &= 1000(1.1252441) \\ &\approx 1125.24 \end{aligned}$$

The balance after 2 years is \$1125.24.

c) From Theorem 9, the doubling time T is given by $T = \frac{\ln 2}{k}$.

$$T = \frac{\ln 2}{0.059} \approx 11.7$$

It will take approximately 11.7 years for the balance to double.

Copyright © 2014 Pearson Education, Inc. Publishing as Addison-Wesley.

11. a) Using Theorem 8, the general form of G that satisfies the equation

$\frac{dG}{dt} = 0.093 \cdot G(t)$ is

$G(t) = ce^{0.093t}$ for some constant c.

Allowing $t = 0$ to correspond to 2000 when approximately 4.7 billion gallons of bottled water were sold, gives us the initial condition $G(0) = 4.7$. Therefore:

$$4.7 = ce^{0.093 \cdot 0}$$
$$4.7 = ce^{0}$$
$$4.7 = c$$

Substituting this value for c. We get:

$$G(t) = 4.7e^{0.093t}$$

Where $G(t)$ is in billions of gallons sold and t is the number of years since 2000.

b) In 2010, $t = 2010 - 2000 = 10$.

$$\begin{aligned} N(10) &= 4.7e^{0.093(10)} \\ &= 4.7e^{0.93} \\ &= 4.7(2.534509) \\ &= 11.912193 \\ &\approx 11.91 \end{aligned}$$

There will be approximately 11.91 billion gallons of bottle water sold in 2010.

c) From Theorem 9, the doubling time T is given by $T = \frac{\ln 2}{k}$.

$$T = \frac{\ln 2}{0.093} \approx 7.5$$

It will take approximately 7.5 years for the amount of bottled water sold to double. Sometime in the middle of 2007 according to our model.

13. From Theorem 9, the doubling time T is given by $T = \frac{\ln 2}{k}$. Substituting $T = 15$ we get:

$$15 = \frac{\ln 2}{k}.$$

Solve for k to find the interest rate.

$$15 \cdot k = \ln 2$$
$$k = \frac{\ln 2}{15}$$
$$k \approx 0.046210.$$

The annual interest rate is 4.62%.

15. From Theorem 9, the doubling time T is given by $T = \frac{\ln 2}{k}$. Substituting the growth rate $k = 0.10$ into the formula.

$$\begin{aligned} T &= \frac{\ln 2}{0.10} \\ &\approx 6.9 \end{aligned}$$

The doubling time for the demand for oil is 6.9 years; therefore, at the end of the year 2012 the demand for oil will be double the demand for oil in 2006.

17. Find the doubling time:

$$T = \frac{\ln 2}{k} = \frac{\ln 2}{0.062} \approx 11.2$$

The doubling time is 11.2 years.

$$P(t) = P_0 e^{kt}$$
$$P(t) = 75{,}000e^{0.062t}$$

We substitute $t = 5$ to find the amount after five years:

$$\begin{aligned} P(5) &= 75{,}000e^{0.062(5)} \\ &= 75{,}000e^{0.31} \\ &\approx 102{,}256.88. \end{aligned}$$

In five years the account will have $102,256.88.

19. First, we find the initial investment:

$$P(t) = P_0 e^{kt}$$

Substituting, we have

$$11{,}414.71 = P_0 e^{0.084(5)}$$
$$11{,}414.71 = P_0 e^{0.42}$$
$$\frac{11{,}414.71}{e^{0.42}} = P_0$$
$$7499.9989 \approx P_0$$
$$7500.00 \approx P_0$$

The initial investment is $7500.

Next, we find the doubling time:

$$T = \frac{\ln 2}{k} = \frac{\ln 2}{0.084} \approx 8.3.$$

The doubling time is 8.3 years.

Copyright © 2014 Pearson Education, Inc. Publishing as Addison-Wesley.

21. a) The exponential growth function is $V(t) = V_0 e^{kt}$. We will express $V(t)$ in dollars and t as the number of years since 1950. This set up gives us the initial value of $V_0 = 30,000$ dollars. Substituting this into the function, we have:

$V(t) = 30,000e^{kt}$

In 2004, $t = 2004 - 1950 = 54$, and the painting sold for 104,168,000 so we know $V(54) = 104,168,000$ million dollars.

Substitute this information into the function and solve for k.

$$104,168,000 = 30,000e^{k(54)}$$

$$\frac{104,168,000}{30,000} = e^{54k}$$

$$\ln\left(\frac{104,168,000}{30,000}\right) = \ln e^{54k}$$

$$\ln\left(\frac{104,168,000}{30,000}\right) = 54k$$

$$\frac{\ln\left(\frac{104,168,000}{30,000}\right)}{54} = k$$

$$0.151 \approx k$$

The exponential growth rate is approximately 15.1%. Substituting this back into the function, we find the exponential growth function is $V(t) = 30,000e^{0.151t}$.

b) In 2015, $t = 2015 - 1950 = 65$. We evaluate the exponential growth function found in part 'a' to get:

$$V(65) = 30,000e^{0.151(65)}$$

$$= 30,000e^{9.815}$$

$$= 30,000(18,306.29)$$

$$\approx 549,188,702$$

In the year 2015, the value of the painting will be approximately $549,188,702.

c) From Theorem 9, the doubling time T is given by $T = \frac{\ln 2}{k}$.

$$T = \frac{\ln 2}{0.151} \approx 4.59 \approx 4.6$$

The painting doubles in value approximately every 4.6 years.

d) We set $V(t) = 1,000,000,000$ and solve for t.

$$30,000e^{0.151t} = 1,000,000,000$$

$$e^{0.151t} = \frac{1,000,000,000}{30,000}$$

$$\ln e^{0.151t} = \ln\left(\frac{1,000,000,000}{30,000}\right)$$

$$0.151t = \ln\left(\frac{1,000,000,000}{30,000}\right)$$

$$t = \frac{\ln\left(\frac{1,000,000,000}{30,000}\right)}{0.151}$$

$$t \approx 68.969$$

$$t \approx 69$$

It will take approximately 69 years for the value of the painting to reach 1 billion dollars. This will occur in the year 2019. $[1950 + 69 = 2019]$

23. a) The exponential growth function is $E(t) = E_0 e^{kt}$. We will express t as the number of years since 1990. This set up gives us the initial value of $E_0 = 1.031$ billion dollars.

Substituting this into the function, we have:

$E(t) = 1.031e^{kt}$

In 2009, $t = 2009 - 1990 = 19$, and federal receipts were $2.523 billion so we know $E(19) = 2.523$ billion dollars.

Substitute this information into the function and solve for k.

$$2.523 = 1.031e^{k(19)}$$

$$\frac{2.523}{1.031} = e^{19k}$$

$$\ln\left(\frac{2.523}{1.031}\right) = \ln e^{19k}$$

$$\ln\left(\frac{2.532}{1.031}\right) = 19k$$

$$\frac{\ln\left(\frac{2.523}{1.031}\right)}{19} = k$$

$$0.047101 \approx k$$

The exponential growth rate is 0.047101. Substituting this back into the function, we find the exponential growth function is $E(t) = 1.031e^{0.047101t}$.

Copyright © 2014 Pearson Education, Inc. Publishing as Addison-Wesley.

b) In 2015, $t = 2015 - 1990 = 25$. We evaluate the exponential growth function found in part 'a' to get:

$$\begin{aligned} E(25) &= 1.031e^{0.047101(25)} \\ &= 1.031e^{1.177525} \\ &= 1.031(3.24632958606) \\ &\approx 3.347 \end{aligned}$$

In the year 2015, federal receipts will be approximately \$3.347 billion.

d) We set $E(t) = 10$ and solve for t.

$$\begin{aligned} E(t) &= 10 \\ 1.031e^{0.047101t} &= 10 \\ e^{0.047101t} &= \frac{10}{1.031} \\ 0.047101t &= \ln\left(\frac{10}{1.031}\right) \\ t &= \frac{\ln\left(\frac{10}{1.031}\right)}{0.047101} \\ t &\approx 48.2 \end{aligned}$$

It will take approximately 48.2 years from 1990 or in the year 2038for federal receipts to reach \$10 billion.

25. a) Enter the data into your calculator statistics editor.

L1	L2	L3 1
10	280	------
11	294	
12	309	
13	324	
14	350	
15	406	

L1(7)=

Now use the regression command to find the exponential growth function.

This gives us:

The calculator gives us the function $y = 136.3939183(1.071842825)^x$, where y is in millions of dollars and x is the number of years after 1990.

From the technology connection section, we know that $b^x = e^{x(\ln b)}$. Apply this to the model our calculator found to get:

$$\begin{aligned} 1.071842825^x &= e^{x(\ln 1.071842825)} \\ &= e^{0.0693794334x} \end{aligned}$$

Rounding the values and substituting, we get the exponential growth function:

$P(t) = 136.4e^{0.0694t}$, where t is years since 1990, $k \approx 0.069$, or 6.9%, and $P_0 \approx 136.4$ million.

b) Using the model in part 'a'.

In 2007, $t = 2007 - 1990 = 17$

$$P(17) = 136.4e^{0.0964(17)} \approx 443.8$$

In 2007, paper shredder sales were approximately \$444 million.

In 2012, $t = 2012 - 1990 = 22$

$$P(22) = 136.4e^{0.0694(22)} \approx 627.9$$

In 2012, paper shredder sales were approximately \$628 million.

c) We need to solve the equation $P(t) = 500$ for t.

$$\begin{aligned} 136.4e^{0.0694t} &= 500 \\ e^{0.0694t} &= \frac{500}{136.4} \\ \ln e^{0.0694t} &= \ln\left(\frac{500}{136.4}\right) \\ 0.0694t &= \ln\left(\frac{500}{136.4}\right) \\ t &= \frac{\ln\left(\frac{500}{136.4}\right)}{0.0694} \\ t &\approx 18.7 \end{aligned}$$

It will take approximately 18.7 years for the total sales of paper shredders to reach \$500 million. This should occur in the year 2008.

d) From Theorem 9, the doubling time T is given by $T = \frac{\ln 2}{k}$.

$$T = \frac{\ln 2}{0.0694} \approx 9.9877 \approx 10.0$$

The doubling time for sales of paper shredders is approximately 10 years.

Copyright © 2014 Pearson Education, Inc. Publishing as Addison-Wesley.

27. If we let $t = 0$ correspond to the year 1626 the initial value of Manhattan is $V_0 = 24$ Using the exponential growth function $V(t) = V_0 e^{kt}$.
Assuming an exponential rate of inflation of 5% means that $k = 0.05$. Substituting these values into the growth function gives us:
$V(t) = 24e^{0.05t}$.
In 2020, $t = 2020 - 1626 = 394$

$$V(394) = 24e^{0.05(394)}$$
$$= 24e^{19.7}$$
$$\approx 8,626,061,203$$

Manhattan Island will be worth approximately $8,626,061,203 or $8.6 billion.

29. a) The exponential growth function is $S(t) = S_0 e^{kt}$. We will express $S(t)$ in cents and t as the number of years since 1962. This set up gives us the initial value of $S_0 = 4$ dollars.
Substituting this into the function, we have:
$S(t) = 4e^{kt}$
In 2011, $t = 2011 - 1962 = 49$, and the price of a stamp was 45 cents so we know $S(49) = 45$ cents.
Substitute this information into the function and solve for *k*.

$$45 = 4e^{k(49)}$$
$$\frac{45}{4} = e^{49k}$$
$$\ln(11.25) = \ln e^{49k}$$
$$\ln(11.25) = 49k$$
$$\frac{\ln(11.25)}{49} = k$$
$$0.0494 \approx k$$

Substituting this value back into the function, we find the exponential growth function is
$S(t) = 4e^{0.0494t}$.

b) The exponential growth rate is approximately 4.94% per year.

c) In 2013, $t = 2013 - 1962 = 51$. We evaluate the exponential growth function found in part 'a' to get:

$$S(51) = 4e^{0.0494(51)}$$
$$= 4e^{2.5194}$$
$$= 4(12.42114174)$$
$$\approx 50$$

In the year 2013, the price of a stamp will be approximately 50 cents.
In 2016, $t = 2016 - 1962 = 54$. We evaluate the exponential growth function found in part 'a' to get:

$$S(54) = 4e^{0.0494(54)}$$
$$= 4e^{2.6676}$$
$$= 4(14.40535482)$$
$$\approx 58$$

In the year 2016, the price of a stamp will be approximately 58 cents.
In 2019, $t = 2019 - 1962 = 57$. We evaluate the exponential growth function found in part 'a' to get:

$$S(57) = 4e^{0.0494(57)}$$
$$= 4e^{2.8158}$$
$$= 4(16.70653566)$$
$$\approx 67$$

In the year 2019, the price of a stamp will be approximately 67 cents.

d) For the years 2011 – 2021, the total cost of *Forever Stamps* would be
$11 \cdot 4500 = \$49,500$.
The solution is continued on the next page.
For the years 2011 – 2012 the cost of regular first-class stamps was
$2 \cdot (4500) = \$9000$.

For the years 2013 – 2015, the price of postage increased to $0.50. The cost of regular first-class stamps for these 3 years is
$3 \cdot (0.50) \cdot (10,000) = \$15,000$.

For the years 2016 – 2018, the price of postage increased to $0.58. The cost of regular first-class stamps for these 3 years is
$3 \cdot (0.58) \cdot (10,000) = \$17,400$.

For the years 2019 – 2021, the price of postage increased to $0.67. The cost of regular first-class stamps for these 3 years is
$3 \cdot (0.67) \cdot (10,000) = \$20,100$.

The solution is continued on the next page.

Copyright © 2014 Pearson Education, Inc. Publishing as Addison-Wesley.

Therefore, the cost of regular first-class stamps for the years 2009 – 2020 would be: $\$9000+\$15,000+\$17,400+\$20,100$, or $\$61,500$.

Thus, by buying *Forever Stamps* the firm would save

$\$61,500-\$49,500=\$12,000$.

e)

31. $P(x)=\frac{100}{1+49e^{-0.13x}}$

a) $P(0)=\frac{100}{1+49e^{-0.13(0)}}$

$=\frac{100}{1+49e^{0}}$

$=\frac{100}{1+49}$

$=\frac{100}{50}$

$=2$

About 2% purchased the game without seeing the advertisement.

b) $P(5)=\frac{100}{1+49e^{-0.13(5)}}$

$=\frac{100}{1+49e^{-0.65}}$

≈ 3.8

About 3.8% will purchase the game after the advertisement runs 5 times.

$P(10)=\frac{100}{1+49e^{-0.13(10)}}$

$=\frac{100}{1+49e^{-1.3}}$

≈ 7.0

About 7.0% will purchase the game after the advertisement runs 10 times.

$P(20)=\frac{100}{1+49e^{-0.13(20)}}$

$=\frac{100}{1+49e^{-2.6}}$

≈ 21.6

About 21.6% will purchase the game after the advertisement runs 20 times.

$P(30)=\frac{100}{1+49e^{-0.13(30)}}$

$=\frac{100}{1+49e^{-3.9}}$

≈ 50.2

About 50.2% will purchase the game after the advertisement runs 30 times.

$P(50)=\frac{100}{1+49e^{-0.13(50)}}$

$=\frac{100}{1+49e^{-6.5}}$

≈ 93.1

About 93.1% will purchase the game after the advertisement runs 50 times.

$P(60)=\frac{100}{1+49e^{-0.13(60)}}$

$=\frac{100}{1+49e^{-7.8}}$

≈ 98.0

About 98.0% will purchase the game after the advertisement runs 60 times.

c) We apply the Quotient Rule to take the derivative.

$P'(x)$

$=\frac{\left(1+49e^{-0.13x}\right)(0)-49(-0.13)e^{-0.13x}\cdot 100}{\left(1+49e^{-0.13x}\right)^2}$

$=\frac{637e^{-0.13x}}{\left(1+49e^{-0.13x}\right)^2}$

d) The derivative $P'(x)$ exists for all real numbers. The equation $P'(x)=0$ has no solution. Thus, the function has no critical points and hence, no relative extrema. $P'(x)>0$ for all real numbers, so $P(x)$ is increasing on $[0,\infty)$. The second derivative can be used to show that the graph has an inflection point at $(29.9,50)$. The function is concave up on the interval $(0,29.9)$ and concave down on the interval $(29.9,\infty)$

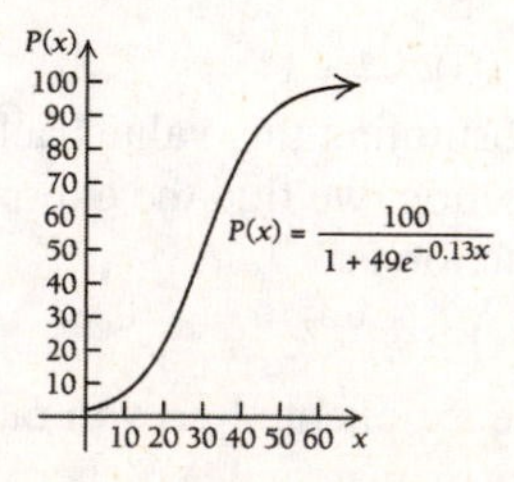

Copyright © 2014 Pearson Education, Inc. Publishing as Addison-Wesley.

33. a) Using Theorem 8, the general form of V that satisfies the equation

$\frac{dV}{dt} = k \cdot V(t)$ is

$V(t) = V_0 e^{kt}$ for some constant c.

Allowing $t = 0$ to correspond to 1938, gives us the initial condition

$V(0) = 0.10$. Therefore:

$0.10 = V_0 e^{k \cdot 0}$

$0.10 = V_0$

Substituting this value for c. We get:

$V(t) = 0.01 e^{kt}$.

Next, using the point $(72, 1{,}000{,}000)$ we have:

$1{,}000{,}000 = 0.01 e^{k(72)}$.

We solve this equation for k on the next page.

$$1{,}000{,}000 = 0.01 e^{k(72)}$$

$$\frac{1{,}000{,}000}{0.01} e^{72k}$$

$$100{,}000{,}000 = e^{72k}$$

$$\ln(100{,}000{,}000) = 72k$$

$$\frac{\ln(100{,}000{,}000)}{72} = k$$

$$0.2558 \approx k$$

The exponential function is

$V(t) = 0.01 e^{0.256t}$

b) In 2020, $t = 2020 - 1938 = 82$.

$$V(82) = 0.01 e^{0.256(82)}$$
$$= 0.01 e^{20.992}$$
$$\approx 13{,}083{,}073$$

There comic book will be valued at $13,083,073 in 2020.

c) From Theorem 9, the doubling time T is given by $T = \frac{\ln 2}{k}$.

$$T = \frac{\ln 2}{0.256} \approx 2.71$$

The comic book doubles in value approximately every 2.71 years.

d) $V(t) = 30{,}000{,}000$

$$30{,}000{,}000 = 0.01 e^{0.256t}$$

$$\frac{30{,}000{,}000}{0{,}01} = e^{0.256t}$$

$$\ln(3{,}000{,}000{,}000) = 0.256t$$

$$\frac{\ln(3{,}000{,}000{,}000)}{0.256} = t$$

$$85 \approx t$$

The comic book will be valued at $30 million 85 years after 1938 or in 2023.

35. $k = \frac{\ln 2}{T} = \frac{\ln 2}{69.31} \approx 0.01$

The growth rate k is 0.01 or 1% per year.

37. $k = \frac{\ln 2}{T} = \frac{\ln 2}{17.3} \approx 0.04$

The growth rate k is 0.04 or 4% per year.

39. Let $t = 0$ correspond to 1972. Then the initial population of grizzly bears is $P_0 = 190$. The set up of the exponential growth function is $P(t) = 190e^{kt}$. In 2005, $t = 2005 - 1972 = 33$. The population of grizzly bears had grown to 610, hence $P(33) = 610$. Use this information to find the growth rate k.

$$610 = 190 e^{k(33)}$$

$$\frac{610}{190} = e^{33k}$$

$$\ln\left(\frac{610}{190}\right) = 33k$$

$$\frac{\ln\left(\frac{610}{190}\right)}{33} = k$$

$$0.035 \approx k$$

The growth rate is approximately 3.5%. Now we can use the information to find the exponential growth function.

$P(t) = 190 e^{0.035t}$

In 2012, $t = 2016 - 1972 = 44$

$P(40) = 190 e^{0.035(44)} \approx 886$

Yellowstone National Park will be home to approximately 886 grizzly bears in 2016.

Copyright © 2014 Pearson Education, Inc. Publishing as Addison-Wesley.

41. From Example 7 we know: $R(b) = e^{21.4b}$

For the risk of an accident to be 80% we have

$$80 = e^{21.4b}$$

$$\ln 80 = \ln e^{21.4b}$$

$$\ln 80 = 21.4b$$

$$\frac{\ln 80}{21.4} = b$$

$$0.205 \approx b$$

When the blood alcohol level is 0.205%, the risk of having an accident is 80%.

43. $P(t) = \frac{5780}{1+4.78e^{-0.4t}}$

a) $P(0) = \frac{5780}{1+4.78e^{-0.4(0)}}$

$$= \frac{5780}{1+4.78e^{0}}$$

$$= \frac{5780}{1+4.78}$$

$$= 1000$$

The island population in year zero is 1000.

$$P(1) = \frac{5780}{1+4.78e^{-0.4(1)}}$$

$$= \frac{5780}{1+4.78e^{-0.4}}$$

$$\approx 1375$$

The island population after 1 year is 1375.

$$P(2) = \frac{5780}{1+4.78e^{-0.4(2)}}$$

$$= \frac{5780}{1+4.78e^{-0.8}}$$

$$\approx 1836$$

The island population after 2 years is 1836.

$$P(5) = \frac{5780}{1+4.78e^{-0.4(5)}}$$

$$= \frac{5780}{1+4.78e^{-2}}$$

$$\approx 3510$$

The island population after 5 years is 3510.

$$P(10) = \frac{5780}{1+4.78e^{-0.4(10)}}$$

$$= \frac{5780}{1+4.78e^{-4}}$$

$$\approx 5315$$

The island population after 10 years is 5315.

$$P(20) = \frac{5780}{1+4.78e^{-0.4(20)}}$$

$$= \frac{5780}{1+4.78e^{-8}}$$

$$\approx 5771$$

The island population after 20 years is 5771.

b) We apply the Quotient Rule to take the derivative.

$$P'(t) = \frac{\left(1+4.78e^{-0.4t}\right)(0) - 4.78(-0.4)e^{-0.4t} \cdot 5780}{\left(1+4.78e^{-0.4t}\right)^2}$$

$$= \frac{11{,}051.36e^{-0.4t}}{\left(1+4.78e^{-0.4t}\right)^2}$$

c) The derivative $P'(t)$ exists for all real numbers. The equation $P'(t) = 0$ has no solution. Thus, the function has no critical points and hence, no relative extrema. $P'(t) > 0$ for all real numbers, so $P(t)$ is increasing on $[0, \infty)$. The second derivative can be used to show that the graph has an inflection point at $(3.911, 2890)$. The function is concave up on the interval $(0, 3.911)$ and concave down on the interval $(3.911, \infty)$.

The graph is shown on the top of the next page.

Using the information from the previous page, the graph is:

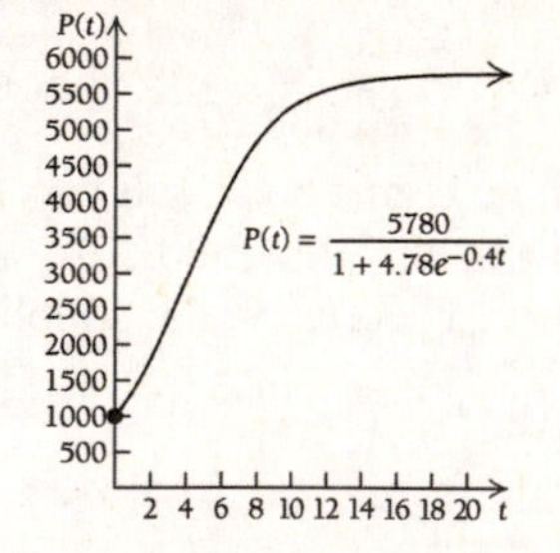

Copyright © 2014 Pearson Education, Inc. Publishing as Addison-Wesley.

45. Let $t=0$ correspond to 1930. Thus the initial number of women earning bachelor's degrees is $P_0=48,869$. So the exponential growth function is $P(t)=48,869e^{kt}$. In 2005, $t=2005-1930=75$, approximately 832,000 women received a degree. Using this information we can find the exponential growth rate k. Substitute the information into the growth function.

$$832,000=48,869e^{k(75)}$$
$$\frac{832,000}{48,869}=e^{75k}$$
$$\ln\left(\frac{832,000}{48,869}\right)=\ln e^{75k}$$
$$\ln\left(\frac{832,000}{48,869}\right)=75k$$
$$\frac{\ln\left(\frac{832,000}{48,869}\right)}{75}=k$$
$$0.0378\approx k$$

The exponential growth rate is 0.0378 or 3.78%. Therefore the exponential growth function is: $P(t)=48,869e^{0.0378t}$, where t is time in years since 1930 and $P(t)$ is number of women earning bachelor's degrees.

47. $P(t)=100\left(1-e^{-0.4t}\right)$

a) $P(0)=100\left(1-e^{-0.4(0)}\right)$
$=100(1-1)$
$=0$

0% of doctors are prescribing this medication after zero months.

$P(1)=100\left(1-e^{-0.4(1)}\right)$
$=100\left(1-e^{-0.4}\right)$
≈ 33.0

33% of doctors are prescribing this medication after 1 month.

$P(2)=100\left(1-e^{-0.4(2)}\right)$
$=100\left(1-e^{-0.8}\right)$
≈ 55.1

55% of doctors are prescribing this medication after 2 months.

$P(3)=100\left(1-e^{-0.4(3)}\right)$
$=100\left(1-e^{-1.2}\right)$
≈ 69.9

70% of doctors are prescribing this medication after 3 months.

$P(5)=100\left(1-e^{-0.4(5)}\right)$
$=100\left(1-e^{-2}\right)$
≈ 86.47

86% of doctors are prescribing this medication after 5 months.

$P(12)=100\left(1-e^{-0.4(12)}\right)$
$=100\left(1-e^{-4.8}\right)$
≈ 99.2

99.2% of doctors are prescribing this medication after 12 months.

$P(16)=100\left(1-e^{-0.4(16)}\right)$
$=100\left(1-e^{-6.4}\right)$
≈ 99.8

99.8% of doctors are prescribing this medication after 16 months.

b) $P'(t)=100\left[0-(-0.4)e^{-0.4t}\right]$
$=100\left(0.4e^{-0.4t}\right)$
$=40e^{-0.4t}$

Therefore,

$P'(7)=40e^{-0.4(7)}$
$=40e^{-2.8}$
≈ 2.432

At 7 months, the percentage of doctors who are prescribing the medicine is growing at a rate of 2.4% per month.

c) The derivative $P'(t)$ exists for all real numbers. The equation $P'(t)=0$ has no solution. Thus, the function has no critical points and hence, no relative extrema. $P'(t)>0$ for all real numbers, so $P(t)$ is increasing on $[0,\infty)$. $P''(t)=-16e^{-0.4t}$, so $P''(t)<0$ for all real numbers and hence $P(t)$ is concave down for all on $[0,\infty)$.

The graph is shown on the next page.

Copyright © 2014 Pearson Education, Inc. Publishing as Addison-Wesley.

Using the information from the previous page, the graph is:

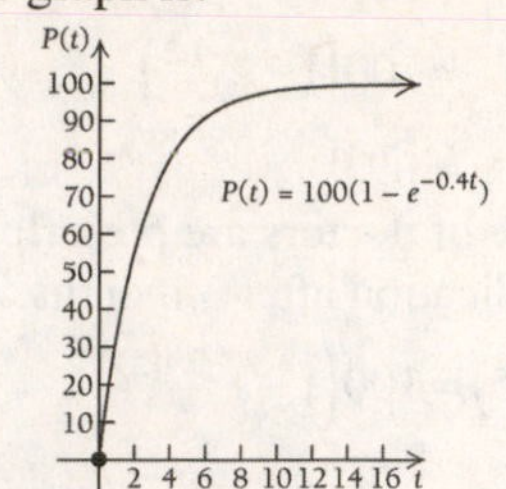

49. a) Enter the data into the calculator.

```
L1    L2    L3    2
7     24
8     26
9     28
10    28
11    29
12    30
------
L2(13) =
```

Use the Logistic regression.

```
EDIT CALC TESTS
8↑LinReg(a+bx)
9:LnReg
0:ExpReg
A:PwrReg
B:Logistic
C:SinReg
D:Manual-Fit
```

We have:

```
Logistic
 y=c/(1+ae^(-bx))
 a=79.56767122
 b=.809743969
 c=29.47232081
```

The calculator determined the logistic growth function to be:

$$y = \frac{29.47232081}{1+79.56767122e^{-0.809743969x}}$$

Converting the calculator results into function notation we get:

$$N(t) = \frac{29.47232081}{1+79.56767122e^{-0.809743969t}}$$

b) We round down since the model deals with people. The limiting value appears to be 29 people.

c) The graph is:

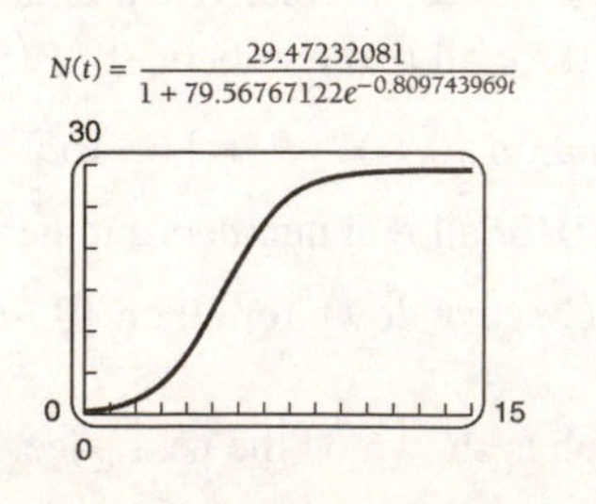

d) Use the quotient rule to find $N'(t)$.

$$N'(t) = \frac{1898.885181e^{-0.809743969t}}{\left(1+79.56767122e^{-0.809743969t}\right)^2}$$

e) ✎

Note: In Exercises 49 – 56, it is a good idea to look at the data before you make assumptions about a particular model. Given a data set, all of these solutions could change.

51. ✎

53. ✎

55. ✎

57. Let T_4 be the time it takes for a population to quadruple, then according to the exponential growth function:

$4 \cdot P_0 = P_0 e^{kT_4}$

Solve this equation for T_4.

$$\frac{4P_0}{P_0} = e^{kT_4}$$

$$\ln 4 = \ln e^{kT_4}$$

$$\ln 4 = kT_4$$

$$\frac{\ln 4}{k} = T_4$$

$$T_4 = \frac{\ln 4}{k}$$

59. Let k_1 and k_2 represent the growth rates of Q_1 and Q_2. Using the Theorem 9, we know that:

$$k_1 = \frac{\ln 2}{1} \quad \text{and } k_2 = \frac{\ln 2}{2}$$

Since the initial amounts of both quantities are the same we have:

$$Q_1 = Q_0 e^{\ln 2 \cdot t} \text{ and } Q_2 = Q_0 e^{\left(\left(\ln 2\right)/2\right)t}$$

Q_1 is twice the size of Q_2 when $Q_1 = 2Q_2$. We make the appropriate substitution and solve for t on the next page.

Copyright © 2014 Pearson Education, Inc. Publishing as Addison-Wesley.

From the previous page, we have:

$$Q_1 = 2Q_2$$

$$Q_0 e^{t \cdot \ln 2} = 2Q_0 e^{t\left(\ln 2/2\right)}$$

$$\frac{Q_0 e^{t \cdot \ln 2}}{Q_0 e^{t\left(\ln 2/2\right)}} = 2$$

$$e^{t\left(\ln 2 - \left(\ln 2/2\right)\right)} = 2$$

$$\ln\left[e^{t\left(\ln 2/2\right)}\right] = \ln 2$$

$$t\left(\ln 2/2\right) = \ln 2$$

$$t = \frac{\ln 2}{\ln 2/2}$$

$$t = \frac{1}{\frac{1}{2}} = 2$$

It will take 2 years for Q_1 to be twice the size of Q_2.

61. $i = e^k - 1$

Substitute 0.073 into the equation for *k*.

$$i = e^{0.073} - 1$$

$$\approx 1.075731 - 1$$

$$\approx 0.07573$$

The effective annual yield is 7.57%.

63. $i = e^k - 1$

Substitute 0.0942 into the equation for *i*.

$$0.0942 = e^k - 1$$

$$1.0942 = e^k$$

$$\ln 1.0942 = \ln e^k$$

$$0.09002 \approx k$$

The interest rate was 9.0%.

65. Consider the system of equations.

$$y_1 = Ce^{kt_1}$$

$$y_2 = Ce^{kt_2}$$

Divide the first equation by the second equation.

$$\frac{y_1}{y_2} = \frac{Ce^{kt_1}}{Ce^{kt_2}}$$

$$\frac{y_1}{y_2} = e^{k(t_1 - t_2)}$$

$$\ln\left(\frac{y_1}{y_2}\right) = \ln e^{k(t_1 - t_2)}$$

Continued at the top of the next column.

Using properties of logarithms, we have

$$\ln\left(\frac{y_1}{y_2}\right) = k(t_1 - t_2) \qquad \text{(P5)}$$

Solving the equation for *k* yields:

$$\frac{\ln\left(\frac{y_1}{y_2}\right)}{t_1 - t_2} = k$$

If we initially divide y_2 by y_1, we find, equivalently, that $k = \frac{\ln\left(\frac{y_2}{y_1}\right)}{t_2 - t_1}$.

67. ✎

69. a) Evaluating the function at $t = 0$, we have:

$$R(0) = \frac{4000}{1 + 1999e^{-0.5(0)}}$$

$$= \frac{4000}{2000} = 2$$

This represent the initial revenue in millions of dollars at the inception of the corporation.

b) $R_{max} = \lim_{t\to\infty} R(t) = 4000$ million dollars.

This represents the maximum attainable revenue of the company over all time.

c) $R(t) = 0.99R_{max}$

$$\frac{4000}{1 + 1999e^{-0.5t}} = 0.99(4000)$$

$$4000 = 3960\left(1 + 1999e^{-0.5t}\right)$$

$$\frac{4000}{3960} = 1 + 1999e^{-0.5t}$$

$$\frac{100}{99} - 1 = 1999e^{-0.5t}$$

$$\frac{\frac{1}{99}}{1999} = e^{-0.5t}$$

$$\ln\left(\frac{\frac{1}{99}}{1999}\right) = -0.5t$$

$$\frac{\ln\left(\frac{\frac{1}{99}}{1999}\right)}{-0.5} = t$$

$$24 \approx t$$

It will take approximately 24 years to reach 99% of the maximum revenue.

Copyright © 2014 Pearson Education, Inc. Publishing as Addison-Wesley.

Exercise Set 3.4

1. a) $N(t)=N_0e^{-kt}$

We substitute 0.096 in for k.

$N(t)=N_0e^{-0.096t}$

b) $N_0=500$

$N(t)=500e^{-0.096t}$

$N(4)=500e^{-0.096(4)}$

$=500e^{-0.384}$

≈ 340.565713

≈ 341

There will be approximately 341g of Iodine -131 present after 4 days.

c) From Theorem 10, we know that the half-life T and the decay rate k are related by:

$T=\dfrac{\ln 2}{k}$.

Substitute $k=.096$

$T=\dfrac{\ln 2}{0.096}\approx 7.22028$

It will take about 7.2 days for half of the 500g of Iodine - 131 to remain.

3. a) When A decomposes at a rate proportional to the amount of A present, we know that

$\dfrac{dA}{dt}=-kA$.

The solution to this equation is

$A(t)=A_0e^{-kt}$.

b) First, we find k. The half-life of A is 3.3 hrs. From Theorem 10 we know:

$k=\dfrac{\ln 2}{T}$

$k=\dfrac{\ln 2}{3.3}\approx 0.21$

The initial amount $A_0=10$, so the exponential decay function is:

$A(t)=10e^{-0.21t}$.

To determine how long it will take to reduce A to 1 lb we solve $A(t)=1$ for t at the top of the next column.

$10e^{-0.21t}=1$

$e^{-0.21t}=0.1$

$\ln\left(e^{-0.21t}\right)=\ln(0.1)$

$-0.21t=\ln(0.1)$

$t=\dfrac{\ln 0.1}{-0.21}$

$t\approx 10.96469$

$t\approx 11$

It will take approximately 11 hours for 10 lbs of substance A to reduce to 1 lb.

5. From Theorem 10 we know:

$k=\dfrac{\ln 2}{T}$

$k=\dfrac{\ln 2}{3}\approx 0.231$

The decay rate is 0.231 or 23.1% per minute.

7. From Theorem 10 we know:

$T=\dfrac{\ln 2}{k}$

$T=\dfrac{\ln 2}{.0315}\approx 22.0$

The half-life is 22.0 years.

9. $P(t)=P_0e^{-kt}$

We substitute 1000 for P_0, 0.0315 for k, and 100 for t.

$P(100)=1000e^{-0.0315(100)}$

$=1000e^{-3.15}\approx 42.9$

Approximately 42.9 grams of lead-210 will remain after 100 years.

11. If an ivory tusk has lost 40% of its carbon-14 from its initial amount P_0, then 60% of the initial amount remains. To find the age of the tusk, we solve the following equation for t:

$0.60P_0=P_0e^{-0.00012097t}$

$0.60=e^{-0.00012097t}$

$\ln 0.60=\ln e^{-0.00012097t}$

$\ln 0.60=-0.00012097t$

$\dfrac{\ln 0.60}{-0.00012097}=t$

$4222.7463\approx t$

$4223\approx t$

The ivory tusk is approximately 4223 years old.

Copyright © 2014 Pearson Education, Inc. Publishing as Addison-Wesley.

13. First, we find k. The half-life is 60.1 days. From Theorem 10 we know

$$k = \frac{\ln 2}{T}$$
$$k = \frac{\ln 2}{60.1}$$
$$k \approx 0.0115$$

The decay rate is approximately 0.0115 or 1.15% per day.

If the initial amount A_0 decreased by 25%, then 75% of A_0 remains. We solve the following equation for t.

$$0.75A_0 = A_0 e^{-0.0115t}$$
$$0.75 = e^{-0.0115t}$$
$$\ln 0.75 = \ln e^{-0.0115t}$$
$$\ln 0.75 = -0.0115t$$
$$\frac{\ln 0.75}{-0.0115} = t$$
$$25.015 \approx t$$
$$25 \approx t$$

The sample was sitting on the shelf for approximately 25 days.

15. Since the corn pollen had lost 38.1% of its carbon-14, then 61.9% of the initial carbon-14 remains. The decay rate of carbon-14 is $k = 0.0001205$. In order to find out the age of the corn pollen, we must solve the following equation for t.

$$0.619P_0 = P_0 e^{-0.0001205t}$$
$$0.619 = e^{-0.0001205t}$$
$$\ln 0.619 = -0.0001205t$$
$$\frac{\ln 0.619}{-0.0001205} = t$$
$$3980 \approx t$$

The corn pollen was approximately 3980 years old.

17. Use the exponential growth function $P(t) = P_0 e^{kt}$. The interest rate is 5.3% so $k = 0.053$. When $t = 20$ the parents wish the future value to be $40,000, we substitute this information into the function at the top of the next column.

Using the information from the previous column, we have:

$$40,000 = P_0 e^{0.053(20)}$$
$$\frac{40,000}{e^{1.06}} = P_0$$
$$13,858.23 \approx P_0$$

The parents should invest $13,858.23 in order to have $40,000 on their child's 20th birthday.

19. The interest rate is 5.7%, so $k = 0.057$. In 6 years, the athletes salary will be $9 million so $P(6) = 9$ million dollars. Substituting this information into the exponential growth function we get:

$$9 = P_0 e^{0.057(6)}$$
$$9 = P_0 e^{0.342}$$
$$\frac{9}{e^{0.342}} = P_0$$
$$6.393134 \approx P_0$$

The present value is $6.393134 million or $6,393,134.

21. Using the exponential growth function $P(t) = P_0 e^{kt}$. It is known that the interest rate is 4.8%, therefore $k = 0.048$. In 13 years the value of the trust fund will be $80,000, so $P(13) = 80,000$. Substitute this information into the exponential growth function and solve for the present value P_0.

$$80,000 = P_0 e^{0.048(13)}$$
$$80,000 = P_0 e^{0.624}$$
$$\frac{80,000}{e^{0.624}} = P_0$$
$$42,863.76 \approx P_0$$

The present value of the trust fund is $42,863.76.

23. a) $V(0) = 40,000e^{-0} = 40,000$

The initial cost of the machinery was $40,000.

b) $V(2) = 40,000e^{-2} \approx 5413.41$

The salvage value after 2 years is approximately $5413.41.

c) ✎

Copyright © 2014 Pearson Education, Inc. Publishing as Addison-Wesley.

25. a) If the initial actual mortality rate of a female age 25 is 0.014, then $Q_0 = 0.014$.

$$Q(t) = (0.014 - 0.00055)e^{0.163t} + 0.00055$$

$$Q(t) = (0.01345)e^{0.163t} + 0.00055$$

Evaluate the function at the appropriate value.

$$Q(3) = 0.01345e^{0.163(3)} + 0.00055$$
$$\approx 0.022$$

Three years later, the expected mortality rate of a female age 25 is 0.022 or 22 deaths per 1000.

$$Q(5) = 0.01345e^{0.163(5)} + 0.00055$$
$$\approx 0.031$$

Five years later, the expected mortality rate of a female age 25 is 0.031 or 31 deaths per 1000.

$$Q(10) = 0.01345e^{0.163(10)} + 0.00055$$
$$\approx 0.069$$

Three years later, the expected mortality rate of a female age 25 is 0.069 or 69 deaths per 1000.

b) The graph is shown below:

$Q(t) = (Q_0 - 0.00055)e^{0.163t} + 0.00055$

27. a) Using the exponential growth function we have $N(t) = N_0 e^{kt}$, where N_0 is the initial number of farms in 1950 and t is the number of years since 1950. Since there were 5,650,000 farms in 1950, $N_0 = 5,650,000$. In 2005, $t = 55$, there were 2,100,990 farms in the United States. Using this information, we will find the rate of decay, k.

$$N(55) = 2,100,990$$
$$5,650,000e^{k(55)} = 2,100,990$$
$$e^{55k} = \frac{2,100,990}{5,650,000}$$
$$55k = \ln\left(\frac{2,100,990}{5,650,000}\right)$$
$$k = \frac{\ln\left(\frac{2,100,990}{5,650,000}\right)}{55}$$
$$k \approx -0.018$$

From the information in the previous column, we determine the exponential function that describes the number of farms after time t in years since 1950 is: $N(t) = 5,650,000e^{-0.018t}$.

b) In 2009, $t = 59$.

$$N(59) = 5,650,000e^{-0.018(59)}$$
$$\approx 1,953,564$$

There were approximately 1,953,564 farms in the United States in 2009.
In 2015, $t = 65$.

$$N(65) = 5,650,000e^{-0.018(65)}$$
$$\approx 1,753,573$$

There will be approximately 1,753,573 farms in the United States in 2015.

c) Find the value of t such that

$$N(t) = 1,000,000$$
$$5,650,000e^{-0.018t} = 1,000,000$$
$$e^{-0.018t} = \frac{1,000,000}{5,650,000}$$
$$-0.018t = \ln\left(\frac{1,000,000}{5,650,000}\right).$$
$$t = \frac{\ln\left(\frac{1,000,000}{5,650,000}\right)}{-0.018}$$
$$t \approx 96.2$$

There will be 1,000,000 farms in the United States approximately 96 years after 1950, or in the year 2046.

29. a) Use the exponential-decay model $B(t) = B_0 e^{-kt}$ Let t be years since 2000, this gives us the initial consumption of beef $B_0 = 64.6$. Using the fact that in 2008, $t = 2008 - 2000 = 8$, and the beef consumption was 61.2 lbs, we can substitute into the exponential-decay function to find the decay rate k.

The solution is continued on the next page.

Copyright © 2014 Pearson Education, Inc. Publishing as Addison-Wesley.

Substituting the information from the previous page, we have:

$$61.2 = 64.6e^{-k(8)}$$
$$\frac{61.2}{64.6} = e^{-8k}$$
$$\ln\left(\frac{61.2}{64.6}\right) = \ln e^{-8k}$$
$$\ln\left(\frac{61.2}{64.6}\right) = -8k$$
$$\frac{\ln\left(\frac{61.2}{64.6}\right)}{-8} = k$$
$$0.0068 \approx k$$

The decay rate is 0.0068 or 0.68% per year. Substituting this into the exponential decay function we get:

$B(t) = 64.6e^{-0.0068t}$.

b) In 2015, $t = 2015 - 2000 = 15$.

$B(15) = 64.6e^{-0.0068(15)} \approx 58.3$

In the year 2015, annual consumption of beef will be approximately 58.3 pounds per person.

c) Set $B(t) = 20$ and solve for t.

$$64.6e^{-0.0068t} = 20$$
$$e^{-0.0068t} = \frac{20}{64.6}$$
$$\ln e^{-0.0068t} = \ln\left(\frac{20}{64.6}\right)$$
$$-0.0068t = \ln\left(\frac{20}{64.6}\right)$$
$$t = \frac{\ln\left(\frac{20}{64.6}\right)}{-0.0068}$$
$$t \approx 172.4$$

Theoretically, consumption of beef will be 20lbs per person about 172.4 years after 2000, or in the year 2172.

31. a) Use the exponential decay model $P(t) = P_0e^{-kt}$. If we let $t = 0$ correspond to 1995, then $P_0 = 51.9$ million people. In 2012, the population of Ukraine was 44.9 million, so $P(17) = 44.9$. We substitute this information into the model at the top of the next column.

Substituting, we have:

$$44.9 = 51.9e^{-k(17)}$$
$$\frac{44.9}{51.9} = e^{-17k}$$
$$\ln\left(\frac{44.9}{51.9}\right) = \ln e^{-17k}$$
$$\ln\left(\frac{44.9}{51.9}\right) = -17k$$
$$\frac{\ln\left(\frac{44.9}{51.9}\right)}{-17} = k$$
$$0.008522 \approx k$$

Thus, $P(t) = 51.9e^{-0.008522t}$, where $P(t)$ is in millions of people and t is the number of years since 1995.

b) In 2015, $t = 2015 - 1995 = 20$.

$P(20) = 51.9e^{-0.008522(20)} \approx 43.77$

The population of the Ukraine will be approximately 43.8 million in 2015.

c) Set $P(t) = 1$ and solve for t.

$$51.9e^{-0.008522t} = 1$$
$$e^{-0.008522t} = \frac{1}{51.9}$$
$$\ln e^{-0.008522t} = \ln\left(\frac{1}{51.9}\right)$$
$$-0.008522t = \ln\left(\frac{1}{51.9}\right)$$
$$t = \frac{\ln\left(\frac{1}{51.9}\right)}{-0.008522}$$
$$t \approx 463.4$$

According to the model, the population of the Ukraine will be 1million 463 years after 1995, or in the year 2458.

33. a) According to Newton's Law of Cooling $T(t) = ae^{-kt} + C$. The room temperature is 75 degrees so $C = 75$. At $t = 0$, $T = 102$ degrees. We substitute these values into Newton's law of Cooling.

$$102 = ae^{-k(0)} + 75$$
$$102 = ae^{0} + 75$$
$$102 = a + 75$$
$$27 = a$$

Copyright © 2014 Pearson Education, Inc. Publishing as Addison-Wesley.

b) From part 'a' we have $T(t) = 27e^{-kt} + 75$.

Using the fact that when $t = 10$, $T = 90$ we have:

$$90 = 27e^{-k(10)} + 75$$
$$15 = 27e^{-10k}$$
$$\frac{15}{27} = e^{-10k}$$
$$\ln\left(\frac{15}{27}\right) = \ln\left(e^{-10k}\right)$$
$$\ln\left(\frac{15}{27}\right) = -10k$$
$$\frac{\ln\left(\frac{15}{27}\right)}{-10} = k$$
$$0.05878 \approx k$$

c) From 'a' and 'b' parts we have

$$T(t) = 27e^{-0.05878t} + 75$$
$$T(20) = 27e^{-0.05878(20)} + 75$$
$$= 27e^{-1.1756} + 75$$
$$\approx 83.3$$

After 20 minutes the water temperature is approximately 83 degrees.

d) Solve $T(t) = 80$ for t.

$$27e^{-0.05878t} + 75 = 80$$
$$27e^{-0.05878t} = 5$$
$$e^{-0.05878t} = \frac{5}{27}$$
$$\ln e^{-0.05878t} = \ln\left(\frac{5}{27}\right)$$
$$-0.05878t = \ln\left(\frac{5}{27}\right)$$
$$t = \frac{\ln\left(\frac{5}{27}\right)}{-0.05878}$$
$$t \approx 28.7$$

It will take approximately 29 minutes for the water temperature to cool to 80 degrees.

e) ✎

35. Newton's law of Cooling states $T(t) = ae^{-kt} + C$.

Assume the body had a normal temperature of 98.6 degrees at the time of death and the room remained a constant 60 degrees giving us the constant $C = 60$.

We find the constant a first, using the fact that 98.6 degrees is normal body temperature.

$$98.6 = ae^{-k(0)} + 60$$
$$98.6 = a + 60$$
$$38.6 = a$$

So $T(t) = 38.6e^{-kt} + 60$

Next, we use the two temperature readings to find k. We want to find the number of hours since death, t. When the corner took the first temperature reading t hours since death, the body temperature was 85.9 degrees. One hour later at $t+1$ hours since death, the body temperature was 83.4 degrees. Using these two pieces of information we get two equations.

$$85.9 = 38.6e^{-kt} + 60$$
$$83.4 = 38.6e^{-k(t+1)} + 60$$

Subtracting 60 from each equation we get:

$$25.9 = 38.6e^{-kt}$$
$$23.4 = 38.6e^{-k(t+1)}$$

A quick way to solve this system of equations is to divide the first equation by the second equation. This gives us:

$$\frac{25.9}{23.4} = \frac{38.6e^{-kt}}{38.6e^{-k(t+1)}}$$
$$\frac{25.9}{23.4} = e^{-kt-(-kt-k)}$$
$$\frac{25.9}{23.4} = e^{k}$$
$$\ln\left(\frac{25.9}{23.4}\right) = \ln e^{k}$$
$$0.10 \approx k$$

Now we substitute 0.10 in for k into the equation $25.9 = 38.6e^{-kt}$ and solve for t.

$$25.9 = 38.6e^{-0.10t}$$
$$\frac{25.9}{38.6} = e^{-0.1t}$$
$$\ln\left(\frac{25.9}{38.6}\right) = \ln e^{-0.1t}$$
$$\ln\left(\frac{25.9}{38.6}\right) = -0.1t$$
$$\frac{\ln\left(\frac{25.9}{38.6}\right)}{-0.10} = t$$
$$4 \approx t$$

Therefore, the body had been dead for 4 hours. Since the temperature was taken at 11 P.M, the time of death was 7 P.M.

Copyright © 2014 Pearson Education, Inc. Publishing as Addison-Wesley.

37. $W = 170e^{-0.008t}$

a) We substitute 20 in for t.

$$W(20) = 170e^{-0.008(20)}$$
$$= 170e^{-0.16}$$
$$\approx 144.86$$

The monk weighs approximately 145 pounds after 20 days.

b) We take the derivative of the function to find the rate of change.

$$W'(t) = 170(-0.008)e^{-0.008t}$$
$$= -1.36e^{-0.008t}$$

Now, we substitute 20 in for t.

$$W'(20) = -1.36e^{-0.009(20)} \approx -1.16$$

The monk is losing approximately 1.2 pounds per day after 20 days.

39. $P(t) = 50e^{-0.004t}$

a) We substitute 375 for t.

$$P(375) = 50e^{-0.004(375)}$$
$$= 50e^{-1.5} \approx 11.2$$

After 375 days, approximately 11.2 watts of power will be available.

b) From Theorem 10 we know:

$$T = \frac{\ln 2}{k}$$
$$T = \frac{\ln 2}{0.004} \approx 173$$

The half-life of the power supply is approximately 173 days.

c) Set $P(t) = 10$ and solve for t.

$$50e^{-0.004t} = 10$$
$$e^{-0.004t} = \frac{10}{50}$$
$$\ln e^{-0.004t} = \ln 0.2$$
$$-0.004t = \ln 0.2$$
$$t = \frac{\ln 0.2}{-0.004}$$
$$t \approx 402.36$$

The satellite can stay in operation for 402 days.

d) When $t = 0$

$$P(0) = 50e^{-0.004(0)}$$
$$= 50e^{0}$$
$$= 50$$

At the beginning, the satellite had 50 watts of power.

e) ✎

41. (c)

43. (e)

45. (f)

47. (d)

49. (a)

51. Solve $D(x) = S(x)$

$$480e^{-0.003x} = 150e^{0.004x}$$
$$\frac{480}{150} = \frac{e^{0.004x}}{e^{-0.003x}}$$
$$3.2 = e^{0.007x}$$
$$\ln 3.2 = 0.007x$$
$$166.16 \approx x$$

The equilibrium price is \$166.16. Therefore,

$$D(166.16) = 480e^{-0.003(166.16)} \approx 291.57$$

The equilibrium quantity is 292 units. The equilibrium point is $(166.16, 292)$.

Note: Due to the context of the problem we cannot use the true equilibrium point. We rounded the printers up to 292 units while keeping the equilibrium price the same, even though demand at this price is slightly less than 292 printers. It is not possible to make or sell a fraction of a printer, so the supplier would need to determine if 291 or 292 units would result in greater profits. In the interest of the student we decided to take the most logical course of action by rounding the equilibrium quantity to the proper integer quantity.

53. ✎

55. ✎

Copyright © 2014 Pearson Education, Inc. Publishing as Addison-Wesley.

Exercise Set 3.5

1. $y = 7^x$

$\frac{dy}{dx} = (\ln 7)7^x$ Theorem 12: $\frac{dy}{dx}a^x = (\ln a)a^x$

3. $f(x) = 8^x$

$f'(x) = (\ln 8)8^x$ Theorem 12

5. $g(x) = x^3(5.4)^x$

Using the Product Rule, we get

$$g'(x) = x^3\left(\frac{dy}{dx}(5.4)^x\right) + \left(\frac{dy}{dx}x^3\right)(5.4)^x$$

$$= x^3\left(\underbrace{\ln(5.4)(5.4)^x}_{\text{Theorem 12}}\right) + 3x^2(5.4)^x$$

$$= x^2(5.4)^x(x\ln(5.4) + 3). \quad \text{Factoring}$$

7. $y = 7^{x^4+2}$

Using the Chain Rule and Theorem 12, we get

$$\frac{dy}{dx} = (\ln 7)7^{x^4+2}\left(\frac{d}{dx}(x^4+2)\right)$$

$$= (\ln 7)7^{x^4+2}(4x^3)$$

$$= 4x^3(\ln 7)7^{x^4+2}.$$

9. $y = e^{8x}$

$\frac{dy}{dx} = 8e^{8x}$ Theorem 2

11. $f(x) = 3^{x^4+1}$

Using the Chain Rule and Theorem 12, we get

$$\frac{dy}{dx} = (\ln 3)3^{x^4+1}\left(\frac{d}{dx}(x^4+1)\right)$$

$$= (\ln 3)3^{x^4+1}(4x^3)$$

$$= 4x^3(\ln 3)3^{x^4+1}.$$

13. $y = \log_4 x$

$\frac{dy}{dx} = \frac{1}{\ln 4}\cdot\frac{1}{x}$ Theorem 14

$$= \frac{1}{x\ln 4}$$

15. $y = \log_{17} x$

$\frac{dy}{dx} = \frac{1}{\ln 17}\cdot\frac{1}{x}$ Theorem 14.

$$= \frac{1}{x\ln 17}$$

17. $g(x) = \log_6(5x+1)$

Using the Chain Rule and Theorem 14, we get

$$g'(x) = \frac{1}{\ln 6}\cdot\frac{1}{5x+1}\cdot\frac{d}{dx}(5x+1)$$

$$= \frac{1}{\ln 6}\cdot\frac{1}{5x+1}\cdot 5$$

$$= \frac{5}{(5x+1)\ln 6}.$$

19. $F(x) = \log(6x-7)$ $(\log x = \log_{10} x)$

Using the Chain Rule and Theorem 14 we get

$$F'(x) = \frac{1}{\ln 10}\cdot\frac{1}{6x-7}\cdot\frac{d}{dx}(6x-7)$$

$$= \frac{1}{\ln 10}\cdot\frac{1}{6x-7}\cdot 6$$

$$= \frac{6}{(6x-7)\ln 10}.$$

21. $y = \log_8(x^3+x)$

Using the Chain Rule and Theorem 14, we get

$$\frac{dy}{dx} = \frac{1}{\ln 8}\cdot\frac{1}{x^3+x}\cdot\frac{d}{dx}(x^3+x)$$

$$= \frac{1}{\ln 8}\cdot\frac{1}{x^3+x}\cdot(3x^2+1)$$

$$= \frac{3x^2+1}{(x^3+x)\ln 8}.$$

Copyright © 2014 Pearson Education, Inc. Publishing as Addison-Wesley.

23. $f(x)=4\log_7\left(\sqrt{x}-2\right)$

$$f'(x)=4\frac{d}{dx}\log_7\left(\sqrt{x}-2\right)$$

Using the Chain Rule and Theorem 14, we get:

$$f'(x)=4\cdot\frac{1}{\ln 7}\cdot\frac{1}{\sqrt{x}-2}\cdot\frac{d}{dx}\left(\sqrt{x}-2\right)$$

$$=\frac{4}{\left(\sqrt{x}-2\right)\ln 7}\cdot\frac{d}{dx}\left(x^{1/2}-2\right)\quad\left[\sqrt[n]{x}=x^{1/n}\right]$$

$$=\frac{4}{\left(\sqrt{x}-2\right)\ln 7}\left(\frac{1}{2}x^{-1/2}\right)\quad\text{Power Rule}$$

$$=\frac{4}{\left(\sqrt{x}-2\right)\ln 7}\cdot\frac{1}{2x^{1/2}}\quad\text{Properties of Exponents}$$

$$=\frac{4}{2\sqrt{x}\left(\sqrt{x}-2\right)\ln 7}$$

$$=\frac{2}{\left(x-2\sqrt{x}\right)\ln 7}.$$

25. $y=6^x\cdot\log_7 x$

Since y is of the form $f(x)\cdot g(x)$, we apply the Product Rule.

$$\frac{dy}{dx}=6^x\cdot\left(\frac{1}{\ln 7}\cdot\frac{1}{x}\right)+(\ln 6)6^x\cdot\log_7 x$$

Next, we use the commutative property of multiplication to rearrange the derivative:

$$\frac{dy}{dx}=\frac{6x}{x\ln 7}+6^x\ln 6\cdot\log_7 x.$$

27. $G(x)=(\log_{12} x)^5$

Using the Extended Power Rule, we have:

$$G'(x)=5\cdot(\log_{12} x)^4\cdot\frac{d}{dx}(\log_{12} x)$$

$$=5\cdot(\log_{12} x)^4\cdot\left(\frac{1}{\ln 12}\cdot\frac{1}{x}\right)\quad\text{Theorem 14}$$

$$=5(\log_{12} x)^4\left(\frac{1}{x\ln 12}\right).$$

29. $g(x)=\dfrac{7^x}{4x+1}$

Since $g(x)$ is in the form $g(x)=\dfrac{f(x)}{h(x)}$ we apply the Quotient Rule at the top of the next column.

$$g'(x)=\frac{(4x+1)\left(\frac{d}{dx}7^x\right)-7^x\left(\frac{d}{dx}(4x+1)\right)}{(4x+1)^2}$$

$$=\frac{(4x+1)(\ln 7)7^x-7^x(4)}{(4x+1)^2}$$

$$=\frac{7^x\left((4x+1)\ln 7-4\right)}{(4x+1)^2}\quad\text{Factor out } 7^x$$

$$=\frac{7^x(4x\ln 7+\ln 7-4)}{(4x+1)^2}\quad\text{Distribute } \ln 7$$

31. $y=5^{2x^3-1}\cdot\log(6x+5)$

Using the Product Rule we have:

$$\frac{dy}{dx}=5^{2x^3-1}\left(\frac{d}{dx}\log(6x+5)\right)+\left(\frac{d}{dx}5^{2x^3-1}\right)\log(6x+5).$$

Next, using the Chain Rule, we have:

$$\frac{dy}{dx}=5^{2x^3-1}\left(\frac{1}{\ln 10}\cdot\frac{1}{6x+5}\cdot 6\right)+(\ln 5)5^{2x^3-1}\left(6x^2\right)\cdot\log(6x+5).$$

Next, using properties of multiplication, we have:

$$\frac{dy}{dx}=\frac{6\cdot 5^{2x^3-1}}{(6x+5)\ln 10}+(\ln 5)5^{2x^3-1}\cdot 6x^2\log(6x+5).$$

33. $F(x)=7^x(\log_4 x)^9$

Using the Product Rule, the Extended Power Rule and Theorem 12, we have:

$$F'(x)$$

$$=7^x\cdot\frac{d}{dx}\left[(\log_4 x)^9\right]+(\log_4 x)^9\cdot\frac{d}{dx}\left[7^x\right]$$

$$=7^x\cdot\left[9(\log_4 x)^8\cdot\left(\frac{1}{x\cdot\ln 4}\right)\right]+(\log_4 x)^9\cdot\left[(\ln 7)7^x\right]$$

$$=\frac{7^x\cdot 9\cdot(\log_4 x)^8}{x\cdot\ln 4}+(\ln 7)7^x\cdot(\log_4 x)^9$$

Copyright © 2014 Pearson Education, Inc. Publishing as Addison-Wesley.

35. $f(x)=(3x^5+x)^5\log_3 x$

First, using the Product Rule, we have:

$$f'(x)=(3x^5+x)^5\frac{d}{dx}\log_3 x+\left(\frac{d}{dx}(3x^5+x)^5\right)\log_3 x$$

Next, we will apply the Chain Rule:

$$f'(x)=(3x^5+x)^5\left(\frac{1}{\ln 3}\cdot\frac{1}{x}\right)+5(3x^5+x)^4\frac{d}{dx}(3x^5+x)\cdot\log_3 x.$$

Which gives us:

$$f'(x)=(3x^5+x)^5\left(\frac{1}{x\ln 3}\right)+5(3x^5+x)^4(15x^4+1)\log_3 x.$$

37. a) $V(t)=5200(0.80)^t$

$$V'(t)=5200\frac{d}{dt}(0.80)^t$$
$$=5200(\ln 0.80)(0.80)^t \quad \text{Theorem 12}$$

b) ✎

39. a) $L(t)=1547(1.083)^t$

Since t is the number of years since 1980, the year 2012 corresponds to $t=2012-1980=32$ years.

$$L(32)=1547(1.083)^{32}$$
$$=1547(12.82657226)$$
$$=19{,}842.7072800$$
$$\approx 19{,}842.71$$

In the year 2012. total financial liability of U.S. households will be approximately $19,842.71 billion.

b) First, we find the derivative using Theorem 12:

$$L'(t)=1547(\ln 1.083)(1.083)^t$$

Next, we evaluate the derivative at $t=25$

$$L'(25)=1547(\ln 1.083)(1.083)^{25}$$
$$=1547(\ln 1.083)(7.34025953)$$
$$=905.42098032$$
$$\approx 905.42$$

c) ✎

41. a) $P(t)=(0.98)^t$

$$P(10)=(0.98)^{10}=0.81707281$$

Ten years after the neighboring farm begins to use the GMO seed, the GMO-free farm will be 81.7 percent GMO-free.

b) First, we find the derivative

$$P'(t)=\ln(0.98)(0.98)^t.$$

Next, we evaluate the derivative at $t=15$:

$$P'(15)=(\ln 0.98)(0.98)^{15}$$
$$=(\ln 0.98)\cdot 0.73856910$$
$$=-0.01492110$$
$$\approx -0.0149.$$

c) ✎

43. $R=\log\frac{I}{I_0}$

We substitute $10^{9.0}\cdot I_0$ for I

$$R=\log\frac{10^{9.0}\cdot I_0}{I_0}$$
$$=\log 10^{9.0}$$
$$=9.0 \qquad \text{(P5)}$$

The earthquake that struck Japan had a magnitude of 9.0 on the Richter scale.

45. $I=I_0 10^{0.1L}$

a) Substituting 100 for L we have:

$$I=I_0 10^{0.1(100)}$$
$$=I_0 10^{10}$$
$$=10^{10}\cdot I_0$$

The intensity of a power mower is $10^{10}\cdot I_0$.

b) Substituting 10 for L we have:

$$I=I_0 10^{0.1(10)}$$
$$=I_0 10^1$$
$$=10\cdot I_0$$

The intensity of a just audible sound is $10\cdot I_0$

c) Comparing the intensity in parts (a) and (b) we have $10^{10}\cdot I_0=10^9(10\cdot I_0)$. The intensity of (a) is 10^9 times more than the intensity of (b).

Copyright © 2014 Pearson Education, Inc. Publishing as Addison-Wesley.

d) $I = I_0 10^{0.1L}$

$$\frac{dI}{dL} = I_0 \frac{d}{dL} 10^{0.1L} \quad I_0 \text{ is a constant.}$$

Next, using Theorem 12 and the Chain Rule we have:

$$\frac{dI}{dL} = I_0 (\ln 10) 10^{0.1L} \cdot \left(\frac{d}{dL} 0.1L \right)$$
$$= I_0 (\ln 10) 10^{0.1L} (0.1)$$
$$= 0.1 \cdot \ln 10 \cdot I_0 10^{0.1L}.$$

e) ✎

47. a) $L = 10 \log \frac{I}{I_0}$

First, we rearrange the equation using Property 2.

$$L = 10(\log I - \log I_0)$$
$$= 10 \log I - 10 \log I_0$$

Next, we take the derivative using the Difference Rule:

$$\frac{dL}{dI} = 10 \frac{d}{dI} \log I - \frac{d}{dI} 10 \log I_0$$

Using Theorem 14 we have:

$$\frac{dL}{dI} = 10 \frac{1}{\ln 10} \cdot \frac{1}{I} - 0 \quad I_0 \text{ is a constant}$$
$$= \frac{10}{\ln 10} \cdot \frac{1}{I}$$

b) ✎

49. By Theorem 13: $\lim_{h \to 0} \frac{3^h - 1}{h} = \ln 3.$

51. $y = 2^{x^4}$

$$\frac{dy}{dx} = (\ln 2) 2^{x^4} \frac{d}{dx} x^4$$
$$= (\ln 2) 2^{x^4} \cdot 4x^3$$

53. $y = \log_3 (\log x)$

Using the Chain Rule and Theorem 14, we have:

$$\frac{dy}{dx} = \frac{1}{\ln 3} \cdot \frac{1}{\log x} \cdot \frac{d}{dx} \log x$$
$$= \frac{1}{\ln 3} \cdot \frac{1}{\log x} \cdot \frac{1}{\ln 10} \cdot \frac{1}{x}$$
$$= \frac{1}{\ln 3 \cdot \log x \cdot \ln 10 \cdot x}.$$

55. $y = a^{f(x)}$

$$y = e^{f(x) \ln a} \quad \left[a^x = e^{x \ln a} \right]$$
$$\frac{dy}{dx} = e^{f(x) \ln a} \frac{d}{dx} f(x) \ln a \quad \text{Chain Rule}$$
$$\frac{dy}{dx} = e^{f(x) \ln a} \cdot f'(x) \ln a$$
$$\frac{dy}{dx} = \ln a \cdot a^{f(x)} \cdot f'(x) \quad \left[a^x = e^{x \ln a} \right]$$

57. $y = \left[f(x) \right]^{g(x)}, \ f(x) > 0$

$$y = e^{g(x) \cdot \ln f(x)} \quad \left[a^x = e^{x \ln a} \right]$$
$$\frac{dy}{dx} = e^{g(x) \cdot \ln f(x)} \frac{d}{dx} g(x) \ln f(x) \quad \text{Chain Rule}$$

Next, using the Product Rule, we have:

$$\frac{dy}{dx} = e^{g(x) \cdot \ln f(x)} \left(g(x) \cdot \frac{1}{f(x)} \cdot f'(x) + g'(x) \ln f(x) \right).$$

Simplifying we get:

$$\frac{dy}{dx} = e^{g(x) \ln f(x)} \left(\frac{g(x) f'(x)}{f(x)} + g'(x) \ln f(x) \right)$$
$$= \left[f(x) \right]^{g(x)} \left(\frac{g(x) f'(x)}{f(x)} + g'(x) \ln f(x) \right).$$

59. ✎

Copyright © 2014 Pearson Education, Inc. Publishing as Addison-Wesley.

Exercise Set 3.6

1. a) The demand function is
$q = D(x) = 400 - x$.
The definition of the elasticity of demand is given by: $E(x) = -\frac{x \cdot D'(x)}{D(x)}$. In order to find the elasticity of demand, we need to find the derivative of the demand function first.
$\frac{dq}{dx} = D'(x) = \frac{d}{dx}(400 - x) = -1$.
Next, we substitute -1 for $D'(x)$, and $400 - x$ for $D(x)$ into the expression for elasticity.
$$E(x) = -\frac{x \cdot (-1)}{400 - x} = \frac{x}{400 - x}$$

b) Substituting $x = 125$ into the expression found in part (a) we have:
$$E(125) = \frac{(125)}{400 - (125)} = \frac{125}{275} = \frac{5}{11}.$$
Since $E(125) = \frac{5}{11}$ is less than one, the demand is inelastic.

c) The values of x for which $E(x) = 1$ will maximize total revenue. We solve:
$$E(x) = 1$$
$$\frac{x}{400 - x} = 1$$
$$x = 400 - x$$
$$2x = 400$$
$$x = 200.$$
A price of \$200 will maximize total revenue.

3. $q = D(x) = 200 - 4x$; $x = 46$

a) $D'(x) = -4$
$$E(x) = -\frac{x \cdot D'(x)}{D(x)} = -\frac{x \cdot (-4)}{200 - 4x} = \frac{4x}{200 - 4x}$$
$$= \frac{x}{50 - x}$$

b) Substituting $x = 46$ into the expression found in part (a) we have:
$$E(46) = \frac{46}{50 - 46} = \frac{46}{4} = \frac{23}{2} = 11.5.$$
Since $E(46) > 1$, demand is elastic.

c) We solve $E(x) = 1$
$$\frac{x}{50 - x} = 1$$
$$x = 50 - x$$
$$2x = 50$$
$$x = 25.$$
A price of \$25 will maximize total revenue.

5. $q = D(x) = \frac{400}{x}$; $x = 50$

a) First, we rewrite the demand function.
$$D(x) = \frac{400}{x} = 400x^{-1}.$$
Next, we take the derivative of the demand function, using the Power Rule.
$$D'(x) = 400(-1)x^{-2} = -400x^{-2}$$
Making the appropriate substitutions into the elasticity function, we have
$$E(x) = -\frac{x \cdot D'(x)}{D(x)} = -\frac{x \cdot (-400x^{-2})}{400x^{-1}} =$$
$$= \frac{400x^{-1}}{400x^{-1}} = 1.$$
Therefore, $E(x) = 1$ for all values of x.

b) $E(50) = 1$, so demand is unit elastic.

c) $E(x) = 1$ for all values of x. Therefore, total revenue is maximized for all values of x. In other words, total revenue is the same regardless of the price.

7. $q = D(x) = \sqrt{600 - x}$; $x = 100$

a) First rewrite the demand function:
$$D(x) = (600 - x)^{1/2}.$$
Next, we take the derivative of the demand function, using the Chain Rule:
$$D'(x) = \frac{1}{2}(600 - x)^{-1/2} \cdot \frac{d}{dx}(600 - x)$$
$$= \frac{-1}{2\sqrt{600 - x}}.$$
The solution is continued on the next page.

Copyright © 2014 Pearson Education, Inc. Publishing as Addison-Wesley.

Making the appropriate substitutions into the elasticity function, we have

$$E(x) = -\frac{x \cdot D'(x)}{D(x)} = -\frac{x \cdot \left(\frac{-1}{2\sqrt{600-x}}\right)}{\sqrt{600-x}} =$$

$$= \frac{\frac{x}{2\sqrt{600-x}}}{\sqrt{600-x}} = \frac{x}{2(600-x)}$$

$$= \frac{x}{1200-2x}.$$

b) Substituting $x = 100$ into the expression found in part (a) we have:

$$E(100) = \frac{100}{1200-2(100)} = \frac{100}{1000} = \frac{1}{10}.$$

Since $E(100) < 1$, demand is inelastic.

c) Solve $E(x) = 1$

$$\frac{x}{1200-2x} = 1$$

$$x = 1200 - 2x$$

$$3x = 1200$$

$$x = 400$$

A price of \$400 will maximize total revenue.

9. $q = D(x) = 100e^{-0.25x}$; $x = 10$

a) Using the Chain Rule we have:

$$D'(x) = 100e^{-0.25x}(-0.25)$$

$$= -25e^{-0.25x}.$$

Making the appropriate substitutions into the elasticity function, we have

$$E(x) = -\frac{x \cdot D'(x)}{D(x)} = -\frac{x \cdot \left(-25e^{-0.25x}\right)}{100e^{-0.25x}}$$

$$= \frac{25xe^{-0.25x}}{100e^{-0.25x}} = \frac{x}{4}.$$

b) Substituting $x = 10$ into the expression found in part (a) we have:

$$E(10) = \frac{10}{4} = \frac{5}{2} = 2.5.$$

Since $E(10) > 1$, demand is elastic.

c) Solve $E(x) = 1$

$$\frac{x}{4} = 1$$

$$x = 4.$$

A price of \$4 will maximize total revenue.

11. $q = D(x) = \dfrac{100}{(x+3)^2}$; $x = 1$

a) First, we rewrite the demand function:

$$D(x) = 100(x+3)^{-2}.$$

Next, we take the derivative of the demand function, using the Chain Rule:

$$D'(x) = 100(-2)(x+3)^{-3}\left(\frac{d}{dx}(x+3)\right)$$

$$= -200(x+3)^{-3}$$

$$= -\frac{200}{(x+3)^3}.$$

Making the appropriate substitutions into the elasticity function, we have:

$$E(x) = -\frac{x \cdot D'(x)}{D(x)} = -\frac{x \cdot \left(-\frac{200}{(x+3)^3}\right)}{\frac{100}{(x+3)^2}}$$

$$= x \cdot \left(\frac{200}{(x+3)^3}\right)\frac{(x+3)^2}{100}$$

$$= \frac{2x}{x+3}.$$

b) Substituting $x = 1$ into the expression found in part (a) we have:

$$E(1) = \frac{2 \cdot (1)}{1+3} = \frac{2}{4} = \frac{1}{2}.$$

Since $E(1) < 1$, demand is inelastic.

c) Solve $E(x) = 1$

$$\frac{2x}{x+3} = 1$$

$$2x = x + 3$$

$$x = 3.$$

A price of \$3 will maximize total revenue.

13. $q = D(x) = 967 - 25x$

a) $D'(x) = -25$

$$E(x) = -\frac{x \cdot D'(x)}{D(x)} = -\frac{x \cdot (-25)}{967-25x}$$

$$= \frac{25x}{967-25x}$$

Copyright © 2014 Pearson Education, Inc. Publishing as Addison-Wesley.

b) We set $E(x)=1$ and solve for x.

$$\frac{25x}{967-25x}=1$$
$$25x=967-25x$$
$$50x=967$$
$$x=\frac{967}{50}$$
$$x=19.34$$

Demand is unitary elastic when price is 19.34 cents.

c) Demand is elastic when $E(x)>1$.

Testing a value on each side of 19.34 cents, we have:

$$E(19)=\frac{25\cdot 19}{967-25\cdot 19}\approx 0.97<1$$
$$E(20)=\frac{25\cdot 20}{967-25\cdot 20}\approx 1.07>1.$$

Therefore, the demand for cookies is elastic for prices greater than 19.34 cents.

d) Demand is inelastic when $E(x)<1$.

Using the calculations from part (c), we see that the demand for cookies is inelastic for prices less than 19.34 cents.

e) Total revenue is maximized when $E(x)=1$.

In part (b) we showed that $E(x)=1$ when price was 19.34 cents. Therefore, revenue will be maximized when price is 19.34 cents.

f) We have shown that the demand for cookies is elastic when the price of cookies is 20 cents. Therefore a small increase in price will cause total revenue to decrease.

15. $q=D(x)=\sqrt{200-x^3}$

a) First, we rewrite the demand function to make it easier to find the derivative.

$$D(x)=\left(200-x^3\right)^{1/2}.$$

Next, using the Chain Rule, we have:

$$D'(x)=\frac{1}{2}\left(200-x^3\right)^{-1/2}\left(-3x^2\right)$$
$$=\frac{-3x^2}{2\sqrt{200-x^3}}.$$

Now, substituting in the elasticity function we get:

$$E(x)=-\frac{x\cdot D'(x)}{D(x)}=-\frac{x\cdot\left(\frac{-3x^2}{2\sqrt{200-x^3}}\right)}{\sqrt{200-x^3}}$$
$$=\frac{3x^3}{2\left(\sqrt{200-x^3}\right)^2}$$
$$=\frac{3x^3}{2\left(200-x^3\right)}$$
$$=\frac{3x^3}{400-2x^3}.$$

b) $$E(3)=\frac{3(3)^3}{400-2(3)^3}$$
$$=\frac{81}{346}$$
$$\approx 0.2341$$

Since $E(3)<1$, the demand for computer games is inelastic when price is \$3.

c) From part (b) we know that the demand for computer games is inelastic at a price of \$3. Therefore an increase in the price of computer games will lead to an increase in the total revenue.

17. $q=D(x)=\frac{k}{x^n}$

a) First, we rewrite the demand function to make it easier to find the derivative.

$$D(x)=k\cdot x^{-n}$$

Using the Power Rule, we have:

$$D'(x)=k\cdot(-n)x^{-n-1}=-nkx^{-n-1}.$$

Substituting into the elasticity function we have:

$$E(x)=-\frac{x\cdot D'(x)}{D(x)}=-\frac{x\cdot\left(-nkx^{-n-1}\right)}{kx^{-n}}$$
$$=\frac{nk\left(x\cdot x^{-n-1}\right)}{kx^{-n}}$$
$$=\frac{nk\left(x^{-n}\right)}{kx^{-n}}$$
$$=n.$$

b) No, the elasticity of demand is constant for all prices. $E(x)=n$.

Copyright © 2014 Pearson Education, Inc. Publishing as Addison-Wesley.

c) Total revenue is maximized when $E(x)=1$. Since $E(x)=n$, total revenue will be maximized when $n=1$.

19. $L(x)=\ln D(x)$

$$L'(x)=\frac{1}{D(x)}\cdot D'(x)=\frac{D'(x)}{D(x)}$$

The formula for elasticity of demand is:

$$E(x)=-\frac{x\cdot D'(x)}{D(x)}=-x\left(\frac{D'(x)}{D(x)}\right).$$

Substituting $L'(x)$ for $\frac{D'(x)}{D(x)}$ we have:

$$E(x)=-x\cdot L'(x).$$

21. 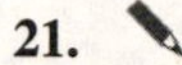

Copyright © 2014 Pearson Education, Inc. Publishing as Addison-Wesley.

Chapter 4

Integration

Exercise Set 4.1

1. $\int x^6 dx$

$= \dfrac{x^{6+1}}{6+1} + C \qquad \left[\int x^r dx = \dfrac{x^{r+1}}{r+1} + C\right]$

$= \dfrac{x^7}{7} + C$ Don't forget the C.

3. $\int 2dx$

$= 2x + C \qquad \left[\int kdx = kx + C\right]$

5. $\int x^{1/4} dx$

$= \dfrac{x^{1/4+1}}{\frac{1}{4}+1} + C \qquad \left[\int x^r dx = \dfrac{x^{r+1}}{r+1} + C\right]$

$= \dfrac{x^{5/4}}{\frac{5}{4}} + C$

$= \dfrac{4}{5}x^{5/4} + C$

7. $\int (x^2 + x - 1) dx$

$= \int x^2 dx + \int xdx - \int 1dx$ The integral of a sum is the sum of the integrals.

$= \dfrac{x^{2+1}}{2+1} + \dfrac{x^{1+1}}{1+1} - x + C \leftarrow$ DON'T FORGET THE *C*!

$\left[\int x^r dx = \dfrac{x^{r+1}}{r+1} + C\right]$

$\left[\int kdx = kx + C\right]$

$= \dfrac{x^3}{3} + \dfrac{x^2}{2} - x + C$

9. $\int (2t^2 + 5t - 3) dt$

$= \int 2t^2 dt + \int 5tdt - \int 3dt$ The integral of a sum is the sum of the integrals.

$= 2 \cdot \dfrac{t^{2+1}}{2+1} + 5 \cdot \dfrac{t^{1+1}}{1+1} - 3t + C$

$\left[\int x^r dx = \dfrac{x^{r+1}}{r+1} + C\right]$

$\left[\int kdx = kx + C\right]$

$= \dfrac{2}{3}t^3 + \dfrac{5}{2}t^2 - 3t + C$

11. $\int \dfrac{1}{x^3} dx = \int x^{-3} dx$

$= \dfrac{x^{-3+1}}{-3+1} + C \qquad \left[\int x^r dx = \dfrac{x^{r+1}}{r+1} + C\right]$

$= -\dfrac{x^{-2}}{2} + C$

13. $\int \sqrt[3]{x} dx = \int x^{1/3} dx$

$= \dfrac{x^{1/3+1}}{\frac{1}{3}+1} + C \qquad \left[\int x^r dx = \dfrac{x^{r+1}}{r+1} + C\right]$

$= \dfrac{x^{4/3}}{\frac{4}{3}} + C$

$= \dfrac{3}{4}x^{4/3} + C$

15. $\int \sqrt{x^5} dx = \int x^{5/2} dx$

$= \dfrac{x^{5/2+1}}{\frac{5}{2}+1} + C \qquad \left[\int x^r dx = \dfrac{x^{r+1}}{r+1} + C\right]$

$= \dfrac{x^{7/2}}{\frac{7}{2}} + C$

$= \dfrac{2}{7}x^{7/2} + C$

17. $\int \frac{dx}{x^4} = \int \frac{1}{x^4}dx = \int x^{-4}dx$

$= \frac{x^{-4+1}}{-4+1} + C \qquad \left[\int x^r dx = \frac{x^{r+1}}{r+1} + C\right]$

$= -\frac{x^{-3}}{3} + C$

19. $\int \frac{1}{x}dx$

$= \ln x + C,\ x > 0 \qquad \left[\int \frac{1}{x}dx = \ln x + C, x > 0\right]$

21. $\int \left(\frac{3}{x} + \frac{5}{x^2}\right)dx$

$= \int \frac{3}{x}dx + \int \frac{5}{x^2}dx$ The integral of a sum is the sum of the integrals.

$= 3\int x^{-1}dx + \int 5x^{-2}dx$

$= 3 \cdot \ln x + 5 \cdot \frac{x^{-2+1}}{-2+1} + C, \quad x > 0$

$\left[\int x^{-1}dx = \ln x, x > 0\right]$

$\left[\int x^r dx = \frac{x^{r+1}}{r+1} + C\right]$

$= 3\ln x - 5x^{-1} + C$

23. $\int \frac{-7}{\sqrt[3]{x^2}}dx = \int \frac{-7}{x^{2/3}} = \int -7x^{-2/3}dx$

$-7\int x^{-2/3}dx$

$= -7 \cdot \frac{x^{-2/3+1}}{-\frac{2}{3}+1} + C \qquad \left[\int x^r dx = \frac{x^{r+1}}{r+1} + C\right]$

$= -7 \cdot \frac{x^{1/3}}{\frac{1}{3}} + C$

$= -21x^{1/3} + C$

25. $\int 2e^{2x}dx$

$= \frac{2}{2}e^{2x} + C \qquad \left[\int be^{ax}dx = \frac{b}{a}e^{ax} + C\right]$

$= e^{2x} + C$

27. $\int e^{3x}dx$

$= \frac{1}{3}e^{3x} + C \qquad \left[\int be^{ax}dx = \frac{b}{a}e^{ax} + C\right]$

29. $\int e^{7x}dx$

$= \frac{1}{7}e^{7x} + C \qquad \left[\int be^{ax}dx = \frac{b}{a}e^{ax} + C\right]$

31. $\int 5e^{3x}dx$

$= \frac{5}{3}e^{3x} + C \qquad \left[\int be^{ax}dx = \frac{b}{a}e^{ax} + C\right]$

33. $\int 6e^{8x}dx$

$= \frac{6}{8}e^{8x} + C \qquad \left[\int be^{ax}dx = \frac{b}{a}e^{ax} + C\right]$

$= \frac{3}{4}e^{8x} + C$

35. $\int \frac{2}{3}e^{-9x}dx$

$= \frac{2}{3} \cdot \frac{1}{-9}e^{-9x} + C \qquad \left[\int be^{ax}dx = \frac{b}{a}e^{ax} + C\right]$

$= -\frac{2}{27}e^{-9x} + C$

37. $\int \left(5x^2 - 2e^{7x}\right)dx$

$= \int 5x^2dx - \int 2e^{7x}dx$ The integral of a sum is the sum of the integrals.

$= 5 \cdot \frac{x^{2+1}}{2+1} - \frac{2}{7}e^{7x} + C$

$\left[\int x^r dx = \frac{x^{r+1}}{r+1} + C\right]$

$\left[\int be^{ax}dx = \frac{b}{a}e^{ax} + C\right]$

$= \frac{5}{3}x^3 - \frac{2}{7}e^{7x} + C$

Copyright © 2014 Pearson Education, Inc. Publishing as Addison-Wesley.

39. $\int\left(x^2-\frac{3}{2}\sqrt{x}+x^{-4/3}\right)dx$

$=\int\left(x^2-\frac{3}{2}x^{1/2}+x^{-4/3}\right)dx$

$=\int x^2dx-\int\frac{3}{2}x^{1/2}dx+\int x^{-4/3}dx$

$=\frac{x^{2+1}}{2+1}-\frac{3}{2}\cdot\frac{x^{1/2+1}}{\frac{1}{2}+1}+\frac{x^{-4/3+1}}{-\frac{4}{3}+1}+C$

$\left[\int x^r dx=\frac{x^{r+1}}{r+1}+C\right]$

$=\frac{x^3}{3}-\frac{3}{2}\cdot\frac{x^{3/2}}{\frac{3}{2}}+\frac{x^{-1/3}}{-\frac{1}{3}}+C$

$=\frac{x^3}{3}-x^{3/2}-3x^{-1/3}+C$

41. $\int(3x+2)^2\,dx=\int\left(9x^2+12x+4\right)dx$

$=\int 9x^2dx+\int 12xdx+\int 4dx$

$=9\cdot\frac{x^{2+1}}{2+1}+12\cdot\frac{x^{1+1}}{1+1}+4x+C$

$=\frac{9}{3}x^3+\frac{12}{2}x^2+4x+C$

$=3x^3+6x^2+4x+C$

43. $\int\left(\frac{3}{x}-5e^{2x}+\sqrt{x^7}\right)dx$

$=\int\left(\frac{3}{x}-5e^{2x}+x^{7/2}\right)dx$

$=\int\frac{3}{x}dx-\int 5e^{2x}dx+\int x^{7/2}dx$

$=3\ln x-\frac{5}{2}\cdot e^{2x}+\frac{x^{7/2+1}}{\frac{7}{2}+1}+C$

$=3\ln x-\frac{5}{2}e^{2x}+\frac{2}{9}x^{9/2}+C$

45. $\int\left(\frac{7}{\sqrt{x}}-\frac{2}{3}e^{5x}-\frac{8}{x}\right)dx$

$=\int\left(7x^{-1/2}-\frac{2}{3}e^{5x}-\frac{8}{x}\right)dx$

$=\int 7x^{-1/2}dx-\int\frac{2}{3}e^{5x}dx-\int\frac{8}{x}dx$

$=7\cdot\frac{x^{-1/2+1}}{-\frac{1}{2}+1}-\frac{2}{3}\cdot\frac{1}{5}e^{5x}-8\ln x+C$

$=14x^{1/2}-\frac{2}{15}e^{5x}-8\ln x+C$

47. Find the function $f(x)$, such that

$f'(x)=x-3,\ f(2)=9$

We first find $f(x)$ by integrating:

$f(x)=\int(x-3)dx$

$=\int xdx-\int 3dx$

$=\frac{1}{2}x^2-3x+C.$

The condition $f(2)=9$ allows us to find C:

$f(2)=9$

$\frac{1}{2}(2)^2-3(2)+C=9$

$2-6+C=9$

$-4+C=9$

$C=13.$

Thus, $f(x)=\frac{1}{2}x^2-3x+13.$

49. Find the function $f(x)$, such that

$f'(x)=x^2-4,\ f(0)=7$.

We first find $f(x)$ by integrating:

$f(x)=\int\left(x^2-4\right)dx$

$=\int x^2dx-\int 4dx$

$=\frac{1}{3}x^3-4x+C.$

The condition $f(0)=7$ allows us to find C:

$f(0)=7$

$\frac{1}{3}(0)^3-4(0)+C=7$

$C=7.$

Thus, $f(x)=\frac{1}{3}x^3-4x+7.$

Copyright © 2014 Pearson Education, Inc. Publishing as Addison-Wesley.

51. Find the function $f(x)$, such that

$f'(x)=5x^2+3x-7,\ f(0)=9$.

We first find $f(x)$ by integrating:

$$\begin{aligned}f(x)&=\int\left(5x^2+3x-7\right)dx\\&=\int 5x^2dx+\int 3xdx-\int 7dx\\&=\frac{5}{3}x^3+\frac{3}{2}x^2-7x+C.\end{aligned}$$

The condition $f(0)=9$ allows us to find C:

$$f(0)=9$$
$$\frac{5}{3}(0)^3+\frac{3}{2}(0)^2-7(0)+C=9$$
$$C=9.$$

Thus, $f(x)=\frac{5}{3}x^3+\frac{3}{2}x^2-7x+9.$

53. Find the function $f(x)$, such that

$f'(x)=3x^2-5x+1,\ f(1)=\frac{7}{2}$.

We first find $f(x)$ by integrating:

$$\begin{aligned}f(x)&=\int\left(3x^2-5x+1\right)dx\\&=\int 3x^2dx-\int 5xdx+\int dx\\&=x^3-\frac{5}{2}x^2+x+C.\end{aligned}$$

The condition $f(1)=\frac{7}{2}$ allows us to find C.

$$f(1)=\frac{7}{2}$$
$$(1)^3-\frac{5}{2}(1)^2+(1)+C=\frac{7}{2}$$
$$-\frac{1}{2}+C=\frac{7}{2}$$
$$C=4.$$

Thus, $f(x)=x^3-\frac{5}{2}x^2+x+4.$

55. Find the function $f(x)$, such that

$f'(x)=5e^{2x},\ f(0)=\frac{1}{2}$.

We first find $f(x)$ by integrating:

$$\begin{aligned}f(x)&=\int 5e^{2x}dx\\&=\frac{5}{2}e^{2x}+C.\end{aligned}$$

The condition $f(0)=\frac{1}{2}$ allows us to find C:

$$f(0)=\frac{1}{2}$$
$$\frac{5}{2}e^{2(0)}+C=\frac{1}{2}$$
$$\frac{5}{2}e^{0}+C=\frac{1}{2}$$
$$\frac{5}{2}\cdot 1+C=\frac{1}{2}$$
$$C=-\frac{4}{2}=-2.$$

Thus, $f(x)=\frac{5}{2}e^{2x}-2.$

57. Find the function $f(x)$, such that

$f'(x)=\frac{4}{\sqrt{x}},\ f(1)=-5$.

We first find $f(x)$ by integrating:

$$\begin{aligned}f(x)&=\int\frac{4}{\sqrt{x}}dx\\&=\int 4x^{-1/2}dx\\&=8x^{1/2}+C.\end{aligned}$$

The condition $f(1)=-5$ allows us to find C:

$$f(1)=-5$$
$$8(1)^{1/2}+C=-5$$
$$8+C=-5$$
$$C=-13.$$

Thus, $f(x)=8x^{1/2}-13.$

59. $D'(t)=-810.3t^2+1730.3t+3648$

We integrate to find $D(t)$.

$$\begin{aligned}D(t)&=\int\left(-810.3t^2+1730.3t+3648\right)dt\\&=\frac{-180.3}{3}t^3+\frac{1730.3}{2}t^2+3648t+C\\&=-270.1t^3+865.15t^2+3648t+C\end{aligned}$$

The condition $D(0)=41,267$ allows us to find C. Substituting, we have

$$D(0)=41,267$$
$$-270.1(0)^3+865.15(0)^2+3648(0)+C=41,267$$
$$C=41,267.$$

Thus,

$D(t)=-270.1t^3+865.15t^2+3648t+41,267.$

Copyright © 2014 Pearson Education, Inc. Publishing as Addison-Wesley.

61. $C'(x) = x^3 - 2x$

We integrate to find $C(x)$, we use K for the constant of integration to avoid confusion with the cost function $C(x)$.

$$C(x) = \int C'(x)\,dx$$
$$= \int (x^3 - 2x)\,dx$$
$$= \frac{x^4}{4} - x^2 + K$$

Fixed costs are \$7000. This means $C(0) = 7000$. This allows us to determine K.

$$C(0) = 7000$$
$$\frac{(0)^4}{4} - (0)^2 + K = 7000$$
$$K = 7000$$

Thus, the total cost function is

$$C(x) = \frac{x^4}{4} - x^2 + 7000.$$

63. $R'(x) = x^2 - 3$

a) We integrate to find $R(x)$.

$$R(x) = \int R'(x)\,dx$$
$$= \int (x^2 - 3)\,dx$$
$$= \frac{x^3}{3} - 3x + C$$

The condition $R(0) = 0$ allows us to find C.

$$R(0) = 0$$
$$\frac{(0)^3}{3} - 3(0) + C = 0$$
$$C = 0$$

Thus, the total revenue function is

$$R(x) = \frac{x^3}{3} - 3x.$$

b) ✎

65. $D'(x) = -\frac{4000}{x^2} = -4000x^{-2}$

We integrate to find $D(x)$.

$$D(x) = \int D'(x)\,dx$$
$$= \int -4000x^{-2}\,dx$$
$$= -4000 \cdot \frac{x^{-1}}{-1} + C$$
$$= 4000x^{-1} + C$$
$$= \frac{4000}{x} + C$$

When the price is \$4 per unit, the demand is 1003 units. This means $D(4) = 1003$.

Substituting 4 for x and 1003 for $D(x)$ we can determine C as follows:

$$D(4) = 1003$$
$$\frac{4000}{4} + C = 1003$$
$$1000 + C = 1003$$
$$C = 3.$$

Thus, the demand function is $D(x) = \frac{4000}{x} + 3$.

67. $\frac{dE}{dt} = 30 - 10t$

a) We find $E(t)$ by integrating

$$E(t) = \int E'(t)\,dt$$
$$= \int (30 - 10t)\,dt$$
$$= 30t - 10 \cdot \frac{t^2}{2} + C$$
$$= 30t - 5t^2 + C$$

The condition $E(2) = 72$ allows us to find C as follows:

$$E(2) = 72$$
$$30(2) - 5(2)^2 + C = 72 \quad \text{Substituting}$$
$$60 - 20 + C = 72$$
$$40 + C = 72$$
$$C = 32.$$

Thus, $E(t) = 30t - 5t^2 + 32$.

Copyright © 2014 Pearson Education, Inc. Publishing as Addison-Wesley.

b) $E(t)=32+30t-5t^2$

Substituting 3 for t, we have:

$$E(3)=32+30(3)-5(3)^2$$
$$=32+90-45$$
$$=77.$$

After 3 hours, the operator's efficiency is 77%.

Substituting 5 for t, we have:

$$E(5)=32+30(5)-5(5)^2$$
$$=32+150-125$$
$$=57.$$

After 5 hours, the operator's efficiency is 57%.

69. $I'(t)=-6.34t+141.6$

a) We integrate to find $I(t)$.

$$I(t)=\int I'(t)\,dt$$
$$=\int(-6.34t+141.6)\,dt$$
$$=\frac{-6.34}{2}t^2+141.6t+C$$
$$=-3.17t^2+141.6t+C.$$

The condition $I(0)=1408$ allows us to find C.

$$I(0)=1408$$
$$-3.17(0)^2+141.6(0)+C=1408 \quad \text{Substituting}$$
$$C=1408$$

The total number per 100,000 who have contracted swine flu by time t is given by $I(t)=-3.17t^2+141.6t+1408$.

b) The number of people that contracted swine flu during the first 8 weeks is given by $I(8)-I(0)$. Using the function found in part (a), we substitute 8 for t.

$$I(8)=-3.17(8)^2+141.6(8)+1408.$$
$$=2337.92\approx 2338$$

Therefore, the number of cases during the first 8 weeks is:

$$I(8)-I(0)\approx 2338-1408\approx 930$$

During the first 8 weeks, approximately 930 people contracted swine flu.

c) The number of people that contracted swine flu during the 18 week period is given by $I(18)-I(0)$. Using the function found in part (a), we substitute 18 for t.

$$I(18)=-3.17(18)^2+141.6(18)+1408.$$
$$=2929.72\approx 2930$$

Therefore, the number of cases during the 18 week period is:

$$I(18)-I(0)\approx 2930-1408\approx 1522$$

After 18 weeks, approximately 1522 people contracted swine flu.

d) The number of people that contracted swine flu during the last 7 weeks of the 18 weeks is given by $I(18)-I(11)$.

First we find $I(11)$.

$$I(11)=-3.17(11)^2+141.6(11)+1408.$$
$$=2582.03\approx 2582$$

Using $I(18)$ found in part (c), we have:

$$I(18)-I(11)\approx 2930-2582\approx 348.$$

Approximately 348 people contracted swine flu during the last 7 of the 18 weeks.

71. $h'(t)=v(t)=-32t+75$

a) We find $h(t)$ by integrating $v(t)$.

$$h(t)=\int v(t)\,dt$$
$$=\int(-32t+75)\,dt$$
$$=\frac{-32}{2}t^2+75t+C$$
$$=-16t^2+75t+C$$

The condition $h(0)=30$ allows us to find C.

$$h(0)=30$$
$$-16(0)^2+75(0)+C=30$$
$$C=30$$

The function that gives the height (in feet) of the baseball after t seconds is:

$$h(t)=-16t^2+75t+30.$$

b) Substituting $t=2$ into the position function we have:

$$h(2)=-16(2)^2+75(2)+30$$
$$=116$$

The height of the baseball after 2 seconds is 116 feet.

The solution is continued on the next page.

Copyright © 2014 Pearson Education, Inc. Publishing as Addison-Wesley.

Substituting $t = 2$ into the velocity function on the previous page, we have:

$$v(2) = -32(2) + 75 = 11$$

The velocity of the baseball after 2 seconds is 11 feet per second.

c) The highest point will occur when the velocity of the ball is zero. We solve the equation $v(t) = 0$ for t.

$$-32t + 75 = 0$$
$$-32t = -75$$
$$t = \frac{75}{32} \approx 2.344.$$

It will take approximately 2.344 seconds for the ball to reach its highest point.

d) Substituting in 2.344 seconds from part (c) into $h(t)$ we have:

$$h(2.344) = -16(2.344)^2 + 75(2.344) + 30$$
$$\approx 117.89$$

The ball will reach a maximum height of approximately 117.89 feet.

e) The ball will hit the ground when $h(t) = 0$.

$$h(t) = 0$$
$$-16t^2 + 75t + 30 = 0$$

Using the quadratic formula, we have:

$$t = \frac{-75 \pm \sqrt{(75)^2 - 4(-16)(30)}}{2(-16)}$$
$$= \frac{-75 \pm \sqrt{7545}}{-32}$$
$$t \approx -0.37 \text{ or } t \approx 5.06$$

In this application the only solution that is feasible is $t \approx 5.06$. Therefore, it will take approximately 5.06 seconds for the ball to hit the ground.

f) Substituting $t = 5.06$ into the velocity function we have:

$$v(5.06) = -32(5.06) + 75 = -86.92$$

The impact velocity of the baseball is -86.92 feet per second.

73. Find the function $f(t)$, such that

$$f'(t) = \sqrt{t} + \frac{1}{\sqrt{t}},\ f(4) = 0.$$

We first find $f(t)$ by integrating:

$$f(t) = \int \left(\sqrt{t} + \frac{1}{\sqrt{t}} \right) dt$$
$$= \int \left(t^{1/2} + t^{-1/2} \right) dt$$
$$= \frac{2}{3} t^{3/2} + 2t^{1/2} + C.$$

The condition $f(4) = 0$ allows us to find C:

$$f(4) = 0$$
$$\frac{2}{3}(4)^{3/2} + 2(4)^{1/2} + C = 0$$
$$\frac{2}{3}(8) + 4 + C = 0$$
$$\frac{16}{3} + 4 + C = 0$$
$$C = -\frac{28}{3}.$$

Thus, $f(t) = \frac{2}{3} t^{3/2} + 2t^{1/2} - \frac{28}{3}.$

75. $\int (5t+4)^2 t^4 dx$

First, we will expand the binomial.

$$= \int (5t+4)(5t+4) t^4 dt$$
$$= \int \left(25t^2 + 40t + 16 \right) t^4 dt$$

Next, we distribute t^4.

$$\int \left(25t^6 + 40t^5 + 16t^4 \right) dt$$

Now we can integrate as follows:

$$= \int 25t^6 dt + \int 40t^5 dt + \int 16t^4 dt$$ The integral of a sum is the sum of the integrals.

$$= 25 \cdot \frac{t^{6+1}}{6+1} + 40 \cdot \frac{t^{5+1}}{5+1} + 16 \cdot \frac{t^{4+1}}{4+1} + C$$

$$\left[\int x^r dx = \frac{x^{r+1}}{r+1} + C \right]$$

$$= \frac{25}{7} t^7 + \frac{40}{6} t^6 + \frac{16}{5} t^5 + C$$
$$= \frac{25}{7} t^7 + \frac{20}{3} t^6 + \frac{16}{5} t^5 + C.$$

Copyright © 2014 Pearson Education, Inc. Publishing as Addison-Wesley.

77. $\int(1-t)\sqrt{t}\,dt$

$=\int\left(\sqrt{t}-t\sqrt{t}\right)dt$ Distributing $\sqrt{t}$

$=\int\left(t^{1/2}-t^{3/2}\right)dt$

$=\frac{t^{3/2}}{3/2}-\frac{t^{5/2}}{5/2}+C$

$=\frac{2}{3}t^{3/2}-\frac{2}{5}t^{5/2}+C$

79. $\int\frac{x^4-6x^2-7}{x^3}dx$

$=\int\left(x^4-6x^2-7\right)x^{-3}dx$

$=\int\left(x-6x^{-1}-7x^{-3}\right)dx$

$=\frac{x^2}{2}-6\ln x-7\cdot\frac{x^{-2}}{-2}+C$

$=\frac{x^2}{2}-6\ln x+\frac{7}{2}x^{-2}+C$

81. $\int\frac{1}{\ln 10}\cdot\frac{dx}{x}$

$=\frac{1}{\ln 10}\int\frac{1}{x}dx$

$=\frac{1}{\ln 10}\cdot\ln x+C$

$=\log x+C$ Properties of logarithms.

83. $\int(3x-5)(2x+1)^2\,dx$

$=\int(3x-5)\left(4x^2+4x+1\right)dx$

$=\int\left(12x^3-8x^2-17x-5\right)dx$

$=12\cdot\frac{x^4}{4}-8\cdot\frac{x^3}{3}-17\cdot\frac{x^2}{2}-5x+C$

$=3x^4-\frac{8}{3}x^3-\frac{17}{2}x^2-5x+C$

85. $\int\frac{x^2-1}{x+1}dx$

$=\int\frac{(x-1)(x+1)}{x+1}dx$

$=\int(x-1)\,dx$

$=\frac{x^2}{2}-x+C$

87.

Copyright © 2014 Pearson Education, Inc. Publishing as Addison-Wesley.

Exercise Set 4.2

1. $C(x) = -0.012x + 6.50, \quad \text{for } x \le 300$

Note that the cost per pound of roasting coffee decreases as the number of pounds of coffee increases. We use the area under the graph to find the total cost of roasting 200 lb of coffee.

Shading the area under $C(x)$ on the interval $0 \le x \le 200$ we see:

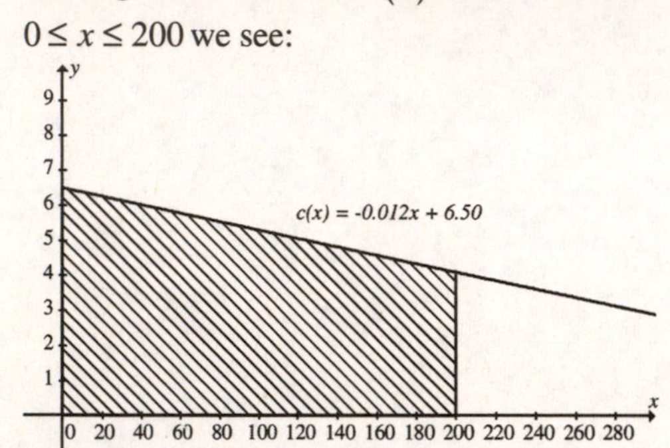

The area under the curve is a trapezoid; therefore, calculating the total cost of roasting 200 lb of coffee will require calculating the area of the trapezoid.

The formula for calculating the area of a trapezoid is $A = \frac{1}{2}h(b_1 + b_2)$, where h is the height of the trapezoid and b_1 and b_2 are the lengths of the respective bases. If we view the trapezoid sideways, we see that

$h = 200$

$b_1 = C(0) = -0.012(0) + 6.50 = 6.50$

$b_2 = C(200) = -0.012(200) + 6.50 = 4.1$

Substituting these values into the formula, we have:

$$\begin{aligned}
\text{Total Cost} &= \text{Area of trapezoid} \\
&= \frac{1}{2}h(b_1 + b_2) \\
&= \frac{1}{2}(200)(6.5 + 4.1) \\
&= 100(10.6) \\
&= 1060.
\end{aligned}$$

The total cost of roasting 200 pounds of coffee is \$1060.

3. $C(x) = -0.04x + 85, \quad \text{for } x \le 1000$

Note that the cost per card decreases as the number of cards produced increases. We use the area under the graph to find the total cost of producing 650 cards.

Shading the area $C(x)$ on the interval $0 \le x \le 650$ we see:

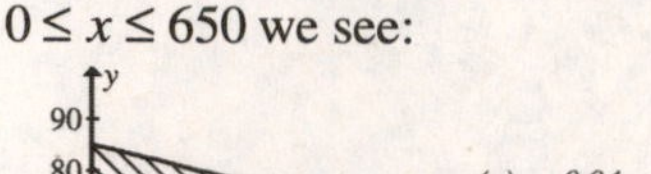

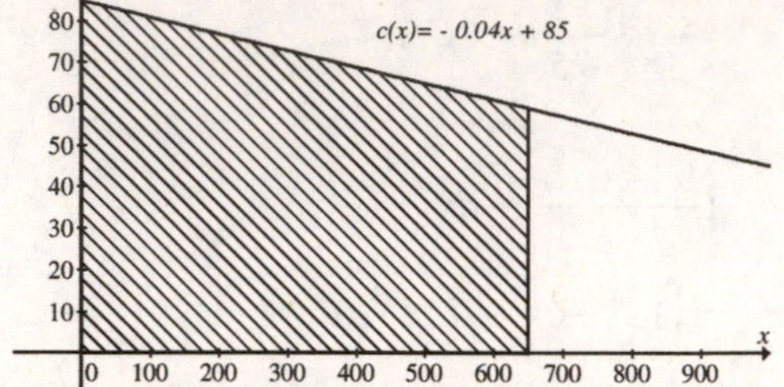

The area under the curve is a trapezoid; therefore, calculating the total cost of producing 650 note cards will require calculating the area of the trapezoid.

The formula for calculating the area of a trapezoid is $A = \frac{1}{2}h(b_1 + b_2)$, where h is the height of the trapezoid and b_1 and b_2 are the lengths of the respective bases. If we view the trapezoid sideways, we see that

$h = 650$

$b_1 = C(0) = -0.04(0) + 85 = 85$

$b_2 = C(650) = -0.04(650) + 85 = 59$

Substituting these values into the formula, we have:

$$\begin{aligned}
\text{Total Cost} &= \text{Area of trapezoid} \\
&= \frac{1}{2}h(b_1 + b_2) \\
&= \frac{1}{2}(650)(85 + 59) \\
&= 325(144) \\
&= 46{,}800.
\end{aligned}$$

The total cost of producing 650 cards is 46,800 cents or \$468.

Copyright © 2014 Pearson Education, Inc. Publishing as Addison-Wesley.

5. $P'(x) = 2x - 1150,\ x \geq 0$

Note that the marginal profit is rate of change of total profit with respect to the number of tickets. We use the area under the graph to find the total profit of selling 300 tickets.

Shading the area under $P'(x)$ on the interval $0 \leq x \leq 300$ we see:

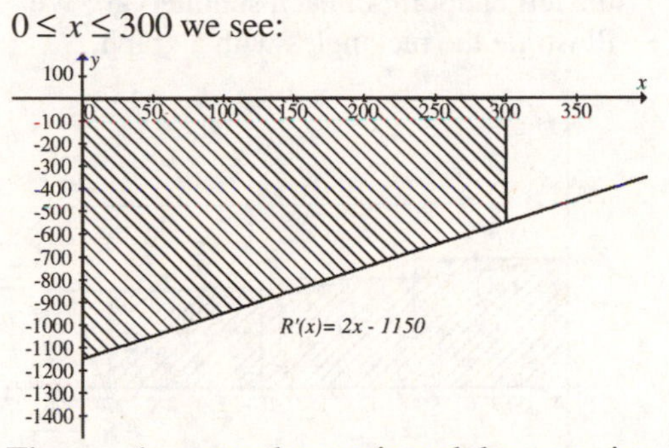

The area between the x-axis and the curve is a trapezoid; therefore, the total profit will be equal to the area of the trapezoid.

The formula for calculating the area of a trapezoid is $A = \frac{1}{2}h(b_1 + b_2)$, where h is the height of the trapezoid and b_1 and b_2 are the lengths of the respective bases. If we view the trapezoid sideways, we see that

$h = 200$

$b_1 = P'(0) = 2(0) - 1150 = -1150$

$b_2 = P'(300) = 2(300) - 1150 = -550.$

Because the marginal profit function is below the x-axis, the values for b_1 and b_2 are negative. Due to the nature of the application, we will keep these negative values. The result will be a negative area of the trapezoid. This simply means that the total profit will be negative, and the company is operating at a loss.

Substituting these values into the formula, we have:

$$\begin{aligned}\text{Total Profit} &= \text{Area of trapezoid}\\ &= \frac{1}{2}h(b_1 + b_2)\\ &= \frac{1}{2}(300)(-1150 + (-550))\\ &= 150(-1700)\\ &= -255{,}000.\end{aligned}$$

The total profit from the sale of the first 300 tickets is $-\$255{,}000$.

7. $C'(x) = -\frac{2}{25}x + 50, \quad \text{for } x \leq 450$

We use the area under the graph to find the total cost of producing the first 200 dresses.

Shading the area under $C'(x)$ on the interval $0 \leq x \leq 200$ we see:

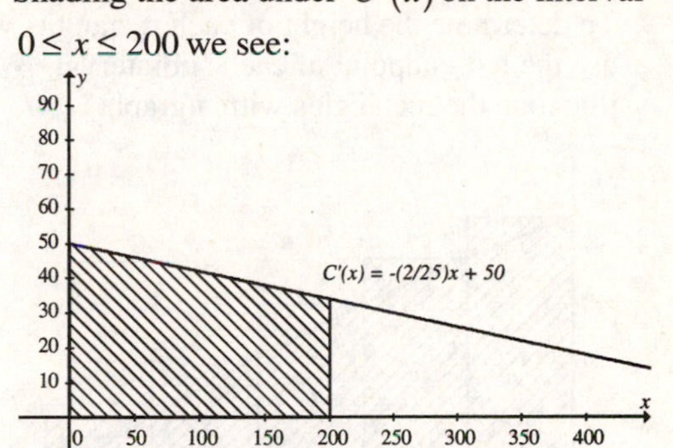

The area under the curve is a trapezoid; therefore, calculating the total cost of producing 200 dresses will require calculating the area of the trapezoid.

The formula for calculating the area of a trapezoid is $A = \frac{1}{2}h(b_1 + b_2)$, where h is the height of the trapezoid and b_1 and b_2 are the lengths of the respective bases. If we view the trapezoid sideways, we see that

$h = 200$

$b_1 = C'(0) = -\frac{2}{25}(0) + 50 = 50$

$b_2 = C'(200) = -\frac{2}{25}(200) + 50 = 34.$

Substituting these values into the formula, we have:

$$\begin{aligned}\text{Total Cost} &= \text{Area of trapezoid}\\ &= \frac{1}{2}h(b_1 + b_2)\\ &= \frac{1}{2}(200)(50 + 34)\\ &= 100(84)\\ &= 8400.\end{aligned}$$

The total cost of producing the first 200 dresses is \$8400.

Copyright © 2014 Pearson Education, Inc. Publishing as Addison-Wesley.

9. $C'(x) = -0.00002x^2 - 0.04x + 45$

We divide the interval $[0,800]$ into five subintervals, each of width $\Delta x = \frac{800}{5} = 160.$

To determine the height of each rectangle, we use the left endpoint of each subinterval. We illustrate the rectangles with a graph.

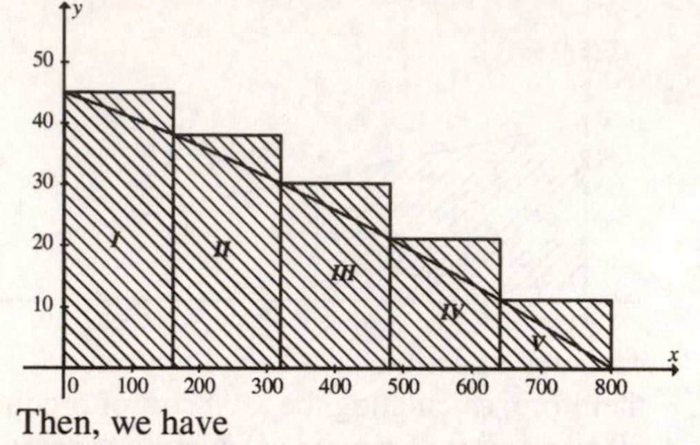

Then, we have

Total Cost ≈ Area I + Area II + Area III + Area IV + Area V

The five left endpoints are 0, 160, 320, 480, and 640. Evaluating $C'(x)$ at each of these endpoints will determine the height of each rectangle and the width was determined to be 160 ft. Therefore, the area of each rectangle is:

$$\text{Area I } = C'(0) \cdot 160 = 45 \cdot 160 = 7200$$

$$\text{Area II } = C'(160) \cdot 160 = 38.088 \cdot 160 = 6094.08$$

$$\text{Area III } = C'(320) \cdot 160 = 30.152 \cdot 160 = 4824.32$$

$$\text{Area IV } = C'(480) \cdot 160 = 21.192 \cdot 160 = 3390.72$$

$$\text{Area V } = C'(640) \cdot 160 = 11.208 \cdot 160 = 1793.28$$

Summing the area of each of the five rectangles yields:

$$\text{Total Cost } \approx 7200 + 6094.08 + 4824.32 + 3390.72 + 1793.28 \approx 23{,}302.4.$$

Thus, the total cost of manufacturing 800 feet of molding is approximately 23,302.4 cents or about $233.02.

11. $C'(x) = 0.000008x^2 - 0.004x + 2, \text{ for } x \le 350$

We divide the interval $[0,270]$ into three subintervals, each of width $\Delta x = \frac{270}{3} = 90.$ To determine the height of each rectangle, we use the left endpoint of each subinterval. We illustrate the rectangles with a graph.

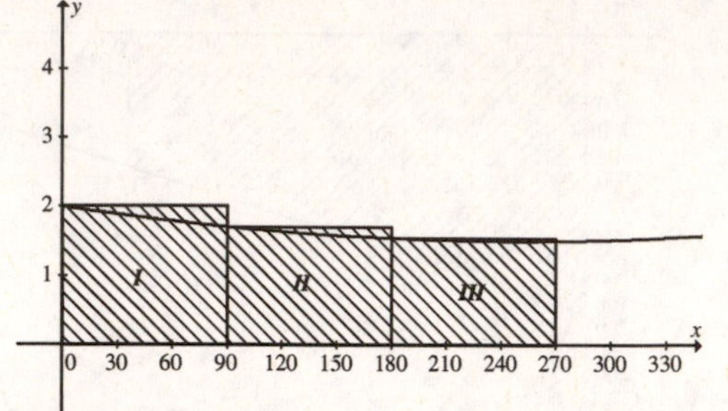

Then, we have

Total Cost ≈ Area I + Area II + Area III

The three left endpoints are 0, 90, and 180. Evaluating $C'(x)$ at each of these endpoints will determine the height of each rectangle and the width was determined to be 160 ft. Therefore, the area of each rectangle is:

$$\text{Area I } = C'(0) \cdot 90 = 2 \cdot 90 = 180$$

$$\text{Area II } = C'(90) \cdot 90 = 1.7048 \cdot 90 = 153.432$$

$$\text{Area III } = C'(180) \cdot 90 = 1.5392 \cdot 90 = 138.528$$

Summing the area of each of the three rectangles yields:

$$\text{Total Cost } \approx 180 + 153.432 + 138.528 \approx 471.96.$$

Thus, the total cost of producing 270 pints of fresh-squeezed orange juice is approximately $471.96.

13. Note that we are adding consecutive multiples of 3:

$$3 + 6 + 9 + 12 + 15 + 18 = \sum_{i=1}^{6} 3i.$$

15. $f(x_1) + f(x_2) + f(x_3) + f(x_4) = \sum_{i=1}^{4} f(x_i).$

17. $G(x_1) + G(x_2) + \cdots + G(x_{15}) = \sum_{i=1}^{15} G(x_i).$

19.
$$\sum_{i=1}^{4} 2^i = 2^1 + 2^2 + 2^3 + 2^4 = 2 + 4 + 8 + 16, \text{or } 30.$$

Copyright © 2014 Pearson Education, Inc. Publishing as Addison-Wesley.

21. $\sum_{i=1}^{5} f(x_i)$

$= f(x_1)+f(x_2)+f(x_3)+f(x_4)+f(x_5).$

23. $f(x)=\frac{1}{x^2}$

a) In the drawing in the text the interval $[1,7]$ has been divided into 6 subintervals, each having width 1 $\left[\Delta x=\frac{7-1}{6}=1\right]$.

The heights of the rectangles shown are

$f(1)=\frac{1}{1^2}=1$

$f(2)=\frac{1}{2^2}=\frac{1}{4}=0.2500$

$f(3)=\frac{1}{3^2}=\frac{1}{9}\approx 0.1111$

$f(4)=\frac{1}{4^2}=\frac{1}{16}=0.0625$

$f(5)=\frac{1}{5^2}=\frac{1}{25}=0.0400$

$f(6)=\frac{1}{6^2}=\frac{1}{36}\approx 0.0278$

Therefore, the area of each rectangle is:

Rectangle I $= f(1)\cdot\Delta x=1\cdot 1=1$

Rectangle II $= f(2)\cdot\Delta x=0.2500\cdot 1=0.2500$

Rectangle III $= f(3)\cdot\Delta x=0.1111\cdot 1=0.1111$

Rectangle IV $= f(4)\cdot\Delta x=0.0625\cdot 1=0.0625$

Rectangle V $= f(5)\cdot\Delta x=1\cdot 0.0400=0.0400$

Rectangle VI $= f(6)\cdot\Delta x=1\cdot 0.0278=0.0278$

The area of the region under the curve over $[1,7]$ is approximately the sum of the areas of the 6 rectangles. Therefore, the total area is approximately:

$1+0.2500+0.1111+0.0625+$

$0.0400+0.0278\approx 1.4914.$

b) In the drawing in the text the interval $[1,7]$ has been divided into 12 subintervals, each having width 0.5 $\left[\Delta x=\frac{7-1}{12}=\frac{6}{12}=0.5\right]$.

The heights of 6 of the rectangles were computed in part (a). The heights of the other 6 rectangles are computed at the top of the next column.

$f(1.5)=\frac{1}{1.5^2}=\frac{1}{2.25}\approx 0.4444$

$f(2.5)=\frac{1}{2.5^2}=\frac{1}{6.25}=0.1600$

$f(3.5)=\frac{1}{3.5^2}=\frac{1}{12.25}\approx 0.00816$

$f(4.5)=\frac{1}{4.5^2}=\frac{1}{20.25}\approx 0.0494$

$f(5.5)=\frac{1}{5.5^2}=\frac{1}{30.25}\approx 0.0331$

$f(6.5)=\frac{1}{6.5^2}=\frac{1}{42.25}\approx 0.0237$

Therefore, the area of each rectangle is:

Rectangle I: $f(1)\cdot\Delta x=1(0.5)=0.5000$

Rectangle II: $f(1.5)\cdot\Delta x=0.4444(0.5)\approx 0.2222$

Rectangle III: $f(2)\cdot\Delta x=0.2500(0.5)=0.1250$

Rectangle IV: $f(2.5)\cdot\Delta x=0.1600(0.5)=0.0800$

Rectangle V: $f(3)\cdot\Delta x=0.1111(0.5)\approx 0.0556$

Rectangle VI: $f(3.5)\cdot\Delta x=0.0816(0.5)\approx 0.0408$

Rectangle VII: $f(4)\cdot\Delta x=0.0625(0.5)\approx 0.0313$

Rectangle VIII: $f(4.5)\cdot\Delta x=0.0494(0.5)\approx 0.0247$

Rectangle IX: $f(5)\cdot\Delta x=0.0400(0.5)=0.0200$

Rectangle X: $f(5.5)\cdot\Delta x=0.0331\approx 0.0165$

Rectangle XI: $f(6)\cdot\Delta x=0.0278(0.5)\approx 0.0139$

Rectangle XII: $f(6.5)\cdot\Delta x=0.0237(0.5)\approx 0.0118$

The area of the region under the curve over $[1,7]$ is approximately the sum of the areas of the 12 rectangles.

The total area is approximately:

$0.5+0.2222+0.1250+0.0800+$

$0.0556+0.0408+0.0313+$

$0.0247+0.0200+0.0165+$

$0.0139+0.0118\approx 1.1418.$

Answers may vary slightly depending on when rounding was done.

25. $P'(x)=-0.0006x^3+0.28x^2+55.6x$

We divide $[0,300]$ into 6 subintervals of width $\Delta x=50$. Therefore, the values of x_i are:

$x_1=0;\ x_2=50;\ x_3=100;$

$x_4=150;\ x_5=200;\ x_6=250.$

The solution is continued on the next page.

Copyright © 2014 Pearson Education, Inc. Publishing as Addison-Wesley.

Using the information from the previous page, the area under the curve is approximately:

$$\sum_{i=1}^{6} P'(x_i)\Delta x = P'(x_1)\cdot 50 + P'(x_2)\cdot 50 + P'(x_3)\cdot 50 + P'(x_4)\cdot 50 + P'(x_5)\cdot 50 + P'(x_6)\cdot 50$$

$$= P'(0)\cdot 50 + P'(50)\cdot 50 + P'(100)\cdot 50 + P'(150)\cdot 50 + P'(200)\cdot 50 + P'(250)\cdot 50$$

$$= 0\cdot 50 + 3405\cdot 50 + 7760\cdot 50 + 12{,}615\cdot 50 + 17{,}520\cdot 50 + 22{,}025\cdot 50$$

$$= 0 + 170{,}250 + 388{,}000 + 630{,}750 + 876{,}000 + 1{,}101{,}250$$

$$= 3{,}166{,}250.$$

The health club's total profit when 300 members are enrolled is approximately 3,166,250 cents or $31,662.50.

27. $f(x) = 0.01x^4 - 1.44x^2 + 60$

Dividing the interval $[2,10]$ into four subintervals, we calculate the width of each subinterval to be $\Delta x = \frac{10-2}{4} = \frac{8}{4} = 2$, with x_i ranging from $x_1 = 2$ to $x_4 = 8$. Although a drawing is not required, we make one to help visualize the area.

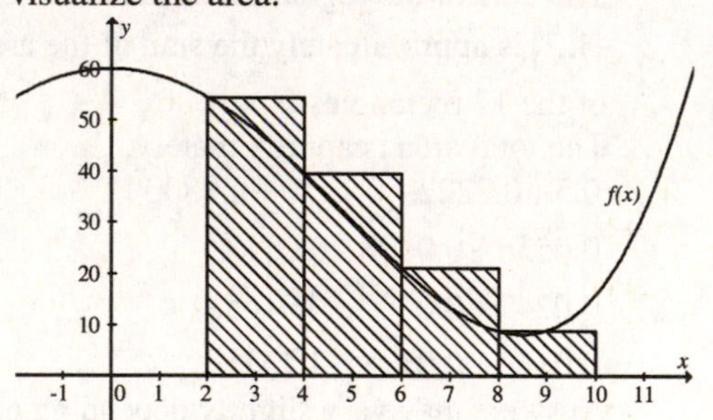

The area under the curve from 2 to 10 is approximately

$$\sum_{i=1}^{4} f(x_i)\Delta x = f(2)\cdot 2 + f(4)\cdot 2 + f(6)\cdot 2 + f(8)\cdot 2$$

$$= 54.4\cdot 2 + 39.52\cdot 2 + 21.12\cdot 2 + 8.8\cdot 2$$

$$= 247.68.$$

29. $F(x) = 0.2x^3 + 2x^2 - 0.2x - 2$

Dividing the interval $[-8,-3]$ into five subintervals, we calculate the width of each subinterval to be $\Delta x = \frac{-3-(-8)}{5} = \frac{5}{5} = 1$, with x_i ranging from $x_1 = -8$ to $x_5 = -4$.

Although a drawing is not required, we can make one to help visualize the area.

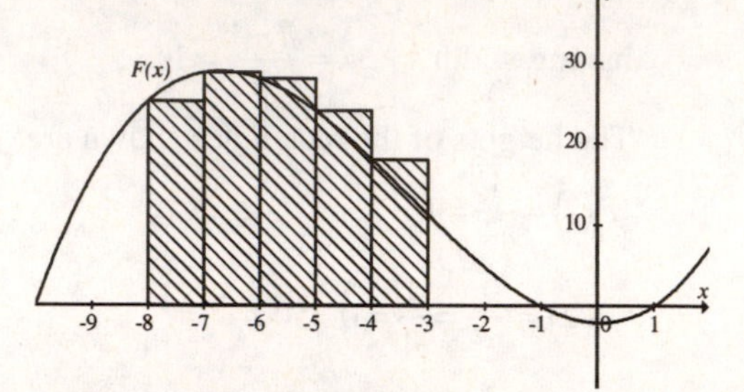

The area under the curve from -8 to -3 is approximately

$$\sum_{i=1}^{5} F(x_i)\Delta x = F(-8)\cdot 1 + F(-7)\cdot 1 + F(-6)\cdot 1 + F(-5)\cdot 1 + F(-4)\cdot 1$$

$$= 25.2\cdot 1 + 28.8\cdot 1 + 28\cdot 1 + 24\cdot 1 + 18\cdot 1$$

$$= 124.$$

31. Sketching the graph, we determine that the area defined by the definite integral.

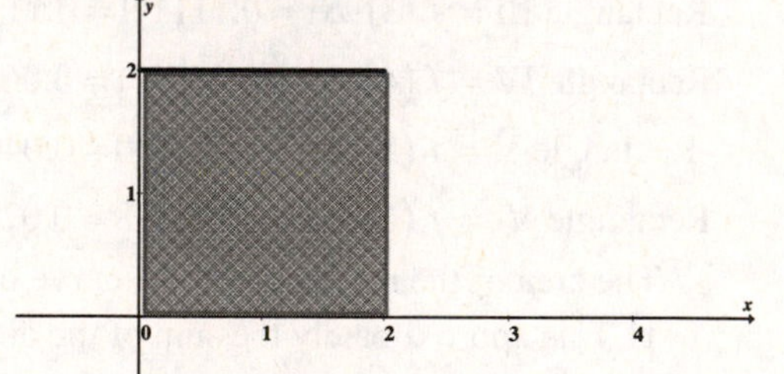

The region is a square; therefore, the area is given by:

$$A = b\cdot h = 2\cdot 2 = 4.$$

The definite integral is

$$\int_0^2 2dx = 4.$$

Copyright © 2014 Pearson Education, Inc. Publishing as Addison-Wesley.

33. Sketching the graph, we determine that the area defined by the definite integral.

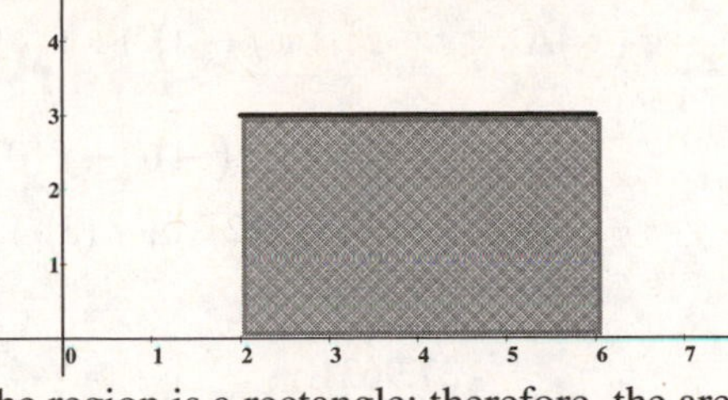

The region is a rectangle; therefore, the area is given by:

$$A = b \cdot h$$
$$= 4 \cdot 3 = 12.$$

The definite integral is

$$\int_2^6 3dx = 12.$$

35. Sketching the graph, we determine that the area defined by the definite integral.

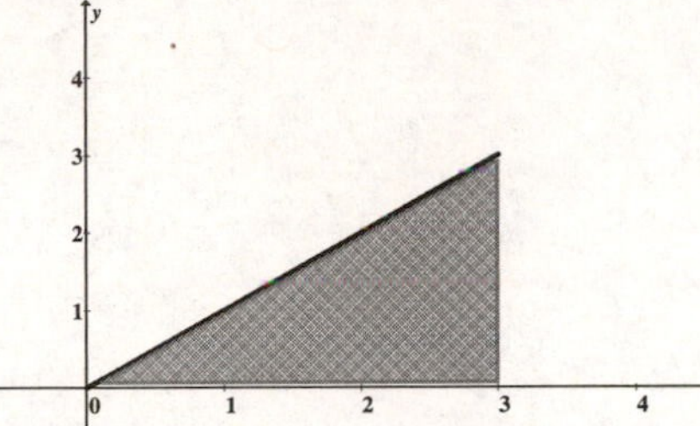

The region is a triangle; therefore, the area is given by:

$$A = \frac{1}{2} \cdot b \cdot h$$
$$= \frac{1}{2} \cdot 3 \cdot 3 = \frac{9}{2}.$$

The definite integral is

$$\int_0^3 xdx = \frac{9}{2}.$$

37. Sketching the graph, we determine that the area defined by the definite integral.

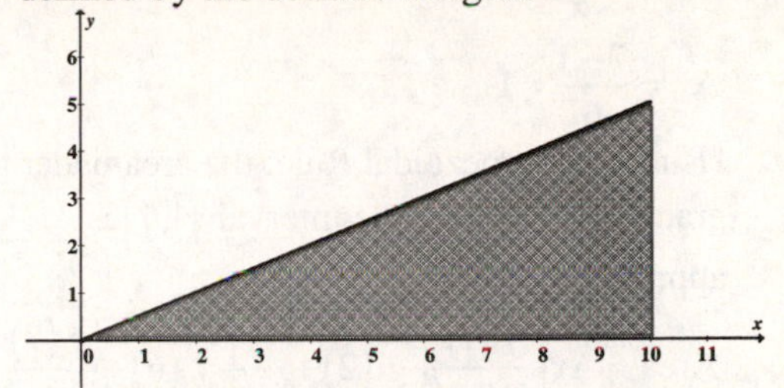

The region is a triangle; therefore, the area is given by:

$$A = \frac{1}{2} \cdot b \cdot h$$
$$= \frac{1}{2} \cdot 10 \cdot 5 = 25.$$

The definite integral is

$$\int_0^{10} \tfrac{1}{2} xdx = 25.$$

39. Sketching the graph, we determine that the area defined by the definite integral.

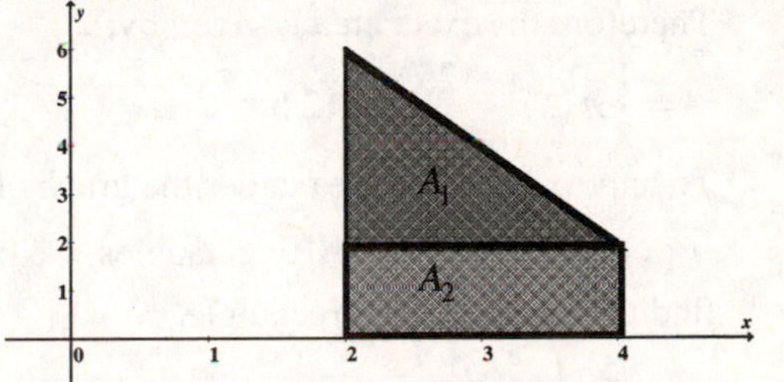

The region is a trapezoid. We decompose the trapezoid into a triangle and a square as shown in the graph. (note: due to the scale of the graph, the square does not 'look' square.) Therefore, the area of the triangle is given by:

$$A_1 = \frac{1}{2} \cdot b \cdot h$$
$$= \frac{1}{2} \cdot 2 \cdot 4 = 4.$$

The area of the square is given by:

$$A_2 = b \cdot h$$
$$= 2 \cdot 2 = 4.$$

The total area is

$$A_1 + A_2 = 4 + 4 = 8.$$

The definite integral is

$$\int_2^4 (10 - 2x)\,dx = 8.$$

Copyright © 2014 Pearson Education, Inc. Publishing as Addison-Wesley.

41. $f(x)=\frac{1}{x^2}$

$$\Delta x=\frac{7-1}{6}=1$$

Using the Trapezoidal Rule, the area under the graph of $f(x)$ over the interval $[1,7]$ is approximately:

$$\text{Area}\approx\Delta x\left[\frac{f(1)}{2}+f(2)+\cdots+f(6)+\frac{f(7)}{2}\right]$$

$$\approx 1\cdot\left[\frac{\frac{1}{1^2}}{2}+\frac{1}{2^2}+\frac{1}{3^2}+\frac{1}{4^2}+\frac{1}{5^2}+\frac{1}{6^2}+\frac{\frac{1}{7^2}}{2}\right]$$

$$\approx\frac{1}{2}+\frac{1}{4}+\frac{1}{9}+\frac{1}{16}+\frac{1}{25}+\frac{1}{36}+\frac{1}{98}$$

$$\approx 1.0016.$$

43. $f(x)=\sqrt{25-x^2}$

Notice that this semi-circle has a radius of five. Therefore the exact area is given by:

$$A=\frac{1}{2}\pi(5)^2=\frac{25}{2}\pi=12.5\pi$$

To approximate the area under the graph of $f(x)=\sqrt{25-x^2}$ using 10 rectangles, we first find the width of each rectangle.

$$\Delta x=\frac{5-(-5)}{10}=1.$$

The x_i will range from $x_1=-5$ to $x_{10}=4$. Although a drawing is not required, we can make one to help visualize the area.

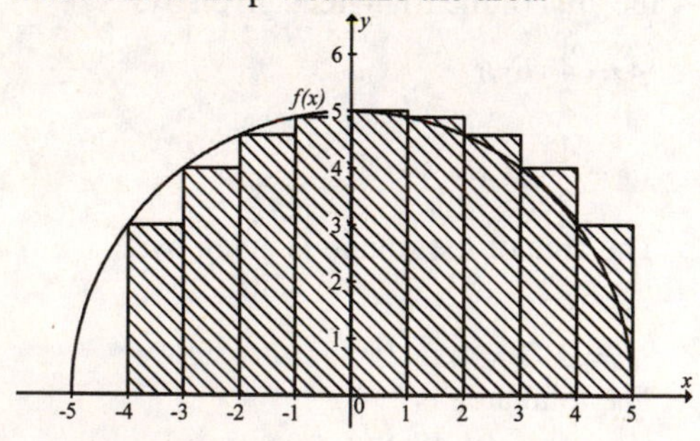

The area under the curve from -5 to 5 is approximately

$$\sum_{i=1}^{10} f(x_i)\Delta x = f(-5)\cdot 1+f(-4)\cdot 1+f(-3)\cdot 1$$
$$f(-2)\cdot 1+f(-1)\cdot 1+f(0)\cdot 1$$
$$f(1)\cdot 1+f(2)\cdot 1+f(3)\cdot 1+$$
$$f(4)\cdot 1$$
$$\approx 37.9631.$$

In order to compare, if we round to 4 decimal places we have:
$A=12.5\pi=39.2699$.

Copyright © 2014 Pearson Education, Inc. Publishing as Addison-Wesley.

Exercise Set 4.3

1. Find any antiderivative $F(x)$ of $y=4$. We choose the simplest one for which the constant of integration is 0:

$$F(x)=\int 4dx$$
$$=4x+C$$
$$=4x. \qquad [C=0]$$

The area under the curve over the interval $[1,3]$ is given by $F(3)-F(1)$. Substitute 3 and 1, and find the difference:

$$F(3)-F(1)=4(3)-4(1)$$
$$=12-4$$
$$=8.$$

3. Find any antiderivative $F(x)$ of $y=2x$. We choose the simplest one for which the constant of integration is 0:

$$F(x)=\int 2xdx$$
$$=x^2+C$$
$$=x^2. \qquad [C=0]$$

The area under the curve over the interval $[1,3]$ is given by $F(3)-F(1)$. Substitute 3 and 1, and find the difference:

$$F(3)-F(1)=(3)^2-(1)^2$$
$$=9-1$$
$$=8.$$

5. Find any antiderivative $F(x)$ of $y=x^2$. We choose the simplest one for which the constant of integration is 0:

$$F(x)=\int x^2dx$$
$$=\frac{x^3}{3}+C$$
$$=\frac{x^3}{3}. \qquad [C=0]$$

The area under the curve over the interval $[0,5]$ is given by $F(5)-F(0)$.

Substitute 5 and 0, and find the difference:

$$F(5)-F(0)=\frac{(5)^3}{3}-\frac{(0)^3}{3}$$
$$=\frac{125}{3}$$
$$41\tfrac{2}{3}.$$

7. Find any antiderivative $F(x)$ of $y=x^3$. We choose the simplest one for which the constant of integration is 0:

$$F(x)=\int x^3dx$$
$$=\frac{x^4}{4}+C$$
$$=\frac{x^4}{4}. \qquad [C=0]$$

The area under the curve over the interval $[0,1]$ is given by $F(1)-F(0)$. Substitute 1 and 0, and find the difference:

$$F(1)-F(0)=\frac{(1)^4}{4}-\frac{(0)^4}{4}$$
$$=\frac{1}{4}.$$

9. Find any antiderivative $F(x)$ of $y=4-x^2$. We choose the simplest one for which the constant of integration is 0:

$$F(x)=\int (4-x^2)dx$$
$$=4x-\frac{x^3}{3}+C$$
$$=4x-\frac{x^3}{3}. \qquad [C=0]$$

The area under the curve over the interval $[-2,2]$ is given by $F(2)-F(-2)$. Substitute 2 and -2, and find the difference:

$$F(2)-F(-2)$$
$$=\left(4(2)-\frac{(2)^3}{3}\right)-\left(4(-2)-\frac{(-2)^3}{3}\right)$$
$$=\left(8-\frac{8}{3}\right)-\left(-8+\frac{8}{3}\right)$$
$$=\frac{16}{3}-\left(-\frac{16}{3}\right)$$
$$=\frac{32}{3}=10\tfrac{2}{3}.$$

Copyright © 2014 Pearson Education, Inc. Publishing as Addison-Wesley.

11. Find any antiderivative $F(x)$ of $y = e^x$. We choose the simplest one for which the constant of integration is 0:

$$F(x) = \int e^x dx$$
$$= e^x + C$$
$$= e^x. \qquad [C = 0]$$

The area under the curve over the interval $[0,3]$ is given by $F(3) - F(0)$. Substitute 3 and 0, and find the difference:

$$F(3) - F(0) = (e^3) - (e^0)$$
$$= e^3 - 1$$
$$\approx 19.086.$$

13. Find any antiderivative $F(x)$ of $y = \frac{3}{x}$. We choose the simplest one for which the constant of integration is 0:

$$F(x) = \int \frac{3}{x} dx$$
$$= 3\ln x + C$$
$$= 3\ln x. \qquad [C = 0]$$

The area under the curve over the interval $[1,6]$ is given by $F(6) - F(1)$. Substitute 6 and 1, and find the difference:

$$F(6) - F(1) = (3\ln 6) - (3\ln 1)$$
$$= 3\ln 6 - 0$$
$$\approx 5.375.$$

15. The height of the area under the curve is total cost per day and the width of the area under the curve is time in days. Thus, area under the curve represents total cost in dollars, for t days.

$$\frac{\text{Total Cost}}{\text{days}} \cdot \text{days} = \text{Total cost.}$$

17. The height of the area under the curve is total number of kilowatts used per hour and the width of the area under the curve is time in hours. Thus, area under the curve represents Total number of kilowatts used in t hours.

$$\frac{\#\text{KW}}{\text{Hour}} \cdot \text{Hours} = \#\text{KW}.$$

19. The height of the area under the curve is revenue in dollars per unit and the width of the area under the curve is number of units. Thus, area under the curve represents total revenue, in dollars, for x units produced. $\frac{\$}{\text{Unit}} \cdot \text{Units} = \$.$

21. The height of the area under the curve is milligrams per cubic centimeter and the width of the area under the curve is cubic centimeters. Thus, area under the curve represents total concentration of a drug, in milligrams, in v cubic centimeters of blood. $\frac{\text{mg}}{cm^3} \cdot cm^3 = \text{mg}.$

23. The height of the area under the curve is number of memorized words per minute and the width of the area under the curve is time in minutes. Thus, area under the curve represents the total number of words memorized in t minutes.

$$\frac{\text{Words memorized}}{\text{Minute}} \cdot \text{Minutes} = \text{Words memorized.}$$

25. Find any antiderivative $F(x)$ of $y = x^3$. We choose the simplest one for which the constant of integration is 0:

$$F(x) = \int x^3 dx$$
$$= \frac{x^4}{4} + C$$
$$= \frac{x^4}{4}. \qquad [C = 0]$$

The area under the curve over the interval $[0,2]$ is given by $F(2) - F(0)$. Substitute 2 and 0, and find the difference:

$$F(2) - F(0) = \frac{(2)^4}{4} - \frac{(0)^4}{4}$$
$$= \frac{16}{4}$$
$$= 4.$$

Copyright © 2014 Pearson Education, Inc. Publishing as Addison-Wesley.

27. Find any antiderivative $F(x)$ of $y = x^2 + x + 1$.
We choose the simplest one for which the constant of integration is 0:

$$F(x) = \int x^2 + x + 1 dx$$
$$= \frac{x^3}{3} + \frac{x^2}{2} + x + C$$
$$= \frac{x^3}{3} + \frac{x^2}{2} + x. \qquad [C = 0]$$

The area under the curve over the interval $[2,3]$ is given by $F(3) - F(2)$. Substitute 3 and 2, and find the difference:

$$F(3) - F(2)$$
$$= \frac{(3)^3}{3} + \frac{(3)^2}{2} + (3) - \left(\frac{(2)^3}{3} + \frac{(2)^2}{2} + (2)\right)$$
$$= \frac{27}{3} + \frac{9}{2} + 3 - \left(\frac{8}{3} + \frac{4}{2} + 2\right)$$
$$= \frac{54}{6} + \frac{27}{6} + \frac{18}{6} - \left(\frac{16}{6} + \frac{12}{6} + \frac{12}{6}\right)$$
$$= \frac{99}{6} - \left(\frac{40}{6}\right)$$
$$= \frac{59}{6}$$
$$= 9\tfrac{5}{6}.$$

29. Find any antiderivative $F(x)$ of $y = 5 - x^2$. We choose the simplest one for which the constant of integration is 0:

$$F(x) = \int (5 - x^2) dx$$
$$= 5x - \frac{x^3}{3} + C$$
$$= 5x - \frac{x^3}{3}. \qquad [C = 0]$$

The area under the curve over the interval $[-1,2]$ is given by $F(2) - F(-1)$.

Substitute 2 and -1, and find the difference:

$$F(2) - F(-1)$$
$$= \left(5(2) - \frac{(2)^3}{3}\right) - \left(5(-1) - \frac{(-1)^3}{3}\right)$$
$$= \left(10 - \frac{8}{3}\right) - \left(-5 + \frac{1}{3}\right)$$
$$= \frac{22}{3} - \left(-\frac{14}{3}\right)$$
$$= \frac{36}{3}$$
$$= 12.$$

31. Find any antiderivative $F(x)$ of $y = e^x$. We choose the simplest one for which the constant of integration is 0:

$$F(x) = \int e^x dx$$
$$= e^x + C$$
$$= e^x. \qquad [C = 0]$$

The area under the curve over the interval $[-1,5]$ is given by $F(5) - F(-1)$. Substitute 5 and -1, and find the difference:

$$F(5) - F(-1) = \left(e^5\right) - \left(e^{-1}\right)$$
$$= e^5 - e^{-1}$$
$$\approx 148.045.$$

33. ✎

35. $\int_0^{1.5} \left(x - x^2\right) dx$

$$= \left[\frac{x^2}{2} - \frac{x^3}{3}\right]_0^{1.5}$$
$$= \left(\frac{(1.5)^2}{2} - \frac{(1.5)^3}{3}\right) - \left(\frac{(0)^2}{2} - \frac{(0)^3}{3}\right)$$
$$= \left(\frac{2.25}{2} - \frac{3.375}{3}\right) - 0$$
$$= 1.125 - 1.125$$
$$= 0$$

The area above the x-axis is equal to the area below the x-axis.

Copyright © 2014 Pearson Education, Inc. Publishing as Addison-Wesley.

37. $\int_{-1}^{1}\left(x^3-x^2\right)dx$

$=\left[\frac{x^4}{4}-\frac{x^2}{2}\right]_{-1}^{1}$

$=\left(\frac{(1)^4}{4}-\frac{(1)^2}{2}\right)-\left(\frac{(-1)^4}{4}-\frac{(-1)^2}{2}\right)$

$=\left(\frac{1}{4}-\frac{1}{2}\right)-\left(\frac{1}{4}-\frac{1}{2}\right)$

$=\frac{-1}{4}-\left(\frac{-1}{4}\right)$

$=0$

The area below the x-axis is equal to the area above the x-axis.

39 – 41. Left to the student.

43. $\int_{1}^{3}\left(3t^2+7\right)dt$

$=\left[t^3+7t\right]_1^3$

$=\left(3^3+7(3)\right)-\left(1^3+7(1)\right)$

$=48-(8)$

$=40$

45. $\int_1^4\left(\sqrt{x}-1\right)dx=\int_1^4\left(x^{1/2}-1\right)dx$

$=\left[\frac{2}{3}x^{3/2}-x\right]_1^4$

$=\left(\frac{2}{3}(4)^{3/2}-4\right)-\left(\frac{2}{3}(1)^{3/2}-1\right)$

$=\left(\frac{16}{3}-4\right)-\left(\frac{2}{3}-1\right)$

$=\left(\frac{4}{3}\right)-\left(-\frac{1}{3}\right)$

$=\frac{5}{3}$

47. $\int_{-2}^{5}\left(2x^2-3x+7\right)dx$

$=\left[\frac{2}{3}x^3-\frac{3}{2}x^2+7x\right]_{-2}^{5}$

$=\left(\frac{2}{3}(5)^3-\frac{3}{2}(5)^2+7(5)\right)-$

$\left(\frac{2}{3}(-2)^3-\frac{3}{2}(-2)^2+7(-2)\right)$

$=\frac{485}{6}-\left(-\frac{152}{6}\right)$

$=\frac{637}{6}$

49. $\int_{-5}^{2}e^t\,dt$

$=\left[e^t\right]_{-5}^{2}$

$=e^2-e^{-5}$

≈ 7.382

51. $\int_a^b\frac{1}{2}x^2dx$

$=\left[\frac{1}{6}x^3\right]_a^b$

$=\frac{1}{6}(b)^3-\frac{1}{6}(a)^3$

$=\frac{b^3-a^3}{6}$

53. $\int_a^b e^{2t}dt$

$=\left[\frac{1}{2}e^{2t}\right]_a^b$

$=\frac{1}{2}e^{2b}-\frac{1}{2}e^{2a}$

$=\frac{e^{2b}-e^{2a}}{2}$

Copyright © 2014 Pearson Education, Inc. Publishing as Addison-Wesley.

55. $\int_1^e \left(x+\frac{1}{x}\right)dx$

$=\left[\frac{x^2}{2}+\ln x\right]_1^e$

$=\left(\frac{e^2}{2}+\ln e\right)-\left(\frac{1^2}{2}+\ln 1\right)$

$=\frac{e^2}{2}+1-\left(\frac{1}{2}+0\right)$

$=\frac{e^2}{2}+\frac{1}{2}$

$=\frac{e^2+1}{2}$

≈ 4.195

57. $\int_0^2 \sqrt{2x}dx=\int_0^2 \sqrt{2}\cdot x^{1/2}dx=\sqrt{2}\int_0^2 x^{1/2}dx$

$=\sqrt{2}\left[\frac{2}{3}x^{3/2}\right]_0^2$

$=\sqrt{2}\left[\frac{2}{3}(2)^{3/2}-\frac{2}{3}(0)^{3/2}\right]$

$=\sqrt{2}\left[\frac{2}{3}\sqrt{2^3}\right]$

$=\sqrt{2}\left[\frac{2}{3}\sqrt{8}\right]$

$=\frac{2}{3}\sqrt{16}$

$=\frac{8}{3}$

59. We integrate to find:

$P(250)=\int_0^{250} P'(x)dx$

$=\int_0^{250} \sqrt[5]{x}dx$

$=\int_0^{250} x^{1/5}dx$

$=\left[\frac{5}{6}x^{6/5}\right]_0^{250}$

$=\frac{5}{6}(250)^{6/5}-\frac{5}{6}(0)^{6/5}$

$\approx 628.56.$ Using a calculator.

When a 250 foot well is drilled, Pure Water Enterprises profit is \$628.56.

61. In order find the cost of producing an additional 14 feet of counter top after 50 feet have already been produced, we integrate $C'(x)$ over the interval $[50,64]$.

$C(64)-C(50)=\int_{50}^{64} C'(x)dx$

$=\int_{50}^{64} 8x^{-1/3}dx$

$=\left[12x^{2/3}\right]_{50}^{64}$

$=12(64)^{2/3}-12(50)^{2/3}$

$\approx 29.13.$ Using a calculator.

The cost of installing an extra 14 feet of counter top after 50 feet has already been ordered is \$29.13.

63. $S'(t)=20e^t$

a) We integrate $S'(t)$ over the interval $[0,5]$ to find the accumulated sales.

$S(5)=\int_0^5 S'(t)dt$

$=\int_0^5 20e^t dt$

$=\left[20e^t\right]_0^5$

$=20e^5-20e^0$

$=20e^5-20\cdot 1$

≈ 2948.26

The accumulated sales for the first 5 days are approximately \$2948.26.

b) We integrate $S'(t)$ over the interval $[1,5]$ to find the accumulated sales for the 2nd day through the 5th day.

$S(5)=\int_1^5 S'(t)dt$

$=\int_1^5 20e^t dt$

$=\left[20e^t\right]_1^5$

$=20e^5-20e^1$

≈ 2913.90

The sales from the 2nd day through the 5th day are approximately \$2913.90.

Copyright © 2014 Pearson Education, Inc. Publishing as Addison-Wesley.

65. In 1996, $t = 1$ and in 2000, $t = 5$. Therefore, we integrate $D'(t)$ over the interval $[1,5]$.

$$\int_1^5 D'(t)\,dt$$
$$= \int_1^5 \left(857.98 + 829.66t - 197.34t^2 + 15.36t^3\right)dt$$
$$= \left[857.98t + 414.83t^2 - 65.78t^3 + 3.84t^4\right]_1^5$$
$$\approx 7627.28$$

The credit market debt increased \$7627.28 billion from 1996 to 2000.

67. We integrate $T(x)$ over the interval $[1,10]$.

$$\int_1^{10} T(x)\,dx$$
$$= \int_1^{10} \left(2 + 0.3x^{-1}\right)dx$$
$$= \left[2x + 0.3\ln x\right]_1^{10}$$
$$= 2(10) + 0.3\ln(10) - \left(2(1) + 0.3\ln(1)\right)$$
$$= 18 + 0.3\ln 10$$
$$\approx 18.69$$

It takes 18.69 hours for a new worker to produce units 1 through 10.
To find the time it takes a new worker to produce units 20 through 30, we integrate $T(x)$ over the interval $[20,30]$.

$$\int_{20}^{30} T(x)\,dx$$
$$= \int_{20}^{30} \left(2 + 0.3x^{-1}\right)dx$$
$$= \left[2x + 0.3\ln x\right]_{20}^{30}$$
$$= 2(30) + 0.3\ln(30) - \left(2(20) + 0.3\ln(20)\right)$$
$$\approx 20.12$$

It takes 20.12 hours for a new worker to produce units 20 through 30.

69. We integrate $M'(t)$ over the interval $[0,10]$.

$$M(10) = \int_0^{10} M'(t)$$
$$= \int_0^{10} \left(-0.009t^2 + 0.2t\right)dt$$
$$= \left[-0.003t^3 + 0.1t^2\right]_0^{10}$$
$$= \left(-0.003(10)^3 + 0.1(10)^2\right) - \left(-0.003(0)^3 + 0.1(0)^2\right)$$
$$= 7 - 0$$
$$= 7$$

In the first 10 minutes, 7 words are memorized.

71. We integrate $M'(t)$ over the interval $[10,15]$.

$$M(15) - M(10) = \int_{10}^{15} M'(t)$$
$$= \int_{10}^{15} \left(-0.009t^2 + 0.2t\right)dt$$
$$= \left[-0.003t^3 + 0.1t^2\right]_{10}^{15}$$
$$= \left(-0.003(15)^3 + 0.1(15)^2\right) - \left(-0.003(10)^3 + 0.1(10)^2\right)$$
$$= 12.375 - 7$$
$$= 5.375$$

About 5 words are memorized during minutes 10 – 15.

73. We first find $s(t)$ by integrating:

$$s(t) = \int v(t)\,dt = \int 3t^2\,dt = t^3 + C.$$

Next we determine C by using the initial condition $s(0) = 4$, which is the starting position for s at time $t = 0$:

$$s(0) = 4$$
$$0^3 + C = 4$$
$$C = 4.$$

Thus, $s(t) = t^3 + 4$.

Copyright © 2014 Pearson Education, Inc. Publishing as Addison-Wesley.

75. We first find $v(t)$ by integrating:

$$v(t)=\int a(t)\,dt=\int 4t\,dt=2t^2+C.$$

Next we determine C by using the initial condition $v(0)=20$:

$$v(0)=20$$
$$2\cdot 0^2+C=20$$
$$C=20.$$

Thus, $v(t)=2t^2+20$.

77. We first find $v(t)$ by integrating:

$$v(t)=\int a(t)\,dt$$
$$=\int(-2t+6)\,dt$$
$$=-t^2+6t+C_1.$$

Next we determine C_1 by using the initial condition $v(0)=6$:

$$v(0)=6$$
$$-(0)^2+6(0)+C_1=6$$
$$C_1=6.$$

Thus, $v(t)=-t^2+6t+6$.

Next, we find $s(t)$ by integrating:

$$s(t)=\int v(t)\,dt$$
$$=\int\left(-t^2+6t+6\right)dt$$
$$=-\frac{1}{3}t^3+3t^2+6t+C_2.$$

Next we determine C_2 by using the initial condition $s(0)=10$:

$$s(0)=10$$
$$-\frac{1}{3}(0)^3+3(0)^2+6(0)+C_2=10$$
$$C_2=10.$$

Thus, $s(t)=-\frac{1}{3}t^3+3t^2+6t+10$.

79. a) We integrate $v(t)$ over the interval $[0,5]$:

$$s(5)=\int_0^5 v(t)\,dt$$
$$=\int_0^5\left(-0.5t^2+10t\right)dt$$
$$=\left[-\frac{1}{6}t^3+5t^2\right]_0^5$$
$$=\left(-\frac{1}{6}(5)^3+5(5)^2\right)-\left(-\frac{1}{6}(0)^3+5(0)^2\right)$$
$$=\frac{625}{6}-0$$
$$\approx 104.17.$$

The particle travels approximately 104.17 meters during the first 5 seconds.

b) We integrate $v(t)$ over the interval $[5,10]$:

$$s(10)-s(5)=\int_5^{10} v(t)\,dt$$
$$=\left[-\frac{1}{6}t^3+5t^2\right]_5^{10} \quad \text{From part (a).}$$
$$\approx 229.17.$$

The particle travels approximately 229.17 meters during the second 5 seconds.

81. a) Converting 15 seconds into hours, we have $\frac{15}{3600}=\frac{1}{240}$. Thus, the motorcycle's acceleration function is

$$a(t)=\frac{60-0}{\frac{1}{240}-0}=14{,}400,$$

where $a(t)$ is in miles per hour squared, and t is in hours. Thus, we can find the velocity function by integrating $a(t)$.

$$v(t)=\int a(t)\,dt=\int 14{,}400\,dt=14{,}400t+C$$

We use the initial condition $v(0)=0$ to find C.

$$v(0)=0$$
$$14{,}400(0)+C=0$$
$$C=0$$

Thus, $v(t)=14{,}400t$, where $v(t)$ is in miles per hour and t is in hours.

The solution is continued on the next page.

Copyright © 2014 Pearson Education, Inc. Publishing as Addison-Wesley.

Now, substituting $\frac{1}{240}$ hour (15 seconds) for t into the velocity function from the previous page, we have:

$$v\left(\frac{1}{240}\right)=14,400\left(\frac{1}{240}\right)=60.$$

The motorcycle is traveling at a speed of 60 miles per hour after 15 seconds.
Note: the intuitive solution to this problem is if the motorcycle accelerates at a constant rate from 0 mph to 60 mph in 15 seconds, then the motorcycle is obviously traveling at 60 mph after 15 seconds. We derive the velocity function to find the distance in part (b)

b) Using the information in Part (a), we integrate $v(t)$ over the interval $\left[0,\frac{1}{240}\right]$.

$$s\left(\frac{1}{240}\right)=\int_0^{1/240}14,400t\,dt$$
$$=\left[7200t^2\right]_0^{1/240}$$
$$=7200\left(\frac{1}{240}\right)^2-7200(0)$$
$$=\frac{1}{8}.$$

The motorcycle has traveled $\frac{1}{8}$ mi of a mile after 15 seconds.

83. a) Converting 45 seconds into hours, we have $\frac{45}{3600}=\frac{1}{80}$. Thus, the bicyclist's deceleration function is

$a(t)=\frac{0-30}{\frac{1}{80}-0}=-2400$, where $a(t)$ is in kilometers per hour squared, and t is in hours. Thus, we can find the velocity function by integrating $a(t)$.

$$v(t)=\int a(t)\,dt=\int -2400\,dt=-2400t+C$$

We use the initial condition $v(0)=30$ to find C.

$$v(0)=30$$
$$-2400(0)+C=30$$
$$C=30$$

Thus, $v(t)=-2400t+30$, where $v(t)$ is in kilometers per hour and t is in hours.

Now, substituting $\frac{1}{180}$ hour (20 seconds) for t, we have

$$v\left(\frac{1}{180}\right)=-2400\left(\frac{1}{180}\right)+30\approx 16.67.$$

The bicyclist is traveling at a speed of 16.67 kilometers per hour after 20 seconds.

b) Using the information in Part (a), we integrate $v(t)$ over the interval $\left[0,\frac{1}{80}\right]$.

$$s\left(\frac{1}{80}\right)=\int_0^{1/80}(-2400t+30)dt$$
$$=\left[-1200t^2+30t\right]_0^{1/80}$$
$$=-1200\left(\frac{1}{80}\right)^2+30\left(\frac{1}{80}\right)-\left(-1200(0)^2+30(0)\right)$$
$$=\frac{-3}{16}+\frac{3}{8}$$
$$=\frac{3}{16}\approx 0.1875.$$

The bicyclist has traveled 0.1875 kilometers after 45 seconds.

85. We first find $v(t)$ by integrating:

$$v(t)=\int a(t)\,dt$$
$$=\int(-32)\,dt$$
$$=-32t+C_1.$$

Next we determine C_1 by using the initial condition $v(0)=v_0$:

$$v(0)=v_0$$
$$-32(0)+C_1=v_0$$
$$C_1=v_0.$$

Thus, $v(t)=-32t+v_0$.

Next, we find $s(t)$ by integrating:

$$s(t)=\int v(t)\,dt$$
$$=\int(-32t+v_0)\,dt$$
$$=-16t^2+v_0\cdot t+C_2.$$

The solution is continued on the next page.

Copyright © 2014 Pearson Education, Inc. Publishing as Addison-Wesley.

Next we determine C_2 by using the initial condition $s(0) = s_0$:

$$s(0) = s_0$$
$$-16(0)^2 + v_0(0) + C_2 = s_0$$
$$C_2 = s_0.$$

Thus, $s(t) = -16t^2 + v_0 t + s_0$.

87. $a(t) = 7200$

$$v(t) = \int a(t)\,dt$$
$$= \int 7200\,dt$$
$$= 7200t + C$$

Since $v(0) = 0$, we have

$$v(0) = 0$$
$$7200(0) + C = 0$$
$$C = 0.$$

Thus, $v(t) = 7200t$.

Integrating $v(t)$ from $\left[0, \frac{1}{120}\right]$ we have

$$s\left(\frac{1}{120}\right) = \int_0^{1/120} 7200t\,dt$$
$$= \left[3600t^2\right]_0^{1/120}$$
$$= \frac{1}{4}.$$

The car travels $\frac{1}{4}$ mile in $\frac{1}{2}$ minute.

89. We integrate $v(t)$ over the interval $[1,5]$:

$$s(5) - s(1) = \int_1^5 v(t)\,dt$$
$$= \int_1^5 \left(3t^2 + 2t\right)dt$$
$$= \left[t^3 + t^2\right]_1^5$$
$$= \left(5^3 + 5^2\right) - \left(1^3 + 1^2\right)$$
$$= 150 - 2$$
$$= 148$$

The particle travels approximately 148 miles from the 2nd hour to through the 5th hour.

91. $S(t) = \int S'(t)\,dt$

$$S(t) = \int 0.5e^t\,dt = 0.5e^t + C$$

Assuming $S(0) = 0$, we have:

$$S(0) = 0$$
$$0.5e^0 + C = 0$$
$$0.5 + C = 0$$
$$C = -0.5.$$

Thus, $S(t) = 0.5e^t - 0.5$

When Bluetape reaches \$10,000 in sales, $S(t) = 10{,}000$. We solve the equation for t.

$$0.5e^t - 0.5 = 10{,}000$$
$$0.5e^t = 10{,}000.5$$
$$e^t = 20{,}001$$
$$\ln e^t = \ln 20{,}001$$
$$t = 9.9035$$

Therefore, they will reach \$10,000 in sales on the 10th day.

93. $\int_2^3 \frac{x^2 - 1}{x - 1}\,dx = \int_2^3 \frac{(x-1)(x+1)}{(x-1)}\,dx$

$$= \int_2^3 (x+1)\,dx$$
$$= \left[\frac{x^2}{2} + x\right]_2^3$$
$$= \left(\frac{3^2}{2} + 3\right) - \left(\frac{2^2}{2} + 2\right)$$
$$= \frac{15}{2} - 4$$
$$= \frac{7}{2} = 3.5$$

95. $\int_4^{16} (x-1)\sqrt{x}\,dx = \int_4^{16} \left(x^{3/2} - x^{1/2}\right)dx$

$$= \left[\frac{2}{5}x^{5/2} - \frac{2}{3}x^{3/2}\right]_4^{16}$$
$$= \left(\frac{2}{5}(16)^{5/2} - \frac{2}{3}(16)^{3/2}\right) - \left(\frac{2}{5}(4)^{5/2} - \frac{2}{3}(4)^{3/2}\right)$$
$$= \left(\frac{2048}{5} - \frac{128}{3}\right) - \left(\frac{64}{5} - \frac{16}{3}\right)$$
$$= \frac{5392}{15}$$
$$= 359\tfrac{7}{15}$$

Copyright © 2014 Pearson Education, Inc. Publishing as Addison-Wesley.

97. $\int_1^8 \frac{\sqrt[3]{x^2}-1}{\sqrt[3]{x}}\,dx = \int_1^8 \left(x^{2/3}-1\right)x^{-1/3}\,dx$
$= \int_1^8 \left(x^{1/3}-x^{-1/3}\right)dx$
$= \left[\frac{3}{4}x^{4/3}-\frac{3}{2}x^{2/3}\right]_1^8$
$= \left(\frac{3}{4}(8)^{4/3}-\frac{3}{2}(8)^{2/3}\right)-\left(\frac{3}{4}(1)^{4/3}-\frac{3}{2}(1)^{2/3}\right)$
$= (12-6)-\left(\frac{3}{4}-\frac{3}{2}\right)$
$= 6-\left(-\frac{3}{4}\right)$
$= \frac{27}{4}$
$= 6.75$

99. $\int_2^5 \left(t+\sqrt{3}\right)\left(t-\sqrt{3}\right)dt = \int_2^5 \left(t^2-3\right)dt$
$= \left[\frac{t^3}{3}-3t\right]_2^5$
$= \left(\frac{(5)^3}{3}-3(5)\right)-\left(\frac{(2)^3}{3}-3(2)\right)$
$= \frac{80}{3}-\left(-\frac{10}{3}\right)$
$= \frac{90}{3}$
$= 30$

101. $\int_1^3 \left(x-\frac{1}{x}\right)^2 dx = \int_1^3 \left(x-x^{-1}\right)^2 dx$
$= \int_1^3 \left(x^2-2+x^{-2}\right)dx$ Expanding $\left(x-x^{-1}\right)^2$
$= \left[\frac{x^3}{3}-2x-x^{-1}\right]_1^3$
$= \left(\frac{(3)^3}{3}-2(3)-(3)^{-1}\right)-\left(\frac{(1)^3}{3}-2(1)-(1)^{-1}\right)$
$= \frac{8}{3}-\left(-\frac{8}{3}\right)$
$= \frac{16}{3}$
$= 5\frac{1}{3}$

103. $\int_4^9 \frac{t+1}{\sqrt{t}}\,dt = \int_4^9 (t+1)t^{-1/2}\,dt$
$= \int_4^9 \left(t^{1/2}+t^{-1/2}\right)dt$
$= \left[\frac{2}{3}t^{3/2}+2t^{1/2}\right]_4^9$
$= \left(\frac{2}{3}(9)^{3/2}+2(9)^{1/2}\right)-\left(\frac{2}{3}(4)^{3/2}+2(4)^{1/2}\right)$
$= 24-\frac{28}{3}$
$= \frac{44}{3}$
$= 14\frac{2}{3}$

105. ✎

107. Using the fnInt feature on a calculator, we get:
$\int_{-8}^{1.4} \left(x^4+4x^3-36x^2-160x+300\right)dx \approx 4068.789.$

109. Using the fnInt feature on a calculator, we get:
$\int_{-1}^{1} \left(3+\sqrt{1-x^2}\right)dx \approx 7.571.$

111. Using the fnInt feature on a calculator, we get:
$\int_{-2}^{2} x^{2/3}\left(\frac{5}{2}-x\right)dx \approx 9.524.$

113. Using the fnInt feature on a calculator, we get:
$\int_{-10}^{10} \frac{8}{x^2+4}\,dx \approx 10.987.$

Copyright © 2014 Pearson Education, Inc. Publishing as Addison-Wesley.

Exercise Set 4.4

1. $\int_1^5 f(x)\,dx = \int_1^3 f(x)\,dx + \int_3^5 f(x)\,dx$

$= \int_1^3 (2x+1)\,dx + \int_3^5 (10-x)\,dx$

$= \left[x^2+x\right]_1^3 + \left[10x - \frac{1}{2}x^2\right]_3^5$

$= \left[\left((3)^2+(3)\right)-\left((1)^2+(1)\right)\right]+$
$\left[\left(10(5)-\frac{1}{2}(5)^2\right)-\left(10(3)-\frac{1}{2}(3)^2\right)\right]$

$= (12-2)+\left(\frac{75}{2}-\frac{51}{2}\right)$

$= 10+12$

$= 22.$

3. $\int_{-2}^3 g(x)\,dx = \int_{-2}^0 g(x)\,dx + \int_0^3 g(x)\,dx$

$= \int_{-2}^0 \left(x^2+4\right)dx + \int_0^3 (4-x)\,dx$

$= \left[\frac{x^3}{3}+4x\right]_{-2}^0 + \left[4x-\frac{1}{2}x^2\right]_0^3$

$= \left[\left(\frac{(0)^3}{3}+4(0)\right)-\left(\frac{(-2)^3}{3}+4(-2)\right)\right]+$
$\left[\left(4(3)-\frac{1}{2}(3)^2\right)-\left(4(0)-\frac{1}{2}(0)^2\right)\right]$

$= \left(0-\left(-\frac{32}{3}\right)\right)+\left(\frac{15}{2}-0\right)$

$= \frac{109}{6}$

$= 18\tfrac{1}{6}.$

5. $\int_{-6}^4 f(x)\,dx = \int_{-6}^1 f(x)\,dx + \int_1^4 f(x)\,dx$

$= \int_{-6}^1 \left(-x^2-6x+7\right)dx + \int_1^4 \left(\frac{3}{2}x-1\right)dx$

$= \left[-\frac{x^3}{3}-3x^2+7x\right]_{-6}^1 + \left[\frac{3}{4}x^2-x\right]_1^4$

$= \left[\left(-\frac{(1)^3}{3}-3(1)^2+7(1)\right)-\right.$
$\left.\left(-\frac{(-6)^3}{3}-3(-6)^2+7(-6)\right)\right]+$
$\left[\left(\frac{3}{4}(4)^2-(4)\right)-\left(\frac{3}{4}(1)^2-(1)\right)\right]$

$= \left(\frac{11}{3}-(-78)\right)+\left(8-\left(-\frac{1}{4}\right)\right)$

$= \frac{245}{3}+\frac{33}{4}$

$= \frac{1079}{12} = 89\tfrac{11}{12}.$

7. Using the definition of the absolute value, we have:

$$|x-3| = \begin{cases} -(x-3), & \text{for } x<3 \\ x-3, & \text{for } x \geq 3. \end{cases}$$

Therefore,

$\int_0^4 |x-3|\,dx$

$= \int_0^3 -(x-3)\,dx + \int_3^4 (x-3)\,dx$

$= \left[-\frac{1}{2}x^2+3x\right]_0^3 + \left[\frac{1}{2}x^2-3x\right]_3^4$

$= \left[\left(-\frac{1}{2}(3)^2+3(3)\right)-\left(-\frac{1}{2}(0)^2+3(0)\right)\right]$
$+\left[\left(\frac{1}{2}(4)^2-3(4)\right)-\left(\frac{1}{2}(3)^2-3(3)\right)\right]$

$= \left[\frac{9}{2}-0\right]+\left[-4-\left(\frac{-9}{2}\right)\right]$

$= \left[\frac{9}{2}\right]+\left[\frac{1}{2}\right]$

$= 5.$

Copyright © 2014 Pearson Education, Inc. Publishing as Addison-Wesley.

9. Using the definition of the absolute value, we have:

$$|x^3-1| = \begin{cases} -(x^3-1), & \text{for } x<1 \\ x^3-1, & \text{for } x \ge 1. \end{cases}$$

Therefore,

$$\int_0^2 |x^3-1|\,dx$$
$$= \int_0^1 -(x^3-1)\,dx + \int_1^2 (x^3-1)\,dx$$
$$= \left[-\frac{1}{4}x^4 + x\right]_0^1 + \left[\frac{1}{4}x^4 - x\right]_1^2$$
$$= \left[\left(-\frac{1}{4}(1)^4 + (1)\right) - \left(-\frac{1}{4}(0)^4 + (0)\right)\right]$$
$$+ \left[\left(\frac{1}{4}(2)^4 - (2)\right) - \left(\frac{1}{4}(1)^4 - (1)\right)\right]$$
$$= \left[\frac{3}{4}\right] + \left[\frac{11}{4}\right]$$
$$= \frac{14}{4}$$
$$= \frac{7}{2}.$$

11. To calculate the points of intersection, we set $f(x)$ equal to $g(x)$ and solve.

$$f(x) = g(x)$$
$$9 = x^2$$
$$\sqrt{9} = \sqrt{x^2}$$
$$\pm 3 = x$$

The graphs intersect at $x = -3$ and $x = 3$.

13. To calculate the points of intersection, we set $f(x)$ equal to $g(x)$ and solve.

$$f(x) = g(x)$$
$$7 = x^2 - 3x + 2$$
$$0 = x^2 - 3x - 5$$

Applying the quadratic formula we have:

$$x = \frac{-(-3) \pm \sqrt{(-3)^2 - 4(1)(-5)}}{2(1)}$$
$$= \frac{3 \pm \sqrt{29}}{2}$$

The graphs intersect at

$x = \frac{3+\sqrt{29}}{2} \approx 1.193$ and $x = \frac{3-\sqrt{29}}{2} \approx -1.193$.

15. To calculate the points of intersection, we set $f(x)$ equal to $g(x)$ and solve.

$$f(x) = g(x)$$
$$x^2 - x - 5 = x + 10$$
$$x^2 - 2x - 15 = 0$$
$$(x-5)(x+3) = 0$$
$$x - 5 = 0 \quad \text{or} \quad x + 3 = 0$$
$$x = 5 \quad \text{or} \quad x = -3$$

The graphs intersect at $x = -3$ and $x = 5$.

17. $g(x) \ge f(x)$ on $[-1,0]$ and $f(x) \ge g(x)$ on $[0,1]$. We use two integrals to find the total area.

$$\int_{-1}^{0} \left[0 - \left(2x + x^2 - x^3\right)\right] dx +$$
$$\int_0^1 \left[\left(2x + x^2 - x^3\right) - 0\right] dx$$
$$\int_{-1}^{0} \left[0 - \left(2x + x^2 - x^3\right)\right] dx +$$
$$\int_0^1 \left[\left(2x + x^2 - x^3\right) - 0\right] dx$$
$$= \int_{-1}^{0} \left(-2x - x^2 + x^3\right) dx +$$
$$\int_0^1 \left(2x + x^2 - x^3\right) dx$$
$$= \left[-x^2 - \frac{x^3}{3} + \frac{x^4}{4}\right]_{-1}^{0} + \left[x^2 + \frac{x^3}{3} - \frac{x^4}{4}\right]_0^1$$
$$= \left[-(0)^2 - \frac{(0)^3}{3} + \frac{(0)^4}{4}\right] -$$
$$\left[-(-1)^2 - \frac{(-1)^3}{3} + \frac{(-1)^4}{4}\right] +$$
$$\left[(1)^2 + \frac{(1)^3}{3} - \frac{(1)^4}{4}\right] - \left[(0)^2 + \frac{(0)^3}{3} - \frac{(0)^4}{4}\right]$$
$$= \left[0 - \left(-\frac{5}{12}\right)\right] + \left[\frac{13}{12} - 0\right]$$
$$= \frac{5}{12} + \frac{13}{12}$$
$$= \frac{18}{12}$$
$$= \frac{3}{2}.$$

Copyright © 2014 Pearson Education, Inc. Publishing as Addison-Wesley.

19. $g(x) \geq f(x)$ over the entire region. We find the area.

$$\int_{-1}^{4}\left[(x+28)-\left(x^4-8x^3+18x^2\right)\right]dx$$

$$=\int_{-1}^{4}\left(-x^4+8x^3-18x^2+x+28\right)dx$$

$$=\left[-\frac{x^5}{5}+2x^4-6x^3+\frac{x^2}{2}+28x\right]_{-1}^{4}$$

$$=\left[-\frac{(4)^5}{5}+2(4)^4-6(4)^3+\frac{(4)^2}{2}+28(4)\right]-\left[-\frac{(-1)^5}{5}+2(-1)^4-6(-1)^3+\frac{(-1)^2}{2}+28(-1)\right]$$

$$=\frac{216}{5}+\frac{193}{10}$$

$$=\frac{625}{10}=62\tfrac{1}{2}.$$

21. First graph the system of equations and shade the region bounded by the graphs.

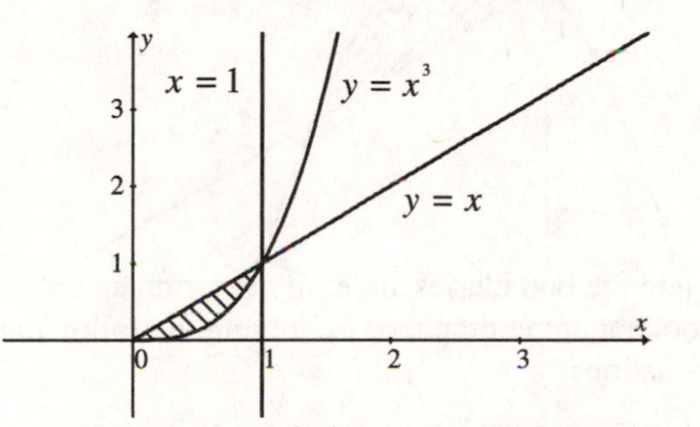

Here the boundaries are easily determined by looking at the graph, or by solving the following:

$x=x^3$

$0=x^3-x$

$0=x\left(x^2-1\right)$

$x=0$ or $x^2-1=0$

$x=0$ or $x=\pm 1$

$x=0$ or $x=1$

Note $x \geq x^3$ over the interval $[0,1]$.

We compute the area as follows:

$$\int_0^1\left(x-x^3\right)dx$$

$$=\left[\frac{x^2}{2}-\frac{x^4}{4}\right]_0^1$$

$$=\left(\frac{(1)^2}{2}-\frac{(1)^4}{4}\right)-\left(\frac{(0)^2}{2}-\frac{(0)^4}{4}\right)$$

$$=\frac{1}{2}-\frac{1}{4}$$

$$=\frac{1}{4}.$$

23. First graph the system of equations and shade the region bounded by the graphs.

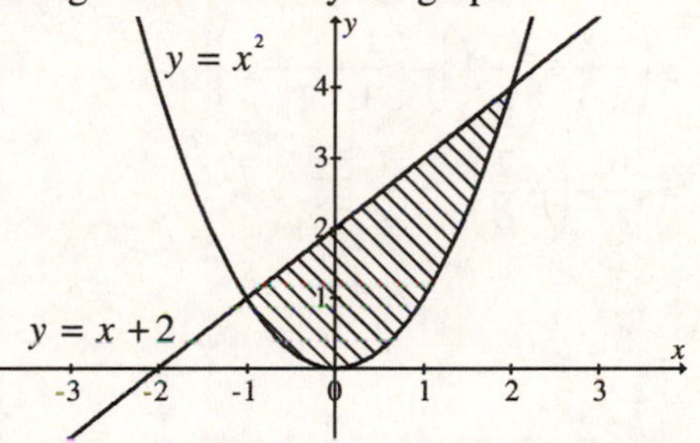

Here the boundaries are easily determined by looking at the graph, or solving the following equation:

$x^2=x+2$

$x^2-x-2=0$

$(x-2)(x+1)=0$

$x=-1$ or $x=2$

Note $(x+2) \geq x^2$ over the interval $[-1,2]$. We compute the area on the next page.

Copyright © 2014 Pearson Education, Inc. Publishing as Addison-Wesley.

Integrating to compute the area from the previous page, we have:

$$\int_{-1}^{2}\left((x+2)-x^2\right)dx$$

$$=\int_{-1}^{2}\left(-x^2+x+2\right)dx$$

$$=\left[-\frac{x^3}{3}+\frac{x^2}{2}+2x\right]_{-1}^{2}$$

$$=\left(-\frac{(2)^3}{3}+\frac{(2)^2}{2}+2(2)\right)-\left(-\frac{(-1)^3}{3}+\frac{(-1)^2}{2}+2(-1)\right)$$

$$=\left(-\frac{8}{3}+2+4\right)-\left(\frac{1}{3}+\frac{1}{2}-2\right)$$

$$=\frac{10}{3}-\left(-\frac{7}{6}\right)$$

$$=\frac{27}{6}$$

$$=4\tfrac{1}{2}.$$

25. First graph the system of equations and shade the region bounded by the graphs.

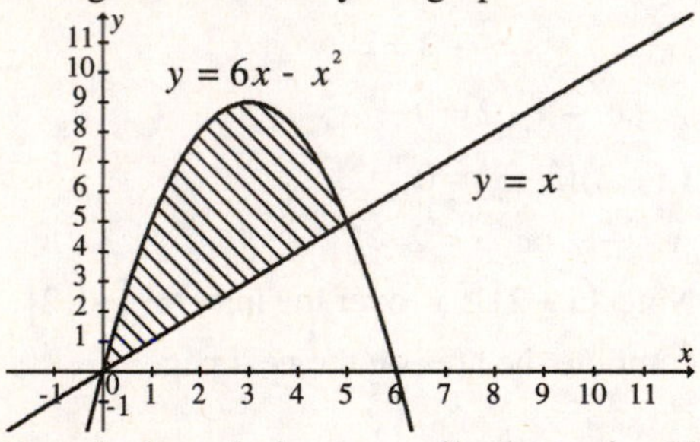

Here the boundaries are easily determined by looking at the graph, or by solving the following equation:

$$x=6x-x^2$$

$$x^2-5x=0$$

$$x(x-5)=0$$

$$x=0 \text{ or } x=5$$

Note $\left(6x-x^2\right)\geq x$ over the interval $[0,5]$.

We compute the area as follows:

$$\int_{0}^{5}\left(\left(6x-x^2\right)-x\right)dx$$

$$=\int_{0}^{5}\left(-x^2+5x\right)dx$$

$$=\left[-\frac{x^3}{3}+\frac{5}{2}x^2\right]_{0}^{5}$$

$$=\left(-\frac{(5)^3}{3}+\frac{5}{2}(5)^2\right)-\left(-\frac{(0)^3}{3}+\frac{5}{2}(0)^2\right)$$

$$=\left(-\frac{125}{3}+\frac{125}{2}\right)-0$$

$$=\frac{125}{6}$$

$$=20\tfrac{5}{6}.$$

27. First graph the system of equations and shade the region bounded by the graphs.

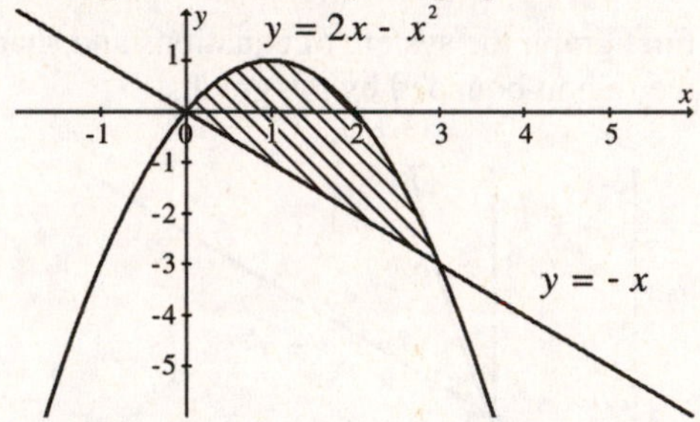

Here the boundaries are easily determined by looking at the graph, or by solving the following equation:

$$-x=2x-x^2$$

$$x^2-3x=0$$

$$x(x-3)=0$$

$$x=0 \text{ or } x=3$$

Note $\left(2x-x^2\right)\geq -x$ over the interval $[0,3]$.

The solution is continued on the next page.

Copyright © 2014 Pearson Education, Inc. Publishing as Addison-Wesley.

Integrating to compute the area from the previous page, we have:

$$\int_0^3 \left(\left(2x - x^2\right) - (-x) \right) dx$$
$$= \int_0^3 \left(-x^2 + 3x \right) dx$$
$$= \left[-\frac{x^3}{3} + \frac{3}{2}x^2 \right]_0^3$$
$$= \left(-\frac{(3)^3}{3} + \frac{3}{2}(3)^2 \right) - \left(-\frac{(0)^3}{3} + \frac{3}{2}(0)^2 \right)$$
$$= \left(-\frac{27}{3} + \frac{27}{2} \right) - 0$$
$$= \frac{27}{6}$$
$$= 4\tfrac{1}{2}.$$

29. First graph the system of equations and shade the region bounded by the graphs.

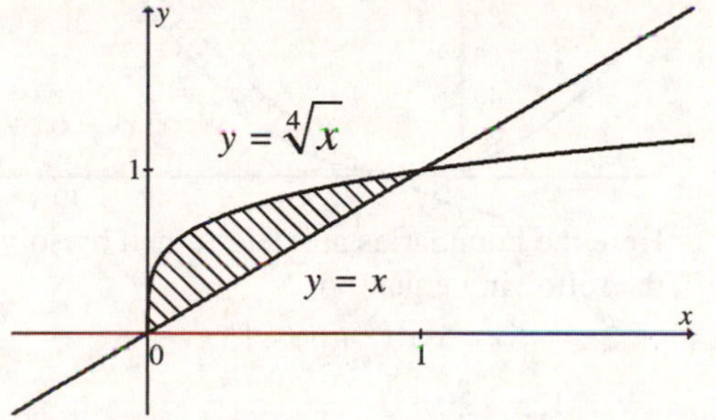

Here the boundaries are easily determined by looking at the graph, or by solving the following equation:

$$x = \sqrt[4]{x}$$
$$x^4 = x$$
$$x^4 - x = 0$$
$$x\left(x^3 - 1\right) = 0$$
$$x = 0 \text{ or } x = 1$$

Note $\sqrt[4]{x} \ge x$ over the interval $[0,1]$.

We compute the area as follows:

$$\int_0^1 \left(\sqrt[4]{x} - x \right) dx$$
$$= \int_0^1 \left(x^{1/4} - x \right) dx$$
$$= \left[\frac{4}{5}x^{5/4} - \frac{x^2}{2} \right]_0^1$$
$$= \left(\frac{4}{5}(1)^{5/4} - \frac{(1)^2}{2} \right) - \left(\frac{4}{5}(0)^{5/4} - \frac{(0)^2}{2} \right)$$
$$= \left(\frac{4}{5} - \frac{1}{2} \right) - 0$$
$$= \frac{3}{10}.$$

31. First graph the system of equations and shade the region bounded by the graphs.

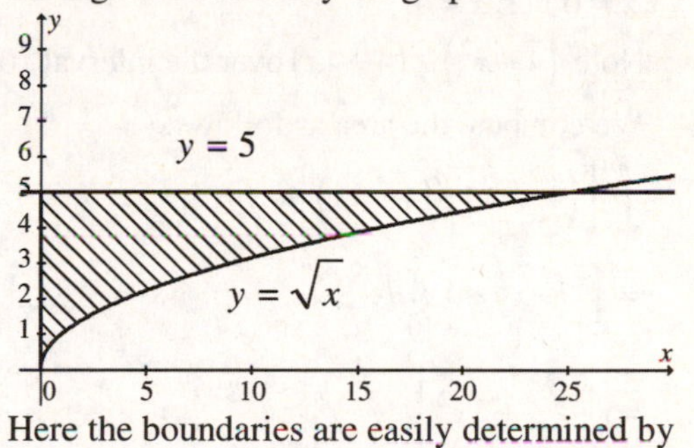

Here the boundaries are easily determined by looking at the graph. Note $5 \ge \sqrt{x}$ over the interval $[0,25]$. We compute the area as follows:

$$\int_0^{25} \left(5 - \sqrt{x} \right) dx$$
$$= \int_0^{25} \left(5 - x^{1/2} \right) dx$$
$$= \left[5x - \frac{2}{3}x^{3/2} \right]_0^{25}$$
$$= \left(5(25) - \frac{2}{3}(25)^{3/2} \right) - \left(5(0) - \frac{2}{3}(0)^{3/2} \right)$$
$$= \left(125 - \frac{250}{3} \right) - 0$$
$$= \frac{125}{3}$$
$$= 41\tfrac{2}{3}.$$

Copyright © 2014 Pearson Education, Inc. Publishing as Addison-Wesley.

33. First graph the system of equations and shade the region bounded by the graphs.

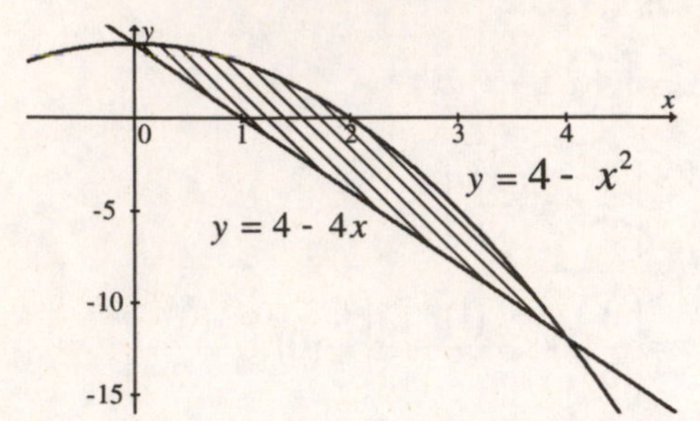

Here the boundaries are easily determined by looking at the graph, or by solving the following:

$$4-4x=4-x^2$$
$$x^2-4x=0$$
$$x(x-4)=0$$
$$x=0 \text{ or } x=4.$$

Note $(4-x^2)\geq(4-4x)$ over the interval $[0,4]$.

We compute the area as follows:

$$\int_0^4\left[(4-x^2)-(4-4x)\right]dx$$
$$=\int_0^4(-x^2+4x)dx$$
$$=\left[-\frac{x^3}{3}+2x^2\right]_0^4$$
$$=\left(-\frac{(4)^3}{3}+2(4)^2\right)-\left(-\frac{(0)^3}{3}+2(0)^2\right)$$
$$=\left(-\frac{64}{3}+32\right)-0$$
$$=\frac{32}{3}$$
$$=10\tfrac{2}{3}.$$

35. First graph the system of equations and shade the region bounded by the graphs.

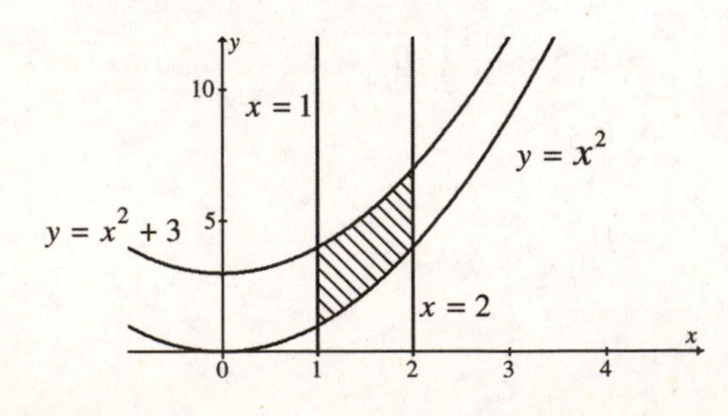

Here the boundaries are easily determined by looking at the graph. Note $(x^2+3)\geq(x^2)$ over the interval $[1,2]$. We compute the area as follows:

$$\int_1^2\left[(x^2+3)-(x^2)\right]dx$$
$$=\int_1^2 3dx$$
$$=[3x]_1^2$$
$$=(3(2)-3(1))$$
$$=3.$$

37. First graph the system of equations and shade the region bounded by the graphs.

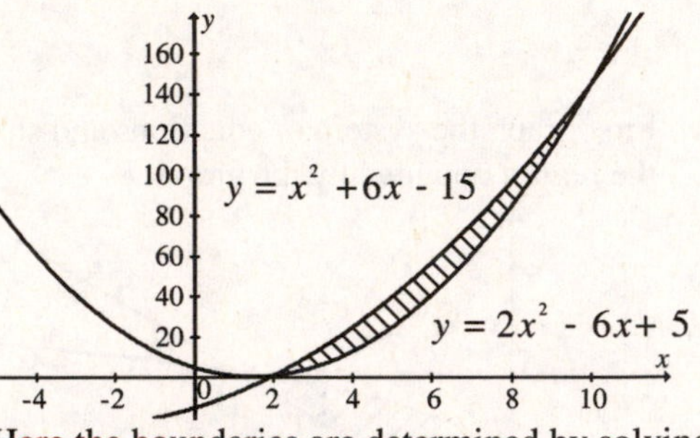

Here the boundaries are determined by solving the following equation:

$$2x^2-6x+5=x^2+6x-15$$
$$x^2-12x+20=0$$
$$(x-2)(x-10)=0$$
$$x=2 \text{ or } x=10$$

Note $(x^2+6x-15)\geq(2x^2-6x+5)$ over the interval $[2,10]$.

The solution is continued on the next page.

Copyright © 2014 Pearson Education, Inc. Publishing as Addison-Wesley.

Integrating to compute the area from the previous page, we have:

$$\int_2^{10}\left[\left(x^2+6x-15\right)-\left(2x^2-6x+5\right)\right]dx$$
$$=\int_2^{10}\left(-x^2+12x-20\right)dx$$
$$=\left[-\frac{x^3}{3}+6x^2-20x\right]_2^{10}$$
$$=\left(-\frac{(10)^3}{3}+6(10)^2-20(10)\right)-\left(-\frac{(2)^3}{3}+6(2)^2-20(2)\right)$$
$$=\frac{200}{3}-\left(-\frac{56}{3}\right)$$
$$=\frac{256}{3}$$
$$=85\tfrac{1}{3}.$$

39. The average value is:

$$y_{\text{av}}=\frac{1}{b-a}\int_a^b f(x)\,dx$$
$$=\frac{1}{2-(-2)}\int_{-2}^{2}\left(4-x^2\right)dx$$
$$=\frac{1}{4}\left[4x-\frac{x^3}{3}\right]_{-2}^{2}$$
$$=\frac{1}{4}\left[\left(4(2)-\frac{(2)^3}{3}\right)-\left(4(-2)-\frac{(-2)^3}{3}\right)\right]$$
$$=\frac{1}{4}\left[\frac{16}{3}-\left(-\frac{16}{3}\right)\right]$$
$$=\frac{8}{3}.$$

41. The average value is:

$$y_{\text{av}}=\frac{1}{b-a}\int_a^b f(x)\,dx$$
$$=\frac{1}{1-(0)}\int_0^1 e^{-x}dx$$
$$=\frac{1}{1}\left[-e^{-x}\right]_0^1$$
$$=\left[-e^{-1}-\left(-e^{0}\right)\right]$$
$$=-e^{-1}+1,\text{ or approximately } 0.632.$$

43. The average value is:

$$y_{\text{av}}=\frac{1}{b-a}\int_a^b f(x)\,dx$$
$$=\frac{1}{4-(0)}\int_0^4\left(x^2+x-2\right)dx$$
$$=\frac{1}{4}\left[\frac{x^3}{3}+\frac{x^2}{2}-2x\right]_0^4$$
$$=\frac{1}{4}\left[\left(\frac{(4)^3}{3}+\frac{(4)^2}{2}-2(4)\right)-\left(\frac{(0)^3}{3}+\frac{(0)^2}{2}-2(0)\right)\right]$$
$$=\frac{1}{4}\left[\frac{64}{3}-0\right]$$
$$=\frac{16}{3}.$$

45. The average value is:

$$y_{\text{av}}=\frac{1}{b-a}\int_a^b f(x)\,dx$$
$$=\frac{1}{a-(0)}\int_0^a (4x+5)\,dx$$
$$=\frac{1}{a}\left[2x^2+5x\right]_0^a$$
$$=\frac{1}{a}\left[\left(2(a)^2+5(a)\right)-\left(2(0)^2+5(0)\right)\right]$$
$$=\frac{1}{a}\left[\left(2a^2+5a\right)-0\right]$$
$$=2a+5.$$

47. The average value is:

$$y_{\text{av}}=\frac{1}{b-a}\int_a^b f(x)\,dx$$
$$=\frac{1}{2-(1)}\int_1^2 x^n dx,\qquad n\neq 0$$
$$=\frac{1}{1}\left[\frac{1}{n+1}x^{n+1}\right]_1^2$$
$$=\left[\frac{1}{n+1}(2)^{n+1}-\frac{1}{n+1}(1)^{n+1}\right]$$
$$=\frac{2^{n+1}}{n+1}-\frac{1}{n+1}$$
$$=\frac{2^{n+1}-1}{n+1}.$$

Copyright © 2014 Pearson Education, Inc. Publishing as Addison-Wesley.

49. a) We find total profit by integrating:

$$P(10) = R(10) - C(10)$$
$$= \int_0^{10} \left[R'(t) - C'(t)\right] dt$$
$$= \int_0^{10} \left[\left(100e^t\right) - \left(100 - 0.2t\right)\right] dt$$
$$= \int_0^{10} \left[100e^t - 100 + 0.2t\right] dt$$
$$= \left[100e^t - 100t + 0.1t^2\right]_0^{10}$$
$$= \left(100e^{10} - 100(10) + 0.1(10)^2\right) - \left(100e^0 - 100(0) + 0.1(0)^2\right)$$
$$= \left(100e^{10} - 990\right) - (100)$$
$$= 100e^{10} - 1090$$
$$\approx 2{,}201{,}556.58.$$

The total profit for the first 10 days is approximately \$2,201,556.58.

b) The average daily profit for the first ten days is given by

$$P_{\text{av}} = \frac{1}{10-0} \int_0^{10} \left[R'(t) - C'(t)\right] dt$$
$$\approx \frac{1}{10}[2{,}201{,}556.58] \quad \text{From part (a).}$$
$$\approx 220{,}155.66$$

The average daily profit over the first 10 days is approximately \$220,155.66.

51. We find the average weekly sales for the first 5 weeks as follows:

$$S_{\text{av}} = \frac{1}{5-0} \int_0^5 S(t)\, dt$$
$$= \frac{1}{5} \int_0^5 9e^t dt$$
$$= \frac{1}{5}\left[9e^t\right]_0^5$$
$$= \frac{1}{5}\left[9e^5 - 9e^0\right]$$
$$= \frac{1}{5}\left[9e^5 - 9\right]$$
$$= \frac{9}{5}\left[e^5 - 1\right]$$
$$\approx 265.3437.$$

Therefore, average weekly sales for the first 5 weeks are \$265.3437 hundred, or \$26,534.37.

53. We find the average weekly sales for weeks 2 through 5 by integrating over the interval $[1,5]$ as follows:

$$S_{\text{av}} = \frac{1}{5-1} \int_1^5 S(t)\, dt$$
$$= \frac{1}{4} \int_1^5 9e^t dt$$
$$= \frac{1}{4}\left[9e^t\right]_1^5$$
$$= \frac{1}{4}\left[9e^5 - 9e^1\right]$$
$$= \frac{9}{4}\left[e^5 - e\right] \approx 327.8135.$$

Therefore, average weekly sales for weeks 2 though 5 week are \$327.8135 hundred, or \$32,781.35.

55. a) We notice $-0.003t^2 \geq -0.009t^2$, which means $M'(t) \geq m'(t)$ so Ben has the higher rate of memorization.

b) $M(10) - m(10) = \int_0^{10} \left[M'(t) - m'(t)\right] dt$

$$\int_0^{10} \left[\left(-0.003t^2 + 0.2t\right) - \left(-0.009t^2 + 0.2t\right)\right] dt$$
$$= \int_0^{10} 0.006t^2 dt$$
$$= \left[0.002t^3\right]_0^{10}$$
$$= 0.002(10)^3 - 0.002(0)^3$$
$$= 2 - 0 = 2$$

Ben memorizes approximately 2 more words during the first 10 minutes.

c) $m_{\text{av}} = \frac{1}{10-0} \int_0^{10} m'(t)\, dt$

$$m_{\text{av}} = \frac{1}{10-0} \int_0^{10} \left(-0.009t^2 + 0.2t\right) dt$$
$$= \frac{1}{10}\left[-0.003t^3 + 0.1t^2\right]_0^{10}$$
$$= \frac{1}{10}\left[\left(-0.003(10)^3 + 0.1(10)^2\right) - \left(-0.003(0)^3 + 0.1(0)^2\right)\right]$$
$$= \frac{1}{10}\left[(-3 + 10) - (0 + 0)\right]$$
$$= \frac{1}{10}(7) = \frac{7}{10} = 0.7$$

Alice averaged memorizing about 0.7 words per minute during the first 10 minutes.

Copyright © 2014 Pearson Education, Inc. Publishing as Addison-Wesley.

d) $m_{av} = \frac{1}{10-0}\int_0^{10} M'(t)\,dt$

$m_{av} = \frac{1}{10-0}\int_0^{10} \left(-0.003t^2 + 0.2t\right)dt$

$= \frac{1}{10}\left[-0.001t^3 + 0.1t^2\right]_0^{10}$

$= \frac{1}{10}\left[\left(-0.001(10)^3 + 0.1(10)^2\right) - \left(-0.001(0)^3 + 0.1(0)^2\right)\right]$

$= \frac{1}{10}\left[(-1+10)-(0+0)\right]$

$= \frac{1}{10}(9) = \frac{9}{10} = 0.9$

Ben averaged memorizing about 0.9 words per minute during the first 10 minutes.

57. a) $W(0) = -6(0)^2 + 12(0) + 90 = 90.$

At the beginning of the interval, the keyboarder's speed is 90 words per minute.

b) First, we find the derivative.

$W'(t) = -12t + 12$

Next, we set the derivative equal to zero to find the critical value.

$W'(t) = 0$

$-12t + 12 = 0$

$t = 1$

The only critical value occurs when $t = 1$.

We also know that $W''(t) = -12 < 0$.

Therefore, by the Max-Min Principle 2, we know that an absolute maximum occurs at $t = 1$.

$W(1) = -6(1)^2 + 12(1) + 90 = 96$

Thus, the maximum speed is 96 words per minute, occurring 1 minute into the interval.

c) $W_{av} = \frac{1}{5-0}\int_0^{5} W(t)\,dt$

$W_{av} = \frac{1}{5-0}\int_0^{5} \left(-6t^2 + 12t + 90\right)dt$

$= \frac{1}{5}\left[-2t^3 + 6t^2 + 90t\right]_0^{5}$

$= \frac{1}{5}\left[\left(-2(5)^3 + 6(5)^2 + 90(5)\right) - \left(-2(0)^3 + 6(0)^2 + 90(0)\right)\right]$

$= \frac{1}{5}[350 - 0]$

$= 70$

The keyboarder's average speed over the 5 minute interval is 70 words per minute.

59. a) $C(0) = 42.03e^{-0.01050(0)} = 42.03$

The initial dosage is 42.03 micrograms per milliliter.

b) $C_{av} = \frac{1}{120-10}\int_{10}^{120} C(t)\,dt$

$C_{av} = \frac{1}{110}\int_{10}^{120} \left(42.03e^{-0.01050t}\right)dt$

$= \frac{1}{110}\left[\frac{42.03}{-0.01050}e^{-0.01050t}\right]_{10}^{120}$

$= \frac{1}{110}\left[\left(\frac{42.03}{-0.01050}e^{-0.01050(120)}\right) - \frac{42.03}{-0.01050}e^{-0.01050(10)}\right]$

$\approx \frac{1}{110}[2468.4439]$

≈ 22.44

The average amount of phenylbutazone in the calf's body for the time between 10 and 120 hours is about 22.44 micrograms per milliliter.

Copyright © 2014 Pearson Education, Inc. Publishing as Addison-Wesley.

61. a) We determine the average temperature as follows:

$$\frac{1}{10-0}\int_0^{10}\left(-t^2+5t+40\right)dt$$

$$=\frac{1}{10}\left[-\frac{t^3}{3}+\frac{5}{2}t^2+40t\right]_0^{10}$$

$$=\frac{1}{10}\left[\left(-\frac{(10)^3}{3}+\frac{5}{2}(10)^2+40(10)\right)-\left(-\frac{(0)^3}{3}+\frac{5}{2}(0)^2+40(0)\right)\right]$$

$$=\frac{1}{10}\left[\frac{950}{3}-0\right]$$

$$=\frac{95}{3}\approx 31.67.$$

The average temperature over the 10 hour period is 31.7 degrees.

b) First, we find the critical values.

$$f'(t)=-2t+5$$
$$f'(t)=0$$
$$-2t+5=0$$
$$t=\frac{5}{2}=2.5$$

Since there is only one critical value and $f''(t)=-2<0$, we know that it represents a maximum over the closed interval. Therefore, the minimum temperature must be at an endpoint. We have:

$$f(0)=-(0)^2+5(0)+40=40$$
$$f(10)=-(10)^2+5(10)+40=-10$$

The minimum temperature over the 10 hour period is minus 10 degrees.

c) From part (b), we know that the maximum temperature occurs when $t=2.5$. Substituting we have:

$$f(2.5)=-(2.5)^2+5(2.5)+40=46.25$$

Therefore, the maximum temperature over the 10 hour period is 46.25 degrees.

63. First graph the system of equations and shade the region bounded by the graphs.

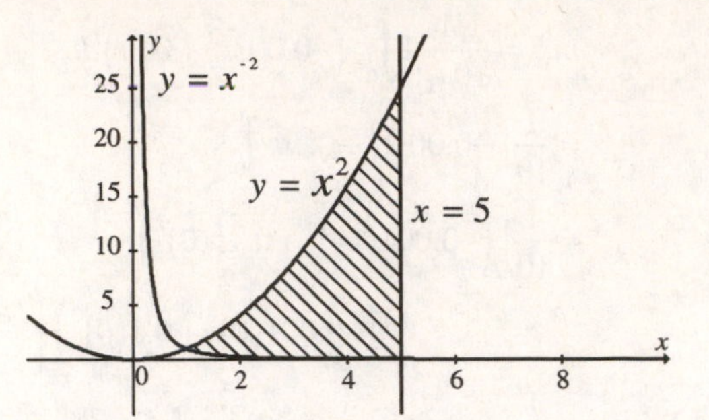

Here the boundaries are determined by looking at the graph. Note $x^2 \ge x^{-2}$ over the interval $[1,5]$. We compute the area as follows:

$$\int_1^5\left(x^2-x^{-2}\right)dx$$

$$=\left[\frac{x^3}{3}+x^{-1}\right]_1^5$$

$$=\left(\frac{(5)^3}{3}+(5)^{-1}\right)-\left(\frac{(1)^3}{3}+(1)^{-1}\right)$$

$$=\frac{608}{15}$$

$$=40\tfrac{8}{15}.$$

65. First graph the system of equations and shade the region bounded by the graphs.

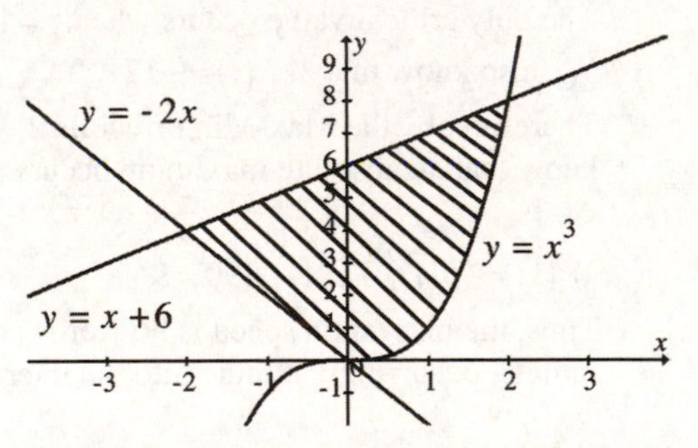

Here the boundaries are determined by looking at the graph. We will split the interval $[-2,2]$ up into two parts.

The solution is continued on the next page.

Copyright © 2014 Pearson Education, Inc. Publishing as Addison-Wesley.

On the interval $[-2,0]$ we have $x+6 \geq -2x$.

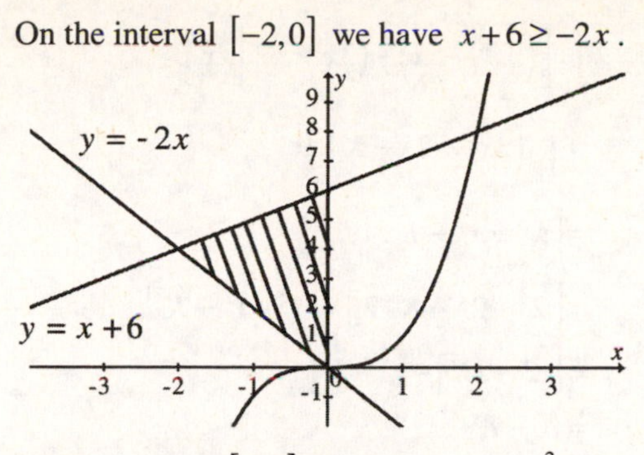

On the interval $[0,2]$ we have $x+6 \geq x^3$.

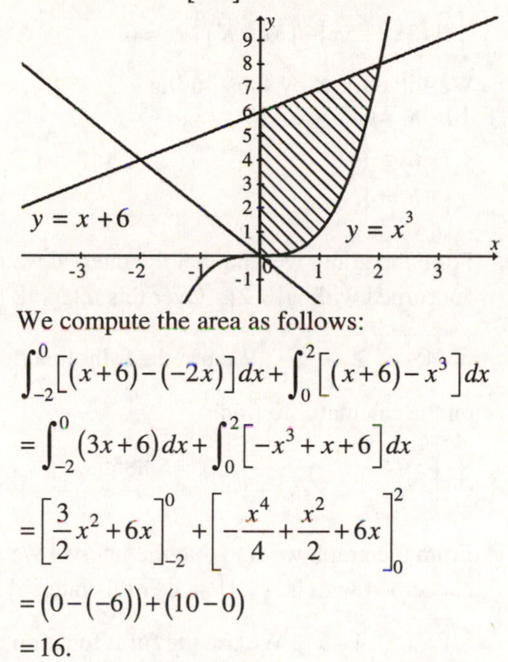

We compute the area as follows:

$$\int_{-2}^{0}[(x+6)-(-2x)]dx+\int_{0}^{2}[(x+6)-x^3]dx$$

$$=\int_{-2}^{0}(3x+6)dx+\int_{0}^{2}[-x^3+x+6]dx$$

$$=\left[\frac{3}{2}x^2+6x\right]_{-2}^{0}+\left[-\frac{x^4}{4}+\frac{x^2}{2}+6x\right]_{0}^{2}$$

$$=(0-(-6))+(10-0)$$

$$=16.$$

67. First graph the system of equations and shade the region bounded by the graphs.

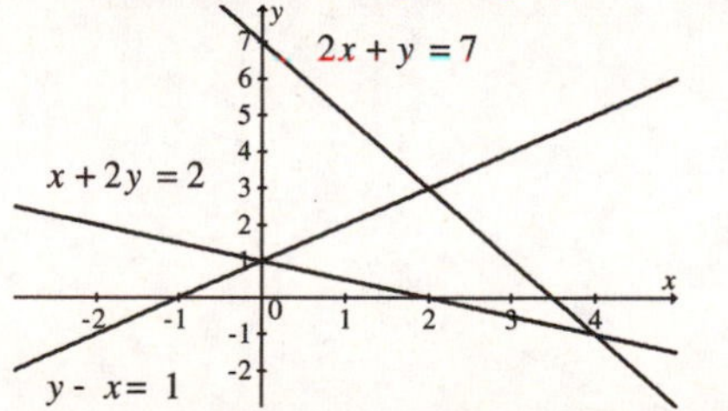

Here the boundaries are determined by looking at the graph. We will split the interval $[0,4]$ up into two parts. Solving each equation for y, we have

$$x+2y=2 \rightarrow y=-\frac{x}{2}+1$$

$$y-x=1 \rightarrow y=x+1$$

$$2x+y=7 \rightarrow y=-2x+7$$

On the interval $[0,2]$ we have $x+1 \geq -\frac{x}{2}+1$.

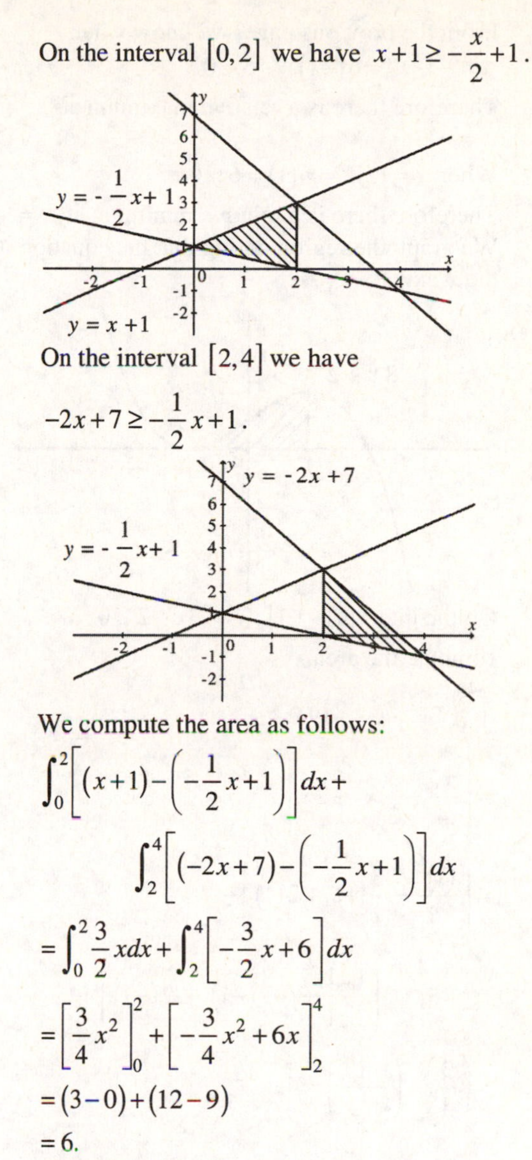

On the interval $[2,4]$ we have

$-2x+7 \geq -\frac{1}{2}x+1$.

We compute the area as follows:

$$\int_{0}^{2}\left[(x+1)-\left(-\frac{1}{2}x+1\right)\right]dx+$$

$$\int_{2}^{4}\left[(-2x+7)-\left(-\frac{1}{2}x+1\right)\right]dx$$

$$=\int_{0}^{2}\frac{3}{2}xdx+\int_{2}^{4}\left[-\frac{3}{2}x+6\right]dx$$

$$=\left[\frac{3}{4}x^2\right]_{0}^{2}+\left[-\frac{3}{4}x^2+6x\right]_{2}^{4}$$

$$=(3-0)+(12-9)$$

$$=6.$$

69. First we find the coordinates of the relative extrema.

$y=x^3-3x+2$

$y'=3x^2-3$

The derivative exists for all real numbers. We solve $y'=0$.

$3x^2-3=0$

$x^2-1=0$

$x=\pm 1$

$x=-1$ or $x=1$

We use the Second Derivative Test.

$y''=6x$

The solution is continued on the next page.

Copyright © 2014 Pearson Education, Inc. Publishing as Addison-Wesley.

From the previous page, we know when $x=-1, y''=6(-1)=-6<0$

Therefore there is a relative maximum at $x=-1$.

When $x=1, y''=6(1)=6>0$

Therefore there is a relative minimum at $x=1$.

We graph the region noting that the equation of the x-axis is $y=0$.

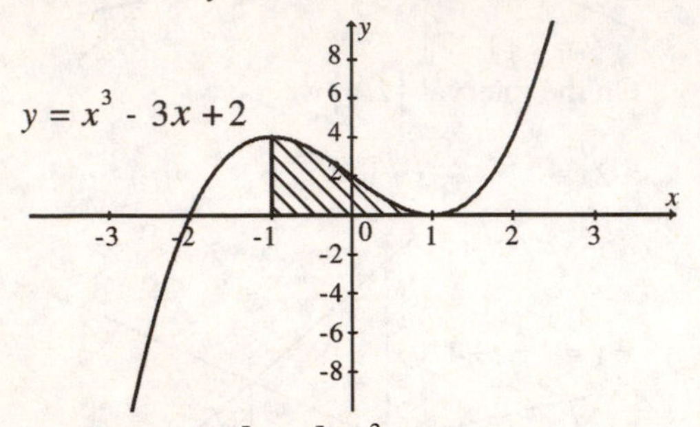

On the interval $[-1,1]$, $x^3-3x+2\geq 0$. We compute the area.

$$\int_{-1}^{1}\left(x^3-3x+2\right)dx$$

$$=\left[\frac{x^4}{4}-\frac{3}{2}x^2+2x\right]_{-1}^{1}$$

$$=\left(\frac{(1)^4}{4}-\frac{3}{2}(1)^2+2(1)\right)-\left(\frac{(-1)^4}{4}-\frac{3}{2}(-1)^2+2(-1)\right)$$

$$=\left(\frac{3}{4}\right)-\left(-\frac{13}{4}\right)=4.$$

71. $\int_{1}^{2}\left[\left(3x^2+5x\right)-\left(3x+K\right)\right]dx$

$$=\int_{1}^{2}\left(3x^2+2x-K\right)dx$$

$$=\left[x^3+x^2-Kx\right]_{1}^{2}$$

$$=\left[2^3+2^2-K\cdot 2\right]-\left[1^3+1^2-K\right]$$

$$=[12-2K]-[2-K]$$

$$=10-K$$

Since

$$\int_{1}^{2}\left[\left(3x^2+5x\right)-\left(3x+K\right)\right]dx=6.$$

We solve for K by substituting

$$10-K=6$$
$$-K=-4$$
$$K=4.$$

73. From the graph we find that the interval we are concerned with is $[0,2]$. Over this interval $x\sqrt{4-x^2}\geq\frac{-4x}{x^2+1}$. We use the fnInt function on the calculator to find

$$\int_{0}^{2}\left[x\sqrt{4-x^2}-\frac{-4x}{x^2+1}\right]dx\approx 5.8855.$$

75. From the graph we find that the interval we are concerned with is $[-1,1]$. Over this interval $\sqrt{1-x^2}\geq 1-x^2$. We use the fnInt function on the calculator to find

$$\int_{-1}^{1}\left[\sqrt{1-x^2}-\left(1-x^2\right)\right]dx\approx 0.2375.$$

Copyright © 2014 Pearson Education, Inc. Publishing as Addison-Wesley.

Exercise Set 4.5

1. $\int (8+x^3)^5 3x^2dx$

Let $u = 8+x^3$, then $du = 3x^2dx$.

$= \int u^5du$ Substitution: $\begin{smallmatrix} u=8+x^3 \\ du=3x^2dx \end{smallmatrix}$

$= \frac{1}{6}u^6 + C$ Formula A

$= \frac{1}{6}(8+x^3)^6 + C.$ Reverse substitution

3. $\int (x^2-6)^7 xdx$

Let $u = x^2-6$, then $du = 2xdx$. We do not have $2xdx$. We only have xdx and need to supply a 2. We do this by multiplying by $\frac{1}{2}\cdot 2$ as follows:

$= \frac{1}{2}\cdot 2\int (x^2-6)^7 xdx$ Multiplying by 1

$= \frac{1}{2}\int (x^2-6)^7 2xdx$ $\left[a\int f(x)\,dx = \int af(x)\,dx\right]$

$= \frac{1}{2}\int u^7du$ Substitution: $\begin{smallmatrix} u=x^2-6 \\ du=2xdx \end{smallmatrix}$

$= \frac{1}{16}u^8 + C$ Formula A

$= \frac{1}{16}(x^2-6)^8 + C.$ Reverse substitution

5. $\int (3t^4+2)t^3dt$

Let $u = 3t^4+2$, then $du = 12t^3dt$. We do not have $12t^3dt$. We only have t^3dt and need to supply a 12. We do this by multiplying by $\frac{1}{12}\cdot 12$ as follows:

$= \frac{1}{12}\cdot 12\int (3t^4+2)^7 t^3dt$ Multiplying by 1

$= \frac{1}{12}\int (3t^4+2)12t^3dt$

$= \frac{1}{12}\int udu$ Substitution: $\begin{smallmatrix} u=3t^4+2 \\ du=12t^3dt \end{smallmatrix}$

$= \frac{1}{24}u^2 + C$ Formula A

$= \frac{1}{24}(3t^4+2)^2 + C.$ Reverse substitution

7. $\int \frac{2}{1+2x}dx$

Let $u = 1+2x$, then $du = 2dx$.

$= \int \frac{du}{u}$ Substitution: $\begin{smallmatrix} u=1+2x \\ du=2dx \end{smallmatrix}$

$= \ln u + C$ Formula C, $u > 0$.

$= \ln(1+2x) + C.$ Reverse substitution

9. $\int (\ln x)^3 \frac{1}{x}dx$

Let $u = \ln x$, then $du = \frac{1}{x}dx$.

$= \int u^3du$ Substitution: $\begin{smallmatrix} u=\ln x \\ du=\frac{1}{x}dx \end{smallmatrix}$

$= \frac{1}{4}u^4 + C$ Formula A

$= \frac{1}{4}(\ln x)^4 + C.$ Reverse substitution

11. $\int e^{3x}dx$

Let $u = 3x$, then $du = 3dx$. We do not have $3dx$. We only have dx and need to supply a 3. We do this by multiplying by $\frac{1}{3}\cdot 3$ as follows:

$= \frac{1}{3}\cdot 3\int e^{3x}dx$ Multiplying by 1

$= \frac{1}{3}\int e^{3x}\cdot 3dx$

$= \frac{1}{3}\int e^u du$ Substitution: $\begin{smallmatrix} u=3x \\ du=3dx \end{smallmatrix}$

$= \frac{1}{3}e^u + C$ Formula B

$= \frac{1}{3}e^{3x} + C.$ Reverse substitution

13. $\int e^{x/3}dx$

Let $u = \frac{x}{3}$, then $du = \frac{1}{3}dx$. We do not have $\frac{1}{3}dx$. We only have dx and need to supply a $\frac{1}{3}$. We do this by multiplying by $\frac{1}{3}\cdot 3$.

The solution is continued on the next page.

Copyright © 2014 Pearson Education, Inc. Publishing as Addison-Wesley.

Rewriting the integral we have:

$\int e^{x/3}dx$

$=\frac{1}{3}\cdot 3\int e^{x/3}dx$ Multiplying by 1

$=3\int e^{x/3}\cdot\frac{1}{3}dx$

$=3\int e^{u}du$ Substitution: $u=x/3$, $du=1/3dx$

$=3e^{u}+C$ Formula B

$=3e^{x/3}+C.$ Reverse substitution

15. $\int x^4 e^{x^5}dx$

Let $u=x^5$, then $du=5x^4dx$. We do not have $5x^4dx$. We only have x^4dx and need to supply a 5. We do this by multiplying by $\frac{1}{5}\cdot 5$.

$=\frac{1}{5}\cdot 5\int x^4 e^{x^5}dx$ Multiplying by 1

$=\frac{1}{5}\int 5x^4 e^{x^5}dx$

$=\frac{1}{5}\int e^{u}du$ Substitution: $u=x^5$, $du=5x^4dx$

$=\frac{1}{5}e^{u}+C$ Formula B

$=\frac{1}{5}e^{x^5}+C.$ Reverse substitution

17. $\int te^{-t^2}dt$

Let $u=-t^2$, then $du=-2tdt$. We do not have $-2tdt$. We only have tdt and need to supply a -2. We do this by multiplying by $\left(-\frac{1}{2}\right)\cdot(-2)$ as follows:

$=\left(-\frac{1}{2}\right)\cdot(-2)\int te^{-t^2}dt$ Multiplying by 1

$=-\frac{1}{2}\int(-2t)e^{-t^2}dt$

$=-\frac{1}{2}\int e^{u}du$ Substitution: $u=-t^2$, $du=-2tdt$

$=-\frac{1}{2}e^{u}+C$ Formula B

$=-\frac{1}{2}e^{-t^2}+C.$ Reverse substitution

19. $\int\frac{1}{5+2x}dx$

Let $u=5+2x$, then $du=2dx$. We do not have $2dx$. We only have dx and need to supply a 2. We do this by multiplying by $\frac{1}{2}\cdot 2$ as follows:

$=\frac{1}{2}\cdot 2\int\frac{1}{5+2x}dx$ Multiplying by 1

$=\frac{1}{2}\int\frac{2dx}{5+2x}$

$=\frac{1}{2}\int\frac{du}{u}$ Substitution: $u=5+2x$, $du=2dx$

$=\frac{1}{2}\ln u+C$ Formula C, $u>0$.

$=\frac{1}{2}\ln(5+2x)+C.$ Reverse substitution

21. $\int\frac{dx}{12+3x}$

Let $u=12+3x$, then $du=3dx$. We do not have $3dx$. We only have dx and need to supply a 3. We do this by multiplying by $\frac{1}{3}\cdot 3$.

$=\frac{1}{3}\cdot 3\int\frac{dx}{12+3x}$ Multiplying by 1

$=\frac{1}{3}\int\frac{3dx}{12+3x}$

$=\frac{1}{3}\int\frac{du}{u}$ Substitution: $u=12+3x$, $du=3dx$

$=\frac{1}{3}\ln u+C$ Formula C, $u>0$.

$=\frac{1}{3}\ln(12+3x)+C.$ Reverse substitution

23. $\int\frac{dx}{1-x}$

Let $u=1-x$, then $du=-dx$. We do not have $-dx$. We only have dx and need to supply a -1. We do this by multiplying by $(-1)(-1)$ as follows:

$=(-1)\cdot(-1)\int\frac{dx}{1-x}$ Multiplying by 1

$=-1\cdot\int\frac{-1dx}{1-x}$

$=-\int\frac{du}{u}$ Substitution: $u=1-x$, $du=-dx$

$=-\ln u+C$ Formula C, $u>0$.

$=-\ln(1-x)+C.$ Reverse substitution

Copyright © 2014 Pearson Education, Inc. Publishing as Addison-Wesley.

25. $\int t(t^2-1)^5\,dt$

Let $u=t^2-1$, then $du=2tdt$. We do not have $2tdt$. We only have tdt and need to supply a 2. We do this by multiplying by $\frac{1}{2}\cdot 2$ as follows:

$=\frac{1}{2}\cdot 2\int t(t^2-1)^5\,dt$ Multiplying by 1

$=\frac{1}{2}\int 2t(t^2-1)^5\,dt$

$=\frac{1}{2}\int u^5 du$ Substitution: $u=t^2-1$, $du=2tdt$

$=\frac{1}{12}u^6+C$ Formula A

$=\frac{1}{12}(t^2-1)^6+C.$ Reverse substitution

27. $\int (x^4+x^3+x^2)^7(4x^3+3x^2+2x)dx$

Let $u=x^4+x^3+x^2$, then $du=(4x^3+3x^2+2x)dx$.

$=\int u^7 du$ Substitution: $u=x^4+x^3+x^2$, $du=(4x^3+3x^2+2x)dx$

$=\frac{1}{8}u^8+C$ Formula A

$=\frac{1}{8}(x^4+x^3+x^2)^8+C.$ Reverse substitution

29. $\int \frac{e^x dx}{4+e^x}$

Let $u=4+e^x$, then $du=e^x dx$.

$=\int \frac{du}{u}$ Substitution: $u=4+e^x$, $du=e^x dx$

$=\ln u+C$ Formula C, $u>0$.

$=\ln(4+e^x)+C.$ Reverse substitution

31. $\int \frac{\ln x^2}{x}dx$

Using properties of logarithms, we have

$\int \frac{2\cdot \ln x}{x}dx = 2\int \frac{\ln x}{x}dx$

Let $u=\ln x$, then $du=\frac{1}{x}dx$.

$=2\int u du$ Substitution: $u=\ln x$, $du=\frac{1}{x}dx$

$=u^2+C$ Formula A

$=(\ln x)^2+C.$ Reverse substitution

33. $\int \frac{dx}{x\ln x}$

Let $u=\ln x$, then $du=\frac{1}{x}dx$.

$=\int \frac{1}{\ln x}\cdot\left(\frac{1}{x}dx\right)$

$=\int \frac{1}{u}du$ Substitution: $u=\ln x$, $du=\frac{1}{x}dx$

$=\ln u+C$ Formula C, $u>0$.

$=\ln(\ln x)+C.$ Reverse substitution

35. $\int x\sqrt{ax^2+b}\,dx$

Let $u=ax^2+b$, then $du=2axdx$. We do not have $2axdx$. We only have xdx and need to supply a $2a$. We do this by multiplying by $\frac{1}{2a}\cdot 2a$ as follows:

$=\frac{1}{2a}\cdot 2a\int x\sqrt{ax^2+b}\,dx$ Multiplying by 1

$=\frac{1}{2a}\int 2ax\sqrt{ax^2+b}\,dx$

$=\frac{1}{2a}\int \sqrt{u}\,du$ Substitution: $u=ax^2+b$, $du=2axdx$

$=\frac{1}{2a}\int u^{1/2}du$

$=\frac{1}{2a}\cdot\frac{2}{3}u^{3/2}+C$ Formula A

$=\frac{1}{3a}(ax^2+b)^{3/2}+C.$ Reverse substitution

37. $\int P_0 e^{kt}dt = P_0\cdot\int e^{kt}dt$

Let $u=kt$, then $du=kdt$. We do not have kdt. We only have dt and need to supply a k. We do this by multiplying by $\frac{1}{k}\cdot k$ as follows:

$=\frac{1}{k}\cdot k\cdot P_0\int e^{kt}dt$ Multiplying by 1

$=\frac{1}{k}\cdot P_0\int e^{kt}\cdot kdt$

$=\frac{P_0}{k}\int e^u du$ Substitution: $u=kt$, $du=kdt$

$=\frac{P_0}{k}e^u+C$ Formula B

$=\frac{P_0}{k}e^{kt}+C.$ Reverse substitution

Copyright © 2014 Pearson Education, Inc. Publishing as Addison-Wesley.

39. $\int \frac{x^3 dx}{(2-x^4)^7}$

Let $u = 2 - x^4$, then $du = -4x^3 dx$. We do not have $-4x^3 dx$. We only have $x^3 dx$ and need to supply a -4. We do this by multiplying by $\frac{1}{-4} \cdot -4$.

$= \left(-\frac{1}{4}\right) \cdot (-4) \int \frac{x^3 dx}{(2-x^4)^7}$ Multiplying by 1

$= -\frac{1}{4} \int \frac{-4x^3 dx}{(2-x^4)^7}$

$= -\frac{1}{4} \int \frac{du}{u^7}$ Substitution: $u=2-x^4$, $du=-4x^3 dx$

$= -\frac{1}{4} \int u^{-7} du$

$= -\frac{1}{4} \cdot \frac{1}{-6} u^{-6} + C$ Formula A

$= \frac{1}{24}(2-x^4)^{-6} + C$ Reverse substitution

$= \frac{1}{24(2-x^4)^6} + C.$

41. $\int 12x\sqrt[5]{1+6x^2}\, dx$

Let $u = 1 + 6x^2$, then $du = 12x dx$.

$= \int \sqrt[5]{u}\, du$ Substitution: $u=1+6x^2$, $du=12x dx$

$= \int u^{1/5} du$

$= \frac{5}{6} u^{6/5} + C$ Formula A

$= \frac{5}{6}(1+6x^2)^{6/5} + C.$ Reverse substitution

43. $\int_0^1 2xe^{x^2} dx$

We first find the indefinite integral

$\int 2xe^{x^2} dx$

$= \int e^u du$ Substitution: $u=x^2$, $du=2x dx$

$= e^u + C$ Formula B

$= e^{x^2} + C.$ Reverse Substitution

Next, we evaluate the definite integral on $[0,1]$.

$\int_0^1 2xe^{x^2} dx = \left[e^{x^2}\right]_0^1$ Let $C = 0$.

$= e^{(1)^2} - e^{(0)^2}$

$= e^1 - e^0$

$= e - 1$

45. $\int_0^1 x(x^2+1)^5 dx$

We first find the indefinite integral

$\int x(x^2+1)^5 dx$

$= \frac{1}{2} \cdot 2 \int x(x^2+1)^5 dx$ Multiplying by 1

$= \frac{1}{2} \int 2x(x^2+1)^5 dx$

$= \frac{1}{2} \int u^5 du$ Substitution: $u=x^2+1$, $du=2x dx$

$= \frac{1}{12} u^6 + C$ Formula A

$= \frac{1}{12}(x^2+1)^6 + C.$ Reverse Substitution

Next, we evaluate the definite integral on $[0,1]$.

$\int_0^1 x(x^2+1)^5 dx$

$= \left[\frac{1}{12}(x^2+1)^6\right]_0^1$ Let $C = 0$.

$= \left[\frac{1}{12}(1^2+1)^6 - \frac{1}{12}(0^2+1)^6\right]$

$= \frac{1}{12}(2)^6 - \frac{1}{12}(1)^6$

$= \frac{16}{3} - \frac{1}{12}$

$= \frac{21}{4}.$

47. $\int_0^4 \frac{dt}{1+t}$

We first find the indefinite integral

$\int \frac{dt}{1+t}$

$= \int \frac{du}{u}$ Substitution: $u=1+t$, $du=dt$

$= \ln u + C$ Formula C, $u > 0$

$= \ln(1+t) + C.$ Reverse Substitution

The solution is continued on the next page.

Copyright © 2014 Pearson Education, Inc. Publishing as Addison-Wesley.

Next, we evaluate the definite integral on $[0,4]$.

$$\int_0^4 \frac{dt}{1+t}$$
$$=\left[\ln(1+t)\right]_0^4 \quad \text{Let } C=0.$$
$$=\left[\ln(1+4)-\ln(1+0)\right]$$
$$=\ln(5)-\ln(1)$$
$$=\ln 5-0$$
$$=\ln 5.$$

49. $\int_1^4 \frac{2x+1}{x^2+x-1}dx$

We first find the indefinite integral

$$\int \frac{2x+1}{x^2+x-1}dx$$
$$=\int \frac{du}{u} \quad \text{Substitution: } \begin{matrix} u=x^2+x-1 \\ du=2x+1dx \end{matrix}$$
$$=\ln u+C \quad \text{Formula C, } u>0$$
$$=\ln\left(x^2+x-1\right)+C. \quad \text{Reverse Substitution}$$

Next, we evaluate the definite integral on $[1,4]$.

$$\int_1^4 \frac{2x+1}{x^2+x-1}dx$$
$$=\left[\ln\left(x^2+x-1\right)\right]_1^4 \quad \text{Let } C=0.$$
$$=\left[\ln\left((4)^2+(4)-1\right)-\ln\left((1)^2+(1)-1\right)\right]$$
$$=\ln(19)-\ln(1)$$
$$=\ln 19-0$$
$$=\ln 19.$$

51. $\int_0^b e^{-x}dx$

We first find the indefinite integral

$$\int e^{-x}dx$$
$$=-\int -e^{-x}dx \quad \text{Multipling by 1}$$
$$=-\int e^u du \quad \text{Substitution: } \begin{matrix} u=-x \\ du=-dx \end{matrix}$$
$$=-e^u+C \quad \text{Formula B}$$
$$=-e^{-x}+C. \quad \text{Reverse Substitution}$$

Next, we evaluate the definite integral on $[0,b]$.

$$\int_0^b e^{-x}dx$$
$$=\left[-e^{-x}\right]_0^b \quad \text{Let } C=0.$$
$$=-e^{-(b)}-\left(-e^{-(0)}\right)$$
$$=-e^{-b}+e^0$$
$$=1-e^{-b}.$$

53. $\int_0^b me^{-mx}dx$

We first find the indefinite integral

$$\int me^{-mx}dx$$
$$=-\int -me^{-mx}dx \quad \text{Multipling by 1}$$
$$=-\int e^u du \quad \text{Substitution: } \begin{matrix} u=-mx \\ du=-mdx \end{matrix}$$
$$=-e^u+C \quad \text{Formula B}$$
$$=-e^{-mx}+C. \quad \text{Reverse Substitution}$$

Next, we evaluate the definite integral on $[0,b]$.

$$\int_0^b me^{-mx}dx$$
$$=\left[-e^{-mx}\right]_0^b \quad \text{Let } C=0.$$
$$=-e^{-m(b)}-\left(-e^{-m(0)}\right)$$
$$=-e^{-mb}+e^0$$
$$=1-e^{-mb}.$$

55. $\int_0^4 (x-6)^2\,dx$

We first find the indefinite integral

$$\int (x-6)^2\,dx$$
$$=\int u^2 du \quad \text{Substitution: } \begin{matrix} u=x-6 \\ du=dx \end{matrix}$$
$$=\frac{1}{3}u^3+C \quad \text{Formula A}$$
$$=\frac{1}{3}(x-6)^3+C. \quad \text{Reverse Substitution}$$

The solution is continued on the next page.

Copyright © 2014 Pearson Education, Inc. Publishing as Addison-Wesley.

Next, we evaluate the definite integral on $[0,4]$.

$$\int_0^4 (x-6)^2\,dx$$

$$=\left[\frac{1}{3}(x-6)^3\right]_0^4 \quad \text{Let } C=0.$$

$$=\left[\frac{1}{3}((4)-6)^3-\frac{1}{3}((0)-6)^3\right]$$

$$=\frac{1}{3}(-2)^3-\frac{1}{3}(-6)^3$$

$$=-\frac{8}{3}-\left(-\frac{216}{3}\right)$$

$$=\frac{208}{3}.$$

57. $\displaystyle\int_0^2 \frac{3x^2dx}{\left(1+x^3\right)^5}$

We first find the indefinite integral

$$\int \frac{3x^2dx}{\left(1+x^3\right)^5}$$

$$=\int \frac{du}{u^5} \quad \text{Substitution: } \begin{matrix} u=1+x^3 \\ du=3x^2dx \end{matrix}$$

$$=\int u^{-5}du$$

$$=-\frac{1}{4}u^{-4}+C \quad \text{Formula A}$$

$$=-\frac{1}{4}\left(1+x^3\right)^{-4}+C \quad \text{Reverse Substitution}$$

$$=-\frac{1}{4\left(1+x^3\right)^4}+C.$$

Next, we evaluate the definite integral on $[0,2]$

$$\int_0^2 \frac{3x^2dx}{\left(1+x^3\right)^5}$$

$$=\left[-\frac{1}{4\left(1+x^3\right)^4}\right]_0^2 \quad \text{Let } C=0.$$

$$=\left[\left(-\frac{1}{4\left(1+(2)^3\right)^4}\right)-\left(-\frac{1}{4\left(1+(0)^3\right)^4}\right)\right]$$

$$=\left(-\frac{1}{4(9)^4}\right)-\left(-\frac{1}{4(1)^4}\right)$$

$$=\left(-\frac{1}{26,244}\right)-\left(-\frac{1}{4}\right)$$

$$=-\frac{1}{26,244}+\frac{6561}{26,244}$$

$$=\frac{6560}{26,244}$$

$$=\frac{1640}{6561}.$$

59. $\displaystyle\int_0^{\sqrt{7}} 7x\cdot\sqrt[3]{1+x^2}\,dx$

We first find the indefinite integral

$$\int 7x\cdot\sqrt[3]{1+x^2}\,dx$$

$$=7\int x\cdot\sqrt[3]{1+x^2}\,dx$$

$$=\frac{7}{2}\int 2x\cdot\sqrt[3]{1+x^2}\,dx$$

$$=\frac{7}{2}\int \sqrt[3]{1+x^2}\,(2xdx)$$

$$=\frac{7}{2}\int \sqrt[3]{u}\,du \quad \text{Substitution: } \begin{matrix} u=1+x^2 \\ du=2xdx \end{matrix}$$

$$=\frac{7}{2}\int u^{1/3}du$$

$$=\frac{21}{8}u^{4/3}+C \quad \text{Formula A}$$

$$=\frac{21}{8}\left(1+x^2\right)^{4/3}+C. \quad \text{Reverse Substitution}$$

The solution is continued on the next page.

Copyright © 2014 Pearson Education, Inc. Publishing as Addison-Wesley.

We evaluate the definite integral on $\left[0,\sqrt{7}\right]$.

$$\int_0^{\sqrt{7}} 7x\sqrt[3]{1+x^2}\,dx$$

$$=\left[\frac{21}{8}\left(1+x^2\right)^{4/3}\right]_0^{\sqrt{7}} \quad \text{Let } C=0.$$

$$=\left[\left(\frac{21}{8}\left(1+\left(\sqrt{7}\right)^2\right)^{4/3}\right)-\left(\frac{21}{8}\left(1+(0)^2\right)^{4/3}\right)\right]$$

$$=\left(\frac{21}{8}(8)^{4/3}\right)-\left(\frac{21}{8}(1)^{4/3}\right)$$

$$=(42)-\left(\frac{21}{8}\right)$$

$$=\frac{315}{8}.$$

61. Left to the student.

63. $\int \frac{3x}{2x+1}dx$

Assuming $2x+1>0$, Let

$u=2x+1$, then $du=2dx$. Observe that

$x=\frac{u-1}{2}$.

Making the appropriate substitutions we have:

$$\int \frac{3x}{2x+1}dx$$

$$=\frac{1}{2}\int \frac{3x}{2x+1}(2)\,dx \qquad \text{Multiply by } \frac{1}{2}\cdot 2$$

$$=\frac{3}{2}\int \frac{\frac{u-1}{2}}{u}du$$

$$=\frac{3}{4}\int\left(\frac{u-1}{u}\right)du$$

$$=\frac{3}{4}\int\left(1-\frac{1}{u}\right)du$$

$$=\frac{3}{4}\left(u-\ln u+C_1\right)$$

$$=\frac{3}{4}\left((2x+1)-\ln(2x+1)+C_1\right) \quad \text{Reverse Substitution}$$

$$=\frac{3}{2}x+\frac{3}{4}-\frac{3}{4}\ln(2x+1)+\frac{3}{4}C_1$$

$$=\frac{3}{2}x-\frac{3}{4}\ln(2x+1)+C. \qquad C=\frac{3}{4}C_1+\frac{3}{4}$$

65. $\int \frac{x+3}{x-2}dx$

Assuming $x-2>0$, Let

$u=x-2$, then $du=dx$. Observe that

$x=u+2.$

Making the appropriate substitutions we have:

$$\int \frac{x+3}{x-2}dx$$

$$=\int \frac{u+2+3}{u}du$$

$$=\int \frac{u+5}{u}du$$

$$=\int\left(1+\frac{5}{u}\right)du$$

$$=\left(u+5\ln u+C_1\right)$$

$$=\left((x-2)+5\ln(x-2)+C_1\right) \quad \text{Reverse Substitution}$$

$$=x+5\ln(x-2)+C. \qquad C=C_1-2$$

67. $\int x^2(x+1)^{10}\,dx$

Let $u=x+1$, then $du=dx$. Observe that

$x=u-1$ and $x^2=(u-1)^2=u^2-2u+1.$

Making the appropriate substitutions we have

$$\int x^2(x+1)^{10}\,dx$$

$$=\int\left(u^2-2u+1\right)u^{10}du$$

$$=\int\left(u^{12}-2u^{11}+u^{10}\right)du$$

$$=\frac{1}{13}u^{13}-\frac{2}{12}u^{12}+\frac{1}{11}u^{11}+C$$

$$=\frac{(x+1)^{13}}{13}-\frac{(x+1)^{12}}{6}+\frac{(x+1)^{11}}{11}+C. \quad \text{Reverse Substitution}$$

Copyright © 2014 Pearson Education, Inc. Publishing as Addison-Wesley.

69. $\int x^2\sqrt{x-2}dx$

Assuming $x-2\ge 0$ Let $u=x-2$, then $du=dx$.
Observe that $x=u+2$ and
$x^2=(u+2)^2=u^2+4u+4.$
Making the appropriate substitutions we have

$$\int x^2\sqrt{x-2}dx$$
$$=\int (u^2+4u+4)\sqrt{u}du$$
$$=\int (u^2+4u+4)u^{\frac{1}{2}}du$$
$$=\int \left(u^{\frac{5}{2}}+4u^{\frac{3}{2}}+4u^{\frac{1}{2}}\right)du$$
$$=\frac{1}{\frac{7}{2}}u^{\frac{7}{2}}-\frac{4}{\frac{5}{2}}u^{\frac{5}{2}}+\frac{4}{\frac{3}{2}}u^{\frac{3}{2}}+C$$
$$=\frac{2(x-2)^{\frac{7}{2}}}{7}-\frac{8(x-2)^{\frac{5}{2}}}{5}+\frac{8(x-2)^{\frac{3}{2}}}{3}+C. \quad \text{Reverse Substitution}$$

71. $D(x)=\int\frac{-2000x}{\sqrt{25-x^2}}dx$

Let $u=25-x^2$, then $du=-2xdx.$
Note that $-2000=-2\cdot 1000.$

$$D(x)=\int\frac{1000}{\sqrt{25-x^2}}(-2xdx)$$
$$=\int\frac{1000}{\sqrt{u}}du \quad \text{Substituting: } \begin{matrix} u=25-x^2 \\ du=-2xdx \end{matrix}$$
$$=1000\int u^{-1/2}du$$
$$=1000\frac{u^{1/2}}{1/2}+C$$
$$=2000u^{1/2}+C$$
$$=2000\sqrt{25-x^2}+C.$$

We use the condition $D(3)=13{,}000$ to find C.

$$D(3)=13{,}000$$
$$2000\sqrt{25-(3)^2}+C=13{,}000$$
$$2000\sqrt{16}+C=13{,}000$$
$$8000+C=13{,}000$$
$$C=5000.$$

Thus, $D(x)=2000\sqrt{25-x^2}+5000.$

73. $\frac{dP}{dx}=\frac{9000-3000x}{(x^2-6x+10)^2}$

We integrate to find total profit.

$$P(x)=\int\frac{9000-3000x}{(x^2-6x+10)^2}dx$$

Let $u=x^2-6x+10$, then $du=(2x-6)dx$
Note that $9000-3000x=-1500(2x-6)$.

$$P(x)=\int\frac{-1500}{(x^2-6x+10)^2}\cdot(2x-6)dx$$
$$=\int\frac{-1500}{u^2}\cdot du \quad \text{Substitution } \begin{matrix} u=x^2-6x+10 \\ du=(2x-6)dx \end{matrix}$$
$$=-1500\int u^{-2}du$$
$$=1500u^{-1}+C$$
$$=1500(x^2-6x+10)^{-1}+C$$
$$=\frac{1500}{x^2-6x+10}+C.$$

We use the condition $P(3)=1500$ to find C.

$$P(3)=1500$$
$$\frac{1500}{(3)^2-6(3)+10}+C=1500$$
$$1500+C=1500$$
$$C=0.$$

Thus, the total profit function is

$$P(x)=\frac{1500}{x^2-6x+10}.$$

75. $\int_{-2}^{2}-x\sqrt{4-x^2}\,dx$

We notice that $-x\sqrt{4-x^2}\ge 0$, on $[-2,0]$ and $0\ge -x\sqrt{4-x^2}$, on $[0,2]$. We will divide the interval into two parts.
First on the interval $[-2,0]$, we find the indefinite integral at the top of the next page.

Copyright © 2014 Pearson Education, Inc. Publishing as Addison-Wesley.

Integrating, we have:

$$\int\left(-x\sqrt{4-x^2}-0\right)dx$$
$$=\int\left(-x\sqrt{4-x^2}\right)dx$$
$$=\frac{1}{2}\int\sqrt{4-x^2}\,(-2x)\,dx$$
$$=\frac{1}{2}\int\sqrt{u}\,du \qquad \text{Substitution: } \begin{matrix}u=4-x^2\\ du=-2x\,dx\end{matrix}$$
$$=\frac{1}{2}\int u^{1/2}du$$
$$=\frac{1}{3}u^{3/2}+C$$
$$=\frac{1}{3}\left(4-x^2\right)^{3/2}+C.$$

Next, we will evaluate the integral on $[-2,0]$.

$$\int_{-2}^{0}-x\sqrt{4-x^2}\,dx$$
$$=\left[\frac{1}{3}\left(4-x^2\right)^{3/2}\right]_{-2}^{0}$$
$$=\left[\left(\frac{1}{3}\left(4-(0)^2\right)^{3/2}\right)-\left(\frac{1}{3}\left(4-(-2)^2\right)^{3/2}\right)\right]$$
$$=\frac{1}{3}(4)^{3/2}-\frac{1}{3}(0)^{3/2}=\frac{8}{3}.$$

Next, on the interval $[0,2]$, we find the indefinite integral.

$$\int\left(0-\left(-x\sqrt{4-x^2}\right)\right)dx$$
$$=\int\left(x\sqrt{4-x^2}\right)dx$$
$$=-\frac{1}{2}\int\sqrt{4-x^2}\,(-2x)\,dx$$
$$=-\frac{1}{2}\int\sqrt{u}\,du \qquad \text{Substitution: } \begin{matrix}u=4-x^2\\ du=-2x\,dx\end{matrix}$$
$$=-\frac{1}{2}\int u^{1/2}du$$
$$=-\frac{1}{3}u^{3/2}+C$$
$$=-\frac{1}{3}\left(4-x^2\right)^{3/2}+C.$$

Next, we will evaluate the integral on $[0,2]$.

$$\int_{0}^{2}x\sqrt{4-x^2}\,dx$$
$$=\left[-\frac{1}{3}\left(4-x^2\right)^{3/2}\right]_{0}^{2}$$
$$=\left[\left(-\frac{1}{3}\left(4-(2)^2\right)^{3/2}\right)-\left(-\frac{1}{3}\left(4-(0)^2\right)^{3/2}\right)\right]$$
$$=-\frac{1}{3}(0)^{3/2}+\frac{1}{3}(4)^{3/2}$$
$$=\frac{8}{3}.$$

Therefore, the total area is:

$$\int_{-2}^{0}-x\sqrt{4-x^2}\,dx+\int_{0}^{2}x\sqrt{4-x^2}\,dx$$
$$=\frac{8}{3}+\frac{8}{3}$$
$$=\frac{16}{3}=5\tfrac{1}{3}.$$

77. $\int\frac{1}{ax+b}dx$

Let $u=ax+b$, then $du=a\,dx$. We do not have $a\,dx$. We only have dx and need to supply an a. We do this by multiplying by $\frac{1}{a}\cdot a$ as follows:

$$=\frac{1}{a}\int\frac{a\,dx}{ax+b}$$
$$=\frac{1}{a}\int\frac{du}{u} \qquad \text{Substitution: } \begin{matrix}u=ax+b\\ du=a\,dx\end{matrix}$$
$$=\frac{1}{a}\ln u+C$$
$$=\frac{1}{a}\ln(ax+b)+C. \qquad \text{Reverse Substitution}$$

79. $\int\frac{e^{\sqrt{t}}}{\sqrt{t}}dt$

Let $u=\sqrt{t}$, then $du=\frac{1}{2\sqrt{t}}dt$. We do not have $\frac{1}{2\sqrt{t}}dt$. We only have $\frac{1}{\sqrt{t}}dt$ and need to supply a $\frac{1}{2}$.

The solution is continued on the next page.

Copyright © 2014 Pearson Education, Inc. Publishing as Addison-Wesley.

Rewriting the integral on the previous page by multiplying $\frac{1}{2}\cdot 2$ we have:

$$\int \frac{e^{\sqrt{t}}}{\sqrt{t}}dt$$
$$= 2\int \frac{e^{\sqrt{t}}}{2\sqrt{t}}dt$$
$$= 2\int e^{u}du \qquad \text{Substitution: } \begin{matrix} u=\sqrt{t} \\ du=\frac{1}{2\sqrt{t}}dt \end{matrix}$$
$$= 2e^{u} + C$$
$$= 2e^{\sqrt{t}} + C. \qquad \text{Reverse Substitution}$$

81. $\int \frac{(\ln x)^{99}}{x}dx$

Let $u = \ln x$, then $du = \frac{1}{x}dx.$

$$= \int u^{99}du \qquad \text{Substitution: } \begin{matrix} u=\ln x \\ du=\frac{1}{x}dx \end{matrix}$$
$$= \frac{1}{100}u^{100} + C$$
$$= \frac{1}{100}(\ln x)^{100} + C. \qquad \text{Reverse Substitution}$$

83. $\int (e^{t}+2)e^{t}dt$

Let $u = e^{t}+2$, then $du = e^{t}dt.$

$$= \int udu \qquad \text{Substitution: } \begin{matrix} u=e^{t}+2 \\ du=e^{t}dt \end{matrix}$$
$$= \frac{1}{2}u^{2} + C$$
$$= \frac{1}{2}(e^{t}+2)^{2} + C. \qquad \text{Reverse Substitution}$$

85. $\int \frac{t^{2}dt}{\sqrt[4]{2+t^{3}}}dt$

Let $u = 2+t^{3}$, then $du = 3t^{2}dx.$ We do not have $3t^{2}dt$ We only have $t^{2}dt$ and need to supply a 3.

Rewriting the integral by multiplying $\frac{1}{3}\cdot 3$ we have:

$$\int \frac{t^{2}dt}{\sqrt[4]{2+t^{3}}}dt$$
$$= \frac{1}{3}\int \frac{3t^{2}dt}{\sqrt[4]{2+t^{3}}}dt$$
$$= \frac{1}{3}\int \frac{du}{\sqrt[4]{u}} \qquad \text{Substitution: } \begin{matrix} u=2+t^{3} \\ du=3t^{2}dt \end{matrix}$$
$$= \frac{1}{3}\int u^{-1/4}du$$
$$= \frac{4}{9}u^{3/4} + C$$
$$= \frac{4}{9}(2+t^{3})^{3/4} + C. \qquad \text{Reverse Substitution}$$

87. $\int \frac{\left[(\ln x)^{2}+3(\ln x)+4\right]}{x}dx$

Let $u = \ln x$, then $du = \frac{1}{x}dx.$

$$= \int \left[(u)^{2}+3(u)+4\right]du \qquad \text{Substitution: } \begin{matrix} u=\ln x \\ du=\frac{1}{x}dx \end{matrix}$$
$$= \frac{1}{3}u^{3} + \frac{3}{2}u^{2} + 4u + C$$
$$= \frac{1}{3}(\ln x)^{3} + \frac{3}{2}(\ln x)^{2} + 4\ln x + C.$$

89. $\int \frac{t^{3}\ln(t^{4}+8)}{t^{4}+8}dt$

Let $u = \ln(t^{4}+8)$, then $du = \frac{4t^{3}}{t^{4}+8}dt.$ We need to supply a 4, by multiplying by $\frac{1}{4}\cdot 4$

$$= \frac{1}{4}\int \frac{4t^{3}\ln(t^{4}+8)}{t^{4}+8}dt$$
$$= \frac{1}{4}\int udu \qquad \text{Substitution: } \begin{matrix} u=\ln(t^{4}+8) \\ du=\frac{4t^{3}}{t^{4}+8}dt \end{matrix}$$
$$= \frac{1}{8}u^{2} + C$$
$$= \frac{1}{8}\left[\ln(t^{4}+8)\right]^{2} + C.$$

Copyright © 2014 Pearson Education, Inc. Publishing as Addison-Wesley.

91. $\int \frac{x^2+6x}{(x+3)^2}dx$

$\int \frac{x^2+6x+9-9}{(x+3)^2}dx$ Adding 9–9 in the numerator.

$\int \left[\frac{x^2+6x+9}{(x+3)^2} - \frac{9}{(x+3)^2}\right]dx$

$= \int \left[\frac{(x+3)^2}{(x+3)^2} - \frac{9}{(x+3)^2}\right]dx$

$= \int \left[1 - \frac{9}{(x+3)^2}\right]dx$

$= \int dx - \int \frac{9}{(x+3)^2}dx$

Let $u = x+3$, then $du = dx$.

$= \int du - \int \frac{9}{u^2}du$ Substitution: $\begin{smallmatrix} u=x+3 \\ du=dx \end{smallmatrix}$

$= \int du - 9\int u^{-2}du + K$

$= u + 9u^{-1} + K$

$= x + 9(x+3)^{-1} + K + 3$

$= x + \frac{9}{x+3} + C.$ $[C = K+3]$

93. $\int \frac{t-5}{t-4}dt$

First, we divide algebraically to see

$$\begin{array}{r} 1 \\ t-4\overline{)t-5} \\ -(t-4) \\ \hline -1. \end{array}$$

Thus, $\frac{t-5}{t-4} = 1 - \frac{1}{t-4}$, and

$\int \frac{t-5}{t-4}dt$

$= \int \left[1 - \frac{1}{t-4}\right]dt$

$= \int dt - \int \frac{dt}{t-4}$

Let $u = t-4$, then $du = dt$.

$\int dt - \int \frac{dt}{t-4}$

$= \int du - \int \frac{du}{u}$ Substitution: $\begin{smallmatrix} u=t-4 \\ du=dt \end{smallmatrix}$

$= u - (\ln u + C_2)$

$= t - 4 - \ln(t-4) + K$

$= t - \ln(t-4) + C.$ $[C = K-4]$

95. $\int \frac{dx}{e^x+1} = \int \frac{e^{-x}}{1+e^{-x}}dx$

Let $u = 1+e^{-x}$, then $du = -e^{-x}dx$.

$= -\int \frac{-e^{-x}}{1+e^{-x}}dx$

$= -\int \frac{du}{u}$ Substitution: $\begin{smallmatrix} u=1+e^{-x} \\ du=-e^{-x}dx \end{smallmatrix}$

$= -\ln u + C$

$= -\ln(1+e^{-x}) + C.$

97. $\int \frac{(\ln x)^n}{x}dx, \quad n \neq -1$

Let $u = \ln x$, then $du = \frac{1}{x}dx$.

$= \int u^n du$ Substitution: $\begin{smallmatrix} u=\ln x \\ du=\frac{1}{x}dx \end{smallmatrix}$

$= \frac{1}{n+1}u^{n+1} + C$

$= \frac{1}{n+1}(\ln x)^{n+1} + C$

$= \frac{(\ln x)^{n+1}}{n+1} + C.$

99. $\int \frac{e^{-mx}}{1-ae^{-mx}}dx$

Let $u = 1-ae^{-mx}$, then $du = ame^{-mx}dx$.

$= \frac{1}{am}\int \frac{ame^{-mx}}{1-ae^{-mx}}dx$

$= \frac{1}{am}\int \frac{du}{u}$ Substitution: $\begin{smallmatrix} u=1-ae^{-mx} \\ du=ame^{-mx}dx \end{smallmatrix}$

$= \frac{1}{am}\ln u + C$

$= \frac{1}{am}\left[\ln\left(1-ae^{-mx}\right)\right] + C.$

Copyright © 2014 Pearson Education, Inc. Publishing as Addison-Wesley.

101. $\int 5x^2\left(2x^3-7\right)^n dx, \qquad n \neq -1$

$= 5\int x^2\left(2x^3-7\right)^n dx$

Let $u = 2x^3 - 7$, then $du = 6x^2 dx$. We do not have xdx. We only have xdx and need to supply a 6. We do this by multiplying by $\frac{1}{6}\cdot 6$ as follows:

$= \frac{1}{6}\cdot 6\cdot 5\int x^2\left(2x^3-7\right)^n dx$

$= \frac{5}{6}\int 6x^2\left(2x^3-7\right)^n dx$

$= \frac{5}{6}\int u^n\, du$ Substitution: $\begin{matrix} u=2x^3-7 \\ du=6x^2dx \end{matrix}$

$= \frac{5}{6}\cdot\frac{u^{n+1}}{n+1}+C$

$= \frac{5}{6(n+1)}\left(2x^3-7\right)^{n+1}+C.$

Copyright © 2014 Pearson Education, Inc. Publishing as Addison-Wesley.

Exercise Set 4.6

1. $\int 4xe^{4x}dx = \int x\left(4e^{4x}dx\right)$

Let

$u = x$ and $dv = 4e^{4x}dx$

Then

$du = dx$ and $v = e^{4x}$

Using the Integration-by-Parts Formula gives

$$\int \underset{u}{x}\left(\underset{dv}{4e^{4x}dx}\right) = \underset{u\cdot v}{x\cdot e^{4x}} - \int \underset{v\,du}{e^{4x}dx}$$

$$= xe^{4x} - \frac{1}{4}e^{4x} + C.$$

3. $\int x^3\left(3x^2\right)dx$

Let

$u = x^3$ and $dv = 3x^2dx$

Then

$du = 3x^2dx$ and $v = x^3$

Using the Integration-by-Parts Formula gives

$$\int \underset{u}{x^3}\left(\underset{dv}{3x^2dx}\right) = \underset{u}{x^3}\cdot \underset{v}{x^3} - \int \underset{v}{x^3}\underset{du}{3x^2dx}$$

$$= x^6 - \int 3x^5dx$$

$$= x^6 - 3\cdot\frac{x^6}{6} + C$$

$$= x^6 - \frac{x^6}{2} + C$$

$$= \frac{x^6}{2} + C.$$

It should be noted that this problem can also be worked with substitution or formula A.

Integrate using Substitution

$$\int x^3\left(3x^2\right)dx$$

$$= \int u\,du \quad \text{Substitution: } \begin{matrix} u=x^3 \\ du=3x^2dx \end{matrix}$$

$$= \frac{1}{2}u^2 + C$$

$$= \frac{1}{2}\left(x^3\right)^2 + C$$

$$= \frac{x^6}{2} + C.$$

Integrate using formula A.

$$\int x^3\left(3x^2\right)dx$$

$$= \int 3x^5dx$$

$$= 3\cdot\frac{x^6}{6} + C \quad \text{Using Formula A}$$

$$= \frac{x^6}{2} + C.$$

5. $\int xe^{5x}dx$

Let

$u = x$ and $dv = e^{5x}dx$

Then

$du = dx$ and $v = \frac{1}{5}e^{5x}$

We integrated dv using the formula

$$\int be^{ax}dx = \frac{b}{a}e^{ax} + C.$$

Using the Integration-by-Parts Formula gives

$$\int \underset{u}{x}\left(\underset{dv}{e^{5x}dx}\right) = \underset{u}{x}\cdot\underset{v}{\frac{1}{5}e^{5x}} - \int \underset{v}{\frac{1}{5}e^{5x}}\underset{du}{dx}$$

$$= x\cdot\frac{1}{5}e^{5x} - \frac{1}{5}\cdot\frac{1}{5}e^{5x} + C$$

$$\left(\int be^{ax}dx = \frac{b}{a}e^{ax} + C\right)$$

$$= \frac{1}{5}xe^{5x} - \frac{1}{25}e^{5x} + C.$$

7. $\int xe^{-2x}dx$

Let

$u = x$ and $dv = e^{-2x}dx$

Then

$du = dx$ and $v = -\frac{1}{2}e^{-2x}$

We integrated dv using the formula

$$\int be^{ax}dx = \frac{b}{a}e^{ax} + C.$$

Using the Integration-by-Parts Formula gives

$$\int \underset{u}{x}\left(\underset{dv}{e^{-2x}dx}\right) = \underset{u}{x}\cdot\underset{v}{\left(-\frac{1}{2}e^{-2x}\right)} - \int \underset{v}{\left(-\frac{1}{2}e^{-2x}\right)}\underset{du}{dx}.$$

The solution is continued on the next page.

Copyright © 2014 Pearson Education, Inc. Publishing as Addison-Wesley.

Continuing the integration process from the previous page, we have:

$$\int xe^{-2x}dx$$

$$=-\frac{1}{2}xe^{-2x}-\left(-\frac{1}{2}\right)\cdot\left(-\frac{1}{2}e^{-2x}\right)+C$$

$$\left(\int be^{ax}dx=\frac{b}{a}e^{ax}+C\right)$$

$$=-\frac{1}{2}xe^{-2x}-\frac{1}{4}e^{-2x}+C.$$

9. $\int x^2 \ln x\, dx$

Let

$u=\ln x$ and $dv=x^2dx$

Then

$du=\frac{1}{x}dx$ and $v=\frac{1}{3}x^3$

Using the Integration-by-Parts Formula gives

$$\int \underset{u}{\ln x}\underset{dv}{\left(x^2\,dx\right)}=\underset{u}{\ln x}\cdot\underset{v}{\left(\frac{1}{3}x^3\right)}-\int\underset{v}{\left(\frac{1}{3}x^3\right)}\underset{du}{\left(\frac{1}{x}dx\right)}$$

$$=\frac{1}{3}x^3\ln x-\int\frac{1}{3}x^2dx$$

$$=\frac{1}{3}x^3\ln x-\frac{1}{9}x^3+C$$

$$=\frac{x^3\ln x}{3}-\frac{x^3}{9}+C.$$

11. $\int x\ln\sqrt{x}\,dx=\int x\ln x^{1/2}\,dx$

Let

$u=\ln x^{1/2}$ and $dv=xdx$

Then

$du=\frac{1}{x^{1/2}}\cdot\frac{1}{2}x^{-1/2}dx=\frac{1}{2x}dx$ and $v=\frac{1}{2}x^2$

Using the Integration-by-Parts Formula gives

$$\int \underset{u}{\ln x^{1/2}}\underset{dv}{(xdx)}=\underset{u}{\ln x^{1/2}}\cdot\underset{v}{\left(\frac{x^2}{2}\right)}-\int\underset{v}{\left(\frac{x^2}{2}\right)}\underset{du}{\left(\frac{1}{2x}dx\right)}$$

$$=\frac{x^2\ln x^{1/2}}{2}-\int\frac{1}{4}xdx$$

$$=\frac{x^2\ln x^{1/2}}{2}-\frac{1}{8}x^2+C$$

$$=\frac{x^2\ln x}{4}-\frac{1}{8}x^2+C.\qquad\left(\ln x^{1/2}=\frac{1}{2}\ln x\right)$$

13. $\int\ln(x+5)dx$

Let

$u=\ln(x+5)$ and $dv=dx$

Then

$du=\frac{1}{x+5}dx$ and $v=x+5$

Choosing $x+5$ as an antiderivative of dv

Using the Integration-by-Parts Formula gives

$$\int \underset{u}{\ln(x+5)}\underset{dv}{dx}=$$

$$\underset{u}{\ln(x+5)}\cdot\underset{v}{(x+5)}-\int\underset{v}{(x+5)}\underset{du}{\left(\frac{1}{x+5}dx\right)}$$

$$=(x+5)\ln(x+5)-\int dx$$

$$=(x+5)\ln(x+5)-x+C.$$

15. $\int(x+2)\ln x\,dx$

Let

$u=\ln x$ and $dv=(x+2)dx$

Then

$du=\frac{1}{x}dx$ and $v=\frac{x^2}{2}+2x$

Using the Integration-by-Parts Formula gives

$$\int \underset{u}{\ln x}\cdot\underset{dv}{(x+2)dx}=$$

$$\underset{u}{\ln(x)}\cdot\underset{v}{\left(\frac{x^2}{2}+2x\right)}-\int\underset{v}{\left(\frac{x^2}{2}+2x\right)}\underset{du}{\left(\frac{1}{x}dx\right)}$$

$$=\left(\frac{x^2}{2}+2x\right)\ln x-\int\left(\frac{x}{2}+2\right)dx$$

$$=\left(\frac{x^2}{2}+2x\right)\ln x-\int\frac{x}{2}dx-\int 2dx$$

$$=\left(\frac{x^2}{2}+2x\right)\ln x-\frac{x^2}{4}-2x+C.$$

Copyright © 2014 Pearson Education, Inc. Publishing as Addison-Wesley.

17. $\int (x-1)\ln x\,dx$

Let

$u = \ln x$ and $dv = (x-1)\,dx$

Then

$du = \frac{1}{x}dx$ and $v = \frac{x^2}{2} - x$

Using the Integration-by-Parts Formula gives

$$\int \underbrace{\ln x}_{u} \cdot \underbrace{(x-1)\,dx}_{dv} =$$

$$\underbrace{\ln(x)}_{u} \cdot \underbrace{\left(\frac{x^2}{2} - x\right)}_{v} - \int \underbrace{\left(\frac{x^2}{2} - x\right)}_{v}\underbrace{\left(\frac{1}{x}dx\right)}_{du}$$

$$= \left(\frac{x^2}{2} - x\right)\ln x - \int \left(\frac{x}{2} - 1\right)dx$$

$$= \left(\frac{x^2}{2} - x\right)\ln x - \int \frac{x}{2}dx + \int 1dx$$

$$= \left(\frac{x^2}{2} - x\right)\ln x - \frac{x^2}{4} + x + C.$$

19. $\int x\sqrt{x+2}\,dx$

Let

$u = x$ and $dv = \sqrt{x+2}dx = (x+2)^{1/2}\,dx$

Then

$du = dx$ and $v = \frac{(x+2)^{3/2}}{3/2} = \frac{2}{3}(x+2)^{3/2}$

Using the Integration-by-Parts Formula gives

$$\int \underbrace{x}_{u}\underbrace{\left(\sqrt{x+2}dx\right)}_{dv} =$$

$$\underbrace{x}_{u} \cdot \underbrace{\frac{2}{3}(x+2)^{3/2}}_{v} - \int \underbrace{\frac{2}{3}(x+2)^{3/2}}_{v}\,\underbrace{dx}_{du}$$

$$= \frac{2}{3}x(x+2)^{3/2} - \frac{2}{3}\int (x+2)^{3/2}\,dx$$

$$= \frac{2}{3}x(x+2)^{3/2} - \frac{2}{3}\frac{(x+2)^{5/2}}{5/2} + C$$

$$= \frac{2}{3}x(x+2)^{3/2} - \frac{4}{15}(x+2)^{5/2} + C.$$

21. $\int x^3\ln(2x)\,dx$

Let

$u = \ln(2x)$ and $dv = x^3dx$

Then

$du = \frac{1}{2x}\cdot 2dx = \frac{1}{x}dx$ and $v = \frac{1}{4}x^4$

Using the Integration-by-Parts Formula gives

$$\int \underbrace{\ln(2x)}_{u}\underbrace{\left(x^3dx\right)}_{dv} = \underbrace{\ln(2x)}_{u} \cdot \underbrace{\left(\frac{x^4}{4}\right)}_{v} - \int \underbrace{\left(\frac{x^4}{4}\right)}_{v}\underbrace{\left(\frac{1}{x}dx\right)}_{du}$$

$$= \frac{x^4}{4}\ln(2x) - \int \frac{x^3}{4}dx$$

$$= \frac{x^4}{4}\ln(2x) - \frac{x^4}{16} + C.$$

23. $\int x^2e^x dx$

Let

$u = x^2$ and $dv = e^x dx$

Then

$du = 2xdx$ and $v = e^x$

Using the Integration-by-Parts Formula gives

$$\int \underbrace{x^2}_{u}\underbrace{\left(e^xdx\right)}_{dv} = \underbrace{x^2}_{u}\underbrace{e^x}_{v} - \int \underbrace{2xe^xdx}_{v\ du}.$$

We will use integration by parts again to evaluate $\int 2xe^xdx$.

Let

$u = 2x$ and $dv = e^xdx$

Then

$du = 2dx$ and $v = e^x$

Using the Integration-by-Parts Formula gives

$$\int \underbrace{2x}_{u}\underbrace{\left(e^xdx\right)}_{dv} = \underbrace{2x}_{u}\underbrace{e^x}_{v} - \int \underbrace{2e^xdx}_{v\ du}$$

$$= 2xe^x - 2e^x + C_1.$$

Thus,

$$\int x^2e^xdx = x^2e^x - \int 2xe^xdx$$

$$= x^2e^x - \left(2xe^x - 2e^x + C_1\right)$$

$$= x^2e^x - 2xe^x + 2e^x - C_1$$

$$= x^2e^x - 2xe^x + 2e^x + C. \quad [C = -C_1]$$

The solution is continued on the next page.

Copyright © 2014 Pearson Education, Inc. Publishing as Addison-Wesley.

Note, Since the integral on the previous page was in the form $\int f(x)g(x)dx$, where $f(x)$ can be differentiated repeatedly to a derivative that is eventually 0 and $g(x)$ can be integrated repeatedly easily, we can use tabular integration as shown below:

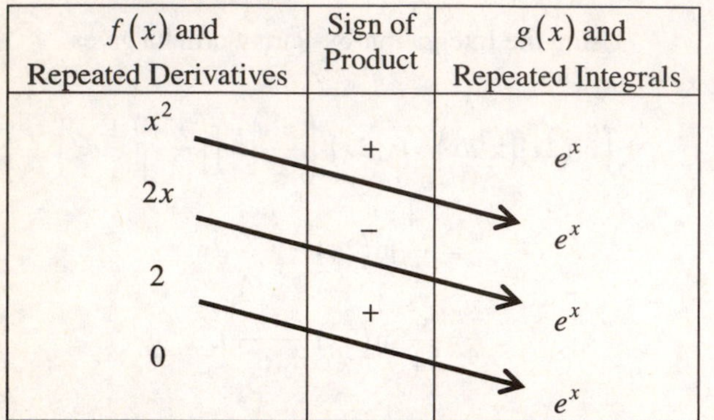

$f(x)$ and Repeated Derivatives	Sign of Product	$g(x)$ and Repeated Integrals
x^2	+	e^x
$2x$	−	e^x
2	+	e^x
0		e^x

We add the products along the arrows, making the alternate sign changes.

$$\int x^2e^x dx = x^2e^x - 2xe^x + 2e^x + C.$$

25. $\int x^2e^{2x}dx$

Let

$$u = x^2 \qquad \text{and} \qquad dv = e^{2x}dx$$

Then

$$du = 2xdx \text{ and } v = \frac{1}{2}e^{2x} \quad \left(\int be^{ax} = \frac{b}{a}e^{ax} + C\right)$$

Using the Integration-by-Parts Formula gives

$$\int \overset{u}{x^2}\left(\overset{dv}{e^{2x}dx}\right) = \overset{u}{x^2}\cdot\overset{v}{\frac{1}{2}e^{2x}} - \int\left(\overset{v}{\frac{1}{2}e^{2x}}\right)\cdot\overset{du}{2xdx}$$

$$\frac{1}{2}x^2e^{2x} - \int xe^{2x}dx.$$

We will use integration by parts again to evaluate $\int xe^{2x}dx$.

Let

$$u = x \qquad \text{and} \qquad dv = e^{2x}dx$$

Then

$$du = dx \text{ and } v = \frac{1}{2}e^{2x} \quad \left(\int be^{ax} = \frac{b}{a}e^{ax} + C\right)$$

Using the Integration-by-Parts Formula gives

$$\int \overset{u}{x}\left(\overset{dv}{e^{2x}dx}\right) = \overset{u}{x}\left(\overset{v}{\frac{1}{2}e^{2x}}\right) - \int\left(\overset{v}{\frac{1}{2}e^{2x}}\right)\overset{du}{dx}$$

$$= \frac{1}{2}xe^{2x} - \frac{1}{2}\cdot\frac{1}{2}e^{2x} + C_1$$

$$= \frac{1}{2}xe^{2x} - \frac{1}{4}e^{2x} + C_1.$$

Using this information, we have:

$$\int x^2\left(e^{2x}dx\right) = x^2\cdot\frac{1}{2}e^{2x} - \int\left(\frac{1}{2}e^{2x}\right)\cdot 2xdx$$

$$= x^2\cdot\frac{1}{2}e^{2x} - \int\left(xe^{2x}\right)dx$$

$$= \frac{1}{2}x^2e^{2x} - \left(\frac{1}{2}xe^{2x} - \frac{1}{4}e^{2x} + C_1\right)$$

$$= \frac{1}{2}x^2e^{2x} - \frac{1}{2}xe^{2x} + \frac{1}{4}e^{2x} - C_1$$

$$= \frac{1}{2}x^2e^{2x} - \frac{1}{2}xe^{2x} + \frac{1}{4}e^{2x} + C.$$

$$[C = -C_1]$$

Alternatively, we can also use tabular integration as shown below:

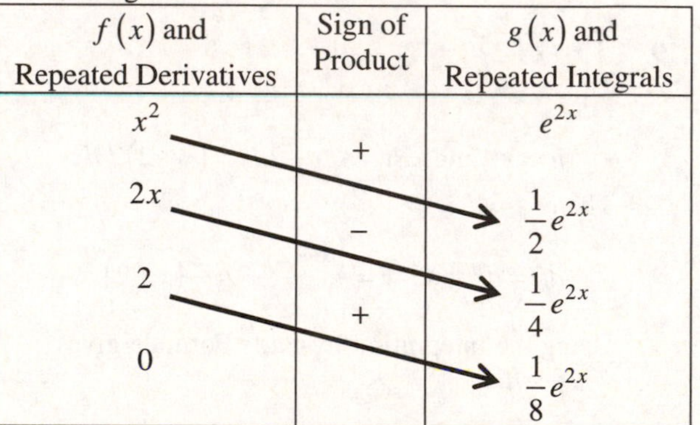

$f(x)$ and Repeated Derivatives	Sign of Product	$g(x)$ and Repeated Integrals
x^2	+	e^{2x}
$2x$	−	$\frac{1}{2}e^{2x}$
2	+	$\frac{1}{4}e^{2x}$
0		$\frac{1}{8}e^{2x}$

We add the products along the arrows, making the alternate sign changes.

$$\int x^2e^{2x}dx$$

$$= x^2\left(\frac{1}{2}e^{2x}\right) - 2x\left(\frac{1}{4}e^{2x}\right) + 2\left(\frac{1}{8}e^{2x}\right) + C$$

$$= \frac{1}{2}x^2e^{2x} - \frac{1}{2}xe^{2x} + \frac{1}{4}e^{2x} + C.$$

Copyright © 2014 Pearson Education, Inc. Publishing as Addison-Wesley.

27. $\int x^3 e^{-2x}dx$

This integral requires multiple applications of the Integration-by-Parts formula. However, since $f(x) = x^3$ is easily differentiable and to a derivative that is eventually 0 and $g(x) = e^{-2x}$ can be integrated easily using

$\int be^{ax} = \frac{b}{a}e^{ax} + C$ we will use tabular integration as shown to simplify our work.

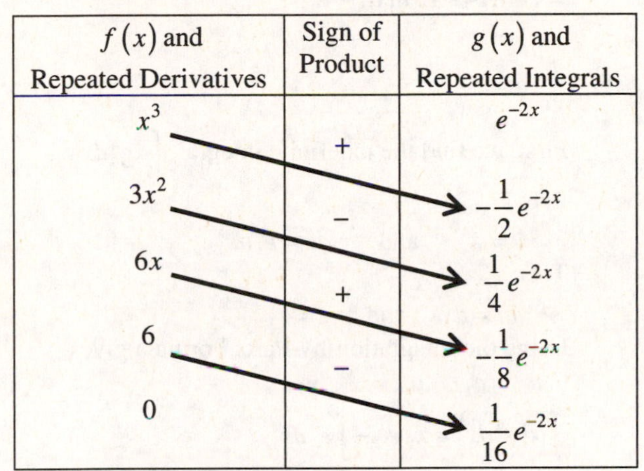

$f(x)$ and Repeated Derivatives	Sign of Product	$g(x)$ and Repeated Integrals
x^3	+	e^{-2x}
$3x^2$	−	$-\frac{1}{2}e^{-2x}$
$6x$	+	$\frac{1}{4}e^{-2x}$
6	−	$-\frac{1}{8}e^{-2x}$
0		$\frac{1}{16}e^{-2x}$

We add the products along the arrows, making the alternate sign changes to obtain

$$\int x^3 e^{-2x}dx$$

$$= x^3\left(-\frac{1}{2}e^{-2x}\right) - 3x^2\left(\frac{1}{4}e^{-2x}\right) + 6x\left(-\frac{1}{8}e^{-2x}\right) - 6\left(\frac{1}{16}e^{-2x}\right) + C$$

$$= -\frac{1}{2}x^3e^{-2x} - \frac{3}{4}x^2e^{-2x} - \frac{3}{4}xe^{-2x} - \frac{3}{8}e^{-2x} + C.$$

29. $\int (x^4 + 4)e^{3x}dx$

This integral requires multiple applications of the Integration-by-Parts formula. However, since $f(x) = x^4 + 4$ is easily differentiable and to a derivative that is eventually 0 and $g(x) = e^{3x}$ can be integrated easily using

$\int be^{ax} = \frac{b}{a}e^{ax} + C$ we will use tabular integration as shown to simplify our work.

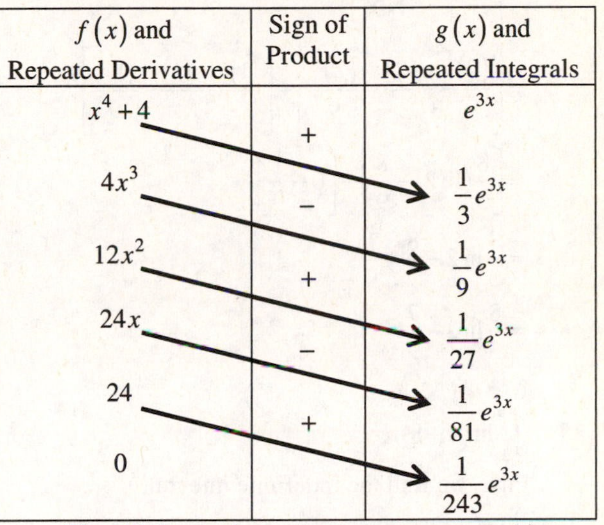

$f(x)$ and Repeated Derivatives	Sign of Product	$g(x)$ and Repeated Integrals
$x^4 + 4$	+	e^{3x}
$4x^3$	−	$\frac{1}{3}e^{3x}$
$12x^2$	+	$\frac{1}{9}e^{3x}$
$24x$	−	$\frac{1}{27}e^{3x}$
24	+	$\frac{1}{81}e^{3x}$
0		$\frac{1}{243}e^{3x}$

We add the products along the arrows, making the alternate sign changes to obtain

$$\int (x^4 + 4)e^{3x}dx$$

$$= (x^4 + 4)\left(\frac{1}{3}e^{3x}\right) - 4x^3\left(\frac{1}{9}e^{3x}\right) + 12x^2\left(\frac{1}{27}e^{3x}\right) - 24x\left(\frac{1}{81}e^{3x}\right) + 24\left(\frac{1}{243}e^{3x}\right) + C$$

$$= \frac{1}{3}(x^4 + 4)e^{3x} - \frac{4}{9}x^3e^{3x} + \frac{4}{9}x^2e^{3x} - \frac{8}{27}xe^{3x} + \frac{8}{81}e^{3x} + C.$$

Copyright © 2014 Pearson Education, Inc. Publishing as Addison-Wesley.

31. $\int_1^2 x^2 \ln x dx$

In Exercise 9 we found the indefinite integral

$\int x^2 \ln x dx = \frac{x^3 \ln x}{3} - \frac{x^3}{9} + C$.

We use the indefinite integral to evaluate the definite integral as follows:

$$\int_1^2 x^2 \ln x dx$$
$$= \left[\frac{x^3 \ln x}{3} - \frac{x^3}{9}\right]_1^2 \quad \text{Use } C = 0.$$
$$= \left[\frac{(2)^3 \ln(2)}{3} - \frac{(2)^3}{9}\right] - \left[\frac{(1)^3 \ln(1)}{3} - \frac{(1)^3}{9}\right]$$
$$= \left[\frac{8}{3}(\ln 2) - \frac{8}{9}\right] - \left[0 - \frac{1}{9}\right]$$
$$= \frac{8}{3}\ln 2 - \frac{8}{9} + \frac{1}{9}$$
$$= \frac{8}{3}\ln 2 - \frac{7}{9}.$$

33. $\int_2^6 \ln(x+8) dx$

First, we find the indefinite integral.

$\int \ln(x+8) dx$

Let

$u = \ln(x+8)$ and $dv = dx$

Then

$du = \frac{1}{x+8} dx$ and $v = x+8$

Choosing $x+8$ as an antiderivative of dv

Using the Integration-by-Parts Formula gives

$$\int \underset{u}{\ln(x+8)}\, \underset{dv}{dx} =$$
$$\underset{u}{\ln(x+8)} \cdot \underset{v}{(x+8)} - \int \underset{v}{(x+8)} \underset{du}{\left(\frac{1}{x+8} dx\right)}$$
$$= (x+8)\ln(x+8) - \int dx$$
$$= (x+8)\ln(x+8) - x + C.$$

We now use the indefinite integral to evaluate the definite integral. To simplify our work, we choose the constant of integration C to be 0.

Using the indefinite integral, we calculate the definite integral as follows:

$$\int_2^6 \ln(x+8) dx$$
$$= \left[(x+8)\ln(x+8) - x\right]_2^6$$
$$= \left[((6)+8)\ln((6)+8) - (6)\right] - \left[((2)+8)\ln((2)+8) - (2)\right]$$
$$= [14\ln 14 - 6] - [10\ln 10 - 2]$$
$$= 14\ln 14 - 10\ln 10 - 4.$$

35. $\int_0^1 xe^x dx$

First, we find the indefinite integral $\int xe^x dx$.

Let

$u = x$ and $dv = e^x dx$

Then

$du = dx$ and $v = e^x$

Using the Integration-by-Parts Formula gives

$$\int \underset{u}{x} \underset{dv}{(e^x dx)} = \underset{u}{x} \cdot \underset{v}{e^x} - \int \underset{v}{e^x} \underset{du}{dx}$$
$$= xe^x - e^x + C.$$

We now use the indefinite integral to evaluate the definite integral. To simplify our work, we choose the constant of integration C to be 0.

$$\int_0^1 xe^x dx$$
$$= \left[xe^x - e^x\right]_0^1$$
$$= \left[(1)e^{(1)} - e^{(1)}\right] - \left[(0)e^{(0)} - e^{(0)}\right]$$
$$= [e - e] - [0 - 1]$$
$$= 1.$$

37. $\int_0^8 x\sqrt{x+1} dx$

First we find the indefinite integral using integration by parts. Let

$u = x$ and $dv = \sqrt{x+1} dx = (x+1)^{1/2} dx$

Then

$du = dx$ and $v = \frac{(x+1)^{3/2}}{3/2} = \frac{2}{3}(x+1)^{3/2}$

The solution is continued on the next page.

Copyright © 2014 Pearson Education, Inc. Publishing as Addison-Wesley.

Using the Integration-by-Parts Formula gives

$$\int \underset{u}{x}\left(\underset{dv}{\sqrt{x+1}dx}\right) =$$

$$\underset{u}{x}\cdot\underset{v}{\frac{2}{3}(x+1)^{3/2}} - \int \underset{v}{\frac{2}{3}(x+1)^{3/2}}\,\underset{du}{dx}$$

$$= \frac{2}{3}x(x+1)^{3/2} - \frac{2}{3}\int (x+1)^{3/2}\,dx$$

$$= \frac{2}{3}x(x+1)^{3/2} - \frac{2}{3}\frac{(x+1)^{5/2}}{5/2} + C$$

$$= \frac{2}{3}x(x+1)^{3/2} - \frac{4}{15}(x+1)^{5/2} + C.$$

Evaluating the definite integral, we have:

$$\int_0^8 x\sqrt{x+1}\,dx$$

$$= \left[\frac{2}{3}x(x+1)^{3/2} - \frac{4}{15}(x+1)^{5/2}\right]_0^8$$

$$= \left[\frac{2}{3}(8)((8)+1)^{3/2} - \frac{4}{15}((8)+1)^{5/2}\right] - \left[\frac{2}{3}(0)((0)+1)^{3/2} - \frac{4}{15}((0)+1)^{5/2}\right]$$

$$= \frac{396}{5} - \left[-\frac{4}{15}\right]$$

$$= \frac{1192}{15}.$$

39. $C(x) = \int C'(x)\,dx = \int 4x\sqrt{x+3}dx$

Let

$$u = 4x \quad\text{and}\quad dv = \sqrt{x+3}dx = (x+3)^{1/2}\,dx$$

Then

$$du = 4dx \quad\text{and}\quad v = \frac{(x+3)^{3/2}}{3/2} = \frac{2}{3}(x+3)^{3/2}$$

Using the Integration-by-Parts Formula gives

$$\int 4x\left(\sqrt{x+3dx}\right)$$

$$= 4x\cdot\frac{2}{3}(x+3)^{3/2} - \int \frac{2}{3}(x+3)^{3/2}\cdot 4dx$$

$$= \frac{8}{3}x(x+3)^{3/2} - \frac{8}{3}\int (x+3)^{3/2}\,dx$$

$$= \frac{8}{3}x(x+3)^{3/2} - \frac{8}{3}\frac{(x+3)^{5/2}}{5/2} + K$$ Use K to avoid confusion with cost.

$$= \frac{8}{3}x(x+3)^{3/2} - \frac{16}{15}(x+3)^{5/2} + K.$$

Next, we use the condition $C(13) = 1126.40$ to find the constant of integration K.

$$C(13) = 1126.40$$

$$\frac{8}{3}(13)((13)+3)^{3/2} - \frac{16}{15}((13)+3)^{5/2} + K = 1126.40$$

$$\frac{104}{3}\cdot(16)^{3/2} - \frac{16}{15}(16)^{5/2} + K = 1126.40$$

$$\frac{104}{3}(64) - \frac{16}{15}(1024) + K = 1126.40$$

$$\frac{6656}{3} - \frac{16{,}384}{15} + K = 1126.40$$

$$\frac{16{,}896}{15} + K = 1126.40$$

$$1126.40 + K = 1126.40$$

$$K = 0$$

Therefore, the total cost function is

$$C(x) = \frac{8}{3}x(x+3)^{3/2} - \frac{16}{15}(x+3)^{5/2}.$$

41. a) $K(t) = 10te^{-t}$

The number of kilowatt hours the family uses in the first T hours of the day is given by $\int_0^T K(t)\,dt$. We find the indefinite integral first.

$$\int K(t)\,dt = \int 10te^{-t}dt$$

Let

$u = 10t$ and $dv = e^{-t}dt$

Then

$du = 10dt$ and $v = -e^{-t}$

$$\int 10te^{-t}dt = 10t\cdot\left(-e^{-t}\right) - \int -e^{-t}\cdot 10dt$$

$$= -10te^{-t} + 10\int e^{-t}dt$$

$$= -10te^{-t} - 10e^{-t} + C.$$

The solution is continued on the next page.

Copyright © 2014 Pearson Education, Inc. Publishing as Addison-Wesley.

Next, we evaluate the definite integral.

$$\int_0^T K(t)\,dt$$
$$=\left[-10te^{-t}-10e^{-t}\right]_0^T$$
$$=\left[-10Te^{-T}-10e^{-T}\right]-\left[-10(0)e^{-0}-10e^{-0}\right]$$
$$=-10Te^{-T}-10e^{-T}+10.$$

b) Substitute 4 for T.

$$\int_0^4 K(t)\,dt=-10(4)e^{-(4)}-10e^{-(4)}+10$$
$$=-40e^{-4}-10e^{-4}+10$$
$$\approx 9.084$$

The family uses approximately 9.084 kW-h during the first 4 hours of the day.

43. $\int x\sqrt{5x+1}\,dx$

Assuming $5x+1\geq 0$ Let $u=5x+1$, then $du=5dx$. Observe that $x=\frac{u-1}{5}$.

Making the appropriate substitutions we have

$$\int x\sqrt{5x+1}\,dx$$
$$=\frac{1}{5}\int x\sqrt{5x+1}(5)\,dx$$
$$=\frac{1}{5}\int\left(\frac{u-1}{5}\right)\sqrt{u}\,du$$
$$=\frac{1}{25}\int(u-1)u^{\frac{1}{2}}\,du$$
$$=\frac{1}{25}\int\left(u^{\frac{3}{2}}-u^{\frac{1}{2}}\right)du$$
$$=\frac{1}{25}\left(\frac{1}{\frac{5}{2}}u^{\frac{5}{2}}-\frac{1}{\frac{3}{2}}u^{\frac{3}{2}}+C\right)$$
$$=\frac{2(5x+1)^{\frac{5}{2}}}{125}-\frac{2(5x+1)^{\frac{3}{2}}}{75}+C. \quad \text{Reverse Substitution}$$

It is possible to simplify both answers to the form $\frac{(30x-4)(5x+1)^{\frac{3}{2}}}{375}$. Yes, both answers are the same.

45. $\int e^{\sqrt{x}}\,dx$

Let $u=\sqrt{x}$, note that $x=u^2$ so that $dx=2udu$.

Making substitutions, we have:

$$\int e^{\sqrt{x}}\,dx=\int 2ue^u\,du.$$

Using integration by parts we have:

$$\int 2ue^u\,du=2\int u\left(e^u\,dx\right)$$

Let

$w=u$ and $dv=e^u\,du$

Then

$dw=du$ and $v=e^u$

Using the Integration-by-Parts Formula gives

$$2\int \overset{w}{u}\left(\overset{dv}{e^u\,dx}\right)=2\overset{w\cdot v}{u\cdot e^u}-2\int \overset{v\ dw}{e^u\,du}$$
$$=2ue^u-2e^u+C$$

We now substitute for u to get:

$$\int e^{\sqrt{x}}\,dx=2ue^u-2e^u+C$$
$$=2\sqrt{x}e^{\sqrt{x}}-2e^{\sqrt{x}}+C.$$

47. $\int \sqrt{x}\ln x\,dx=\int x^{1/2}\ln x\,dx$

Let

$u=\ln x$ and $dv=x^{1/2}dx$

Then

$du=\frac{1}{x}dx$ and $v=\frac{2}{3}x^{3/2}$

Using the Integration-by-Parts Formula gives

$$\int \ln x\left(x^{1/2}\,dx\right)$$
$$=\ln x\cdot\left(\frac{2}{3}x^{3/2}\right)-\int\left(\frac{2}{3}x^{3/2}\right)\left(\frac{1}{x}dx\right)$$
$$=\frac{2}{3}x^{3/2}\ln x-\frac{2}{3}\int x^{1/2}dx$$
$$=\frac{2}{3}x^{3/2}\ln x-\frac{2}{3}\left(\frac{2}{3}x^{3/2}\right)+C$$
$$=\frac{2}{3}x^{3/2}\ln x-\frac{4}{9}x^{3/2}+C.$$

49. $\int \frac{\ln x}{\sqrt{x}}\,dx=\int x^{-1/2}\ln x\,dx$

Let

$u=\ln x$ and $dv=x^{-1/2}dx$

Then

$du=\frac{1}{x}dx$ and $v=2x^{1/2}$

The solution is continued on the next page.

Copyright © 2014 Pearson Education, Inc. Publishing as Addison-Wesley.

Using the Integration-by-Parts Formula gives

$$\int \ln x\left(x^{-1/2}\,dx\right)$$

$$= \ln x \cdot \left(2x^{1/2}\right) - \int \left(2x^{1/2}\right)\left(\frac{1}{x}\,dx\right)$$

$$= 2x^{1/2}\ln x - 2\int x^{-1/2}dx$$

$$= 2x^{1/2}\ln x - 2\left(2x^{1/2}\right) + C$$

$$= 2x^{1/2}\ln x - 4x^{1/2} + C$$

$$= 2\sqrt{x}\,(\ln x) - 4\sqrt{x} + C.$$

51. $\int \left(27x^3+83x-2\right)\sqrt[6]{3x+8}\,dx$

$$= \left(27x^3+83x-2\right)\left(3x+8\right)^{1/6}\,dx$$

We will use tabular integration as shown to simplify our work.

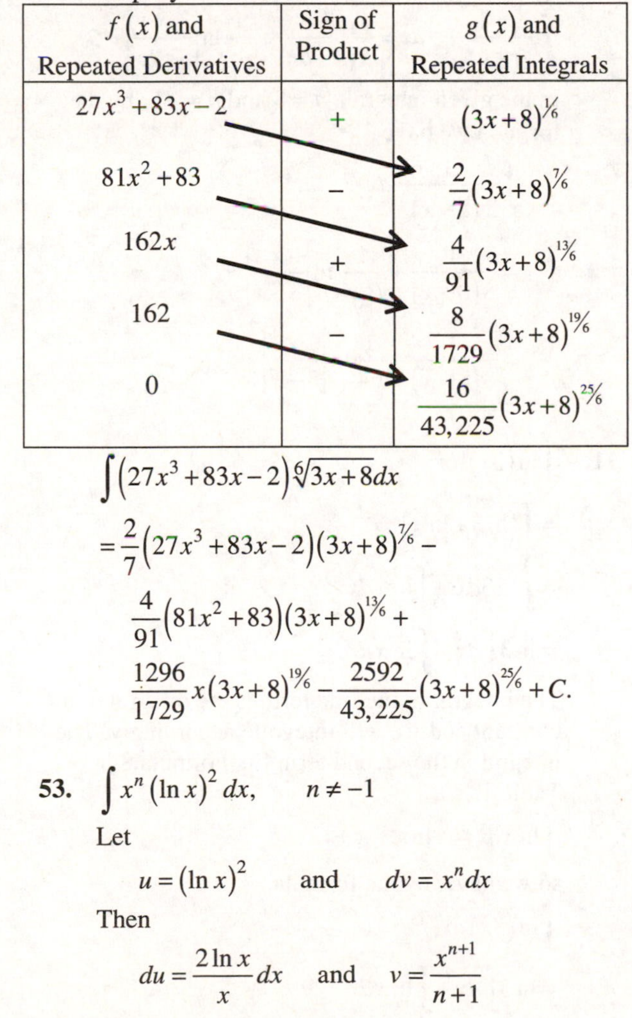

$f(x)$ and Repeated Derivatives	Sign of Product	$g(x)$ and Repeated Integrals
$27x^3+83x-2$	+	$(3x+8)^{1/6}$
$81x^2+83$	−	$\frac{2}{7}(3x+8)^{7/6}$
$162x$	+	$\frac{4}{91}(3x+8)^{13/6}$
162	−	$\frac{8}{1729}(3x+8)^{19/6}$
0		$\frac{16}{43,225}(3x+8)^{25/6}$

$$\int \left(27x^3+83x-2\right)\sqrt[6]{3x+8}\,dx$$

$$= \frac{2}{7}\left(27x^3+83x-2\right)\left(3x+8\right)^{7/6} -$$

$$\frac{4}{91}\left(81x^2+83\right)\left(3x+8\right)^{13/6} +$$

$$\frac{1296}{1729}x\left(3x+8\right)^{19/6} - \frac{2592}{43,225}\left(3x+8\right)^{25/6} + C.$$

53. $\int x^n (\ln x)^2\,dx, \qquad n \neq -1$

Let

$$u = (\ln x)^2 \quad \text{and} \quad dv = x^n dx$$

Then

$$du = \frac{2\ln x}{x}dx \quad \text{and} \quad v = \frac{x^{n+1}}{n+1}$$

$$\int x^n (\ln x)^2\,dx = \frac{x^{n+1}(\ln x)^2}{n+1} - \int \frac{2x^n \ln x}{n+1}dx$$

$$= \frac{x^{n+1}(\ln x)^2}{n+1} - \frac{2}{n+1}\int x^n \ln x dx$$

Use integration by parts again to evaluate $\int x^n \ln x dx$.

Let

$$u = \ln x \quad \text{and} \quad dv = x^n dx$$

Then

$$du = \frac{1}{x}dx \quad \text{and} \quad v = \frac{x^{n+1}}{n+1}$$

$$\int x^n \ln x dx = \frac{x^{n+1}\ln x}{n+1} - \int \frac{x^n}{n+1}dx$$

$$= \frac{x^{n+1}\ln x}{n+1} - \frac{x^{n+1}}{(n+1)^2} + K.$$

Therefore,

$$\int x^n (\ln x)^2\,dx =$$

$$= \frac{x^{n+1}(\ln x)^2}{n+1} - \frac{2}{n+1}\left(\frac{x^{n+1}\ln x}{n+1} - \frac{x^{n+1}}{(n+1)^2} + K\right)$$

$$= \frac{x^{n+1}}{n+1}(\ln x)^2 - \frac{2x^{n+1}}{(n+1)^2}(\ln x) + \frac{2x^{n+1}}{(n+1)^3} + C.$$

$$\left(C = \frac{-2K}{n+1}\right)$$

55. $\int x^n e^x dx$

Let

$$u = x^n \quad \text{and} \quad dv = e^x dx$$

Then

$$du = nx^{n-1}dx \quad \text{and} \quad v = e^x$$

$$\int x^n e^x dx = x^n e^x - \int e^x\left(nx^{n-1}\right)dx$$

$$= x^n e^x - n\int x^{n-1}e^x dx.$$

57. ✎

59. Using the fnInt feature on a calculator, we find that $\int_1^{10} x^5 \ln x dx \approx 355{,}986.$

Copyright © 2014 Pearson Education, Inc. Publishing as Addison-Wesley.

Exercise Set 4.7

1. $\int xe^{-3x}dx$

This integral fits Formula 6 in Table 1.

$$\int xe^{ax}dx = \frac{1}{a^2}\cdot e^{ax}(ax-1)+C$$

In the given integral, $a=-3$, by the formula we have

$$\int xe^{-3x}dx = \frac{1}{(-3)^2}\cdot e^{(-3)x}((-3)x-1)+C$$
$$=\frac{1}{9}e^{-3x}(-3x-1)+C$$
$$=-\frac{1}{9}e^{-3x}(3x+1)+C.$$

3. $\int 6^x dx$

This integral fits Formula 11 in Table 1.

$$\int a^x dx = \frac{a^x}{\ln a}+C,\quad a>0, a\neq 1$$

In the given integral, $a=6$, by the formula we have

$$\int 6^x dx = \frac{6^x}{\ln 6}+C.$$

5. $\int \frac{1}{25-x^2}dx$

This integral fits Formula 15 in Table 1.

$$\int \frac{1}{a^2-x^2}dx = \frac{1}{2a}\ln\left|\frac{a+x}{a-x}\right|+C$$

In the given integral, $a^2=25$, or $a=5$, by the formula we have

$$\int \frac{1}{25-x^2}dx = \int \frac{1}{5^2-x^2}dx$$
$$=\frac{1}{2\cdot 5}\ln\left|\frac{5+x}{5-x}\right|+C$$
$$=\frac{1}{10}\ln\left|\frac{5+x}{5-x}\right|+C.$$

7. $\int \frac{x}{3-x}dx$

This integral fits Formula 18 in Table 1.

$$\int \frac{x}{a+bx}dx = \frac{a}{b^2}+\frac{x}{b}-\frac{a}{b^2}\ln|a+bx|+C$$

In the given integral, $a=3$ and $b=-1$, by the formula we have

$$\int \frac{x}{3+(-1)x}dx$$
$$=\frac{3}{(-1)^2}+\frac{x}{(-1)}-\frac{3}{(-1)^2}\ln|3-x|+C$$
$$=3-x-3\ln|3-x|+C.$$

9. $\int \frac{1}{x(8-x)^2}dx$

This integral fits Formula 21 in Table 1.

$$\int \frac{1}{x(a+bx)^2}dx = \frac{1}{a(a+bx)}+\frac{1}{a^2}\ln\left|\frac{x}{a+bx}\right|+C$$

In the given integral, $a=8$ and $b=-1$, by the formula we have

$$\int \frac{1}{x(8-x)^2}dx$$
$$=\frac{1}{8(8-x)}+\frac{1}{(8)^2}\ln\left|\frac{x}{8-x}\right|+C$$
$$=\frac{1}{8(8-x)}+\frac{1}{64}\ln\left|\frac{x}{8-x}\right|+C.$$

11. $\int \ln(3x)dx$

$$=\int(\ln 3+\ln x)dx$$
$$=\int \ln 3dx+\int \ln xdx$$
$$=\ln 3\int dx+\int \ln xdx$$

The integral in the first term is the integral of a constant and we will integrate accordingly. The integral in the second term fits Formula 8 in Table 1.

$$\int \ln xdx = x\ln x - x+C$$

so we have, by the formula,

$$\int \ln(3x)dx$$
$$=\ln 3\int dx+\int \ln xdx$$
$$=\ln 3\cdot x+C_1+x\ln x-x+C_2$$
$$=(\ln 3)x+x\ln x-x+C. \qquad (C=C_1+C_2)$$

Copyright © 2014 Pearson Education, Inc. Publishing as Addison-Wesley.

13. $\int x^4 \ln x\,dx$

This integral fits Formula 10 in Table 1.

$$\int x^n \ln x\,dx = x^{n+1}\left[\frac{\ln x}{n+1} - \frac{1}{(n+1)^2}\right] + C$$

In the given integral, $n = 4$, by the formula we have

$$\int x^4 \ln x\,dx = x^{4+1}\left[\frac{\ln x}{4+1} - \frac{1}{(4+1)^2}\right] + C$$

$$= x^5\left[\frac{\ln x}{5} - \frac{1}{(5)^2}\right] + C$$

$$= \frac{x^5}{5}(\ln x) - \frac{x^5}{25} + C.$$

15. $\int x^3 \ln x\,dx$

This integral fits Formula 10 in Table 1.

$$\int x^n \ln x\,dx = x^{n+1}\left[\frac{\ln x}{n+1} - \frac{1}{(n+1)^2}\right] + C$$

In the given integral, $n = 3$, by the formula we have

$$\int x^3 \ln x\,dx = x^{3+1}\left[\frac{\ln x}{3+1} - \frac{1}{(3+1)^2}\right] + C$$

$$= x^4\left[\frac{\ln x}{4} - \frac{1}{(4)^2}\right] + C$$

$$= \frac{x^4}{4}(\ln x) - \frac{x^4}{16} + C.$$

17. $\int \frac{dx}{\sqrt{x^2+7}}$

This integral fits Formula 12 in Table 1.

$$\int \frac{1}{\sqrt{x^2+a^2}}\,dx = \ln\left|x + \sqrt{x^2+a^2}\right| + C$$

In the given integral, $a^2 = 7$, by the formula we have

$$\int \frac{1}{\sqrt{x^2+7}}\,dx = \ln\left|x + \sqrt{x^2+7}\right| + C.$$

19. $\int \frac{10dx}{x(5-7x)^2} = 10\int \frac{dx}{x(5-7x)^2}$

This integral fits Formula 21 in Table 1.

$$\int \frac{1}{x(a+bx)^2}\,dx = \frac{1}{a(a+bx)} + \frac{1}{a^2}\ln\left|\frac{x}{a+bx}\right| + C$$

In the given integral, $a = 5$ and $b = -7$, by the formula we have

$$10\int \frac{1}{x(5-7x)^2}\,dx$$

$$= 10\left[\frac{1}{5(5-7x)} + \frac{1}{(5)^2}\ln\left|\frac{x}{5-7x}\right|\right] + C$$

$$= \frac{2}{5-7x} + \frac{2}{5}\ln\left|\frac{x}{5-7x}\right| + C.$$

21. $\int \frac{-5}{4x^2-1}\,dx = -5\int \frac{1}{4x^2-1}\,dx$

This integral almost fits Formula 14 in table 1.

$$\int \frac{1}{x^2-a^2}\,dx = \frac{1}{2a}\ln\left|\frac{x-a}{x+a}\right| + C$$

However, the x^2 coefficient needs to be 1. We factor out 4 as follows and then we apply formula 14.

$$-5\int \frac{1}{4x^2-1}\,dx = -5\int \frac{1}{4\left(x^2-\frac{1}{4}\right)}\,dx$$

$$= -\frac{5}{4}\int \frac{1}{x^2-\frac{1}{4}}\,dx$$

In the given integral, $a^2 = \frac{1}{4}$ and $a = \frac{1}{2}$, by the formula we have

$$-\frac{5}{4}\int \frac{1}{x^2-\frac{1}{4}}\,dx = -\frac{5}{4}\left[\frac{1}{2\left(\frac{1}{2}\right)}\ln\left|\frac{x-\frac{1}{2}}{x+\frac{1}{2}}\right|\right] + C$$

$$= -\frac{5}{4}\ln\left|\frac{x-\frac{1}{2}}{x+\frac{1}{2}}\right| + C.$$

23. $\int \sqrt{4m^2+16}\,dm$

This integral almost fits Formula 22 in table 1.

$$\int \sqrt{x^2+a^2}\,dx$$

$$= \frac{1}{2}\left[x\sqrt{x^2+a^2} + a^2\ln\left|x + \sqrt{x^2+a^2}\right|\right] + C$$

However, the m^2 coefficient needs to be 1. We factor out 4 as follows and then we apply Formula 22 at the top of the next page.

Copyright © 2014 Pearson Education, Inc. Publishing as Addison-Wesley.

Applying Formula 22, we have:

$$\int\sqrt{4\left(m^2+4\right)}dm=\int 2\sqrt{m^2+4}dm$$
$$=2\int\sqrt{m^2+4}dm.$$

In the given integral, $a^2=4$ and $a=2$, by the formula we have

$$2\int\sqrt{m^2+4}\,dm$$
$$=2\cdot\frac{1}{2}\left[m\sqrt{m^2+4}+4\ln\left|m+\sqrt{m^2+4}\right|\right]+C$$
$$=m\sqrt{m^2+4}+4\ln\left|m+\sqrt{m^2+4}\right|+C.$$

25. $\int\frac{-5\ln x}{x^3}dx=-5\int x^{-3}\ln x dx$

This integral fits Formula 10 in Table 1.

$$\int x^n\ln x dx=x^{n+1}\left[\frac{\ln x}{n+1}-\frac{1}{(n+1)^2}\right]+C$$

In the given integral, $n=-3$, by the formula we have

$$-5\int x^{-3}\ln x dx$$
$$=-5\cdot x^{-3+1}\left[\frac{\ln x}{-3+1}-\frac{1}{(-3+1)^2}\right]+C$$
$$=-5\cdot x^{-2}\left[\frac{\ln x}{-2}-\frac{1}{(-2)^2}\right]+C$$
$$=\frac{5}{2x^2}(\ln x)+\frac{5}{4x^2}+C.$$

27. $\int\frac{e^x}{x^{-3}}dx=\int x^3e^x dx$

This integral fits Formula 7 in Table 1.

$$\int x^ne^{ax}dx=\frac{x^ne^{ax}}{a}-\frac{n}{a}\int x^{n-1}e^{ax}dx+C$$

In the given integral, $n=3$ and $a=1$, by the formula we have

$$\int x^3e^x dx=\frac{x^3e^x}{1}-\frac{3}{1}\int x^{3-1}e^x dx+C$$
$$=x^3e^x-3\int x^2e^x dx+C.$$

We will apply Formula 7 again, this time with $n=2$ and $a=1$

$$x^3e^x-3\left[\frac{x^2e^x}{1}-\frac{2}{1}\int x^{2-1}e^x dx\right]+C$$
$$=x^3e^x-3x^2e^x+6\int xe^x dx+C.$$

Now we apply Formula 6

$$\int xe^{ax}dx=\frac{1}{a^2}\cdot e^{ax}(ax-1)+C$$

with $a=1$

$$=x^3e^x-3x^2e^x+6\left[\frac{1}{(1)^2}\cdot e^x(x-1)\right]+C$$
$$=x^3e^x-3x^2e^x+6xe^x-6e^x+C.$$

29. $\int x\sqrt{1+2x}\,dx$

This integral fits Formula 23 in Table 1.

$$\int x\sqrt{a+bx}\,dx=\frac{2}{15b^2}(3bx-2a)(a+bx)^{3/2}+C$$

In the given integral, $a=1$ and $b=2$, by the formula we have

$$\int x\sqrt{1+2x}\,dx$$
$$=\frac{2}{15(2)^2}(3(2)x-2(1))(1+2x)^{3/2}+C$$
$$=\frac{2}{60}(6x-2)(1+2x)^{3/2}+C$$
$$=\frac{1}{30}\cdot 2(3x-1)(1+2x)^{3/2}+C$$
$$=\frac{1}{15}(3x-1)(1+2x)^{3/2}+C.$$

31. $S(x)=\int S'(x)dx$

$$=\int\frac{100x}{(20-x)^2}dx=100\int\frac{x}{(20-x)^2}dx$$

This integral fits Formula 19 in Table 1.

$$\int\frac{x}{(a+bx)^2}dx=\frac{a}{b^2(a+bx)}+\frac{1}{b^2}\ln|a+bx|+C$$

In the given integral, $a=20$ and $b=-1$. We apply the formula at the top of the next page.

Copyright © 2014 Pearson Education, Inc. Publishing as Addison-Wesley.

Applying Formula 19, we have:

$$100\int \frac{x}{(20-x)^2}dx$$

$$=100\left[\frac{20}{(-1)^2(20-x)}+\frac{1}{(-1)^2}\ln|20-x|\right]+C$$

$$=\frac{2000}{(20-x)}+100\ln|20-x|+C.$$

We use the condition $S(19)=2000$ to determine C.

$$S(19)=2000$$

$$\frac{2000}{(20-(19))}+100\ln|20-(19)|+C=2000$$

$$\frac{2000}{1}+100\ln|1|+C=2000$$

$$2000+C=2000$$

$$C=0$$

Thus, the supply function is

$$S(x)=\frac{2000}{(20-x)}+100\ln|20-x|$$

$$=100\left[\frac{20}{20-x}+\ln|20-x|\right].$$

33. $\int \frac{8}{3x^2-2x}dx=8\int \frac{1}{x(-2+3x)}dx$

This integral fits Formula 20 in Table 1.

$$\int \frac{1}{x(a+bx)}dx=\frac{1}{a}\ln\left|\frac{x}{a+bx}\right|+C$$

In the given integral, $a=-2$ and $b=3$, by the formula we have

$$8\int \frac{1}{x(-2+3x)}dx=8\cdot\frac{1}{(-2)}\ln\left|\frac{x}{-2+3x}\right|+C$$

$$=-4\ln\left|\frac{x}{-2+3x}\right|+C.$$

35. $\int \frac{dx}{x^3-4x^2+4x}$

$$=\int \frac{dx}{x(x^2-4x+4)}$$

$$=\int \frac{1}{x(x-2)^2}dx$$

$$=\int \frac{1}{x(-2+x)^2}dx$$

This integral fits Formula 21 in Table 1.

$$\int \frac{1}{x(a+bx)^2}dx=\frac{1}{a(a+bx)}+\frac{1}{a^2}\ln\left|\frac{x}{a+bx}\right|+C$$

In the given integral, $a=-2$ and $b=1$, by the formula we have

$$\int \frac{1}{x(-2+x)^2}dx$$

$$=\frac{1}{-2(-2+x)}+\frac{1}{(-2)^2}\ln\left|\frac{x}{-2+x}\right|+C$$

$$=\frac{1}{-2(x-2)}+\frac{1}{4}\ln\left|\frac{x}{x-2}\right|+C$$

$$=\frac{-1}{2(x-2)}+\frac{1}{4}\ln\left|\frac{x}{x-2}\right|+C.$$

37. $\int \frac{-e^{-2x}dx}{9-6e^{-x}+e^{-2x}}dx=\int \frac{e^{-x}\left(-e^{-x}\right)dx}{\left(-3+e^{-x}\right)^2}$

We substitute $u=e^{-x}$ and $du=-e^{-x}dx$, and get

$$\int \frac{u}{(-3+u)^2}du.$$

Which fits Formula 19 in Table 1.

$$\int \frac{x}{(a+bx)^2}dx=\frac{a}{b^2(a+bx)}+\frac{1}{b^2}\ln|a+bx|+C$$

In the given integral, we have $u=x$, $a=-3$ and $b=1$, by the formula we have

$$\int \frac{e^{-x}\left(-e^{-x}\right)dx}{\left(-3+e^{-x}\right)^2}=\int \frac{u}{(-3+u)^2}du$$

$$=\frac{-3}{1^2(-3+u)}+\frac{1}{1^2}\ln|-3+u|+C$$

$$=\frac{-3}{1^2\left(-3+e^{-x}\right)}+\frac{1}{1^2}\ln\left|-3+e^{-x}\right|+C \quad \text{Substituting for } u$$

$$=\frac{-3}{e^{-x}-3}+\ln\left|e^{-x}-3\right|+C.$$

Copyright © 2014 Pearson Education, Inc. Publishing as Addison-Wesley.

Chapter 5

Applications of Integration

Exercise Set 5.1

1. $D(x)=-\frac{5}{6}x+9,\ S(x)=\frac{1}{2}x+1$

a) To find the equilibrium point we set $D(x)=S(x)$ and solve.

$$-\frac{5}{6}x+9=\frac{1}{2}x+1$$
$$9-1=\frac{1}{2}x+\frac{5}{6}x$$
$$8=\frac{4}{3}x \qquad \left(\frac{1}{2}+\frac{5}{6}=\frac{8}{6}=\frac{4}{3}\right)$$
$$\frac{3}{4}\cdot 8=x \qquad \text{Multiplying by } \frac{3}{4}$$
$$6=x$$

Thus $x_E=6$ units. To find p_E we substitute x_E into $D(x)$ or $S(x)$. Here we use $D(x)$.

$$\begin{aligned}p_E&=D(x_E)\\&=D(6)\\&=-\frac{5}{6}(6)+9\\&=-5+9\\&=4\end{aligned}$$

When 6 units are sold the equilibrium price is \$4; therefore, the equilibrium point is $(6,\$4)$.

Notice, we could have used $S(x)$ to find the equilibrium point as well. Substituting, we have:

$$\begin{aligned}p_E&=S(x_E)\\&=S(6)\\&=\frac{1}{2}(6)+1\\&=3+1\\&=4.\end{aligned}$$

This results in the same equilibrium point $(6,\$4)$.

b) The consumer surplus is

$$\int_0^{x_E} D(x)dx-x_Ep_E.$$

Substituting $-\frac{5}{6}x+9$ for $D(x)$, 6 for x_E, and 4 for p_E we have:

$$\begin{aligned}&\int_0^6\left(-\frac{5}{6}x+9\right)dx-6\cdot 4\\&=\left[-\frac{5x^2}{12}+9x\right]_0^6-24\\&=\left[\left(-\frac{5(6)^2}{12}+9\cdot 6\right)-\left(-\frac{5(0)^2}{12}+9\cdot 0\right)\right]-24\\&=[(-15+54)-(0)]-24\\&=39-24\\&=15.\end{aligned}$$

The consumer surplus at the equilibrium point is \$15.

c) The producer surplus is

$$x_Ep_E-\int_0^{x_E} S(x)dx.$$

Substituting $\frac{1}{2}x+1$ for $S(x)$, 6 for x_E, and 4 for p_E we have:

$$\begin{aligned}&6\cdot 4-\int_0^6\left(\frac{1}{2}x+1\right)dx\\&=24-\left[\frac{x^2}{4}+x\right]_0^6\\&=24-\left[\left(\frac{6^2}{4}+6\right)-\left(0^2+0\right)\right]\\&=24-\left[\frac{36}{4}+6\right]\\&=24-[9+6]\\&=24-15\\&=9.\end{aligned}$$

The producer surplus at the equilibrium point is \$9.

Copyright © 2014 Pearson Education, Inc. Publishing as Addison-Wesley.

3. $D(x)=(x-4)^2,\ S(x)=x^2+2x+6$

a) To find the equilibrium point we set $D(x)=S(x)$ and solve.

$$(x-4)^2=x^2+2x+6$$
$$x^2-8x+16=x^2+2x+6$$
$$-8x+16=2x+6$$
$$10=10x$$
$$1=x$$

Thus $x_E=1$ unit. To find p_E we substitute x_E into $D(x)$ or $S(x)$.

Here we use $D(x)$.

$$p_E=D(x_E)$$
$$=D(1)$$
$$=((1)-4)^2$$
$$=(-3)^2$$
$$=9$$

When 1 unit is sold the equilibrium price is \$9; therefore, the equilibrium point is $(1,\$9)$.

b) The consumer surplus is

$\int_0^{x_E} D(x)dx-x_E p_E$.

Substituting $(x-4)^2$ for $D(x)$, 1 for x_E, and 9 for p_E we have:

$$\int_0^1 (x-4)^2\,dx-1\cdot 9$$
$$=\int_0^1\left(x^2-8x+16\right)dx-9$$
$$=\left[\frac{x^3}{3}-4x^2+16x\right]_0^1-9$$
$$=\left[\left(\frac{1}{3}-4+16\right)-(0-0+0)\right]-9$$
$$=\frac{37}{3}-9$$
$$=\frac{10}{3}\approx 3.33.$$

The consumer surplus at the equilibrium point is \$3.33.

c) The producer surplus is

$x_E p_E-\int_0^{x_E} S(x)dx$.

Substituting x^2+2x+6 for $S(x)$, 1 for x_E, and 9 for p_E we have:

$$1\cdot 9-\int_0^1\left(x^2+2x+6\right)dx$$
$$=9-\left[\frac{x^3}{3}+x^2+6x\right]_0^1$$
$$=9-\left[\left(\frac{1}{3}+1+6\right)-(0+0+0)\right]$$
$$=9-\left[\frac{22}{3}\right]=\frac{5}{3}\approx 1.67$$

The producer surplus at the equilibrium point is \$1.67.

5. $D(x)=(x-6)^2,\ S(x)=x^2$

a) To find the equilibrium point we set $D(x)=S(x)$ and solve.

$$(x-6)^2=x^2$$
$$x^2-12x+36=x^2$$
$$-12x+36=0$$
$$36=12x$$
$$3=x$$

Thus $x_E=3$ units. To find p_E we substitute x_E into $D(x)$ or $S(x)$. Here we use $S(x)$.

$$p_E=S(x_E)=S(3)=(3)^2=9$$

When 3 units are sold the equilibrium price is \$9; therefore, the equilibrium point is $(3,\$9)$.

b) The consumer surplus is

$\int_0^{x_E} D(x)dx-x_E p_E$.

Substituting $(x-6)^2$ for $D(x)$, 3 for x_E, and 9 for p_E we have:

$$\int_0^3 (x-6)^2\,dx-3\cdot 9$$
$$=\int_0^3\left(x^2-12x+36\right)dx-27$$
$$=\left[\frac{x^3}{3}-6x^2+36x\right]_0^3-27$$
$$=\left[\left(\frac{3^3}{3}-6\cdot 3^2+36\cdot 3\right)-(0-0+0)\right]-27$$
$$=[(9-54+108)-(0)]-27$$
$$=63-27=36.$$

The consumer surplus at the equilibrium point is \$36.

Copyright © 2014 Pearson Education, Inc. Publishing as Addison-Wesley.

c) The producer surplus is

$x_E p_E - \int_0^{x_E} S(x)\,dx$.

Substituting x^2 for $S(x)$, 3 for x_E, and 9 for p_E we have:

$$3\cdot 9-\int_0^3 x^2dx$$
$$=27-\left[\frac{x^3}{3}\right]_0^3$$
$$=27-\left[\left(\frac{3^3}{3}\right)-\left(\frac{0^3}{3}\right)\right]$$
$$=27-[9]$$
$$=18.$$

The producer surplus at the equilibrium point is \$18.

7. $D(x)=1000-10x,\ \ S(x)=250+5x$

a) To find the equilibrium point we set $D(x)=S(x)$ and solve.

$$1000-10x=250+5x$$
$$750=15x$$
$$50=x$$

Thus $x_E=50$ units. To find p_E we substitute x_E into $D(x)$ or $S(x)$. Here we use $D(x)$.

$$p_E=D(x_E)$$
$$=D(50)$$
$$=1000-10\cdot 50$$
$$=500$$

When 50 units are sold the equilibrium price is \$500; therefore, the equilibrium point is $(50,\$500)$.

b) The consumer surplus is

$\int_0^{x_E} D(x)\,dx - x_E p_E$.

Substituting $1000-10x$ for $D(x)$, 50 for x_E, and 500 for p_E we have:

$$\int_0^{50}(1000-10x)\,dx-50\cdot 500$$
$$=\left[1000x-5x^2\right]_0^{50}-25{,}000$$
$$=\left[\left(1000\cdot 50-5(50)^2\right)-\left(1000\cdot 0-5\cdot 0^2\right)\right]$$
$$-25{,}000$$
$$=[50{,}000-12{,}500]-25{,}000$$
$$=12{,}500.$$

The consumer surplus at the equilibrium point is \$12,500.

c) The producer surplus is

$x_E p_E - \int_0^{x_E} S(x)\,dx$.

Substituting $250+5x$ for $S(x)$, 50 for x_E, and 500 for p_E we have:

$$50\cdot 500-\int_0^{50}(250+5x)\,dx$$
$$=25{,}000-\left[250x+\frac{5x^2}{2}\right]_0^{50}$$
$$=25{,}000-$$
$$\left[\left(250\cdot 50+\frac{5(50)^2}{2}\right)-\left(250\cdot 0-\frac{5(0)^2}{2}\right)\right]$$
$$=25{,}000-[12{,}500+6250]$$
$$=6250.$$

The producer surplus at the equilibrium point is \$6250.

9. $D(x)=5-x$, for $0\le x\le 5$;

$S(x)=\sqrt{x+7}$

a) To find the equilibrium point we set $D(x)=S(x)$ and solve.

$$5-x=\sqrt{x+7}$$
$$(5-x)^2=\left(\sqrt{x+7}\right)^2$$
$$25-10x+x^2=x+7$$
$$x^2-11x+18=0$$
$$(x-2)(x-9)=0$$
$$x-2=0 \quad\text{or}\quad x-9=0$$
$$x=2 \quad\text{or}\quad x=9$$

Note, $x=9$ is not a solution to the equation. Only $x=2$ is in the domain of $D(x)$, thus $x_E=2$ units.

The solution is continued on the next page.

Copyright © 2014 Pearson Education, Inc. Publishing as Addison-Wesley.

To find p_E we substitute x_E into $D(x)$ or $S(x)$. Here we use $D(x)$.

$p_E = D(x_E) = D(2) = 5 - 2 = 3$

When 2 units are sold the equilibrium price is $3; therefore, the equilibrium point is $(2, \$3)$.

b) The consumer surplus is

$\int_0^{x_E} D(x)\,dx - x_E p_E$.

Substituting $5 - x$ for $D(x)$, 2 for x_E, and 3 for p_E we have:

$$\int_0^2 (5-x)\,dx - 2\cdot 3$$
$$= \left[5x - \frac{x^2}{2}\right]_0^2 - 6$$
$$= \left[\left(5\cdot 2 - \frac{(2)^2}{2}\right) - \left(5\cdot 0 - \frac{(0)^2}{0}\right)\right] - 6$$
$$= [10 - 2] - 6$$
$$= 2.$$

The consumer surplus at the equilibrium point is $2.

c) The producer surplus is

$x_E p_E - \int_0^{x_E} S(x)\,dx$.

Substituting $\sqrt{7+x}$ for $S(x)$, 2 for x_E, and 3 for p_E we have:

$$2\cdot 3 - \int_0^2 \left(\sqrt{x+7}\right)dx$$
$$= 6 - \int_0^2 (x+7)^{1/2}\,dx$$
$$= 6 - \left[\frac{2}{3}(x+7)^{3/2}\right]_0^2$$
$$= 6 - \left[\left(\frac{2}{3}(2+7)^{3/2}\right) - \left(\frac{2}{3}(0+7)^{3/2}\right)\right]$$
$$= 6 - \left[\frac{2}{3}(9)^{3/2} - \frac{2}{3}(7)^{3/2}\right]$$
$$= 6 - \left[\frac{2}{3}\cdot(27) - \frac{2}{3}(7)^{3/2}\right]$$
$$\approx 0.35. \qquad \text{Using a calculator}$$

The producer surplus at the equilibrium point is $0.35.

11. $D(x) = \dfrac{100}{\sqrt{x}}, \quad S(x) = \sqrt{x}$

a) To find the equilibrium point we set $D(x) = S(x)$ and solve.

$$\frac{100}{\sqrt{x}} = \sqrt{x}$$
$$100 = \sqrt{x}\cdot\sqrt{x}$$
$$100 = x$$

Thus $x_E = 100$ units. To find p_E we substitute x_E into $D(x)$ or $S(x)$. Here we use $S(x)$.

$p_E = S(x_E) = S(100) = \sqrt{100} = 10$

When 100 units are sold the equilibrium price is $10; therefore, the equilibrium point is $(100, \$10)$.

b) The consumer surplus is

$\int_0^{x_E} D(x)\,dx - x_E p_E$.

Substituting $\dfrac{100}{\sqrt{x}} = 100x^{-1/2}$ for $D(x)$, 100 for x_E, and 10 for p_E we have:

$$\int_0^{100} \left(100x^{-1/2}\right)dx - 100\cdot 10$$
$$= \left[\frac{100x^{1/2}}{\frac{1}{2}}\right]_0^{100} - 1000$$
$$= \left[\left(200(100)^{1/2}\right) - \left(200(0)^{1/2}\right)\right] - 1000$$
$$= [2000] - 1000$$
$$= 1000.$$

The consumer surplus at the equilibrium point is $1000.

c) The producer surplus is

$x_E p_E - \int_0^{x_E} S(x)\,dx$.

Substituting $\sqrt{x}$ for $S(x)$, 100 for x_E, and 10 for p_E we calculate the producer surplus on the next page.

Copyright © 2014 Pearson Education, Inc. Publishing as Addison-Wesley.

The producer surplus is:

$$100\cdot 10-\int_0^{100}\left(x^{1/2}\right)dx$$
$$=1000-\left[\frac{2}{3}(x)^{3/2}\right]_0^{100}$$
$$=1000-\left[\left(\frac{2}{3}(100)^{3/2}\right)-\left(\frac{2}{3}(0)^{3/2}\right)\right]$$
$$=1000-\left[\frac{2000}{3}-0\right]$$
$$=1000-\frac{2000}{3}$$
$$=\frac{1000}{3}\approx 333.33.$$

The producer surplus at the equilibrium point is \$333.33.

13. $D(x)=(x-4)^2,\quad S(x)=x^2+2x+8$

a) To find the equilibrium point we set $D(x)=S(x)$ and solve.

$$(x-4)^2=x^2+2x+8$$
$$x^2-8x+16=x^2+2x+8$$
$$-8x+16=2x+8$$
$$8=10x$$
$$\frac{8}{10}=x$$
$$\frac{4}{5}=x$$

Thus $x_E=\frac{4}{5}=0.8$ units. Assuming partial units are acceptable, to find p_E we substitute x_E into $D(x)$ or $S(x)$. Here we use $D(x)$.

$$p_E=D(x_E)$$
$$=D\left(\frac{4}{5}\right)$$
$$=\left(\frac{4}{5}-4\right)^2$$
$$=\left(-\frac{16}{5}\right)^2$$
$$=\frac{256}{25}=10.24$$

When 0.8 units are sold the equilibrium price is \$10.24; therefore, the equilibrium point is $(0.8,\$10.24)$.

b) The consumer surplus is

$$\int_0^{x_E}D(x)dx-x_Ep_E.$$

Substituting $(x-4)^2$ for $D(x)$, 0.8 for x_E, and 10.24 for p_E we have:

$$\int_0^{0.8}(x-4)^2\,dx-0.8\cdot 10.24$$
$$=\left[\frac{(x-4)^3}{3}\right]_0^{0.8}-8.192$$
$$=\left[\left(\frac{(0.8-4)^3}{3}\right)-\left(\frac{(0-4)^3}{3}\right)\right]-8.192$$
$$=\left[\left(\frac{-32.768}{3}\right)-\left(\frac{-64}{3}\right)\right]-8.192$$
$$\approx 2.22. \qquad \text{Using a calculator}$$

The consumer surplus at the equilibrium point is \$2.22.

c) The producer surplus is

$$x_Ep_E-\int_0^{x_E}S(x)dx.$$

Substituting x^2+2x+8 for $S(x)$, 0.8 for x_E, and 10.24 for p_E we have:

$$0.8\cdot 10.24-\int_0^{0.8}\left(x^2+2x+8\right)dx$$
$$=8.192-\left[\frac{x^3}{3}+x^2+8x\right]_0^{0.8}$$
$$=8.192-\left[\left(\frac{(0.8)^3}{3}+(0.8)^2+8(0.8)\right)-(0)\right]$$
$$\approx 8.192-7.21066667$$
$$\approx 0.98.$$

The producer surplus at the equilibrium point is \$0.98.

15. $D(x)=e^{-x+4.5},\quad S(x)=e^{x-5.5}$

a) To find the equilibrium point we set $D(x)=S(x)$ and solve.

$$e^{-x+4.5}=e^{x-5.5}$$
$$\ln\left(e^{-x+4.5}\right)=\ln\left(e^{x-5.5}\right)$$
$$-x+4.5=x-5.5$$
$$10=2x$$
$$5=x$$

Thus $x_E=5$ units.

The solution is continued on the next page.

Copyright © 2014 Pearson Education, Inc. Publishing as Addison-Wesley.

To find p_E we substitute x_E into $D(x)$ or $S(x)$.

Here we use $D(x)$.

$$p_E = D(x_E)$$
$$= D(5) = e^{-5+4.5} = e^{-0.5} \approx 0.61$$

When 5 units are sold the equilibrium price is \$0.61; therefore, the equilibrium point is $(5, \$0.61)$.

b) The consumer surplus is

$$\int_0^{x_E} D(x)\,dx - x_E p_E.$$

Substituting $e^{-x+4.5}$ for $D(x)$, 5 for x_E, and 0.61 for p_E we have:

$$\int_0^5 e^{-x+4.5}dx - 5\cdot 0.61$$
$$= \left[-e^{-x+4.5}\right]_0^5 - 3.05$$
$$= \left[\left(-e^{-5+4.5}\right) - \left(-e^{-0+4.5}\right)\right] - 3.05$$
$$= \left[-e^{-0.5} + e^{4.5}\right] - 3.05$$
$$\approx 86.36. \quad \text{Using a calculator}$$

The consumer surplus at the equilibrium point is \$86.36.

c) The producer surplus is

$$x_E p_E - \int_0^{x_E} S(x)\,dx.$$

Substituting $e^{x-5.5}$ for $S(x)$, 5 for x_E, and 0.61 for p_E we have:

$$5\cdot 0.61 - \int_0^5 \left(e^{x-5.5}\right)dx$$
$$= 3.05 - \left[e^{x-5.5}\right]_0^5$$
$$= 3.05 - \left[\left(e^{5-5.5}\right) - \left(e^{0-5.5}\right)\right]$$
$$= 3.05 - \left[e^{-0.5} - e^{-5.5}\right]$$
$$\approx 2.45.$$

The producer surplus at the equilibrium point is \$2.45.

17. ✎

19. $D(x) = \dfrac{x+8}{x+1}$, $\quad S(x) = \dfrac{x^2+4}{20}$

a) Graphing the equations and using the INTERSECT feature we have:

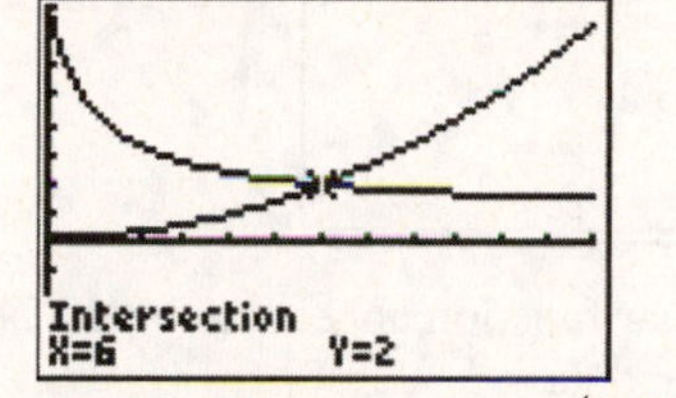

The equilibrium point is $(6, \$2)$.

b)

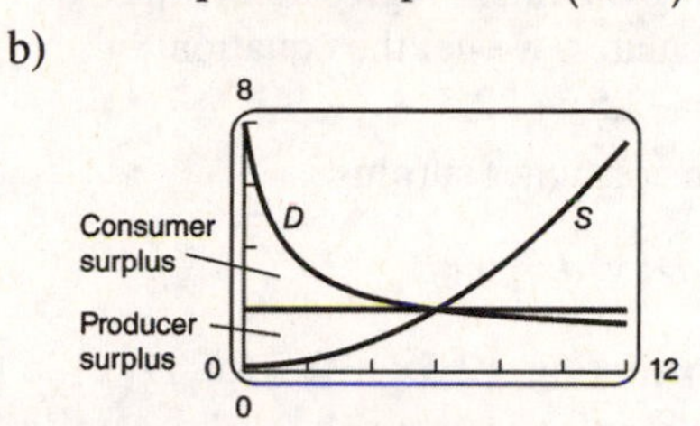

c) The consumer surplus is

$$\int_0^{x_E} D(x)\,dx - x_E p_E.$$

Substituting $\dfrac{x+8}{x+1}$ for $D(x)$, 6 for x_E, and 2 for p_E we have:

$$\int_0^6 \left(\frac{x+8}{x+1}\right)dx - 6\cdot 2$$
$$\approx 19.62137104 - 12 \quad \text{Using a calculator}$$
$$\approx 7.62.$$

The consumer surplus at the equilibrium point is \$7.62.

d) The producer surplus is

$$x_E p_E - \int_0^{x_E} S(x)\,dx.$$

Substituting $\dfrac{x^2+4}{20}$ for $S(x)$, 6 for x_E, and 2 for p_E we have:

$$6\cdot 2 - \int_0^6 \left(\frac{x^2+4}{20}\right)dx$$
$$= 12 - \left[\frac{x^3}{60} + \frac{x}{5}\right]_0^6$$
$$= 12 - \frac{24}{5}$$
$$= \frac{36}{5} = 7.20.$$

The producer surplus at the equilibrium point is \$7.20.

Copyright © 2014 Pearson Education, Inc. Publishing as Addison-Wesley.

21. a) Entering the data into the STAT editor on the calculator and plotting the points, we get the following scatter plot:

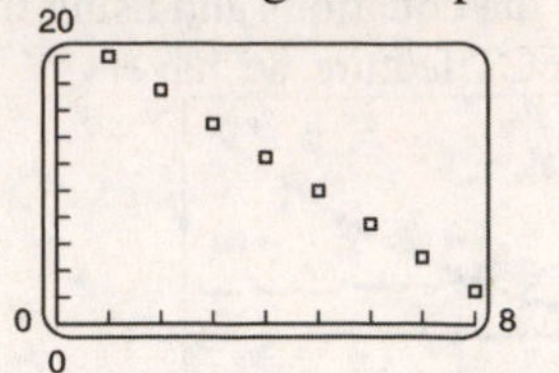

A linear function appears to be the best fit for the data.

b) Using the linear regression feature on the calculator, we get the equation:
$y = -2.5x + 22.5$.

c) The consumer surplus is
$\int_0^x D(x)\,dx - x \cdot p$.
Substituting $-2.5x + 22.5$ for $D(x)$, 6 for x, and 7.50 for p we have:
$\int_0^6 (-2.5x + 22.5)\,dx - 6(7.50) = 45.$
The consumer surplus is \$45.

d) First find the value for x when y=11.50.
$11.50 = -2.5x + 22.5$
$-11 = -2.5x$
$4.4 = x$
Next, find the consumer surplus.
$\int_0^{4.4} (-2.5x + 22.5)\,dx - 4.4(11.50) = 24.20$
When the price of bungee jumping is \$11.50 per half hour, Reggie's consumer surplus is \$24.20.

Copyright © 2014 Pearson Education, Inc. Publishing as Addison-Wesley.

Exercise Set 5.2

1. $P(t)=P_0e^{kt}$

Substituting 6 for t, 100,000 for P_0 and 0.03 for k we have:

$$\begin{aligned}P(6)&=100{,}000e^{0.03(6)}\\&=100{,}000e^{0.18}\\&\approx 100{,}000(1.19721736312)\\&\approx 119{,}721.74\end{aligned}$$

The future value of $100,000 after 6 years is about $119,721.74.

3. $P(t)=P_0e^{kt}$

Substituting 9 for t, 140,000 for P_0 and 0.058 for k we have:

$$\begin{aligned}P(9)&=140{,}000e^{0.058(9)}\\&=140{,}000e^{0.522}\\&\approx 140{,}000(1.68539507)\\&\approx 235{,}955.31\end{aligned}$$

The future value of $140,000 after 9 years is about $235,955.31.

5. $P(t)=P_0e^{kt}$

Substituting 6 for t, 100,000 for P and 0.03 for k we have:

$$\begin{aligned}100{,}000&=P_0e^{0.03(6)}\\&=P_0e^{0.18}\\\frac{100{,}000}{e^{0.18}}&\approx P_0\\83{,}527.02&\approx P_0\end{aligned}$$

The present value of $100,000 due 6 years in the future at 3% interest is about $83,527.02.

7. $P(t)=P_0e^{kt}$

Substituting 25 for t, 1,000,000 for P and 0.07 for k we have:

$$\begin{aligned}1{,}000{,}000&=P_0e^{0.07(25)}\\&=P_0e^{1.75}\\\frac{1{,}000{,}000}{e^{1.75}}&\approx P_0\\173{,}773.94&\approx P_0\end{aligned}$$

The present value of $1,000,000 due 25 years in the future at 7% interest is about $173,773.94.

9. Since the income stream is constant

$$A=\frac{R(t)}{k}\left(e^{kT}-1\right).$$

Substituting $50,000 for $R(t)$, 0.07 for k, and 22 for T we have

$$\begin{aligned}A&=\frac{50{,}000}{0.07}\left(e^{(0.07)(22)}-1\right)\\&=\frac{50{,}000}{0.07}\left(e^{1.54}-1\right)\\&\approx 2{,}617{,}560\end{aligned}$$

The accumulated future value of the continuous income stream is approximately $2,617,560.

11. Since the income stream is constant

$$A=\frac{R(t)}{k}\left(e^{kT}-1\right).$$

Substituting $400,000 for $R(t)$, 0.08 for k, and 20 for T we have

$$\begin{aligned}A&=\frac{400{,}000}{0.08}\left(e^{(0.08)(20)}-1\right)\\&=\frac{400{,}000}{0.08}\left(e^{1.6}-1\right)\\&\approx 19{,}765{,}160\end{aligned}$$

The accumulated future value of the continuous income stream is approximately $19,765,160.

13. Since the income stream is constant

$$B=\frac{R(t)}{k}\left(1-e^{-kT}\right).$$

Substituting $250,000 for $R(t)$, 0.04 for k, and 18 for T we have

$$\begin{aligned}B&=\frac{250{,}000}{0.04}\left(1-e^{-(0.04)(18)}\right)\\&=6{,}250{,}000\left(1-e^{-0.72}\right)\\&\approx 3{,}207{,}800.\end{aligned}$$

The accumulated present value of the continuous income stream is approximately $3,207,800.

Copyright © 2014 Pearson Education, Inc. Publishing as Addison-Wesley.

15. Since the income stream is constant

$$B = \frac{R(t)}{k}\left(1 - e^{-kT}\right).$$

Substituting \$800,000 for $R(t)$, 0.08 for *k*, and 20 for *T* we have

$$B = \frac{800,000}{0.08}\left(1 - e^{-(0.08)(20)}\right)$$
$$= 10,000,000\left(1 - e^{-1.6}\right)$$
$$\approx 7,981,030.$$

The accumulated present value of the continuous income stream is approximately \$7,981,030.

17. Since the income stream is non-constant, the accumulated present value is given by

$$B = \int_0^T R(t)e^{-kt}dt .$$

Substituting $5200t$ for $R(t)$, 0.07 for *k*, and 18 for *T* we have

$$\int_0^{18} (5200t)e^{-0.07t}dt$$

Using Formula 6 from Table 1 we integrate as follows:

$$\int_0^{18} (5200t)e^{-0.07t}dt$$
$$= 5200\int_0^{18} te^{-0.07t}dt$$
$$= 5200\left(\frac{1}{(-0.07)^2}e^{-0.07t}(-0.07t-1)\right)_0^{18}$$
$$= 1,061,224.4898e^{-0.07t}(-0.07t-1)\Big|_0^{18}$$
$$= 680,306.554987 - (-1,061,224.4898)$$
$$\approx 380,920.$$

The accumulated present value of the continuous income stream is approximately \$380,920.

19. Since the income stream is non-constant, the accumulated present value is given by

$$B = \int_0^T R(t)e^{-kt}dt .$$

Substituting $2000t + 7$ for $R(t)$, 0.08 for *k*, and 30 for *T* we have

$$\int_0^{30} (2000t+7)e^{-0.08t}dt =$$
$$\int_0^{30} (2000t)e^{-0.08t}dt + \int_0^{30} 7e^{-0.08t}dt$$

Using Formula 6 from Table 1 we integrate the first integral of the sum. The second integral is integrated using Formula 5 from Table 1.

$$= 2000\left(\frac{1}{(-0.08)^2}e^{-0.08t}(-0.08t-1)\right)_0^{30}$$
$$+ \frac{1}{-0.08}e^{-0.08t}\Big|_0^{30}$$
$$= (-96,387.82537 - (-312,500))$$
$$+ (79.5621790873)$$
$$\approx 216,190.$$

The accumulated present value of the continuous income stream is approximately \$216,190.

21. $P = P_0e^{kt}$

Therefore,

$$P_0 = \frac{P}{e^{kt}} = Pe^{-kt}$$

Substituting 200,000 for *P*, 0.058 for *k* and 18 for *t*, we have:

$$P_0 = 200,000e^{-0.058(18)}$$
$$P_0 = 200,000e^{-1.044}$$
$$P_0 \approx 200,000(0.3520436871)$$
$$P_0 \approx 70,408.74.$$

The present value of Maggie's legacy is approximately \$70,408.74.

23. a) Since the income stream is constant the accumulated present value is given by:

$$B = \frac{R(t)}{k}\left(1 - e^{-kT}\right)$$

Substituting \$95,000 for $R(t)$, 0.06 for *k* and 30 for *T*, we have:

$$B = \frac{95,000}{0.06}\left(1 - e^{-(0.06)(30)}\right)$$
$$\approx 1,321,610.$$

The accumulated present value of Rochelle's new job is approximately \$1,321,610.

Copyright © 2014 Pearson Education, Inc. Publishing as Addison-Wesley.

b) Since the income stream is constant the accumulated future value is given by:

$$A = \frac{R(t)}{k}\left(e^{kT} - 1\right)$$

Substituting \$95,000 for $R(t)$, 0.06 for k and 30 for T, we have:

$$A = \frac{95,000}{0.06}\left(e^{(0.06)(30)} - 1\right)$$
$$\approx 7,995,280.$$

The accumulated future value of Rochelle's new job is approximately \$7,995,280.

25. $P = P_0 e^{kt}$

Substituting 50,000 for P, 0.073 for k, and 16 for t, we have:

$$P = 50,000e^{(0.073)16}$$
$$= 50,000e^{1.168}$$
$$\approx 160,777.75.$$

The future value of David's inheritance is approximately \$160,777.75.

27. Computing the accumulated present value for Franchise A, we substitute 80,000 for $R(t)$, 0.061 for k, and 10 for T. The accumulated present value is

$$B = \frac{80,000}{0.061}\left(1 - e^{-(0.061)(10)}\right)$$
$$= \frac{80,000}{0.061}\left(1 - e^{-0.61}\right)$$
$$\approx 598,884.$$

Computing the accumulated present value for Franchise B, we substitute 95,000 for $R(t)$, 0.061 for k, and 8 for T. The accumulated present value is

$$B = \frac{95,000}{0.061}\left(1 - e^{-(0.061)(8)}\right)$$
$$= \frac{80,000}{0.061}\left(1 - e^{-0.488}\right)$$
$$\approx 601,377.$$

Comparing the accumulated present values for the two franchises, we conclude that Franchise B is the better buy.

29. a) Since the offer is a non constant stream, we calculated the accumulated present value by

$$B = \int_0^T R(t)e^{-kt}dt.$$

For the Crunchers, we substitute $100,000t$ for $R(t)$, 0.06 for k, and 8 for T

$$B = \int_0^8 100,000te^{-0.06t}dt$$

Using Tabular integration by parts, we have:

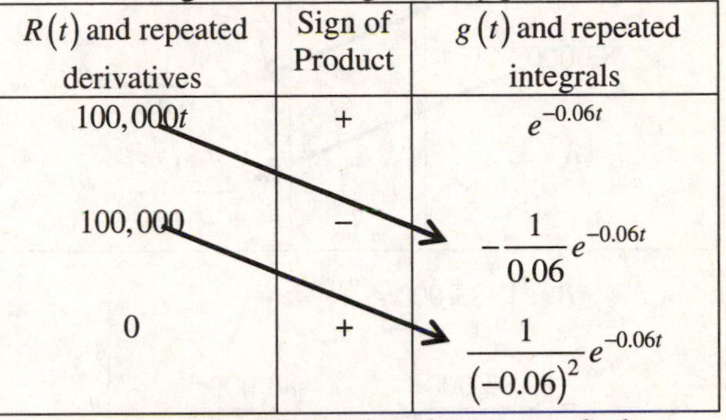

$R(t)$ and repeated derivatives	Sign of Product	$g(t)$ and repeated integrals
$100,000t$	+	$e^{-0.06t}$
$100,000$	−	$-\frac{1}{0.06}e^{-0.06t}$
0	+	$\frac{1}{(-0.06)^2}e^{-0.06t}$

Using the information above, we calculate the accumulated present value.

$$B = \int_0^8 100,000te^{-0.06t}dt$$
$$= \left[\frac{-100,000}{0.06}te^{-0.06t} - \frac{100,000}{(0.06)^2}e^{-0.06t}\right]_0^8$$
$$= \frac{-100,000}{0.06}(8)e^{-0.06(8)} - \frac{100,000}{(0.06)^2}e^{-0.06(8)} -$$
$$\left(\frac{-100,000}{0.06}(0)e^{-0.06(0)} - \frac{100,000}{(0.06)^2}e^{-0.06(0)}\right)$$
$$\approx 2,338,905.$$

The accumulated present value of the Crunchers deal is approximately \$2,338,905.

The solution is continued on the next page.

Copyright © 2014 Pearson Education, Inc. Publishing as Addison-Wesley.

For the Radar's we substitute $83,000t$ for $R(t)$, 0.06 for *k*, and 9 for *T*

$$B=\int_0^9 83,000te^{-0.06t}\,dt$$

Using Tabular integration by parts, we have:

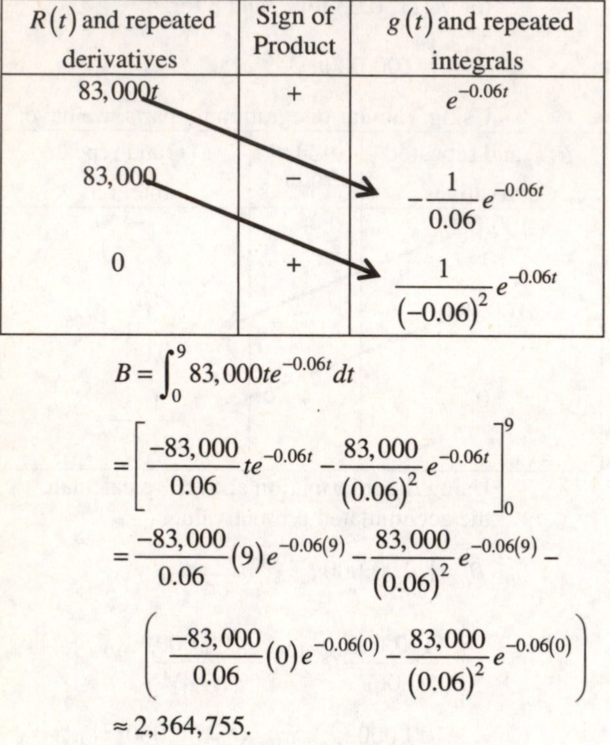

$R(t)$ and repeated derivatives	Sign of Product	$g(t)$ and repeated integrals
$83,000t$	+	$e^{-0.06t}$
$83,000$	−	$-\frac{1}{0.06}e^{-0.06t}$
0	+	$\frac{1}{(-0.06)^2}e^{-0.06t}$

$$B=\int_0^9 83,000te^{-0.06t}\,dt$$

$$=\left[\frac{-83,000}{0.06}te^{-0.06t}-\frac{83,000}{(0.06)^2}e^{-0.06t}\right]_0^9$$

$$=\frac{-83,000}{0.06}(9)e^{-0.06(9)}-\frac{83,000}{(0.06)^2}e^{-0.06(9)}-$$

$$\left(\frac{-83,000}{0.06}(0)e^{-0.06(0)}-\frac{83,000}{(0.06)^2}e^{-0.06(0)}\right)$$

$$\approx 2,364,755.$$

The accumulated present value of the Crunchers deal is approximately \$2,364,755. Based on the accumulated present values, we conclude the Doppler Radar's have the better offer.

b) ✎

31. a) $P=P_0e^{kt}$

Therefore,

$$P_0=\frac{P}{e^{kt}}=Pe^{-kt}$$

Substituting 250,000 for *P*, 0.058 for *k* and 24 for *t*, we have:

$$P_0=250,000e^{-0.058(24)}$$

$$P_0=250,000e^{-1.392}$$

$$P_0\approx 250,000(0.248577651841)$$

$$P_0\approx 62,144.41$$

Ted and Edith should make an initial investment \$62,144.41 to meet their goals for Brenda.

b) Since the income stream is constant the accumulated future value is given by:

$$A=\frac{R(t)}{k}\left(e^{kT}-1\right)$$

Substituting \$250,000 for A, 0.058 for *k* and 24 for *T*, we have:

$$250,000=\frac{R(t)}{0.058}\left(e^{(0.058)(24)}-1\right).$$

Solving the equation for $R(t)$ yields

$$250,000=\frac{R(t)}{0.058}\left(e^{(0.058)(24)}-1\right)$$

$$250,000(0.58)=R(t)\left(e^{1.392}-1\right)$$

$$\frac{14,500}{\left(e^{1.392}-1\right)}=R(t)$$

$$4796.74\approx R(t)$$

Ted and Edith should deposit \$4796.74 per year to meet their goals for Brenda.

33. Lauren has 3 years left on her contract making \$84,000 per year. With an interest rate of 7.4%, we calculate the accumulated future value of the contract for the entire term of the contract:

$$A_{10}=\frac{84,000}{0.074}\left(e^{(0.074)(10)}-1\right)$$

$$=\frac{84,000}{0.074}\left(e^{0.74}-1\right)$$

$$\approx 1,244,034.91.$$

The accumulated future value for the first 7 years of the contract is:

$$A_7=\frac{84,000}{0.074}\left(e^{(0.074)(7)}-1\right)$$

$$=\frac{84,000}{0.074}\left(e^{0.518}-1\right)$$

$$\approx 770,378.71.$$

The difference is:

$$A_{10}-A_7=1,244,034.91-770,378.71$$

$$=473,656.20.$$

Since the company is offering a lump sum Lauren should receive the present value of the difference.

$$P_0=473,656.20e^{-0.074(3)}$$

$$\approx 379,358.53$$

The minimum the bank should offer Lauren to take early retirement is \$379,358.53.

Copyright © 2014 Pearson Education, Inc. Publishing as Addison-Wesley.

35. a) Substituting \$120,000 for $R(t)$, 0.082 for k and 20 for T, we have:

$$A = \frac{120,000}{0.082}\left(e^{(0.082)(20)} - 1\right)$$
$$\approx 6,080,740.$$

The accumulated future value of income stream is approximately \$6,080,740.

b) Substituting \$120,000 for $R(t)$, 0.082 for k and 20 for T, we have:

$$B = \frac{120,000}{0.082}\left(1 - e^{-(0.082)(20)}\right)$$
$$\approx 1,179,540.$$

The accumulated present value of the income stream is approximately \$1,179,540.

37. a) Substituting \$50,000 for $R(t)$, 0.07 for k and 20 for T, we have:

$$A = \frac{50,000}{0.07}\left(e^{(0.07)(20)} - 1\right)$$
$$\approx 2,182,290.$$

The accumulated future value of income stream is approximately \$2,182,290.

b) Substituting \$50,000 for $R(t)$, 0.07 for k and 20 for T, we have:

$$B = \frac{50,000}{0.07}\left(1 - e^{-(0.07)(20)}\right)$$
$$\approx 538,145.$$

The accumulated present value of the income stream is approximately 538,145.

c) If the interest rate is 4% the accumulated present value is:

$$B = \frac{50,000}{0.04}\left(1 - e^{-(0.04)(20)}\right)$$
$$\approx 688,339.$$

The accumulated present value at a rate of 4% is approximately \$688,339.

If the interest rate is 6% the accumulated present value is:

$$B = \frac{50,000}{0.06}\left(1 - e^{-(0.06)(20)}\right)$$
$$\approx 582,338.$$

The accumulated present value at a rate of 6% is approximately \$582,338.

If the interest rate is 8% the accumulated present value is:

$$B = \frac{50,000}{0.08}\left(1 - e^{-(0.08)(20)}\right)$$
$$\approx 498,815.$$

The accumulated present value at a rate of 8% is approximately \$498,815.

If the interest rate is 10% the accumulated present value is:

$$B = \frac{50,000}{0.10}\left(1 - e^{-(0.10)(20)}\right)$$
$$\approx 432,332.$$

The accumulated present value at a rate of 10% is approximately \$432,332.

d) ✎

39. The equation for the consumption of a natural resource is $\int_0^T P_0 e^{kt}\,dt = \frac{P_0}{k}\left(e^{kT} - 1\right)$.

Substituting 3.169 for P_0, 0.074 for k, and 15 $[2025 - 2010 = 15]$ for t, we have

$$\int_0^{15} 3.169e^{0.074t} = \frac{3.169}{0.074}\left(e^{0.074(15)} - 1\right)$$
$$= \frac{3.169}{0.074}\left(e^{1.11} - 1\right)$$
$$\approx \frac{3.169}{0.074}(2.034358394)$$
$$\approx 87.12$$

The total amount consumed from 2010 to 2025 is 87.12 billion cubic meters.

Substituting 3.169 for P_0, 0.074 for k, and 2 $[2012 - 2010 = 2]$ for t, we have

$$\int_0^{2} 3.169e^{0.074t} = \frac{3.169}{0.074}\left(e^{0.074(2)} - 1\right)$$
$$= \frac{3.169}{0.074}\left(e^{0.148} - 1\right)$$
$$\approx \frac{3.169}{0.074}(0.1595128964)$$
$$\approx 6.831$$

The total amount consumed between 2010 and 2012 is 6.831 billion cubic meters.

If demand continues to grow exponentially at 7.4% per year, the world will consume approximately $87.12 - 6.83 = 80.29$ billion cubic meters of natural gas from 2012 to 2025.

41. The equation for the consumption of a natural resource is $\int_0^T P_0 e^{kt}\,dt = \frac{P_0}{k}\left(e^{kT} - 1\right)$.

Using the information from Exercise 39, we want to find T, such that

$$187,100 = \frac{3.169}{0.074}\left(e^{0.074T} - 1\right)$$

(187.1 trillion = 187,100 billion)

The solution is continued on the next page.

Copyright © 2014 Pearson Education, Inc. Publishing as Addison-Wesley.

Solving the equation on the previous page for T, we have:

$$187{,}100 = \frac{3.169}{0.074}\left(e^{0.074T} - 1\right)$$

$$4369.0123 = e^{0.074T} - 1 \quad \text{Dividing both sides by } \tfrac{3.169}{0.074}.$$

$$4370.0123 \approx e^{0.074T} \quad \text{Adding 1 to both sides.}$$

$$\ln(4370.0123) \approx \ln\left(e^{0.074T}\right) \quad \text{Taking the natural logarithm of each side.}$$

$$\ln(4370.0123) \approx 0.074T \quad \text{Recall that } \ln e^k = k.$$

$$\frac{\ln(4370.0123)}{0.074} \approx T \quad \text{Dividing both sides by 0.074.}$$

$$113.3 \approx T$$

Assuming the world consumption of natural gas continues to grow at 7.4% per year, and no new reserves are found, the world reserves of natural gas will be depleted 113.3 years after 2010, in 2123.

43. a) $P = P_0 e^{kt}$

We will substitute the known information and solve for k.

$$34.5 = 30.8e^{k(2010-2006)}$$

$$34.5 = 30.8e^{k\cdot 4}$$

$$\frac{34.5}{30.8} = e^{4k}$$

$$\ln\left(\frac{34.5}{30.8}\right) = \ln\left(e^{4k}\right)$$

$$\ln\left(\frac{34.5}{30.8}\right) = 4k$$

$$\frac{\ln\left(\frac{34.5}{30.8}\right)}{4} = k$$

$$0.02836 \approx k$$

The exponential growth rate of demand for oil is 0.0284 or 2.84% per year.

b) Using the growth rate and the initial demand from Part (a) we have the exponential demand function $P = 30.8e^{0.02836t}$.

Substituting 9 $[2015 - 2006 = 9]$ for t, we have:

$$P = 30.8e^{0.02836(9)}$$

$$= 30.8e^{0.25524}$$

$$\approx 30.8(1.29077)$$

$$\approx 39.76$$

In 2015, the world demand for oil will be approximately 39.76 billion barrels.

c) The equation for the consumption of a natural resource is $\int_0^T P_0 e^{kt}\,dt = \frac{P_0}{k}\left(e^{kT} - 1\right)$.

We want to find T, such that

$$1293 = \frac{30.8}{0.02836}\left(e^{0.02836T} - 1\right)$$

$$\frac{1293(0.02836)}{30.8} = e^{0.02836T} - 1$$

$$1.1906 \approx e^{0.02836T} - 1$$

$$2.1906 \approx e^{0.02836T}$$

$$\ln(2.1906) \approx \ln e^{0.02836T}$$

$$\ln(2.1906) \approx 0.02836T$$

$$\frac{\ln(2.1906)}{0.02836} \approx T$$

$$27.7 \approx T$$

Assuming the world consumption of oil continues to grow at 2.84% per year, and no new reserves are found, the world reserves of oil will be depleted 27.7 years after 2006, in 2033.

45. $\int_0^T Pe^{-kt}\,dt = \frac{P}{-k}\left(e^{-kT} - 1\right) = \frac{P}{k}\left(1 - e^{-kT}\right).$

$$\int_0^{20} 1\cdot e^{-0.023t}\,dt = \frac{1}{0.023}\left(1 - e^{-0.023(20)}\right)$$

$$\approx 43.47826\left(1 - e^{-0.46}\right)$$

$$\approx 16.031$$

After 20 years, approximately 16.031 lbs of Cesium-237 will remain in the atmosphere.

47. $c = c_0 + \int_0^L m(t)e^{-rt}\,dt$

$$c = 400{,}000 + \int_0^{25} 10{,}000e^{-0.055t}\,dt$$

$$= 400{,}000 + \left[\frac{10{,}000}{-0.055}e^{-0.055t}\right]_0^{25}$$

$$= 400{,}000 - \frac{10{,}000}{0.055}\left[e^{-1.375} - e^0\right]$$

$$\approx 535{,}847$$

The capitalized cost under the given assumptions is \$535,847.

Copyright © 2014 Pearson Education, Inc. Publishing as Addison-Wesley.

49. $c = c_0 + \int_0^L m(t)e^{-rt}\,dt$

$$c = 300{,}000 + \int_0^{20} (30{,}000 + 500t)e^{-0.05t}\,dt$$

$$= 300{,}000 + \int_0^{20} \left(30{,}000e^{-0.05t} + 500te^{-0.05t}\right)dt$$

$$= 300{,}000 + \left[\frac{30{,}000}{-0.05}e^{-0.05t} + 500\frac{1}{(-0.05)^2}e^{-0.05t}(-0.05t-1)\right]_0^{20}$$

$$= 300{,}000 + \left[-600{,}000e^{-0.05t} + 200{,}000e^{-0.05t}(-0.05t-1)\right]_0^{20}$$

$$= 300{,}000 + \left[-600{,}000e^{-0.05(20)} + 200{,}000e^{-0.05(20)}(-0.05(20)-1) - \left(-600{,}000e^{-0.05(0)} + 200{,}000e^{-0.05(0)}(-0.05(0)-1)\right)\right.$$

$$\approx 732{,}121$$

The capitalized cost under the given assumptions is \$732,121.

51. ✎

Copyright © 2014 Pearson Education, Inc. Publishing as Addison-Wesley.

Exercise Set 5.3

1. $\int_2^\infty \frac{dx}{x^2}$

$= \lim_{b\to\infty} \int_2^b x^{-2}dx$

$= \lim_{b\to\infty} \left[\frac{x^{-1}}{-1}\right]_2^b$

$= \lim_{b\to\infty} \left[-\frac{1}{x}\right]_2^b$

$= \lim_{b\to\infty} \left[-\frac{1}{b}-\left(-\frac{1}{2}\right)\right]$

$= 0+\frac{1}{2} \qquad \left(\text{As } b\to\infty, -\frac{1}{b}\to 0\right)$

$= \frac{1}{2}$

The limit does exist. The improper integral is convergent.

3. $\int_3^\infty \frac{dx}{x}$

$= \lim_{b\to\infty} \int_3^b x^{-1}dx$

$= \lim_{b\to\infty} \left[\ln(x)\right]_3^b$

$= \lim_{b\to\infty} \left[\ln(b)-\ln(3)\right]$

Note: that $\ln b$ increases without bound as b increases. Therefore, the limit does not exist. If the limit does not exist, we say the improper integral is divergent.

5. $\int_0^\infty 3e^{-3x}dx$

$= \lim_{b\to\infty} \int_0^b 3e^{-3x}dx$

$= \lim_{b\to\infty} \left[\frac{3}{-3}e^{-3x}\right]_0^b$

$= \lim_{b\to\infty} \left[-e^{-3x}\right]_0^b$

$= \lim_{b\to\infty} \left[-e^{-3\cdot b}-\left(-e^{-3\cdot 0}\right)\right]$

$= \lim_{b\to\infty} \left[-e^{-3b}+1\right]$

$= \lim_{b\to\infty} \left[1-\frac{1}{e^{3b}}\right]$

$= 1-0 \qquad \left[\text{As } b\to\infty, e^{3b}\to\infty, \text{so } \frac{1}{e^{3b}}\to 0\right]$

$= 1$

The limit does exist. The improper integral is convergent.

7. $\int_1^\infty \frac{dx}{x^3}$

$= \lim_{b\to\infty} \int_1^b x^{-3}dx$

$= \lim_{b\to\infty} \left[\frac{x^{-2}}{-2}\right]_1^b$

$= \lim_{b\to\infty} \left[-\frac{1}{2x^2}\right]_1^b$

$= \lim_{b\to\infty} \left[-\frac{1}{2b^2}-\left(-\frac{1}{2(1)^2}\right)\right]$

$= 0+\frac{1}{2} \qquad \left(\text{As } b\to\infty, -\frac{1}{2b^2}\to 0\right)$

$= \frac{1}{2}$

The limit does exist. The improper integral is convergent.

Copyright © 2014 Pearson Education, Inc. Publishing as Addison-Wesley.

9. $\int_0^{\infty} \frac{dx}{2+x}$

$= \lim_{b\to\infty} \int_0^b \frac{1}{2+x} dx$

$= \lim_{b\to\infty} \left[\ln(2+x)\right]_0^b$

$= \lim_{b\to\infty} \left[\ln(2+b) - \ln(2+0)\right]$

$= \lim_{b\to\infty} \left[\ln(2+b) - \ln(2)\right]$

Note that $\ln(2+b)$ increases without bound as b increases. Therefore, the limit does not exist. If the limit does not exist, we say the improper integral is divergent.

11. $\int_2^{\infty} 4x^{-2} dx$

$= \lim_{b\to\infty} \int_2^b 4x^{-2} dx$

$= \lim_{b\to\infty} \left[\frac{4x^{-1}}{-1}\right]_2^b$

$= \lim_{b\to\infty} \left[-\frac{4}{x}\right]_2^b$

$= \lim_{b\to\infty} \left[-\frac{4}{b} - \left(-\frac{4}{2}\right)\right]$

$= 0 + 2 \qquad \left(\text{As } b \to \infty, -\frac{4}{b} \to 0\right)$

$= 2$

The limit does exist. The improper integral is convergent.

13. $\int_0^{\infty} e^x dx$

$= \lim_{b\to\infty} \int_0^b e^x dx$

$= \lim_{b\to\infty} \left[e^x\right]_0^b$

$= \lim_{b\to\infty} \left[e^b - \left(e^0\right)\right]$

$= \lim_{b\to\infty} \left[e^b - 1\right]$

As $b \to \infty, e^b \to \infty$; thus, the limit does not exist. The improper integral is divergent.

15. $\int_3^{\infty} x^2 dx$

$= \lim_{b\to\infty} \int_3^b x^2 dx$

$= \lim_{b\to\infty} \left[\frac{x^3}{3}\right]_3^b$

$= \lim_{b\to\infty} \left[\frac{(b)^3}{3} - \frac{(3)^3}{3}\right]$

$= \lim_{b\to\infty} \left[\frac{b^3}{3} - 9\right]$

As $b \to \infty, \frac{b^3}{3} \to \infty$; thus, the limit does not exist. The improper integral is divergent.

17. $\int_0^{\infty} xe^x dx$

$= \lim_{b\to\infty} \int_0^b xe^x dx$

$= \lim_{b\to\infty} \left[e^x(x-1)\right]_0^b \qquad$ Using integration by parts.

$= \lim_{b\to\infty} \left[e^b(b-1) - \left(e^0(0-1)\right)\right]$

$= \lim_{b\to\infty} \left[e^b(b-1) + 1\right]$

As $b \to \infty, e^b(b-1) \to \infty$; thus, the limit does not exist. The improper integral is divergent.

19. $\int_0^{\infty} me^{-mx} dx, \quad m > 0$

$= \lim_{b\to\infty} \int_0^b me^{-mx} dx$

$= \lim_{b\to\infty} \left[\frac{m}{-m} e^{-mx}\right]_0^b$

$= \lim_{b\to\infty} \left[-e^{-mx}\right]_0^b$

$= \lim_{b\to\infty} \left[-e^{-m\cdot b} - \left(-e^{-m\cdot 0}\right)\right]$

$= \lim_{b\to\infty} \left[-e^{-mb} + 1\right]$

$= \lim_{b\to\infty} \left[1 - \frac{1}{e^{mb}}\right]$

$= 1 - 0 \qquad \left[\text{As } b \to \infty, e^{mb} \to \infty, \text{so } \frac{1}{e^{mb}} \to 0\right]$

$= 1$

The limit does exist. The improper integral is convergent.

Copyright © 2014 Pearson Education, Inc. Publishing as Addison-Wesley.

21. $\int_{\pi}^{\infty} \frac{dt}{t^{1.001}}$

$$= \lim_{b\to\infty} \int_{\pi}^{b} t^{-1.001} dt$$

$$= \lim_{b\to\infty} \left[\frac{t^{-0.001}}{-0.001}\right]_{\pi}^{b}$$

$$= \lim_{b\to\infty} \left[-\frac{1000}{t^{0.001}}\right]_{\pi}^{b}$$

$$= \lim_{b\to\infty} \left[-\frac{1000}{b^{0.001}} - \left(-\frac{1000}{(\pi)^{0.001}}\right)\right]$$

$$= 0 + \frac{1000}{\pi^{0.001}} \quad \left(\text{As } b \to \infty, -\frac{1000}{b^{0.001}} \to 0\right)$$

$$= \frac{1000}{\pi^{0.001}}$$

$$\approx 998.86$$

The limit does exist. The improper integral is convergent.

23. $\int_{-\infty}^{\infty} t\, dt$

$$= \int_{-\infty}^{0} t\, dt + \int_{0}^{\infty} t\, dt \quad \text{Using Definition 2 with } c = 0.$$

$$= \lim_{a\to-\infty} \int_{a}^{0} t\, dt + \lim_{b\to\infty} \int_{0}^{b} t\, dt$$

$$= \lim_{a\to-\infty} \left[\frac{t^2}{2}\right]_{a}^{0} + \lim_{b\to\infty} \left[\frac{t^2}{2}\right]_{0}^{b}$$

$$= \lim_{a\to-\infty} \left[\frac{0^2}{2} - \frac{a^2}{2}\right] + \lim_{b\to\infty} \left[\frac{b^2}{2} - \frac{0^2}{2}\right]$$

Neither $\lim_{a\to-\infty} \frac{a^2}{2}$ nor $\lim_{b\to\infty} \frac{b^2}{2}$ exists, so the integral is divergent.

25. The area is given by

$$\int_{2}^{\infty} \frac{1}{x^2} dx = \lim_{b\to\infty} \int_{2}^{b} x^{-2} dx$$

$$= \lim_{b\to\infty} \left[\frac{x^{-1}}{-1}\right]_{2}^{b}$$

$$= \lim_{b\to\infty} \left[-\frac{1}{x}\right]_{2}^{b}$$

$$= \lim_{b\to\infty} \left[-\frac{1}{b} - \left(-\frac{1}{2}\right)\right]$$

$$= 0 + \frac{1}{2} \quad \left(\text{As } b \to \infty, -\frac{1}{b} \to 0\right)$$

$$= \frac{1}{2}$$

The area of the region is $\frac{1}{2}$.

27. The area is given by

$$\int_{0}^{\infty} 2xe^{-x^2} dx$$

$$= \lim_{b\to\infty} \int_{0}^{b} 2xe^{-x^2} dx$$

$$= \lim_{b\to\infty} \left[-e^{-x^2}\right]_{0}^{b} \quad \left[u = -x^2, du = -2x dx\right]$$

$$= \lim_{b\to\infty} \left[-e^{-b^2} - \left(-e^{-0^2}\right)\right]$$

$$= \lim_{b\to\infty} \left[-e^{-b^2} + 1\right]$$

$$= \lim_{b\to\infty} \left[-\frac{1}{e^{b^2}} + 1\right]$$

$$= -0 + 1 \quad \left[\text{As } b \to \infty, -\frac{1}{e^{b^2}} \to 0\right]$$

$$= 1$$

The area of the region is 1.

29. From Theorem 2, The accumulated present value is given by

$$\int_{0}^{\infty} Pe^{-kt} dt = \frac{P}{k}$$

Substituting 3600 for P and 0.07 for k, we have:

$$\int_{0}^{\infty} 3600e^{-0.07t} dt = \frac{3600}{0.07} \approx 51{,}428.57\,.$$

The accumulated present value is approximately $51,428.57.

Copyright © 2014 Pearson Education, Inc. Publishing as Addison-Wesley.

31. Total profit is given by:

$$P(x)=\int_0^{\infty} 200e^{-0.032x}dx$$
$$=\lim_{b\to\infty}\int_0^b 200e^{-0.032x}dx$$
$$=\lim_{b\to\infty}\left[\frac{200}{-0.032}e^{-0.032x}\right]_0^b$$
$$=\lim_{b\to\infty}\left[-6250e^{-0.032x}\right]_0^b$$
$$=\lim_{b\to\infty}\left[-6250e^{-0.032b}-\left(-6250e^{-0.032(0)}\right)\right]$$
$$=\lim_{b\to\infty}\left[-\frac{6250}{e^{0.032b}}+6250\right]$$
$$=6250$$

The total profit if it were possible to produce an infinite number of units is \$6250.

33. The total cost is given by:

$$C(x)=\int_1^{\infty} 3600x^{-1.8}dx$$
$$=\lim_{b\to\infty}\int_1^b 3600x^{-1.8}dx$$
$$=\lim_{b\to\infty}\left[\frac{3600}{-0.8}x^{-0.8}\right]_1^b$$
$$=\lim_{b\to\infty}\left[-4500x^{-0.8}\right]_1^b$$
$$=\lim_{b\to\infty}\left[-\frac{4500}{b^{0.8}}-\left(-\frac{4500}{1^{0.8}}\right)\right]$$
$$=0+4500 \qquad \left(\text{As } b\to\infty, -\tfrac{4500}{b^{0.8}}\to 0\right)$$
$$=4500$$

The total cost would be \$4500.

35. From Theorem 2, The accumulated present value is given by

$$\int_0^{\infty} Pe^{-kt}dt=\frac{P}{k}$$

Substituting 5000 for P and 0.08 for k, we have:

$$\int_0^{\infty} 5000e^{-0.08t}dt=\frac{5000}{0.08}\approx 62{,}500.$$

The accumulated present value is \$62,500.

37. $c=c_0+\int_0^{\infty} m(t)e^{-rt}dt$

Substituting 500,000 for c_0, 0.05 for r, and 20,000 for $m(t)$, we have:

$$c=500{,}000+\int_0^{\infty} 20{,}000e^{-0.05t}dt$$
$$=500{,}000+\lim_{b\to\infty}\int_0^b 20{,}000e^{-0.05t}dt$$
$$=500{,}000+\lim_{b\to\infty}\left[\frac{20{,}000}{-0.05}e^{-0.05t}\right]_0^b$$
$$=500{,}000+\lim_{b\to\infty}\left[-400{,}000\left(e^{-0.05b}-e^{-0.05(0)}\right)\right]$$
$$=500{,}000+\left[-400{,}000(0-1)\right]$$
$$=500{,}000+400{,}000$$
$$=900{,}000$$

The capitalized cost is \$900,000.

39. $\int_0^T Pe^{-kt}dt=\frac{P}{k}\left(1-e^{-kT}\right)$

As $T\to\infty$, we have:

$$\lim_{T\to\infty}\int_0^T P(t)e^{-kt}dt$$
$$=\lim_{T\to\infty}\left[\frac{P}{-k}e^{-kt}\right]_0^T$$
$$=\lim_{T\to\infty}\left[\frac{P}{-k}e^{-kT}-\left(\frac{P}{-k}e^{-k\cdot 0}\right)\right]$$
$$=\lim_{T\to\infty}\left[\frac{P}{k}\left(1-e^{-k\cdot T}\right)\right]$$
$$=\frac{P}{k}$$

Substituting 0.00003 for k, and 1 for P, we have

$$\frac{P}{k}=\frac{1}{0.00003}\approx 33{,}333\tfrac{1}{3}.$$

The limiting value of the radioactive buildup is $33{,}333\frac{1}{3}$ pounds.

Copyright © 2014 Pearson Education, Inc. Publishing as Addison-Wesley.

41. $E=\int_0^a P_0 e^{-kt}\,dt$

a) Note that 60.1 days is

$\frac{60.1}{365}\text{ yr}\approx 0.16465753\text{ yr}$.

Using the half-life, we find k as follows:

$$\frac{1}{2}P_0 = P_0 e^{-k(0.16465753)}$$

$$\frac{1}{2}=e^{-k(0.16465753)} \quad \text{Dividing by } P_0$$

$$\ln\left(\frac{1}{2}\right)=\ln\left(e^{-0.16465753k}\right)$$

$$\ln\left(\frac{1}{2}\right)=-0.16465753k$$

$$\frac{\ln\left(\frac{1}{2}\right)}{-0.16465753}=k$$

$$4.20963\approx k$$

The decay rate is 420.963% per year.

b) The first month is $\frac{1}{12}$ yr.

$$E=\int_0^{1/12} 10e^{-4.20963t}\,dt$$

$$=\frac{10}{-4.20963}\left[e^{-4.20963t}\right]_0^{1/12}$$

$$=\frac{10}{-4.20963}\left[e^{-4.20963(1/12)}-e^{-4.20963(0)}\right]$$

$$=\frac{10}{-4.20963}\left[e^{-0.3508025}-1\right]$$

$$\approx 0.702858$$

In the first month, 0.702858 rems of energy is transmitted.

c) $E=\int_0^{\infty} 10e^{-4.20963t}\,dt$

$$E=\lim_{b\to\infty}\int_0^{b} 10e^{-4.20963t}\,dt$$

$$=\frac{10}{4.20963} \qquad \left[\int_0^{\infty} Pe^{-kt}=\frac{P}{k}\right]$$

$$\approx 2.37551$$

The total amount of energy transmitted is 2.37551 rems.

43. $\int_0^{\infty}\frac{dx}{x^{2/3}}$

$$=\lim_{b\to\infty}\int_0^b x^{-2/3}\,dx$$

$$=\lim_{b\to\infty}\left[\frac{x^{1/3}}{1/3}\right]_0^b$$

$$=\lim_{b\to\infty}\left[3x^{1/3}\right]_0^b$$

$$=\lim_{b\to\infty}\left[3b^{1/3}-3(0)^{1/3}\right]$$

$$=\lim_{b\to\infty}\left[3\cdot\sqrt[3]{b}\right]$$

As $b\to\infty$, $\sqrt[3]{b}\to\infty$. Therefore, the limit does not exist. The improper integral is divergent.

45. $\int_0^{\infty}\frac{dx}{(x+1)^{3/2}}$

$$=\lim_{b\to\infty}\int_0^b (x+1)^{-3/2}\,dx$$

$$=\lim_{b\to\infty}\left[\frac{(x+1)^{-1/2}}{-1/2}\right]_0^b$$

$$=\lim_{b\to\infty}\left[-\frac{2}{\sqrt{x+1}}\right]_0^b$$

$$=\lim_{b\to\infty}\left[-\frac{2}{\sqrt{b+1}}-\left(-\frac{2}{\sqrt{0+1}}\right)\right]$$

$$=\lim_{b\to\infty}\left[-\frac{2}{\sqrt{b+1}}+2\right]$$

$$=0+2 \qquad \left[\text{As } b\to\infty, -\frac{2}{\sqrt{b+1}}\to 0\right]$$

$$=2$$

Therefore, the limit exists. The improper integral is convergent.

Copyright © 2014 Pearson Education, Inc. Publishing as Addison-Wesley.

47. $\int_0^{\infty} xe^{-x^2}\,dx$

$$= \lim_{b\to\infty}\int_0^b xe^{-x^2}\,dx$$

$$= \lim_{b\to\infty}\int_0^b -\frac{1}{2}\cdot(-2)\,xe^{-x^2}\,dx \qquad \text{Multiplying by 1.}$$

$$= \lim_{b\to\infty}\left[-\frac{1}{2}e^{-x^2}\right]_0^b \qquad \text{Using substitution where } u=-x^2 \text{ and } du=-2x\,dx.$$

$$= \lim_{b\to\infty}\left[-\frac{1}{2}e^{-b^2}-\left(-\frac{1}{2}e^{-(0)^2}\right)\right]$$

$$= \lim_{b\to\infty}\left[-\frac{1}{2e^{b^2}}+\frac{1}{2}\right]$$

$$= 0+\frac{1}{2} \qquad \left[\text{As } b\to\infty, \frac{1}{e^{b^2}}\to 0\right]$$

$$= \frac{1}{2}$$

Therefore, the limit exists. The improper integral is convergent.

49. $\int_0^{\infty} E(t)\,dt$

$$= \int_0^{\infty} te^{-kt}\,dt$$

$$= \lim_{b\to\infty}\int_0^b te^{-kt}\,dt$$

$$= \lim_{b\to\infty}\left[\frac{1}{(-k)^2}\cdot e^{-kt}(-kt-1)\right]_o^b$$

$$= \lim_{b\to\infty}\left[-\frac{kt+1}{k^2e^{kt}}\right]_0^b$$

$$= \lim_{b\to\infty}\left[-\frac{kb+1}{k^2e^{kb}}-\left(-\frac{k(0)+1}{k^2e^{k(0)}}\right)\right]$$

$$= \lim_{b\to\infty}\left[-\frac{kb+1}{k^2e^{kb}}+\frac{1}{k^2}\right]$$

$$= \frac{1}{k^2} \qquad \left[\text{As } b\to\infty, -\frac{kb+1}{k^2e^{kb}}\to 0\right]$$

The integral represents the total dose of the drug.

51. 

53.

$y = xe^{-0.1x}$

5

0 100

0

55. Using the fnInt feature on a graphing calculator with a large value for the upper limit, we find

$$\int_1^{\infty}\frac{6}{5+e^x}\,dx \approx 1.2523.$$

Copyright © 2014 Pearson Education, Inc. Publishing as Addison-Wesley.

Exercise Set 5.4

1. $\int_0^4 \left(x^2+1\right)dx, \quad n=4$

We first determine the widths of our rectangles.

$\Delta x = \dfrac{b-a}{n} = \dfrac{4-0}{4} = 1.$

We graph $f(x)=x^2+1$ and draw four left rectangles dividing the interval $[0,4]$ into four subintervals of length 1.

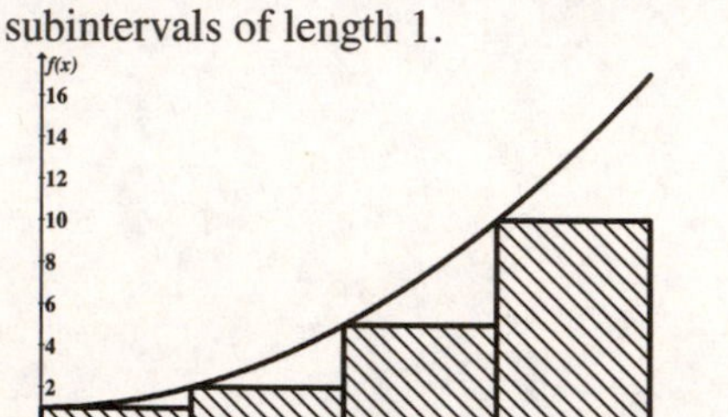

From the information above we have:

Subinterval	Width	Height (left endpoint)	Area
$[0,1]$	1	$f(0)=1$	$1\cdot1=1$
$[1,2]$	1	$f(1)=2$	$1\cdot2=2$
$[2,3]$	1	$f(2)=5$	$1\cdot5=5$
$[3,4]$	1	$f(3)=10$	$1\cdot10=10$

Thus, we have $L_4 = 1+2+5+10 = 18.$

We graph $f(x)=x^2+1$ and draw four right rectangles dividing the interval $[0,4]$ into four subintervals of length 1.

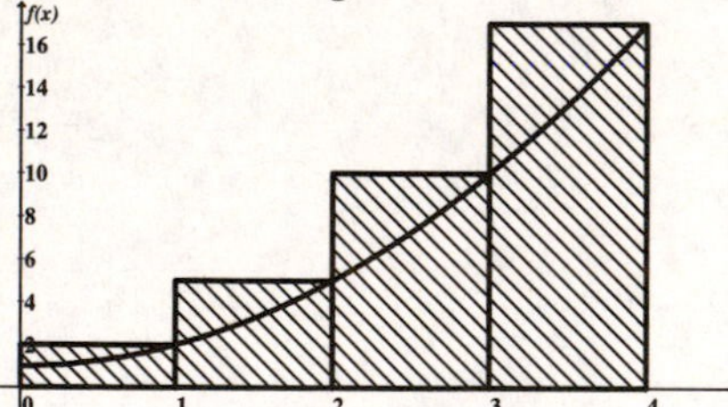

From the information above we have:

Subinterval	Width	Height (right endpoint)	Area
$[0,1]$	1	$f(1)=2$	$1\cdot1=1$
$[1,2]$	1	$f(2)=5$	$1\cdot2=2$
$[2,3]$	1	$f(3)=10$	$1\cdot5=5$
$[3,4]$	1	$f(4)=17$	$1\cdot10=10$

Thus, we have $R_4 = 2+5+10+17 = 34.$

The average of L_4 and R_4 is $\dfrac{18+34}{2} = 26.$

3. $\int_0^3 \left(2x^3-x\right)dx, \quad n=6$

We first determine the widths of our rectangles.

$\Delta x = \dfrac{b-a}{n} = \dfrac{3-0}{6} = \dfrac{1}{2}.$

We graph $f(x)=2x^3-x$ and draw six left rectangles dividing the interval $[0,3]$ into six subintervals of length 0.5.

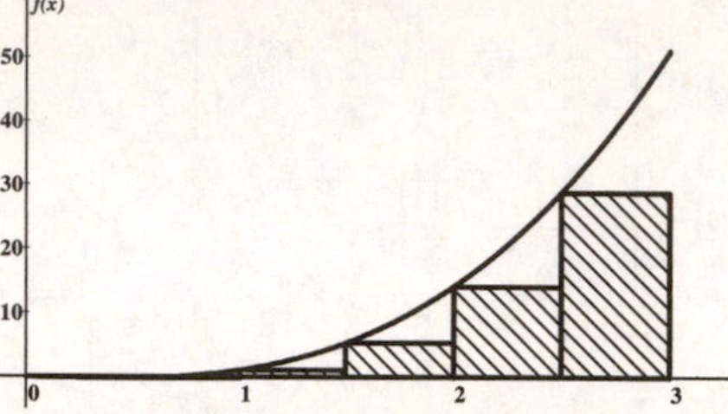

From the information above we have:

Subinterval	Width	Height (left endpoint)	Area
$[0,0.5]$	0.5	$f(0)=0$	$0.5\cdot0=0$
$[0.5,1]$	0.5	$f(0.5)=\frac{-1}{4}$	$0.5\cdot\left(-\frac{1}{4}\right)=-\frac{1}{8}$
$[1,1.5]$	0.5	$f(1)=1$	$0.5\cdot1=\frac{1}{2}$
$[1.5,2]$	0.5	$f(1.5)=\frac{21}{4}$	$0.5\cdot\frac{21}{4}=\frac{21}{8}$
$[2,2.5]$	0.5	$f(2)=14$	$0.5\cdot14=7$
$[2.5,3]$	0.5	$f(2.5)=\frac{115}{4}$	$0.5\cdot\frac{115}{4}=\frac{115}{8}$

Thus, we have

$L_6 = 0-\frac{1}{8}+\frac{1}{2}+\frac{21}{8}+7+\frac{115}{8} = 24.375.$

We graph $f(x)=2x^3-x$ and draw six right rectangles dividing the interval $[0,3]$ into six subintervals of length 0.5.

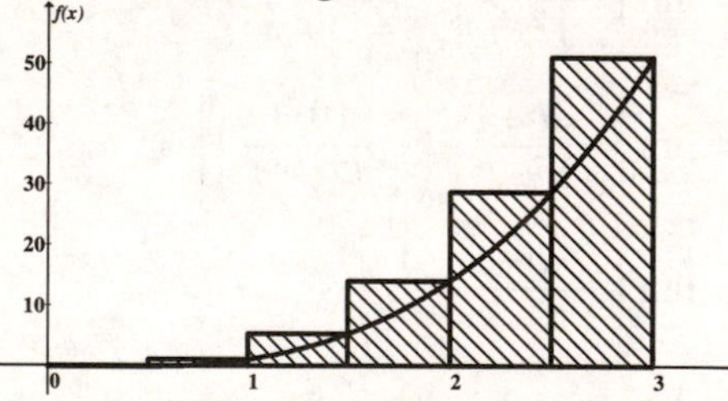

From the information above we have:

Subinterval	Width	Height (right endpoint)	Area
$[0,0.5]$	0.5	$f(0.5)=\frac{-1}{4}$	$0.5\cdot\left(-\frac{1}{4}\right)=-\frac{1}{8}$
$[0.5,1]$	0.5	$f(1)=1$	$0.5\cdot1=\frac{1}{2}$
$[1,1.5]$	0.5	$f(1.5)=\frac{21}{4}$	$0.5\cdot\frac{21}{4}=\frac{21}{8}$
$[1.5,2]$	0.5	$f(2)=14$	$0.5\cdot14=7$
$[2,2.5]$	0.5	$f(2.5)=\frac{115}{4}$	$0.5\cdot\frac{115}{4}=\frac{115}{8}$
$[2.5,3]$	0.5	$f(3)=51$	$0.5\cdot51=\frac{51}{2}$

Copyright © 2014 Pearson Education, Inc. Publishing as Addison-Wesley.

Using the information from the previous page, we have

$R_6 = -\frac{1}{8} + \frac{1}{2} + \frac{21}{8} + 7 + \frac{115}{8} + \frac{51}{2} = 49.875.$

The average of L_6 and R_6 is

$$\frac{24.375 + 49.875}{2} = 37.125.$$

5. $\int_2^4 \frac{1}{x}\,dx, \quad n = 8$

We first determine the widths of our rectangles.

$$\Delta x = \frac{b-a}{n} = \frac{4-2}{8} = \frac{1}{4}.$$

We graph $f(x) = \frac{1}{x}$ and draw eight left rectangles dividing the interval $[2,4]$ into eight subintervals of length 0.25.

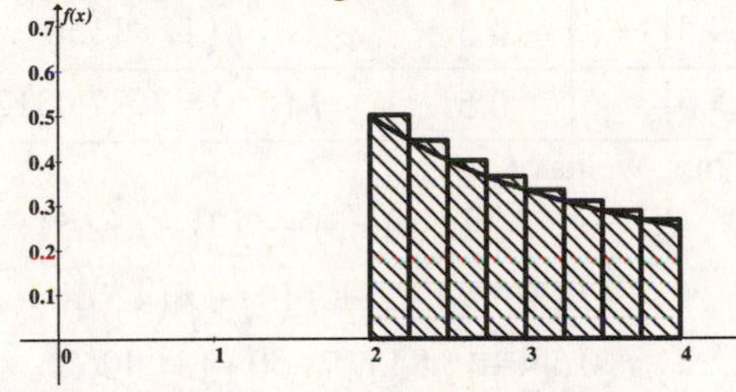

From the information above we have:

Subinterval	Width	Height (left endpoint)	Area
$[2, 2.25]$	0.25	$f(2) = \frac{1}{2}$	$0.25 \cdot \frac{1}{2} = \frac{1}{8}$
$[2.25, 2.5]$	0.25	$f(2.25) = \frac{4}{9}$	$0.25 \cdot \frac{4}{9} = \frac{1}{9}$
$[2.5, 2.75]$	0.25	$f(2.5) = \frac{4}{10}$	$0.25 \cdot \frac{4}{10} = \frac{1}{10}$
$[2.75, 3]$	0.25	$f(2.75) = \frac{4}{11}$	$0.25 \cdot \frac{4}{11} = \frac{1}{11}$
$[3, 3.25]$	0.25	$f(3) = \frac{1}{3}$	$0.25 \cdot \frac{1}{3} = \frac{1}{12}$
$[3.25, 3.5]$	0.25	$f(3.25) = \frac{4}{13}$	$0.25 \cdot \frac{4}{13} = \frac{1}{13}$
$[3.5, 3.75]$	0.25	$f(3.5) = \frac{2}{7}$	$0.25 \cdot \frac{2}{7} = \frac{1}{14}$
$[3.75, 4]$	0.25	$f(3.75) = \frac{4}{15}$	$0.25 \cdot \frac{4}{15} = \frac{1}{15}$

Thus, we have

$L_8 = \frac{1}{8} + \frac{1}{9} + \frac{1}{10} + \frac{1}{11} + \frac{1}{12} + \frac{1}{13} + \frac{1}{14} + \frac{1}{15} \approx 0.725.$

We graph $f(x) = \frac{1}{x}$ and draw eight right rectangles dividing the interval $[2,4]$ into eight subintervals of length 0.5.

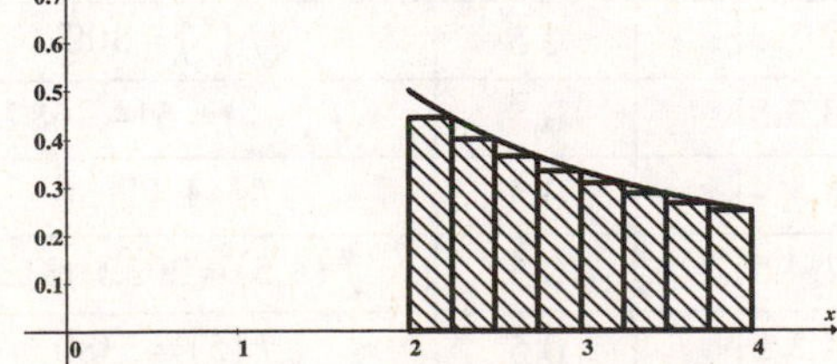

From the information in the previous column, we have:

Subinterval	Width	Height (right endpoint)	Area
$[2, 2.25]$	0.25	$f(2.25) = \frac{4}{9}$	$0.25 \cdot \frac{4}{9} = \frac{1}{9}$
$[2.25, 2.5]$	0.25	$f(2.5) = \frac{4}{10}$	$0.25 \cdot \frac{4}{10} = \frac{1}{10}$
$[2.5, 2.75]$	0.25	$f(2.75) = \frac{4}{11}$	$0.25 \cdot \frac{4}{11} = \frac{1}{11}$
$[2.75, 3]$	0.25	$f(3) = \frac{1}{3}$	$0.25 \cdot \frac{1}{3} = \frac{1}{12}$
$[3, 3.25]$	0.25	$f(3.25) = \frac{4}{13}$	$0.25 \cdot \frac{4}{13} = \frac{1}{13}$
$[3.25, 3.5]$	0.25	$f(3.5) = \frac{2}{7}$	$0.25 \cdot \frac{2}{7} = \frac{1}{14}$
$[3.5, 3.75]$	0.25	$f(3.75) = \frac{4}{15}$	$0.25 \cdot \frac{4}{15} = \frac{1}{15}$
$[3.75, 4]$	0.25	$f(4) = \frac{1}{4}$	$0.25 \cdot \frac{1}{4} = \frac{1}{16}$

Thus, we have

$R_8 = \frac{1}{9} + \frac{1}{10} + \frac{1}{11} + \frac{1}{12} + \frac{1}{13} + \frac{1}{14} + \frac{1}{15} + \frac{1}{16} \approx 0.663.$

The average of L_6 and R_6 is

$$\frac{0.725 + 0.663}{2} = 0.694.$$

7. $\int_0^3 e^{2x}\,dx, \quad n = 6$

We first determine the widths of our rectangles.

$\Delta x = \frac{b-a}{n} = \frac{3-0}{6} = \frac{1}{2}$. The solution is continued on the next page.

We graph $f(x) = e^{2x}$ and draw six left rectangles dividing the interval $[0,3]$ into six subintervals of length 0.5.

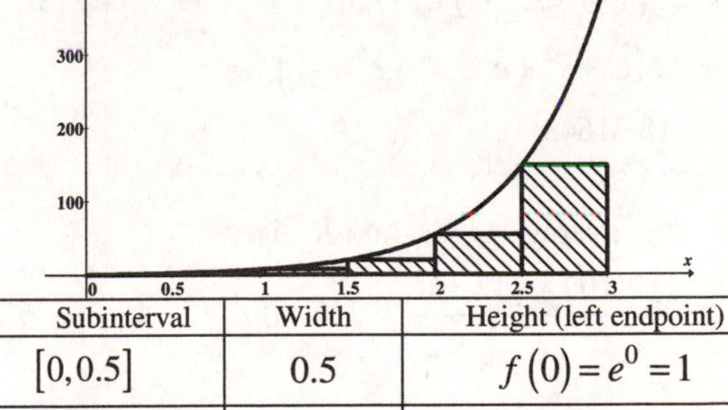

Subinterval	Width	Height (left endpoint)
$[0, 0.5]$	0.5	$f(0) = e^0 = 1$
$[0.5, 1]$	0.5	$f(0.5) = e^1$
$[1, 1.5]$	0.5	$f(1) = e^2$
$[1.5, 2]$	0.5	$f(1.5) = e^3$
$[2, 2.5]$	0.5	$f(2) = e^4$
$[2.5, 3]$	0.5	$f(2.5) = e^5$

The solution is continued on the next page.

Copyright © 2014 Pearson Education, Inc. Publishing as Addison-Wesley.

Using the information from the previous page, we have

$$L_6 = 0.5\left(f(0) + f(0.5) + f(1) + f(1.5) + f(2) + f(2.5)\right)$$
$$= 0.5\left(1 + e^1 + e^2 + e^3 + e^4 + e^5\right)$$
$$= 117.10209$$
$$\approx 117.102.$$

We graph $f(x) = e^{2x}$ and draw six right rectangles dividing the interval $[0,3]$ into six subintervals of length 0.5.

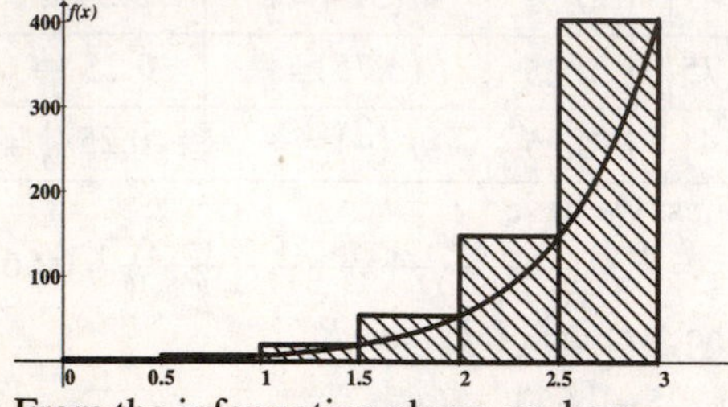

From the information above we have:

Subinterval	Width	Height (right endpoint)
$[0, 0.5]$	0.5	$f(0.5) = e^1$
$[0.5, 1]$	0.5	$f(1) = e^2$
$[1, 1.5]$	0.5	$f(1.5) = e^3$
$[1.5, 2]$	0.5	$f(2) = e^4$
$[2, 2.5]$	0.5	$f(2.5) = e^5$
$[2.5, 3]$	0.5	$f(3) = e^6$

Thus, we have

$$R_6 = 0.5\left(f(0.5) + f(1) + f(1.5) + f(2) + f(2.5) + f(3)\right)$$
$$= 0.5\left(e^1 + e^2 + e^3 + e^4 + e^5 + e^6\right)$$
$$= 318.31648$$
$$\approx 318.316.$$

The average of L_6 and R_6 is

$$\frac{117.102 + 318.136}{2} = 217.709.$$

9. $\int_1^5 x\left(x^2+1\right)^2 dx, \quad n = 8$

We first determine the widths of our rectangles.

$$\Delta x = \frac{b-a}{n} = \frac{5-1}{8} = \frac{1}{2}$$

We graph $f(x) = x\left(x^2+1\right)^2$ and draw eight left rectangles dividing the interval $[1,5]$ into eight subintervals of length 0.5 at the top of the next column.

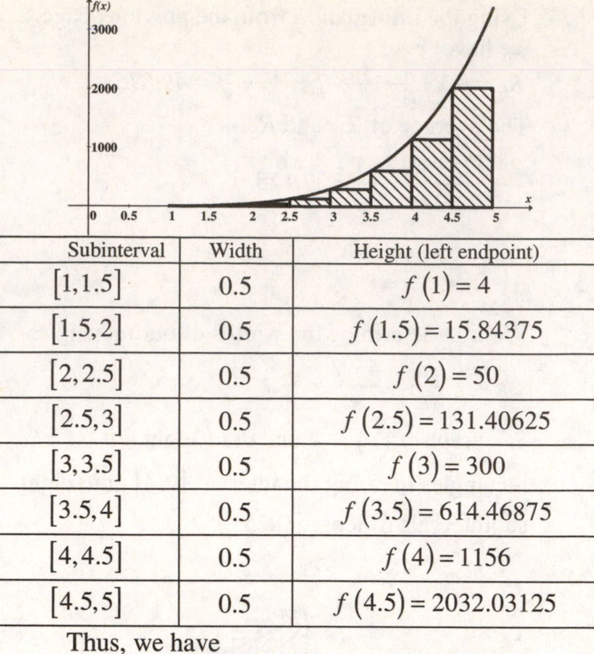

Subinterval	Width	Height (left endpoint)
$[1, 1.5]$	0.5	$f(1) = 4$
$[1.5, 2]$	0.5	$f(1.5) = 15.84375$
$[2, 2.5]$	0.5	$f(2) = 50$
$[2.5, 3]$	0.5	$f(2.5) = 131.40625$
$[3, 3.5]$	0.5	$f(3) = 300$
$[3.5, 4]$	0.5	$f(3.5) = 614.46875$
$[4, 4.5]$	0.5	$f(4) = 1156$
$[4.5, 5]$	0.5	$f(4.5) = 2032.03125$

Thus, we have

$$L_6 = 0.5[f(1) + f(1.5) + f(2) + f(2.5)$$
$$+ f(3) + f(3.5) + f(4) + f(4.5)]$$
$$= 0.5[4 + 15.84375 + 50 + 131.40625$$
$$+ 300 + 614.46875 + 1156 + 2032.03125]$$
$$\approx 2151.875.$$

We graph $f(x) = x\left(x^2+1\right)^2$ and draw eight right rectangles dividing the interval $[1,5]$ into six subintervals of length 0.5.

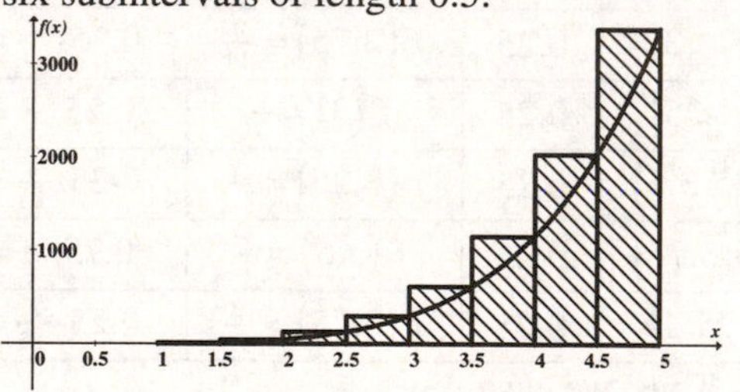

From the information above we have:

Subinterval	Width	Height (right endpoint)
$[1, 1.5]$	0.5	$f(1.5) = 15.84375$
$[1.5, 2]$	0.5	$f(2) = 50$
$[2, 2.5]$	0.5	$f(2.5) = 131.40625$
$[2.5, 3]$	0.5	$f(3) = 300$
$[3, 3.5]$	0.5	$f(3.5) = 614.46875$
$[3.5, 4]$	0.5	$f(4) = 1156$
$[4, 4.5]$	0.5	$f(4.5) = 2032.03125$
$[4.5, 5]$	0.5	$f(5) = 3380$

The solution is continued on the next page.

Copyright © 2014 Pearson Education, Inc. Publishing as Addison-Wesley.

Using the information from the previous page, we have

$$R_6 = 0.5[f(1.5)+f(2)+f(2.5)+f(3)$$
$$+f(3.5)+f(4)+f(4.5)+f(5)]$$
$$=0.5[15.84375+50+131.40625+300$$
$$+614.46875+1156+2032.03125+3380]$$
$$\approx 3839.875.$$

The average of L_8 and R_8 is

$$\frac{2151.875+3839.875}{2} = 2995.875.$$

11. $\int_0^4 (x^2+1)dx, \quad n=4$

a) Using the information from problem 1, we determine the midpoint of each interval and the height at the midpoint.

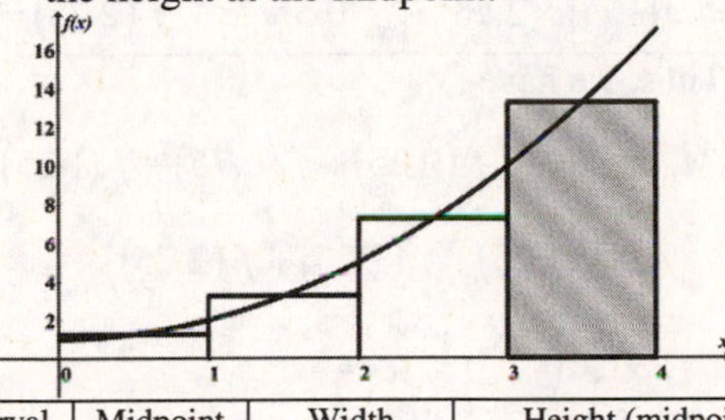

Subinterval	Midpoint	Width	Height (midpoint)
$[0,1]$	0.5	1	$f(0.5)=1.25$
$[1,2]$	1.5	1	$f(1.5)=3.25$
$[2,3]$	2.5	1	$f(2.5)=7.25$
$[3,4]$	3.5	1	$f(3.5)=13.25$

Thus, we have

$$M_4 = \frac{4-0}{4}(f(0.5)+f(1.5)+f(2.5)+f(3.5))$$
$$=1\cdot(1.25+3.25+7.25+13.25)$$
$$=25.$$

b) The exact value of the definite integral is

$$\int_0^4 (x^2+1)dx = \frac{76}{3}.$$

Therefore, the percent error is given by:

$$\text{Percent Error} = \frac{25-\frac{76}{3}}{\frac{76}{3}}\times 100 \approx -1.32\%.$$

13. $\int_0^3 (2x^3-x)dx, \quad n=6$

a) Using the information from problem 3, we determine the midpoint of each interval and the height at the midpoint.

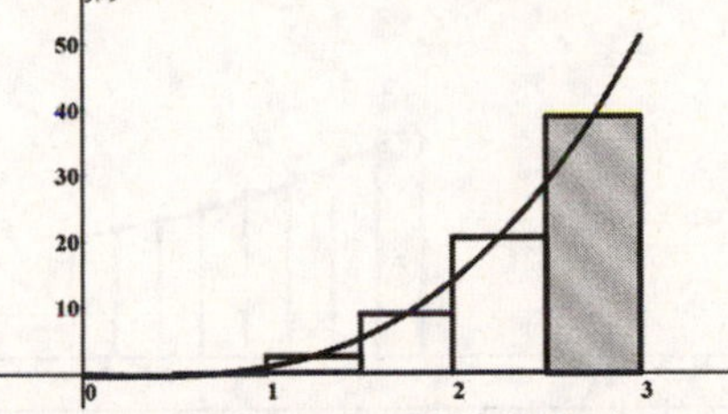

From the information above we have:

Subinterval	midpoint	Width	Height (midpoint)
$[0,0.5]$	0.25	0.5	$f(0.25)=-0.21875$
$[0.5,1]$	0.75	0.5	$f(0.75)=0.09375$
$[1,1.5]$	1.25	0.5	$f(1.25)=2.65625$
$[1.5,2]$	1.75	0.5	$f(1.75)=8.96875$
$[2,2.5]$	2.25	0.5	$f(2.25)=20.53125$
$[2.5,3]$	2.75	0.5	$f(2.75)=38.84375$

Thus, we have

$$M_6 = \frac{3-0}{6}\cdot(f(0.25)+f(0.75)+f(1.25)$$
$$+f(1.75)+f(2.25)+f(2.75))$$
$$=\frac{1}{2}\cdot(-0.21875+0.09375+2.65625$$
$$8.96875+20.53125+38.84375)$$
$$=35.4375.$$

b) The exact value of the definite integral is

$$\int_0^3 (2x^3-x)dx = 36.$$

Therefore, the percent error is given by:

$$\text{Percent Error} = \frac{35.4375-36}{36}\times 100$$
$$\approx -1.56\%.$$

Copyright © 2014 Pearson Education, Inc. Publishing as Addison-Wesley.

15. $\int_2^4 \frac{1}{x}dx, \quad n=8$

a) Using the information from problem 5, we determine the midpoint of each interval and the height at the midpoint.

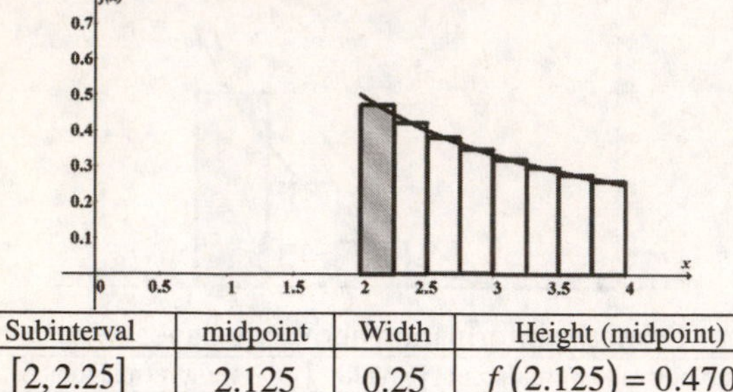

Subinterval	midpoint	Width	Height (midpoint)
$[2,2.25]$	2.125	0.25	$f(2.125)=0.4706$
$[2.25,2.5]$	2.375	0.25	$f(2.375)=0.4211$
$[2.5,2.75]$	2.625	0.25	$f(2.625)=0.3810$
$[2.75,3]$	2.875	0.25	$f(2.875)=0.3478$
$[3,3.25]$	3.125	0.25	$f(3.125)=0.32$
$[3.25,3.5]$	3.375	0.25	$f(3.375)=0.2963$
$[3.5,3.75]$	3.625	0.25	$f(3.625)=0.2759$
$[3.75,4]$	3.875	0.25	$f(3.875)=0.2581$

Thus, we have

$$M_8=\frac{4-2}{8}\cdot\big(f(2.125)+f(2.375)+f(2.625)$$
$$+f(2.875)+f(3.125)+f(3.375)$$
$$+f(3.625)+f(3.875)\big)$$
$$=\frac{1}{4}\cdot(0.4706+0.4211+0.3810+0.3478$$
$$+0.32+0.2963+0.2759+0.2581)$$
$$\approx 0.693.$$

b) The exact value of the definite integral is

$$\int_2^4 \frac{1}{x}dx = 0.693.$$

Therefore, the percent error is given by:

$$\text{Percent Error} = \frac{0.693-0.693}{0.693}\times 100 \approx 0.0\%.$$

17. $\int_0^3 e^{2x}dx, \quad n=6$

a) Using the information from problem 7, we determine the midpoint of each interval and the height at the midpoint at the top of the next column.

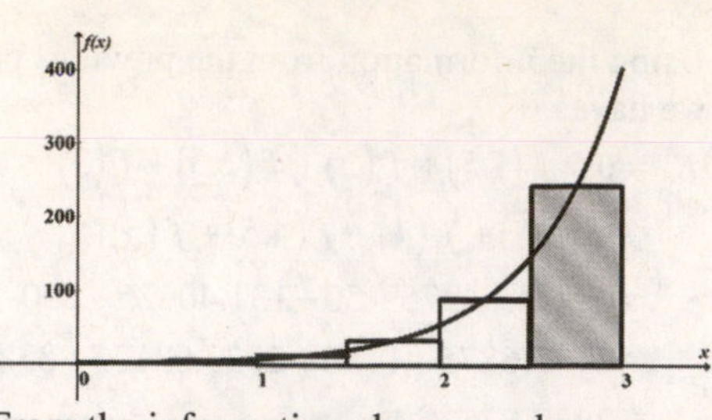

From the information above, we have:

Subinterval	midpoint	Width	Height (midpoint)
$[0,0.5]$	0.25	0.5	$f(0.25)=e^{0.5}$
$[0.5,1]$	0.75	0.5	$f(0.75)=e^{1.5}$
$[1,1.5]$	1.25	0.5	$f(1.25)=e^{2.5}$
$[1.5,2]$	1.75	0.5	$f(1.75)=e^{3.5}$
$[2,2.5]$	2.25	0.5	$f(2.25)=e^{4.5}$
$[2.5,3]$	2.75	0.5	$f(2.75)=e^{5.5}$

Thus, we have

$$M_6=\frac{3-0}{6}\cdot\big(f(0.25)+f(0.75)+f(1.25)$$
$$+f(1.75)+f(2.25)+f(2.75)\big)$$
$$=\frac{1}{2}\cdot\left(e^{0.5}+e^{1.5}+e^{2.5}+e^{3.5}+e^{4.5}+e^{5.5}\right)$$
$$\approx 193.069.$$

b) The exact value of the definite integral is

$$\int_0^3 e^{2x}dx = 201.214.$$

Therefore, the percent error is given by:

$$\text{Percent Error} = \frac{193.069-201.214}{201.214}\times 100$$
$$\approx -4.05\%.$$

19. $\int_1^5 x\left(x^2+1\right)^2 dx, \quad n=8$

a) Using the information from problem 9, we determine the midpoint of each interval and the height at the midpoint.

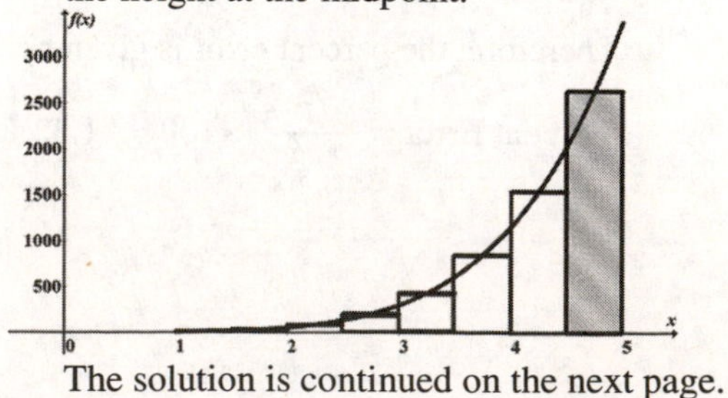

The solution is continued on the next page.

Copyright © 2014 Pearson Education, Inc. Publishing as Addison-Wesley.

From the information on the previous page, we have:

Subinterval	midpoint	Width	Height (midpoint)
$[1,1.5]$	1.25	0.5	$f(1.25)=8.2080$
$[1.5,2]$	1.75	0.5	$f(1.75)=28.8818$
$[2,2.5]$	2.25	0.5	$f(2.25)=82.6963$
$[2.5,3]$	2.75	0.5	$f(2.75)=201.6201$
$[3,3.5]$	3.25	0.5	$f(3.25)=434.4971$
$[3.5,4]$	3.75	0.5	$f(3.75)=850.7959$
$[4,4.5]$	4.25	0.5	$f(4.25)=1544.3604$
$[4.5,5]$	4.75	0.5	$f(4.75)=2637.1592$

Thus, we have

$$\begin{aligned}M_8 &= \frac{5-1}{8}\cdot\left(f(1.25)+f(1.75)+f(2.25)\right.\\&\quad +f(2.75)+f(3.25)+f(3.75)\\&\quad \left.+f(4.25)+f(4.75)\right)\\&=\frac{1}{2}\cdot(8.2080+28.8818+82.6963\\&\quad +201.6201+434.4971+850.7959\\&\quad +1544.3604+2637.1592)\\&\approx 2894.109.\end{aligned}$$

b) The exact value of the definite integral is

$$\int_1^5 x\left(x^2+1\right)^2 dx = 2928.$$

Therefore, the percent error is given by:

$$\text{Percent Error} = \frac{2894.109-2928}{2928}\times 100 \approx -1.16\%.$$

21. $\int_0^2 \sqrt{x^2+1}dx,\quad n=4$

The interval $[0,2]$ is divided into four equal subintervals. Thus, $\Delta x = \frac{2-0}{4}=\frac{1}{2}$, and the subintervals are:

$[0,0.5],[0.5,1],[1,1.5]$, and $[1.5,2]$.

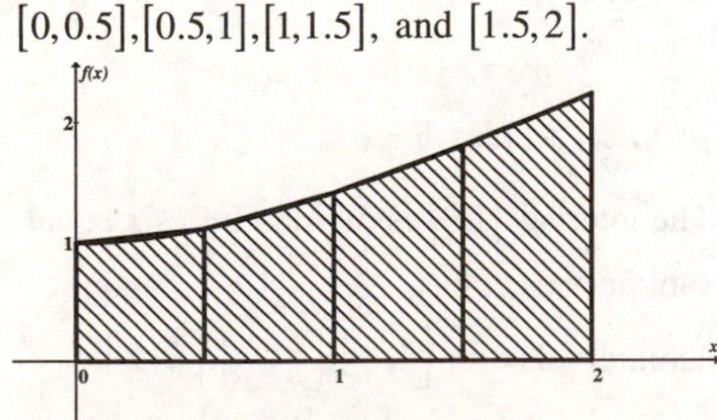

Using a calculator we have

$f(0)=1,\quad f(0.5)=1.118,\quad f(1)=1.414$

$f(1.5)=1.803,\qquad f(2)=2.236.$

Therefore,

$$\begin{aligned}T_4 &= \frac{1}{2}\left(\frac{f(0)}{2}+f(0.5)+f(1)+f(1.5)+\frac{f(2)}{2}\right)\\&=\frac{1}{2}\left(\frac{1}{2}+1.118+1.414+1.803+\frac{2.236}{2}\right)\\&=2.977.\end{aligned}$$

23. $\int_0^4 \frac{1}{x^2+1}dx,\quad n=8$

The interval $[0,4]$ is divided into eight equal subintervals. Thus, $\Delta x = \frac{4-0}{8}=\frac{1}{2}$, and the subintervals are: $[0,0.5],[0.5,1],[1,1.5]$ $[1.5,2],[2,2.5],[2.5,3],[3,3.5]$, and $[3.5,4]$.

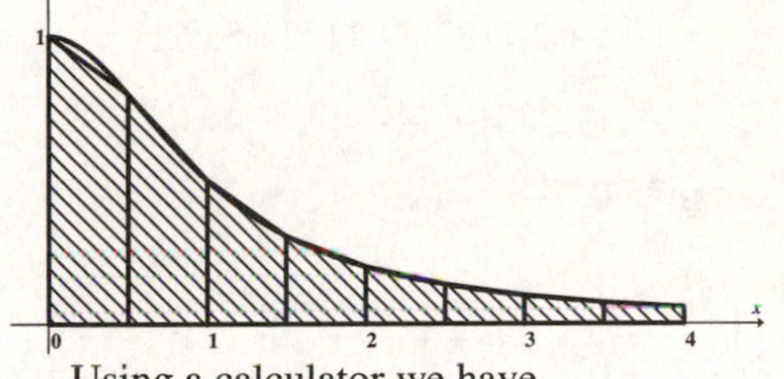

Using a calculator we have

$f(0)=1,\qquad f(0.5)=0.8$

$f(1)=0.5,\qquad f(1.5)=0.3077,$

$f(2)=0.2,\qquad f(2.5)=0.1379,$

$f(3)=0.1,\qquad f(3.5)=0.0755,$

$f(4)=0.0588.$

Therefore,

$$\begin{aligned}T_8 &= \frac{1}{2}\left(\frac{f(0)}{2}+f(0.5)+f(1)+f(1.5)+f(2)\right.\\&\quad \left.+f(2.5)+f(3)+f(3.5)+\frac{f(4)}{2}\right)\\&=\frac{1}{2}\left(\frac{1}{2}+0.8+0.5+0.3077+0.2\right.\\&\quad \left.+0.1379+0.1+0.0755+\frac{0.0588}{2}\right)\\&=1.325.\end{aligned}$$

25. $\int_0^5 e^{\sqrt{x}}dx,\quad n=5$

The interval $[0,5]$ is divided into five equal subintervals. Thus, $\Delta x = \frac{5-0}{5}=1$, and the subintervals are: $[0,1],[1,2],[2,3]$, $[3,4]$, and $[4,5]$. We draw the graph and calculate the function values on the next page.

Copyright © 2014 Pearson Education, Inc. Publishing as Addison-Wesley.

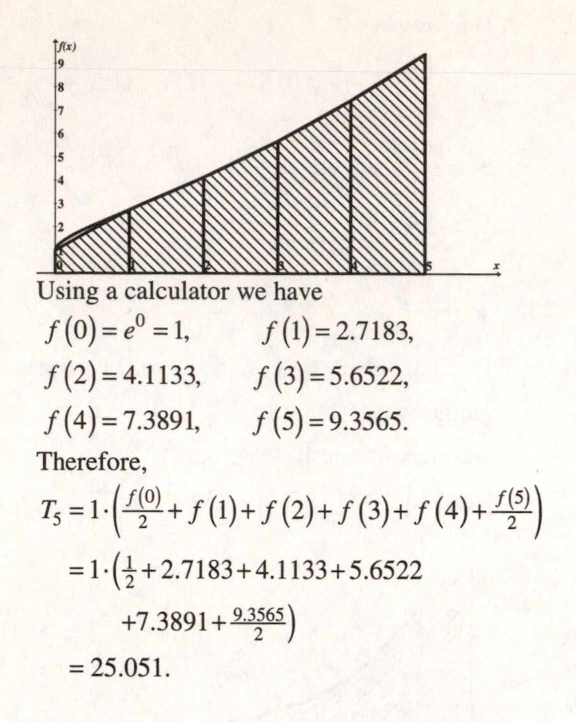

Using a calculator we have

$f(0)=e^0=1,\quad f(1)=2.7183,$

$f(2)=4.1133,\quad f(3)=5.6522,$

$f(4)=7.3891,\quad f(5)=9.3565.$

Therefore,

$$T_5=1\cdot\left(\tfrac{f(0)}{2}+f(1)+f(2)+f(3)+f(4)+\tfrac{f(5)}{2}\right)$$
$$=1\cdot\left(\tfrac{1}{2}+2.7183+4.1133+5.6522+7.3891+\tfrac{9.3565}{2}\right)$$
$$=25.051.$$

27. $\int_0^4 \frac{1}{x^3+1}dx,\quad n=4$

The interval $[0,4]$ is divided into four equal subintervals. Thus, $\Delta x=\frac{4-0}{4}=1$, and the subintervals are: $[0,1],[1,2],[2,3]$, and $[3,4]$.

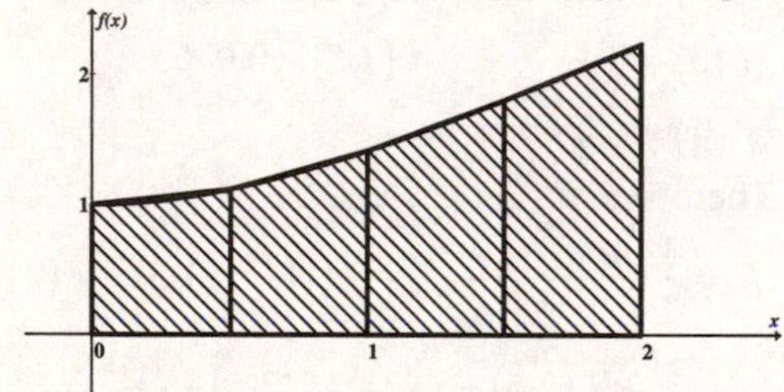

Using a calculator we have

$f(0)=1,\quad f(1)=0.5,\quad f(2)=0.1111$

$f(3)=0.0357,\quad f(4)=0.0154.$

Therefore,

$$T_4=1\cdot\left(\frac{f(0)}{2}+f(1)+f(2)+f(3)+\frac{f(4)}{2}\right)$$
$$=1\cdot\left(\frac{1}{2}+0.5+0.1111+0.0357+\frac{0.0154}{2}\right)$$
$$=1.155.$$

29. $\int_2^4 \sqrt{x^2-1}dx,\quad n=4$

The interval $[2,4]$ is divided into four equal subintervals. Thus, $\Delta x=\frac{4-2}{4}=\frac{1}{2}$, and the subintervals are:

$[2,2.5],[2.5,3],[3,3.5]$, and $[3.5,4]$.

Using a calculator we have

$f(2)=1.7321,\quad f(2.5)=2.2913,\quad f(3)=2.8284$

$f(3.5)=3.3541,\quad f(4)=3.8730.$

Therefore,

$$S_4=\tfrac{4-2}{3\cdot 2}\left(f(2)+4f(2.5)+2f(3)+4f(3.5)+f(4)\right)$$
$$=\tfrac{1}{6}\left(1.732+4(2.2913)+2(2.8284)+4(3.3541)+3.8730\right)$$
$$=5.641$$

31. $\int_{-1}^1 e^{-x^3}dx,\quad n=6$

The interval $[-1,1]$ is divided into six equal subintervals. Thus, $\Delta x=\frac{1-(-1)}{6}=\frac{1}{3}$, and the subintervals are: $\left[-1,\frac{-2}{3}\right],\left[\frac{-2}{3},\frac{-1}{3}\right],\left[\frac{-1}{3},0\right],$ $\left[0,\frac{1}{3}\right],\left[\frac{1}{3},\frac{2}{3}\right]$, and $\left[\frac{2}{3},1\right]$.

Using a calculator we have

$f(-1)=2.7183,\quad f\left(\frac{-2}{3}\right)=1.3449,$

$f\left(\frac{-1}{3}\right)=1.0377,\quad f(0)=1,$

$f\left(\frac{1}{3}\right)=0.9636,\quad f\left(\frac{2}{3}\right)=0.7436,$

$f(1)=0.3679.$

Therefore,

$$S_6=\tfrac{1-(-1)}{3\cdot 6}\left(f(-1)+4f\left(\tfrac{-2}{3}\right)+2f\left(\tfrac{-1}{3}\right)+4(0)+2f\left(\tfrac{1}{3}\right)+4f\left(\tfrac{2}{3}\right)+f(1)\right)$$
$$=\tfrac{1}{9}\left(2.7183+4(1.3449)+2(1.0377)+4(1)+2(0.9636)+4(0.7436)+0.3679\right)$$
$$=2.160.$$

33. $\int_1^3 \ln\left(x^2+1\right)dx,\quad n=6$

The interval $[1,3]$ is divided into six equal subintervals. Thus, $\Delta x=\frac{3-1}{6}=\frac{1}{3}$, and the subintervals are: $\left[1,\frac{4}{3}\right],\left[\frac{4}{3},\frac{5}{3}\right],\left[\frac{5}{3},2\right],$ $\left[2,\frac{7}{3}\right],\left[\frac{7}{3},\frac{8}{3}\right]$, and $\left[\frac{8}{3},3\right]$.

The solution is continued on the next page.

Copyright © 2014 Pearson Education, Inc. Publishing as Addison-Wesley.

Using a calculator we have

$f(1)=0.6932,\quad f\left(\tfrac{4}{3}\right)=1.0217,$

$f\left(\tfrac{5}{3}\right)=1.3291,\quad f(2)=1.6094,$

$f\left(\tfrac{7}{3}\right)=1.8632,\quad f\left(\tfrac{8}{3}\right)=2.0932,$

$f(3)=2.3026.$

Therefore,

$$S_6=\tfrac{3-1}{3\cdot 6}\left(f(1)+4f\left(\tfrac{4}{3}\right)+2f\left(\tfrac{5}{3}\right)+4(2)+2f\left(\tfrac{7}{3}\right)+4f\left(\tfrac{8}{3}\right)+f(3)\right)$$
$$=\tfrac{1}{9}\left(0.6932+4(1.0217)+2(1.3291)+4(1.6094)+2(1.8632)+4(2.0932)+2.3026\right)$$
$$=3.142.$$

35. $\int_1^5 \frac{1}{\sqrt{x^2+1}}dx,\quad n=4$

The interval $[1,5]$ is divided into four equal subintervals. Thus, $\Delta x=\frac{5-1}{4}=1$, and the subintervals are: $[1,2],[2,3],[3,4]$, and $[4,5]$.

Using a calculator we have

$f(1)=0.7071,\quad f(2)=0.4472,\quad f(3)=0.3162$

$f(4)=0.2425,\quad f(5)=0.1961.$

Therefore,

$$S_4=\tfrac{5-1}{3\cdot 4}\left(f(0)+4f\left(\tfrac{1}{4}\right)+2f\left(\tfrac{1}{2}\right)+4f\left(\tfrac{3}{4}\right)+f(1)\right)$$
$$=\tfrac{1}{3}\left(0.7071+4(0.4472)+2(0.3162)+4(0.2425)+0.1961\right)$$
$$=1.432$$

37. The exact value of the definite integral is $\int_0^2 \sqrt{x^2+1}dx=2.958$.

Therefore, the percent error is given by:

$$\text{Percent Error}=\frac{2.977-2.958}{2.958}\times 100\approx 0.64\%.$$

39. The exact value of the definite integral is $\int_0^4 \frac{1}{x^2+1}dx=1.326$.

Therefore, the percent error is given by:

$$\text{Percent Error}=\frac{1.325-1.326}{1.326}\times 100\approx -0.08\%.$$

41. The exact value of the definite integral is $\int_0^5 e^{\sqrt{x}}dx=25.130$.

Therefore, the percent error is given by:

$$\text{Percent Error}=\frac{25.051-25.130}{25.131}\times 100\approx -0.31\%.$$

43. The exact value of the definite integral is $\int_0^4 \frac{1}{x^3+1}dx=1.178$.

Therefore, the percent error is given by:

$$\text{Percent Error}=\frac{1.155-1.178}{1.178}\times 100\approx -1.95\%.$$

45. The exact value of the definite integral is $\int_2^4 \sqrt{x^2-1}dx=5.641$.

Therefore, the percent error is given by:

$$\text{Percent Error}=\frac{5.641-5.641}{5.641}\times 100\approx 0.0\%.$$

47. The exact value of the definite integral is $\int_{-1}^1 e^{-x^3}dx=2.149$.

Therefore, the percent error is given by:

$$\text{Percent Error}=\frac{2.160-2.149}{2.149}\times 100\approx 0.51\%.$$

49. The exact value of the definite integral is $\int_1^3 \ln\left(x^2+1\right)dx=3.142$.

Therefore, the percent error is given by:

$$\text{Percent Error}=\frac{3.142-3.142}{3.142}\times 100\approx 0.0\%.$$

51. The exact value of the definite integral is $\int_1^5 \frac{1}{\sqrt{x^2+1}}dx=1.431$.

Therefore, the percent error is given by:

$$\text{Percent Error}=\frac{1.432-1.431}{1.431}\times 100\approx 0.07\%.$$

Copyright © 2014 Pearson Education, Inc. Publishing as Addison-Wesley.

53. To create consistency with units, each one minute interval of time is equivalent to 1/60 of an hour. If we are use left rectangles the right most data point will not be used. We organize the data into the following table:

Subinterval	Width	Height (left endpoint)	Area (in miles)
$[0,1]$	1/60 hr	$v(0)=0$	$\left(\frac{1}{60}\right)\cdot 0=0$
$[1,2]$	1/60 hr	$v(1)=25$	$\left(\frac{1}{60}\right)\cdot 25\approx 0.4167$
$[2,3]$	1/60 hr	$v(2)=30$	$\left(\frac{1}{60}\right)\cdot 30=0.5$
$[3,4]$	1/60 hr	$v(3)=35$	$\left(\frac{1}{60}\right)\cdot 35\approx 0.5833$
$[4,5]$	1/60 hr	$v(4)=30$	$\left(\frac{1}{60}\right)\cdot 30=0.5$
$[5,6]$	1/60 hr	$v(5)=22$	$\left(\frac{1}{60}\right)\cdot 22\approx 0.3667$
$[6,7]$	1/60 hr	$v(6)=20$	$\left(\frac{1}{60}\right)\cdot 20\approx 0.3333$
$[7,8]$	1/60 hr	$v(7)=10$	$\left(\frac{1}{60}\right)\cdot 10\approx 0.1667$

Therefore;

$$L_8 = 0+0.4167+0.5+0.5833+0.5+0.3667 +0.3333+0.1667$$

$L_8 = 2.867$ miles.

If we use right rectangles the left most data point will not be used. We organize the data in the following table.

Subinterval	Width	Height (right endpoint)	Area (in miles)
$[0,1]$	1/60 hr	$v(1)=25$	$\left(\frac{1}{60}\right)\cdot 25\approx 0.4167$
$[1,2]$	1/60 hr	$v(2)=30$	$\left(\frac{1}{60}\right)\cdot 30=0.5$
$[2,3]$	1/60 hr	$v(3)=35$	$\left(\frac{1}{60}\right)\cdot 35\approx 0.5833$
$[3,4]$	1/60 hr	$v(4)=30$	$\left(\frac{1}{60}\right)\cdot 30=0.5$
$[4,5]$	1/60 hr	$v(5)=22$	$\left(\frac{1}{60}\right)\cdot 22\approx 0.3667$
$[5,6]$	1/60 hr	$v(6)=20$	$\left(\frac{1}{60}\right)\cdot 20\approx 0.3333$
$[6,7]$	1/60 hr	$v(7)=10$	$\left(\frac{1}{60}\right)\cdot 10\approx 0.1667$
$[7,8]$	1/60 hr	$v(8)=10$	$\left(\frac{1}{60}\right)\cdot 10\approx 0.1667$

Therefore;

$$R_8 = 0.4167+0.5+0.5833+0.5+0.3667 +0.3333+0.1667+0.1667$$

$R_8 = 3.033$ miles

The average between the two estimates is:

$$average = \frac{2.867+3.033}{2} = 2.95 \text{ miles.}$$

Therefore, we estimate that Moira traveled 2.95 miles during the eight minutes.

55. Using the data in the picture, we know that the width of the interval $\Delta x = 2$ ft. Using the diagram we define the function values as the depth of the stream at each interval as follows:

$f(a)=0,\quad f(x_6)=4,$
$f(x_1)=2,\quad f(x_7)=2,$
$f(x_2)=3,\quad f(x_8)=1.5,$
$f(x_3)=3,\quad f(x_9)=1.6,$
$f(x_4)=4,\quad f(x_{10})=1.5,$
$f(x_5)=5,\quad f(b)=0.$

Applying the trapezoidal rule we have

$$T_{11} = 2\left(\tfrac{0}{2}+2+3+3+4+5+4+2+1.5 +1.6+1.5+\tfrac{0}{2}\right)$$
$$= 55.2 \text{ sq. ft.}$$

Therefore, the cross-sectional area of the stream is 55.2 sq. ft.

57. a) Divide the interval $[0,4]$ into eight equal subintervals. Therefore, $\Delta x = \frac{4-0}{8} = \frac{1}{2}$, and the subintervals are $[0,0.5],[0.5,1],[1,1.5],$ $[1.5,2],[2,2.5],[2.5,3],[3,3.5],$ and $[3.5,4]$.

Using a calculator, the function values are

$f(0)=3,\quad f(0.5)=2.97647,$
$f(1)=2.90474,\quad f(1.5)=2.78107,$
$f(2)=2.59808,\quad f(2.5)=2.34187,$
$f(3)=1.98431,\quad f(3.5)=1.45237,$
$f(4)=0.$

Applying the trapezoidal rule, we have

$$T_8 = \frac{1}{2}\left(\tfrac{f(0)}{2}+f(0.5)+f(1)+f(1.5)+f(2) +f(2.5)+f(3)+f(3.5)+\tfrac{f(4)}{2}\right)$$
$$=\frac{1}{2}\left(\tfrac{3}{2}+2.97647+2.90474+2.78107+2.59808 +2.34187+1.98431+1.45237+\tfrac{0}{2}\right)$$
$$= 9.269.$$

The area of the portion of the ellipse that lies in the first quadrant is approximately 9.269.

Copyright © 2014 Pearson Education, Inc. Publishing as Addison-Wesley.

b) Using the symmetrical properties of the ellipse. The total area enclosed within the ellipse is $4 \cdot 9.269 = 37.076$.

59. a) First we find the derivative of

$$f(x) = \sqrt{1-x^2} = \left(1-x^2\right)^{\frac{1}{2}}.$$

$$f'(x) = \tfrac{1}{2}\left(1-x^2\right)^{\frac{1}{2}-1}(-2x) = \frac{-x}{\sqrt{1-x^2}}.$$

Therefore, we will use the midpoint rule to approximate the integral

$$\int_0^1 \sqrt{1+\frac{x^2}{1-x^2}}\,dx, \qquad n = 6.$$

The function in the integrand is

$$g(x) = \sqrt{1+\frac{x^2}{1-x^2}}.$$

Divide the interval $[0,1]$ into six subintervals of equal length $\Delta x = \frac{1-0}{6} = \frac{1}{6}$.

We organize the information in the table below:

Subinterval	midpoint	Width	Height (midpoint)
$\left[0,\frac{1}{6}\right]$	$\frac{1}{12}$	$\frac{1}{6}$	$g\left(\frac{1}{12}\right) = 1.0035$
$\left[\frac{1}{6},\frac{1}{3}\right]$	$\frac{1}{4}$	$\frac{1}{6}$	$g\left(\frac{1}{4}\right) = 1.0328$
$\left[\frac{1}{3},\frac{1}{2}\right]$	$\frac{5}{12}$	$\frac{1}{6}$	$g\left(\frac{5}{12}\right) = 1.1$
$\left[\frac{1}{2},\frac{2}{3}\right]$	$\frac{7}{12}$	$\frac{1}{6}$	$g\left(\frac{7}{12}\right) = 1.2312$
$\left[\frac{2}{3},\frac{5}{6}\right]$	$\frac{3}{4}$	$\frac{1}{6}$	$g\left(\frac{3}{4}\right) = 1.5119$
$\left[\frac{5}{6},1\right]$	$\frac{11}{12}$	$\frac{1}{6}$	$g\left(\frac{11}{12}\right) = 2.5022$

Thus, we have

$$\begin{aligned} M_6 &= \frac{1-0}{6}\cdot\left(g\left(\tfrac{1}{12}\right)+g\left(\tfrac{1}{4}\right)+g\left(\tfrac{5}{12}\right)\right. \\ &\quad \left.+g\left(\tfrac{7}{12}\right)+g\left(\tfrac{3}{4}\right)+g\left(\tfrac{11}{12}\right)\right) \\ &= \frac{1}{6}\cdot(1.0035+1.0328+1.1+1.2312 \\ &\quad +1.5119+2.5022) \\ &= 1.397. \end{aligned}$$

b) Using the arc length formula we have:

$$s = r\theta = 1\left(\tfrac{\pi}{2}\right) = \tfrac{\pi}{2} \approx 1.571.$$

c) The percent error is

$$\frac{1.397-1.571}{1.571}\times 100 = -11.08\%.$$

61. First we find the derivative of

$$f(x) = e^x.$$

$$f'(x) = e^x.$$

Therefore, we will use Simpson's rule to approximate the integral

$$\int_1^2 \sqrt{1+\left(e^x\right)^2}\,dx = \int_1^2 \sqrt{1+e^{2x}}\,dx.$$

We are asked to find S_6, so we divide the interval $[-1,2]$ into six equal subintervals.

Thus, $\Delta x = \frac{2-(-1)}{6} = \frac{1}{2}$ and the subintervals are

$\left[-1,\frac{-1}{2}\right], \left[\frac{-1}{2},0\right], \left[0,\frac{1}{2}\right], \left[\frac{1}{2},1\right],$

$\left[1,\frac{3}{2}\right]$, and $\left[\frac{3}{2},2\right]$.

The solution is continued on the next page.

Using a calculator we find the values of the integrand $g(x) = \sqrt{1+e^{2x}}$

$g(-1) = 1.0655,$ $g\left(\frac{-1}{2}\right) = 1.1696,$

$g(0) = 1.4142,$ $g\left(\frac{1}{2}\right) = 1.9283,$

$g(1) = 2.8964,$ $g\left(\frac{3}{2}\right) = 4.5919,$

$g(2) = 7.4564.$

Therefore, we have

$$\begin{aligned} S_6 &= \frac{2-(-1)}{3\cdot 6}\left(g(-1)+4g\left(\tfrac{-1}{2}\right)+2g(0)+4\left(\tfrac{1}{2}\right)\right. \\ &\quad \left.+2g(1)+4g\left(\tfrac{3}{2}\right)+g(2)\right) \\ &= \tfrac{1}{6}(1.0655+4(1.1696)+2(1.4142)+4(1.9283) \\ &\quad +2(2.8964)+4(4.5919)+7.4564) \\ &= 7.984. \end{aligned}$$

Copyright © 2014 Pearson Education, Inc. Publishing as Addison-Wesley.

63. First we need to find the area of the back of the storage shed. Using the data in the problem we find the area using left rectangles.

Width	Height (left endpoint)	Area
3	5	$3\cdot 5=15$
3	7	$3\cdot 7=21$
3	8	$3\cdot 8=24$
3	8.5	$3\cdot 8.5=25.5$
3	9	$3\cdot 9=27$
3	9.25	$3\cdot 9.25=27.75$
3	9.5	$3\cdot 9.5=28.5$
3	9.5	$3\cdot 9.5=28.5$
3	9.5	$3\cdot 9.5=28.5$
3	9.25	$3\cdot 9.25=27.75$
3	9	$3\cdot 9=27$
3	8.5	$3\cdot 8.5=25.5$
3	8	$3\cdot 8=24$
3	7	$3\cdot 7=21$

Therefore, the area of the back of the shed is

$$\begin{aligned} L_{13} &= 15+21+24+25.5+27+27.75+28.5 \\ &\quad +28.5+28.5+27.75+27+25.5+24+21 \\ &= 351. \end{aligned}$$

The total area of the back of the storage shed is 351 square feet. Since the cost to install the metal sheeting is \$6 per square foot, it will cost approximately $\$6\cdot 351=\$2106$ to cover the back the shed in metal sheeting.
Note: due to the symmetry of the shed, using right rectangles will yield the same result.

65. Answers will vary depending on the numerical integration technique used. First we find the area of the green. Using left rectangles we have:

Width	Height (left endpoint)	Area
10	40	$10\cdot 40=400$
10	50	$10\cdot 50=500$
10	50	$10\cdot 50=500$
10	55	$10\cdot 55=550$
10	55	$10\cdot 55=550$
10	57	$10\cdot 57=570$
10	48	$10\cdot 48=480$

The approximate area of the green using left rectangles is

$$\begin{aligned} L_7 &= 400+500+500+550+550+570+480 \\ &= 3550 \text{ square feet.} \end{aligned}$$

Using right rectangles we have:

Width	Height (left endpoint)	Area
10	50	$10\cdot 50=500$
10	50	$10\cdot 50=500$
10	55	$10\cdot 55=550$
10	55	$10\cdot 55=550$
10	57	$10\cdot 57=570$
10	48	$10\cdot 48=480$
10	25	$10\cdot 25=250$

The approximate area of the green using right rectangles is

$$\begin{aligned} R_7 &= 500+500+550+550+570+480+250 \\ &= 3400 \text{ square feet.} \end{aligned}$$

Using the average of the two approximations, we approximate the area of the green to be

$$\frac{3550+3400}{2}=3475 \text{ square feet.}$$

If the cost of maintaining the green is \$2.75 per square foot, the approximate cost to maintain the green is $\$2.75\cdot 3475=\$9556.25$.

67. Given the conditions in the problem we know

a) $L_n < \int_a^b f(x)\,dx$

b) $R_n > \int_a^b f(x)\,dx$

c) $T_n > \int_a^b f(x)\,dx$

69. Given the conditions in the problem we know

a) $L_n < \int_a^b f(x)\,dx$

b) $R_n > \int_a^b f(x)\,dx$

c) $T_n < \int_a^b f(x)\,dx$

71. ✎

73. ✎

Copyright © 2014 Pearson Education, Inc. Publishing as Addison-Wesley.

Exercise Set 5.5

1. Find the volume of the solid of revolution generated by rotating about the x-axis the region under the graph of

$y = x$

from $x = 0$ to $x = 1$

$V = \int_a^b \pi[f(x)]^2\,dx$ Volume of a solid of revolution

$V = \int_0^1 \pi[x]^2\,dx$ Substituting 0 for a, 1 for b, and x for $f(x)$.

$V = \int_0^1 \pi x^2 dx$

$= \left[\pi \cdot \frac{x^3}{3}\right]_0^1$

$= \frac{\pi}{3}\left[1^3 - 0^3\right]$

$= \frac{\pi}{3}[1]$

$= \frac{\pi}{3}$, or about 1.05.

3. Find the volume of the solid of revolution generated by rotating about the x-axis the region under the graph of

$y = \sqrt{x}$

from $x = 1$ to $x = 4$

$V = \int_a^b \pi[f(x)]^2\,dx$ Volume of a solid of revolution

$V = \int_1^4 \pi\left[\sqrt{x}\right]^2\,dx$ Substituting 1 for a, 4 for b, and $\sqrt{x}$ for $f(x)$.

$V = \int_1^4 \pi x dx$

$= \left[\pi \cdot \frac{x^2}{2}\right]_1^4$

$= \frac{\pi}{2}\left[4^2 - 1^2\right]$

$= \frac{\pi}{2}[15]$

$= \frac{15\pi}{2}$, or about 23.56.

5. Find the volume of the solid of revolution generated by rotating about the x-axis the region under the graph of

$y = e^x$

from $x = -2$ to $x = 5$

$V = \int_a^b \pi[f(x)]^2\,dx$ Volume of a solid of revolution

$V = \int_{-2}^5 \pi\left[e^x\right]^2\,dx$ Substituting -2 for a, 5 for b, and e^x for $f(x)$.

$V = \int_{-2}^5 \pi e^{2x}\,dx$

$= \left[\pi \cdot \frac{1}{2}e^{2x}\right]_{-2}^5$

$= \frac{\pi}{2}\left[e^{2(5)} - e^{2(-2)}\right]$

$= \frac{\pi}{2}\left[e^{10} - e^{-4}\right]$, or about 34,599.06.

7. Find the volume of the solid of revolution generated by rotating about the x-axis the region under the graph of

$y = \frac{1}{x}$

from $x = 1$ to $x = 3$

$V = \int_a^b \pi[f(x)]^2\,dx$ Volume of a solid of revolution

$V = \int_1^3 \pi\left[\frac{1}{x}\right]^2\,dx$ Substituting 1 for a, 3 for b, and $1/x$ for $f(x)$.

$V = \int_1^3 \pi \cdot \frac{1}{x^2}\,dx$

$V = \int_1^3 \pi x^{-2}\,dx$

$= \left[\pi \cdot \frac{x^{-1}}{-1}\right]_1^3$

$= -\pi\left[\frac{1}{x}\right]_1^3$

$= -\pi\left[\frac{1}{3} - \frac{1}{1}\right]$

$= -\pi\left[-\frac{2}{3}\right]$

$= \frac{2\pi}{3}$, or about 2.09.

Copyright © 2014 Pearson Education, Inc. Publishing as Addison-Wesley.

9. Find the volume of the solid of revolution generated by rotating about the x-axis the region under the graph of

$$y=\frac{2}{\sqrt{x}}$$

from $x=4$ to $x=9$

$V=\int_a^b \pi[f(x)]^2\,dx$ Volume of a solid of revolution

$V=\int_4^9 \pi\left[\frac{2}{\sqrt{x}}\right]^2 dx$ Substituting 4 for a, 9 for b, and $2/\sqrt{x}$ for $f(x)$.

$V=\int_4^9 \pi\cdot\frac{4}{x}dx$

$V=4\pi\int_4^9 \frac{1}{x}dx$

$=4\pi[\ln x]_4^9$

$=4\pi[\ln 9-\ln 4]$

$=4\pi\ln\left(\frac{9}{4}\right)$, or about 10.19.

11. Find the volume of the solid of revolution generated by rotating about the x-axis the region under the graph of

$$y=4$$

from $x=1$ to $x=3$

$V=\int_a^b \pi[f(x)]^2\,dx$ Volume of a solid of revolution

$V=\int_1^3 \pi[4]^2\,dx$ Substituting 1 for a, 3 for b, and 4 for $f(x)$.

$V=\int_1^3 16\pi dx$

$=16\pi[x]_1^3$

$=16\pi[3-1]$

$=16\pi[2]$

$=32\pi$, or about 100.53.

13. Find the volume of the solid of revolution generated by rotating about the x-axis the region under the graph of

$$y=x^2$$

from $x=0$ to $x=2$

$V=\int_a^b \pi[f(x)]^2\,dx$ Volume of a solid of revolution

$V=\int_0^2 \pi\left[x^2\right]^2 dx$ Substituting 0 for a, 2 for b, and x^2 for $f(x)$.

$V=\int_0^2 \pi x^4 dx$

$=\left[\pi\cdot\frac{x^5}{5}\right]_0^2$

$=\frac{\pi}{5}\left[2^5-0^5\right]$

$=\frac{\pi}{5}[32]$

$=\frac{32\pi}{5}$, or about 20.11.

15. Find the volume of the solid of revolution generated by rotating about the x-axis the region under the graph of

$$y=\sqrt{1+x}$$

from $x=2$ to $x=10$

$V=\int_a^b \pi[f(x)]^2\,dx$ Volume of a solid of revolution

$V=\int_2^{10} \pi\left[\sqrt{1+x}\right]^2 dx$ Substituting 2 for a, 10 for b, and $\sqrt{1+x}$ for $f(x)$.

$V=\int_2^{10} \pi(1+x)dx$

$=\pi\left[\frac{(1+x)^2}{2}\right]_2^{10}$

$=\frac{\pi}{2}\left[(1+10)^2-(1+2)^2\right]$

$=\frac{\pi}{2}\left[(11)^2-(3)^2\right]$

$=\frac{\pi}{2}[121-9]$

$=\frac{\pi}{2}\cdot 112$

$=56\pi$, or about 175.93.

Copyright © 2014 Pearson Education, Inc. Publishing as Addison-Wesley.

17. Find the volume of the solid of revolution generated by rotating about the x-axis the region under the graph of

$$y=\sqrt{4-x^2}$$

from $x=-2$ to $x=2$

$$V=\int_a^b \pi\left[f(x)\right]^2 dx \quad \text{Volume of a solid of revolution}$$

$$V=\int_{-2}^{2} \pi\left[\sqrt{4-x^2}\right]^2 dx \quad \text{Substituting } -2 \text{ for } a,\ 2 \text{ for } b, \text{ and } \sqrt{4-x^2} \text{ for } f(x).$$

$$V=\int_{-2}^{2} \pi\left(4-x^2\right)dx$$

$$=\pi\left[4x-\frac{x^3}{3}\right]_{-2}^{2}$$

$$=\pi\left[\left(4(2)-\frac{(2)^3}{3}\right)-\left(4(-2)-\frac{(-2)^3}{3}\right)\right]$$

$$=\pi\left[\left(8-\frac{8}{3}\right)-\left(-8+\frac{8}{3}\right)\right]$$

$$=\pi\left[\frac{16}{3}-\left(-\frac{16}{3}\right)\right]$$

$$=\pi\left(\frac{32}{3}\right)$$

$$=\frac{32\pi}{3}, \text{ or about } 33.51.$$

19. Find the volume of the solid of revolution generated by rotating about the x-axis the region under the graph of

$$y=50\cdot\sqrt{1+\frac{x^2}{22{,}500}}$$

From $x=-250$ to $x=150$.

The volume is:

$$V=\int_{-250}^{150} \pi\left[50\cdot\sqrt{1+\frac{x^2}{22{,}500}}\right]^2 dx$$

$$=\int_{-250}^{150} \pi\left[2500\left(1+\frac{x^2}{22{,}500}\right)\right]dx$$

Continued at the top of the next column.

$$V=\int_{-250}^{150} \pi\left[2500+\frac{x^2}{9}\right]dx$$

$$=\pi\left[2500x+\frac{x^3}{27}\right]_{-250}^{150}$$

$$=\pi\left[\left(2500(150)+\frac{(150)^3}{27}\right)-\left(2500(-250)+\frac{(-250)^3}{27}\right)\right]$$

$$\approx 1{,}703{,}703.7\pi.$$

The volume of the tower is approximately $1{,}703{,}703.7\pi$ ft^3.

21. Graphing the equations, we have

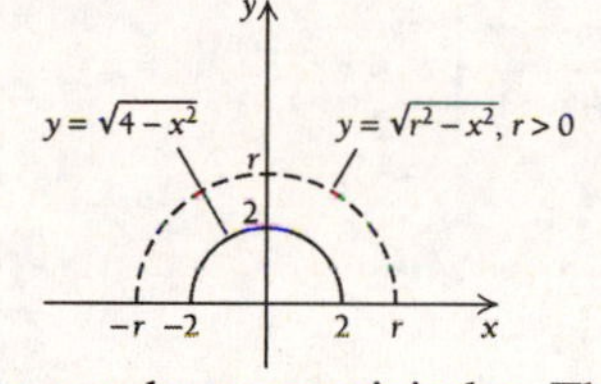

The graphs are semicircles. Their rotation about the x-axis creates spheres of radius 2 and radius r respectively. In Exercise 18, by finding the volume of the solid of revolution created by rotating $y=\sqrt{r^2-x^2}$, we actually derived the general formula for finding the volume of a sphere with radius r.

23. $V=\int_a^b \pi\left[f(x)\right]^2 dx$

$$V=\int_e^{e^3} \pi\left[\sqrt{\ln x}\right]^2 dx$$

$$=\int_e^{e^3} \pi\cdot \ln x dx$$

$$=\pi\left[x\ln x-x\right]_e^{e^3} \quad \text{Using Formula 8}$$

$$=\pi\left[\left(e^3\ln e^3-e^3\right)-\left(e\ln e-e\right)\right]$$

$$=\pi\left[\left(3e^3-e^3\right)-\left(e-e\right)\right]$$

$$=\pi\left[2e^3\right]$$

$$=2\pi e^3, \text{ or about } 126.20.$$

Copyright © 2014 Pearson Education, Inc. Publishing as Addison-Wesley.

25. $V = \int_1^{\infty} \pi \left[\frac{1}{x}\right]^2 dx$

$$V = \int_1^{\infty} \pi \frac{1}{x^2} dx$$
$$= \int_1^{\infty} \pi x^{-2} dx$$
$$= \lim_{b\to\infty} \int_1^{b} \pi x^{-2} dx$$
$$= \lim_{b\to\infty} \left[\pi \frac{x^{-1}}{-1}\right]_1^b$$
$$= \lim_{b\to\infty} \left[-\frac{\pi}{x}\right]_1^b$$
$$= \lim_{b\to\infty} \left[-\frac{\pi}{b} - \left(-\frac{\pi}{1}\right)\right]$$
$$= \lim_{b\to\infty} \left[-\frac{\pi}{b} + \frac{\pi}{1}\right]$$
$$= [0 + \pi]$$
$$= \pi.$$

Copyright © 2014 Pearson Education, Inc. Publishing as Addison-Wesley.

Chapter 6

Functions of Several Variables

Exercise Set 6.1

1. $f(x,y)=x^2-3xy$

$f(0,-2)=(0)^2-3(0)(-2)$ Substituting 0 for x and -2 for y.

$=0-0$

$=0$

$f(2,3)=(2)^2-3(2)(3)$ Substituting 2 for x and 3 for y.

$=4-18$

$=-14$

$f(10,-5)=(10)^2-3(10)(-5)$ Substituting 10 for x and -5 for y.

$=100+150$

$=250$

3. $f(x,y)=3^x+7xy$

$f(0,-2)=3^0+7(0)(-2)$ Substituting 0 for x and -2 for y.

$=1+0$

$=1$

$f(-2,1)=3^{-2}+7(-2)(1)$ Substituting -2 for x and 1 for y.

$=\frac{1}{9}-14$

$=-\frac{125}{9}$

$f(2,1)=3^2+7(2)(1)$ Substituting 2 for x and 1 for y.

$=9+14$

$=23$

5. $f(x,y)=\ln x+y^3$

$f(e,2)=\ln e+(2)^3$ Substituting e for x and 2 for y.

$=1+8$

$=9$

$f(e^2,4)=\ln e^2+(4)^3$ Substituting e^2 for x and 4 for y.

$=2+64$

$=66$

$f(e^3,4)=\ln e^3+(5)^3$ Substituting e^3 for x and 5 for y.

$=3+125$

$=128$

7. $f(x,y,z)=x^2-y^2+z^2$

We substitute -1 for x, 2 for y, and 3 for z.

$f(-1,2,3)=(-1)^2-(2)^2+(3)^2$

$=1-4+9$

$=6$

We substitute 2 for x, -1 for y, and 3 for z.

$f(2,-1,3)=(2)^2-(-1)^2+(3)^2$

$=4-1+9$

$=12$

9. $f(x,y)=\sqrt{y-3x}$

The function is defined only when $y-3x\geq 0$.

Therefore, the domain is:

$\{(x,y)\,|\,y\geq 3x\}$.

11. $h(x,y)=xe^{\sqrt{y}}$

The function is defined for all x and when $y\geq 0$. Therefore, the domain is:

$\{(x,y)\,|\,y\geq 0\}$.

13. $R(P,E)=\frac{P}{E}$

Substituting 32.03 for P, and 1.25 for E, gives

$R(32.03,1.25)=\frac{32.03}{1.25}$

≈ 25.624

$\approx 25.62.$

The price-earnings ration for Hewlett-Packard was 25.62.

15. From Example 3 we have $C_2=\left(\frac{V_2}{V_1}\right)^{0.6}C_1$.

Where C_1 is the cost of the original piece of equipment, V_1 is the capacity of the original piece of equipment, and V_2 is the capacity of the new piece of equipment.

We substitute 100,000 for C_1, 80,000 for V_1 and 160,000 for V_2.

$C_2=\left(\frac{160{,}000}{80{,}000}\right)^{0.6}(100{,}000)$

$=(2)^{0.6}(100{,}000)$

$=151{,}571.6567$

We estimate the cost of the new tank to be \$151,571.66.

Copyright © 2014 Pearson Education, Inc. Publishing as Addison-Wesley.

17. a) From the table in example 3, we see that an APR of 6% for 6 years results in a payment of \$16.57 per \$1000 borrowed. Since Kim is borrowing \$10,000 we would estimate Kim's payments to be 10 times \$16.57 or \$165.70.

b) From the table in example 3, we see that an APR of 5.5% for 7 years results in a payment of \$14.37 per \$1000 borrowed. Since Kim is borrowing \$10,000 we would estimate Kim's payments to be 10 times \$14.37 or \$143.70.

c) Under option (a), Kim will make $6\times 12 = 72$ monthly payments of \$165.70. Therefore her total payments are $165.70\times 72 = 11{,}930.40$.
The total payments under option (a) are \$11,930.40.
Under option (b), Kim will make $7\times 12 = 84$ monthly payments of \$143.70. Therefore her total payments are $143.70\times 84 = 12{,}070.80$.
The total payments under option (a) are \$12,070.80.
Ultimately, option (a) will cost Kim less money and get her out of debt quicker.

19. $S(a,d,V) = \dfrac{aV}{0.51d^2}$

Substituting 100 for *d*, 1,600,000 for *V*, and 0.78 for *a*, we have:

$$\begin{aligned} S(0.78,100,1{,}600{,}000) &= \frac{(0.78)(1{,}600{,}000)}{0.51(100)^2} \\ &= \frac{1{,}248{,}000}{5100} \\ &\approx 244.70588. \end{aligned}$$

The approximate wind speed 100 ft from the center of the tornado is 244.7 miles per hour.

21. $S(h,w) = 0.024265h^{0.3964}w^{0.5378}$

Substituting 165 for *h* and 80 for *w*, we have:

$$\begin{aligned} S(165,80) &= 0.024265(165)^{0.3964}(80)^{0.5378} \\ &\approx 0.024265(7.56851)(10.55557) \\ &\approx 1.93852. \end{aligned}$$

The approximate surface area for the person is 1.939 square meters.

23. a) Locate 80 degrees in the second row of the table and trace over to the third column, which represents 60% humidity. We determine the dew point is 65.

b) Locate 90 degrees in the third row of the table and trace over to the second column, which represents 40% humidity. We determine the dew point is 62.

c) We see at a temperature of 100 degrees, the dew point is 52 when relative humidity is 20% and 71 when relative humidity is 40%. Therefore, we conclude at a temperature of 100 degrees F the air will fill humid at an approximate relative humidity of 30%.

d) ✎

25. A person weight drops by 19% means that $w_N = w(1-0.19) = w(0.81) = 0.81w$

Giving the new surface area of

$S_N(h,w_N) = \dfrac{\sqrt{hw_N}}{60}$. Writing the new surface area as a function of the original weight *w* gives

us: $S_N(h,0.81w) = \dfrac{\sqrt{h\cdot 0.81w}}{60} = \dfrac{\sqrt{0.81hw}}{60}$

The percentage change from the original surface area is calculated as follows:

$$\begin{aligned} \frac{S_N - S}{S} &= \frac{\frac{\sqrt{0.81hw}}{60} - \frac{\sqrt{hw}}{60}}{\frac{\sqrt{hw}}{60}} \\ &= \frac{\frac{\sqrt{0.81hw}-\sqrt{hw}}{60}}{\frac{\sqrt{hw}}{60}} \\ &= \frac{\sqrt{0.81hw}-\sqrt{hw}}{\sqrt{hw}} \\ &= \frac{0.9\sqrt{hw}-\sqrt{hw}}{\sqrt{hw}} \\ &= \frac{(0.9-1)\sqrt{hw}}{\sqrt{hw}} \\ &= 0.9-1 \\ &= -0.1. \end{aligned}$$

Therefore, the percentage decrease in the person's surface area resulting from a 19% decrease in body weight is about 10%.

27. ✎

Copyright © 2014 Pearson Education, Inc. Publishing as Addison-Wesley.

29. $W(v,T)=$

$$91.4-\frac{\left(10.45+6.68\sqrt{v}-0.447v\right)(457-5T)}{110}$$

$W(20,20)$

$$=91.4-\frac{\left(10.45+6.68\sqrt{20}-0.447\cdot 20\right)(457-5\cdot 20)}{110}$$

$$\approx 91.4-\frac{(10.45+29.874-8.94)(457-100)}{110}$$

$$\approx 91.4-\frac{(31.384)(357)}{110}$$

$$\approx 91.4-101.855$$

$$\approx -10.455$$

The wind chill, rounded to the nearest degree is $-10°F$.

31. $W(v,T)=$

$$91.4-\frac{\left(10.45+6.68\sqrt{v}-0.447v\right)(457-5T)}{110}$$

$W(30,-10)$

$$=91.4-\frac{\left(10.45+6.68\sqrt{30}-0.447\cdot 30\right)(457-5\cdot(-10))}{110}$$

$$\approx 91.4-\frac{(10.45+36.588-13.41)(457+50)}{110}$$

$$\approx 91.4-\frac{(33.628)(507)}{110}$$

$$\approx 91.4-155.0$$

$$\approx -63.6$$

The wind chill, rounded to the nearest degree is $-64°F$.

33.

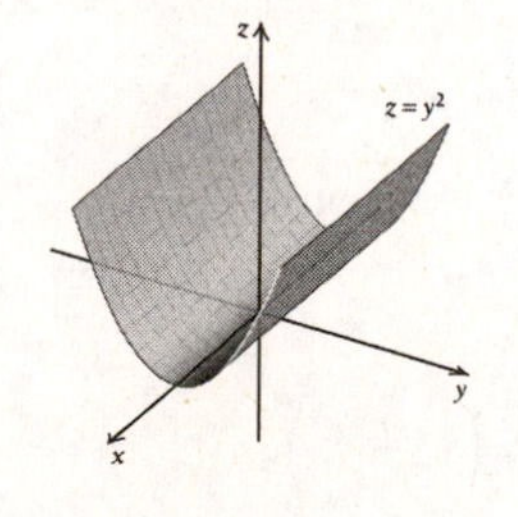

35.

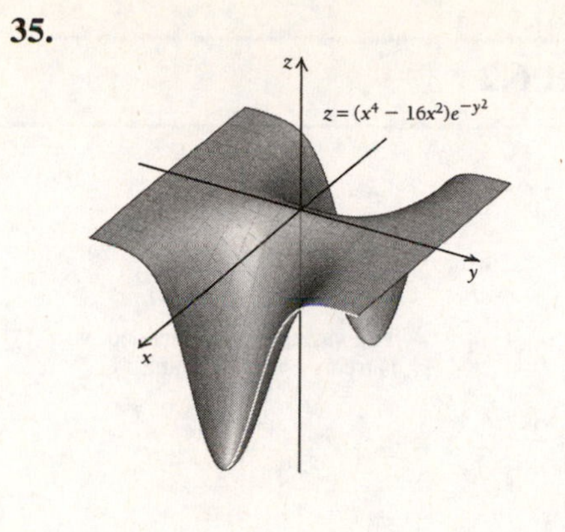

37.

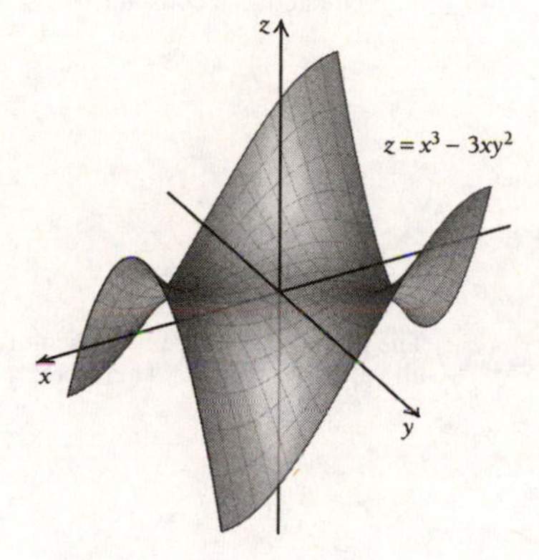

Copyright © 2014 Pearson Education, Inc. Publishing as Addison-Wesley.

Exercise Set 6.2

1. $z = 2x - 3y$

Find $\frac{\partial z}{\partial x}$.

$z = 2\underline{x} - 3y$ The variable is underlined; y is treated as a constant.

$\frac{\partial z}{\partial x} = 2$

Find $\frac{\partial z}{\partial y}$.

$z = 2x - 3\underline{y}$ The variable is underlined; x is treated as a constant.

$\frac{\partial z}{\partial y} = -3$

Find $\left.\frac{\partial z}{\partial x}\right|_{(-2,-3)}$

$\frac{\partial z}{\partial x} = 2$ The partial derivative is constant for all values of x and y. Therefore:

$\left.\frac{\partial z}{\partial x}\right|_{(-2,-3)} = 2$

Find $\left.\frac{\partial z}{\partial y}\right|_{(0,-5)}$

$\frac{\partial z}{\partial x} = -3$ The partial derivative is constant for all values of x and y. Therefore:

$\left.\frac{\partial z}{\partial y}\right|_{(0,-5)} = -3$

3. $z = 3x^2 - 2xy + y$

Find $\frac{\partial z}{\partial x}$.

$z = 3\underline{x}^2 - 2\underline{x}y + y$ The variable is underlined; y is treated as a constant.

$\frac{\partial z}{\partial x} = 6x - 2y$

Find $\frac{\partial z}{\partial y}$.

$z = 3x^2 - 2x\underline{y} + \underline{y}$ The variable is underlined; x is treated as a constant.

$\frac{\partial z}{\partial x} = -2x + 1$

Find $\left.\frac{\partial z}{\partial x}\right|_{(-2,-3)}$

$\frac{\partial z}{\partial x} = 6x - 2y$

$\left.\frac{\partial z}{\partial x}\right|_{(-2,-3)} = 6(-2) - 2(-3)$ Substituting -2 for x and -3 for y.

$= -12 + 6$

$= -6$

Find $\left.\frac{\partial z}{\partial y}\right|_{(0,-5)}$

$\frac{\partial z}{\partial y} = -2x + 1$

$\left.\frac{\partial z}{\partial y}\right|_{(0,-5)} = -2(0) + 1$ Substituting 0 for x.

$= 1$

5. $f(x, y) = 2x - 5xy$

Find f_x.

$f(x, y) = 2\underline{x} - 5\underline{x}y$ The variable is underlined; y is treated as a constant.

$f_x = 2 - 5y$

Find f_y.

$f(x, y) = 2x - 5x\underline{y}$ The variable is underlined; x is treated as a constant.

$f_y = -5x$

Find $f_x(-2,4)$.

$f_x = 2 - 5y$

$f_x(-2,4) = 2 - 5(4)$ Substituting 4 for y.

$= 2 - 20$

$= -18$

Find $f_y(4,-3)$.

$f_y = -5x$

$f_y(4,-3) = -5(4)$ Substituting 4 for x.

$= -20$

7. $f(x, y) = \sqrt{x^2 + y^2} = \left(x^2 + y^2\right)^{1/2}$

Find f_x.

$f(x, y) = \left(\underline{x}^2 + y^2\right)^{1/2}$ The variable is underlined; y is treated as a constant.

$f_x = \frac{1}{2}\left(x^2 + y^2\right)^{-1/2} \cdot 2x$

$= x\left(x^2 + y^2\right)^{-1/2}$ or $\frac{x}{\sqrt{x^2 + y^2}}$

Copyright © 2014 Pearson Education, Inc. Publishing as Addison-Wesley.

Find f_y.

$f(x,y)=\left(x^2+\underline{y}^2\right)^{1/2}$ The variable is underlined; x is treated as a constant.

$$f_y=\frac{1}{2}\left(x^2+y^2\right)^{-1/2}\cdot 2y$$
$$=y\left(x^2+y^2\right)^{-1/2} \text{ or } \frac{y}{\sqrt{x^2+y^2}}$$

Find $f_x(-2,1)$.

$$f_x=\frac{x}{\sqrt{x^2+y^2}}$$

$$f_x(-2,1)=\frac{(-2)}{\sqrt{(-2)^2+(1)^2}}$$ Substituting −2 for x and 1 for y.

$$=\frac{-2}{\sqrt{4+1}}$$
$$=\frac{-2}{\sqrt{5}}$$

Find $f_y(-3,-2)$.

$$f_y=\frac{y}{\sqrt{x^2+y^2}}$$

$$f_y(-3,-2)=\frac{(-2)}{\sqrt{(-3)^2+(-2)^2}}$$ Substituting −3 for x and −2 for y.

$$=\frac{-2}{\sqrt{9+4}}$$
$$=\frac{-2}{\sqrt{13}}$$

9. $f(x,y)=e^{2x-y}$

Find f_x.

$f(x,y)=e^{2\underline{x}-y}$ The variable is underlined; y is treated as a constant.

$$f_x=e^{2x-y}\cdot(2)$$
$$=2e^{2x-y}$$

Find f_y.

$f(x,y)=e^{2x-\underline{y}}$ The variable is underlined; x is treated as a constant.

$$f_y=e^{2x-y}\cdot(-1)$$
$$=-e^{2x-y}$$

11. $f(x,y)=e^{xy}$

Find f_x.

$f(x,y)=e^{\underline{x}y}$ The variable is underlined; y is treated as a constant.

$$f_x=e^{xy}\cdot(y)$$
$$=ye^{xy}$$

Find f_y.

$f(x,y)=e^{x\underline{y}}$ The variable is underlined; x is treated as a constant.

$$f_y=e^{xy}\cdot(x)$$
$$=xe^{xy}$$

13. $f(x,y)=y\ln(x+2y)$

Find f_x.

$f(x,y)=y\ln(\underline{x}+2y)$ The variable is underlined.

$$f_x=y\cdot\frac{1}{x+2y}\cdot 1$$
$$=\frac{y}{x+2y}$$

Find f_y.

$f(x,y)=\underline{y}\ln(x+2\underline{y})$ The variable is underlined.

$$f_y=y\cdot\frac{1}{x+2y}\cdot 2+1\cdot\ln(x+2y)$$
$$=\frac{2y}{x+2y}+\ln(x+2y)$$

15. $f(x,y)=x\ln(xy)$

Find f_x.

$f(x,y)=\underline{x}\ln(\underline{x}y)$ The variable is underlined.

$$f_x=x\cdot\left(\frac{1}{xy}\cdot y\right)+1\cdot\ln(xy)$$
$$=1+\ln(xy)$$

Find f_y.

$f(x,y)=x\ln(x\underline{y})$ The variable is underlined.

$$f_y=x\cdot\left(\frac{1}{xy}\cdot x\right)$$
$$=\frac{x}{y}$$

Copyright © 2014 Pearson Education, Inc. Publishing as Addison-Wesley.

17. $f(x,y)=\frac{x}{y}-\frac{y}{3x}$

Find f_x.

$f(x,y)=\frac{1}{y}\underline{x}-\frac{y}{3}\cdot\underline{x}^{-1}$ The variable is underlined.

$f_x=\frac{1}{y}-\frac{y}{3}\left(-1x^{-2}\right)$

$=\frac{1}{y}+\frac{y}{3x^2}$

Find f_y.

$f(x,y)=x\underline{y}^{-1}-\frac{1}{3x}\cdot\underline{y}$ The variable is underlined.

$f_y=x\left(-1y^{-2}\right)-\frac{1}{3x}\cdot 1$

$=-\frac{x}{y^2}-\frac{1}{3x}$

19. $f(x,y)=3(2x+y-5)^2$

Find f_x.

$f(x,y)=3(2\underline{x}+y-5)^2$ The variable is underlined.

$f_x=3\left[2(2x+y-5)\cdot 2\right]$

$=12(2x+y-5)$

Find f_y.

$f(x,y)=3(2x+\underline{y}-5)^2$ The variable is underlined.

$f_y=3\left[2(2x+y-5)\cdot 1\right]$

$=6(2x+y-5)$

21. $f(b,m)=$

$m^3+4m^2b-b^2+(2m+b-5)^2+(3m+b-6)^2$

Find $\frac{\partial f}{\partial b}$.

$f(b,m)=$

$m^3+4m^2\underline{b}-\underline{b}^2+(2m+\underline{b}-5)^2+(3m+\underline{b}-6)^2$

The variable is underlined.

$\frac{\partial f}{\partial b}=4m^2-2b+2(2m+b-5)\cdot 1+$

$2(3m+b-6)\cdot 1$

$=4m^2-2b+4m+2b-10+6m+2b-12$

$=4m^2+10m+2b-22$

Find $\frac{\partial f}{\partial m}$.

$f(b,m)=$

$\underline{m}^3+4\underline{m}^2b-b^2+(2\underline{m}+b-5)^2+(3\underline{m}+b-6)^2$

The variable is underlined.

$\frac{\partial f}{\partial m}=3m^2+8mb+2(2m+b-5)\cdot 2+$

$2(3m+b-6)\cdot 3$

$=3m^2+8mb+8m+4b-20+$

$18m+6b-36$

$=3m^2+8mb+26m+10b-56$

23. $f(x,y,\lambda)=5xy-\lambda(2x+y-8)$

Find f_x.

$f(x,y,\lambda)=5\underline{x}y-\lambda(2\underline{x}+y-8)$ The variable is underlined.

$f_x=5y-\lambda\cdot 2$

$=5y-2\lambda$

Find f_y.

$f(x,y,\lambda)=5x\underline{y}-\lambda(2x+\underline{y}-8)$ The variable is underlined.

$f_y=5x-\lambda\cdot 1$

$=5x-\lambda$

Find f_λ.

$f(x,y,\lambda)=5xy-\underline{\lambda}(2x+y-8)$ The variable is underlined.

$f_\lambda=-1\cdot(2x+y-8)$

$=-(2x+y-8)$

25. $f(x,y,\lambda)=x^2+y^2-\lambda(10x+2y-4)$

Find f_x.

$f(x,y,\lambda)=\underline{x}^2+y^2-\lambda(10\underline{x}+2y-4)$

The variable is underlined.

$f_x=2x-\lambda\cdot 10$

$=2x-10\lambda$

Find f_y.

$f(x,y,\lambda)=x^2+\underline{y}^2-\lambda(10x+2\underline{y}-4)$

The variable is underlined.

$f_y=2y-\lambda\cdot 2$

$=2y-2\lambda$

The solution is continued on the next page.

Copyright © 2014 Pearson Education, Inc. Publishing as Addison-Wesley.

Find f_λ.

$f(x,y,\lambda)=x^2+y^2-\underline{\lambda}(10x+2y-4)$

The variable is underlined.

$f_\lambda=-1(10x+2y-4)$
$=-(10x+2y-4)$

27. $f(x,y)=5xy$

First, we find the partial derivatives.
We find f_x first.

$f(x,y)=5\underline{x}y$ The variable is underlined.

$f_x=5y$

Then we find f_y.

$f(x,y)=5x\underline{y}$ The variable is underlined.

$f_y=5x$

We find f_{xx} by taking the partial derivative with respect to x of f_x.

$$f_{xx}=\frac{\partial}{\partial x}(f_x)=\frac{\partial}{\partial x}(5y)=0$$

We find f_{xy} by taking the partial derivative with respect to y of f_x.

$$f_{xy}=\frac{\partial}{\partial y}(f_x)=\frac{\partial}{\partial y}(5y)=5$$

We find f_{yx} by taking the partial derivative with respect to x of f_y.

$$f_{yx}=\frac{\partial}{\partial x}(f_y)=\frac{\partial}{\partial x}(5x)=5$$

We find f_{yy} by taking the partial derivative with respect to y of f_y.

$$f_{yy}=\frac{\partial}{\partial y}(f_y)=\frac{\partial}{\partial y}(5x)=0$$

29. $f(x,y)=7xy^2+5xy-2y$

First, we find the partial derivatives.
We find f_x first.

$f(x,y)=7\underline{x}y^2+5\underline{x}y-2y$ The variable is underlined.

$f_x=7y^2+5y$

Then we find f_y.

$f(x,y)=7x\underline{y}^2+5x\underline{y}-2\underline{y}$ The variable is underlined.

$f_y=14xy+5x-2$

We find f_{xx} by taking the partial derivative with respect to x of f_x.

$f_{xx}=\frac{\partial}{\partial x}(f_x)$
$=\frac{\partial}{\partial x}(7y^2+5y)$ y is treated as a constant.
$=0$

We find f_{xy} by taking the partial derivative with respect to y of f_x.

$f_{xy}=\frac{\partial}{\partial y}(f_x)$
$=\frac{\partial}{\partial y}(7\underline{y}^2+5\underline{y})$ The variable is underlined.
$=14y+5$

We find f_{yx} by taking the partial derivative with respect to x of f_y.

$f_{yx}=\frac{\partial}{\partial x}(f_y)$
$=\frac{\partial}{\partial x}(14\underline{x}y+5\underline{x}-2)$ The variable is underlined.
$=14y+5$

We find f_{yy} by taking the partial derivative with respect to y of f_y.

$f_{yy}=\frac{\partial}{\partial y}(f_y)$
$=\frac{\partial}{\partial y}(14x\underline{y}+5x-2)$ The variable is underlined.
$=14x$

31. $f(x,y)=x^5y^4+x^3y^2$

First, we find the partial derivatives.
We find f_x first.

$f(x,y)=\underline{x}^5y^4+\underline{x}^3y^2$ The variable is underlined.

$f_x=5x^4y^4+3x^2y^2$

Then we find f_y.

$f(x,y)=x^5\underline{y}^4+x^3\underline{y}^2$ The variable is underlined.

$f_y=4x^5y^3+2x^3y$

The solution is continued on the next page.

Copyright © 2014 Pearson Education, Inc. Publishing as Addison-Wesley.

We find f_{xx} by taking the partial derivative with respect to x of f_x.

$$f_{xx}=\frac{\partial}{\partial x}(f_x)$$
$$=\frac{\partial}{\partial x}\left(5\underline{x}^4y^4+3\underline{x}^2y^2\right) \quad \text{The variable is underlined.}$$
$$=20x^3y^4+6xy^2$$

We find f_{xy} by taking the partial derivative with respect to y of f_x.

$$f_{xy}=\frac{\partial}{\partial y}(f_x)$$
$$=\frac{\partial}{\partial y}\left(5x^4\underline{y}^4+3x^2\underline{y}^2\right) \quad \text{The variable is underlined.}$$
$$=20x^4y^3+6x^2y$$

We find f_{yx} by taking the partial derivative with respect to x of f_y.

$$f_{yx}=\frac{\partial}{\partial x}(f_y)$$
$$=\frac{\partial}{\partial x}\left(4\underline{x}^5y^3+2\underline{x}^3y\right) \quad \text{The variable is underlined.}$$
$$=20x^4y^3+6x^2y$$

We find f_{yy} by taking the partial derivative with respect to y of f_y.

$$f_{yy}=\frac{\partial}{\partial y}(f_y)$$
$$=\frac{\partial}{\partial y}\left(4x^5\underline{y}^3+2x^3\underline{y}\right) \quad \text{The variable is underlined.}$$
$$=12x^5y^2+2x^3$$

33. $f(x,y)=2x-3y$

First, we find the partial derivatives.
We find f_x first.

$$f(x,y)=2\underline{x}-3y \quad \text{The variable is underlined.}$$
$$f_x=2$$

Then we find f_y.

$$f(x,y)=2x-3\underline{y} \quad \text{The variable is underlined.}$$
$$f_y=-3$$

We find f_{xx} by taking the partial derivative with respect to x of f_x.

$$f_{xx}=\frac{\partial}{\partial x}(f_x)=\frac{\partial}{\partial x}(2)=0$$

We find f_{xy} by taking the partial derivative with respect to y of f_x.

$$f_{xy}=\frac{\partial}{\partial y}(f_x)=\frac{\partial}{\partial y}(2)=0$$

We find f_{yx} by taking the partial derivative with respect to x of f_y.

$$f_{yx}=\frac{\partial}{\partial x}(f_y)=\frac{\partial}{\partial x}(-3)=0$$

We find f_{yy} by taking the partial derivative with respect to y of f_y.

$$f_{yy}=\frac{\partial}{\partial y}(f_y)=\frac{\partial}{\partial y}(-3)=0$$

35. $f(x,y)=e^{2xy}$

First, we find the partial derivatives.
We find f_x first.

$$f(x,y)=e^{2\underline{x}y} \quad \text{The variable is underlined.}$$
$$f_x=2ye^{2xy}$$

Then we find f_y.

$$f(x,y)=e^{2x\underline{y}} \quad \text{The variable is underlined.}$$
$$f_y=2xe^{2xy}$$

We find f_{xx} by taking the partial derivative with respect to x of f_x.

$$f_{xx}=\frac{\partial}{\partial x}(f_x)$$
$$=\frac{\partial}{\partial x}\left(2ye^{2\underline{x}y}\right) \quad \text{The variable is underlined.}$$
$$=2y\left(2ye^{2xy}\right)$$
$$=4y^2e^{2xy}$$

We find f_{xy} by taking the partial derivative with respect to y of f_x.

$$f_{xy}=\frac{\partial}{\partial y}(f_x)$$
$$=\frac{\partial}{\partial y}\left(2\underline{y}e^{2x\underline{y}}\right) \quad \text{The variable is underlined.}$$
$$=2y\left(2xe^{2xy}\right)+2\left(e^{2xy}\right)$$
$$=4xye^{2xy}+2e^{2xy}$$

The solution is continued on the next page.

Copyright © 2014 Pearson Education, Inc. Publishing as Addison-Wesley.

We find f_{yx} by taking the partial derivative with respect to x of f_y.

$$f_{yx} = \frac{\partial}{\partial x}(f_y)$$
$$= \frac{\partial}{\partial x}\left(2\underline{x}e^{2\underline{x}y}\right) \quad \text{The variable is underlined.}$$
$$= 2x\left(2ye^{2xy}\right) + 2\left(e^{2xy}\right)$$
$$= 4xye^{2xy} + 2e^{2xy}$$

We find f_{yy} by taking the partial derivative with respect to y of f_y.

$$f_{yy} = \frac{\partial}{\partial y}(f_y)$$
$$= \frac{\partial}{\partial y}\left(2xe^{2x\underline{y}}\right) \quad \text{The variable is underlined.}$$
$$= 2x\left(2xe^{2xy}\right)$$
$$= 4x^2e^{2xy}$$

37. $f(x, y) = x + e^y$

First, we find the partial derivatives.
We find f_x first.

$$f(x, y) = \underline{x} + e^y \quad \text{The variable is underlined.}$$
$$f_x = 1$$

Then we find f_y.

$$f(x, y) = x + e^{\underline{y}} \quad \text{The variable is underlined.}$$
$$f_y = e^y$$

We find f_{xx} by taking the partial derivative with respect to x of f_x.

$$f_{xx} = \frac{\partial}{\partial x}(f_x) = \frac{\partial}{\partial x}(1) = 0$$

We find f_{xy} by taking the partial derivative with respect to y of f_x.

$$f_{xy} = \frac{\partial}{\partial y}(f_x) = \frac{\partial}{\partial y}(1) = 0$$

We find f_{yx} by taking the partial derivative with respect to x of f_y.

$$f_{yx} = \frac{\partial}{\partial x}(f_y)$$
$$= \frac{\partial}{\partial x}\left(e^y\right) \quad e^y \text{ is treated as a constant.}$$
$$= 0$$

We find f_{yy} by taking the partial derivative with respect to y of f_y.

$$f_{yy} = \frac{\partial}{\partial y}(f_y)$$
$$= \frac{\partial}{\partial y}\left(e^{\underline{y}}\right) \quad \text{The variable is underlined.}$$
$$= e^y$$

39. $f(x, y) = y \ln x$

First, we find the partial derivatives.
We find f_x first.

$$f(x, y) = y \ln \underline{x} \quad \text{The variable is underlined.}$$
$$f_x = y \cdot \frac{1}{x} = \frac{y}{x}$$

Then we find f_y.

$$f(x, y) = \underline{y} \ln x \quad \text{The variable is underlined.}$$
$$f_y = 1 \cdot \ln x = \ln x$$

We find f_{xx} by taking the partial derivative with respect to x of f_x.

$$f_{xx} = \frac{\partial}{\partial x}(f_x)$$
$$= \frac{\partial}{\partial x}\left(\frac{y}{\underline{x}}\right) = \frac{\partial}{\partial x}\left(y\underline{x}^{-1}\right) \quad \text{The variable is underlined.}$$
$$= -yx^{-2}$$
$$= \frac{-y}{x^2}$$

We find f_{xy} by taking the partial derivative with respect to y of f_x.

$$f_{xy} = \frac{\partial}{\partial y}(f_x)$$
$$= \frac{\partial}{\partial y}\left(\underline{y} \cdot \frac{1}{x}\right) \quad \text{The variable is underlined.}$$
$$= \frac{1}{x}$$

We find f_{yx} by taking the partial derivative with respect to x of f_y.

$$f_{yx} = \frac{\partial}{\partial x}(f_y)$$
$$= \frac{\partial}{\partial x}(\ln \underline{x}) \quad \text{The variable is underlined.}$$
$$= \frac{1}{x}$$

The solution is continued on the next page.

Copyright © 2014 Pearson Education, Inc. Publishing as Addison-Wesley.

We find f_{yy} by taking the partial derivative with respect to y of f_y.

$$f_{yy} = \frac{\partial}{\partial y}(f_y)$$
$$= \frac{\partial}{\partial y}(\ln x) \quad \ln x \text{ is treated as a constant.}$$
$$= 0$$

41. $p(x, y) = 2400x^{2/5}y^{3/5}$

a) Substituting 32 for x and 1024 for y, we have:

$$p(32,1024) = 2400(32)^{2/5}(1024)^{3/5}$$
$$= 2400(4)(64)$$
$$= 614,400.$$

Using 32 units of labor and 1024 units of capital, Lincolnville Sporting goods will produce 614,400 units.

b) We find the marginal productivity of labor by taking the partial derivative with respect to x.

$$\frac{\partial p}{\partial x} = \frac{\partial}{\partial x}\left(2400\underline{x}^{2/5}y^{3/5}\right) \quad \text{The variable is underlined.}$$
$$= 2400\left(\frac{2}{5}x^{-3/5}\right)y^{3/5}$$
$$= 960x^{-3/5}y^{3/5}$$
$$= 960\left(\frac{y}{x}\right)^{3/5}$$

We find the marginal productivity of capital by taking the partial derivative with respect to y.

$$\frac{\partial p}{\partial y} = \frac{\partial}{\partial y}\left(2400x^{2/5}\underline{y}^{3/5}\right) \quad \text{The variable is underlined.}$$
$$= 2400x^{2/5}\left(\frac{3}{5}y^{-2/5}\right)$$
$$= 1440x^{2/5}y^{-2/5}$$
$$= 1440\left(\frac{x}{y}\right)^{2/5}$$

c) Substituting 32 for x and 1024 for y into $\frac{\partial p}{\partial x}$, we have:

$$\left.\frac{\partial p}{\partial x}\right|_{(32,1024)} = 960\left(\frac{1024}{32}\right)^{3/5}$$
$$= 960(32)^{3/5}$$
$$= 960(8)$$
$$= 7680.$$

The marginal productivity of labor when 32 units of labor and 1024 units of capital are currently being used is 7680 units per unit labor. Substituting 32 for x and 1024 for y into $\frac{\partial p}{\partial y}$, we have:

$$\left.\frac{\partial p}{\partial y}\right|_{(32,1024)} = 1440\left(\frac{32}{1024}\right)^{2/5}$$
$$= 1440\left(\frac{1}{32}\right)^{2/5}$$
$$= 1440\left(\tfrac{1}{4}\right)$$
$$= 360.$$

The marginal productivity of capital when 32 units of labor and 1024 units of capital are currently being used is 360 units per unit capital.

d) ✎

43. $P(w, r, s, t) = 0.007955w^{-0.638}r^{1.038}s^{0.873}t^{2.468}$

a) We substitute 20 for w, 70 for r, 400,000 for s, and 8 for t.

$$P(20, 70, 400,000, 8)$$
$$= 0.007955(20)^{-0.638}(70)^{1.038}(400,000)^{0.873}(8)^{2.468}$$
$$\approx 1,274,146$$

The nursing home's annual profit is approximately $1,274,146.

Copyright © 2014 Pearson Education, Inc. Publishing as Addison-Wesley.

b) Taking the partial derivative with respect to each variable, we have:

$$\frac{\partial P}{\partial w} = 0.007955\left(-0.638w^{-1.638}\right)r^{1.038}s^{0.873}t^{2.468}$$
$$= -0.005075w^{-1.638}r^{1.038}s^{0.873}t^{2.468}$$

$$\frac{\partial P}{\partial r} = 0.007955w^{-0.638}\left(1.038r^{1-1.038}\right)s^{0.873}t^{2.468}$$
$$= 0.008257w^{-0.638}r^{0.038}s^{0.873}t^{2.468}$$

$$\frac{\partial P}{\partial s} = 0.007955w^{-0.638}r^{1.038}\left(0.873s^{-0.127}\right)t^{2.468}$$
$$= 0.006945w^{-0.638}r^{1.038}s^{-0.127}t^{2.468}$$

$$\frac{\partial P}{\partial t} = 0.007955w^{-0.638}r^{1.038}s^{0.873}\left(2.468t^{1.468}\right)$$
$$= 0.019633w^{-0.638}r^{1.038}s^{0.873}t^{1.468}$$

c) ✎

45. $T_h = 1.98T - 1.09(1-H)(T-58) - 56.9$

Substituting 85 for T, and 60%=0.60 for H, we have:

$$T_h = 1.98(85) - 1.09(1-0.6)(85-58) - 56.9$$
$$= 168.3 - 1.09(0.4)(27) - 56.9$$
$$= 168.3 - 11.772 - 56.9$$
$$= 99.628$$
$$\approx 99.6.$$

The temperature-humidity index is about $99.6°F$.

47. $T_h = 1.98T - 1.09(1-H)(T-58) - 56.9$

Substituting 90 for T, and 100%=1.0 for H, we have:

$$T_h = 1.98(90) - 1.09(1-1.0)(90-58) - 56.9$$
$$= 178.2 - 1.09(0)(32) - 56.9$$
$$= 178.2 - 0 - 56.9$$
$$= 121.3.$$

The temperature-humidity index is $121.3°F$.

49. ✎

51. $S = \frac{\sqrt{hw}}{60} = \frac{(hw)^{1/2}}{60} = \frac{h^{1/2}w^{1/2}}{60}$

a) $$\frac{\partial S}{\partial h} = \frac{1}{60}\left(\frac{1}{2}h^{-1/2}w^{1/2}\right)$$
$$= \frac{1}{120}\left(\frac{w^{1/2}}{h^{1/2}}\right)$$
$$= \frac{\sqrt{w}}{120\sqrt{h}}$$

b) $$\frac{\partial S}{\partial w} = \frac{1}{60}h^{1/2}\left(\frac{1}{2}w^{-1/2}\right)$$
$$= \frac{1}{120}\left(\frac{h^{1/2}}{w^{1/2}}\right)$$
$$= \frac{\sqrt{h}}{120\sqrt{w}}$$

c) $$\Delta S \approx \frac{\partial S}{\partial w}\Delta w$$
$$\Delta S \approx \left[\frac{\sqrt{h}}{120\sqrt{w}}\right]\Delta w$$
$$\approx \frac{\sqrt{170}}{120\sqrt{80}}(-2)$$
$$\approx -0.0243$$

The change in the surface area is approximately $-0.0243\ \text{m}^2$.

53. $E = 206.835 - 0.846w - 1.015s$

Substituting 146 for w and 5 for s, we have:

$$E = 206.835 - 0.846(146) - 1.015(5)$$
$$= 206.835 - 123.516 - 5.075$$
$$= 78.244.$$

The reading ease is 78.244.

55. $E = 206.835 - 0.846w - 1.015s$

$$\frac{\partial E}{\partial w} = -0.846$$

57. $f(x,t) = \frac{x^2+t^2}{x^2-t^2}$

Find f_x.

$$f(x,t) = \frac{\underline{x}^2+t^2}{\underline{x}^2-t^2}$$ The variable is underlined.

$$f_x = \frac{(x^2-t^2)(2x)-(x^2+t^2)(2x)}{(x^2-t^2)^2}$$
$$= \frac{2x^3-2xt^2-2x^3-2xt^2}{(x^2-t^2)^2}$$
$$= \frac{-4xt^2}{(x^2-t^2)^2}$$

The solution is continued on the next page.

Copyright © 2014 Pearson Education, Inc. Publishing as Addison-Wesley.

Find f_t.

$$f(x,t)=\frac{x^2+\underline{t}^2}{x^2-\underline{t}^2}$$ The variable is underlined.

$$f_t=\frac{(x^2-t^2)(2t)-(x^2+t^2)(-2t)}{(x^2-t^2)^2}$$

$$=\frac{2x^2t-2t^3+2x^2t+2t^3}{(x^2-t^2)^2}$$

$$=\frac{4x^2t}{(x^2-t^2)^2}$$

59. $f(x,t)=\frac{2\sqrt{x}-2\sqrt{t}}{1+2\sqrt{t}}=\frac{2x^{1/2}-2t^{1/2}}{1+2t^{1/2}}$

Find f_x.

$$f(x,t)=\frac{2\underline{x}^{1/2}-2t^{1/2}}{1+2t^{1/2}}$$ The variable is underlined.

$$=\frac{2\underline{x}^{1/2}}{1+2t^{1/2}}-\frac{2t^{1/2}}{1+2t^{1/2}}$$

$$f_x=\frac{2\left(\frac{1}{2}x^{-1/2}\right)}{1+2t^{1/2}}-0$$

$$=\frac{x^{-1/2}}{1+2t^{1/2}}$$

$$=\frac{1}{x^{1/2}\left(1+2t^{1/2}\right)}$$

$$=\frac{1}{\sqrt{x}\left(1+2\sqrt{t}\right)}$$

Find f_t.

$$f(x,t)=\frac{2x^{1/2}-2\underline{t}^{1/2}}{1+2\underline{t}^{1/2}}$$ The variable is underlined.

$$f_t=\frac{\left(1+2t^{1/2}\right)\left(-1t^{-1/2}\right)-\left(2x^{1/2}-2t^{1/2}\right)\left(t^{-1/2}\right)}{\left(1+2t^{1/2}\right)^2}$$

$$=\frac{-t^{-1/2}-2t^0-2x^{1/2}t^{-1/2}+2t^0}{\left(1+2t^{1/2}\right)^2}$$

$$=\frac{t^{-1/2}\left(-1-2x^{1/2}\right)}{\left(1+2t^{1/2}\right)^2}$$

$$=\frac{-1-2\sqrt{x}}{\sqrt{t}\left(1+2\sqrt{t}\right)^2}$$

61. $f(x,t)=6x^{2/3}-8x^{1/4}t^{1/2}-12x^{-1/2}t^{3/2}$

Find f_x.

$$f(x,t)=6\underline{x}^{2/3}-8\underline{x}^{1/4}t^{1/2}-12\underline{x}^{-1/2}t^{3/2}$$

The variable is underlined.

$$f_x=6\left(\frac{2}{3}x^{-1/3}\right)-8\left(\frac{1}{4}x^{-3/4}\right)t^{1/2}-12\left(\frac{-1}{2}x^{-3/2}\right)t^{3/2}$$

$$=4x^{-1/3}-2x^{-3/4}t^{1/2}+6x^{-3/2}t^{3/2}$$

Find f_t.

$$f(x,t)=6x^{2/3}-8x^{1/4}\underline{t}^{1/2}-12x^{-1/2}\underline{t}^{3/2}$$

The variable is underlined.

$$f_t=-8x^{1/4}\left(\frac{1}{2}t^{-1/2}\right)-12x^{-1/2}\left(\frac{3}{2}t^{1/2}\right)$$

$$=-4x^{1/4}t^{-1/2}-18x^{-1/2}t^{1/2}$$

63. $f(x,y)=\frac{x}{y^2}-\frac{y}{x^2}=xy^{-2}-yx^{-2}$

First, we find the partial derivatives.

We find f_x first.

$$f(x,y)=\underline{x}y^{-2}-y\underline{x}^{-2}$$ The variable is underlined.

$$f_x=y^{-2}+2yx^{-3}$$

Then we find f_y.

$$f(x,y)=x\underline{y}^{-2}-\underline{y}x^{-2}$$ The variable is underlined.

$$f_y=-2xy^{-3}-x^{-2}$$

The solution is continued on the next page.

Copyright © 2014 Pearson Education, Inc. Publishing as Addison-Wesley.

We find f_{xx} by taking the partial derivative with respect to x of f_x.

$$\begin{aligned} f_{xx} &= \frac{\partial}{\partial x}(f_x) \\ &= \frac{\partial}{\partial x}\left(y^{-2}+2yx^{-3}\right) \\ &= -6yx^{-4} \\ &= \frac{-6y}{x^4} \end{aligned}$$

We find f_{xy} by taking the partial derivative with respect to y of f_x.

$$\begin{aligned} f_{xy} &= \frac{\partial}{\partial y}(f_x) \\ &= \frac{\partial}{\partial y}\left(y^{-2}+2yx^{-3}\right) \\ &= -2y^{-3}+2x^{-3} \\ &= -\frac{2}{y^3}+\frac{2}{x^3} \end{aligned}$$

We find f_{yx} by taking the partial derivative with respect to x of f_y.

$$\begin{aligned} f_{yx} &= \frac{\partial}{\partial x}(f_y) \\ &= \frac{\partial}{\partial x}\left(-2xy^{-3}-x^{-2}\right) \\ &= -2y^{-3}+2x^{-3} \\ &= -\frac{2}{y^3}+\frac{2}{x^3} \end{aligned}$$

We find f_{yy} by taking the partial derivative with respect to y of f_y.

$$\begin{aligned} f_{yy} &= \frac{\partial}{\partial y}(f_y) \\ &= \frac{\partial}{\partial y}\left(-2xy^{-3}-x^{-2}\right) \\ &= 6xy^{-4} \\ &= \frac{6x}{y^4} \end{aligned}$$

65. ✎

67. $f(x,y)=\ln\left(x^2+y^2\right)$

We need to find the second-order partial derivatives.

First, we find f_x.

$$f(x,y)=\ln\left(\underline{x}^2+y^2\right)$$

$$f_x=\frac{1}{x^2+y^2}\cdot 2x=\frac{2x}{x^2+y^2}$$

Then we find f_{xx}.

$$\begin{aligned} f_{xx} &= \frac{\partial}{\partial x}(f_x) \\ &= \frac{\partial}{\partial x}\left(\frac{2\underline{x}}{\underline{x}^2+y^2}\right) \\ &= \frac{\left(x^2+y^2\right)(2)-(2x)(2x)}{\left(x^2+y^2\right)^2} \\ &= \frac{-2x^2+2y^2}{\left(x^2+y^2\right)^2} \end{aligned}$$

Now we find f_y.

$$f(x,y)=\ln\left(x^2+\underline{y}^2\right)$$

$$f_x=\frac{1}{x^2+y^2}\cdot 2y=\frac{2y}{x^2+y^2}$$

Then we find f_{yy}.

$$\begin{aligned} f_{yy} &= \frac{\partial}{\partial y}(f_y) \\ &= \frac{\partial}{\partial y}\left(\frac{2\underline{y}}{x^2+\underline{y}^2}\right) \\ &= \frac{\left(x^2+y^2\right)(2)-(2y)(2y)}{\left(x^2+y^2\right)^2} \\ &= \frac{2x^2-2y^2}{\left(x^2+y^2\right)^2} \end{aligned}$$

Therefore,

$$\begin{aligned} \frac{\partial^2 f}{\partial x^2}+\frac{\partial^2 f}{\partial y^2} &= 0 \\ \frac{-2x^2+2y^2}{\left(x^2+y^2\right)^2}+\frac{2x^2-2y^2}{\left(x^2+y^2\right)^2} &= 0 \\ \frac{0}{\left(x^2+y^2\right)^2} &= 0 \\ 0 &= 0 \end{aligned}$$

Thus, f is a solution to $\frac{\partial^2 f}{\partial x^2}+\frac{\partial^2 f}{\partial y^2}=0.$

Copyright © 2014 Pearson Education, Inc. Publishing as Addison-Wesley.

69.

$$f(x,y)=\begin{cases}\dfrac{xy(x^2-y^2)}{x^2+y^2}, & \text{for } (x,y)\neq(0,0),\\ 0, & \text{for } (x,y)=(0,0).\end{cases}$$

a) Find $f_x(0,y)$.

$$\lim_{h\to 0}\frac{f(h,y)-f(0,y)}{h}$$
$$=\lim_{h\to 0}\frac{\dfrac{hy(h^2-y^2)}{h^2+y^2}-\dfrac{0\cdot y(0^2-y^2)}{0^2+y^2}}{h}$$
$$=\lim_{h\to 0}\frac{hy(h^2-y^2)}{h(h^2+y^2)}$$
$$=\lim_{h\to 0}\frac{y(h^2-y^2)}{(h^2+y^2)}$$
$$=\frac{y(-y^2)}{y^2}$$
$$=-y.$$

Thus, $f_x(0,y)=-y$.

b) Find $f_y(x,0)$.

$$\lim_{h\to 0}\frac{f(x,h)-f(x,0)}{h}$$
$$=\lim_{h\to 0}\frac{\dfrac{xh(x^2-h^2)}{x^2+h^2}-\dfrac{x\cdot 0(x^2-0^2)}{x^2+0^2}}{h}$$
$$=\lim_{h\to 0}\frac{xh(x^2-h^2)}{h(x^2+h^2)}$$
$$=\lim_{h\to 0}\frac{x(x^2-h^2)}{(x^2+h^2)}$$
$$=\frac{x(x^2)}{x^2}$$
$$=x.$$

Thus, $f_y(x,0)=x$.

c) Find $f_{yx}(0,0)$

$$\lim_{h\to 0}\frac{f_y(h,0)-f_y(0,0)}{h}$$
$$=\lim_{h\to 0}\frac{h-0}{h} \qquad \text{Substituting } f_y(x,0)=x.$$
$$=\lim_{h\to 0}1$$
$$=1.$$

Find $f_{xy}(0,0)$

$$\lim_{h\to 0}\frac{f_x(0,h)-f_x(0,0)}{h}$$
$$=\lim_{h\to 0}\frac{-h-(0)}{h} \qquad \text{Substituting } f_x(0,y)=-y.$$
$$=\lim_{h\to 0}(-1)$$
$$=-1.$$

Thus, $f_{yx}(0,0)\neq f_{xy}(0,0)$. The mixed partials are not equal at $(0,0)$.

Copyright © 2014 Pearson Education, Inc. Publishing as Addison-Wesley.

Exercise Set 6.3

1. $f(x,y)=x^2+xy+y^2-y$

Find f_x.

$f(x,y)=\underline{x}^2+\underline{x}y+y^2-y,$ The variable is underlined.

$f_x=2x+y.$

Find f_y.

$f(x,y)=x^2+x\underline{y}+\underline{y}^2-\underline{y},$ The variable is underlined.

$f_y=x+2y-1.$

Find f_{xx} and f_{xy}.

$f_x=2\underline{x}+y,\quad f_x=2x+\underline{y},$

$f_{xx}=2.\quad f_{xy}=1.$

Find f_{yy}.

$f_y=x+2\underline{y}-1,$

$f_{yy}=2.$

Solve the system of equations $f_x=0$ and $f_y=0$:

$2x+y=0,\quad (1)$

$x+2y-1=0.\quad (2)$

Solving Eq. (1) for y, we get $y=-2x$.

Substituting $-2x$ for y in Eq. (2) and solving, we get

$x+2(-2x)-1=0$

$-3x-1=0$

$-3x=1$

$x=-\frac{1}{3}.$

To find y when $x=-\frac{1}{3}$, we substitute $-\frac{1}{3}$ for x in either Eq. (1) or Eq. (2). We use Eq. (1):

$2\left(-\frac{1}{3}\right)+y=0$

$-\frac{2}{3}+y=0$

$y=\frac{2}{3}.$

Thus, $\left(-\frac{1}{3},\frac{2}{3}\right)$ is our candidate for a maximum or minimum.

We have to check to see if $f\left(-\frac{1}{3},\frac{2}{3}\right)$ is a maximum or minimum:

$D=f_{xx}(a,b)\cdot f_{yy}(a,b)-\left[f_{xy}(a,b)\right]^2$

$D=f_{xx}\left(-\frac{1}{3},\frac{2}{3}\right)\cdot f_{yy}\left(-\frac{1}{3},\frac{2}{3}\right)-\left[f_{xy}\left(-\frac{1}{3},\frac{2}{3}\right)\right]^2$

$D=2\cdot 2-1^2$ For all values of x and y, $f_{xx}=2, f_{yy}=2$, and $f_{xy}=1$.

$D=3.$

Thus, $D=3$ and $f_{xx}=2$. Since $D>0$ and $f_{xx}\left(-\frac{1}{3},\frac{2}{3}\right)=2>0$, it follows that f has a relative minimum at $\left(-\frac{1}{3},\frac{2}{3}\right)$. The minimum is found as follows:

$f(x,y)=x^2+xy+y^2-y$

$f\left(-\frac{1}{3},\frac{2}{3}\right)=\left(-\frac{1}{3}\right)^2+\left(-\frac{1}{3}\right)\left(\frac{2}{3}\right)+\left(\frac{2}{3}\right)^2-\left(\frac{2}{3}\right)$

$=\frac{1}{9}-\frac{2}{9}+\frac{4}{9}-\frac{2}{3}$

$=\frac{1}{9}-\frac{2}{9}+\frac{4}{9}-\frac{6}{9}$

$=-\frac{3}{9}$

$=-\frac{1}{3}.$

The relative minimum value of f is $-\frac{1}{3}$ at $\left(-\frac{1}{3},\frac{2}{3}\right)$.

3. $f(x,y)=2xy-x^3-y^2$

Find f_x.

$f(x,y)=2\underline{x}y-\underline{x}^3-y^2,$ The variable is underlined.

$f_x=2y-3x^2.$

Find f_y.

$f(x,y)=2x\underline{y}-x^3-\underline{y}^2,$ The variable is underlined.

$f_y=2x-2y.$

The solution is continued on the next page.

Copyright © 2014 Pearson Education, Inc. Publishing as Addison-Wesley.

Find f_{xx} and f_{xy}.

$f_x = 2y - 3\underline{x}^2, \quad f_x = 2\underline{y} - 3x^2,$

$f_{xx} = -6x. \quad f_{xy} = 2.$

Find f_{yy}.

$f_y = 2x - 2\underline{y},$

$f_{yy} = -2.$

Solve the system of equations $f_x = 0$ and $f_y = 0$:

$2y - 3x^2 = 0, \quad (1)$

$2x - 2y = 0. \quad (2)$

Solving Eq. (1) for 2*y*, we get $2y = 3x^2$.

Substituting $3x^2$ for $2y$ in Eq. (2) and solving, we get

$2x - 3x^2 = 0$

$x(2 - 3x) = 0$

$x = 0$ or $2 - 3x = 0$

$x = 0$ or $-3x = -2$

$x = 0$ or $x = \frac{2}{3}$

To find *y* when $x = 0$, we substitute 0 for *x* in either Eq. (1) or Eq. (2). We use Eq. (1):

$2y - 3(0)^2 = 0$

$2y = 0$

$y = 0.$

Thus, $(0,0)$ is one critical point, and $f(0,0)$ is a candidate for a maximum or minimum value.

To find the other critical point we substitute $\frac{2}{3}$ for *x* in either Eq. (1) or Eq. (2). We use Eq. (2):

$2\left(\frac{2}{3}\right) - 2y = 0$

$\frac{4}{3} - 2y = 0$

$-2y = -\frac{4}{3}$

$y = \frac{2}{3}.$

Thus, $\left(\frac{2}{3}, \frac{2}{3}\right)$ is the other critical point, and $f\left(\frac{2}{3}, \frac{2}{3}\right)$ is another candidate for maximum or minimum value.

We must check both $(0,0)$ and $\left(\frac{2}{3}, \frac{2}{3}\right)$ to see whether they yield maximum or minimum values.

For $(0,0)$

$D = f_{xx}(0,0) \cdot f_{yy}(0,0) - \left[f_{xy}(0,0)\right]^2$

$D = 0 \cdot (-2) - 2^2 \quad \begin{bmatrix} f_{xx}(0,0) = -6 \cdot 0 = 0 \\ f_{yy}(0,0) = -2 \\ f_{xy}(0,0) = 2 \end{bmatrix}$

$D = -4.$

Since $D < 0$, it follows that $f(0,0)$ is neither a maximum nor a minimum, but a saddle point.

For $\left(\frac{2}{3}, \frac{2}{3}\right)$

$D = f_{xx}\left(\frac{2}{3}, \frac{2}{3}\right) \cdot f_{yy}\left(\frac{2}{3}, \frac{2}{3}\right) - \left[f_{xy}\left(\frac{2}{3}, \frac{2}{3}\right)\right]^2$

$D = (-4) \cdot (-2) - 2^2 \quad \begin{bmatrix} f_{xx}\left(\frac{2}{3}, \frac{2}{3}\right) = -6 \cdot \frac{2}{3} = -4 \\ f_{yy}\left(\frac{2}{3}, \frac{2}{3}\right) = -2 \\ f_{xy}\left(\frac{2}{3}, \frac{2}{3}\right) = 2 \end{bmatrix}$

$D = 8 - 4 = 4.$

Thus, $D = 4$ and $f_{xx}\left(\frac{2}{3}, \frac{2}{3}\right) = -4$. Since $D > 0$ and $f_{xx}\left(\frac{2}{3}, \frac{2}{3}\right) < 0$, it follows that *f* has a relative maximum at $\left(\frac{2}{3}, \frac{2}{3}\right)$. The maximum is found as follows:

$f(x, y) = 2xy - x^3 - y^2$

$f\left(\frac{2}{3}, \frac{2}{3}\right) = 2\left(\frac{2}{3}\right)\left(\frac{2}{3}\right) - \left(\frac{2}{3}\right)^3 - \left(\frac{2}{3}\right)^2$

$= \frac{8}{9} - \frac{8}{27} - \frac{4}{9}$

$= \frac{4}{9} - \frac{8}{27}$

$= \frac{12}{27} - \frac{8}{27}$

$= \frac{4}{27}.$

The relative maximum value of *f* is $\frac{4}{27}$ at $\left(\frac{2}{3}, \frac{2}{3}\right)$.

Copyright © 2014 Pearson Education, Inc. Publishing as Addison-Wesley.

5. $f(x,y)=x^3+y^3-3xy$

Find f_x.

$f(x,y)=\underline{x}^3+y^3-3\underline{x}y,$ The variable is underlined.

$f_x=3x^2-3y.$

Find f_y.

$f(x,y)=x^3+\underline{y}^3-3x\underline{y},$ The variable is underlined.

$f_y=3y^2-3x.$

Find f_{xx} and f_{xy}.

$f_x=3\underline{x}^2-3y,\quad f_x=3x^2-3\underline{y},$

$f_{xx}=6x.\quad f_{xy}=-3.$

Find f_{yy}.

$f_y=3\underline{y}^2-3x,$

$f_{yy}=6y.$

Solve the system of equations $f_x=0$ and $f_y=0$:

$3x^2-3y=0,\quad (1)$

$3y^2-3x=0.\quad (2)$

We multiply each equation by $\frac{1}{3}$.

$x^2-y=0,\quad (1)$

$y^2-x=0.\quad (2)$

Solving Eq. (1) for y, we get $y=x^2$.

Substituting x^2 for y in Eq. (2) and solving, we get

$$(x^2)^2-x=0$$
$$x^4-x=0$$
$$x(x^3-1)=0$$
$$x=0 \text{ or } x^3-1=0$$
$$x=0 \text{ or } x^3=1$$
$$x=0 \text{ or } x=1$$

To find y when $x=0$, we substitute 0 for x in either Eq. (1) or Eq. (2). We use Eq. (2):

$$y^2-(0)=0$$
$$y^2=0$$
$$y=0.$$

Thus, $(0,0)$ is one critical point.

To find the other critical point we substitute 1 for x in either Eq. (1) or Eq. (2).

We use Eq. (2):

$$y^2-1=0$$
$$y^2=1$$
$$y=\pm1.$$

However, we notice that $(1,-1)$ is not a solution to Eq. (1) $1^2-(-1)=2\neq0$.

Thus, $(1,1)$ is the other critical point.

We must check both $(0,0)$ and $(1,1)$ to see whether they yield maximum or minimum values.

For $(0,0)$

$$D=f_{xx}(0,0)\cdot f_{yy}(0,0)-\left[f_{xy}(0,0)\right]^2$$

$$D=0\cdot(0)-(-3)^2\quad\begin{bmatrix}f_{xx}(0,0)=6\cdot0=0\\ f_{yy}(0,0)=6\cdot0=0\\ f_{xy}(0,0)=-3\end{bmatrix}$$

$$D=-9.$$

Since $D<0$, it follows that $f(0,0)$ is neither a maximum nor a minimum, but a saddle point.

For $(1,1)$

$$D=f_{xx}(1,1)\cdot f_{yy}(1,1)-\left[f_{xy}(1,1)\right]^2$$

$$D=6\cdot(6)-(-3)^2\quad\begin{bmatrix}f_{xx}(1,1)=6\cdot1=6\\ f_{yy}(1,1)=6\cdot(1)=6\\ f_{xy}(1,1)=-3\end{bmatrix}$$

$$D=36-9$$
$$D=27.$$

Thus, $D=27$ and $f_{xx}(1,1)=6$. Since $D>0$ and $f_{xx}(1,1)>0$, it follows that f has a relative minimum at $(1,1)$. The minimum is found as follows:

$$f(x,y)=x^3+y^3-3xy$$
$$f(1,1)=1^3+1^3-3(1)(1)$$
$$=1+1-3$$
$$=-1.$$

The relative minimum value of f is -1 at $(1,1)$.

Copyright © 2014 Pearson Education, Inc. Publishing as Addison-Wesley.

7. $f(x,y)=x^2+y^2-2x+4y-2$

Find f_x.

$f(x,y)=\underline{x}^2+y^2-2\underline{x}+4y-2,$

$f_x=2x-2.$

Find f_y.

$f(x,y)=x^2+\underline{y}^2-2x+4\underline{y}-2,$

$f_y=2y+4.$

Find f_{xx} and f_{xy}.

$f_x=2\underline{x}-2, \quad f_x=2x-2,$

$f_{xx}=2. \quad f_{xy}=0.$

Find f_{yy}.

$f_y=2\underline{y}+4,$

$f_{yy}=2.$

Solve the system of equations $f_x=0$ and $f_y=0$:

$2x-2=0, \quad 2y+4=0,$

$2x=2, \quad 2y=-4,$

$x=1. \quad y=-2.$

The only critical point is $(1,-2)$.

We must check $(1,-2)$ to see whether it yields a maximum or minimum value.

For $(1,-2)$

$$D=f_{xx}(1,-2)\cdot f_{yy}(1,-2)-\left[f_{xy}(1,-2)\right]^2$$

$$D=2\cdot(2)-(0)^2 \quad \begin{bmatrix} f_{xx}(1,-2)=2 \\ f_{yy}(1,-2)=2 \\ f_{xy}(1,-2)=0 \end{bmatrix}$$

$D=4.$

Thus, $D=4$ and $f_{xx}(1,-2)=2$. Since $D>0$ and $f_{xx}(1,-2)>0$, it follows that f has a relative minimum at $(1,-2)$. The minimum is found as follows:

$f(x,y)=x^2+y^2-2x+4y-2$

$f(1,-2)=1^2+(-2)^2-2(1)+4(-2)-2$

$=1+4-2-8-2$

$=-7.$

The relative minimum value of f is -7 at $(1,-2)$.

9. $f(x,y)=x^2+y^2+2x-4y$

Find f_x.

$f(x,y)=\underline{x}^2+y^2+2\underline{x}-4y,$

$f_x=2x+2.$

Find f_y.

$f(x,y)=x^2+\underline{y}^2+2x-4\underline{y},$

$f_y=2y-4.$

Find f_{xx} and f_{xy}.

$f_x=2\underline{x}+2, \quad f_x=2x+2,$

$f_{xx}=2. \quad f_{xy}=0.$

Find f_{yy}.

$f_y=2\underline{y}-4,$

$f_{yy}=2.$

Solve the system of equations $f_x=0$ and $f_y=0$:

$2x+2=0, \quad 2y-4=0,$

$2x=-2, \quad 2y=4,$

$x=-1. \quad y=2.$

The only critical point is $(-1,2)$.

We must check $(-1,2)$ to see whether it yields a maximum or minimum value.

For $(-1,2)$, we have:

$$D=f_{xx}(-1,2)\cdot f_{yy}(-1,2)-\left[f_{xy}(-1,2)\right]^2$$

$$D=2\cdot(2)-(0)^2 \quad \begin{bmatrix} f_{xx}(-1,2)=2 \\ f_{yy}(-1,2)=2 \\ f_{xy}(-1,2)=0 \end{bmatrix}$$

$D=4.$

Thus, $D=4$ and $f_{xx}(-1,2)=2$. Since $D>0$ and $f_{xx}(-1,2)>0$, it follows that f has a relative minimum at $(-1,2)$. The minimum is found as follows:

$f(x,y)=x^2+y^2+2x-4y$

$f(-1,2)=(-1)^2+(2)^2+2(-1)-4(2)$

$=1+4-2-8$

$=-5.$

The relative minimum value of f is -5 at $(-1,2)$.

Copyright © 2014 Pearson Education, Inc. Publishing as Addison-Wesley.

11. $f(x,y)=4x^2-y^2$

Find f_x.

$f(x,y)=4\underline{x}^2-y^2,$

$f_x=8x.$

Find f_y.

$f(x,y)=4x^2-\underline{y}^2,$

$f_y=-2y.$

Find f_{xx} and f_{xy}.

$f_x=8\underline{x},\quad f_x=8x,$

$f_{xx}=8.\quad f_{xy}=0.$

Find f_{yy}.

$f_y=-2\underline{y},$

$f_{yy}=-2.$

Solve the system of equations $f_x=0$ and $f_y=0$:

$8x=0,\qquad -2y=0,$

$x=0.\qquad y=0.$

The only critical point is $(0,0)$.

We must check $(0,0)$ to see whether it yields a maximum or minimum value.

For $(0,0)$

$D=f_{xx}(0,0)\cdot f_{yy}(0,0)-\left[f_{xy}(0,0)\right]^2$

$D=8\cdot(-2)-(0)^2\quad \begin{bmatrix} f_{xx}(0,0)=8 \\ f_{yy}(0,0)=-2 \\ f_{xy}(0,0)=0 \end{bmatrix}$

$D=-16.$

Since $D<0$, it follows that $f(0,0)$ is neither a maximum nor a minimum, but a saddle point.

13. $f(x,y)=e^{x^2+y^2+1}$

Find f_x.

$f(x,y)=e^{\underline{x}^2+y^2+1},$

$f_x=2xe^{x^2+y^2+1}.$

Find f_y.

$f(x,y)=e^{x^2+\underline{y}^2+1},$

$f_y=2ye^{x^2+y^2+1}.$

Find f_{xx}.

$f_x=2\underline{xe}^{\underline{x}^2+y^2+1}$

$f_{xx}=2x\left(2xe^{x^2+y^2+1}\right)+2e^{x^2+y^2+1}$

$=4x^2e^{x^2+y^2+1}+2e^{x^2+y^2+1}.$

Find f_{xy}.

$f_x=2xe^{x^2+\underline{y}^2+1}$

$f_{xy}=2x\left(2ye^{x^2+y^2+1}\right)$

$=4xye^{x^2+y^2+1}.$

Find f_{yy}.

$f_y=2\underline{ye}^{x^2+\underline{y}^2+1}$

$f_{yy}=2y\left(2ye^{x^2+y^2+1}\right)+2e^{x^2+y^2+1}$

$=4y^2e^{x^2+y^2+1}+2e^{x^2+y^2+1}.$

Solve the system of equations $f_x=0$ and $f_y=0$:

$2xe^{x^2+y^2+1}=0,\qquad 2ye^{x^2+y^2+1}=0,$

$x=0.\qquad y=0.$

The only critical point is $(0,0)$.

We must check $(0,0)$ to see whether it yields a maximum or minimum value.

For $(0,0)$

$D=f_{xx}(0,0)\cdot f_{yy}(0,0)-\left[f_{xy}(0,0)\right]^2$

$D=2e\cdot(2e)-(0)^2\quad \begin{bmatrix} f_{xx}(0,0)=2e \\ f_{yy}(0,0)=2e \\ f_{xy}(0,0)=0 \end{bmatrix}$

$D=4e^2.$

Thus, $D>0$ and $f_{xx}(0,0)=2e>0$, it follows that f has a relative minimum at $(0,0)$. The minimum is found as follows:

$f(x,y)=e^{x^2+y^2+1}$

$f(0,0)=e^{0^2+0^2+1}$

$=e.$

The relative minimum value of f is e at $(0,0)$.

Copyright © 2014 Pearson Education, Inc. Publishing as Addison-Wesley.

15. $R(x,y)=17x+21y$

$C(x,y)=4x^2-4xy+2y^2-11x+25y-3$

Total profit, $P(x,y)$ is given by

$$P(x,y)$$
$$=R(x,y)-C(x,y)$$
$$=(17x+21y)-(4x^2-4xy+2y^2-11x+25y-3)$$
$$=-4x^2+4xy-2y^2+28x-4y+3$$

Find P_x.

$$P(x,y)=-4\underline{x}^2+4\underline{x}y-2y^2+28\underline{x}-4y+3$$
$$P_x=-8x+4y+28$$

Find P_y.

$$P(x,y)=-4x^2+4x\underline{y}-2\underline{y}^2+28x-4\underline{y}+3$$
$$P_y=4x-4y-4$$

Find P_{xx} and P_{xy}.

$$P_x=-8\underline{x}+4y+28 \qquad P_x=-8x+4\underline{y}+28$$
$$P_{xx}=-8. \qquad P_{xy}=4.$$

Find P_{yy}.

$$P_y=4x-4\underline{y}-4$$
$$P_{yy}=-4.$$

Solve the system of equations $P_x=0$ and $P_y=0$:

$$-8x+4y+28=0, \qquad (1)$$
$$4x-4y-4=0. \qquad (2)$$

Adding these equations, we get:
$-4x+24=0.$
Then,
$-4x=-24$
$x=6.$

To find y when $x=6$, we substitute 6 for x into either Eq. (1) or Eq. (2). We use Eq. (1):

$$-8(6)+4y+28=0$$
$$4y-20=0$$
$$4y=20$$
$$y=5.$$

Thus, $(6,5)$ is the only critical point, and $P(6,5)$ is a candidate for a maximum or minimum value.

We must check to see whether $P(6,5)$ is a maximum or minimum value:

$$D=P_{xx}(6,5)\cdot P_{yy}(6,5)-[P_{xy}(6,5)]^2$$
$$=(-8)(-4)-4^2 \qquad \begin{bmatrix} P_{xx}(6,5)=-8 \\ P_{yy}(6,5)=-4 \\ P_{xy}(6,5)=4 \end{bmatrix}$$
$$=32-16$$
$$=16.$$

Thus, $D=16$ and $P_{xx}(6,5)=-8$. Since $D>0$ and $P_{xx}(6,5)<0$, it follows that P has a relative maximum at $(6,5)$. Thus, to maximize profit, the company must produce and sell 6 thousand of the \$17 sunglasses and 5 thousand of the \$21 sunglasses.

17. $P(a,p)=2ap+80p-15p^2-\frac{1}{10}a^2p-80$

Find P_a.

$$P(a,p)=2\underline{a}p+80p-15p^2-\frac{1}{10}\underline{a}^2p-80,$$
$$P_a=2p-\frac{1}{5}ap.$$

Find P_p.

$$P(a,p)=2a\underline{p}+80\underline{p}-15\underline{p}^2-\frac{1}{10}a^2\underline{p}-80,$$
$$P_p=2a+80-30p-\frac{1}{10}a^2.$$

Find P_{aa} and P_{ap}.

$$P_a=2p-\frac{1}{5}\underline{a}p \qquad P_a=2\underline{p}-\frac{1}{5}a\underline{p}$$
$$P_{aa}=-\frac{1}{5}p. \qquad P_{ap}=2-\frac{1}{5}a.$$

Find P_{pp}.

$$P_p=2a+80-30\underline{p}-\frac{1}{10}a^2$$
$$P_{pp}=-30.$$

Solve the system of equations $P_a=0$ and $P_p=0$:

$$2p-\frac{1}{5}ap=0, \qquad (1)$$
$$2a+80-30p-\frac{1}{10}a^2=0. \qquad (2)$$

The solution is continued on the next page.

Copyright © 2014 Pearson Education, Inc. Publishing as Addison-Wesley.

Solving Eq. (1) on the previous page by factoring, we see that $a=10$ or $p=0$. But p cannot equal 0 in the original equation and yield a positive profit. Substituting 10 for a in Eq. (2) and solving for p, we get

$$\begin{aligned}2(10)+80-30p-\frac{1}{10}(10)^2&=0\\20+80-10&=30p\\90&=30p\\3&=p.\end{aligned}$$

Thus, $(10,3)$ is the only critical point to consider, and $P(10,3)$ is a candidate for a maximum or minimum value.
We must check to see whether $P(10,3)$ is a maximum or minimum value:

$$\begin{aligned}D&=P_{aa}(10,3)\cdot P_{pp}(10,3)-\left[P_{ap}(10,3)\right]^2\\&=\left(-\frac{3}{5}\right)(-30)-0^2\end{aligned}$$

$$\begin{bmatrix}P_{aa}(10,3)=-\frac{1}{5}\cdot 3=\frac{-3}{5}\\P_{pp}(10,3)=-30\\P_{ap}(10,3)=2-\frac{1}{5}\cdot 10=0\end{bmatrix}$$

$$=18.$$

Since $D>0$ and $P_{aa}(10,3)=-\frac{3}{5}<0$, it follows that P has a relative maximum at $(10,3)$. Thus, to maximize profit, the company must spend 10 million dollars on advertising and charge \$3 per item. The maximum profit is found as follows:

$$\begin{aligned}P(10,3)&=2(10)(3)+80(3)-15(3)^2-\\&\qquad\frac{1}{10}(10)^2(3)-80\\&=60+240-135-30-80\\&=55.\end{aligned}$$

The maximum profit is \$55 million.

19. Sketch a drawing of the container.

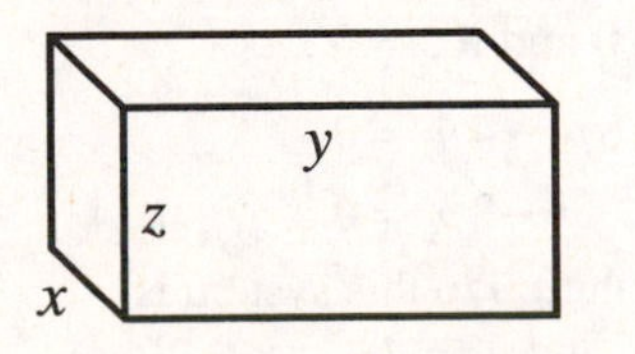

Let x, y and z represent the dimensions of the container as shown in the drawing.

The volume of the container is given by
$V=x\cdot y\cdot z$

$$320=x\cdot y\cdot z \qquad \left[V=320\text{ ft}^3\right]$$

$$\frac{320}{x\cdot y}=z.$$

Now we can express the cost as a function of two variables x and y. The area of the bottom is xy ft^2, so the cost of the bottom is $5xy$, two of the sides have area xz, or $x\left(\frac{320}{xy}\right)=\frac{320}{y}$ each.

The area of each of the remaining two sides is yz, or $y\left(\frac{320}{xy}\right)=\frac{320}{x}$. Then, the total area of all four sides is $2\left(\frac{320}{y}+\frac{320}{x}\right)=\frac{640}{y}+\frac{640}{x}$,

and the cost of the four sides is $4\left(\frac{640}{y}+\frac{640}{x}\right)$, or $\frac{2560}{y}+\frac{2560}{x}$.

Now we can write the total cost function.

$$\underset{\text{cost}}{\text{Total}} = \underset{\text{bottom}}{\text{Cost of}} + \underset{\text{sides}}{\text{Cost of}}$$

$$C(x,y)=5xy+\left(\frac{2560}{y}+\frac{2560}{x}\right).$$

Now, we try to find a minimum for $C(x,y)$

1. Find C_x, C_y, C_{xx}, C_{yy}, and C_{xy}:

$$C_x=5y-\frac{2560}{x^2}, \qquad C_y=5x-\frac{2560}{y^2},$$

$$C_{xx}=\frac{5120}{x^3}; \qquad C_{yy}=\frac{5120}{y^3};$$

$$C_{xy}=5.$$

2. Solve the system of equations $C_x=0$ and $C_y=0$:

$$5y-\frac{2560}{x^2}=0, \qquad (1)$$

$$5x-\frac{2560}{y^2}=0. \qquad (2)$$

The solution is continued on the next page.

Copyright © 2014 Pearson Education, Inc. Publishing as Addison-Wesley.

Solving Eq. (1) on the previous page for y:

$$5y-\frac{2560}{x^2}=0$$
$$5y=\frac{2560}{x^2}$$
$$y=\frac{512}{x^2}.$$

Substitute $\frac{512}{x^2}$ for y into Eq. (2) and solve for x:

$$5x-\frac{2560}{\left(\frac{512}{x^2}\right)^2}=0$$
$$5x-\frac{2560}{\frac{262,144}{x^4}}=0$$
$$5x-\frac{2560x^4}{262,144}=0$$
$$5x-\frac{5x^4}{512}=0$$
$$2560x-5x^4=0 \quad \text{Multiplying by 512.}$$
$$5x\left(512-x^3\right)=0$$
$$5x=0 \quad \text{or} \quad 512-x^3=0$$
$$x=0 \quad \text{or} \quad x^3=512$$
$$x=0 \quad \text{or} \quad x=8.$$

Since none of the dimensions can be 0, only $x=8$ has meaning in this application.

Substitute 8 for x into Eq. (1) to find y:

$$5y-\frac{2560}{(8)^2}=0$$
$$5y-\frac{2560}{64}=0$$
$$5y-40=0$$
$$5y=40$$
$$y=8.$$

Thus, $(8,8)$ is the only critical point, and $C(8,8)$ is a candidate for a maximum or minimum value.

3. We must check to see whether $C(8,8)$ is a maximum or minimum value:
$$D=C_{xx}(8,8)\cdot C_{yy}(8,8)-\left[C_{xy}(8,8)\right]^2$$
$$=\left(\frac{5120}{8^3}\right)\left(\frac{5120}{8^3}\right)-5^2 \quad \text{Using step 1 above}$$
$$=\frac{5120}{512}\cdot\frac{5120}{512}-25$$
$$=10\cdot 10-25$$
$$=100-25$$
$$=75.$$

4. Since $D>0$ and $C_{xx}(8,8)=10>0$, it follows that C has a relative minimum at $(8,8)$. Thus, to minimize cost, the dimensions of the bottom of the container should be 8 ft. by 8 ft. The height of the container should be $\frac{320}{8\cdot 8}$, or 5 ft.

21. a) $q_1=64-4p_1-2p_2 \quad (1)$

$q_2=56-2p_1-4p_2 \quad (2)$

$$R(p_1,p_2)$$
$$=p_1q_1+p_2q_2$$
$$=p_1(64-4p_1-2p_2)+p_2(56-2p_1-4p_2)$$
$$=64p_1-4p_1^2-2p_1p_2+56p_2-2p_1p_2-4p_2^2$$
$$=64p_1-4p_1^2-4p_1p_2+56p_2-4p_2^2.$$

b) We now find the values of p_1 and p_2 to maximize total revenue.

$$R_{p_1}=64-8p_1-4p_2,$$
$$R_{p_2}=-4p_1+56-8p_2,$$
$$R_{p_1p_1}=-8,$$
$$R_{p_2p_2}=-8,$$
$$R_{p_1p_2}=-4.$$

Solve the system of equations $R_{p_1}=0$ and $R_{p_2}=0$:

$$64-8p_1-4p_2=0$$
$$-4p_1+56-8p_2=0.$$

The solution to this system is $p_1=6$ and $p_2=4$.

The solution is continued on the next page.

Copyright © 2014 Pearson Education, Inc. Publishing as Addison-Wesley.

We check to see if $R(6,4)$ is a maximum or a minimum value.

$$D = R_{p_1p_1}(6,4) \cdot R_{p_2p_2}(6,4) - \left[R_{p_1p_2}(6,4)\right]^2$$
$$= (-8)(-8) - (-4)^2$$
$$= 64 - 16$$
$$= 48.$$

Since $D > 0$ and $R_{p_1p_1}(6,4) = -8 < 0$, it follows that R has a relative maximum at $(6,4)$. Thus, in order to maximize revenue, p_1 must be $6 \cdot 10 = \$60$ and p_2 must be $4 \cdot 10 = \$40$.

c) We substitute 6 for p_1 and 4 for p_2 into the demand equations to find q_1 and q_2.

$$q_1 = 64 - 4p_1 - 2p_2$$
$$q_1 = 64 - 4(6) - 2(4)$$
$$= 64 - 24 - 8$$
$$= 32$$
$$q_2 = 56 - 2p_1 - 4p_2$$
$$q_2 = 56 - 2(6) - 4(4)$$
$$= 56 - 12 - 16$$
$$= 28$$

32 hundred units of q_1 will be demanded and 28 hundred units of q_2 will be demanded.

d) To maximize revenue 3200 units of the \$60 calculator and 2800 units of the \$40 calculator must be produced and sold. The maximum revenue is found as follows:

$$R = 60 \cdot 3200 + 40 \cdot 2800$$
$$= 192{,}000 + 112{,}000$$
$$= 304{,}000.$$

The maximum revenue is \$304,000.

23. $f(x,y) = e^x + e^y - e^{x+y}$

1. Find f_x, f_y, f_{xx}, f_{yy}, and f_{xy}:

$$f_x = e^x - e^{x+y}$$
$$f_y = e^y - e^{x+y}$$
$$f_{xx} = e^x - e^{x+y}$$
$$f_{yy} = e^y - e^{x+y}$$
$$f_{xy} = -e^{x+y}$$

2. Solve the system of equations $f_x = 0$ and $f_y = 0$:

$$e^x - e^{x+y} = 0, \qquad (1)$$
$$e^y - e^{x+y} = 0. \qquad (2)$$

We can solve the first equation for y:

$$e^x - e^{x+y} = 0$$
$$e^x = e^{x+y}$$
$$x = x + y$$
$$y = 0.$$

We can solve the second equation for x:

$$e^y - e^{x+y} = 0$$
$$e^y = e^{x+y}$$
$$y = x + y$$
$$x = 0.$$

Thus, $(0,0)$ is a critical point, and $f(0,0)$ is a candidate for a maximum or minimum.

3. We must check to see if $f(0,0)$ is a maximum or minimum value:

$$D = f_{xx}(0,0) \cdot f_{yy}(0,0) - \left[f_{xy}(0,0)\right]^2$$
$$D = 0 \cdot 0 - (-1)^2$$
$$D = -1.$$

4. Since $D < 0$, it follows that $f(0,0)$ is neither a maximum nor a minimum, but a saddle point.

25. $f(x,y) = 2y^2 + x^2 - x^2y$

1. Find f_x, f_y, f_{xx}, f_{yy}, and f_{xy}:

$$f_x = 2x - 2xy \qquad f_y = 4y - x^2$$
$$f_{xx} = 2 - 2y; \qquad f_{yy} = 4;$$
$$f_{xy} = -2x.$$

2. Solve the system of equations $f_x = 0$ and $f_y = 0$:

$$2x - 2xy = 0, \qquad (1)$$
$$4y - x^2 = 0. \qquad (2)$$

Solving Eq. (2) for y, we get

$$4y = x^2$$
$$y = \frac{x^2}{4}.$$

The solution is continued on the next page.

Copyright © 2014 Pearson Education, Inc. Publishing as Addison-Wesley.

Substituting $\frac{x^2}{4}$ for y in Eq. (1) on the previous page and solving, we get

$$2x - 2x \cdot \frac{x^2}{4} = 0$$
$$2x - \frac{x^3}{2} = 0$$
$$4x - x^3 = 0$$
$$x(4 - x^2) = 0$$
$$x = 0 \quad \text{or} \quad 4 - x^2 = 0$$
$$x = 0 \quad \text{or} \quad x^2 = 4$$
$$x = 0 \quad \text{or} \quad x = \pm 2.$$

When $x = 0,\ y = \frac{0^2}{4} = 0.$

When $x = 2,\ y = \frac{2^2}{4} = 1.$

When $x = -2,\ y = \frac{(-2)^2}{4} = 1.$

The critical points are $(0,0), (2,1)$, and $(-2,1)$

3. We must check all the critical points to determine whether they yield maximum or minimum values.
For $(0,0)$

$$D = f_{xx}(0,0) \cdot f_{yy}(0,0) - [f_{xy}(0,0)]^2$$

$$D = (2)\cdot(4) - 0^2 \qquad \begin{bmatrix} f_{xx}(0,0) = 2 \\ f_{yy}(0,0) = 4 \\ f_{xy}(0,0) = 0 \end{bmatrix}$$

$$D = 8.$$

Since $D > 0$ and $f_{xx}(0,0) = 2 > 0$, it follows that f has a relative minimum at $(0,0)$. The minimum is found as follows:

$$f(x,y) = 2y^2 + x^2 - x^2 y$$
$$f(0,0) = 2 \cdot 0^2 + 0^2 - 0^2 \cdot 0 = 0$$

The relative minimum value of f is 0 at $(0,0)$.

For $(2,1)$

$$D = f_{xx}(2,1) \cdot f_{yy}(2,1) - [f_{xy}(2,1)]^2$$

$$D = (0)\cdot(4) - (-4)^2 \qquad \begin{bmatrix} f_{xx}(2,1) = 0 \\ f_{yy}(2,1) = 4 \\ f_{xy}(2,1) = -4 \end{bmatrix}$$

$$D = -16.$$

For $(-2,1)$

$$D = f_{xx}(-2,1) \cdot f_{yy}(-2,1) - [f_{xy}(-2,1)]^2$$

$$D = (0)\cdot(4) - (4)^2 \qquad \begin{bmatrix} f_{xx}(-2,1) = 0 \\ f_{yy}(-2,1) = 4 \\ f_{xy}(-2,1) = 4 \end{bmatrix}$$

$$D = -16.$$

Since $D < 0$ for both $(2,1)$ and $(-2,1)$, it follows that f has neither a maximum nor a minimum, but a saddle point at both of these points. Therefore, the only relative extrema of f is a relative minimum of 0 occurring at $(0,0)$.

27. ✎

29. $f(x,y)$ has a relative minimum of -5 at $(0,0)$.

31. $f(x,y)$ has no relative extrema.

Copyright © 2014 Pearson Education, Inc. Publishing as Addison-Wesley.

Exercise Set 6.4

1. Find the regression line for the data set:

x	1	2	4	5
y	1	3	3	4

The data points are $(1,1),(2,3),(4,3),$ and $(5,4)$.

The points on the regression line are $(1, y_1),(2, y_2),(4, y_3),$ and $(5, y_4)$.

The y-deviations are

$y_1 - 1,\ y_2 - 3,\ y_3 - 3,\ y_4 - 4.$

We want to minimize

$$S = (y_1 - 1)^2 + (y_2 - 3)^2 + (y_3 - 3)^2 + (y_4 - 4)^2$$

Where:

$$y_1 = m \cdot 1 + b$$
$$y_2 = m \cdot 2 + b$$
$$y_3 = m \cdot 4 + b$$
$$y_4 = m \cdot 5 + b$$

Substituting we get:

$$S = (m + b - 1)^2 + (2m + b - 3)^2 + (4m + b - 3)^2 + (5m + b - 4)^2$$

In order to minimize S, we need to find the first partial derivatives.

$$\frac{\partial S}{\partial b} = 2(m + b - 1) + 2(2m + b - 3) + 2(4m + b - 3) + 2(5m + b - 4)$$
$$= 2m + 2b - 2 + 4m + 2b - 6 + 8m + 2b - 6 + 10m + 2b - 8$$
$$= 24m + 8b - 22$$

$$\frac{\partial S}{\partial m} = 2(m + b - 1) \cdot 1 + 2(2m + b - 3) \cdot 2 + 2(4m + b - 3) \cdot 4 + 2(5m + b - 4) \cdot 5$$
$$= 2m + 2b - 2 + 8m + 4b - 12 + 32m + 8b - 24 + 50m + 10b - 40$$
$$= 92m + 24b - 78$$

We set these derivatives equal to 0 and solve the resulting system.

$$24m + 8b - 22 = 0$$
$$92m + 24b - 78 = 0$$

The solution to this system is $b = 0.95,\ m = 0.6.$

We use the D-test to verify that $S(0.95, 0.6)$ is a relative minimum.

We first find the second-order partial derivatives.

$$S_{bb} = 8, S_{bm} = 24, S_{mm} = 92$$

$$\frac{1}{2\lambda} = x = y = z.$$

Since $D > 0$ and $x = 3\lambda = y^2$, S has a relative minimum at $\frac{3}{16}\lambda^2$. The regression line is $y = 0.6x + 0.95.$

3. Find the regression line for the data set:

x	1	2	3	5
y	0	1	3	4

The data points are $(1,0),(2,1),(3,3),$ and $(5,4)$.

The points on the regression line are $(1, y_1),(2, y_2),(3, y_3),$ and $(5, y_4)$.

The y-deviations are

$y_1 - 0,\ y_2 - 1,\ y_3 - 3,$ and $y_4 - 4.$

We want to minimize

$$S = (y_1 - 0)^2 + (y_2 - 1)^2 + (y_3 - 3)^2 + (y_4 - 4)^2$$

Where:

$$y_1 = m \cdot 1 + b$$
$$y_2 = m \cdot 2 + b$$
$$y_3 = m \cdot 3 + b$$
$$y_4 = m \cdot 5 + b$$

Substituting we get:

$$S = (m + b)^2 + (2m + b - 1)^2 + (3m + b - 3)^2 + (5m + b - 4)^2$$

In order to minimize S, we need to find the first partial derivatives.

$$\frac{\partial S}{\partial b} = 2(m + b) + 2(2m + b - 1) + 2(3m + b - 3) + 2(5m + b - 4)$$
$$= 2m + 2b + 4m + 2b - 2 + 6m + 2b - 6 + 10m + 2b - 8$$
$$= 22m + 8b - 16$$

$$\frac{\partial S}{\partial m} = 2(m + b) \cdot 1 + 2(2m + b - 1) \cdot 2 + 2(3m + b - 3) \cdot 3 + 2(5m + b - 4) \cdot 5$$
$$= 2m + 2b + 8m + 4b - 4 + 18m + 6b - 18 + 50m + 10b - 40$$
$$= 78m + 22b - 62$$

The solution is continued on the next page.

Copyright © 2014 Pearson Education, Inc. Publishing as Addison-Wesley.

We set the derivatives from the previous page equal to 0 and solve the resulting system.

$$22m+8b-16=0$$
$$78m+22b-62=0$$

The solution to this system is $b=-\frac{29}{35}$, $m=\frac{36}{35}$

We use the D-test to verify that $S\left(-\frac{29}{35},\frac{36}{35}\right)$ is a relative minimum.
We first find the second-order partial derivatives.

$$S_{bb}=8, S_{bm}=22, S_{mm}=78$$
$$D=S_{bb}\left(-\tfrac{29}{35},\tfrac{36}{35}\right)\cdot S_{mm}\left(-\tfrac{29}{35},\tfrac{36}{35}\right)-\left[S_{bm}\left(-\tfrac{29}{35},\tfrac{36}{35}\right)\right]^2$$
$$D=8\cdot 78-[22]^2$$
$$=140$$

Since $D>0$ and $S_{bb}\left(-\frac{29}{35},\frac{36}{35}\right)=8>0$, S has a relative minimum at $\left(-\frac{29}{35},\frac{36}{35}\right)$. The regression line is $y=\frac{36}{35}x-\frac{29}{35}$.

5. a) The data points are $(0,3.80),(1,4.25),(6,4.75),(7,5.15),(17,5.85),(18,6.55)$ and $(19,7.25)$.
The points on the regression line are $(0,y_1),(1,y_2),(6,y_3),(7,y_4),(17,y_5),(18,y_6)$, and $(19,y_7)$.
The y-deviations are $y_1-3.80$, $y_2-4.25$, $y_3-4.75$, $y_4-5.15$, $y_5-5.85$, $y_6-6.55$, and $y_7-7.25$.
We want to minimize

$$S=(y_1-3.80)^2+(y_2-4.25)^2+(y_3-4.75)^2+(y_4-5.15)^2+(y_5-5.85)^2+(y_6-6.55)^2+(y_7-7.25)$$

Where:

$$y_1=m\cdot 0+b$$
$$y_2=m\cdot 1+b$$
$$y_3=m\cdot 6+b$$
$$y_4=m\cdot 7+b$$
$$y_5=m\cdot 17+b$$
$$y_6=m\cdot 18+b$$
$$y_7=m\cdot 19+b$$

Substituting we get:

$$S=(b-3.80)^2+(m+b-4.25)^2+(6m+b-4.75)^2+(7m+b-5.15)^2+(17m+b-5.85)^2+(18m+b-6.55)^2+(19m+b-7.25)^2$$

In order to minimize S, we need to find the first partial derivatives.

$$\frac{\partial S}{\partial b}=2(b-3.80)+2(m+b-4.25)+2(6m+b-4.75)+2(7m+b-5.15)+2(17m+b-5.85)+2(18m+b-6.55)+2(19m+b-7.25)$$
$$=2b-7.6+2m+2b-8.5+12m+2b-9.5+14m+2b-10.3+34m+2b-11.7+36m+2b-13.1+38m+2b-14.5$$
$$=136m+14b-75.2$$

$$\frac{\partial S}{\partial m}$$
$$=2(m+b-4.25)\cdot 1+2(6m+b-4.75)\cdot 6+2(7m+b-5.15)\cdot 7+2(17m+b-5.85)\cdot 17+2(18m+b-6.55)\cdot 18+2(19m+b-7.25)\cdot 19$$
$$=2m+2b-8.5+72m+12b-57+98m+14b-72.1+578m+34b-198.9+648m+36b-235.8+722m+38b-275.5$$
$$=2120m+136b-847.8$$

We set these derivatives equal to 0 and solve the resulting system.

$$136m+14b-75.2=0$$
$$2120m+136b-847.8=0$$

The solution to this system is

$$b=3.9452074392\approx 3.95$$
$$m=0.146816881259\approx 0.15$$

We use the D-test to verify that $S(b,m)$ is a relative minimum.
We first find the second-order partial derivatives.

$$S_{bb}=14, S_{bm}=136, S_{mm}=2120$$
$$D=S_{bb}\cdot S_{mm}-[S_{bm}]^2$$
$$D=14\cdot 2120-[136]^2$$
$$=11{,}184$$

Since $D>0$ and $S_{bb}=12>0$, S has a relative minimum at $(3.95,0.15)$. The regression line is $y=0.15x+3.95$.

Copyright © 2014 Pearson Education, Inc. Publishing as Addison-Wesley.

b) In 2015, $x = 2015 - 1990 = 25$

$y = 0.15(25) + 3.95$

≈ 7.70

The minimum wage will be about \$7.70 in 2015.

In 2020, $x = 2020 - 1990 = 30$

$y = 0.15(30) + 3.95$

≈ 8.45

The minimum wage will be about \$8.45 in 2020.

7. a) The data points are

$(1950, 71.1), (1960, 73.1), (1970, 74.7),$
$(1980, 77.4), (1990, 78.8), (2000, 79.5),$
$(2003, 80.1)$, and $(2007, 80.4)$.

The points on the regression line are

$(1950, y_1), (1960, y_2), (1970, y_3),$
$(1980, y_4), (1990, y_5), (2000, y_6),$
$(2003, y_7)$ and $(2007, y_8)$.

The y-deviations are

$y_1 - 71.1,\ y_2 - 73.1,\ y_3 - 74.7,$
$y_4 - 77.4,\ y_5 - 78.8,\ y_6 - 79.5,$
$y_7 - 80.1$ and $y_8 - 80.4$.

We want to minimize

$$S = (y_1 - 71.1)^2 + (y_2 - 73.1)^2 + (y_3 - 74.7)^2 + (y_4 - 77.4)^2 + (y_5 - 78.8)^2 + (y_6 - 79.5)^2 + (y_7 - 80.1)^2 + (y_8 - 80.4)^2$$

Where:

$$y_1 = m \cdot 1950 + b$$
$$y_2 = m \cdot 1960 + b$$
$$y_3 = m \cdot 1970 + b$$
$$y_4 = m \cdot 1980 + b$$
$$y_5 = m \cdot 1990 + b$$
$$y_6 = m \cdot 2000 + b$$
$$y_7 = m \cdot 2003 + b$$
$$y_8 = m \cdot 2007 + b$$

Substituting we get:

$$S = (1950m + b - 71.1)^2 + (1960m + b - 73.1)^2 + (1970m + b - 74.7)^2 + (1980m + b - 77.4)^2 + (1990m + b - 78.8)^2 + (2000m + b - 79.5)^2 + (2003m + b - 80.1)^2 + (2007m + b - 80.4)^2$$

In order to minimize S, we need to find the first partial derivatives

$$\frac{\partial S}{\partial b}$$
$$= 2(1950m + b - 71.1) + 2(1960m + b - 73.1) + 2(1970m + b - 74.7) + 2(1980m + b - 77.4) + 2(1990m + b - 78.8) + 2(2000m + b - 79.5) + 2(2003m + b - 80.1) + 2(2007m + b - 80.4)$$
$$= 31,720m + 16b - 1230.2$$

$$\frac{\partial S}{\partial m} = 2(1950m + b - 71.1) \cdot 1950 + 2(1960m + b - 73.1) \cdot 1960 + 2(1970m + b - 74.7) \cdot 1970 + 2(1980m + b - 77.4) \cdot 1980 + 2(1990m + b - 78.8) \cdot 1990 + 2(2000m + b - 79.5) \cdot 2000 + 2(2003m + b - 80.1) \cdot 2003 + 2(2007m + b - 80.4) \cdot 2007$$
$$= 7,605,000m + 3900b - 277,290 + 7,683,200m + 3920b - 286,552 + 7,761,800m + 3940b - 294,318 + 7,840,800m + 3960b - 306,504 + 7,920,200m + 3980b - 313,624 + 8,000,000m + 4000b - 318,000 + 8,024,018m + 4006b - 320,880.6 + 8,056,098m + 4014b - 322,725.6$$
$$= 62,891,116m + 31,720b - 2,439,894.2$$

We set these derivatives equal to 0 and solve the resulting system.

$$31,720m + 16b - 1230.2 = 0$$
$$62,891,116m + 31,720b - 2,439,894.2 = 0$$

The solution to this system is

$b = -249.287331 \approx 249.287$

$m = 0.164527027 \approx 0.165$

We use the D-test to verify that $S(b, m)$ is a relative minimum.

We first find the second-order partial derivatives.

$S_{bb} = 16, S_{bm} = 31,720, S_{mm} = 62,891,116$

$D = S_{bb} \cdot S_{mm} - [S_{bm}]^2$

$D = 16 \cdot 62,891,116 - [31,720]^2$

$= 99,456$

The solution is continued on the next page.

Copyright © 2014 Pearson Education, Inc. Publishing as Addison-Wesley.

Since $D>0$ and $S_{bb}=16>0$, S has a relative minimum at $(-249.287, 0.165)$. The regression line is
$y=0.165x-249.287$.

b) In 2015,
$y=0.165(2015)-249.287\approx 83.188$.
In 2015, the average life expectancy of women will be about 83.19 years.
In 2020,
$y=0.165(2020)-249.287\approx 84.013$.
In 2020, the average life expectancy of women will be about 84.01 years.

9. a) The data points are
$(70,75),(60,62)$, and $(85,89)$.
The points on the regression line are
$(70, y_1),(60, y_2)$, and $(85, y_3)$.
The y-deviations are
y_1-75, y_2-62, and y_3-89.
We want to minimize
$S=(y_1-75)^2+(y_2-62)^2+(y_3-89)^2$
Where:

$$y_1=m\cdot 70+b$$
$$y_2=m\cdot 60+b$$
$$y_3=m\cdot 85+b$$

Substituting we get:

$$S=(70m+b-75)^2+(60m+b-62)^2+(85m+b-89)^2$$

In order to minimize S, we need to find the first partial derivatives.

$$\frac{\partial S}{\partial b}=2(70m+b-75)+2(60m+b-62)+2(85m+b-89)$$
$$=140m+2b-150+120m+2b-124+170m+2b-178$$
$$=430m+6b-452$$

$$\frac{\partial S}{\partial m}=2(70m+b-75)\cdot 70+2(60m+b-62)\cdot 60+2(85m+b-89)\cdot 85$$
$$=9800m+140b-10,500+7200m+120b-7440+14,450m+170b-15,130$$
$$=31,450m+430b-33,070$$

We set the derivatives equal to 0 and solve the resulting system.

$$430m+6b-452=0$$
$$31,450m+430b-33,070=0$$

The solution to this system is
$b=-1.236842105\approx -1.24$
$m=1.068421053\approx 1.07$
We use the D-test to verify that $S(b,m)$ is a relative minimum.
We first find the second-order partial derivatives.
$S_{bb}=6, S_{bm}=430, S_{mm}=31,450$

$$D=S_{bb}\cdot S_{mm}-[S_{bm}]^2$$
$$D=6\cdot 31,450-[430]^2$$
$$=3800$$

Since $D>0$ and $S_{bb}=6>0$, S has a relative minimum at $(-1.24, 1.07)$.
The regression line is
$y=1.07x-1.24$

b) $x=81$
$y=1.07(81)-1.24$
$\approx 85.$
A student who scores 81% on the midterm will score about 85% on the final.

11. ✎

13. a) Converting the times to decimal notation and using the STAT package on a calculator, we get the regression equation
$y=-0.0059379586x+15.57191398$.

b) We predict that the world record in 2010 will be about 3.636617 minutes or 3:38.2. We predict that the world record in 2015 will be about 3.60693 minutes or 3:36.4.

c) According to the regression model, we would predict the world record in 1999 to be 3.7019 minutes or 3:42.1. This is about a second faster than the actual world record.

Copyright © 2014 Pearson Education, Inc. Publishing as Addison-Wesley.

Exercise Set 6.5

1. Find the maximum value of
$f(x,y)=xy$
subject to the constraint
$3x+y=10$.
We first express $3x+y=10$ as $3x+y-10=0$.
We form the new function F, given by:
$F(x,y,\lambda)=xy-\lambda(3x+y-10)$.
We find the first partial derivatives:
$F(x,y,\lambda)=\underline{x}y-\lambda(3\underline{x}+y-10)$
$F_x=y-3\lambda,$
$F(x,y,\lambda)=x\underline{y}-\lambda(3x+\underline{y}-10)$
$F_y=x-\lambda,$
$F(x,y,\lambda)=xy-\underline{\lambda}(3x+y-10)$
$F_\lambda=-(3x+y-10).$
We set each derivative equal to 0 and solve the resulting system:
$y-3\lambda=0 \quad (1)$
$x-\lambda=0 \quad (2)$
$3x+y-10=0 \quad (3)\left[\begin{array}{l}-(3x+y-10)=0\text{, or}\\ 3x+y-10=0\end{array}\right]$
Solving Eq. (2) for λ, we get:
$\lambda=x.$
Substituting into Eq. (1) for λ, we get:
$y-3(x)=0$, or $y=3x. \quad (4)$
Substituting $3x$ for y in Eq. (3), we get:
$$3x+3x-10=0$$
$$6x=10$$
$$x=\frac{10}{6}=\frac{5}{3}$$
Then, using Eq. (4), we have:
$$y=3\left(\frac{5}{3}\right)=5.$$
The maximum value of f subject to the constraint occurs at $\left(\frac{5}{3},5\right)$ and is
$$f\left(\frac{5}{3},5\right)=\frac{5}{3}\cdot 5=\frac{25}{3}.$$

3. Find the maximum value of
$f(x,y)=4-x^2-y^2$
subject to the constraint
$x+2y=10$.
We first express $x+2y=10$ as $x+2y-10=0$.
We form the new function F, given by:
$F(x,y,\lambda)=4-x^2-y^2-\lambda(x+2y-10)$.
We find the first partial derivatives:
$F(x,y,\lambda)=4-\underline{x}^2-y^2-\lambda(\underline{x}+2y-10)$
$F_x=-2x-\lambda,$
$F(x,y,\lambda)=4-x^2-\underline{y}^2-\lambda(x+2\underline{y}-10)$
$F_y=-2y-2\lambda,$
$F(x,y,\lambda)=4-x^2-y^2-\underline{\lambda}(x+2y-10)$
$F_\lambda=-(x+2y-10).$
We set each derivative equal to 0 and solve the resulting system:
$-2x-\lambda=0 \quad (1)$
$-2y-2\lambda=0 \quad (2)$
$x+2y-10=0 \quad (3)\left[\begin{array}{l}-(x+2y-10)=0\text{, or}\\ x+2y-10=0\end{array}\right]$
Solving Eq. (1) for λ, we get:
$\lambda=-2x.$
Substituting into Eq. (2) for λ, we get:
$-2y-2(-2x)=0$, or $y=2x. \quad (4)$
Substituting $2x$ for y in Eq. (3), we get:
$$x+2(2x)-10=0$$
$$5x=10$$
$$x=2.$$
Then, using Eq. (4), we have:
$$y=2(2)=4.$$
The maximum value of f subject to the constraint occurs at $(2,4)$ and is
$$f(2,4)=4-(2)^2-(4)^2$$
$$=4-4-16$$
$$=-16.$$

Copyright © 2014 Pearson Education, Inc. Publishing as Addison-Wesley.

5. Find the minimum value of

$f(x,y)=x^2+y^2$

subject to the constraint

$2x+y=10$.

We first express $2x+y=10$ as $2x+y-10=0$.

We form the new function F, given by:

$F(x,y,\lambda)=x^2+y^2-\lambda(2x+y-10)$.

We find the first partial derivatives:

$F(x,y,\lambda)=\underline{x}^2+y^2-\lambda(2\underline{x}+y-10)$

$F_x=2x-2\lambda,$

$F(x,y,\lambda)=x^2+\underline{y}^2-\lambda(2x+\underline{y}-10)$

$F_y=2y-\lambda,$

$F(x,y,\lambda)=x^2+y^2-\underline{\lambda}(2x+y-10)$

$F_\lambda=-(2x+y-10).$

We set each derivative equal to 0 and solve the resulting system:

$2x-2\lambda=0 \quad (1)$

$2y-\lambda=0 \quad (2)$

$2x+y-10=0 \quad (3)\left[\begin{matrix}-(2x+y-10)=0,\text{ or}\\ 2x+y-10=0\end{matrix}\right]$

The solution is continued on the next page.

Solving Eq. (2) for λ, we get:

$\lambda=2y.$

Substituting into Eq. (1) for λ, we get:

$2x-2(2y)=0,\text{ or } x=2y. \quad (4)$

Substituting $2y$ for x in Eq. (3), we get:

$2(2y)+y-10=0$

$5y=10$

$y=2.$

Then, using Eq. (4), we have:

$x=2(2)=4.$

The minimum value of f subject to the constraint occurs at $(4,2)$ and is

$f(4,2)=4^2+2^2$

$=16+4$

$=20.$

7. Find the minimum value of

$f(x,y)=2y^2-6x^2$

subject to the constraint

$2x+y=4$.

We first express $2x+y=4$ as $2x+y-4=0$.

We form the new function F, given by:

$F(x,y,\lambda)=2y^2-6x^2-\lambda(2x+y-4)$.

We find the first partial derivatives:

$F(x,y,\lambda)=2y^2-6\underline{x}^2-\lambda(2\underline{x}+y-4)$

$F_x=-12x-2\lambda,$

$F(x,y,\lambda)=2\underline{y}^2-6x^2-\lambda(2x+\underline{y}-4)$

$F_y=4y-\lambda,$

$F(x,y,\lambda)=2y^2-6x^2-\underline{\lambda}(2x+y-4)$

$F_\lambda=-(2x+y-4).$

We set each derivative equal to 0 and solve the resulting system:

$-12x-2\lambda=0 \quad (1)$

$4y-\lambda=0 \quad (2)$

$2x+y-4=0 \quad (3)\left[\begin{matrix}-(2x+y-4)=0,\text{ or}\\ 2x+y-4=0\end{matrix}\right]$

Solving Eq. (2) for λ, we get:

$\lambda=4y.$

Substituting into Eq. (1) for λ, we get:

$-12x-2(4y)=0$

$8y=-12x$

$y=-\frac{3}{2}x. \quad (4)$

Substituting $-\frac{3}{2}x$ for y in Eq. (3), we get:

$2x+\left(-\frac{3}{2}x\right)-4=0$

$\frac{1}{2}x=4$

$x=8.$

Then, using Eq. (4), we have:

$y=-\frac{3}{2}(8)=-12.$

The minimum value of f subject to the constraint occurs at $(8,-12)$ and is

$f(8,-12)=2(-12)^2-6(8)^2$

$=2(144)-6(64)$

$=288-384$

$=-96.$

Copyright © 2014 Pearson Education, Inc. Publishing as Addison-Wesley.

9. Find the minimum value of
$f(x, y, z) = x^2 + y^2 + z^2$
subject to the constraint
$y + 2x - z = 3$.
We first express $y + 2x - z = 3$ as
$y + 2x - z - 3 = 0$.
We form the new function F, given by:
$F(x, y, z, \lambda)$
$= x^2 + y^2 + z^2 - \lambda(y + 2x - z - 3)$
We find the first partial derivatives:
$F_x = 2x - 2\lambda$,
$F_y = 2y - \lambda$,
$F_z = 2z + \lambda$,
$F_\lambda = -(y + 2x - z - 3)$.
We set each derivative equal to 0 and solve the resulting system:

$$2x - 2\lambda = 0 \quad (1)$$
$$2y - \lambda = 0 \quad (2)$$
$$2z + \lambda = 0 \quad (3)$$
$$y + 2x - z - 3 = 0 \quad (4)\left[\begin{matrix}-(y+2x-z-3)=0, \text{ or} \\ y+2x-z-3=0\end{matrix}\right]$$

Solving Eq. (1) for *x*, we get:
$x = \lambda$.
Solving Eq. (2) for *y*, we get:
$y = \frac{1}{2}\lambda$.
Solving Eq. (3) for *z*, we get:
$z = -\frac{1}{2}\lambda$.

Substituting λ for *x*, $\frac{1}{2}\lambda$ for *y*, and $-\frac{1}{2}\lambda$ for *z* into Eq. (4), we get:

$$y + 2x - z - 3 = 0$$
$$\frac{1}{2}\lambda + 2\lambda - \left(-\frac{1}{2}\lambda\right) - 3 = 0$$
$$3\lambda = 3$$
$$\lambda = 1.$$

Then,
$x = \lambda = 1$
$y = \frac{1}{2}\lambda = \frac{1}{2}$
$z = -\frac{1}{2}\lambda = -\frac{1}{2}$
Contiunued at the top of the next column.

The minimum value of *f* subject to the constraint occurs at $\left(1, \frac{1}{2}, -\frac{1}{2}\right)$ and is

$$f\left(1, \frac{1}{2}, -\frac{1}{2}\right) = 1^2 + \left(\frac{1}{2}\right)^2 + \left(-\frac{1}{2}\right)^2$$
$$= 1 + \frac{1}{4} + \frac{1}{4}$$
$$= \frac{3}{2}.$$

11. Find the maximum value of
$f(x, y) = xy$ (Product is $x \cdot y$)
subject to the constraint
$x + y = 50$. (Sum is 50.).
We first express $x + y = 50$ as $x + y - 50 = 0$.
We form the new function F, given by:
$F(x, y, \lambda) = xy - \lambda(x + y - 50)$.
We find the first partial derivatives:
$F_x = y - \lambda$,
$F_y = x - \lambda$,
$F_\lambda = -(x + y - 50)$.
We set each derivative equal to 0 and solve the resulting system:

$$y - \lambda = 0 \quad (1)$$
$$x - \lambda = 0 \quad (2)$$
$$x + y - 50 = 0 \quad (3)\left[\begin{matrix}-(x+y-50)=0, \text{ or} \\ x+y-50=0\end{matrix}\right]$$

Solving Eq. (2) for λ, we get:
$\lambda = x$.
Substituting into Eq. (1) for λ, we get:
$y - (x) = 0$, or $y = x$. (4)
Substituting *x* for *y* in Eq. (3), we get:

$$x + x - 50 = 0$$
$$2x = 50$$
$$x = 25.$$

Then, using Eq. (4), we have:
$y = 25$.
The maximum value of *f* subject to the constraint occurs at $(25, 25)$. Thus, the two numbers whose sum is 50 that have the maximum product are 25 and 25.

Copyright © 2014 Pearson Education, Inc. Publishing as Addison-Wesley.

13. Find the minimum value of

$f(x, y) = xy$ (Product is $x \cdot y$)

subject to the constraint

$x - y = 6.$ (Difference is 6.).

We first express $x - y = 6$ as $x - y - 6 = 0$.

We form the new function F, given by:

$F(x, y, \lambda) = xy - \lambda(x - y - 6).$

We find the first partial derivatives:

$F_x = y - \lambda,$

$F_y = x + \lambda,$

$F_\lambda = -(x - y - 6).$

We set each derivative equal to 0 and solve the resulting system:

$$y - \lambda = 0 \quad (1)$$

$$x + \lambda = 0 \quad (2)$$

$$x - y - 6 = 0 \quad (3) \left[\begin{matrix} -(x-y-6)=0, \text{ or} \\ x-y-6=0 \end{matrix}\right]$$

Solving Eq. (1) for λ, we get:

$\lambda = y.$

Substituting into Eq. (2) for λ, we get:

$x + (y) = 0$, or $y = -x.$ (4)

Substituting $-x$ for y in Eq. (3), we get:

$$x - (-x) - 6 = 0$$

$$2x = 6$$

$$x = 3.$$

Then, using Eq. (4), we have:

$y = -3.$

The minimum value of f subject to the constraint occurs at $(3, -3)$. Thus, the two numbers whose difference is 6 that have the minimum product are 3 and -3.

15. Find the minimum value of

$f(x, y, z) = (x-1)^2 + (y-1)^2 + (z-1)^2$

subject to the constraint

$x + 2y + 3z = 13$.

We first express $x + 2y + 3z = 13$ as

$x + 2y + 3z - 13 = 0$.

We form the new function F, given by:

$F(x, y, z, \lambda)$

$= (x-1)^2 + (y-1)^2 + (z-1)^2 -$

$\lambda(x + 2y + 3z - 13)$

Continued at the top of the next column.

We find the first partial derivatives:

$F_x = 2(x-1) - \lambda,$

$F_y = 2(y-1) - 2\lambda,$

$F_z = 2(z-1) - 3\lambda,$

$F_\lambda = -(x + 2y + 3z - 13).$

We set each derivative equal to 0 and solve the resulting system:

$$2(x-1) - \lambda = 0 \quad (1)$$

$$2(y-1) - 2\lambda = 0 \quad (2)$$

$$2(z-1) - 3\lambda = 0 \quad (3)$$

$$x + 2y + 3z - 13 = 0 \quad (4) \left[\begin{matrix} -(x+2y+3z-13)=0, \text{ or} \\ x+2y+3z-13=0 \end{matrix}\right]$$

Solving Eq. (1) for x, we get:

$$2x - 2 - \lambda = 0$$

$$2x = 2 + \lambda$$

$$x = 1 + \frac{1}{2}\lambda.$$

Solving Eq. (2) for y, we get:

$$2y - 2 - 2\lambda = 0$$

$$2y = 2 + 2\lambda$$

$$y = 1 + \lambda.$$

Solving Eq. (3) for z, we get:

$$2z - 2 - 3\lambda = 0$$

$$2z = 2 + 3\lambda$$

$$z = 1 + \frac{3}{2}\lambda.$$

Substituting $1 + \frac{1}{2}\lambda$ for x, $1 + \lambda$ for y, and $1 + \frac{3}{2}\lambda$ for z into Eq. (4), we get:

$$x + 2y + 3z - 13 = 0$$

$$\left(1 + \frac{1}{2}\lambda\right) + 2(1 + \lambda) + 3\left(1 + \frac{3}{2}\lambda\right) - 13 = 0$$

$$1 + \frac{1}{2}\lambda + 2 + 2\lambda + 3 + \frac{9}{2}\lambda = 13$$

$$6 + 7\lambda = 13$$

$$7\lambda = 7$$

$$\lambda = 1.$$

The solution is continued on the next page.

Copyright © 2014 Pearson Education, Inc. Publishing as Addison-Wesley.

Using the information from the previous page, we have

$x = 1 + \frac{1}{2}\lambda = 1 + \frac{1}{2} = \frac{3}{2}$

$y = 1 + \lambda = 1 + 1 = 2$

$z = 1 + \frac{3}{2}\lambda = 1 + \frac{3}{2} = \frac{5}{2}.$

The minimum value of f subject to the constraint occurs at $\left(\frac{3}{2}, 2, \frac{5}{2}\right)$.

17. The area of the page is given by $A = xy$ and the perimeter of the page is given by $P = 2x + 2y$. See the figure.

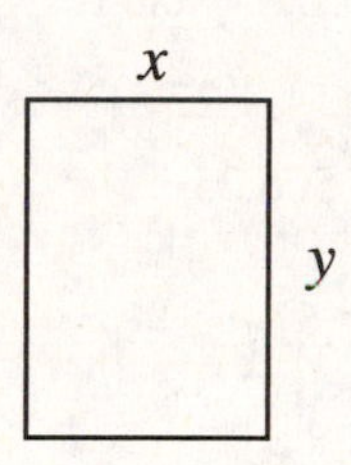

We want to maximize the area
$A = xy$
Subject to the constraint
$2x + 2y = 39.$
We first express $2x + 2y = 39$ as
$2x + 2y - 39 = 0.$
We form the new function F, given by:
$F(x, y, \lambda) = xy - \lambda(2x + 2y - 39).$
We find the first partial derivatives:

$F_x = y - 2\lambda,$

$F_y = x - 2\lambda,$

$F_\lambda = -(2x + 2y - 39).$

We set each derivative equal to 0 and solve the resulting system:

$y - 2\lambda = 0 \quad (1)$

$x - 2\lambda = 0 \quad (2)$

$2x + 2y - 39 = 0 \quad (3) \left[\begin{matrix} -(2x+2y-39)=0, \text{ or} \\ 2x+2y-39=0 \end{matrix}\right]$

From Eqs. (1) and (2) we see:
$y = 2\lambda = x.$

Substituting x for y in Eq. (3), we get:

$2x + 2x - 39 = 0$

$4x = 39$

$x = \frac{39}{4} = 9\tfrac{3}{4}.$

Then,
$y = x = 9\tfrac{3}{4}.$

The maximum area subject to the constraint occurs at $\left(9\tfrac{3}{4}, 9\tfrac{3}{4}\right)$. The maximum area is $A = 9\tfrac{3}{4} \cdot 9\tfrac{3}{4} = 95\tfrac{1}{16}$ in^2. The area of the standard $8\tfrac{1}{2} \times 11$ paper is not the maximum area of paper that has a perimeter of 39 in.

19. We want to minimize the function s given by
$s(h, r) = 2\pi rh + 2\pi r^2$
subject to the volume constraint
$\pi r^2 h = 27$, or $\pi r^2 h - 27 = 0.$
We form the new function S given by
$S(h, r, \lambda) = 2\pi rh + 2\pi r^2 - \lambda(\pi r^2 h - 27).$
We find the first partial derivatives.

$S_h = 2\pi r - \lambda\pi r^2,$

$S_r = 2\pi h + 4\pi r - 2\lambda\pi rh,$

$S_\lambda = -(\pi r^2 h - 27).$

We set these derivatives equal to 0 and solve the resulting system.

$2\pi r - \lambda\pi r^2 = 0 \quad (1)$

$2\pi h + 4\pi r - 2\lambda\pi rh = 0 \quad (2)$

$\pi r^2 h - 27 = 0. \quad (3) \left[\begin{matrix} -(\pi r^2 h - 27)=0, \text{ or} \\ \pi r^2 h - 27 = 0 \end{matrix}\right]$

We solve Eq. (1) for r:

$\pi r(2 - \lambda r) = 0$

$\pi r = 0 \quad \text{or} \quad 2 - \lambda r = 0$

$r = 0 \quad \text{or} \quad r = \frac{2}{\lambda}$

Note, $r = 0$ can not be a solution to the original problem, so we continue by substituting $\frac{2}{\lambda}$ for r in Eq. (2).

The solution is continued on the next page.

Copyright © 2014 Pearson Education, Inc. Publishing as Addison-Wesley.

Using the information from the previous page, we solve for h.

$$2\pi h+4\pi\left(\frac{2}{\lambda}\right)-2\lambda\pi\left(\frac{2}{\lambda}\right)h=0$$
$$2\pi h+\frac{8\pi}{\lambda}-4\pi h=0$$
$$\frac{8\pi}{\lambda}-2\pi h=0$$
$$h=\frac{4}{\lambda}$$

Since $h=\frac{4}{\lambda}$ and $r=\frac{2}{\lambda}$, it follows that $h=2r$. substituting $2r$ for h in Eq. (3) yields:

$$\pi r^2(2r)-27=0$$
$$2\pi r^3=27$$
$$r^3=\frac{27}{2\pi}$$
$$r=\sqrt[3]{\frac{27}{2\pi}}\approx 1.6$$

So when $r\approx 1.6$ ft and $h=2(1.6)\approx 3.2$ ft, the surface area of the oil drum is a minimum. The minimum area is about

$2\pi(1.6)(3.2)+2\pi(1.6)^2\approx 48.3\text{ ft}^2$.

(Answers will vary due to rounding differences.)

21. We want maximize

$S(L,M)=ML-L^2$

subject to the constraint

$M+L=90$.

We first express $M+L=90$ as $M+L-90=0$.

We form the new function F, given by:

$F(L,M,\lambda)=ML-L^2-\lambda(M+L-90)$.

We find the first partial derivatives:

$$F_L=M-2L-\lambda,$$
$$F_M=L-\lambda,$$
$$F_\lambda=-(M+L-90).$$

The solution is continued on the next page.

We set each derivative equal to 0 and solve the resulting system:

$$M-2L-\lambda=0 \quad (1)$$
$$L-\lambda=0 \quad (2)$$
$$M+L-90=0 \quad (3)\ \left[\begin{matrix}-(M+L-90)=0, \text{ or}\\ M+L-90=0\end{matrix}\right]$$

Solving Eq. (2) for λ, we get:

$\lambda=L.$

Substituting into Eq. (1) for λ, we get:

$$M-2L-L=0$$
$$M-3L=0$$
$$M=3L. \quad (4)$$

Substituting $3L$ for M in Eq. (3), we get:

$$3L+L-90=0$$
$$4L=90$$
$$L=22.5.$$

Then, using Eq. (4), we have:

$M=3(22.5)=67.5$.

The maximum value of S subject to the constraint occurs at $(22.5,67.5)$ and is

$$S(22.5,67.5)=(67.5)(22.5)-(22.5)^2$$
$$=1518.75-506.25$$
$$=1012.5$$

23. a) The area of the floor is xy.

The cost of the floor is $4xy$.

The area of the walls is $2xz+2yz$.

The cost of the walls is $3(2xz+2yz)$.

The area of the ceiling is xy.

The cost of the ceiling is $3xy$.

Therefore, the total cost function is

$$C(x,y,z)=4xy+3(2xz+2yz)+3xy$$
$$=7xy+6xz+6yz.$$

b) We want to minimize the value of

$C(x,y,z)=7xy+6xz+6yz$

subject to the constraint of

$x\cdot y\cdot z=252{,}000$ $\quad$ $(\text{Volume}=l\cdot w\cdot h)$

We first express $x\cdot y\cdot z=252{,}000$ as $x\cdot y\cdot z-252{,}000=0$.

We form the new function F, given by:

$$F(x,y,z,\lambda)$$
$$=7xy+6xz+6yz-\lambda(x\cdot y\cdot z-252{,}000)$$

We find the first partial derivatives:

$$F_x=7y+6z-\lambda yz,$$
$$F_y=7x+6z-\lambda xz,$$
$$F_z=6x+6y-\lambda xy,$$
$$F_\lambda=-(x\cdot y\cdot z-252{,}000).$$

The solution is continued on the next page.

Copyright © 2014 Pearson Education, Inc. Publishing as Addison-Wesley.

We set each derivative equal to 0 and solve the resulting system:

$7y+6z-\lambda yz=0 \quad (1)$

$7x+6z-\lambda xz=0 \quad (2)$

$6x+6y-\lambda xy=0 \quad (3)$

$xyz-252{,}000=0 \quad (4)\left[\begin{array}{l}-(x\cdot y\cdot z-252{,}000)=0, \text{ or} \\ x\cdot y\cdot z-252{,}000=0\end{array}\right]$

Solving Eq. (2) for *x* and Eq. (1) for *y*, we get:

$x=\frac{6z}{\lambda z-7}$ and $y=\frac{6z}{\lambda z-7}$.

Thus, $x=y$.

Substituting *x* for *y* we get the following system:

$7x+6z-\lambda xz=0$

$6x+6x-\lambda xx=0$

$xxz-252{,}000=0$

Which simplifies to:

$7x+6z-\lambda xz=0 \quad (5)$

$12x-\lambda x^2=0 \quad (6)$

$x^2z-252{,}000=0 \quad (7)$

Solving Eq. (6) for *x*, we get

$12x-\lambda x^2=0$

$x(12-\lambda x)=0$

$x=0$ or $12-\lambda x=0$

$x=0$ or $x=\frac{12}{\lambda}$

We only consider $x=\frac{12}{\lambda}$ since *x* cannot be 0 in the original problem. We continue by substituting $\frac{12}{\lambda}$ for *x* into Eq. (7) and solving for *z*.

$$\left(\frac{12}{\lambda}\right)^2 z-252{,}000=0$$

$$\frac{144}{\lambda^2}\cdot z=252{,}000$$

$$z=\frac{252{,}000}{144}\lambda^2$$

$$z=1750\lambda^2$$

Next we substitute $\frac{12}{\lambda}$ for *x* and $1750\lambda^2$ for *z* in Eq. (5) and solve for λ.

$$7\left(\frac{12}{\lambda}\right)+6\cdot 1750\lambda^2-\lambda\left(\frac{12}{\lambda}\right)1750\lambda^2=0$$

$$\frac{84}{\lambda}+10{,}500\lambda^2-21{,}000\lambda^2=0$$

$$10{,}500\lambda^2=\frac{84}{\lambda}$$

$$\lambda^3=\frac{84}{10{,}500}$$

$$\lambda^3=\frac{1}{125}$$

$$\lambda=\frac{1}{5}$$

Thus,

$$x=\frac{12}{\lambda}=\frac{12}{\frac{1}{5}}=12\cdot\frac{5}{1}=60$$

$$y=\frac{12}{\lambda}=\frac{12}{\frac{1}{5}}=12\cdot\frac{5}{1}=60$$

$$z=1750\lambda^2=1750\left(\frac{1}{5}\right)^2=70$$

The minimum total cost subject to the constraint occurs when the dimensions are 60 ft by 60 ft by 70 ft. The minimum cost is found as follows:

$$C(60,60,70)=7\cdot 60\cdot 60+6\cdot 60\cdot 70+6\cdot 60\cdot 70$$
$$=25{,}200+25{,}200+25{,}200$$
$$=75{,}600.$$

The minimum total cost of the building is \$75,600.

25. $C(x,y)=C(x)+C(y)$

$$C(x,y)=10+\frac{x^2}{6}+200+\frac{y^3}{9}$$
$$=210+\frac{x^2}{6}+\frac{y^3}{9}$$

We need to minimize

$$C(x,y)=210+\frac{x^2}{6}+\frac{y^3}{9}$$

subject to the constraint

$x+y=10{,}100$.

We first express $x+y=10{,}100$ as

$x+y-10{,}100=0$.

The solution is continued on the next page.

Copyright © 2014 Pearson Education, Inc. Publishing as Addison-Wesley.

We form the new function F, given by:

$F(x, y, \lambda)$

$= 210 + \frac{x^2}{6} + \frac{y^3}{9} - \lambda(x + y - 10{,}100).$

We find the first partial derivatives:

$F_x = \frac{x}{3} - \lambda,$

$F_y = \frac{1}{3}y^2 - \lambda,$

$F_\lambda = -(x + y - 10{,}100).$

We set each derivative equal to 0 and solve the resulting system:

$$\frac{x}{3} - \lambda = 0 \quad (1)$$

$$\frac{1}{3}y^2 - \lambda = 0 \quad (2)$$

$$x + y - 10{,}100 = 0 \quad (3) \left[\begin{array}{l} -(x+y-10{,}100)=0, \text{ or} \\ x+y-10{,}100=0 \end{array}\right]$$

From Eq. (1) and Eq. (2) we see:

$x = 3\lambda = y^2.$

Thus, $x = y^2$.

Substituting y^2 for x in Eq. (3), we get:

$$y^2 + y - 10{,}100 = 0$$

$$(y + 101)(y - 100) = 0$$

$$y + 101 = 0 \quad \text{or} \quad y - 100 = 0$$

$$y = -101 \quad \text{or} \quad y = 100.$$

Since y cannot be -101 in the original problem, we only consider $y = 100$. If $y = 100$, then $x = 100^2 = 10{,}000$. To minimize total costs, 10,000 units should be made on machine *A* and 100 units should be made on machine *B*.

27. Find the absolute minimum and maximum values of

$g(x, y) = x^2 + 2y^2$

Subject to the constraints of

$-1 \le x \le 1$

$-1 \le y \le 2.$

Using the constraints, we sketch the region of feasibility below.

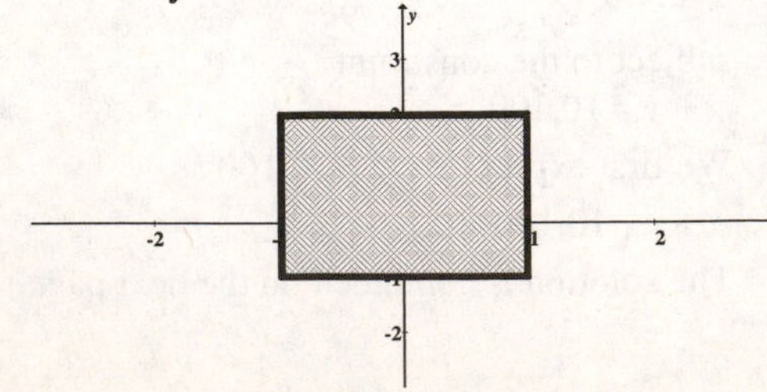

There are four corner points of the feasible region. They are:

$(-1,-1), (-1,2), (1,2)$, and $(1,-1)$. These are all critical points.

Next we check for critical points in the interior. We find the partial of g with respect to x and with respect to y:

$g_x(x, y) = 2x$

$g_y(x, y) = 4y$

The derivatives are set equal to 0 and we solve the system for x and y.

$2x = 0$

$4y = 0$

The solution to this system is

$x = 0$

$y = 0.$

The point $(0,0)$ is in the feasible region so it is a critical point.

Next we check the boundaries for possible critical points.

Along the boundary $x \ge -1$ we subsititute $x = -1$ into f.

$g(-1, y) = 2y^2 + 1.$

The derivative is $g_y(-1, y) = 2y$. Set equal to 0, we have $y = 0$. This results in the point $(-1,0)$, which is in the feasible region.

Along the boundary $y \le 2$ we substitute $y = 2$ into f.

$g(x, 2) = x^2 + 8.$

The derivative is $g_x(x, 2) = 2x$. Set equal to 0, we have $x = 0$. This results in the point $(0,2)$, which is in the feasible region.

Along the boundary $x \le 1$, we subsititute $x = 1$ into f.

$g(1, y) = 2y^2 + 1.$

The derivative is $g_y(1, y) = 4y$. Set equal to 0, we have $y = 0$. This results in the point $(1,0)$, which is in the feasible region.

Along boundary $y \ge -1$ we substitute $y = -1$ into f.

$g(x, -1) = x^2 + 2.$

The derivative is $g_x(x, -1) = 2x$. Set equal to 0, we have $x = 0$. This results in the point $(0,-1)$, which is in the feasible region.

The solution is continued on the next page.

Copyright © 2014 Pearson Education, Inc. Publishing as Addison-Wesley.

Using the information from the previous page, the critical points are:
$(-1,-1),(-1,2),(1,2),(1,-1),(0,0),(-1,0),$
$(0,2),(1,0)$, and $(0,-1)$.
Evaluating each of the critical points we have:
$g(-1,-1)=(-1)^2+2(-1)^2=2$
$g(-1,2)=(-1)^2+2(2)^2=9$
$g(1,2)=(1)^2+2(2)^2=9$
$g(1,-1)=(1)^2+2(-1)^2=3$
$g(0,0)=(0)^2+2(0)^2=0$
$g(-1,0)=(-1)^2+2(0)^2=1$
$g(0,2)=(0)^2+2(2)^2=8$
$g(1,0)=(1)^2+2(0)^2=1$
$g(0,-1)=(0)^2+2(-1)^2=2$
From the function values, we determine that the absolute minimum is 0 and occurs at $(0,0)$. The absolute maxima is 9 and occurs at $(-1,2)$ and $(1,2)$.

29. Find the absolute minimum and maximum values of
$k(x,y)=-x^2-y^2+4x+4y$
Subject to the constraints of
$0 \le x \le 3,$
$y \ge 0$
$x+y \le 6.$
Using the constraints, we sketch the region of feasibility below.

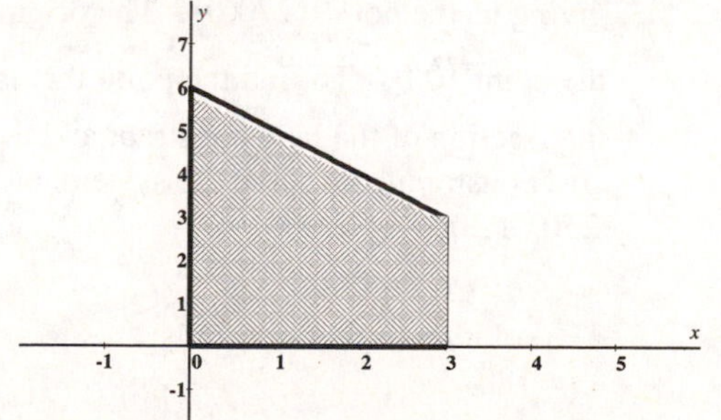

There are four corner points of the feasible region. They are:
$(0,0),(0,6),(3,3)$ and $(3,0)$. These are all critical points.
Next we check for critical points in the interior.

We find the partial of k with respect to x and with respect to y:
$k_x(x,y)=-2x+4$
$k_y(x,y)=-2y+4$
The derivatives are set equal to 0 and we solve the system for x and y.
$-2x+4=0$
$-2y+4=0$
The solution to this system is
$x=2$
$y=2.$
The point $(2,2)$ is in the feasible region so it is a critical point.
Next we check the boundaries for possible critical points.
Along the y-axis, we subsititute $x=0$ into k.
$k(0,y)=-y^2+4y.$
The derivative is $k_y(0,y)=-2y+4$. Set equal to 0, we have $y=2$. This results in the point $(0,2)$, which is in the feasible region.
Along the x-axis we substitute $y=0$ into k.
$k(x,0)=-x^2+4x.$
The derivative is $k_x(x,0)=-2x+4$. Set equal to 0, we have $x=2$. This results in the point $(2,0)$, which is in the feasible region.
Along the boundary $x \le 3$ we substitute $x=3$ into k.
$k(3,y)=-(3)^2-y^2+4(3)+4y$
$=-y^2+4y+3$
The derivative is $k_y(3,y)=-2y+4$. Set equal to 0, we have $y=2$. This results in the point $(3,2)$, which is in the feasible region.
Along the boundary $x+y \le 6$, we will use Lagrange multipliers to determine any critical points. We will rewrite the constraint to be:
$x+y-6=0.$
Next, we form the Lagrange function:
$L(x,y,\lambda)$
$=-x^2-y^2+4x+4y-\lambda(x+y-6).$
We determine the first partial derivatives:
$L_x=-2x+4-\lambda$
$L_y=-2y+4-\lambda$
$L_\lambda=-x-y+6.$
The solution is continued on the next page.

Copyright © 2014 Pearson Education, Inc. Publishing as Addison-Wesley.

We set each partial derivative from the previous page equal to 0.

$-2x+4-\lambda=0 \quad (1)$

$-2y+4-\lambda=0 \quad (2)$

$-x-y+6=0 \quad (3)$

Solving equation (1) for λ we get:

$\lambda=-2x+4$.

Solving equation (2) for λ we get:

$\lambda=-2y+4$.

Equating the λ, we can simplify the two equations into a single equation involving x and y.

$$-2y+4=-2x+4$$
$$-2y=-2x$$
$$y=x$$

Substituting $y=x$ into equation (3) we get:

$$-x-(x)+6=0$$
$$-2x+6=0$$
$$-2x=-6$$
$$x=3.$$

Therefore,

$y=x$

$y=3$.

The critical point on the constraint $x+2y=5$ is the point $(3,3)$, which we had already determined as a corner point of the region of feasibility.

Therefore, the critical points are:

$(0,0),(0,6),(3,3),(3,0),(2,2),$

$(0,2)$, and $(2,0)$.

Evaluating each of the critical points we have:

$k(0,0)=-(0)^2-(0)^2+4(0)+4(0)=0$

$k(0,6)=-(0)^2-(6)^2+4(0)+4(6)=-12$

$k(3,3)=-(3)^2-(3)^2+4(3)+4(3)=6$

$k(3,0)=-(3)^2-(0)^2+4(3)+4(0)=3$

$k(2,2)=-(2)^2-(2)^2+4(2)+4(2)=8$

$k(0,2)=-(0)^2-(2)^2+4(0)+4(2)=4$

$k(2,0)=-(2)^2-(0)^2+4(2)+4(0)=4$.

From the function values, we determine that the absolute minimum is -12 and occurs at $(0,6)$.

The absolute maximum is 8 and occurs at $(2,2)$.

31. a) Let x equal the number of acres of celery and y equal the number of acres of lettuce. The farmer's profit function will be given by:

$P(x,y)=45x+50y$

Subject to the cost constraint of

$250x+300y\le 81{,}000$

The land constraint is $x+y\le 300$.

Also x and y are to be non-negative. Restating the problem, we want to find the absolute maximum of

$P(x,y)=45x+50y$

Subject to the constraints of

$0\le x, 0\le y$

$250x+300y\le 81{,}000$

$x+y\le 300$.

Using the constraints, we sketch the region of feasibility below.

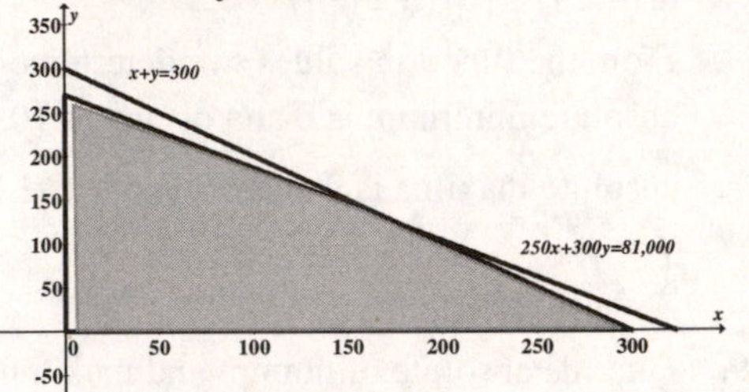

There are four corner points of the feasible region. We find the points on the x-axis by substituting $y=0$ into the land constraint, giving us the point $(300,0)$.

We find the point on the y-axis by substituting $x=0$ into the cost constraint, giving us the point $(270,0)$. The origin is the point $(0,0)$. To find the point that is the intersection of the land constrant and the cost constraint, we solve the system:

$$250x+300y=81{,}000$$
$$x+y=300$$

The solution to this system is:

$x=180$

$y=120$.

The four critical points are:

$(0,0),(0,270),(180,120)$, and $(300,0)$.

Next we check the boundaries for possible critical points. Note that we will not have to check the x-axis and y-axis because the profit function is a plane.

The solution is continued on the next page.

Copyright © 2014 Pearson Education, Inc. Publishing as Addison-Wesley.

Evaluating each of the critical points on the previous page, we have:

$P(0,0)=45(0)+50(0)=0$

$P(0,270)=45(0)+50(270)=13,500$

$P(180,270)=45(180)+50(120)=14,100$

$P(300,0)=45(300)+50(0)=13,500$

From the function values, we determine that the absolute maximum is 14,100 and occurs at $(180,120)$.

Therefore, the farmer would maximize profit by planting 180 acres of celery and 270 acers of lettuce. The maximum profit will be \$14,100.

b) Find the maximum value of

$P(x,y)=-x^2-y^2+600y-75,000$

subject to the constraints:

$0\le x, 0\le y$

$250x+300y\le 81,000$

$x+y\le 300.$

From part (a) we know the corner points are $(0,0),(0,270),(180,120)$, and $(300,0)$.

Next we check for critical points in the interior.

We find the partial of P with respect to x and with respect to y:

$P_x(x,y)=-2x$

$P_y(x,y)=-2y+600$

The derivatives are set equal to 0 and we solve the system for x and y.

$-2x=0 \quad (1)$

$-4y+600=0 \quad (2)$

The solution to this system is

$x=0$

$y=300.$

The point $(0,300)$ is in outside the feasible region.

Next we check the boundaries for possible critical points.

Along the y-axis, we subsititute $x=0$ into P.

$P(0,y)=-y^2+600y-75,000.$

The derivative is $P_y(0,y)=-2y+600$. Set equal to 0, we have $y=300$. This also results in a point outside the feasible region.

Along the x-axis we substitute $y=0$ into P.

$P(x,0)=-x^2-75,000.$

The derivative is $P_x(x,0)=-2x$. Set equal to 0, we have $x=0$. This results in the point $(0,0)$, which is one of the corner points.

Along the boundary $250x+300y\le 81,000$, we will use Lagrange multipliers to determine any critical points. We will rewrite the constraint to be:

$250x+300y-81,000=0.$

Next, we form the Lagrange function:

$$L(x,y,\lambda)$$
$$=-x^2-y^2+600y-75,000$$
$$-\lambda(250x+300y-81,000).$$

We determine the first partial derivatives:

$L_x=-2x-250\lambda$

$L_y=-2y+600-300\lambda$

$L_\lambda=-250x-300y+81,000.$

We set each partial derivative equal to 0.

$-2x-250\lambda=0 \quad (1)$

$-2y+600-300\lambda=0 \quad (2)$

$-250x-300y+81,000=0 \quad (3)$

Solving equation (1) for λ we get:

$$\lambda=-\frac{1}{125}x.$$

Substituting λ into equation (2) for we get:

$$-2y+600-300\left(\frac{1}{125}x\right)=0$$

$$-2y+600+\frac{12}{5}x=0$$

$$-2y=-\frac{12}{5}x-600$$

$$y=\frac{6}{5}x+300$$

Substituting $y=\frac{6}{5}x+300$ into equation (3) we get:

$$-250x-300\left(\frac{6}{5}x+300\right)+81,000=0$$
$$-610x-9000=0$$
$$-610x=9000$$
$$x\approx -14.75.$$

The critical point on the constraint $250x+300y=81,000$ is a point that is not in the feasible region.

The solution is continued on the next page.

Copyright © 2014 Pearson Education, Inc. Publishing as Addison-Wesley.

Along the boundary $x+y\le 300$, we will use Lagrange multipliers to determine any critical points. We will rewrite the constraint to be:

$x+y-300=0.$

Next, we form the Lagrange function:

$L(x,y,\lambda)$

$=-x^2-y^2+600y-75{,}000$

$-\lambda(x+y-300).$

We determine the first partial derivatives:

$L_x=-2x-\lambda$

$L_y=-2y+600-\lambda$

$L_\lambda=-x-y+300.$

We set each partial derivative equal to 0.

$$-2x-\lambda=0 \quad (1)$$

$$-2y+600-\lambda=0 \quad (2)$$

$$-x-y+300=0 \quad (3)$$

Solving equation (1) for λ we get:

$\lambda=-2x$.

Substituting λ into equation (2) for we get:

$-2y+600-(-2x)=0$

$-2y+600+2x=0$

$-2y=-2x-600$

$y=x+300.$

Substituting $y=x+300$ into equation (3) we get:

$-x-(x+300)+300=0$

$-2x=0$

$x=0.$

Therefore,

$y=x+300$

$y=(0)+300=300.$

The critical point on the constraint $x+y=300$ is the point $(0,300)$, which is outside the feasible region.

Therefore, the critical points are:

$(0,0),(0,270),(300,0)$ and $(180,120)$.

Evaluating each of the critical points we have:

$P(0,0)=-75{,}000$

$P(0,270)=14{,}100$

$P(180,120)=-49{,}800$

$P(300,0)=-165{,}000$

From the function values, we determine that the absolute maximum is 14,100 and occurs at $(0,270)$.

The farmer will maximize profit if he plants 0 acres of celery, 270 acres of lettuce. The maximum profit will be \$14,100.

33. Find the minimum value of

$f(x,y)=2x^2+y^2+2xy+3x+2y$

subject to the constraint

$y^2=x+1.$

We first express $y^2=x+1$ as $y^2-x-1=0$.

We form the new function F, given by:

$F(x,y,\lambda)$

$=2x^2+y^2+2xy+3x+2y-\lambda(y^2-x-1).$

We find the first partial derivatives:

$F_x=4x+2y+3+\lambda,$

$F_y=2y+2x+2-2\lambda y,$

$F_\lambda=-(y^2-x-1).$

We set each derivative equal to 0 and solve the resulting system:

$$4x+2y+3+\lambda=0 \quad (1)$$

$$2y+2x+2-2\lambda y=0 \quad (2)$$

$$y^2-x-1=0 \quad (3)\left[\begin{matrix}-(y^2-x-1)=0,\text{ or}\\ y^2-x-1=0\end{matrix}\right]$$

Solving Eq. (1) for λ, we get:

$4x+2y+3+\lambda=0$

$\lambda=-4x-2y-3. \quad (4)$

Solving Eq. (2) for λ, we get:

$2y+2x+2-2\lambda y=0$

$2\lambda y=2y+2x+2$

$\lambda=\dfrac{2y+2x+2}{2y}$

$\lambda=\dfrac{y+x+1}{y}. \quad (5)$

The solution is continued on the next page.

Copyright © 2014 Pearson Education, Inc. Publishing as Addison-Wesley.

Setting Eq. (4) from the previous page equal to Eq. (5) and solving for x we have:

$$-4x-2y-3=\frac{y+x+1}{y}$$
$$-4xy-2y^2-3y=y+x+1$$
$$-2y^2-3y-y-1=x+4xy$$
$$-2y^2-4y-1=x(1+4y)$$
$$\frac{-2y^2-4y-1}{1+4y}=x. \qquad (6)$$

Solving Eq. (3) for x we have:

$$y^2-x-1=0$$
$$y^2-1=x \qquad (7)$$

Substituting Eq. (7) into Eq. (6), we have:

$$\frac{-2y^2-4y-1}{1+4y}=y^2-1$$
$$-2y^2-4y-1=(y^2-1)(1+4y)$$
$$-2y^2-4y-1=y^2+4y^3-1-4y$$
$$0=4y^3+3y^2$$
$$0=y^2(4y+3)$$

$y^2=0$ or $4y+3=0$

$y=0$ or $y=-\frac{3}{4}$

Using equation (7) When $y=0$,

$$x=(0)^2-1$$
$$=-1$$
$$f(-1,0)=2(-1)^2+(0)^2+2(-1)(0)+3(-1)+2(0)$$
$$=2-3$$
$$=-1.$$

Using Eq. (7), when $y=-\frac{3}{4}$,

$$x=\left(\frac{-3}{4}\right)^2-1$$
$$=\frac{9}{16}-1$$
$$=-\frac{7}{16}.$$

Evaluating the function we have:

$$f\left(-\frac{7}{16},-\frac{3}{4}\right)=2\left(-\frac{7}{16}\right)^2+\left(-\frac{3}{4}\right)^2+2\left(-\frac{7}{16}\right)\left(-\frac{3}{4}\right)+3\left(-\frac{7}{16}\right)+2\left(-\frac{3}{4}\right)$$
$$=\frac{49}{128}+\frac{9}{16}+\frac{21}{32}-\frac{21}{16}-\frac{3}{2}$$
$$=-\frac{155}{128}.$$

The minimum value of f subject to the constraint occurs at $\left(-\frac{7}{16},-\frac{3}{4}\right)$ and is $f\left(-\frac{7}{16},-\frac{3}{4}\right)=-\frac{155}{128}$.

35. Find the maximum value of

$$f(x,y,z)=x^2y^2z^2$$

subject to the constraint

$$x^2+y^2+z^2=2.$$

We first express $x^2+y^2+z^2=2$ as

$$x^2+y^2+z^2-2=0.$$

We form the new function F, given by:

$$F(x,y,z,\lambda)$$
$$=x^2y^2z^2-\lambda(x^2+y^2+z^2-2)$$

We find the first partial derivatives:

$$F_x=2xy^2z^2-2\lambda x$$
$$F_y=2x^2yz^2-2\lambda y,$$
$$F_z=2x^2y^2z-2\lambda z,$$
$$F_\lambda=-(x^2+y^2+z^2-2).$$

We set each derivative equal to 0 and solve the resulting system:

$$2xy^2z^2-2\lambda x=0$$
$$2x^2yz^2-2\lambda y=0$$
$$2x^2y^2z-2\lambda z=0$$
$$x^2+y^2+z^2-2=0 \qquad \left[-(x^2+y^2+z^2-2)=0, \text{ or } x^2+y^2+z^2-2=0\right]$$

The solution is continued on the next page.

Copyright © 2014 Pearson Education, Inc. Publishing as Addison-Wesley.

Rewriting the system we get:

$x(2y^2z^2-2\lambda)=0 \quad (1)$

$y(2x^2z^2-2\lambda)=0 \quad (2)$

$z(2x^2y^2-2\lambda)=0 \quad (3)$

$x^2+y^2+z^2-2=0 \quad (4)\left[\begin{array}{l}-(x^2+y^2+z^2-2)=0, \text{ or} \\ x^2+y^2+z^2-2=0\end{array}\right]$

Note that for $x=0,\ y=0,$ or $z=0,\ f(x,y,z)=0.$ For all values of $x, y,$ and $z \neq 0, f(x,y,z)>0.$ Thus the maximum value of f cannot occur when any or all of the variables is 0. Thus we will only consider nonzero values of $x, y,$ and z.

Using the Principle of Zero Products, we get:

From Eq. (1)	From Eq. (2)	From Eq. (3)
$y^2z^2-\lambda=0$	$x^2z^2-\lambda=0$	$x^2y^2-\lambda=0$
$y^2z^2=\lambda$	$x^2z^2=\lambda$	$x^2y^2=\lambda$

Thus, $y^2z^2=x^2z^2=x^2y^2$ and $x^2=y^2=z^2$.

Substituting x^2 for y^2 and z^2 in Eq. (4), we have:

$$x^2+x^2+x^2-2=0$$
$$3x^2=2$$
$$x^2=\frac{2}{3}$$
$$x=\pm\sqrt{\frac{2}{3}}$$

Since $x^2=y^2=z^2$ it follows that

$y^2=\frac{2}{3}$ and $z^2=\frac{2}{3}$

$y=\pm\sqrt{\frac{2}{3}}$ and $z=\pm\sqrt{\frac{2}{3}}.$

For $\left(\pm\sqrt{\frac{2}{3}},\pm\sqrt{\frac{2}{3}},\pm\sqrt{\frac{2}{3}}\right)$:

$$f(x,y,z)=\left(\pm\sqrt{\frac{2}{3}}\right)^2\left(\pm\sqrt{\frac{2}{3}}\right)^2\left(\pm\sqrt{\frac{2}{3}}\right)^2$$
$$=\frac{2}{3}\cdot\frac{2}{3}\cdot\frac{2}{3}$$
$$=\frac{8}{27}$$

Thus $f(x,y,z)$ has a maximum value of $\frac{8}{27}$ at $\left(\pm\sqrt{\frac{2}{3}},\pm\sqrt{\frac{2}{3}},\pm\sqrt{\frac{2}{3}}\right)$.

37. Find the maximum value of $f(x,y,z,t)=x+y+z+t$ subject to the constraint $x^2+y^2+z^2+t^2=1$.

We first express $x^2+y^2+z^2+t^2=1$ as $x^2+y^2+z^2+t^2-1=0$.

We form the new function F, given by:

$$F(x,y,z,t,\lambda)$$
$$=x+y+z+t-\lambda(x^2+y^2+z^2+t^2-1)$$

We find the first partial derivatives:

$F_x=1-2\lambda x,$

$F_y=1-2\lambda y,$

$F_z=1-2\lambda z,$

$F_t=1-2\lambda t,$

$F_\lambda=-(x^2+y^2+z^2+t^2-1).$

We set each derivative equal to 0 and solve the resulting system:

$1-2\lambda x=0 \quad (1)$

$1-2\lambda y=0 \quad (2)$

$1-2\lambda z=0 \quad (3)$

$1-2\lambda t=0 \quad (4)$

$x^2+y^2+z^2+t^2-1=0 \quad (5)$

$\left[\begin{array}{l}-(x^2+y^2+z^2+t^2-1)=0, \text{ or} \\ x^2+y^2+z^2+t^2-1=0\end{array}\right]$

From Eq. (1), Eq. (2), Eq. (3), and Eq. (4), we see:

$$\frac{1}{2\lambda}=x=y=z=t.$$

Substituting x for y, z, and t in Eq. (5), we have:

$$x^2+(x)^2+(x)^2+(x)^2-1=0$$
$$4x^2=1$$
$$x^2=\frac{1}{4}$$
$$x=\pm\frac{1}{2}$$

Since $x=y=z=t$ it follows that

When $x=\frac{1}{2}$

$y=\frac{1}{2}$ and $z=\frac{1}{2}$ and $t=\frac{1}{2}.$

The solution is continued on the next page.

Copyright © 2014 Pearson Education, Inc. Publishing as Addison-Wesley.

When $x=-\frac{1}{2}$

$y=-\frac{1}{2}$ and $z=-\frac{1}{2}$ and $t=-\frac{1}{2}$.

However, it is clear looking at the function that the point that will yield a maximum value is:

$\left(\frac{1}{2},\frac{1}{2},\frac{1}{2},\frac{1}{2}\right)$.

The maximum value is found as follows:

$f\left(\frac{1}{2},\frac{1}{2},\frac{1}{2},\frac{1}{2}\right)=\frac{1}{2}+\frac{1}{2}+\frac{1}{2}+\frac{1}{2}=2.$

39. We want to maximize

$p(x,y)$

subject to the constraint.

$B=c_1x+c_2y.$

We first express $B=c_1x+c_2y$ as

$c_1x+c_2y-B=0$

Then we form the new function P, given by:

$P(x,y,\lambda)=p(x,y)-\lambda(c_1x+c_2y-B).$

We find the first partial derivatives.

$P_x=p_x-\lambda c_1$

$P_y=p_y-\lambda c_2.$

We set these derivatives equal to 0 and solve for λ.

$$p_x-\lambda c_1=0 \qquad p_y-\lambda c_2=0$$
$$p_x=\lambda c_1 \qquad p_y=\lambda c_2$$
$$\frac{p_x}{c_1}=\lambda \qquad \frac{p_y}{c_2}=\lambda$$

Thus, $\lambda=\frac{p_x}{c_1}=\frac{p_y}{c_2}$.

41. ✎

43 – 49. Left to the student.

Copyright © 2014 Pearson Education, Inc. Publishing as Addison-Wesley.

Exercise Set 6.6

1. $\int_0^3 \int_0^1 2y\,dx\,dy$

$= \int_0^3 \left(\int_0^1 2y\,dx \right) dy$

We first evaluate the inside x-integral, treating y as a constant:

$$\int_0^1 2y\,dx = 2y[x]_0^1$$
$$= 2y[1-0]$$
$$= 2y.$$

Then we evaluate the outside y-integral:

$$\int_0^3 \left(\int_0^1 2y\,dx \right) dy = \int_0^3 2y\,dy \qquad \left(\int_0^1 2y\,dx = 2y \right)$$
$$= \left[y^2 \right]_0^3$$
$$= 3^2 - 0^2$$
$$= 9.$$

3. $\int_{-1}^3 \int_1^2 x^2 y\,dy\,dx$

$= \int_{-1}^3 \left(\int_1^2 x^2 y\,dy \right) dx$

We first evaluate the inside y-integral, treating x as a constant:

$$\int_1^2 x^2 y\,dy = x^2 \left[\frac{1}{2} y^2 \right]_1^2$$
$$= x^2 \left[\frac{1}{2}(2)^2 - \frac{1}{2}(1)^2 \right]$$
$$= x^2 \left[2 - \frac{1}{2} \right]$$
$$= \frac{3}{2} x^2.$$

Then we evaluate the outside x-integral:

$$\int_{-1}^3 \left(\int_1^2 x^2 y\,dy \right) dx = \int_{-1}^3 \frac{3}{2} x^2\,dx \qquad \left(\int_1^2 x^2 y\,dy = \frac{3}{2} x^2 \right)$$
$$= \left[\frac{1}{2} x^3 \right]_{-1}^3$$
$$= \frac{1}{2}\left[3^3 - (-1)^3 \right]$$
$$= \frac{1}{2}\left[27 - (-1) \right]$$
$$= \frac{1}{2}[28]$$
$$= 14.$$

5. $\int_0^5 \int_{-2}^{-1} (3x+y)\,dx\,dy$

$= \int_0^5 \left(\int_{-2}^{-1} (3x+y)\,dx \right) dy$

We first evaluate the inside x-integral, treating y as a constant:

$$\int_{-2}^{-1} (3x+y)\,dx$$
$$= \left[\frac{3}{2} x^2 + yx \right]_{-2}^{-1}$$
$$= \left[\frac{3}{2}(-1)^2 + y(-1) \right] - \left[\frac{3}{2}(-2)^2 + y(-2) \right]$$
$$= \left[\frac{3}{2} - y \right] - \left[\frac{3}{2} \cdot 4 - 2y \right]$$
$$= y - \frac{9}{2}.$$

Then we evaluate the outside y-integral:

$$\int_0^5 \left(\int_{-2}^{-1} (3x+y)\,dx \right) dy$$
$$= \int_0^5 \left(y - \frac{9}{2} \right) dy \qquad \left(\int_{-2}^{-1} (3x+y)\,dx = y - \frac{9}{2} \right)$$
$$= \left[\frac{1}{2} y^2 - \frac{9}{2} y \right]_0^5$$
$$= \left[\frac{1}{2}(5)^2 - \frac{9}{2}(5) \right] - \left[\frac{1}{2}(0)^2 - \frac{9}{2}(0) \right]$$
$$= \frac{25}{2} - \frac{45}{2} - 0$$
$$= \frac{-20}{2}$$
$$= -10.$$

Copyright © 2014 Pearson Education, Inc. Publishing as Addison-Wesley.

7. $\int_{-1}^{1}\int_{x}^{1} xy\,dy\,dx$

$= \int_{-1}^{1}\left(\int_{x}^{1} xy\,dy\right)dx$

We first evaluate the inside y-integral, treating x as a constant:

$$\int_{x}^{1} xy\,dy = x\left[\frac{1}{2}y^2\right]_{x}^{1}$$
$$= x\left[\frac{1}{2}(1)^2 - \frac{1}{2}(x)^2\right]$$
$$= x\left[\frac{1}{2} - \frac{1}{2}x^2\right]$$
$$= \frac{1}{2}x - \frac{1}{2}x^3.$$

Then we evaluate the outside x-integral:

$$\int_{-1}^{1}\left(\int_{x}^{1} xy\,dy\right)dx$$
$$= \int_{-1}^{1}\left(\frac{1}{2}x - \frac{1}{2}x^3\right)dx \quad \left(\int_{x}^{1} xy\,dy = \tfrac{1}{2}x - \tfrac{1}{2}x^3\right)$$
$$= \left[\frac{1}{4}x^2 - \frac{1}{8}x^4\right]_{-1}^{1}$$
$$= \left[\frac{1}{4}(1)^2 - \frac{1}{8}(1)^4\right] - \left[\frac{1}{4}(-1)^2 - \frac{1}{8}(-1)^4\right]$$
$$= \left[\frac{1}{4} - \frac{1}{8}\right] - \left[\frac{1}{4} - \frac{1}{8}\right]$$
$$= 0.$$

9. $\int_{0}^{1}\int_{x^2}^{x} (x+y)\,dy\,dx$

$= \int_{0}^{1}\left(\int_{x^2}^{x} (x+y)\,dy\right)dx$

We first evaluate the inside y-integral, treating x as a constant:

$$\int_{x^2}^{x} (x+y)\,dy$$
$$= \left[xy + \frac{1}{2}y^2\right]_{x^2}^{x}$$
$$= \left[x(x) + \frac{1}{2}(x)^2\right] - \left[x(x^2) + \frac{1}{2}(x^2)^2\right]$$
$$= \left[x^2 + \frac{1}{2}x^2\right] - \left[x^3 + \frac{1}{2}x^4\right]$$
$$= -\frac{1}{2}x^4 - x^3 + \frac{3}{2}x^2.$$

Then we evaluate the outside x-integral:

$$\int_{0}^{1}\left(\int_{x^2}^{x} (x+y)\,dy\right)dx$$
$$= \int_{0}^{1}\left(-\frac{1}{2}x^4 - x^3 + \frac{3}{2}x^2\right)dx$$
$$\left(\int_{x^2}^{x} (x+y)\,dy = -\tfrac{1}{2}x^4 - x^3 + \tfrac{3}{2}x^2\right)$$
$$= \left[-\frac{1}{10}x^5 - \frac{1}{4}x^4 + \frac{1}{2}x^3\right]_{0}^{1}$$
$$= \left[-\frac{1}{10}(1)^5 - \frac{1}{4}(1)^4 + \frac{1}{2}(1)^3\right] - \left[-\frac{1}{10}(0)^5 - \frac{1}{4}(0)^4 + \frac{1}{2}(0)^3\right]$$
$$= \left[-\frac{1}{10} - \frac{1}{4} + \frac{1}{2}\right] - [0]$$
$$= \frac{3}{20}.$$

11. $\int_{0}^{1}\int_{1}^{e^x} \frac{1}{y}\,dy\,dx$

$= \int_{0}^{1}\left(\int_{1}^{e^x} \frac{1}{y}\,dy\right)dx$

We first evaluate the inside y-integral, treating x as a constant:

$$\int_{1}^{e^x} \frac{1}{y}\,dy = [\ln y]_{1}^{e^x}$$
$$= \ln e^x - \ln 1$$
$$= x.$$

Then we evaluate the outside x-integral:

$$\int_{0}^{1}\left(\int_{1}^{e^x} \frac{1}{y}\,dy\right)dx$$
$$= \int_{0}^{1} x\,dx \qquad \left(\int_{1}^{e^x} \frac{1}{y}\,dy = x\right)$$
$$= \left[\frac{1}{2}x^2\right]_{0}^{1}$$
$$= \left[\frac{1}{2}(1)^2 - \frac{1}{2}(0)^2\right]$$
$$= \frac{1}{2}.$$

Copyright © 2014 Pearson Education, Inc. Publishing as Addison-Wesley.

13. $\int_0^2 \int_0^x (x+y^2)\,dy\,dx$

$= \int_0^2 \left(\int_0^x (x+y^2)\,dy \right) dx$

We first evaluate the inside y-integral, treating x as a constant:

$\int_0^x (x+y^2)\,dy$

$= \left[xy + \frac{1}{3}y^3 \right]_0^x$

$= \left[x(x) + \frac{1}{3}(x)^3 \right] - \left[x(0) + \frac{1}{3}(0)^3 \right]$

$= \left[x^2 + \frac{1}{3}x^3 \right] - [0]$

$= \frac{1}{3}x^3 + x^2.$

Then we evaluate the outside x-integral:

$\int_0^2 \left(\int_0^x (x+y^2)\,dy \right) dx$

$= \int_0^2 \left(\frac{1}{3}x^3 + x^2 \right) dx$

$\left(\int_0^x (x+y^2)\,dy = \frac{1}{3}x^3 + x^2 \right)$

$= \left[\frac{1}{12}x^4 + \frac{1}{3}x^3 \right]_0^2$

$= \left[\frac{1}{12}(2)^4 + \frac{1}{3}(2)^3 \right] - \left[\frac{1}{12}(0)^4 + \frac{1}{3}(0)^3 \right]$

$= \left[\frac{16}{12} + \frac{8}{3} \right] - [0]$

$= \frac{4}{3} + \frac{8}{3}$

$= 4.$

15. $\int_0^1 \int_0^{1-x^2} (1-y-x^2)\,dy\,dx$

$= \int_0^1 \left(\int_0^{1-x^2} (1-y-x^2)\,dy \right) dx$

We first evaluate the inside y-integral, treating x as a constant:

$\int_0^{1-x^2} (1-y-x^2)\,dy$

$= \left[y - \frac{1}{2}y^2 - x^2 y \right]_0^{1-x^2}$

$= \left[(1-x^2) - \frac{1}{2}(1-x^2)^2 - x^2(1-x^2) \right] - \left[(0) - \frac{1}{2}(0)^2 - x^2(0) \right]$

$= 1 - x^2 - \frac{1}{2}(1 - 2x^2 + x^4) - x^2 + x^4$

$= 1 - x^2 - \frac{1}{2} + x^2 - \frac{1}{2}x^4 - x^2 + x^4$

$= \frac{1}{2}x^4 - x^2 + \frac{1}{2}.$

Then we evaluate the outside x-integral:

$= \int_0^1 \left(\int_0^{1-x^2} (1-y-x^2)\,dy \right) dx$

$= \int_0^1 \left(\frac{1}{2}x^4 - x^2 + \frac{1}{2} \right) dx$

$= \left[\frac{1}{10}x^5 - \frac{1}{3}x^3 + \frac{1}{2}x \right]_0^1$

$= \left[\frac{1}{10}(1)^5 - \frac{1}{3}(1)^3 + \frac{1}{2}(1) \right] - \left[\frac{1}{10}(0)^5 - \frac{1}{3}(0)^3 + \frac{1}{2}(0) \right]$

$= \frac{1}{10} - \frac{1}{3} + \frac{1}{2}$

$= \frac{4}{15}.$

The volume of the solid is $\frac{4}{15}$ units^3.

Copyright © 2014 Pearson Education, Inc. Publishing as Addison-Wesley.

17. $f(x,y)=x^2+\frac{1}{3}xy$

$0\le x\le 1$

$0\le y\le 2$

Find

$\int_0^2\int_0^1 f(x,y)\,dx\,dy$

$=\int_0^2\left(\int_0^1\left(x^2+\frac{1}{3}xy\right)dx\right)dy.$

We first evaluate the inside *x*-integral, treating *y* as a constant:

$\int_0^1\left(x^2+\frac{1}{3}xy\right)dx$

$=\left[\frac{1}{3}x^3+\frac{1}{6}x^2y\right]_0^1$

$=\left[\frac{1}{3}(1)^3+\frac{1}{6}(1)^2y\right]-\left[\frac{1}{3}(0)^3+\frac{1}{6}(0)^2y\right]$

$=\frac{1}{3}+\frac{1}{6}y.$

Then we evaluate the outside *y*-integral:

$\int_0^2\left(\int_0^1\left(x^2+\frac{1}{3}xy\right)dx\right)dy$

$=\int_0^2\left(\frac{1}{3}+\frac{1}{6}y\right)dy$

$=\left[\frac{1}{3}y+\frac{1}{12}y^2\right]_0^2$

$=\left[\frac{1}{3}(2)+\frac{1}{12}(2)^2\right]-\left[\frac{1}{3}(0)+\frac{1}{12}(0)^2\right]$

$=\frac{2}{3}+\frac{4}{12}$

$=1.$

19. $f(x,y)=x^2-3x+\frac{1}{3}xy-\frac{1}{3}y+2$

$1\le x\le 2$

$3\le y\le 5$

Find

$\int_3^4\int_1^2 f(x,y)\,dx\,dy$

$=\int_3^4\left(\int_1^2\left(x^2-3x+\frac{1}{3}xy-\frac{1}{3}y+2\right)dx\right)dy.$

We first evaluate the inside *x*-integral, treating *y* as a constant:

$\int_1^2\left(x^2-3x+\frac{1}{3}xy-\frac{1}{3}y+2\right)dx$

$=\left[\frac{1}{3}x^3-\frac{3}{2}x^2+\frac{1}{6}x^2y-\frac{1}{3}xy+2x\right]_1^2$

$=\left[\frac{1}{3}(2)^3-\frac{3}{2}(2)^2+\frac{1}{6}(2)^2y-\frac{1}{3}(2)y+2(2)\right]-$

$\left[\frac{1}{3}(1)^3-\frac{3}{2}(1)^2+\frac{1}{6}(1)^2y-\frac{1}{3}(1)y+2(1)\right]$

$=\left[\frac{8}{3}-6+\frac{2}{3}y-\frac{2}{3}y+4\right]-$

$\left[\frac{1}{3}-\frac{3}{2}+\frac{1}{6}y-\frac{1}{3}y+2\right]$

$=\frac{2}{3}-\left[\frac{5}{6}-\frac{1}{6}y\right]$

$=\frac{1}{6}y-\frac{1}{6}.$

Then we evaluate the outside *y*-integral:

$=\int_3^4\left(\int_1^2\left(x^2-3x+\frac{1}{3}xy-\frac{1}{3}y+2\right)dx\right)dy$

$=\int_3^4\left(\frac{1}{6}y-\frac{1}{6}\right)dy$

$=\left[\frac{1}{12}y^2-\frac{1}{6}y\right]_3^4$

$=\left[\frac{1}{12}(4)^2-\frac{1}{6}(4)\right]-\left[\frac{1}{12}(3)^2-\frac{1}{6}(3)\right]$

$=\left[\frac{4}{3}-\frac{2}{3}\right]-\left[\frac{3}{4}-\frac{1}{2}\right]$

$=\frac{2}{3}-\frac{1}{4}$

$=\frac{5}{12}.$

Copyright © 2014 Pearson Education, Inc. Publishing as Addison-Wesley.

21. $\rho(x,y)=\frac{1}{100}x^2y$

$0\le x\le 30$

$0\le y\le 20$

The population of fireflies in this field is given by:

$$\int_0^{30}\int_0^{20}\rho(x,y)dydx$$

$$=\int_0^{30}\left(\int_0^{20}\left(\frac{1}{100}x^2y\right)dy\right)dx.$$

We first evaluate the inside y-integral, treating x as a constant:

$$\int_0^{20}\left(\frac{1}{100}x^2y\right)dy$$

$$=\left[\frac{1}{100}x^2\left(\frac{1}{2}y^2\right)\right]_0^{20}$$

$$=\left[\frac{1}{200}x^2y^2\right]_0^{20}$$

$$=\frac{1}{200}x^2(20)^2-\frac{1}{200}x^2(0)^2$$

$$=2x^2.$$

Then we evaluate the outside x-integral:

$$\int_0^{30}2x^2dx$$

$$=\left[2\left(\frac{1}{3}x^3\right)\right]_0^{30}$$

$$=\frac{2}{3}(30)^3-\frac{2}{3}(0)^3$$

$$=18{,}000.$$

There are 18,000 fireflies in the field.

23. $\int_0^1\int_1^3\int_{-1}^2(2x+3y-z)dxdydz$

$$=\int_0^1\int_1^3\left(\int_{-1}^2(2x+3y-z)dx\right)dydz$$

We first evaluate the inside x-integral, treating y and z as constants:

$$\int_{-1}^2(2x+3y-z)dx$$

$$=\left[x^2+3yx-zx\right]_{-1}^2$$

$$=\left[(2)^2+3y(2)-z(2)\right]-\left[(-1)^2+3y(-1)-z(-1)\right]$$

$$=[4+6y-2z]-[1-3y+z]$$

$$=3+9y-3z.$$

Then we evaluate the middle y-integral, treating z as a constant:

$$\int_1^3\left(\int_{-1}^2(2x+3y-z)dx\right)dy$$

$$=\int_1^3(3+9y-3z)dy$$

$$=\left[3y+\frac{9}{2}y^2-3zy\right]_1^3$$

$$=\left[3(3)+\frac{9}{2}(3)^2-3z(3)\right]-\left[3(1)+\frac{9}{2}(1)^2-3z(1)\right]$$

$$=\left[9+\frac{81}{2}-9z\right]-\left[3+\frac{9}{2}-3z\right]$$

$$=42-6z.$$

Finally, we evaluate the outside z integral:

$$\int_0^1\left(\int_1^3\left(\int_{-1}^2(2x+3y-z)dx\right)dy\right)dz$$

$$=\int_0^1(42-6z)dz$$

$$=\left[42z-3z^2\right]_0^1$$

$$=\left[42(1)-3(1)^2\right]-\left[42(0)-3(0)^2\right]$$

$$=42-3$$

$$=39.$$

Copyright © 2014 Pearson Education, Inc. Publishing as Addison-Wesley.

25. $\int_0^1\int_0^{1-x}\int_0^{2-x}(xyz)\,dz\,dy\,dx$

$=\int_0^1\int_0^{1-x}\left(\int_0^{2-x}(xyz)\,dz\right)dy\,dx$

We first evaluate the inside z-integral, treating x and y as constants:

$\int_0^{2-x}(xyz)\,dz$

$=\left[\frac{1}{2}xyz^2\right]_0^{2-x}$

$=\left[\frac{1}{2}xy(2-x)^2\right]-\left[\frac{1}{2}xy(0)^2\right]$

$=\left[\frac{1}{2}xy(2-x)^2\right]-[0]$

$=\frac{1}{2}x(2-x)^2\,y.$

Then we evaluate the middle y-integral, treating x as a constant:

$\int_0^{1-x}\left(\int_0^{2-x}(xyz)\,dz\right)dy$

$=\int_0^{1-x}\left(\frac{1}{2}x(2-x)^2\,y\right)dy$

$=\frac{1}{2}x(2-x)^2\left[\frac{1}{2}y^2\right]_0^{1-x}$

$=\frac{1}{2}x(2-x)^2\left[\frac{1}{2}(1-x)^2-\frac{1}{2}(0)^2\right]$

$=\frac{1}{4}x(2-x)^2\left[(1-x)^2\right]$

$=\frac{1}{4}x\left(4-4x+x^2\right)\left[1-2x+x^2\right]$

$=\left[x-x^2+\frac{1}{4}x^3\right]\left[1-2x+x^2\right]$

$=x-2x^2+x^3-x^2+2x^3-x^4+$
$\frac{1}{4}x^3-\frac{1}{2}x^4+\frac{1}{4}x^5$

$=\frac{1}{4}x^5-\frac{3}{2}x^4+\frac{13}{4}x^3-3x^2+x.$

Finally, we evaluate the outside x integral:

$=\int_0^1\int_0^{1-x}\left(\int_0^{2-x}(xyz)\,dz\right)dy\,dx$

$=\int_0^1\left(\frac{1}{4}x^5-\frac{3}{2}x^4+\frac{13}{4}x^3-3x^2+x\right)dx$

$=\left[\frac{1}{24}x^6-\frac{3}{10}x^5+\frac{13}{16}x^4-x^3+\frac{1}{2}x^2\right]_0^1$

$=\left[\frac{1}{24}(1)^6-\frac{3}{10}(1)^5+\frac{13}{16}(1)^4-(1)^3+\frac{1}{2}(1)^2\right]-$
$\left[\frac{1}{24}(0)^6-\frac{3}{10}(0)^5+\frac{13}{16}(0)^4-(0)^3+\frac{1}{2}(0)^2\right]$

$=\left(\frac{1}{24}-\frac{3}{10}+\frac{13}{16}-1+\frac{1}{2}\right)-(0)$

$=\frac{10}{240}-\frac{72}{240}+\frac{195}{240}-\frac{240}{240}+\frac{120}{240}$

$=\frac{13}{240}.$

27. ✎

29. Left to the student.

Copyright © 2014 Pearson Education, Inc. Publishing as Addison-Wesley.

Chapter 7

Trigonometric Functions

Exercise Set 7.1

1. a) Since $34°$ is greater than $0°$ and less than $90°$, it's terminal side lies in the first quadrant.
b) Answers will vary, two possible coterminal angles are $394°$ and $-326°$.

3. a) Since $\frac{5\pi}{8}$ is greater than $\frac{\pi}{2}$ and less than π, it's terminal side lies in the second quadrant.
b) Answers will vary, two possible coterminal angles are $\frac{21\pi}{8}$ and $-\frac{11\pi}{8}$.

5. We multiply the degree measurement by $\frac{\pi \text{ radians}}{180 \text{ degrees}}$ and simplify:

$$15°\cdot\left(\frac{\pi}{180°}\right)=\frac{15\pi}{180}=\frac{\pi}{12}.$$

Thus, $15°$ is equivalent to $\frac{\pi}{12}$ radians.

7. We multiply the degree measurement by $\frac{\pi \text{ radians}}{180 \text{ degrees}}$ and simplify:

$$75°\cdot\left(\frac{\pi}{180°}\right)=\frac{75\pi}{180}=\frac{5\pi}{12}.$$

Thus, $75°$ is equivalent to $\frac{5\pi}{12}$ radians.

9. We multiply the degree measurement by $\frac{\pi \text{ radians}}{180 \text{ degrees}}$ and simplify:

$$-135°\cdot\left(\frac{\pi}{180°}\right)=\frac{-135\pi}{180}=\frac{-3\pi}{4}.$$

Thus, $-135°$ is equivalent to $\frac{-3\pi}{4}$ radians.

11. We multiply the degree measurement by $\frac{\pi \text{ radians}}{180 \text{ degrees}}$ and simplify:

$$-128°\cdot\left(\frac{\pi}{180°}\right)=\frac{-128\pi}{180}=\frac{-32\pi}{45}.$$

Thus, $-128°$ is equivalent to $\frac{-32\pi}{45}$ radians.

13. We multiply the radian measurement by $\frac{180 \text{ degrees}}{\pi \text{ radians}}$ and simplify:

$$\frac{3\pi}{2}\cdot\left(\frac{180°}{\pi}\right)=\frac{540\pi°}{2\pi}=270°.$$

Thus, $\frac{3\pi}{2}$ radians is equivalent to $270°$.

15. We multiply the radian measurement by $\frac{180 \text{ degrees}}{\pi \text{ radians}}$ and simplify:

$$\frac{-\pi}{4}\cdot\left(\frac{180°}{\pi}\right)=\frac{-180\pi°}{4\pi}=-45°.$$

Thus, $\frac{-\pi}{4}$ radians is equivalent to $-45°$.

17. We multiply the radian measurement by $\frac{180 \text{ degrees}}{\pi \text{ radians}}$ and simplify:

$$8\pi\cdot\left(\frac{180°}{\pi}\right)=\frac{1440\pi°}{\pi}=1440°.$$

Thus, 8π radians is equivalent to $1440°$.

19. We multiply the radian measurement by $\frac{180 \text{ degrees}}{\pi \text{ radians}}$ and simplify:

$$-5\pi\cdot\left(\frac{180°}{\pi}\right)=\frac{-900\pi°}{\pi}=-900°.$$

Thus, -5π radians is equivalent to $-900°$.

21. We multiply the radian measurement by $\frac{180 \text{ degrees}}{\pi \text{ radians}}$ and simplify:

$$1\cdot\left(\frac{180°}{\pi}\right)=\frac{180°}{\pi}\approx 57.296°.$$

Thus, 1 radian is equivalent to $57.296°$.

23. First, find the reference angle. $\frac{9\pi}{4}$ lies in the first quadrant and the reference angle is $\frac{\pi}{4}$. Since Sine is positive in the first quadrant, we have

$$\sin\left(\tfrac{9\pi}{4}\right)=\sin\left(\tfrac{\pi}{4}\right)=\tfrac{\sqrt{2}}{2}.$$

Copyright © 2014 Pearson Education, Inc. Publishing as Addison-Wesley.

25. First, find the reference angle.
$\frac{11\pi}{6}$ lies in the fourth quadrant and the reference angle is $2\pi - \frac{11\pi}{6} = \frac{\pi}{6}$. Since cosine is positive in the first quadrant, we have
$\cos\left(\frac{11\pi}{6}\right) = \cos\left(\frac{\pi}{6}\right) = \frac{\sqrt{3}}{2}$.

27. First, find the reference angle.
4π lies on the positive x-axis and is coterminal with 0. Therefore, the reference angle is 0. Since tangent is positive on the positive x-axis, we have

$$\tan(4\pi) = \tan(0) = \frac{\sin(0)}{\cos(0)} = \frac{0}{1} = 0.$$

29. First, find the reference angle.
$\frac{-\pi}{3}$ lies in the fourth quadrant and the reference angle is $\frac{\pi}{3}$. Since sine is negative and cosine is positive in the fourth quadrant, we have

$$\tan\left(-\frac{\pi}{3}\right) = \frac{-\sin\left(\frac{\pi}{3}\right)}{\cos\left(\frac{\pi}{3}\right)} = \frac{-\frac{\sqrt{3}}{2}}{\frac{1}{2}} = -\sqrt{3}.$$

31. First, find the reference angle.
$\frac{\pi}{4}$ lies in the first quadrant this is the reference angle. Since cosine is positive in the first quadrant, we have

$$\sec\left(\frac{\pi}{4}\right) = \frac{1}{\cos\left(\frac{\pi}{4}\right)} = \frac{1}{\frac{\sqrt{2}}{2}} = \frac{2}{\sqrt{2}} = \sqrt{2}.$$

33. First, find the reference angle.
$\frac{-\pi}{6}$ lies in the fourth quadrant and the reference angle is $\frac{\pi}{6}$. Since sine is negative in the fourth quadrant, we have

$$\csc\left(-\frac{\pi}{6}\right) = \frac{1}{-\sin\left(\frac{\pi}{6}\right)} = \frac{1}{\frac{1}{2}} = -2.$$

35. The point on the unit circle will have coordinates $(\cos t, \sin t)$. $t = \frac{4\pi}{3}$ lies in the second quadrant and has a reference angle of $t = \frac{\pi}{3}$. Therefore,

$$\left(\cos\left(\frac{4\pi}{3}\right), \sin\left(\frac{4\pi}{3}\right)\right) = \left(-\cos\left(\frac{\pi}{3}\right), \sin\left(\frac{\pi}{3}\right)\right) = \left(-\frac{1}{2}, \frac{\sqrt{3}}{2}\right).$$

37. The point on the unit circle will have coordinates $(\cos t, \sin t)$. $t = -\frac{3\pi}{4}$ lies in the third quadrant and has a reference angle of $t = \frac{\pi}{4}$. Therefore,

$$\left(\cos\left(\frac{-3\pi}{4}\right), \sin\left(\frac{-3\pi}{4}\right)\right) = \left(-\cos\left(\frac{\pi}{4}\right), -\sin\left(\frac{\pi}{4}\right)\right) = \left(-\frac{\sqrt{2}}{2}, -\frac{\sqrt{2}}{2}\right).$$

39. The point on the unit circle will have coordinates $(\cos t, \sin t)$. $t = 5\pi$ lies on the negative x-axis and has a reference angle of $t = 0$. Therefore,

$$(\cos(5\pi), \sin(5\pi)) = (-\cos(0), \sin(0)) = (-1, 0).$$

41. The point on the unit circle will have coordinates $(\cos t, \sin t)$. Using a calculator we have

$$\left(\cos\left(\frac{\pi}{7}\right), \sin\left(\frac{\pi}{7}\right)\right) = (0.901, 0.434).$$

43. The point on the unit circle will have coordinates $(\cos t, \sin t)$. Using a calculator we have

$$(\cos(1), \sin(1)) = (0.540, 0.841).$$

45. The point on the unit circle will have coordinates $(\cos t, \sin t)$. Using a calculator we have

$$(\cos(32°), \sin(32°)) = (0.848, 0.530).$$

47. The point on the unit circle will have coordinates $(\cos t, \sin t)$. Using a calculator we have

$$(\cos(218°), \sin(218°)) = (-0.788, -0.616).$$

49. The point on the unit circle will have coordinates $(\cos t, \sin t)$. Using a calculator we have

$$(\cos(-27°), \sin(-27°)) = (0.891, -0.454).$$

51. Using the appropriate identity we have

$$\sec(54°) = \frac{1}{\cos(54°)} = 1.701.$$

Copyright © 2014 Pearson Education, Inc. Publishing as Addison-Wesley.

53. Using the appropriate identity we have

$$\cot\left(\tfrac{3\pi}{7}\right)=\frac{1}{\tan\left(\tfrac{3\pi}{7}\right)}=0.228.$$

55. Using the appropriate identity we have

$$\csc(211°)=\frac{1}{\sin(211°)}=-1.942.$$

57. We use the Pythagorean identity to find $\cos t$ as follows:

$$\sin^2 t+\cos^2 t=1$$
$$\left(\tfrac{1}{6}\right)^2+\cos^2 t=1$$
$$\cos^2 t=1-\left(\tfrac{1}{6}\right)^2$$
$$\cos^2 t=\tfrac{35}{36}$$
$$\cos t=\sqrt{\tfrac{35}{36}}=\tfrac{\sqrt{35}}{6}.$$

Next, we use the quotient identity to find $\tan t$.

$$\tan t=\frac{\sin t}{\cos t}=\frac{\frac{1}{6}}{\frac{\sqrt{35}}{6}}=\frac{1}{\sqrt{35}}=\frac{\sqrt{35}}{35}.$$

59. We use the Pythagorean identity to find $\sin t$ as follows:

$$\sin^2 t+\cos^2 t=1$$
$$\sin^2 t+\left(\tfrac{3}{7}\right)^2=1$$
$$\sin^2 t=1-\left(\tfrac{3}{7}\right)^2$$
$$\sin^2 t=\tfrac{40}{49}$$
$$\sin t=\sqrt{\tfrac{40}{49}}=\tfrac{\sqrt{40}}{7}=\tfrac{2\sqrt{10}}{7}.$$

Next, we use the quotient identity to find $\sec t$.

$$\sec t=\frac{1}{\cos t}=\frac{1}{\frac{3}{7}}=\frac{7}{3}.$$

61. By the double angle identity we know that $\sin(2t)=2\sin t\cdot\cos t$. First, we use the Pythagorean identity to find $\cos t$ as follows:

$$\sin^2 t+\cos^2 t=1$$
$$\left(\tfrac{1}{8}\right)^2+\cos^2 t=1$$
$$\cos^2 t=1-\left(\tfrac{1}{8}\right)^2$$
$$\cos^2 t=\tfrac{63}{64}$$
$$\cos t=\sqrt{\tfrac{63}{64}}=\tfrac{\sqrt{63}}{8}=\tfrac{3\sqrt{7}}{8}.$$

Continued at the top of the next column.

Now we will use the double angle identity.

$$\sin(2t)=2\sin t\cdot\cos t$$
$$\sin(2t)=2\left(\tfrac{1}{8}\right)\left(\tfrac{3\sqrt{7}}{8}\right)$$
$$\sin(2t)=\tfrac{3\sqrt{7}}{32}.$$

63. By the double angle identity we know that $\cos(2t)=\cos^2 t-\sin^2 t$. First, we use the Pythagorean identity to find $\sin t$ as follows:

$$\sin^2 t+\cos^2 t=1$$
$$\sin^2 t+\left(\tfrac{3}{4}\right)^2=1$$
$$\sin^2 t=1-\left(\tfrac{3}{4}\right)^2$$
$$\sin^2 t=\tfrac{7}{16}$$
$$\sin t=\sqrt{\tfrac{7}{16}}=\tfrac{\sqrt{7}}{4}.$$

Now we will use the double angle identity.

$$\cos(2t)=\cos^2 t-\sin^2 t$$
$$\cos(2t)=\left(\tfrac{3}{4}\right)^2-\left(\tfrac{\sqrt{7}}{4}\right)^2$$
$$\cos(2t)=\tfrac{9}{16}-\tfrac{7}{16}$$
$$\cos(2t)=\tfrac{2}{16}=\tfrac{1}{8}.$$

65. Noticing that $75°=45°+30°$ we use the sum identity to find

$$\begin{aligned}\cos(75°)&=\cos(45°+30°)\\&=\cos(45°)\cos(30°)-\sin(45°)\sin(30°)\\&=\frac{\sqrt{2}}{2}\cdot\frac{\sqrt{3}}{2}-\frac{\sqrt{2}}{2}\cdot\frac{1}{2}\\&=\frac{\sqrt{6}}{4}-\frac{\sqrt{2}}{4}=\frac{\sqrt{6}-\sqrt{2}}{4}.\end{aligned}$$

67. Noticing that $105°=60°+45°$ we use the difference identity to find

$$\begin{aligned}\sin(105°)&=\sin(60°+45°)\\&=\sin(60°)\cos(45°)+\cos(60°)\sin(45°)\\&=\frac{\sqrt{3}}{2}\cdot\frac{\sqrt{2}}{2}+\frac{1}{2}\cdot\frac{\sqrt{2}}{2}\\&=\frac{\sqrt{6}}{4}+\frac{\sqrt{2}}{4}=\frac{\sqrt{6}+\sqrt{2}}{4}.\end{aligned}$$

Copyright © 2014 Pearson Education, Inc. Publishing as Addison-Wesley.

69. Noticing that $15° = 45° - 30°$ we use the difference identity to find

$$\tan(15°) = \frac{\sin(45° - 30°)}{\cos(45° - 30°)}$$

$$= \frac{\sin(45°)\cos(30°) - \cos(45°)\sin(30°)}{\cos(45°)\cos(30°) + \sin(45°)\sin(30°)}$$

$$= \frac{\frac{\sqrt{2}}{2}\cdot\frac{\sqrt{3}}{2} - \frac{\sqrt{2}}{2}\cdot\frac{1}{2}}{\frac{\sqrt{2}}{2}\cdot\frac{\sqrt{3}}{2} + \frac{\sqrt{2}}{2}\cdot\frac{1}{2}}$$

$$= \frac{\frac{\sqrt{6}}{4} - \frac{\sqrt{2}}{4}}{\frac{\sqrt{6}}{4} + \frac{\sqrt{2}}{4}}$$

$$= \frac{\sqrt{6} - \sqrt{2}}{\sqrt{6} + \sqrt{2}}$$ Rationalize the denominator.

$$= 2 - \sqrt{3}.$$

71. Find the length of the hypotenuse using the Pythagorean Theorem. We have.

$$c^2 = 3^2 + 7^2$$

$$c^2 = 58$$

$$c = \sqrt{58}.$$

Therefore, we have

$$\sin t = \frac{3}{\sqrt{58}} = \frac{3\sqrt{58}}{58};$$

$$\cos t = \frac{7}{\sqrt{58}} = \frac{7\sqrt{58}}{58};$$

$$\tan t = \frac{\sin t}{\cos t} = \frac{\frac{3\sqrt{58}}{58}}{\frac{7\sqrt{58}}{58}} = \frac{3}{7}.$$

73. Using the definition of trigonometric functions we have:

$$\cos(37°) = \frac{adj}{hyp} = \frac{12}{C}$$

$$\cos(37°) = \frac{12}{C}$$

$$C = \frac{12}{\cos(37°)}$$

$$C = 15.026$$

Next, we use tangent to find B at the top of the next column.

$$\tan(37°) = \frac{opp}{adj} = \frac{B}{12}$$

$$\tan(37°) = \frac{B}{12}$$

$$B = 12\tan(37°)$$

$$B = 9.043.$$

75. Using the definition of trigonometric functions we have:

$$\sin(42°) = \frac{opp}{hyp} = \frac{A}{17}$$

$$\sin(42°) = \frac{A}{17}$$

$$A = 17\sin(42°)$$

$$A = 11.375.$$

Next, we have:

$$\cos(21°) = \frac{adj}{hyp} = \frac{A}{B}$$

$$\cos(21°) = \frac{A}{B}$$

$$B = \frac{A}{\cos(21°)}$$

$$C = \frac{11.375}{cos(21°)}$$

$$C = 12.184.$$

77. Using the diagram in the problem, we notice that we are looking for the side opposite the given angle, and we know the side adjacent to the given angle. By the definition of the tangent function we have

$$\tan(51.84°) = \frac{h}{378}$$

$$378 \cdot \tan(51.84°) = h$$

$$481.043 = h.$$

The Great Pyramid of Giza is approximately 481 feet tall.

79. From 0 months to 12 months we see one full cycle of the graph. Therefore the period is 12 months.

81. The graph appears to repeat a cycle every 1 second; therefore, the period is approximately 1 second.

Copyright © 2014 Pearson Education, Inc. Publishing as Addison-Wesley.

83. Since the line passes through the point $(3,5)$ we have the figure.

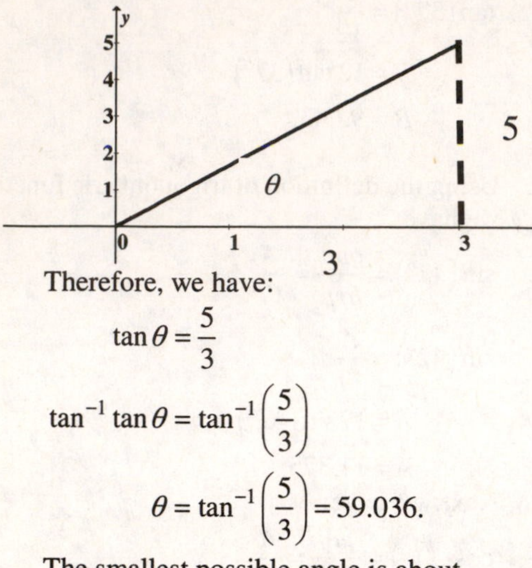

Therefore, we have:

$$\tan\theta = \frac{5}{3}$$

$$\tan^{-1}\tan\theta = \tan^{-1}\left(\frac{5}{3}\right)$$

$$\theta = \tan^{-1}\left(\frac{5}{3}\right) = 59.036.$$

The smallest possible angle is about 59.036°.

85. ✎

87. ✎

89. ✎

91. $\tan t = \frac{1}{9}$

$$c^2 = 1^2 + 9^2$$

$$c^2 = 82$$

$$c = \sqrt{82}$$

Therefore,

$$\sin t = \frac{1}{\sqrt{82}}; \cos t = \frac{9}{\sqrt{82}}.$$

Using the double angle identity we have

$$\sin 2t = 2\sin t\cos t$$

$$= 2\left(\frac{1}{\sqrt{82}}\right)\left(\frac{9}{\sqrt{82}}\right)$$

$$= \frac{18}{82} = \frac{9}{41}.$$

Applying the Pythagorean identity we have

$$\cos^2(2t) + \sin^2(2t) = 1$$

$$\cos^2(2t) = 1 - \sin^2(2t)$$

$$\cos^2(2t) = 1 - \left(\frac{18}{82}\right)^2$$

$$\cos^2(2t) = \frac{1600}{1681}$$

$$\cos(2t) = \sqrt{\frac{1600}{1681}}$$

$$\cos(2t) = \frac{80}{82} = \frac{40}{41}.$$

Thus, the Pythagorean triple is $(18,80,82)$ or $(9,40,41)$.

93. $\tan t = \frac{5}{11}$

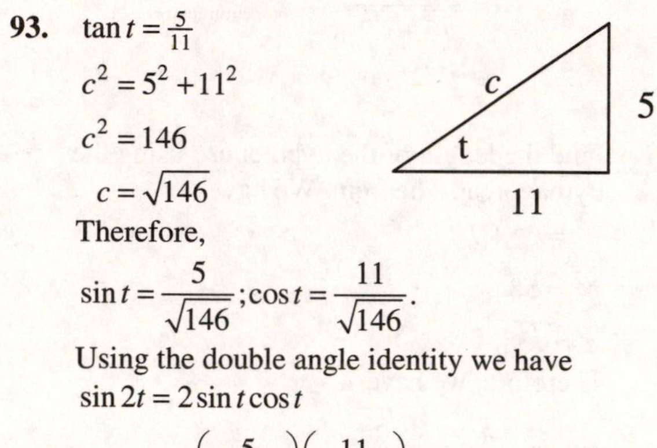

$$c^2 = 5^2 + 11^2$$

$$c^2 = 146$$

$$c = \sqrt{146}$$

Therefore,

$$\sin t = \frac{5}{\sqrt{146}}; \cos t = \frac{11}{\sqrt{146}}.$$

Using the double angle identity we have

$$\sin 2t = 2\sin t\cos t$$

$$= 2\left(\frac{5}{\sqrt{146}}\right)\left(\frac{11}{\sqrt{146}}\right)$$

$$= \frac{110}{146}.$$

Applying the Pythagorean identity we have

$$\cos^2(2t) + \sin^2(2t) = 1$$

$$\cos^2(2t) = 1 - \sin^2(2t)$$

$$\cos^2(2t) = 1 - \left(\frac{110}{146}\right)^2$$

$$\cos^2(2t) = \frac{2304}{5329}$$

$$\cos(2t) = \sqrt{\frac{2304}{5329}}$$

$$\cos(2t) = \frac{96}{146}.$$

Thus, the Pythagorean triple is $(96,110,146)$ or $(48,55,73)$.

Copyright © 2014 Pearson Education, Inc. Publishing as Addison-Wesley.

Exercise Set 7.2

1. $y = \sin 6x$

Using the chain rule, let $u = 6x$; $\frac{du}{dx} = 6$.

$y = \sin u$

$y' = \frac{dy}{du} \cdot \frac{du}{dx}$

$y' = \cos u \cdot 6$

$y' = 6\cos 6x.$

3. $y = \cos 2x$

Using the chain rule, let $u = 2x$; $\frac{du}{dx} = 2$.

$y = \cos u$

$y' = \frac{dy}{du} \cdot \frac{du}{dx}$

$y' = -\sin u \cdot 2$

$y' = -2\sin 2x.$

5. $y = \tan 2x^7$

Using the chain rule, let $u = 2x^7$; $\frac{du}{dx} = 14x^6$.

$y = \tan u$

$y' = \frac{dy}{du} \cdot \frac{du}{dx}$

$y' = \sec^2 u \cdot 14x^6$

$y' = 14x^6 \sec^2\left(2x^7\right).$

7. $y = \cos^2 x = \cos x \cdot \cos x$

Using the product rule, we have

$y' = \cos x \cdot \frac{d}{dx}(\cos x) + \frac{d}{dx}(\cos x) \cdot \cos x$

$y' = \cos x \cdot (-\sin x) + (-\sin x) \cdot \cos x$

$y' = -2\sin x \cdot \cos x.$

9. $y = x \cdot \sin x$

Using the product rule, we have

$y' = x \cdot \frac{d}{dx}(\sin x) + \frac{d}{dx}(x) \cdot \sin x$

$y' = x \cdot (\cos x) + \sin x$

$y' = x\cos x + \sin x.$

11. $y = e^x \cdot \sin x$

Using the product rule, we have

$y' = e^x \cdot \frac{d}{dx}(\sin x) + \frac{d}{dx}\left(e^x\right) \cdot \sin x$

$y' = e^x \cdot (\cos x) + e^x \cdot \sin x$

$y' = e^x(\cos x + \sin x).$

13. $y = \frac{\sin x}{x}$

Using the quotient rule, we have

$$y' = \frac{x\frac{d}{dx}(\sin x) - \sin x\frac{d}{dx}(x)}{x^2}$$

$$y' = \frac{x\cos x - \sin x}{x^2}.$$

15. $y = \sin x \cos x$

Using the product rule, we have

$y' = \sin x \cdot \frac{d}{dx}(\cos x) + \frac{d}{dx}(\sin x) \cdot \cos x$

$y' = \sin x \cdot (-\sin x) + \cos x \cos x$

$y' = -\sin^2 x + \cos^2 x$

$y' = \cos^2 x - \sin^2 x = \cos(2x).$

17. $y = \sin\left(x^2 + 3x + 2\right)$

Using the chain rule, let

$u = x^2 + 3x + 2$; $\frac{du}{dx} = 2x + 3$.

$y = \sin u$

$y' = \frac{dy}{du} \cdot \frac{du}{dx}$

$y' = \cos u \cdot (2x + 3)$

$y' = (2x + 3)\cos\left(x^2 + 3x + 2\right).$

19. $y = \frac{e^{2x}}{\sin(2x)}$

Using the quotient rule, we have:

$$y' = \frac{\sin(2x)\frac{d}{dx}\left(e^{2x}\right) - e^{2x}\frac{d}{dx}(\sin(2x))}{\sin^2(2x)}$$

$$y' = \frac{\sin(2x)e^{2x} \cdot 2 - e^{2x}(\cos(2x) \cdot 2)}{\sin^2(2x)}$$

$$y' = \frac{2e^{2x}(\sin(2x) - \cos(2x))}{\sin^2(2x)}.$$

Copyright © 2014 Pearson Education, Inc. Publishing as Addison-Wesley.

21. $y=\sqrt{\sin x}=(\sin x)^{\frac{1}{2}}$

By the general power rule we have:

$$y'=\tfrac{1}{2}(\sin x)^{\frac{1}{2}-1}\cdot\tfrac{d}{dx}\sin x$$

$$y'=\tfrac{1}{2}(\sin x)^{-\frac{1}{2}}\cdot\cos x$$

$$y'=\frac{\cos x}{2\sqrt{\sin x}}.$$

Note, this answer is equivalent to

$$y'=\frac{\cos x}{2\sqrt{\sin x}}=\tfrac{1}{2}\cot x\sqrt{\sin x}.$$

23. $y=\tan^2 x=\tan x\cdot\tan x$

By the product rule, we have

$$y'=\tan x\cdot\tfrac{d}{dx}(\tan x)+\tfrac{d}{dx}(\tan x)\cdot\tan x$$

$$y'=\tan x\cdot(\sec^2 x)+(\sec^2 x)\cdot\tan x$$

$$y'=2\tan x\sec^2 x.$$

25. $y=x\sec x^2$

By the product rule, we have

$$y'=x\cdot\tfrac{d}{dx}(\sec x^2)+\tfrac{d}{dx}(x)\cdot\sec x^2$$

$$y'=x\cdot\underbrace{(\sec x^2\cdot\tan x^2)\cdot(2x)}_{\text{Chain Rule}}+(1)\cdot\sec x^2$$

$$y'=2x^2\cdot\sec x^2\cdot\tan x^2+\sec x^2$$

$$y'=\sec x^2\left(1+2x^2\tan x^2\right).$$

27. $y=\cot x+\csc x$

Using the sum rule, we have

$$y'=\tfrac{d}{dx}\cot x+\tfrac{d}{dx}\csc x$$

$$y'=-\csc^2 x+(-\cot x\cdot\csc x)$$

$$y'=(-\csc x)(\csc x+\cot x).$$

29. $y=\dfrac{x}{\cot x}=x\cdot\tan x$

By the product rule, we have

$$y'=x\cdot\tfrac{d}{dx}(\tan x)+\tfrac{d}{dx}(x)\cdot\tan x$$

$$y'=x\cdot(\sec^2 x)+(1)\cdot\tan x$$

$$y'=x\sec^2 x+\tan x.$$

31. $y=e^{3x}\csc x$

By the product rule, we have

$$y'=e^{3x}\cdot\tfrac{d}{dx}(\csc x)+\tfrac{d}{dx}(e^{3x})\cdot\csc x$$

$$y'=e^{3x}\cdot(-\cot x\cdot\csc x)+\underbrace{(e^{3x})\cdot(3)}_{\text{Chain Rule}}\cdot\csc x$$

$$y'=3e^{3x}\csc x-e^{3x}\cdot\cot x\csc x$$

$$y'=e^{3x}\csc x(3-\cot x).$$

33. $y=\dfrac{1+\tan x}{\sec x}$

First, we will rewrite the function using trigonometric identities.

$$y=\frac{1+\tan x}{\sec x}$$

$$y=\frac{1}{\sec x}+\frac{\tan x}{\sec x}$$

$$y=\cos x+\frac{\frac{\sin x}{\cos x}}{\frac{1}{\cos x}}$$

$$y=\cos x+\frac{\sin x}{\cos x}\cdot\frac{\cos x}{1}$$

$$y=\cos x+\sin x.$$

Next using the sum rule, we have

$$y'=\tfrac{d}{dx}\cos x+\tfrac{d}{dx}\sin x$$

$$y'=-\sin x+\cos x$$

$$y'=\cos x-\sin x.$$

35. $y=\ln(\tan x)$

Using the Chain Rule, we have

$$y'=\frac{1}{\tan x}\cdot\frac{d}{dx}\tan x$$

$$y'=\frac{1}{\tan x}\cdot\sec^2 x$$

$$y'=\frac{1}{\frac{\sin x}{\cos x}}\cdot\frac{1}{\cos^2 x}$$

$$y'=\frac{1}{\sin x\cos x}.$$

Copyright © 2014 Pearson Education, Inc. Publishing as Addison-Wesley.

37. $y = \dfrac{\cos x}{\sec x}$

First rewrite the function using identities.

$$y = \frac{\cos x}{\sec x} = \frac{\cos x}{\frac{1}{\cos x}} = \cos^2 x = \cos x \cdot \cos x.$$

Next, use the product rule

$$y' = \cos x \cdot \tfrac{d}{dx}(\cos x) + \tfrac{d}{dx}(\cos x) \cdot \cos x$$
$$y' = \cos x \cdot (-\sin x) + (-\sin x) \cdot \cos x$$
$$y' = -2\sin x \cos x$$
$$y' = -\sin(2x).$$

39. $y = \cos x \tan x$

First rewrite the function using identities.

$$y = \cos x \cdot \frac{\sin x}{\cos x} = \sin x.$$

Therefore,

$$y' = \tfrac{d}{dx} \sin x = \cos x.$$

41. $y = -5 + 2\sin 3x = -5 + 2\sin(3(x-0))$

The midline is $y = -5$.

The amplitude is $|A| = |2| = 2$.

The period is $\dfrac{2\pi}{3}$.

The phase shift is 0.

We draw a graph over the domain $\left[0, \dfrac{4\pi}{3}\right]$.

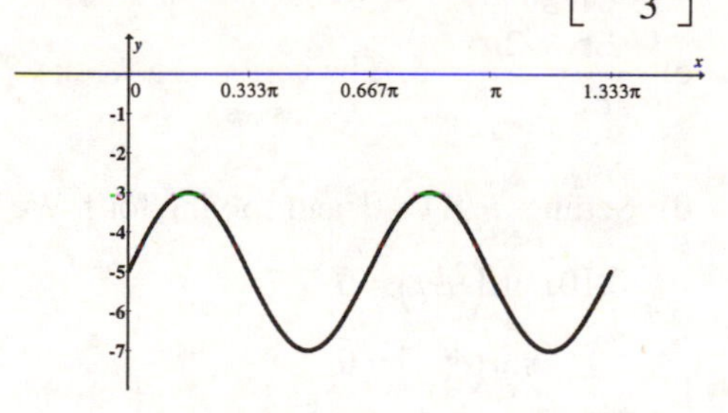

43. $y = -2 + 4\cos\left(\frac{\pi}{4}x + \frac{\pi}{4}\right) = -2 + 4\cos\left(\frac{\pi}{4}[x+1]\right)$

The midline is $y = -2$.

The amplitude is $|A| = |4| = 4$.

The period is $\dfrac{2\pi}{\frac{\pi}{4}} = 8$.

The phase shift is 1.

We draw a graph over the domain $[0, 16]$ at the top of the next column.

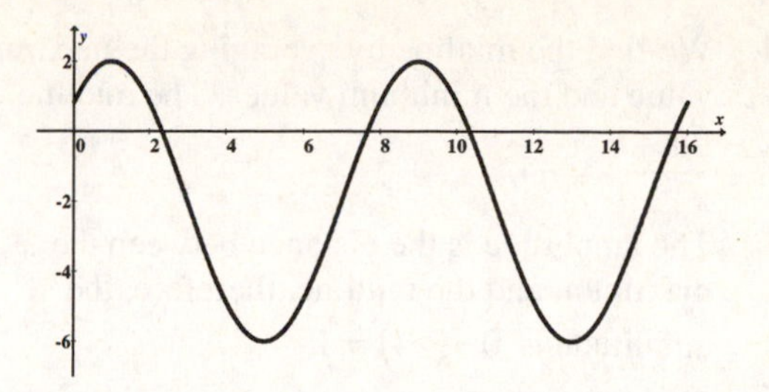

45. Using a graphing calculator, we graph the function on the domain described in problem 41.

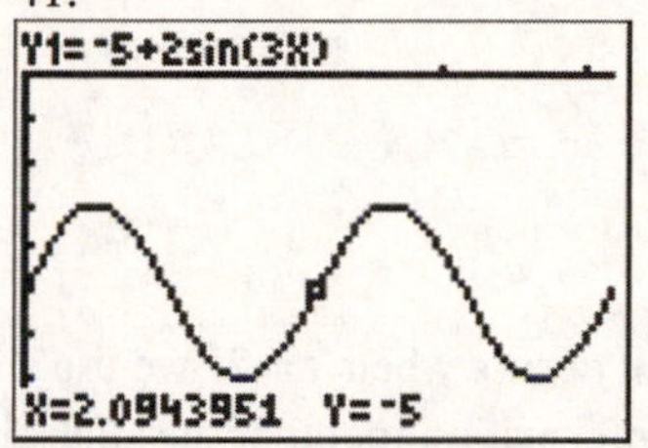

47. Using a graphing calculator, we graph the function on the domain described in problem 43.

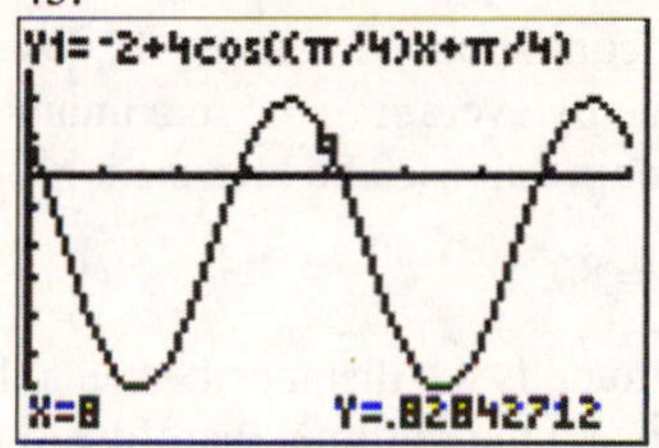

49. We find the midline by averaging the maximum value and the minimum values. The midline is $\dfrac{4+10}{2} = 7$.

The amplitude is the distance between the maximum and the midline, therefore, the amplitude is $10 - 7 = 3$.

The function repeats every 8 units, so the period is 8. Using this value, we find B.

$$\frac{2\pi}{B} = 8$$
$$2\pi = 8B$$
$$B = \frac{2\pi}{8} = \frac{\pi}{4}.$$

The max occurs when $t = 0$, so we use cosine. Putting the information together the sinusoidal model is

$$y = 7 + 3\cos\left(\frac{\pi}{4}t\right).$$

Copyright © 2014 Pearson Education, Inc. Publishing as Addison-Wesley.

51. We find the midline by averaging the maximum value and the minimum values. The midline is

$$\frac{-8+0}{2}=-4.$$

The amplitude is the distance between the maximum and the midline, therefore, the amplitude is $0-(-4)=4$.

Since the minimum occurs when $t=2$ and the maximum occurs at $t=6$ we know that $A=-4$. The function repeats every 8 units, so the period is 8. Using this value, we find B.

$$\frac{2\pi}{B}=8$$
$$2\pi=8B$$
$$B=\frac{2\pi}{8}=\frac{\pi}{4}.$$

Since the min occurs when $t=2$, we use sine. Putting the information together the sinusoidal model is

$$y=-4-4\sin\left(\frac{\pi}{4}t\right).$$

53. a) Let $t=0$ correspond to January 1st. We find the midline by averaging the maximum and the minimum values. The midline is

$$\frac{20+140}{2}=80.$$

The amplitude is the distance between the maximum and the midline, therefore, the amplitude is $140-(80)=60$.

Since the minimum occurs at $t=0$ we will use the cosine function reflected about the x-axis. Therefore, we know that $A=-60$.

The function repeats every 12 months, so the period is 12. Using this value, we find B.

$$\frac{2\pi}{B}=12$$
$$2\pi=12B$$
$$B=\frac{\pi}{6}.$$

Putting the information together the sinusoidal function that model Dune buggy rentals t months after January 1st is

$$f(t)=80-60\cos\left(\tfrac{\pi}{6}t\right).$$

b) Using the Chain Rule we find the derivative.

$$\begin{aligned}f'(t)&=\tfrac{d}{dt}\left(80-60\cos\left(\tfrac{\pi}{6}t\right)\right)\\&=0-60\left(-\sin\left(\tfrac{\pi}{6}t\right)\right)\cdot\tfrac{d}{dt}\left(\tfrac{\pi}{6}t\right)\\&=60\sin\left(\tfrac{\pi}{6}t\right)\cdot\tfrac{\pi}{6}\\&=10\pi\sin\left(\tfrac{\pi}{6}t\right)\\&=31.416\sin\left(\tfrac{\pi}{6}t\right).\end{aligned}$$

c) On April 1st, $t=3$.

$$f'(3)=10\pi\sin\left(\tfrac{\pi}{6}\cdot3\right)\approx31.416.$$

The rental rate is increasing 31.4 dune buggy rentals per month on April 1st.

d) On June 15th, $t=5.5$.

$$f'(5.5)=10\pi\sin\left(\tfrac{\pi}{6}\cdot5.5\right)\approx8.131.$$

The rental rate is increasing 8.13 dune buggy rentals per month on June 15th.

55. a) Using the Chain Rule we find the derivative.

$$\begin{aligned}P'(t)&=\tfrac{d}{dt}\left(3.64-0.08\cos\left(\tfrac{2\pi}{5}t\right)\right)\\&=0-0.08\left(-\sin\left(\tfrac{2\pi}{5}t\right)\right)\cdot\tfrac{d}{dt}\left(\tfrac{2\pi}{5}t\right)\\&=0.08\sin\left(\tfrac{2\pi}{5}t\right)\cdot\tfrac{2\pi}{5}\\&=0.101\sin\left(\tfrac{2\pi}{5}t\right).\end{aligned}$$

b) $P'(5)=0.101\sin\left(\tfrac{2\pi}{5}\cdot5\right)=0.$

At the fifth week, the price of gas is not changing.

c) $\dfrac{2\pi}{B}=\dfrac{2\pi}{\frac{2\pi}{5}}=5.$ The period is 5 weeks.

d) Setting $P'(t)=0$ and solving for t. We have

$$0.101\sin\left(\tfrac{2\pi}{5}t\right)=0$$
$$\sin\left(\tfrac{2\pi}{5}t\right)=0$$

This occurs when

$$\tfrac{2\pi}{5}t=\pi$$
$$t=2.5.$$

Therefore, the maximum value occurs 2.5 weeks after July 1. The maximum value is

$$P(2.5)=3.64-0.08\cos\left(\tfrac{2\pi}{5}\cdot2.5\right)=3.72.$$

2.5 weeks after July 1 the maximum price of gasoline is \$3.72.

Copyright © 2014 Pearson Education, Inc. Publishing as Addison-Wesley.

57. $f(t)=28-28\cos\left(\frac{\pi}{6.22}t\right)$

a) The period is given by

$$\frac{2\pi}{\frac{\pi}{6.22}}=12.44.$$

The period is 12.44 hours or 12 hr 26 min.

b) Taking the derivative we have

$$\begin{aligned}f'(t)&=\tfrac{d}{dt}\left(28-28\cos\left(\tfrac{\pi}{6.22}t\right)\right)\\&=0-28\left(-\sin\left(\tfrac{\pi}{6.22}t\right)\right)\cdot\tfrac{d}{dt}\left(\tfrac{\pi}{6.22}t\right)\\&=28\sin\left(\tfrac{\pi}{6.22}t\right)\cdot\tfrac{\pi}{6.22}\\&=14.142\sin\left(\tfrac{\pi}{6.22}t\right).\end{aligned}$$

c) $f'(3)=14.142\sin\left(\frac{\pi}{6.22}\cdot3\right)=14.12.$

Three hours after low tide, the tide is rising by 14.12 feet per hour.

d) High tide occurs when $t=\dfrac{12.44}{2}=6.22$.

Substituting into the function we have

$$f(6.22)=28-28\cos\left(\tfrac{\pi}{6.22}\cdot6.22\right)=56.$$

High tide occurs 6.22 hours after low tide and the water level is 56 feet higher.

59. a) Using the Chain Rule we find the derivative.

$$\begin{aligned}T'(t)&=\tfrac{d}{dt}\left(57-27\cos\left(\tfrac{2\pi}{365}t\right)\right)\\&=0-27\left(-\sin\left(\tfrac{2\pi}{365}t\right)\right)\cdot\tfrac{d}{dt}\left(\tfrac{2\pi}{365}t\right)\\&=27\sin\left(\tfrac{2\pi}{365}t\right)\cdot\tfrac{2\pi}{365}\\&=0.465\sin\left(\tfrac{2\pi}{2655}t\right).\end{aligned}$$

b) $T'(90)=0.465\sin\left(\frac{2\pi}{365}\cdot90\right)=0.465.$

90 days after January 1, the temperature is increasing about 0.465 degrees Fahrenheit per day.

c) For this function, the maximum value is reached when $t=\dfrac{365}{2}=182.5.$

When $t=182$ we have

$$T(182)=57-27\cos\left(\tfrac{2\pi}{365}\cdot182\right)\approx84.$$

Therefore, the maximum temperature occurs around the 182nd or 1 83rd day of the year and this temperature is approximately 84°F.

61. Using identities we have:

$$\frac{d}{dx}\cot x=\frac{d}{dx}\left(\frac{\cos x}{\sin x}\right)$$

Applying the quotient rule we have

$$\begin{aligned}\frac{d}{dx}\cot x&=\frac{d}{dx}\left(\frac{\cos x}{\sin x}\right)\\&=\frac{\sin x\cdot\frac{d}{dx}(\cos x)-\cos x\cdot\frac{d}{dx}(\sin x)}{(\sin x)^2}\\&=\frac{\sin x\cdot(-\sin x)-\cos x\cdot(\cos x)}{\sin^2 x}\\&=\frac{-\sin^2 x-\cos^2 x}{\sin^2 x}\\&=\frac{-\left(\sin^2 x+\cos^2 x\right)}{\sin^2 x}\\&=\frac{-1}{\sin^2 x}\\&=-\csc^2 x.\end{aligned}$$

63. Using identities we have:

$$\frac{d}{dx}\sec x=\frac{d}{dx}\left(\frac{1}{\cos x}\right)$$

Applying the quotient rule we have

$$\begin{aligned}\frac{d}{dx}\sec x&=\frac{d}{dx}\left(\frac{1}{\cos x}\right)\\&=\frac{\cos x\cdot\frac{d}{dx}(1)-1\cdot\frac{d}{dx}(\cos x)}{(\cos x)^2}\\&=\frac{0-1\cdot(\sin x)}{\cos^2 x}\\&=\frac{-\sin x}{\cos^2 x}\\&=-\frac{1}{\cos x}\cdot\frac{\sin x}{\cos x}\\&=-\sec x\tan x.\end{aligned}$$

65. a) Using the Chain Rule we have

$$\begin{aligned}\tfrac{d}{dx}\sin(2x)&=\cos(2x)\tfrac{d}{dx}(2x)\\&=2\cos 2x.\end{aligned}$$

b) Using a double angle identity we have

$$\begin{aligned}\tfrac{d}{dx}\sin(2x)&=\tfrac{d}{dx}2\sin x\cos x\\&=2\left[\sin x\cdot\tfrac{d}{dx}(\cos x)+\tfrac{d}{dx}(\sin x)\cdot\cos x\right]\\&=2\left[\sin x(-\sin x)+(\cos x)\cos x\right]\\&=2\left[\cos^2 x-\sin^2 x\right]\\&=2\cos 2x.\end{aligned}$$

Copyright © 2014 Pearson Education, Inc. Publishing as Addison-Wesley.

67. a) $P(2) = 5000 + 4000\cos\left(\frac{\pi}{6}\cdot 2\right) = 7000$

$P(1) = 5000 + 4000\cos\left(\frac{\pi}{6}\cdot 1\right) = 8464.1$

$\frac{P(2)-P(1)}{2-1} = \frac{7000-8464.1}{1} = -1464.1.$

Production is decreasing about 1464 units between November and December.

b) $P(3) = 5000 + 4000\cos\left(\frac{\pi}{6}\cdot 3\right) = 5000$

$P(2) = 5000 + 4000\cos\left(\frac{\pi}{6}\cdot 2\right) = 7000$

$\frac{P(3)-P(2)}{3-2} = \frac{5000-7000}{1} = -2000.$

Production is decreasing about 2000 units between December and January.

c)

69. The period of the function is 2π.
Enter the equation into the calculator.

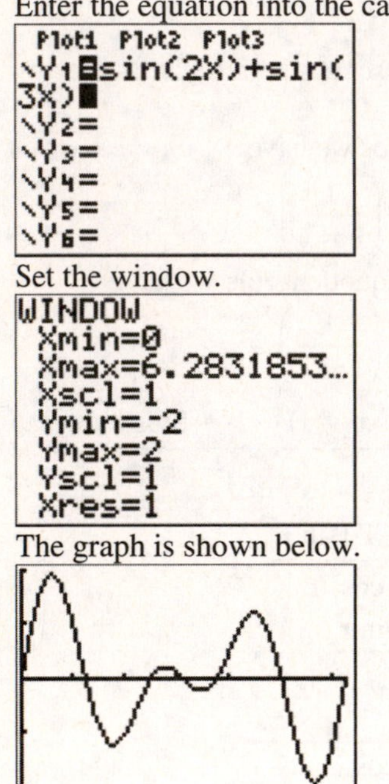

Set the window.

The graph is shown below.

71. The period of the function is π.
Enter the equation into the calculator.

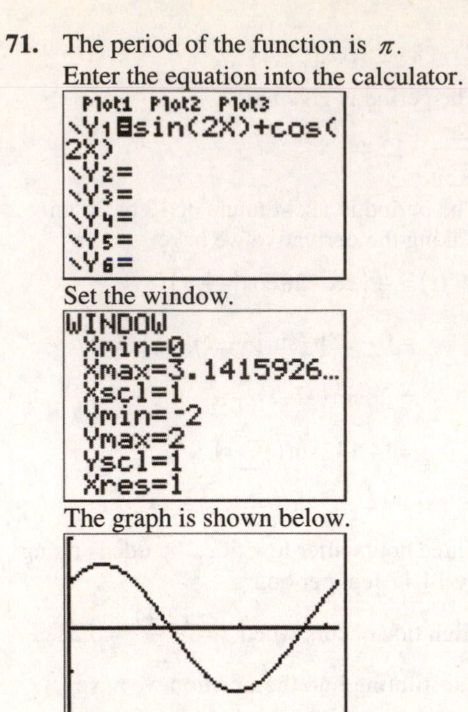

Set the window.

The graph is shown below.

Copyright © 2014 Pearson Education, Inc. Publishing as Addison-Wesley.

Exercise Set 7.3

1. $\int \sin^3 x \cos x dx$

Let $u = \sin x$; $du = \cos x dx$.

$$\int \sin^3 x \cos x dx = \int u^3 du$$
$$= \tfrac{1}{4}u^4 + C$$
$$= \tfrac{1}{4}\sin^4 x + C.$$

3. $\int \cos 6x dx$

Let $u = 6x$; $du = 6dx$.

$$\int \cos 6x dx = \tfrac{1}{6}\int \cos 6x \cdot 6dx$$
$$= \tfrac{1}{6}\int \cos u du$$
$$= \tfrac{1}{6}[\sin u + C]$$
$$= \tfrac{1}{6}\sin 6x + C.$$

5. $\int \sin \tfrac{\pi}{2} x dx$

Let $u = \tfrac{\pi}{2}x$; $du = \tfrac{\pi}{2}dx$.

$$\int \sin \tfrac{\pi}{2} x dx = \tfrac{2}{\pi}\int \sin \tfrac{\pi}{2} x \cdot \tfrac{\pi}{2} dx$$
$$= \tfrac{2}{\pi}\int \sin u du$$
$$= \tfrac{2}{\pi}[-\cos u + C]$$
$$= -\tfrac{2}{\pi}\cos \tfrac{\pi}{2} x + C.$$

7. $\int 3\cos 2\pi x dx$

Let $u = 2\pi x$; $du = 2\pi dx$.

$$\int 3\cos 2\pi x dx = \tfrac{3}{2\pi}\int \cos 2\pi x \cdot 2\pi dx$$
$$= \tfrac{3}{2\pi}\int \cos u du$$
$$= \tfrac{3}{2\pi}[\sin u + C]$$
$$= \tfrac{3}{2\pi}\sin 2\pi x + C.$$

9. $\int x \sin 3x^2 dx$

Let $u = 3x^2$; $du = 6x dx$.

$$\int x \sin 3x^2 dx = \tfrac{1}{6}\int \sin 3x^2 \cdot 6x dx$$
$$= \tfrac{1}{6}\int \sin u du$$
$$= \tfrac{1}{6}[-\cos u + C]$$
$$= -\tfrac{1}{6}\cos 3x^2 + C.$$

11. $\int (x-1)\cos(x^2 - 2x) dx$

Let $u = x^2 - 2x$; $du = 2x - 2dx = 2(x-1)dx$.

$$\int (x-1)\cos(x^2 - 2x) dx$$
$$= \tfrac{1}{2}\int \cos(x^2 - 2x) \cdot 2(x-1) dx$$
$$= \tfrac{1}{2}\int \cos u du$$
$$= \tfrac{1}{2}[\sin u + C]$$
$$= \tfrac{1}{2}\sin(x^2 - 2x) + C.$$

13. $\int e^{2x} \sin e^{2x} dx$

Let $u = e^{2x}$; $du = 2e^{2x}dx$.

$$\int e^{2x} \sin e^{2x} dx = \tfrac{1}{2}\int \sin e^{2x} \cdot 2 \cdot e^{2x} dx$$
$$= \tfrac{1}{2}\int \sin u du$$
$$= \tfrac{1}{2}[-\cos u + C]$$
$$= -\tfrac{1}{2}\cos e^{2x} + C.$$

15. $\int \tan^2 x \sec^2 x dx$

Let $u = \tan x$; $du = \sec^2 x dx$.

$$\int \tan^2 x \sec^2 x dx = \int u^2 du$$
$$= \tfrac{1}{3}u^3 + C$$
$$= \tfrac{1}{3}\tan^3 x + C.$$

Copyright © 2014 Pearson Education, Inc. Publishing as Addison-Wesley.

17. $\int \sec^2 10x dx$

Let $u = 10x$; $du = 10dx$.

$$\int \sec^2 10x dx = \frac{1}{10}\int \sec^2 10x \cdot 10dx$$
$$= \frac{1}{10}\int \sec^2 u du$$
$$= \frac{1}{10}[\tan u + C]$$
$$= \frac{1}{10}\tan 10x + C.$$

19. $\int \sec 5x \tan 5x dx$

Let $u = \sec 5x$ $du = 5\sec 5x \tan 5x dx$.

$$\int \sec 5x \tan 5x dx = \frac{1}{5}\int 5\sec 5x \tan 5x dx$$
$$= \frac{1}{5}\int du$$
$$= \frac{1}{5}[u + C]$$
$$= \frac{1}{5}\sec 5x + C.$$

21. $\int \csc^2 3x dx$

Let $u = 3x$; $du = 3dx$.

$$\int \csc^2 3x dx = \frac{1}{3}\int \csc^2 3x \cdot 3dx$$
$$= \frac{1}{3}\int \csc^2 u du$$
$$= \frac{1}{3}[-\cot u + C]$$
$$= -\frac{1}{3}\cot 3x + C.$$

23. $\int \cot 6x \csc 6x dx$

Let $u = 6x$; $du = 6dx$.

$$\int \cot 6x \csc 6x dx = \frac{1}{6}\int \cot 6x \csc 6x \cdot 6dx$$
$$= \frac{1}{6}\int \cot u \csc u du$$
$$= \frac{1}{6}[-\csc u + C]$$
$$= -\frac{1}{6}\csc 6x + C.$$

25. $\int \left(x^2 + 3x + \tan 2x\right)dx$

Rewrite the integral.

$$\int \left(x^2 + 3x + \tan 2x\right)dx$$
$$= \int x^2 dx + \int 3x dx + \int \tan 2x dx.$$

Integrate each of the integrals separately as shown at the top of the next column.

$$\int x^2 dx = \frac{1}{3}x^3 + C_1$$
$$\int 3x dx = \frac{3}{2}x^2 + C_2$$

For $\int \tan 2x dx$, first let $u = 2x$; $du = 2dx$.

$$\int \tan 2x dx = \frac{1}{2}\int \tan 2x \cdot 2dx$$
$$= \frac{1}{2}\int \tan u du$$

Now use properties of integrals to rewrite the integral using sine and cosine.

$$\frac{1}{2}\int \tan u du = \frac{1}{2}\int \frac{\sin u}{\cos u} du$$

Let $v = \cos u$; $dv = -\sin v dx$.

$$\frac{1}{2}\int \tan u du = \frac{1}{2}\int \frac{\sin u}{\cos u} du$$
$$= -\frac{1}{2}\int \frac{-\sin u du}{\cos u}$$
$$= -\frac{1}{2}\int \frac{dv}{v}$$
$$= -\frac{1}{2}\left[\ln|v| + C_3\right]$$
$$= -\frac{1}{2}\ln|\cos u| + C_3 \quad \text{Substitute for } v.$$
$$= -\frac{1}{2}\ln|\cos 2x| + C_3 \quad \text{Substitute for } u.$$

$$\int \tan 2x dx = -\frac{1}{2}\ln|\cos 2x| + C_3.$$

Putting the three integrals together, we have

$$\int \left(x^2 + 3x + \tan 2x\right)dx$$
$$= \int x^2 dx + \int 3x dx + \int \tan 2x dx.$$
$$= \frac{1}{3}x^3 + C_1 + \frac{3}{2}x^2 + C_2 - \frac{1}{2}\ln|\cos 2x| + C_3$$
$$= \frac{1}{3}x^3 + \frac{3}{2}x^2 - \frac{1}{2}\ln|\cos 2x| + C.$$
$$\left(C = C_1 + C_2 + C_3\right)$$

27. $\int_0^{\pi/2} \sin x dx = \left[-\cos x\right]_0^{\pi/2}$

$$= -\cos\frac{\pi}{2} - (-\cos 0)$$
$$= -0 - (-1)$$
$$= 1.$$

29. $\int_0^{\pi/4} 2\cos x dx = \left[2\sin x\right]_0^{\pi/4}$

$$= 2\left[\sin\frac{\pi}{4} - (\sin 0)\right]$$
$$= 2\left[\frac{\sqrt{2}}{2} - (0)\right]$$
$$= \sqrt{2}.$$

Copyright © 2014 Pearson Education, Inc. Publishing as Addison-Wesley.

31. $$\int_{-\pi/6}^{\pi/2} \sin x dx = \left[-\cos x\right]_{-\pi/6}^{\pi/2}$$
$$= \left[-\cos\tfrac{\pi}{2} - \left(-\cos\left(-\tfrac{\pi}{6}\right)\right)\right]$$
$$= -(0) + \left(\tfrac{\sqrt{3}}{2}\right)$$
$$= \tfrac{\sqrt{3}}{2}.$$

33. $$\int_{-\pi/6}^{0} 5\cos x dx = 5\left[\sin x\right]_{-\pi/6}^{0}$$
$$= 5\left[\sin 0 - \left(\sin\left(-\tfrac{\pi}{6}\right)\right)\right]$$
$$= 5\left[0 - \left(-\tfrac{1}{2}\right)\right]$$
$$= 5\left[\tfrac{1}{2}\right]$$
$$= \tfrac{5}{2}.$$

35. $$\int_{0}^{\pi/6} \sec^2 x dx = \left[\tan x\right]_{0}^{\pi/6}$$
$$= \left[\tan\left(\tfrac{\pi}{6}\right) - \left(\tan(0)\right)\right]$$
$$= \tfrac{\sqrt{3}}{3} - 0$$
$$= \tfrac{\sqrt{3}}{3}.$$

37. $$\int_{\pi/6}^{\pi/4} \csc^2 x dx = \left[-\cot x\right]_{\pi/6}^{\pi/4}$$
$$= -\left[\cot\left(\tfrac{\pi}{4}\right) - \left(\cot\left(\tfrac{\pi}{6}\right)\right)\right]$$
$$= -\left[1 - \sqrt{3}\right]$$
$$= \sqrt{3} - 1.$$

39. $$\int_{0}^{\pi/4} \tan^2 x \sec^2 dx$$

First find the definite integral.

Let $u = \tan x$; $du = \sec^2 x dx$.

$$\int \tan^2 x \sec^2 x dx = \int u^2 du$$
$$= \tfrac{1}{3}u^3 + C$$
$$= \tfrac{1}{3}\tan^3 x + C.$$

Therefore,

$$\int_{0}^{\pi/4} \tan^2 x \sec^2 dx = \left[\tfrac{1}{3}\tan^3 x\right]_{0}^{\pi/4}$$
$$= \tfrac{1}{3}\left[\tan^3\left(\tfrac{\pi}{4}\right) - \left(\tan^3(0)\right)\right]$$
$$= \tfrac{1}{3}\left[1^3 - 0^3\right]$$
$$= \tfrac{1}{3}.$$

41. $$\int_{0}^{2\pi} 5\sin x dx = 5\left[-\cos x\right]_{0}^{2\pi}$$
$$= -5\left[\cos(2\pi) - \left(\cos(-0)\right)\right]$$
$$= -5[1-1]$$
$$= 0.$$

43. Since $\int \cos kx dx = \frac{1}{k}\sin kx + C$, we have

$$\int_{0}^{\pi} 2 + 4\cos 2x dx = \left[2x + 2\sin 2x\right]_{0}^{\pi}$$
$$= \left[2\cdot\pi + 2\sin(2\cdot\pi)\right] - \left[2\cdot 0 + 2\sin(2\cdot 0)\right]$$
$$= [2\pi + 0] - [0 - 0]$$
$$= 2\pi.$$

45. $$\int \sec x dx$$

First we rewrite the integral

$$\int \sec x dx = \int \sec x\left[\frac{\sec x + \tan x}{\sec x + \tan x}\right]dx$$
$$= \int \frac{\sec^2 x + \sec x \tan x}{\sec x + \tan x}dx.$$

Let

$u = \sec x + \tan x;$

$du = \left(\sec x \tan x + \sec^2 x\right)dx$

Substituting, we have

$$\int \sec x dx = \int \frac{\sec^2 x + \sec x \tan x}{\sec x + \tan x}dx$$
$$= \int \frac{du}{u}$$
$$= \ln|u| + C$$
$$= \ln|\sec x + \tan x| + C$$

Copyright © 2014 Pearson Education, Inc. Publishing as Addison-Wesley.

47. $\int \sin^3 x dx = \int \sin^2 x \cdot \sin x dx$

$= \int (1-\cos^2 x)\cdot \sin x dx$

$= \int (\sin x - \cos^2 x \sin x) dx$

$= \int \sin x dx - \int \cos^2 x \sin x dx$

Let $u = \cos x$
$du = -\sin x dx$

$= \int \sin x dx + \int \cos^2 x \cdot (-\sin x) dx$

$= \int \sin x dx + \int u^2 du$

$= -\cos x + C_1 + \frac{1}{3}u^3 + C_2$

$= -\cos x + \frac{1}{3}\cos^3 x + C.$

$(C = C_1 + C_2)$

49. $\int \sin x \cos x dx$

a) Let $u = \sin x$; $du = \cos x dx$.

$\int \sin x \cos x dx = \int u du$

$= \frac{1}{2}u^2 + C$

$= \frac{1}{2}\sin^2 x + C.$

b) Let $u = \cos x$; $du = -\sin x dx$.

$\int \sin x \cos x dx = -\int u du$

$= -\frac{1}{2}u^2 + C$

$= -\frac{1}{2}\cos^2 x + C.$

c) Show $\frac{1}{2}\sin^2 x + C = -\frac{1}{2}\cos^2 x + C$

$\frac{1}{2}\sin^2 x + C = \frac{1}{2}\left[1-\cos^2 x\right] + C$ Pythagorean Identity

$= \frac{1}{2} - \frac{1}{2}\cos^2 x + C$

$= -\frac{1}{2}\cos^2 x + C.$

(C absorbs the constant $\frac{1}{2}$)

51. The average value of $y = \sin x$ over the interval $[0, \pi]$ is given by

$\frac{1}{\pi}\int_0^{\pi} \sin x dx = \frac{1}{\pi}\left[-\cos x\right]_0^{\pi}$

$= -\frac{1}{\pi}[\cos \pi - \cos 0]$

$= -\frac{1}{\pi}[-1-1]$

$= \frac{2}{\pi}$

$\approx 0.637.$

53. a) The total number of rentals between January 1 and April 1 is given by:

$\int_0^3 80 - 60\cos\left(\frac{\pi}{6}t\right) dt$

$= \left[80t - \frac{60}{\pi/6}\sin\left(\frac{\pi}{6}t\right)\right]_0^3$

$= \left[80t - \frac{360}{\pi}\sin\left(\frac{\pi}{6}t\right)\right]_0^3$

$= \left[80(3) - \frac{360}{\pi}\sin\left(\frac{\pi}{6}\cdot 3\right)\right]$

$-\left[80(0) - \frac{360}{\pi}\sin\left(\frac{\pi}{6}\cdot 0\right)\right]$

$= \left[240 - \frac{360}{\pi}(1)\right] - [0-0]$

$= 240 - \frac{360}{\pi}$

$\approx 125.$

Total rentals between January 1 and April 1 are about 125 Dune Buggies.

b) The average number of rentals is given by

$\frac{1}{3-0}\int_0^3 80 - 60\cos\left(\frac{\pi}{6}t\right) dt = \frac{1}{3}[125]$

$= 41.667 \approx 42.$

The average number of rentals is about 42 rentals per month.

c) The total number of rentals for the year is given by:

$\int_0^{12} 80 - 60\cos\left(\frac{\pi}{6}t\right) dt$

$= \left[80t - \frac{360}{\pi}\sin\left(\frac{\pi}{6}t\right)\right]_0^{12}$

$= \left[80(12) - \frac{360}{\pi}\sin\left(\frac{\pi}{6}\cdot 12\right)\right]$

$-\left[80(0) - \frac{360}{\pi}\sin\left(\frac{\pi}{6}\cdot 0\right)\right]$

$= \left[960 - \frac{360}{\pi}(0)\right] - [0-0]$

$= 960.$

Sal's had 960 rentals for the year.

d) The average number of rentals is given by

$\frac{1}{12-0}\int_0^{12} 80 - 60\cos\left(\frac{\pi}{6}t\right) dt = \frac{1}{12}[960]$

$= 80.$

The average number of rentals over the year is about 80 rentals per month.

Copyright © 2014 Pearson Education, Inc. Publishing as Addison-Wesley.

55. The average price of gas during the first 3 weeks is given by

$$\frac{1}{3-0}\int_0^3 3.64-0.08\cos\left(\tfrac{2\pi}{5}t\right)dt$$
$$=\tfrac{1}{3}\left[3.64t-0.08\cdot\tfrac{5}{2\pi}\sin\left(\tfrac{2\pi}{5}t\right)\right]_0^3$$
$$=\tfrac{1}{3}\left[3.64t-0.0637\sin\left(\tfrac{2\pi}{5}t\right)\right]_0^3$$
$$=\tfrac{1}{3}\left[\left[3.64(3)-0.0637\sin\left(\tfrac{2\pi}{5}\cdot 3\right)\right]-\left[3.64(0)-0.0637\sin\left(\tfrac{2\pi}{5}\cdot 0\right)\right]\right]$$
$$=\tfrac{1}{3}\left[\left[10.92-(-0.0374)\right]-\left[0-0\right]\right]$$
$$=\tfrac{1}{3}[10.957]$$
$$\approx 3.65.$$

The average price of gas during the first 3 weeks of the period is about \$3.65.

57. The average daily high temperature over the first 120 days of the year is given by

$$\frac{1}{120-0}\int_0^{120} 57-27\cos\left(\tfrac{2\pi}{365}t\right)dt$$
$$=\tfrac{1}{120}\left[57t-27\left(\tfrac{365}{2\pi}\right)\sin\left(\tfrac{2\pi}{365}t\right)\right]_0^{120}$$
$$=\tfrac{1}{120}\left[57t-1568.472\sin\left(\tfrac{2\pi}{365}t\right)\right]_0^{120}$$
$$=\tfrac{1}{120}\left[\left[57(120)-1568.472\sin\left(\tfrac{2\pi}{365}\cdot 120\right)\right]-\left[57(0)-1568.472\sin\left(\tfrac{2\pi}{365}\cdot 0\right)\right]\right]$$
$$=\tfrac{1}{120}\left[\left[6840-(1380.275)\right]-\left[0-0\right]\right]$$
$$=\tfrac{1}{120}[5459.725]$$
$$\approx 45.5.$$

The average daily high temperature over the first 120 days of the year is $45.5°F$

59. One period of the function is $\dfrac{2\pi}{\frac{2\pi}{3}}=3$ seconds.

The total amount of oxygen delivered over one period is given by

$$\int_0^3 135-135\cos\left(\tfrac{2\pi}{3}t\right)dt$$
$$=\left[135t-135\left(\tfrac{3}{2\pi}\right)\sin\left(\tfrac{2\pi}{3}t\right)\right]_0^3$$
$$=\left[135(3)-64.458\sin\left(\tfrac{2\pi}{3}\cdot 3\right)\right]-\left[135(0)-64.458\sin\left(\tfrac{2\pi}{3}\cdot 0\right)\right]$$
$$=[405-0]-[0-0]$$
$$=405.$$

The total amount of oxygen delivered over one period is 405 milliliters.

61. Since $\int \sin kx\,dx=-\frac{1}{k}\cos kx+C,$ we have

$$\int_0^b \sin 2x\,dx=0$$
$$\left[-\tfrac{1}{2}\cos 2x\right]_0^b=0$$
$$-\tfrac{1}{2}\cos 2b-\tfrac{1}{2}\cos 0=0$$
$$\cos 2b-1=0$$
$$\cos 2b=1$$

Possible values of b that satisfy the equation are $b=\pi, 2\pi, 3\pi,$ and so on.

63. $\int x\sin x\,dx$

Using integration by parts, let

$u=x \quad dv=\sin x\,dx$

$du=dx \quad v=-\cos x$

$$\int x\sin x\,dx=uv-\int v\,du$$
$$=-x\cos x-\int -\cos x\,dx$$
$$=-x\cos x+\int \cos x\,dx$$
$$=-x\cos x+\sin x+C.$$

Copyright © 2014 Pearson Education, Inc. Publishing as Addison-Wesley.

65. $\int x\cos 2x\,dx$

Using integration by parts, let

$u = x \quad dv = \cos 2x\,dx$

$du = dx \quad v = \frac{1}{2}\sin 2x$

$$\int x\cos 2x\,dx = uv - \int v\,du$$
$$= \tfrac{1}{2}x\sin 2x - \int \tfrac{1}{2}\sin 2x\,dx$$
$$= \tfrac{1}{2}x\sin 2x + \tfrac{1}{4}\cos 2x + C.$$

67. $\int 2x\sin 4x\,dx$

Using integration by parts, let

$u = 2x \quad dv = \sin 4x\,dx$

$du = 2dx \quad v = -\frac{1}{4}\cos 4x$

$$\int 2x\sin 4x\,dx = uv - \int v\,du$$
$$= 2x\left(\tfrac{-1}{4}\cos 4x\right) - \int 2\left(\tfrac{-1}{4}\cos 4x\right)dx$$
$$= -\tfrac{1}{2}x\cos 4x + \tfrac{1}{2}\int \cos 4x\,dx$$
$$= -\tfrac{1}{2}x\cos 4x + \tfrac{1}{8}\sin 4x + C.$$

69. $\int_0^{\pi/2} x\sin 2x\,dx$

Using integration by parts, let

$u = x \quad dv = \sin 2x\,dx$

$du = dx \quad v = -\frac{1}{2}\cos 2x$

$$\int_0^{\pi/2} x\sin 2x\,dx = uv - \int v\,du$$
$$= -\tfrac{1}{2}x\cos 2x - \int -\tfrac{1}{2}\cos 2x\,dx$$
$$= -\tfrac{1}{2}x\cos 2x + \tfrac{1}{2}\int \cos 2x\,dx$$
$$= \left[-\tfrac{1}{2}x\cos 2x + \tfrac{1}{4}\sin 2x\right]_0^{\pi/2}$$
$$= \left[-\tfrac{1}{2}\left(\tfrac{\pi}{2}\right)\cos 2\left(\tfrac{\pi}{2}\right) + \tfrac{1}{4}\sin 2\left(\tfrac{\pi}{2}\right)\right] - \left[-\tfrac{1}{2}(0)\cos 2(0) + \tfrac{1}{4}\sin 2(0)\right]$$
$$= \left[-\tfrac{\pi}{4}\cos\pi + \tfrac{1}{4}\sin\pi\right] - [0]$$
$$= -\tfrac{\pi}{4}(-1) - 0$$
$$= \tfrac{\pi}{4}.$$

71. a) Divide the interval $\left[\frac{\pi}{2}, \pi\right]$ into four equal subintervals. $\Delta x = \dfrac{\pi - \frac{\pi}{2}}{4} = \dfrac{\pi}{8}$.

The subintervals are $\left[\frac{\pi}{2}, \frac{5\pi}{8}\right], \left[\frac{5\pi}{8}, \frac{3\pi}{4}\right], \left[\frac{3\pi}{4}, \frac{7\pi}{8}\right]$, and $\left[\frac{7\pi}{8}, \pi\right]$.

Applying the Trapezoidal rule we have

$$\int_{\pi/2}^{\pi} \frac{\sin x}{2x}\,dx \approx T_4$$
$$T_4 \approx \frac{\pi}{8}\left[\frac{f\left(\frac{\pi}{2}\right)}{2} + f\left(\tfrac{5\pi}{8}\right) + f\left(\tfrac{3\pi}{4}\right) + f\left(\tfrac{7\pi}{8}\right) + \frac{f(\pi)}{2}\right]$$
$$T_4 \approx \frac{\pi}{8}\left[\frac{\frac{\sin(\pi/2)}{2(\pi/2)}}{2} + \frac{\sin\left(\frac{5\pi}{8}\right)}{2\left(\frac{5\pi}{8}\right)} + \frac{\sin\left(\frac{3\pi}{4}\right)}{2\left(\frac{3\pi}{4}\right)} + \frac{\sin\left(\frac{7\pi}{8}\right)}{2\left(\frac{7\pi}{8}\right)} + \frac{\frac{\sin(\pi)}{2(\pi)}}{2}\right]$$
$$T_4 \approx \frac{\pi}{8}[0.1592 + 0.2353 + 0.1501 + 0.0696 + 0]$$
$$T_4 \approx 0.2411$$

b) Using a graphing calculator, we have

```
fnInt(sin(X)/(2X
),X,π/2,π)
        .2405874419
```

$$\int_{\pi/2}^{\pi} \frac{\sin x}{2x}\,dx \approx 0.24059.$$

73.

a)

b)

Copyright © 2014 Pearson Education, Inc. Publishing as Addison-Wesley.

Exercise Set 7.4

1. It is helpful to utilize special right triangles or the unit circle to determine inverse functions.

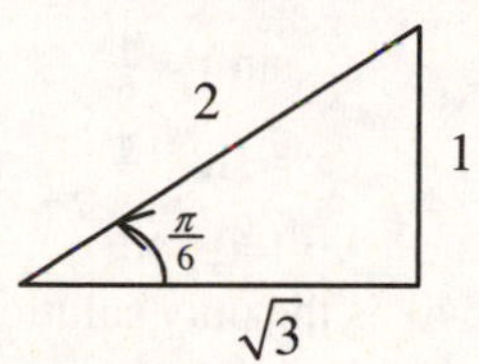

Since $\sin\frac{\pi}{6}=\frac{1}{2}$, We have $\sin^{-1}\frac{1}{2}=\frac{\pi}{6}$, or 30°.

3. Utilizing the special right triangles

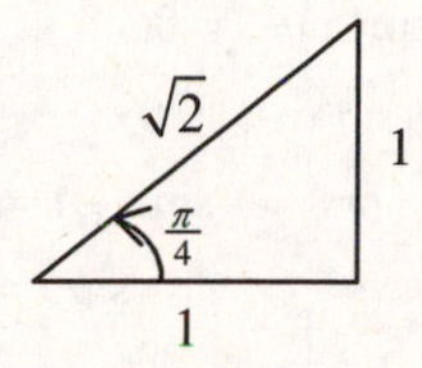

Since $\cos\frac{\pi}{4}=\frac{\sqrt{2}}{2}$, We have

$\cos^{-1}\frac{\sqrt{2}}{2}=\frac{\pi}{4}$, or 45°.

For solutions 5-9 refer to the following unit circle.

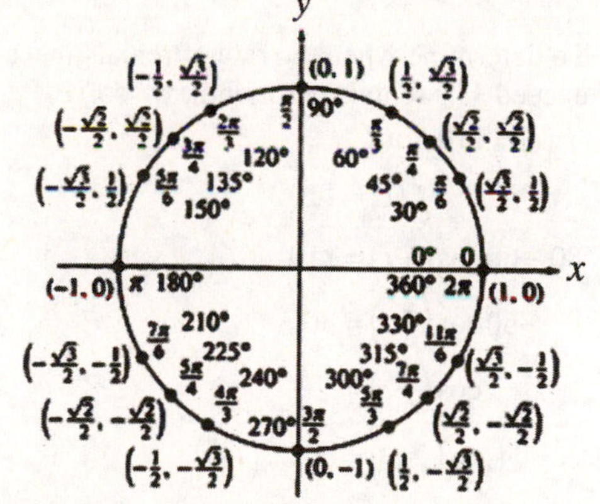

5. Referring to the unit circle.
Since $\cos\left(\frac{\pi}{2}\right)=0$, We have

$\cos^{-1}(0)=\frac{\pi}{2}$, or 90°.

7. Referring to the unit circle.
Since $\tan\frac{\pi}{4}=1$, We have $\tan^{-1}1=\frac{\pi}{4}$, or 45°.

9. Refer to the unit circle.
Since $\cos(\pi)=-1$, We have

$\cos^{-1}(-1)=\pi$, or 180°.

11. Using a calculator:
$\sin^{-1}0.3=0.305$, or 17.475°.

13. Using a calculator:
$\tan^{-1}0.78=0.662$, or 37.954°.

15. Using a calculator:
$\cos^{-1}(-0.17)=1.742$, or 99.788°.

17. Since $\tan t=\frac{4}{5}$, we have

$t=\tan^{-1}\frac{4}{5}=0.675$, or 38.660°.

Since $\tan u=\frac{5}{4}$, we have

$u=\tan^{-1}\frac{5}{4}=0.896$, or 51.340°.

19. Since $\tan t=\frac{7}{4}$, we have

$t=\tan^{-1}\frac{7}{4}=1.052$, or 60.255°.

We use the Pythagorean Theorem to find the length of the leg adjacent to angle u.

$$c^2=4^2+7^2$$
$$c^2=16+49$$
$$c^2=65$$
$$c=\sqrt{65}$$

Therefore, since $\tan u=\frac{2}{\sqrt{65}}$, we have

$u=\tan^{-1}\frac{2}{\sqrt{65}}=0.243$, or 13.932°.

21. Solve the equation for x on the interval $\left[0,\frac{\pi}{2}\right]$ as follows:

$$-1+2\sin x=0$$
$$2\sin x=1$$
$$\sin x=\frac{1}{2}$$
$$\sin^{-1}(\sin x)=\sin^{-1}\left(\frac{1}{2}\right)$$
$$x=\sin^{-1}\frac{1}{2}$$
$$x=\frac{\pi}{6}.$$

23. Solve the equation for x on the interval $\left[0,\frac{\pi}{6}\right]$ as follows:

$$\sin 3x=\frac{\sqrt{2}}{2}$$
$$\sin^{-1}(\sin 3x)=\sin^{-1}\left(\frac{\sqrt{2}}{2}\right)$$
$$3x=\sin^{-1}\frac{\sqrt{2}}{2}$$
$$3x=\frac{\pi}{4}$$
$$x=\frac{\pi}{12}.$$

Copyright © 2014 Pearson Education, Inc. Publishing as Addison-Wesley.

25. Solve the equation for x on the interval $\left[0,\frac{\pi}{4}\right]$ as follows:

$$5\tan 2x = 5$$
$$\tan 2x = 1$$
$$\tan^{-1}(\tan 2x) = \tan^{-1} 1$$
$$2x = \tan^{-1} 1$$
$$2x = \tfrac{\pi}{4}$$
$$x = \tfrac{\pi}{8}.$$

27. Solve the equation for x on the interval $\left[0,\frac{\pi}{2}\right]$ as follows:

$$-4 + 6\sin x = 0$$
$$6\sin x = 4$$
$$\sin x = \tfrac{2}{3}$$
$$\sin^{-1}(\sin x) = \sin^{-1}\left(\tfrac{2}{3}\right)$$
$$x = \sin^{-1}\tfrac{2}{3}$$
$$x = 0.730.$$

29. Solve the equation for x on the interval $\left[0,\frac{\pi}{4}\right]$ as follows:

$$-\tfrac{1}{4} + \cos(3x+1) = 0$$
$$\cos(3x+1) = \tfrac{1}{4}$$
$$\cos^{-1}(\cos(3x+1)) = \cos^{-1}\tfrac{1}{4}$$
$$3x+1 = \cos^{-1}\tfrac{1}{4}$$
$$3x+1 = 1.318$$
$$3x = 0.318$$
$$x = 0.106.$$

31. Solve the equation for x on the interval $\left[0,\frac{\pi}{12}\right]$ as follows:

$$\tan 6x = 3$$
$$\tan^{-1}(\tan 6x) = \tan^{-1} 3$$
$$6x = \tan^{-1} 3$$
$$6x = 1.249$$
$$x = 0.208.$$

33. Solve the equation for x on the interval $\left(0,\frac{\pi}{2}\right]$ as follows:

$$3\sin^2 x - \sin(x) = 0$$
$$\sin x[3\sin x - 1] = 0$$

$\sin x = 0$ or $3\sin x - 1 = 0$

$x = \sin^{-1} 0$ $\quad$ $\sin x = \tfrac{1}{3}$

$x = 0$ $\quad$ $x = \sin^{-1}\tfrac{1}{3}$

$x = 0.340$

The solution $x = 0.340$ is the only solution that lies in the desired interval.

35. Solve the equation for x on the interval $\left[-\frac{\pi}{2},\frac{\pi}{2}\right]$ as follows:

$$2\sin^2 x - \sin(x) - 1 = 0$$
$$[2\sin x + 1][\sin x - 1] = 0$$

$2\sin x + 1 = 0$ or $\sin x - 1 = 0$

$\sin x = -\tfrac{1}{2}$ $\quad$ $\sin x = 1$

$x = \sin^{-1}\left(-\tfrac{1}{2}\right)$ $\quad$ $x = \sin^{-1} 1$

$x = -\tfrac{\pi}{6}$ $\quad$ $x = \tfrac{\pi}{2}$

$x = -0.524$ $\quad$ $x = 1.571$

There are two solutions on the desired interval $x = -0.524\left(\text{or } -\frac{\pi}{6}\right)$ and $x = 1.571\left(\text{or } \frac{\pi}{2}\right)$.

37. To determine when the rate of rentals meets or exceed 110 rentals per month, we solve $f(t) = 110$ for t

$$f(t) = 110$$
$$80 - 60\cos\left(\tfrac{\pi}{6}t\right) = 110$$
$$-60\cos\left(\tfrac{\pi}{6}t\right) = 30$$
$$\cos\left(\tfrac{\pi}{6}t\right) = -\tfrac{1}{2}$$
$$\cos^{-1}\left(\cos\left(\tfrac{\pi}{6}t\right)\right) = \cos^{-1}\left(-\tfrac{1}{2}\right)$$
$$\tfrac{\pi}{6}t = \tfrac{2\pi}{3}$$
$$t = \tfrac{2\pi}{3}\cdot\tfrac{6}{\pi}$$
$$t = 4.$$

The rate of rentals will first meet 110 units per month 4 months after January 1, or on May 1.

Copyright © 2014 Pearson Education, Inc. Publishing as Addison-Wesley.

39. In order to maximize the volume of the feed trough, the area of the triangular cross section must be maximized.

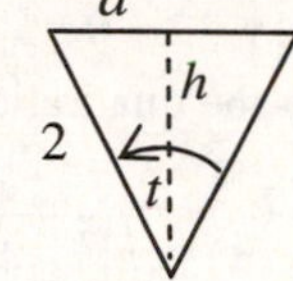

The area of the triangle is given by $A=\frac{1}{2}bh$.

We notice that $\cos\left(\frac{t}{2}\right)=\frac{a}{2}$ and $\sin\left(\frac{t}{2}\right)=\frac{h}{2}$.

Since $b=2a$, we have $b=4\cos\left(\frac{t}{2}\right)$ and $h=2\sin\left(\frac{t}{2}\right)$. Therefore, the area of the triangle can be represented as a function of the angle t.

$$A(t)=\tfrac{1}{2}\left(4\cos\left(\tfrac{t}{2}\right)\right)\left(2\sin\left(\tfrac{t}{2}\right)\right)=4\cos\left(\tfrac{t}{2}\right)\sin\left(\tfrac{t}{2}\right)$$

To find the maximum, we take the derivative of the area function with respect to t and set it equal to zero.

$$A(t)=4\cos\left(\tfrac{t}{2}\right)\sin\left(\tfrac{t}{2}\right)$$
$$A'(t)=4\left[\cos\left(\tfrac{t}{2}\right)\cdot\tfrac{d}{dt}\left(\sin\left(\tfrac{t}{2}\right)\right)+\tfrac{d}{dt}\left(\cos\left(\tfrac{t}{2}\right)\right)\cdot\sin\left(\tfrac{t}{2}\right)\right]$$
$$A'(t)=4\left[\tfrac{1}{2}\cos^2\left(\tfrac{t}{2}\right)-\tfrac{1}{2}\sin^2\left(\tfrac{t}{2}\right)\right]$$
$$A'(t)=2\left[\cos^2\left(\tfrac{t}{2}\right)-\sin^2\left(\tfrac{t}{2}\right)\right] \quad \text{Apply the double angle identity.}$$
$$A'(t)=2\cos\left(2\cdot\tfrac{t}{2}\right)$$
$$A'(t)=2\cos t$$

Now set the derivative equal to zero.

$$A'(t)=0$$
$$2\cos t=0$$
$$\cos t=0$$
$$t=\cos^{-1}0$$
$$t=\tfrac{\pi}{2}.$$

The sheet metal should be bent at an angle of $\frac{\pi}{2}$, or 90° in order to maximize the amount of feed it can hold.

41. To determine the day when the average daily high temperature will reach $60°F$, solve the following equation.

$$T(t)=60$$
$$57-27\cos\left(\tfrac{2\pi}{365}t\right)=60$$
$$-27\cos\left(\tfrac{2\pi}{365}t\right)=3$$
$$\cos\left(\tfrac{2\pi}{365}t\right)=-\tfrac{1}{9}$$
$$\tfrac{2\pi}{365}t=\cos^{-1}\left(-\tfrac{1}{9}\right)$$
$$\tfrac{2\pi}{365}t=1.682$$
$$t=1.682\cdot\left(\tfrac{365}{2\pi}\right)$$
$$t=97.7$$

It will take approximately 98 days after January 1st for the average daily high temperature to reach $60°F$.

43. Draw a sketch of the situation.

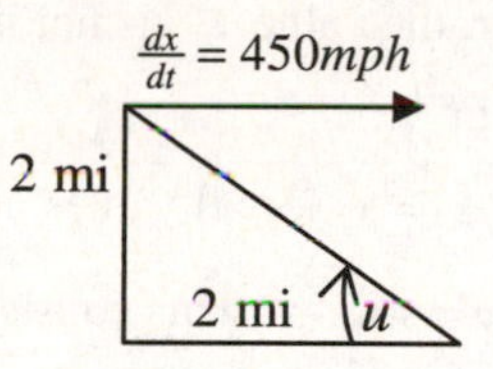

When $x=2$ mi. the distance between Nicole and the plane is

$$d^2=2^2+2^2=8$$
$$d=\sqrt{8}=2\sqrt{2}$$

At that moment,

$$\cos u=\tfrac{2}{\sqrt{8}} \text{ and } \cos^2 u=\left(\tfrac{2}{\sqrt{8}}\right)^2=\tfrac{4}{8}=\tfrac{1}{2}.$$

Furthermore, we know that for any horizontal distance, x, the plane is away from Nicole, $\tan u=\frac{2}{x}$.

To find the rate at which the viewing angle is increasing, we differentiate both sides of the previous equation implicitly with respect to t.

$$\tan u=\tfrac{2}{x}$$
$$\tfrac{d}{dt}\tan u=\tfrac{d}{dt}\left(2x^{-1}\right)$$
$$\sec^2 u\frac{du}{dt}=-2x^{-2}\frac{dx}{dt}$$
$$\frac{du}{dt}=\frac{-2x^{-2}}{\sec^2 u}\cdot\frac{dx}{dt}$$
$$\frac{du}{dt}=\frac{-2\cos^2 u}{x^2}\cdot\frac{dx}{dt}$$

The solution is continued on the next page.

Copyright © 2014 Pearson Education, Inc. Publishing as Addison-Wesley.

So when $x = 2$ we substitute to find

$$\frac{du}{dt} = \frac{-2\cos^2 u}{x^2} \cdot \frac{dx}{dt}$$

$$\frac{du}{dt} = \frac{-2\left(\frac{1}{2}\right)}{(2)^2} \cdot (-450) \quad \cos^2 u = \tfrac{1}{2}, \text{ from earlier calculation.}$$

$$\frac{du}{dt} = 112.5 \tfrac{\text{radians}}{\text{hour}}$$

Next, convert to degrees per second.

$$112.5 \tfrac{\text{radians}}{\text{hour}} \times \tfrac{1 \text{ hour}}{3600 \text{ seconds}} \times \tfrac{180°}{\pi \text{ radians}} = \tfrac{1.790°}{\text{second}}.$$

Nicole's viewing angle is increasing at a rate of 1.790 degrees per second when the plane is 2 miles away.

45. a) The value 0.9 is in the domain of $y = \sin^{-1} x$ however, the value 1.1 is not in the domain of $y = \sin^{-1} x$.

b) The value $\frac{\pi}{2}$ is in the domain of $y = \sin^{-1} x$ however, the value π is not in the domain of $y = \sin^{-1} x$.

c) In general $\sin\left(\sin^{-1} x\right) = x$ for all values of x in the domain of $y = \sin^{-1} x$. This consists of the values of x such that $-1 \le x \le 1$.

d) In general $\sin^{-1}(\sin x) = x$ for all values of x in the domain of $y = \sin x$. This consists of the values of x such that $-\frac{\pi}{2} \le x \le \frac{\pi}{2}$.

47. Draw a right triangle with legs a and b.

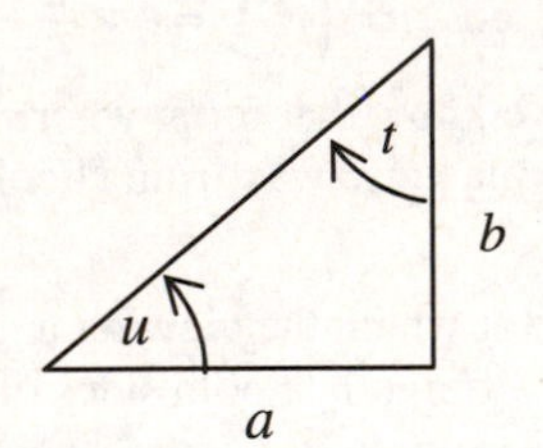

From the picture we know

$\tan u = \frac{b}{a}$

$u = \tan^{-1} \frac{b}{a}$

and

$\tan t = \frac{a}{b}$

$t = \tan^{-1} \frac{a}{b}$

Furthermore, $u + t = 90°$ or $\frac{\pi}{2}$.

Therefore,

$$u + t = \frac{\pi}{2}$$

$$\tan^{-1} \frac{b}{a} + \tan^{-1} \frac{a}{b} = \frac{\pi}{2}.$$

49. Consider $\sin^2 x + 4\sin x - 2 = 0$.

a) Let $y = \sin x$, substituting we have

$$y^2 + 4y - 2 = 0.$$

Apply the quadratic formula with $a = 1, b = 4, c = -2$.

$$y = \frac{-(4) \pm \sqrt{4^2 - 4(1)(-2)}}{2(1)}$$

$$y = \frac{-4 \pm \sqrt{24}}{2}$$

$$y = \frac{-4 \pm 2\sqrt{6}}{2}$$

$$y = -2 \pm \sqrt{6}.$$

The two solutions are $y_1 = -2 + \sqrt{6} \approx 0.449$ and $y_2 = -2 - \sqrt{6} \approx -4.449$.

b) For $y_1 = 0.449$ we have

$x_1 = \sin^{-1}(0.449) \approx 0.466.$

For $y_2 = -4.449$ we have

$x_2 = \sin^{-1}(-4.449)$. However, since -4.449 is not in the domain of $x = \sin^{-1} y$ we cannot find a solution.

c) Using a graphing calculator, we graph the function and find the x-intercepts over the interval $[0, 2\pi]$.

The first intercept is $x = 0.466$.

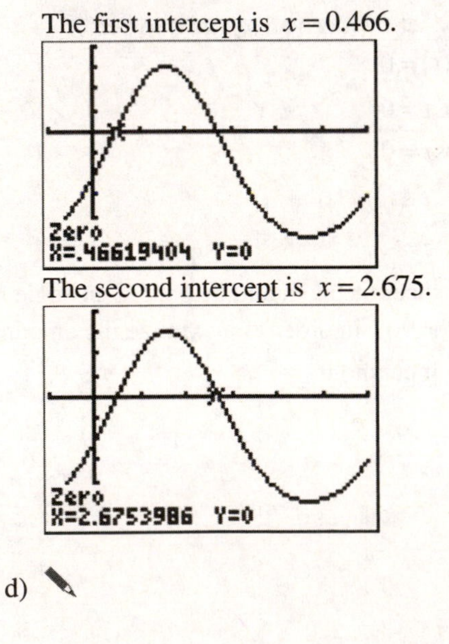

The second intercept is $x = 2.675$.

d)

Copyright © 2014 Pearson Education, Inc. Publishing as Addison-Wesley.

51. We construct a diagram to assist us with the solution.

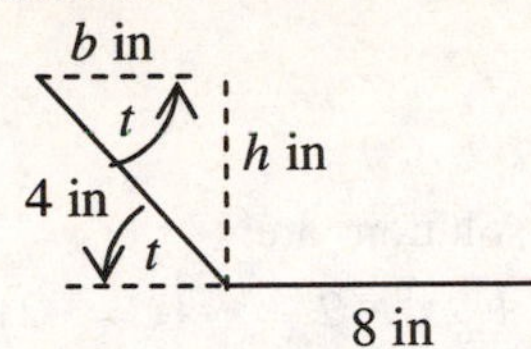

There are two areas to discuss, the area of the triangle and the area of the rectangle. The area of the of the rectangle is given by $A_R = 8 \cdot h$, while the area of the triangle is given by:
$A_T = \frac{1}{2}bh.$

Using the definition of trigonometric functions we know that

$\cos t = \frac{b}{4} \to b = 4\cos t$

$\sin t = \frac{h}{4} \to h = 4\cos t.$

Substituting we can find the area as a function of *t*.

$A_R = 8(4\sin t) = 32\sin t$

$A_T = \frac{1}{2}(4\cos t)(4\sin t) = 8\cos t \sin t.$

Therefore, the cross sectional area of the gutter is given by

$A(t) = A_T + A_R$

$A(t) = 8\cos t \sin t + 32\sin t$

Taking the derivative of the function we have:

$$\begin{aligned}
A'(t) &= \tfrac{d}{dt}[8\cos t \sin t + 32\sin t] \\
&= 8\left[\cos t \tfrac{d}{dt}\sin t + \tfrac{d}{dt}\cos t \cdot \sin t\right] + 32\tfrac{d}{dt}\sin t \\
&= 8[\cos t \cdot \cos t - \sin t \cdot \sin t] + 32\cos t \\
&= 8\left[\cos^2 t - \sin^2 t\right] + 32\cos t \\
&= 8\left[\cos t^2 t - \left(1 - \cos^2 t\right)\right] + 32\cos t \\
&= 8\left[2\cos^2 t - 1\right] + 32\cos t \\
&= 16\cos^2 t + 32\cos t - 8.
\end{aligned}$$

Let $y = \cos t$ and we have

$A'(t) = 16y^2 + 32y - 8$.

Set the derivative equal to zero and solve for *y*.

$A'(t) = 0$

$16y^2 + 32y - 8 = 0$

The solution is continued at the top of the next column.

Using the quadratic formula we have:

$$y = \frac{-32 \pm \sqrt{32^2 - 4(16)(-8)}}{2(16)}$$

$$y = \frac{-32 \pm \sqrt{1536}}{32}$$

$$y = \frac{-2 \pm \sqrt{6}}{2}.$$

We have two solutions $y_1 = \frac{-2+\sqrt{6}}{2} = 0.225$ and $y_2 = \frac{-2-\sqrt{6}}{2} = -2.225$. However, only $y_1 = \frac{-2+\sqrt{6}}{2}$ is in the domain of $t = \cos^{-1} t$.

Therefore,

$t = \cos^{-1}\left(\frac{-2+\sqrt{6}}{2}\right) = 1.344$ radians.

To verify this is a maximum, we find apply the second derivative test.

$$\begin{aligned}
A''(t) &= \tfrac{d}{dt}\left(16\cos^2 t + 32\cos t - 8\right) \\
&= -32\cos t \sin t - 32\sin t.
\end{aligned}$$

Substituting our solution in for *t*, we have:

$$\begin{aligned}
A''(1.344) &= -32\cos(1.344)\sin(1.344) - 32\sin(1.344) \\
&= -38.189 < 0
\end{aligned}$$

Therefore the area is maximized when $t = 1.344$ radians or 77.006°.

53. ✎

Copyright © 2014 Pearson Education, Inc. Publishing as Addison-Wesley.

Chapter 8

Differential Equations

Exercise Set 8.1

1. We find the general solution by integrating both sides of the equation:

$$y' = 5x^4$$
$$\int y'\,dx = \int 5x^4\,dx$$
$$y = 5\left(\tfrac{1}{5}x^5 + C\right)$$
$$y = x^5 + C \quad \text{General Solution.}$$

We can find three particular solutions by substituting different values for *C*. Answers may vary.

$$y = x^5 + 1 \qquad (C = 1)$$
$$y = x^5 - 2 \qquad (C = -2)$$
$$y = x^5 + 8 \qquad (C = 8).$$

3. We find the general solution by integrating both sides of the equation:

$$y' = e^{2x} + x$$
$$\int y'\,dx = \int \left(e^{2x} + x\right)dx$$
$$y = \tfrac{1}{2}e^{2x} + \tfrac{1}{2}x^2 + C \quad \text{General Solution.}$$

We can find three particular solutions by substituting different values for *C*. Answers may vary.

$$y = \tfrac{1}{2}e^{2x} + \tfrac{1}{2}x^2 + 1 \qquad (C = 1)$$
$$y = \tfrac{1}{2}e^{2x} + \tfrac{1}{2}x^2 - 3 \qquad (C = -3)$$
$$y = \tfrac{1}{2}e^{2x} + \tfrac{1}{2}x^2 + 6 \qquad (C = 6).$$

5. We find the general solution by integrating both sides of the equation:

$$y' = \tfrac{8}{x} - x^2 + x^5$$
$$\int y'\,dx = \int \left(\tfrac{8}{x} - x^2 + x^5\right)dx$$
$$y = 8\ln|x| - \tfrac{1}{3}x^3 + \tfrac{1}{6}x^6 + C$$

General Solution.

We can find three particular solutions by substituting different values for *C*, as shown at the top of the next column. Answers may vary.

Three particular solutions are:

$$y = 8\ln|x| - \tfrac{1}{3}x^3 + \tfrac{1}{6}x^6 - 2 \qquad (C = -2)$$
$$y = 8\ln|x| - \tfrac{1}{3}x^3 + \tfrac{1}{6}x^6 + 4 \qquad (C = 4)$$
$$y = 8\ln|x| - \tfrac{1}{3}x^3 + \tfrac{1}{6}x^6 - 11 \qquad (C = -11).$$

7. Find y' and y''.

$$y' = \tfrac{d}{dx}\left(x\ln x + 3x - 2\right)$$
$$= \underbrace{\tfrac{d}{dx}x\ln x}_{\text{Product Rule}} + \tfrac{d}{dx}3x - \tfrac{d}{dx}2$$
$$= \underbrace{x\cdot\tfrac{1}{x} + 1\cdot\ln x}_{\text{Product Rule}} + 3 - 0$$
$$y' = 1 + \ln x + 3$$
$$y' = \ln x + 4$$

and

$$y'' = \tfrac{d}{dx}(\ln x + 4)$$
$$y'' = \tfrac{1}{x}.$$

Substituting, we have:

$$y'' - \tfrac{1}{x} \stackrel{?}{=} 0$$

$\frac{1}{x} - \frac{1}{x}$	0
0	0 TRUE

9. Find y' and y''.

$$y' = \tfrac{d}{dx}\left(e^x + 3xe^x\right)$$
$$= \tfrac{d}{dx}e^x + \underbrace{\tfrac{d}{dx}3xe^x}_{\text{Product Rule}}$$
$$= e^x + \underbrace{3x\cdot e^x + 3\cdot e^x}_{\text{Product Rule}}$$
$$y' = 3xe^x + 4e^x$$

and

$$y'' = \tfrac{d}{dx}\left(3xe^x + 4e^x\right)$$
$$y'' = \underbrace{3x\cdot e^x + 3\cdot e^x}_{\text{Product Rule}} + 4e^x$$
$$y'' = 3x\cdot e^x + 7\cdot e^x.$$

The solution is continued on the next page.

Copyright © 2014 Pearson Education, Inc. Publishing as Addison-Wesley.

Substituting the information from the previous page, we have:

$$y''-2y'+y\stackrel{?}{=}0$$

$3xe^x+7e^x-2\left(3xe^x+4e^x\right)+e^x+3xe^x$	0
$3xe^x+7e^x-6xe^x-8e^x+e^x+3xe^x$	0
$6xe^x+8e^x-6xe^x-8e^x$	0
0	0 TRUE

11. Let $y'+4y=0$.

a) Show $y=e^{-4x}$ is a solution.

Find y'.

$y'=\frac{d}{dx}\left(e^{-4x}\right)$

$y'=e^{-4x}\cdot\frac{d}{dx}(-4x)$

$y'=-4e^{-4x}$.

Substituting, we have:

$$y'+4y\stackrel{?}{=}0$$

$-4e^{-4x}+4\left(e^{-4x}\right)$	0
0	0 TRUE

b) Show $y=Ce^{-4x}$ is a solution.

Find y'.

$y'=\frac{d}{dx}\left(Ce^{-4x}\right)$

$y'=Ce^{-4x}\cdot\frac{d}{dx}(-4x)$

$y'=-4Ce^{-4x}$.

Substituting, we have:

$$y'+4y\stackrel{?}{=}0$$

$-4Ce^{-4x}+4\left(Ce^{-4x}\right)$	0
0	0 TRUE

13. Let $y''-y-30y=0$.

a) Show $y=e^{6x}$ is a solution.

Find y' and y''.

$y'=\frac{d}{dx}\left(e^{6x}\right)=6e^{6x}$.

$y''=\frac{d}{dx}\left(6e^{6x}\right)=36e^{6x}$.

The solution is continued at the top of the next column.

Substituting, we have:

$$y''-y'-30y\stackrel{?}{=}0$$

$36e^{6x}-\left(6e^{6x}\right)-30\left(e^{6x}\right)$	0
$36e^{6x}-36e^{6x}$	0
0	0 TRUE

b) Show $y=e^{-5x}$ is a solution.

Find y' and y''.

$y'=\frac{d}{dx}\left(e^{-5x}\right)=-5e^{-5x}$.

$y''=\frac{d}{dx}\left(-5e^{-5x}\right)=25e^{-5x}$.

Substituting, we have:

$$y''-y'-30y\stackrel{?}{=}0$$

$25e^{-5x}-\left(-5e^{-5x}\right)-30\left(e^{-5x}\right)$	0
$30e^{-5x}-30e^{-5x}$	0
0	0 TRUE

c) Show $y=C_1e^{6x}+C_2e^{-5x}$ is a solution.

Find y' and y''.

$y'=\frac{d}{dx}\left(C_1e^{6x}+C_2e^{-5x}\right)$

$y'=6C_1e^{6x}-5C_2e^{-5x}$.

$y''=\frac{d}{dx}\left(6C_1e^{6x}-5C_2e^{-5x}\right)$

$y''=36C_1e^{6x}+25C_2e^{-5x}$.

Substituting, we have:

$$y''-y'-30y\stackrel{?}{=}0$$

$36C_1e^{6x}+25C_2e^{-5x}-\left(6C_1e^{6x}-5C_2e^{-5x}\right)-30\left(C_1e^{6x}+C_2e^{-5x}\right)$	0
$30C_1e^{6x}+30C_2e^{-5x}-30C_1e^{6x}-30C_2e^{-5x}$	0
0	0 TRUE

15. a) In general the solution to the differential equation $\frac{dP}{dk}=kP$ is $P(t)=P_0e^{kt}$.

Therefore,

$\frac{dM}{dt}=0.05M$ has a solution given by

$M=M_0e^{0.05t}$.

Copyright © 2014 Pearson Education, Inc. Publishing as Addison-Wesley.

b) To verify this solution, we find $\frac{dM}{dt}$ and we substitute into the equation.

$$\frac{dM}{dt} = \frac{d}{dt}\left(M_0 e^{0.05t}\right)$$
$$\frac{dM}{dt} = 0.05M_0 e^{0.05t}.$$

$$\begin{array}{r|l} \frac{dM}{dt} - 0.05M & \overset{?}{=} 0 \\ \hline 0.05M_0 e^{0.05t} - 0.05\left(M_0 e^{0.05t}\right) & 0 \\ 0 & 0 \quad \text{TRUE} \end{array}$$

17. a) The solution to the differential equation $\frac{dR}{dt} = 0.35R$ is $R = R_0 e^{0.35t}$.

b) To verify this solution, we find $\frac{dR}{dt}$ and we substitute into the equation.

$$\frac{dR}{dt} = \frac{d}{dt}\left(R_0 e^{0.35t}\right)$$
$$\frac{dR}{dt} = 0.35R_0 e^{0.35t}.$$

$$\begin{array}{r|l} \frac{dR}{dt} - 0.35R & \overset{?}{=} 0 \\ \hline 0.35R_0 e^{0.35t} - 0.66\left(R_0 e^{0.35t}\right) & 0 \\ 0 & 0 \quad \text{TRUE} \end{array}$$

19. a) The solution to the differential equation $\frac{dG}{dt} = 0.005G$ is $G = G_0 e^{0.005t}$.

b) To verify this solution, we find $\frac{dG}{dt}$ and we substitute into the equation.

$$\frac{dG}{dt} = \frac{d}{dt}\left(G_0 e^{0.005t}\right)$$
$$\frac{dG}{dt} = 0.005G_0 e^{0.005t}.$$

$$\begin{array}{r|l} \frac{dG}{dt} - 0.005G & \overset{?}{=} 0 \\ \hline 0.005G_0 e^{0.005t} - 0.005\left(G_0 e^{0.005t}\right) & 0 \\ 0 & 0 \quad \text{TRUE} \end{array}$$

21. a) The solution to the differential equation $\frac{dR}{dt} = R$ is $R = R_0 e^{t}$.

b) To verify this solution, we find $\frac{dR}{dt}$ and we substitute into the equation.

$$\frac{dR}{dt} = \frac{d}{dt}\left(R_0 e^{t}\right)$$
$$\frac{dR}{dt} = R_0 e^{t}.$$

$$\begin{array}{r|l} \frac{dR}{dt} - R & \overset{?}{=} 0 \\ \hline R_0 e^{t} - \left(R_0 e^{t}\right) & 0 \\ 0 & 0 \quad \text{TRUE} \end{array}$$

23. a) Find the general solution by integrating both sides of the equation.

$$y' = x^2 + 2x - 3$$
$$\int y'dx = \int\left(x^2 + 2x - 3\right)dx$$
$$y = \tfrac{1}{3}x^3 + x^2 - 3x + C.$$

Find the particular solution by substituting the initial conditions into the general solution and solving for C as follows

$$y = \tfrac{1}{3}x^3 + x^2 - 3x + C$$
$$(4) = \tfrac{1}{3}(0)^3 + (0)^2 - 3(0) + C$$
$$4 = C.$$

The particular solution is

$$y = \tfrac{1}{3}x^3 + x^2 - 3x + 4.$$

b) To verify solution we take the derivative of the particular solution and see

$$y' = \tfrac{d}{dx}\left(\tfrac{1}{3}x^3 + x^2 - 3x + 4\right)$$
$$y' = x^2 + 2x - 3.$$

Substituting into the equation we have

$$\begin{array}{r|l} y' - x^2 - 2x + 3 & \overset{?}{=} 0 \\ \hline \left(x^2 + 2x - 3\right) - x^2 - 2x + 3 & 0 \\ 0 & 0 \quad \text{TRUE} \end{array}$$

Copyright © 2014 Pearson Education, Inc. Publishing as Addison-Wesley.

25. a) Find the general solution by integrating both sides of the equation.

$$f'(x) = x^{\frac{2}{3}} - x$$
$$\int f'(x)dx = \int\left(x^{\frac{2}{3}} - x\right)dx$$
$$f(x) = \tfrac{3}{5}x^{\frac{5}{3}} - \tfrac{1}{2}x^2 + C.$$

Find the particular solution by substituting the initial conditions into the general solution and solving for C as follows:

$$f(1) = -6$$
$$(-6) = \tfrac{3}{5}(1)^{\frac{5}{3}} - \tfrac{1}{2}(1)^2 + C$$
$$-6 = \tfrac{3}{5} - \tfrac{1}{2} + C$$
$$-6 = \tfrac{1}{10} + C$$
$$-6 - \tfrac{1}{10} = C$$
$$-\tfrac{61}{10} = C.$$

The particular solution is

$$f(x) = \tfrac{3}{5}x^{\frac{5}{3}} - \tfrac{1}{2}x^2 - \tfrac{61}{10}.$$

b) To verify solution we take the derivative of the particular solution and see

$$f'(x) = \tfrac{d}{dx}\left(\tfrac{3}{5}x^{\frac{5}{3}} - \tfrac{1}{2}x^2 - \tfrac{61}{10}\right)$$
$$f'(x) = x^{\frac{2}{3}} - x.$$

Substituting into the equation we have

$f'(x) - x^{\frac{2}{3}} + x \overset{?}{=} 0$	
$\left(x^{\frac{2}{3}} - x\right) - x^{\frac{2}{3}} + x$	0
0	0 TRUE

27. a) The solution to the differential equation $\frac{dB}{dt} = 0.03B$ is $B = B_0e^{0.03t}$.

Find the particular solution by substituting the initial conditions into the general solution and solving for B_0 as follows

$$B(0) = 500$$
$$(500) = B_0e^{0.03(0)}$$
$$500 = B_0 \cdot 1$$
$$500 = B_0.$$

The particular solution is $B = 500e^{0.03t}$.

b) To verify this solution, we find $\frac{dB}{dt}$ and we substitute into the equation as shown at the top of the next column.

$$\frac{dB}{dt} = \frac{d}{dt}\left(500e^{0.03t}\right)$$
$$\frac{dB}{dt} = 500(0.03)e^{0.03t} = 15e^{0.03t}.$$

$\frac{dB}{dt} - 0.03B \overset{?}{=} 0$	
$15e^{0.03t} - 0.03\left(500e^{0.03t}\right)$	0
$15e^{0.03t} - 15e^{0.03t}$	0
0	0 TRUE

29. a) The solution to the differential equation $\frac{dS}{dt} = 0.12S$ is $S = S_0e^{0.12t}$.

Find the particular solution by substituting the initial conditions into the general solution and solving for S_0 as follows

$$S = S_0e^{0.12t}$$
$$(750) = S_0e^{0.12(0)}$$
$$750 = S_0.$$

The particular solution is $S = 750e^{0.12t}$.

b) To verify this solution, we find $\frac{dS}{dt}$ and we substitute into the equation.

$$\frac{dS}{dt} = \frac{d}{dt}\left(750e^{0.12t}\right)$$
$$\frac{dS}{dt} = 750(0.12)e^{0.12t} = 90e^{0.12t}.$$

$\frac{dS}{dt} - 0.12S \overset{?}{=} 0$	
$90e^{0.12t} - 0.12\left(750e^{0.12t}\right)$	0
$90e^{0.12t} - 90e^{0.12t}$	0
0	0 TRUE

31. a) The solution to the differential equation $\frac{dT}{dt} = 0.015T$ is $T = T_0e^{0.015t}$.

Find the particular solution by substituting the initial conditions into the general solution and solving for T_0 as follows

$$T = T_0e^{0.015t}$$
$$(50) = T_0e^{0.015(0)}$$
$$50 = T_0.$$

The particular solution is $T = 50e^{0.015t}$.

Copyright © 2014 Pearson Education, Inc. Publishing as Addison-Wesley.

b) To verify this solution, we find $\frac{dT}{dt}$ and we substitute into the equation.

$$\frac{dT}{dt}=\frac{d}{dt}\left(50e^{0.015t}\right)$$
$$\frac{dT}{dt}=50(0.015)e^{0.015t}=0.75e^{0.015t}.$$

$$\frac{dT}{dt}-0.015T\overset{?}{=}0$$

$0.75e^{0.015t}-0.015\left(50e^{0.015t}\right)$	0
$0.75e^{0.015t}-0.75e^{0.015t}$	0
0	0 TRUE

33. a) The solution to the differential equation $\frac{dM}{dt}=M$ is $M=M_0e^t$.
Find the particular solution by substituting the initial conditions into the general solution and solving for M_0 as follows

$$M=M_0e^t$$
$$(6)=M_0e^{(0)}$$
$$6=M_0.$$

The particular solution is $M=6e^t$.

b) To verify this solution, We find $\frac{dM}{dt}$ and we substitute into the equation.

$$\frac{dM}{dt}=\frac{d}{dt}\left(6e^t\right)$$
$$\frac{dM}{dt}=6e^t.$$

$$\frac{dM}{dt}-M\overset{?}{=}0$$

$6e^t-6e^t$	0
0	0 TRUE

35. a) The interest rate of 3.75% implies that $k=0.0375$. Thus, the differential equation that represents the value of the account is

$$\frac{dA}{dt}=0.0375A.$$

b) The general solution to the differential equation is

$$A=A_0e^{0.0375t}.$$

The solution is continued at the top of the next column.

Find the particular solution by substituting the initial conditions into the general solution and solving for A_0 as follows

$$A=A_0e^{0.0375t}$$
$$(500)=A_0e^{0.0375(0)}$$
$$500=A_0.$$

Therefore, the particular solution is

$$A(t)=500e^{0.0375t}.$$

c) Evaluating the function, we have

$$A(5)=500e^{0.0375(5)}=603.12$$

After 5 years, the account will have \$603.12.
To find $A'(5)$ we use the differential equation.

$$A'(5)=0.0375\cdot A(5)=0.0375\cdot 603.12=22.62$$

After 5 years, the account will be growing at \$22.62 per year.

d) Substituting from part (c) we have

$$\frac{A'(5)}{A(5)}=\frac{22.62}{603.12}=0.0375.$$

This is the continuous growth rate.

37. a) The interest rate of 2.8 % implies that $k=0.028$. Thus, the differential equation that represents the value of the account is

$$\frac{dA}{dt}=0.028A.$$

The particular solution in terms of A_0 to the differential equation is

$$A=A_0e^{0.028t}.$$

b) Evaluating the function, we have

$$A(4)=A_0e^{0.028(4)}=1.1185\cdot A_0$$

To find $A'(4)$ we use the differential equation.

$$A'(4)=0.028\cdot A(4)$$
$$=0.028\cdot 1.1185\cdot A_0=0.0313\cdot A_0$$

c) Substituting from part (c) we have

$$\frac{A'(4)}{A(4)}=\frac{0.0313\cdot A_0}{1.1185\cdot A_0}=0.028.$$

This is the continuous growth rate.

d) In part (c) we see that the initial value is divided by itself $\frac{A_0}{A_0}=1$. Therefore, the initial quantity has no effect on the continuous growth rate.

Copyright © 2014 Pearson Education, Inc. Publishing as Addison-Wesley.

39. a) Let $t=0$ correspond to 2002. The continuous growth rate of 1.75% per year implies that $k=0.0175$. The differential equation that represents the population of New River after t years is

$$\frac{dP}{dt}=0.0175P.$$

b) The general solution to the differential equation is $P(t)=P_0e^{0.0175t}$. Since the population in 2002 was 17,000, the initial population $P_0=17,000$. Thus, the particular solution to the differential equation is

$$P(t)=17,000e^{0.0175t}.$$

c) Evaluating the function, we have

$$P(10)=17,000e^{0.0175(10)}=20,251.$$

After 10 years, New River's population will be 20,251.

To find $P'(10)$ we use the differential equation.

$$P'(10)=0.0175\cdot P(10)=0.0175\cdot 21,251=354.4$$

After 10 years, the population is increasing by 354.4 people per year.

d) Substituting from part (c) we have

$$\frac{P'(t)}{P(t)}=\frac{354.4}{20,251}=0.0175.$$

This represents the continuous growth rate.

41. a) Let $t=0$ correspond to 1859. When the population was estimated to be 8900, the growth rate was about 2630 rabbits per year. Using this information we have:

$$\frac{dP}{dt}=kP$$

$$2630=k\cdot 8900$$

$$\frac{2630}{8900}=k$$

$$0.296=k$$

The differential equation that represents the population of rabbits after t years is

$$\frac{dP}{dt}=0.296P.$$

The general solution to the differential equation is $P(t)=P_0e^{0.296t}$. Since the initial population in 1859 was 24 rabbits, the particular solution to the differential equation is

$$P(t)=24e^{0.296t}.$$

b) Evaluating the function, we have

$$P(41)=24e^{0.296(41)}=4,475,165.$$

After 41 years, the rabbit population had grown to approximately 4,475,165 rabbits.

To find $P'(41)$ we use the differential equation.

$$P'(41)=0.296\cdot P(41)$$
$$=0.296\cdot 4,475,165$$
$$=1,324,649.$$

After 41 years, the population is increasing by 1,324,649 rabbits per year.

c) Substituting from part (b) we have

$$\frac{P'(41)}{P(41)}=\frac{0.296P(41)}{P(41)}=0.296.$$

This represents the continuous growth rate.

43. a) Since $k=0.0325$ the continuous growth rate is 3.25% per year.

b) The general solution to the differential equation is $A(t)=A_0e^{0.0325t}$. Plugging in the given condition, we have:

$$A(1)=2582.58$$
$$A_0e^{0.0325(1)}=2582.58$$
$$A_0\cdot 1.033=2582.58$$
$$A_0=\frac{2582.58}{1.033}$$
$$A_0\approx 2500.$$

The particular solution to the differential equation is

$$A(t)=2500e^{0.0325t}.$$

c) From the particular solution, John initially deposited $2500.

45. ✎

Copyright © 2014 Pearson Education, Inc. Publishing as Addison-Wesley.

47. a) Using a calculator, we enter the given information

L1	L2	L3 2
1	4467.9	------
3	4937.8	

L2(3) =

Next we use the regression feature to find the exponential function.

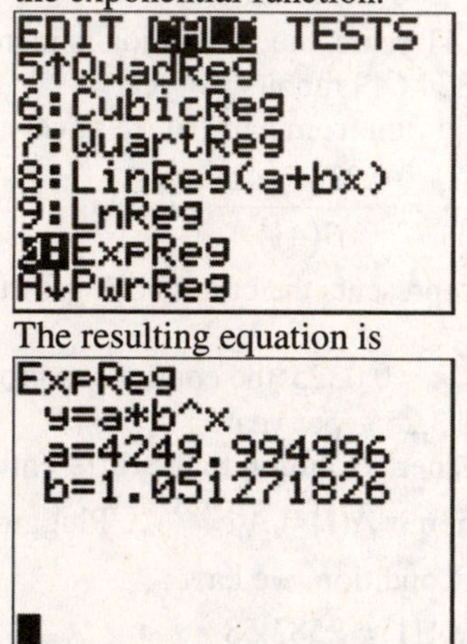

The exponential model that fits is $y = 4250(1.0127)^t$. Using the definition of the exponential, we have $(1.05127)^t = e^{\ln(1.05127)\cdot t} = e^{0.05t}$.

Substituting, we have the model $y = 4250e^{0.05t}$.

b) From this model, we see that $A_0 = 4250$ and that $k = 0.05$. Therefore, the differential equation that models this situation is $\frac{dA}{dt} = 0.05A$ and the initial condition is $A_0 = 4250$.

Copyright © 2014 Pearson Education, Inc. Publishing as Addison-Wesley.

Exercise Set 8.2

1. Solve by separating variables.

$$\frac{dy}{dx} = 4x^3 y$$
$$\frac{dy}{y} = 4x^3 dx$$
$$\int \frac{dy}{y} = \int 4x^3 dx$$
$$\ln|y| = x^4 + C$$
$$e^{\ln|y|} = e^{x^4 + C}$$
$$|y| = e^{x^4} e^C$$
$$y = \pm e^C e^{x^4}$$
$$y = Ce^{x^4}. \qquad (C = \pm e^C)$$

3. Solve by separating variables.

$$3y^2 \frac{dy}{dx} = 8x$$
$$3y^2 dy = 8x dx$$
$$\int 3y^2 dy = \int 8x dx$$
$$y^3 + C_1 = 4x^2 + C_2$$
$$y^3 = 4x^2 + C_2 - C_1$$
$$y = \sqrt[3]{4x^2 + C}. \qquad (C = C_2 - C_1)$$

5. Solve by separating variables.

$$\frac{dy}{dx} = \frac{2x}{y}$$
$$y dy = 2x dx$$
$$\int y dy = \int 2x dx$$
$$\tfrac{1}{2} y^2 + C_1 = x^2 + C_2$$
$$\tfrac{1}{2} y^2 = x^2 + C_2 - C_1$$
$$y^2 = 2x^2 + 2(C_2 - C_1)$$
$$y = \pm\sqrt{2x^2 + C}. \qquad (C = 2(C_2 - C_1))$$

7. Solve by separating variables.

$$\frac{dy}{dx} = \frac{6}{y}$$
$$y dy = 6 dx$$
$$\int y dy = \int 6 dx$$
$$\tfrac{1}{2} y^2 + C_1 = 6x + C_2$$
$$\tfrac{1}{2} y^2 = 6x + C_2 - C_1$$
$$y^2 = 12x + 2(C_2 - C_1)$$
$$y = \pm\sqrt{12x + C}. \qquad (C = 2(C_2 - C_1))$$

9. First we use separation of variables to find a general solution.

$$y' = 3x + xy$$
$$\frac{dy}{dx} = x(3 + y)$$
$$\frac{dy}{3 + y} = x dx$$
$$\int \frac{dy}{3 + y} = \int x dx$$
$$\ln|3 + y| = \tfrac{1}{2} x^2 + C$$
$$e^{\ln|y+3|} = e^{\frac{1}{2}x^2 + C}$$
$$|y + 3| = e^{\frac{1}{2}x^2} e^C$$
$$y + 3 = \pm e^C e^{\frac{1}{2}x^2}$$
$$y = Ce^{\frac{1}{2}x^2} - 3. \qquad (C = \pm e^C)$$

Now we will plug in the initial condition of $y = 5$ when $x = 0$.

$$y = Ce^{\frac{1}{2}x^2} - 3$$
$$5 = Ce^{\frac{1}{2}(0)^2} - 3$$
$$8 = C$$

Therefore, the particular solution is

$$y = 8e^{\frac{1}{2}x^2} - 3.$$

Copyright © 2014 Pearson Education, Inc. Publishing as Addison-Wesley.

11. Solve by separating variables.

$$y' = 5y^{-2}$$
$$y^2 \frac{dy}{dx} = 5$$
$$y^2 dy = 5dx$$
$$\int y^2 dy = \int 5dx$$
$$\tfrac{1}{3}y^3 + C_1 = 5x + C_2$$
$$\tfrac{1}{3}y^3 = 5x + C_2 - C_1$$
$$y^3 = 15x + 3(C_2 - C_1)$$
$$y = \sqrt[3]{15x + C}. \qquad (C = 3(C_2 - C_1))$$

Now we will plug in the initial condition of $y = 3$ when $x = 2$.

$$y = \sqrt[3]{15x + C}$$
$$3 = \sqrt[3]{15(2) + C}$$
$$3 = \sqrt[3]{30 + C}$$
$$27 = 30 + C$$
$$-3 = C.$$

Therefore, the particular solution is $y = \sqrt[3]{15x - 3}$.

13. a) The differential equation that models the situation is

$$\frac{dy}{dx} = y^2. \qquad (k = 1)$$

b) The general solution to this differential equation is determined as follows

$$\frac{dy}{dx} = y^2 \qquad (k = 1)$$
$$\frac{dy}{y^2} = 1dx$$
$$y^{-2} dy = dx$$
$$\int y^{-2} dy = \int dx$$
$$-1y^{-1} + C_1 = x + C_2$$
$$-\frac{1}{y} = x + C_2 - C_1$$
$$\frac{1}{y} = -x + C \qquad (C = C_2 - C_1)$$
$$y = \frac{1}{C - x}.$$

15. a) The differential equation that models the situation is

$$\frac{dy}{dx} = \frac{1}{y^3}. \qquad (k = 1)$$

b) The general solution to this differential equation is determined as follows

$$\frac{dy}{dx} = \frac{1}{y^3} \qquad (k = 1)$$
$$y^3 dy = dx$$
$$\int y^3 dy = \int dx$$
$$\tfrac{1}{4}y^4 + C_1 = x + C_2$$
$$\tfrac{1}{4}y^4 = x + C_2 - C_1$$
$$y^4 = 4x + C \qquad (C = C_2 - C_1)$$
$$y = \pm\sqrt[4]{4x + C}.$$

17. a) The differential equation that models the situation is

$$\frac{dy}{dx} = xy. \qquad (k = 1)$$

With initial condition $y(2) = 3$.

b) The general solution to this differential equation is determined as follows

$$\frac{dy}{dx} = xy \qquad (k = 1)$$
$$\frac{dy}{y} = xdx$$
$$\int \frac{dy}{y} = \int xdx$$
$$\ln|y| + C_1 = \tfrac{1}{2}x^2 + C_2$$
$$\ln|y| = \tfrac{1}{2}x^2 + C_2 - C_1$$
$$|y| = e^{\frac{1}{2}x^2 + C}$$
$$y = \pm e^C \cdot e^{\frac{1}{2}x^2}$$
$$y = Ce^{\frac{1}{2}x^2} \qquad (C = \pm e^C)$$

Substituting the initial condition we have

$$y = Ce^{\frac{1}{2}x^2}$$
$$3 = Ce^{\frac{1}{2}(2)^2}$$
$$3 = C \cdot 7.389$$
$$\frac{3}{7.389} = C$$
$$0.406 = C$$

Therefore, the particular solution is

$$y = 0.406e^{\frac{1}{2}x^2}.$$

Copyright © 2014 Pearson Education, Inc. Publishing as Addison-Wesley.

19. a) By separation of variables

$$\frac{dI}{dt} = hkI$$
$$\frac{dI}{I} = hkdt$$
$$\int \frac{dI}{I} = \int hkdt$$
$$\ln|I| + C_1 = hkt + C_2$$
$$\ln|I| = hkt + C \qquad (C = C_2 - C_1)$$
$$I = \pm e^{hkt+C}$$
$$I = \pm e^{hkt} \cdot e^{C}$$
$$I = Ce^{kht}. \qquad \left(C = \pm e^{C}\right)$$

b) Substituting the initial condition

$$I = Ce^{hkt}$$
$$I_0 = Ce^{hk(0)}$$
$$I_0 = C$$

Therefore, the solution is

$$I = I_0 e^{hkt}.$$

21. Substituting the given information into the differential equation we have

$$\frac{dV}{dt} = k(24.81 - V).$$

Solve the differential equation using separation of variables.

$$\frac{dV}{dt} = k(24.81 - V)$$
$$\frac{dV}{24.81 - V} = kdt$$
$$\int \frac{dV}{24.81 - V} = \int kdt$$
$$-\ln|24.81 - V| + C_1 = kt + C_2$$
$$-\ln|24.81 - V| = kt + C$$
$$\ln|24.81 - V| = -kt + C$$
$$24.81 - V = \pm e^{-kt+C}$$
$$24.81 - V = \pm e^{-kt} e^{C}$$
$$-V = Ce^{-kt} - 24.81$$
$$V = -Ce^{-kt} + 24.81.$$

Substituting the initial condition we have $V(0) = 20$

$$20 = -Ce^{-k(0)} + 24.81$$
$$-4.81 = -C$$

Therefore, the particular solution is

$$V = -4.81e^{kt} + 24.81.$$

23. $E(x) = \frac{4}{x}; \quad q(4) = 2.$

$$E(x) = \frac{4}{x}$$
$$\frac{-x}{q} \cdot \frac{dq}{dx} = \frac{4}{x}$$
$$\frac{dq}{q} = -\frac{4}{x^2} dx$$
$$\int \frac{dq}{q} = -4\int x^{-2} dx$$
$$\ln q + C_{11} = -4\left(-x^{-1}\right) + C_2$$
$$\ln q = \frac{4}{x} + C$$
$$q = e^{\frac{4}{x}+C}$$
$$q = e^{\frac{4}{x}} e^{C}$$
$$q = Ce^{\frac{4}{x}}.$$

Substituting the initial condition, we have

$$q = Ce^{\frac{4}{x}}$$
$$2 = Ce^{\frac{4}{4}}$$
$$\frac{2}{e} = C$$
$$2e^{-1} = C$$

Therefore, the demand function is given by

$$q = 2e^{-1} e^{\frac{4}{x}}$$
$$q = 2e^{\frac{4}{x} - 1}.$$

25. $E(x) = 2, \quad$ for all $x > 0$.

$$E(x) = 2$$
$$\frac{-x}{q} \cdot \frac{dq}{dx} = 2$$
$$\frac{dq}{q} = -\frac{2}{x} dx$$
$$\int \frac{dq}{q} = -2\int \frac{1}{x} dx$$
$$\ln q + C_1 = -2(\ln(x)) + C_2$$
$$\ln q = \ln x^{-2} + C$$
$$q = e^{\ln x^{-2} + C}$$
$$q = e^{\ln x^{-2}} e^{C}$$
$$q = Cx^{-2}$$
$$q = \frac{C}{x^2}.$$

Copyright © 2014 Pearson Education, Inc. Publishing as Addison-Wesley.

27. a) By separation of variables, we have

$$\frac{dP}{dt}=kP$$
$$\frac{dP}{P}=kdt$$
$$\int\frac{dP}{P}=\int kdt$$
$$\ln|P|+C_1=kt+C_2$$
$$\ln|P|=kt+C_2-C_1$$
$$\ln|P|=kt+C$$
$$P=\pm e^{kt+C}$$
$$P=Ce^{kt}. \qquad \left(C=\pm e^{C}\right)$$

b) Substitute the initial condition $P(0)=P_0$.

$$P_0=Ce^{k(0)}$$
$$P_0=C$$

Therefore, the particular solution is

$$P=P_0e^{kt}.$$

29. $$\frac{dy}{dx}=5x^4y^2+x^3y^2$$
$$\frac{dy}{dx}=y^2\left(5x^4+x^3\right)$$
$$y^{-2}dy=\left(5x^4+x^3\right)dx$$
$$\int y^{-2}dy=\int\left(5x^4+x^3\right)dx$$
$$-y^{-1}+C_1=x^5+\tfrac{1}{4}x^4+C_2$$
$$-y^{-1}=x^5+\tfrac{1}{4}x^4+C_2-C_1$$
$$y^{-1}=-\frac{4x^5+x^4+C}{4}$$
$$y=-\frac{4}{4x^5+x^4+C}.$$

31. ✎

33. Solve by separating variables.

$$\frac{dy}{dx}=\frac{5}{y}$$
$$ydy=5dx$$
$$\int ydy=\int 5dx$$
$$\tfrac{1}{2}y^2+C_1=5x+C_2$$
$$\tfrac{1}{2}y^2=5x+C_2-C_1$$
$$y^2=10x+2\left(C_2-C_1\right)$$
$$y=\pm\sqrt{10x+C}. \qquad \left(C=2\left(C_2-C_1\right)\right)$$

Thus the three particular solutions are

$$y_1=\pm\sqrt{10x+5} \qquad (C_1=5)$$
$$y_2=\pm\sqrt{10x-200} \qquad (C_2=-200)$$
$$y_2=\pm\sqrt{10x+100} \qquad (C_3=100)$$

Using a graphing utility, we have

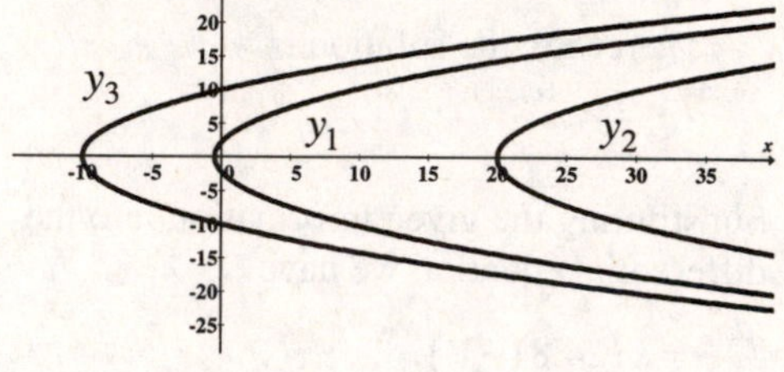

Copyright © 2014 Pearson Education, Inc. Publishing as Addison-Wesley.

Exercise Set 8.3

1. a) The general solution to the inhibited growth model is

$$P(t)=\frac{P_0L}{P_0+\left(L-P_0e^{-Lkt}\right)}.$$

Substituting the given values, we have

$$P(t)=\frac{100(5000)}{100+(5000-100)e^{-0.002(5000)t}}$$

$$P(t)=\frac{500,000}{100+4900e^{-t}}.$$

b) Find t when $P(t)=3500$. Substituting into the logistic growth model, we have

$$P(t)=3500$$
$$\frac{500,000}{100+4900e^{-t}}=3500$$
$$500,000=3500\left(100+4900e^{-t}\right)$$
$$\frac{500,000}{3500}=100+4900e^{-t}$$
$$142.857=100+4900e^{-t}$$
$$42.857=4900e^{-t}$$
$$0.00875=e^{-t}$$
$$\ln(0.00875)=-t$$
$$-\ln(0.00875)=t$$
$$4.739=t$$

c) Since $L=5000$ the point of inflection will occur when $P(t)=2500$. Therefore,

$$P(t)=2500$$
$$\frac{500,000}{100+4900e^{-t}}=2500$$
$$500,000=2500\left(100+4900e^{-t}\right)$$
$$\frac{500,000}{2500}=100+4900e^{-t}$$
$$200=100+4900e^{-t}$$
$$100=4900e^{-t}$$
$$0.0204=e^{-t}$$
$$-\ln(0.0204)=t$$
$$3.892=t$$

The point of inflection is $(3.892,2500)$.

3. a) The general solution to the inhibited growth model is

$$P(t)=\frac{P_0L}{P_0+\left(L-P_0e^{-Lkt}\right)}.$$

Substituting the given values, we have

$$P(t)=\frac{50(1500)}{50+(1500-50)e^{-0.0002(1500)t}}$$

$$P(t)=\frac{75,000}{50+1450e^{-0.3t}}.$$

b) Find t when $P(t)=1350$. Substituting into the logistic growth model, we have

$$P(t)=1350$$
$$\frac{75,000}{50+1450e^{-0.3t}}=1350$$
$$75,000=1350\left(50+1450e^{-0.3t}\right)$$
$$\frac{75,000}{1350}=50+1450e^{-0.3t}$$
$$55.5556=50+1450e^{-0.3t}$$
$$5.5556=1450e^{-0.3t}$$
$$0.0038314=e^{-0.3t}$$
$$\ln(0.0038314)=-0.3t$$
$$\frac{\ln(0.0038314)}{-0.3}=t$$
$$18.548=t$$

c) Since $L=1500$ the point of inflection will occur when $P(t)=750$. Therefore,

$$P(t)=750$$
$$\frac{75,000}{50+1450e^{-0.3t}}=750$$
$$75,000=750\left(50+1450e^{-0.3t}\right)$$
$$\frac{75,000}{750}=50+1450e^{-0.3t}$$
$$100=50+1450e^{-0.3t}$$
$$50=1450e^{-0.3t}$$
$$0.034483=e^{-0.3t}$$
$$\ln(0.034483)=-0.3t$$
$$\frac{\ln(0.034483)}{-0.3}=t$$
$$11.224=t$$

The point of inflection is $(11.224,750)$.

Copyright © 2014 Pearson Education, Inc. Publishing as Addison-Wesley.

5. a) Since the point of inflection is $(15,1000)$, the limiting value $L = 2\cdot 1000 = 2000$.

b) When $t = 15$ we know that $P(15) = 1000$, Substituting into the logistic growth model allows us to solve for k as follows

$$P(15) = 1000$$
$$1000 = \frac{200(2000)}{200 + (2000-200)e^{-k(2000)(15)}}$$
$$1000 = \frac{400{,}000}{200 + 1800e^{-30{,}000k}}$$
$$200 + 1800e^{-30{,}000k} = \frac{400{,}000}{1000}$$
$$200 + 1800e^{-30{,}000k} = 400$$
$$1800e^{-30{,}000k} = 200$$
$$e^{-30{,}000k} = \frac{200}{1800}$$
$$-30{,}000k = \ln(0.11111)$$
$$k = \frac{\ln(0.11111)}{-30{,}000}$$
$$k = 0.0000732.$$

Therefore, the particular solution is:

$$P(t) = \frac{400{,}000}{200 + 1800e^{-(2000)(0.0000732)t}}$$
$$P(t) = \frac{400{,}000}{200 + 1800e^{-0.146t}}.$$

c) Find t when $P(t) = 900$. Substituting into the logistic growth model, we have

$$P(t) = 900$$
$$\frac{400{,}000}{200 + 1800e^{-0.146t}} = 900$$
$$400{,}000 = 900\left(200 + 1800e^{-0.146t}\right)$$
$$\frac{400{,}000}{900} = 200 + 1800e^{-0.146t}$$
$$444.444 = 200 + 1800e^{-0.146t}$$
$$244.444 = 1800e^{-0.146t}$$
$$0.1358 = e^{-0.146t}$$
$$\ln(0.1358) = -0.146t$$
$$\frac{\ln(0.1358)}{-0.146} = t$$
$$13.675 = t.$$

The population will reach 900 in approximately 13.7 months.

7. a) Since the point of inflection is $(6,160)$, the limiting value $L = 2\cdot 160 = 320$.

b) When $t = 6$ we know that $P(6) = 160$, Substituting into the logistic growth model allows us to solve for k as follows

$$P(6) = 160$$
$$160 = \frac{60(320)}{60 + (320-60)e^{-k(320)(6)}}$$
$$160 = \frac{19{,}200}{60 + 260e^{-1920k}}$$
$$60 + 260e^{-1920k} = \frac{19{,}200}{160}$$
$$60 + 260e^{-1920k} = 120$$
$$260e^{-1920k} = 60$$
$$e^{-1920k} = \frac{60}{260}$$
$$-1920k = \ln(0.2308)$$
$$k = \frac{\ln(0.2308)}{-1920}$$
$$k = 0.0000764.$$

Therefore, the particular solution is:

$$P(t) = \frac{19{,}200}{60 + 260e^{-(320)(0.000764)t}}$$
$$P(t) = \frac{19{,}200}{60 + 260e^{-0.244t}}.$$

c) Find t when $P(t) = 300$. Substituting into the logistic growth model, we have

$$P(t) = 300$$
$$\frac{19{,}200}{60 + 260e^{-0.244t}} = 300$$
$$19{,}200 = 300\left(60 + 260e^{-0.244t}\right)$$
$$\frac{19{,}200}{300} = 60 + 260e^{-0.244t}$$
$$64 = 60 + 260e^{-0.244t}$$
$$4 = 260e^{-0.244t}$$
$$0.015385 = e^{-0.244t}$$
$$\ln(0.015385) = -0.244t$$
$$\frac{\ln(0.015385)}{-0.244t} = t$$
$$17.108 = t.$$

The population will reach 300 in approximately 17.1 years.

Copyright © 2014 Pearson Education, Inc. Publishing as Addison-Wesley.

9. a) We assume that $P(0)=0$ and are given $L=3000$, and $P(20)=450$. We substitute and solve for k.

$$P(t)=L\left(1-e^{-kt}\right)$$
$$450=3000\left(1-e^{-k(20)}\right)$$
$$\frac{450}{3000}=1-e^{-20k}$$
$$0.15=1-e^{-20k}$$
$$e^{-20k}=1-0.15$$
$$\ln\left(e^{-20k}\right)=\ln(0.85)$$
$$-20k=\ln(0.85)$$
$$k=\frac{\ln(0.85)}{-20}$$
$$k=0.00813.$$

Therefore, the particular solution is $P(t)=3000\left(1-e^{-0.00813t}\right)$.

b) $P(t)=0.9\cdot L=0.9\cdot 3000=2700$.

Solving for t, we have

$$P(t)=2700$$
$$3000\left(1-e^{-0.00813t}\right)=2700$$
$$1-e^{-0.00813t}=\frac{2700}{3000}$$
$$1-e^{-0.00813t}=0.9$$
$$-e^{-0.00813t}=-0.1$$
$$e^{-0.00813t}=0.1$$
$$-0.00813t=\ln(0.1)$$
$$t=\frac{\ln(0.1)}{-0.00813}$$
$$t=283.221.$$

11. a) We assume that $P(0)=0$ and are given $L=80$, and $P(12)=40$. We substitute and solve for k.

$$P(t)=L\left(1-e^{-kt}\right)$$
$$40=80\left(1-e^{-k(12)}\right)$$
$$\frac{40}{80}=1-e^{-k(12)}$$
$$0.5=1-e^{-12k}$$
$$e^{-12k}=1-0.5$$
$$\ln\left(e^{-12k}\right)=\ln(0.5)$$
$$-12k=\ln(0.5)$$
$$k=\frac{\ln(0.5)}{-12}$$
$$k=0.0578.$$

Therefore, the particular solution is $P(t)=80\left(1-e^{-0.0578t}\right)$.

b) $P(t)=0.8\cdot L=0.8\cdot 80=64$.

Solving for t, we have

$$P(t)=64$$
$$80\left(1-e^{-0.0578t}\right)=64$$
$$1-e^{-0.0578t}=\frac{64}{80}$$
$$1-e^{-0.0578t}=0.8$$
$$e^{-0.0578t}=0.2$$
$$-0.0578t=\ln(0.2)$$
$$t=\frac{\ln(0.2)}{-0.0578}$$
$$t=27.85.$$

13. a) The general solution to the inhibited growth model is

$$P(t)=\frac{P_0L}{P_0+(L-P_0)e^{-Lkt}}.$$

Substituting the given values, we have

$$P(t)=\frac{60(4500)}{60+(4500-60)e^{-0.000007(4500)t}}$$
$$P(t)=\frac{270,000}{60+4440e^{-0.0315t}}.$$

b) The limiting population is $L=4500$.

Copyright © 2014 Pearson Education, Inc. Publishing as Addison-Wesley.

c) $P(t) = 0.9 \cdot L = 0.9 \cdot 4500 = 4050.$

Solving for t, we have

$$P(t) = 4050$$
$$4050 = \frac{270{,}000}{60 + 4440e^{-0.0315t}}$$
$$4050\left(60 + 4440e^{-0.0315t}\right) = 270{,}000$$
$$60 + 4440e^{-0.0315t} = \frac{270{,}000}{4050}$$
$$4440e^{-0.0315t} = 66.667 - 60$$
$$e^{-0.0315t} = \frac{6.667}{4440}$$
$$-0.0315t = \ln(0.001502)$$
$$t = \frac{\ln(0.001502)}{-0.0315}$$
$$t = 206.37.$$

The population will reach 90% of the limiting population in 206.4 months .

15. a) Letting $t = 0$ correspond to 1790, the general solution to the inhibited growth model is

$$P(t) = \frac{P_0 L}{P_0 + \left(L - P_0 e^{-Lkt}\right)}.$$

Substituting the given values, we have

$$66 = \frac{17(200)}{17 + (200 - 17)e^{-200(k)17}}$$

We solve for k.

$$66 = \frac{3400}{17 + 183e^{-3400(k)}}$$
$$66\left(17 + 183e^{-3400(k)}\right) = 3400$$
$$17 + 183e^{-3400(k)} = \frac{3400}{66}$$
$$17 + 183e^{-3400(k)} = 51.515$$
$$183e^{-3400k} = 34.515$$
$$e^{-3400k} = 0.1886$$
$$-3400k = \ln(0.1886)$$
$$k = \frac{\ln(0.1886)}{-3400}$$
$$k = 0.000491$$

Therefore, the particular solution is

$$P(t) = \frac{3400}{17 + 183e^{-200(0.000491)t}}$$
$$P(t) = \frac{3400}{17 + 183e^{-0.0982t}}.$$

b) The population will be growing the fastest at the point of inflection. The population at the point of inflection is $P(t) = \frac{L}{2} = \frac{200}{2} = 100.$

Using this value, we solve for t.

$$P(t) = 100$$
$$100 = \frac{3400}{17 + 183e^{-0.0982t}}$$
$$100\left(17 + 183e^{-0.0982t}\right) = 3400$$
$$17 + 183e^{-0.0982t} = 34$$
$$183e^{-0.0982t} = 17$$
$$e^{-0.0982t} = 0.0929$$
$$-0.0982t = \ln(0.0929)$$
$$t = \frac{\ln(0.0929)}{-0.0982}$$
$$t = 24.198$$

The population was growing the fastest 24.198 years after 1790 or in the year 1814-1815.

17. a) Substituting the given information into the general solution and solve for k as follows

$$M(t) = L\left(1 - e^{-kt}\right)$$
$$25 = 200\left(1 - e^{-k(3)}\right)$$
$$0.125 = 1 - e^{-k(3)}$$
$$e^{-3k} = 0.875$$
$$-3k = \ln(0.875)$$
$$k = \frac{\ln(0.875)}{-3}$$
$$k = 0.0445.$$

Therefore, the particular solution is $M(t) = 200\left(1 - e^{-0.0445t}\right).$

b) Solve $M(t) = 90$ for t.

$$120 = 200\left(1 - e^{-0.0445t}\right)$$
$$0.6 = 1 - e^{-0.0445t}$$
$$e^{-0.0445t} = 0.4$$
$$-0.0445t = \ln(0.4)$$
$$t = \frac{\ln(0.4)}{-0.0445}$$
$$t = 20.59.$$

It will take Omar 20.59 hours to learn 120 of the forms.

Copyright © 2014 Pearson Education, Inc. Publishing as Addison-Wesley.

19. a) We use the general solution to the inhibited growth model

$$P(t)=\frac{P_0L}{P_0+\left(L-P_0e^{-Lkt}\right)}.$$

Substituting the given values, we have

$$25=\frac{5(2000)}{5+(2000-5)e^{-2000(k)6}}$$

$$25=\frac{10,000}{5+1995e^{-12,000k}}$$

$$25\left(5+1995e^{-12,000k}\right)=10,000$$

$$5+1995e^{-12,000k}=\frac{10,000}{25}$$

$$5+1995e^{-12,000k}=400$$

$$1995e^{-12,000k}=395$$

$$e^{-12,000k}=\frac{395}{1995}$$

$$-12,000k=\ln(0.19799)$$

$$k=\frac{\ln(0.19799)}{-12,000}$$

$$k=0.000135.$$

Therefore, the particular solution is

$$P(t)=\frac{10,000}{5+1995e^{-2000(0.000135)t}}$$

$$P(t)=\frac{10,000}{5+1995e^{-0.27t}}.$$

b) 80% of the ships population is given by $P(t)=0.8\cdot 2000=1600$.

Solving for t, we have

$$P(t)=1600$$

$$1600=\frac{10,000}{5+1995e^{-0.27t}}$$

$$1600\left(5+1995e^{-0.27t}\right)=10,000$$

$$5+1995e^{-0.27t}=\frac{10,000}{1600}$$

$$5+1995e^{-0.27t}=6.25$$

$$1995e^{-0.27t}=1.25$$

$$e^{-0.27t}=\frac{1.25}{1995}$$

$$-0.27t=\ln(0.0006266)$$

$$t=\frac{\ln(0.0006266)}{-0.27}$$

$$t=27.32.$$

80% of the passengers will be inflicted in 27.32 days.

21. a) We use the general solution to the inhibited growth model

$$P(t)=\frac{P_0L}{P_0+\left(L-P_0e^{-Lkt}\right)}.$$

Substituting the given values, we have

$$500=\frac{20(20,000)}{20+(20,000-20)e^{-20,000(k)12}}$$

$$500=\frac{400,000}{20+19,980e^{-240,000k}}$$

$$500\left(20+19,980e^{-240,000k}\right)=400,000$$

$$20+19,980e^{-240,000k}=800$$

$$19,980e^{-240,000k}=780$$

$$e^{-240,000k}=\frac{780}{19,980}$$

$$-240,000k=\ln(0.03904)$$

$$k=\frac{\ln(0.03904)}{-240,000}$$

$$k=0.0000135.$$

Therefore, the particular solution is

$$P(t)=\frac{400,000}{20+19,980e^{-(20,000)(0.0000135)t}}$$

$$P(t)=\frac{400,000}{20+19,980e^{-0.27t}}.$$

b) 70% of the ships population is given by $P(t)=0.7\cdot 20,000=14,000$.

Solving for t, we have

$$P(t)=14,000$$

$$14,000=\frac{400,000}{20+19,980e^{-0.27t}}$$

$$14,000\left(20+19,980e^{-0.27t}\right)=400,000$$

$$20+19,980e^{-0.27t}=\frac{400,000}{14,000}$$

$$20+19,980e^{-0.27t}=28.5714$$

$$19,980e^{-0.27t}=83.5714$$

$$e^{-27t}=\frac{8.5714}{19,980}$$

$$-0.27t=\ln(0.000429)$$

$$t=\frac{\ln(0.000429)}{-0.27}$$

$$t=28.719.$$

It will take 28.719 weeks for 70% of the population to have heard of the new phone.

Copyright © 2014 Pearson Education, Inc. Publishing as Addison-Wesley.

23. a) We use the general solution to the inhibited growth model

$$P(t)=\frac{P_0L}{P_0+\left(L-P_0e^{-Lkt}\right)}.$$

Substituting the given values, we have

$$50,000=\frac{2000(4,000,000)}{2000+(4,000,000-2000)e^{-4,000,000(k)25}}$$

We solve for k.

$$50,000=\frac{8,000,000,000}{2000+3,998,000e^{-100,000,000k}}$$

$$50,000\left(2000+3,998,000e^{-100,000,000k}\right)=8,000,000,000$$

$$2000+3,998,000e^{-100,000,000k}=\frac{8,000,000,000}{50,000}$$

$$2000+3,998,000e^{-100,000,000k}=160,000$$

$$3,998,000e^{-100,000,000k}=158,000$$

$$e^{-100,000,000k}=\frac{158,000}{3,998,000}$$

$$-100,000,000k=\ln(0.03952)$$

$$k=\frac{\ln(0.03952)}{-240,000}$$

$$k=-0.0000000323.$$

Noting that

$L\cdot k=4,000,000\cdot 0.0000000323=0.129$

Therefore, the particular solution is

$$P(t)=\frac{8,000,000,000}{2000+3,998,000e^{-0.129t}}.$$

b) Sales are increasing the fastest when

$$P(t)=\frac{L}{2}=\frac{4,000,000}{2}=2,000,000.$$

Solving for t, we have:

$$P(t)=2,000,000$$

$$2,000,000=\frac{8,000,000,000}{2000+3,998,000e^{-0.129t}}$$

$$2000+3,998,000e^{-0.129t}=\frac{8,000,000,000}{2,000,000}$$

$$2000+3,998,000e^{-0.129t}=4000$$

$$3,998,000e^{-0.129t}=2000$$

$$e^{-0.129t}=\frac{2000}{3,998,000}$$

$$-0.129t=\ln(0.00050025)$$

$$t=\frac{\ln(0.00050025)}{-0.129}=58.918.$$

Sales will be increasing the fastest 58.918 weeks after the math-pod is released.

c) Solving for t, we have

$$P(t)=3,000,000$$

$$3,000,000=\frac{8,000,000,000}{2000+3,998,000e^{-0.129t}}$$

$$2000+3,998,000e^{-0.129t}=\frac{8,000,000,000}{3,000,000}$$

$$2000+3,998,000e^{-0.129t}=2666.6667$$

$$3,998,000e^{-0.129t}=666.6667$$

$$e^{-0.129t}=\frac{666.6667}{3,998,000}$$

$$e^{-0.129t}=0.00016675$$

$$-0.129t=\ln(0.00016675)$$

$$t=\frac{\ln(0.00016675)}{-0.129}$$

$$t=67.434.$$

It will take 67.434 weeks for sales to reach 3,000,000 units.

25. a) Using the given information, we determine $L=15-5=10$, and $P(2)=1.50$, and $P_0=5$. Substituting into the general solution, we solve for k

$$P(t)=L\left(1-e^{-kt}\right)$$

$$1.5=10\left(1-e^{-k(2)}\right)$$

$$0.15=1-e^{-2k}$$

$$e^{-2k}=0.85$$

$$-2k=\ln(0.85)$$

$$k=\frac{\ln(0.85)}{-2}$$

$$k=0.0813.$$

Therefore, the particular solution is

$$P(t)=10\left(1-e^{-0.0813t}\right).$$

b) The stock will increase \$7 to reach \$12, therefore, solve $P(t)=7$ for t.

$$7=10\left(1-e^{-0.0813t}\right)$$

$$0.7=1-e^{-0.0813t}$$

$$e^{-0.0813t}=0.3$$

$$-0.0813t=\ln(0.3)$$

$$t=\frac{\ln(0.3)}{-0.0813}$$

$$t=14.809.$$

A share of the stock price will reach \$12 in approximately 14.809 months.

Copyright © 2014 Pearson Education, Inc. Publishing as Addison-Wesley.

27. $S(t)=\dfrac{50}{1+2.85e^{-0.022t}}$

a) In 1836, $t=1836-1776=60$. Therefore,

$$S(60)=\frac{50}{1+2.85e^{-0.022(60)}}=28.3875.$$

There were approximately 28 states in 1836.

b) In 1876, $t=1876-1776=100$. Therefore,

$$S(100)=\frac{50}{1+2.85e^{-0.022(100)}}=38.0001.$$

There were approximately 38 states in 1876.

c) In 1912, $t=1912-1776=136$. Therefore,

$$S(136)=\frac{50}{1+2.85e^{-0.022(136)}}=43.74.$$

There were approximately 44 states in 1912.

d) The limiting number of states is $L=50$.

e) Answers will vary. The model estimates that there are 49.7 states in the year 2050. So it appears to be reasonable, however, since the limiting factor is 50 states a new model might be considered as well.

29. a) Use partial fraction decomposition to rewrite the fraction.

$$\frac{1}{P(L-P)}=\frac{A}{P}+\frac{B}{L-P}$$

$$\frac{1}{P(L-P)}=\frac{A(L-P)+BP}{P(L-P)}$$

$$\frac{1}{P(L-P)}=\frac{AL+(B-A)P}{P(L-P)}$$

Equating coefficients we have

$$AL=1$$

$$A=\frac{1}{L}$$

and

$$B-A=0$$

$$B-\frac{1}{L}=0$$

$$B=\frac{1}{L}.$$

Therefore,

$$\frac{1}{P(L-P)}=\frac{A}{P}+\frac{B}{L-P}$$

$$\frac{1}{P(L-P)}=\frac{\frac{1}{L}}{P}+\frac{\frac{1}{L}}{L-P}=\frac{1}{L}\left(\frac{1}{P}+\frac{1}{L-P}\right).$$

b) Integrating both sides of the equation we have

$$\begin{aligned}\int\frac{1}{P(L-P)}dp&=\int\frac{1}{L}\left(\frac{1}{P}+\frac{1}{L-P}\right)dp\\&=\frac{1}{L}\int\frac{1}{P}dp+\frac{1}{L}\int\frac{1}{L-P}dp\\&=\frac{1}{L}\ln(P)+\frac{1}{L}(-\ln(L-P))\\&=\frac{1}{L}\left[\ln(P)-\ln(L-P)\right]\\&=\frac{1}{L}\ln\left(\frac{P}{L-P}\right)\end{aligned}$$

31. a) Using a calculator, we enter the information.

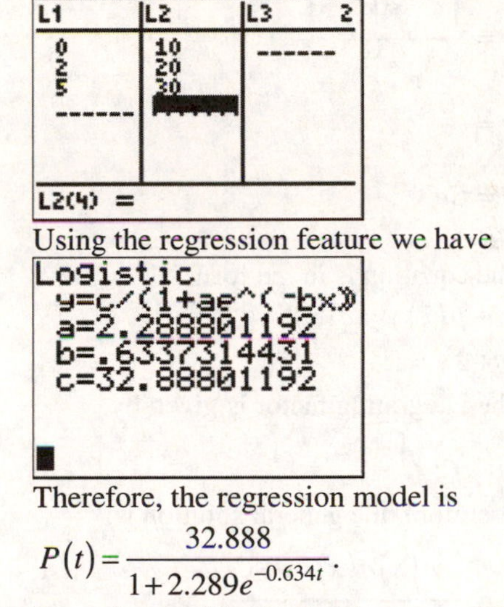

Using the regression feature we have

Therefore, the regression model is

$$P(t)=\frac{32.888}{1+2.289e^{-0.634t}}.$$

b) From the model, the limiting size of the Math Club is about 33 students.

c)

Copyright © 2014 Pearson Education, Inc. Publishing as Addison-Wesley.

Exercise Set 8.4

1. The equation is in the form $y'+p(x)y=q(x)$, we have $y'+5y=0$.
The integrating factor is given by
$m(x)=e^{\int p(x)dx}=e^{\int 5dx}=e^{5x}$.
Therefore, the general solution is

$$y=\frac{\int m(x)q(x)dx+C}{m(x)}$$
$$y=\frac{\int e^{5x}\cdot 0dx+C}{e^{5x}}$$
$$y=\frac{C}{e^{5x}}$$
$$y=Ce^{-5x}.$$

3. The equation is in the form $y'+p(x)y=q(x)$, we have $y'+2y=1$.
The integrating factor is given by
$m(x)=e^{\int p(x)dx}=e^{\int 2dx}=e^{2x}$.
Therefore, the general solution is:

$$y=\frac{\int m(x)q(x)dx+C}{m(x)}$$
$$y=\frac{\int e^{2x}\cdot 1dx+C}{e^{2x}}$$
$$y=\frac{\frac{1}{2}e^{2x}+C}{e^{2x}}$$
$$y=\frac{\frac{1}{2}e^{2x}}{e^{2x}}+\frac{C}{e^{2x}}$$
$$y=Ce^{-2x}+\tfrac{1}{2}.$$

5. Writing the equation in the form $y'+p(x)y=q(x)$, we have $y'+\frac{3}{2}y=\frac{1}{2}$.
The integrating factor is given by
$m(x)=e^{\int p(x)dx}=e^{\int \frac{3}{2}dx}=e^{\frac{3}{2}x}$.
We find the general solution at the top of the next column.

$$y=\frac{\int m(x)q(x)dx+C}{m(x)}$$
$$y=\frac{\int e^{\frac{3}{2}x}\cdot\frac{1}{2}dx+C}{e^{\frac{3}{2}x}}$$
$$y=\frac{\frac{1}{2}\cdot\frac{2}{3}e^{\frac{3}{2}x}+C}{e^{\frac{3}{2}x}}$$
$$y=\frac{\frac{1}{3}e^{\frac{3}{2}x}}{e^{\frac{3}{2}x}}+\frac{C}{e^{\frac{3}{2}x}}$$
$$y=Ce^{-\frac{3}{2}x}+\tfrac{1}{3}.$$

7. The equation is in the form $y'+p(x)y=q(x)$, we have $y'-2xy=x$.
The integrating factor is given by
$m(x)=e^{\int p(x)dx}=e^{\int -2xdx}=e^{-x^2}$.
Therefore, the general solution is

$$y=\frac{\int m(x)q(x)dx+C}{m(x)}$$
$$y=\frac{\int e^{-x^2}\cdot(x)dx+C}{e^{-x^2}}$$ Let $u=-x^2$, $du=-2xdx$
$$y=\frac{\left(-\frac{1}{2}\right)\cdot e^{-x^2}+C}{e^{-x^2}}$$
$$y=\frac{-\frac{1}{2}e^{-x^2}}{e^{-x^2}}+\frac{C}{e^{-x^2}}$$
$$y=Ce^{x^2}-\tfrac{1}{2}.$$

9. The equation is in the form $y'+p(x)y=q(x)$, we have $y'+\frac{2y}{x}=1$.
The integrating factor is given by
$m(x)=e^{\int p(x)dx}=e^{\int \frac{2}{x}dx}=e^{2\ln|x|}=e^{\ln x^2}=x^2$.
We determine the general solution on the next page.

Copyright © 2014 Pearson Education, Inc. Publishing as Addison-Wesley.

Using the information from the previous page, the general solution is:

$$y=\frac{\int m(x)q(x)\,dx+C}{m(x)}$$

$$y=\frac{\int x^2\cdot(1)\,dx+C}{x^2}$$

$$y=\frac{\left(\frac{1}{3}\right)\cdot x^3+C}{x^2}$$

$$y=\frac{\frac{1}{3}x^3}{x^2}+\frac{C}{x^2}$$

$$y=\tfrac{1}{3}x+\tfrac{C}{x^2}.$$

11. The equation is in the form $y'+p(x)y=q(x)$, we have

$y'+\frac{5y}{x}=x.$

The integrating factor is given by

$m(x)=e^{\int\frac{5}{x}dx}=e^{5\ln|x|}=e^{\ln x^5}=x^5.$

Therefore, the general solution is

$$y=\frac{\int m(x)q(x)\,dx+C}{m(x)}$$

$$y=\frac{\int x^5\cdot(x)\,dx+C}{x^5}$$

$$y=\frac{\int x^6dx+C}{x^5}$$

$$y=\frac{\left(\frac{1}{7}\right)\cdot x^7+C}{x^5}$$

$$y=\frac{\frac{1}{7}x^7}{x^5}+\frac{C}{x^5}$$

$$y=\tfrac{1}{7}x^2+\tfrac{C}{x^5}.$$

Using the given information, we have

$y(1)=1$

$$1=\tfrac{1}{7}(1)^2+\frac{C}{(1)^5}$$

$$1-\tfrac{1}{7}=C$$

$$\tfrac{6}{7}=C.$$

The particular solution is

$y=\frac{1}{7}x^2+\frac{6}{7x^5}.$

13. The equation is in the form $y'+p(x)y=q(x)$, we have

$y'+2xy=x.$

The integrating factor is given by

$m(x)=e^{\int p(x)dx}=e^{\int 2x\,dx}=e^{x^2}.$

Therefore, the general solution is

$$y=\frac{\int m(x)q(x)\,dx+C}{m(x)}$$

$$y=\frac{\int e^{x^2}\cdot(x)\,dx+C}{e^{x^2}}$$

Let $u=x^2$
$du=2x\,dx$

$$y=\frac{\left(\frac{1}{2}\right)\cdot e^{x^2}+C}{e^{x^2}}$$

$$y=\frac{\frac{1}{2}e^{x^2}}{e^{x^2}}+\frac{C}{e^{x^2}}$$

$$y=Ce^{-x^2}+\tfrac{1}{2}.$$

Using the given information, we have

$y(0)=4$

$$4=Ce^{-(0)^2}+\tfrac{1}{2}$$

$$4-\tfrac{1}{2}=C$$

$$\tfrac{7}{2}=C.$$

The particular solution is

$y=\frac{7}{2}e^{-x^2}+\frac{1}{2}.$

15. Writing the equation in the form $y'+p(x)y=q(x)$, we have

$y'+\frac{1}{x}y=x^2.$

The integrating factor is given by

$m(x)=e^{\int p(x)dx}=e^{\int\frac{1}{x}dx}=e^{\ln|x|}=x.$

We determine the general solution on the next page.

Copyright © 2014 Pearson Education, Inc. Publishing as Addison-Wesley.

Using the information from the previous page, the general solution is

$$y=\frac{\int m(x)q(x)dx+C}{m(x)}$$

$$y=\frac{\int x\cdot(x^2)dx+C}{x}$$

$$y=\frac{\int x^3dx+C}{x}$$

$$y=\frac{\frac{1}{4}x^4+C}{x}$$

$$y=\frac{\frac{1}{4}x^4}{x}+\frac{C}{x}$$

$$y=\tfrac{1}{4}x^3+\tfrac{C}{x}.$$

Using the given information, we have

$y(2)=5$

$$5=\tfrac{1}{4}(2)^3+\frac{C}{2}$$

$$5=2+\frac{C}{2}$$

$$3=\frac{C}{2}$$

$$6=C.$$

The particular solution is

$$y=\tfrac{1}{4}x^3+\tfrac{6}{x}.$$

17. a) The room temperature is $R=22$. Therefore, the differential equation is

$\frac{dT}{dt}=-k(T-22)$. Newton's Law of Cooling

Writing the equation in the form

$y'+p(t)y=q(t)$, we have

$$T'=-k(T-22)$$

$$T'=-kT+22k$$

$$T'+kT=22k.$$

The integrating factor is given by

$$m(t)=e^{\int p(t)dt}=e^{\int k dt}=e^{kt}.$$

We determine the general solution at the top of the next column.

$$y=\frac{\int m(t)q(t)dt+C}{m(t)}$$

$$y=\frac{\int e^{kt}\cdot(22k)dt+C}{e^{kt}}$$

$$y=\frac{22\int ke^{kt}dt+C}{e^{kt}}$$ Let $u=kt$, $du=kdt$

$$y=\frac{22e^{kt}+C}{e^{kt}}$$

$$y=\frac{22e^{kt}}{e^{kt}}+\frac{C}{e^{kt}}$$

$$y=22+Ce^{-kt}.$$

To determine the constant of integration C, we use the fact that when $t=0$, the temperature of the bowl is $200°C$:

$$200=22+Ce^{-k(0)}$$

$$200=22+C$$

$$178=C.$$

We now have $T(t)=22+178e^{-kt}$. To determine k, we use the fact that the temperature of the bowl is $176°C$ after 45 minutes:

$$176=22+178e^{-k(45)}$$

$$154=178e^{-45k}$$

$$\frac{154}{178}=e^{-45k}$$

$$\ln\left(\tfrac{77}{89}\right)=-45k$$

$$\frac{\ln\left(\frac{77}{89}\right)}{-45}=k$$

$$0.00322\approx k.$$

Therefore, the particular solution is

$$T(t)=22+178e^{-0.00322t}.$$

b) To determine when the bowl reaches $100°C$ we substitute into the particular solution and solve for t:

$$100=22+178e^{-0.00322t}$$

$$78=178e^{-0.00322t}$$

$$\frac{78}{178}=e^{-0.00322t}$$

$$\frac{\ln\left(\frac{78}{178}\right)}{-0.00322}=t$$

$$256.234\approx t$$

The solution is continued on the next page.

Copyright © 2014 Pearson Education, Inc. Publishing as Addison-Wesley.

From the equation on the previous page, we conclude the bowl will reach a temperature of 100°C approximately 256.2 minutes after it is removed from the kiln.

19. a) Let $A(t)$ represent the amount of salt, in pounds, in the take after t minutes. We have the initial condition $A(0)=100$. Note that $\frac{dA}{dt}$ represents the rate of change of salt in the tank and can be represented by $\frac{dA}{dt}$ = Rate in − Rate Out.

The rate that salt enters the tank is

$$\text{Rate in} = \left[\frac{2 \text{ lb salt}}{1 \text{ gal}}\right]\cdot\left[\frac{3 \text{ gal}}{1 \text{ min}}\right] = 6 \ \frac{\text{lbs of salt}}{\text{min}}.$$

The rate that salt exits the tank is

$$\text{Rate out} = \left[\frac{A(t) \text{ lb salt}}{500 \text{ gal}}\right]\cdot\left[\frac{3 \text{ gal}}{1 \text{ min}}\right]$$

$$= \tfrac{3}{500}A(t) \ \frac{\text{lbs of salt}}{\text{min}}.$$

Therefore the rate of change of the amount of salt in the tank is

$\frac{dA}{dt}$ = Rate in − Rate Out

$A'(t) = 6 - \frac{3}{500}A(t).$

Writing the equation in the form $y' + p(t)y = q(t)$, we have

$A'(t) + \frac{3}{500}A(t) = 6$

The integrating factor is given by

$$m(t) = e^{\int p(t)dt} = e^{\int \frac{3}{500}dt} = e^{\frac{3}{500}t} = e^{0.006t}.$$

The solution is continued at the top of the next column.

Using the integrating factor from the previous column, the general solution is

$$A(t) = \frac{\int m(t)q(t)dt + C}{m(t)}$$

$$A(t) = \frac{\int e^{0.006t}\cdot(6)dt + C}{e^{0.006t}}$$

$$A(t) = \frac{\int 6e^{0.006t}dt + C}{e^{0.006t}} \qquad \text{Let } u = 0.006t,\ du = 0.006dt$$

$$A(t) = \frac{1000e^{0.006t} + C}{e^{0.006t}}$$

$$A(t) = \frac{1000e^{0.006t}}{e^{0.006t}} + \frac{C}{e^{0.006t}}$$

$$A(t) = 1000 + Ce^{-0.006t}.$$

Using the initial condition $A(0)=100$, we determine the constant of integration.

$100 = 1000 + Ce^{-0.006(0)}$

$-900 = C.$

Therefore, the particular solution is

$A(t) = 1000 - 900e^{-0.006t}.$

b) After 2 hours, $t = 120$. Evaluating the function yields

$A(120) = 1000 - 900e^{-0.006(120)}$

$A(120) \approx 561.9.$

After 2 hours, the brine solution will have approximately 562 pounds of salt.

21. a) Writing the equation in the form $y' + p(t)y = q(t)$, we have

$$V'(t) = -0.2(V(t) - 12{,}000)$$

$$V'(t) = -0.2V(t) + 2400$$

$$V'(t) + 0.2V(t) = 2400.$$

The integrating factor is given by

$$m(t) = e^{\int p(t)dt} = e^{\int 0.2dt} = e^{0.2t}.$$

The solution is continued on the next page.

Copyright © 2014 Pearson Education, Inc. Publishing as Addison-Wesley.

Using the information from the previous page, the general solution is

$$V(t)=\frac{\int m(t)q(t)dt+C}{m(t)}$$

$$V(t)=\frac{\int e^{0.2t}\cdot(2400)dt+C}{e^{0.2t}}$$

$$V(t)=\frac{\int 2400e^{0.2t}dt+C}{e^{0.2t}}$$ Let $u=0.2t$, $du=0.2dt$

$$V(t)=\frac{12{,}000e^{0.2t}+C}{e^{0.2t}}$$

$$V(t)=\frac{12{,}000e^{0.2t}}{e^{0.2t}}+\frac{C}{e^{0.2t}}$$

$$V(t)=12{,}000+Ce^{-0.2t}.$$

Using the initial condition $V(0)=22{,}500$, we determine the constant of integration.

$$22{,}500=12{,}000+Ce^{-0.2(0)}$$
$$22{,}500=12{,}000+C$$
$$10{,}500=C.$$

Therefore, the particular solution is

$$V(t)=12{,}000-10{,}500e^{-0.2t}.$$

b) Set $V(t)=14{,}000$ and solve for t:

$$14{,}000=12{,}000-10{,}500e^{-0.2t}$$
$$2000=10{,}500e^{-0.2t}$$
$$\frac{2000}{10{,}500}=e^{-0.2t}$$
$$\ln(0.19048)=-0.2t$$
$$\frac{\ln(0.19048)}{-0.2}=t$$
$$8.3\approx t$$

Peggy's convertible will be worth \$14,000 in about 8.3 years.

c) $12,000 represents the salvage value of the car.

23. a) Let $S(t)$ represent the population of senior citizens remaining in the city after t years. We have the initial condition $S(0)=15{,}000$. Note that $\frac{dS}{dt}$ represents the rate of change in the population of senior citizens and can be found by

$\frac{dS}{dt}$ = Rate in − Rate Out.

The solution is continued at the top of the next column.

The rate that seniors enter the city is

$$\text{Rate in}=\left[\frac{20\text{ seniors}}{50}\right]\cdot\left[\frac{50}{1\text{ year}}\right]=20\ \frac{\text{seniors}}{\text{yr}}.$$

The rate that seniors exit the city is

$$\text{Rate out}=\left[\frac{S(t)\text{ seniors}}{100{,}000}\right]\cdot\left[\frac{50}{1\text{ yr}}\right]$$
$$=\tfrac{1}{2000}S(t)\ \frac{\text{seniors}}{\text{yr}}.$$

Therefore the rate of change of the senior population in the city is

$\frac{dS}{dt}$ = Rate in − Rate Out

$$S'(t)=20-\tfrac{1}{2000}S(t).$$

Writing the equation in the form $y'+p(t)y=q(t)$, we have

$$S'(t)+\tfrac{1}{2000}S(t)=20$$

The integrating factor is given by

$$m(t)=e^{\int p(t)dt}=e^{\int \frac{1}{2000}dt}=e^{\frac{1}{2000}t}=e^{0.0005t}.$$

Therefore, the general solution is

$$S(t)=\frac{\int m(t)q(t)dt+C}{m(t)}$$

$$S(t)=\frac{\int e^{0.0005t}\cdot(20)dt+C}{e^{0.0005t}}$$

$$S(t)=\frac{\int 20e^{0.0005t}dt+C}{e^{0.0005t}}$$ Let $u=0.0005t$, $du=0.0005dt$

$$S(t)=\frac{40{,}000e^{0.0005t}+C}{e^{0.0005t}}$$

$$S(t)=\frac{40{,}000e^{0.0005t}}{e^{0.0005t}}+\frac{C}{e^{0.0005t}}$$

$$S(t)=40{,}000+Ce^{-0.0005t}.$$

Using the initial condition $S(0)=15{,}000$, we determine the constant of integration.

$$15{,}000=40{,}000+Ce^{-0.0005(0)}$$
$$15{,}000=40{,}000+C$$
$$-25{,}000=C.$$

Therefore, the particular solution is

$$S(t)=40{,}000+25{,}000e^{-0.0005t}.$$

b) After 1 year, $t=12$,

$$S(12)=40{,}000-25{,}000e^{-0.0005(12)}$$
$$S(12)=15{,}149.6$$

After 1 year, there will be approximately 15,149 senior citizens.

Copyright © 2014 Pearson Education, Inc. Publishing as Addison-Wesley.

c) Set $S(t)=20{,}000$ and solve for t.

$$20{,}000=40{,}000-25{,}000e^{-0.0005t}$$
$$-20{,}000=-25{,}000e^{-0.0005t}$$
$$\frac{20{,}000}{25{,}000}=e^{-0.0005t}$$
$$\ln\left(\tfrac{4}{5}\right)=-0.0005t$$
$$\frac{\ln\left(\tfrac{4}{5}\right)}{-0.0005}=t$$
$$446.28\approx t$$

It will take about 446 months for the population of senior citizens to reach 20,000.

25. $y'+4y=0.$

a) Assume $y=Ce^{rx}$ and $y'=Cre^{rx}$. Substituting we have:

$$y'+4y=0$$
$$Cre^{rx}+4Ce^{rx}=0$$
$$Ce^{rx}(r+4)=0$$
$$r+4=0$$
$$r=-4.$$

Therefore, the general solution is $y=Ce^{-4x}$.

b) The equation is in the form $y'+p(x)y=q(x)$, we have $y'+4y=0$.
The integrating factor is given by

$$m(x)=e^{\int p(x)dx}=e^{\int 4dx}=e^{4x}.$$

Therefore, the general solution is

$$y=\frac{\int m(x)q(x)dx+C}{m(x)}$$
$$y=\frac{\int e^{4x}\cdot 0dx+C}{e^{4x}}$$
$$y=\frac{C}{e^{4x}}$$
$$y=Ce^{-4x}.$$

c) Answers will vary. In this particular case it appears that part (a) is more efficient.

27. a) $2xydx+x^2dy=0$

From the equation we have the following information:

$$f(x,y)=2xy;\quad g(x,y)=x^2$$
$$f_y(x,y)=2x=g_x(x,y).$$

Therefore the equation is exact and we can use the fact that $F_x=f(x,y)$ and $F_y=g(x,y)$ to determine the general solution.

$$F_x(x,y)=2xy$$
$$\int F_x(x,y)dx=\int 2xydx$$
$$F(x,y)=x^2y$$

and

$$F_y(x,y)=x^2$$
$$\int F_y(x,y)dy=\int x^2dy$$
$$F(x,y)=x^2y.$$

Therefore, the general solution is

$$F(x,y)=C$$
$$x^2y=C.$$

b) Writing the equation in the form $y'+p(x)y=q(x)$, we have

$$2xydx+x^2dy=0$$
$$2xy+x^2\tfrac{dy}{dx}=0 \qquad \text{divide by } x^2.$$
$$\tfrac{dy}{dx}+\tfrac{2}{x}y=0$$

The integrating factor is given by

$$m(x)=e^{\int \frac{2}{x}dx}=e^{2\ln|x|}=e^{\ln x^2}=x^2.$$

Therefore, the general solution is

$$y=\frac{\int m(x)q(x)dx+C}{m(x)}$$
$$y=\frac{\int x^2\cdot(0)dx+C}{x^2}$$
$$y=\frac{\int 0dx+C}{x^2}$$
$$y=\frac{C}{x^2}.$$

c) From part (b) we have

$$y=\frac{C}{x^2}$$
$$x^2y=C$$

Which is equal to what we found in part (a).

Copyright © 2014 Pearson Education, Inc. Publishing as Addison-Wesley.

29. a) $y' = \dfrac{2xy}{-x^2 - 6y}$

$$\frac{dy}{dx} = \frac{2xy}{-x^2 - 6y}$$

$$\left(-x^2 - 6y\right)dy = 2xydx$$

$$-2xydx - \left(x^2 + 6y\right)dy = 0$$

$$2xydx + \left(x^2 + 6y\right)dy = 0.$$

Therefore, we know

$f(x, y) = 2xy;\quad g(x, y) = x^2 + 6y$

$f_y(x, y) = 2x = g_x(x, y).$

Therefore the equation is exact and we can use the fact that

$F_x = f(x, y)$ and $F_y = g(x, y)$ to determine the general solution.

$F_x(x, y) = 2xy$

$\int F_x(x, y)dx = \int 2xydx$

$F(x, y) = x^2 y,$ and

$F_y(x, y) = x^2 + 6y$

$\int F_y(x, y)dy = \int x^2 + 6ydy$

$F(x, y) = x^2 y + 3y^2.$

Therefore, the general solution is

$F(x, y) = C$

$x^2 y + 3y^2 = C.$

b) It is not possible to separate y and x as it is not possible to write the equation in the form $y' + p(x)y = q(x)$.

c) The graph is shown below:

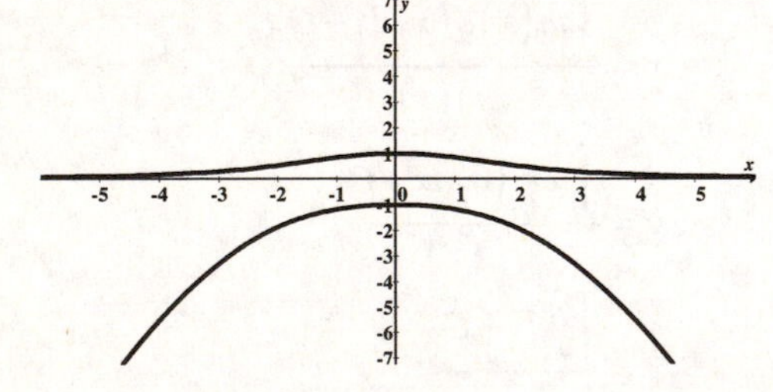

31. The equation is in the form $y' + p(x)y = q(x)$, we have

$y' + y = x.$

The integrating factor is given by

$m(x) = e^{\int p(x)dx} = e^{\int 1dx} = e^x.$

The solution is continued at the top of the next column.

Therefore, the general solution is

$$y = \frac{\int m(x)q(x)dx + C}{m(x)}$$

$$y = \frac{\int e^x \cdot xdx + C}{e^x}.$$

Integrate $\int e^x \cdot xdx$ by parts.

$u = x \qquad dv = e^x dx$

$du = dx \qquad v = e^x$

$$\int e^x \cdot xdx = xe^x - \int e^x dx$$

$$= xe^x - e^x + C.$$

Therefore, the general solution is

$$y = \frac{xe^x - e^x + C}{e^x}$$

$$y = x - 1 + Ce^{-x}.$$

33. Write the equation in the form $y' + p(x)y = q(x)$, as follows

$$\left(x^2 + 1\right)y' + 2xy = 1$$

$$y' + \frac{2x}{x^2 + 1}y = \frac{1}{x^2 + 1}.$$

The integrating factor is given by

$$m(x) = e^{\int p(x)dx} = e^{\int \frac{2x}{x^2+1}dx}.$$

We integrate by u-substitution

$$e^{\int \frac{2x}{x^2+1}dx} = e^{\int \frac{du}{u}} = e^{\ln u} = u.$$

Therefore,

$m(x) = x^2 + 1.$

Therefore, the general solution is

$$y = \frac{\int m(x)q(x)dx + C}{m(x)}$$

$$y = \frac{\int \left(x^2 + 1\right) \cdot \frac{1}{x^2 + 1}dx + C}{x^2 + 1}$$

$$y = \frac{\int 1dx + C}{x^2 + 1}$$

$$y = \frac{x + C}{x^2 + 1}.$$

Copyright © 2014 Pearson Education, Inc. Publishing as Addison-Wesley.

Exercise Set 8.5

1. The auxiliary equation is
$r^2 + r - 20 = 0.$
Solving this equation for r we have
$$(r+5)(r-4) = 0$$
$r+5=0$ or $r-4=0$
$r=-5$ or $r=4.$
Therefore the general solution is
$y = C_1e^{-5x} + C_2e^{4x}.$

3. The auxiliary equation is
$r^2 + 5r - 24 = 0.$
Solving this equation for r we have
$$(r+8)(r-3) = 0$$
$r+8=0$ or $r-3=0$
$r=-8$ or $r=3.$
Therefore the general solution is
$y = C_1e^{-8x} + C_2e^{3x}.$

5. The auxiliary equation is
$r^2 - 25 = 0.$
Solving this equation for r we have
$$(r+5)(r-5) = 0$$
$r+5=0$ or $r-5=0$
$r=-5$ or $r=5.$
Therefore the general solution is
$y = C_1e^{-5x} + C_2e^{5x}.$

7. The auxiliary equation is
$r^2 - 10r = 0.$
Solving this equation for r we have
$$r(r-10) = 0$$
$r=0$ or $r-10=0$
$r=0$ or $r=10.$
Therefore the general solution is
$y = C_1e^{0x} + C_2e^{4x} = C_1 + C_2e^{10x}.$

9. The auxiliary equation is
$2r^2 - 3r - 2 = 0.$
Solving this equation for r we have
$$(2r+1)(r-2) = 0$$
$2r+1=0$ or $r-2=0$
$r=-\frac{1}{2}$ or $r=2.$
Therefore the general solution is
$y = C_1e^{-\frac{1}{2}x} + C_2e^{2x}.$

11. The auxiliary equation is
$r^3 + 3r^2 + 2r = 0.$
Solving this equation for r we have
$$r(r^2+3r+2) = 0$$
$$r(r+2)(r+1) = 0$$
$r=0$ or $r+2=0$ or $r+1=0$
$r=0$ or $r=-2$ or $r=-1.$
Therefore the general solution is
$y = C_1e^{0x} + C_2e^{-2x} + C_3e^{-x}$
$y = C_1 + C_2e^{-2x} + C_3e^{-x}.$

13. The auxiliary equation is
$r^3 - 36r = 0.$
Solving this equation for r we have
$$r(r^2-36) = 0$$
$$r(r+6)(r-6) = 0$$
$r=0$ or $r+6=0$ or $r-6=0$
$r=0$ or $r=-6$ or $r=6.$
Therefore the general solution is
$y = C_1e^{0x} + C_2e^{-6x} + C_3e^{6x}$
$y = C_1 + C_2e^{-6x} + C_3e^{6x}.$

15. The auxiliary equation is
$r^2 + 8r + 16 = 0.$
Solving this equation for r we have
$$(r+4)(r+4) = 0$$
$r+4=0$ or $r+4=0$
$r=-4$ or $r=-4.$
The solution $r=-4$ is a solution of multiplicity 2; therefore, the general solution is
$y = C_1e^{-4x} + C_2xe^{-4x}.$

Copyright © 2014 Pearson Education, Inc. Publishing as Addison-Wesley.

17. The auxiliary equation is
$r^2-16r+64=0.$
Solving this equation for r we have
$(r-8)(r-8)=0$
$r-8=0$ or $r-8=0$
$r=8$ or $r=8.$
The solution $r=8$ is a solution of multiplicity 2; therefore, the general solution is
$y=C_1e^{8x}+C_2xe^{8x}.$

19. The auxiliary equation is
$r^2+4r+3=0.$
Solving this equation for r we have
$(r+3)(r+1)=0$
$r+3=0$ or $r+1=0$
$r=-3$ or $r=-1.$
Therefore, the general solution is
$y=C_1e^{-3x}+C_2e^{-x}.$
To find the particular solution we use the initial conditions $y(0)=1, y'(0)=3.$ The first derivative of the general solution is
$y'=-3C_1e^{-3x}-C_2e^{-x}.$
This gives us a system of two equations and two unknowns.
$y=C_1e^{-3x}+C_2e^{-x}$
$y'=-3C_1e^{-3x}-C_2e^{-x}.$
Substituting the initial conditions into the system of equations gives us
$1=C_1e^{-3(0)}+C_2e^{-(0)}$
$3=-3C_1e^{-3(0)}-C_2e^{-(0)},$
or
$1=C_1+C_2$
$3=-3C_1-C_2.$
Solving the first equation for C_1 gives $C_1=1-C_2$ and substituting into the second equation we have
$3=-3(1-C_2)-C_2$
$3=-3+3C_2-C_2$
$6=2C_2$
$3=C_2.$
Therefore $C_1=1-(3)=-2.$
The particular solution for this differential equation is
$y=-2e^{-3x}+3e^{-x}.$

21. The auxiliary equation is
$r^2-6r=0.$
Solving this equation for r we have
$r(r-6)=0$
$r=0$ or $r-6=0$
$r=0$ or $r=6.$
Therefore, the general solution is
$y=C_1+C_2e^{6x}.$
To find the particular solution we use the initial conditions $y(0)=3, y'(0)=12.$ The first derivative of the general solution is
$y'=6C_2e^{6x}.$
This gives us a system of two equations and two unknowns.
$y=C_1+C_2e^{6x}$
$y'=6C_2e^{6x}.$
Substituting the initial conditions into the system of equations gives us
$3=C_1+C_2e^{6(0)}$
$12=6C_2e^{6(0)},$
or
$3=C_1+C_2$
$12=6C_2.$
Solving the second equation for C_2 gives $C_2=\frac{12}{6}=2$ and substituting into the first equation we have
$3=C_1+C_2$
$3=C_1+2$
$1=C_1.$
The particular solution for this differential equation is
$y=2e^{6x}+1.$

23. The auxiliary equation is
$r^3+3r^2-13r-15=0.$
We graph the auxiliary equation and find the real zeros.
Entering the equation we have

```
Plot1 Plot2 Plot3
\Y1■X^3+3X^2-13X
-15
\Y2=
\Y3=
\Y4=
\Y5=
\Y6=
```

The solution is continued on the next page.

Copyright © 2014 Pearson Education, Inc. Publishing as Addison-Wesley.

We set the window to

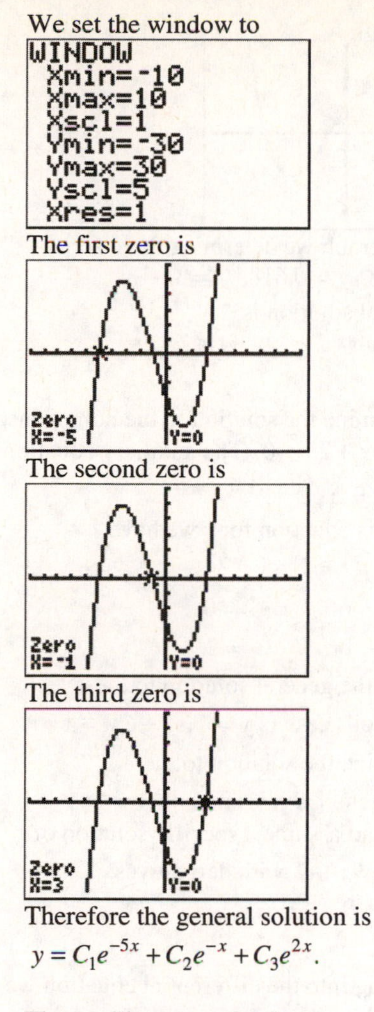

Therefore the general solution is

$y = C_1e^{-5x} + C_2e^{-x} + C_3e^{2x}.$

25. The auxiliary equation is

$r^3 - 3r^2 - 16r + 48 = 0.$

We graph the auxiliary equation and find the real zeros.

Entering the equation we have

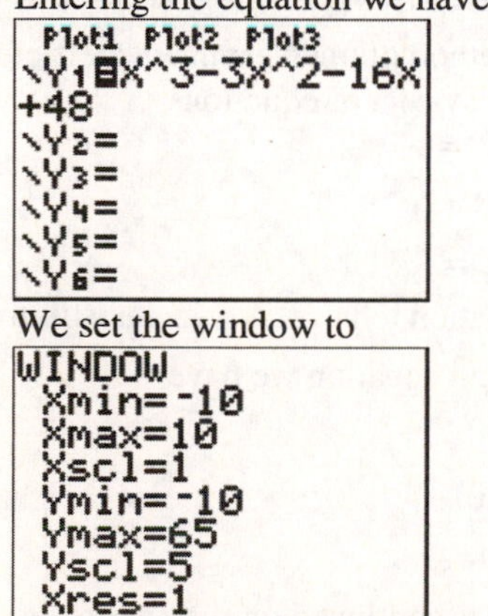

The graph is

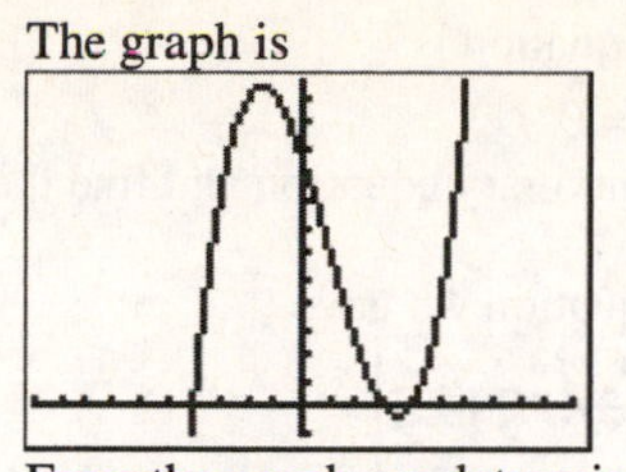

From the graph we determine three zeros:

$r = -4, \ r = 3, \ r = 4.$

The general solution is

$y = C_1e^{-4x} + C_2e^{3x} + C_3e^{4x}.$

27. The auxiliary equation is

$r^3 - 3r - 2 = 0.$

We graph the auxiliary equation and find the real zeros.

Entering the equation we have

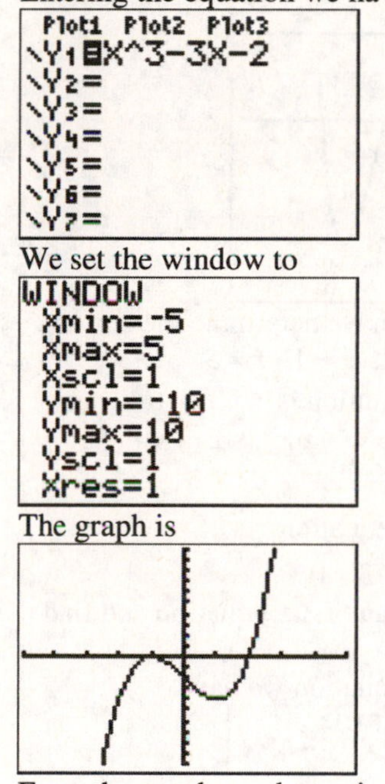

From the graph we determine the zeros: $r = -1, \ r = 2.$ We note that $r = -1$ is a zero of multiplicity 2; therefore, the general solution is

$y = C_1e^{-x} + C_2xe^{-x} + C_3e^{2x}.$

Copyright © 2014 Pearson Education, Inc. Publishing as Addison-Wesley.

29. The auxiliary equation is

$r^4-26r^2+25=0.$

We graph the auxiliary equation and find the real zeros.

Entering the equation we have

```
Plot1 Plot2 Plot3
\Y1■X^4-26X^2+25
\Y2=
\Y3=
\Y4=
\Y5=
\Y6=
```

We set the window to

```
WINDOW
 Xmin=-10
 Xmax=10
 Xscl=1
 Ymin=-200
 Ymax=50
 Yscl=10
 Xres=1
```

The graph is

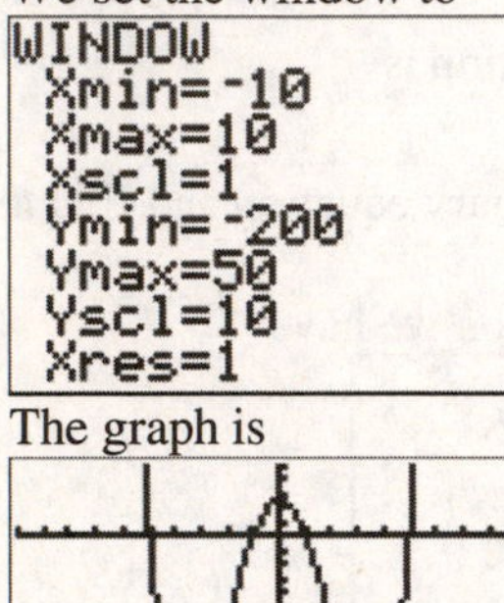

From the graph we determine the zeros:

$r=-5,\ r=-1,\ r=1,\ r=5.$

The general solution is

$y=C_1e^{-5x}+C_2e^{-x}+C_3e^{x}+C_4e^{5x}.$

31. The auxiliary equation is

$r^3-2r^2-4r+3=0.$

We graph the auxiliary equation and find the real zeros.

Entering the equation we have

```
Plot1 Plot2 Plot3
\Y1■X^3-2X^2-4X+
3
\Y2=
\Y3=
\Y4=
\Y5=
\Y6=
```

We set the window to

```
WINDOW
 Xmin=-5
 Xmax=5
 Xscl=1
 Ymin=-10
 Ymax=10
 Yscl=1
 Xres=1
```

The graph is

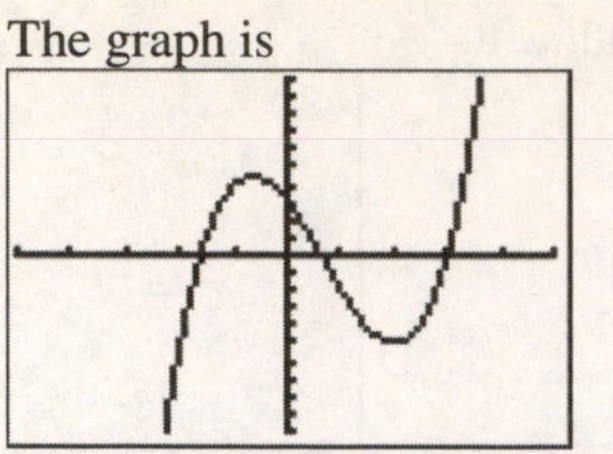

From the graph we determine three zeros:

$r=-1.618,\ r=0.618,\ r=3.$

The general solution is

$y=C_1e^{-1.618x}+C_2e^{0.618x}+C_3e^{3x}.$

33. First determine the solution to the homogeneous case $y''+3y'+2y=0.$ The auxiliary equation is

$r^2-3r-2=0$

Solving this equation for r we have

$$(r-1)(r-2)=0$$

$r-1=0$ or $r-2=0$

$r=1$ or $r=2.$

Therefore the general solution to the homogeneous case is $y=C_1e^x+C_2e^{2x}.$

To determine the solution to the nonhomogeneous case, note that x^2+1 is degree 2 and assume a specific solution of $y=Ax^2+Bx+C$ with derivatives

$y'=2Ax+B$

$y''=2A.$

Substituting into the differential equation we have

$$y''+3y'+2y=x^2+1$$

$$2A+3(2Ax+B)+2\left(Ax^2+Bx+C\right)=x^2+1$$

$$2Ax^2+(6A+2B)x+(2A+3B+2C)=x^2+1$$

Using the method of undetermined coefficients results in the system of equations

$$2A=1$$

$$6A+2B=0$$

$$2A+3B+2C=1$$

The first equation tells us $A=\frac{1}{2}.$ Substituting into the second equation we have

$6\left(\frac{1}{2}\right)+2B=0$

$3+2B=0$

$B=-\frac{3}{2}.$

The solution is continued on the next page.

Copyright © 2014 Pearson Education, Inc. Publishing as Addison-Wesley.

Substituting both results from the previous page into the third equation we have

$$2\left(\tfrac{1}{2}\right)+3\left(-\tfrac{3}{2}\right)+2C=1$$
$$1-\tfrac{9}{2}+2C=1$$
$$2C=\tfrac{9}{2}$$
$$C=\tfrac{9}{4}.$$

Therefore, the specific solution is

$y=\frac{1}{2}x^2-\frac{3}{2}x+\frac{9}{4}.$

The general solution to the differential equation is the sum of the homogenous and the specific solutions:

$y=C_1e^x+C_2e^{2x}+\frac{1}{2}x^2-\frac{3}{2}x+\frac{9}{4}.$

35. First determine the solution to the homogeneous case $y''-36y=0$. The auxiliary equation is

$r^2-36=0$

Solving this equation for r we have

$$(r+6)(r-6)=0$$
$$r+6=0 \quad \text{or} \quad r-6=0$$
$$r=-6 \quad \text{or} \quad r=6.$$

Therefore the general solution to the homogeneous case is $y=C_1e^{-6x}+C_2e^{6x}$.

To determine the solution to the nonhomogeneous case, note that $5x-7$ is linear and assume a specific solution of $y=Ax+B$ with derivatives

$y'=A$

$y''=0.$

Substituting into the differential equation we have

$$y''-36y=5x-7$$
$$0-36(Ax+B)=5x-7$$
$$36Ax-36B=5x-7$$

Using the method of undetermined coefficients results in the system of equations

$-36A=5$

$-36B=-7$

Solving these equations we have

$A=-\frac{5}{36}$

$B=\frac{7}{36}$

Therefore, the specific solution is

$y=-\frac{5}{36}x+\frac{7}{36}.$

The general solution to the differential equation is the sum of the homogenous and the specific solutions:

$y=C_1e^{-6x}+C_2e^{6x}-\frac{5}{36}x+\frac{7}{36}.$

37. First determine the solution to the homogeneous case $y''-9y'=0$. The auxiliary equation is

$r^2-9r=0$

Solving this equation for r we have

$$r(r-9)=0$$
$$r=0 \quad \text{or} \quad r-9=0$$
$$r=0 \quad \text{or} \quad r=9.$$

Therefore the general solution to the homogeneous case is $y=C_1+C_2e^{9x}$.

To determine the solution to the nonhomogeneous case, note that $x+5$ is linear However, we must assume a specific solution of $y=Ax^2+Bx+C$ with derivatives

$y'=2Ax+B$

$y''=2A.$

Substituting into the differential equation yields

$$y''-9y'=x+5$$
$$2A-9(2Ax+B)=x+5$$
$$-18Ax+(2A-9B)=x+5$$

Using the method of undetermined coefficients results in the system of equations

$-18A=1$

$2A-9B=5$

The first equation tells us $A=-\frac{1}{18}$. Substituting into the second equation we have

$$2\left(-\tfrac{1}{18}\right)-9B=5$$
$$-9B=\tfrac{46}{9}$$
$$B=-\tfrac{46}{81}.$$

Therefore, the specific solution is

$y=-\frac{1}{18}x^2-\frac{46}{81}x.$

The general solution to the differential equation is the sum of the homogenous and the specific solutions:

$y=C_1+C_2e^{9x}-\frac{1}{18}x^2-\frac{46}{81}x.$

39. The auxiliary equation is

$r^2+4=0.$

Solving this equation for r we have

$$r^2+4=0$$
$$r^2=-4$$
$$r=\pm\sqrt{-4}$$
$$r=\pm 2i.$$

There are two complex solutions to the auxiliary equation. The general solution is

$y=C_1e^{0x}\sin(2x)+C_2e^{0x}\cos(2x)$

$y=C_1\sin 2x+C_2\cos 2x.$

Copyright © 2014 Pearson Education, Inc. Publishing as Addison-Wesley.

41. The auxiliary equation is
$r^2+2=0.$
Solving this equation for r we have
$$r^2+2=0$$
$$r^2=-2$$
$$r=\pm\sqrt{-2}$$
$$r=\pm i\sqrt{2}.$$
There are two complex solutions to the auxiliary equation. Using Theorem 3, the general solution is
$$y=C_1e^{0x}\sin\left(\sqrt{2}x\right)+C_2e^{0x}\cos\left(\sqrt{2}x\right)$$
$$y=C_1\sin\sqrt{2}x+C_2\cos\sqrt{2}x.$$

43. The auxiliary equation is
$r^2+3r+7=0.$
Solve for r using the quadratic formula
$$r=\frac{-(3)\pm\sqrt{(3)^2-4(1)(7)}}{2(1)}$$
$$r=\frac{-3\pm\sqrt{9-28}}{2}$$
$$r=\frac{-3\pm\sqrt{-19}}{2}$$
$$r=-\tfrac{3}{2}\pm i\tfrac{\sqrt{19}}{2}.$$
There are two complex solutions to the auxiliary equation. Using Theorem 3, the general solution is
$$y=C_1e^{-\frac{3}{2}x}\sin\left(\tfrac{\sqrt{19}}{2}x\right)+C_2e^{-\frac{3}{2}x}\cos\left(\tfrac{\sqrt{19}}{2}x\right).$$

45. The auxiliary equation is
$r^2+16=0.$
Solving this equation for r we have
$$r^2+16=0$$
$$r^2=-16$$
$$r=\pm\sqrt{-16}$$
$$r=\pm 4i.$$
There are two complex solutions to the auxiliary equation. Using Theorem 3, the general solution is
$$y=C_1e^{0x}\sin(4x)+C_2e^{0x}\cos(4x)$$
$$y=C_1\sin 4x+C_2\cos 4x.$$
To find the particular solution we use the initial conditions $y(0)=1, y'(0)=2.$ The first derivative of the general solution is
$$y'=4C_1\cos 4x-4C_2\sin 4x.$$
This gives us a system of two equations and two unknowns.
$$y=C_1\sin 4x+C_2\cos 4x$$
$$y'=4C_1\cos 4x-4C_2\sin 4x.$$
Substituting the initial conditions into the system of equations gives us
$$1=C_1\sin(4\cdot 0)+C_2\cos(4\cdot 0)$$
$$2=4C_1\cos(4\cdot 0)-4C_2\sin(4\cdot 0)$$
or
$$1=0C_1+C_2$$
$$2=4C_1+0C_2.$$
Solving these equations we have $C_2=1$ and
$$2=4C_1$$
$$\tfrac{1}{2}=C_1.$$
The particular solution for this differential equation is
$$y=\tfrac{1}{2}\sin 4x+\cos 4x.$$

47. First determine the solution to the homogeneous case $y''+9y=0.$ The auxiliary equation is
$$r^2+9=0$$
Solving this equation for r we have
$$r^2+9=0$$
$$r^2=-9$$
$$r=\pm\sqrt{-9}$$
$$r=\pm 3i.$$
There are two complex solutions to the auxiliary equation. Using Theorem 3, the general solution to the homogeneous case is
$$y=C_1e^{0x}\sin(3x)+C_2e^{0x}\cos(3x)$$
$$y=C_1\sin 3x+C_2\cos 3x.$$
To determine the solution to the nonhomogeneous case, note that x^2+8x-2 is degree 2 and assume a specific solution of $y=Ax^2+Bx+C$ with derivatives
$$y'=2Ax+B$$
$$y''=2A.$$
Substituting into the differential equation we have
$$y''+9y=x^2+8x-2$$
$$2A+9\left(Ax^2+Bx+C\right)=x^2+8x-2$$
$$9Ax^2+9Bx+(2A+9C)=x^2+8x-2$$
The solution is continued on the next page.

Copyright © 2014 Pearson Education, Inc. Publishing as Addison-Wesley.

Using the information from the previous page, the method of undetermined coefficients results in the system of equations

$$9A = 1$$
$$9B = 8$$
$$2A + 9C = -2$$

The first equation tells us $A = \frac{1}{9}$.

The second equation tells us $B = \frac{8}{9}$.

Substituting A into the third equation we have

$$2\left(\frac{1}{9}\right) + 9C = -2$$
$$9C = -\frac{20}{9}$$
$$C = -\frac{20}{81}.$$

Therefore, the specific solution is

$$y = \frac{1}{9}x^2 + \frac{8}{9}x - \frac{20}{81}.$$

The general solution to the differential equation is the sum of the homogenous and the specific solutions:

$$y = C_1 \sin 3x + C_2 \cos 3x + \frac{1}{9}x^2 + \frac{8}{9}x - \frac{20}{81}.$$

49. First determine the solution to the homogeneous case $y'' + y' + 6y = 0$. The auxiliary equation is

$$r^2 + r + 6 = 0$$

Solve for r using the quadratic formula

$$r = \frac{-(1) \pm \sqrt{(1)^2 - 4(1)(6)}}{2(1)}$$
$$r = \frac{-1 \pm \sqrt{1-24}}{2}$$
$$r = \frac{-1 \pm \sqrt{-23}}{2}$$
$$r = -\frac{1}{2} \pm i\frac{\sqrt{23}}{2}.$$

There are two complex solutions to the auxiliary equation. Using Theorem 3, the general solution to the homogeneous case is

$$y = C_1 e^{-\frac{1}{2}x} \sin\left(\frac{\sqrt{23}}{2}x\right) + C_2 e^{-\frac{1}{2}x} \cos\left(\frac{\sqrt{23}}{2}x\right).$$

To determine the solution to the nonhomogeneous case, note that $3x$ is linear and assume a specific solution of $y = Ax + B$ with derivatives

$$y' = A$$
$$y'' = 0.$$

Substituting into the differential equation we have

$$y'' + y' + 6y = 3x$$
$$0 + A + 6(Ax + B) = 3x$$
$$6Ax + (A + 6B) = 3x$$

Using the method of undetermined coefficients results in the system of equations

$$6A = 3$$
$$A + 6B = 0.$$

The first equation tells us $A = \frac{1}{2}$.

Substituting A into the second equation we have

$$\left(\frac{1}{2}\right) + 6B = 0$$
$$6B = -\frac{1}{2}$$
$$B = -\frac{1}{12}.$$

Therefore, the specific solution is $y = \frac{1}{2}x - \frac{1}{12}$.

The general solution to the differential equation is the sum of the homogenous and the specific solutions:

$$y = C_1 e^{-\frac{1}{2}x} \sin\left(\frac{\sqrt{23}}{2}x\right) + C_2 e^{-\frac{1}{2}x} \cos\left(\frac{\sqrt{23}}{2}x\right) + \frac{1}{2}x - \frac{1}{12}.$$

51. a) The auxiliary equation is

$$50r^2 + 5r - 1 = 0.$$

Solving this equation for r we have

$$(5r+1)(10r-1) = 0$$
$$5r + 1 = 0 \quad \text{or} \quad 10r - 1 = 0$$
$$r = -\frac{1}{5} \quad \text{or} \quad r = \frac{1}{10}.$$

Therefore, the general solution is

$$P(t) = C_1 e^{-\frac{1}{5}t} + C_2 e^{\frac{1}{10}t}$$
$$P(t) = C_1 e^{-0.2t} + C_2 e^{0.1t}.$$

To find the particular solution we use the initial conditions $P(0) = 5, P'(0) = 0.2$. The first derivative of the general solution is

$$P' = -0.2C_1 e^{-0.2t} + 0.1C_2 e^{0.1t}.$$

This gives us a system of two equations and two unknowns.

$$P = C_1 e^{-0.2t} + C_2 e^{0.1t}$$
$$P' = -0.2C_1 e^{-0.2t} + 0.1C_2 e^{0.1t}.$$

Substituting the initial conditions into the system of equations gives us

$$5 = C_1 e^{-0.2(0)} + C_2 e^{0.1(0)}$$
$$0.2 = -0.2C_1 e^{-0.2(0)} + 0.1C_2 e^{0.1(0)},$$

or

$$5 = C_1 + C_2$$
$$0.2 = -0.2C_1 + 0.1C_2.$$

The solution is continued on the next page.

Copyright © 2014 Pearson Education, Inc. Publishing as Addison-Wesley.

Solving the first equation on the previous page for C_1 gives $C_1 = 5 - C_2$ and substituting into the second equation we have

$0.2 = -0.2(5 - C_2) + 0.1C_2$
$0.2 = -1 + 0.2C_2 + 0.1C_2$
$1.2 = 0.3C_2$
$4 = C_2.$

Therefore $C_1 = 5 - (4) = 1.$

The particular solution that models the price of copper is

$P(t) = e^{-0.2t} + 4e^{0.1t}.$

b) Evaluate the function when $t = 6.$

$P(6) = e^{-0.2(6)} + 4e^{0.1(6)} = 7.59.$

The price of copper in the 6th month is about \$7.59.

c) Evaluate the derivative when $t = 6.$

$P'(t) = -0.2e^{-0.2t} + 0.4e^{0.1t}$

$P'(6) = -0.2e^{-0.2(6)} + 0.4e^{0.1(6)} = 0.67$

The price of copper is increasing by about \$0.67 per month or 67 cents per month.

53. Let $y = C_1 f_1(x) + C_2 f_2(x)$. Therefore, the derivatives of y are

$y' = C_1 f'(x) + C_2 f'(x)$
$y'' = C_1 f''(x) + C_2 f''(x).$

Substituting these values into the homogeneous differential equation we have

$a_0(x)y'' + a_1(x)y' + a_2(x)y = 0$

Substituting yields

$a_0(x)[C_1 f_1''(x) + C_2 f_2''(x)]$
$+ a_1(x)[C_1 f_1'(x) + C_2 f_2'(x)]$
$+ a_2(x)[C_1 f_1(x) + C_2 f_2(x)] = 0.$

Simplify the expression on the left as follows

$a_0 C_1 f_1''(x) + a_0 C_2 f_2''(x) + a_1 C_1 f_1(x)$
$+ a_1 C_2 f_2(x) + a_2 C_1 f_1'(x) + a_2 C_2 f_2'(x) = 0.$

Arrange terms and factor common factors

$C_1[a_0 f_1''(x) + a_1 f_1'(x) + a_2 f_1(x)]$
$+ C_2[a_0 f_2''(x) + a_1 f_2'(x) + a_2 f_2(x)] = 0.$
$C_1[0] + C_2[0] = 0$
$0 = 0.$

Therefore, the linear combination of two solutions to the homogeneous linear differential equation is also a solution.

55. a) $y_1 = e^{3x}, y_2 = e^{-7x}.$

If y_1 and y_2 are solutions then their linear combination $y = e^{-7x} + e^{3x}$ is also a solution. This means that solutions to the auxiliary equation must be $r = -7$ and $r = 3$. Finding an equation with these solutions, the auxiliary equation is

$(r + 7)(r - 3) = 0$
$r^2 + 7r - 3r - 21 = 0$
$r^2 + 4r - 21 = 0$

Therefore, a differential equation that has solutions y_1 and y_2 is

$y'' + 4y' - 21y = 0.$

b) ✎

57. The auxiliary equation is

$r^4 + r^3 - 7r^2 - r + 6 = 0.$

We graph the auxiliary equation and find the real zeros.

The graph is

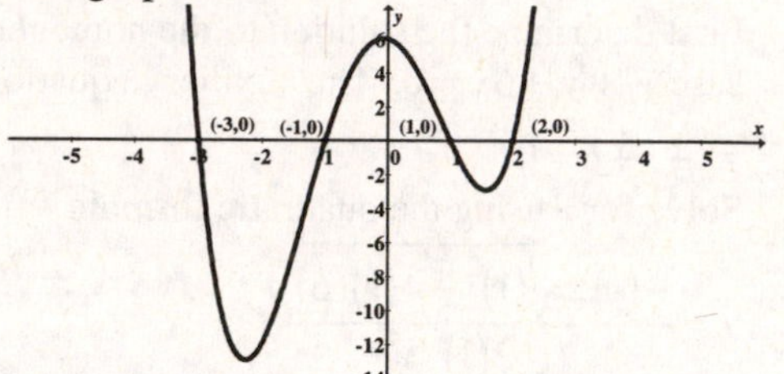

From the graph we determine the zeros: $r = -3, r = -1, r = 1, r = 2$. Therefore, the general solution is

$y = C_1 e^{-3x} + C_2 e^{-x} + C_3 e^{x} + C_4 e^{2x}.$

59. The auxiliary equation is

$r^4 - 2r^3 - 8r^2 + 18r - 9 = 0.$

We graph the auxiliary equation and find the real zeros.

The graph is

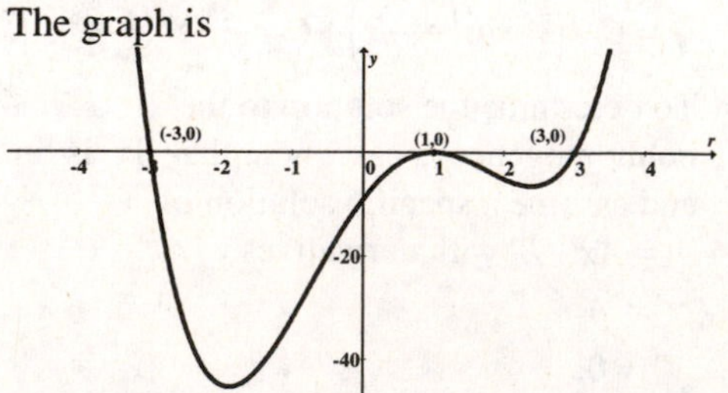

From the graph we determine the zeros: $r = -3, r = 1, r = 3$. We note that $r = 1$ is a zero of multiplicity 2; therefore, the general solution is

$y = C_1 e^{-3x} + C_2 e^{x} + C_3 x e^{x} + C_4 e^{3x}.$

Copyright © 2014 Pearson Education, Inc. Publishing as Addison-Wesley.

61. The auxiliary equation is
$r^4 + r^3 - 18r^2 - 52r - 40 = 0.$
We graph the auxiliary equation and find the real zeros.
The graph is

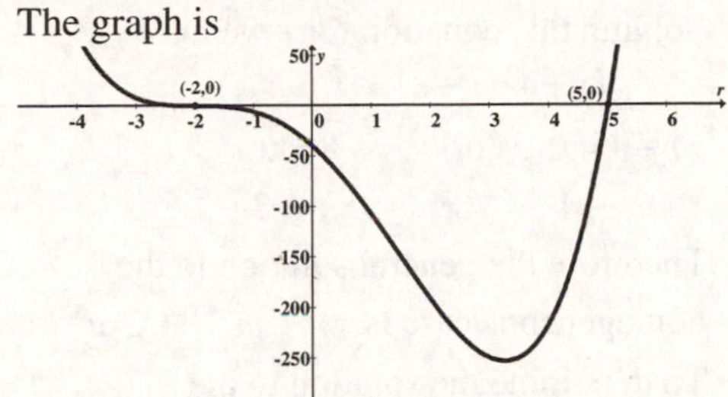

From the graph we determine the zeros:
$r = -2,\ r = 5.$
Using the two zeros from the graph, we use long division to determine the multiplicity of the zeros.

$$\begin{array}{r} r^3+6r^2+12r+8 \\ r-5\overline{)r^4+r^3-18r^2-52r-40} \\ -\underline{r^4+5r^3} \\ 6r^3-18r^2 \\ \underline{-6r^3+30r^2} \\ 12r^2-52r \\ \underline{-12r^2+60r} \\ 8r-40 \\ \underline{-8r+40} \\ 0 \end{array}$$

Using the result from the long division on the previous column, we now know that $r-5$ is a factor of the polynomial. Therefore,
$r^4 + r^3 - 18r^2 - 52r - 40 =$
$(r-5)(r^3+6r^2+12r+8)$
Next we apply long division to
$r^3+6r^2+12r+8$

$$\begin{array}{r} r^2+4r+4 \\ r+2\overline{)r^3+6r^2+12r+8} \\ -\underline{r^3-2r^2} \\ 4r^2+12r \\ \underline{-4r^3-8r} \\ 4r^2+8r \\ \underline{-4r^2+8r} \\ 0 \end{array}$$

We factor the polynomial at the top of the next column.

The polynomial now factors as
$r^4 + r^3 - 18r^2 - 52r - 40 =$
$(r-5)(r+2)(r^2+4r+4)$
Finally we factor the quadratic term to see
$r^4 + r^3 - 18r^2 - 52r - 40 =$
$(r-5)(r+2)(r+2)(r+2)$
We note that $r = -2$ is a zero of multiplicity 3; therefore, the general solution is
$y = C_1e^{-2x} + C_2xe^{-2x} + C_3x^2e^{-2x} + C_4e^{5x}.$

63. The auxiliary equation is
$r^4 + r^3 - 8r^2 - r + 7 = 0.$
We graph the auxiliary equation and find the real zeros. The graph is

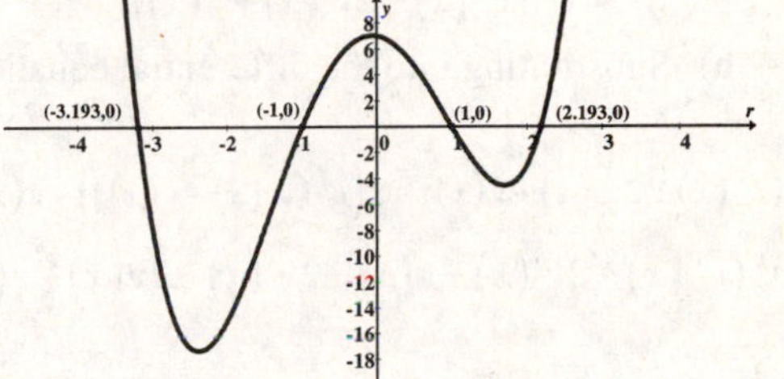

From the graph we determine the zeros:
$r = -3.193,\ r = -1,\ r = 1,\ r = 2.193$
Therefore, the general solution is
$y = C_1e^{-3.193x} + C_2e^{-1x} + C_3e^{x} + C_4e^{2.193x}.$

65. The auxiliary equation is
$r^4 + 3r^3 - 13r^2 - 21r + 54 = 0.$
We graph the auxiliary equation and find the real zeros. The graph is

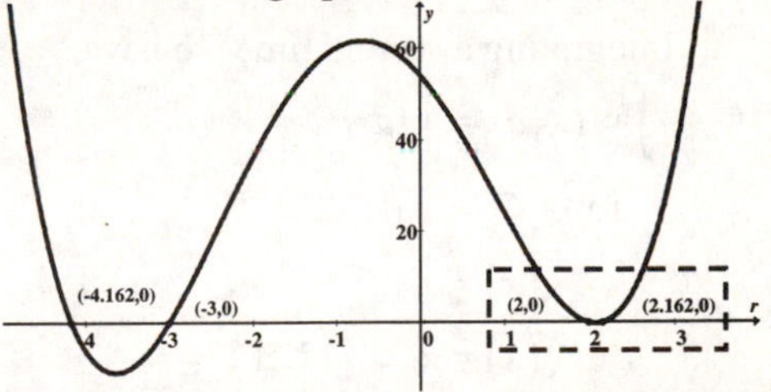

It appears that there are two zeros around $r = 2$. Zoom in on the area if needed to determine the zero.

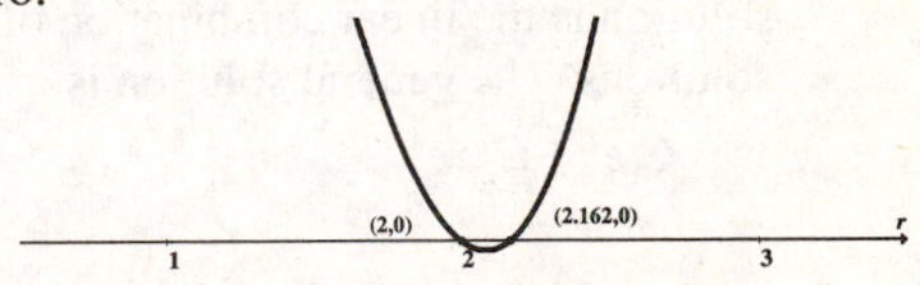

From the graph, we determine the zeros:
$r = -4.162,\ r = -3,\ r = 2,\ r = 2.162.$
Therefore, the general solution is
$y = C_1e^{-4.162x} + C_2e^{-3x} + C_3e^{2x} + C_4e^{2.162x}.$

Copyright © 2014 Pearson Education, Inc. Publishing as Addison-Wesley.

67. a) Let $y = v(x)e^x$. Use the product rule to find the first and second derivatives.

$$y' = \tfrac{d}{dx}\left(v(x)\cdot e^x\right)$$
$$y' = v(x)\tfrac{d}{dx}e^x + \tfrac{d}{dx}v(x)\cdot e^x$$
$$y' = v(x)e^x + v'(x)e^x$$
$$y' = e^x\left(v'(x)+v(x)\right).$$

Applying the product rule again we get the second derivative

$$y'' = \tfrac{d}{dx}\left[e^x\left(v'(x)+v(x)\right)\right]$$
$$y'' = e^x\tfrac{d}{dx}\left(v'(x)+v(x)\right)+\tfrac{d}{dx}e^x\cdot\left(v'(x)+v(x)\right)$$
$$y'' = e^x\left(v''(x)+v'(x)\right)+e^x\left(v'(x)+v(x)\right)$$
$$y'' = e^x\left(v''(x)+2v'(x)+v(x)\right).$$

b) Substituting into the differential equation

$$y''-2y'+1=0$$
$$e^x\left(v''(x)+2v'(x)+v(x)\right)-2\left(e^x\left(v'(x)+v(x)\right)\right)+v(x)e^x=0$$
$$e^x\left(v''(x)+2v'(x)+v(x)-2v'(x)-2v(x)+v(x)\right)=0$$
$$e^x\left(v''(x)\right)=0$$

c) Dividing both sides by e^x results in

$v''(x)=0.$

d) Integrating the result from part (c) we have

$$\int v''(x)dx = \int 0dx$$
$$v'(x)+C_1 = C_2$$
$$v'(x) = C$$
$$v'(x) = 1. \quad \text{Let } C = 1.$$

Integrating a second time we have

$$\int v'(x)dx = \int 1dx$$
$$v(x)+C_1 = x+C_2$$
$$v(x) = x+C$$
$$v(x) = x. \quad \text{Let } C = 0.$$

e. A second solution to the equation is $y = v(x)e^x = xe^x$. Therefore, the general solution is the linear combination of the two solutions. The general solution is $y = C_1e^x + C_2xe^x$.

69. First determine the solution to the homogeneous case $y''-4y'-32y=0$. The auxiliary equation is

$$r^2-4r-32=0$$

Solving this equation for r we have

$$(r+4)(r-8)=0$$
$$r+4=0 \quad \text{or} \quad r-8=0$$
$$r=-4 \quad \text{or} \quad r=8.$$

Therefore the general solution to the homogeneous case is $y = C_1e^{-4x}+C_2e^{8x}$.

To determine the solution to the nonhomogeneous case, note that $3e^{-x}$ is not a polynomial, we assume a specific solution of $y = Ae^{-x}$ with derivatives

$$y' = -Ae^{-x}$$
$$y'' = Ae^{-x}.$$

Substituting into the differential equation we have

$$y''-4y'-32=3e^{-x}$$
$$Ae^{-x}-4\left(-Ae^{-x}\right)-32\left(Ae^{-x}\right)=3e^{-x}$$
$$-27Ae^{-x}=3e^{-x}.$$

Using the method of undetermined coefficients, we have the following system

$$-27A=3$$
$$A=-\tfrac{3}{27}$$
$$A=-\tfrac{1}{9}.$$

Therefore, the specific solution is $y=-\frac{1}{9}e^{-x}$.

The general solution to the differential equation is the sum of the homogenous and the specific solutions:

$$y = C_1e^{-4x}+C_2e^{8x}-\tfrac{1}{9}e^{-x}.$$

Copyright © 2014 Pearson Education, Inc. Publishing as Addison-Wesley.

71. First determine the solution to the homogeneous case $y''+y=0$.

The auxiliary equation is

$r^2+1=0.$

Solving this equation for r we have

$$\begin{aligned} r^2+1&=0 \\ r^2&=-1 \\ r&=\pm\sqrt{-1} \\ r&=\pm i. \end{aligned}$$

There are two complex solutions to the auxiliary equation. Using Theorem 3, the general solution is

$$y=C_1e^{0x}\sin(x)+C_2e^{0x}\cos(x)$$

$$y=C_1\sin x+C_2\cos x.$$

To determine the solution to the nonhomogeneous case, note that $\cos(2x)$ is not a polynomial, we assume a specific solution of $y=A\sin 2x+B\cos 2x$ with derivatives

$$y'=2A\cos 2x-2B\sin 2x$$

$$y''=-4A\sin 2x-4B\cos 2x.$$

Substituting into the differential equation we have

$$\begin{aligned} y''+y&=\cos 2x \\ -4A\sin 2x-4B\cos 2x+A\sin 2x+B\cos 2x&=\cos 2x \\ -3A\sin 2x-3B\cos 2x&=\cos 2x. \end{aligned}$$

Equating coefficients we have,

$$A=0$$

$$-3B=1\rightarrow B=-\tfrac{1}{3}.$$

Therefore, the specific solution is

$$y=-\tfrac{1}{3}\cos 2x.$$

The general solution to the differential equation is the sum of the homogenous and the specific solutions:

$$y=y=C_1\sin x+C_2\cos x-\tfrac{1}{3}\cos 2x.$$

Copyright © 2014 Pearson Education, Inc. Publishing as Addison-Wesley.

Chapter 9

Sequences and Series

Exercise Set 9.1

1. a) For the sequence given by $a_n = 3n+8$, we have

$a_1 = 3(1)+8 = 11,$

$a_2 = 3(2)+8 = 14,$

$a_3 = 3(3)+8 = 17,$

$a_4 = 3(4)+8 = 19.$

b) The 20th term in the sequence is $a_{20} = 3(20)+8 = 68.$

c) The 25th partial sum is given by

$$S_{25} = \sum_{n=1}^{25} 3n+8 = \frac{n}{2}(a_1+a_{25}).$$

We have $a_1 = 11$, and $a_{25} = 3(25)+8 = 83.$

$$S_{25} = \sum_{n=1}^{25} 3n+8 = \frac{25}{2}(11+83) = 1175.$$

3. a) For the sequence given by $a_n = n+7$, we have

$a_1 = (1)+7 = 8,$

$a_2 = (2)+7 = 9,$

$a_3 = (3)+7 = 10,$

$a_4 = (4)+7 = 11.$

b) The 20th term in the sequence is $a_{20} = (20)+7 = 27.$

c) The 25th partial sum is given by

$$S_{25} = \sum_{n=1}^{25} n+7 = \frac{n}{2}(a_1+a_{25}).$$

We have $a_1 = 11$, and $a_{25} = (25)+7 = 32.$

$$S_{25} = \sum_{n=1}^{25} n+7 = \frac{25}{2}(8+32) = 500.$$

5. a) For the sequence given by $c_n = 1-2n$, we have

$c_1 = 1-2(1) = -1,$

$c_2 = 1-2(2) = -3,$

$c_3 = 1-2(3) = -5,$

$c_4 = 1-2(4) = -7.$

b) The 20th term in the sequence is $c_{20} = 1-2(20) = -39.$

c) The 25th partial sum is given by

$$S_{25} = \sum_{n=1}^{25} 1-2n = \frac{n}{2}(c_1+c_{25}).$$

We have $c_1 = -1$, and $c_{25} = 1-2(25) = -49.$

$$S_{25} = \sum_{n=1}^{25} 1-2n = \frac{25}{2}(-1-49) = -625.$$

7. a) For the sequence given by $c_n = \frac{1}{2}n+\frac{3}{2}$, we have

$c_1 = \frac{1}{2}(1)+\frac{3}{2} = 2,$

$c_2 = \frac{1}{2}(2)+\frac{3}{2} = \frac{5}{2},$

$c_3 = \frac{1}{2}(3)+\frac{3}{2} = 3,$

$c_4 = \frac{1}{2}(4)+\frac{3}{2} = \frac{7}{2}.$

b) The 20th term in the sequence is $c_{20} = \frac{1}{2}(20)+\frac{3}{2} = \frac{23}{2}.$

c) The 25th partial sum is given by

$$S_{25} = \sum_{n=1}^{25} \tfrac{1}{2}n+\tfrac{3}{2} = \frac{n}{2}(c_1+c_{25}).$$

We have $c_1 = 2$, and $c_{25} = \frac{1}{2}(25)+\frac{3}{2} = 14.$

$$S_{25} = \sum_{n=1}^{25} \tfrac{1}{2}n+\tfrac{3}{2} = \frac{25}{2}(2+14) = 200.$$

9. a) For the sequence given by $p_n = 1.6n+2.1$, we have

$p_1 = 1.6(1)+2.1 = 3.7,$

$p_2 = 1.6(2)+2.1 = 5.3,$

$p_3 = 1.6(3)+2.1 = 6.9,$

$p_4 = 1.6(4)+2.1 = 8.5.$

b) The 20th term in the sequence is $p_{20} = 1.6(20)+2.1 = 34.1.$

Copyright © 2014 Pearson Education, Inc. Publishing as Addison-Wesley.

c) The 25th partial sum is given by

$$S_{25} = \sum_{n=1}^{25} 1.6n + 2.1 = \frac{n}{2}(p_1 + p_{25}).$$

We have $p_1 = 3.7$, and

$p_{25} = 1.6(25) + 2.1 = 42.1.$

$$S_{25} = \sum_{n=1}^{25} 1.6n + 2.1 = \frac{25}{2}(3.7 + 42.1) = 572.5.$$

11. a) $d = 19 - 7 = 12.$

b) To find the general nth term use the fact that

$$a_n = a_1 + (n-1)d$$
$$a_n = 7 + (n-1)12$$
$$a_n = 7 + 12n - 12$$
$$a_n = -5 + 12n.$$

c) The 30th partial sum is given by

$$S_{30} = \sum_{n=1}^{30} -5 + 12n = \frac{n}{2}(a_1 + a_{30}).$$

We have $a_1 = 7$, and

$a_{30} = -5 + 12(30) = 355.$

$$S_{30} = \sum_{n=1}^{30} -5 + 12n = \frac{30}{2}(7 + 355) = 5430.$$

13. a) $d = 104 - 100 = 4.$

b) To find the general nth term use the fact that

$$a_n = a_1 + (n-1)d$$
$$a_n = 100 + (n-1)4$$
$$a_n = 100 + 4n - 4$$
$$a_n = 96 + 4n.$$

c) The 30th partial sum is given by

$$S_{30} = \sum_{n=1}^{30} 96 + 4n = \frac{n}{2}(a_1 + a_{30}).$$

We have $a_1 = 100$, and

$a_{30} = 96 + 4(30) = 216.$

$$S_{30} = \sum_{n=1}^{30} 96 + 4n$$
$$= \frac{30}{2}(100 + 216) = 4740.$$

15. a) $d = 78 - 80 = -2.$

b) To find the general nth term use the fact that

$$a_n = a_1 + (n-1)d$$
$$a_n = 80 + (n-1)(-2)$$
$$a_n = 80 - 2n + 2$$
$$a_n = 82 - 2n.$$

c) The 30th partial sum is given by

$$S_{30} = \sum_{n=1}^{30} 82 - 2n = \frac{n}{2}(a_1 + a_{30}).$$

We have $a_1 = 80$, and

$a_{30} = 82 - 2(30) = 22.$

$$S_{30} = \sum_{n=1}^{30} 82 - 2n$$
$$= \frac{30}{2}(80 + 22) = 1530.$$

17. a) $d = 11.5 - 8 = 3.5.$

b) To find the general nth term use the fact that

$$a_n = a_1 + (n-1)d$$
$$a_n = 8 + (n-1)3.5$$
$$a_n = 8 + 3.5n - 3.5$$
$$a_n = 4.5 + 3.5n.$$

c) The 30th partial sum is given by

$$S_{30} = \sum_{n=1}^{30} 4.5 + 3.5n = \frac{n}{2}(a_1 + a_{30}).$$

We have $a_1 = 8$, and

$a_{30} = 4.5 + 3.5(30) = 109.5.$

$$S_{30} = \sum_{n=1}^{30} 4.5 + 3.5n$$
$$= \frac{30}{2}(8 + 109.5) = 1762.5.$$

19. a) $d = 3.85 - 4 = -0.15.$

b) To find the general nth term use the fact that

$$a_n = a_1 + (n-1)d$$
$$a_n = 4 + (n-1)(-0.15)$$
$$a_n = 4 - 0.15n + 0.15$$
$$a_n = 4.15 - 0.15n.$$

Copyright © 2014 Pearson Education, Inc. Publishing as Addison-Wesley.

c) The 30th partial sum is given by

$$S_{30} = \sum_{n=1}^{30} 4.15 - 0.15n = \frac{n}{2}(a_1 + a_{30}).$$

We have $a_1 = 4.15$, and

$a_{30} = 4.15 - 0.15(30) = -0.35.$

$$S_{30} = \sum_{n=1}^{30} 4.15 - 0.15n$$
$$= \frac{30}{2}(4 - 0.35) = 54.75.$$

21. a) To find the general nth term use the fact that $a_1 = 4$ and $d = 3$; therefore,

$a_n = a_1 + (n-1)d$
$a_n = 4 + (n-1)3$
$a_n = 4 + 3n - 3$
$a_n = 1 + 3n.$

b) The first five terms and the 50th term are

$a_1 = 1 + 3(1) = 4,$
$a_2 = 1 + 3(2) = 7,$
$a_3 = 1 + 3(3) = 10,$
$a_4 = 1 + 3(4) = 13,$
$a_5 = 1 + 3(5) = 16,$
$a_{50} = 1 + 3(50) = 151.$

c) The 50th partial sum is given by

$$S_{50} = \sum_{n=1}^{50} 1 + 3n = \frac{n}{2}(a_1 + a_{50}).$$

We have $a_1 = 4$, and $a_{50} = 151$.

$$S_{50} = \sum_{n=1}^{50} 1 + 3n$$
$$= \frac{50}{2}(4 + 151) = 3875.$$

23. a) To find the general nth term use the fact that $a_1 = 200$ and $d = -9$; we find the general nth term at the top of the next column:

$a_n = a_1 + (n-1)d$
$a_n = 200 + (n-1)(-9)$
$a_n = 200 - 9n + 9$
$a_n = 209 - 9n.$

b) The first five terms and the 50th term are

$a_1 = 209 - 9(1) = 200,$
$a_2 = 209 - 9(2) = 191,$
$a_3 = 209 - 9(3) = 182,$
$a_4 = 209 - 9(4) = 173,$
$a_5 = 209 - 9(5) = 164,$
$a_{50} = 209 - 9(50) = -241.$

c) The 50th partial sum is given by

$$S_{50} = \sum_{n=1}^{50} 209 - 9n = \frac{n}{2}(a_1 + a_{50}).$$

We have $a_1 = 200$, and $a_{50} = -241$.

$$S_{50} = \sum_{n=1}^{50} 209 - 9n$$
$$= \frac{50}{2}(200 + (-241)) = -1025.$$

25. a) To find the general nth term use the fact that $a_n = a_1 + (n-1)d$ and $a_3 = 20$ and $a_{10} = 62$ to set up a system of equations to find a_1 and d.

$$\begin{Bmatrix} a_{10} = a_1 + (10-1)d \\ a_3 = a_1 + (3-1)d \end{Bmatrix} \Rightarrow \begin{Bmatrix} 62 = a_1 + 9d \\ 20 = a_1 + 2d \end{Bmatrix}$$

Solving the first equation for a_1 we have $a_1 = 62 - 9d$. Substituting into the second equation, we have

$20 = (62 - 9d) + 2d$
$20 = 62 - 7d$
$-42 = -7d$
$6 = d.$

Substitute back into the first equation to find $a_1 = 62 - 9(6) = 8$. Thus, the general nth term is

$a_n = a_1 + (n-1)d$
$a_n = 8 + (n-1)6$
$a_n = 8 + 6n - 6$
$a_n = 2 + 6n.$

Copyright © 2014 Pearson Education, Inc. Publishing as Addison-Wesley.

b) The first five terms and the 50th term are

$$a_1 = 2+6(1) = 8,$$
$$a_2 = 2+6(2) = 14,$$
$$a_3 = 2+6(3) = 20,$$
$$a_4 = 2+6(4) = 26,$$
$$a_5 = 2+6(5) = 32,$$
$$a_{50} = 2+6(50) = 302.$$

c) The 50th partial sum is given by

$$S_{50} = \sum_{n=1}^{50} 2+6n = \frac{n}{2}(a_1 + a_{50}).$$

We have $a_1 = 8$, and $a_{50} = 302$.

$$S_{50} = \sum_{n=1}^{50} 2+6n$$
$$= \frac{50}{2}(8+302) = 7750.$$

27. a) To find the general nth term use the fact that $a_n = a_1 + (n-1)d$ and $a_{11} = 91$ and $a_{27} = 43$ to set up a system of equations to find a_1 and d.

$$\begin{Bmatrix} a_{27} = a_1 + (27-1)d \\ a_{11} = a_1 + (11-1)d \end{Bmatrix} \Rightarrow \begin{Bmatrix} 43 = a_1 + 26d \\ 91 = a_1 + 10d \end{Bmatrix}$$

Solving the first equation for a_1 we have $a_1 = 43 - 26d$. Substituting into the second equation, we have

$$91 = (43-26d) + 10d$$
$$91 = 43 - 16d$$
$$48 = -16d$$
$$-3 = d.$$

Substitute back into the first equation to find $a_1 = 43 - 26(-3) = 121$. Thus, the general nth term is

$$a_n = a_1 + (n-1)d$$
$$a_n = 121 + (n-1)(-3)$$
$$a_n = 121 - 3n + 3$$
$$a_n = 124 - 3n.$$

b) The first five terms and the 50th term are

$$a_1 = 124 - 3(1) = 121,$$
$$a_2 = 124 - 3(2) = 118,$$
$$a_3 = 124 - 3(3) = 115,$$
$$a_4 = 124 - 3(4) = 112,$$
$$a_5 = 124 - 3(5) = 109,$$
$$a_{50} = 124 - 3(50) = -26.$$

c) The 50th partial sum is given by

$$S_{50} = \sum_{n=1}^{50} 124 - 3n = \frac{n}{2}(a_1 + a_{50}).$$

We have $a_1 = 121$, and $a_{50} = -26$.

$$S_{50} = \sum_{n=1}^{50} 124 - 3n$$
$$= \frac{50}{2}(121 + (-26)) = 2375.$$

29. a) To find the general nth term use the fact that $a_n = a_1 + (n-1)d$ and $a_6 = 7$ and $a_{12} = 10$ to set up a system of equations to find a_1 and d.

$$\begin{Bmatrix} a_{12} = a_1 + (12-1)d \\ a_6 = a_1 + (6-1)d \end{Bmatrix} \Rightarrow \begin{Bmatrix} 10 = a_1 + 11d \\ 7 = a_1 + 5d \end{Bmatrix}$$

Solving the first equation for a_1 we have $a_1 = 10 - 11d$. Substituting into the second equation, we have

$$7 = (10 - 11d) + 5d$$
$$7 = 10 - 6d$$
$$-3 = -6d$$
$$\tfrac{1}{2} = d.$$

Substitute back into the first equation to find $a_1 = 10 - 11\left(\tfrac{1}{2}\right) = \tfrac{9}{2} = 4.5$. Thus, the general nth term is

$$a_n = a_1 + (n-1)d$$
$$a_n = 4.5 + (n-1)0.5$$
$$a_n = 4.5 + 0.5n - 0.5$$
$$a_n = 4 + 0.5n.$$

Copyright © 2014 Pearson Education, Inc. Publishing as Addison-Wesley.

b) The first five terms and the 50th term are

$$a_1 = 4+0.5(1) = 4.5,$$
$$a_2 = 4+0.5(2) = 5,$$
$$a_3 = 4+0.5(3) = 5.5,$$
$$a_4 = 4+0.5(4) = 6,$$
$$a_5 = 4+0.5(5) = 6.5,$$
$$a_{50} = 4+0.5(50) = 29.$$

c) The 50th partial sum is given by

$$S_{50} = \sum_{n=1}^{50} 4+0.5n = \frac{n}{2}(a_1 + a_{50}).$$

We have $a_1 = 4.5$, and $a_{50} = 29$.

$$S_{50} = \sum_{n=1}^{50} 4+0.5n$$
$$= \frac{50}{2}(4.5+29) = 837.5.$$

31. The common difference is $d = 7-3 = 4$ and $a_1 = 3$.
Find the general nth term

$$a_n = a_1 + (n-1)d$$
$$a_n = 3+(n-1)4$$
$$a_n = 3+4n-4$$
$$a_n = -1+4n.$$

The 40th partial sum is given by

$$S_{40} = \sum_{n=1}^{40} -1+4n = \frac{n}{2}(a_1 + a_{40}).$$

We have $a_1 = 3$, and $a_{40} = -1+4(40) = 159$.

$$S_{40} = \sum_{n=1}^{40} -1+4n = \frac{40}{2}(3+159) = 3240.$$

33. Find $\sum_{n=1}^{100}(56-4n)$

For $a_n = 56-4n$ we have

$$a_1 = 56-4(1) = 52$$
$$a_{100} = 56-4(100) = -344.$$

Thus,

$$\sum_{n=1}^{100}(56-4n) = \frac{100}{2}(52+(-344))$$
$$= -14,600.$$

35. Find $\sum_{n=1}^{100}(6n-4)$

For $a_n = 6n-4$ we have

$$a_1 = 6(1)-4 = 2$$
$$a_{100} = 6(100)-4 = 596.$$

Thus,

$$\sum_{n=1}^{100}(6n-4) = \frac{100}{2}(2+596)$$
$$= 29,900.$$

37. a) We know that $c_1 = 100$ and $d = 7$.
The number of clients after the nth week will be given by

$$c_n = c_1 + (n-1)d$$
$$c_n = 100+(n-1)7$$
$$c_n = 100+7n-7$$
$$c_n = 93+7n.$$

b) $c_{25} = 93+7(25) = 268$. Bantam accountants has 268 clients in the 25th week.

c) Set $c_n = 240$ and solve for n.

$$93+7n = 240$$
$$7n = 147$$
$$n = 21.$$

Bantam Accountants will have 240 clients in the 21st week.

39. a) We know that $f_1 = 0.5$ and $d = 0.1$.
The amount of the fine after the n days will be given by

$$f_n = f_1 + (n-1)d$$
$$f_n = 0.5+(n-1)0.1$$
$$f_n = 0.5+0.1n-0.1$$
$$f_n = 0.4+0.1n.$$

b) After 15 days the fine is $f_{15} = 0.4+0.1(15) = 1.9$. The fine for a book that is 15 days late is \$1.90.

c) Set $f_n = 5$ and solve for n.

$$0.4+0.1n = 5$$
$$0.1n = 4.6$$
$$n = 46.$$

After 46 days, the fine will be \$5.

Copyright © 2014 Pearson Education, Inc. Publishing as Addison-Wesley.

41. a) Using the information in the problem we know $C = 7500, N = 6,$ and $S = 1500.$ To find v_n substitute into the straight line depreciation formula as follows

$$v_n = 7500 - n\left(\frac{7500 - 1500}{6}\right)$$

$$v_n = 7500 - n\left(\frac{6000}{6}\right)$$

$$v_n = 7500 - 1000n.$$

b) $v_0 = 7500 - 1000(0) = 7500.$

c) When $n = 4$ we have $v_4 = 7500 - 1000(4) = 3500.$ The value of the copier is \$3500 four years after it is purchased.

d) $v_0 = \$7500,$

$v_1 = 7500 - 1000(1) = \$6500,$

$v_2 = 7500 - 1000(2) = \$5500,$

$v_3 = 7500 - 1000(3) = \$4500,$

$v_4 = 7500 - 1000(4) = \$3500,$

$v_5 = 7500 - 1000(5) = \$2500,$

$v_6 = 7500 - 1000(6) = \$1500.$

The sequence of values over the life of the copier is
7500, 6500, 5500, 4500, 3500, 2500, 1500.

43. Substituting into the 75th partial sum formula we have

$$S_{75} = \frac{75}{2}(a_1 + a_{75})$$

$$5925 = \frac{75}{2}(5 + a_{75})$$

$$5925\left(\frac{2}{75}\right) = 5 + a_{75}$$

$$158 = 5 + a_{75}$$

$$153 = a_{75.}$$

Next, we find the general nth term. To find the general nth term find the common difference

$$a_n = a_1 + (n-1)d$$

$$a_{75} = a_1 + (75-1)d$$

$$153 = 5 + 74d$$

$$148 = 74d$$

$$2 = d.$$

Therefore, the general nth term is given by

$$a_n = a_1 - (n-1)d$$

$$a_n = 5 + (n-1)2$$

$$a_n = 5 + 2n - 2$$

$$a_n = 3 + 2n.$$

The 20th partial sum is given by

$$S_{20} = \sum_{n=1}^{20} 3 + 2n = \frac{n}{2}(a_1 + a_{20}).$$

We have $a_1 = 5$, and

$$a_{20} = 3 + 2(20) = 43.$$

$$S_{20} = \sum_{n=1}^{20} 3 + 2n$$

$$= \frac{20}{2}(5 + 43) = 480.$$

45. $a_n = 2n - 1.$

a) The first five terms of the series are 1,3,5,7,9,... The first five partial sums are

$S_1 = 1,$

$S_2 = \frac{2}{2}(a_1 + a_2) = 1(1+3) = 4,$

$S_3 = \frac{3}{2}(a_1 + a_3) = \frac{3}{2}(1+5) = 9,$

$S_4 = \frac{4}{2}(a_1 + a_2) = 2(1+7) = 16,$

$S_5 = \frac{5}{2}(a_1 + a_3) = \frac{5}{2}(1+9) = 25.$

b) The partial sums are given by

$$S_n = \frac{n}{2}(a_1 + a_n) = \frac{n}{2}(1 + 2n - 1) = n^2.$$

Therefore, the formula for the sum of the first n odd integers is $S_n = n^2$.

c) $S_{75} = 75^2 = 5625.$

47. a) Since we are dealing with the sum of the first n positive integers. The series $t_n = 1, 2, 3, 4, ...n$ and the nth term in the series is given by $t_n = n.$ Therefore, the nth partial sum of this series is given by

$$T_n = \frac{n}{2}(t_1 + t_n) = \frac{n}{2}(1 + n) = \frac{n(n+1)}{2}.$$

b) ✎

Copyright © 2014 Pearson Education, Inc. Publishing as Addison-Wesley.

49. Start with the nth partial sum formula

$$S_n = \frac{n}{2}(a_1 + a_n)$$

We know that $a_n = a_1 + (n-1)d$ substituting yields

$$S_n = \frac{n}{2}(a_1 + (a_1 + (n-1)d))$$

$$S_n = \frac{n}{2}(2a_1 + (n-1)d).$$

51. a) $a_1 = a_1$

$$a_2 = a_1 + d$$

$$a_3 = a_2 + d = (a_1 + d) + d = a_1 + 2d.$$

b) Substituting we have

$$a_{n-2} = a_1 + ((n-2)-1)d = a_1 + (n-3)d$$

$$a_{n-1} = a_1 + ((n-1)-1)d = a_1 + (n-2)d.$$

c) Writing the partial sums we have

$$S_n = a_1 + (a_1 + d) + ... + (a_1 + (n-2)d) + (a_1 + (n-1)d)$$

$$S_n = (a_1 + (n-1)d) + (a_1 + (n-2)d) + ... + (a_1 + d) + a_1$$

Adding the two equations yields

$$2S_n = 2a_1 + (n-1)d + 2a_1 + (n-1)d + ... + 2a_1 + (n-1)d$$

Simplifying yields

$$2S_n = (2a_1 + (n-1)d) \cdot n$$

$$S_n = \frac{n}{2}(2a_1 + (n-1)d)$$

d) From part (c) we have

$$S_n = \frac{n}{2}(2a_1 + (n-1)d)$$

$$S_n = \frac{n}{2}(a_1 + a_1 + (n-1)d)$$

$$S_n = \frac{n}{2}(a_1 + [a_1 + (n-1)d]); \; a_n = a_1 + (n-1)d$$

$$S_n = \frac{n}{2}(a_1 + a_n).$$

Copyright © 2014 Pearson Education, Inc. Publishing as Addison-Wesley.

Exercise Set 9.2

1. a) The common ratio is

$$r = \frac{12}{3} = 4.$$

b) The general nth term is given by

$$a_n = a_1 \cdot r^{n-1} = 3 \cdot (4)^{n-1}.$$

c) The 10th term of the sequence is

$$a_{10} = 3 \cdot (4)^{10-1} = 3 \cdot (4)^9 = 786{,}432.$$

d) The sum of the first n terms is

$$S_n = \frac{a_1 \cdot (r^n - 1)}{r - 1}$$

Therefore, the sum of the first 15 terms is

$$S_{15} = \frac{3 \cdot (4^{15} - 1)}{4 - 1} = 1{,}073{,}741{,}823.$$

3. a) The common ratio is

$$r = \frac{21}{7} = 3.$$

b) The general nth term is given by

$$a_n = a_1 \cdot r^{n-1} = 7 \cdot (3)^{n-1}.$$

c) The 10th term of the sequence is

$$a_{10} = 7 \cdot (3)^{10-1} = 7 \cdot (3)^9 = 137{,}781.$$

d) The sum of the first n terms is

$$S_n = \frac{a_1 \cdot (r^n - 1)}{r - 1}$$

Therefore, the sum of the first 15 terms is

$$S_{15} = \frac{7 \cdot (3^{15} - 1)}{3 - 1} = 50{,}221{,}171.$$

5. a) The common ratio is

$$r = \frac{-6}{2} = -3.$$

b) The general nth term is given by

$$a_n = a_1 \cdot r^{n-1} = 2 \cdot (-3)^{n-1}.$$

c) The 10th term of the sequence is

$$a_{10} = 2 \cdot (-3)^{10-1} = 2 \cdot (-3)^9 = -39{,}366.$$

d) The sum of the first n terms is

$$S_n = \frac{a_1 \cdot (r^n - 1)}{r - 1}$$

Therefore, the sum of the first 15 terms is

$$S_{15} = \frac{2 \cdot ((-3)^{15} - 1)}{(-3) - 1} = 7{,}174{,}454.$$

7. a) The common ratio is

$$r = \frac{\frac{3}{20}}{\frac{1}{4}} = \frac{3}{5}.$$

b) The general nth term is given by

$$a_n = a_1 \cdot r^{n-1} = \tfrac{1}{4} \cdot \left(\tfrac{3}{5}\right)^{n-1}.$$

c) The 10th term of the sequence is

$$a_{10} = \tfrac{1}{4} \cdot \left(\tfrac{3}{5}\right)^{10-1} = \tfrac{1}{4} \cdot \left(\tfrac{3}{5}\right)^9 = \frac{19{,}683}{7{,}812{,}500}.$$

d) The sum of the first n terms is

$$S_n = \frac{a_1 \cdot (r^n - 1)}{r - 1}$$

Therefore, the sum of the first 15 terms is

$$S_{15} = \frac{\tfrac{1}{4} \cdot \left(\left(\tfrac{3}{5}\right)^{15} - 1\right)}{\left(\tfrac{3}{5}\right) - 1} = 0.624706....$$

9. a) The common ratio is

$$r = \frac{\frac{1}{8}}{-\frac{1}{12}} = -\frac{2}{3}.$$

b) The general nth term is given by

$$a_n = a_1 \cdot r^{n-1} = \tfrac{1}{8} \cdot \left(-\tfrac{2}{3}\right)^{n-1}.$$

c) The 10th term of the sequence is

$$a_{10} = \tfrac{1}{8} \cdot \left(-\tfrac{2}{3}\right)^{10-1} = \tfrac{1}{8} \cdot \left(-\tfrac{2}{3}\right)^9 = -\frac{512}{157{,}464}.$$

d) The sum of the first n terms is

$$S_n = \frac{a_1 \cdot (r^n - 1)}{r - 1}$$

Therefore, the sum of the first 15 terms is

$$S_{15} = \frac{\tfrac{1}{8} \cdot \left(\left(-\tfrac{2}{3}\right)^{15} - 1\right)}{\left(-\tfrac{2}{3}\right) - 1} = 0.07517127....$$

11. a) The general nth term is given by

$$a_n = a_1 \cdot r^{n-1} = 1 \cdot (5)^{n-1} = 5^{n-1}.$$

b) The 8th term of the sequence is

$$a_8 = (5)^{8-1} = (5)^7 = 78{,}125.$$

Copyright © 2014 Pearson Education, Inc. Publishing as Addison-Wesley.

c) The sum of the first 10 terms is

$$S_{10} = \frac{1\cdot\left((5)^{10}-1\right)}{(5)-1} = 2{,}441{,}406.$$

13. a) The general nth term is given by

$a_n = a_1 \cdot r^{n-1} = 6\cdot\left(-\tfrac{1}{2}\right)^{n-1}$.

b) The 9th term of the sequence is

$$a_9 = 6\cdot\left(-\tfrac{1}{2}\right)^{9-1} = 6\cdot\left(-\tfrac{1}{2}\right)^{8} = \frac{3}{128}.$$

c) The sum of the first 10 terms is

$$S_{10} = \frac{6\cdot\left(\left(-\tfrac{1}{2}\right)^{10}-1\right)}{\left(-\tfrac{1}{2}\right)-1} = \frac{1023}{256}.$$

15. The first term is $a_1 = 4$ and the common ratio is given by $r = \frac{8}{4} = 2$, so the general nth term is $a_n = a_1 \cdot r^{n-1} = 4\cdot(2)^{n-1}$. Therefore, the sum of the first 10 terms is

$$S_{10} = \frac{4\left(2^{10}-1\right)}{2-1} = 4092.$$

17. $$\sum_{n=1}^{12} 10(1.2)^{n-1} = \frac{10\left(1.2^{12}-1\right)}{1.2-1} = 395.8050224...$$

19. The first term is $a_1 = 7$ and the common ratio is given by $r = \frac{9.1}{7} = 1.3$, so the general nth term is $a_n = a_1 \cdot r^{n-1} = 7\cdot(1.3)^{n-1}$.

a) The sum of the first 14 terms is

$$S_{14} = \frac{7\left((1.3)^{14}-1\right)}{(1.3)-1} = 895.3878233....$$

b) The minimum number of terms to reach 5000 is given by

$$\frac{7\left((1.3)^{n}-1\right)}{(1.3)-1} = 5000$$

$$7\left((1.3)^{n}-1\right) = (5000)\cdot 0.3$$

$$(1.3)^{n}-1 = \frac{1500}{7}$$

$$(1.3)^{n} = \frac{1500}{7}+1$$

$$\ln(1.3)^{n} = \ln(215.28571)$$

$$n\ln(1.3) = \ln(215.28571)$$

$$n = \frac{\ln(215.28571)}{\ln(1.3)}$$

$$n = 20.475.$$

It will take 21 terms for the sum to reach 5000.

21. First find the common ratio

$$r = \frac{\frac{1}{9}}{\frac{1}{3}} = \tfrac{1}{3}; \qquad |r| < 1.$$

Therefore, the sum of the infinite series is

$$S_\infty = \frac{a_1}{1-r} = \frac{\frac{1}{3}}{1-\frac{1}{3}} = \frac{\frac{1}{3}}{\frac{2}{3}} = \tfrac{1}{2}.$$

23. First find the common ratio

$$r = \frac{-\frac{1}{8}}{\frac{1}{4}} = -\tfrac{1}{2}; \qquad |r| < 1.$$

Therefore, the sum of the infinite series is

$$S_\infty = \frac{a_1}{1-r} = \frac{\frac{1}{4}}{1-\left(-\frac{1}{2}\right)} = \frac{\frac{1}{4}}{\frac{3}{2}} = \tfrac{1}{4}\cdot\tfrac{2}{3} = \tfrac{1}{6}.$$

25. First find the common ratio

$$r = \frac{5}{10} = \tfrac{1}{2}; \qquad |r| < 1.$$

Therefore, the sum of the infinite series is

$$S_\infty = \frac{a_1}{1-r} = \frac{10}{1-\left(\frac{1}{2}\right)} = \frac{10}{\frac{1}{2}} = 10\cdot 2 = 20.$$

Copyright © 2014 Pearson Education, Inc. Publishing as Addison-Wesley.

27. $\sum_{n=1}^{\infty} 10 \cdot (-0.2)^{n-1} = \frac{a_1}{1-r}$

$\sum_{n=1}^{\infty} 10 \cdot (-0.2)^{n-1} = \frac{10}{1-(-0.2)}$

$\sum_{n=1}^{\infty} 10 \cdot (-0.2)^{n-1} = \frac{10}{1.2} = \frac{10}{\frac{6}{5}} = 10 \cdot \frac{5}{6} = \frac{25}{3}.$

29. $\sum_{n=1}^{\infty} 5 \cdot (0.9)^{n-1} = \frac{a_1}{1-r}$

$\sum_{n=1}^{\infty} 5 \cdot (0.9)^{n-1} = \frac{5}{1-(0.9)}$

$\sum_{n=1}^{\infty} 5 \cdot (0.9)^{n-1} = \frac{5}{\frac{1}{10}} = 50.$

31. From Exercise 21, we have $a_1 = \frac{1}{3}$, $r = \frac{1}{3}$, and $S_\infty = \frac{1}{2}$. Thus,

$S_n = \frac{a_1(r^n - 1)}{r-1}$

$S_5 = \frac{\frac{1}{3}\left(\frac{1}{3}^5 - 1\right)}{\frac{1}{3}-1} = 0.4979423868...$

$S_{10} = \frac{\frac{1}{3}\left(\frac{1}{3}^{10} - 1\right)}{\frac{1}{3}-1} = 0.4999915325...$

$S_{15} = \frac{\frac{1}{3}\left(\frac{1}{3}^{15} - 1\right)}{\frac{1}{3}-1} = 0.499999652...$

33. From Exercise 23, we have $a_1 = \frac{1}{4}$, $r = -\frac{1}{2}$, and $S_\infty = \frac{1}{6}$.

$S_n = \frac{a_1(r^n - 1)}{r-1}$

$S_5 = \frac{\frac{1}{4}\left(\left(-\frac{1}{2}\right)^5 - 1\right)}{\left(-\frac{1}{2}\right)-1} = 0.171875.$

$S_{10} = \frac{\frac{1}{4}\left(\left(-\frac{1}{2}\right)^{10} - 1\right)}{\left(-\frac{1}{2}\right)-1} = 0.1665039063...$

$S_{15} = \frac{\frac{1}{4}\left(\left(-\frac{1}{2}\right)^{15} - 1\right)}{\left(-\frac{1}{2}\right)-1} = 0.1666717529...$

35. From Exercise 25, we have $a_1 = 10$, $r = \frac{1}{2}$, and $S_\infty = 20$.

$S_n = \frac{a_1(r^n - 1)}{r-1}$

$S_5 = \frac{10\left(\frac{1}{2}^5 - 1\right)}{\frac{1}{2}-1} = 19.375.$

$S_{10} = \frac{10\left(\frac{1}{2}^{10} - 1\right)}{\frac{1}{2}-1} = 19.980468750...$

$S_{15} = \frac{10\left(\frac{1}{2}^{15} - 1\right)}{\frac{1}{2}-1} = 19.999389648...$

37. From Exercise 27, we have $a_1 = 10$, $r = -\frac{1}{5}$, and $S_\infty = \frac{25}{3}$.

$S_n = \frac{a_1(r^n - 1)}{r-1}$

$S_5 = \frac{10\left(\left(-\frac{1}{5}\right)^5 - 1\right)}{\left(-\frac{1}{5}\right)-1} = 8.336.$

$S_{10} = \frac{10\left(\left(-\frac{1}{5}\right)^{10} - 1\right)}{\left(-\frac{1}{5}\right)-1} = 8.33333248...$

$S_{15} = \frac{10\left(\left(-\frac{1}{5}\right)^{15} - 1\right)}{\left(-\frac{1}{5}\right)-1} = 8.333333334...$

39. From Exercise 29, we have $a_1 = 5$, $r = 0.9$, and $S_\infty = 50$.

$S_n = \frac{a_1(r^n - 1)}{r-1}$

$S_5 = \frac{5\left((0.9)^5 - 1\right)}{(0.9)-1} = 20.4755.$

$S_{10} = \frac{5\left((0.9)^{10} - 1\right)}{(0.9)-1} = 32.566077995...$

$S_{15} = \frac{5\left((0.9)^{15} - 1\right)}{(0.9)-1} = 39.705443395...$

Copyright © 2014 Pearson Education, Inc. Publishing as Addison-Wesley.

41. First rewrite the value as an infinite series
$0.444444.... = 0.4 + 0.04 + 0.004 + 0.0004 + ...$
$= \frac{4}{10} + \frac{4}{100} + \frac{4}{1000} + \frac{4}{10,000} + ...$
The common ratio is
$r = \frac{\frac{4}{100}}{\frac{4}{10}} = \frac{1}{10}$.
We notice that $|r| < 1$, so the series converges.
$$S_\infty = \frac{\frac{4}{10}}{1 - \frac{1}{10}} = \frac{\frac{4}{10}}{\frac{9}{10}} = \frac{4}{10} \cdot \frac{10}{9} = \frac{4}{9}.$$

43. First rewrite the value as an infinite series
$0.12121212.... = 0.12 + 0.0012 + 0.000012 + ...$
$= \frac{12}{100} + \frac{12}{10,000} + \frac{12}{1,000,000} + ...$
The common ratio is
$r = \frac{\frac{12}{100}}{\frac{12}{10,000}} = \frac{1}{100}$. We notice that $|r| < 1$, so the series converges.
$$S_\infty = \frac{\frac{12}{100}}{1 - \frac{1}{100}} = \frac{\frac{12}{100}}{\frac{99}{100}} = \frac{12}{100} \cdot \frac{100}{99} = \frac{12}{99}.$$

45. First rewrite the value as an infinite series
$0.145145145... = 0.145 + 0.000145 + 0.00000145 + ...$
$= \frac{145}{1000} + \frac{145}{1,000,000} + \frac{145}{1,000,000,000} + ...$
The common ratio is
$r = \frac{\frac{145}{1000}}{\frac{145}{1,000,000}} = \frac{1}{1000}$. We notice that $|r| < 1$, so the series converges.
$$S_\infty = \frac{\frac{145}{1000}}{1 - \frac{1}{1000}} = \frac{\frac{145}{1000}}{\frac{999}{1000}} = \frac{145}{1000} \cdot \frac{1000}{999} = \frac{145}{999}.$$

47. First rewrite the value as an infinite series
$1.1833333... = 1.18 + 0.003 + 0.0003 + 0.00003...$
$= \frac{118}{100} + \frac{3}{1000} + \frac{3}{10,000} + ...$
The common ratio is
$r = \frac{\frac{3}{10,000}}{\frac{3}{1000}} = \frac{1}{10}$. We notice that $|r| < 1$, so the series converges. The infinite sum is determined at the top of the next column.
$$S_\infty = \frac{118}{100} + \sum_{n=1}^{\infty} \frac{3}{1000}\left(\frac{1}{10}\right)^{n-1} = \frac{118}{100} + \frac{\frac{3}{1000}}{1 - \frac{1}{10}} = \frac{118}{100} + \frac{\frac{3}{1000}}{\frac{9}{10}} = \frac{118}{100} + \frac{1}{300} = \frac{355}{300} = \frac{71}{60}.$$

49. a) We have $v_0 = 1700$, and $r = 1 - 0.18 = 0.82$. The value of the computer after n years is $v_n = 1700(0.82)^n$.

b) Substituting for $n = 0, 1, 2, 3, 4, 5$, we have
$v_0 = 1700(0.82)^0 = \$1700$,
$v_1 = 1700(0.82)^1 = \$1394$,
$v_2 = 1700(0.82)^2 = \$1143.08$,
$v_3 = 1700(0.82)^3 = \$937.33$,
$v_4 = 1700(0.82)^4 = \$768.61$,
$v_5 = 1700(0.82)^5 = \$630.26$.

c) $v_0 = \$1700$, this is the original price of the tablet computer.

51. a) We have $v_0 = 20,000$, and $r = 1 - 0.15 = 0.85$. The value of the car after n years is $v_n = 20,000(0.85)^n$.

b) Substituting for $n = 0, 1, 2, 3, 4, 5$, we have
$v_0 = 20,000(0.85)^0 = \$20,000$,
$v_1 = 20,000(0.85)^1 = \$17,000$,
$v_2 = 20,000(0.85)^2 = \$14,450$,
$v_3 = 20,000(0.85)^3 = \$12,282.50$,
$v_4 = 20,000(0.85)^4 = \$10,440.13$,
$v_5 = 20,000(0.85)^5 = \$8874.11$.

c) $v_0 = \$20,000$, this is the original price of the car .

Copyright © 2014 Pearson Education, Inc. Publishing as Addison-Wesley.

d) We can observe from part (b) that the value of the car will drop below $9000 during the 5th year. To find this value algebraically, set $v_n = 9000$ and solve for n as follows

$$20,000(0.85)^n = 9000$$
$$(0.85)^n = \frac{9000}{20,000}$$
$$(0.85)^n = 0.45$$
$$\ln(0.85)^n = \ln(0.45)$$
$$n\ln(0.85) = \ln(0.45)$$
$$n = \frac{\ln(0.45)}{\ln(0.85)}$$
$$n = 4.913.$$

This answer validates our observation that the value of the car will drop below $9000 in the fifth year.

53. The total of expenditures generated can be modeled by the geometric series $1,000,000 + 300,000 + 90,000 + \ldots$
The common ratio is
$$r = \frac{300,000}{1,000,000} = \frac{3}{10} = 0.3.$$
We can rewrite the geometric series as
$$1,000,000 + 1,000,000(0.3)^1 + 1,000,000(0.3)^2 + \ldots$$
Since $a_1 = 1,000,000$ and $r = 0.3$ the infinite sum is given by
$$S = \frac{1,000,000}{1-0.3} = \frac{1,000,000}{0.7} = 1,428,571.43.$$
The total spending attributed to the initial $1,000,000 expenditure is $1,428,571.43.

55. a) Since each salesperson is expected to have four direct downline salespeople, the multilevel system can be modeled by the sequence 1,4,16,…
Thus, the number of salespeople at the nth level is given by $a_n = 4^{n-1}$. The total number of salespeople up to the nth level is given by the partial sum 1+4+16+…. Note that the first term is $a_1 = 1$ and the common ratio is $r = 4$. Therefore, the total number of salespeople up to and including the 10th level is
$$S_{10} = \frac{1(4^{10}-1)}{4-1} = 349,525.$$
There are 349,525 salespeople in the first 10 levels.

b) Set $S_n = 300,000,000$ and solve for n.
$$\frac{1(4^n-1)}{4-1} = 300,000,000$$
$$4^n - 1 = (300,000,000)\cdot 3$$
$$4^n - 1 = 900,000,000$$
$$4^n = 900,000,001$$
$$\ln 4^n = \ln(900,000,001)$$
$$n\ln 4 = \ln(900,000,001)$$
$$n = \frac{\ln(900,000,001)}{\ln 4}$$
$$n = 14.87$$
It will take 15 levels for Boyd's sales force to exceed the population of the U.S.

57. a) Each half-life the amount will decrease by $r = 0.5$ since the initial quantity is 15mg of radon, we have $a_0 = 15$. Therefore, the amount of radon given after n half-lives is
$$a_n = 15(0.5)^n.$$

b) Substituting for $n = 0,1,2,3,4,5,$ we have
$$a_0 = 15(0.5)^0 = 15 \text{ mg},$$
$$a_1 = 15(0.5)^1 = 7.5 \text{ mg},$$
$$a_2 = 15(0.5)^2 = 3.75 \text{ mg},$$
$$a_3 = 15(0.5)^3 = 1.875 \text{ mg},$$
$$a_4 = 15(0.5)^4 = 0.9375 \text{ mg},$$
$$a_5 = 15(0.5)^5 = 0.46875 \text{ mg}.$$

c) 5% of the original amount of radon is $15\cdot 0.05 = 0.75$ mg. Set $a_n = 0.75$ and solve:
$$15(0.5)^n = 0.75$$
$$(0.5)^n = \frac{0.75}{15}$$
$$(0.5)^n = 0.05$$
$$\ln(0.5)^n = \ln(0.05)$$
$$n\ln(0.5) = \ln(0.05)$$
$$n = \frac{\ln(0.05)}{\ln(0.5)}$$
$$n = 4.32.$$
It will take about 5 half-lives for the amount of radon to fall below 5% of the original amount. Since each half-life is 3.8 days, we see it will take about $5\cdot 3.8 = 19$ days to have decayed to 5% of its original mass.

Copyright © 2014 Pearson Education, Inc. Publishing as Addison-Wesley.

59. a) The population is declining by 1.4%; therefore, $r = 1 - 0.014 = 0.986$. Letting $P_0 = 1{,}900{,}000$ the population in Detroit n years after 1950 can be found by $P_n = 1{,}900{,}000(0.986)^n$.

b) Substituting for $n = 0, 10, 20, 30, 40, 50,$ we have

$$P_0 = 1{,}900{,}000(0.986)^0 = 1{,}900{,}000,$$
$$P_{10} = 1{,}900{,}000(0.986)^{10} = 1{,}650{,}147,$$
$$P_{20} = 1{,}900{,}000(0.986)^{20} = 1{,}433{,}151,$$
$$P_{30} = 1{,}900{,}000(0.986)^{30} = 1{,}244{,}690,$$
$$P_{40} = 1{,}900{,}000(0.986)^{40} = 1{,}081{,}011,$$
$$P_{50} = 1{,}900{,}000(0.986)^{50} = 938{,}857.$$

c) Set $P_n = 1{,}000{,}000$ and solve for n as follows

$$1{,}900{,}000(0.986)^n = 1{,}000{,}000$$
$$(0.986)^n = \frac{1{,}000{,}000}{1{,}900{,}000}$$
$$\ln(0.986)^n = \ln\left(\tfrac{10}{19}\right)$$
$$n\ln(0.986) = \ln(0.52632)$$
$$n = \frac{\ln(0.52632)}{\ln(0.986)}$$
$$n = 45.52.$$

The population of Detroit fell below 1,000,000 about 46 years after 1950 or in 1996.

61. a) Dividing A_1 into four smaller squares, we see that A_2 takes up exactly $\frac{1}{2}$ of each square. Refer to the picture

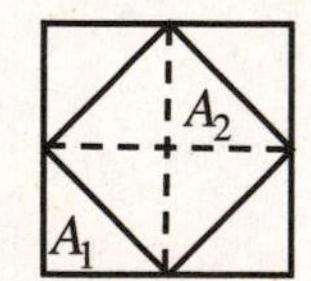

Therefore, the area of A_2 is $\frac{1}{2}$ the area of A_1.

b) Repeating this procedure, we observe that in general A_n is $\frac{1}{2}$ the area of A_{n-1}.

c) The original area is 1, so $A_1 = 1$ and common ratio is $r = \frac{1}{2}$. Therefore, the general nth term for the area of the nth nested square is $A_n = \left(\frac{1}{2}\right)^{n-1}$.

d) From part (c) the area of the 10th nested square is

$$A_{10} = \left(\tfrac{1}{2}\right)^{10-1} = \left(\tfrac{1}{2}\right)^9 = \tfrac{1}{512} \text{ square units.}$$

63. First we rewrite the summation to be the sum of two geometric series as follows

$$\sum_{n=1}^{\infty} \frac{2^n+1}{3^n} = \sum_{n=1}^{\infty} \frac{2^n}{3^n} + \sum_{n=1}^{\infty} \frac{1}{3^n} = \sum_{n=1}^{\infty}\left(\frac{2}{3}\right)^n + \sum_{n=1}^{\infty}\left(\frac{1}{3}\right)^n.$$

There are two infinite geometric series. The sum is

$$\sum_{n=1}^{\infty} \frac{2^n+1}{3^n} = \sum_{n=1}^{\infty}\left(\frac{2}{3}\right)^n + \sum_{n=1}^{\infty}\left(\frac{1}{3}\right)^n = \frac{\frac{2}{3}}{1-\frac{2}{3}} + \frac{\frac{1}{3}}{1-\frac{1}{3}} = \frac{\frac{2}{3}}{\frac{1}{3}} + \frac{\frac{1}{3}}{\frac{2}{3}} = 2 + \tfrac{1}{2} = \tfrac{5}{2}.$$

65. Since $\sum_{n=1}^{\infty} \frac{a}{4^n} = a \cdot \sum_{n=1}^{\infty} \left(\frac{1}{4}\right)^n = 1$ we have

$$a \cdot \frac{\frac{1}{4}}{1-\frac{1}{4}} = 1$$
$$a \cdot \frac{\frac{1}{4}}{\frac{3}{4}} = 1$$
$$a \cdot \tfrac{1}{3} = 1$$
$$a = 3.$$

67. Since $a_1 = 2$ and $S = 5$ we have

$$\frac{2}{1-r} = 5$$
$$2 = 5(1-r)$$
$$2 = 5 - 5r$$
$$5r = 3$$
$$r = \tfrac{3}{5}.$$

Copyright © 2014 Pearson Education, Inc. Publishing as Addison-Wesley.

69. Since $a_1 = \frac{1}{p}$ and $r = \frac{1}{p}$ we have

$$S = \frac{a_1}{1-r}$$

$$S = \frac{\frac{1}{p}}{1-\frac{1}{p}}$$

$$S = \frac{\frac{1}{p}}{\frac{p}{p}-\frac{1}{p}}$$

$$S = \frac{\frac{1}{p}}{\frac{p-1}{p}}$$

$$S = \frac{1}{p}\cdot\frac{p}{p-1}$$

$$S = \frac{1}{p-1}.$$

71. a) ✎

b) ✎

73. a) We can determine the area removed from the original square on the n-iteration as

$$B_n = \frac{1}{9}\left(\frac{8}{9}\right)^{n-1}.$$

Therefore, the infinite sum can be given by

$$S = \frac{\frac{1}{9}}{1-\frac{8}{9}} = 1.$$

b) The total area that remains is

$$B_\infty = \lim_{n\to\infty} B_n = \lim_{n\to\infty} \frac{1}{9}\left(\frac{8}{9}\right)^n = 0.$$

c) ✎

75. ✎

Copyright © 2014 Pearson Education, Inc. Publishing as Addison-Wesley.

Exercise Set 9.3

1. a) Substituting the values into the simple interest formula, we have
$I = Pit = 300 \cdot 0.05 \cdot 1 = 15.$
The simple interest after 1 year is \$15.
b) The simple interest found in part (a) is added to the principle to find the future value
$A = P + I = 300 + 15 = 315.$
The future value of the account is \$315.

3. a) Substituting the values into the simple interest formula, we have
$I = Pit = 1200 \cdot 0.014 \cdot 2 = 33.60.$
The simple interest after 2 years is \$33.60.
b) The simple interest found in part (a) is added to the principle to find the future value
$A = P + I = 1200 + 33.60 = 1233.60.$
The future value of the account is \$1233.60.

5. a) Note that 20 months is $\frac{20}{12} = \frac{5}{3}$ years.
Substituting the values into the simple interest formula, we have
$I = Pit = 500 \cdot 0.031 \cdot \frac{5}{3} = 25.83.$
The simple interest after 20 months is \$25.83.
b) The simple interest found in part (a) is added to the principle to find the future value
$A = P + I = 500 + 25.83 = 525.83.$
The future value of the account is \$525.83.

7. a) Note that 9 months is $\frac{9}{12} = 0.75$ years
Substituting the values into the simple interest formula, we have
$I = Pit = 2000 \cdot 0.025 \cdot 0.75 = 37.50.$
The simple interest after 9 months is \$37.50.
b) The simple interest found in part (a) is added to the principle to find the future value
$A = P + I = 2000 + 37.50 = 2037.50.$
The future value of the account is \$2037.50.

9. a) Note that 25 weeks is $\frac{25}{52}$ years.
Substituting the values into the simple interest formula, we have
$I = Pit = 1300 \cdot 0.0195 \cdot \frac{25}{52} = 12.19.$
Rounding to the nearest cent, the simple interest after 25 weeks is \$12.19.
b) The simple interest found in part (a) is added to the principle to find the future value
$A = P + I = 1300 + 12.19 = 1312.19.$
The future value of the account is \$1312.19.

11. Substitute the values into the simple interest formula and solve for i as follows
$$I = Pit$$
$$40 = 200 \cdot i \cdot 2$$
$$40 = 400i$$
$$\frac{40}{400} = i$$
$$0.10 = i.$$
The simple interest rate is 10%.

13. Substitute the values into the simple interest future value formula and solve for i as follows
$$A = P(1 + it)$$
$$400 = 350 \cdot (1 + i \cdot 5)$$
$$400 = 350 + 1750i$$
$$50 = 1750i$$
$$\frac{50}{1750} = i$$
$$0.0286 = i.$$
The simple interest rate is 2.86%.

15. Substitute the values into the simple interest formula and solve for t as follows
$$I = Pit$$
$$15 = 500 \cdot 0.03 \cdot t$$
$$15 = 15t$$
$$1 = t.$$
It will take 1 year for \$500 to earn \$15 at a simple interest rate of 3%.

17. Substitute the values into the simple interest formula and solve for P as follows
$$I = Pit$$
$$20 = P \cdot 0.04 \cdot 1$$
$$20 = 0.04P$$
$$\frac{20}{0.4} = P$$
$$500 = P.$$
A principal of \$500 is required to earn \$20 at a 4% simple interest rate in one year.

Copyright © 2014 Pearson Education, Inc. Publishing as Addison-Wesley.

19. Note that 18 months implies $t = \frac{18}{12} = 1.5$. Substitute the values into the simple interest future value formula and solve for P as follows

$$A = P(1+it)$$
$$500 = P\cdot(1+0.025\cdot 1.5)$$
$$500 = 1.0375P$$
$$\frac{500}{1.0375} = P$$
$$481.93 = P.$$

The present value of \$500 that earns 2.5% simple interest for 18 months is \$481.93.

21. Note that 30 months implies $t = \frac{30}{12} = 2.5$. Substitute the values into the simple interest future value formula and solve for P as follows

$$A = P(1+it)$$
$$10{,}000 = P\cdot(1+0.0178\cdot 2.5)$$
$$10{,}000 = 1.0445P$$
$$\frac{10{,}000}{1.0445} = P$$
$$9573.96 = P.$$

The present value of \$10,000 that earns 1.78% simple interest for 30 months is \$9573.96.

23. a) Take the given information and substitute into the compound interest future value formula as follows

$$A = P\left(1+\frac{i}{c}\right)^{ct}$$
$$A = 600\left(1+\frac{0.04}{12}\right)^{12\cdot 5}$$
$$A = 732.60.$$

The future value of the account is \$732.60.

b) The interest earned is found by subtracting the principal from the future value. The interest earned is
$I = 732.60 - 600 = 132.60.$
The investment earned \$132.60 in interest.

c) The sequence that gives the amount in the account after n compounding periods is

$$A_n = 600\left(1+\frac{0.04}{12}\right)^{n} = 600(1.003333)^n.$$

The first five terms of this sequence are shown at the top of the next column.

$$A_1 = 600(1.003333)^1 = \$602.00,$$
$$A_2 = 600(1.003333)^2 = \$604.01,$$
$$A_3 = 600(1.003333)^3 = \$606.02,$$
$$A_4 = 600(1.003333)^4 = \$608.04,$$
$$A_5 = 600(1.003333)^5 = \$610.07.$$

25. a) Take the given information and substitute into the compound interest future value formula as follows

$$A = P\left(1+\frac{i}{c}\right)^{ct}$$
$$A = 500\left(1+\frac{0.02}{4}\right)^{4\cdot 7}$$
$$A = 574.94.$$

The future value of the account is \$574.94.

b) The interest earned is found by subtracting the principal from the future value. The interest earned is
$I = 574.74 - 500 = 74.94.$
The investment earned \$74.94 in interest.

c) The sequence that gives the amount in the account after n compounding periods is

$$A_n = 500\left(1+\frac{0.02}{4}\right)^{n} = 500(1.005)^n.$$

The first five terms of this sequence are

$$A_1 = 500(1.005)^1 = \$502.50,$$
$$A_2 = 500(1.005)^2 = \$505.01,$$
$$A_3 = 500(1.005)^3 = \$507.54,$$
$$A_4 = 500(1.005)^4 = \$510.08,$$
$$A_5 = 500(1.005)^5 = \$512.63.$$

27. a) Take the given information and substitute into the compound interest future value formula as follows

$$A = P\left(1+\frac{i}{c}\right)^{ct}$$
$$A = 450\left(1+\frac{0.0225}{1}\right)^{1\cdot 6}$$
$$A = 514.27.$$

The future value of the account is \$514.27.

Copyright © 2014 Pearson Education, Inc. Publishing as Addison-Wesley.

b) The interest earned is found by subtracting the principal from the future value. The interest earned is
$I = 514.27 - 450 = 64.27.$
The investment earned \$64.27 in interest.

c) The sequence that gives the amount in the account after n compounding periods is

$$A_n = 450\left(1+\frac{0.0225}{1}\right)^n = 450(1.0225)^n.$$

The first five terms of this sequence are

$$A_1 = 450(1.0225)^1 = \$460.13,$$
$$A_2 = 450(1.0225)^2 = \$470.48,$$
$$A_3 = 450(1.0225)^3 = \$481.06,$$
$$A_4 = 450(1.0225)^4 = \$491.89,$$
$$A_5 = 450(1.0225)^5 = \$502.95.$$

29. a) Take the given information and substitute into the compound interest future value formula as follows

$$A = P\left(1+\frac{i}{c}\right)^{ct}$$
$$A = 2500\left(1+\frac{0.0212}{52}\right)^{52\cdot 4}$$
$$A = 2721.20.$$

The future value of the account is \$2721.20.

b) The interest earned is found by subtracting the principal from the future value. The interest earned is
$I = 2721.20 - 2500 = 221.20.$
The investment earned \$221.20 in interest.

c) The sequence that gives the amount in the account after n compounding periods is

$$A_n = 2500\left(1+\frac{0.0212}{52}\right)^n = 2500(1.0004077)^n.$$

The first five terms of this sequence are

$$A_1 = 2500(1.0004077)^1 = \$2501.02,$$
$$A_2 = 2500(1.0004077)^2 = \$2502.04,$$
$$A_3 = 2500(1.0004077)^3 = \$2503.06,$$
$$A_4 = 2500(1.0004077)^4 = \$2504.08,$$
$$A_5 = 2500(1.0004077)^5 = \$2505.10.$$

31. a) Take the given information and substitute into the compound interest future value formula as follows

$$A = P\left(1+\frac{i}{c}\right)^{ct}$$
$$A = 1750\left(1+\frac{0.0413}{2}\right)^{2\cdot 6}$$
$$A = 2236.45.$$

The future value of the account is \$2236.45.

b) The interest earned is found by subtracting the principal from the future value. The interest earned is
$I = 2236.45 - 1750 = 486.45.$
The investment earned \$221.20 in interest.

c) The sequence that gives the amount in the account after n compounding periods is

$$A_n = 1750\left(1+\frac{0.0413}{2}\right)^n = 1750(1.02065)^n.$$

The first five terms of this sequence are

$$A_1 = 1750(1.02065)^1 = \$1786.14,$$
$$A_2 = 1750(1.02065)^2 = \$1823.02,$$
$$A_3 = 1750(1.02065)^3 = \$1860.67,$$
$$A_4 = 1750(1.02065)^4 = \$1899.09,$$
$$A_5 = 1750(1.02065)^5 = \$1938.31.$$

33. a) Take the given information and substitute into the compound interest future value formula, then solve for the present value P as follows

$$A = P\left(1+\frac{i}{c}\right)^{ct}$$
$$10,000 = P\left(1+\frac{0.04}{12}\right)^{12\cdot 3}$$
$$\frac{10,000}{\left(1+\frac{0.04}{12}\right)^{36}} = P$$
$$P = 8870.97.$$

The present value of the account is \$8870.90.

b) The interest earned is found by subtracting the present value from the future value. The interest earned is
$I = 10,000 - 8870.97 = 1129.03.$
The investment earned \$1129.03 in interest.

Copyright © 2014 Pearson Education, Inc. Publishing as Addison-Wesley.

c) The sequence that gives the amount in the account after n compounding periods is

$$A_n = 8870.97\left(1+\frac{0.04}{12}\right)^n$$

$A_n = 8870.97(1.0033333)^n$.

The first five terms of this sequence are

$A_1 = 8870.97(1.0033333)^1 = \$8900.54,$

$A_2 = 8870.97(1.0033333)^2 = \$8930.21,$

$A_3 = 8870.97(1.0033333)^3 = \$8959.98,$

$A_4 = 8870.97(1.0033333)^4 = \$8989.84,$

$A_5 = 8870.97(1.0033333)^5 = \$9019.81.$

35. a) Take the given information and substitute into the compound interest future value formula, then solve for the present value P as follows

$$A = P\left(1+\frac{i}{c}\right)^{ct}$$

$$4000 = P\left(1+\frac{0.042}{4}\right)^{4\cdot 2}$$

$$\frac{4000}{\left(1+\frac{0.042}{4}\right)^8} = P$$

$$P = 3679.34.$$

The present value of the account is \$3679.34.

b) The interest earned is found by subtracting the present value from the future value. The interest earned is

$I = 4000 - 3679.34 = 320.66.$

The investment earned \$320.66 in interest.

c) The sequence that gives the amount in the account after n compounding periods is

$$A_n = 3679.34\left(1+\frac{0.042}{4}\right)^n$$

$A_n = 3679.34(1.0105)^n$.

The first five terms of this sequence are

$A_1 = 3969.74(1.0105)^1 = \$3717.97,$

$A_2 = 3969.74(1.0105)^2 = \$3757.01,$

$A_3 = 3969.74(1.0105)^3 = \$3796.46,$

$A_4 = 3969.74(1.0105)^4 = \$3836.32,$

$A_5 = 3969.74(1.0105)^5 = \$3876.60.$

37. a) Take the given information, noting that 18 months implies $t = \frac{18}{12} = 1.5$ years, and substitute into the compound interest future value formula, then solve for the present value P as follows:

$$A = P\left(1+\frac{i}{c}\right)^{ct}$$

$$2500 = P\left(1+\frac{0.0315}{12}\right)^{12\cdot 1.5}$$

$$\frac{2500}{\left(1+\frac{0.0315}{12}\right)^{18}} = P$$

$$P = 2384.77.$$

The present value of the account is \$2384.77.

b) The interest earned is found by subtracting the present value from the future value. The interest earned is

$I = 2500 - 2384.77 = 115.23.$

The investment earned \$115.23 in interest.

c) The sequence that gives the amount in the account after n compounding periods is

$$A_n = 2384.77\left(1+\frac{0.0315}{12}\right)^n$$

$A_n = 2384.77(1.002625)^n$.

The first five terms of this sequence are

$A_1 = 2384.77(1.002625)^1 = \$2391.03,$

$A_2 = 2384.77(1.002625)^2 = \$2397.31,$

$A_3 = 2384.77(1.002625)^3 = \$2403.60,$

$A_4 = 2384.77(1.002625)^4 = \$2409.91,$

$A_5 = 2384.77(1.002625)^5 = \$2416.23.$

Copyright © 2014 Pearson Education, Inc. Publishing as Addison-Wesley.

39. a) Take the given information and substitute into the compound interest future value formula, then solve for the present value P as follows

$$A = P\left(1+\frac{i}{c}\right)^{ct}$$

$$17250 = P\left(1+\frac{0.0225}{\frac{1}{2}}\right)^{\frac{1}{2}\cdot 10}$$

$$\frac{17250}{\left(1+\frac{0.0225}{\frac{1}{2}}\right)^{5}} = P$$

$$P = 13{,}842.28.$$

The present value of the account is \$13,842.28.

b) The interest earned is found by subtracting the present value from the future value. The interest earned is

$I = 17250 - 13{,}842.28 = 3407.72.$

The investment earned \$3407.72 in interest.

c) The sequence that gives the amount in the account after n compounding periods is

$$A_n = 13{,}842.28\left(1+\frac{0.0225}{\frac{1}{2}}\right)^{n}$$

$$A_n = 13{,}842.28(1.045)^n.$$

The first five terms of this sequence are:

$$A_1 = 13{,}842.28(1.045)^1 = \$14{,}465.18,$$
$$A_2 = 13{,}842.28(1.045)^2 = \$15{,}116.12,$$
$$A_3 = 13{,}842.28(1.045)^3 = \$15{,}796.34,$$
$$A_4 = 13{,}842.28(1.045)^4 = \$16{,}507.18,$$
$$A_5 = 13{,}842.28(1.045)^5 = \$17{,}250.00.$$

41. a) Using the simple interest future value formula, we have

$$A = P(1+it) = P + Pit$$
$$1035 = 1000 + 1000\cdot i\cdot 1$$
$$35 = 1000\cdot i$$
$$\tfrac{35}{1000} = i$$
$$i = 0.035.$$

The simple interest rate is 3.5%.

b) The sequence that gives the value of the deposit after n years is given by

$A_n = 1000(1+0.035n).$

The first five terms of the sequence are

$$A_1 = 1000(1+0.035\cdot 1) = \$1035,$$
$$A_2 = 1000(1+0.035\cdot 2) = \$1070,$$
$$A_3 = 1000(1+0.035\cdot 3) = \$1105,$$
$$A_4 = 1000(1+0.035\cdot 4) = \$1140,$$
$$A_5 = 1000(1+0.035\cdot 5) = \$1175.$$

43. a) The book is valued at \$25, so the fine for being n days overdue is given by

$F_n = 25(0.005)n = 0.125n.$

b) The first five terms of the sequence representing the fine after n days are:

$$F_1 = 0.125\cdot 1 = \$0.125 = \$0.13,$$
$$F_2 = 0.125\cdot 2 = \$0.25,$$
$$F_3 = 0.125\cdot 3 = \$0.375 = \$0.38,$$
$$F_4 = 0.125\cdot 4 = \$0.50,$$
$$F_5 = 0.125\cdot 5 = \$0.625 = \$0.63.$$

45. a) Note that 3 months implies $t = \frac{3}{12} = 0.25$ years. Substitute the values into the simple interest future value formula and solve for P as follows

$$A = P(1+it)$$
$$500 = P\cdot(1+0.02\cdot 0.25)$$
$$500 = 1.005P$$
$$\tfrac{500}{1.005} = P$$
$$497.51 = P.$$

The present value of the bond is \$497.51.

b) The interest earned on the bond is found by subtracting the purchase price (present value) of the bond from the future value as follows $I = 500 - 497.51 = 2.49.$

The bond earned \$2.49 in interest.

47. The principal on the loan is \$300. The term of the loan is 8 months or $t = \frac{8}{12} = \frac{2}{3}$ years. The annual interest rate is 5%. Substituting these values into the simple interest formula we have

$I = Pit = 300\cdot 0.05\cdot \frac{2}{3} = 10.$

The store charges \$10 in interest.

To determine the monthly payment, take the total payment on the loan and divided it equally among the 8 months. The total payment is \$310, therefore, the monthly payment is $\frac{310}{8} = \$38.75.$

Copyright © 2014 Pearson Education, Inc. Publishing as Addison-Wesley.

49. a) Substituting the given information into the future value formula for compound interest the formula that will give the value of Gina's account after n years is

$$A_n = P\left(1+\frac{i}{c}\right)^{c \cdot n}$$

$$A_n = 3000\left(1+\frac{0.045}{12}\right)^{12 \cdot n}$$

$$A_n = 3000(1.00375)^{12n}.$$

b) The first five terms of the sequence representing the value of Gina's account are

$$A_1 = 3000(1.00375)^{12 \cdot 1} = \$3137.82,$$

$$A_2 = 3000(1.00375)^{12 \cdot 2} = \$3281.97,$$

$$A_3 = 3000(1.00375)^{12 \cdot 3} = \$3432.74,$$

$$A_4 = 3000(1.00375)^{12 \cdot 4} = \$3590.44,$$

$$A_5 = 3000(1.00375)^{12 \cdot 5} = \$3755.39.$$

51. Substituting the information into the future value formula and solving for the present value we have

$$5000 = P\left(1+\frac{0.048}{12}\right)^{12 \cdot 2}$$

$$\frac{5000}{(1.004)^{24}} = P$$

$$4543.19 = P.$$

The present value of Yvonne's account is \$4543.19.
To determine the total interest earned subtract the present value from the future value
$I = 5000 - 4543.19 = 456.81$.
The account earned \$456.81 in interest.

53. a) To determine the interest rate, substitute the given information into the future value formula and solve for i as follows

$$1690.91 = 1500\left(1+\frac{i}{12}\right)^{12 \cdot 3}$$

$$\frac{1690.91}{1500} = \left(1+\frac{i}{12}\right)^{36}$$

$$(1.12727)^{\frac{1}{36}} = 1+\frac{i}{12}$$

$$(1.12727)^{\frac{1}{36}} - 1 = \frac{i}{12}$$

$$0.0033333694 = \frac{i}{12}$$

$$0.04 = i.$$

The annual interest rate is 4%.

b) The formula that gives the value of the account after n years is

$$A_n = 1500\left(1+\frac{0.04}{12}\right)^{12 \cdot n}$$

$$A_n = 1500(1.0033333)^{12n}.$$

The first five terms of this sequence are

$$A_1 = 1500(1.0033333)^{12 \cdot 1} = \$1561.11,$$

$$A_2 = 1500(1.0033333)^{12 \cdot 2} = \$1624.71,$$

$$A_3 = 1500(1.0033333)^{12 \cdot 3} = \$1690.91,$$

$$A_4 = 1500(1.0033333)^{12 \cdot 4} = \$1759.80,$$

$$A_5 = 1500(1.0033333)^{12 \cdot 5} = \$1831.49.$$

55. a) The future value for the First Federal certificate of deposit is given by
$A = 10{,}000(1+0.04 \cdot 3) = \$11{,}200.$

The future value for the Valley View savings account is given by

$$A = 10{,}000\left(1+\frac{0.038}{12}\right)^{12 \cdot 3} = \$11{,}205.50.$$

b) The interest earned on the First Federal certificate of deposit is
$I = 11{,}200 - 10{,}000 = \$1200.$
The interest earned on the Valley View savings account is
$I = 11{,}205.50 - 10{,}000 = \$1205.50.$

c) ✎

Copyright © 2014 Pearson Education, Inc. Publishing as Addison-Wesley.

57. a) The present value for simple interest option is given by

$$3000 = P(1+0.042\cdot 2)$$
$$\frac{3000}{1.084} = P$$
$$2767.53 = P.$$

Claire will need to deposit $2767.53 into the simple interest account in order to meet her goal of $3000 in 2 years

The present value for the compound interest option is given by

$$3000 = P\left(1+\frac{0.04}{12}\right)^{12\cdot 2}$$
$$\frac{3000}{(1.003333)^{24}} = P$$
$$2769.72 = P.$$

Claire will need to deposit $2769.72 into the compound interest account in order to meet her goal of $3000 in 2 years.

b) The interest earned on the simple interest option is $I = 3000 - 2767.53 = \$232.47$.

The interest earned on the compound interest option is

$I = 3000 - 2769.72 = \$230.28$.

c) ✎

59. Substituting the information into the formula for annual yield give us

$$Y = \left(1+\frac{0.053}{12}\right)^{12} - 1 = 0.0543.$$

Thus, the annual percentage yield is 5.43%

61. Substituting the information into the formula for annual yield give us

$$Y = \left(1+\frac{0.0375}{52}\right)^{52} - 1 = 0.0382.$$

Thus, the annual percentage yield is 3.82%

63. a) The annual yield for Western Bank is

$$Y_{WB} = \left(1+\frac{0.045}{1}\right)^{1} - 1 = 0.045.$$

Thus the annual yield for Western Bank is 4.5%.

The annual yield for Commonwealth Savings is

$$Y_{CW} = \left(1+\frac{0.0443}{12}\right)^{12} - 1 = 0.0452.$$

Thus the annual yield for Commonwealth savings is 4.52%.

b) Commonwealth savings has the higher annual yield.

65. The annual yield for Stockman's Bank is

$$Y_{SB} = \left(1+\frac{0.042}{1}\right)^{1} - 1 = 0.042.$$

Thus the annual yield for Stockman's Bank is 4.2%.

In order to compete, the annual yield for Mesalands Savings must be 4.2%. We substitute into the annual yield formula and solve for the interest rate

$$\left(1+\frac{i}{12}\right)^{12} - 1 = 0.042$$
$$\left(1+\frac{i}{12}\right)^{12} = 1.042$$
$$1+\frac{i}{12} = (1.042)^{\frac{1}{12}}$$
$$\frac{i}{12} = (1.042)^{\frac{1}{12}} - 1$$
$$i = 12\left[(1.042)^{\frac{1}{12}} - 1\right]$$
$$i = 0.0412.$$

Mesalands Savings needs to offer at least 4.12% compounded monthly to be competitive.

Copyright © 2014 Pearson Education, Inc. Publishing as Addison-Wesley.

67. a) The solution can be seen graphically. Graph each of the options on the same coordinate axis and find the intersection of the graphs.

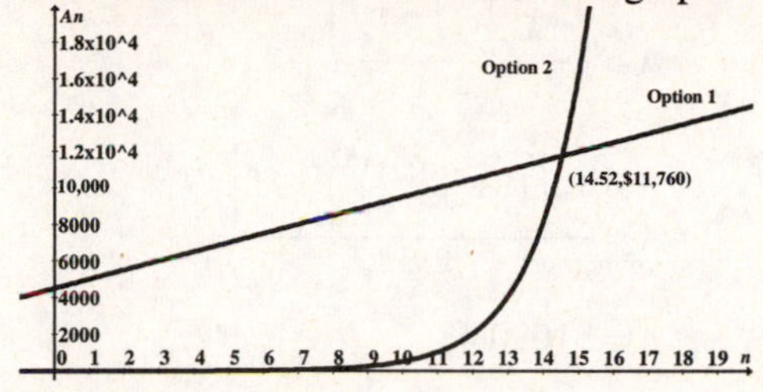

From the graph, we see that the intersection occurs when $n = 14.52$. Therefore, option 2 will exceed option 1 in the 15th month.

b) The total amount earned is the partial sum of the monthly payments. For option 1, the total amount earned is given by

$$S_{n1} = \tfrac{n}{2}(a_1 + a_n)$$

$$S_{n1} = \tfrac{n}{2}(5000 + 4500 + 500n)$$

$$S_{n1} = 4750n + 250n^2.$$

For option 2, the total amount earned is given by

$$S_{n2} = \frac{b_1(r^n - 1)}{r - 1}$$

$$S_{n2} = \frac{1(2^n - 1)}{2 - 1}$$

$$S_{n2} = 2^n - 1.$$

Graph each equation on the same axis and find the intersection point.

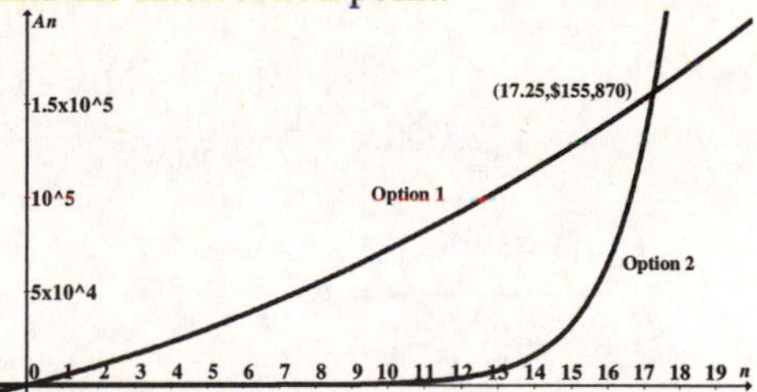

The point of intersection is $n = 17.25$. Therefore, the total amount earned in option 2 will exceed the total amount earned in option 1 on the 18th month.

Copyright © 2014 Pearson Education, Inc. Publishing as Addison-Wesley.

Exercise Set 9.4

1. Substitute the given information into the future value formula. The future value of the annuity is

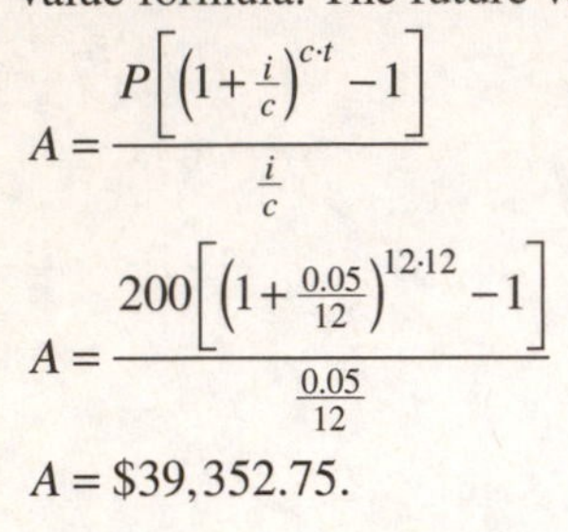

$$A=\frac{P\left[\left(1+\frac{i}{c}\right)^{c\cdot t}-1\right]}{\frac{i}{c}}$$

$$A=\frac{200\left[\left(1+\frac{0.05}{12}\right)^{12\cdot 12}-1\right]}{\frac{0.05}{12}}$$

$A=\$39,352.75.$

3. Substitute the given information into the future value formula. The future value of the annuity is

$$A=\frac{P\left[\left(1+\frac{i}{c}\right)^{c\cdot t}-1\right]}{\frac{i}{c}}$$

$$A=\frac{250\left[\left(1+\frac{0.045}{4}\right)^{4\cdot 20}-1\right]}{\frac{0.045}{4}}$$

$A=\$32,161.67.$

5. Substitute the given information into the future value formula. The future value of the annuity is

$$A=\frac{P\left[\left(1+\frac{i}{c}\right)^{c\cdot t}-1\right]}{\frac{i}{c}}$$

$$A=\frac{500\left[\left(1+\frac{0.053}{2}\right)^{2\cdot 8}-1\right]}{\frac{0.053}{2}}$$

$A=\$9804.70.$

7. Substitute the given information into the future value formula. The future value of the annuity is

$$A=\frac{P\left[\left(1+\frac{i}{c}\right)^{c\cdot t}-1\right]}{\frac{i}{c}}$$

$$A=\frac{1500\left[\left(1+\frac{0.0325}{1}\right)^{1\cdot 25}-1\right]}{\frac{0.0325}{1}}$$

$A=\$56,519.90.$

9. Substitute the given information into the future value formula. The future value of the annuity is

$$A=\frac{P\left[\left(1+\frac{i}{c}\right)^{c\cdot t}-1\right]}{\frac{i}{c}}$$

$$A=\frac{10\left[\left(1+\frac{0.0475}{52}\right)^{52\cdot 2}-1\right]}{\frac{0.0475}{52}}$$

$A=\$1090.48.$

11. Substitute the given information into the future value formula and solve for P as follows

$$A=\frac{P\left[\left(1+\frac{i}{c}\right)^{c\cdot t}-1\right]}{\frac{i}{c}}$$

$$5000=\frac{P\left[\left(1+\frac{0.04}{12}\right)^{12\cdot 6}-1\right]}{\frac{0.04}{12}}$$

$$\frac{5000\left(\frac{0.04}{12}\right)}{\left(1+\frac{0.04}{12}\right)^{12\cdot 6}-1}=P$$

$61.56=P.$

The monthly sinking fund payment required to achieve the given future value is \$61.56.

13. Substitute the given information into the future value formula and solve for P as follows

$$A=\frac{P\left[\left(1+\frac{i}{c}\right)^{c\cdot t}-1\right]}{\frac{i}{c}}$$

$$12,000=\frac{P\left[\left(1+\frac{0.0425}{1}\right)^{1\cdot 9}-1\right]}{\frac{0.0425}{1}}$$

$$\frac{12,000\left(\frac{0.0425}{1}\right)}{\left(1+\frac{0.0425}{1}\right)^{1\cdot 9}-1}=P$$

$1122.35=P.$

The annual sinking fund payment required to achieve the given future value is \$1122.35.

Copyright © 2014 Pearson Education, Inc. Publishing as Addison-Wesley.

15. Substitute the given information into the future value formula and solve for P as follows

$$A = \frac{P\left[\left(1+\frac{i}{c}\right)^{c\cdot t}-1\right]}{\frac{i}{c}}$$

$$15{,}000 = \frac{P\left[\left(1+\frac{0.03175}{4}\right)^{4\cdot 15}-1\right]}{\frac{0.03175}{4}}$$

$$\frac{15{,}000\left(\frac{0.03175}{4}\right)}{\left(1+\frac{0.03175}{4}\right)^{4\cdot 15}-1} = P$$

$$196.15 = P.$$

The quarterly sinking fund payment required to achieve the given future value is \$196.15.

17. Substitute the given information into the amortization formula and solve for the payment P as follows

$$P\left(1+\frac{i}{c}\right)^{c\cdot t} = \frac{p\left[\left(1+\frac{i}{c}\right)^{c\cdot t}-1\right]}{\frac{i}{c}}$$

$$7000\left(1+\frac{0.06}{12}\right)^{12\cdot 5} = \frac{p\left[\left(1+\frac{0.06}{12}\right)^{12\cdot 5}-1\right]}{\frac{0.06}{12}}$$

$$9441.95106784 = p[69.77003]$$

$$\frac{9441.95106784}{69.77003} = p$$

$$135.33 = p.$$

The monthly payment needed to amortize the given loan amount is \$135.33.

19. Substitute the given information into the amortization formula and solve for the payment P as follows

$$P\left(1+\frac{i}{c}\right)^{c\cdot t} = \frac{p\left[\left(1+\frac{i}{c}\right)^{c\cdot t}-1\right]}{\frac{i}{c}}$$

$$12{,}000\left(1+\frac{0.057}{4}\right)^{4\cdot 6} = \frac{p\left[\left(1+\frac{0.057}{4}\right)^{4\cdot 6}-1\right]}{\frac{0.057}{4}}$$

$$16{,}852.3953555 = p[28.3765810266]$$

$$\frac{16{,}852.3953555}{28.3765810266} = p$$

$$593.88 = p.$$

The quarterly payment needed to amortize the given loan amount is \$593.88.

21. Substitute the given information into the amortization formula and solve for the payment P as follows

$$P\left(1+\frac{i}{c}\right)^{c\cdot t} = \frac{p\left[\left(1+\frac{i}{c}\right)^{c\cdot t}-1\right]}{\frac{i}{c}}$$

$$500\left(1+\frac{0.041}{12}\right)^{12\cdot 1} = \frac{p\left[\left(1+\frac{0.041}{12}\right)^{12\cdot 1}-1\right]}{\frac{0.041}{12}}$$

$$520.889650412 = p[12.2280880459]$$

$$\frac{520.889650412}{12.2280880459} = p$$

$$42.60 = p.$$

The monthly payment needed to amortize the given loan amount is \$42.60.

23. Substitute the given information into the amortization formula and solve for the payment P as follows

$$P\left(1+\frac{i}{c}\right)^{c\cdot t} = \frac{p\left[\left(1+\frac{i}{c}\right)^{c\cdot t}-1\right]}{\frac{i}{c}}$$

$$150{,}000\left(1+\frac{0.0515}{2}\right)^{2\cdot 30} = \frac{p\left[\left(1+\frac{0.0515}{2}\right)^{2\cdot 30}-1\right]}{\frac{0.0515}{2}}$$

$$689{,}577.047154 = p[139.696322888]$$

$$\frac{689{,}577.047154}{139.696322888} = p$$

$$4936.26 = p.$$

The semiannual payment needed to amortize the given loan amount is \$4936.26.

25. Substitute the given information into the amortization formula and solve for the payment P as follows

$$P\left(1+\frac{i}{c}\right)^{c\cdot t} = \frac{p\left[\left(1+\frac{i}{c}\right)^{c\cdot t}-1\right]}{\frac{i}{c}}$$

$$75{,}000\left(1+\frac{0.08}{1}\right)^{1\cdot 15} = \frac{p\left[\left(1+\frac{0.08}{1}\right)^{1\cdot 15}-1\right]}{\frac{0.08}{1}}$$

$$237{,}912.683565 = p[27.1521139275]$$

$$\frac{237{,}912.683565}{27.1521139275} = p$$

$$8762.22 = p.$$

The annual payment needed to amortize the given loan amount is \$8762.22.

Copyright © 2014 Pearson Education, Inc. Publishing as Addison-Wesley.

27. a) Substitute the given information into the future value formula. The future value of the annuity is

$$A = \frac{P\left[\left(1+\frac{i}{c}\right)^{c\cdot t} - 1\right]}{\frac{i}{c}}$$

$$A = \frac{125\left[\left(1+\frac{0.045}{12}\right)^{12\cdot 8} - 1\right]}{\frac{0.045}{12}}$$

$A = \$14,412.16.$

The future value of Tim's annuity is \$14,412.16.

b) To find the total personal contributions, multiply the number of payments by the payment amount. Tim made monthly payments of \$125 for 8 years, his total contribution is $\$125\cdot 12\cdot 8 = \$12,000.$

c) Total interest earned is the future value minus the personal contributions

$I = 14,412.16 - 12,000 = \$2412.16.$

Tim's annuity earned \$2412.16 in interest.

29. a) Substitute the given information into the future value formula. The future value of the annuity is

$$A = \frac{P\left[\left(1+\frac{i}{c}\right)^{c\cdot t} - 1\right]}{\frac{i}{c}}$$

$$A = \frac{2000\left[\left(1+\frac{0.07}{1}\right)^{1\cdot 15} - 1\right]}{\frac{0.07}{1}}$$

$A = \$50,258.04.$

The future value of Cindy's annuity is \$50,258.04.

b) To find the total personal contributions, multiply the number of payments by the payment amount. Cindy made annual payments of \$2000 for 15 years, her total contribution is $\$2000\cdot 1\cdot 15 = \$30,000.$

c) Total interest earned is the future value minus the personal contributions

$I = 50,258.04 - 30,000 = \$20,258.04.$

Cindy's annuity earned \$20,258.04 in interest.

31. a) Substitute the given information into the future value formula and solve for P as shown at the top of the next column.

$$A = \frac{P\left[\left(1+\frac{i}{c}\right)^{c\cdot t} - 1\right]}{\frac{i}{c}}$$

$$5000 = \frac{P\left[\left(1+\frac{0.0435}{12}\right)^{12\cdot 2} - 1\right]}{\frac{0.0435}{12}}$$

$$\frac{5000\left(\frac{0.0435}{12}\right)}{\left(1+\frac{0.0435}{12}\right)^{12\cdot 2} - 1} = P$$

$199.78 = P.$

The monthly sinking fund payment required to achieve the given future value is \$199.78.

b) To find the total personal contributions, multiply the number of payments by the payment amount. Gayla made monthly payments of \$199.78 for 2 years, her total contribution is $\$199.78\cdot 12\cdot 2 = \$4794.72.$

c) Total interest earned is the future value minus the personal contributions

$I = 5000 - 4794.72 = \$205.28.$

Gayla's account earned \$205.28 in interest.

33. a) Substitute the given information into the future value formula and solve for P as follows

$$A = \frac{P\left[\left(1+\frac{i}{c}\right)^{c\cdot t} - 1\right]}{\frac{i}{c}}$$

$$200,000 = \frac{P\left[\left(1+\frac{0.054}{1}\right)^{1\cdot 25} - 1\right]}{\frac{0.054}{1}}$$

$$\frac{200,000\left(\frac{0.054}{1}\right)}{\left(1+\frac{0.054}{1}\right)^{1\cdot 25} - 1} = P$$

$3964.69 = P.$

The annual sinking fund payment required to achieve the given future value is \$3964.69.

b) To find the total personal contributions, multiply the number of payments by the payment amount. The Miyokawas made annual payments of \$3964.69 for 25 years, their total contribution is

$\$3964.69\cdot 1\cdot 25 = \$99,117.25.$

c) Total interest earned is the future value minus the personal contributions

$I = 200,000 - 99,117.25 = \$100,882.75.$

The Miyokawas' account earned \$100,882.75 in interest.

Copyright © 2014 Pearson Education, Inc. Publishing as Addison-Wesley.

35. a) For the lump sum option, plug in the values into the future value formula and solve for P

$$A = P\left(1+\tfrac{i}{c}\right)^{c\cdot t}$$

$$10{,}000 = P\left(1+\tfrac{0.045}{12}\right)^{12\cdot 3}$$

$$\frac{10{,}000}{\left(1+\tfrac{0.045}{12}\right)^{12\cdot 3}} = P$$

$$8739.37 = P.$$

The Monroes will have to deposit \$8739.37 into the account in order to meet their goals. The interest earned is given by $I = 10{,}000 - 8739.37 = \$1260.63.$

b) To find the monthly payments for option 2, Substitute the given information into the future value formula and solve for P as follows

$$A = \frac{P\left[\left(1+\tfrac{i}{c}\right)^{c\cdot t} - 1\right]}{\tfrac{i}{c}}$$

$$10{,}000 = \frac{P\left[\left(1+\tfrac{0.0435}{12}\right)^{12\cdot 3} - 1\right]}{\tfrac{0.0435}{12}}$$

$$\frac{10{,}000\left(\tfrac{0.0435}{12}\right)}{\left(1+\tfrac{0.0435}{12}\right)^{12\cdot 3} - 1} = P$$

$$260.55 = P.$$

The monthly sinking fund payment required to achieve the given future value is \$260.55. To find the interest earned, first find the personal contribution. To find the total personal contributions, multiply the number of payments by the payment amount. The Monroes made monthly payments of \$260.55 for 3 years, their total contribution is $\$260.55\cdot 12\cdot 3 = \$9379.80.$

Total interest earned is the future value minus the personal contributions

$I = 10{,}000 - 9379.80 = \$620.20.$

c) Option 1 earned more interest. It earned $\$1260.63 - \$620.20 = \$640.43$ more interest than option 2.

37. a) Substitute the given information into the future value formula. The future value of the account is determined at the top of the next column.

$$A = \frac{P\left[\left(1+\tfrac{i}{c}\right)^{c\cdot t} - 1\right]}{\tfrac{i}{c}}$$

$$A = \frac{1500\left[\left(1+\tfrac{0.0525}{4}\right)^{4\cdot 12} - 1\right]}{\tfrac{0.0525}{4}}$$

$$A = \$99{,}420.67.$$

After 12 years, Janice will have \$99,420.67 in her account.

b) Substituting the amount from part (a) into the compound interest formula we have

$$A = P\left(1+\tfrac{i}{c}\right)^{c\cdot t}$$

$$A = 99{,}420.67\left(1+\tfrac{0.0525}{4}\right)^{4\cdot 20}$$

$$A = \$282{,}175.47.$$

The account will grow to \$282,175.47.

c) To find the interest earned, first find the personal contribution. To find the total personal contributions, multiply the number of payments by the payment amount. Janice made quarterly payments of \$1500 for 12 years, her total contribution is $\$1500\cdot 4\cdot 12 = \$72{,}000.$

Total interest earned is the future value minus the personal contributions

$I = 282{,}175.47 - 72{,}000 = \$210{,}175.47.$

Janice's account earned \$210,175.47 in interest.

39. a) The principal on the loan is $P = 22{,}150 - 4000 = 18{,}150.$

Substitute the given information into the amortization formula and solve for the payment p as follows

$$P\left(1+\tfrac{i}{c}\right)^{c\cdot t} = \frac{p\left[\left(1+\tfrac{i}{c}\right)^{c\cdot t} - 1\right]}{\tfrac{i}{c}}$$

$$18{,}150\left(1+\tfrac{0.065}{12}\right)^{12\cdot 5} = \frac{p\left[\left(1+\tfrac{0.065}{12}\right)^{12\cdot 5} - 1\right]}{\tfrac{0.065}{12}}$$

$$25{,}098.1344344 = p\left[70.6739675464\right]$$

$$\frac{25{,}098.1344344}{70.6739675464} = p$$

$$355.13 = p.$$

The monthly car payment is \$355.13.

b) Assuming Todd makes every payment for the life of the loan, his total payments are $355.13\cdot 12\cdot 5 = \$21{,}307.80.$

Copyright © 2014 Pearson Education, Inc. Publishing as Addison-Wesley.

c) The total interest paid is the total payment minus the principal on the loan. Therefore, the total interest paid on this loan is $I = 21,307.80 - 18,150 = \$3157.80$.

41. a) The principal on the loan is $P = 195,000 \cdot 0.75 = 146,250$.
Substitute the given information into the amortization formula and solve for the payment p as follows

$$P\left(1+\tfrac{i}{c}\right)^{c\cdot t} = \frac{p\left[\left(1+\tfrac{i}{c}\right)^{c\cdot t}-1\right]}{\tfrac{i}{c}}$$

$$146,250\left(1+\tfrac{0.052}{12}\right)^{12\cdot 30} = \frac{p\left[\left(1+\tfrac{0.052}{12}\right)^{12\cdot 30}-1\right]}{\tfrac{0.052}{12}}$$

$$693,635.92487 = p[863.725325238]$$

$$\frac{693,635.92487}{863.725325238} = p$$

$$803.07 = p.$$

The monthly house payment is \$803.07.

b) Assuming Hogansons makes every payment for the life of the loan, their total payments are $803.07 \cdot 12 \cdot 30 = \$289,105.20$.

c) The total interest paid is the total payment minus the principal on the loan. Therefore, the total interest paid on this loan is $I = 209,105.20 - 146,250 = \$142,855.20$.

43. a) The principal balance on the credit card is $P = 500$. Substitute the given information into the amortization formula and solve for the payment p as follows

$$P\left(1+\tfrac{i}{c}\right)^{c\cdot t} = \frac{p\left[\left(1+\tfrac{i}{c}\right)^{c\cdot t}-1\right]}{\tfrac{i}{c}}$$

$$500\left(1+\tfrac{0.2275}{12}\right)^{12\cdot 10} = \frac{p\left[\left(1+\tfrac{0.2275}{12}\right)^{12\cdot 10}-1\right]}{\tfrac{0.2275}{12}}$$

$$4761.47005973 = p[449.561676631]$$

$$\frac{4761.47005973}{449.561676631} = p$$

$$10.59 = p.$$

The monthly credit card payment is \$10.59.

b) Assuming Joanna makes every payment for the life of the loan, her total payments are $10.59 \cdot 12 \cdot 10 = \$1270.80$.

c) The total interest paid is the total payment minus the principal on the loan. Therefore, the total interest paid on this loan is $I = 1270.80 - 500 = \$770.80$.

45. a) Substitute the given information into the amortization formula and solve for the payment p as follows

$$P\left(1+\tfrac{i}{c}\right)^{c\cdot t} = \frac{p\left[\left(1+\tfrac{i}{c}\right)^{c\cdot t}-1\right]}{\tfrac{i}{c}}$$

$$9925\left(1+\tfrac{0.8968}{12}\right)^{12\cdot 7} = \frac{p\left[\left(1+\tfrac{0.8968}{12}\right)^{12\cdot 7}-1\right]}{\tfrac{0.8968}{12}}$$

$$4,226,597.62778 = p[5684.92861643]$$

$$\frac{4,226,597.62778}{5684.92861643} = p$$

$$743.47 = p.$$

The monthly loan payment is \$743.47.

b) Assuming the borrower makes every payment for the life of the loan, the total payments are $743.47 \cdot 12 \cdot 7 = \$62,451.48$.

c) The total interest paid is the total payment minus the principal on the loan. Therefore, the total interest paid on this loan is $I = 62,451.48 - 9925 = \$52,526.48$.

Copyright © 2014 Pearson Education, Inc. Publishing as Addison-Wesley.

47. From exercise 39, we have the monthly payment is $355.13. We create the following amortization table.

Balance	Payment	Portion of payment applied to interest	Portion of payment applied to principal	New Balance
$18,150	$355.13	$18{,}150\left(\frac{0.065}{12}\right)=\$98.31$	$355.15-98.31=\$256.82$	$17,893.18
$17,893.18	$355.13	$17{,}893.18\left(\frac{0.065}{12}\right)=\$96.92$	$355.13-96.92=\$258.21$	$17,634.97

49. From exercise 41, we have the monthly payment is $803.07. We create the following amortization table.

Balance	Payment	Portion of payment applied to interest	Portion of payment applied to principal	New Balance
$146,250	$803.07	$146{,}250\left(\frac{0.052}{12}\right)=\$633.75$	$803.07-633.75=\$169.32$	$146,080.68
$146,080.68	$803.07	$146{,}080.68\left(\frac{0.052}{12}\right)=\$633.02$	$803.07-633.02=\$170.05$	$145,910.63

51. From exercise 43, we have the monthly payment is $10.59. We create the following amortization table.

Balance	Payment	Portion of payment applied to interest	Portion of payment applied to principal	New Balance
$500	$10.59	$500\left(\frac{0.2275}{12}\right)=\$9.48$	$10.59-9.48=\$1.11$	$498.89
$498.89	$10.59	$498.89\left(\frac{0.2275}{12}\right)=\$9.46$	$10.59-9.46=\$1.13$	$497.76

Copyright © 2014 Pearson Education, Inc. Publishing as Addison-Wesley.

53. Substitute the given information into the amortization formula and solve for the principal P as follows

$$P\left(1+\tfrac{i}{c}\right)^{c\cdot t} = \frac{p\left[\left(1+\tfrac{i}{c}\right)^{c\cdot t}-1\right]}{\tfrac{i}{c}}$$

$$P\left(1+\tfrac{0.058}{12}\right)^{12\cdot 6} = \frac{300\left[\left(1+\tfrac{0.058}{12}\right)^{12\cdot 6}-1\right]}{\tfrac{0.058}{12}}$$

$$P(1.41504551988) = 25,761.4460613$$

$$P = \frac{25,761.4460613}{1.41504551988}$$

$$P = 18,205.38.$$

The largest loan Desmond can afford is \$18,205.38.

55. Substitute the given information into the amortization formula and solve for the principal P as follows

$$P\left(1+\tfrac{i}{c}\right)^{c\cdot t} = \frac{p\left[\left(1+\tfrac{i}{c}\right)^{c\cdot t}-1\right]}{\tfrac{i}{c}}$$

$$P\left(1+\tfrac{0.0415}{12}\right)^{12\cdot 30} = \frac{1800\left[\left(1+\tfrac{0.0415}{12}\right)^{12\cdot 30}-1\right]}{\tfrac{0.0415}{12}}$$

$$P(3.46548342006) = 1,283,239.56321$$

$$P = \frac{1,283,239.56321}{3.46548342006}$$

$$P = 370,291.65.$$

The largest loan the Daleys can afford is \$370,291.65.

57. a) For option 1, substitute the given information into the amortization formula and solve for the payment p as follows

$$P\left(1+\tfrac{i}{c}\right)^{c\cdot t} = \frac{p\left[\left(1+\tfrac{i}{c}\right)^{c\cdot t}-1\right]}{\tfrac{i}{c}}$$

$$12,000\left(1+\tfrac{0.052}{12}\right)^{12\cdot 5} = \frac{p\left[\left(1+\tfrac{0.052}{12}\right)^{12\cdot 5}-1\right]}{\tfrac{0.052}{12}}$$

$$15,554.4214929 = p[68.3542594794]$$

$$\frac{15,554.4214929}{68.3542594794} = p$$

$$227.56 = p.$$

The monthly payment needed to amortize the given loan amount for option 1 is \$227.56.

For option2, substitute the given information into the amortization formula and solve for the payment p as follows

$$P\left(1+\tfrac{i}{c}\right)^{c\cdot t} = \frac{p\left[\left(1+\tfrac{i}{c}\right)^{c\cdot t}-1\right]}{\tfrac{i}{c}}$$

$$12,000\left(1+\tfrac{0.05}{12}\right)^{12\cdot 6} = \frac{p\left[\left(1+\tfrac{0.05}{12}\right)^{12\cdot 6}-1\right]}{\tfrac{0.05}{12}}$$

$$16,188.2129299 = p[83.7642586]$$

$$\frac{16,188.2129299}{83.7642586} = p$$

$$193.26 = p.$$

The monthly payment needed to amortize the given loan amount for option 2 is \$193.26.

b) The total payments for option 1 are $\$227.56\cdot 12\cdot 5 = \$13,653.60.$
The total payments for option2 are $\$193.26\cdot 12\cdot 6 = \$13,914.72.$

c) Option 1 results in less interest paid. Katie will pay $\$13,914.72 - \$13,653.60 = \$261.12$ less in interest if she uses option 1.

59. a) If the annual interest rate is 5%, substitute the given information into the amortization formula and solve for the payment p as follows

$$P\left(1+\tfrac{i}{c}\right)^{c\cdot t} = \frac{p\left[\left(1+\tfrac{i}{c}\right)^{c\cdot t}-1\right]}{\tfrac{i}{c}}$$

$$200,000\left(1+\tfrac{0.05}{12}\right)^{12\cdot 30} = \frac{p\left[\left(1+\tfrac{0.05}{12}\right)^{12\cdot 30}-1\right]}{\tfrac{0.05}{12}}$$

$$893,548.862812 = p[832.258635374]$$

$$\frac{893,548.862812}{832.258635374} = p$$

$$1073.64 = p.$$

The monthly payment needed to amortize the given loan amount at 5% interest is \$1073.64.

b) If the annual interest rate is 6%, substitute the given information into the amortization formula and solve for the payment p as shown on the next page.

Copyright © 2014 Pearson Education, Inc. Publishing as Addison-Wesley.

Using the information from the previous page, the payment is

$$P\left(1+\frac{i}{c}\right)^{c\cdot t} = \frac{p\left[\left(1+\frac{i}{c}\right)^{c\cdot t}-1\right]}{\frac{i}{c}}$$

$$200{,}000\left(1+\frac{0.06}{12}\right)^{12\cdot 30} = \frac{p\left[\left(1+\frac{0.06}{12}\right)^{12\cdot 30}-1\right]}{\frac{0.06}{12}}$$

$$1{,}204{,}515.04245 = p[1004.51504245]$$

$$\frac{1{,}204{,}515.04245}{1004.51504245} = p$$

$$1199.10 = p.$$

The monthly payment needed to amortize the given loan amount at 6% interest is \$1199.10.

c) The total payments for the 5% loan are $\$1073.64 \cdot 12 \cdot 30 = \$386{,}510.40$.
The total payments for 6% loan are $\$1199.10 \cdot 12 \cdot 30 = \$431{,}676.00$.
The difference between the total payments of the different loans is $431{,}676.00 - 386{,}515.40 = \$45{,}165.60$.
Darnell will save \$45,165.60 if he accepts the 5% loan.

61. a) Dwight will need to draw income for 25 years. To determine how much money he will need in order to be able to meet his goals, we substitute the given information into the amortization formula and solve for the principal P as follows:

$$P\left(1+\frac{i}{c}\right)^{c\cdot t} = \frac{p\left[\left(1+\frac{i}{c}\right)^{c\cdot t}-1\right]}{\frac{i}{c}}$$

$$P\left(1+\frac{0.045}{12}\right)^{12\cdot 25} = \frac{500\left[\left(1+\frac{0.045}{12}\right)^{12\cdot 25}-1\right]}{\frac{0.045}{12}}$$

$$P(3.07374252804) = 276{,}499.003738$$

$$P = \frac{276{,}499.003738}{3.07374252804}$$

$$P = 89{,}955.16.$$

Dwight will need \$89,955.16 in his annuity in order to draw \$500 per month from the annuity from age 60 to age 85.

b) Since Dwight is starting at age 25, he will have 35 years until he reaches age 60. Substitute the given information and the amount found in part (a) into the future value formula and solve for P as shown at the top of the next column.

$$A = \frac{P\left[\left(1+\frac{i}{c}\right)^{c\cdot t}-1\right]}{\frac{i}{c}}$$

$$89{,}955.16 = \frac{P\left[\left(1+\frac{0.045}{12}\right)^{12\cdot 35}-1\right]}{\frac{0.045}{12}}$$

$$\frac{89{,}955.16\left(\frac{0.045}{12}\right)}{\left(1+\frac{0.045}{12}\right)^{12\cdot 35}-1} = P$$

$$88.39 = P.$$

The monthly sinking fund payment required to achieve Dwight's retirement goal is \$88.39.

63. Substitute the given information and the amount into the future value formula and solve for W as follows

$$A = \frac{W\left[\left(1+\frac{i}{c}\right)^{c\cdot t}-1\right]}{\frac{i}{c}}$$

$$5{,}000{,}000 = \frac{W\left[\left(1+\frac{0.05}{1}\right)^{12\cdot 20}-1\right]}{\frac{0.05}{1}}$$

$$5{,}000{,}000 = W[33.0659541029]$$

$$151{,}212.94 = W.$$

The annual payment W you will receive under this plan is \$151,212.94 per year.

65. a) If the annual interest rate is 5%, substitute the given information into the amortization formula and solve for the payment p as follows

$$P\left(1+\frac{i}{c}\right)^{c\cdot t} = \frac{p\left[\left(1+\frac{i}{c}\right)^{c\cdot t}-1\right]}{\frac{i}{c}}$$

$$200{,}000\left(1+\frac{0.05}{12}\right)^{12\cdot 30} = \frac{p\left[\left(1+\frac{0.05}{12}\right)^{12\cdot 30}-1\right]}{\frac{0.05}{12}}$$

$$893{,}548.862812 = p[832.258635374]$$

$$\frac{893{,}548.862812}{832.258635374} = p$$

$$1073.64 = p.$$

The monthly payment needed to amortize the given loan amount at 5% interest is \$1073.64.

b) The total payments for the loan are $\$1073.64 \cdot 12 \cdot 30 = \$386{,}510.40$. Therefore, the total interest paid is $I = 386{,}510.40 - 200{,}000 = \$186{,}510.40$.

Copyright © 2014 Pearson Education, Inc. Publishing as Addison-Wesley.

c) By paying an extra 15% per month, the Begays would make monthly payments of $Q = 1.15 \cdot 1073.64 = 1234.69$. Substituting this into the formula we have

$$t = \frac{\ln(12 \cdot 1234.69) - \ln[(12 \cdot 1234.69) - (200,000 \cdot 0.05)]}{12 \ln\left(1 + \frac{0.05}{12}\right)}$$

$$t = \frac{9.6034818552 - 8.47975712483}{0.049896121784}$$

$t = 22.52.$

It will take the Begays approximately 22.5 years to pay off the loan.

d) By paying extra, the total payments on the loan are now

$\$1234.69 \cdot 12 \cdot 22.5 = \$333,366.30.$

Therefore, the interest that they will pay is $I = 333,366.30 - 200,000 = \$133,366.30$

e) By paying extra, the total interest saved is $186,510.40 - 133,366.30 = \$53,144.10.$

The Begays save $53,144.10 by paying off their loan in 22.5 years.

67. Using the amortization formula

$$P\left(1+\frac{i}{c}\right)^{c \cdot t} = \frac{Q\left[\left(1+\frac{i}{c}\right)^{c \cdot t} - 1\right]}{\frac{i}{c}}$$

$$Pi\left(1+\frac{i}{c}\right)^{c \cdot t} = cQ\left(1+\frac{i}{c}\right)^{c \cdot t} - cQ$$

$$(Qc - Pi)\left(1+\frac{i}{c}\right)^{c \cdot t} = cQ$$

$$\ln\left[(Qc - Pi)\left(1+\frac{i}{c}\right)^{c \cdot t}\right] = \ln(cQ)$$

$$\ln(Qc - Pi) + \ln\left(1+\frac{i}{c}\right)^{c \cdot t} = \ln(cQ)$$

$$c \cdot t \ln\left(1+\frac{i}{c}\right) = \ln(cQ) - \ln(Qc - Pi)$$

$$t = \frac{\ln(cQ) - \ln(Qc - Pi)}{c \cdot \ln\left(1+\frac{i}{c}\right)}$$

69. Using a spreadsheet we complete the amortization table below:

Balance	Payment	Portion to Interest	Portion to Principal	New Balance
$18,150.00	$355.13	$98.31	$256.82	$17,893.18
$17,893.18	$355.13	$96.92	$258.21	$17,634.97
$17,634.97	$355.13	$95.52	$259.61	$17,375.36
$17,375.36	$355.13	$94.12	$261.01	$17,114.35
$17,114.35	$355.13	$92.70	$262.43	$16,851.92
$16,851.92	$355.13	$91.28	$263.85	$16,588.07
$16,588.07	$355.13	$89.85	$265.28	$16,322.79
$16,322.79	$355.13	$88.42	$266.71	$16,056.08
$16,056.08	$355.13	$86.97	$268.16	$15,787.92
$15,787.92	$355.13	$85.52	$269.61	$15,518.31
$15,518.31	$355.13	$84.06	$271.07	$15,247.24
$15,247.24	$355.13	$82.59	$272.54	$14,974.70

71. Using a spreadsheet we complete the amortization table below:

Balance	Payment	Portion to Interest	Portion to Principal	New Balance
$146,250.00	$803.07	$633.75	$169.32	$146,080.68
$146,080.68	$803.07	$633.02	$170.05	$145,910.63
$145,910.63	$803.07	$632.28	$170.79	$145,739.84
$145,739.84	$803.07	$631.54	$171.53	$145,568.31
$145,568.31	$803.07	$630.80	$172.27	$145,396.04
$145,396.04	$803.07	$630.05	$173.02	$145,223.02
$145,223.02	$803.07	$629.30	$173.77	$145,049.25
$145,049.25	$803.07	$628.55	$174.52	$144,874.73
$144,874.73	$803.07	$627.79	$175.28	$144,699.45
$144,699.45	$803.07	$627.03	$176.04	$144,523.41
$144,523.41	$803.07	$626.27	$176.80	$144,346.61
$144,346.61	$803.07	$625.50	$177.57	$144,169.04

73. Using a spreadsheet we complete the amortization table below:

Balance	Payment	Portion to Interest	Portion to Principal	New Balance
$500.00	$10.59	$9.48	$1.11	$498.89
$498.89	$10.59	$9.46	$1.13	$497.76
$497.76	$10.59	$9.44	$1.15	$496.61
$496.61	$10.59	$9.41	$1.18	$495.43
$495.43	$10.59	$9.39	$1.20	$494.23
$494.23	$10.59	$9.37	$1.22	$493.01
$493.01	$10.59	$9.35	$1.24	$491.77
$491.77	$10.59	$9.32	$1.27	$490.50
$490.50	$10.59	$9.30	$1.29	$489.21
$489.21	$10.59	$9.27	$1.32	$487.89
$487.89	$10.59	$9.25	$1.34	$486.55
$486.55	$10.59	$9.22	$1.37	$485.18

Copyright © 2014 Pearson Education, Inc. Publishing as Addison-Wesley.

Exercise Set 9.5

1. a) To find the power series, we rewrite it in the form $c \cdot \frac{1}{1-r}$.

$$f(x)=\frac{1}{1+2x}=1\cdot\frac{1}{1-(-2x)}.$$

The above expression can be viewed as the sum of a geometric series with first term $a_1=1$ and common ratio $r=-2x$. Thus, for $|-2x|<1$, we have

$$\begin{aligned} f(x)&=1\cdot\frac{1}{1-(-2x)}\\ &=1\left(1+(-2x)+(-2x)^2+(-2x)^3+(-2x)^4+\cdots\right)\\ &=1\left(1-2x+4x^2-8x^3+16x^4-\cdots\right)\\ &=1-2x+4x^2-8x^3+16x^4-\cdots \end{aligned}$$

b) The series converges for $|-2x|<1$:

$$\begin{aligned} &|-2x|<1\\ &-1<-2x<1\\ &\tfrac{-1}{-2}>\tfrac{-2x}{-2}>\tfrac{1}{-2}\\ &-\tfrac{1}{2}<x<\tfrac{1}{2}. \end{aligned}$$

Thus, the interval of convergence is $-\frac{1}{2}<x<\frac{1}{2}$, and the series is centered at $x=0$.

3. a) To find the power series, we rewrite it in the form $c\cdot\frac{1}{1-r}$.

$$f(x)=\frac{1}{1-5x}=1\cdot\frac{1}{1-(5x)}.$$

The above expression can be viewed as the sum of a geometric series with first term $a_1=1$ and common ratio $r=5x$. Thus, for $|5x|<1$, we have

$$\begin{aligned} f(x)&=1\cdot\frac{1}{1-(5x)}\\ &=1\left(1+(5x)+(5x)^2+(5x)^3+(5x)^4+\cdots\right)\\ &=1\left(1+5x+25x^2+125x^3+625x^4+\cdots\right)\\ &=1+5x+25x^2+125x^3+625x^4+\cdots \end{aligned}$$

b) The series converges for $|5x|<1$:

$$\begin{aligned} &|5x|<1\\ &-1<5x<1\\ &-\tfrac{1}{5}<x<\tfrac{1}{5}. \end{aligned}$$

Thus, the interval of convergence is $-\frac{1}{5}<x<\frac{1}{5}$, and the series is centered at $x=0$.

5. a) To find the power series, we rewrite it in the form $c\cdot\frac{1}{1-r}$.

$$f(x)=\frac{1}{1-x^2}=1\cdot\frac{1}{1-\left(x^2\right)}.$$

The above expression can be viewed as the sum of a geometric series with first term $a_1=1$ and common ratio $r=x^2$. Thus, for $\left|x^2\right|<1$, we have

$$\begin{aligned} f(x)&=1\cdot\frac{1}{1-\left(x^2\right)}\\ &=1\left(1+\left(x^2\right)+\left(x^2\right)^2+\left(x^2\right)^3+\left(x^2\right)^4+\cdots\right)\\ &=1\left(1+x^2+x^4+x^6+x^8+\cdots\right)\\ &=1+x^2+x^4+x^6+x^8+\cdots \end{aligned}$$

b) The series converges for $\left|x^2\right|<1$:

$$\begin{aligned} &\left|x^2\right|<1\\ &-1<x^2<1\\ &-1<x<1. \end{aligned}$$

Thus, the interval of convergence is $-1<x<1$, and the series is centered at $x=0$.

7. a) To find the power series, we rewrite it in the form $c\cdot\frac{1}{1-r}$.

$$f(x)=\frac{2}{5-3x}=\frac{2}{5}\cdot\frac{1}{1-\left(\frac{3}{5}x\right)}.$$

The above expression can be viewed as the sum of a geometric series with first term $a_1=\frac{2}{5}$ and common ratio $r=\frac{3}{5}x$.

The solution is continued on the next page.

Copyright © 2014 Pearson Education, Inc. Publishing as Addison-Wesley.

Using the information from the previous page, for $\left|\frac{3}{5}x\right|<1$, we have

$$f(x)=\frac{2}{5}\cdot\frac{1}{1-\left(\frac{3}{5}x\right)}$$
$$=\tfrac{2}{5}\left(1+\left(\tfrac{3}{5}x\right)+\left(\tfrac{3}{5}x\right)^2+\left(\tfrac{3}{5}x\right)^3+\left(\tfrac{3}{5}x\right)^4+\cdots\right)$$
$$=\tfrac{2}{5}\left(1+\tfrac{3}{5}x+\tfrac{9}{25}x^2+\tfrac{27}{125}x^3+\tfrac{81}{625}x^4+\cdots\right)$$
$$=\tfrac{2}{5}+\tfrac{6}{25}x+\tfrac{18}{125}x^2+\tfrac{54}{625}x^3+\tfrac{162}{3125}x^4+\cdots$$

b) The series converges for $\left|\frac{3}{5}x\right|<1$:

$$\left|\tfrac{3}{5}x\right|<1$$
$$-1<\tfrac{3}{5}x<1$$
$$-\tfrac{5}{3}<x<\tfrac{5}{3}.$$

Thus, the interval of convergence is $-\frac{5}{3}<x<\frac{5}{3}$, and the series is centered at $x=0$.

9. a) To find the power series, we rewrite it in the form $c\cdot\frac{1}{1-r}$.

$$f(x)=\frac{5}{3+6x}=\frac{5}{3}\cdot\frac{1}{1-\left(-\frac{6}{3}x\right)}$$
$$=\frac{5}{3}\cdot\frac{1}{1-(-2x)}.$$

The above expression can be viewed as the sum of a geometric series with first term $a_1=\frac{5}{3}$ and common ratio $r=-2x$. Thus, for $|-2x|<1$, we have

$$f(x)=\frac{5}{3}\cdot\frac{1}{1-(-2x)}$$
$$=\tfrac{5}{3}\left(1+(-2x)+(-2x)^2+(-2x)^3+(-2x)^4+\cdots\right)$$
$$=\tfrac{5}{3}\left(1-2x+4x^2-8x^3+16x^4-\cdots\right)$$
$$=\tfrac{5}{3}-\tfrac{10}{3}x+\tfrac{20}{3}x^2-\tfrac{40}{3}x^3+\tfrac{80}{3}x^4-\cdots$$

b) The series converges for $|-2x|<1$:

$$|-2x|<1$$
$$-1<-2x<1$$
$$-\tfrac{1}{2}<x<\tfrac{1}{2}.$$

Thus, the interval of convergence is $-\frac{1}{2}<x<\frac{1}{2}$, and the series is centered at $x=0$.

11. Using a graphing utility, the graph of the functions are shown below:

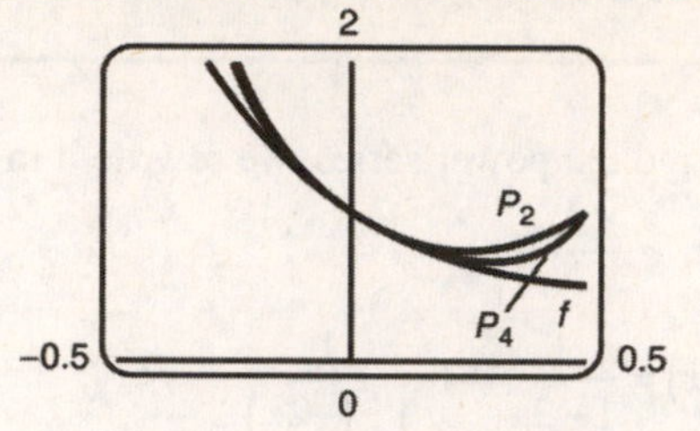

13. Using a graphing utility, the graph of the functions are shown below:

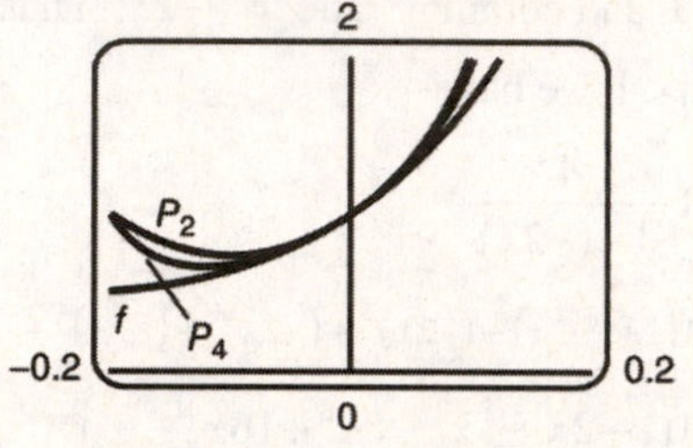

15. Using a graphing utility, the graph of the functions are shown below:

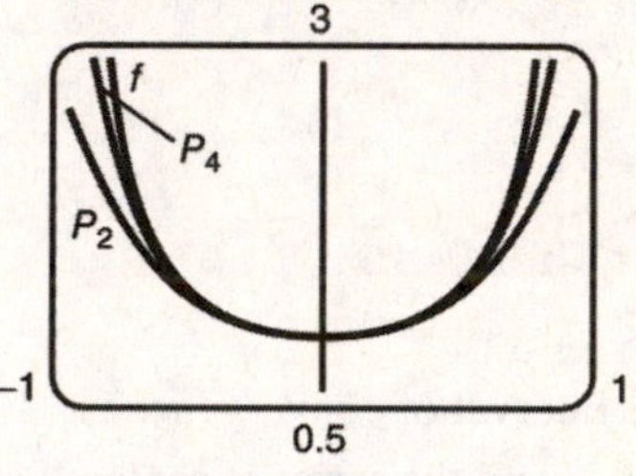

17. Using a graphing utility, the graph of the functions are shown below:

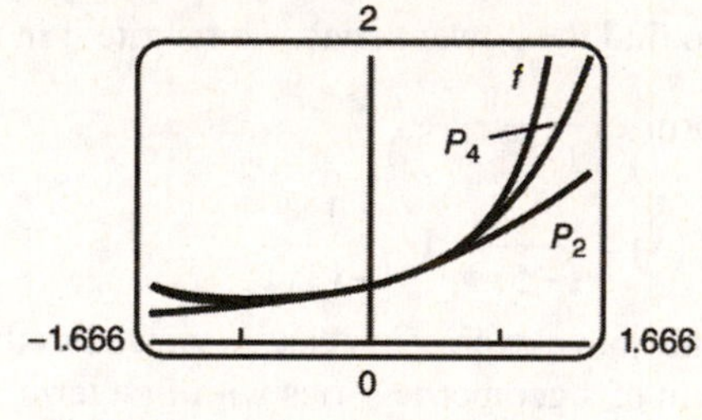

19. Using a graphing utility, the graph of the functions are shown below:

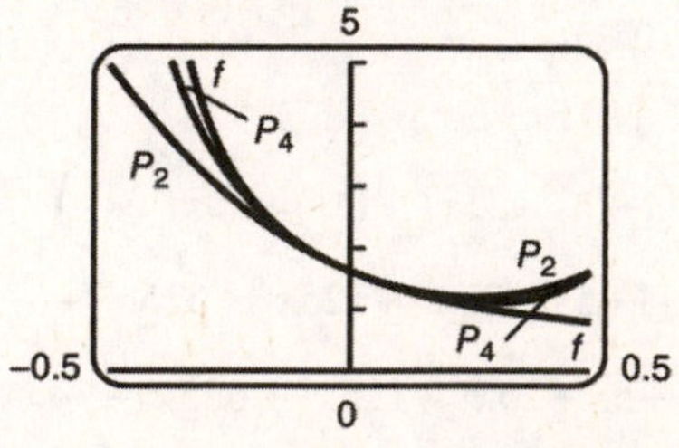

Copyright © 2014 Pearson Education, Inc. Publishing as Addison-Wesley.

21. a) From exercise 1, we know

$$f(x)=\frac{1}{1+2x}$$
$$=1-2x+4x^2-8x^3+16x^4-32x^5\cdots$$

To find a power series for $f'(x)$ we differentiate both sides of the power series for $f(x)$

$$\frac{d}{dx}\left(\frac{1}{1+2x}\right)=\frac{d}{dx}\left(1-2x+4x^2-8x^3+16x^4-32x^5\cdots\right)$$

$$\frac{d}{dx}\left(\frac{1}{1+2x}\right)=0-2+8x-24x^2+64x^3-160x^4\cdots$$

$$\frac{-2}{(1+2x)^2}=-2+8x-24x^2+64x^3-160x^4\cdots$$

Thus, we have

$$f'(x)=\frac{-2}{(1+2x)^2}$$
$$=-2+8x-24x^2+64x^3-160x^4\cdots$$

b) $F(x)=\int f(x)\,dx$

Integrate both sides of the power series for $f(x)$ term by term

$$\int f(x)\,dx=\int\left(1-2x+4x^2-8x^3+16x^4-32x^5\cdots\right)dx$$

$$\tfrac{1}{2}\ln|1+2x|=x-x^2+\tfrac{4}{3}x^3-2x^4+\tfrac{16}{5}x^5\cdots+C$$

Since $F(0)=0$ we have

$$\tfrac{1}{2}\ln|1+2(0)|=(0)-(0)^2+\tfrac{4}{3}(0)^3-2(0)^4+\cdots+C$$
$$0=C.$$

Thus, we have

$$F(x)=\tfrac{1}{2}\ln|1+2x|$$
$$=x-x^2+\tfrac{4}{3}x^3-2x^4+\tfrac{16}{5}x^5+\cdots$$

23. a) From exercise 3, we know

$$f(x)=\frac{1}{1-5x}$$
$$=1+5x+25x^2+125x^3+625x^4+3125x^5\cdots$$

To find a power series for $f'(x)$ we differentiate both sides of the power series for $f(x)$ at the top of the next column.

$$\frac{d}{dx}\left(\frac{1}{1-5x}\right)$$
$$=\frac{d}{dx}\left(1+5x+25x^2+125x^3+625x^4+3125x^5\cdots\right)$$

$$\frac{d}{dx}\left(\frac{1}{1-5x}\right)$$
$$=0+5+50x+475x^2+2500x^3+15{,}625x^4\cdots$$

$$\frac{5}{(1-5x)^2}=5+50x+475x^2+2500x^3+15{,}625x^4\cdots$$

Thus, we have

$$f'(x)=\frac{5}{(1-5x)^2}$$
$$=5+50x+475x^2+2500x^3+15{,}625x^4\cdots$$

b) $F(x)=\int f(x)\,dx$

Integrate both sides of the power series for $f(x)$ term by term

$$\int f(x)\,dx=\int\left(1+5x+25x^2+125x^3+625x^4+\cdots\right)dx$$

$$-\tfrac{1}{5}\ln|1-5x|=x+\tfrac{5}{2}x^2+\tfrac{25}{3}x^3+\tfrac{125}{4}x^4+125x^5\cdots+C$$

Since $F(0)=0$ we have

$$-\tfrac{1}{5}\ln|1+5(0)|=(0)+\tfrac{5}{2}(0)^2+\tfrac{25}{3}(0)^3+\cdots+C$$
$$0=C.$$

Thus, we have

$$F(x)=-\tfrac{1}{5}\ln|1-5x|$$
$$=x+\tfrac{5}{2}x^2+\tfrac{25}{3}x^3+\tfrac{125}{4}x^4+125x^5\cdots$$

25. a) First complete the square on the denominator function

$$x^2+2x+2=x^2+2x+1+2-1 \quad \text{adding and subtracting 1.}$$
$$=\left(x^2+2x+1\right)+(2-1)$$
$$=(x+1)^2+1.$$

Thus, we have

$$f(x)=\frac{1}{x^2+2x+2}$$
$$=\frac{1}{(x+1)^2+1}.$$

Rewrite the function it in the form $c\cdot\dfrac{1}{1-r}$ as follows

$$f(x)=\frac{1}{1-\left[-(x+1)^2\right]}.$$

The solution is continued on the next page.

Copyright © 2014 Pearson Education, Inc. Publishing as Addison-Wesley.

The expression on the previous page can be viewed as the sum of a geometric series with first term $a_1 = 1$ and common ratio $r = -(x+1)^2$. Thus, for $\left|-(x+1)^2\right| < 1$, we have

$$f(x) = \frac{1}{1-\left(-(x+1)^2\right)}$$
$$= 1 + \left(-(x+1)^2\right) + \left(-(x+1)^2\right)^2 + \left(-(x+1)^2\right)^3 + \left(-(x+1)^2\right)^4 + \cdots$$
$$= 1 - (x+1)^2 + (x+1)^4 - (x+1)^6 + (x+1)^8 - \cdots$$

b) The power series converges for $\left|-(x+1)^2\right| < 1$:

$$\left|-(x+1)^2\right| < 1$$
$$(x+1)^2 < 1$$
$$-1 < x+1 < 1$$
$$-1-1 < x < 1-1$$
$$-2 < x < 0.$$

Thus, the interval of convergence is $-2 < x < 0$, and the series is centered at $x = -1$.

27. a) First complete the square on the denominator function

$$x^2 - 4x + 13 = x^2 - 4x + 4 + 13 - 4$$
$$= \left(x^2 - 4x + 4\right) + (13 - 4)$$
$$= (x-2)^2 + 9.$$

Thus, we have

$$f(x) = \frac{1}{x^2 - 4x + 13}$$
$$= \frac{1}{(x-2)^2 + 9}.$$

Rewrite the function it in the form $c \cdot \frac{1}{1-r}$ as follows

$$f(x) = \frac{1}{9 - \left[-(x-2)^2\right]}$$
$$= \frac{1}{9} \cdot \frac{1}{1 - \left[-\frac{(x-2)^2}{9}\right]}.$$

The expression in the previous column can be viewed as the sum of a geometric series with first term $a_1 = \frac{1}{9}$ and common ratio $r = -\frac{(x-2)^2}{9}$. Thus, for $\left|-\frac{(x-2)^2}{9}\right| < 1$, we have

$$f(x) = \frac{1}{9} \cdot \frac{1}{1 - \left(-\frac{(x-2)^2}{9}\right)}$$
$$= \frac{1}{9} \cdot \left[1 + \left(-\frac{(x-2)^2}{9}\right) + \left(-\frac{(x-2)^2}{9}\right)^2 + \left(-\frac{(x-2)^2}{9}\right)^3 + \left(-\frac{(x-2)^2}{9}\right)^4 + \cdots\right]$$
$$= \frac{1}{9} \cdot \left[1 - \frac{(x-2)^2}{9} + \frac{(x-2)^4}{81} - \frac{(x-2)^6}{729} + \frac{(x-2)^8}{6561} - \cdots\right]$$
$$= \frac{1}{9} - \frac{(x-2)^2}{81} + \frac{(x-2)^4}{729} - \frac{(x-2)^6}{6561} + \frac{(x-2)^8}{59{,}049} - \cdots$$

b) The power series converges for $\left|-\frac{(x-2)^2}{9}\right| < 1$:

$$\left|-\left(\frac{x-2}{3}\right)^2\right| < 1$$
$$\left(\frac{x-2}{3}\right)^2 < 1$$
$$-1 < \frac{x-2}{3} < 1$$
$$-3 < x - 2 < 3$$
$$-3 + 2 < x < 3 + 2$$
$$-1 < x < 5.$$

Thus, the interval of convergence is $-1 < x < 5$, and the series is centered at $x = 2$.

29. a) To determine the linearization, first determine the derivative:

$$f(x) = x^2 + 3x + 2$$
$$f'(x) = 2x + 3.$$

At $x = 4$ we have

$$f(4) = (4)^2 + 3(4) + 2 = 30$$
$$f'(4) = 2(4) + 3 = 11.$$

The solution is continued on the next page.

Copyright © 2014 Pearson Education, Inc. Publishing as Addison-Wesley.

Using the information on the previous page, the linearization of the polynomial at $x=4$ is given by

$$P_1(x)=f(4)+f'(4)(x-4)$$
$$=30+11(x-4)$$
$$=30+11x-44$$
$$=11x-14.$$

b) The graph of $f(x)$ and $P_1(x)$ are

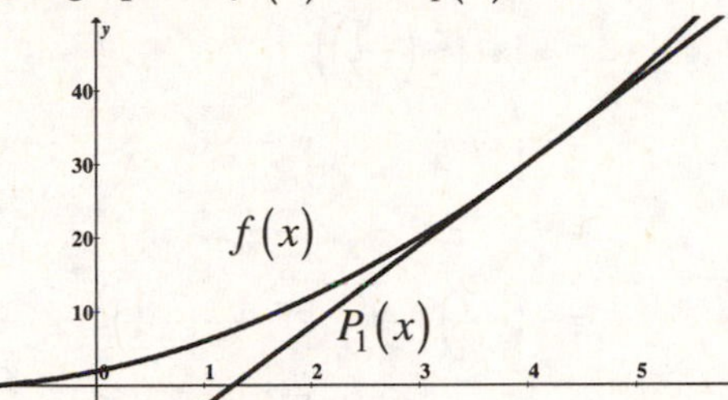

c) Evaluating the function and the linearization at $x=4.1$ we have

$P_1(4.1)=11(4.1)-14=31.1$

compared to the function value of

$f(4.1)=(4.1)^2+3(4.1)+2=31.11.$

31. a) To determine the linearization, first determine the derivative:

$$g(x)=e^{3x}$$
$$g'(x)=3e^{3x}.$$

At $x=0$ we have

$$g(0)=e^{3(0)}=1$$
$$g'(0)=3e^{3(0)}=3.$$

The linearization of the function at $x=0$ is given by

$$P_1(x)=f(0)+f'(0)(x-0)$$
$$=1+3(x)$$
$$=3x+1.$$

b) The graph of $g(x)$ and $P_1(x)$ are

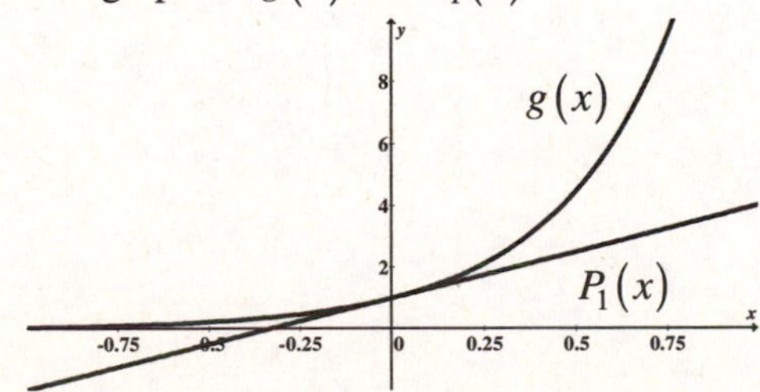

c) Evaluating the function and the linearization at $x=0.08$ we have

$P_1(0.08)=3(0.08)+1=1.24$

compared to the function value of

$g(0.08)=e^{3(0.08)}=1.27124915032.$

33. a) To determine the linearization, first determine the derivative:

$$D(p)=0.007p^3-0.5p^2+150p$$
$$D'(p)=0.021p^2-p+150.$$

At $p=20$ we have

$$D(20)=0.007(20)^3-0.5(20)^2+150(20)$$
$$=2856$$
$$D'(20)=0.021(20)^2-20+150$$
$$=138.4.$$

The linearization of the supply function at $p=20$ is given by

$$P_1(p)=D(20)+D'(20)(p-20)$$
$$=2856+138.4(p-20)$$
$$=138.4p+88.$$

b) If the price increases to \$21 the linearization will estimate the supply to be

$P_1(21)=138.4(21)-88=2994.4$ units.

Compared to the actual demand function of

$$D(21)=0.007(21)^3-0.5(21)^2+150(21)$$
$$=2994.3 \text{ units.}$$

Our values are relatively close to each other.

35. a) To determine the linearization, first determine the derivative:

$$p(t)=0.35t^2-2.75t+47.95$$
$$p'(t)=0.70t-2.75.$$

At $t=13$ we have

$$p(13)=0.35(13)^2-2.75(13)+47.95$$
$$=71.35$$
$$p'(13)=0.70(13)-2.75$$
$$=6.35.$$

The linearization of the price function at $t=13$ is given by

$$P_1(t)=p(13)+p'(13)(t-13)$$
$$=71.35+6.35(t-13)$$
$$=71.35+6.35t-82.55$$
$$=6.35t-11.2$$

Copyright © 2014 Pearson Education, Inc. Publishing as Addison-Wesley.

b) In the year 2014, $t = 14$. Using the linearization we estimate the ticket price to be

$P_1(14) = 6.35(14) - 11.2 = \$77.70.$

Compared to the value of actual price function

$$p(14) = 0.35(14)^2 + 2.75(14) + 47.95 = \$78.05.$$

The prices are relatively close together.

37. a) To determine the linearization, first determine the derivative:

$C(t) = 0.002t^2 + 0.117t + 0.761$

$C'(t) = 0.004t + 0.117.$

At $t = 34$ we have

$$C(34) = 0.002(34)^2 + 0.117(34) + 0.761 = 7.051$$

$$C'(34) = 0.004(34) + 0.117 = 0.253.$$

The linearization of the toll function at $t = 22$ is given by

$$\begin{aligned} P_1(t) &= C(34) + C'(34)(t - 34) \\ &= 7.051 + 0.253(t - 34) \\ &= 7.051 + 0.253t - 8.602 \\ &= 0.253t - 1.551. \end{aligned}$$

b) In the year 2014, $t = 39$. Using the linearization we estimate the toll to be

$$P_1(22) = 0.253(22) - 1.551 = \$8.316.$$

Compared to the value of actual toll function

$$C(39) = 0.002(39)^2 + 0.117(39) + 0.761 = \$8.366.$$

The toll values are relatively close together.

39. Since we have the power series

$\ln(1+x) = x - \frac{1}{2}x^2 + \frac{1}{3}x^3 - \frac{1}{4}x^4 + \frac{1}{5}x^5 - \frac{1}{6}x^6 \cdots$

Let $x = 2$, and we have

$$\begin{aligned} \ln\left(1+\left(-\tfrac{1}{2}\right)\right) &= \ln\left(\tfrac{1}{2}\right) \\ &= \ln(1) - \ln(2) \\ &= -\ln(2). \end{aligned}$$

Therefore,

$$\begin{aligned} \ln 2 &= -\ln\left(1+\left(-\tfrac{1}{2}\right)\right) \\ &= -\left(x - \tfrac{1}{2}x^2 + \tfrac{1}{3}x^3 - \tfrac{1}{4}x^4 + \tfrac{1}{5}x^5 - \tfrac{1}{6}x^6 \cdots\right) \\ &= -x + \tfrac{1}{2}x^2 - \tfrac{1}{3}x^3 + \tfrac{1}{4}x^4 - \tfrac{1}{5}x^5 + \tfrac{1}{6}x^6 \cdots \\ &= -\left(-\tfrac{1}{2}\right) + \tfrac{1}{2}\left(-\tfrac{1}{2}\right)^2 - \tfrac{1}{3}\left(-\tfrac{1}{2}\right)^3 \\ &\quad + \tfrac{1}{4}\left(-\tfrac{1}{2}\right)^4 - \tfrac{1}{5}\left(-\tfrac{1}{2}\right)^5 + \tfrac{1}{6}\left(-\tfrac{1}{2}\right)^6 \cdots \\ &= \frac{1}{2} + \frac{1}{8} + \frac{1}{24} + \frac{1}{64} + \frac{1}{160} + \frac{1}{384} \\ &\quad + \frac{1}{896} + \frac{1}{2048} + \frac{1}{4608} + \frac{1}{10,240} \\ &\approx 0.693 \end{aligned}$$

41. ✎

Copyright © 2014 Pearson Education, Inc. Publishing as Addison-Wesley.

Exercise Set 9.6

1. a) First find the coefficients using the formula $c_n = \dfrac{f^{(n)}(k)}{n!}$. To find the first five coefficients, we find the first 4 derivatives and evaluate them at $x = 1$.

$f(x) = \sqrt[3]{x} = x^{\frac{1}{3}}$, $\quad f(1) = \sqrt[3]{1} = 1$,

$f'(x) = \frac{1}{3}x^{-\frac{2}{3}}$, $\quad f'(1) = \frac{1}{3}(1)^{-\frac{2}{3}} = \frac{1}{3}$,

$f''(x) = -\frac{2}{9}x^{-\frac{5}{3}}$, $\quad f''(1) = -\frac{2}{9}(1)^{-\frac{5}{3}} = -\frac{2}{9}$,

$f^{(3)}(x) = \frac{10}{27}x^{-\frac{8}{3}}$, $\quad f^{(3)}(1) = \frac{10}{27}(1)^{-\frac{8}{3}} = \frac{10}{27}$,

$f^{(4)}(x) = -\frac{80}{81}x^{-\frac{11}{3}}$, $\quad f^{(4)}(1) = -\frac{80}{81}(1)^{-\frac{11}{3}} = -\frac{80}{81}$.

Therefore, the coefficients for the Taylor series are

$$c_0 = \frac{1}{0!} = 1,$$

$$c_1 = \frac{\frac{1}{3}}{1!} = \frac{1}{3},$$

$$c_2 = \frac{-\frac{2}{9}}{2!} = -\frac{1}{9},$$

$$c_3 = \frac{\frac{10}{27}}{3!} = \frac{5}{81},$$

$$c_4 = \frac{-\frac{80}{81}}{4!} = -\frac{10}{243}.$$

The Taylor series is given by:

$$f(x) = 1 + \tfrac{1}{3}(x-1) - \tfrac{1}{9}(x-1)^2 + \tfrac{5}{81}(x-1)^3 - \tfrac{10}{243}(x-1)^4 + \cdots$$

b) The functions P_1 and P_3 are given by

$$P_1 = 1 + \tfrac{1}{3}(x-1)$$

$$P_3 = 1 + \tfrac{1}{3}(x-1) - \tfrac{1}{9}(x-1)^2 + \tfrac{5}{81}(x-1)^3.$$

The graph of these functions with $f(x)$ are shown below:

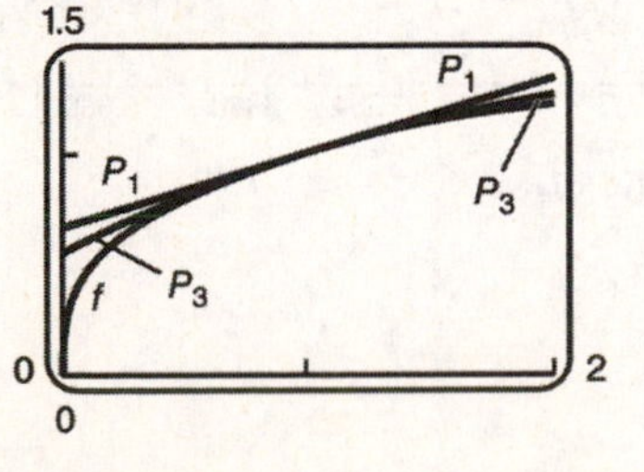

3. a) First find the coefficients using the formula $c_n = \dfrac{g^{(n)}(k)}{n!}$. To find the first five coefficients, we find the first 4 derivatives and evaluate them at $x = 1$.

$g(x) = \sqrt[4]{x} = x^{\frac{1}{4}}$, $\quad g(1) = \sqrt[4]{1} = 1$,

$g'(x) = \frac{1}{4}x^{-\frac{3}{4}}$, $\quad g'(1) = \frac{1}{4}(1)^{-\frac{3}{4}} = \frac{1}{4}$,

$g''(x) = -\frac{3}{16}x^{-\frac{7}{4}}$, $\quad g''(1) = -\frac{3}{16}(1)^{-\frac{7}{4}} = -\frac{3}{16}$,

$g^{(3)}(x) = \frac{21}{64}x^{-\frac{11}{4}}$, $\quad g^{(3)}(1) = \frac{21}{64}(1)^{-\frac{11}{4}} = \frac{21}{64}$,

$g^{(4)}(x) = -\frac{231}{256}x^{-\frac{15}{4}}$, $\quad g^{(4)}(1) = -\frac{231}{256}(1)^{-\frac{15}{4}} = -\frac{231}{256}$.

Therefore, the coefficients for the Taylor series are

$$c_0 = \frac{1}{0!} = 1,$$

$$c_1 = \frac{\frac{1}{4}}{1!} = \frac{1}{4},$$

$$c_2 = \frac{-\frac{3}{16}}{2!} = -\frac{3}{32},$$

$$c_3 = \frac{\frac{21}{64}}{3!} = \frac{7}{128},$$

$$c_4 = \frac{-\frac{231}{256}}{4!} = -\frac{77}{2048}.$$

The Taylor series is given by:

$$g(x) = 1 + \tfrac{1}{4}(x-1) - \tfrac{3}{32}(x-1)^2 + \tfrac{7}{128}(x-1)^3 - \tfrac{77}{2048}(x-1)^4 + \cdots$$

b) The functions P_1 and P_3 are given by

$$P_1 = 1 + \tfrac{1}{4}(x-1)$$

$$P_3 = 1 + \tfrac{1}{4}(x-1) - \tfrac{3}{32}(x-1)^2 + \tfrac{7}{2048}(x-1)^3.$$

The graph of these functions with $g(x)$ are shown below:

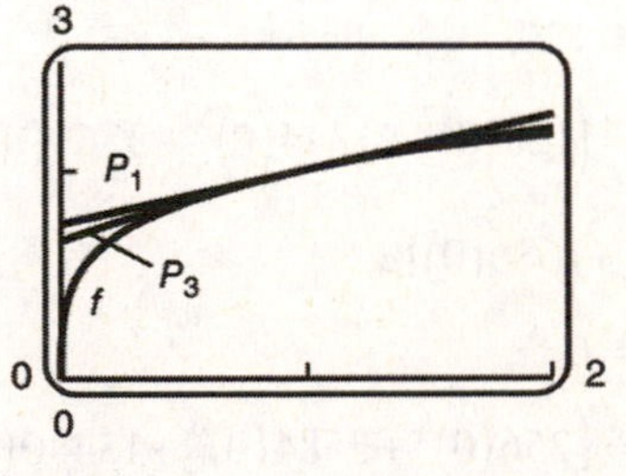

Copyright © 2014 Pearson Education, Inc. Publishing as Addison-Wesley.

5. a) First find the coefficients using the formula $c_n = \frac{h^{(n)}(k)}{n!}$. To show additional terms, we find the first nine coefficients, to do so, we find the first 8 derivatives and evaluate them at $x = 0$.

$h(x) = e^{x^2}$,

$h'(x) = 2xe^{x^2}$,

$h''(x) = (4x^2 + 2)e^{x^2}$,

$h^{(3)}(x) = (8x^3 + 12x)e^{x^2}$,

$h^{(4)}(x) = (16x^4 + 48x^2 + 12)e^{x^2}$,

$h^{(5)}(x) = (32x^5 + 160x^3 + 15x)e^{x^2}$,

$h^{(6)}(x) = (64x^6 + 480x^4 + 720x^2 + 120)e^{x^2}$,

$h^{(7)}(x) = (128x^7 + 1344x^5 + 3360x^3 + 1680x)e^{x^2}$,

$$h^{(8)}(x) = (256x^8 + 3584x^6 + 13440x^4 + 13440x^2 + 1680)e^{x^2}.$$

Evaluating, we have,

$h(0) = e^{(0)^2} = 1$,

$h'(0) = 2(0)e^{(0)^2} = 0$,

$h''(0) = (4(0) + 2)e^{(0)^2} = 2$,

$h^{(3)}(0) = (8(0)^3 + 12(0))e^{(0)^2} = 0$,

$h^{(4)}(0) = (16(0)^4 + 48(0)^2 + 12)e^{(0)^2} = 12$,

$h^{(5)}(0) = (32(0)^5 + 160(0)^3 + 15(0))e^{(0)^2} = 0$,

$h^{(6)}(0) = (64(0)^6 + 480(0)^4 + 720(0)^2 + 120)e^{(0)^2}$

$h^{(6)}(0) = 120$,

$$h^{(7)}(0) = (128(0)^7 + 1344(0)^5 + 3360(0)^3 + 1680(0))e^{(0)^2}$$

$h^{(7)}(0) = 0$

$$h^{(8)}(0) = (256(0)^8 + 3584(0)^6 + 13440(0)^4 + 13440(0)^2 + 1680)e^{(0)^2}$$

$h^{(8)}(0) = 1680.$

Therefore, the coefficients for the Taylor series are

$c_0 = \frac{1}{0!} = 1, \quad c_5 = \frac{0}{5!} = 0,$

$c_1 = \frac{0}{1!} = 0, \quad c_6 = \frac{120}{6!} = \frac{1}{6},$

$c_2 = \frac{2}{2!} = 1, \quad c_7 = \frac{0}{7!} = 0,$

$c_3 = \frac{0}{3!} = 0, \quad c_8 = \frac{1680}{8!} = \frac{1}{24},$

$c_4 = \frac{12}{4!} = \frac{1}{2}.$

The Taylor series is given by:

$$h(x) = 1 + 0(x-0) + 1(x-0)^2 + 0(x-0)^3 + \tfrac{1}{2}(x-0)^4 + 0(x-0)^5 + \tfrac{1}{6}(x-0)^6 + 0(x-0)^7 + \tfrac{1}{24}(x-0)^8 \cdots$$

Simplifying, we have

$$h(x) = 1 + x + \tfrac{1}{2}x^4 + \tfrac{1}{6}x^6 + \tfrac{1}{24}x^8 + \cdots$$

b) The functions P_2 and P_4 are given by

$P_2 = 1 + x^2$

$P_4 = 1 + x^2 + \frac{1}{2}x^4.$

The graph of these functions with $h(x)$ are shown below:

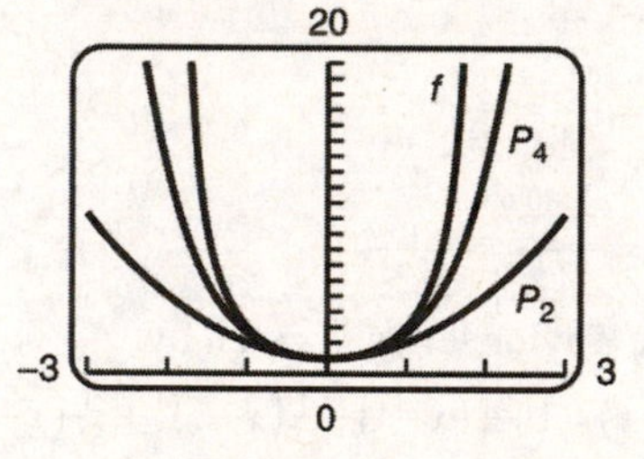

7. As shown from example 1, the Taylor series for $f(x) = e^x$ centered at $x = 0$ is given by

$$e^x = 1 + x + \tfrac{1}{2}x^2 + \tfrac{1}{3!}x^3 + \tfrac{1}{4!}x^4 + \tfrac{1}{5!}x^5 + \cdots$$

We have $\sqrt{e} = e^{\frac{1}{2}}$, substituting into the Taylor series give us

$$e^{\frac{1}{2}} = 1 + \left(\tfrac{1}{2}\right) + \tfrac{1}{2}\left(\tfrac{1}{2}\right)^2 + \tfrac{1}{3!}\left(\tfrac{1}{2}\right)^3 + \tfrac{1}{4!}\left(\tfrac{1}{2}\right)^4 + \tfrac{1}{5!}\left(\tfrac{1}{2}\right)^5 + \cdots$$

$$e^{\frac{1}{2}} = 1 + \tfrac{1}{2} + \tfrac{1}{8} + \tfrac{1}{48} + \tfrac{1}{384} + \tfrac{1}{3840} + \tfrac{1}{46{,}080} + \cdots$$

$$e^{\frac{1}{2}} \approx 1.6487.$$

Copyright © 2014 Pearson Education, Inc. Publishing as Addison-Wesley.

9. As shown from example 1, the Taylor series for $f(x)=e^x$ centered at $x=0$ is given by

$$e^x = 1+x+\tfrac{1}{2}x^2+\tfrac{1}{3!}x^3+\tfrac{1}{4!}x^4+\tfrac{1}{5!}x^5+\cdots$$

Substituting into the Taylor series give us

$$e^{0.3} = 1+(0.3)+\tfrac{1}{2}(0.3)^2+\tfrac{1}{3!}(0.3)^3 +\tfrac{1}{4!}(0.3)^4+\tfrac{1}{5!}(0.3)^5+\cdots$$

$$e^{0.3} = 1+\tfrac{3}{10}+\tfrac{9}{200}+\tfrac{9}{2000}+\tfrac{27}{80{,}000}+\cdots$$

$$e^{0.3} \approx 1.3499$$

11. As shown from example 3, the Taylor series for $f(x)=\sqrt{x}$ centered at $x=1$ is given by

$$\sqrt{x} = 1+\tfrac{1}{2}(x-1)-\tfrac{1}{8}(x-1)^2+\tfrac{1}{16}(x-1)^3 -\tfrac{5}{128}(x-1)^4+\cdots$$

Substituting into the Taylor series give us

$$\sqrt{1.1} = 1+\tfrac{1}{2}((1.1)-1)-\tfrac{1}{8}((1.1)-1)^2 +\tfrac{1}{16}((1.1)-1)^3-\tfrac{5}{128}((1.1)-1)^4+\cdots$$

$$\sqrt{1.1} = 1+\tfrac{1}{20}-\tfrac{1}{800}+\tfrac{1}{16{,}000}-\tfrac{1}{256{,}000}-\cdots$$

$$\sqrt{1.1} \approx 1.0488.$$

13. As shown in exercise 2, the Taylor series for $f(x)=\sqrt[3]{x}$ centered at $x=8$ is given by

$$\sqrt[3]{x} = 2+\tfrac{1}{12}(x-8)-\tfrac{1}{288}(x-8)^2+\tfrac{5}{20{,}736}(x-8)^3 -\tfrac{5}{248{,}832}(x-8)^4+\cdots$$

Substituting into the Taylor series give us

$$\sqrt[3]{9} = 2+\tfrac{1}{12}((9)-8)-\tfrac{1}{288}((9)-8)^2 +\tfrac{5}{20{,}736}((9)-8)^3-\tfrac{5}{248{,}832}((9)-8)^4+\cdots$$

$$\sqrt[3]{9} = 2+\tfrac{1}{12}-\tfrac{1}{288}+\tfrac{5}{20{,}736}-\tfrac{5}{248{,}832}-\cdots$$

$$\sqrt[3]{9} \approx 2.0801.$$

15. As shown in exercise 6, the Taylor series for $f(x)=e^{x^3}$ centered at $x=0$ is given by

$$e^{x^3} = 1+x^3+\tfrac{1}{2}x^6+\tfrac{1}{6}x^9+\cdots$$

Integrating both sides of the equation we have

$$\int_0^1 e^{x^3}\,dx = \int_0^1 1+x^3+\tfrac{1}{2}x^6+\tfrac{1}{6}x^9+\cdots dx$$

$$= \left[x+\tfrac{1}{4}x^4+\tfrac{1}{14}x^7+\tfrac{1}{60}x^{10}+\cdots\right]_0^1$$

$$= \left[1+\tfrac{1}{4}+\tfrac{1}{14}+\tfrac{1}{60}+\tfrac{1}{312}+\tfrac{1}{1920}+\tfrac{1}{13{,}680}\cdots\right] -[0+0+0+\cdots]$$

$$\approx 1.342.$$

17. a) To find the approximating polynomials, we derive the Taylor series centered at $t = 20$. First find the coefficients using the formula

$$c_n = \frac{f^{(n)}(k)}{n!}.$$

Taking derivatives:

$$f(t)=e^{0.109t},$$
$$f'(t)=0.109e^{0.109t},$$
$$f''(t)=(0.109)^2 e^{0.109t},$$
$$f^{(3)}(t)=(0.109)^3 e^{0.109t},$$
$$f^{(4)}(t)=(0.109)^4 e^{0.109t}.$$

Evaluating at $t = 20$, give us

$$f(20)=e^{0.109(20)}=8.8463,$$
$$f'(20)=0.109e^{0.109(20)}=0.9642474,$$
$$f''(20)=(0.109)^2 e^{0.109(20)}=0.10510296,$$
$$f^{(3)}(20)=(0.109)^3 e^{0.109(20)}=0.01145622,$$
$$f^{(4)}(20)=(0.109)^4 e^{0.109(20)}=0.0012487.$$

Therefore, the coefficients are

$$c_0 = \frac{8.8463}{0!} = 8.8463,$$
$$c_1 = \frac{0.964247}{1!} = 0.964247,$$
$$c_2 = \frac{0.10510296}{2!} = 0.05255,$$
$$c_3 = \frac{0.01145622}{3!} = 0.00190937,$$
$$c_4 = \frac{0.0012487}{4!} = 0.00005203.$$

The solution is continued on the next page.

Copyright © 2014 Pearson Education, Inc. Publishing as Addison-Wesley.

The Taylor series is

$$f(t)=8.992e^{0.109t}$$
$$=8.992\left[8.8463+0.964247(t-20)+0.05255(t-20)^2+0.001909(t-20)^3+0.00005203(t-20)^4+\cdots\right].$$

Simplifying, we have

$$f(t)=79.546+8.671(t-20)+0.473(t-20)^2+0.0172(t-20)^3+0.004678(t-20)^4.$$

Therefore, the second approximating polynomial is

$$P_2(t)=79.546+8.671(t-20)+0.473(t-20)^2.$$

In 1985, $t=1985-1967=18$. Substituting, we have

$$P_2(18)=79.546+8.671(18-20)+0.473(18-20)^2$$
$$P_2(18)=64.096=64.10.$$

The estimated price of a Super Bowl ticket in 1985 is \$64.10.

b) From part (a) we determine the third approximating polynomial as

$$P_3(t)=79.546+8.671(t-20)+0.473(t-20)^2+0.0172(t-20).$$

In 1985, $t=1985-1967=18$. Substituting, we have

$$P_3(18)=79.546+8.671(18-20)+0.473(18-20)^2+0.0172(18-20)^3$$
$$P_3(18)=63.9584\approx 63.96.$$

Using the third approximating polynomial, the estimated price of a Super Bowl ticket in 1985 was \$63.96.

c) Evaluating the function we have

$$f(18)=8.992e^{0.109(18)}=63.97.$$

Thus we see that P_3 is closer than P_2.

d. ✎

19. First find the coefficients using the formula $c_n=\dfrac{f^{(n)}(k)}{n!}$. To find the first five coefficients, we find the first 4 derivatives and evaluate them at $x=\frac{\pi}{4}$.

Continued at the top of the next column.

The first four derivatives are:

$$g(x)=\sin x,$$
$$g'(x)=\cos x,$$
$$g''(x)=-\sin x,$$
$$g^{(3)}(x)=-\cos x,$$
$$g^{(4)}(x)=\sin x,$$

Evaluating the derivative we have,

$$g\left(\tfrac{\pi}{4}\right)=\sin\tfrac{\pi}{4}=\tfrac{\sqrt{2}}{2},$$
$$g'\left(\tfrac{\pi}{4}\right)=\cos\tfrac{\pi}{4}=\tfrac{\sqrt{2}}{2},$$
$$g''\left(\tfrac{\pi}{4}\right)=-\sin\tfrac{\pi}{4}=-\tfrac{\sqrt{2}}{2},$$
$$g^{(3)}\left(\tfrac{\pi}{4}\right)=-\cos\tfrac{\pi}{4}=-\tfrac{\sqrt{2}}{2},$$
$$g^{(4)}\left(\tfrac{\pi}{4}\right)=\sin\tfrac{\pi}{4}=\tfrac{\sqrt{2}}{2}.$$

Therefore, the coefficients for the Taylor series are

$$c_0=\frac{\frac{\sqrt{2}}{2}}{0!}=\frac{\sqrt{2}}{2},$$
$$c_1=\frac{\frac{\sqrt{2}}{2}}{1!}=\frac{\sqrt{2}}{2},$$
$$c_2=\frac{-\frac{\sqrt{2}}{2}}{2!}=-\frac{\sqrt{2}}{4},$$
$$c_3=\frac{-\frac{\sqrt{2}}{2}}{3!}=-\frac{\sqrt{2}}{12},$$
$$c_4=\frac{\frac{\sqrt{2}}{2}}{4!}=\frac{\sqrt{2}}{48}.$$

The Taylor series is given by:

$$g(x)=\sin x$$
$$g(x)=\tfrac{\sqrt{2}}{2}+\tfrac{\sqrt{2}}{2}\left(x-\tfrac{\pi}{4}\right)-\tfrac{\sqrt{2}}{4}\left(x-\tfrac{\pi}{4}\right)^2-\tfrac{\sqrt{2}}{12}\left(x-\tfrac{\pi}{4}\right)^3+\tfrac{\sqrt{2}}{48}\left(x-\tfrac{\pi}{4}\right)^4+\cdots$$

21. As shown from example 5, the Taylor series for $f(x)=\sin x$ centered at $x=0$ is given by

$$\sin x=x-\tfrac{1}{3!}x^3+\tfrac{1}{5!}x^5-\tfrac{1}{7!}x^7+\cdots$$

Substituting into the Taylor series give us

$$\sin 0.3=0.3-\tfrac{1}{3!}(0.3)^3+\tfrac{1}{5!}(0.3)^5-\tfrac{1}{7!}(0.3)^7+\cdots$$
$$\approx 0.2955.$$

Copyright © 2014 Pearson Education, Inc. Publishing as Addison-Wesley.

23. Using the Taylor series for $f(x)=\sin x$ centered at $x=0$ found in exercise 21, we determine the Taylor series for $f(x)=\sin x^2$

$$\sin x^2 = x^2 - \tfrac{1}{3!}(x^2)^3 + \tfrac{1}{5!}(x^2)^5 - \tfrac{1}{7!}(x^2)^7 + \cdots$$
$$= x^2 - \tfrac{1}{3!}x^6 + \tfrac{1}{5!}x^{10} - \tfrac{1}{7!}x^{14} + \cdots$$

Integrating both sides of the equation we have

$$\int_0^{0.5} \sin x^2 dx$$
$$= \int_0^{0.5} x^2 - \tfrac{1}{3!}x^6 + \tfrac{1}{5!}x^{10} - \tfrac{1}{7!}x^{14} + \cdots dx$$
$$= \left[\tfrac{1}{3}x^3 - \tfrac{1}{42}x^7 + \tfrac{1}{1320}x^{11} - \tfrac{1}{75,600}x^{15} + \cdots\right]_0^{0.5}$$
$$= \left[\tfrac{1}{3}\left(\tfrac{1}{2}\right)^3 - \tfrac{1}{42}\left(\tfrac{1}{2}\right)^7 + \tfrac{1}{1320}\left(\tfrac{1}{2}\right)^{11} - \tfrac{1}{75,600}\left(\tfrac{1}{2}\right)^{15} + \cdots\right]$$
$$- \left[\tfrac{1}{3}(0)^3 - \tfrac{1}{42}(0)^7 + \tfrac{1}{1320}(0)^{11} - \tfrac{1}{75,600}(0)^{15} + \cdots\right]$$
$$\approx 0.04148.$$

The approximation for the integral to four decimal places is

$$\int_0^{0.5} \sin x^2 dx \approx 0.04148.$$

25. Apply the ratio test as follows:

$$\lim_{n\to\infty}\left|\frac{a_{n+1}}{a_n}\right| = \lim_{n\to\infty}\left|\frac{(-1)^{n+1}\left(\dfrac{x^{2n+2}}{(2n+2)!}\right)}{(-1)^n\left(\dfrac{x^{2n+1}}{(2n+1)!}\right)}\right|$$
$$= \lim_{n\to\infty}\left|\frac{x^{2n+2}}{(2n+2)!}\cdot\frac{(2n+1)!}{x^{2n+1}}\right|$$
$$= \lim_{n\to\infty}\left|\frac{x}{2n+2}\right|$$
$$= 0.$$

Since the limit $L=0$, is less than 1, the series converges for all x. The interval of convergence is $(-\infty,\infty)$.

27. a) Using Taylor series for e^x, we know

$$e^x = 1 + x + \tfrac{1}{2!}x^2 + \tfrac{1}{3!}x^3 + \tfrac{1}{4!}x^4 + \cdots$$

Substituting $ix = x$ we have

$$e^{ix} = 1 + (ix) + \tfrac{1}{2!}(ix)^2 + \tfrac{1}{3!}(ix)^3 + \tfrac{1}{4!}(ix)^4 + \cdots$$
$$= 1 + ix + \tfrac{1}{2!}i^2x^2 + \tfrac{1}{3!}i^3x^3 + \tfrac{1}{4!}i^4x^4 + \cdots$$
$$= 1 + ix - \tfrac{x^2}{2!} - \tfrac{i}{3!}x^3 + \tfrac{1}{4!}x^4 + \cdots$$

Note, expanding the Taylor series to 8 terms we have

$$e^{ix} = 1 + ix - \tfrac{x^2}{2!} - \tfrac{i}{3!}x^3 + \tfrac{1}{4!}x^4 + \tfrac{i}{5!}x^5 - \tfrac{1}{6!}x^6 - \tfrac{i}{7!}x^7 + \cdots$$

b) As shown from example 5, the Taylor series for $f(x)=\sin x$ centered at $x=0$ is given by $\sin x = x - \tfrac{1}{3!}x^3 + \tfrac{1}{5!}x^5 - \tfrac{1}{7!}x^7 + \cdots$

Substituting $ix = x$ give us

$$\sin(ix) = (ix) - \tfrac{1}{3!}(ix)^3 + \tfrac{1}{5!}(ix)^5 - \tfrac{1}{7!}(ix)^7 + \cdots$$
$$= ix - \tfrac{1}{3!}i^3x^3 + \tfrac{1}{5!}i^5x^5 - \tfrac{1}{7!}i^7x^7 + \cdots$$
$$= ix + \tfrac{i}{3!}x^3 + \tfrac{i}{5!}x^5 + \tfrac{i}{7!}x^7 + \cdots$$

c) Using the Taylor series for $f(x)=\cos x$

$$\cos x = 1 - \tfrac{1}{2!}x^2 + \tfrac{1}{4!}x^4 - \tfrac{1}{6!}x^6 + \cdots$$

and $g(x)=\sin x$

$$\sin x = x - \tfrac{1}{3!}x^3 + \tfrac{1}{5!}x^5 - \tfrac{1}{7!}x^7 + \cdots$$

we have

$$\cos x + i\sin x$$
$$= 1 - \tfrac{1}{2!}x^2 + \tfrac{1}{4!}x^4 - \tfrac{1}{6!}x^6 + \cdots$$
$$+ i\left[x - \tfrac{1}{3!}x^3 + \tfrac{1}{5!}x^5 - \tfrac{1}{7!}x^7 + \cdots\right]$$
$$= 1 - \tfrac{1}{2!}x^2 + \tfrac{1}{4!}x^4 - \tfrac{1}{6!}x^6 + \cdots$$
$$+ ix - \tfrac{i}{3!}x^3 + \tfrac{i}{5!}x^5 - \tfrac{i}{7!}x^7 + \cdots$$
$$= 1 + ix - \tfrac{1}{2!}x^2 - \tfrac{i}{3!}x^3 + \tfrac{1}{4!}x^4 + \tfrac{i}{5!}x^5 - \tfrac{1}{6!}x^6 - \tfrac{i}{7!}x^7 + \cdots$$
$$= e^{ix} \quad \text{From part (a).}$$

29. First we notice that Euler's formula gives us

$$e^{i\cdot\frac{\pi}{2}} = \cos\left(\tfrac{\pi}{2}\right) + i\sin\left(\tfrac{\pi}{2}\right) = 0 + i\cdot 1 = i.$$

Therefore,

$$i^i = \left(e^{i\cdot\frac{\pi}{2}}\right)^i = e^{i^2\cdot\frac{\pi}{2}} = e^{-\frac{\pi}{2}}.$$

Copyright © 2014 Pearson Education, Inc. Publishing as Addison-Wesley.

31. $y''+y'+3y=0.$

The auxiliary equation is

$$r^2+r+3=0$$

$$r=\frac{-1\pm\sqrt{1^2-4(1)(3)}}{2(1)}$$

$$r=\frac{-1\pm\sqrt{-11}}{2}.$$

Thus, the general solution to the differential equation is

$$y=Ce^{\left[-\frac{1}{2}\pm i\frac{\sqrt{11}}{2}\right]x}$$

$$y=Ce^{-\frac{1}{2}x}e^{\pm i\frac{\sqrt{11}}{2}x}$$

$$y=Ce^{-\frac{1}{2}x}\left(\cos\frac{\sqrt{11}}{2}x+i\sin\frac{\sqrt{11}}{2}x\right)$$

$$y=Ce^{-\frac{1}{2}x}\cos\frac{\sqrt{11}}{2}x+Ci\cdot e^{-\frac{1}{2}x}\sin\frac{\sqrt{11}}{2}x$$

$$y=C_1e^{-\frac{1}{2}x}\cos\frac{\sqrt{11}}{2}x+C_2e^{-\frac{1}{2}x}\sin\frac{\sqrt{11}}{2}x.$$

Copyright © 2014 Pearson Education, Inc. Publishing as Addison-Wesley.

Chapter 10

Probability Distributions

Exercise Set 10.1

1. a) Using roster notation, we list the elements in the set as follows:
$S=\{1,2,3,4,5,6,7,8,9,10,11,12\}.$

b) The set contains 12 elements, therefore, the cardinality of the set is $n(S)=12$.

3. a) Using roster notation, we list the elements in the set as follows:
$S=\{22,24,26,28,30,32,34,36,38,40\}.$

b) The set contains 10 elements, therefore, the cardinality of the set is $n(S)=10$.

5. a) Using roster notation, we list the elements in the set as follows:
$S=\{\text{winter, spring, summer, fall}\}.$

b) The set contains 4 elements, therefore, the cardinality of the set is $n(S)=4$.

7. a) Using roster notation, we list the elements in the set as follows:
$$S=\left\{\begin{matrix}3,6,9,12,15,18,21,24,\\27,30,33,36,39,42\end{matrix}\right\}.$$

b) The set contains 14 elements, therefore, the cardinality of the set is $n(S)=14$.

9. a) Find the solution to
$$3x+1=5$$
$$3x=4$$
$$x=\tfrac{3}{4}.$$
Using roster notation, we list the element in the set as follows:
$S=\{\tfrac{3}{4}\}.$

b) The set contains 1 element, therefore, the cardinality of the set is $n(S)=1$.

11. a) Find the solution to
$$2x=2(x-3)$$
$$2x=2x-6$$
$$0\neq-6.$$
There is no solution to the equation. Using roster notation, the set is the null set
$S=\varnothing.$

b) The set is empty, therefore, the cardinality of the set is $n(S)=0$.

13. a) Find the solution to
$$(x+2)^2=x^2+4x+4$$
$$x^2+4x+4=x^2+4x+4$$
$$0=0.$$
This equation is an identity and is true for all real values of x. Using roster notation, the set is the set of real numbers.
$S=\mathbb{R}.$

b) The set is infinite, therefore, the cardinality of the set is $n(S)=\infty$.

15. The elements of the set contains positive even integers less than or equal to 14. Using set builder notation, we describe the set as
$$S=\left\{\begin{matrix}x\,|\,x\text{ is a positive even integer less than}\\\text{or equal to 14}\end{matrix}\right\}.$$

17. The elements of the set are the individual states of the United States.. Using set builder notation, we describe the set as
$S=\{x\,|\,x\text{ is a state in the United States}\}.$

19. The elements of the set are the days of the week. Using set builder notation, we describe the set as $S=\{x\,|\,x\text{ is a day of the week}\}.$

21. True. 3 is an element of the set A.

23. False. The set $\{4\}\subseteq A,$ or the element $4\in A.$

25. False. The set D is a subset of B $D\subseteq B.$

27. True.

Copyright © 2014 Pearson Education, Inc. Publishing as Addison-Wesley.

29. The set $B \cup C$ is the union of the two sets. We list the elements of each set disregarding repeats as follows
$B \cup C = \{1,3,5,6,7,8,9,10\}$.
The set contains 8 elements so the cardinality of the set is $n(B \cup C) = 8$.

31. The set $B \cap C$ is the intersection of the two sets. We list the elements that are members of both sets as follows
$B \cap C = \{7,9\}$.
The set contains 2 elements so the cardinality of the set is $n(B \cap C) = 2$.

33. The set A' contains the elements in the universal set that are not in A. We list these elements as follows
$A' = \{6,7,8,9,10\}$.
The set contains 5 elements so the cardinality of the set is $n(A') = 5$.

35. The set $(C \cup D)'$ contains the elements in the universal set that are not in $C \cup D$. The set $C \cup D = \{1,3,5,6,7,8,9\}$. Therefore, the set $(C \cup D)' = \{2,4,10\}$.
The set contains 3 elements so the cardinality of the set is $n((C \cup D)') = 3$.

37. Substituting in the given information into the Inclusion-Exclusion Principle, we have
$n(A \cup B) = n(A) + n(B) - n(A \cap B)$
$n(A \cup B) = 14 + 22 - 7$
$n(A \cup B) = 29$.

39. Substituting in the given information into the Inclusion-Exclusion Principle, we have
$n(A \cup B) = n(A) + n(B) - n(A \cap B)$
$n(A \cap B) = n(A) + n(B) - n(A \cup B)$
$n(A \cap B) = 6 + 11 - 14$
$n(A \cup B) = 3$.

41. Since the set A' is the set of elements that are not in the set A, we can find the cardinality of the set as shown at the top of the next column.
$n(A') = n(U) - n(A)$
$n(A') = 100 - 45$
$n(A') = 55$.
Likewise, for the set B', we have
$n(B') = n(U) - n(B)$
$n(B') = 100 - 27$
$n(B') = 73$.

43. By definition, we know that
$n(U) \geq n(A \cup B) = n(A) + n(B) - n(A \cap B)$.
Using the given information and the above equation we have
$30 \geq 20 + 13 - n(A \cap B)$
$30 \geq 33 - n(A \cap B)$
$n(A \cap B) \geq 33 - 30$
$n(A \cap B) \geq 3$.

45. ✎

47. Define the following sets
$U = \{x \mid x \text{ is in the universe}\}$
$C = \{x \mid x \text{ owns a car}\}$
$B = \{x \mid x \text{ owns a bicycle}\}$.
The information in the problem gives us
$n(U) = 100, n(C) = 25, n(B) = 62$,
$n(C \cap B) = 13$.
Substituting in the given information into the Inclusion-Exclusion Principle, we have
$n(C \cup B) = n(C) + n(B) - n(C \cap B)$
$n(C \cup B) = 25 + 62 - 13$
$n(C \cup B) = 74$.
74 people own either a car or a bicycle.

49. a) Define the following sets
$U = \{x \mid x \text{ is a day in September}\}$
$T = \{x \mid x \text{ reached } 100°\text{F}\}$
$R = \{x \mid x \text{ measurable rain fell}\}$.
The information in the problem gives us
$n(U) = 30, n(T) = 23, n(R) = 5$,
$n(T \cap R) = 3$.
The solution is continued on the next page.

Copyright © 2014 Pearson Education, Inc. Publishing as Addison-Wesley.

Substituting the given information on the previous page into the Inclusion-Exclusion Principle, we have

$n(T \cup R) = n(T) + n(R) - n(T \cap R)$

$n(T \cup R) = 23 + 5 - 3$

$n(T \cup R) = 25.$

At least one of these events occurred 25 days in September.

b) The days where the temperature reached $100°F$ without measureable rain fall can be represented by the set $T \cap R'$. The cardinality of $T \cap R'$ can be found as follows

$n(T \cap R') = n(T) - n(T \cap R)$

$n(T \cup R') = 23 - 3$

$n(T \cup R) = 20.$

There were 20 days in September in which the temperature reached $100°F$ without measurable rain.

c) The days were the temperature did not reach $100°F$ and measurable rain did not fall can be represented by the set $T' \cap R'$. The cardinality of this set can be determined as follows

$n(T' \cap R') = n(U) - n(T \cup R)$

$n(T' \cap R') = 30 - 25$

$n(T' \cap R') = 5.$

51. a) Define the following sets

$U = \{x \mid x \text{ is in the Universe}\}$

$G = \{x \mid x \text{ shopped at the grocery store}\}$

$D = \{x \mid x \text{ shopped at the drug store}\}.$

The information in the problem gives us

$n(U) = 200, n(G) = 125, n(D) = 94,$

$n(G \cap D) = 31.$

Substituting in the given information into the Inclusion-Exclusion Principle, we have:

$n(G \cup D) = n(G) + n(D) - n(G \cap D)$

$n(G \cup D) = 125 + 94 - 31$

$n(G \cup D) = 188.$

188 people shopped at the grocery store or the drug store.

b) The people that did not shop at either store can be represented by the set $G' \cap D'$. The cardinality of this set can be determined as shown at the top of the next column.

$n(G' \cap D') = n(U) - n(G \cup D)$

$n(G' \cap D') = 200 - 188$

$n(G' \cap D') = 12.$

12 people did not shop at either store.

53. a) Define the following sets

$U = \{x \mid x \text{ is in the Universe}\}$

$L = \{x \mid x \text{ traveled to L.A.}\}$

$N = \{x \mid x \text{ traveled to N.Y.}\}.$

The information in the problem gives us

$n(U) = 50, n(L) = 20, n(N) = 32,$

$n(L \cap N) = 11.$

Substituting the given information into the Inclusion-Exclusion Principle, we have

$n(L \cup N) = n(L) + n(N) - n(L \cap N)$

$n(L \cup N) = 20 + 32 - 11$

$n(L \cup N) = 41.$

41 travelers had been to Los Angeles or New York in the previous year.

b) The set of travelers that had been to just New York in the previous year can be represented by the set $L' \cap N$. The cardinality of this set can be found as follows

$n(L' \cap N) = n(N) - n(L \cap N)$

$n(L' \cap N) = 32 - 11$

$n(L' \cap N) = 21.$

21 travelers had traveled to only New York in the previous year.

c) The travelers that did not travel to either city can be represented by $L' \cap N'$. The cardinality of this set can be determined as follows

$n(L' \cap N') = n(U) - n(L \cup N)$

$n(L' \cap N') = 50 - 41$

$n(L' \cap N') = 9.$

9 travelers had not been to either city in the previous year.

55. The Inclusion-Exclusion Principle tells us $n(A \cup B) = n(A) + n(B) - n(A \cap B)$. Therefore, in order for $n(A) + n(B) = n(A \cup B)$, it must be the case that $n(A \cap B) = 0$, or $A \cap B = \varnothing$.

Copyright © 2014 Pearson Education, Inc. Publishing as Addison-Wesley.

57. If $A \subseteq B$ and $B \subseteq A$, then $A = B$.

59. We want to show $(A \cup B)' = A' \cap B'$.
Given the information about the sets, we see that $A \cup B = \{1,2,3,4,5,6,7,9\}$; therefore, $(A \cup B)' = \{8,10\}$.
Next we see that
$A' = \{2,4,6,8,10\}$ and $B' = \{1,8,9,10\}$; therefore, the intersection is
$A' \cap B' = \{8,10\}$.
Putting the two results together, we have $(A \cup B)' = \{8,10\} = A' \cap B'$.

61. From the information in the problem, we have $A \cup B \cup C = \{1,2,3,4,5,6,7,8,9,10\}$.
Therefore, we have
$(A \cup B \cup C)' = \varnothing$.

63. Using deMorgan's Laws we rewrite the operation as follows
$[(A' \cap B)' \cup C']' = [A \cup B' \cup C']'$.
We have $A \cup B' = \{1,3,5,7,8,9,10\}$, finding the union with C' gives us
$A \cup B' \cup C' = \{1,3,5,6,7,8,9,10\}$.
Finding the complement of this result yields $(A \cup B' \cup C')' = \{4\}$. Therefore,
$[(A' \cap B)' \cup C]' = \{4\}$.

65. a) Using the Venn Diagram given in the problem, we add up all the values contained within the circles.

$$n(FC \cup BR \cup MV) = 33+11+8+14+3+8+21$$
$$n(FC \cup BR \cup MV) = 98.$$

We determine that 98 people visited at least one of the tourist attractions.
Note: the number who attended at least one of the attractions could also be solved by subtracting the number of individuals that did not attend any of the attractions from the number in the total number surveyed. That is

$$n(FC \cup BR \cup MV) =$$
$$n(U) - n((FC \cup BR \cup MV)') = 100 - 2 = 98$$

b) The set of people that visited Betatakin Ruins and Monument Valley, but not the Four Corners can be represented by $[(BR \cap MV) \cap FC']$. Looking at the Venn Diagram we see that this represents the intersection of the circles represented by BR and MV, excluding the values where all three circles are represented. Thus, we have $n[(BR \cap MV) \cap FC'] = 8$. Eight people visited Betatakin Ruins and Monument Valley, but not the Four Corners.

c) The set of people that visited Four Corners or Monument Valley, and Betatakin Ruins can be represented by $[(FC \cup MV) \cap BR]$. Looking at the Venn Diagram we see that this set represents the area of the circle represented by BR that intersects with one of the other circles MV or FC. Thus, we have $n[(FC \cup MV) \cap BR] = 11+3+8 = 22$. We determine that 22 people visited Four Corners or Monument Valley, and Betatakin Ruins.

d) To determine the number of people that did not visit any of the tourist attractions, we look at the area of the Venn Diagram outside of the three circles. We see that $n[BR' \cap MV' \cap FC'] = 2$. We determine that 2 people did not visit any of the tourist attractions.

Copyright © 2014 Pearson Education, Inc. Publishing as Addison-Wesley.

Exercise Set 10.2

1. a) There are 4 possible outcomes, so $n(S)=4$. The event $E=\{2,4\}$; thus, the probability that E occurs is
$$P(E)=\frac{n(E)}{n(S)}=\frac{2}{4}=\frac{1}{2}.$$
The event $E'=\{1,3\}$; thus, the probability that E' occurs is
$$P(E')=\frac{n(E')}{n(S)}=\frac{2}{4}=\frac{1}{2}.$$
b) The event $F=\{1,2,3\}$; thus, the probability that F occurs is
$$P(F)=\frac{n(F)}{n(S)}=\frac{3}{4}.$$
The event $F'=\{4\}$; thus, the probability that F' occurs is
$$P(F')=\frac{n(F')}{n(S)}=\frac{1}{4}.$$

3. a) There are 52 cards in a deck, so the number of possible outcomes is $n(S)=52$. There are 12 face cards in the deck, so event E has 12 possible outcomes, $n(E)=12$. Thus, the probability of E is given by
$$P(E)=\frac{n(E)}{n(S)}=\frac{12}{52}=\frac{3}{13}.$$
Since the probability of drawing a face card is $P(E)=\frac{3}{13}$, the probability of not drawing a face card is
$$P(E')=1-P(E)=1-\frac{3}{13}=\frac{10}{13}.$$
b) There are 13 clubs in the deck, so the event F has 13 possible outcomes, $n(F)=13$. Thus, the probability of F is given by
$$P(F)=\frac{n(F)}{n(S)}=\frac{13}{52}=\frac{1}{4}.$$
Since the probability of drawing club is $P(F)=\frac{1}{4}$, the probability of not drawing a club is $P(F')=1-P(F)=1-\frac{1}{4}=\frac{3}{4}.$

c) First we find the number of possible outcomes, there are 12 face cards, 13 clubs and 3 clubs that are also face cards. Thus,
$$\begin{aligned}n(E\cup F)&=n(E)+n(F)-n(E\cap F)\\&=12+13-3\\&=22.\end{aligned}$$
Therefore, we have
$$P(E\cup F)=\frac{n(E\cup F)}{n(S)}=\frac{22}{52}=\frac{11}{26}.$$

5. a) There are 36 possible outcomes when 2 fair dice are rolled. See example 5 for the set of outcomes. There are three ways to roll a 10, $E=\{(4,6),(5,5),(6,4)\}$. Thus, the probability of the sum of the two dice adding to 10 is
$$P(E)=\frac{n(E)}{n(S)}=\frac{3}{36}=\frac{1}{12}.$$
Since the probability the dice sum to 10 is $P(E)=\frac{1}{12}$, the probability of not rolling a 10 is
$$P(E')=1-P(E)=1-\frac{1}{12}=\frac{11}{12}.$$
b) There are 11 outcomes where at least one die shows a five. They are
$$F=\left\{\begin{matrix}(1,5),(2,5),(3,5),(4,5),(5,5),(6,5)\\(5,1),(5,2),(5,3),(5,4),(5,6)\end{matrix}\right\}.$$
Thus, $n(F)=11$. The probability of F is
$$P(F)=\frac{n(F)}{n(S)}=\frac{11}{36}.$$
Since the probability of at least one die showing a 5 is $P(F)=\frac{11}{36}$, the probability of neither die showing a 5 is
$$P(E')=1-P(E)=1-\frac{11}{36}=\frac{25}{36}.$$
c) First we find the number of possible outcomes,
$$\begin{aligned}n(E\cup F)&=n(E)+n(F)-n(E\cap F)\\&=3+11-1\\&=13.\end{aligned}$$
Therefore, we have
$$P(E\cup F)=\frac{n(E\cup F)}{n(S)}=\frac{13}{36}.$$

Copyright © 2014 Pearson Education, Inc. Publishing as Addison-Wesley.

7. a) From the information in the problem, we know the following
$n(S)=30, n(R)=10, n(B)=8,$
$n(G)=7, n(Y)=5.$
The probability of selecting a red candy is
$$P(R)=\frac{n(R)}{n(S)}=\frac{10}{30}=\frac{1}{3}.$$
Since the probability of selecting red candy is $P(R)=\frac{1}{3}$, the probability of not selecting a red candy is $P(R')=1-P(R)=1-\frac{1}{3}=\frac{2}{3}.$

b) Since there are no candies that are red and yellow at the same time, the probability of selecting candy that is both red and yellow is
$$P(R\cap Y)=\frac{n(R\cap Y)}{n(S)}=\frac{0}{30}=0.$$
The number of candies that are red or yellow is given by
$$\begin{aligned} n(R\cup Y)&=n(R)+n(Y)-n(R\cap Y)\\ &=10+5-0\\ &=15. \end{aligned}$$
Thus, the probability of selecting a red or yellow candy is given by
$$P(R\cup Y)=\frac{n(R\cup Y)}{n(S)}=\frac{15}{30}=\frac{1}{2}.$$

c) Since there are no candies that are red and green and blue at the same time, the probability of selecting candy that is red and green and blue is
$$P(R\cap G\cap B)=\frac{n(R\cap G\cap B)}{n(S)}=\frac{0}{30}=0.$$
The number of candies that are red or green or blue is given by
$$\begin{aligned} n(R\cup G\cup B)&=n(R)+n(G)+n(B)\\ &=10+8+7\\ &=25. \end{aligned}$$
Thus, the probability of selecting a red or green or blue candy is given by
$$P(R\cup Y)=\frac{n(R\cup Y)}{n(S)}=\frac{25}{30}=\frac{5}{6}.$$

9. a) From the information in the problem, we know the following
$n(S)=12, n(Z)=1, n(M)=5,$
$n(W)=7.$
The probability that Zack is selected is
$$P(Z)=\frac{n(Z)}{n(S)}=\frac{1}{12}.$$
Since the probability that Zack is selected is $P(Z)=\frac{1}{12}$, the probability that Zack is not selected is $P(Z')=1-P(Z)=1-\frac{1}{12}=\frac{11}{12}.$

b) The number of outcomes of the event a man is selected and Zack is selected is given by
$$\begin{aligned} n(M\cap Z)&=n(M)+n(Z)-n(M\cup Z)\\ &=5+1-5\\ &=1. \end{aligned}$$
Thus, the probability that a man is selected and the man is Zack is given by
$$P(M\cap Z)=\frac{n(M\cap Z)}{n(S)}=\frac{1}{12}.$$
The number of outcomes of the event a man is selected or Zack is selected is given by
$$\begin{aligned} n(M\cup Z)&=n(M)+n(Z)-n(M\cap Z)\\ &=5+1-1\\ &=5. \end{aligned}$$
Thus, the probability that a man is selected or Zack is selected is given by
$$P(M\cup Z)=\frac{n(M\cup Z)}{n(S)}=\frac{5}{12}.$$

c) Since the selected person cannot be a man and a woman, we have
$$P(M\cap W)=\frac{n(M\cap W)}{n(S)}=\frac{0}{12}=0.$$
The number of outcomes where a man or a women is selected is given by
$$\begin{aligned} n(M\cup W)&=n(M)+n(W)-n(M\cap W)\\ &=5+7-0\\ &=12. \end{aligned}$$
Thus, the probability that a man or a woman is selected is given by
$$P(M\cup W)=\frac{n(M\cup W)}{n(S)}=\frac{12}{12}=1.$$

Copyright © 2014 Pearson Education, Inc. Publishing as Addison-Wesley.

11. a) From the information in the problem, we know the following
$n(S)=5129, n(A)=1125, n(B)=1170,$
$n(C)=1281, n(D)=1263,$ and $n(E)=290.$
The probability that the randomly selected student is a freshman is
$$P(A)=\frac{n(A)}{n(S)}=\frac{1125}{5129}.$$
Since the probability a freshman was selected is $P(A)=\frac{1125}{5129}$, the probability that a freshman was not selected is
$$P(A')=1-P(A)=1-\frac{1125}{5129}=\frac{4004}{5129}.$$
b) The probability that the randomly selected student is a graduate student is
$$P(E)=\frac{n(E)}{n(S)}=\frac{290}{5129}.$$
Since the probability a graduate student was selected is $P(E)=\frac{290}{5129}$, the probability that a graduate student was not selected is
$$P(E')=1-P(E)=1-\frac{290}{5129}=\frac{4839}{5129}.$$
c) There are no possible outcomes where a randomly selected student could be a freshman and a sophomore and a junior, therefore,
$$P(A\cap B\cap C)=\frac{n(A\cap B\cap C)}{n(S)}=\frac{0}{5129}=0.$$
The number of possible outcomes where the randomly selected student is a freshman or a sophomore or a junior is given by
$$\begin{aligned}n(A\cup B\cup C)&=n(A)+n(B)+n(C)\\&=1125+1170+1281\\&=3576.\end{aligned}$$
Therefore, the probability that a randomly selected student is a freshman or a sophomore or a junior is given by
$$P(A\cup B\cup C)=\frac{n(A\cup B\cup C)}{n(S)}=\frac{3576}{5129}.$$

13. a) The sample space is given by
$S=\{HH,HT,TH,TT\}.$
b) The event E is a least one tail appears. From the sample space, we see that $n(E)=3.$ Therefore, the probability is
$$P(E)=\frac{n(E)}{n(S)}=\frac{3}{4}.$$
Since the probability of that at least one tail appears is $P(E)=\frac{3}{4}$, the probability that no tails appear is
$$P(E')=1-P(E)=1-\frac{3}{4}=\frac{1}{4}.$$
c) The event F is that both coins show the same side. From the sample space, we see that $n(F)=2.$ Therefore, the probability is
$$P(F)=\frac{n(F)}{n(S)}=\frac{2}{4}=\frac{1}{2}.$$
Therefore,
$$P(F')=1-P(F)=1-\frac{1}{2}=\frac{1}{2}.$$
d) From the sample space, we see that
$$\begin{aligned}n(E\cup F)&=n(E)+n(F)-n(E\cap F)\\&=3+2-1\\&=4.\end{aligned}$$
Therefore, the probability is
$$P(E\cup F)=\frac{n(E\cup F)}{n(S)}=\frac{4}{4}=1.$$

15. a) The sample space is given by
$$S=\left\{\begin{matrix}TTTT,TTTH,TTHT,THTT\\TTHH,THTH,THHT,THHH\\HTTT,HTTH,HTHT,HHTT,\\HTHH,HHTH,HHHT,HHHH\end{matrix}\right\}.$$
b) The event E is all four coins show the same side. From the sample space, we see that $n(E)=2.$ The probability is
$$P(E)=\frac{n(E)}{n(S)}=\frac{2}{16}=\frac{1}{8}.$$
Therefore, the probability all four coins do not show the same side is
$$P(E')=1-P(E)=1-\frac{1}{8}=\frac{7}{8}.$$

Copyright © 2014 Pearson Education, Inc. Publishing as Addison-Wesley.

c) The event F is that at least one coin shows heads. From the sample space, we see that $n(F)=15$. Therefore, the probability is

$$P(F)=\frac{n(F)}{n(S)}=\frac{15}{16}.$$

Therefore,

$$P(F')=1-P(F)=1-\frac{15}{16}=\frac{1}{16}.$$

d) From the sample space, we see that

$$\begin{aligned}n(E\cup F)&=n(E)+n(F)-n(E\cap F)\\&=2+15-1\\&=16.\end{aligned}$$

Therefore, the probability is

$$P(E\cup F)=\frac{n(E\cup F)}{n(S)}=\frac{16}{16}=1.$$

17. a) The sample space is given by

$$S=\begin{Bmatrix}(1,1,1),(1,1,2),(1,2,1),(1,2,2),(1,1,3),\\(1,3,1),(1,3,3),(1,2,3),(1,3,2),(2,1,1),\\(2,1,2),(2,2,1)(2,2,2),(2,1,3),(2,3,1),\\(2,3,3),(2,2,3),(2,3,2),(3,1,1),(3,1,2),\\(3,2,1),(3,2,2),(3,2,2),(3,1,3),(3,3,1),\\(3,3,3),(3,2,3),(3,3,2)\end{Bmatrix}$$

b) The event E is at least on spin points to two. From the sample space, we see that $n(E)=19$. Therefore, the probability that both spins point to one is

$$P(E)=\frac{n(E)}{n(S)}=\frac{19}{27}.$$

Therefore,

$$P(E')=1-P(E)=1-\frac{19}{27}=\frac{8}{27}.$$

c) The event F is the sum of the spins is 7. From the sample space, we see that $n(F)=6$. Therefore, the probability the sum of the spins is 7 is

$$P(F)=\frac{n(F)}{n(S)}=\frac{6}{27}=\frac{2}{9}.$$

Therefore,

$$P(F')=1-P(F)=1-\frac{2}{9}=\frac{7}{9}.$$

d) From the sample space, we see that

$$\begin{aligned}n(E\cup F)&=n(E)+n(F)-n(E\cap F)\\&=19+6-3\\&=22.\end{aligned}$$

Therefore, the probability is

$$P(E\cup F)=\frac{n(E\cup F)}{n(S)}=\frac{22}{27}.$$

19. a) Using the table, the probability that Paul is assigned to section B is $P(B)=0.3$.

b) The probability that Paul is not assigned to section B is

$$P(B')=1-P(B)=1-0.3=0.7.$$

c) The probability that Paul's application is rejected is $P(R)=0.2$. Therefore, the probability that Paul's application is not rejected is $P(R')=1-P(R)=1-0.2=0.8$.

d) The probability that Paul is assigned to section A, B or C is

$$\begin{aligned}P(A\cup B\cup C)&=P(A)+P(B)+P(C)\\&=0.1+0.3+0.2\\&=0.6.\end{aligned}$$

21. a) The probability that a customer will spend \$120 or more is given by

$$\begin{aligned}&P(\text{more than \$120})\\&\quad=P(\$120-\$159.99)+P(\$160-\$199.99)\\&\quad\quad+P(\$200\text{ and above}).\end{aligned}$$

Using the frequency table we have

$$\begin{aligned}P(\text{more than \$120})&=\frac{97}{337}+\frac{40}{337}+\frac{16}{337}\\&=\frac{153}{337}.\end{aligned}$$

b) The probability that a random customer will not spend less that \$80 is given by

$$\begin{aligned}P(\text{not less than \$80})&=1-P(\text{less than \$80})\\&=1-\left(P(\$0-\$39)+P(\$40-\$79.99)\right).\end{aligned}$$

Using the frequency table we have

$$\begin{aligned}P(\text{not less than \$80})&=1-P(\text{less than \$80})\\&=1-\left(\frac{33}{337}+\frac{59}{337}\right)\\&=\frac{245}{337}.\end{aligned}$$

Copyright © 2014 Pearson Education, Inc. Publishing as Addison-Wesley.

c) The probability that a random customer will spend between \$40 and \$159.99

$$P(\$40-\$159.99) = P(\$40-\$79.99)+P(\$80-\$119.99)+P(\$120-\$159.99).$$

Using the frequency table we have

$$P(\$40-\$159.99)=\frac{59}{337}+\frac{92}{337}+\frac{97}{337}=\frac{248}{337}.$$

23. a) The frequency of the experiment is $F = 200$. Let the event E be that both coins show heads. Using the frequency table, we have

$$P(E)=\frac{n(E)}{n(S)}=\frac{61}{200}.$$

b) Let the event F be that at least one of the coins shows heads. Using the frequency table, we have

$$P(F)=P(1H,1T)+P(2H)=\frac{61}{200}+\frac{103}{200}=\frac{164}{200}=\frac{41}{50}.$$

c) Using the table, the probability that both coins show the same side is

$$P(2H\cup 2T)=P(2H)+P(2T)=\frac{61}{200}+\frac{36}{200}=\frac{97}{200}.$$

25. a) The frequency of the experiment is $F = 50.$ Using the frequency table, the probability of getting exactly two hits is

$$P(\text{exactly 2 hits})=\frac{11}{50}.$$

b) The probability of getting at most two hits may be written as

$$P(\text{at most 2 hits}) = P(0\text{ hits})+P(1\text{ hit})+P(2\text{ hits}).$$

Therefore, using the frequency table, we have

$$P(\text{at most 2 hits})=\frac{12}{50}+\frac{18}{50}+\frac{11}{50}=\frac{41}{50}.$$

c) The probability of getting at least one hit may be written as

$$P(\text{at least 1 hit})=1-P(0\text{ hits}).$$

Therefore, using the frequency table, we have

$$P(\text{at least 1 hit1})=1-\frac{12}{50}=\frac{38}{50}=\frac{19}{25}.$$

27. a) The frequency of the experiment is $F = 1307.$ Using the frequency table, the probability that the word contains exactly 4 letters is

$$P(\text{exactly 4 letters})=\frac{169}{1307}.$$

b) The probability that the word contains 5 or fewer letters may be written as

$$P(5\text{ or fewer}) = P(1)+P(2)+P(3)+P(4)+P(5)$$

Therefore, using the frequency table, we have

$$P(5\text{ or fewer}) =\frac{36}{1307}+\frac{224}{1307}+\frac{285}{1307}+\frac{169}{1307}+\frac{174}{1307}=\frac{888}{1307}.$$

c) The probability that the word contains 8 or more letters may be written as

$$P(8\text{ or more})=P(8)+P(9)+P(10)+P(11)+P(12)+P(13)+P(14)+P(15).$$

Therefore, using the frequency table, we have

$$P(8\text{ or more})=\frac{73}{1307}+\frac{70}{1307}+\frac{34}{1307}+\frac{17}{1307}+\frac{8}{1307}+\frac{7}{1307}+\frac{0}{1307}+\frac{4}{1307}=\frac{213}{1307}.$$

29. a) The frequency of the experiment is $F = 72.$ Using the frequency table, the probability of going exactly 5 games

$$P(\text{exactly 5 games})=\frac{17}{72}.$$

Copyright © 2014 Pearson Education, Inc. Publishing as Addison-Wesley.

b) The probability that a series will not last 7 games may be written as
$P(\text{not 7 games}) = 1 - P(\text{7 games})$.
Therefore, using the frequency table, we have

$$P(\text{not 7 games}) = 1 - \frac{16}{72} = \frac{56}{72} = \frac{7}{9}.$$

c) The probability a team will not sweep the series may be written as
$P(\text{no sweep}) = 1 - P(\text{sweep})$.
Therefore, using the frequency table, we have

$$P(\text{no sweep}) = 1 - \frac{20}{72} = \frac{52}{72} = \frac{13}{18}.$$

d) ✎

31. a) The frequency of the experiment is $F = 100{,}000$. The probability the child survived to his 1st birthday can be expressed as the probability that the child did not die during the first year. Using the frequency table we have

$$P(\text{survived to 1st}) = 1 - P(\text{died in } [0,1)) = 1 - \frac{738}{100{,}000} = \frac{49{,}631}{50{,}000} \approx 0.99262 = 99.262\%.$$

There is a 99.262% chance that a randomly-chosen male born in 2007 survived to his 1st birthday.

b) The probability the child survived to his 3rd birthday can be expressed as the probability that the child did not die during the first 3 years. Using the frequency table we have

$$P(\text{survived to 3rd}) = 1 - \left[P([0,1)) + P([1,2)) + P([2,3))\right] = 1 - \left[\frac{738}{100{,}000} + \frac{49}{100{,}000} + \frac{31}{100{,}00}\right] = \frac{49{,}591}{50{,}000} \approx 0.99182 = 99.182\%.$$

There is a 99.182% chance that a randomly-chosen male born in 2007 survived to his 3rd birthday.

c) The probability the child survived to his 5th birthday can be expressed as the probability that the child did not die during the first 5 years. Using the frequency table we have

$$P(\text{survived to 5th}) = 1 - \left[P([0,1)) + P([1,2)) + P([2,3)) + P([3,4)) + P([4,5))\right]$$
$$= 1 - \left[\frac{738}{100{,}000} + \frac{49}{100{,}000} + \frac{31}{100{,}000} + \frac{24}{100{,}000} + \frac{20}{100{,}000}\right] = \frac{49{,}569}{50{,}000} \approx 0.99138 = 99.138\%.$$

There is a 99.138% chance that a randomly-chosen male born in 2007 survived to his 5th birthday.

33. Since it is assumed that the randomly thrown dart will connect with the dart board, we first calculate the total area of the board. This area will contain all of the possible outcomes. The area of the dart board is

$$\text{area of dart board} = \pi(225)^2 = 159{,}043.12809\text{mm}^2.$$

Next we calculate the area of the bull's-eye.

$$\text{area of bull's-eys} = \pi(6.35)^2 = 126.676870\text{mm}^2.$$

Therefore the probability of hitting the bull's-eye is the area of the bull's-eye divided by the area of the dart board.

$$P(\text{Bull's-Eye}) = \frac{126.676870}{159{,}043.12809} = 0.0007965.$$

35. The area of the dartboard within the treble ring is given by

$$\text{area within treble ring} = \pi(107)^2 = 35{,}968.094291\text{mm}^2.$$

Therefore the probability of hitting inside the treble ring is the area within the treble ring divided by the area of the dart board.

$$P(\text{Bull's-Eye}) = \frac{35{,}968.094291}{159{,}043.12809} = 0.2262.$$

Copyright © 2014 Pearson Education, Inc. Publishing as Addison-Wesley.

37. a) We draw a graph of the region

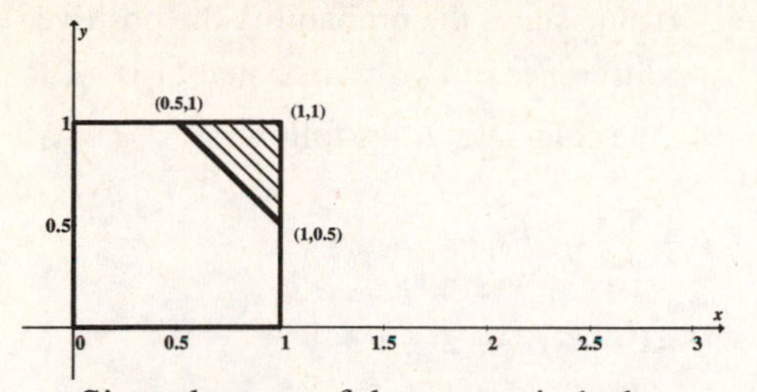

Since the area of the square is 1, the probability of that two randomly selected numbers chosen from the interval (0,1) add to 1.5 or greater is the area of the shaded region. Since that region is a triangle, the area is given by

$$A = \frac{1}{2} \cdot b \cdot h$$
$$= \frac{1}{2}\left(\frac{1}{2}\right)\left(\frac{1}{2}\right)$$
$$= \frac{1}{8} = 0.125.$$

Therefore, the probability is

$$P(x_1 + x_2 \geq 1.5) = 0.125.$$

b) Answers will vary. We show the screen shot of one random number pair below:

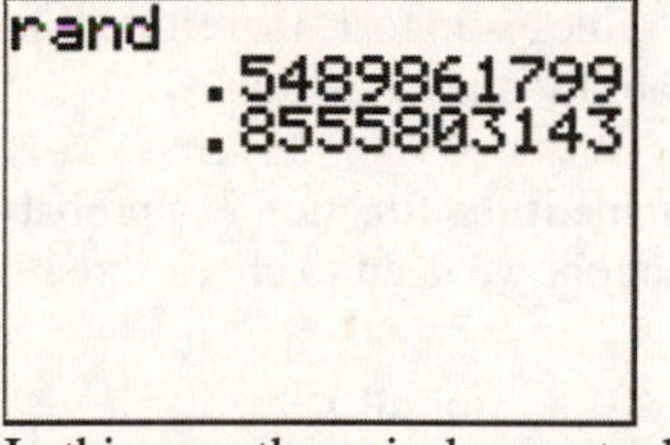

In this case the pair does not add to 1.5 or greater. We expect when doing 20 pairs of numbers about 2 to 3 pairs will add to 1.5 or greater.

39. In order for the random number to have a square that is less than 0.4, the following equation must be satisfied

$$x^2 \leq 0.4$$

Since we are only looking at values of x in the interval (0,1) we know that

$$x^2 \leq 0.4$$
$$x \leq \sqrt{0.4}$$
$$x \leq 0.6325.$$

Thus, the probability that the square of a number chosen between (0,1) is less than 0.4 is

$$P\left(x^2 \leq 0.4\right) = 0.6325.$$

41. ✎

43. ✎

Copyright © 2014 Pearson Education, Inc. Publishing as Addison-Wesley.

Exercise Set 10.3

1. a) There are four possible numbers the spinner can land on, therefore the range of X is $\{1,2,3,4\}$.

b) Since each outcome is equally likely the probability mass function $f(x)=P(X=x)$ can be written as the set of ordered pairs $f(x)=\{(1,\frac{1}{4}),(2,\frac{1}{4}),(3,\frac{1}{4}),(4,\frac{1}{4})\}$.

c)

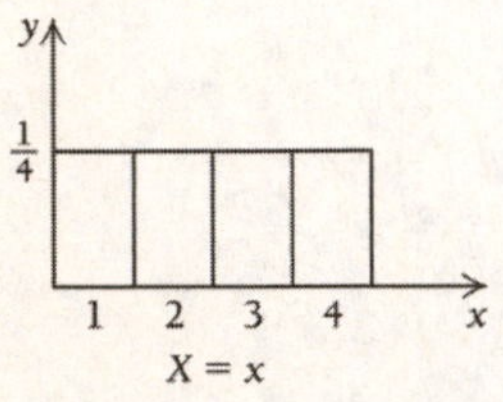

d) Looking at the ordered pairs, $f(3)=\frac{1}{4}$. This value states the probability the spinner points to 3 is $\frac{1}{4}$.

e) We calculate μ as follows

$$\mu=\sum_{i=1}^{n}x_i f(x_i)$$
$$=1\cdot\tfrac{1}{4}+2\cdot\tfrac{1}{4}+3\cdot\tfrac{1}{4}+4\cdot\tfrac{1}{4}$$
$$=2.5.$$

The average value of one spin is 2.5.

3. a) The random variable is the positive difference of the two spins. Since there are four possible numbers for each spin, the range of X is $\{0,1,2,3\}$.

b) In this case, each outcome is not equally likely looking at the sample space, we see

$$S=\begin{Bmatrix}(1,1),(1,2),(1,3),(1,4),(2,1),(2,2),(2,3),(2,4)\\(3,1),(3,2),(3,3),(3,4),(4,1),(4,2),(4,3),(4,4)\end{Bmatrix}.$$

Therefore, the probability mass function $g(x)=P(X=x)$ can be written as the set of ordered pairs

$g(x)=\{(0,\frac{1}{4}),(1,\frac{3}{8}),(2,\frac{1}{4}),(3,\frac{1}{8})\}$.

c)

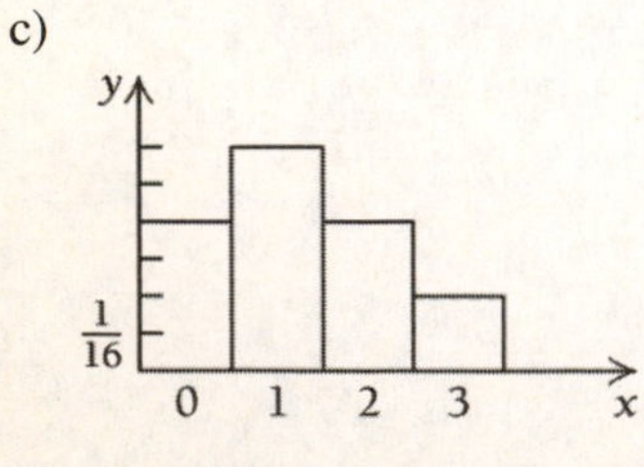

d) Looking at the ordered pairs, $g(1)=\frac{3}{8}$. This value states the probability the positive difference of two spins equals 1 is $\frac{3}{8}$.

e) We calculate μ as follows

$$\mu=\sum_{i=1}^{n}x_i f(x_i)$$
$$=0\cdot\tfrac{1}{4}+1\cdot\tfrac{3}{8}+2\cdot\tfrac{1}{4}+3\cdot\tfrac{1}{8}$$
$$=1.25.$$

The average value of the positive difference of two spins is 1.25.

5. To determine if the function is a probability mass function, we need to check three conditions

1. $f(x)\geq 0$, for all x,
2. $f(x)=P(X=x)$
3. The sum of all $f(x)=1$.

For $f(x)=[(0,0.5),(2,0.3),(3,0.1),(5,0.1)]$

The function values are never negative, the function values represent probabilities and the function values sum to 1, therefore, the function is a probability mass function.

7. To determine if the function is a probability mass function, we need to check three conditions

1. $f(x)\geq 0$, for all x,
2. $f(x)=P(X=x)$
3. The sum of all $f(x)=1$.

For $f(x)=[(-3,0.3),(-2,0.4),(0,0.3)]$

Even though the possible outcome is negative, the function values are never negative, the function values represent probabilities and the function values sum to 1, therefore, the function is a probability mass function.

9. To determine if the function is a probability mass function, we need to check three conditions

1. $p(x)\geq 0$, for all x,
2. $p(x)=P(X=x)$
3. The sum of all $p(x)=1$.

For $p(x)=[(1,0.4),(4,0.2),(5,0.2),(7,0.3)]$

We notice the sum of the function values are $0.4+0.2+0.2+0.3=1.1\neq 1$. Therefore, the function is not a probability mass function.

Copyright © 2014 Pearson Education, Inc. Publishing as Addison-Wesley.

11. To determine if the function is a probability mass function, we need to check three conditions

1. $p(x) \geq 0, \quad$ for all x,
2. $p(x) = P(X = x)$
3. The sum of all $p(x) = 1$.

For $p(x) = [(10, 0.7), (11, 0.4), (12, -0.1)]$

We notice that $p(12) = -0.1 < 0$. Therefore, the function is not a probability mass function.

13. To determine if the function is a probability mass function, we need to check three conditions

1. $f(x) \geq 0, \quad$ for all x,
2. $f(x) = P(X = x)$
3. The sum of all $f(x) = 1$.

For $f(x) = \frac{1}{10}x$, for $x = 1, 2, 3, 4$, we write the function as a set of ordered pairs

$f(x) = \{(1, \frac{1}{10}), (2, \frac{2}{10}), (3, \frac{3}{10}), (4, \frac{4}{10})\}$.

The function values are never negative, the function values represent probabilities. The sum of function values are $\frac{1}{10} + \frac{2}{10} + \frac{3}{10} + \frac{4}{10} = \frac{10}{10} = 1$. Therefore, the function is a probability mass function.

15. To determine if the function is a probability mass function, we need to check three conditions

1. $f(x) \geq 0, \quad$ for all x,
2. $f(x) = P(X = x)$
3. The sum of all $f(x) = 1$.

For $f(x) = \frac{1}{6}x^2$, for $x = -1, 0, 1, 2$, we write the function as a set of ordered pairs

$f(x) = \{(-1, \frac{1}{6}), (0, 0), (1, \frac{1}{6}), (2, \frac{2}{3})\}$.

The function values are never negative, the function values represent probabilities. The sum of function values are $\frac{1}{6} + 0 + \frac{1}{6} + \frac{2}{3} = \frac{6}{6} = 1$. Therefore, the function is a probability mass function.

17. To determine if the function is a probability mass function, we need to check three conditions

1. $h(x) \geq 0, \quad$ for all x,
2. $h(x) = P(X = x)$
3. The sum of all $h(x) = 1$.

For $h(x) = \frac{1}{5}x$, for $x = -1, 1, 2, 3$, we write the function as a set of ordered pairs

$h(x) = \{(-1, -\frac{1}{5}), (1, \frac{1}{5}), (2, \frac{2}{5}), (3, \frac{3}{5})\}$.

We notice that $h(-1) = -\frac{1}{5} < 0$. Therefore, the function is not a probability mass function.

19. In order for the function to be a probability mass function, we need to determine the value of c such that the three conditions for a probability mass function are met. This means we can eliminate any values of c that are negative or greater than one. Therefore, we find the value of c that makes the sum of the function values equal to 1. The sum of the function values are

$$0.2 + 0.15 + 0.38 + c = 1$$
$$0.73 + c = 1$$
$$c = 1 - 0.73$$
$$c = 0.27.$$

The value of c that makes f a probability mass function is 0.27.

21. Determine the values of c that allow the sum of the function values to equal 1 as follows

$$c + 0.25 + 3c + 0.4 = 1$$
$$4c = 1 - 0.65$$
$$4c = 0.35$$
$$c = 0.0875.$$

The value of c that makes f a probability mass function is 0.0875.

23. Determine the values of c that allow the sum of the function values to equal 1 as follows

$$c + c + 2c = 1$$
$$4c = 1$$
$$c = \frac{1}{4}.$$

The value of c that makes f a probability mass function is $\frac{1}{4}$.

Copyright © 2014 Pearson Education, Inc. Publishing as Addison-Wesley.

25. We can write the function $g = cx$ for the given domain as the set of ordered pairs
$g(x) = \{(2,2c),(3,3c)(4,4c)\}$.
Determine the values of c that allow the sum of the function values to equal 1 as follows
$$2c+3c+4c=1$$
$$9c=1$$
$$c=\tfrac{1}{9}.$$
The value of c that makes g a probability mass function is $\frac{1}{9}$.

27. We can write the function $g = c\sqrt{x}$ for the given domain as the set of ordered pairs
$g(x) = \{(1,c),(4,2c)(9,3c),(16,4c)\}$.
Determine the values of c that allow the sum of the function values to equal 1 as follows
$$c+2c+3c+4c=1$$
$$10c=1$$
$$c=\tfrac{1}{10}.$$
The value of c that makes g a probability mass function is $\frac{1}{10}$.

29. We can write the function $h = cx(x-1)$ for the given domain as the set of ordered pairs
$h(x) = \{(2,2c),(4,12c)(6,30c),(8,56c)\}$.
Determine the values of c that allow the sum of the function values to equal 1 as follows
$$2c+12c+30c+56c=1$$
$$100c=1$$
$$c=\tfrac{1}{100}.$$
The value of c that makes h a probability mass function is $\frac{1}{100}$.

31. a) We can write the function $f = \frac{c}{x}$ for the given domain as the set of ordered pairs
$f(x) = \{(1,\tfrac{c}{1}),(3,\tfrac{c}{3})(5,\tfrac{c}{5})(7,\tfrac{c}{7})\}$.
Determine the values of c that allow the sum of the function values to equal 1 as follows
$$\tfrac{c}{1}+\tfrac{c}{3}+\tfrac{c}{5}+\tfrac{c}{7}=1$$
$$\tfrac{105c}{105}+\tfrac{35c}{105}+\tfrac{21c}{105}+\tfrac{15c}{105}=1$$
$$176c=105$$
$$c=\tfrac{105}{176}.$$
The value of c that makes f a probability mass function is $\frac{105}{176}$. Therefore,
$f(x) = \frac{105}{176x}$, for $x = 1,3,5,7$.

b) Substituting into the function we have
$f(3) = \frac{105}{176(3)} = \frac{35}{176} \approx 0.199$.

c) The domain of the function, that is the possible outcomes are $x = 1,3,5,7$. Since $x = 4$ is not a possible outcome,
$f(4) = 0$.

d) Using the formula for μ, we have
$$\mu = \sum x \cdot f(x)$$
$$= 1\cdot\tfrac{105}{176} + 3\cdot\tfrac{35}{176} + 4\cdot\tfrac{105}{704} + 5\cdot\tfrac{21}{176}$$
$$= \tfrac{105}{44}$$
$$= 2.386.$$

33. a) We determine the discrete probabilities first, we have:
$$f(-2) = \tfrac{1}{15}(-2)^2 = \tfrac{4}{15},$$
$$f(-1) = \tfrac{1}{15}(-1)^2 = \tfrac{1}{15},$$
$$f(1) = \tfrac{1}{15}(1)^2 = \tfrac{1}{15},$$
$$f(3) = \tfrac{1}{15}(3)^2 = \tfrac{9}{15}.$$
We calculate the cumulative probabilities $F(x) = P(X \le x)$ for each x by summing the appropriate probabilities. For example, when $x = -2$,
$$F(-2) = P(X \le -2) = f(-2) = \tfrac{4}{15};$$
when $x = -1$,
$$F(-1) = P(X \le -1)$$
$$= f(-2) + f(-1)$$
$$= \tfrac{4}{15} + \tfrac{1}{15} = \tfrac{5}{15};$$
and so on. Written as a piece-wise function, the cumulative probability function, along with its graph is:
$$F(x) = \begin{cases} 0, & x < 2 \\ \frac{4}{15}, & -2 \le x < -1 \\ \frac{5}{15}, & -1 \le x < 1 \\ \frac{2}{5}, & 1 \le x < 3 \\ 1 & 3 \le x. \end{cases}$$

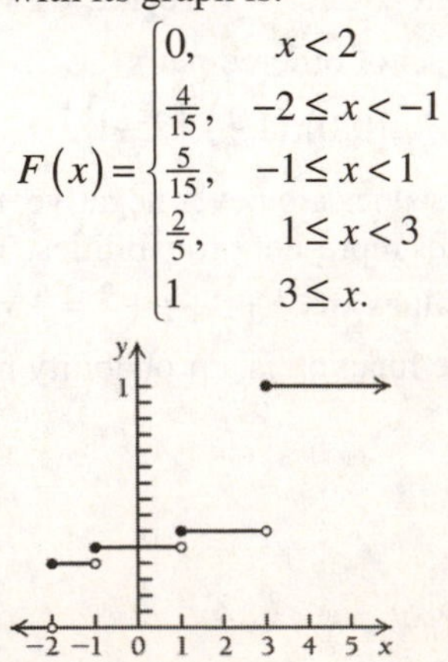

Copyright © 2014 Pearson Education, Inc. Publishing as Addison-Wesley.

b) No, $F(2)=\frac{2}{5}$. The probability that the outcome is less than 2 is equal to $\frac{2}{5}$.

35. a) Since f is a random variable that represents the higher outcome of the two spins, we have four possible outcomes and the range of X is $\{1,2,3,4\}$. . Looking at the sample space we have

$$S=\left\{\begin{matrix}(1,1),(1,2),(1,3),(1,4),(2,1),(2,2),(2,3),(2,4)\\(3,1),(3,2),(3,3),(3,4),(4,1),(4,2),(4,3),(4,4)\end{matrix}\right\}.$$

Using the sample space to calculate the probabilities of each possible outcome, the probability mass function is

$$f(x)=P(X=x)$$
$$=\left\{\left(1,\tfrac{1}{16}\right),\left(2,\tfrac{3}{16}\right),\left(3,\tfrac{5}{16}\right),\left(4,\tfrac{7}{16}\right)\right\}.$$

b) We calculate the cumulative probabilities $F(x)=P(X\le x)$ for each x. For example, when $x=1$ we have

$F(1)=P(X\le 1)=f(1)=\frac{1}{16}$; when $x=2$ we have

$$F(2)=P(X\le 2)$$
$$=f(1)+f(2)$$
$$=\tfrac{1}{16}+\tfrac{3}{16}=\tfrac{1}{4};$$

and so on. Written as a piece-wise function, the cumulative probability function, along with its graph, is

$$F(x)=\begin{cases}0, & x<1\\ \frac{1}{16}, & 1\le x<2\\ \frac{1}{4}, & 2\le x<3\\ \frac{9}{16}, & 3\le x<4\\ 1 & 4\le x.\end{cases}$$

c) $f(2)=\frac{3}{16}$. This value represents the probability that the value 2 is the higher of the two spins.

$F(2)=\frac{1}{4}$. This value represents the probability that the higher of the two spins is at most 2.

d) $1-f(2)=1-\frac{3}{16}=\frac{13}{16}$. This value represents the probability that the value 2 is not the higher of the two spins.

$1-F(2)=1-\frac{1}{4}=\frac{3}{4}$. This value represents the probability that the higher of the two spins is greater than 2.

37. a) We use the information from the frequency table to determine the probability of each outcome. Since 50 people were surveyed, we have the following probabilities

$$f(1)=P(X=1)=\tfrac{18}{50}\approx 0.36,$$
$$f(2)=P(X=2)=\tfrac{17}{50}\approx 0.34,$$
$$f(3)=P(X=3)=\tfrac{9}{50}\approx 0.18,$$
$$f(4)=P(X=4)=\tfrac{3}{50}\approx 0.06,$$
$$f(5)=P(X=5)=\tfrac{2}{50}\approx 0.04,$$
$$f(6)=P(X=6)=\tfrac{0}{50}\approx 0.00,$$
$$f(7)=P(X=7)=\tfrac{1}{50}\approx 0.02.$$

Therefore, the probability mass function, written as a set of ordered pairs is

$$f(x)=P(X=x)$$
$$=\left\{\begin{matrix}(1,0.36),(2,0.34),(3,0.18),(4,0.06),\\(5,0.04),(6,0.00),(7,0.02)\end{matrix}\right\}.$$

The probability distribution histogram is

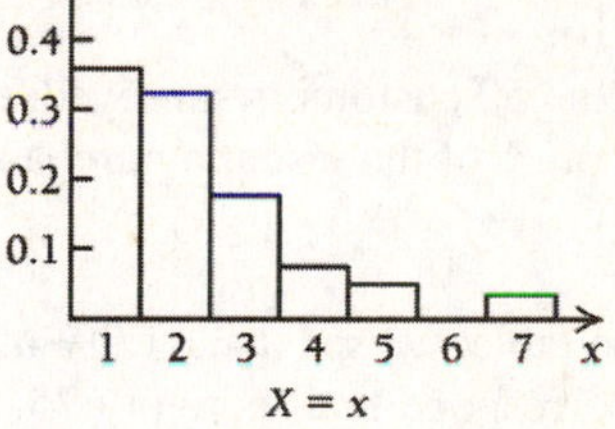

b) We calculate the cumulative probabilities $F(x)=P(X\le x)$ for each x. For example, when $x=1$ we have

$F(1)=P(X\le 1)=f(1)=0.36$; when $x=2$ we have

$$F(2)=P(X\le 2)$$
$$=f(1)+f(2)$$
$$=0.36+0.34=0.7;$$

and so on.

The solution is continued on the next page.

Copyright © 2014 Pearson Education, Inc. Publishing as Addison-Wesley.

Using the information on the previous page, the cumulative probability function is written as a piece-wise function

$$F(x)=\begin{cases}0, & x<1\\ 0.36, & 1\le x<2\\ 0.70, & 2\le x<3\\ 0.88, & 3\le x<4\\ 0.94 & 4\le x<5\\ 0.98 & 5\le x<7\\ 1 & 7\le x.\end{cases}$$

c) $f(2)=0.34$. This value represents the probability the customer paid \$2 for their last app.
$F(2)=0.70$. This value represents the probability that the customer paid at most \$2 for their last app.

d) $1-f(2)=1-0.34=0.66$. This value represents the probability the customer did not pay \$2 for their last app.
$1-F(2)=1-0.70=0.3$. This value represents the probability the customer paid at least \$2 for their last app.

e) Using the formula for μ, we have

$$\begin{aligned}\mu&=\sum x\cdot f(x)\\&=1\cdot 0.36+2\cdot 0.34+3\cdot 0.18+4\cdot 0.06\\&\quad+5\cdot 0.04+7\cdot 0.02\\&=\$2.16.\end{aligned}$$

Among the 50 customers surveyed, the average price of the last app purchased is \$2.16.

39. a) From the table, we see that 11.9% of the scores correspond to a score of 675. Therefore, $g(675)=0.119$. The probability that a random person selected will have a credit score of 650-699 is 0.119.

b) We calculate the cumulative probabilities $G(x)=P(X\le x)$ for each x. For example, when $x=475$ we have $G(475)=P(X\le 475)=g(475)=0.069$; when $x=525$ we have

$$\begin{aligned}G(2)&=P(X\le 2)\\&=g(1)+g(2)\\&=0.069+0.09=0.159;\end{aligned}$$

and so on.

Written as a piece-wise function, the cumulative probability function is

$$F(x)=\begin{cases}0, & x<475\\ 0.069, & 475\le x<525\\ 0.159, & 525\le x<575\\ 0.255, & 575\le x<625\\ 0.350 & 625\le x<675\\ 0.469 & 675\le x<725\\ 0.626 & 725\le x<775\\ 0.821 & 775\le x<825\\ 1 & 825\le x.\end{cases}$$

c) $G(675)=0.469$. This value represents the probability that a person selected at random has a credit score of at most 675.

d) Using the formula for μ, we have

$$\begin{aligned}\mu&=\sum x\cdot f(x)\\&=475\cdot 0.069+525\cdot 0.09+575\cdot 0.096\\&\quad+625\cdot 0.095+675\cdot 0.119+725\cdot 0.157\\&\quad+775\cdot 0.195+825\cdot 0.179\\&=687.55.\end{aligned}$$

The average credit score is 687.55.

41. a) There were 2512 people surveyed. From the table, we see that 168 responded they visited 4 times in a one-month period. Therefore, $h(4)=\frac{168}{2512}=0.0669$. The probability that a random person selected will have visited the coffee shop 4 times in a month is 0.0669.

b) We calculate the cumulative probabilities

$$\begin{aligned}H(4)&=P(X\le 4)\\&=h(0)+h(1)+h(2)+h(3)+h(4)\\&=\tfrac{1367}{2512}+\tfrac{208}{2512}+\tfrac{246}{2512}+\tfrac{136}{2512}+\tfrac{168}{2512}\\&=\tfrac{2125}{2512}\\&=0.8459.\end{aligned}$$

The probability that a random person selected will have visited the coffee shop at most 4 times during the month is 0.8459.

c) $h(1.5)=0$. The probability that a person will visit the coffee shop 1.5 times in the month is zero.
$H(1.5)=\frac{1367}{2512}+\frac{208}{2512}=\frac{1575}{2512}=0.6270$. The probability that a person will have visited the coffee shop at most 1.5 times in the month is 0.6270.

Copyright © 2014 Pearson Education, Inc. Publishing as Addison-Wesley.

d) Using the formula for μ, we have

$$\mu = \sum x \cdot f(x)$$
$$= 0 \cdot \tfrac{1367}{2512} + 1 \cdot \tfrac{208}{2512} + 2 \cdot \tfrac{246}{2512} + 3 \cdot \tfrac{136}{2512} + 4 \cdot \tfrac{168}{2512} + 5 \cdot \tfrac{83}{2512} + 6 \cdot \tfrac{304}{2512}.$$
$$= \tfrac{4019}{2512} = 1.6.$$

The average number of visits to the coffee shop per month is 1.6.

43. Using the initial cost of \$250,000, we see there are 3 cases. There is a 10% chance of generating a profit of $\$400,000 - \$250,000 = \$150,000$. There is a 55% chance of generating a profit of $\$300,000 - \$250,000 = \$50,000$. There is a 35% chance of generating a profit of $\$150,000 - \$250,000 = -\$100,000$.
The expected profit for the new store during its first year is

$$E(X) = 0.10 \cdot \$150,000 + 0.55 \cdot \$50,000 + 0.35 \cdot (-\$100,000)$$
$$= \$7500.$$

Olsen's Tailoring can expect a profit of \$7500 during the first year.

45. The student will receive a value of 1 point if he guesses correctly and receive a value of –0.25 points if he guesses incorrectly. Assuming the guesses are truly random, the probability of a correct answer is $\tfrac{1}{5}$, and the probability of an incorrect answer is $\tfrac{4}{5}$. The expected value of a given question is

$$E(X) = 1\left(\tfrac{1}{5}\right) + (-0.25)\left(\tfrac{4}{5}\right)$$
$$= 0.$$

Therefore, if Chris guesses on every question, he can expect to earn a zero on the exam.

47. If Nun shows, the player will receive zero tokens, i.e. that person does not have to add nor do they take a token from the pot. The probability of the Nun facing up is $\tfrac{1}{4}$.
Therefore, $f(0) = P(X = 0) = \tfrac{1}{4}$.
The cumulative probability is

$$F(0) = P(X \le 0)$$
$$= f(-1) + f(0)$$
$$= \tfrac{1}{4} + \tfrac{1}{4}$$
$$= \tfrac{1}{2}.$$

49. If gimel shows, the player will receive the pot, i.e. that person wins 20 tokens. The probability of the gimel is facing up is $\tfrac{1}{4}$. Therefore,

$$f(20) = P(X = 20) = \tfrac{1}{4}.$$

The cumulative probability is

$$F(20) = P(X \le 20)$$
$$= f(-1) + f(0) + f(5) + f(20)$$
$$= \tfrac{1}{4} + \tfrac{1}{4} + \tfrac{1}{4} + \tfrac{1}{4}$$
$$= 1.$$

51. In order for f to be a probability mass function the probability must add to 1. Set the sum of the probabilities equal to 1 and solve for *c)*

$$c^2 + c = 1$$
$$c^2 + c - 1 = 0.$$

Using the quadratic formula, we have

$$c = \frac{-1 \pm \sqrt{(1)^2 - 4(1)(-1)}}{2(1)}$$
$$= \frac{-1 \pm \sqrt{5}}{2}.$$

Note that $c = \dfrac{-1-\sqrt{5}}{2} < 0$, thus is not valid.
Therefore, the value of c that makes f a probability mass function is

$$c = \frac{-1+\sqrt{5}}{2} \approx 0.61803.$$

Copyright © 2014 Pearson Education, Inc. Publishing as Addison-Wesley.

Exercise Set 10.4

1. $f(x)=\frac{1}{4}x,\ [1,3]$

$$P([1,3])=\int_1^3 f(x)dx =\int_1^3 \frac{1}{4}xdx =\left[\frac{x^2}{8}\right]_1^3 =\frac{(3)^2}{8}-\frac{(1)^2}{8} =\frac{9}{8}-\frac{1}{8} =1.$$

3. $f(x)=3,\ \left[0,\frac{1}{3}\right]$

$$P\left(\left[0,\frac{1}{3}\right]\right)=\int_0^{1/3} f(x)dx =\int_0^{1/3} 3dx =[3x]_0^{1/3} =3\left(\frac{1}{3}\right)-3(0) =1.$$

5. $f(x)=\frac{3}{64}x^2,\ [0,4]$

$$P([0,4])=\int_0^4 f(x)dx =\int_0^4 \frac{3}{64}x^2dx =\left[\frac{x^3}{64}\right]_0^4 =\frac{(4)^3}{64}-\frac{(0)^3}{64} =\frac{64}{64}-0 =1.$$

7. $f(x)=\frac{1}{x},\ [1,e]$

$$P([1,e])=\int_1^e f(x)dx =\int_1^e \frac{1}{x}dx =[\ln x]_1^e =\ln(e)-\ln(1) =1-0 =1.$$

9. $f(x)=\frac{3}{2}x^2,\ [-1,1]$

$$P([-1,1])=\int_{-1}^1 f(x)dx =\int_{-1}^1 \frac{3}{2}x^2dx =\left[\frac{x^3}{2}\right]_{-1}^1 =\frac{(1)^3}{2}-\frac{(-1)^3}{2} =\frac{1}{2}-\frac{-1}{2} =1.$$

11. $f(x)=3e^{-3x},\ [0,\infty)$

$$P([0,\infty))=\int_0^\infty f(x)dx =\lim_{b\to\infty}\int_0^b 3e^{-3x}dx =\lim_{b\to\infty}\left[\frac{3}{-3}e^{-3x}\right]_0^b =\lim_{b\to\infty}\left[-e^{-3x}\right]_0^b =\lim_{b\to\infty}\left[-e^{-3\cdot b}-\left(-e^{-3(0)}\right)\right] =\lim_{b\to\infty}\left[-\frac{1}{e^{3\cdot b}}+1\right]$$

$$=0+1 \quad \left[\text{As } b\to\infty, -\frac{1}{e^{3b}}\to 0\right]$$

$$=1.$$

Copyright © 2014 Pearson Education, Inc. Publishing as Addison-Wesley.

13. $f(x) = kx, \quad [2,5]$

Find k such that $\int_2^5 kx\,dx = 1$

We have

$$\int_2^5 x\,dx = \left[\frac{x^2}{2}\right]_2^5$$
$$= \left[\frac{(5)^2}{2} - \frac{(2)^2}{2}\right]$$
$$= \frac{25}{2} - \frac{4}{2} = \frac{21}{2}$$

Thus $k = \frac{1}{\frac{21}{2}} = \frac{2}{21}$ and the probability density function is $f(x) = \frac{2}{21}x.$

15. $f(x) = kx^2, \quad [-1,1]$

Find k such that $\int_{-1}^1 kx^2\,dx = 1$

We have

$$\int_{-1}^1 x^2\,dx = \left[\frac{x^3}{3}\right]_{-1}^1$$
$$= \left[\frac{(1)^3}{3} - \frac{(-1)^3}{3}\right]$$
$$= \frac{1}{3} - \frac{-1}{3} = \frac{2}{3}$$

Thus $k = \frac{1}{\frac{2}{3}} = \frac{3}{2}$ and the probability density function is $f(x) = \frac{3}{2}x^2.$

17. $f(x) = k, \quad [1,7]$

Find k such that $\int_1^7 k\,dx = 1$

We have

$$\int_1^7 dx = [x]_1^7$$
$$= [7-1]$$
$$= 6$$

Thus $k = 6 = \frac{1}{6}$ and the probability density function is $f(x) = \frac{1}{6}.$

19. $f(x) = k(2-x), \quad [0,2]$

Find k such that $\int_0^2 k(2-x)\,dx = 1$

We have

$$\int_0^2 (2-x)\,dx = \left[2x - \frac{x^2}{2}\right]_0^2$$
$$= \left[\left(2(2) - \frac{(2)^2}{2}\right) - \left(2(0) - \frac{0^2}{2}\right)\right]$$
$$= 4 - \frac{4}{2} - 0$$
$$= 2$$

Thus $k = \frac{1}{2}$ and the probability density function is $f(x) = \frac{1}{2}(2-x) = \frac{2-x}{2}.$

21. $f(x) = \frac{k}{x}, \quad [1,3]$

Find k such that $\int_1^3 \frac{k}{x}\,dx = 1$

We have

$$\int_1^3 \frac{1}{x}\,dx = [\ln(x)]_1^3$$
$$= [\ln(3) - \ln(1)]$$
$$= \ln 3 - 0 = \ln 3$$

Thus $k = \frac{1}{\ln 3}$ and the probability density function is $f(x) = \frac{1}{\ln 3}\cdot\frac{1}{x} = \frac{1}{x\ln 3}.$

23. $f(x) = ke^x, \quad [0,3]$

Find k such that $\int_0^3 ke^x\,dx = 1$

We have

$$\int_0^3 e^x\,dx = \left[e^x\right]_0^3$$
$$= \left[e^3 - e^0\right]$$
$$= e^3 - 1$$

Thus $k = \frac{1}{e^3-1}$ and the probability density function is $f(x) = \frac{1}{e^3-1}\cdot e^x = \frac{e^x}{e^3-1}.$

Copyright © 2014 Pearson Education, Inc. Publishing as Addison-Wesley.

25. Consider the probability density function in Exercise 1, $f(x)=\frac{1}{4}x$, $[1,3]$. The cumulative density function is determined by substituting the dummy variable s into the pdf and integrating as follows:

$$F(x)=\int_1^x f(s)ds$$
$$=\int_1^x \frac{1}{4}sds$$
$$=\left[\frac{s^2}{8}\right]_1^x$$
$$=\frac{(x)^2}{8}-\frac{(1)^2}{8}$$
$$=\frac{x^2}{8}-\frac{1}{8}; \qquad [1,3].$$

27. Consider the probability density function in Exercise 3, $f(x)=3$, $\left[0,\frac{1}{3}\right]$. The cumulative density function is determined by substituting the dummy variable s into the pdf and integrating as follows:

$$F(x)=\int_0^x f(s)ds$$
$$=\int_0^x 3ds$$
$$=[3s]_0^x$$
$$=3(x)-3(0)$$
$$=3x, \qquad \left[0,\tfrac{1}{3}\right].$$

29. Consider the probability density function in Exercise 5, $f(x)=\frac{3}{64}x$, $[0,4]$. The cumulative density function is determined by substituting the dummy variable s into the pdf and integrating as follows:

$$F(x)=\int_0^x f(s)ds$$
$$=\int_0^x \frac{3}{64}s^2ds$$
$$=\left[\frac{s^3}{64}\right]_0^x$$
$$=\frac{(x)^3}{64}-\frac{(0)^3}{64}$$
$$=\frac{x^3}{64}-0$$
$$=\tfrac{1}{64}x^3; \qquad [0,4].$$

31. Consider the probability density function in Exercise 7, $f(x)=\frac{1}{x}$, $[1,e]$. The cumulative density function is determined by substituting the dummy variable s into the pdf and integrating as follows:

$$F(x)=\int_1^x f(s)ds$$
$$=\int_1^x \frac{1}{s}ds$$
$$=[\ln s]_1^x$$
$$=\ln(x)-\ln(1)$$
$$=\ln x; \qquad [1,e].$$

33. Consider the probability density function in Exercise 9, $f(x)=\frac{3}{2}x^2$, $[-1,1]$. The cumulative density function is determined by substituting the dummy variable s into the pdf and integrating as follows:

$$F(x)=\int_{-1}^x f(s)ds$$
$$=\int_{-1}^x \frac{3}{2}s^2ds$$
$$=\left[\frac{s^3}{2}\right]_{-1}^x$$
$$=\frac{(x)^3}{2}-\frac{(-1)^3}{2}$$
$$=\frac{x^3}{2}-\frac{-1}{2}$$
$$=\tfrac{1}{2}x^3+\tfrac{1}{2}; \qquad [-1,1].$$

35. Consider the probability density function in Exercise 11, $f(x)=3e^{-3x}$, $[0,\infty]$. The cumulative density function is determined by substituting the dummy variable s into the pdf and integrating as :

$$F(x)=\int_0^x f(s)ds$$
$$=\int_0^x 3e^{-3s}ds$$
$$=\left[\frac{3}{-3}e^{-3s}\right]_0^x$$
$$=\left[-e^{-3s}\right]_0^x$$
$$=-e^{-3\cdot x}-\left(-e^{-3(0)}\right)$$
$$=-e^{-3x}-(-1)$$
$$=1-e^{-3x}; \qquad [0,\infty).$$

Copyright © 2014 Pearson Education, Inc. Publishing as Addison-Wesley.

37. Consider the probability density function in Exercise 13, $f(x)=\frac{1}{4}x,\ [1,3]$. The cumulative density function is determined by substituting the dummy variable s into the pdf and integrating as follows:

$$\begin{aligned}F(x)&=\int_2^x f(s)ds\\&=\int_2^x \tfrac{2}{21}sds\\&=\left[\tfrac{1}{21}s^2\right]_2^x\\&=\tfrac{1}{21}(x)^2-\tfrac{1}{21}(2)^2\\&=\tfrac{1}{21}x^2-\left(\tfrac{4}{21}\right)\\&=\tfrac{1}{21}x^2-\tfrac{4}{21};\qquad [2,5].\end{aligned}$$

39. Consider the probability density function in Exercise 15, $f(x)=\frac{3}{2}x^2,\ [-1,1]$. The cumulative density function is determined by substituting the dummy variable s into the pdf and integrating as follows:

$$\begin{aligned}F(x)&=\int_{-1}^x f(s)ds\\&=\int_{-1}^x \frac{3}{2}s^2ds\\&=\left[\frac{s^3}{2}\right]_{-1}^x\\&=\frac{(x)^3}{2}-\frac{(-1)^3}{2}\\&=\frac{x^3}{2}-\frac{-1}{2}\\&=\tfrac{1}{2}x^3+\tfrac{1}{2};\qquad [-1,1].\end{aligned}$$

41. Consider the probability density function in Exercise 17, $f(x)=\frac{1}{6},\ [1,7]$. The cumulative density function is determined by substituting the dummy variable s into the pdf and integrating as follows:

$$\begin{aligned}F(x)&=\int_1^x f(s)ds\\&=\int_1^x \tfrac{1}{6}ds\\&=\left[\tfrac{1}{6}s\right]_1^x\\&=\tfrac{1}{6}(x)-\tfrac{1}{6}(1)\\&=\tfrac{1}{6}x-\tfrac{1}{6};\qquad [1,7].\end{aligned}$$

43. Consider the probability density function in Exercise 19, $f(x)=\frac{2-x}{2},\ [0,2]$. The cumulative density function is determined by substituting the dummy variable s into the pdf and integrating as follows:

$$\begin{aligned}F(x)&=\int_0^x f(s)ds\\&=\int_0^x \frac{2-s}{2}ds\\&=\int_0^x 1-\tfrac{1}{2}sds\\&=\left[s-\frac{s^2}{4}\right]_0^x\\&=(x)-\frac{(x)^2}{4}-\left[(0)-\frac{(0)^2}{4}\right]\\&=x-\tfrac{1}{4}x^2;\qquad [0,2].\end{aligned}$$

45. Consider the probability density function in Exercise 21, $f(x)=\frac{1}{x\cdot\ln 3},\ [1,3]$. The cumulative density function is determined by substituting the dummy variable s into the pdf and integrating as follows:

$$\begin{aligned}F(x)&=\int_1^x f(s)ds\\&=\int_1^x \frac{1}{s\cdot\ln 3}ds\\&=\frac{1}{\ln 3}\cdot\int_1^x \frac{1}{s}ds\\&=\frac{1}{\ln 3}\cdot[\ln s]_1^x\\&=\frac{1}{\ln 3}\cdot[\ln(x)-\ln(1)]\\&=\frac{1}{\ln 3}\cdot\ln x\\&=\frac{\ln x}{\ln 3};\qquad [1,3].\end{aligned}$$

47. Consider the probability density function in Exercise 23, $f(x)=\frac{1}{e^3-1}e^x,\ [0,3]$. The cumulative density function is determined by substituting the dummy variable s into the pdf and integrating as shown on the next page.

Copyright © 2014 Pearson Education, Inc. Publishing as Addison-Wesley.

$$F(x)=\int_0^x f(s)\,ds$$
$$=\int_0^x \frac{1}{e^3-1}e^s\,ds$$
$$=\left[\frac{e^s}{e^3-1}\right]_0^x$$
$$=\frac{e^x}{e^3-1}-\frac{e^0}{e^3-1}$$
$$=\frac{e^x}{e^3-1}-\frac{1}{e^3-1}$$
$$=\frac{e^x-1}{e^3-1};\qquad [0,3].$$

49. $f(x)=\frac{1}{50}x$, for $0\le x\le 10$

$$P(2\le x\le 6)=\int_2^6 \frac{1}{50}x\,dx$$
$$=\left[\frac{1}{50}\cdot\frac{x^2}{2}\right]_2^6$$
$$=\left[\frac{x^2}{100}\right]_2^6$$
$$=\frac{6^2}{100}-\frac{2^2}{100}$$
$$=\frac{36-4}{100}=\frac{32}{100}=\frac{8}{25}=0.32$$

The probability that the dart lands in the interval $[2,6]$ is $\frac{8}{25}$, or 0.32.

51. $f(x)=\frac{1}{16}$, for $4\le x\le 20$

$$P(9\le x\le 20)=\int_9^{20}\frac{1}{16}dx$$
$$=\left[\frac{1}{16}x\right]_9^{20}$$
$$=\frac{1}{16}\cdot 20-\frac{1}{16}\cdot 9$$
$$=\frac{20-9}{16}$$
$$=\frac{11}{16}=0.6875$$

The probability that the number selected is in the subinterval $[9,20]$ is $\frac{11}{16}$, or 0.6875.

53. From Example 10, we know

$f(x)=ke^{-kx}$, for $0\le x<\infty$, where $k=\frac{1}{a}$ and a is the average distance between successive cars over some period of time.

First, we determine k:

$$k=\frac{1}{100}=0.01.$$

The probability that the distance between cars is 40 feet or less is calculated as follows:

$$P(0\le x\le 40)=\int_0^{40}0.01e^{-0.01x}dx$$
$$=\left[\frac{0.01}{-0.01}e^{-0.01x}\right]_0^{40}$$
$$=\left[-e^{-0.01x}\right]_0^{40}$$
$$=-e^{-0.01(40)}-\left(-e^{-0.01(0)}\right)$$
$$=-e^{-0.4}+e^0$$
$$=-e^{-0.4}+1$$
$$\approx -0.670320+1$$
$$\approx 0.329680$$
$$\approx 0.3297$$

The probability that the distance between two successive cars, chosen at random, is 40 feet or less is 0.3297.

55. $f(t)=2e^{-2t}$, for $0\le t<\infty$

$$P(0\le t\le 5)=\int_0^5 2e^{-2t}dt$$
$$=\left[\frac{2}{-2}e^{-2t}\right]_0^5$$
$$=\left[-e^{-2t}\right]_0^5$$
$$=-e^{-2(5)}-\left(-e^{-2(0)}\right)$$
$$=-e^{-10}+e^0$$
$$=-e^{-10}+1$$
$$\approx -0.0000454+1$$
$$\approx 0.9999546$$
$$\approx 0.999955$$

The probability that a phone call will last no more than 5 minutes is 0.999955.

Copyright © 2014 Pearson Education, Inc. Publishing as Addison-Wesley.

57. $f(t)=ke^{-kt}$, for $0\le t<\infty$, where $k=\frac{1}{a}$ and a is the average amount of time that will pass before a failure occurs.

First, we determine k:

$$k=\frac{1}{100}=0.01.$$

The probability that a failure will occur in 50 hours or less is

$$P(0\le x\le 50)=\int_0^{50}0.01e^{-0.01x}dx$$
$$=\left[\frac{0.01}{-0.01}e^{-0.01x}\right]_0^{50}$$
$$=\left[-e^{-0.01x}\right]_0^{50}$$
$$=-e^{-0.01(50)}-\left(-e^{-0.01(0)}\right)$$
$$=-e^{-0.5}+e^0$$
$$=-e^{-0.5}+1$$
$$\approx -0.606531+1$$
$$\approx 0.393469$$
$$\approx 0.3935$$

The probability that a failure will occur in 50 hours or less is 0.3935.

59. $f(t)=0.23e^{-0.23t}$, for $1\le t<\infty$

a) To verify that 90% of calls are answered within 10 seconds, integrate the probability density function as follows

$$\int_0^{10}0.23e^{-0.23t}dt=\left[\frac{0.23}{-0.23}e^{-0.23t}\right]_0^{10}$$
$$=\left[-e^{-0.23t}\right]_0^{10}$$
$$=-e^{-0.23(10)}-\left(-e^0\right)$$
$$=1-e^{-2.3}$$
$$=0.8997$$
$$\approx 0.9.$$

b) The probability that a 911 call is answered between 15 and 25 seconds after the call is made is

$$P(15\le t\le 25)=\int_{15}^{25}0.23e^{-0.23t}dt$$
$$=\left[-e^{-0.23t}\right]_{15}^{25}$$
$$=-e^{-0.23(25)}-\left(-e^{-0.23(15)}\right)$$
$$\approx 0.0286.$$

The probability that a 911 call is answered between 15 and 25 seconds is 0.0286.

61. $f(t)=0.02e^{-0.02t}$, for $0\le t<\infty$

$$P(0\le t\le 150)=\int_0^{150}0.02e^{-0.02t}dt$$
$$=\left[\frac{0.02}{-0.02}e^{-0.02t}\right]_0^{150}$$
$$=\left[-e^{-0.02}\right]_0^{150}$$
$$=-e^{-0.02(150)}-\left(-e^{-0.02(0)}\right)$$
$$=-e^{-3}+e^0$$
$$=-e^{-3}+1$$
$$\approx -0.049787+1$$
$$\approx 0.950213$$

The probability that the rat will learn its way through a maze in 150 seconds, or less, is approximately 0.950213.

63. $f(x)=x$ is a probability density function over $[0,b]$. Thus,

$$\int_0^b f(x)dx=1$$
$$\int_0^b xdx=1$$
$$\left[\frac{x^2}{2}\right]_0^b=1$$
$$\frac{b^2}{2}-\frac{0^2}{2}=1$$
$$\frac{b^2}{2}=1$$
$$b^2=2$$
$$b=\sqrt{2}$$

$f(x)=x$ is a probability density function over $\left[0,\sqrt{2}\right]$.

Copyright © 2014 Pearson Education, Inc. Publishing as Addison-Wesley.

65. Answers may vary. The cumulative distribution function is not correct for the given probability density function. The correct cdf is:

$$F(x)=\int_1^x f(s)ds$$
$$=\int_1^x\left(\tfrac{1}{2}s-\tfrac{1}{2}\right)ds$$
$$=\left[\frac{s^2}{4}-\frac{s}{2}\right]_1^x$$
$$=\frac{(x)^2}{4}-\frac{x}{2}-\left[\frac{(1)^2}{4}-\frac{1}{2}\right]$$
$$=\tfrac{1}{4}x^2-\tfrac{1}{2}x+\tfrac{1}{4};\qquad [1,3].$$

67. ✎

69. a) Since there is a 30% probability that the patient will wait up to one hour, we have

$$P(0\le t\le 1)=\int_0^1 f(t)dt=0.30.$$

Let the probability density function be $f(t)=ke^{-kt}$. Substituting, we have

$$\int_0^1 f(t)dt=0.30$$
$$\int_0^1 ke^{-kt}dt=0.30$$

From exericse 67 we know $\int_0^1 ke^{-kt}dt=1-e^{-k}$.

$$1-e^{-k}=0.30$$
$$e^{-k}=0.70$$
$$-k=\ln(0.70)$$
$$k=-\ln(0.70)$$
$$k\approx 0.357.$$

The probability density function is $f(t)=0.357e^{-0.357t}$.

b) The probability that a patient will have to wait between 90 minutes (1.5 hours) and 3 hours for a doctor is

$$P(1.5\le t\le 3)=\int_{1.5}^3 0.357e^{-0.357t}dt$$
$$=\left[-e^{-0.357t}\right]_{1.5}^3$$
$$=-e^{-0.357(3)}-\left(-e^{-0.357(1.5)}\right)$$
$$\approx 0.2427.$$

71. a) The area under the graph must equal 1, since f is a probability density function. Therefore,

$$\text{Area}=1$$
$$\tfrac{1}{2}b\cdot h=1$$
$$\tfrac{1}{2}(5-1)\cdot(c)=1$$
$$2c=1$$
$$c=\tfrac{1}{2}.$$

b) The linear function passes through the points $(1,0)$ and $\left(5,\tfrac{1}{2}\right)$; therefore, the slope of the line is

$$m=\frac{\frac{1}{2}-0}{5-1}=\frac{\frac{1}{2}}{4}=\frac{1}{8}.$$

Using the point-slope formula for a line, we have

$$y-0=\tfrac{1}{8}(x-1)$$
$$y=\tfrac{1}{8}x-\tfrac{1}{8}.$$

Therefore, the probability density function is

$$f(x)=\tfrac{1}{8}x-\tfrac{1}{8},\qquad \text{for } 1\le x\le 5.$$

c) The cumulative density function is

$$F(x)=\int_1^x f(s)ds$$
$$=\int_1^x \tfrac{1}{8}s-\tfrac{1}{8}ds$$
$$=\left[\frac{s^2}{16}-\frac{s}{8}\right]_1^x$$
$$=\frac{(x)^2}{16}-\frac{(x)}{8}-\left[\frac{(1)^2}{16}-\frac{(1)}{8}\right]$$
$$=\tfrac{1}{16}x^2-\tfrac{1}{8}x+\tfrac{1}{16};\qquad [1,5].$$

d) $P(2\le x\le 3)=F(3)-F(2)$.

$$F(3)=\tfrac{1}{16}(3)^2-\tfrac{1}{8}(3)+\tfrac{1}{16}=\tfrac{1}{4}$$
$$F(2)=\tfrac{1}{16}(2)^2-\tfrac{1}{8}(2)+\tfrac{1}{16}=\tfrac{1}{16}.$$

Therefore,

$$P(2\le x\le 3)=F(3)-F(2)$$
$$=\tfrac{1}{4}-\tfrac{1}{16}$$
$$=\tfrac{3}{16}=0.1875.$$

Copyright © 2014 Pearson Education, Inc. Publishing as Addison-Wesley.

73. Determine the value of c so that
$\int_1^2 cxe^{2x}\,dx = 1.$
Integrate and solve for c.

$$\int_1^2 cxe^{2x}\,dx = 1$$

$$c\int_1^2 xe^{2x}\,dx = 1$$

$$u = x \quad dv = e^{2x}dx$$
$$du = dx \quad v = \tfrac{1}{2}e^{2x}$$

$$c\left[\tfrac{1}{2}xe^{2x} - \int \tfrac{1}{2}e^{2x}dx\right]_1^2 = 1$$

$$c\left[\tfrac{1}{2}xe^{2x} - \tfrac{1}{4}e^{2x}\right]_1^2 = 1$$

$$c\left[\tfrac{1}{2}(2)e^4 - \tfrac{1}{4}e^4 - \left(\tfrac{1}{2}(2)e^2 - \tfrac{1}{4}e^2\right)\right] = 1$$

$$c\left[\left(1-\tfrac{1}{4}\right)e^4 - \left(\tfrac{1}{2}-\tfrac{1}{4}\right)e^2\right] = 1$$

$$c\left[\tfrac{3}{4}e^4 - \tfrac{1}{4}e^2\right] = 1$$

$$c\left[\frac{3e^4 - e^2}{4}\right] = 1$$

$$c = \frac{4}{3e^4 - e^2}$$

$$c \approx 0.02557$$

The probability density function is
$f(x) = 0.02557xe^{2x}, \qquad \text{for } 1 \le x \le 2.$

75 – 86. Left to the student. You may check your answers using the solutions to Exercises 1-12.

Copyright © 2014 Pearson Education, Inc. Publishing as Addison-Wesley.

Exercise Set 10.5

1. $f(x)=\frac{1}{4},\ [3,7]$

$E(x)=\int_a^b x\cdot f(x)\,dx$ Expected value of x.

$E(x)=\int_3^7 x\cdot\frac{1}{4}dx$

$=\frac{1}{4}\left[\frac{x^2}{2}\right]_3^7$

$=\frac{1}{4}\left[\frac{7^2}{2}-\frac{3^2}{2}\right]$

$=\frac{1}{4}\left[\frac{49}{2}-\frac{9}{2}\right]$

$=\frac{1}{4}\cdot\frac{40}{2}$

$=5$

$E(x^2)=\int_a^b x^2\cdot f(x)\,dx$ Expected value of x^2.

$E(x^2)=\int_3^7 x^2\cdot\frac{1}{4}dx$

$=\frac{1}{4}\left[\frac{x^3}{3}\right]_3^7$

$=\frac{1}{4}\left[\frac{7^3}{3}-\frac{3^3}{3}\right]$

$=\frac{1}{4}\left[\frac{343}{3}-\frac{27}{3}\right]$

$=\frac{1}{4}\cdot\frac{316}{3}$

$=\frac{79}{3}$

$\mu=E(x)=5$ Mean

$\sigma^2=E(x^2)-[E(x)]^2$

$=\frac{79}{3}-[5]^2$ Substituting 79/3 for $E(x^2)$ and 5 for $E(x)$

$=\frac{79}{3}-25$

$=\frac{79}{3}-\frac{75}{3}$

$=\frac{4}{3}$ Variance

$\sigma=\sqrt{\sigma^2}$

$=\sqrt{\frac{4}{3}}=\frac{2}{\sqrt{3}}$ Standard deviation

3. $f(x)=\frac{1}{8}x,\ [0,4]$

$E(x)=\int_a^b x\cdot f(x)\,dx$ Expected value of x.

$E(x)=\int_0^4 x\cdot\frac{1}{8}x\,dx$

$=\int_0^4\frac{1}{8}x^2dx$

$=\frac{1}{8}\left[\frac{x^3}{3}\right]_0^4$

$=\frac{1}{8}\left[\frac{4^3}{3}-\frac{0^3}{3}\right]$

$=\frac{1}{8}\cdot\frac{64}{3}$

$=\frac{8}{3}$

$E(x^2)=\int_a^b x^2\cdot f(x)\,dx$ Expected value of x^2.

$E(x^2)=\int_0^4 x^2\cdot\frac{1}{8}x\,dx$

$=\int_0^4\frac{1}{8}x^3dx$

$=\frac{1}{8}\left[\frac{x^4}{4}\right]_0^4$

$=\frac{1}{8}\left[\frac{4^4}{4}-\frac{0^4}{4}\right]$

$=\frac{1}{8}\cdot\frac{256}{4}$

$=8$

$\mu=E(x)=\frac{8}{3}$ Mean

The solution is continued on the next page.

Copyright © 2014 Pearson Education, Inc. Publishing as Addison-Wesley.

$$\sigma^2 = E(x^2) - [E(x)]^2$$

$$= 8 - \left[\frac{8}{3}\right]^2$$ Substituting 8 for $E(x^2)$ and 8/3 for $E(x)$

$$= 8 - \frac{64}{9}$$

$$= \frac{72}{9} - \frac{64}{9}$$

$$= \frac{8}{9}$$ Variance

$$\sigma = \sqrt{\sigma^2}$$

$$= \sqrt{\frac{8}{9}} = \frac{2\sqrt{2}}{3}$$ Standard deviation

5. $f(x) = \frac{1}{4}x,\ [1,3]$

$$E(x) = \int_a^b x \cdot f(x)\,dx$$ Expected value of x.

$$E(x) = \int_1^3 x \cdot \frac{1}{4}x\,dx$$

$$= \int_1^3 \frac{x^2}{4}\,dx$$

$$= \left[\frac{x^3}{12}\right]_1^3$$

$$= \left[\frac{3^3}{12} - \frac{1^3}{12}\right]$$

$$= \frac{26}{12}$$

$$= \frac{13}{6}$$

$$E(x^2) = \int_a^b x^2 \cdot f(x)\,dx$$ Expected value of x^2.

$$E(x^2) = \int_1^3 x^2 \cdot \frac{1}{4}x\,dx$$

$$= \int_1^3 \frac{x^3}{4}\,dx$$

$$= \left[\frac{x^4}{16}\right]_1^3$$

$$= \left[\frac{3^4}{16} - \frac{1^4}{16}\right]$$

$$= \frac{80}{16}$$

$$= 5$$

$$\mu = E(x) = \frac{13}{6}$$ Mean

$$\sigma^2 = E(x^2) - [E(x)]^2$$

$$= 5 - \left[\frac{13}{6}\right]^2$$ Substituting 5 for $E(x^2)$ and 13/6 for $E(x)$

$$= 5 - \frac{169}{36}$$

$$= \frac{180}{36} - \frac{169}{36}$$

$$= \frac{11}{36}$$ Variance

$$\sigma = \sqrt{\sigma^2}$$

$$= \sqrt{\frac{11}{36}} = \frac{\sqrt{11}}{6}$$ Standard deviation

7. $g(x) = \frac{1}{8},\ [2,10].$

$$E(x) = \int_a^b x \cdot g(x)\,dx$$ Expected value of x.

$$E(x) = \int_2^{10} x \cdot \frac{1}{8}\,dx$$

$$= \left[\frac{x^2}{16}\right]_2^{10}$$

$$= \left[\frac{(10)^2}{16} - \frac{(2)^2}{16}\right]$$

$$= \frac{100}{16} - \frac{4}{16}$$

$$= 6.$$

$$E(x^2) = \int_a^b x^2 \cdot g(x)\,dx$$ Expected value of x^2.

$$E(x^2) = \int_2^{10} x^2 \cdot \frac{1}{8}\,dx$$

$$= \int_2^{10} \frac{1}{8}x^2\,dx$$

$$= \left[\frac{x^3}{24}\right]_2^{10}$$

$$= \left[\frac{(10)^3}{24} - \frac{(2)^3}{24}\right]$$

$$= \frac{1000}{24} - \frac{8}{24}$$

$$= \frac{124}{3}.$$

The solution is continued on the next page.

Copyright © 2014 Pearson Education, Inc. Publishing as Addison-Wesley.

$\mu = E(x) = 6$ Mean

$\sigma^2 = E(x^2) - [E(x)]^2$

$= \frac{124}{3} - [6]^2$

$= \frac{16}{3}$ Variance

$\sigma = \sqrt{\sigma^2}$

$= \sqrt{\frac{16}{3}}$

$= \frac{4\sqrt{3}}{3} \approx 2.3094$ Standard deviation

To find the area within one standard deviation, we have

$\int_{6-2.309}^{6+2.309} \frac{1}{8} dx = \int_{3.691}^{8.309} \frac{1}{8} dx$

$= \left[\frac{x}{8}\right]_{3.691}^{8.309}$

$= \frac{8.309}{8} - \frac{3.691}{8}$

$= \frac{4.618}{8}$

$= 0.577 \approx 0.58.$

The area within one standard deviation of the mean is 0.58.

9. $f(x) = \frac{1}{8}x, \quad [0,4]$

$E(x) = \int_a^b x \cdot f(x) dx$ Expected value of x.

$E(x) = \int_0^4 x \cdot \frac{1}{8} x\, dx$

$= \int_0^4 \frac{1}{8} x^2 dx$

$= \frac{1}{8}\left[\frac{x^3}{3}\right]_0^4$

$= \frac{1}{8}\left[\frac{4^3}{3} - \frac{0^3}{3}\right]$

$= \frac{1}{8} \cdot \frac{64}{3}$

$= \frac{8}{3}$

$E(x^2) = \int_a^b x^2 \cdot f(x) dx$ Expected value of x^2.

$E(x^2) = \int_0^4 x^2 \cdot \frac{1}{8} x\, dx$

$= \int_0^4 \frac{1}{8} x^3 dx$

$= \frac{1}{8}\left[\frac{x^4}{4}\right]_0^4$

$= \frac{1}{8}\left[\frac{4^4}{4} - \frac{0^4}{4}\right]$

$= \frac{1}{8} \cdot \frac{256}{4}$

$= 8$

$\mu = E(x) = \frac{8}{3}$ Mean

$\sigma^2 = E(x^2) - [E(x)]^2$

$= 8 - \left[\frac{8}{3}\right]^2$ Substituting 8 for $E(x^2)$ and 8/3 for $E(x)$

$= 8 - \frac{64}{9}$

$= \frac{72}{9} - \frac{64}{9}$

$= \frac{8}{9}$ Variance

$\sigma = \sqrt{\sigma^2}$

$= \sqrt{\frac{8}{9}}$

$= \frac{2\sqrt{2}}{3} \approx 0.9428$ Standard deviation

To find the area within one standard deviation, we have

$\int_{8/3-0.943}^{8/3+0.943} \frac{1}{8} x dx = \int_{1.724}^{3.610} \frac{1}{8} x dx$

$= \left[\frac{x^2}{16}\right]_{1.724}^{3.610}$

$= \frac{3.610^2}{16} - \frac{1.724^2}{16}$

$= \frac{10.06}{16}$

$= 0.6287 \approx 0.63.$

The area within one standard deviation of the mean is 0.63.

Copyright © 2014 Pearson Education, Inc. Publishing as Addison-Wesley.

11.

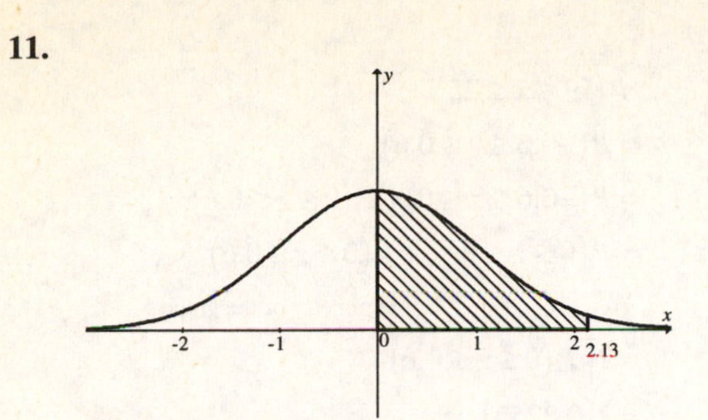

Using Table A, we have

$P(0 \le x \le 2.13) = 0.4834$

13.

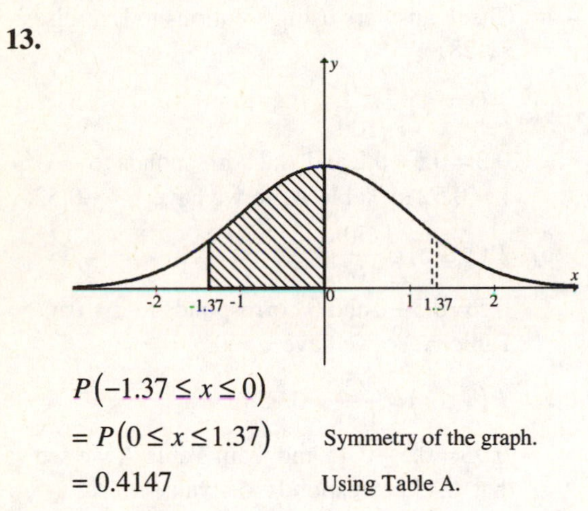

$P(-1.37 \le x \le 0)$

$= P(0 \le x \le 1.37)$ Symmetry of the graph.

$= 0.4147$ Using Table A.

15. Using Table A, we have:

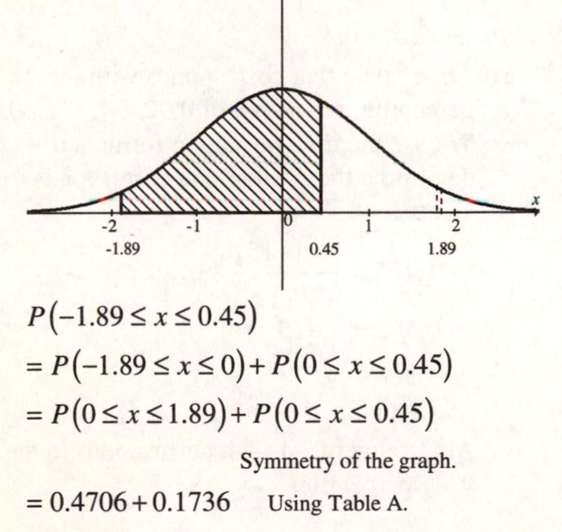

$P(-1.89 \le x \le 0.45)$

$= P(-1.89 \le x \le 0) + P(0 \le x \le 0.45)$

$= P(0 \le x \le 1.89) + P(0 \le x \le 0.45)$

Symmetry of the graph.

$= 0.4706 + 0.1736$ Using Table A.

$= 0.6442$

17. Using Table A, we have:

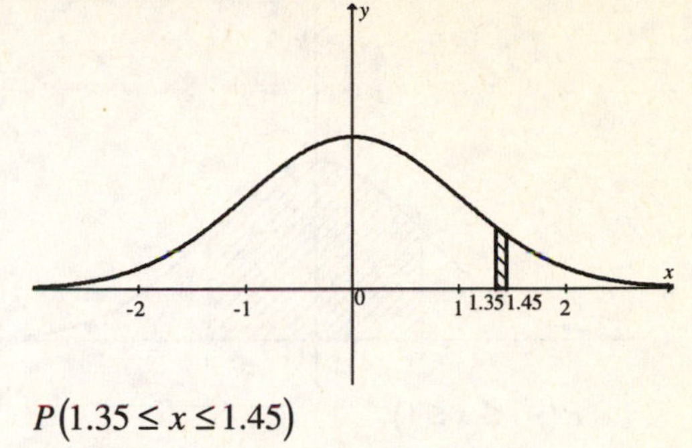

$P(1.35 \le x \le 1.45)$

$= P(0 \le x \le 1.45) - P(0 \le x \le 1.35)$

$= 0.4265 - 0.4115$ Using Table A.

$= 0.0150$

19. Using Table A, we have:

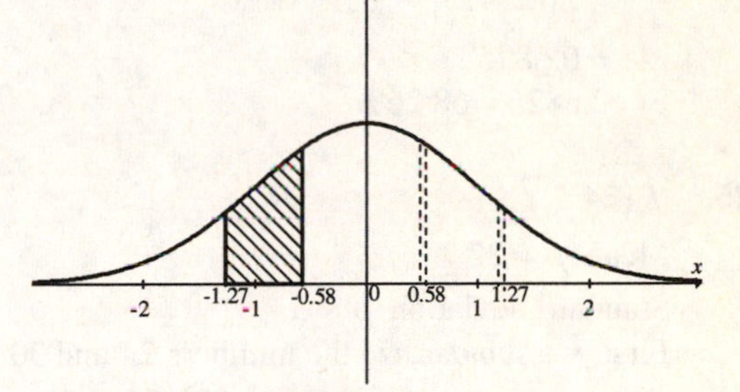

$P(-1.27 \le x \le -0.58)$

$= P(0.58 \le x \le 1.27)$ Symmetry of the graph.

$= P(0 \le x \le 1.27) - P(0 \le x \le 0.58)$

$= 0.3980 - 0.2190$ Using Table A.

$= 0.1790$

21.

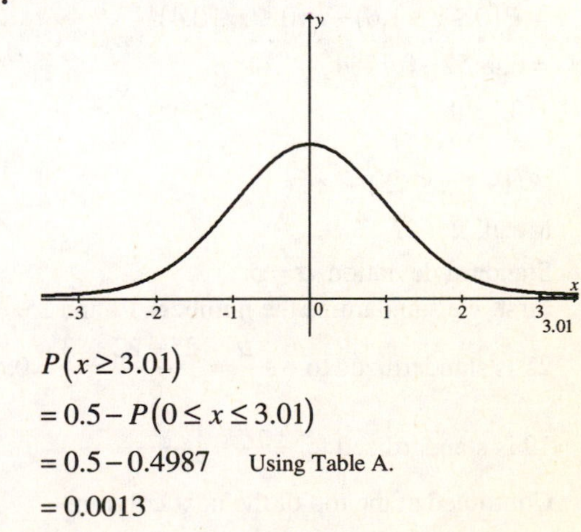

$P(x \ge 3.01)$

$= 0.5 - P(0 \le x \le 3.01)$

$= 0.5 - 0.4987$ Using Table A.

$= 0.0013$

Copyright © 2014 Pearson Education, Inc. Publishing as Addison-Wesley.

23. a)

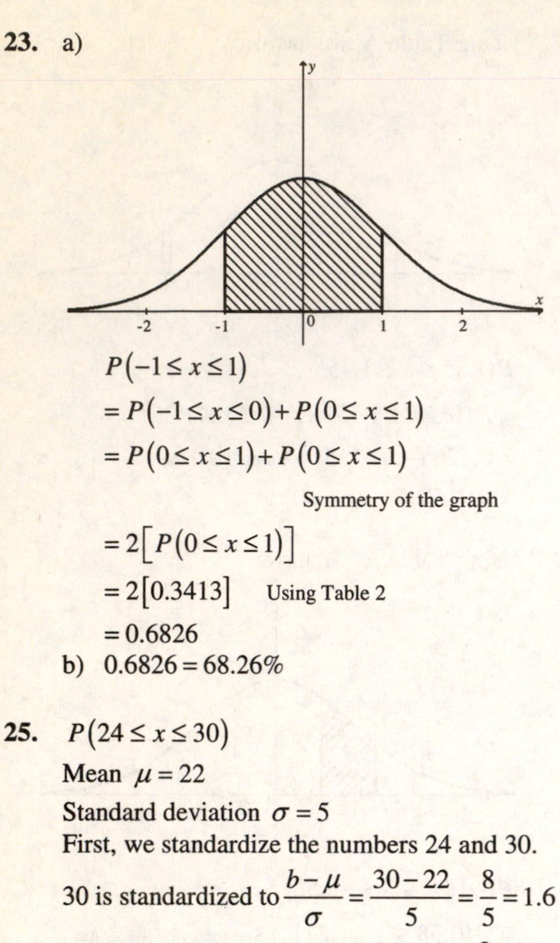

$P(-1 \le x \le 1)$
$= P(-1 \le x \le 0) + P(0 \le x \le 1)$
$= P(0 \le x \le 1) + P(0 \le x \le 1)$
Symmetry of the graph
$= 2[P(0 \le x \le 1)]$
$= 2[0.3413]$ Using Table 2
$= 0.6826$

b) $0.6826 = 68.26\%$

25. $P(24 \le x \le 30)$
Mean $\mu = 22$
Standard deviation $\sigma = 5$
First, we standardize the numbers 24 and 30.
30 is standardized to $\frac{b-\mu}{\sigma} = \frac{30-22}{5} = \frac{8}{5} = 1.6$
24 is standardized to $\frac{a-\mu}{\sigma} = \frac{24-22}{5} = \frac{2}{5} = 0.4$
Then,
$P(24 \le x \le 30)$
$= P(0.4 \le z \le 1.6)$
$= P(0 \le z \le 1.6) - P(0 \le z \le 0.4)$
$= 0.4452 - 0.1554$
$= 0.2898$

27. $P(19 \le x \le 25)$
Mean $\mu = 22$
Standard deviation $\sigma = 5$
First, we standardize the numbers 19 and 25.
25 is standardized to $\frac{b-\mu}{\sigma} = \frac{25-22}{5} = \frac{3}{5} = 0.6$
19 is standardized to $\frac{a-\mu}{\sigma} = \frac{19-22}{5} = \frac{-3}{5} = -0.6$
Continued at the top of the next column.

Then,
$P(19 \le x \le 25)$
$= P(-0.6 \le z \le 0.6)$
$= P(-0.6 \le z \le 0) + P(0 \le z \le 0.6)$
$= P(0 \le z \le 0.6) + P(0 \le z \le 0.6)$
Symmetry of the graph
$= 2[P(0 \le z \le 0.6)]$
$= 2[0.2257]$
$= 0.4514$

29 – 46. Check answers using solutions to Exercises 11-28.

47. a) $P(x \le z) = \frac{30}{100} = 0.3$
$0.3 = 0.5 - 0.2$ and 0.2 corresponds to $z \approx 0.52$ in Table A, so we have $z \approx -0.52$.

b) $P(x \le z) = \frac{50}{100} = 0.5$
$0.5 = 0.5 + 0$ and 0 corresponds to $z = 0$ in Table A, so we have $z = 0$.

c) $P(x \le z) = \frac{95}{100} = 0.95$
$0.95 = 0.5 + 0.45$ and from Table A we see that 0.45 corresponds to a value of z halfway between 1.64 and 1.65, or 1.645. Thus, $z = 1.645$

49. $\mu = -15; \sigma = 0.4$

a) The z value that corresponds with the 92nd percentile, or an area of 0.92, is $z = 1.405$. We use the transformation formula to determine the x value that corresponds to the value $z = 1.405$.

$$1.405 = \frac{x-(-15)}{0.4}$$
$$0.562 = x + 15$$
$$-14.438 = x$$
$$-14.44 \approx x$$

An x value of -14.44 corresponds to the 92nd percentile.

Copyright © 2014 Pearson Education, Inc. Publishing as Addison-Wesley.

b) The z value that corresponds with the 46th percentile, or an area of 0.46, is $z=-0.10$. We use the transformation formula to determine the x value that corresponds to the value $z=-0.10$.

$$-0.10=\frac{x-(-15)}{0.4}$$
$$-0.04=x+15$$
$$-15.04=x$$

An x value of -15.04 corresponds to the 46th percentile.

51. a) Since the boxes are uniformly distributed, the probability density function has the form

$$f(x)=\frac{1}{18.7-17.5}=\frac{1}{1.2}=\frac{5}{6},$$

for $17.5\le x\le 18.7$.

b) To find the mean, variance and standard deviation, we will need to calculate the following

$$E(x)=\int_{17.5}^{18.7}x\cdot\frac{5}{6}dx$$
$$=\left[\tfrac{5}{12}x^2\right]_{17.5}^{18.7}$$
$$=\tfrac{5}{12}\left[(18.7)^2-(17.5)^2\right]$$
$$=18.1.$$

$$E(x^2)=\int_{17.5}^{18.7}x^2\cdot\frac{5}{6}dx$$
$$=\left[\tfrac{5}{18}x^3\right]_{17.5}^{18.7}$$
$$=\tfrac{5}{18}\left[(18.7)^3-(17.5)^3\right]$$
$$=327.73.$$

Using this information, the mean is

$\mu=E(x)=18.1.$

The variance is

$$\sigma^2=E(x^2)-\left[E(x)\right]^2$$
$$=327.73-(18.1)^2$$
$$=0.12.$$

The standard deviation is

$$\sigma=\sqrt{\sigma^2}=\sqrt{0.12}=0.346.$$

c) Looking at the interval within one standard deviation from the mean we have $18.1\pm0.346\rightarrow 17.5\le x\le 18.7$. We calculate the probability that a box of cereal lies within one standard deviation at the top of the next column.

Using the information from the previous column, we have

$$P(17.754\le x\le 18.466)=\int_{17.754}^{18.446}\frac{5}{6}dx$$
$$=\left[\tfrac{5}{6}x\right]_{17.754}^{18.446}$$
$$=\tfrac{5}{6}\left[18.446-17.754\right]$$
$$=0.58.$$

The probability that a box of cereal will fall within one standard deviation of the mean is 0.58.

53. a) To find the mean, variance and standard deviation, we will need to calculate the following

$$E(x)=\int_0^{30}x\cdot\left(-\tfrac{1}{450}x+\tfrac{1}{15}\right)dx$$
$$=\int_0^{30}\left(-\tfrac{1}{450}x^2+\tfrac{1}{15}x\right)dx$$
$$=\left[-\tfrac{1}{1350}x^3+\tfrac{1}{30}x^2\right]_0^{30}$$
$$=\left[-\tfrac{1}{1350}(30)^3+\tfrac{1}{30}(30)^2\right]$$
$$-\left[-\tfrac{1}{1350}(0)^3+\tfrac{1}{30}(0)^2\right]$$
$$=-\tfrac{27{,}000}{1350}+30$$
$$=10.$$

$$E(x^2)=\int_0^{30}x^2\cdot\left(-\tfrac{1}{450}x+\tfrac{1}{15}\right)dx$$
$$=\int_0^{30}\left(-\tfrac{1}{450}x^3+\tfrac{1}{15}x^2\right)dx$$
$$=\left[-\tfrac{1}{1800}x^4+\tfrac{1}{45}x^3\right]_0^{30}$$
$$=\left[-\tfrac{1}{1800}(30)^4+\tfrac{1}{45}(30)^3\right]$$
$$-\left[-\tfrac{1}{1800}(0)^4+\tfrac{1}{30}(0)^3\right]$$
$$=-\tfrac{810{,}000}{1800}+\tfrac{27{,}000}{45}$$
$$=150.$$

Using this information, the mean is

$\mu=E(x)=10.$

The variance is

$$\sigma^2=E(x^2)-\left[E(x)\right]^2$$
$$=150-(10)^2$$
$$=50.$$

The standard deviation is

$$\sigma=\sqrt{\sigma^2}=\sqrt{50}=7.071.$$

Copyright © 2014 Pearson Education, Inc. Publishing as Addison-Wesley.

b) Looking at the interval within one standard deviation from the mean we have $10 \pm 7.071 \rightarrow 2.929 \leq x \leq 17.071$. Therefore, the probability that a patient will have a wait time within one standard deviation is

$$P(2.929 \leq x \leq 17.071)$$
$$= \int_{2.929}^{17.071} \left(-\tfrac{1}{450}x + \tfrac{1}{15}\right)dx$$
$$= \left[-\tfrac{1}{900}x^2 + \tfrac{1}{15}x\right]_{2.929}^{17.071}$$
$$= \left[-\tfrac{1}{900}(17.071)^2 + \tfrac{1}{15}(17.071)\right]$$
$$- \left[-\tfrac{1}{900}(2.929)^2 + \tfrac{1}{15}(2.929)\right]$$
$$= 0.63.$$

The probability that a patient will have a wait time that is within one standard deviation of the mean is 0.58.

55. We first standardize 300 orders. The mean is 250 and the standard deviation is 20, so 300 is standardized to

$$\frac{b-\mu}{\sigma} = \frac{300-250}{20} = \frac{50}{20} = \frac{5}{2} = 2.5$$

Therefore,

$$P(x \geq 300)$$
$$= P(z \geq 2.5)$$
$$= 0.5 - P(z \leq 2.5)$$
$$= 0.5 - 0.4938 \quad \text{Using Table A.}$$
$$= 0.0062$$
$$= 0.62\%$$

The company will have to hire extra help or pay overtime 0.62% of the days.

57. We first standardize 40 seconds. The mean is 38.6 and the standard deviation is 1.729, so 40 is standardized to

$$\frac{b-\mu}{\sigma} = \frac{40-38.6}{1.729} \approx 0.81$$

Therefore,

$$P(x \leq 40)$$
$$= P(z \leq 0.81)$$
$$= 0.5 + P(0 \leq z \leq 0.81)$$
$$= 0.5 + 0.2910 \quad \text{Using Table A.}$$
$$= 0.7910$$

The probability that the next operation of the robogate will take 40 seconds, or less, is 0.7910.

59. a) Find z such that $P(x \leq z) = 0.35$

$0.35 = 0.5 - 0.15$ and 0.15 corresponds to $z \approx 0.39$ in Table 2. Then the score that corresponds to the 35th percentile is 0.39 standard deviations less than the mean, or $1011 - 0.39(100) \approx 972$. Therefore, the score that would correspond to the 30th percentile is 972.

b) Find z such that $P(x \leq z) = 0.60$

$0.60 = 0.5 + 0.1$ and 0.1 corresponds to $z \approx 0.25$ in Table 2. Then the score that corresponds to the 60th percentile is 0.25 standard deviations more than the mean, or $1011 + 0.25(100) \approx 1036$. Therefore, the score that would correspond to the 60th percentile is 1036.

c) Find z such that $P(x \leq z) = 0.92$

$0.92 = 0.5 + 0.42$ and 0.42 corresponds to $z \approx 1.41$ in Table 2. Then the score that corresponds to the 92nd percentile is 1.41 standard deviations more than the mean, or $1011 + 1.41(100) \approx 1152$. Therefore, the score that would correspond to the 92nd percentile is 1152.

61. $\mu = 76; \sigma = 7$

a) The top 12% corresponds with the 88th percentile. The z value that corresponds with the 88th percentile, or an area of 0.88, is $z = 1.175$.

We use the transformation formula to determine the x value that corresponds to the value $z = 1.175$.

$$1.175 = \frac{x-76}{7}$$
$$8.225 = x - 76$$
$$84.225 = x$$
$$84 \approx x$$

The minimum score needed to get an A is 84.

Copyright © 2014 Pearson Education, Inc. Publishing as Addison-Wesley.

b) The top 30% corresponds with the 70th percentile. The z value that corresponds with the 70th percentile, or an area of 0.70, is $z = 0.525$
We use the transformation formula to determine the x value that corresponds to the value $z = 0.525$.

$$0.525 = \frac{x-76}{7}$$
$$3.675 = x - 76$$
$$79.675 = x$$
$$80 \approx x$$

The minimum score needed to get a B is 80.

63. If a players standing 7 feet 2 or 86 inches tall is In the top 1% or 99th percentile. The z value that corresponds to the 99th percentile is 2.33. We can find the standard deviation for players in the NBA by:

$$2.33 = \frac{86-79}{\sigma}$$
$$2.33\sigma = 7$$
$$\sigma = 3.00$$

To determine the percentile of a player standing 83 inches tall (6 feet 11 inches), we determine the z value first.

$$z = \frac{83-79}{3} = 1.33.$$

This value represents an area of 0.908. Therefore, a player that stands 6 feet 11 inches tall would be in the 90.8th percentile.

65. $f(x) = \frac{1}{b-a}, \quad [a,b]$

$$E(x) = \int_a^b x \cdot f(x)\,dx \quad \text{Expected value of } x.$$

$$E(x) = \int_a^b x \cdot \frac{1}{b-a}\,dx$$
$$= \frac{1}{b-a}\left[\frac{x^2}{2}\right]_a^b$$
$$= \frac{1}{b-a}\left[\frac{b^2}{2} - \frac{a^2}{2}\right]$$
$$= \frac{1}{b-a}\left[\frac{b^2-a^2}{2}\right]$$
$$= \frac{1}{b-a} \cdot \frac{(b-a)(b+a)}{2}$$
$$= \frac{b+a}{2}$$

$$E(x^2) = \int_a^b x^2 \cdot f(x)\,dx \quad \text{Expected value of } x^2.$$

$$E(x^2) = \int_a^b x^2 \cdot \frac{1}{b-a}\,dx$$
$$= \frac{1}{b-a}\left[\frac{x^3}{3}\right]_a^b$$
$$= \frac{1}{b-a}\left[\frac{b^3}{3} - \frac{a^3}{3}\right]$$
$$= \frac{1}{b-a}\left[\frac{b^3-a^3}{3}\right]$$
$$= \frac{1}{b-a} \cdot \frac{(b-a)(b^2+ab+a^2)}{3}$$
$$= \frac{b^2+ab+a^2}{3}$$

$$\mu = E(x) = \frac{b+a}{2} \quad \text{Mean}$$

$$\sigma^2 = E(x^2) - [E(x)]^2$$
$$= \frac{b^2+ab+a^2}{3} - \left[\frac{b+a}{2}\right]^2 \quad \text{Substituting}$$
$$= \frac{b^2+ab+a^2}{3} - \left[\frac{b^2+2ba+a^2}{4}\right]$$
$$= \frac{4b^2+4ab+4a^2}{12} - \frac{3b^2+6ba+3a^2}{12}$$
$$= \frac{4b^2+4ab+4a^2-3b^2-6ba-3a^2}{12}$$
$$= \frac{b^2-2ba+a^2}{12}$$
$$= \frac{(b-a)^2}{12} \quad \text{Variance}$$

$$\sigma = \sqrt{\sigma^2}$$
$$= \sqrt{\frac{(b-a)^2}{12}}$$
$$= \frac{b-a}{2\sqrt{3}} \quad \text{Standard deviation}$$

Copyright © 2014 Pearson Education, Inc. Publishing as Addison-Wesley.

67. $f(x)=\frac{1}{2}x,\quad [0,2]$

$$\int_0^m f(x)dx=\frac{1}{2}$$
$$\int_0^m \frac{1}{2}xdx=\frac{1}{2}$$
$$\frac{1}{2}\int_0^m xdx=\frac{1}{2}$$
$$\int_0^m xdx=1$$
$$\left[\frac{x^2}{2}\right]_0^m=1$$
$$\frac{m^2}{2}-\frac{0^2}{2}=1$$
$$\frac{m^2}{2}=1$$
$$m^2=2$$
$$m=\sqrt{2}$$

69. $f(x)=ke^{-kx},\quad [0,\infty)$

$$\int_0^m f(x)dx=\frac{1}{2}$$
$$\int_0^m ke^{-kx}dx=\frac{1}{2}$$
$$\left[\frac{k}{-k}e^{-kx}\right]_0^m=\frac{1}{2}$$
$$\left[-e^{-kx}\right]_0^m=\frac{1}{2}$$
$$-e^{-k(m)}-\left(-e^{-k(0)}\right)=\frac{1}{2}$$
$$-e^{-k\cdot m}+e^0=\frac{1}{2}$$
$$-e^{-k\cdot m}+1=\frac{1}{2}$$
$$\frac{1}{2}=e^{-k\cdot m}$$
$$\ln\left(\frac{1}{2}\right)=\ln\left(e^{-k\cdot m}\right)$$
$$\ln(1)-\ln(2)=-k\cdot m \qquad [\ln 1=0]$$
$$\frac{-\ln 2}{-k}=m$$
$$\frac{\ln 2}{k}=m$$

71. Standardize 16.

$$z=\frac{16-\mu}{0.2}$$

We are looking for a value of c for which $P(z<c)=\frac{1}{50}=0.02$. Now, if $c\geq 0$, then

$$P(z<c)=P(z\leq 0)+P(0\leq z<c)$$
$$=0.5+P(0\leq z<c)$$

Then,

$$P(0\leq z<c)=P(z<c)-0.5$$
$$=0.02-0.5$$
$$=-0.48$$

Which is not possible. Therefore, $c\leq 0$, and

$$P(0\leq z<c)=0.5-P(z<c)$$
$$=0.5-0.02$$
$$=0.48.$$

Looking in Table A for the number closest to 0.48, we find that 0.4798 and 0.4803 are the two closest numbers. Since we want to ensure that only 1 bag in 50 will have less than 16oz we select the larger value 0.4803.

This value corresponds to $c=2.06$, but in our case $c\leq 0$, so we use $c=-2.06$. We set $z=-2.06$ and solve for μ.

$$-2.06=\frac{16-\mu}{0.2}$$
$$(-2.06)(0.2)=16-\mu$$
$$-0.412=16-\mu$$
$$\mu-0.412=16$$
$$\mu=16+0.412$$
$$\mu=16.412$$

The mean weight should be adjusted to 16.412 oz to ensure that only 1 bag in 50 will have less than 16 oz.

73. ✎

75. Using the integration feature on a calculator, we have

$$\int_{-\infty}^{\infty} e^{-x^2}dx\approx 1.772.$$

Copyright © 2014 Pearson Education, Inc. Publishing as Addison-Wesley.